The Urban Book Series

The Urban Book Series is a resource for urban studies and geography research worldwide. It provides a unique and innovative resource for the latest developments in the field, nurturing a comprehensive and encompassing publication venue for urban studies, urban geography, planning and regional development.

The series publishes peer-reviewed volumes related to urbanization, sustainability, urban environments, sustainable urbanism, governance, globalization, urban and sustainable development, spatial and area studies, urban management, transport systems, urban infrastructure, urban dynamics, green cities and urban landscapes. It also invites research which documents urbanization processes and urban dynamics on a national, regional and local level, welcoming case studies, as well as comparative and applied research.

The series will appeal to urbanists, geographers, planners, engineers, architects, policy makers, and to all of those interested in a wide-ranging overview of contemporary urban studies and innovations in the field. It accepts monographs, edited volumes and textbooks.

Now Indexed by Scopus!

More information about this series at http://www.springer.com/series/14773

Wenzhong Shi · Michael F. Goodchild ·
Michael Batty · Mei-Po Kwan · Anshu Zhang
Editors

Urban Informatics

Springer

Editors
Wenzhong Shi
The Hong Kong Polytechnic University
Hong Kong, China

Michael F. Goodchild
University of California
Santa Barbara, USA

Michael Batty
University College London
London, UK

Mei-Po Kwan
The Chinese University of Hong Kong
Hong Kong, China

Anshu Zhang
The Hong Kong Polytechnic University
Hong Kong, China

ISSN 2365-757X ISSN 2365-7588 (electronic)
The Urban Book Series
ISBN 978-981-15-8982-9 ISBN 978-981-15-8983-6 (eBook)
https://doi.org/10.1007/978-981-15-8983-6

This Springer imprint is published by the registered company Springer Nature Singapore Pte Ltd.
The registered company address is: 152 Beach Road, #21-01/04 Gateway East, Singapore 189721, Singapore

Acknowledgements

The publication of this book is supported by The Hong Kong Polytechnic University (1-847V, 1-ZVN6, 1-99YX, 1-99XK).

We would like to thank all the authors for their valuable contributions to this book. Also, we greatly appreciate the comments from anonymous reviewers for the chapters. We would like to thank Prof. Shaowen Wang, Prof. Shih-lung Shaw, and Dr. Yang Xu for their input to the book organization at the early stage and Ms. Mengfei Ma and Ms. Zihan Kan for their editorial assistance in all parts and in Part II of the book, respectively.

Contents

About the Editors

Wenzhong Shi is the Otto Poon Charitable Foundation Professor in Urban Informatics, Chair Professor in GISci and remote sensing, Director of the Smart Cities Research Institute and Head of the Department of Land Surveying and Geo-Informatics, The Hong Kong Polytechnic University. He obtained his doctoral degree from the University of Osnabrück, Germany, in 1994. His current research interests include urban informatics for Smart Cities, GISci and remote sensing, focusing on analytics and quality control for spatial big data, object extraction, and change detection from satellite images and LiDAR data, integrated mobile mapping technology, and 3D and dynamic GISci modeling. He was elected as an academician to the International Eurasian Academy of Sciences in 2019, as a fellow of HKIS in 2019 and a fellow of RICS in 2018. He has published more than 400 scientific articles and 15 books. He has received a number of prestige awards, including an award from the International Society of Photogrammetry and Remote Sensing and the Natural Science Award from the State Council, China. He serves as President of the International Society for Urban Informatics. His full CV is at http://www.lsgi.polyu.edu.hk/academic_staff/John.Shi/index.htm.

Michael F. Goodchild is Professor Emeritus of Geography at the University of California, Santa Barbara. Until 2012, he held the Jack and Laura Dangermond Chair of Geography and was Director of UCSB's Center for Spatial Studies. He received his BA degree from Cambridge University in Physics in 1965 and his Ph.D. in Geography from McMaster University in 1969. His research and teaching interests focus on issues in geographic information science, including uncertainty in geographic information, discrete global grids, and volunteered geographic information. He has directed or co-directed several large funded projects, including for the National Center for Geographic Information and Analysis, the Alexandria Digital Library, and the Center for Spatially Integrated Social Science. He was elected member of the US National Academy of Sciences in 2002, Foreign Member of the Royal Society, and Corresponding Fellow of the British Academy in 2010; and in 2007, he received the Prix Vautrin Lud. He has published over 550 books and articles. He moved to Seattle upon retirement in 2012 and currently holds part-time positions as Research Professor at Arizona State University and as Distinguished Chair Professor at The Hong Kong Polytechnic University. His full CV is at www.geog.ucsb.edu/~good.

Michael Batty is Bartlett Professor of Planning, Chairman of the Center for Advanced Spatial Analysis at University College London, and Distinguished Chair Professor at the Hong Kong Polytechnic University. He received his BA degree from The University of Manchester in Town and Country Planning in 1966 and his PhD in Architecture from the University of Wales in 1984. He has worked on computer models of cities and their visualization since the 1970s and has published several books, such as *Cities and Complexity* (2005), *The New Science of Cities* (2013), and *Inventing Future Cities* (2018) all MIT Press. Prior to his current position, he was Professor of City Planning and Dean of the School of Environmental Design at the University of Wales at Cardiff from 1979 to 1990 and then Director of the National Center for Geographic Information and Analysis at the State University of New York at Buffalo from 1990 to 1995. He was awarded the CBE in the Queen's Birthday Honours in 2004 and was the 2013 recipient of the Lauréat Prix International de Géographie Vautrin Lud. In 2015, he was awarded the Gold Medal of the Royal Geographical Society and in 2016 the Gold Medal of the Royal Town Planning Institute. He is a Fellow of the Royal Society and the British Academy.

Mei-Po Kwan is Choh-Ming Li Professor of Geography and Resource Management and Director of the Institute of Space and Earth Information Science at the Chinese University of Hong Kong. She was a Visiting Distinguished Chair Professor of Geography and Geographic Information Science at the Hong Kong Polytechnic University. She received her MA degree in Urban Planning in 1989 from the University of California, Los Angeles, and her Ph.D. in 1994 in Geography from the University of California, Santa Barbara. Her research interests include environmental health, human mobility, social and transport issues in cities, healthy cities, and GIScience. Kwan is an Elected Fellow of the UK Academy of Social Sciences, the John Simon Guggenheim Memorial Foundation, the American Association for the Advancement of Science, and the American Association of Geographers (AAG). She was included in the 2019 Highly Cited Researchers list of the Web of Science Group and has received many prestigious honors and awards, including the Distinguished Scholarship Honors from the AAG and a Research Award from the US University Consortium for Geographic Information Science. Kwan has published over 310 books, articles and book chapters. She has delivered over 240 keynote addresses and invited lectures in more than 20 countries. More information about Kwan is available at http://meipokwan.org.

Anshu Zhang is a Research Assistant Professor in the Department of Land Surveying and Geo-Informatics, The Hong Kong Polytechnic University. She received her B.Sc. degree in Geo-Information Technology in 2011 and her Ph.D. in Geographic Information Systems in 2017 from The Hong Kong Polytechnic University. Her research interests include spatial data mining, human mobility modeling and prediction, with the emphasis on improving the robustness and reliability of data analytics by approaches such as statistical tests, evolutionary computing, and explainable artificial intelligence. She was the Secretary of Working Group II/1, The International Society for Photogrammetry and Remote Sensing (2012–2016). She has participated in multiple governmental funded projects and is the Principle Investigator of a sub-research program under the State Key R&D Scheme funded by the Ministry of Science and Technology of China. She received the China Science and Technology Progress Award in Surveying and Mapping (Grand Award) in 2017.

Chapter 1
Overall Introduction

Wenzhong Shi, Michael F. Goodchild, Michael Batty, Mei-Po Kwan, and Anshu Zhang

Abstract Urban informatics is an interdisciplinary approach to understanding, managing, and designing the city using systematic theories and methods based on new information technologies. Integrating urban science, geomatics, and informatics, urban informatics is a particularly timely way of fusing many interdisciplinary perspectives in studying city systems. This edited book aims to meet the urgent need for works that systematically introduce the principles and technologies of urban informatics. The book gathers over 40 world-leading research teams from a wide range of disciplines, who provide comprehensive reviews of the state of the art and the latest research achievements in their various areas of urban informatics. The book is organized into six parts, respectively covering the conceptual and theoretical basis of urban informatics, urban systems and applications, urban sensing, urban big data infrastructure, urban computing, and prospects for the future of urban informatics. This introductory chapter provides a definition of urban informatics and an outline of the book's structure and scope.

W. Shi (✉) · A. Zhang
Department of Land Surveying and Geo-Informatics and Smart Cities Research Institute, The Hong Kong Polytechnic University, Hong Kong, China
e-mail: lswzshi@polyu.edu.hk

A. Zhang
e-mail: aszhang@polyu.edu.hk

M. F. Goodchild
University of California, Santa Barbara, USA
e-mail: good@geog.ucsb.edu

M. Batty
Centre for Advanced Spatial Analysis, University College London, London, UK
e-mail: m.batty@ucl.ac.uk

M.-P. Kwan
Department of Geography and Resource Management and Institute of Space and Earth Information Science, The Chinese University of Hong Kong, Hong Kong, China
e-mail: mpk654@gmail.com

W. Shi et al. (eds.), *Urban Informatics*, The Urban Book Series,
https://doi.org/10.1007/978-981-15-8983-6_1

1.1 Defining Urban Informatics

Urban informatics is an interdisciplinary approach to understanding, managing, and designing the city using systematic theories and methods based on new information technologies, and grounded in contemporary developments of computers and communications. It integrates urban science, geomatics, and informatics: urban science provides studies of activities, places, and flows in the urban area; geomatics provides the science and technologies for measuring spatiotemporal and dynamic urban objects in the real world and managing the data obtained from the measurements; informatics provides the science and technologies of information processing, information systems, computer science, and statistics which support the quest to develop applications to cities.

The field covers many sectors that define city systems. Those sectors are often studied in their own right, such as transportation, housing, retail activity, physical infrastructure involving the distribution of waste, water, electricity, and other sources of energy, as well as demographic structure, economic location, urban development, and a host of related perspectives that pertain to cities and urban systems. What makes urban informatics different and complementary to these disciplinary approaches is the fact that computation is central to the way in which methods and models are used to generate a deeper understanding: of many problems that involve working out how cities function, how they generate different forms, how their dynamics reflects the ways in which they grow and decline, and how they mix, segregate, and polarize different populations and activities.

What makes urban informatics a particularly timely way of gathering together and fusing many interdisciplinary perspectives which involve computation is that in the last twenty years, computers have scaled down to the point where they can be used as sensors and embedded in a variety of physical infrastructures as well as being used in a mobile context by the population at large. This has meant that quite suddenly we are now endowed with streams of data about a city's functioning in real time, something that was not generally available hitherto when most of our methods of data collection were not automated through sensors. This has led to what is called big data—data that are generated in real time, with great variety, and hence almost limitless in volume. Such data may be the product of sensors that operate continuously and provide immediate updates to the system of our concern. For these data, we need new methods and models to help our understanding and to interpret old models that still have relevance. This has thrown the 24-hour city onto the agenda, and many of the chapters in this book reflect the fact that temporal dynamics is now a serious feature of this field of informatics. Time is now being deeply reflected in our models, whereas in the past the focus was more on spatial variety.

The field of urban informatics is still developing rapidly in its embrace of new sensing technologies, new kinds of spatial data science, new methods of analysis that range from traditional statistical methods as in spatial econometrics, all the way to new developments in machine learning, and multivariate analysis that enable analysts to explore big data in ways that have not been possible hitherto. In terms of the fields that are distinct within the contributions we have collected here, it is worth noting that new approaches to the structure, form, and dynamics of cities using mainly physical approaches are being used to define a new kind of urban science. New methods of urban analytics are being fashioned using these ideas, and the fact that we are now able to exploit real-time movement data from sensors—either fixed to monitor traffic or mobile to do the same through telephone calls and other social media—means that we have a much richer understanding of cities than anything we have been able to develop so far. Mobility studies have thus become central to urban informatics, while developments in the dynamics of infrastructure, urban pollution, and waste—in short, the metabolism of the city—are coming to the fore through urban analytics. A large part of urban informatics involves sensing at many spatial scales from satellite remote sensing to indoor navigation, while the development of the third dimension in cities in terms of sensing and visualization is now becoming routine. Stitching all these ideas together is another important function of urban informatics, while the development of what was seen as rather disconnected types of urban models—land use and transportation, urban microsimulation, cellular automata, and agent-based models—is now part of the wider agenda. Last but not least, the field also has regard to how its theories, models, and tools relate to wider questions of governance, risk, security, crime, health, and welfare, as well as geodemographics. All these features are encapsulated in our definition of urban informatics here, and we hope readers will thus be able to piece together their own big picture of the field as they navigate many contributions in this book.

1.2 The Background: The Origins of Urban Informatics

The idea of publishing this book is rooted in the fast development of urban informatics in both academia and industry in the big data era. In academia, many universities have established programs to offer both undergraduate and postgraduate degrees related to urban informatics. Examples of such programs include a undergraduate program in Urban Informatics at Shenzhen University, an MSc program in Smart Cities and Urban Analytics at University College London (UCL), a graduate program in Applied Urban Science and Informatics at New York University, an MSc program in Urban Informatics at Northeastern University, an MSc program in Urban Informatics and Analytics at Warwick University, and an MSc program and a PhD research area in Urban Informatics and Smart Cities at The Hong Kong Polytechnic University (HKPU). These kinds of courses are rapidly expanding as different research groups recognize the importance of training and research in the ways in which urban informatics might be applied to contemporary urban problems. The common goal shared by these programs is to promote education and research activities to cope with various challenges in cities under the rapid global urbanization process. In industry, the smart city is a major new trend in urban development and management, and urban informatics is the core technology of smart cities. According to recent reports by Grand View Research and Zion Market Research, the global smart city market accounted for USD 955.3 billion in 2017 and is anticipated to reach USD 2.57 trillion by 2025. Such a huge and increasing market is driven by many factors, such as rapid growth of urban populations around the world and the need to foster sustainable urban development. However, there are very few books systematically introducing the principles and technologies of urban informatics, including urban science, urban systems and applications, urban sensing, urban big data infrastructure, and urban computing. There is an urgent need to edit and publish such books to equip the current and next-generation workforce with the knowledge to tackle the challenges that cities are facing. Our contribution here is to address this urgent need.

The publication of this book is among a series of activities carried out by HKPU for promoting urban informatics internationally. Other activities include initiating and organizing the International Conference on Urban Informatics (ICUI) series, establishing the International Society of Urban Informatics (ISUI) and International Journal of Urban Informatics (IJUI), developing a new MSc program and a PhD research area in Urban Informatics and Smart Cities, and founding the Smart Cities Research Institute for conducting cutting-edge research.

Hosted by the Department of Land Surveying and Geo-Informatics (LSGI), HKPU, ICUI provides a platform for leading scientists, young scholars, and researchers worldwide to share an interest in urban informatics. The first conference in the ICUI series was held in 2017, with around 40 presentations on topics in urban systems, urban sensing, spatiotemporal big data, urban computing, and urban solutions. The second conference was held in 2019 with the theme "Toward Future Smart Cities". Over 280 participants from 18 countries and institutions such as MIT, Harvard University, the University of Cambridge, UCL, ETH, and the Alan Turing Research Institute, joined the conference and delivered over 120 presentations on 18 topics. Also introduced in ICUI 2019 was the International Society of Urban Informatics (ISUI). ISUI aims to promote the international exchange of knowledge and experience in the field of urban informatics, helping its members to succeed in their professions through regional and international academic exchange programs, publications, and networks of cross-disciplinary experts.

A number of other universities in Hong Kong have also contributed to urban informatics and smart city development. For example, the University of Hong Kong has formed the Hong Kong Urban Labs, the Chinese University of Hong Kong has established the Institute of Future Cities, and the Hong Kong University of Science and Technology has developed the GREAT Smart Cities Institute. HKPU has been conducting research on various topics in urban informatics and has accumulated numerous theories, methods, advanced technologies, and successful application cases that provide updated materials for this book.

The book is based on invitations to over 40 world-leading scholars and their teams across a wide range of fields in urban informatics who were asked to write the chapters of this book. In the book, they not only give comprehensive reviews but also share their latest research achievements in various topics within urban informatics, as well as vivid examples of employing emerging urban informatics technologies for solving urban problems. Some of the chapters have been contributed by the participants of the ICUI series, but include new material rather than the presentations at these conferences.

This book is intended for use by researchers and students from a wide range of disciplines related to urban informatics, urban science, urban systems and applications, urban sensing, urban big data infrastructure, and urban computing. It will serve as a textbook for those undergraduate and graduate students majoring in urban informatics, studies in smart cities, transport and civil engineering, geography, geosciences, urban planning, geographic information science, environmental science, resources science, and land use. It can also be used as a reference book for practitioners and professionals in the governmental, commercial, and industrial sectors, such as urban planners, computer scientists, data scientists, geographers, policy makers, architect designers, surveyors, urban governors, and environmental scientists.

1.3 Structure of the Book

This book has six parts that cover the latest developments in a wide range of topics in urban informatics. These topics include the conceptual and theoretical basis of urban informatics, applications of urban informatics in understanding and managing various urban systems, urban sensing, urban big data infrastructure, and urban computing. While the parts are related, they can be read in any order except Part I, which intends to provide an overview of the backgrounds of urban informatics and thus should be read before the other parts.

After the overall introduction, Part I (Dimensions of Urban Science) focuses on the conceptual and theoretical basis of urban science as it has evolved in the examination of the city as a system. It highlights contemporary theories of urban interactions, human dynamics, metabolisms, and the urban economy, and relates these to the wider vision of a new urban science for examining cities in the twenty-first century. The chapters in Part II (Urban Systems and Applications) discuss applications of urban informatics in understanding, analyzing, and managing various urban systems. These include applications in urban travel and human mobility, urban freight systems, crime and security, pollution monitoring, energy systems, health and well-being, risk and resilience, as well as urban governance. The state-of-the art urban informatics are used to identify the problems and provide viable solutions for those problems. The chapters in Part III (Urban Sensing) describe existing and new methods of urban sensing, including remote sensing, ground-based sensors, global navigation satellite systems (GNSS), mobile mapping technologies, indoor positioning technologies, user-generated content, and other developments that have a considerable potential for advancing urban science.

Part IV (Urban Big Data Infrastructure) focuses on issues related to the new developments in urban big data infrastructure, including those concerning big data, geoprivacy, 3D city modeling, 3D cadastre, rule-based modeling, cyber infrastructure, spatial search, and urban IoT. These new developments will likely contribute to significant progress in urban informatics and in urban science more broadly. The chapters in Part V (Urban Computing) cover various topics in urban informatics from the perspectives of computer science and urban modeling. Specific research or application areas examined include visual analytics, cloud and mobile computing, data mining, artificial intelligence (AI) and deep learning, agent-based modeling, microsimulation, Cellular Automata modeling, and transportation modeling. The chapters highlight the development and use of computing technologies, principles, and models for urban contexts and applications. Part VI (The Value of Urban Informatics) concludes the book with a broadly based and forward-looking discussion by Michael F. Goodchild on the goals of urban informatics, the potential for unintended consequences, and possible approaches to accountability.

1.4 Retrospective and Prospective

In the third decade of the twenty-first century, we find ourselves with a well-developed ability to acquire vast amounts of information about the city and with the tools to perform a wide range of analyses. Projects under way in world cities such as Beijing, London, New York, Hong Kong, and Singapore are described at many points in the chapters of this book, and there is every reason to believe that the burgeoning field of urban informatics will continue to grow. But while the reader will find rich detail in the pages that follow, he or she will also recognize that what is being described is a first-world activity, largely confined to the Global North. What all of this means for the Global South remains an issue that is scarcely addressed, and we can only speculate as to what is likely to happen if this omission continues.

Urban informatics is a young field, and not surprisingly it is difficult to organize into self-contained subfields. The reader will become well aware of this issue as he or she navigates the parts of the book and encounters issues such as urban mobility or urban heat islands in different chapters and parts and in different contexts. Hopefully, a better and more robust conceptual model of urban informatics will emerge in time, as the field matures and as its principles become more clearly articulated. We look forward to one or more future textbooks that distill the field into a simple, concise, and theory-based structure. For now, however, the approach has to be more encyclopedic.

What else is missing? First is a sense of history, of how earlier cities dealt with their limited information resources and their lack of the tools to make sense of what they had. John Snow's map of the London cholera outbreak of 1854 was a masterful exercise in inference (Johnson 2007); while the concept of the smart lamppost has a fascinating precursor in the Pluto lamps that were installed in London in the late 1890s (https://www.british-history.ac.uk/survey-london/vol47/pp52-83). We should be able to learn much from a counterfactual approach from earlier times. Second is a sense of what the future may hold in the way of unintended consequences, gaming of technology, and subversion. The history of information technologies is rich in examples of breakthroughs gone astray, finding application for purposes that are malicious and dystopic. Many of the chapters are full of enthusiasm and excitement for the positive potential of urban informatics and understandably do not dwell on the negative. These possibilities are addressed at the end of the book in Part VI. Finally, as in any data-intensive field there will always be a need to address uncertainty, and associated issues of data provenance and measurement error, especially given the spatiotemporal focus of the field. Dealing with uncertainty is not simply a matter of putting a plus or minus on each item of data, given the strong existence of statistical dependence in both spatial and temporal domains. To quote Korszybski (1933), the map is not the territory; the data are only an approximation and representation of reality.

References

Johnson S (2007) The ghost map: the story of London's most terrifying epidemic–and how it changed science, cities, and the modern world. Penguin, London

Korszybski A (1933) Science and sanity. An introduction to non-Aristotelian systems and general semantics. The International Non-Aristotelian Library Pub. *Co.* pp 747–761

Part I
Dimensions of Urban Science

Chapter 2
Introduction to Urban Science

Michael Batty

Abstract This introduction outlines a portfolio of theory and methods in the chapters that develop a basic urban science for urban informatics. Inductive and deductive methods for generating data, analytics, and urban simulation, form the focus. In this first Part of the book, the emphasis is on mobility, space-time theory, energy and infrastructure, the spatial economy, and the role of modelling in understanding and planning the smart city.

There are many different but related disciplinary perspectives underpinning urban informatics, and each of these brings a different science to bear on the tools and techniques which form the core of this new domain. In this introduction, we will not sketch all of these different approaches, for many of these will be developed throughout this book. Here, we will simply outline some of the basic physical theories that pertain to the structure of cities, in particular how the form of the city and its functions influence the location of different activities and the ways in which these activities are linked together. We call this "urban science," which is a little more comprehensive than particular sciences relevant to cities, which relate to ecology, energy, social structure, economic development, and so on, and which develop theories and concepts of these particular subsystems in greater depth. Urban science deals with generic theories of how cities are structured and how they grow and evolve in time, how they change qualitatively with respect to growth, and how their populations organize themselves in space. These features often reveal the kinds of problems that urban planning is designed to alleviate, and in this context, the ways in which urban informatics might progress physical planning can be rooted in some of the theories and principles which urban science is able to elucidate.

Like any science, urban science articulates relationships that define the components of the city using quantitative methods which are generally validated by observations that are drawn from actual cities. In short, the conventional scientific method is key to developing the best tools and techniques that comprise urban informatics. The

M. Batty (✉)
Centre for Advanced Spatial Analysis, University College London, London, UK
e-mail: m.batty@ucl.ac.uk

W. Shi et al. (eds.), *Urban Informatics*, The Urban Book Series,
https://doi.org/10.1007/978-981-15-8983-6_2

tool set that is evolving rapidly is based on the classic distinctions between methods that are used to infer order and pattern in data drawn from the city, as well as testing hypotheses that are framed about this order and pattern with respect to data about the city. In short, these tools are based on generating theory through induction or testing theory through deduction. The scientific method usually involves both induction that generates ideas, often alongside deductions from these ideas which in turn are tested. The loop that defines this method is continuous as new ideas are evolved, improved, or discarded, revealing whether or not they are fit for purpose. But at any point in this cycle, these theories need to be translated into forms that are useful in applying the methods of urban informatics. Indeed, the first substantive chapter by Daniel Zünd and Luis Bettencourt illustrates how we can capture data in real time from various objects in the city and by using machine learning, can generate patterns that define how the form of the city can be interpreted. In a later chapter, Shih Lung Shaw illustrates how a series of models about the dynamics of the city can be defined in terms of how the city changes in space and time, with the models then validated in classic deductive terms. Thus, induction and deduction are both brought to bear on the development of urban informatics.

This entire area is dominated by many new methods emanating from computer science, which in turn have developed as computers have scaled down to the point where we can use them to sense any movement and change in the built environment. These sensors may be fixed or mobile, but they have given rise to new data sets that measure how different components in the city change through time. This has led to very large data volumes that tend to produce highly unstructured data that we can only interpret using new methods of pattern recognition and statistical analysis that search for pattern and order in the data. These data are often called 'big' in that they pertain to individual movements and decisions in real time and are only bounded by the time the sensors are active. In this way, data streams can be continuous, and if they grow to terabyte or petabyte levels, we need new and different techniques to explore them, that is, to find the pattern in such data. This is in stark contrast to traditional data sets in cities that usually do have structure, as they are collected in one-off fashion through interview or census. The focus in this book on techniques that involve machine learning and data search has emerged primarily from the need to find structure in data that in their raw form are often completely unstructured. At the same time, increasing amounts of data which might become big can be fashioned from individuals generating their own data either individually or through crowdsourcing. Crowdsourcing has always been used to collect some data, but the existence of new information technologies to support such sourcing has given a new momentum to this kind of data collection.

The elements of urban science that the chapters in this first part of the book address deal with urban morphology, which defines the form and function of the city in terms of location and interactions. Morphology is developed in terms of a threefold characterization of the size, scale, and shape of the city, and much of urban informatics addresses ways in which we might improve the city by changing and manipulating these dimensions. Mobility is the generic area that has grown to encompass the relations between locations and interactions, and this immediately raises the role of networks at different hierarchical levels in the city, as well as the flows that are directed by these networks. Transportation modeling encompasses the best-developed set of tools in this domain, and many of the chapters here allude to such modeling. The relations that bind all these ideas together and are the essence of urban science are scaling, which formalizes the way the hierarchy of elements of different sizes and scales, such as neighborhoods and districts, function within the city. The classic signature of such scaling is the power law, which is ubiquitous as a measure of nonlinearity in urban systems; and in the next chapter, these ideas are spelt out in more detail. In absorbing the contents of this book, readers will find that they emerge in many different guises.

With respect to what follows in this first part, Daniel Zünd and Luís Bettencourt illustrate how it is possible to sense the most obvious objects in a small town in the Galapagos Islands using a blanket coverage and street-view-like cameras. This produces data that can be mined for the more abstract morphology of the place, showing how a judicious mix of user-generated content can be used to sense the spatial structure of the town. Shih Lung Shaw then provides a detailed review of different dynamic models of cities based on urban systems dynamics, cellular automata, and agent-based simulations, setting this in the wider context of human dynamics at the individual person level, and space-time theory as originally developed by Torsten Hägerstrand. The use of new technologies in unpacking individual movements is explored by Martin Raubal, Dominik Bucher, and Henry Martin, who show how personalized tracking can be scaled to look more generally at mobile decision-making, complementing the two previous chapters, with the focus very much on urban dynamics, spatial structure, and individual mobility.

The argument then changes direction. Sybil Derrible, Lynette Cheah, Mohit Arora, and Lih Wei Yeow explore urban metabolism that they articulate using input–output relations and flows of energy and materials that define linkages between many different components of the urban system. These models are static in that they simulate flow at a cross section in time, and although the authors provide an example based on Singapore, they illustrate how problematic it is to generalize these kinds of models to embrace the fine spatial scale. Ying Jin then explores a simple spatial econometric model which looks at GDP in Guangzhou province in China, where he uses the classic measure of gravitational potential or accessibility to relate this to the way the urban system functions with respect to innovative economic activities. This has important implications for future planning of industrial development in the region. Helen Couclelis then concludes this part by standing back and speculating on how all these trends in digital modeling at different scales pertain to the planning of future cities, particularly smart cities. This serves as gentle closure to the ideas in this first part of the book, which establishes many of the theoretical concepts to be picked up and operationalized in the chapters that follow.

Chapter 3
Defining Urban Science

Michael Batty

Abstract This introductory chapter provides a brief overview of the theories and models that constitute what has come to be called urban science. Explaining and measuring the spatial structure of the city in terms of its form and function is one of the main goals of this science. It provides links between the way various theories about how the city is formed, in terms of its economy and social structure, and how these theories might be transformed into models that constitute the operational tools of urban informatics. First the idea of the city as a system is introduced, and then various models pertaining to the forces that determine what is located where in the city are presented. How these activities are linked to one another through flows and networks are then introduced. These models relate to formal models of spatial interaction, the distribution of the sizes of different cities, and the qualitative changes that take place as cities grow and evolve to different levels. Scaling is one of the major themes uniting these different elements grounding this science within the emerging field of complexity. We then illustrate how we might translate these ideas into operational models which are at the cutting edge of the new tools that are being developed in urban informatics, and which are elaborated in various chapters dealing with modeling and mobility throughout this book.

3.1 A Science of Cities

There are many sciences that encompass our understanding of cities. In this introductory chapter, we seek to define the range of scientific disciplines and perspectives that underpin theories pertaining to urban form, social structure, and the built environment in contemporary cities. The science that we will present is based on abstracting the critical functions that determine processes of change that characterize cities, processes such as the way markets operate; the way goods, people, and information are distributed across networks; the economic rationale for the location of activities in cities; and the way these functions and processes grow and change as cities get

M. Batty (✉)
Centre for Advanced Spatial Analysis, University College London, London, UK
e-mail: m.batty@ucl.ac.uk

W. Shi et al. (eds.), *Urban Informatics*, The Urban Book Series,
https://doi.org/10.1007/978-981-15-8983-6_3

bigger or smaller. There are many sciences of the city that are not included in our remit, such as those involving the physics of the built environment, the ecology of cities, and the way climate impacts on city form and function; and there are many aspects of the social domain such as political actions and social mixing that are not considered in this review. But it is important at the outset to be clear about the limits to this science (Lobo et al. 2020). The purpose of this chapter is to suggest a wide variety of scientific ideas that support the quest for establishing urban informatics. We loosely define it here as the technologies and tools as well as the data that enable our city science to be embodied in the models and simulations that are used to improve the management and planning of cities and regions across many different scales and topic areas (Batty 2019).

Urban informatics has emerged as a coherent field largely due to the scaling down of computers and sensors to the point where they can be embedded at very high densities in every part of the urban environment. This includes mobile devices that people activate and operate, as well as fixed sensors that record data pertaining to their functions, often in real time. Urban informatics thus covers a wide range of digital data, from that which is collected in traditional terms from universal or sample censuses at typically low frequencies such as years or decades, all the way to real-time big data streams that are captured at very high frequencies and which provide a portrait of how the city is changing continuously. This field not only covers data, but it also embraces the tools and models that are collectively referred to as urban analytics. In all these tools, we need good theory, and thus, it is the purpose of this chapter to sketch the rudiments of a city science that covers both low- and high-frequency processes in cities, as well as methods of representing and visualizing the form these processes take when we are able to incorporate them in models, simulations, and predictions.

Accordingly we begin by exploring the nature of the city as a system, which was the dominant way of articulating its structure and dynamics in the middle years of the past century. This will establish the key components of cities and how they function at different levels of organization arranged in hierarchical fashion. This then leads us to extend our knowledge to systems of cities, although in this book we will only occasionally refer to such extended systems when we explore cities at regional and national levels. In reviewing these ideas, we introduce the notion that cities can also be seen as systems that emerge from a multitude of local individual decisions, implemented from the bottom up. These generate order from the apparent chaos of non-coordination, and this grounds the study of cities and this science as one of the main exemplars of complexity theory. The theories that have emerged from this focus on systems and complexity are often referred to loosely as social physics in analogy to mechanical systems, and we review these before we develop two key constructs that define the essence of this science of cities: scale and size. The way a city's spatial form—often through its geometry—is reflected in its functions generates the key properties of cities that are articulated in theories about how cities function economically and socially. We then present these functions, linking these to the networks and flows that form the cement that binds the various subsystems, components, and the city's elements together. Many of these models form the basis of operational applications, and we will note a wide variety of these simulations to give readers some idea of the range of possibilities in using simulation in urban informatics. We will then conclude with some speculations about how these theories, as viewed in terms of urban informatics, influence the distribution of different types of cities world-wide and the way in which they can be used to develop tools to improve the quality of life and sustainability of cities through the development of urban informatics.

3.2 City Systems and Systems of Cities

Up to the beginning of the industrial revolution, all cities evolved from some central location where people came together to trade or to rule. From ancient times, populations clustered around these central places and cities developed in such a way that competition for locating closer to the center depended upon the ability of those who engaged in production to capture sufficient demand for their goods to be able to outbid others with respect to the price of space and proximity. Although this model was distorted in the early industrial revolution with the exploitation of fossil fuels around which cities also grew, the notion of the city having a dominant core with bands of different land-use activities or land uses surrounding it, became the received wisdom for how cities came to be formed . As transportation routes bringing producers and consumers to the center to engage in trade could not be built everywhere, cities also developed in radial fashion, with the dominant model being the radially concentric form that was most clearly articulated by Park and Burgess (1925) in their classic studies of Chicago.

The system underlying this model is much more complex, in that different subsystems exist, each with a radially concentric form at different hierarchical levels. These form neighborhoods, districts, communities, villages, and even small towns within bigger cities, and as the city grows and evolves, these hubs or clusters become ever more differentiated. In short, these subsystems form highly structured networks which in turn mirror a hierarchy of different functions, each serving local areas. The kinds of models that have been developed, and are still widely applied, simulate flows of people and goods between different places within the city, using analogies from gravitation that mirror the increasing deterrence effects that distance imposes on movement. The standard model divides the city into different locations (or zones) which we can label i and j, and we assume that a generic flow between these locations T_{ij} is a direct function of the size of places i, O_i and j, D_j and an inverse function of the distance or some function of spatial impedance d_{ij} between them. The typical model is

$$T_{ij} \sim O_i D_j f(d_{ij}) \tag{3.1}$$

and this is still widely applied to simulate transportation in cities, migration between cities, flows of expenditure to retail centers, and many other flow systems that define how the subsystems of the city engage with one another across many different hierarchical levels. A key element in this new science of cities is that patterns of spatial interaction also reflect underlying networks, and that the activities at different specific locations can be simulated as being proportional to the flows that emanate from all locations. From Eq. (3.1), these accumulations of flow at different locations might be predicted as proportional to the relevant activities as

$$\left.\begin{aligned} P_i &\propto \sum_j T_{ij} \sim O_i \sum_j D_j f(d_{ij}) \\ P_j &\propto \sum_i T_{ij} \sim D_j \sum_i O_i f(d_{ij}) \end{aligned}\right\}, \tag{3.2}$$

where P_i and P_j might be defined as some measure of population size at their respective locations.

The models in Eq. (3.2) are in essence measures of potential—in analogy to gravitation once again—or accessibility, and measure the relative nearness of all places to each place in question (Stewart 1947; Hansen 1959). The models developed by Jin (Chap. 8), which measure hotspots with respect to income and GDP, are in this tradition. In fact, this generic model can be made subject to constraints on locations in various ways. The usual version of the model used for transportation modeling is to make sure the trip distribution produced by the model in Eq. (3.1) meets the constraints on the size of trips generated at origins and attracted to destinations. This is the so-called doubly constrained model. If there are constraints solely on the origins or the destinations, these are singly constrained models, and it is possible to use them to predict the cumulative flow of trips at origins or destinations; in this sense, these are location models. If there are no constraints on either origins or destinations, the model in Eq. (3.1) predicts the location of activities such as the populations given by Eq. (3.2). This is the unconstrained model. This family of models and other variants was introduced by Wilson (1971) and has become the de facto standard in spatial interaction modeling.

This link between location and spatial interaction is key to the science that we are referring to. We can in fact generalize these ideas to many cities—to systems of cities as Berry (1964) first referred to them—in that although functions such as retailing specialize across a hierarchy within individual cities, this same sort of differentiation exists between cities. It was Christaller (1933) who first defined the hierarchy of cities with respect to the different functions different-sized cities have, using the idea that the bigger the city, the more specialist services it could provide—largely through its division of labor. The population would demand more specialist services in the bigger cities, and this would imply that the bigger city would need a much bigger hinterland to capture this demand than smaller cities. This would then be reflected in the area of the hinterland and thus implies a hierarchy of cities based on nested hinterlands associated with different city sizes, and a decreasing number of large cities and their hinterlands as the demand for more and more specialist functions grew. Christaller did two things with these ideas. He first demonstrated that this pattern of nested hinterlands could be observed in the relatively well-developed landscape of Bavaria, while his second contribution was to abstract these hinterlands into a regular hierarchy of hexagonal market areas that could be nested and which reflected a progression of ever fewer but bigger central places. In fact, the model is one of the cornerstones of human geography, and it is consistent with much of location theory (Isard 1956), with spatial interaction models, with network representations of cities, and with the development of urban economics (Alonso 1964).

If we order the cities in such a system by size from the largest to the smallest, we can then rank them, and when we examine this ranking, it is easy to show that these sizes follow an inverse scaling relation which is often assumed to be an inverse-power law. Of course, the frequency of cities of the same size increases with rank in this theoretical central place system based on regular-nested hexagons, but if we consider that some noise is always present in such an evolving system, then it is not difficult to imagine that we get a smoother continuum, and it is this that has been used to demonstrate a strong relationship between city size and rank. It was Zipf (1949) who first popularized this relationship, and we can give some form to this by first thinking about the size of the various neighborhoods within a single city using the model that we introduced in Eqs. (3.1) and (3.2). Let us assume that the destination activity in Eq. (3.2), that is, P_j, can be ordered from largest to smallest. Then, we can use the index $1, 2, \ldots, n$ to define these cities where $P(1)_j = P(\max)_j$ and $P(1)_j > P(2)_k > P(3)_z > \cdots$. We can dispense with the index j because we are now rank-ordering the locations with respect to size, not location. The formal relation which has been demonstrated many times in many places for locations within cities and also between cities themselves—Zipf's Law or the rank-size rule—can thus be stated as:

$$P(r) \propto 1/r^{\alpha} \tag{3.3}$$

where r is the rank of the location or city with population $P(r)$ and α is a parameter which defines the slope of the power law. In fact, the strict form of Zipf's law is where $\alpha = 1$ but most applications suggest that this parameter differs from 1. This is due to the relative stage which particular cities have reached in the evolutionary process, the fact that the distribution of cities is not in a steady state, and the fact that the spatial regions over which the relationship is defined, are not usually closed in any sense.

3.3 Urban Growth: Urbanization from the Bottom Up

The models that define the city in terms of spatial interaction are essentially static, in that they articulate the workings of the city at a cross section in time. There is little concern for process other than developing average relationships that encapsulate the entire historical development of the city at the given point in time, and there is little concern for urban growth and change. As soon as the models from social physics were applied and adapted to urban applications, there was a move to embed and extend them to deal with related dynamic processes. Some of these applications simply used the models to simulate a series of cross sections and to explore the time series that was generated, but some have been used to simulate the actual changes as increments in each time interval, which provides a more basic representations of the dynamics. However, these kinds of application do not embrace the fundamentals of urban dynamics, and other models which are essentially temporal have been adopted.

Many of these models articulate the city not as a mechanism but as an organism, evolving like a biological system rather than being manufactured like a machine. In this sense, cities are represented not as aggregates of populations but as sets of individuals—agents—that act purposively in making decisions pertaining to urban development. Thus cities develop from the bottom up rather than being organized or planned from the top down. There are many models of how city populations grow and change but in aggregate, it looks now as though world population, whose growth until quite recently appeared to be exponential or even super-exponential, is likely to become logistic with the total population stabilizing by the end of the century. This of course is one prediction too far, but it appears currently to be the most likely, and in some respects, the growth of cities is following a similar trend. Big cities are getting bigger, but they are achieving this by fusing with other cities, generating polycentric urban landscapes while still attracting population, but at a decreasing rate. Cities are thus fusing into larger urban agglomerations, but their dynamics is much more mixed than following simple exponential and capacitated-exponential curves. A number of models that illustrate chaotic patterns of urban growth have been suggested, and although none of these have been operationalized for real cities, other than as thought experiments illustrated by stylized facts, they have provided an arsenal of tools for studying nonlinear dynamical systems that underpin many of the tools and techniques presented in the rest of this book.

As cities grow in size, they change qualitatively, generating economies and diseconomies of scale that do not cancel each other out. As cities get bigger, they bring more specialized people together, and as central place theory reveals, the bigger cities are much more specialized and serve a much larger population than the smaller ones. Their economies of scale are reflected in the fact that big cities are more innovative, more creative, and consequently often more wealthy, and there is considerable evidence that as cities grow, they do indeed become more than proportionately richer, creative, and innovative. But at the same time, there are diseconomies of scale which relate to more-than-proportionately increasing levels of crime, lower incomes among the poorest, and increasing inequalities between rich and poor. These relationships are captured in the key relationship between the income of a city $Y(t)$ and its population $P(t)$ that can be written as:

$$Y(t) \sim P(t)^{\beta}, \ \beta > 1 \tag{3.4}$$

where β is a measure of the economies of scale. If $\beta < 1$, then the model in Eq. (3.4) illustrates that income increases less than proportionately with population size. This in fact is unlikely, but if we were to break the population down into different groups, then the poorest group would have to get more than proportionately much greater when cities increase in size for the relationships in Eq. (3.4) to hold. This sort of model was originally developed to look at growth in biological systems, but it presents a good analog of economies of scale, and has been widely applied to examples of ancient and modern city systems as well as firms, individual incomes, and a host of related socio-economic phenomena (West 2017).

In fact, this allometric model has not been developed temporally for individual cities or sets of cities, and there is considerable debate about the effect of scale economies, as the underlying processes which lead to this are defined away by such models; as such they remain implicit in these formulations. In fact, there is still a dearth of dynamic models that represent the way cities evolve, although with the development of complexity theory, there are several key dimensions to the way we now characterize these dynamics. There are no well-worked-out dynamics that coincide with the processes that determine how cities grow and evolve, and this is as much because there are very few good, robust theories that we have been able to discover to date. This is also because of our inability to observe such processes at first hand and compile good data. Urban systems like many social systems are highly resistant to detailed observation and show a degree of invisibility that is much more problematic than in many physical systems where we are able to instrument most features of any relevance.

Complexity theory does, however, reveal certain features of cities that define the limits to our existing models. Cities are always in disequilibrium and this is the new normal, as if it was anything other than that hitherto. In fact, cities are far from equilibrium, in that equilibrium is an abstract concept that in some models represents a long-term steady state, but in most models cannot be defined and probably does not exist. As cities grow from the bottom up, patterns emerge at higher levels. Although there are features of self-similarity at these different levels that we can grasp and sometimes articulate in terms of fractal phenomena, it is often difficult to tie the patterns that we see in cities at different levels to specific bottom-up processes. In this sense, history is all important as we perceive an average randomness in how decisions about urban development are made at the lowest levels. Decisions are for the most part rational if they are unpacked to the level at which they become understandable, but the physical limits of the city and the way we interact socially are such that these constrain what is possible and enable the emergence of order at all levels. In this sense, history matters just as much as geography does. As we implied above, our models and theories need to rapidly reflect the fact that the systems we are dealing vary in space and time. Our abilities to improve the quality of life in cities must take account of such variations which of course reflect underlying human behaviors. In short, in any complex system, there is a degree of historical path dependence that reflects the fact that decisions, although rational, are not necessarily ordered in any obvious way.

There are some processes that are now quite well defined such as those that reveal remarkably clear organization based on decisions that are initially random. For example, the model of segregation first developed by Schelling (1978) demonstrates that if a population system composed of agents are initially randomly distributed, but these agents have distinct preferences to always live with as many of their own kind around them, then if agents begin to move when this is not the case, very quickly an extreme pattern of segregation can evolve. The degree of extremeness—like ghettoization or gentrification in modern cities—appears to be entirely unwarranted, given that the agents have a very mild preference to live side by side with those of their own kind (being quite content to have an equal number of their own kind as well as an equal number of other kinds around them). The reason for this segregation, then, is that there is no coordination at the micro-level. Individuals move of their own accord when they see those around them dominating the neighborhood. It is processes like these that we need to identify in cities because part of our quest to make cities less polarized, more efficient, and to increase the quality of life, are closely bound up with this kind of decision making.

All issues pertaining to complexity influence our current thinking about cities (Batty 2005), but the theories we have about how the city system functions are still quite rudimentary. Many of the models we have hinted at so far are being developed for individual sectors and distinct dynamic processes, and many are being adapted to deal with short- as well as long-term change in the high- as well as the low-frequency city. For example, in this book, there are several chapters that deal with mobility and new data sets that pertain to networks and flows, and the models in this chapter are reflected in these. To an extent, urban informatics is much more about tools, techniques, and models than about theories, although theory is essential to constructing the bigger picture of how this domain can improve our understanding, prediction, and design of future cities. In the next section, we will pull the ideas of the previous two sections together, emphasizing how these models can be consistently linked in terms of what we know about scale and size, networks, and flows.

3.4 Scale and Size, Networks, and Flows

To all intents and purposes, by the end of the century, everyone will be living in cities of one size or another, where the distribution of sizes will follow the rank-size rule. The biggest cities will be up to 100 million in population, but all of these will be urban agglomerations that consist of polycentric hierarchies of smaller cities, towns, and villages that have fused together. But as we have shown in the previous two sections, the size of a city can also be measured with respect to its local morphology, its geometry, and the distances that define the bounds over which people will interact intensively to enact the business of the city. Since the industrial revolution and the invention of new technologies for mobility and interaction, all cities are part of a global urban form where distances, travel costs, travel times, and like measures of impedance condition the interactions and networks that bind all cities together. In short, we can no longer think of cities as being freestanding entities; they are now networked in ways that make it ever more difficult to disentangle them from one another.

The ideas that we have introduced all pertain to different levels of size and scale. A metropolitan area for example has a certain population size, a density which is some measure of size with respect to unit area, and various distances from its core to its boundary. There is a common force which relates scale to size, and this is referred to in statistical physics as scaling. In essence, it means that as a city grows in size, density, in the length of its perimeter, and in the distances travelled within it, we can identify a common scaling that enables us to represent these various properties with respect to size. As we change their size, then the quantities involve scale in a relatively simple way. We can demonstrate this quite easily with respect to the various models that we have introduced. Starting with the standard spatial interaction model in Eq. (3.1), we can now write it in more specific terms using the inverse-power function of distance as follows:

$$T_{ij} \sim O_i D_j d_{ij}^{-\gamma} \tag{3.5}$$

If we increase the scale of the city by a factor λ, which to fix ideas, we might consider being equal to 2, this will change the model to:

$$T_{ij} \sim \lambda^{-\gamma} T_{ij} \sim \lambda^{-\gamma} O_i D_j d_{ij}^{-\gamma} = O_i D_j (\lambda d_{ij})^{-\gamma} \tag{3.6}$$

We have doubled the distance, but the number of trips has not halved, for the nonlinearity applied in the model reduces the number of trips by the factor $\lambda^{-\gamma}$. If we define an inverse square law of distance $\gamma = 2$, then the number of trips reduces by a factor of 4. In the same way, if our model incorporated economies of scale ϑ and μ which we apply to the origin and destination attractors as

$$T_{ij} \sim O_i^{\vartheta} D_j^{\mu} d_{ij}^{-\gamma} \tag{3.7}$$

and if we scale these attractors by $(\xi O_i)^{\vartheta}$ and $\left(\varpi D_j\right)^{\mu}$, then we can easily show that the trips also scale in a nonlinear way, but remain proportionate to the existing flows.

When we look at the distribution of population sizes and any of the cumulative flows that can be predicted from the model in Eqs. (3.5) or (3.6), we have also noted in Eq. (3.3) that these follow an inverse-power law in the form of the rank-size rule. If we scale the rank of the cities by a rate α, then the rank-size relation becomes:

$$\lambda^{-\alpha} P(r) \sim (\lambda r)^{-\alpha} = \lambda^{-\alpha}(r^{-\alpha}) \sim P(r) \tag{3.8}$$

The same kind of self-similar scaling is evident in any power-law relationship such as the urban allometric relationship in Eq. (3.4). If the population in all cities grows by a factor λ, then

$$\lambda^{\beta} Y(t) \sim (\lambda P(t))^{\beta} = \lambda^{\beta} P(t)^{\beta} \sim Y(t) \tag{3.9}$$

It is also worth noting that several key relationships which emerge from urban economics, such as the relationship between the density of population, rents charged, and indeed income itself, vary with respect to distance in the city. The long-standing observation that densities and rents decline inversely with distance from the core of the city has been widely modeled using inverse relationships as either a negative exponential or a power law. The density ρ_i (population P_i divided by area A_i) defined as

$$\rho_i = {P_i}/{A_i} \sim \exp(-\varphi d_i) \quad \text{or} \quad \rho_i = {P_i}/{A_i} \sim d_i^{-\varphi} \tag{3.10}$$

is also scaling, as a simple change in the scale of distance in either of these relationships in Eq. (3.10) would show. These relationships indicate that as size increases in cities, quantities such as income, the numbers of trips, etc., increase or decrease more or less than proportionately, and this indicates that as cities grow or decline, there are qualitative changes that are likely to change the kinds of informatics that are appropriate. This is certainly true of issues concerning economic development, the provision of transportation, and the ability of the city to generate wealth, innovations, and new industries (Bettencourt 2021).

In some senses, what we know about the pattern of locations and interactions in cities is reflected in the underlying networks that support them. There are a multitude of such networks, other than the most obvious and visible systems that transport people and goods using different technologies or modes, but many are hard to observe and measure, particularly those that involve information, such as email, Web access, social media, even telephone, television, and countless other media. All of these networks have scaling properties that suggest that the distribution of their hubs in terms of their indegrees and outdegrees—the number of links that enter or leave the hubs or nodes defining these networks—follow rank-size distributions, and the number of clusters in such networks by size also follow similar inverse-power laws (Barabási 2018). In many of the chapters in this book that deal with mobility, networks form the basis of the various simulations, and the properties introduced here are key to the way such flows are measured and modeled.

3.5 The Development of Operational Urban Models

The theories and models that we have introduced form many of the elements of more comprehensive urban models that deal with various sectors of the urban system. Most models developed so far tend to be those that deal with the low-frequency city, but some of these tools, particularly those dealing with flows and networks which involve transportation, are being developed to deal with movements over short periods of time, focusing on real-time movements, usually on a daily basis. There are at least four classes of model that we can define as the pillars of urban science with respect to urban informatics: first, those that depend on aggregate populations and activities which we call land-use transportation interaction (LUTI models), physical urban-development models using cellular automata (CA models), agent-based models that deal with disaggregate populations of individuals moving and making decisions through time (ABM models), and dynamic models that deal with individual decision-making, focusing largely on mobility and geodemographics such as microsimulation models (Chap. 44).

The generic spatial interaction model in Eq. (3.1) and its derivatives, such as accessibility potentials in Eq. (3.2), lie at the heart of many land-use transportation models that essentially stitch together several such models to replicate the locations and interactions between many population and employment sectors of the urban system. These models were first developed as pure transportation models and then extended to deal with land uses and activities in the 1960s. The problems they encountered were due to limits on computation which have now largely disappeared, but more important were the limitations of good theory and of course data. Data still remain an enormous problem, for data on spatial movements have always been hard to get, notwithstanding new sources from real-time capture on mobile devices. The fact that such models and their variants only simulate the city at a cross section in time spurred the development of more dynamic urban models, and in the later years of the last century, models based not on simulating the dynamics of population and employment location but on urban land use more generally at the physical level were developed. These models were largely based on cellular automata whose roots lie in complexity theory and in physical diffusion processes (such as forest fires). Because they focus literally on the physical development of land-use change, they are not easily linked to the numerical characterization of the city in terms of population, employment, income, and related properties. As such, rather than providing operational applications, CA processes as articulated in this genus of model find their use in more specific processes such as traffic simulation at the level of detailed flows.

In the quest for better representations, much more disaggregate models are being built using two different but complementary approaches: agent-based modeling and microsimulation. In terms of ABM models, urban models formulated in this way at the operational level are highly detailed with large data requirements on the behaviors of individual decision makers, usually households and firms, but most suffer from difficulties over developing good theory for the key urban dynamics processes at work in cities. As such, many models tend to be pilots and demonstrations, prototypes used to illustrate what is possible, and very few reach the level of full operationality. UrbanSim and PECAS are exceptions. The fourth class of model based on microsimulation uses techniques based on constructing synthetic populations which are more tolerant of the lack of data pertaining to individual behaviors. Such simulations reflect probability distributions pertaining to the attributes of individuals in a population, and such profiles are used to construct synthetic estimates of populations according to a series of conditional probabilities. There are two subtypes of model, the first being traditional microsimulation models reflecting population profiles in terms of geodemographics. The second set are rather different in that these have been quite widely developed for transportation modeling. These are loosely referred to as activity models, where households generate decisions about trip-making over the course a day, and the probabilities associated with such decision making translate into trip patterns at a very detailed level, such that these are much more powerful than detailed traffic-flow models. MATSIM is one of the best-known such models, although others such as SimMobility, SimAgent, and so on have been developed. All of these models derive from TRANSIMS, the original Los Alamos microsimulation of traffic flow. There are a number of reviews of all these models, and the reader is referred to Batty (2008), Wegener (2014), and Moeckel et al. (2018) for definitions, theoretical expositions, and applications.

In the rest of this book, these dimensions of urban science map out into many areas of urban informatics, and it is worth noting some of the key chapters that relate to this science before we conclude. In terms of modeling, all four of the areas that we have just defined are covered in detail in the chapters at the end of the book, in Part 5 where Eric Miller deals with transportation modeling (Chap. 47), Anthony Yeh with CA modeling (Chap. 45), Andrew Crooks and his co-authors (Chap. 46) with agent-based modeling, and Mark Birkin (Chap. 44) with microsimulation. Mobility of course runs through all these themes and is dealt with from different perspectives in several parts of the book, particularly by Shih-Lung Shaw (Chap. 5) and Martin Raubal and his co-authors (Chap. 6) in Part 1, by Marta Gonzalez et al. (Chap. 11) linking mobility to urban science in Part 2, Chiang Kai-Wei et al. (Chap. 25) explaining developments in mobile mapping in Part 3, methods for spatial search by Liping Di and Eugent Yu (Chap. 37) in Part 4, and with respect to the visualization of movement data by Gennady Andrienko et al. (Chap. 40) in Part 5. Sybil Derrible et al. (Chap. 7) and Budhendra Bhaduri et al. (Chap. 18) examine energy and infrastructure in their contributions in Parts 1 and 2, respectively. In terms of an overview, urban informatics is such a broad area that many of the authors here develop the big picture from their own perspectives. But in particular, Helen Couclelis (Chap. 9) sets all this in context of the smart city in Part 1, and Michael Goodchild provides the wider perspective for how this whole area of urban informatics is addressing questions of new and big data and geographic information science in Part 6.

3.6 Future Directions in Urban Informatics

There are many aspects of urban systems which we have not addressed in this brief review of what constitutes urban science. There is a general question as to how the tools and techniques of urban informatics apply to different types and sizes of cities in different cultures and societies. Much of urban studies is focused on such comparative analysis from the point of view of social and economic differences, and there are implications for the use of urban informatics in different sizes of city with different social cultures, political regimes, and governance. In particular, the distinction between the Global North and Global South is important, and there are already attempts at extending the ideas of city science to these domains, as in the reports from Acuto et al. (2018) and Lobo et al. (2020). Urban science deals with how we define cities in terms of their spatial scale and their boundaries, and in this sense, the size of the city is all important with respect to the kinds of models and techniques that spin off from the ideas introduced in this chapter and elaborated in the rest of this book.

The theories that we have hinted at in this introductory chapter are by no means complete and never will be. Cities are driven by individuals, and complexity theory tells us that they grow and evolve from the bottom up. If there is a hidden hand in this process, it is in the fact that we appear to be able to produce quite ordered structures from our actions that in many respects are quite independent of each other. How we intervene in such complex systems is highly problematic, and urban informatics is in the front line of how we move toward a planning system that is effective in developing more sustainable, equitable, and efficient cities. This book introduces a very wide range of tools that can be used at many points in the planning and policy process, and a major focus needs to be on developing models and techniques that are able to adapt to new changes that continue to beset cities, as well as new technologies that are being introduced ever more rapidly.

References

Acuto M, Parnell S, Seto KC (eds) (2018) Science and the future of cities. Report of the international expert panel on science and the future of cities. London and Melbourne. https://www.nature.com/documents/Science_and_the_future_of_cites.pdf

Alonso W (1964) Location and land use: toward a general theory of land rent. Harvard University Press, Cambridge, MA

Barabási AL (2018) Network science. Cambridge University Press, Cambridge, UK

Batty M (2005) Cities and complexity: understanding cities with cellular automata, agent-based models, and fractals. The MIT Press, Cambridge, MA

Batty M (2008) Fifty years of urban modelling: macro statics to micro dynamics. In: Albeverio S, Andrey D, Giordano P, Vancheri A (eds) The dynamics of complex urban systems: an interdisciplinary approach. Physica-Verlag, Heidelberg, pp 1–20

Batty M (2019) On the confusion of terminologies. Environ Plan B, Urban Anal City Sci 46:997–998

Berry BJL (1964) Cities as systems within systems of cities. Pap Reg Sci 13:147–163

Bettencourt L (2021) Introduction to urban science. The MIT Press, Cambridge, MA

Christaller W (1933, 1966) Die zentralen orte in Süddeutschland. Gustav Fischer Verlag, Jena. Translated in part by Baskin CW. Central places in southern Germany. Prentice Hall, Englewood Cliffs

Hansen WG (1959) How accessibility shapes land use. J Am Inst Plan 25:73–76

Isard W (1956) Location and the space-economy; a general theory relating to industrial location, market areas, land use, trade, and urban structure. The MIT Press, Cambridge, MA

Lobo J, Alberti M, Allen-Dumas M, Arcaute E, Barthelemy M, Bojorquez Tapia LA, Brail S, Bettencourt L, Beukes A, Chen WQ, Florida R (2020) Urban science: integrated theory from the first cities to sustainable metropolises. Report to the National Science Foundation. Santa Fe Institute, Santa Fe, NM

Moeckel R, Llorca Garcia CM, Chou AT, Okrah MB (2018) Trends in integrated land use/transport modeling: an evaluation of the state of the art. J Transp Land Use 11. https://doi.org/10.5198/jtlu.2018.1205

Park R, Burgess E (1925) The city. University of Chicago Press, Chicago, IL

Schelling TC (1978) Micromotives and macrobehavior. W W Norton Company, New York, NY
Stewart JQ (1947) Empirical mathematical rules concerning the distribution and equilibrium of population. Geogr Rev 37:461–485
Wegener M (2014) Land-use transport interaction models. In: Fischer MM, Nijkamp P (eds) Handbook of regional science. Springer, New York, London, pp 741–758
West G (2017) Scale: the universal laws of life and death in organisms, cities and companies. Weidenfeld and Nicolson, London, UK
Wilson AG (1971) A family of spatial interaction models, and associated developments. Environ Plan 3:1–32
Zipf GK (1949) Human behavior and the principle of least effort. Addison-Wesley, Cambridge, MA

Michael Batty is Bartlett Professor of Planning and Chairman of the Centre for Advanced Spatial Analysis at University College London. He is also a Distinguished Chair Professor at The Hong Kong Polytechnic University. He is a Fellow of the Royal Society and the British Academy.

Chapter 4
Street View Imaging for Automated Assessments of Urban Infrastructure and Services

Daniel Zünd and Luís M. A. Bettencourt

Abstract Many forms of ambient data in cities are starting to become available that allows tracking of short-term urban operations, such as traffic management, trash collections, inspections, or non-emergency maintenance requests. However, arguably the greatest promise of urban analytics is to set up measurable objectives and track progress toward systemic development goals connected to human development and sustainability over the longer term. The challenge for such an approach is the connection between new technological capabilities, such as sensing and machine learning and local knowledge, and operations of residents and city governments. Here, we describe an emerging project for the long-term monitoring of sustainable development in fast-growing towns in the Galapagos Islands through the convergence of these methods. We demonstrate how collaborative mapping and the capture of 360-degree street views can produce a general basis for a broad set of quantitative analytics, when such actions are coupled to mapping and deep-learning characterizations of urban environments. We map and assess the precision of urban assets via automatic object classification and characterize their abundance and spatial heterogeneity. We also discuss how these methods, as they continue to improve, can provide the means to perform an ambient census of urban assets (buildings, vehicles, services) and environmental conditions.

4.1 Introduction

Many forms of ambient data in cities are starting to allow tracking of short-term operations and services (Park et al. 2014; Townsend 2015). Uses of these technologies range from facilitating traffic management to air quality control, or the management of non-emergency requests (Park et al. 2014; O'Brien 2015). However, arguably one

D. Zünd (✉) · L. M. A. Bettencourt
Mansueto Institute for Urban Innovation and Ecology and Evolution, The University of Chicago, Chicago, USA
e-mail: dzuend@uchicago.edu

L. M. A. Bettencourt
e-mail: bettencourt@uchicago.edu

W. Shi et al. (eds.), *Urban Informatics*, The Urban Book Series,
https://doi.org/10.1007/978-981-15-8983-6_4

of the greatest promises of urban analytics is to set up measurable objectives and track progress toward systemic development goals connected to human development and sustainability over the longer term (Brelsford et al. 2017). A main challenge to achieving long-term monitoring of processes in urban settings is the convergence of new technology, local knowledge, and the operations of residents and local governance. Whereas these objectives already constitute challenges for developed cities, they are even more daunting in developing country settings (Praharaj et al. 2017). In rapidly developing cities, data are often far less abundant or even non-existent. Additionally, urban environments often change at a much faster pace and in informal ways (Sarin 2016).This makes it much more difficult to track change, and specifically, to generate statistical progress in development trajectories toward sustainable development goals (Randhawa and Kuma 2015; Komninos 2015).

A good case study to research the potential of new technology in semi-informal settings, and the impact it has on managing and tracking the progress of long-term goals, are the Galapagos Islands. The archipelago, famous for its unique ecosystems, lies about 1000 km off the Pacific coast of Ecuador (the blue square in Fig. 4.1). Though most of the islands remain a natural reserve, the human presence on land and sea is growing very quickly, with four fast-growing towns concentrating most of the immigrant human population. The remote location and the unique coupled urban–natural system of these islands constitute a particularly interesting and poignant setting to study the development trajectories of urbanization (Batty et al. 2019).

Fig. 4.1 The Galapagos Islands are an archipelago in the midst of the Pacific Ocean (blue square). Their secluded location, fast-growing towns, and unique ecosystems offer a particularly interesting and poignant setting for developing models of sustainable development for coupled urban–natural systems. The manageable size of these urban areas makes it possible to study novel methods of collaborative data collection and the convergence of new technology and local knowledge. We exemplify the method on the capital of the islands, Puerto Baquerizo Moreno on San Cristóbal, depicted in the inset. Map designs are from Mapillary (2019) and OpenStreetMap (2019)

From a modeling perspective, the islands provide a unique setting due to their remote location, and the fact that all materials and goods in and out the system are registered upon arrival or departure, just as are people's migration (Bettencourt 2019), provides a good basis for assessing the impact of the island system on its external environment and vice versa.

Together with the emergence of a plan to harmonize tourism with sustainable stewardship of the local charismatic ecosystem (Rousseaud et al. 2017), the towns in the Galapagos Islands provide a unique chance to study novel approaches to urban planning, urban management of resource flows, and tracking of development toward sustainability goals (Batty et al. 2019).

We will focus in this study on the second largest town in the Galapagos, Puerto Baquerizo Moreno, which is also the regional capital and has a population of about eight thousand residents (Andrade and Ferri 2019). The town is located on the eastern part of the Archipelago, on the island of San Cristóbal, as depicted in Fig. 4.1. In terms of materials, the island is relatively independent of the other islands in the archipelago since it has its own harbor and airport that directly connect it to continental Ecuador where most people, construction materials, energy, and consumer goods originate.

Historically, the island of San Cristóbal has not been the archipelago's main tourist hotspot. However, since the airport opened in 1986, the island is increasingly attractive to a growing number of tourists—as can be seen by the number of arrivals at the airport—which shows a higher growth rate than the total growth rate of tourist arrivals across the Galapagos Islands (Izurieta 2017). The annual increase of 3.72% in tourism (about 225 thousand visitors in 2015; Izurieta 2017) creates a growing economy on the islands, but also places pressure on the urban–natural interfaces of the islands. These pressures and possible solutions remain hard to track in detail, therefore precluding a balanced path where economic opportunities may be expanded, while ecosystems in the islands are protected.

Thus, innovative approaches that track the growth and effects of urbanization on the islands are becoming paramount. Here, we exemplify how collaborative data collection and new imaging and artificial intelligence technology can support this process in the context of an emerging project for long-term sustainable development of the Galapagos Islands.

4.2 Data Collection and Object Localization

The rapid development of computer vision and object recognition has opened up efficient ways to process large image datasets (Chen et al. 2016). For urban science and policy, these capabilities have great potential to follow the trajectory of the built infrastructure and to assess the heterogeneity of urban assets and services, including the consumption of energy and materials. However, data about these issues are often lacking, outdated, or too coarse in many developing urban areas. This is even more so the case for remote locations, such as the towns in the Galapagos Islands and specifically, the town of Puerto Baquerizo Moreno. Before we started the project of monitoring the town's built environment, very few data were available online (about a dozen images) of which only a few depicted the island's urban areas.

Monitoring the urban development, however, asks for data that capture the urban fabric as a whole and over time. In the following, we introduce a method that makes it possible to document the whole town within only a few days' work and with only minimal initial investments, thus making collaborative data collection possible. The data pipeline consists of three main steps, of which two are fully automated. The first involves capturing street-level photographs, and the second analyzes single images in order to recognize and segment objects, as depicted on the right panel of Fig. 4.2. The third step consists of identifying the same object in different images and geolocating its position in space and time.

The most time-consuming step is the collection of enough imagery to cover the whole town. The process is entirely parallelizable and can involve a group of people or vehicles. There must be enough overlap in the images so that the geolocation of objects is possible and thus becomes unambiguous. Figure 4.3 depicts an example where a store sign was recognized in six different images.

In this study, we used a 360-degree action camera able to automatically take images with a chosen temporal frequency. The camera is capable of taking images

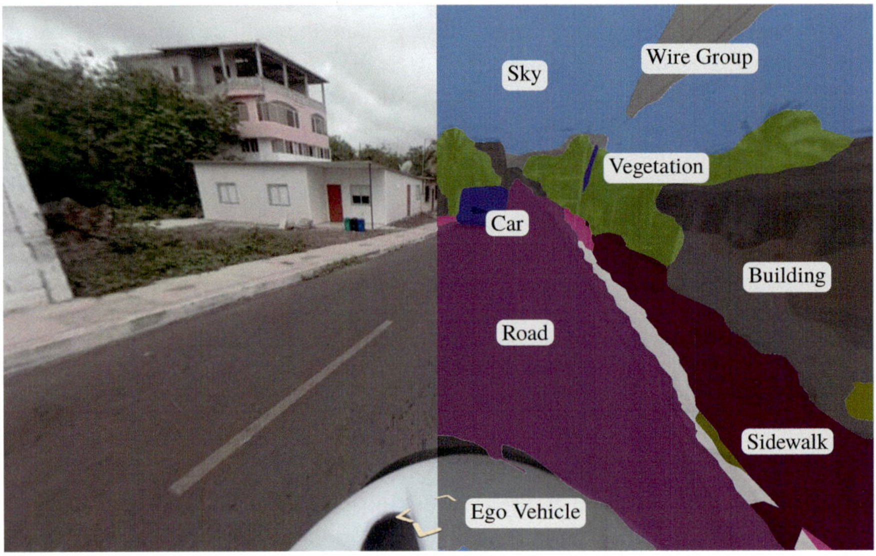

Fig. 4.2 Street-level imagery can be captured with relatively simple tools. For this study, we collected data by attaching a 360-degree sports camera on a helmet and rode a bicycle through the town. The imagery is available through Mapillary's (2019) user interface, as depicted on the left panel. The right panel shows processed and segmented imagery. The automatic object classification identifies structures and objects out of almost three-dozen categories. However, on the island, the algorithms sometimes fail to properly identify certain objects. For example, the sidewalk on the right is classified as ground. Nevertheless, the methods provide a powerful tool to assess urban features in developing towns experiencing rapid change

Fig. 4.3 The imagery covers most of the accessible street network of Puerto Baquerizo Moreno on San Cristóbal, Galapagos. The green dots show the locations of all 360-degree imagery produced by us. When a series of images are available along a street, objects can be identified and geolocated. The inset depicts a situation in which the same store sign is recognized in six different images in the right inset panel, taken from slightly different locations, of which three are shown in the left inset panel. Map designs are from Mapillary (2019) and OpenStreetMap (2019)

that cover the whole surrounding from the current location which, with some post-processing, produced globes at each location. We attached the camera to a helmet and drove around the town with it. Since the camera also added the GPS coordinates to each image's metadata, we were able to cover about 75 km of geotagged image globes within only a couple of days. The collected imagery accounts for more than 10,000 images, of which many overlap and provide a good dataset for the next steps in the data pipeline. Each location of a 360-degree image is depicted by a trace of green dots in Fig. 4.3.

We executed steps two and three in collaboration with Mapillary (2019), a technology company dedicated to creating crowdsourced street view maps. Mapillary provides an engine that automatically processes uploaded images, including a user interface to walk from one image to the next and, thus, ultimately throughout the entire city. The left side of Fig. 4.2 depicts the interface that is accessible to the public. The images are further processed using computer vision and object recognition algorithms, of which many have been developed and optimized by the Mapillary research teams (Bulo and Kontschieder 2016; Bulo et al. 2017; Cariucci et al. 2017; Neuhold et al. 2017). The algorithms segment the images and add semantic information to different parts of the visual field.

The field of computer vision and object recognition has made significant strides in recent years by using deep-learning algorithms to perform image segmentation (Krylov et al. 2018). However, these techniques are not yet perfect and the resulting semantic information extracted from images is often only an approximation to reality. For street-level data, this is especially the case for areas that differ from the data that were used to train the object recognition classifier. Nevertheless, the algorithms are able to recognize core properties in the imagery, as depicted in the right inset panel of Fig. 4.2.

When the same object is recognized in several images, it can be geolocated uniquely in space. Figure 4.3 shows an example where a single store sign is recognized in six different images located in the right inset, three of which are shown in the left inset panel. The task of geolocating objects from different images at street level involves several major technical challenges. Besides aggregating the same object present in several images, the main challenge in processing crowdsourced street-level data is the varying qualities of the imagery, such as blurring or restricted field of view, and variability in camera positions. The latter is important, since high-quality geolocation depends on the camera position relative to the object in the field of view for accurate triangulation and location (Krylov and Dahyot 2018).

Despite these challenges, the engine was able to geolocate almost 12,000 objects in the small town of Puerto Baquerizo Moreno, including 777 trash cans, 343 store signs, 412 advertisement signs, and 224 driveways. These are the classes of objects that we use in the next section to derive the functions of certain parts of the town and to exemplify the conclusions that can be drawn from these methods, as they continue to improve.

4.3 Deriving Urban Functions from Object Statistics

The collection of data and the identification and localization of objects in space provides a basic functional mapping of an urban area. The spatial distribution of different classes of objects makes it possible to study the location and functions of different districts. For example, the density distribution of store signs in Fig. 4.4b shows the areas in Puerto Baquerizo Moreno that provides a range of specific services, typically associated with tourism (Andrade and Ferri 2019).

Figure 4.4 shows two object–class density distributions that are good indicators of residential areas: the distributions of trash cans and driveways (subfigures (a) and (c)). Trash cans in residential areas of Puerto Baquerizo Moreno are standardized vessels with a unique shape and color combination. Each household is required to have their trash cans outside of the building, close to the street for easy access for trash collectors. They additionally serve as public trash bins. The trash bins in tourist areas are different, not as prominently placed, and often obfuscated. The segmentation engine has problems identifying them as such, but this is also a clear sign of a different look and function and of an intentional effort to deal with the issue differently. The waterfront area with the most tourist services is much denser than

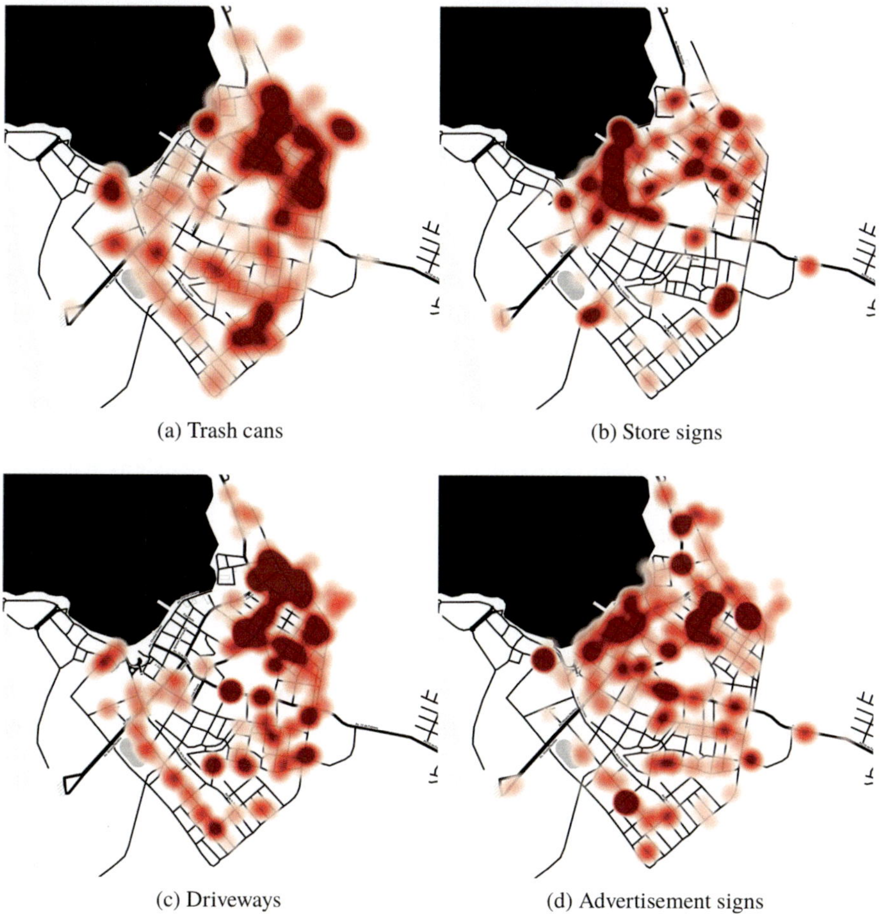

Fig. 4.4 Geolocated objects help to identify and locate different properties of the town. The figures depict the distribution of **a** trash cans, **b** store signs, **c** driveways, and **d** advertisement signs. The distribution of the trash cans shows the importance of local knowledge. The ones identified by the segmentation are private trash cans, whereas the public ones are not recognized and are largely in the business parts of town, close to the sea and indicated by a high volume of shop signs in **b**. The driveways in **c** indicate a lower density of houses in those areas, since they are set back from the street. The advertisement signs in **d** have a similar pattern as the store signs in **b**, but are more uniformly distributed, mainly along principal roads. Map designs are from Stamen Design (2019)

the rest of the town. The buildings are often located next to the street and not set back. This is indicated by the abundance of driveways in the residential area in the northeast and their absence in the denser locations, such as the area central of the town toward the sea. Figure 4.4c depicts this clearly.

The last indicator we want to point out in this study is the distribution of advertisement signs. Their spatial distribution is depicted in Fig. 4.4d. According to the density distributions of advertisement signs, there are three main patterns specific to places with a large accumulation of advertising signs. The first pattern is where most tourists spend their time within the town and also where most restaurants and tourist services are located, corresponding to the highest density of store signs in Fig. 4.4b.

The second area with a high density of advertisements consists of the main thoroughfares that cut through the town from east to west, each a one-way street. Within the town, these are the streets where most shops frequented by locals are located. The main street also connects further to the only other settlement on the island and is the only street that cuts through the San Cristóbal from east to west. This road constitutes the main axis in the town, together with the street that is orthogonal to it and starts at the airport on the left of the map. However, these signals are not as clear as for other indicators.

The third cluster, the one with the highest density of advertising signs according to the data, is located at the international convention center close to the center top of the image. This cluster has to be regarded with care, because many of our data collection trips started here, so that the region is oversampled in terms of imagery. The data-processing engine has some difficulties to cope with this sampling effect, separates advertisement signs that are the same, and geolocates them in very similar locations.

The above interpretations of the different density distributions in Fig. 4.4 are clearly highly reliant on local knowledge. For example, the unique form and shape of the private trash cans are not a general pattern across different urban systems, but a very local feature. There would not have been an obvious conclusion from the extracted data without knowledge of local choices, habits, and rules.

4.4 Discussion

Recent technological advancements are paving the way to novel ways of monitoring, studying, and assessing characteristics and change in urban environments that are closer to the human experience. Our present study shows how collecting street view imagery and identifying and locating associated functional objects require little initial investment. These methods are also suitable for collaborative approaches involving both image collection and interpretation of resulting spatial statistics. Thus, this type of result demonstrates that concepts of smart cities and the collection of extensive and detailed ambient urban data are no longer restricted to large investments and efforts by large corporations or universities, but are also feasible in developing towns by relatively small numbers of people.

It is desirable that local citizens take a greater part in this type of process for a number of different reasons. First, on purely technical grounds, an ongoing data collection effort helps improve the system's evidence pool in terms of coverage and accuracy of object identification statistics. Second, local knowledge is critical for good urban planning and policy, and there have been thus far few systematic strategies that combine data and technology with people's local experiences. Third, and most important, data collections by corporations and governments rarely speak to the perspective and priorities of local communities, who, in the case of sustainable development, have a clear stake in the future of their environment and can act as the best stewards of its well-being (Burke et al. 2006). Fourth, the use of methods such as the ones discussed here provides a number of interesting educational and training opportunities that can contribute to the growth of local human capital and may have spillovers to other innovative local practices.

There are still a number of technical obstacles for turning the pilot described here into an effective system that can speak to these objectives. Object recognition in images of developing cities is far from working perfectly. This is likely due to biases in training of the artificial intelligence algorithms with imagery from more formal environments, such as cities of the Global North. As a result, the present algorithms often fail to extract all semantic information from the images in the Galapagos and thus fail to achieve high levels of accuracy in object recognition and segmentation. Nevertheless, the methods already offer powerful tools in their current state, so that we can reasonably expect that they will improve in the near future as more evidence from informal and variable environments becomes part of training corpora.

Aspects of algorithms that need improvement are likely related to increased knowledge of geographic and cultural contexts. We have seen for example that the recognition of sidewalks remains difficult as these rather irregular spaces are often classified as parts of the streets or simply as ground. Another example is the classification of beaches. In the data, we collected on the Galapagos Islands, sand beaches are often classified as snow. Simple contextual clues would certainly improve this type of classification.

Nevertheless, the methodology provides initial stages of potentially powerful artificial intelligence tools to assess the assets of cities and towns and to study the development trajectory of urban microenvironments. This will become even more powerful in the future, as the algorithms become capable of more fine-grained object classification and segmentation in a ways that can track, for example, construction processes and the materials and costs involved.

A big impact in future studies of urban areas will arise from extracting three-dimensional (3D) city models (Schläpfer et al. 2015) from the type of imagery produced and analyzed in this study. In combination with more traditional aerial and remote sensing (Qin and Fang 2014; Weng et al. 2018) and citizen engagement, high-quality 3D models of whole towns and cities are just now becoming accessible also in fast-changing settings in the developing world (see also Chap. 34). The simplicity and generalizability of data collection demonstrated here provide a way to easily and quickly track these development trajectories in ways that are closer to the experience of individuals and households living and working in these environments, and at the same time allow us to characterize material and information flows through these systems across scales.

References

Andrade JEL, Ferri DQ (2019) The Galapagos urban context. In: Kvan T, Karakiewicz J (eds) Urban Galapagos: transition to sustainability in complex adaptive systems. Springer, Berlin, pp 9–22

Batty M, Bettencourt LMA, Kirley M (2019) Understanding coupled urban natural dynamics as the key to sustainability: the example of the Galapagos. In: Kvan T, Karakiewicz J (eds) Urban Galapagos: transition to sustainability in complex adaptive systems. Springer, Berlin, pp 23–41

Bettencourt LMA (2019) Critical paths to sustainability: the research challenge from island urban systems. In: Kvan T, Karakiewicz J (eds) Urban Galapagos: transition to sustainability in complex adaptive systems. Springer, Berlin, pp 155–160

Brelsford C, Lobo J, Hand J, Bettencourt LM (2017) Heterogeneity and scale of sustainable development in cities. Proc Natl Acad Sci 114(34):8963–8968

Bulo SR, Kontschieder P (2016) Online learning with Bayesian classification trees. In: Proceedings of the IEEE conference on computer vision and pattern recognition, pp 3985–3993

Bulo SR, Neuhold G, Kontschieder P (2017) Loss max-pooling for semantic image segmentation. In: 2017 IEEE conference on computer vision and pattern recognition (CVPR), pp 7082–7091

Burke JA, Estrin D, Hansen M, Parker AP, Ramanathan N, Reddy S, Srivastava MB (2006) Participatory sensing. Research Gate. https://www.researchgate.net/publication/228613934_Participatory_Sensing/link/544e3db20cf29473161a67f4/download

Cariucci FM, Porzi L, Caputo B, Ricci E, Bulò SR (2017) Autodial: automatic domain alignment layers. In: 2017 IEEE international conference on computer vision (ICCV), pp 5077–5085

Chen LC, Yang Y, Wang J, Xu W, Yuille AL (2016) Attention to scale: scale-aware semantic image segmentation. In: Proceedings of the IEEE conference on computer vision and pattern recognition, pp 3640–3649

Izurieta JC (2017) Behavior and trends in tourism in Galapagos between 2007 and 2015. In: Galapagos report 2015–2016. GNPD, GCREG, CDF and GC, 2017

Komninos K (2015) Intelligent cities: variable geometries of spatial intelligence. In: From intelligent to smart cities. Routledge, London, pp 46–62

Krylov VA, Dahyot R (2018) Object geolocation from crowdsourced street level imagery. In: European conference on machine learning (ECML) workshops

Krylov V, Kenny K, Dahyot R (2018) Automatic discovery and geotagging of objects from street view imagery. Remote Sens 10(5):661

Mapillary (2019) www.mapillary.com. Accessed 2019-03-05

Neuhold G, Ollmann T, Bulo SR, Kontschieder P (2017) The Mapillary vistas dataset for semantic understanding of street scenes. In: Proceedings of the IEEE international conference on computer vision, pp 4990–4999

O'Brien DT (2015) Custodians and custodianship in urban neighborhoods: a methodology using reports of public issues received by a city's 311 hotline. Environ Behav 47(3):304–327

OpenStreetMap (2019) www.openstreetmap.org. Accessed 2019-03-05

Park TH, Turner J, Musick M, Lee JH, Jacoby C, Mydlarz C, Salamon J (2014) Sensing urban soundscapes. In: EDBT/ICDT workshops. Citeseer, pp 375–382

Praharaj S, Hoon Han J, Hawken S (2017) Innovative civic engagement and digital urban infrastructure: lessons from 100 smart cities mission in India. Procedia Eng 180:1423–1432

Qin R, Fang W (2014) A hierarchical building detection method for very high resolution remotely sensed images combined with DSM using graph cut optimization. Photogramm Eng Remote Sens 80(9):873–883

Randhawa A, Kuma A (2015) Exploring sustainability of smart development initiatives in India. Int J Sustain Built Environ 6(2):701–710

Rousseaud A, Cruz E, Naula E, Ramos A, Granda M, Calvopiña M, León P, Sanchez D, Zapata F, Guerrero J, Babriel L, Falconi E (2017) Plan Galapagos: an instrument for the holistic sustainable development of the province. In: Galapagos report 2015–2016. GNPD, GCREG, CDF and GC
Sarin G (2016) Developing smart cities using internet of things: an empirical study. In: 3rd international conference on computing for sustainable global development (INDIACom), pp 315–320
Schläpfer M, Lee J, Bettencourt LMA (2015) Urban skylines: building heights and shapes as measures of city size. arXiv preprint arXiv:1512.00946
Stamen Design (2019) maps.stamen.com. Accessed 2019-03-04
Townsend A (2015) Cities of data: examining the new urban science. Public Cult 27(2):201–212
Weng Q, Quattrochi D, Gamba PE (eds) (2018) Urban remote sensing. CRC Press, Boca Raton, FL

Daniel Zünd is a Postdoctoral Fellow at the Mansueto Institute for Urban Innovation and a Postdoctoral researcher in Ecology and Evolution at the University of Chicago. He holds a Ph.D. in Architecture and Urban Planning, as well as a Master's degree in Computer Science.

Luís M. A. Bettencourt is the Inaugural Director of the Mansueto Institute for Urban Innovation and Professor of Ecology and Evolution at the University of Chicago, as well as an External Professor of Complex Systems at the Santa Fe Institute.

Chapter 5
Urban Human Dynamics

Shih-Lung Shaw

Abstract Urban areas are places where people concentrate in a relatively high density built environment to carry out a wide range of activities. Each urban area should provide adequate infrastructure and services to support the needs of its population. Since various resources, services, and facilities are at different locations, urban areas manifest a complex system of flows of people, goods, and information to support the economic, social, cultural, and political systems in human society. These activities, flows, and systems are driven by various processes and exhibit various spatiotemporal patterns that are the outcomes of human dynamics. However, how we investigate the various dynamic processes and complex systems in urban areas has been and continues to be a challenging research topic. Urban human dynamics cover multiple aspects and can be studied from different perspectives. This chapter discusses urban dynamics and human dynamics in terms of their respective approaches and methods, along with some selected examples. It then connects urban human dynamics research with urban informatics to highlight their relationships and how together they could lead to urban areas that can better serve human needs and improve the quality of life.

5.1 Introduction

Urban areas are places where people concentrate in a relatively high density built environment to carry out a wide range of activities. The terms urban area and city are often used interchangeably. The National Geographic Society, for example, indicates that "An urban area is the region surrounding a city" (https://www.nationalgeographic.org/encyclopedia/urban-area/). Each urban area requires adequate infrastructure and services such as electricity, water, sewer, transportation, schools, hospitals, shops, and parks to support the needs of its population. Since various resources, services, and facilities are at different locations, urban areas therefore have a complex system of flows of people, goods, and information to support their economic, social, cultural,

S.-L. Shaw (✉)
Department of Geography, University of Tennessee, Knoxville, USA
e-mail: sshaw@utk.edu

W. Shi et al. (eds.), *Urban Informatics*, The Urban Book Series,
https://doi.org/10.1007/978-981-15-8983-6_5

and political systems. These activities, flows, and systems are driven by various processes and exhibit various spatiotemporal patterns that are the outcomes of urban human dynamics. It should be noted that urban human dynamics also constantly evolve across space and over time with changing technologies, environmental issues, and social values.

According to the United Nations Educational, Scientific and Cultural Organization (UNESCO, https://www.unesco.org/education/tlsf/mods/theme_c/popups/mod13t01s009.html) and Our World in Data (https://ourworldindata.org/urbanization), the trend to global urbanization has dramatically accelerated in the past several decades. Approximately 30% of world population lived in urban areas in 1950 and around 55% in 2019. This urbanization trend is expected to continue, and it is estimated that close to 70% of the world's population are likely to live in urban areas by 2050. With this trend, many existing cities must grow bigger to accommodate the increasing population. Given the fact that many big cities already face significant challenges with respect to their current population size, how to accommodate a continuously increasing urban population without sacrificing our general quality of life has become an important and urgent research topic.

Urban areas have long been considered to be dynamic and complex in nature (Crosby 1983; Batty 2003). Batty (2005) suggested that the emphasis of urban models is no longer on spatial interaction but on development dynamics and local movement. However, how to investigate the various dynamic processes and complex systems in urban areas has been and remains a challenging research topic. Urban human dynamics cover multiple aspects and can be studied with different perspectives. In general, we can divide research in urban human dynamics into two major types, urban dynamics research, and human dynamics research. Urban dynamics research tends to focus on the evolution of an urban area in terms of its growth, change, and decline. In this case, the focus is mainly on the urban area itself, and human activities often are considered implicitly through the outcomes of human activities such as land-use types. For example, we can study how a city evolves spatially through its land-use change patterns over time in terms of its growth, change, and decline. Urban dynamics research also can investigate the dynamics among a system of urban areas such as studying various types of flows among a set of cities. In this case, the focus is mainly on the interactions between cities. Human dynamics research, on the other hand, has a focus on humans per se and studies the dynamics of human activities and interactions that lead to various flows and patterns in an urban area or between urban areas. Although urban dynamics and human dynamics are closely related to each other and should not be treated as two independent types of dynamics in urban areas, this chapter discusses each of these two types of urban human dynamics separately since they tend to use different research approaches and research methods.

5.2 Urban Dynamics

One way of studying complex and dynamic urban areas is to employ general systems theory (von Bertalanffy 1968; Straussfogel 1991; Alfeld 1995; Xie 1996). General systems theory considers a system comprising of a set of interdependent subsystems. A system, which can be more than the sum of its parts, exhibits emerging patterns from the interactions of its parts. Changes in one subsystem can affect other subsystems as well as the system as a whole. Forrester (1969), who is considered the founder of system dynamics, published a book titled *Urban Dynamics* in 1969. He states that "In this book, the nature of the urban problem, its causes, and possible corrections are examined in terms of interactions between components of the urban system" (Forrester 1969, p. ix). Forrester uses computer simulations to study the life cycle of an urban area to reveal its dynamic characteristics. This was an early effort at studying urban dynamics with a computer simulation approach to systematically examine the structure, growth, stagnation, and revival of urban areas.

Due to the influence of Forrester's approach in investigating urban dynamics, two volumes of *Readings in Urban Dynamics* were subsequently published in 1974 and 1975, respectively (Mass 1974; Schroeder et al. 1975). These two volumes include articles that cover conceptual issues, models, and applications of various aspects of urban dynamics as well as responses to the criticisms of the approach presented in Forrester's book. For example, Forrester uses a five-step process to reach his conclusions about the dynamics of a typical inner area of a US city, his example loosely related to Boston. The first step chooses certain basic variables to represent the social and economic composition of an urban area, followed by a second step of using specific equations to describe the development of an urban area. The third step introduces public policies to modify the development expressed in the equations, which then leads to the fourth step of deriving the development outcomes due to the public policies introduced into the equations. The fifth step compares the different development outcomes and recommends the public policy that would generate a desirable development outcome. Kadanoff (1971) pointed out several shortcomings of Forrester's approach, which includes (1) Forrester's model fails to include city–suburban interactions, (2) migration is the only interaction between an urban area and the outside world in Forrester's model, and (3) Forrester's model focuses mainly on predictive methods and does not give sufficient attention to the goals behind the normative approach. Kadanoff (1971, p. 262) then concluded that "I would reject the conclusions, but accept the model as an appropriate basis for further work."

In response to these criticisms, Forrester (1974, p. vii) wrote: "With the publication of *Readings in Urban Dynamics*, it seems important to emphasize that the original *Urban Dynamics* model represented more a viewpoint and a methodology for analyzing urban behavior than a single, finished model. *Urban Dynamics* was a first step in a continuously evolving set of ideas about social systems. The urban dynamics approach has several major distinguishing features. First, it focuses primarily upon the *interrelationships* between economic, political, psychological, and sociological variables rather than analyzing in detail any one subsystem of the urban environment. Second, it deals with the long-term evolution of an urban area; it treats the positive feedback processes that lead to urban growth as well as the nonlinearities and negative feedback processes that arise to limit growth. Finally, it provides a formal means for testing the implications of our collective assumptions about urban behavior." The above statements provide a clear picture of Jay Forrester's approach to studying urban dynamics; it is associated with general systems theory and uses computer simulations to examine the interrelationships among different subsystems of an urban area. More importantly, the computer simulation approach suggested by Forrester has been pursued by many other researchers in their investigation of urban dynamics, although different simulation models have been used in various studies.

5.2.1 Cellular Automata for Urban Dynamics Research

Cellular automata (CA), which were developed in the 1940s by Ulam (1950) and von Neumann (1966), are frequently used to model and simulate urban dynamics. Following these ideas, Tobler (1979) proposed a cellular geography that uses cellular spaces in geographic modeling. A cellular space can be considered as a two-dimensional grid, and each cell in the grid has a *state* that is determined by the states of its neighboring cells. The *neighbor* of a given cell can be defined in different ways, by either the four cells sharing a common side (known as von Neumann neighborhood) or the eight cells that share a common side or a common corner of a given cell (known as the Moore neighborhood). A *transition rule* then determines how the state of a cell changes into a different state from time t to time $t + 1$ based on the specific configurations of the states of its neighboring cells. For example, a transition rule could convert a given cell from the state of non-residential at time t to residential at time $t + 1$ if three of its four neighboring cells have a state of residential at time t. Cells, states, neighbors, and transition rules therefore serve as the foundation of cellular automata models.

There are two characteristics of cellular automata that are attractive to geographical problems (White and Engelen 1993). First, cellular automata divide a study area into a grid that is intrinsically spatial. Second, cellular automata can generate very complex forms from very simple rules that are useful to study complex spatial phenomena. In other words, simple local changes due to interactions among the neighboring cells in a CA model could lead to complex emergent global patterns (Wolfram 1983, 1984). CA models therefore can reflect micro–macro interactions in a simple and direct way, and the key contribution of CA models is to provide insights into how urban systems work rather than offer a simulation tool of urban dynamics (Couclelis 1985). This presents a way of linking the processes operating at different scales to tackle a major research challenge in many fields that attempt to link forms to processes and address local to global structures (Batty and Xie 1994; Emmeche 1994). In fact, Jacobs (1961) suggested that the observed disorders in urban areas could be viewed as organized complexity due to a deeper order reflecting their diversity. Cellular automata models enable us to investigate urban dynamics from local processes in order to understand global complex patterns and to gain insights into the evolution of various aspects of urban dynamics.

Chapin and Weiss (1968) first applied the concepts of cellular automata to an urban land development model, and Tobler (1970) employed the idea of cellular space to simulate urban growth in the Detroit region, although both studies did not use the term cellular automata. Tobler (1970, p. 234) suggested that "the utmost effort must be exercised to avoid writing a complicated model. … Because a process appears complicated is also no reason to assume that it is the result of complicated rules." White and Engelen (1993) argued that most geographic theories, such as central place theory and urban economic models as embodied in the Alonso-Muth land-use theory, are static in nature and assume a state of stable equilibrium, which is contrary to our common sense and experience that all urban areas are undergoing continual growth, change, decline, and restructuring. White and Engelen (1993) consequently developed a CA model that generates fractal patterns of land use from relatively simple rules of spatial behavior in order to address the issue of complexity in urban structure. The objective of this study is to gain insight into the underlying reasons behind the evolution of land-use structures and to demonstrate the existence of a complex fractal order of land-use patterns. Their findings suggest that complexity is a necessary feature of cities. When cities are too simple in their structure, they probably will not evolve successfully and could cease to function effectively. This study is a good example of using a CA model to assess the complexity of urban structure and to establish general guidelines for planning policy.

Couclelis (1985) pointed out that the standard cell-space model has many limitations to its usefulness for tackling real-world geographic problems. These limitations include the infinite plane, neighborhood stationarity, spatial homogeneity, spatial and temporal invariance of transition rules, and closure to external events that are directly related to the basic assumptions of cell-space models. Batty and Xie (1994, p. S46) also suggested that a major problem of applying CA models to urban systems is that "It is most unlikely that urban systems can be simulated entirely at the local scale, but the value of this approach lies in focusing our attention on this scale and the extent to which a hierarchy of processes and scales is essential to understanding how cities work." Xie (1996) discussed improvements to CA models over the years and proposed a generalized model for cellular urban dynamics, named dynamic urban evolutionary modeling (DUEM), to demonstrate the theoretical integrity and technical merit of the CA approach for urban dynamic applications. One major contribution of DUEM is to adopt a hierarchical system of CA spaces consisting of neighborhood, field, and region that can be used to simulate interactions between cell space, model space, and geographic space to overcome some limitations of the conventional cell-space models. DUEM further connects with a geographic information system (GIS) to benefit from GIS data, analysis, and visualization capabilities.

Anthony Yeh, Xia Li and their collaborators have used cellular automata models extensively to study urban dynamics. Li and Yeh (2000) developed a constrained CA model within a raster GIS that includes local, regional, and global constraints to regulate cellular space and defines gray cells as representing the percentages of urban land development at any iteration of the CA model. Yeh and Li (2001) further used a constrained CA model and a raster GIS to simulate seven different types of urban forms and developments ranging from compact-monocentric to very highly dispersed development patterns. Their model considers various criteria such as urban forms, environmental suitability, and land consumption for the purpose of planning sustainable cities. They also combined CA models with computational intelligence methods such as neural networks (Li and Yeh 2001), ant colony optimization (Liu et al. 2008), and artificial immune systems (Liu et al. 2010) to investigate complex urban systems. Santé et al. (2010) offered a helpful review of urban cellular automata models applied to the simulation of real-world urban processes with respect to their capabilities and limitations. They also conclude that the widespread use of CA models is due to their simplicity. In the meantime, the simplicity of CA models is also the main weakness that limits their ability to represent real-world phenomena. Another major shortcoming is the lack of a standard method for the definition of transition rules in urban CA models which represent the complexity of the processes.

5.2.2 Other Urban Dynamics Approaches

Batty (2008) indicated that traditional urban models treated cities as aggregate equilibrium systems and mainly used spatial interaction. The approach changed in the late twentieth century to consider urban dynamics more as evolving complex systems whose structure emerges from the bottom-up. In his book *Cities and Complexity: Understanding Cities with Cellular Automata, Agent-Based Models, and Fractals*, Batty (2007) presented agent-based models as another useful approach to study complex urban dynamics as urban planning moves from a top-down centralized perspective to a bottom-up decentralized perspective. An agent-based model (ABM) consists of autonomous agents, which can be either individual or collective entities, with defined behaviors to simulate the effects on emerging system patterns from the actions and interactions of the autonomous agents. One key difference between cellular automata models and agent-based models is that agents in ABM are free to move and interact with each other and the environment. The goal of agent-based models is mainly to gain insights into the collective behavior of agents that follow simple behavioral rules. Huang et al. (2014) reviewed 51 agent-based residential choice models in three research domains, which are (1) urban land-use models based on classical theories, (2) different stages of the urbanization process, and (3) integrated agent-based and microsimulation models, to offer a retrospective on developments in agent-based models (ABMs) of urban residential choices. This review paid special attention to the progress of the representation of agent heterogeneity, the extent of land-market representation, and the method of measuring the extensive model outputs. They concluded that "Urban land-use models can benefit from agent-based modeling by incorporating heterogeneous intelligent agents and explicit modeling of an institution that stands behind land exchange" (Huang et al. 2014, p. 681).

Xie et al. (2007) applied agent-based modeling to study the development of desakota, which is a mixed urban-rural space adjacent to a metropolitan area, in the Suzhou-Wuxian region in China for the period of 1990–2000. They developed an ABM that links local household reform to global urban reform in order to examine processes of local land developments that are moderated by the higher-level macroeconomy. Benenson et al. (2008), on the other hand, developed an agent-based model to study the complex self-organizing dynamics of parking patterns in a non-homogeneous road space by examining the distributions of search time, walking distance, and parking costs of different driver groups. Hosseinali et al. (2013) introduced an agent-based model with new methods of modeling agent movements and competition among agents to simulate urban land-use development in Qazvin, Iran. After the model is calibrated with existing data, it is used to predict land-use developments under four scenarios of development policies.

There are also studies of urban dynamics of a system of urban areas. For example, Batty (2003) presented an approach to urban dynamics that generalized Zipf's rank-size model to investigate the changing rank-size relationships among cities through time. He used data of the 100 largest towns and cities from 1790 to 2000 at a ten-year interval to examine the volatility of the distributions of individual cities within the rank-size distributions with a measure of the half life of cities. He found that there is considerable volatility in the rank-size relationships which change almost entirely over a 200 year period. This study illustrates the dynamics of how an individual city rises, falls, or holds its position in a system of cities. In addition, Batty's (2013a) book *The New Science of Cities*, which suggested that we must view cities not only as places in space but also as systems of networks and flows, further indicated the need for looking into the connections and interactions both within an individual city and among a system of cities to better understand various aspects of urban dynamics.

5.3 Human Dynamics

Human dynamics are the foundation of human society. All economic, social, cultural, and political systems and all built environments are developed to serve human needs that are dynamic in nature. The focus of human dynamics research therefore is on the dynamics of disaggregate individual behaviors as well as aggregate group behaviors (Shaw et al. 2016; Shaw and Sui 2018a, b, c). Human dynamics has been a research topic in many disciplines ranging from business, geography, planning, psychology, and sociology to physics. A recent surge of research interests in human dynamics is partially due to the work of Albert-László Barabási and his associates on scale-free networks and heavy-tailed distributions of human behavior. Barabási and Bonabeau (2003) suggested that many complex systems share an important characteristic of some nodes having a large number of connections to other nodes in a network while most nodes have just a handful connections. In other words, these networks appear to have no scale or are scale-free. Barabási (2005) further indicated that individuals often execute tasks with bursts of rapidly executed tasks separated by long periods of inactivity that results in heavy-tailed distributions. This line of research identifies some general laws of human dynamics from the perspective of statistical physics.

From an urban planning perspective, we need to go beyond the general laws of human behaviors and gain further insights into human dynamics to facilitate policy making and planning practices. Human dynamics evolve with the changing environment, technology, and society (Shaw and Sui 2018b). The ways that people carried out their activities and interacted with other people and the environment 50 years ago are very different from human dynamics today. It is therefore important to gain a better understanding of evolving human dynamics in order to design and develop smarter cities to better serve human needs in the next 10–20 years, if not longer.

5.3.1 Effects of Information and Communications Technologies on Human Dynamics

Information and communications technologies (ICT) such as the Internet and mobile phones have significantly influenced the ways that people carry out their activities and interactions. The Internet allows us to access a huge amount of information and a wide range of services online through a global system of interconnected computer networks. With Wi-Fi technology, we can connect to the Internet from any locations that have a wireless local area network. Mobile phones and tablets which are equipped with increasingly powerful computing power further free us from the fixed landline phones and bulky computers, to stay connected almost anywhere and at any time. It is now feasible to find a journal article when a library is closed, purchase an item without a physical visit to a store, and stay in touch with friends almost all of the time. In other words, modern technologies have removed many spatial and temporal constraints on human activities and interactions to extend our activity space (Janelle 1973). Human activities and interactions therefore have become more flexible and spontaneous which in turn can change the nature and spatiotemporal patterns of human dynamics.

There have been many studies of the effects of ICT on travel and human activity patterns (e.g., Salomon 1986; Salomon and Koppelman 1988; Mokhtarian and Meenakshisundaram 1999; Townsend 2000; Hjorthol 2002; Ben-Elia et al. 2014). Mokhtarian (2003) suggested that there exist four types of relationships between telecommunications and travel. The first type of relationship is *substitution* such as teleconferencing or e-shopping, where an online activity substitutes for a trip in physical space. The second type of relationship is *complementarity*, which suggests that the use of ICT will increase activities in physical space. For example, sales messages pushed to smart phones could attract more people to visit stores in physical space. The third type of relationship is *modification*, such as when information obtained from an online real-time traffic information service changes the route that a traveler takes to make a trip. This simply modifies a trip pattern in physical space without adding or reducing the number of trips in physical space. The last type of relationship is *neutrality*, which means that an activity using ICT has no effect on activities in physical space. This study illustrates the challenge of identifying specific effects of ICT on human dynamics.

Humans must move between different locations in physical space to carry out their activities (e.g., work, school, shopping, social, recreation). Transportation provides the means for people to move from a location to another location in physical space. Since physical movements take time, humans have to trade time to overcome spatial separation. As transportation technologies improve over time, we can overcome the same distance over a shorter time period, which is known as time-space convergence (Janelle 1968, 1969). With the rapid growth and widespread use of ICT in today's world, an increasing number of human activities and interactions are carried out in virtual space using ICT devices to navigate among different places in virtual space. For example, many people stay in touch with their friends via online social network apps and shop online with their smart phone or computer. These activities in virtual space can have major implications for the activities in physical space. For instance, an online order at Amazon.com triggers a shipment from a distribution center to the customer's location via a courier delivery service (e.g., FedEx or UPS). This delivery replaces a personal trip to a store. When there are many people who engage online shopping, a large number of personal trips are replaced by a few delivery truck trips that normally take different routes and occur at different times from those of personal shopping trips. We therefore need to consider human activities and interactions in both physical and virtual spaces, in order to study their interactions and gain a better understanding of human dynamics in the modern world (Shaw and Yu 2009).

5.3.2 *Time Geography*

Time geography, which was developed by Torsten Hägerstrand (1970), presents a useful framework for studying individual activities in a space-time context. A well-known time-geographic concept is the space-time path that tracks the movements of an individual across space and over time. When there are multiple space-time paths for a group of people, we can analyze their spatiotemporal relationships (Parkes and Thrift 1980; Golledge and Stimson 1997; Janelle 2004; Shaw and Yu 2009). For example, when two or more individuals are at the same location during the same time period, they have a *co-existence* relationship. If two or more individuals visit the same location at different times, they have a *co-location in space* relationship. If two or more people communicate with each other at different locations during the same time period (e.g., online chat), then they have a *co-location in time* relationship. When two or more people interact asynchronously in both space and time (e.g., email communications), it does not require co-existence, co-location in space, or co-location in time. These relationships make it feasible to study human activity patterns at the individual level to understand human dynamics in a space-time context.

Time geography also covers many other useful concepts for human dynamics research. Time geography assumes that every individual faces three types of constraints on their activities. Capability constraints are related to an individual's biological system and ability for utilizing tools. For example, all people must sleep and eat, which take time at certain locations. Also, a person who can drive a car can reach more distant locations than people who do not drive. Coupling constraints require that an individual be coupled with other people or entities to carry out particular activities. For example, a class lecture requires an instructor, and the students to be present at the same location during the same time period. Authority constraints are imposed by a domain. An example is that an individual cannot access a grocery store when it is closed. Our daily activities and interactions are conditioned by these three types of constraints, which in turn influence spatiotemporal human dynamics. Another useful time-geographic concept is the space-time prism, which allows us to identify the maximum feasible space-time extent that an individual could reach under given constraints. A space-time prism can help us understand why an individual exhibits certain space-time activity patterns. Diorama is another critical concept. Hägerstrand puts various time-geographic concepts together in a diorama to emphasize the presence of an individual in an immersive environment, such that the individual appreciates how situations evolve as an aggregate outcome while considering various constraints and situations to achieve the goal of a project (Hägerstrand 1982). In fact, Hägerstrand (1982, p. 338) stated that "without a diorama approach, the revealing power of time geography cannot be fully explored."

Although time geography offers a useful framework for human dynamics research, it has not been widely used in empirical studies, due mainly to two limitations (Shaw 2012). First, time geography requires detailed spatial movement data over time at the individual level that is costly and time-consuming to collect. Most previous time geography studies used data collected from surveys or interviews that had a relatively small sample size. Second, even though many studies collected data of large sample size, it was challenging to conduct time-geographic analyses using a space-time path and a space-time prism due to a lack of computational tools to process, analyze, and visualize the data. These limitations have been overcome to some extent in the big data era, along with the advances in space-time-geographic information systems (GIS).

5.3.3 Big Data and Space-Time GIS for Human Dynamics Research

With advances in sensing, mobile, and information and communications technologies in recent decades, it has become far easier and much cheaper to collect individual data. Mobile phones can constantly track our locations across space and over time at unprecedented spatial and temporal granularity using built-in global positioning system (GPS) capability. Phone companies have records of our phone communications including phone calls, text messages, and websites accessed. Credit card companies know where, when, and what we purchased, and how much we paid for each purchased item. Smart cards used in many cities for public transit know where and when we used public transit, which transit routes we used, and how often we used them. Search engine service providers such as Google know when we have searched online, which websites we visited, and how long we browsed a particular website. Online social network service providers like Facebook, Twitter, Flickr, and LinkedIn know who our friends and connections are, how frequently we communicated with each other, and what we discussed with each other. These tracking data cover not only human activities in physical space but also human activities and interactions in virtual space. They provide extremely useful data sources to conduct empirical studies of human dynamics, although the research community needs to pay close attention to the ethical and privacy issues of using such data (see Chap. 32).

In the meantime, the large amount of data available for human dynamics research demand adequate tools to process, manage, analyze, and visualize the data. GIS was designed to handle spatial data, yet they were not adequate to dealing with space-time data. Efforts extending the conventional GIS to space-time GIS started in the 1990s by developing functions in GIS that support time-geographic concepts. Miller (1991) first implemented the space-time prism concept in GIS to study individual accessibility, followed by many other efforts at expanding time-geographic functions in GIS (e.g., Kwan 2000a, b; Buliung and Kanaroglou 2006; Yu 2006; Chen et al. 2011; Scott and He 2012). One of the major challenges of applying time geography to human dynamics research is that most time-geographic concepts are based on human activities in physical space. Since many human activities and interactions today are taking place in virtual space, it is critical to extend the conventional time-geographic concepts to cover human dynamics in both physical and virtual spaces. Yu and Shaw (2008) developed a space-time GIS that extends the conventional space-time prism concept to support analysis of potential human activities and interactions in both physical and virtual spaces. Shaw and Yu (2009) further extended the time-geographic concepts of space-time path, station, bundle, activity, event, and project into a hybrid physical–virtual space and implement them in a space-time GIS. Yin and Shaw (2015) then developed a method for creating social closeness of space-time paths in a GIS environment, such that we can assess the relationships between any pair of individuals in both physical space and social closeness space. These efforts make it feasible to study human dynamics in a hybrid physical–virtual space based on time-geographic concepts, although many research challenges remain to be addressed.

5.3.4 Some Other Examples Human Dynamics Studies

In addition to human dynamics research based on time-geographic concepts, there exist a large volume of studies investigating human dynamics using a wide range of individual data collected in the Big Data era. Candia et al. (2008) used mobile phone data to study the mean collective behavior and identify the rise, clustering, and decay of anomalous events that can be useful in real-time detection of emergency situations. They also examined calling activities at the individual level and found that they follow a heavy-tailed distribution. Vazquez-Prokopec et al. (2013) employed GPS tracking of residents in Iquitos, Peru to study mobility patterns, infer mobility networks, and model infectious disease transmission within an Iquitos neighborhood. This study demonstrated how to use data collected from location-aware technology to characterize complex social systems in a developing country and then use the identified mobility patterns and networks to address an important health issue of infectious disease dynamics in an urban environment. Zhong et al. (2014) applied methods in network science to identify the spatial structure of city hubs using smartcard transit data collected in Singapore. They illustrated the evolving roles and influences of local areas in the overall spatial structure of urban movements and indicated that collective movement can shape local communities similar to what happens in social networks. Xu et al. (2016), on the other hand, used mobile phone data collected in Shenzhen and Shanghai, China, to compare their human dynamics patterns based on the number of major activity points, activity range, and frequency of movements (for further examples of this kind of research see Chaps. 28 and 29).

Liu et al. (2015) proposed a concept of social sensing, in contrast to remote sensing, to characterize the research that employs individual level Big Geospatial Data to study socioeconomic aspects of human dynamics. They also considered each individual person as a sensor that helps contribute data to human dynamics research. The concept of social sensing is clearly related to human dynamics research. Due to an explosion of research related to urban human dynamics in recent years using crowdsourcing data and other big data, it is not an intention of this chapter to provide a comprehensive review. Instead, readers can find various examples in other chapters of this book.

5.4 Urban Human Dynamics and Urban Informatics

With this brief review of urban human dynamics research, it is important to connect urban human dynamics to the theme of this book: urban informatics. Urban informatics, which is a relatively new field, takes a data-driven approach enabled by modern sensing, mobile, and information and communications technologies to gain insights into how people function in an urban area and how various systems and services operate in an urban area (Kontokosta 2018). Foth et al. (2011, p. 4) define urban informatics as "the study, design, and practice of urban experiences across different urban contexts that are created by new opportunities for real-time, ubiquitous technology, and the augmentation that mediate the physical and digital layers of people networks and urban infrastructures." This definition links place, technology, and people together in an urban environment.

As urban areas continue to grow in their geographic size and population density in order to accommodate the ever-increasing urban population, there is an urgent need for improving our understanding of how urban areas function, what causes urban problems, and how we can address these urban problems in smart and sustainable ways. These challenges are not new at all, and they have been studied for many decades. Unfortunately, it appears that we have not been able to reign in these urban problems, and many urban areas are experiencing worse traffic congestion, air pollution, heat-island effects, housing issues, job mismatches, etc., than ever before. If we accept that human dynamics are the fundamental driving forces of the economic, social, cultural, political, and other systems in urban areas, we must better understand human needs and how they interact with other people and the environment under various constraints imposed by the environment, society, and technology. When infrastructure and services in an urban area cannot adequately accommodate human needs, we run into problems. Since human needs emerge at different locations and different times, they present a challenge of matching supply and demand spatially and temporally. From an urban planning perspective, our goal is to design urban areas that can best meet human needs and improve the quality of life. This is a significant challenge, as evidenced by a wide range of problems facing most urban areas today.

In his article "big data, smart cities, and city planning," Batty (2013b, p. 274) stated that "the growth of big data is shifting the emphasis from longer term strategic planning to short-term thinking about how cities function and can be managed; although with the possibility that over much longer periods of time, this kind of big data will become a source for information about every time horizon." Batty (2013b, p. 276) further indicated that "There is, however, a coincidence between what are now being called smart cities and big data, with smartness in cities pertaining primarily to the ways in which sensors can generate new data streams in real time with precise geo-positioning; of course, it is often pointed out that cities only become smart when people are smart, and this is sine qua non of our argument here." Technologies clearly play an important role in urban informatics and smart cities. However, we must keep in mind that urban informatics and smart cities are developed to better serve human needs and improve quality of life. Whether or not a city or a particular system in a city is smart should be assessed by how well it serves the needs of various population groups to improve the quality of life (Shaw and Sui 2019).

Shared bicycles experienced an amazing rapid growth in many Chinese cities a few years ago and this created a motive for reviving bicycles as a popular travel alternative in Chinese cities. However, the entire business collapsed quickly. As indicated by Huang (2018), "Bike-sharing apps seemed poised to be the solution—and millions of bikes were poured into China's streets by the private sector in the last three years. But today, as the companies fail, unused units pile up in bicycle graveyards, and queues of angry users demand their deposits back, it is obvious just how doomed the idea was from the start." The bike-sharing apps were smart in the sense that users could unlock and lock bicycles and pay rent by smart phones anywhere in a city. Yet, it is not clear to what extent the shared bicycles fit well with human needs with respect to various constraints people face in urban areas to carry out their dynamic activity patterns. This example reminds us that it is critical to keep human dynamics in mind when we pursue urban informatics. In conclusion, it is beneficial to combine urban informatics with urban human dynamics research to better understand human activities and interactions in an increasingly hybrid physical–virtual space; yet we must remember that various systems and services in urban areas are created to better serve and meet the human needs in order to improve the quality of life.

References

Alfeld LE (1995) Urban dynamics—the first fifty years. Syst Dyn Rev 11(3):199–217
Barabási AL (2005) The origin of bursts and heavy tails in human dynamics. Nature 435(7039):207–211
Barabási AL, Bonabeau E (2003) Scale-free networks. Sci Am 288(5):50–59
Batty M (2003) The emergence of cities: complexity and urban dynamics. Center for advanced spatial analysis working paper series, paper 64. University College London
Batty M (2005) Agents, cells, and cities: new representational models for simulating multiscale urban dynamics. Environ Plan A 37(8):1373–1394
Batty M (2007) Cities and complexity: understanding cities with cellular automata, agent-based models, and fractals. MIT Press, Cambridge, MA
Batty M (2008) Cities as complex systems: scaling, interactions, networks, dynamics and urban morphologies. Centre for advanced spatial analysis working paper series, paper 131. University College London
Batty M (2013a) The new science of cities. MIT Press, Cambridge, MA
Batty M (2013b) Big data, smart cities and city planning. Dialogues Hum Geogr 3(3):274–279
Batty M, Xie Y (1994) From cells to cities. Environ Plann B Plann Des 21(7):S31–S48
Ben-Elia E, Alexander B, Hubers C, Ettema D (2014) Activity fragmentation, ICT and travel: an exploratory path analysis of spatiotemporal interrelationships. Transp Res Part A Policy Pract 68:56–74
Benenson I, Martens K, Birfir S (2008) PARKAGENT: an agent-based model of parking in the city. Comput Environ Urban Syst 32(6):431–439
Buliung R, Kanaroglou P (2006) A GIS toolkit for exploring geographies of household activity/travel behavior. J Transp Geogr 14(1):35–51
Candia J, González MC, Wang P, Schoenharl T, Madey G, Barabási AL (2008) Uncovering individual and collective human dynamics from mobile phone records. J Phys A Math Theor 41(22):1–11
Chapin FS, Weiss SF (1968) A probabilistic model for residential growth. Transp Res 2:375–390
Chen J, Shaw SL, Yu H, Lu F, Chai Y, Jia Q (2011) Exploratory data analysis of activity diary data: a space-time GIS approach. J Transp Geogr 19(3):394–404
Couclelis H (1985) Cellular worlds: a framework for modeling micro-macro dynamics. Environ Plan A 17(5):585–596
Crosby RW (1983) Introduction and asking better questions. In: Crosby RW (ed) Cities and regions as nonlinear decision systems. Westview Press, Boulder, CO, pp 1–28
Emmeche C (1994) The garden in the machine: the emerging science of artificial life. Princeton University Press, Princeton
Forrester JW (1969) Urban dynamics. MIT Press, Cambridge, MA

Forrester JW (1974) Preface. In: Mass NJ (ed) Readings in urban dynamics, vol 1. Wright-Allen Press, Cambridge, MA, pp vii–viii
Foth M, Choi JH, Satchell C (2011) Urban informatics. In: Proceedings of the ACM 2011 conference on computer supported cooperative work, Hangzhou, pp 1–8
Golledge R, Stimson R (1997) Spatial behavior: a geographic perspective. Guilford Press, New York, NY
Hägerstrand T (1970) What about people in regional science? Pap Reg Sci Assoc 24(1):7–21
Hägerstrand T (1982) Diorama, path and project. Tijdschr Voor Econ Soc Geogr 73(6):323–339
Hjorthol RJ (2002) The relation between daily travel and use of the home computer. Transp Res A 36(5):437–452
Hosseinali F, Alesheikh AA, Nourian F (2013) Agent-based modeling of urban land-use development, case study: simulating future scenarios of Qazvin city. Cities 31:105–113
Huang F (2018) The rise and fall of China's cycling empires. Foreign Policy, 31 Dec. https://foreignpolicy.com/2018/12/31/a-billion-bicyclists-can-be-wrong-china-business-bikeshare/
Huang Q, Parker DC, Filatova T, Sun S (2014) A review of urban residential choice models using agent-based modeling. Environ Plann B Plann Des 41(4):661–689
Jacobs J (1961) The death and life of great American cities. Vintage Books, New York, NY
Janelle DG (1968) Central place development in a time-space framework. Prof Geogr 20(1):5–10
Janelle DG (1969) Spatial organization: a model and concept. Ann Assoc Am Geogr 59(2):348–364
Janelle DG (1973) Measuring human extensibility in a shrinking world. J Geogr 72(5):8–15
Janelle D (2004) Impact of information technologies. In: Hanson S, Giuliano G (eds) The geography of urban transportation, 3rd edn. Guilford Press, New York, NY, pp 86–112
Kadanoff LP (1971) From simulation model to public policy: an examination of Forrester's "Urban Dynamics." Simulation 16(6):261–268
Kontokosta CE (2018) Urban informatics in the science and practice of planning. J Plan Educ Res 1–14. https://doi.org/10.1177/0739456X18793716
Kwan MP (2000a) Human extensibility and individual hybrid-accessibility in space–time: a multi-scale representation using GIS. In: Janelle D, Hodge D (eds) Information, place, and cyberspace issues in accessibility. Springer, Berlin, pp 241–256
Kwan MP (2000b) Interactive geovisualization of activity-travel patterns using three-dimensional geographical information systems: a methodological exploration with a large data set. Transp Res C 8:185–203
Li X, Yeh AG (2000) Modelling sustainable urban development by the integration of constrained cellular automata and GIS. Int J Geogr Inf Sci 14(2):131–152
Li X, Yeh AG (2001) Calibration of cellular automata by using neural networks for the simulation of complex urban systems. Environ Plan A Econ Space 33(8):1145–1162
Liu X, Li X, Liu L, He J, Ai B (2008) A bottom-up approach to discover transition rules of cellular automata using ant intelligence. Int J Geogr Inf Sci 22(11–12):1247–1269
Liu X, Li X, Shi X, Zhang X, Chen Y (2010) Simulating land-use dynamics under planning policies by integrating artificial immune systems with cellular automata. Int J Geogr Inf Sci 24(5):783–802
Liu Y, Liu X, Gao S, Gong L, Kang C, Zhi Y, Chi G, Shi L (2015) Social sensing: a new approach to understanding our socioeconomic environments. Ann Assoc Am Geogr 105(3):512–530
Mass NJ (ed) (1974) Readings in urban dynamics, vol 1. Wright-Allen Press, Cambridge, MA
Miller H (1991) Modelling accessibility using space-time prism concepts within geographical information systems. Int J Geogr Inf Syst 5(3):287–301
Mokhtarian PL (2003) Telecommunications and travel: the case for complementarity. J Ind Ecol 6(2):43–57
Mokhtarian PL, Meenakshisundaram R (1999) Beyond tele-substitution: disaggregate longitudinal structural equations modeling of communication impacts. Transp Res C 7(1):33–52
Parkes D, Thrift N (1980) Times, spaces, and places: a chronogeographic perspective. Wiley, New York
Salomon I (1986) Telecommunications and travel relationships: a review. Transp Res A 20(3):223–238

Salomon I, Koppelman FS (1988) A framework for studying teleshopping versus store shopping. Transp Res A 22(4):247–255
Santé I, García AM, Miranda D, Crecente R (2010) Cellular automata models for the simulation of real-world urban processes: a review and analysis. Landsc Urban Plan 96(2):108–122
Schroeder III WW, Sweeney RE, Alfeld LE (eds) (1975) Readings in urban dynamics, vol 2. Wright-Allen Press, Cambridge
Scott D, He S (2012) Modeling constrained destination choice for shopping: a GIS based time-geographic approach. J Transp Geogr 23:60–71
Shaw SL (2012) Guest editorial introduction: time geography—its past, present and future. J Transp Geogr 23:1–4
Shaw SL, Sui D (eds) (2018a) Human dynamics research in smart and connected communities. Springer International Publishing AG, Cham, Switzerland
Shaw SL, Sui D (2018b) Introduction: human dynamics in perspective. In: Shaw SL, Sui D (eds) Human dynamics research in smart and connected communities. Springer International Publishing AG, Cham, Switzerland, pp 1–11
Shaw SL, Sui D (2018c) GIScience for human dynamics research in a changing world. Trans GIS 22(4):891–899
Shaw SL, Sui D (2019) Understanding the new human dynamics in smart spaces and places: towards a splatial framework. Ann Am Assoc Geogr. https://doi.org/10.1080/24694452.2019.1631145
Shaw SL, Yu H (2009) A GIS-based time-geographic approach of studying individual activities and interactions in a hybrid physical-virtual space. J Transp Geogr 17(2):141–149
Shaw SL, Tsou MH, Ye X (2016) Editorial: Human dynamics in the mobile and Big Data era. Int J Geogr Inf Sci 30(9):1687–1693
Straussfogel D (1991) Modeling suburbanization as an evolutionary system dynamic. Geogr Anal 23:1–24
Tobler WR (1970) A computer movie simulating urban growth in the Detroit region. Econ Geogr 46:234–240
Tobler WR (1979) Cellular geography. In: Gale S, Olsson G (eds) Philosophy in geography. D. Reidel Publishing Company, Dordrecht, Netherlands, pp 379–386
Townsend A (2000) Life in the real-time city: mobile telephones and urban metabolism. J Urban Technol 7(2):85–104
Ulam S (1950) Random processes and transformations. In: Proceedings of the international congress on mathematics, Cambridge, 30 Aug–6 Sept 1950, pp 264–275
Vazquez-Prokopec GM, Bisanzio D, Stoddard ST, Paz-Soldan V, Morrison AC, Elder JP, Ramirez-Paredes J, Halsey ES, Kochel TJ, Scott TW, Kitron U (2013) Using GPS technology to quantify human mobility, dynamic contacts and infectious disease dynamics in a resource-poor urban environment. PLoS ONE 8(4):e58802
von Bertalanffy L (1968) General system theory: foundations, development, applications. George Braziller, New York, NY
von Neumann J (1966) Theory of self-reproducing automata. In: Burks A (ed). University of Illinois Press, Urbana, IL
White R, Engelen G (1993) Cellular automata and fractal urban form: a cellular modelling approach to the evolution of urban land use patterns. Environ Plan A 25(8):1175–1199
Wolfram S (1983) Statistical mechanics of cellular automata. Rev Mod Phys 55(3):601–644
Wolfram S (1984) Universality and complexity in cellular automata. Phys D 10(1–2):1–35
Xie Y (1996) A generalized model for cellular urban dynamics. Geogr Anal 28(4):350–373
Xie Y, Batty M, Zhao K (2007) Simulating emergent urban form using agent-based modeling: desakota in the Suzhou-Wuxian region in China. Ann Assoc Am Geogr 97(3):477–495
Xu Y, Shaw SL, Zhao Z, Yin L, Lu F, Chen J, Fang Z, Li Q (2016) Another tale of two cities—understanding human activity space using actively tracked cellphone location data. Ann Assoc Am Geogr 106(2):489–502
Yeh AG, Li X (2001) A constrained CA model for the simulation and planning of sustainable urban forms by using GIS. Environ Plann B Plann Des 28(5):733–753

Yin L, Shaw SL (2015) Exploring space-time paths in physical and social closeness spaces: a space-time GIS approach. Int J Geogr Inf Sci 22(4):409–430
Yu H (2006) Spatio-temporal GIS design for exploring interactions of human activities. Cartogr Geogr Inf Sci 33:3–19
Yu H, Shaw SL (2008) Exploring potential human interactions in physical and virtual spaces: a spatiotemporal GIS approach. Int J Geogr Inf Sci 22(4):409–430
Zhong C, Arisona SM, Huang X, Batty M, Schmitt G (2014) Detecting the dynamics of urban structure through spatial network analysis. Int J Geogr Inf Sci 28(11):2178–2199

Shih-Lung Shaw is Chancellor's Professor, Alvin and Sally Beaman Professor, and Arts and Sciences Excellence Professor of Geography at the University of Tennessee, Knoxville. He is an elected fellow of the American Association for the Advancement of Science (AAAS). His research interests cover human dynamics, space-time GIScience, and transportation.

Chapter 6
Geosmartness for Personalized and Sustainable Future Urban Mobility

Martin Raubal, Dominik Bucher, and Henry Martin

Abstract Urban mobility and the transport of people have been increasing in volume inexorably for decades. Despite the advantages and opportunities mobility has brought to our society, there are also severe drawbacks such as the transport sector's role as one of the main contributors to greenhouse-gas emissions and traffic jams. In the future, an increasing number of people will be living in large urban settings, and therefore, these problems must be solved to assure livable environments. The rapid progress of information and communication, and geographic information technologies, has paved the way for urban informatics and smart cities, which allow for large-scale urban analytics as well as supporting people in their complex mobile decision making. This chapter demonstrates how geosmartness, a combination of novel spatial-data sources, computational methods, and geospatial technologies, provides opportunities for scientists to perform large-scale spatio-temporal analyses of mobility patterns as well as to investigate people's mobile decision making. Mobility-pattern analysis is necessary for evaluating real-time situations and for making predictions regarding future states. These analyses can also help detect behavioral changes, such as the impact of people's travel habits or novel travel options, possibly leading to more sustainable forms of transport. Mobile technologies provide novel ways of user support. Examples cover movement-data analysis within the context of multi-modal and energy-efficient mobility, as well as mobile decision-making support through gaze-based interaction.

M. Raubal (✉) · D. Bucher · H. Martin
Institute of Cartography and Geoinformation, ETH Zurich, Zurich, Switzerland
e-mail: mraubal@ethz.ch

D. Bucher
e-mail: dobucher@ethz.ch

H. Martin
e-mail: martinhe@ethz.ch

W. Shi et al. (eds.), *Urban Informatics*, The Urban Book Series,
https://doi.org/10.1007/978-981-15-8983-6_6

6.1 Introduction

Urban mobility and the transport of people have been rising inexorably for decades. Despite the many advantages and opportunities, mobility has brought to our society, there are also severe drawbacks such as the transport sector's role as one of the main contributors to CO_2 emissions, traffic jams, and mass event catastrophes (Elliott and Urry 2010; Taaffe et al. 1996). Forecasts show that by 2030, the world will have 41 megacities each with more than 10 million inhabitants (UN 2014), and by the year 2050, approximately 80% of the European population will be living in urban areas (Caragliu et al. 2011). Therefore, these challenging problems must be solved to assure livable environments for future generations.

The rapid progress of information and communication technologies (ICT) and geographic information technologies has paved the way for urban informatics and smart cities, which allow for large-scale urban analytics as well as supporting people in their complex mobile decision making. This chapter demonstrates how geosmartness, a combination of novel spatial-data sources, computational methods, and geospatial technologies, provides ample opportunities for scientists to perform large-scale spatio-temporal analyses of mobility patterns as well as investigate people's mobile decision making. This application of novel methods and technologies with spatial big data will allow for unprecedented possibilities of evaluating current states of urban systems including their citizens in real time, and making predictions and forecasts of future states.

Mobility-pattern analysis is necessary for evaluating real-time situations but also for making short- and longer-term predictions regarding the transportation network. In addition, these analyses can help detect behavioral changes, such as the impact of people's travel habits or novel travel options, possibly leading to more sustainable forms of transport. Sustainable urban mobility will become ever more important in order to curb greenhouse-gas emissions in the future. Long-term decarbonization of transport will not solely be achievable through new technology, such as vehicle efficiency measures, powertrain technology, and new energy carriers, but will require people's efforts in containing demand and shifting to lower-emission transport modes (Boulouchos et al. 2017).

Mobile technologies help to identify individual-oriented problems and provide novel ways of personalized user support. Spatial Big Data can be utilized to support people in their location-based decision making, in combination with novel technologies and interaction concepts, such as location-based services and gaze-based interaction. This will lead to more effective and efficient spatio-temporal decision making, and, hopefully, contribute to sustainable urban mobility of the future.

This chapter starts by introducing geosmartness and its major enablers, namely geospatial technologies, spatial, big data and spatio-temporal computational methods. We then investigate the analysis of urban-mobility patterns, including data, prediction, and labeling methods. The section is complemented by an overview of mobility studies and a detailed example focusing on multi-modal and energy-efficient mobility. In the next section, we elaborate on the potential of geospatial

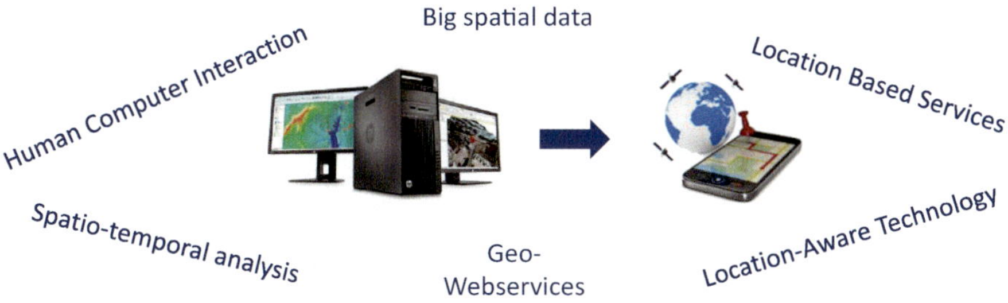

Fig. 6.1 Methods and tools enabling geosmartness

and persuasive technologies to support people in sustainable mobility. This includes motivational aspects and methods for detecting and supporting behavioral change. The section also includes an overview of studies in this area and the description of a recent study targeting the change of mobility behavior. In the penultimate section, we explain the specificities of mobile decision making, introduce the technique of mobile eye-tracking and the concept of gaze-based interaction, and demonstrate how their combination can enable personalized gaze-based decision support. The final section presents conclusions and directions for future work.

6.2 Geosmartness

Geosmartness relates to the vast opportunities of utilizing novel geospatial technologies, spatial big data, and spatio-temporal computational methods for solving many of the world's challenging problems in the domains of mobility, transport, and climate. It has been made possible through the rapid progress of computing, communication, and information technologies, but also by theoretical advancements in fields such as geographic information science (or to be more encompassing, spatial data science including its representations, models, and analysis methods) (Goodchild 1992; Raubal 2019; Reitsma 2012).

Geosmartness is essential for successfully transforming traditional cities and urban areas into smart cities, which are in essence digitally integrated urban spaces based on a real-time sensor-based control system. Such a system comprises technology, people, and community (Nam and Pardo 2011), and its major goal and challenge is to solve key problems of growing cities through integration of technology and environment (Batty et al. 2012). Ratti and Claudel (2016) provide an overview of future smart-city concepts, emphasizing also the value of open data and platforms, and the necessity for smart citizens. Concrete efforts and lessons learned when building a smart city have been demonstrated and described, such as for Barcelona (Gasco-Hernandez 2018).

The various methods and tools enabling geosmartness (Fig. 6.1) cover the traditional stages of a GIS (geographic information system) process, including spatial data modeling, representation, analysis, and presentation (Longley et al. 2011), but on a much wider scale, involving novel interfaces, cutting-edge information technology, and real-time sensor data (not only at the geographic scale; Montello 1993).

Spatial big data results from the ever-increasing progress in computing, communication, and information technologies. They come in the form of massive movement-trajectory datasets, fine-resolution environmental data, or specific user-behavior data (e.g., from eye-tracking), often in real time. Li et al. (2016) characterize geospatial big data by the following dimensions:

- Volume: Exabytes (or more) of imagery, sensor, and location-based social-media data raise both storage and analysis issues.
- Variety: relating to the various types of geospatial data, such as raster, vector, network, structured, and unstructured data and their integration.
- Velocity: Real-time trajectory and social-media data, and other continuous streams of sensor data require data processing at the same speed as data acquisition.
- Veracity: Depending on the sources, geospatial big data vary in accuracy and precision, and impact reliability and trust. Quality assessment may therefore be difficult.
- Visualization: on the one hand providing procedures to impose human reasoning on big data analysis, and on the other hand facilitating the communication of patterns and relationships as the results from such analysis.
- Visibility: Geospatial big data can nowadays be efficiently accessed and processed through cloud-computing technologies.

In order to pursue knowledge discovery from these complex and massive spatial data, traditional spatio-temporal analysis methods are now extended and complemented on a large scale by machine-learning approaches (Raubal et al. 2018). Machine learning is applied to spatial big data in CyberGIS analytics, for spatio-temporal outlier and anomaly detection, and for predicting human spatial behavior. Spatial data science enhances machine learning by proposing methods for spatio-temporal modeling and context integration to achieve better results and higher performance. In the area of mobility and transport, it has recently been demonstrated how graph convolutional neural networks (GCNs) can be used for imputing human activity purposes from GPS trajectory data (Martin et al. 2018). Multiple personalized graphs were utilized to model human mobility behavior and to embed a large variety of spatio-temporal information and structure in the graphs' weights and connections. These graphs served as input to the GCNs, which in turn exploited such structure.

Geographic information technologies encompass systems and services that exploit geoinformation to support people's spatio-temporal decision making (Raubal 2018). They utilize data related to locations in space and time, and process these data with respect to spatial locations, which results in increased complexity during reasoning and data analysis. Nowadays, geographic information technologies not only include desktop GIS for acquiring, representing, analyzing, and visualizing spatio-temporal data, but also location-based services (LBS), which support people in their mobile decision-making by providing spatial information based on their current locations, typically by relying on GPS (Global Positioning System) technology built into them (Brimicombe and Li 2009). LBS can be further enhanced by other context information, such as the user's gaze. This allows taking the user's viewing direction into account (Anagnostopoulos et al. 2017), leading, for example, to personalized audio guides that help users to find objects in the environment, and adapting the audio content to what has previously been looked at (Kwok et al. 2019). This directly relates to geographic human–computer interaction, i.e., people's interaction with geographic information technologies (Hecht et al. 2011). Novel interaction modalities and paradigms, and context-aware user interfaces, are available nowadays. In addition to traditional user interfaces through which people can interact with text-based information or cartographic maps, novel interaction modes, such as audio, gesture, gaze, or vibration (Gkonos et al. 2017), and displays integrating augmented and virtual reality exist (Rudi et al. 2016).

6.3 Analyzing Urban-Mobility Patterns

Mobility has always been a crucial part of urban life. As cities grow larger, moving millions of people for work, errands, or leisure activities becomes increasingly complicated, and when unmanaged, mobility has severe negative effects such as greenhouse-gas emissions, air pollution, health problems (Krzyżanowski et al. 2005), and traffic congestion.

To mitigate these negative effects, system-level actions must be combined with actions that empower mobility behavior change of individuals (Banister 2011). Examples for system-level interventions are the implementation of smart traffic management systems, or adaptive and attractive public transport systems. Individual mobility change may be achieved by enabling new forms of mobility, such as mobility as a service (MaaS), on-the-fly ride sharing or on-demand last-mile buses. These novel mobility concepts are all manifestations of geosmartness as they are ways to optimally allocate spatial resources, for which they require detailed knowledge of individual and aggregated city-wide mobility behavior.

6.3.1 Data

With the proceeding digitalization of our society, cities have become a melting pot for data from many different sources. This development bears new and unprecedented potential of gaining detailed knowledge about people's mobility behavior that can be used to enable sustainable mobility concepts. From the perspective of movement analysis, all available data can be divided into two groups: tracking data and context data.

Quantitative movement analysis is based on tracking data, which can be described as sequentially recorded and time-stamped locations. In the past, the elicitation of these data was based on paper or telephone surveys, but over the past decade the diversity of tracking-data sources multiplied and today, a manifold of different types of tracking data are available. Examples are global navigation satellite system (GNSS) tracking data (Zheng et al. 2008), location data based on the proximity to WiFi hotspots (Sapiezynski et al. 2015), location data from social networks (Hasan et al. 2013), public transport smart card data (Zhong et al. 2016), call detail record (CDR) data (González et al. 2008; Yuan and Raubal 2016b; Yuan et al. 2012), and credit-card transactions (Clemente et al. 2018).

These sources offer new possibilities to analyze movement within cities. However, the many possibilities to record urban movement create a heterogeneous landscape of tracking data sets. Four factors are particularly important when comparing different data sets:

- Tracking style (e.g., fixed versus moving tracking devices as in the Eulerian versus Lagrangian tracking style concept; Laube 2014)
- Spatio-temporal resolution (i.e., sampling rate)
- Spatio-temporal distribution (track point distribution, e.g., regular vs. burst patterns)
- Sample biases (e.g., daily urban mobility vs. mobility of tourists).

Due to these differences, it is difficult to compare results across different data sets and to develop data-agnostic methods. These are still open research challenges to be addressed in the near future in order to ensure the success of urban movement data analytics.

The second part of the data that are available in an urban setting does not describe the movement of people itself but the context in which people are moving. These context data are important for the analysis of human mobility patterns because human movement is always set in and influenced by its spatio-temporal context (Sharif and Alesheikh 2018). For example, when driving, our movement is restricted by the street network, when using public transportation, we depend on fixed schedules; we walk faster when it rains (Knoblauch et al. 1996), and we move differently depending on the urban or suburban setting (Yuan and Raubal 2016a).

In the past, only a few sources of context data, usually with a coarse spatio-temporal resolution, were available. This changed with progress in the digitalization of cities, and today many different context data sources with fine spatio-temporal resolution are available. Among the most important ones, urban movement analytics are volunteered geographic information (VGI) platforms such as OpenStreetMap, which provides easy access to road networks and point-of-interest data. A more recent trend inspired by the success of the open-data community is the open-data movement at the city level. Today many cities have open-data policies and publish their data on open-data platforms. Sensor networks provide another important source for context data, such as temperature, noise, pedestrian counts, or air quality. Examples for sensor networks with publicly available data are VGI-based platforms such as OpenSenseMap or luft-daten.info for air-quality data. There are also sensor networks operated by the cities themselves such as the Array of Things project in Chicago. Other context data include photogrammetry or street imagery data such as Google

Street View. The latter has been used to automatically assess the well-being of neighborhoods (Suel et al. 2019) and to develop image-based navigation systems (Mirowski et al. 2019).

6.3.2 *Computational Methods for Large-Scale Spatio-temporal Mobility-Pattern Analysis*

Movement and context data generated by smart cities offer unprecedented possibilities for analyzing urban-mobility patterns (see also Chaps. 28 and 29). However, the large data volume, the variety of the new urban data sources, and the large bandwidth of tasks require the enhancement of traditional GIS methods known from classical movement analytics (Long and Nelson 2013; Zheng 2015).

6.3.2.1 Data Preparation and Data Fusion

Especially for the preparation of the data and for the combination of different spatial datasets, well-established GIS methods are of great importance. Important preprocessing steps are GPS-trajectory segmentation, map matching, spatial filtering or movement-trajectory compression. In the same way, proven GIS methods can be used to combine different spatial datasets and to enrich trajectories with context data (Jonietz and Bucher 2017).

However, with the growing data volume, manual processing will not be an option in the future. Therefore, scalability of workflows must always be kept in mind. This includes the choice of efficient algorithms, their efficient implementation, and the possibility of processing using distributed frameworks (e.g., big data frameworks).

6.3.2.2 Prediction and Labeling

The following tasks are of great importance when analyzing urban-mobility patterns: adding semantic information to unlabeled data and predicting urban mobility for a short forecast horizon (e.g., hours or days).

Adding semantic information is important because even though digital cities provide large volumes of data, large-scale tracking data sets are usually recorded passively (e.g., without interaction of the user) and are therefore unlabeled (Bauer et al. 2016). In order to interpret and understand urban mobility, these datasets must be enriched with semantic information such as activity labels or mode of transport.

The prediction of movement and mobility is important to optimize future states of the mobility system and to create flexible and personalized mobility offers. Knowing the future mobility demand within a city allows for optimizing the schedule of public transport systems, taxi placements, or timings of traffic lights. On the other hand, knowing about the subsequent places, an individual wants to visit helps in identifying potential ride-sharing partners.

The current state-of-the-art to solve these prediction and labeling tasks is the usage of machine-learning methods (Toch et al. 2018). The usual approach is to extract meaningful features from the available movement and context data, and to use them for training a classifier for label-prediction tasks or a regressor for predicting future mobility demand. Here, the random-forest algorithm (Breiman 2001) is especially worth mentioning, as it is very robust with regard to the distribution of the input data, has generally a very good performance, and does not require extensive hyper-tuning of parameters.

An important research direction is to create spatially aware machine-learning methods (Gilardi and Bengio 2000; Hengl et al. 2018). One problem is that general-purpose machine-learning algorithms do usually not consider spatial dependencies (e.g., spatial autocorrelation present in the input or output data; Cracknell and Reading 2014). Another recent research direction is to avoid the explicit feature extraction step altogether, because it usually implies the assumption of independent and identically distributed data. An alternative is the use of neural networks and learning feature maps directly from the data. However, here, it is often difficult to find a meaningful data representation that is suitable for neural networks. Possible representations are image representations (Chen et al. 2016a) or more recently graph representations (Martin et al. 2018).

6.3.3 Studies

In practice, studies based on tracking data are scarce and usually not publicly available. The most important reason for this is that personal tracking data are extremely privacy sensitive (Keßler and McKenzie 2018). This implies that on the one hand, it is difficult to find participants who are willing to share their geodata due to privacy concerns, and on the other hand, that datasets are unavailable for other research groups once they were collected. Resulting from this situation, there are two types of mobility studies: user studies based on participants that were recruited for the purpose of the study by a research group, and mobility studies based on data that were already collected for different purposes and contained the locations of users as a byproduct. The first type of study are also called active-tracking studies because users in these studies commonly provide feedback that can be used to label the data and to answer the underlying research questions. The second type of study is called passive-tracking studies because users are commonly unaware that they participate in a study and that their location is collected passively in the background without any possibility for the user to provide feedback. Some notable examples of mobility studies based on passive-tracking data sets include:

Brockmann et al. (2006) were among the first to use already-collected data (sightings of dollar bills from www.wheresgeorge.com) that contained information about human mobility as a byproduct. The analysis of this dataset with more than a million displacements uncovered fundamental statistical properties of human movement, such as a power-law distribution of traveling distances.

González et al. (2008) developed an early large mobility study based on CDR data collected for billing purposes by the mobile-phone provider, which also allowed for the reconstruction of human mobility patterns. These data allowed one to analyze the movement of individual persons over a time span of six months and revealed a high spatio-temporal regularity of human movement patterns.

Both studies are early representatives of large-scale empirical studies and are rather descriptive and general. Studies in later years became more specific:

Hasan et al. (2013) used data from smart cards utilized in public transportation systems to specifically analyze human mobility within a city. Among other results, this study reproduced the already known general mobility characteristics in an urban setting.

Yuan and Raubal (2016a) used CDR data that were enriched with demographic information to empirically analyze the spatial distribution of different demographic groups within a city.

Clemente et al. (2018) used credit card records in combination with CDR data from the same users to analyze urban mobility. This allowed them to cluster the users utilizing the semantically rich credit-card data and to interpret these clusters spatially using the CDR data.

The second type of study is significantly different as it involves only a small number of people but with very detailed data about these persons:

Eagle and Pentland (2006) conducted one of the first larger studies using mobile phones as wearable sensors. They collected information such as call logs, Bluetooth proximity data, and the current cell phone tower ID as a proxy for location. The goal of the study was to study not the mobility of the participants but rather their social interactions. This so-called reality-mining dataset is one of the first publicly available datasets that includes tracking data.

Zheng et al. (2008) introduced GeoLife, one of the first large GPS tracking studies, with 65 users being tracked for varying timespans within a ten-month period. These data were used to analyze individual mobility patterns. This dataset is publicly available and can be used for research purposes.

Alessandretti et al. (2018) used different publicly available datasets such as the reality-mining dataset and proprietary datasets such as the CNS dataset from Stopczynski et al. (2014) to show that persons only have a limited number of regularly visited locations and that, while the locations change slowly over time, the total number of locations stays constant.

6.3.4 SBB Green Class (Multi-modal and Energy-Efficient Mobility)

This section presents one case study in greater detail, the SBB Green Class pilot studies. In 2016 and 2017, the Swiss federal railways (SBB) carried out two large, one-year pilot tests of a MaaS concept. In these studies, customers received access to comprehensive mobility options for a fixed yearly fee. The first pilot study had 150 participants from Switzerland, who received a Swiss-wide public transport pass, a battery electric vehicle, a parking space at their local train station, and credit for carsharing and bikesharing services. The second pilot study had 50 participants and included an e-bike instead of the e-car. As part of the pilot study, all participants installed a tracking app on their phone and agreed to label the recorded and segmented GPS tracks with the user mode of transport and a high-level description of the trip purpose. The most interesting characteristic of the SBB Green Class pilot studies is a flat rate for mobility, where almost all costs are covered by the subscription fee, making it the first study of this size that can be used to test the impact of MaaS offers.

To evaluate the mobility behavior of the participants the tracking data had to be prepared using different preprocessing steps, such as the fusion of different data sources, imputation of missing labels, map matching, grouping movement into trips and tours, and the detection of anomalies. Subsequently the participants' mobility behavior could be compared to a pseudo-control group generated from the Swiss mobility and transport microcensus (MTMC). The most important results were:

- Especially the Green Class e-car pilot study participants traveled more than the average Swiss person and were particularly frequently multimodal. These differences can be partially explained by the SBB Green Class offer: on the one hand, there are available parking facilities near the railway station, which clearly promote combined travel, and on the other hand the lower marginal costs for mobility invite passengers to longer and more frequent journeys.
- A comparison with the control group revealed that the electric car primarily replaced journeys with a conventional vehicle; the proportion of train journeys differed only slightly between Green Class customers and the control group.
- The analysis of the longitudinal tracking data showed that the CO_2 emissions of most participants decreased significantly shortly after the start of the project. This can primarily be attributed to the electric vehicle, which has lower average CO_2 emissions than a car with a combustion engine (especially when taking into account the Swiss electricity mix). The overall development of the Green Class e-car users' CO_2 emissions and the possible impact of a MaaS offer can be seen in Fig. 6.2.
- A result that is particularly noteworthy is that the e-car established itself in the mobility mix of the participants in the long term while primarily replacing the conventional car.

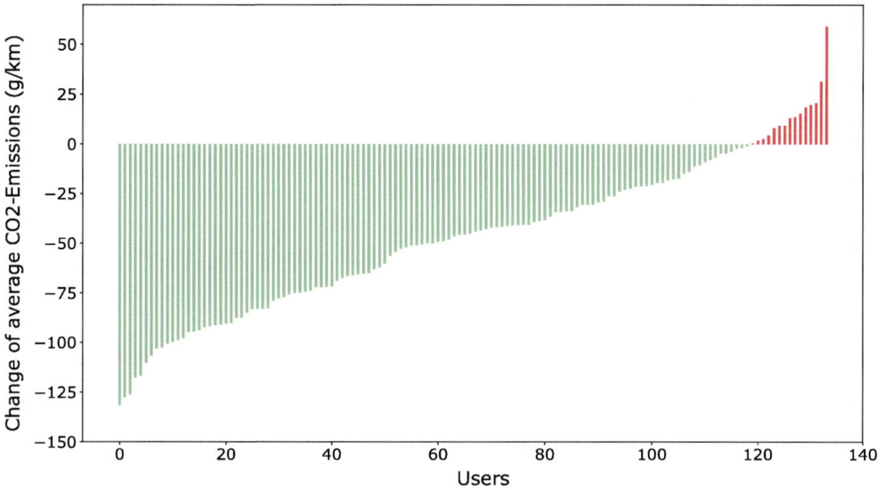

Fig. 6.2 Comparison of SBB Green Class 1 users' average CO_2 emissions during a six-week pre-project tracking phase and their emissions after they got access to the new mobility tools (public transport pass, e-car, etc.). Most participants (indicated in green) were able to reduce their CO_2 emissions significantly and only few participants (indicated in red) increased their average CO_2 emissions compared to before the project

6.4 Behavioral Change and Sustainable Mobility

It is often argued that making mobility ecologically sustainable requires a wide range of technical, institutional, and societal innovations, in particular in the short term (Banister 2008; Holden 2016; Kemp and Rotmans 2004). These innovations are related to the optimization and extension of public transport networks, to the electrification of car fleets alongside an increased renewable energy production, and also to various shifts in our use of mobility, for example from cars to alternative means of transport. The latter is commonly referred to as changing one's mobility behavior, and a substantial body of research concerns the effects of mobility behavior changes on a large scale (Bucher et al. 2019; Taniguchi and Fujii 2007), how ICT impacts people in their mobility planning and choices (Chen et al. 2016b; Cohen-Blankshtain and Rotem-Mindali 2016), how persuasive technologies can be used to nudge people toward certain desired behaviors (Gabrielli et al. 2014; Weiser et al. 2016), and how and where critical support infrastructure should be built to maximize its impact on mobility behavior (Buffat et al. 2018; Huétink et al. 2010). Here, we will focus on the potentials of novel geospatial and persuasive technologies alongside contextualized and personalized computational methods to help people travel sustainably.

6.4.1 Motivation

Behavior is strongly driven by motivation, which in turn arises from two groups of base needs (Deci and Ryan 2004; Reeve 2014): Psychological needs form the most innate group and include the desire for autonomy, competence and relatedness. They describe the facts that humans like to be in control of their actions, that these actions must be challenging yet doable, and that people need to interact with others within meaningful relationships. Social needs are similarly about the cultivation of relationships but are learned over the course of our lives. They encompass achievement, affiliation, intimacy, and a desire for leader- and follower-ship.

Individual actions (such as choosing a particular mode of transport) are usually spurred by either external or internal motivational sources. External sources include monetary incentives, rewards, or simply promises by other people. In stark contrast, intrinsic motivation is generated by one's own goals, expectations, beliefs, and perceptions. At its core is the perception we have of ourselves, subconsciously built by inspecting the effects of our behavior on other people. Based on this, we develop attitudes and beliefs, on which we rely when formulating certain goals or building expectations. Intrinsic motivation correlates with the satisfaction of the above-mentioned base needs (Van den Broeck et al. 2016). If a human does not manage to live up to his or her core beliefs, a state of cognitive dissonance is entered, which forms a strong internal motivational source that can be used to induce behavior change.

Such a change of behavior can be modeled using the trans-theoretical model (Prochaska and Velicer 1997). On a high level, we can classify behavior change into two phases: discovery and maintenance (Li et al. 2011). The trans-theoretical model splits discovery into a pre-contemplation, a contemplation, and a preparation phase, which are characterized by a transition from being unaware of a certain behavior to starting to form plans on how to change it. The transition into maintenance is performed once a person starts taking actions, which are prompted by triggers, for example, receiving a notification about an upcoming appointment (Fogg 2009). After reaching a certain level of competence, people have to be kept from relapsing until the behavior is truly internalized and a new habit is formed. Smart geographic ICT must thus be aware of the different motivational factors and phases that influence individuals in varying ways and provide adapted support for people in different circumstances and contexts.

6.4.2 Detecting and Supporting Behavioral Change

A substantial amount of research focuses on using ICT to detect and identify activities related to movement and mobility (Feng and Timmermans 2013; Gong et al. 2012; Montini et al. 2014), in particular the motives for traveling somewhere as they heavily influence transport-mode choices. This identification of activities and transport modes becomes increasingly accurate as researchers get easier access to large ground-truth datasets that can be effectively used for machine learning and thus automated inference at scale.

Once the activities are known, their change over time can be analyzed to detect sudden or gradual changes in behavior and support users adequately throughout different motivational stages. Jonietz and Bucher (2018) continuously mined trajectories with the aim of identifying behavioral patterns and anomalies. They summarized daily and weekly mobility usage by computing characteristic features; for example, the number of trips taken or the total distance traveled with a certain mode of transport. An anomalous deviation of these features from one week to another can indicate a transition from one phase of behavior change to another and should be reflected within the supporting ICT. Additionally, identifying people in similar behavioral transition phases can be used for analytical purposes or to target individual groups with specific incentives (Zhao et al. 2019).

Depending on the motivational phase, people have different needs for support: someone (pre-)contemplating change is well served by information about the existence of alternative transport options; someone taking action requires external motivators and well-timed and appropriate triggers (Weiser et al. 2015). If a trigger manages to increase our motivation (e.g., by giving additional external rewards) or to decrease the difficulty of the action (e.g., by providing a meaningful sustainable mobility alternative), a user is much more likely to exhibit the desired behavior (Fogg 2009). To provide alternative mobility plans, ICT has to generate and evaluate them, taking into account sustainability as well as the user's context (e.g., the planned activity at the destination, or past and future trips). Based on a wealth of (multi-modal) transport planning systems (Bast et al. 2016), heuristic methods (Bucher et al. 2017), and approaches based on previously recorded movement (Arentze 2013; Campigotto et al. 2016) were developed to generate meaningful routes. The resulting alternatives are scored using the primary feature of interest, e.g., the total CO_2 emissions, the distance, or the duration.

An often employed persuasive method is gamification, i.e., using game design elements in non-game contexts (Deterding et al. 2011). Gamification can be used as an external source of motivation by employing mechanisms such as feedback, rewards, challenges, competition, or cooperation (Weiser et al. 2015). These should follow a set of general design principles, such as offering meaningful suggestions, providing guidance, supporting user choices, or personalizing experiences. It needs to be noted that the use of common gamification elements for feedback on mobility behavior is not as straightforward as in other domains. As mobility is highly individual, simply offering rewards for taking the bicycle to work might be completely unfeasible for some while extremely easy for others. Similarly, rewarding points for taking public transport may lead to people trying to travel more, while the most ecologically friendly choice would likely be not to travel at all (Froehlich et al. 2009).

6.4.3 Studies

Among the well-known early studies on the effects of persuasive ICT on mobility, choices and behavior are applications that feature a combination of movement tracking and technology-assisted feedback, commonly by showing users the impact of the CO_2 emissions caused by their trips (Anagnostopoulou et al. 2018; Gössling 2018). UbiGreen (Froehlich et al. 2009) uses a combination of a mobile sensing platform, GSM cell tower localization, and information entered by users to record mobility patterns. It features a visual representation involving either a tree or an iceberg that indicates the effect of trips taken during a week. While there was no quantitative analysis of behavior change performed (due to the small sample size of 14 people and the short tracking duration of three weeks), interview responses demonstrated the viability of such eco-feedback applications. Similarly, MatkaHupi (Jylhä et al. 2013), tripzoom (Bie et al. 2012), the THELMA project (Bauer et al. 2016), or the Streetlife EU project (Kazhamiakin et al. 2015) featured smartphone applications that were used both as a tracker as well as for providing feedback to the mobility consumer.

Typically, these studies were performed with a smaller sample of participants (approximately 10–50) over the course of up to two months (Anagnostopoulou et al. 2018). Recently, several studies have tried to replicate their results with larger samples over longer periods of time. Research by Semanjski et al. (2016) involved a six-month data collection and intervention period with 3400 participants. During this time, movement data were collected and feedback given via a Web platform. Their results showed that eco-feedback can be used to initiate behavioral changes but the outcomes vary depending on the attitudinal profiles. Ebermann and Brauer (2016) similarly enrolled 248 participants to use a Web site during a three-week period and explored the influence of different goals ("self-exploration," "competition," "climate protection," etc.) on the use of various gamification elements. An additional large body of work emphasizes the use of persuasive technologies to improve personal health—which often leads to more ecologically sustainable travel behavior as well. Consolvo et al. (2008) explored the potential of early smartphones in combination with mobile sensing platforms to promote healthy lifestyles. Similarly, Harries et al. (2013) enrolled 152 participants for their study that used an app to promote walking behavior. They found that the app manages to increase the step count by around 64%, but that comparative social feedback did not improve this value.

The latter also indicates that not all persuasive strategies work well in a mobility context. Gabrielli et al. (2014) summarize these challenges associated with inducing a mobility behavior change for more sustainable future urban mobility. They found that changing mobility behavior is a lengthy process and that it is very difficult to find motivational features that engage a wide range of users. In contrast to the personal health domain, collective mechanisms (i.e., social influence) tend to have a stronger influence on behavior than individual ones. Their findings corresponded to research by Nicholson (2012) and Weiser et al. (2015), who stressed that eco-feedback must be timely and meaningful.

6.4.4 GoEco!

For a more in-depth account of a study targeting the change of mobility behavior, the example of GoEco! is chosen (Cellina et al. 2019). In contrast to previous studies, GoEco! targeted around 200 people from two diverse geographic regions; they were asked to participate in the experiment over the duration of a year. Within this year, three periods were chosen during which participants had to install an application on their smartphone that would simply record their movement in the first phase, give them additional eco-feedback (using gamification elements) in the second phase, and resort back to simple movement tracking for the third one (to determine potential long-term effects of the intervention in the second phase; Cellina et al. 2019).

The application used a naïve Bayes classifier to identify transport modes from several features, such as travel speed, journey distance, or the distance to public transport stops in the vicinity (Bucher et al. 2019). This transport-mode identification was then given to users for verification, after which several potential (and feasible) alternatives were computed for each trip. These alternatives were presented as feedback to people, together with an assessment of potential CO_2 emission reductions stemming from transitions to different transport modes. In addition, the gamified feedback included personal goals, weekly challenges, badges as rewards for desirable behavior (e.g., taking the bicycle to work, or completing a certain challenge), and a leaderboard that ranked people according to the number of badges they collected (Fig. 6.3; Cellina et al. 2019b).

Studying the long-term effects, it was found that people in rural areas changed their behavior on systematic routes. This was partially due to the selection of participants, who came from the city of Zürich (where people are often already eco-friendly travelers due to artificially created impediments for car drivers) and the canton of Ticino (where public transport is less developed, and the private car is the primary means of transport). The fact that people changed their behavior on systematic routes (e.g., from home to work and back) is likely due to having more options on those (as

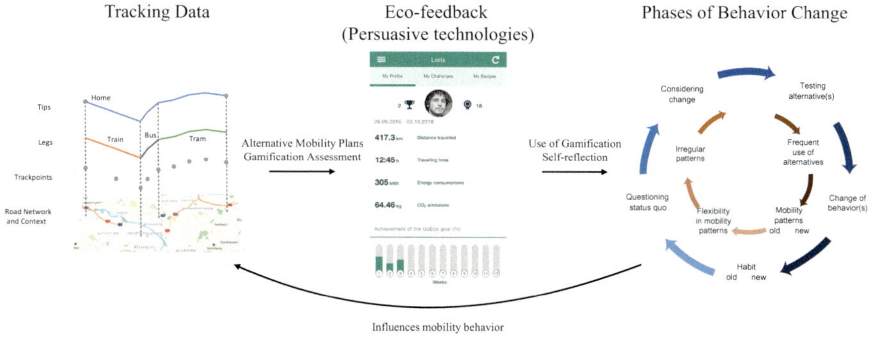

Fig. 6.3 Starting from movement and mobility tracking data, different mobility plans are evaluated, based on which gamified feedback is given. Users interpret and utilize the feedback differently depending on the phase of behavior change, which is reflected in the tracking data again

one is potentially less restricted by context, such as the need to drive the whole family or carry shopping goods) and due to only having to find good alternatives a limited number of times (in contrast to non-systematic routes, where a suitable alternative has to be searched for every time).

6.5 Mobile Decision Making

Mobile geospatial technologies support people in their location-based decision-making, and at the same time acquire spatial big data, which can be utilized for urban planning and the enhancement of urban infrastructure resilience (Heinimann and Hatfield 2017). Mobile location-based decision-making encompasses a variety of spatio-temporal constraints, which relate not only to people's spatio-temporal behavior in large-scale space (Kuipers and Levitt 1988) but also to their interaction with mobile devices, and perceptual, cognitive, and social processes (Raubal 2015). People often need to make fast decisions on the spot, which requires both fast access to spatial memory and immediate system responsiveness. Furthermore, mobile devices such as mobile phones limit the communication process to their users, for example through small screen size, which makes it challenging to present information to someone on the move (Montello and Raubal 2012).

6.5.1 Mobile Eye-Tracking and Gaze-Based Interaction

As described earlier, geosmartness is also enabled by novel interaction modalities and paradigms, and one of these concerns gaze-based interaction. Gaze-based interaction is made possible by eye-tracking technology, and it is regarded as a particularly efficient and intuitive interaction modality (Majaranta and Bulling 2014), especially when interacting with space and visual-spatial representations (Kiefer et al. 2017). In explicit gaze-based interaction, the user deliberately triggers an interaction by looking at a certain position in the stimulus, whereas implicit gaze-based interaction refers to the automatic interpretation of eye movements for recognizing cognitive states, such as search activities on maps.

The ability to track gaze movements with eye-tracking technology allows measuring the current point of regard on a specific stimulus. There exist remote and mobile eye-tracking devices, and nowadays, most of them are video-based corneal reflection systems (Duchowski 2017). Mobile eye trackers measure a person's visual attention on a stimulus in the wild instead of the laboratory. The basic recordings are called gazes, and it is generally assumed that perception takes place only if gaze remains almost still for a minimum amount of time. Gazes are therefore often aggregated spatio-temporally to fixations. A transition between two fixations is called a saccade, which is caused by a rapid movement of the eye. Eye-tracking data can be used for investigating cognitive processes, such as self-localization during wayfinding (Kiefer et al. 2014), for activity recognition (Kiefer et al. 2013), and as input for gaze-based assistants. Many eye-tracking systems allow for real-time data access, which is the principle behind such gaze-assistive systems.

6.5.2 Personalized Gaze-Based Decision Support

Urban mobility and navigation of the future will become more complex for people due to the variety of combined transport modes offered by mobility-as-a-service options, increased environmental complexity (especially in megacities), and the multifaceted decision-making process of how to engage in sustainable mobility. Smart city environments, as described here, in combination with gaze-assistive systems, will allow personalized navigation support for their users.

Nowadays, navigation instructions are typically displayed as turn-by-turn instructions on a digital map presented on small mobile screens (Hirtle and Raubal 2013). Visual attention switches between display and environment can lead to high cognitive load (Bunch and Lloyd 2006) and distraction, such as in busy traffic situations. These problems can be avoided by utilizing gaze-based interaction concepts. An example is GazeNav (Fig. 6.4), which enables gaze-based interaction for pedestrian navigation (Giannopoulos et al. 2015). Gaze is utilized to inform the wayfinder whether the road that he or she is gazing at is the correct one to follow. To use this system, the user wears mobile eye-tracking glasses, which capture the current point of regard. When a decision point with different options is approached, the user starts to examine the possible ones to follow. At the moment when the user's gaze is aligned with the correct street, the system automatically provides feedback to convey this, for example through a vibrotactile belt or, more effectively, its combination with gaze information (Gkonos et al. 2017). Systems for real-time gaze tracking in outdoor environments, which map the gazes from a mobile eye tracker to a georeferenced view using computer vision methods, allow for such personalized gaze-based decision support (Anagnostopoulos et al. 2017).

The example of GazeNav illustrates how novel interaction modalities will impact our spatio-temporal decision-making in the future, leading to more personalized

Fig. 6.4 Gaze-based pedestrian navigation

information that can facilitate and improve people's decision processes. In addition, such technologies will also provide an enormous amount of spatial big data, in this case user-behavior data, which can be utilized by both the private and public sectors to improve old services and offer new ones. This implies that our locations will be shared with a multitude of different services, and therefore, the protection of geoprivacy in combination with other types of personal information will become an even more important issue in smart city environments (Keßler and McKenzie 2018; see Chap. 32).

6.6 Conclusions and Future Work

The ever-increasing urban mobility and transport of people has led to an increase of greenhouse-gas emissions and traffic jams. In this chapter, we demonstrated how geosmartness, a combination of novel spatial-data sources, computational methods, and geospatial technologies made possible through major advances in ICT helps to make urban mobility of the future more sustainable and personalized. On the one hand, novel movement-analytics methods including machine learning can be applied to massive volumes of tracking and context data, in order to make short- and longer-term predictions of transportation network states. This will help to optimize future states of the mobility system and to create flexible and personalized mobility offers. An overview of recent mobility studies and SBB Green Class, a detailed case study of multi-modal and energy-efficient mobility, served as examples. On the other hand, mobility-pattern analysis will help detect people's behavioral changes, and the impact of their travel habits and alternative travel modes, which in turn should pave the way toward more sustainable forms of transport. Sustainable urban mobility will be one contributor to the reduction of CO_2 emissions in the future. We introduced methods for detecting and supporting behavioral change, related studies, and GoEco! as a concrete study targeting the change of mobility behavior through tracking data analysis and eco-feedback. Finally, from a user perspective people must also be directly supported in their complex mobile decision making. We proposed mobile eye tracking as a novel data source, which allows personalized gaze-based decision support in urban navigation. GazNav illustrated how gaze-based pedestrian navigation facilitates people's decision making based on the integration of gaze input, a navigation service, and a representative model of the environment.

Further research is necessary in all three of the discussed aspects of geosmartness, that is, spatial big data, spatio-temporal analysis methods, and geographic information technologies, in order to achieve a fully personalized and sustainable urban mobility of the future. For various states it will be important to have true real-time data from different sources—for example tracking, context, and social-media data—available, in order to evaluate a particular situation comprehensively and to detect the causes of a potential problem. The sheer data volume, and data integration and accuracy issues present obvious challenges. From a data analysis perspective, most machine-learning methods do not account for spatial autocorrelation; therefore, further research on how to make machine-learning methods spatially aware is required. In addition, most machine-learning models come as black boxes, which hinders interpretability and explanation of results. Machine-learning model interpretability is therefore a pressing issue (Hohman et al. 2019). Finally, future advancements in the area of urban informatics will continue to be technology driven. We expect novel geographic information technologies that will enhance both urban system evaluations and predictions, as well as mobile decision-making support for the individual user.

References

Alessandretti L, Sapiezynski P, Sekara V, Lehmann S, Baronchelli A (2018) Evidence for a conserved quantity in human mobility. Nat Hum Behav 2(7):485. https://doi.org/10.1038/s41562-018-0364-x

Anagnostopoulos V, Havlena M, Kiefer P, Giannopoulos I, Schindler K, Raubal M (2017) Gaze-informed location-based services. Int J Geogr Inf Sci 31(9):1770–1797

Anagnostopoulou E, Bothos E, Magoutas B, Schrammel J, Mentzas G (2018) Persuasive technologies for sustainable mobility: state of the art and emerging trends. Sustainability 10(7):2128

Arentze TA (2013) Adaptive personalized travel information systems: a Bayesian method to learn users' personal preferences in multimodal transport networks. IEEE Trans Intell Transp Syst 14(4):1957–1966

Banister D (2008) The sustainable mobility paradigm. Transp Policy 15(2):73–80

Banister D (2011) Cities, mobility and climate change. J Transp Geogr 19(6):1538–1546. https://doi.org/10.1016/j.jtrangeo.2011.03.009

Bast H, Delling D, Goldberg A, Müller-Hannemann M, Pajor T, Sanders P, Wagner D, Wer-neck R (2016) Route planning in transportation networks. In: Algorithm engineering. Springer, pp 19–80

Batty M, Axhausen K, Giannotti F, Pozdnoukhov A, Bazzani A (2012) Smart cities of the future. Eur Phys J Spec Top 214(1):481–518. https://doi.org/10.1140/epjst/e2012-01703-3

Bauer C, Cox B, Heck T, Hirschberg S, Hofer J, Schenler W, Simons A, Del Duce A, Althaus H-J, Georges G (2016) Opportunities and challenges for electric mobility: an interdisciplinary assessment of passenger vehicles. Final report of the THELMA project in co-operation with the Swiss Competence Center for Energy Research "Efficient technologies and systems for mobility". Technical Report. ETH, Zürich

Bie J, Bijlsma M, Broll G, Cao H, Hjalmarsson A, Hodgson F, Holleis P, van Houten Y, Jacob K, Koolwaaij J (2012) Move better with tripzoom. Int J Adv Life Sci 4(3):125–135

Boulouchos K, Cellina F, Ciari F, Cox B, Georges G, Hirschberg S, Hoppe M, Jonietz D, Kannan R, Kovacs N, Küng L, Michl T, Raubal M, Rudel R, Schenler W (2017) Towards an energy efficient and climate compatible future Swiss transportation system. Working paper. SCCER Mobility

Breiman L (2001) Random forests. Mach Learn 45(1):5–32. https://doi.org/10.1023/A:1010933404324

Brimicombe A, Li B (2009) Location-based services and geo-information engineering. Wiley-Blackwell, Chichester, UK

Brockmann D, Hufnagel L, Geisel T (2006) The scaling laws of human travel. Nature 439(7075):462–465. https://doi.org/10.1038/nature04292

Bucher D, Jonietz D, Raubal M (2017) A heuristic for multi-modal route planning. Progress in location-based services 2016. Springer, Berlin, pp 211–229

Bucher D, Mangili F, Cellina F, Bonesana C, Jonietz D, Raubal M (2019) From location tracking to personalized eco-feedback: a framework for geographic information collection, processing and visualization to promote sustainable mobility behaviors. Travel Behav Soc 14:43–56

Buffat R, Bucher D, Raubal M (2018) Using locally produced photovoltaic energy to charge electric vehicles. Comput Sci Res Dev 33(1–2):37–47

Bunch R, Lloyd R (2006) The cognitive load of geographic information. Prof Geogr 58(2):209–220

Campigotto P, Rudloff C, Leodolter M, Baue D (2016) Personalized and situation-aware multimodal route recommendations: the FAVOUR algorithm. IEEE Trans Intell Transp Syst 18(1):92–102

Caragliu A, Del Bo C, Nijkamp P (2011) Smart cities in Europe. J Urban Technol 18(2):65–82. https://doi.org/10.1080/10630732.2011.601117

Cellina F, Bucher D, Mangili F, Simão J, Rudel R, Raubal M (2019) A large scale, app-based behaviour change experiment persuading sustainable mobility patterns: methods, results and lessons learnt. Sustainability 11(9):2674. https://doi.org/10.3390/su11092674

Cellina F, Bucher D, Simão J, Rudel R, Raubal M (2019) Beyond limitations of current behaviour change apps for sustainable mobility: insights from a user-centered design and evaluation process. Sustainability 11(8):2281. https://doi.org/10.3390/su11082281

Chen C, Ma J, Susilo Y, Liu Y, Wang M (2016) The promises of big data and small data for travel behavior aka human mobility analysis. Transp Res Part C: Emerg Technol 68:285–299

Chen Q, Song X, Yamada H, Shibasaki R (2016) Learning deep representation from big and heterogeneous data for traffic accident inference. Paper presented at the Thirtieth AAAI Conference on Artificial Intelligence

Clemente R, Luengo-Oroz M, Travizano M, Xu S, Vaitla B, González M (2018) Sequences of purchases in credit card data reveal lifestyles in urban populations. Nat Commun 9(1):3330. https://doi.org/10.1038/s41467-018-05690-8

Cohen-Blankshtain G, Rotem-Mindali O (2016) Key research themes on ICT and sustainable urban mobility. Int J Sustain Transp 10(1):9–17

Consolvo S, McDonald D, Toscos T, Chen M, Froehlich J, Harrison B, Klasnja P, LaMarca A, LeGrand L, Libby R (2008) Activity sensing in the wild: a field trial of ubifit garden. Paper presented at the SIGCHI Conference on Human Factors in Computing Systems

Cracknell MJ, Reading AM (2014) Geological mapping using remote sensing data: a comparison of five machine learning algorithms, their response to variations in the spatial distribution of training data and the use of explicit spatial information. Comput Geosci 63:22–33. https://doi.org/10.1016/j.cageo.2013.10.008

Deci EL, Ryan RM (2004) Handbook of self-determination research. University of Rochester Press, Rochester, NY

Deterding S, Dixon D, Khaled R, Nacke L (2011) From game design elements to gamefulness: defining "gamification". Paper presented at the 15th International Academic MindTrek Conference: Envisioning Future Media Environments

Duchowski A (2017) Eye tracking methodology: theory and practice, 3rd edn. Springer, London

Eagle N, Pentland A (2006) Reality mining: sensing complex social systems. Pers Ubiquit Comput 10(4):255–268. https://doi.org/10.1007/s00779-005-0046-3

Ebermann C, Brauer B (2016) The role of goal frames regarding the impact of gamified persuasive systems on sustainable mobility behavior. Paper presented at the 24th European Conference on Information Systems

Elliott A, Urry J (2010) Mobile lives London. Routledge, London

Feng T, Timmermans HJ (2013) Transportation mode recognition using GPS and accelerometer data. Transp Res Part C: Emerg Technol 37:118–130

Fogg B (2009) A behavior model for persuasive design. Paper presented at the 4th international conference on persuasive technology

Froehlich J, Dillahunt T, Klasnja P, Mankoff J, Consolvo S, Harrison B, Landay J (2009) UbiGreen: investigating a mobile tool for tracking and supporting green transportation habits. Paper presented at the SIGCHI conference on human factors in computing systems

Gabrielli S, Forbes P, Jylhä A, Wells S, Sirén M, Hemminki S, Nurmi P, Maimone R, Masthoff J, Jacucci G (2014) Design challenges in motivating change for sustainable urban mobility. Comput Hum Behav 41:416–423

Gasco-Hernandez M (2018) Building a smart city: lessons from Barcelona. Commun ACM 61(4):50–57

Giannopoulos I, Kiefer P, Raubal M (2015) GazeNav: Gaze based pedestrian navigation. Paper presented at MobileHCI, 17th International conference on human–computer interaction with mobile devices and services, Copenhagen, Denmark

Gilardi N, Bengio S (2000) Local machine learning models for spatial data analysis. J Geogr Inf Decis Anal 4(1):11–28

Gkonos C, Giannopoulos I, Raubal M (2017) Maps, vibration or gaze? Comparison of novel navigation assistance in indoor and outdoor environments. J Location Based Serv 11(1):29–49. https://doi.org/10.1080/17489725.2017.1323125

Gong H, Chen C, Bialostozky E, Lawson CT (2012) A GPS/GIS method for travel mode detection in New York City. Comput Environ Urban Syst 36(2):131–139

González MC, Hidalgo CA, Barabási AL (2008) Understanding individual human mobility patterns. Nature 453(7196):779–782. https://doi.org/10.1038/nature06958

Goodchild M (1992) Geographical information science. Int J Geogr Inf Sci 6(1):31–45

Gössling S (2018) ICT and transport behavior: a conceptual review. Int J Sustain Transp 12(3):153–164

Harries T, Eslambolchilar P, Stride C, Rettie R, Walton S (2013) Walking in the wild: using an always-on smartphone application to increase physical activity. Paper presented at the IFIP conference on human–computer interaction

Hasan S, Zhan X, Ukkusuri S (2013) Understanding urban human activity and mobility patterns using large-scale location-based data from online social media. In: Proceedings of the 2nd ACM SIGKDD international workshop on urban computing

Hecht B, Schöning J, Erickson T, Priedhorsky R (2011) Geographic human–computer interaction. Paper presented at the CHI 11 Extended Abstracts on human factors in computing systems, Vancouver, BC, Canada

Heinimann H, Hatfield K (2017) Infrastructure resilience assessment, management and governance—state and perspectives. In: Linkov I, Palma-Oliveira J (eds) Resilience and risk—methods and application in environment, cyber and social domains. Springer, Berlin, pp 147–187

Hengl T, Nussbaum M, Wright MN, Heuvelink GB, Gräler B (2018) Random forest as a generic framework for predictive modeling of spatial and spatio-temporal variables. PeerJ 6:e5518

Hirtle S, Raubal M (2013) Many to many mobile maps. In: Raubal M, Mark D, Frank A (eds) Cognitive and linguistic aspects of geographic space: new perspectives on geographic information research. Springer, Berlin, Heidelberg, pp 141–157

Hohman F, Head A, Caruana R, DeLine R, Drucker S (2019) Gamut: a design probe to under-stand how data scientists understand machine learning models. Paper presented at the CHI conference on human factors in computing systems CHI 2019, 4–9 May 2019, Glasgow, UK

Holden E (2016) Achieving sustainable mobility: everyday and leisure-time travel in the EU. Routledge

Huétink FJ, van der Vooren A, Alkemade F (2010) Initial infrastructure development strategies for the transition to sustainable mobility. Technol Forecast Soc Chang 77(8):1270–1281

Jonietz D, Bucher D (2017) Towards an analytical framework for enriching movement trajectories with spatio-temporal context data. Paper presented at the 20th AGILE conference on geographic information science, 9–12 May 2017, Wageningen, The Netherlands

Jonietz D, Bucher D (2018) Continuous trajectory pattern mining for mobility behaviour change detection. In: Proceedings of the 14th international conference on location based services. Springer, Cham, Switzerland, pp 211–230

Jylhä A, Nurmi P, Sirén M, Hemminki S, Jacucci G (2013) Matkahupi: a persuasive mobile application for sustainable mobility. In: Proceedings of the 2013 ACM conference on pervasive and ubiquitous computing, pp 227–230

Kazhamiakin R, Marconi A, Perillo M, Pistore M, Valetto G, Piras L, Avesani F, Perri N (2015) Using gamification to incentivize sustainable urban mobility. Paper presented at the 2015 IEEE first international smart cities conference ISC2.
Kemp R, Rotmans J (2004) Managing the transition to sustainable mobility. In: System innovation and the transition to sustainability: theory, evidence and policy, pp 137–167
Keßler C, McKenzie G (2018) A geoprivacy manifesto. Transactions in GIS 22:3–19. https://doi.org/10.1111/tgis.12305
Kiefer P, Giannopoulos I, Raubal M (2013) Using eye movements to recognize activities on cartographic maps. Paper presented at the 21th SIGSPATIAL international conference on advances in geographic information systems, 9–12 Nov 2013, Orlando, FL, USA
Kiefer P, Giannopoulos I, Raubal M (2014) Where am I? Investigating map matching during self-localization with mobile eye tracking in an urban environment. Trans GIS 18(5):660–686. https://doi.org/10.1111/tgis.12067
Kiefer P, Giannopoulos I, Raubal M, Duchowski A (2017) Eye tracking for spatial research: cognition, computation, challenges. Spatial Cogn Comput 17(1–2):1–19. https://doi.org/10.1080/13875868.2016.1254634
Knoblauch RL, Pietrucha MT, Nitzburg M (1996) Field studies of pedestrian walking speed and start-up time. Transp Res Rec 1538(1):27–38. https://doi.org/10.1177/0361198196153800104
Krzyżanowski M, Kuna-Dibbert B, Schneider J (2005) Health effects of transport-related air pollution. WHO Regional Office Europe
Kuipers B, Levitt T (1988) Navigation and mapping in large-scale space. AI Mag 9(2):25–43
Kwok T, Kiefer P, Schinazi V, Adams B, Raubal M (2019) Gaze-guided narratives: adapting audio guide content to gaze in virtual and real environments. Paper presented at the CHI Conference on Human Factors in Computing systems CHI 2019, 4–9 May 2019, Glasgow, Scotland
Laube P (2014) Computational movement analysis. Springer International Publishing, New York, NY
Li I, Dey AK, Forlizzi J (2011) Understanding my data, myself: supporting self-reflection with ubicomp technologies. In: Proceedings of the 13th international conference on ubiquitous computing, pp 405–414
Li S, Dragicevic S, Castro F, Sester M, Winter S, Coltekin A, Pettit C, Jiang B, Haworth J, Stein A, Cheng T (2016) Geospatial big data handling theory and methods: a review and research challenges. ISPRS J Photogr Remote Sens 115:119–133
Long JA, Nelson TA (2013) A review of quantitative methods for movement data. Int J Geogr Inf Sci 27(2):292–318. https://doi.org/10.1080/13658816.2012.682578
Longley P, Goodchild M, Maguire D, Rhind D (2011) Geographic information systems and science, 3rd edn. Wiley, Chichester, UK
Majaranta P, Bulling A (2014) Eye tracking and eye-based human–computer interaction. In: Fairclough S, Gilleade K (eds) Advances in physiological computing. Springer, pp 39–65
Martin H, Bucher D, Suel E, Zhao P, Perez-Cruz F, Raubal M (2018) Graph convolutional neural networks for human activity purpose imputation. Paper presented at the modeling and decision-making in the spatiotemporal domain workshop at the 32nd conference on neural information processing systems NIPS, Montreal, Canada
Mirowski P, Grimes MK, Malinowski M, Hermann KM, Anderson K, Teplyashin D, Simonyan K, Kavukcuoglu K, Zisserman A, Hadsell R (2019) Learning to navigate in cities without a map. arXiv:180400168[cs]
Montello D (1993) Scale and multiple psychologies of space. In: Frank AU, Campari I (eds) Spatial information theory: theoretical basis for GIS. Springer, Heidelberg-Berlin, pp 312–321
Montello D, Raubal M (2012) Functions and applications of spatial cognition. In: Waller D, Nadel L (eds) Handbook of spatial cognition. American Psychological Association, Washington, DC, pp 249–264
Montini L, Rieser-Schüssler N, Horni A, Axhausen KW (2014) Trip purpose identification from GPS tracks. Transp Res Rec 2405(1):16–23

Nam T, Pardo T (2011) Smart city as urban innovation: Focusing on management, policy, and context. Paper presented at the fifth international conference on theory and practice of electronic governance, Tallinn, Estonia
Nicholson S (2012) A user-centered theoretical framework for meaningful gamification. Paper presented at Games + Learning + Society
Prochaska JO, Velicer WF (1997) The transtheoretical model of health behavior change. Am J Health Promot 12(1):38–48
Ratti C, Claudel M (2016) The city of tomorrow: sensors, networks, hackers, and the future of urban life. Yale University Press, New Haven, CN
Raubal M (2015) Mobile cognition: balancing user support and learning. Paper presented at WMC 2015—workshop on mobile cognition: using mobile devices to enhance human cognition, MobileHCI '15: 17th international conference on human–computer interaction with mobile devices and services adjunct proceedings, Copenhagen, Denmark
Raubal M (2018) Cognition and geographic information technologies. In: Montello D (ed) Handbook of behavioral and cognitive geography. Edward Elgar, Cheltenham, UK, pp 197–210
Raubal M (2019) It's the spatial data science, stupid! Paper presented at the spatial data science symposium—"Setting the spatial data science agenda", Santa Barbara, California. Retrieved from https://spatial.ucsb.edu/eventfiles/SDSS2019/Bio-Position-Raubal.pdf
Raubal M, Wang S, Guo M, Jonietz D, Kiefer P (eds) (2018) Spatial big data and machine learning in GIScience. Workshop at GIScience 2018, Melbourne, Australia
Reeve J (2014) Understanding motivation and emotion. Wiley, New York, NY
Reitsma F (2012) Revisiting the 'Is GIScience a science?' debate or quite possibly scientific gerrymandering. Int J Geogr Inf Sci 27(2):211–221
Rudi D, Giannopoulos I, Kiefer P, Peier C, Raubal M (2016) Interacting with maps on optical head-mounted displays. Paper presented at Spatial User Interaction SUI, 15–16 Oct 2016, Tokyo, Japan
Sapiezynski P, Stopczynski A, Gatej R, Lehmann S (2015) Tracking human mobility using wifi signals. PLoS ONE 10(7):e0130824. https://doi.org/10.1371/journal.pone.0130824
Semanjski I, Lopez Aguirre A, De Mol J, Gautama S (2016) Policy 2.0 platform for mobile sensing and incentivized targeted shifts in mobility behavior. Sensors 16(7):1035
Sharif M, Alesheikh AA (2018) Context-aware movement analytics: implications, taxonomy, and design framework. Wiley Interdisc Rev Data Mining Knowl Discov 8(1):e1233. https://doi.org/10.1002/widm1233
Stopczynski A, Sekara V, Sapiezynski P, Cuttone A, Madsen MM, Larsen JE, Lehmann S (2014) Measuring large-scale social networks with high resolution. PLoS ONE 9(4):e95978. https://doi.org/10.1371/journal.pone.0095978
Suel E, Polak JW, Bennett JE, Ezzati M (2019) Measuring social, environmental and health inequalities using deep learning and street imagery. Sci Rep 9(1):1–10. https://doi.org/10.1038/s41598-019-42036-w
Taaffe E, Gauthier H, O'Kelly M (1996) Geography of transportation. Prentice Hall, Englewood Cliffs, NJ
Taniguchi A, Fujii S (2007) Promoting public transport using marketing techniques in mobility management and verifying their quantitative effects. Transportation 34(1):37
Toch E, Lerner B, Ben-Zion E, Ben-Gal I (2018) Analyzing large-scale human mobility data: a survey of machine learning methods and applications. Knowl Inf Syst. https://doi.org/10.1007/s10115-018-1186-x
UN (2014) World urbanization prospects: 2014 revision, highlights. Retrieved from https://esa.un.org/unpd/wup/publications/files/wup2014-highlights.Pdf
Van den Broeck A, Ferris D, Chang C-H, Rosen C (2016) A review of self-determination theory's basic psychological needs at work. J Manag 42(5):1195–1229
Weiser P, Bucher D, Cellina F, Luca VD (2015) A taxonomy of motivational affordances for meaningful gamified and persuasive technologies. Paper presented at ICT4S 2015

Weiser P, Scheider S, Bucher D, Kiefer P, Raubal M (2016) Towards sustainable mobility behavior: research challenges for location-aware information and communication technology. GeoInformatica 20:213–239. https://doi.org/10.1007/s10707-015-0242-x
Yuan Y, Raubal M (2016a) Analyzing the distribution of human activity space from mobile phone usage—an individual and urban-oriented study. Int J Geogr Inf Sci 30(8):1594–1621. https://doi.org/10.1080/13658816.2016.1143555
Yuan Y, Raubal M (2016b) Exploring georeferenced mobile phone datasets—a survey and reference framework. Geogr Compass 10(6):239–252. https://doi.org/10.1111/gec3.12269
Yuan Y, Raubal M, Liu Y (2012) Correlating mobile phone usage and travel behaviour—a case study of Harbin, China. Comput Environ Urban Syst 36(2):118–130
Zhao P, Bucher D, Martin H, Raubal M (2019) A clustering-based framework for understanding individuals' travel mode choice behavior. In: Kyriakidis P, Hadjimitsis D, Skarlatos D, Mansourian A (eds) Geospatial technologies for local and regional development. In: Pro-ceedings of the 22nd AGILE conference on geographic information science. Springer, Cham, Switzerland, pp 77–94
Zheng Y (2015) Trajectory data mining: an overview. ACM Trans Intel Syst Technol 6:1–41. https://doi.org/10.1145/2743025
Zheng Y, Wang L, Zhang R, Xie X, Ma W (2008) GeoLife: managing and understanding your past life over maps. Paper presented at the ninth international conference on mobile data management
Zhong C, Batty M, Manley E, Wang J, Wang Z, Chen F, Schmitt G (2016) Variability in regularity: mining temporal mobility patterns in London, Singapore and Beijing using smart-card data. PLoS ONE 11(2):e0149222. https://doi.org/10.1371/journal.pone.0149222

Martin Raubal is Professor of Geoinformation Engineering at ETH Zürich, Switzerland. He was a Co-chair of AGILE (Association of Geographic Information Laboratories in Europe) from 2014 to 2019. His research interests include Spatial Data Science, Location Based Services, spatial cognitive engineering, and mobile eye-tracking.

Dominik Bucher is a doctoral student in the group of Geoinformation Engineering at ETH Zürich, Switzerland. Within the Mobility Information Engineering Lab, he is working on methods to process and analyze human movement and mobility data with the aim of supporting people in reaching more sustainable behavior.

Henry Martin is working as a doctoral student in the Mobility Information Engineering Lab (MIE) at the Chair of Geoinformation Engineering at ETH Zürich. He is interested in applying modern data analysis methods to spatio-temporal problems and enable mobility as a driver towards a more sustainable energy system.

Chapter 7
Urban Metabolism

Sybil Derrible, Lynette Cheah, Mohit Arora, and Lih Wei Yeow

Abstract Urban metabolism (UM) is fundamentally an accounting framework whose goal is to quantify the inflows, outflows, and accumulation of resources (such as materials and energy) in a city. The main goal of this chapter is to offer an introduction to UM. First, a brief history of UM is provided. Three different methods to perform an UM are then introduced: the first method takes a bottom-up approach by collecting/estimating individual flows; the second method takes a top-down approach by using nation-wide input–output data; and the third method takes a hybrid approach. Subsequently, to illustrate the process of applying UM, a practical case study is offered using the city-state of Singapore as an exemplar. Finally, current and future opportunities and challenges of UM are discussed. Overall, by the early twenty-first century, the development and application of UM have been relatively slow, but this might change as more and better data sources become available and as the world strives to become more sustainable and resilient.

7.1 Introduction

Water, electricity, gasoline, natural gas, food, concrete, and asphalt are some of the energy and resources that are imported, consumed, stored, or exported to, in, and from cities every day. Keeping track of these exchanges and processes can be extremely challenging and is at the heart of urban metabolism (UM). The term metabolism

S. Derrible (✉)
University of Illinois at Chicago, Chicago, USA
e-mail: derrible@uic.edu

L. Cheah · M. Arora · L. W. Yeow
Singapore University of Technology and Design, Tampines, Singapore
e-mail: lynette@sutd.edu.sg

M. Arora
e-mail: arora_mohit@alumni.sutd.edu.sg

L. W. Yeow
e-mail: lihwei_yeow@alumni.sutd.edu.sg

W. Shi et al. (eds.), *Urban Informatics*, The Urban Book Series,
https://doi.org/10.1007/978-981-15-8983-6_7

relates to how a human body converts nutrient intake into energy. The first attempt at quantitative (human) metabolism accounting was probably developed in the early seventeenth century where, in the first documented experiment, Sanctorius (1561–1636) spent over 30 years weighing his dietary intake and bodily excretions on a weighting chair to create a mass-balance sheet. Understanding that not everything that is consumed is directly excreted, he concluded that a significant portion of his consumption was lost through insensible perspiration via his skin (Eknoyan 1999).

Quantifying the metabolism of a city requires a similar methodological approach. The origins of the modern form of UM date back to 1965 when Abel Wolman wrote a ten-page article in Scientific American titled "The Metabolism of Cities" (Wolman 1965). As a sanitary engineer, Wolman's research interests delved into pollution, recognizing that getting an account of the flows of resources inside and outside of a city was key to solving the problem at its root. The concept then grew in popularity in the early 2000s, notably aided by the rise of the global research agenda toward sustainable development and the need to identify major consumers of energy and emitters of greenhouse gases (GHG). Over the years, UM has grown in its understanding into three main schools: Marxist ecology, industrial ecology, and urban ecology (Newell and Cousins 2014). Marx defined UM as the characterization of complex nature–society relationships that produce uneven outcomes; industrial ecology looks at UM as stocks and flows of materials and energy; and urban ecology looks at it as complex socio-ecological systems. More broadly, UM fits within the realm of sociometabolism defined by Haberl et al. (2019) as "a systems approach to study society–nature interactions at different spatiotemporal scales."

Since its origin, UM has evolved significantly from a methodological point of view, partly due to changes in data format and accessibility. Conceptually, UM remains largely an accounting framework, as illustrated in Fig. 7.1, that includes inputs (I), outputs (O), internal flows (Q), storage (S), and production (P) of water (W), energy (E), material (M), and food (F). With its initial focus on resources and materials, UM has evolved to account for energy (in addition to resources) and for the endogenous processes occurring within cities (e.g., accounting for the production of food in

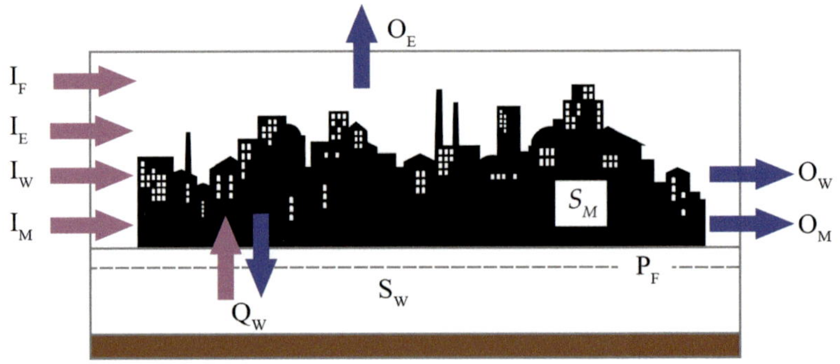

Fig. 7.1 Sketch of UM processes accounting for inputs (I), outputs (O), internal flows (Q), storage (S), and production (P) of water (W), energy (E), material (M), and food (F)

cities and for the internal reuse and recycling of materials), again in line with the global sustainability effort. A commonly adopted definition of UM comes from Kennedy et al. (2007) who defined it as: "the sum total of the technical and socio-economic processes that occur in cities, resulting in growth, production of energy, and elimination of waste."

From a methodological viewpoint, following the industrial ecology way of thinking, UM is largely inspired by material flow analysis (MFA), which for example quantifies the flows of a particular material across industrial sectors. An account of energy flows can then be added to the approach, thus giving material and energy flow analysis (MEFA). Broadly, there are two main methods for studying the UM of a city: the bottom-up method is based on directly collecting flow data from a city (e.g., how much water is consumed), while the top-down method is based on economic input–output data (e.g., from the United Nations International Trade Statistics Database, also known as UN COMTRADE). Both techniques are presented in this chapter. In addition, a hybrid approach combining bottom-up and top-down datasets has facilitated the development of several methods discussed in this chapter and categorized as hybrid methods.

Ultimately, the volume of data available is the main limiting factor to what can be included in an UM study. In spite of the fact that we have entered the era of big data, UM involves such a large number of flows that data availability is arguably the main reason why UM has not been applied more systematically to cities across the world. New datasets and new UM methods might help partly tackle this issue, however, as will be discussed. In fact, when it comes to urban informatics, UM holds a central presence and has the potential to directly inform policies and designs to help cities become more sustainable and resilient (Mohareb et al. 2016; Derrible 2019a).

In line with the general theme of this book, the main goal of this chapter is to give a brief introduction to urban metabolism by:

- Offering a brief review of the history of urban metabolism;
- Introducing two methods to calculate the metabolism of a city;
- Applying UM to a practical case study (Singapore); and
- Discussing the future of urban metabolism.

The structure of the book chapter follows these goals sequentially. To learn more about UM, the reader is referred to several important works (that inspired this chapter), including *Sustainable Urban Metabolism* by Ferrão and Fernández (2013), *Understanding Urban Metabolism: A Tool for Urban Planning* by Chrysoulakis et al. (2014), *Urban Engineering for Sustainability* by Derrible (2019b), and the book chapter "A Mathematical Description of Urban Metabolism" by Kennedy (2012). For quicker references and data on cities, the reader is strongly recommended to look at the Metabolism of Cities online platform accessible at https://metabolismofcities.org/.

7.2 History of Urban Metabolism

As an accounting framework, UM is used to gain an understanding of the flows between a city and its surrounding environment. As cities grew in size and as pollution levels increased significantly because of the Industrial Revolution—that notably spurred the initial push for suburbanization (Hall 2002)—it was only a matter of time before a technique like UM was developed. A first essay titled "Essay on the Metabolism of Berlin" was written by Theodor Weyl in 1894 and quantified the flows of nutrients in and out of Berlin (Lederer and Kral 2015). We can then see some traces of UM in Patrick Geddes's book "Cities in Evolution" (Geddes 1915). It was only when more data started to be collected and become available, however, that UM took its more modern form, and the rise of UM from sanitary engineering and in the twentieth century is, therefore, not surprising. Issues related to data availability have always been central to UM. In fact, even in his original article, Wolman could not calculate the UM of an actual city, and instead estimated the UM of a hypothetical American city of one million inhabitants, focusing on three inputs (water, food, and fuel) and three outputs (sewage, solid waste, and air pollutants). Figure 7.2 shows the original figure used by Wolman, which illustrates the large imports of water and exports of sewage from a typical city.

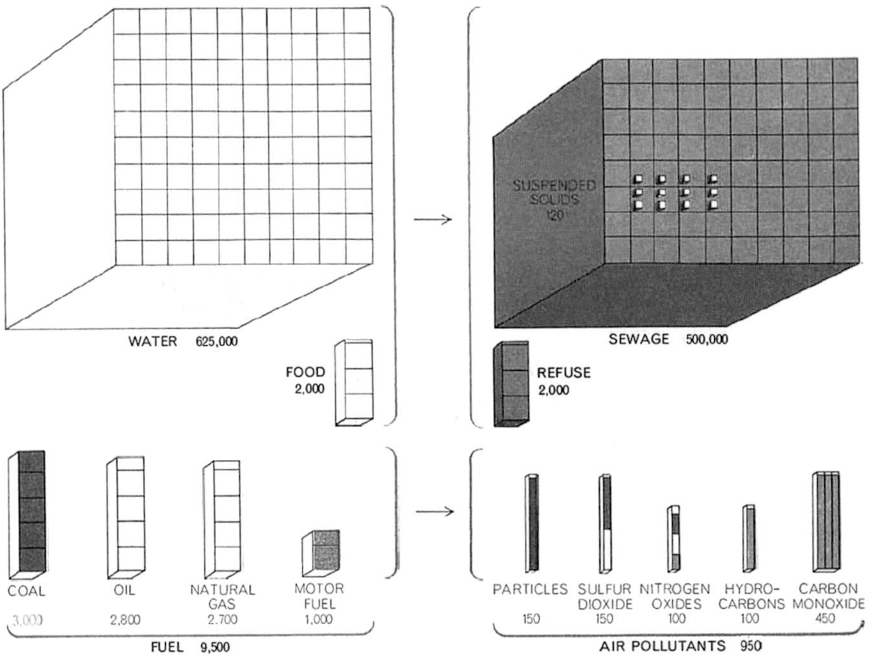

Fig. 7.2 Wolman's 1965 urban metabolism of a hypothetical American city of one million people, focusing on water, food, and fuel as inputs and on sewage, solid waste, and air pollutants as outputs

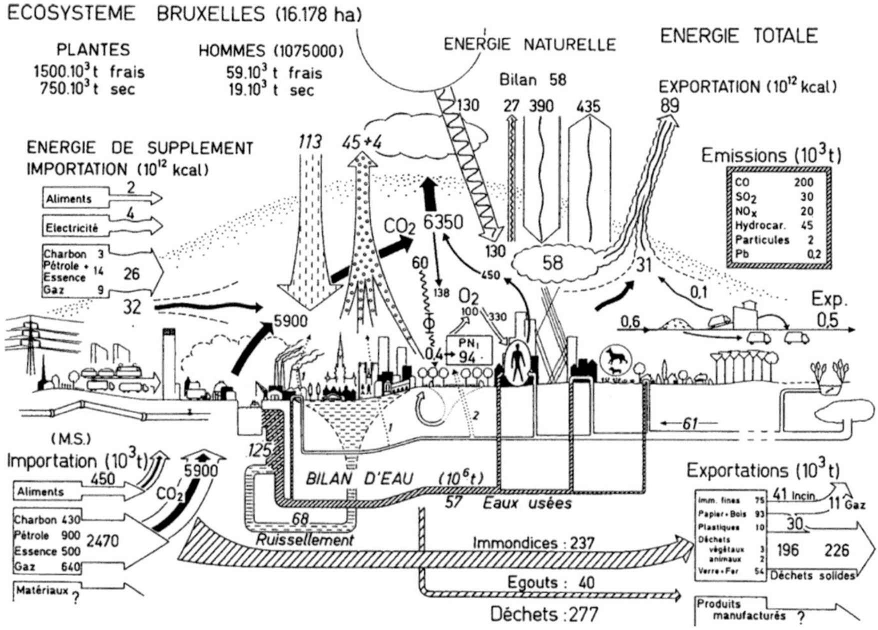

Fig. 7.3 UM of Brussels in the 1970s, Belgium. Adapted from Duvigneaud and Denaeyer-De Smet (1977)

Perhaps the most famous of all early UM studies is the surprisingly exhaustive case study of Brussels in the 1970s by Duvigneaud and Denaeyer-De Smet (1977). The main figure from the study is shown in Fig. 7.3. One year after the Brussels study, in 1978, Newcome et al. (1978) calculated the inflows and outflows of construction materials and finished goods in Hong Kong for 1971, foreseeing the amazing growth in demand for materials and resources for an increasingly wealthy and urban world. In their article, Kennedy et al. (2007) report the UM of nine cities:

- US typical (Wolman's study) in 1965
- Brussels (Belgium) in the 1970s
- Tokyo in 1970
- Hong Kong (China) in 1971 and 1997
- Sydney (Australia) in 1970 and 1990
- Toronto (Canada) in 1987 and 1999
- Vienna (Austria) in the 1990s
- London (United Kingdom) in 2000
- Cape Town (South Africa) in 2000.

Since the early 2000s, many more UM studies have been carried out, from Paris (Barles 2009) to Ho Chi Minh City (ADB 2014), including one particularly large study by Kennedy et al. (2015) that investigated the UM of 27 megacities. Significant data requirements remain a limiting factor to calculate the UM of more cities. In the next section, we will review two standard methods to estimate the metabolism of a city.

7.3 Methods of Urban Metabolism

Estimating the flows in Fig. 7.1 can be done in many different ways. In fact, there is no right technique as long the flows can be identified. Broadly, we can categorize techniques in three groups: bottom-up, top-down, and hybrid methods. From the bottom up, flows are investigated individually, for example, by contacting local water, gas, and electricity utility companies. From the top down, economic input–output (IO) data can be collected, often at the country scale, and then disaggregated to the city scale.

The bottom-up approach is generally preferred because it tends to provide more insights about a city; for example, to investigate differences between residential and commercial consumption patterns. The bottom-up approach tends to be arguably more accurate as well since disaggregating data from the national scale to the urban scale can be challenging. Nevertheless, methodologically, the top-down approach may be easier to apply and thus might be preferred in some instances. Other approaches including using emergy, ecological, or environmental network analysis and other methodological advancements have found lesser momentum but can be powerful tools for UM study. The three groups of approaches are introduced in this section.

7.3.1 Bottom-Up Methods

Identifying the flows in Fig. 7.1 from the bottom up can be done by asking the proper authorities for data or by using some means to estimate them. Flows related to the consumption of water, electricity, gas, and other resources can be collected from local utility companies, for example. Flows related to the amount of water received from precipitation can be collected from local weather stations. Nevertheless, collecting these data can be challenging—local utility companies may not want to share data or they may not have access to data in the first place. This section introduces some of the ways these flows can be estimated.

Primarily, we will use the divide and conquer technique by breaking down a problem into multiple parts; the general approach (not related to UM) is well discussed by Mahajan (2014). This approach is greatly influenced by the IPAT equation, initially developed by Ehrlich and Holdren (1971) and defined as

$$I = P \cdot A \cdot T \tag{7.1}$$

where I, P, A, and T stand for impact, population, affluence, and technology, respectively. Essentially, the end goal is to estimate total energy use or emissions (e.g., in watt-hours or Wh) and the problem is divided cleverly to play with units. For example, if we are looking for the total energy use linked with water consumption in liters [L], we can use the IPAT equation by estimating the average water consumption per person and the average energy use per liter of water; in terms of units, we get: [Wh] = [pers] × [L/pers] × [Wh/L]. In this section, we will cover four sectors: materials, energy, water, and food. The chapter is greatly inspired by Kennedy (2012) and more details can be found in Derrible's (2019b) book.

7.3.1.1 Materials

Cities are physically composed of countless materials. While it is impossible to quantify the flows of every material imported to or exported from a city, certain materials are worth investigating. In particular, for many cities, the two giants are concrete for buildings and asphalt for roads—in terms of weight, concrete production actually tends to be the most produced material in the world, over oil and gas production (Ashby 2013). In this section, we will see two ways to estimate these two materials, but the methods can easily be extended to account for other materials such as steel and other metals.

For buildings, we can try to divide the problem into estimating the floor space available per person, A, in a city in [m^2/pers], and the material intensity M of a building in tons per square meter (i.e., [t/m^2]). Specifically, for building type i, the stock S of material m (e.g., concrete) can be estimated from

$$S_{i,m} = P \cdot A_{i,m} \cdot M_{i,m} \tag{7.2}$$

The units of the three variables on the right-hand side are [pers] × [m^2/pers] × [t/m^2], thus giving us an answer in [t] (i.e., a weight). For roads, we can follow the same procedure or instead try to estimate the proportion of roads space taken by unit area in [km/km^2] for A, using the following equation:

$$S_{i,m} = D \cdot A_{i,m} \cdot M_{i,m} \tag{7.3}$$

where $S_{i,m}$ is the stock of road type i for material m in [t], D is the area of a city in [km^2], A is the affluence of roads in [km/km^2], and M is the material intensity in [t/km].

Results in units of weight can then be multiplied by an energy or carbon conversion factor, for example, in [MWh/t] and [t CO_2/t], respectively. These conversion factors can be found in the literature. For example, the Circular Ecology group offers a fairly extensive and free database accessible at https://www.circularecology.com/. In this database, the energy and carbon conversion factors of concrete are 1.53 MWh/t and

0.95 t CO_2/t, and the same factors for asphalt are 696.95 kWh/m^2 and 99 kg/m^2—note the difference of units between concrete and asphalt.

7.3.1.2 Energy

The UM of energy can include a number of sources since virtually every process requires some kind of energy. Here, we divide total energy use into six sources: buildings, transport, industry, construction, water pumping, and waste, such that:

$$I_E = I_{E,\text{buildings}} + I_{E,\text{transport}} + I_{E,\text{industry}} + I_{E,\text{construction}} + I_{E,\text{water pumping}} + I_{E,\text{waste}} \tag{7.4}$$

where I and E stand for impact and energy, respectively. Quantifying these six sources of energy can be challenging, and other sources might exist depending on the scope of the study. Ideally, data can be collected from local utilities. If not, individual sources can be broken down into quantities that are simpler to estimate.

Energy use in buildings can be broken down into energy use for heating, cooling, water heating, and light and appliances—about 50% of the energy used in buildings is consumed for space conditioning (heating and cooling) and about 20% for water heating, although values vary greatly, especially with climate. In the USA, data for these four subcategories are available from the Department of Energy. Other strategies are available in Derrible's (2019b) book. For transport, we either need to know how much fossil fuel was consumed and convert it into energy/emissions, or we need to estimate the average distance traveled per vehicle type (e.g., car and bus) and multiply it by an energy conversion factor. Local surveys are generally needed to estimate distances traveled per vehicle type, although national surveys can help. In the USA, the National Household Travel Survey offers US-wide travel pattern data, and the Environmental Protection Agency (EPA) offers typical conversion factors for distance traveled to carbon emissions.

For industry and construction, the flows can even be harder to estimate; this is where the top-down approach might offer an alternative. For water pumping, energy uses vary greatly based on several factors, including the topology of a city (i.e., hilly vs. flat terrain). Chini and Stillwell (2018) have gathered and made available a large database for the USA. Other values are available in the literature. We have to be a little bit careful since some values in the literature might take into account the full life cycle of a water distribution system (i.e., including the construction, operation, and disposal of the water treatment plant and water distribution system), while many others will not.

For waste, the quantity of waste generated as a weight must first be estimated (e.g., in [kg/y]). Urban-scale data are rarely available, but many countries offer national per capita estimates that can be sufficient—the World Bank has also compiled a significant database (Kaza et al. 2019). What may be more difficult is to get a breakdown of how much of the waste is recycled versus incinerated versus landfilled. Once achieved, however, the WAste Reduction Model (WARM) of the EPA offers carbon-emission intensity values for different disposal strategies. Finally, some studies also include natural energy inputs, such as the amount of energy received from the sun (that was included in Fig. 7.3). Kennedy (2012) offered an equation which can be referred to if needed. Ultimately, energy uses included in an UM study depend on the scope of the study.

7.3.1.3 Water

As Wolman had already illustrated in his study, water is one of the largest resources imported in a city, and water use is often included in UM studies. Moreover, although energy use and carbon emissions linked to water use tend to be relatively small, water is essential to generate electricity (i.e., Energy–Water Nexus) and for agriculture irrigation (i.e., to produce food), and monitoring water flows within an UM framework is typically desirable.

In general, the overall water balance of a city can be captured by seven variables, following the equation:

$$I_{W,\text{precip}} + I_{W,\text{pipe}} + I_{W,\text{surface}} + I_{W,\text{ground}} = O_{W,\text{evap}} + O_{W,\text{out}} + \Delta S_W \quad (7.5)$$

where $I_{W,\text{ precip}}$ denotes natural inflow from precipitation, $I_{W,\text{ pipe}}$ denotes pipe inflow, $I_{W,\text{ surface}}$ denotes net surface-water inflow (e.g., streams), $I_{W,\text{ ground}}$ denotes net groundwater inflow, $O_{W,\text{ evap}}$ denotes water loss through evapotranspiration, $O_{W,\text{ out}}$ denotes pipe outflow, and ΔS_W denotes annual change in water stored within the city—typically close to 0 unless groundwater levels are changing, for example, because of over pumping.

In Eq. (7.5), four variables are hydrological (precipitation, surface-water inflow, groundwater inflow, and evaporation) and should be available from local weather stations in most places. Pipe inflow relates directly to water use. Pipe outflow relates both to water use and stormwater management. Pipe inflow tends to match water use and accounts for both consumption and losses (e.g., through leaks). Estimating water use can be challenging without adequate data, however. Leakage rates can vary greatly from about 6% in some US cities to 50% in places like Rio de Janeiro (Derrible 2019a). For water consumption, Kennedy (2012) proposed a method that accounts for a base demand and a seasonal demand that was reproduced by Derrible (2019b). Ideally, metered data from water-treatment plants can be collected since it accounts for both consumption and leakage.

Pipe outflow can be broken down into three types: sanitary, stormwater, and infiltrated wastewater (from groundwater aquifers that penetrate the sewer system). Sanitary wastewater comes directly from water use, although the two quantities are not equal since some of the water used is lost through leakage, some evaporates, and some simply does not enter the sanitary sewer system (e.g., lawn watering); Kennedy (2012) found that 20–25% of the water consumed in Toronto did not enter the wastewater system. Here, again, data may be available from local wastewater utilities. Stormwater and wastewater comprise mostly surface runoff that enters the sewer system during heavy precipitation. Local wastewater utilities may have some data here as well, depending on whether the sewer system is combined or separated. Estimates of stormwater flows can also be generated through modeling, for example, by using the Natural Resources Conservation Service curve number model. Infiltrated wastewater flows are harder to estimate and may be negligible.

7.3.1.4 Food

Historically, food, as a specific sector, has rarely been included in UM studies. Nonetheless, UM studies that focus on energy and water often include the amount of energy and water used to prepare and dispose of food. Moreover, it may be more difficult to collect data on food, but we can still think about ways to estimate the UM related to food. First, the term food here includes both solid food and liquid food. Packaged drinks, for example, can be accounted for here. Water use related to food, such as water used in the kitchen, $I_{W,\mathrm{Kit}}$, can be included here, but we should be careful not to double-count it if it was already included in the UM section related to water.

Furthermore, food can be both imported into a city, I_F, as well as produced within a city, P_F. In terms of exports, food waste, $O_{F,\mathrm{FW}}$, can either be disposed of in landfills or it can be recycled (e.g., through composting). We can also account for the carbon and water lost by transpiration and evaporation, $O_{F,\mathrm{MET}}$ (where met stands for metabolism), and for the water disposed of in the sanitary sewer, $O_{F,S}$ (unless it is accounted for in the UM section related to wastewater). Altogether, we get the following equation for the UM of food:

$$I_F + P_F + I_{W,\mathrm{Kit}} = O_{F,\mathrm{FW}} + O_{F,\mathrm{Met}} + O_{F,S} \tag{7.6}$$

All or only some of the variables in Eq. (7.6) may be available depending on the scope of a study. In particular, food imports and exports may be available from freight data sources. It might be more challenging to estimate the other variables. In terms of units, food is generally expressed both as a weight in tons, although it could be expressed as an energy in Wh or Joules with the proper conversion factors. This is all we will cover in this section, but many more methods and techniques can be imagined and applied to study UM from the bottom up. Now, we will switch to a different conceptual approach to UM by estimating flows from the top down.

7.3.2 Top-Down Methods

Bottom-up approaches for UM accounting often tend to be time consuming and data intensive. As an alternative, most countries maintain data for economy-wide import, export, and production of resources, which can be tapped for an UM assessment. A top-down approach primarily benefits from the availability of relevant data in aggregate form. Often generating economy-wide insights on UM can be a powerful tool to influence sustainability efforts at the national or regional scale. In addition, the top-down approach tends to be easier to carry out and relies on international datasets, which helps in making time-series assessments to track progress over time. This section first provides a historical evolution of top-down economy-wide material flow accounting. It also discusses resources categories, data sources, and the accounting methods that can be chosen based on the scope and boundaries of an UM study.

7.3.2.1 General Approach

The MFA in an economy-wide (ew) exercise signifies the socioeconomic metabolism of a territory. Even though this section provides a methodology for an ew-MFA, often only partial accounts are performed, both in terms of materials and commodities as well as inflows and trade, or outflows in some combinations. As illustrated in Fig. 7.4, ew-MFA aims to assess the overall material inputs into a national economy, material stock changes within the economic system, and the material outputs to the external environment and economies (Krausmann et al. 2018). Such an exercise aims to describe the total scale of socio-economic activities in physical quantities. While initial efforts for ew-MFA were initiated in the 1990s in Austria, Japan, and Germany, credit for leading the global comparative ew-MFA methodology has often been assigned to a seminal study by Matthews et al. (2000). They assessed five countries, namely Austria, Netherlands, Germany, Japan and the USA, for their comprehensively mass-balanced material flows from 1975 to 1996, and they developed material flow indicators.

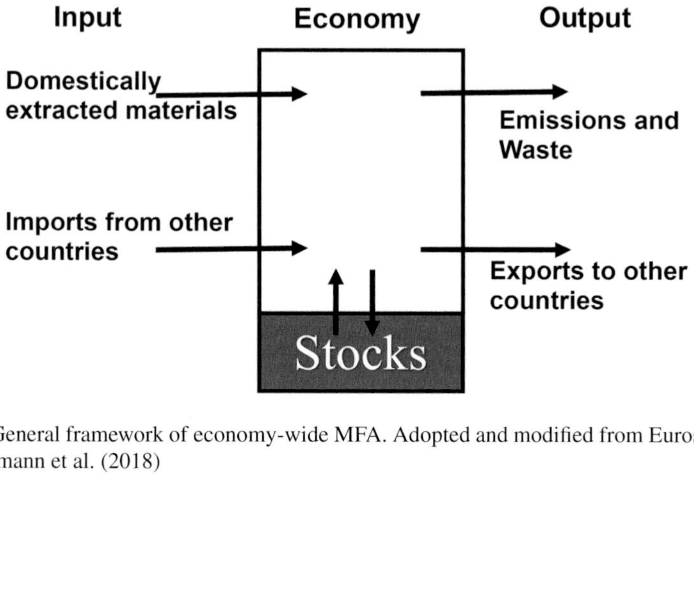

Fig. 7.4 General framework of economy-wide MFA. Adopted and modified from Eurostat (2001) and Krausmann et al. (2018)

In the same fashion, and to harmonize methodological details and indicators, Eurostat published its 2001 report "Economy-wide material flow accounts and derived indicators: A methodological guide" (Eurostat 2001), which has evolved over the years (Eurostat 2018) and which remains widely adopted for ew-MFA. For a step-by-step procedure to perform ew-MFA, the reader can refer to the comprehensive guide developed by Krausmann et al. (2018).

The basic concept of ew-MFA follows the mass-balance principle with a unit of metric tons per year (i.e., [t/y]) where:

$$\begin{aligned}\text{Input} &= \text{Output} + \text{Additions to Stock} - \text{Removals from Stock} \\ &= \text{Output} + \text{Net Stock Changes}\end{aligned} \tag{7.7}$$

Covering over 70 material groups, a typical MFA approach aggregates four material categories, namely biomass, metal ores, non-metallic minerals, and fossil energy carriers. In terms of biophysical bases for society, these four major material categories fulfill all the material and energy requirements for socio-economic metabolism such as food, feed, energy, housing, and infrastructure, including all man-made artifacts. Water and air are typically not accounted along with these four major groups of materials, excluding the mass balancing items such as moisture.

Table 7.1 defines the main MFA parameters for input and output into the economy, as well as for societal stocks. Most commonly, ew-MFA considers direct flows, which are defined as flows crossing the system (national) boundary. Major direct material flow categories include domestic extraction (DE) and imports on the input side, with

Table 7.1 MFA parameters and definition

Parameter	Definition
Domestic extraction (DE)	Used extraction of materials including solid, liquid, and gaseous raw materials from the natural environment (excluding water and air)
Imports, exports	All imported or exported commodities as weights (e.g., metric tons). Traded commodities comprise of goods at all stages of processing from basic commodities to highly processed products
Stocks	Physical structures of society: humans, livestock, and manufactured capital
Manufactured capital	All in-use artifacts (buildings, infrastructures, and durable goods)
NAS	Net additions to stock; year to year change of stocks
DPO	Domestic processed output of wastes and emissions including deliberately applied materials (e.g., fertilizers)
DPO*	DPO excluding balancing flows of oxygen and water (i.e., the fraction of DPO contained in DE)
Balancing flows	Oxygen taken up during combustion and respiration and water uptake by humans and livestock
Metabolic rate	Material consumption per capita of population
Material intensity	Material consumption per unit of GDP

exports and domestic processed outputs (DPO) of waste and emissions on the output side. DPO includes all waste and emissions from processing, manufacturing, use, and final disposal of materials. Unused or indirect flows that do not become an input for production or consumption are ignored. Because of the direct flows into and out of an economy, there are net changes in the stocks, which are taken into consideration to assess the physical growth. All accumulated materials in the form of manufactured capital and discarded or demolished artefacts lead to a net addition to stock (NAS) that can be positive or negative based on the overall balance. Negative NAS is rare in growing cities and national economies.

Considering the mass balance nature of ew-MFA, it is important to account for the water and air flows required in the processing and transformation of materials. Such flows are categorized as balancing items on the input and output sides. These may include water vapors for respiration, oxygen required for combustion of fossil fuels, and atmospheric gases captured or transformed into commodities such as fertilizers. These balancing items can be calculated using stoichiometric equations. Based on these material flow categories, a national material balance for a given year can be given by:

$$\begin{aligned} \text{DE} + \text{Imports} + \text{Input Balancing Items} = {} & \text{Exports} + \text{DPO} \\ & + \text{Output Balancing Items} + \text{NAS} \end{aligned} \quad (7.8)$$

In socioeconomic metabolism, material flows represent the pressure on the environment from an economy. These pressures can be measured through aggregated material flow indicators, which capture the socioeconomic sustainability of the system being studied. Direct material input (DMI) measures the direct input of all materials with an economic value and used in production and consumption activities. Domestic material consumption (DMC) provides all material inputs into an economy that are destined to be consumed and eventually released into the environment as waste, representing domestic waste potential. Physical trade balance (PTB) represents the balance of imports minus exports. These indicators are mathematically defined by:

$$\text{DMI} = \text{DE} + \text{Imports} \quad (7.9)$$

$$\text{DMC} = \text{DE} + \text{Imports} - \text{Exports} \quad (7.10)$$

$$\text{PTB} = \text{Imports} - \text{Exports} \quad (7.11)$$

For cross-country comparisons, material flow indicators require appropriate measures to account for differences in size. Overall, material efficiency is assessed by relating DMC to GDP. The ratio of DMC to GDP is defined as material intensity while the ratio of GDP to DMC is defined as material productivity. The ratio of material flows to total land area measures the scale of the physical economy to its natural environment. The DE to DMC ratio measures the dependence of the physical economy on domestic raw material supply. The proportion of import or export with DMI measures the trade intensity for import or export for a physical economy.

7.3.2.2 Data Sources

Several data sources exist to meet the data requirements needed to carry out an ew-MFA; for example, to collect inflow, outflow, or domestic extraction. National statistics and databases serve as the primary and most reliable data sources due to their direct collection mechanisms. Multiple international databases with harmonized values across countries and commodities also exist. In particular, the United Nations International Trade Statistics Database (UN COMTRADE) remains one of the most comprehensive datasets for international trade that provides monetary as well as quantity data for import and export commodities. This dataset can be aligned with the MFA computation tables based on the focus of the UM exercise for biomass, metals, fossils or non-metallic minerals. In addition, the Food and Agriculture Organization (FAO) maintains the FAOSTAT database for all biomass production and trade, which is more detailed and reliable.

Table 7.2 provides major data sources for various material categories. It is important to highlight that both the time scale (1917–2018) and the geographical coverage

Table 7.2 Major data sources for material flows in world economies

Material	Flows	Main source
Biomass (food, paper, wood, timber, and products, etc.)	Production, import, export, consumption	FAOSTAT, UN COMTRADE
Metals (steel, aluminum, copper, etc.)	Production, import, export, consumption	World Steel Association, The Aluminum Association, British Geological Survey, US Geological Survey, UN COMTRADE, UN Industrial Commodity Statistics Yearbook
Non-metallic minerals (sand, gravel, etc.)	Production, import, export	UN COMTRADE, UN Industrial Commodity Statistics Yearbook, United States Geological Survey
Cement	Production, import, export, consumption	CEMBUREAU, UN COMTRADE, UN Industrial Commodity Statistics Yearbook
Asphalt	Production, import, export, consumption	International Energy Agency (IEA)
Fossil materials and petroleum products (coal, crude and refined oil, gas, etc.)	Import, export, consumption	IEA, UN COMTRADE

(from a few countries to worldwide) of these data sources vary significantly. Additional sources of data include scientific studies, reports, and surveys, which can be very useful in certain cases.

For countries with limited datasets, several academic studies over the years have led to a comprehensive understanding of socio-economic metabolism, leading to significant datasets. Ongoing efforts in UM and industrial ecology communities have resulted in data repositories such as the industrial ecology database at the University of Freiburg Germany (https://www.database.industrialecology.uni-freiburg.de/), the UNEP MFA database (https://www.resourcepanel.org/global-material-flows-database, https://www.materialflows.net/), and the Eurostat MFA database (https://ec.europa.eu/eurostat/web/environment/data/database).

In case of poor data quality for certain commodities or countries, various datasets can be combined. When combining datasets for UM assessment, proper validation processes should be followed. For instance, data for domestic extraction of primary resources such as mining activities and food and vegetable production should ideally be validated with national statistics. Data for consumption of non-metallic minerals can be validated with consumption data for cement and asphalt. Likewise, gross metal ore production can be estimated from metal production and ore grades data in mining. Such exercises help in ensuring the mass balance of material flow. We now move on to hybrid methods to perform a UM study.

7.3.3 Hybrid Methods

Based on the scope and boundary of an MFA study, raw material equivalents (all materials used in the production of a commodity) for traded commodities can be calculated based on life-cycle assessment (LCA), environmentally extended input–output models, or by combining both. This is particularly useful for estimating consumption-based indicators such as the material footprint of an economy. Multiregional input–output (MRIO) models have been most widely used for sectoral resolution of physical flows based on monetary inputs and outputs. Allocating physical amounts of material extraction to products of final consumption can be carried out based on monetary information about the economics and structure of a sector while considering global processing chains and trade; however, challenges also exist (Krausmann et al. 2017a).

To estimate material and substance stocks, several extensions have been developed with varied temporal, sectoral, and spatial resolutions. Methodologically, it includes top-down and bottom-up static or dynamic stock assessment models. The basic concept of stock assessment depends on the service life of built-up stock and stock renewal rates, which are estimated for stock building artifacts such as infrastructure, buildings, road networks, and vehicles (Fishman et al. 2014; Krausmann et al. 2017b). Techniques such as geographical information systems and satellite-based imaging have allowed for various advances in the measurement of stocks and resource flows. In addition, hybrid approaches combine both the bottom-up and top-down approaches for assessing the UM of a city. From an ecological system's point of view, the use of emergy and ecological network analysis (ENA) has found greater interest.

The use of emergy originated in the 1950s through the pioneering work of the Odum brothers on the energetic basis of ecology on Earth. Hau and Bakshi (2004) suggest that emergy analysis "provides an ecocentric view of ecological and human activities, which can be used for evaluating and improving industrial activities." This approach is fundamentally based on the principle that the sun is the primary source of energy for all ecological and economic activities on earth. It considers tidal energy and deep earth heat as additional non-solar sources of energy on Earth and converts them into an objective matrix of energy quality that can be added altogether. As a result, all direct or indirect energy required to manufacture or deliver any or all products and services can be characterized in terms of solar energy equivalents. Emergy, hence, is estimated based on energy required to perform a function or service, with solar energy as the only source of energy (Odum 1996). As a scientific unit, emergy is represented in terms of solar embodied joules, abbreviated as [sej]. To account for energy transformations from high to low quality or into heat, the concept of solar transformity has been developed. Solar transformity, as a measure of energy quality or transformations, is defined as the solar emergy required to make one J of a service or product (measured in [sej/J]). Mathematically,

$$M = \tau \cdot B \tag{7.12}$$

where M is emergy, τ is transformity, and B is available energy.

This equation provides a convenient way of estimating the emergy of commodities, resources, and services. Odum pioneered the estimation of transformity for most inputs and, at the time of this writing, research still relies on Odum's matrix to estimate emergy. Total emergy input to the Earth can be derived from the sum of emergy of solar exposure, tidal energy, and deep Earth heat. To estimate ecological and metabolic pressures, emergy estimations can be carried out from the planetary level to the product or city level. To integrate economic and ecosystems activities, it is possible to estimate emergy of economic inputs based on the total emergy of a country and its gross national economic product, thus allowing for an objective comparison. The thermodynamic rigor behind this approach, the inclusion of ecological contributions in economic activities, and the ease of objective comparison based on a single measurement unit are some of its major advantages. The reader should refer to Odum (1996) for a detailed methodology.

As a different approach, modeling the complexity of nature–societal interactions has been carried out in some studies through ecological network analysis and its variations. This approach develops urban metabolic networks between different actors and assigns possible transformative processes to the flows (Fath et al. 2007). In comparison to linear relationships, network analysis captures more realistic interactions between various stakeholders and flows. However, complexity and assumptions involved in network simulations are primarily data limited. The methodology has evolved to capture the complete dynamics of urban metabolic activities. The scope and boundary of an urban metabolic network varies according to carbon emissions, pollutants, energy, materials, nutrients, and other substances. Finally, several studies have combined network analysis with emergy and MFA to provide robust comparable results for cities such as Beijing and Vienna (Chen and Chen 2012; Zhang et al. 2009). As a practical case study, we will now turn to the UM of Singapore.

7.4 A Case Study: The Metabolism of Singapore

Singapore has unique characteristics that makes it a good case study for showcasing the methodologies of UM. In 2016, the small and dense city-state in Southeast Asia housed 5.6 million people on a total land area of 720 km^2 and imported most of its material, food, and energy requirements. Unlike many other cities, the city-state has clear national and urban boundaries that coincide with each other (Abou-Abdo et al. 2011). Thus, all flows in and out of the city are classified as international trade and are well documented at Singapore's highly regulated ports of entry. Moreover, water flows in Singapore are highly managed by the Public Utilities Board (PUB), making for relatively easy accounting. Stormwater and used water are collected in "separate storm and sanitary sewer systems" (Irvine et al. 2014), which channel stormwater and surface runoff to rivers and reservoirs, and used water to water treatment plants (Tortajada et al. 2013). The water distribution network is robust, with "[no] illegal connections, and all water connections are metered" (Tortajada and Buurman 2017).

The study of Singapore's UM from the perspective of material flows began with Schulz (2007), who used physical trade flows and other data sources to conduct an ew-MFA, as described in the previous section. The flows of biomass, construction materials, industrial minerals, fossil fuels, and semi- and final products were analyzed over a 41-year period from 1962 to 2003. The study found that DMC "remained closely coupled to economic activity," rising in tandem with Singapore's massive economic growth since independence. Chertow et al. (2011) continued this work into the years 2000, 2004, and 2008, and have expanded the scope of flows to include emissions, waste, and recycling. The authors found large variations in DMC of between 14 and 55 metric tons per capita, which is mainly explained by variations in the import of construction minerals. Other UM studies in Singapore include an analysis of phosphorus flows (Pearce and Chertow 2017), and stocks and flows of concrete and steel in residential buildings (Arora et al. 2019). Beyond the analysis of material flows, system dynamics have been used to study urban resource flows (Abou-Abdo et al. 2011) and water (Welling 2011), while Tan et al. (2019) use exergy and ecological network analysis to study Singapore's resource effectiveness.

As an illustration of UM methods, this section adopts the simpler top-down approach to estimate the UM of Singapore in 2016, owing to the fact that as a city-state, national data do not need to be disaggregated to the urban scale. A wide range of data sources was used, such as international trade statistics from UN COMTRADE, data from the Food and Agriculture Organization (FAO), the International Energy Agency (IEA), and Singapore's Department of Statistics. The physical flows reported by these data sources are combined and adjusted to achieve mass balance. From these balanced flows, the key metabolism indicators, such as DMI and DMC (Eurostat 2001), are calculated and compared with the same indicators during Singapore's independence in 1965 (Schulz 2007).

Figure 7.5 shows the material flows of Singapore's economy in 2016. In total, 270.3 million metric tons of material were imported, with a large majority being fossil fuels (187.2 Mt, 69%) followed by non-metallic minerals (65 Mt, 24%), which are mainly used for constructing buildings and infrastructure, such as the 9,308 lane-kilometer long road network (Government of Singapore 2019). As a major oil trading and refining hub, most of the fossil fuels it imports are in the form of crude oil, which is traded or refined into other petroleum products for export (160.8 Mt). As a small island with no natural resources and limited options for renewable energy (NCCS 2019), 95% of Singapore's electricity is generated from the combustion of imported natural gas. A small proportion of energy is also produced from solar power and waste-to-energy facilities that produce energy from incinerating waste (MEWR 2019). Of the 48.6 TWh of electricity consumed in 2016, the largest share was by the manufacturing industry (38%), followed by businesses in the commerce and services sector (36%), and households (16%) (Singstat 2019). Altogether, oil refining, electricity generation and the 956,430 motor vehicles (Land Transport Authority, 2018)—most of which run on fossil fuels—contributed 51.5 Mt of greenhouse gases (CO_2 equivalent) emitted into the air in 2016 (MEWR 2019).

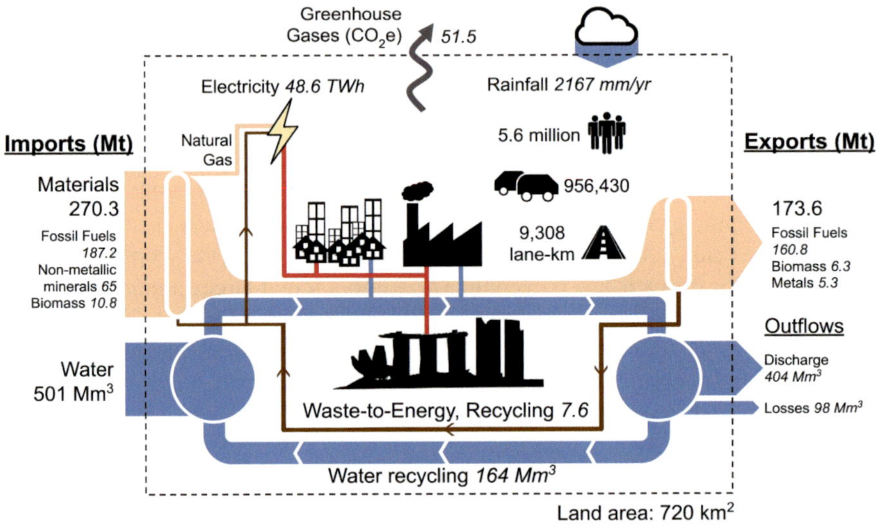

Fig. 7.5 Metabolism of Singapore in 2016. Major flows of materials (in million metric tons, Mt), water, and energy are displayed, along with several key statistics. Data on water flows, recycling, and greenhouse gas emissions obtained from MEWR (2019). Singapore skyline by Kiraan on VectorStock

With a total renewable water resource (TRWR) per capita of 105.1 m^3/year, Singapore is considered to be facing absolute water scarcity (Food and Agriculture Organization 2014, 2019). Even though Singapore is located just one degree north of the equator and receives more than two meters of rainfall per year (weather.gov.sg 2019), its small size gives little room for water catchment sufficient to meet its water demand. Historically reliant on its closest neighbor for water imports, Singapore has invested heavily in water recycling (locally branded as NEWater) and desalination to "close the water loop" (PUB 2016) and achieve self-sufficiency in water resources. Investments in water recycling have resulted in the significant secondary flow of water that makes up more than 25% of all the water sent to the end-users.

Table 7.3 shows how Singapore's UM has grown since independence from 1965 to 2016. Except for DE, which has virtually disappeared relative to the other indicators, all other indicators in 2016 have increased by 5–7 times their values in 1965, with imports growing the most from 6.8 to 48.2 metric tons per capita. Fossil fuels have always made up the bulk of Singapore's imports and exports, although the share of fossil fuels in total exports has increased while the opposite is true for imports. These metabolic indicators show the phenomenal growth of the material flows of Singapore, which occurred in tandem with Singapore's rise from a predominantly agricultural economy to a global one with manufacturing, oil refining, and service industries.

Nonetheless, Singapore is not alone in its trajectory. Other cities have also experienced great increases in material consumption per capita in the past century (Kennedy et al. 2007). For example, the total material consumption per capita in Hong Kong increased by 141% from 2.9 metric tons in 1971 to 7.0 metric tons in 1997 (Warren-Rhodes and Koenig 2001). While cities around the world are growing and reaching new economic heights, will the trend of increasing material consumption and intensity continue without bounds? If the theory of the Environmental Kuznets

Table 7.3 Comparison of Singapore's UM indicators from 1965 to 2016

Indicators (metric tons per capita)	1965[a] (Schulz 2007)	2016[b]	%-change
Imports	6.8	48.2	612
Fossil fuels (% of total)	4.9 (72%)	33.4 (69%)	582
Domestic extraction (DE)	1.4	0.07	−95
Direct material input (DMI)	8.2	48.3	489
Exports	5.0	31.0	522
Fossil fuels (% of total)	4.3 (87%)	28.7 (93%)	561
Domestic material consumption (DMC)	3.2	17.3	441
Population (million)	1.89	5.6	196
GDP per capita (S$, 2015)	5804	77,754	1240

[a]Values estimated from figures published by Schulz (2007)
[b]This study

Curve (EKC) holds, environmental impacts would decline as societies become more affluent. Empirical support for the theory is mixed. DMI, DMC, and DPO were found to correlate poorly with GDP per capita for affluent industrial economies (Fischer-Kowalski and Amann 2001), with similarly poor correlations for water use and solid waste production in megacities from 2001 to 2011 (Kennedy et al. 2015). On the other hand, the latter found that energy use is growing at half the rate of economic growth, with London even reducing its electricity consumption per capita while its GDP grew. Returning to the case of Singapore, DMC grew at less than half the rate of GDP growth from 1965 to 2016 (Table 7.3). Furthermore, Abou-Abdo et al. (2011) presented evidence of per capita water consumption for Singapore following the EKC, reaching a peak in the early 1990s with water consumption at 115 m^3 per capita and a gross urban income of about S$34,000.

The material footprints of cities are direct consequences of their metabolism; to recall the definition of Kennedy et al. (2007): "the sum total of the technical and socio-economic processes." Analyzing the flows of material and energy into, within, and out of cities provides us with a glimpse under the hood of the engine that keeps our cities running. These flows also serve as fingerprints of our cities, reflecting the unique circumstances—past and present—that drive their continuing growth and adaptation.

7.5 Urban Metabolism Applications, Challenges, and Opportunities

The study of UM has been considered for the purposes of urban planning and urban infrastructure planning. The study of resource stocks and flow exchanges in cities offers a perspective for urban systems analysis, and a potential to understand self-sufficiency, efficiency, and resilience. The merit of UM lies in examining resource requirements, availability, rates of change, and accumulation. It offers an understanding of sources (inflows) required to sustain growth, or the abilities of the city to regulate flows, assimilate or treat waste, and capture emissions. As a communications tool, UM can also be used to convey the consumption of resources within cities and allude to limits to growth. Many cities are in fact resource sinks, often accumulating material stocks, and requiring continuous inflows. While UM studies help profile the past and current status of urban systems, many UM studies have not led to actionable recommendations beyond the initial assessment. One main criticism of UM is that since it fundamentally offers a retrospective view of resource stocks and flows, it has to be coupled with other approaches in order to consider opportunities for achieving resource efficiency. UM studies therefore provide diagnosis but are missing a prescription to follow. John et al. (2019) found that two-thirds of 221 UM studies followed a problem-oriented approach to characterize the metabolism of the system and understand risks, as opposed to seeking ways to solve the challenges uncovered.

This limitation of UM is partly due to its systems perspective, which masks many complex interactions that take place within cities and cannot yet be adequately captured. It, therefore, lacks visibility about which actors are driving the flows, where the flows occur, and the underlying usage and consumption patterns. Without a view on the causes and drivers for resource flows, this makes it difficult to extract details on specific infrastructure systems, levers of control, and to consider how to manage, let alone optimize. Many UM scholars have, therefore, highlighted the need to advance the field of practice beyond accounting, assessment, and reporting, to guidance for designing, optimizing, and decision making.

A number of studies have suggested options to couple UM with notions of sustainable design, in order to translate the assessment into practical urban design and planning. Examples include:

- The European BRIDGE research project (2011) developed a GIS-based decision-support UM assessment tool that evaluates urban planning alternatives. The research team emphasized a need for UM to focus on the local scale.
- González et al. (2013) used UM to assess the sustainability impact of urban planning alternatives, such as building types or the location of transportation and infrastructural developments.
- Thomson and Newman (2018) explored the influence of different urban forms on resource inflows, and waste and emissions outflows, for the city of Perth, Australia.
- In a comparative study of UM of different megacities, Han et al. (2018) considered the industrial structures of cities and suggested that the pursuit of service industries instead of manufacturing can allow cities to achieve green growth.

As the field advances, we see four challenges in the further application of UM:

1. As mentioned, unless the internal flows within cities are adequately portrayed in UM, it will be difficult to translate the findings into intervention options. Pincetl et al. (2012) suggested to connect metabolism studies with the actors driving their dynamics. They also highlighted a need to consider the internal political, economic, and social processes within cities, to better understand the complexities of possible change. The aim is to better understand "socioeconomic and policy drivers that govern the flows and patterns."
2. The quantities or qualities of energy and material flowing through cities may not always be the right metric of concern, nor are they all that matter. The forces driving resource consumption are the demands for services derived from these resources, or the utility obtained. There is a need to capture the value of the services derived, and not just the amounts of resources. Carreón and Worrell (2018) argue for the consideration of energy services, and drivers of them, in UM research.
3. The study of UM remains highly constrained by the availability of quality data. Most existing UM studies cover a limited set of resources—materials (particularly metals), energy, water, and nutrients. Analyses are also usually limited to a single time period (e.g., a single year). Moreover, Currie and Musango (2017)

highlighted that UM studies have generally been limited to the cities in the Global North, given the lack of data elsewhere.
4. While there have been attempts to carry out comparative UM studies across cities (including those by Currie and Musango 2017; Han et al. 2018), it is generally difficult to compare UM studies without a standard approach. Beloin-Saint-Pierre et al. (2017) reported on the lack of consistency on assessment methods. Zhang et al. (2015) recommended the establishment of "a multilevel, unified, and standardized system of categories to support the creation of consistent inventory databases," which can guide comparative analysis. Even so, the harmonization of efforts will likely remain highly challenging given disparate and often missing datasets.

Despite these challenges, we see related opportunities to advance the field in several ways. Most essentially, new data sources are becoming more available to better examine urban systems. This allows for disaggregated UM that (i) operates at finer temporal resolutions, (ii) is spatially explicit, and (iii) integrates relevant sources of information. Enabled by pervasive sensing and improved communications technologies, time-series data on the building-, district- and even city-level are increasingly available, such as real-time electricity use, individual mobility patterns, water use, and management tools. With the shortening of the timescale of analysis, it is possible to monitor and track resource consumption more carefully. This also allows for understanding rates of change, to better understand the timescale of impacts and potential interventions. In this direction, Shahrokni et al. (2015) proposed what they termed smart urban metabolism, which is capable of integrating UM concepts with information and communication technologies (ICT) and smart-city technologies, thus enabling user-generated automated data collection, real-time analytics, and feedback for city planners.

The mapping of resource flows for a more spatially explicit UM analysis is another potential area of development. By moving beyond scalar quantities, this allows for an understanding of the direction and distribution of internal flows within the city. Impact arises from the distributed nature of activities that drive the demand for resources, resulting in flows. Planners can then consider the resource efficiency implications of land use or infrastructure location decisions. Voskamp et al. (2018) also recommended finer spatio-temporal resolution for monitoring energy and water flows, arguing that this is required in order to develop interventions to optimize resource flows. There is also the opportunity to integrate different types of information at the disaggregated level to evaluate UM. Related sources of information and tools include supply chain data (e.g., transaction data from enterprise resource planning systems) or building information modeling (BIM) data. Researchers have even used satellite and night-light imagery (Xie and Weng 2016), GIS tools (Li and Kwan 2018), and freight transportation surveys (Yeow and Cheah 2019) to better examine UM.

Furthermore, data concerning different resources can be fused or integrated to allow analysts a better understanding of the interdependencies and relationships between different resource flows, as opposed to examining individual resources separately. Exploring the interactions between water consumption and energy use

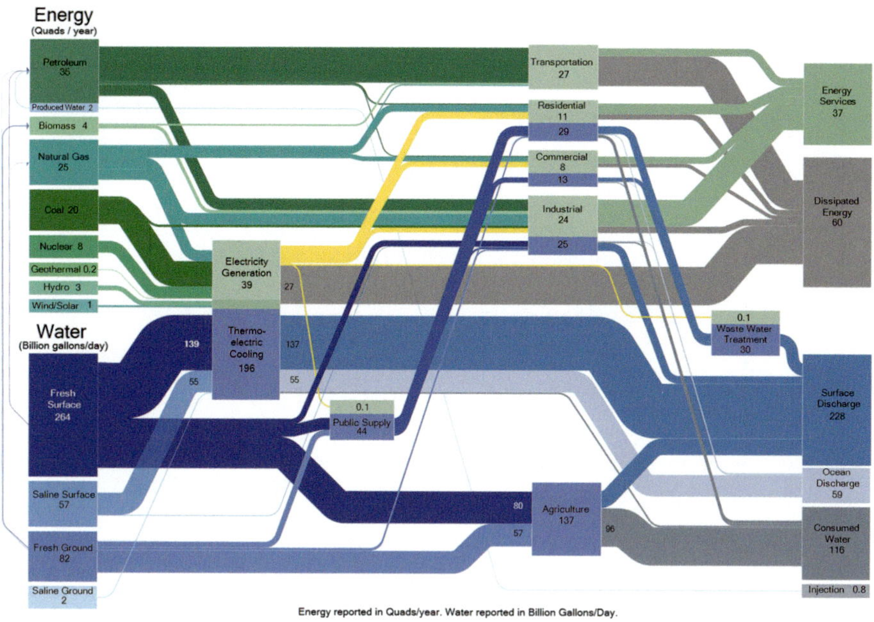

Fig. 7.6 Hybrid Sankey diagram of 2011 U.S. water and energy flows. *Source* U.S. Department of Energy

(water–energy nexus), or linking resource demand with urban activities can aid with holistic policy decision-making and integrated resource management. Hamiche et al. (2016) conducted a review of the water–energy nexus to reveal the complex links between water and electricity generation. Movahedi and Derrible (2020) studied the interrelationships between water, electricity, and gas consumption in large-scale buildings in New York City. Figure 7.6 shows a hybrid Sankey diagram depicting interconnected water and energy flows in the United States in 2011, developed by the US Department of Energy (Bauer et al. 2014).

Finally, UM analysis may progress from a descriptive approach toward a more prescriptive one, when it is considered in simulations of resource flows through cities, allowing the analyst an opportunity to test potential interventions. Figure 7.7 shows the potential evolution of the field, advancing toward more disaggregated analysis with finer temporal and spatial resolution, and eventually using real-time data to offer predictions on the state of the system. With live data streams, one can monitor demand and regulate resource flows in or near real time. This would be analogous to real-time system monitoring, even with the possibility of feedback and control. Such advances are already becoming available at the scale of individual buildings and even neighborhoods, with the possibility of scaling up to virtual city representations in the form of the city's digital twin, albeit with greater complexity. For instance, in the *Virtual Singapore* project, a digital twin of the city has been developed with the intention for urban planners to simulate alternative policies (Wall 2019). When

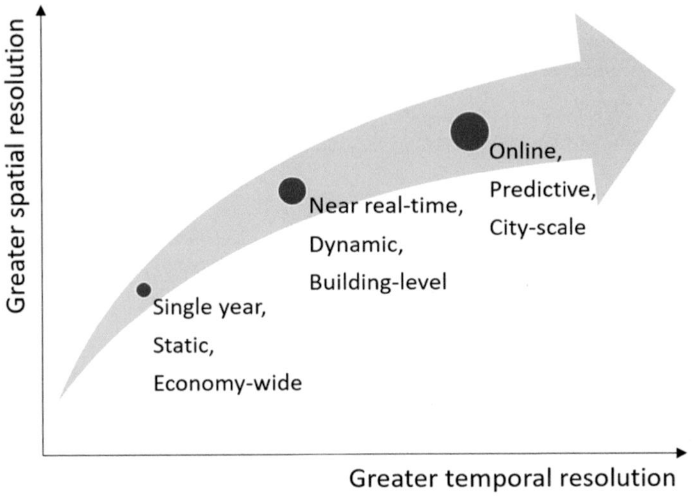

Fig. 7.7 Envisioned developments in the field of urban metabolism

available, such virtual representations of a city's metabolism allow for an opportunity to better monitor, manage, and optimize resource use. In the future, the metabolism of cities can even be predicted and self-regulated.

Ultimately, the coupling of urban metabolism portrayal with sustainable urban planning and design can provide both a comprehensive diagnosis, as well as the capabilities to consider solutions. This allows stakeholders to explore impact mitigation pathways, and consider strategies to achieve sustainable urban renewal and growth. Cities and their metabolism are an outcome of the agglomeration of the complex behaviors of their residents. The study of UM monitors the pulse of the city, allowing insights and actions toward greater urban sustainability.

7.6 Conclusions

From its humble beginnings in quantifying flows of nutrients in and out of Berlin and in sanitary engineering, UM has evolved to become an established field whose main goal is to quantify the inflows, outflows, and production of energy and resources to, from, and in cities. In this chapter, a short history of UM was first offered, notably recalling Wolman's findings from his 1965 study. Because of the significant number of flows that need to be estimated, carrying out a UM is not necessarily straightforward. Methodologically, the goal is primarily to perform a Material and Energy Flow Analysis (MEFA) of a city. In this chapter, two main families of UM approaches were described. The first family attempts to calculate UM from the bottom up by either collecting or estimating individual flows, such as quantifying the amount of water consumed. The second family takes a top-down approach by leveraging and disaggregating nation-wide economic input-output data sources. Finally, some hybrid methods exist to pursue UM studies, including one that utilizes concepts of emergy and another that utilizes concepts of ecological network analysis.

As a practical case study, the UM of Singapore was then studied. As a city-state, Singapore is particularly interesting since both bottom-up and top-down approaches can be adopted. The exercise led to the development of Fig. 7.5 that offers an interesting and insightful snapshot of the material and energy flows that entered or exited Singapore in 2016. Subsequently, the applications, opportunities, and challenges of UM were reviewed. In particular, one main challenge of UM resides in the fact that it is purely an accounting method and it does not directly lead to the development of appropriate designs and policies to tackle specific problems. In contrast, as more numerous and larger data sources are becoming available, it is becoming increasingly possible to perform UM in much finer spatiotemporal resolutions.

Overall, the development and use of UM have evolved relatively slowly in the past century, but significant advances are likely to emerge in the future. On the one hand, more and better data sources are becoming available; on the other hand, cities around the world are striving to become more sustainable and resilient. UM, therefore, offers significant opportunities to help understand how energy and resources are being consumed and, therefore, can contribute to inform better designs and policies to radically change how people live in cities in the twenty-first century.

Acknowledgements This research was supported, in part, by the United States National Science Foundation (NSF) CAREER Award 1551731 and by the Singapore University of Technology and Design (SUTD) graduate research fellowship from the Ministry of Education, Singapore. The authors would also like to acknowledge the OeAD Ernst Mach Grant (Award# ICM-2018-09903) and thank Professor Fridolin Krausmann from the Institute for Social Ecology Vienna.

References

Abou-Abdo T, Davis NR, Krones JS, Welling KN, Fernández JE (2011) Dynamic modeling of Singapore's urban resource flows: historical trends and sustainable scenario development. Paper presented at the 2011 IEEE International symposium on sustainable systems and technology

ADB (2014) Urban metabolism of six Asian cities. Asian Development Bank, Metro Manila, Philippines

Arora M, Raspall F, Cheah L, Silva A (2019) Residential building material stocks and component-level circularity: the case of Singapore. J Cleaner Prod 216:239–248

Ashby MF (2013) Chapter 2—Resource consumption and its drivers. In: Materials and the environment, 2nd edn. Butterworth-Heinemann, Boston, MA, pp 15–48

Barles S (2009) Urban metabolism of Paris and its region. J Ind Ecol 13:898–913

Bauer D, Philbrick M, Vallario B (2014) The water–energy nexus: challenges and opportunities. US Department of Energy, Washington DC

Beloin-Saint-Pierre D, Rugani B, Lasvaux S, Mailhac A, Popovici E, Sibiude G, Benetto E, Schiopu N (2017) A review of urban metabolism studies to identify key methodological choices for future harmonization and implementation. J Cleaner Prod 163:S223–S240

BRIDGE (2011) About BRIDGE project. Retrieved from https://www.bridge-fp7.eu/. Accessed 3 Oct 2019

Carreón JR, Worrell E (2018) Urban energy systems within the transition to sustainable development. A research agenda for urban metabolism. Resour Conserv Recycl 132:258–266
Chen S, Chen B (2012) Network environ perspective for urban metabolism and carbon emissions: a case study of Vienna, Austrai. Environ Sci Technol 46(8):4498–4506
Chertow M, Choi E, Lee K (2011) The material consumption of Singapore's economy: an industrial ecology approach. In: Environment and climate change in Asia: ecological footprints and green prospects. Prentice Hall, Englewood Cliffs, NJ
Chini CM, Stillwell AS (2018) The state of US urban water: data and the energy–water nexus. Water Resour Res 54(3):1796–1811
Chrysoulakis N, de Castro EA, Moors EJ (2014) Understanding urban metabolism: a tool for urban planning. Routledge, London, UK
Currie PK, Musango JK (2017) African urbanization: assimilating urban metabolism into sustainability discourse and practice. J Ind Ecol 21:1262–1276
Derrible S (2019a) An approach to designing sustainable urban infrastructure. MRS Energy Sustain 5:E15
Derrible S (2019b) Urban engineering for sustainability. MIT Press, Cambridge, MA
Duvigneaud P, Denaeyer-De Smet S (1977) L'ecosystéme urbain Bruxellois. In: Duvigneaud P, Kestemont P (eds) Productivité biologique en Belgique. Duculot, Paris, pp 608–613
Ehrlich PR, Holdren JP (1971) Impact of population growth. Science 171(3977):1212–1217
Eknoyan G (1999) Santorio Sanctorius (1561–1636)—founding father of metabolic balance studies. Am J Nephrol 19(2):226–233
Eurostat (2001) Economy-wide material flow accounts and balances with derived resource use indicators: a methodological guide. Office for Official Publications of the European Communities, Luxembourg
Eurostat (2018) Economy-wide material flow accounts and balances with derived resource use indicators: a methodological guide. Office for Official Publications of the European Communities, Luxembourg
Food and Agriculture Organization (2019) Country fact sheet—Singapore. Retrieved from https://www.fao.org/nr/water/aquastat/countries_regions/index.stm. Accessed 6 Oct 2019
Food and Agriculture Organization (2014) Water stress. Retrieved from https://www.fao.org/nr/water/aquastat/infographics/Stress_eng.pdf. Accessed 6 Oct 2019
Fath BD, Scharler UM, Ulanowicz RE, Hannon B (2007) Ecological network analysis: network construction. Ecol Model 208(1):49–55
Ferrão P, Fernández JE (2013) Sustainable urban metabolism. MIT Press, Cambridge, MA
Fischer-Kowalski M, Amann C (2001) Beyond IPAT and Kuznets curves: globalization as a vital factor in analyzing the environmental impact of socio-economic metabolism. Popul Environ 23(1):7–47
Fishman T, Schandl H, Tanikawa H, Walker P, Krausmann F (2014) Accounting for the material stock of nations. J Ind Ecol 18(3):407–420
González A, Donnelly A, Jones M, Chrysoulakis N, Lopes M (2013) A decision-support system for sustainable urban metabolism in Europe. Environ Impact Assess Rev 38:109–119
Government of Singapore (2019) Length of roads maintained by LTA. Retrieved from https://data.gov.sg/dataset/length-of-road-maintained-by-lta. Accessed 11 Sept 2019
Geddes P (1915) Cities in evolution: an introduction to the town planning movement and to the study of civics. Williams and Norgate, London, UK
Haberl H, Wiedenhofer D, Pauliuk S, Krausmann F, Müller DB, Fischer-Kowalski M (2019) Contributions of sociometabolic research to sustainability science. Nat Sustain 2(3):173–184
Hall P (2002) Cities of tomorrow, 3rd edn. Blackwell Publishers, Oxford, UK and Malden, MA
Hamiche AM, Stambouli AB, Flazi S (2016) A review of the water-energy nexus. Renew Sustain Energy Rev 65:319–331
Han W, Geng Y, Lu Y, Wilson J, Sun L, Satoshi O, Geldron A, Qian Y (2018) Urban metabolism of megacities: a comparative analysis of Shanghai, Tokyo, London and Paris to inform low carbon and sustainable development pathways. Energy 155:887–898

Hau JL, Bakshi BR (2004) Promise and problems of emergy analysis. Ecol Model 178(1):215–225
Irvine K, Chua L, Eikass HS (2014) The four national taps of Singapore: a holistic approach to water resources management from drainage to drinking water. J Water Manag Model 22:1–11
John B, Luederitz C, Lang DJ, von Wehrden H (2019) Toward sustainable urban metabolisms. From system understanding to system transformation. Ecol Econ 157:402–414
Kaza S, Yao L, Bhada-Tata P, Van Woerden F (2019) What a waste 2.0: a global snapshot of solid waste management to 2050. World Bank, Washington, DC
Kennedy C (2012) A mathematical description of urban metabolism. In: Weinstein MP, Turner RE (eds) Sustainability science. Springer, New York, NY, pp 275–291
Kennedy C, Cuddihy J, Engel-Yan J (2007) The changing metabolism of cities. J Ind Ecol 11(2):43–59
Kennedy C, Stewart I, Facchini A, Cersosimo I, Mele R, Chen B, Uda M (2015) Energy and material flows of megacities. Proc Natl Acad Sci 112(19):5985–5990
Krausmann F, Schandl H, Eisenmenger N, Giljum S, Jackson T (2017) Material flow accounting: measuring global material use for sustainable development. Annu Rev Environ Resour 42(1):647–675
Krausmann F, Wiedenhofer D, Lauk C, Haas W, Tanikawa H, Fishman T, Miatto A, Schandl H, Haberl H (2017) Global socioeconomic material stocks rise 23-fold over the 20th century and require half of annual resource use. Proc Natl Acad Sci 114(8):1880–1885
Krausmann F, Weisz H, Eisenmenger N, Schütz H, Haas W, Schaffartzik A (2018) Economy-wide material flow accounting: introduction and guide. Social Ecology Working Paper 151v1.2. Institute of Social Ecology, Vienna
Lederer J, Kral U (2015) Theodor Weyl: a pioneer of urban metabolism studies. J Ind Ecol 19:695–702
Li H, Kwan M-P (2018) Advancing analytical methods for urban metabolism studies. Resour Conserv Recycl 132:239–245
Land Transport Authority (2018) Annual vehicle statistics 2018: motor vehicle population by vehicle type. Land Transport Authority, Singapore
Mahajan S (2014) The art of insight in science and engineering: mastering complexity. MIT Press, Cambridge, MA
Matthews E, Amann C, Bringezu S, Fischer-Kowalski M, Huettler W, Kleijn R, Moriguchi Y (2000) The weight of nations: material outflows from industrial economies. World Resources Institute, Washington DC
MEWR (2019) Key environmental statistics 2019. Retrieved from https://www.mewr.gov.sg/docs/default-source/default-document-library/grab-our-research/-kes-2019.pdf. Accessed 7 Oct 2019
Mohareb E, Derrible S, Peiravian F (2016) Intersections of Jane Jacobs' conditions for diversity and low-carbon urban systems: a look at four global cities. J Urban Plann Dev 142(2):05015004
Movahedi A, Derrible S (2020) Interrelated patterns of electricity, gas, and water consumption in large-scale buildings. engr Xiv. https://doi.org/10.31224/osf.io/ahn3e.
NCCS (2019) Singapore's approach to alternative energy. Retrieved from https://www.nccs.gov.sg/climate-change-and-singapore/national-circumstances/singapore's-approach-to-alternative-energy. Accessed 7 Oct 2019
Newcombe K, Kalina JD, Aston AR (1978) The metabolism of a city: the case of Hong Kong. Ambio 7:3–15
Newell JP, Cousins JJ (2014) The boundaries of urban metabolism: towards a political–industrial ecology. Prog Hum Geogr 39(6):702–728
Odum HT (1996) Environmental accounting: EMERGY and environmental decision making. Wiley, New York, NY
Pearce BJ, Chertow M (2017) Scenarios for achieving absolute reductions in phosphorus consumption in Singapore. J Cleaner Prod 140:1587–1601
Pincetl S, Bunje P, Holmes T (2012) An expanded urban metabolism method: toward a systems approach for assessing urban energy processes and causes. Landscape Urban Plann 107(3):193–202

PUB (2016) Our water, our future. Retrieved from https://www.pub.gov.sg/Documents/PUBOurWaterOurFuture.pdf. Accessed 6 Oct 2019
Schulz NB (2007) The direct material inputs into Singapore's development. J Ind Ecol 11(2):117–131
Shahrokni H, Lazarevic D, Brandt N (2015) Smart urban metabolism: towards a real-time understanding of the energy and material flows of a city and its citizens. J Urban Technol 22(1):65–86
Singstat (2019) M890841—Electricity generation and consumption, Annual. Retrieved from https://www.tablebuilder.singstat.gov.sg/publicfacing/createDataTable.action?refId=14602, accessed 7 October 2019
Tan LM, Arbabi H, Brockway PE, Densley Tingley D, Mayfield M (2019) An ecological-thermodynamic approach to urban metabolism: measuring resource utilization with open system network effectiveness analysis. Appl Energy 254:113618
Thomson G, Newman P (2018) Urban fabrics and urban metabolism—from sustainable to regenerative cities. Resour Conserv Recycl 132:218–229
Tortajada C, Joshi Y, Biswas A (2013) The Singapore water story. Routledge, London, UK
Tortajada C, Buurman J (2017) [Singapore Handbook of Public Policy] Water policy in Singapore. Retrieved from https://lkyspp.nus.edu.sg/gia/article/water-policy-in-singapore. Accessed 7 Oct 2019
Voskamp IM, Spiller M, Stremke S, Bregt AK, Vreugdenhil C, Rijnaarts HHM (2018) Space-time information analysis for resource-conscious urban planning and design: a stakeholder based identification of urban metabolism data gaps. Resour Conserv Recycl 128:516–525
Wall M (2019) Virtual cities: designing the metropolises of the future. BBC.com, 18 January 2019. Retrieved from https://www.bbc.com/news/business-46880468. Accessed 12 Oct 2019
Warren-Rhodes K, Koenig A (2001) Escalating trends in the urban metabolism of Hong Kong: 1971–1997. AMBIO J Hum Environ 30(7):429–438
weather.gov.sg (2019) Climate of Singapore. Retrieved from https://www.weather.gov.sg/climate-climate-of-singapore. Accessed 4 Oct 2019
Welling KN (2011) Modeling the water consumption of Singapore using system dynamics. Master of Science, Massachusetts Institute of Technology, Cambridge, MA
Wolman A (1965) The metabolism of cities. Sci Am 213:179–190
Xie Y, Weng Q (2016) Detecting urban-scale dynamics of electricity consumption at Chinese cities using time-series DMSP-OLS (Defense Meteorological Satellite Program-Operational Linescan System) nighttime light imageries. Energy 100:177–189
Yeow L, Cheah L (2019) Using spatially explicit commodity flow and truck activity data to map urban material flows. J Ind Ecol 2019:1–12
Zhang Y, Yang Z, Yu X (2009) Ecological network and emergy analysis of urban metabolic systems: model development, and a case study of four Chinese cities. Ecol Model 220(11):1431–1442
Zhang Y, Yang Z, Yu X (2015) Urban metabolism: a review of current knowledge and directions for future study. Environ Sci Technol 49(19):11247–11263

Sybil Derrible is an Associate Professor in Civil Engineering at the University of Illinois at Chicago and Director of the Complex and Sustainable Urban Networks Laboratory. He is the author of Urban Engineering for Sustainability (MIT Press, 2019) and an Associate Editor for the ASCE Journal of Infrastructure Systems.

Lynette Cheah is an Associate Professor of Engineering Systems at the Singapore University of Technology and Design. She leads the Sustainable Urban Mobility research group, which develops data-driven models and tools to reduce the environmental impacts of passenger and urban freight transport.

Mohit Arora is a Research Associate in Sustainable Built Environment at the University of Edinburgh and Imperial College London. His research combines Circular Economy Strategies with Development Engineering for a low carbon future.

Lih Wei Yeow is a Senior Research Assistant at the Singapore University of Technology and Design. He works with the Sustainable Urban Mobility research group and is interested in urban systems and their interactions.

Chapter 8
Spatial Economics, Urban Informatics, and Transport Accessibility

Ying Jin

Abstract One central pillar in the development of urban science which is key to the development of simulation of models of urban structure is spatial econometrics. In this chapter, we outline the way in which ideas pertaining to accessibility which we define conventionally, as in transport economics, as the relative nearness and size of locations to one another, can be embedded in a wider econometric framework. We are thus able to explore how GDP (gross domestic product) of different locations is influenced by different spatial investments. To illustrate this, we first outline the intellectual context, followed by a review of the most relevant econometric models. We examine the data required for such models and look at various quantifications in terms of elasticities of business productivity with respect to transport accessibility, using ordinary least squares, time-series fixed effects, and a range of dynamic panel-data models which narrow down the valid range of estimates. We then show how the model is applied to Guangdong province (with its connections to Hong Kong and Macau), which is one of the three major mega-city regions and a leading adopter of new technologies in China.

8.1 Introduction

In a nutshell, the contributions of spatial economics to urban informatics relate to the measurement, design, and interpretation of urban data that supports economic, social, and technological decisions regarding the locations, distributions, and layouts of urban activities, buildings, and infrastructure. In past decades, research at the frontier between spatial economics and urban informatics has largely been commissioned by governments, major banks, and businesses. Since new civic groups are playing an increasingly prominent role in investigating alternative options for spatial development (for a recent example in the UK, see the UK2070 Commission 2019), a full range of societal stakeholders have now been actively engaging with this area of interdisciplinary research. Students of urban informatics need an understanding

Y. Jin (✉)
Martin Centre for Architectural and Urban Studies, University of Cambridge, Cambridge, UK
e-mail: yj242@cam.ac.uk

W. Shi et al. (eds.), *Urban Informatics*, The Urban Book Series,
https://doi.org/10.1007/978-981-15-8983-6_8

of spatial economics if they wish to influence the real decisions underpinning the planning, designing, funding, regulating, and maintaining of these spaces in cities and their hinterlands.

Spatial economics has a historic root as deep as all other main branches of modern economics. In particular, it can be traced back to the seminal works of von Thünen (1826). Since then, spatial economics has grown into a vast field of learning, which is sometimes referred to as the new economic geography (although this latter name does not have the consent of all geographers). Comprehensive handbooks on spatial economics have been compiled, for instance, see Duranton et al. (1987, 2018), and the higher-level overview by Redding and Rossi-Hansberg (2017). Somewhat paradoxically, this vastness of learning has often become a formidable barrier for those who work in urban informatics and wish to understand more about how spatial economies actually work.

This chapter adopts an approach that is complementary to the handbooks such as those referred to above—it aims to give students of urban informatics a feel for how spatial economics must tackle one of the critical issues that often confront them, that is, the measurement and interpretation of the contribution of inter-city transport accessibility improvements to the economy. According to Lakshmanan (2011), this is one of the most persistent spatial-economic issues in urban and regional transport studies. This approach which is an introduction by example is meant to encourage students of urban informatics to start with the quantitative skills that they may already have (e.g., simple ordinary least square or OLS regression models) and then engage with a cross section of advanced spatial-economics literature that is cogent to the topic.

The quantification of the economic contribution of transport accessibility improvements is particularly important for infrastructure investment. Significant progress has been made in recent years in spatial economics (see, for example, comprehensive reviews by Rosenthal and Strange 2004; Melo et al. 2009, 2013; Laird and Venables 2017). Nevertheless, in contrast to the considerable volume of research on the relationship between transport investment and productivity in the OECD countries, there are to date very few quantifications in this regard in emerging economies which are suitable for investment and loan decisions.

The complex, slow-evolving, and cumulative nature of the transport infrastructure investment makes the quantification of its impact one of the most challenging. Econometric modeling is the mainstay in current quantification of such impacts. Different types of regression and modeling methods have been developed over the years in this field, which started with OLS and time-series models that tested solely the effects of transport investment, and progressed with the introduction of a series of control variables, instrumental variables, and extended functional forms which are better able to deal with the heterogeneity and endogeneity issues of cumulative causation. This progression has led to more robust econometric models for such analysis.

In econometrics and only until recently, models have tended to be used in isolation rather than jointly. The quantification exercise tends to be carried out using the most advanced functional forms each time and this applies to the transport-related studies. However, using the alternative models jointly can offer valuable new insights into the quantification results. Bond et al. (2001) and Brülhart and Mathys (2008) point out that a comparison of the results of the alternative models with the theoretical, prior expectations may serve as an important bound test. Melo et al. (2013) have recently highlighted the empirical differences of the alternative model forms across different studies through a comprehensive meta-analysis on the effects of investment in transport infrastructure.

In this chapter, we show how a new approach to spatial-economic quantification of the transport effects can be developed using a series of regression models in the assessment of inter-city transport improvements. The econometric models are not only examined on their individual functional forms and estimation diagnostics, but also through a comparison of the outturn coefficient values with the prior theoretical expectations. Through this method, we aim to identify more precisely the transport effects on the real economy, while not substantially increasing the analytical work for practical studies designed, for example, for loan-project assessment.

We report an econometric analyses for Guangdong province, one of the three major mega-city regions and a leading adopter of new technologies in China. The analyses include Hong Kong and Macau as appropriate for the regional economic activities. Although we first started working on this quantification because of World Bank loan projects, we soon realized that Guangdong may be among the best case-study locations for such an investigation. Although the province has contributed to the highest provincial share of national GDP in China for more than two decades, its economic development is polarized, with a prosperous center and an underdeveloped periphery; its ways of doing business are being widely emulated by other provinces in China, thus are likely to represent what is to come in the rest of the country; and its land boundaries consist primarily of mountain chains which makes it straightforward to delineate a study-area boundary. This is in stark contrast to the amorphous limits of the other two main mega-city regions centered upon Beijing and Shanghai.

The chapter is organized accordingly in seven sections: Sect. 8.2 outlines the intellectual context, which is followed by Sect. 8.3 on the alternative econometric models. Section 8.4 presents the data. Section 8.5 presents the various quantifications in terms of elasticities of business productivity with respect to transport accessibility, using ordinary least squares, time-series fixed-effects and various dynamic panel-data models to narrow down the valid range of estimates. Section 8.6 discusses the wider implications of the findings and the extent of corroborations. Section 8.7 concludes with a short summary and considerations for future research directions.

8.2 Intellectual Context

Recent years have seen a growing body of research on the relationship between transport investment and productivity. The arguments are primarily built upon the spatial-economics literature, which gives due recognition to (1) consumers' and producers' love of variety in their use of products and services, (2) increasing returns to scale in production, and (3) the importance of transport costs in shaping the economic landscape. This has led to theoretical models that identify reasons why modern firms tend to be more productive when they either concentrate in or have low cost links to large markets. Empirical studies have so far built up a substantial body of evidence which suggests that production and income are correlated with spatial proximity in the way suggested by the theories. Ciccone and Hall (1996), Rosenthal and Strange (2004), Redding and Venables (2004) and Melo et al. (2009, 2013) provide systematic surveys of the empirical evidence.

Inter-regional and city-scale theoretical models emerged about a decade after the initial trade models (see Fujita et al. 1999). Empirical studies followed. Rice et al. (2006) outlined an analytical framework within which interactions between the different aspects of regional inequality in per-employee productivity can be investigated econometrically using aggregate data. Kopp (2007) used a panel-data model to address the issue of endogeneity and identified contribution from transport investment to productivity, showing that doubling road stock in a country will lead to about 10% growth in total factor productivity in Western Europe. Combes et al. (2008) developed a general framework to investigate, respectively, the sources and mechanisms that lead to wage disparities across regional labor markets through sorting and self-selection. Graham and Kim (2008) investigated the relationship between spatial proximity and productivity using a large sample of financial accounting information from individual firms in the UK.

For emerging economies, Deichmann et al. (2005) distinguished between natural advantage, including infrastructure endowments, wage rates, and natural resource endowments, and production externalities that arise from the co-location of firms in the same or complementary industries, in their examination of the aggregate and sectoral geographic concentration of manufacturing industries for Indonesia. Lall et al. (2010) differentiated local and national infrastructure supply in India, and found that a city's proximity to international ports and highways connecting large domestic markets has the largest effect on its attractiveness for private investment.

In China, there has been a growing volume of literature that associates productivity benefits with agglomeration in Chinese cities and city regions (e.g., IBRD 2006, p. 145; Lu et al. 2007, p. 163). Using two nation-wide Censuses of Establishments of 1996 and 2001, Lu (2010) outlined the spatial distribution of economic activities across China and found through multivariate analysis that, during that period, the micro-economic explanations of agglomeration do not work well with publicly owned institutions, although they do work well with non-publicly owned institutions. Roberts and Goh (2012) showed that distance has a significant role in determining spatial productivity disparities in Chongqing municipality. Roberts et al. (2012) used counterfactual analysis based on a general equilibrium model to show that China's national expressway network has brought sizeable aggregate benefits to the Chinese economy, although its impact on regional disparities may be contingent upon factors such as migration.

These studies have shed an important light both on the statistical relationship between spatial proximity and productivity, and on a variety of complex issues of empirical modeling. Nevertheless, the studies have also shown that such statistical relationships may be highly context-specific.

At the heart of the difficulties of empirical measurements is the very nature of agglomeration as a process of circular, cumulative causation, which has become known since the work of Gunnar Myrdal: agglomeration propels endogenous growth—higher productivity leads to higher wages, which attracts employees of a higher caliber, which in turn draws in new investment, more productive technologies and so on; these lead to a new round of productivity growth. Conventionally, instrumental variables are used to overcome endogeneity issues in regressions; but by its very nature, agglomeration studies rarely have good instrumental variables for dealing with cumulative causation (Redding 2010).

8.3 Econometric Models

The underlying empirical model can thus be presented in a general form:

$$y_i = f(M_i, X_i) \tag{8.1}$$

where y_i is a measure of per-worker income or productivity in zone i, and $f(M_i, X_i)$ is a measure of transport accessibility of zone i, denoted by M_i, and a set of control variables X_i that reflect other zone-specific characteristics that may affect per-worker income or productivity. We define accessibility as measured by an aggregate economic mass (EM) that is accessible from a given location:

$$M_i = \sum_j \left(\frac{P_j}{g_{ij}^{\alpha}}\right), \text{ for all zones } j \text{ including } j = i \tag{8.2}$$

where

- i Location of the 'home' zone, for which the EM is computed as measuring accessibility from this location.
- j All relevant zones in the study area for market access, including $j = i$.
- g_{ij} Cost of travel between i and j, which may include time and monetary costs.
- P_j A measure of economic activity in zone j.
- α A parameter that controls the distance-decay effect; e.g., it was set to 1 by Graham and Kim (2008) and UK DfT (2006).

It goes without saying that the EM of location i increases if there is an increase in the level of economic activity in i, or there are decreases in the generalized costs of travel between i and j (e.g., through some transport intervention). By the same token, increased level of traffic congestion or dispersion of economic activity around a zone will reduce its EM.

We note that with this measure, the calculation of EM includes the contribution from the home zone (i.e., for $j = i$). This is the average travel cost for journeys within each zone, such as defined in transport studies.

A second popular functional form for the EM uses an exponential function to represent the effects of travel costs, in line with travel demand models:

$$M_i = \sum_j \left(P_j e^{-\theta g_{ij}}\right) \tag{8.3}$$

where P, i, j, and g_{ij} are defined as previously, and θ is a parameter for the exponential function that controls the distance-decay effect. θ may be calibrated through observed travel demand, and empirically, for inter-city travel, θ tends to reduce in value as the economic cost of travel increases. Rice et al. (2006) tested a variation of this exponential function as well as the Hansen function in their analyses of productivity effects.

8.3.1 Isotropic Versus Hierarchical Market Linkages for Economic Mass (EM) Computation

The two EM functions above may be used to cover market access to all destinations, or only a subset of the destinations which are relevant to the home zone in question. In the former case, the measurement is said to be isotropic in the sense that economic linkages between any cities, towns, and so on are considered in an identical way. This has been a common approach in the wider New Economic Geography literature.

In developing economies with limited technical specialization across locations, a hierarchical approach to covering the true market area (as originally defined by Christaller 1933) may be more realistic. This means that the cities and towns are central places of different orders in a regional hierarchy, and the linkages between different orders often tend to be stronger than those among centers of the same order. This is particularly true for learning new skills and transferring technology.

This is not a criticism of the existing EM measures in the literature, because they have largely been defined for regions of developed countries where the inter-city and inter-regional transport networks today are so well connected that they enable nearby central places at the same level of hierarchy to specialize and cross-trade to an extent that was not seen in Christaller's time. Extensive analyses of inter-city and inter-regional travel in Europe and Australia during the 1960s and 1970s indicated that the spatial patterns of travel in that era still exhibited features of the central place hierarchies (Bullock 1980). Our field work in Guangdong has also shown that regional hierarchies are important when firms consider their suppliers, markets, and linkages for technology transfer.

8.3.2 Control Variables

Other than transport accessibility that is represented by the EM, per-employee earnings in a given zone are influenced by a range of factors such as the number of hours worked, capital investment, level of skills, industry composition, and so on. If workers in a given zone work longer hours (e.g., through routine overtime working), they get higher nominal total pay. All being equal, better capital endowment enables higher output. Higher-skilled workers are paid more, and a high proportion of skilled workers in zonal employment would raise the level of average earnings. Similarly, employees working in some industries, such as finance, business services, IT, and research and development are often seen to be paid more than in other industries. These influences on per-worker earnings must be tested, and if significant, controlled for.

Here, we control the effect of working hours by modeling the average hourly earnings per employee as the dependent variable, that is, the annual average per-employee earnings are divided by the average number of working weeks and the average working hours per week. Similarly, we control for employee skills using as a proxy the proportions of those who achieved college, university, and post-graduate qualifications among the employees. In addition, we include control variables to represent industry composition and capital investment.

The regression analyses have been conducted using time-series data for 1999–2008, consisting of assembled economic data at the county or urban-district level and the economic mass (EM) data estimated by the study team using car travel times at the inter-county or urban-district level and a real GDP, as discussed above.

8.3.3 Representing Spatial Spillover Effects

The spatial econometrics literature suggests that there can be significant spillover effects between neighboring counties or urban districts. A formal way to deal with such spillover effects is to construct a spatial-weights matrix such that the lagged dependent and independent variables of all the near and distant neighbors are tested as explanatory variables, in addition to the independent variables of each county or urban district. Given that the EM variable has by definition already accounted for spatial proximity to each employment center, a weights matrix containing the influences of both near and distant neighbors would make the regression model over-complicated if used simultaneously with the dynamic panel-data models. We have therefore adopted here a simplified approach of only including as additional control variables the nearest neighbor of each county or urban district for such spillover effects. As a rule, including the nearest neighbor in the spatial spillover, analysis should take account of 70–80% of the spillover effects (LaSage 2012).

In line with our field-survey findings, in the main regression models, we have assumed a lag of up to three years for the EM, capital stock, and education level in each county or urban district to take effect. This is implemented through producing composite independent variables for any year t through producing a moving average of the same variable for t, $t-1$, and $t-2$. For the spillover effects, the main regression models that use spatial-lag variables take variables of the nearest neighbor from one year earlier.

In terms of the regression models, we exploit what is known in theory about the nature of the OLS, fixed-effects (FE) panel-data models, and dynamic panel-data models, in terms of coefficient estimation bias when used with a dataset such as ours which is autoregressive in nature and has a relatively short time-span. On the one hand, the pooled OLS estimation is likely to bias the coefficient upwards, because of potential endogeneity of the EM variable: if there exist un-measured zonal features that impact on per-employee productivity that would attract the businesses and output and thus impact upon the EM variable over time. The corresponding FE model which is intended for use with a long time series will bias the coefficients downwards if the time series is fairly short, which is often the case with the panel-data series assembled for transport impact studies.

Since our aim is to identify causal effects that run from the economic mass to per-employee hourly earnings, we have to account for the fact that all explanatory variables may be potentially endogenous. In this context, the dynamic panel-data model based on a linearized generalized method of moments (GMM) technique (Arellano and Bond 1991; Arellano and Bover 1995; Blundell and Bond 1998) would in theory be more appropriate than the pooled OLS and FE methods above. The idea of the dynamic panel-data model is to use the past realizations of the model variables as internal instrument variables, based on the assumptions that (1) past levels of a variable may have an influence on its current change, but not the opposite, and (2) past changes of a variable may have an influence on its current level, but not the opposite. The method suits well our requirements because truly exogenous instrumental variables are hard to find in investigations of urban agglomeration effects.

In large samples and given some weak assumptions, GMM models can be free of some of the estimation bias inherent in the OLS and FE models. However, the two variants of the GMM methods, namely DIFF-GMM and SYS-GMM, have different properties when used with small samples. While the DIFF-GMM technique may be unreliable under small samples (Bond et al. 2001), the SYS-GMM technique is expected to yield considerable improvements in such situations (Blundell and Bond 1998). As a rule, data samples of transport impact analyses are unlikely to be very big ones, especially in developing economies. It is therefore necessary to test all the above models in order to clarify the robustness of the models. In turn, a comparison with the theoretical, prior expectations may also serve as a robustness test (Brülhart and Mathys 2008).

8.4 Data

The bulk of the Guangdong economy consists of manufacturing and local commerce. Despite being one of the richest provinces in China, Guangdong had a per-capita GDP of US$6500 in 2008, which in real terms is equivalent to the level of the US per-capita output in the 1930s. The primary and manufacturing industries, mostly low-tech and labor intensive, account for over 70% of the provincial output, and the high-end R&D and business services are a small, unknown fraction of the tertiary sector output. Empirical evidence for the developed economies may not therefore be transferrable to Guangdong or elsewhere in China.

Data from Guangdong are available at two different spatial scales: the province is first divided into 21 municipalities, and the municipalities are in turn subdivided into 67 counties or county-level cities and 21 urban districts of the municipalities (therefore, 88 county-level units in total). This is the most detailed spatial level currently reachable.

The earnings data are for fully employed staff and workers in urban establishments. This definition excludes farmers and other workers in rural areas. Compared with other employment and earning data available, these are the most suitable, as the employees in urban establishments are the most relevant to the agglomeration effects on productivity.

The data for calculating the economic mass (EM) consist of the level of economic activity and travel costs. For economic activity, we chose zonal GDP as the main variable, and retained the zonal size of employment as a sensitivity test. The travel costs and times are those of business travel, because these trips are most directly related to business linkages, technology transfer, commercial transactions, and negotiations. Because our regression models presuppose that the EM variable is correlated with the control variables and respective error terms (see choice of regression modeling strategy below), we have opted to using business travel time as the main travel-cost variable, while retaining travel cost and general travel cost as sensitivity tests.

Road construction data have been assembled over the period of 1999–2008 from a variety of provincial sources. Road links from the 2008 road network are then modified backward in time. For time-series analysis, a road network has been produced for each year of 1999–2008 within the GIS tool. The resulting travel distance, cost, and time matrices at the county or urban-district level for 1999–2008 are checked using our transport modeling experience. Up to 2008, the use of rail for business travel was minimal within the province, and thus, it is not necessary to include rail costs and times in the travel data.

In order to carry out comparisons of different EM measures, both the Hansen and exponential EM function forms are calculated for both the isotropic and hierarchical market areas. For the hierarchical market-area computation, we assume that (1) a county or urban district always interacts with itself, with constant business travel times through all years 1999–2008, and (2) a county or urban district interacts with all component counties or urban districts within its own municipality, as well as the provincial-level centers of Guangzhou, Shenzhen, Zhuhai, and Hong Kong. The only exceptions are Guangzhou and Foshan, which are effectively coalesced into the same metropolitan area—the two urban areas are allowed to interact with each other.

For the control variables, we use the percentage of workers with college degree and above as a proxy for labor skills from the statistical yearbooks at the county or urban-district level. The statistical yearbooks report the levels of fixed asset investment per year. The Economic Census of 2004 also reports the total capital stock for production purposes per municipality. We estimate the county or urban-district level capital stock through these sources and build up the yearly capital stock for the entire time series that incorporates a standard capital stock depreciation rate of 5% per year. Investment in residential properties is excluded. We divide the zonal total capital stock by the total of full-time workers and staff in that zone to obtain the per-employee capital endowment. According to the National Labour Statistics Yearbook 2009, finance, information technology, and R&D industries are ranked as the top three high-earning sectors in Guangdong Province. We use the number of employees by region in these three sectors to control for the effects that can potentially arise from such differences in industrial composition. Specifically, we construct the index of sectoral composition following the definition of location quotient (LQ).

8.5 Model Test Results

The regression analyses have been conducted using time-series data for 1999–2008, consisting of assembled economic data at the county or urban-district level and the economic mass (EM) data estimated by the study team using inter-county or urban-district level business car travel times and level of economic activity, as discussed above.

To recap, on the left-hand side of the regression equations, the dependent variable is a vector of zonal data representing per-employee productivity levels: the average nominal hourly earnings at the county or urban-district level is used as the main test variable, with per-employee average GDP as a sensitivity test variable. On the right-hand side of the equations, the list of independent zonal variables at the county or urban-district level includes the EM representing transport accessibility, a range of variables representing zonal capital investment, skills, and industrial composition, and spatial-lag variables from the nearest neighbor zones. The independent variables are tested as appropriate for each specific functional form. In addition, the GMM models use time-lagged independent variables as instruments as specified.

Through the regressions, we have tested different measures of productivity (i.e., hourly earnings and per-employee GDP), different EM terms (i.e., using distance, travel time, and generalized travel cost for isotropic and hierarchical market areas), and different measurements of capital endowment and labor skills. All regression models have retuned consistent results, among which we have found that the equations using hourly nominal earnings, hierarchical EM using time to measure travel cost, accumulated and depreciated capital stock, and parentage of college and above graduates to measure labor skills, have an overall best fit. This is in line with our field-survey findings. Both the Hansen-type and exponential functional forms of the EM variable are tested. Owing to the limit of space, we report the core estimation results in Table 8.1. The other tests are available upon request.

In Table 8.1, Model (1) is a pooled OLS model which returns an EM coefficient of 0.24, with the EM and the control variables (for capital stock and education level) being statistically significant and a relatively high R-squared = 0.69. However, we have good theoretical reasons to suspect that the coefficients are biased upwards and this model result embodies an absolute upper bound of the productivity elasticities.

By contrast, with Model (2) which is the time-series fixed-effect (FE) model, the EM coefficient drops to 0.115 when the period dummies (representing the period-specific effects) are included for the Hansen EM formulation. The EM coefficient further drops to 0.052 in Model (3) when the exponential EM variable is used. Our theoretical expectations are that these are biased downwards for respective EM functional forms, and thus could be considered as a lower bound to the EM coefficient.

This is reflected in the DIFF-GMM model in Column (4). The EM coefficient output from this model is at 0.151, between the upper and lower bounds as we expect, although the coefficients are not statistically significant. The SYS-GMM model (5) gives a similar EM coefficient at 0.141: Both the EM and the capital stock coefficients are now significant; note that this model includes additional explanatory variables that represent the spillover effects from the nearest neighbor zones in terms of capital stock endowment and education level of the employees.

The GMM-SYS Model (6) is a standard test to assess the robustness of the model by reducing the number of instrument variables (from 115 to 69), which has raised somewhat the significance of the education-level variable but has not altered the nature of the model results nor the magnitude of the coefficients. The standard tests of the GMM models suggest that there are no apparent misspecification problems. The Hansen test for over identification restrictions, and the difference Hansen tests for the validity of the GMM and IV instruments, indicate that the instruments are valid. The Arellano-Bond AR2 test suggests that no second-order residual auto-correlation is present.

Model (7) presents the SYS-GMM results for the exponential functional form of EM, which returns an EM coefficient of 0.087. The estimation diagnostics are similarly good. A test to reduce the number of instruments (from 103 to 75) has also been carried out as Model (8) and has confirmed that the instruments are valid.

Given that the exponential form of the EM variable embodies the distance-decay parameters that are consistent with the travel-behavior model calibrated in China, it would seem sensible to consider Model (7) as the preferred estimate of the productivity elasticity (i.e., 0.087 with a standard error of 0.03 and robust t statistic 2.89) with respect to transport accessibility.

In summary, the econometric results show that transport accessibility as represented by the EM is statistically significant after controlling for control-variable endogeneity and spatial spillover effects. Our preferred estimate comes from Model (7) in Table 8.1, which adopts a SYS-GMM formulation and exponential EM formula and returns a productivity elasticity of 0.087, with a robust standard error of 0.030. The

Table 8.1 Time-series model results

Dependent variable: ln(hrly earnings [t])	(1)	(2)	(3)	(4)	(5)	(6)	(7)	(8)
	Pooled OLS (Hansen)	Fixed effects (Hansen)	Fixed effects (Exp)	DIFF-GMM (Hansen)	SYS-GMM (Hansen)	SYS-GMM (Hansen)	SYS-GMM (Exp)	SYS-GMM (Exp)
ln (economic mass)	0.240***	0.115***	0.052	0.151	0.141***	0.162***	0.087***	0.101***
	[0.016]	[0.043]	[0.017]	[0.152]	[0.067]	[0.077]	[0.030]	[0.037]
ln (per-worker capital stock)	0.155***	0.007	0.007	0.026	0.133***	0.151***	0.127***	0.129***
	[0.013]	[0.011]	[0.011]	[0.035]	[0.032]	[0.032]	[0.035]	[0.035]
ln (% of college graduates)	0.193***	−0.021	−0.026	−0.046	0.031	0.062	0.119***	0.110***
	[0.011]	[0.016]	[0.016]	[0.048]	[0.049]	[0.040]	[0.039]	[0.037]
Period dummies ln (per-worker capital stock)—nearest county	No	Yes	Yes	Yes	Yes	Yes	Yes	Yes
					0.008		0.017	
					[0.033]		[0.028]	
ln (% of college graduates)—nearest county					0.071		0.006	
					[0.041]		[0.041]	
Constant	−0.569***	1.975			0.170	0.104	0.387	0.491
	[0.127]	[0.181]			[0.478]	[0.364]	[0.423]	0.407
Observations	704	704		616	704	704	616	616
R-squared	0.69	0.57						
Number of instruments				64	115	69	103	75
AR(2) test				0.430	0.170	0.790	0.370	0.411
Hansen test of over identification restrictions				0.274	0.980	0.352	0.818	0.245

(continued)

Table 8.1 (continued)

Dependent variable: ln(hrly earnings [t])	(1)	(2)	(3)	(4)	(5)	(6)	(7)	(8)
	Pooled OLS (Hansen)	Fixed effects (Hansen)	Fixed effects (Exp)	DIFF-GMM (Hansen)	SYS-GMM (Hansen)	SYS-GMM (Hansen)	SYS-GMM (Exp)	SYS-GMM (Exp)
Difference in Hansen tests of exogeneity of instrument subsets:								
For GMM instruments (*p*-value)					0.713	0.367	0.516	0.249
For IV instruments (*p*-value)				0.115	0.909	0.365	0.744	0.330

Notes (1) For the GMM models, the variables for the economic mass and average percentage of college graduates are treated as GMM instrument variables while the year dummies and average per-worker capital stock are treated as IV variables. For GMM models they are based on two-step Windmeijer finite-sample adjustment. For the FE and GMM models, the fixed effects are modeled but not reported here. (2) Robust standard errors are in square parentheses: ***$p < 0.01$, **$p < 0.05$, *$p < 0.1$

model diagnostics suggest that all the SYS-GMM model results are robust. Furthermore, the GMM model results fit our prior expectations regarding the upper bounds established by the pooled OLS models and the lower bounds by the time-series fixed-effects models.

8.6 Discussions

An extensive series of regression model tests show a consistent pattern for a statistically robust relationship between transport accessibility and business productivity. In particular:

(a) As expected, the pooled OLS regressions produced high elasticity estimates while the time-series fixed-effects (FE) regressions produced low estimates. The dynamic panel-data models using the linearized generalized method of moments (GMM) tend to return intermediate elasticity values.
(b) Our understanding of the regression models and the development process in Guangdong, China gives grounds to prefer the GMM model estimates (particularly the SYS variant which corrects for relatively small samples). This is because the SYS-GMM models are capable of making a sound use of the short panel dataset.
(c) Our preferred estimate of the productivity elasticity with respect to transport accessibility is 0.087 (with robust standard error 0.03 and t statistic 2.89). This comes from the SYS-GMM model which uses the exponential formula in measuring transport accessibility. This positive relationship remains robust after controlling for a range of control variables, endogeneity, and nearest neighbor spillover effects. The robustness of this estimate is confirmed through both the regression diagnostics and a comparison with results from the alternative models.

This central productivity elasticity estimate of 0.087 implies that a 10% improvement in transport accessibility would give rise to an increase of per-worker productivity of 0.83% (i.e., $(1 + 10\%)^{0.087} - 1 = 0.0083$), and a doubling in transport accessibility would imply an increase of per-worker productivity of 6.2% (i.e., $(1 + 100\%)^{0.087} - 1 = 0.0622$). This is well within the consensus range of productivity elasticities from a comprehensive review of such evidence in predominantly developed economies that "doubling city size seems to increase productivity by an amount that ranges from… roughly 5–8%" (Rosenthal and Strange 2004), and is comparable with the elasticity range from the latest meta-analysis of productivity elasticities published by Melo et al. (2013), who suggests the central elasticity value is around 0.05.

In assessing the estimates we may also compare them with our prior expectations: transport accessibility and agglomeration are thought to play an important role in knowledge spillover and technological improvements in China (IBRD 2006).

The empirical findings in this chapter are to an extent supported by emerging estimates for China, although our estimates are considerably lower. For instance, Au and Henderson (2006), using data of 1990 and 1997 from 205 Chinese cities, suggested that there are significant urban agglomeration benefits: for example, moving from a city of 635,000 to one of 1.27 m increases the real output per worker by 14%, after controlling a range of other influences. More recently, Zhang's analysis (Zhang 2008) using the 1993–2004 data put the mean elasticity value at 0.106 in China after controlling for spatial spillover effects.

Our field studies in Guangdong (see EASCS 2014a, b) have also started to investigate the actual mechanisms through which businesses benefit from transport accessibility improvements in terms of employee productivity. It indicates that the agglomeration benefits accrued by transport improvements are well understood by the businesses and individuals, and the extent to which they exploit such benefits is comparable with those observed in developed economies. This provides a degree of corroboration at the micro-level. Of course, further work is still needed to quantify such effects at the level of individual businesses and employees.

8.7 Conclusions

This chapter aims to introduce the theories and methods of spatial economics through one specific example of quantifying the economic contribution of transport accessibility improvements, which may well be a research question that often confronts the students of urban informatics. The chapter starts with simple OLS regression models that are commonly used in urban-informatics research and then extends the models step by step using a cross section of spatial analytical and economic theories. The resulting models reach the current frontier of the field, and they serve to fill a gap in current literature. In developing the models, there is also an ethos of developing a methodology which is theoretically rigorous but can be made operational with a level of data availability that is generally achievable in the emerging economies. In the low- and middle-income developing countries such as China, such empirical evidence for spatial-economic effects of transport is currently poor and the practical needs for them are urgent, for example for assessing major investment initiatives.

Of course, the current econometric models may not yet fully control for other differences between zones, for example, the spatial self-selection and sorting of employees within and among the counties and urban districts. Clearly, spatial proximity resulting from transport improvements plays an enabling role in spatial self-selection and sorting. Nevertheless, it is yet difficult to discern the precise contribution of transport improvements to such mechanisms within the available data sources.

Also, it is not for econometric studies alone to establish causality between transport accessibility and productivity where there is a process of significant cumulative causation; that task should be supported by an in-depth understanding of the actual mechanisms at work, for example through field studies as discussed above.

Additional future work may further improve the robustness of the findings presented here; the list below would serve to indicate the scope of further research on this topic:

First, it may be possible to expand the time series under consideration both in years covered and the range of explanatory variables, which is likely to make the model more robust and improve the precision of the coefficient estimates.

Secondly, similar econometric models can be estimated for the economically less-developed regions in China (e.g., inland regions such as Sichuan), as well as other affluent regions along the Eastern Coast (e.g., the Yangtze River Delta centered upon Shanghai and the Bohai Bay Metropolitan Area centered upon Beijing). This would clarify whether there are significant differences among regions of different levels of development.

Thirdly, if and when the disaggregate Economic Census data become available from the Chinese statistics bureaux, enterprise-level production functions (e.g., of the Translog type) can be estimated, which would provide more precise estimates of the agglomeration effects including possible spatial sorting effects. The Economic Census data were collected by enterprise, although so far they have not been released for use in research in China.

Fourthly, micro-level case studies of firms and institutions will help us understand how firms actually respond to transport improvements, and through what mechanisms they gain from agglomeration effects or otherwise.

The cumulative evidence through the above could eventually provide a fuller understanding of economic development in terms of dynamic general equilibrium processes, for example as suggested by Au and Henderson (2006) and Lakshmanan (2011). Such understanding would in turn enable us to better plan transport projects, particularly to promote shared prosperity and poverty alleviation in under-developed regions.

Acknowledgements and Disclamer The example reported in this chapter is primarily based on a working paper sponsored by the Technical Assistance of the World Bank project "Regional economic impact analysis of high-speed rail in China". The funding support from the World Bank is gratefully acknowledged. The author also acknowledges additional funding support from the UK EPSRC Centre for Smart Infrastructure and Construction Phase 2 (EP/N021614/1) on the theoretical extensions and testing of the quantification methods. The contributions of World Bank colleagues during the project, particularly John Scales, Gerald Ollivier, Richard G. Bullock, and Wanli Fang, are gratefully acknowledged. The views and any remaining errors are the responsibility of the author alone.

References

Arellano M, Bover O (1995) Another look at the instrumental variable estimation of error-components models. J Economet 68:29–51

Arellano M, Bond S (1991) Some tests of specification for panel data: Monte Carlo evidence and an application to employment equations. Rev Econ Stud 58:277–297

Au CC, Henderson JV (2006) Are Chinese cities too small? Rev Econ Stud 73:549–576. https://doi.org/10.1111/j.1467-937X.2006.00387.x
Blundell R, Bond S (1998) Initial conditions and moment restrictions in dynamic panel data models. J Economet 87:115–143
Bond S, Hoeffler A, Temple J (2001) GMM estimation of empirical growth models. CEPR Discussion Paper 3048
Brülhart M, Mathys NA (2008) Sectoral agglomeration economies in a panel of European regions. Reg Sci Urban Econ 38:348–362
Bullock RG (1980) Regional travel patterns: review and empirical model estimation. Mimeo
Christaller (1933) Central places in Southern Germany. English edition 1966. Prentice-Hall, Engelwood Cliffs, NJ
Ciccone A, Hall R (1996) Productivity and the density of economic activity. Am Econ Rev 86:54–70
Combes P, Duranton G, Gobillon L (2008) Spatial wage disparities: sorting matters! J Urban Econ 63:723–742
Deichmann U, Kaiser K, Lall SV, Shalizi Z (2005) Agglomeration, transport, and regional development in Indonesia. World Bank Policy Research Working Paper No. 3477. https://papers.ssrn.com/sol3/papers.cfm?abstract_id=647682
Duranton G et al (1987–2018) Handbook of urban and regional economics, 1–5. https://www-sciencedirect-com.ezp.lib.cam.ac.uk/journal/handbook-of-regional-and-urban-economics/issues
EASCS (2014a) Regional economic impact analysis of high speed rail in China: main report. Report No.: ACS9734. The World Bank, Washington DC
EASCS (2014b) Regional economic impact analysis of high speed rail in China: background Working Paper 2. Report No: ACS9734. The World Bank, Washington DC
Fujita M, Krugman P, Venables AJ (1999) The spatial economy: cities, regions and international trade. MIT Press, Cambridge, MA
Graham DJ, Kim HY (2008) An empirical analytical framework for agglomeration economies. Ann Reg Sci 42:267–289
IBRD (2006) Governance, investment climate, and harmonious society: competitiveness enhancements for 120 cities in China. The World Bank and IBRD, Washington DC
Kopp A (2007) Aggregate productivity effects of road investment: a reassessment for Western Europe. In: Karlsson C, Anderson WP, Johansson B, Kobayashi K (eds) The management and measurement of infrastructure. Edward Elgar, Cheltenham
Laird JJ, Venables AJ (2017) Transport investment and economic performance: a framework for project appraisal. Transp Policy 56:1–11
Lakshmanan TR (2011) The broader economic consequences of transport infrastructure investments. J Transp Geogr 19:1–12
Lall SV, Gun WH, Deichmann U (2010) Infrastructure and city competitiveness in India. UNU-WIDER Working Paper W, World Institute for Development Economic Research
LaSage J (2012) Understanding spill-overs. ERSA 2012 Congress keynote speech, Bratislava, August
Lu D, Yao S, Liu H, Gao X, Li G, Duan J et al (2007) China regional development report: the urbanisation process and spatial expansion. The Commercial Press, Beijing ([In Chinese])
Lu J (2010). Agglomeration of economic activities in China: evidence from Establishment Censuses. Regional Stud 44:281–297
Melo PC, Graham DJ, Noland RB (2009) A meta-analysis of estimates of urban agglomeration economies. Reg Sci Urban Econ 39:332–342
Melo P, Graham DJ, Brage-Ardao R (2013) The productivity of transport infrastructure investment: a meta-analysis of empirical evidence. Reg Sci Urban Econ 43:695–706
Redding SJ (2010) The empirics of new economic geography. J Reg Sci 50:297–311
Redding SJ, Venables A (2004) Economic geography and international inequality. J Int Econ 62:53–82
Redding SJ, Rossi-Hansberg E (2017) Quantitative spatial economics. https://www.princeton.edu/~erossi/QSE.pdf

Rice PG, Venables AJ, Patacchini E (2006) Spatial determinants of productivity: analysis for the regions of Great Britain. Reg Sci Urban Econ 36:727–752
Roberts M, Goh C (2012) Density, distance and division: the case of Chongqing municipality, China. Cambridge J Reg Econ Soc 4:175–188
Roberts M, Deichmann U, Fingleton B, Shi T (2012) Evaluating China's road to prosperity: a new economic geography approach. Reg Sci Urban Econ 42:580–594
Rosenthal S, Strange WC (2004) Evidence on the nature and sources of agglomeration. Rev Econ Stat 85:377–393
UK DfT (2006) Transport, wider economic benefits, and impacts on GDP. UK Department for Transport, London. https://www.dft.gov.uk/pgr/economics/rdg/webia/transportwidereconomicbenefi3137
UK2070 Commission (2019) Fairer and stronger: first report of the UK2070 Commission. https://uk2070.org.uk/publications/
von Thünen JH (1826) Der isolierte staat in beziehung auf landwirthschaft und nationalökonomie. Perthes, Hamburg. English translation: Pergamon Press, Oxford
Zhang X (2008) Transport infrastructure, spatial spillover and economic growth: evidence from China. Front Econ China 3:585–597

Ying Jin is Director of the Martin Centre for Architectural and Urban Studies at University of Cambridge and a Fellow of Robinson College, Cambridge. He was elected a Fellow of the Academy of Urbanism, London in 2011, and his research interests are investigating the causes of urban change through data science and interdisciplinary models.

Chapter 9
Conceptualizing the City of the Information Age

Helen Couclelis

Abstract Cities are among humanity's most important and most complex creations, and they have been steadily increasing in complexity since the advent of the digital age. Informatics, the science of information, has by now advanced to a point where high expectations of improved understanding and evidence-based actionable knowledge for urban researchers, managers, and planners appear justified. But while there is more information than ever before, many kinds of theories, models, approaches, and tools that we have relied on thus far may no longer be of much use in the city of the information age. This chapter provides an overview of the state of affairs in urban science and planning, pointing out the limitations of formerly reliable methods and tools in the face of dramatic developments in the life and function of cities in the developed world. The chapter closes with suggestions for data-oriented strategies that might replace the ways we have used urban data up until recently.

9.1 Introduction

9.1.1 Urban Complexity in the Age of Information and Communication Technologies

A defining characteristic of a complex system is that it can be seen from any number of different, even contradictory angles (Casti 1984). Cities are complex systems by this as well as by many other possible definitions. They are made of asphalt and concrete, but they grow and change; they are places, but also networks; they are spatiotemporal objects, but they are about people; they are physical structures, but also abstract institutions connected with the notion of *citi*zenship; they may fit within a square mile, or they are larger than many small countries; and more recently, they are also both actual and virtual.

H. Couclelis (✉)
University of California, Santa Barbara, USA
e-mail: cook@geog.ucsb.edu

W. Shi et al. (eds.), *Urban Informatics*, The Urban Book Series,
https://doi.org/10.1007/978-981-15-8983-6_9

For some time now, cities have been responding to information and communication technologies (ICTs) while also helping define them, with or without the help of urban analysts, managers, or planners. Years of publications on the topic have shown that the results of mostly piecemeal urban applications of ICT have so far been mixed, with few spectacular achievements or transferable best practices. There are also many questions of time–space perspective, as we transition from the city of yesterday to the city of tomorrow. For example: the repurposing of urban structures and infrastructures for new uses at new times; the anticipation of new divisions of labor, of new forms of urban management, and of new urban decision-making pathways, whereby technology companies increasingly call the shots; the role of supra-local and global agents, and of new political alliances at any scale. And of course, also, the appearance of new technologies not yet on the horizon. Issues such as these are highly likely to arise within the next twenty to thirty years, most of them supported by the unrelenting spread of ICTs across the globe. How does one even begin to grasp what is really going on? But there is hope: this may be the moment when the data, tools, infrastructure, and analytic approaches of the informatics revolution are becoming mature enough to forge unprecedented opportunities for the betterment of cities.

9.1.2 A Different Kind of City

There is no question that ICTs significantly add to the fundamental complexity of cities. Also, the piecemeal nature of most urban applications of ICT to date is antithetical to the notion of complexity, which entails interdependence and interplay. One everyday example of the interdependent complexity contributed by ICTs is captured by the related notions of the disconnect of urban form from function (Batty 2018) and the fragmentation of activity (Couclelis 2009; McBride et al. 2019), which affect the macro- and micro-levels of the city. The former notion concerns the relationship between, on the one hand, the classic urban activities of residing, working, shopping, learning, recreating, etc., and on the other, the urban places where these activities take place. In the traditional pre-ICT city, there is a close correspondence between each kind of urban activity and the urban spaces adapted to support it. The correspondence used to be so reliable that knowing where someone was at some point in time made it relatively easy to guess what they might be doing—and conversely ("if working then at the workplace, if shopping then at the shopping mall, if getting an education then at school"). This match between activity and place was also at the heart of traditional urban land-use and transportation models and planning, since people's movement from place to place was largely dictated by the daily schedule of predictable activities, and urban form and function were tightly linked. In much of today's industrialized world, these close connections between urban activities and spaces are disintegrating, and as a result, model predictions of urban growth and change are becoming less reliable.

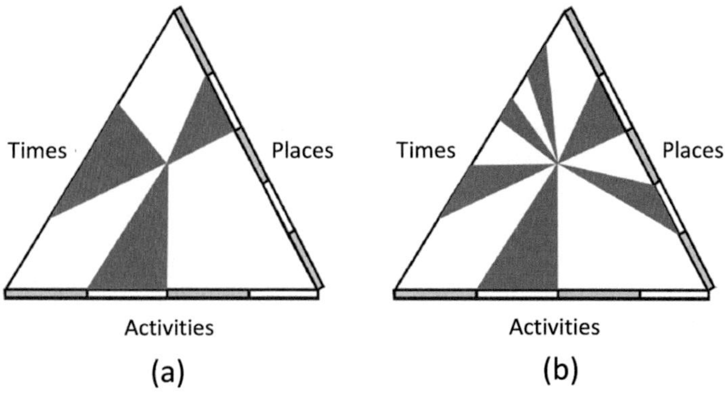

Fig. 9.1 Fragmentation of activity and ICT. **a** Before ICT: one of four activities is carried out at one place, during one time interval; **b** After ICT: that same activity is carried out at two different places, at three distinct time intervals (from Couclelis 2009)

The notion of the fragmentation of activity sheds light on the micro-level of this phenomenon. Indeed, for some time now, thanks to ICTs, increasing numbers of daily activities can be broken down into tasks and carried out consecutively at several different places and several different time intervals during the day (Fig. 9.1). For increasing numbers of people, gone is the compulsive Monday to Friday 8–5 at the office, or the family Saturday trip to the shopping mall. These traditional specialized places still exist, but we can also shop from home after a visit to the drug store, watch movies on our workplace computer during breaks from work, close an extra business deal from our car after the martini lunch at the fancy hotel, follow university lectures on our smartphone while in bed, before cycling to campus, or monitor our real-time health indicators on our smart watch at the gym to expedite the check-up at the clinic later.

9.1.3 The Smart City

The broadening international conversation about the coming smart city is certain to add several more layers of complexity to urban research and management. While the smart-city concept remains ill-defined and open-ended, and few, if any, generally accepted examples exist today, there is agreement on several of the anticipated (or desired) defining characteristics: smart cities will be sustainable, livable, equitable, innovative, and creative. Above all, they will be able to capitalize on the extraordinary possibilities that technology, especially ICTs, artificial intelligence (AI), and big data, are already unraveling before our eyes. Are all these hopes, assumptions, and anticipated characteristics realistic, or even mutually compatible? There is also the unavoidable gap between intention and reality. As Goh (2015, p. 169) asks: "What happens when intelligent plans encounter messy politics, social systems, and divergent scales of urban governance?" San Francisco, USA comes to mind. There, the world's most famous breeding ground of new information technologies coexists with sky-high property values and with some of the worst levels of homelessness and street squalor to be found in any city of the industrialized world.

Further: smart is not quite the same as intelligent. Smart, much like clever, has connotations of something playful, a bit superficial, not terribly serious, of no great consequence. A smart child. A smart dog. A smart answer. Street smarts. A clever trick. The smart city could easily be smart in this sense, with bright flashes of brilliance here and there (and then), but also with much that is technology for technology's sake, unhelpful, unneeded, wasteful, discriminatory, retrograde, ephemeral, or downright damaging—now, or a few years down the road. How can our cities be not just smart, but truly intelligent?

The smart cities phenomenon thus encapsulates many of the major new challenges of current urban research and management. At the one end of a spectrum, smart (urban) growth only recently meant wisely managed urban development, socially, fiscally, and environmentally sustainable, mindful of resource constraints, prepared to capitalize on comparative advantages and to seize opportunities as they arise, while attentive to community input, fairness, and the planners' recommendations. At the other end of the spectrum, a smart city is the bionic city of science fiction. At the moderate middle, we find mixed approaches of some of this and some of that, or even coexisting views that appear incompatible at first sight. As an example, the European Commission's website begins by defining the smart city as "a place where traditional networks and services are made more efficient with the use of digital and telecommunication technologies for the benefit of its inhabitants and business," but a few lines later, the European Partnership on Smart Cities and Communities is introduced as being primarily about governance, citizenship, wise regulation, and other such traditional soft imperatives going back to the Athens of Pericles (European Commission 2020).

Different authors also provide many contrasting definitions and descriptions of the smart city. Thus Caragliu et al. (2009, p. 50) consider "a city to be smart when investments in human and social capital and traditional (transport) and modern (ICT) infrastructure fuel sustainable economic growth and a high quality of life, with a wise management of natural resources, through participatory governance," whereas Batty (2018, p. 178) emphasizes the technological aspects: "The nature of the smart city then lies in the very technology that defines it." Geertman et al. (2015) take a different approach, attempting a classification of smart cities into four categories, as follows: (a) Smart machines and informated [sic] organizations; (b) Partnerships and collaboration; (c) Learning and adaptation; and (d) Investing for the future. These categories are discussed in the above chapter, and they are interesting and plausible, but are meant more as alternative abstract types than as descriptions of possible, actual kinds of cities.

9.1.4 Urban Informatics

Informatics, the increasingly preferred term for information science, has been defined as "the study of the behavior and structure of any system that generates, stores, processes and then presents information; it is basically the science of information. The field takes into consideration the interaction between the information systems and the user, as well as the construction of the interfaces between the two." (Techno-pedia). The smart city is but one application area for urban informatics, albeit one of fundamental importance, considering the ever-increasing significance of the urban in the present world and in any conceivable near future. Not coincidentally, Batty's (2018, p. 176) notion that "Smart cities essentially enable computers and communications to be embedded in the very fabric of the city" is very close to the use of the term "system" in the above definition. But informatics is needed just as much in the still-traditional city, where so many taken-for-granted regularities are being increasingly challenged by ICTs.

The next section provides a broad overview of current approaches to urban research and planning, seeking to identify areas where modern informatics may have a key role to play.

9.2 Urban Research and Planning, Yesterday, and Tomorrow

9.2.1 The City as Place

A direct consequence of the complexity of the urban is the multitude of possible ways of approaching the study of the city. On the one hand, there is the vast range of disciplinary perspectives, whereby the word "urban" may be added as a qualifier to almost any empirical discipline. We thus have urban economics, urban sociology, urban history, urban geography, urban ecology, urban transportation, urban health, urban anthropology, urban planning, etc., and now also urban informatics. In addition, there are numerous cross-disciplinary and methodological viewpoints and approaches applicable to cities, such as post-Marxism, post-structuralism, gender studies, science and technology studies, quantitative social science, spatial analysis, computer simulation and modeling, the networks perspective, the design perspective, and so on. In "Key Thinkers on Cities," Koch and Latham (2017) collected 40 profiles of scholars who in one way or another have made significant contributions to the study of cities, the stress being on "one way or another," as the diversity of approaches represented is quite stunning. While there are significant affinities among cognate disciplines or approaches (urban sociology and urban anthropology, say, or spatial analysis, mathematical modeling, and computer simulation), others are so distant intellectually from one another that they hardly seem to be about the same general topic. One may say that the universe of perspectives and theories on cities is locally coherent but globally not coherent. The most creative new work on cities might be that which discovers and establishes important connections among intellectually or methodologically remote areas of urban research. An example is the work by Reades et al. (2018) on gentrification, which combined spatial analysis, qualitative research, and machine learning to show that it is possible to analyze existing patterns and processes of neighborhood change to identify areas likely to experience change in the future.

Theories of the city have existed since antiquity, but have flourished since World War II along with the establishment of academic units and journals dedicated to their study, and the fast-increasing number, size, complexity, and importance of cities in the modern world. At the same time, the quantitative and computational turns in the social sciences and urban planning have enabled more thorough and empirically relevant work, while also stimulating theory development, motivated by the newly available observations and informed discussions.

This trend toward more realistic empirical theory may now be reversing. We saw earlier how gravity-based spatial interaction modeling, one of the mainstays of quantitative urban theory and planning, risks becoming less and less relevant as urban activity becomes more fragmented in space and time, and as urban form is getting disconnected from function. The same seems true of urban cellular-automata-based modeling, another popular approach that also relies on assumptions of proximal relations among cognate places and land uses. It is true that the principle of distance decay, which underlies these kinds of models, is too fundamental to become obsolete as long as people and cities inhabit the physical world; but having to coexist with principles of the virtual world makes its theoretical utility more elusive.

Other ways of looking at the city, such as those involving cognition (think space syntax, the legibility of urban environments, finding one's way in an unfamiliar area, recognizing place in space) may be more resilient in principle. But faced with ubiquitous digital aids for navigation, point-of-interest (POI) location, place-related information, and environmental problem-solving in general, it is questionable whether human spatial abilities might not degrade over time. More optimistically, spatial abilities should improve in tasks involving ICTs, just as they degenerate where no longer needed.

Economy, demography, and technology remain among the handful of key drivers of urban growth and change, especially in the vast megalopolises of the world that are not yet steeped in ICTs. Increasingly, ecological conditions such as water availability and climate are added to the key drivers of urbanization. Most of these factors are slow-moving and can be accounted for relatively well with traditional data and methods. But the more a city becomes part of the information society, the more its study requires indicators on fleeting phenomena that vary during the course of the day, the hour, or the minute. Many of these may be local quality-of-life factors (noise levels, air pollution, traffic conditions, disturbances due to special events or incidents), while others, such as threats to community health and safety, or to the integrity of energy and information networks at any scale, may be of broader import.

9.2.2 The City as Node on a Network

The vast majority of urban research has approached the city as a kind of place, but an alternative, increasingly relevant way of thinking about cities is as nodes in a network. This idea has been around for some time, and is reflected, among others, in Christaller's widely known Central Place Theory, which views individual settlements as elements in a recursive regional hierarchy of population sizes centered on the largest settlement. The idealized model of the resulting spatial arrangement is a hierarchy of nested hexagons, the vertices of which are the smaller settlements that depend on the central larger one. While Central Place Theory emphasizes the notions of trade and distance, it also clearly describes systems of settlements bound together by networks of relations.

Christaller's notion of networks of interdependent cities also appears, at a much grander scale, in Doxiadis's (1968) vision of Ecumenopolis. This is the author's term for the coming network of cities of all different sizes that spans the entire globe, and which becomes, at the limit, a mesh of continuous corridors of urbanization ('Ecumene' is Greek for the inhabited world). Megalopolis—literally, the big city—is a more modest and better-known version of the same idea, of which there are multiple actual instances around the world. While the term had appeared in earlier writings of the twentieth century, it was popularized by Gottman's (1961) work on the north-eastern seaboard of the USA. The catchy name BosWash, for the urban agglomeration reaching from Boston, MA to Washington, DC is the best-remembered part of Gottman's ground-breaking study.

The most systematic contemporary approach to the notion of the city as node in a network of cities is quite likely represented by the work of the international research network on Global and World Cities (GaWC 2020). Scholars affiliated with the GaWC network sometimes describe their work as metageography—a geography of geographies—to emphasize the global-scale perspective on cities that they adopt. The group's focus is the world-wide hierarchy of cities of different degrees of importance and size (world, global, peripheral, and specialized cities), with an emphasis on the mutual dependencies and other relations that make up the international network of urban interactions. The socioeconomic, political, and physical characteristics of individual cities are examined to the extent that they reflect or promote the forces that bind the world's cities together, such as the global phenomena of capital flight, industrial dislocation, labor migration, trade and resource flows, innovation and technology diffusion, and so on. To study these networks of mostly intangible long-distance flows and their local implications, GaWC researchers must ask novel questions requiring new kinds of data and new forms of visualization—in other words, define a new agenda for urban research. The network's website provides a wealth of information about the work of the close to three hundred affiliated members, who include several prominent names in geography, urban studies, and a number of other fields contributing to research on the information society (e.g., Latham and Sassen 2005; Hoyler et al. 2018).

9.2.3 Planning the City

Urban planning—professional as well as academic—is another field that is being substantially affected by developments in the city of the information age. Like urban studies, planning deals with the city at several different scales, from that of the neighborhood park to that of the megalopolis. Unlike urban studies, the planners' approach is more that of the engineer than of the scientist, more synthetic than analytic, more action-oriented than knowledge-oriented. The major difference between these two fields, however, is the fact that planning is inherently and fundamentally about the future, whereas urban research and data are at best about the very recent past. Predictive models developed by urban researchers still go some way toward meeting the current needs of planning, but the assumptions, generalizations, and rules of thumb built into them may soon become obsolete. It is ironic that deep qualitative uncertainty, the kind that matters most to future-oriented endeavors like planning, might be substantially increasing at a time when the quantity and quality of available data are also increasing dramatically.

Urban management is also a form of planning, operating over shorter time frames and handling more specific sets of problems. Both professional planning and management directly contribute to urban governance, and their errors have consequences well beyond the threat of a research paper rejection. Despite the considerable overlap with urban studies, planning and management thus involve a very different take on the city, and information needs that are as complex but different from those of the urban researcher. For example, planning must now (by law, in many countries) take into account the often vague or conflicting input of the public, while also accommodating political interventions and juggling a myriad of local and regional regulations that may include mutually contradictory, obsolete, or otherwise unhelpful restrictions.

Things were not always as complicated for urban planning. In the modern era, planning was at first a straightforward engineering profession focused on urban sanitation and other infrastructure development, before embracing the systems approach and operations-research methodologies in the 1950s and 60s, and later also additional perspectives by the names of comprehensive, integrated, or strategic planning. It is only with the social movements of the 1970s, when the participatory era began, that the planners' tidy office spilled onto the streets. Planning was no longer carried out for the people but with the people. Opinion surveys, public hearings, story-telling, and politicking increasingly replaced computer models, especially in countries such as the USA that lack a strong planning tradition. However, geographic information systems (GIS) eventually came along to fill the technical void, and there was no way back.

The adoption of GIS in planning was at first not without problems. Critics were concerned about the possibility of disenfranchising those lacking the requisite digital literacy, of affecting societal priorities by focusing on what is easily measurable, of imposing a technocratic view of the world on other people's perspectives, of introducing new issues of privacy and surveillance, and so on. These concerns have been to a large extent resolved, to the point where most of those who used to be the critics are now often using GIS themselves.

In response to the critique, academic planners developed methodologies largely based on GIS for the age of public participation, creating the subfields of public participation GIS (PPGIS) and, for well-defined groups of stakeholders, participatory GIS (PGIS; Jankowski and Nyerges 2001). Planning support systems (PSS) emerged in the early 1990s as a response to the increasing complexity of planning in societies that value both the diversity of opinions and the scientific grounding of public decision making (Brail and Klosterman 2001; Geertman and Stillwell 2009; Geertman et al. 2015). PSS were enabled by major improvements in computational resources and geospatial data availability, and relied heavily on the rapid expansion and increasing sophistication of GIS. The main purpose of PSS is to integrate the societal and technical aspects of planning with the computational bonanza of our age, and are thus, at least in concept, one of the best incarnations of the idea of geodesign to date. Current forms of PSS successfully support public participation, allowing the collection and processing of a wide range of relevant data through crowd-sourcing methods. The adoption of PSS has been slow, but the field continues to attract considerable interest, now also from scholars and practitioners from beyond traditional urban planning.

9.3 Speculations

9.3.1 The Robotic Era?

Humanity spent millennia in the pre-industrial age, then the industrial age lasted some two hundred years, the post-industrial age has been with us for just a few decades, and already the term information age that followed appears too limited. Yes, this is the age of big data, but it is also the dawn of a still nameless era (let's call it the robotic era) where big data become embodied in machines. There is now talk about the second machine age (Brynjolfsson and McAffee 2014), of systems that privilege information over energy as input, and which output intelligence as well as physical objects and physical work: brains added to brawn, thinking built into inert matter. The coming world of sentient machines—the autonomous vehicles, the Internet of things, the drones delivering our packages or fighting our wars, the satellites deciding which information to transmit to which city of the global urban network, and so much else we cannot yet imagine (let's not talk yet about machines built around synthetic biology, or quantum computers)—define a reality that challenges ordinary theoretical treatment. Indeed, the Greek word theory literally means contemplation, viewing, looking at something from the outside. It will eventually be futile to try to develop theories of the traditional kind by "looking from the outside" at cities run at least in part by emergent networks of heterogeneous, interacting smart systems.

We are not there yet, and we still need to figure out how best to use the big data bonanza. It is not likely that data mining alone will ever give the answers that urban research, management, or planning need, especially when it comes to helping prepare for the future. But there might exist certain basic principles at the core of current quantitative theories that can be relied on to remain valid even if the superstructure of the theory (dealing with socioeconomic or other empirical processes) is no longer helpful. Batty and March (1976) called these effects residues, and Couclelis (1984) developed the related idea of prior structure. These principles owe their resilience to the fact that they are formal rather than empirical: they are abstract properties of systems qua systems, or of the formal languages used in their derivation, which constrain what a model can represent. In spatial systems, it is properties of particular forms of abstract space that get transferred to the model. Here are some candidates of such principles that are well-established in the urban and geographic literature: distance decay; spatial heterogeneity; spatial autocorrelation; scaling laws; the rank-size rule; network properties; possibly fractal growth. And so on. There may be additional effects deriving from properties of cyberspace that could be added to the list. One can imagine appropriate combinations of these principles forming the backbone of analysis in hybrid approaches to data mining and any other strongly data-oriented techniques. But this is another discussion, for another kind of book.

9.3.2 The City's Epistemic Planes

The speculations in this section continue, but more realistically now: how could we best capitalize on the wealth and promise of urban informatics—not in a few years, but today? If data do not speak for themselves, what elements of order, what structured approach could make the data sing? Here is a tentative suggestion.

Cities—and even more so, cities of the information age—are not only highly complex but are also made up of many highly complex parts. Moreover, these parts are so qualitatively different from one another that they may be viewed as different realities, partially incompatible. Consider: The smart city as technological achievement versus as home of humanity; the smart city as place versus as node on a global network of urban linkages; the smart city as integration of actual and virtual dimensions.

It is increasingly unlikely that the whole of today's urban reality can be tackled with current notions of modeling. No comprehensive theory or framework may be able to do justice to the growing information-age complexity of the city. What might be possible instead is the development of strategies to guide the selection of data, tools, and methods, so that, depending on the objectives of the research or decision problem, the relevant critical aspects of contrasting views of the city are integrated in the analysis.

To give a sense of what such an informatics strategy might entail, here is an illustrative framework for merging disparate views of the city in response to specific questions or problems. It is based on the notion of a sequence of epistemic planes, each of which would support data and methods for a qualitatively different part of urban reality, and for qualitatively different kinds of knowledge. As a quick example, for any reasonably well-defined problem, one might need to systematically glean and weave together specific relevant information of the following kind from four or five different epistemic planes, e.g.:

- Measurements of the physical, social, and demographic **spatial structure** of the city, including information from and about distributed sensors and associated physical infrastructure;
- Information on social, business, financial, government, etc. **ICT networks**, both local and long-distance, including data on the supporting physical infrastructure;
- Measurements and qualitative information on the level of **functioning** of key aspects of the city (local and long-distance), including transportation, energy production and distribution, commerce, business services, health and human services, government, etc. (local and long-distance);
- Information on the **agents and forces** (local and global) affecting or likely to soon affect city functioning directly or indirectly, including recent technological breakthroughs such as autonomous vehicles and the Internet of things, and political changes such as the power of private companies over personal data.

For each problem or objective (to do with efficiency, growth, social justice, sustainability, quality of life, public safety, governance, etc.), appropriate analytical methods, models, and tools should be selected or developed to allow the problem-specific integration of the highly heterogeneous kinds of knowledge that aspects of the truly smart city demand. Only the most tentative indications of what these tools might look like can be suggested here. Possibilities include some type of information-filtering system (similar to recommender engines) for traversing the set of epistemic planes, artificial intelligence (AI) techniques for formalizing the objective or research question motivating the search, semantic networks and ontologies, to provide structure and help guide the selection of variables from among semantically heterogeneous planes of urban reality. Indeed, the systematic decomposition of urban-system information tentatively sketched above is loosely based on the information ontology proposed by Couclelis (2010).

9.4 Conclusion

This chapter has presented several of the reasons why business as usual in urban research, management, and planning cannot continue for much longer in the information-age city. We will miss the traditional kinds of theories, models, approaches, and methods that have served us well in the past century when these can no longer be relied on, as long as operational new approaches and tools do not yet exist to help us get the most out of ubiquitous, high-quality urban data. As an example of what may be lost along with a good traditional theory or model is its role in restricting the space of possibilities, so that not everything can be the case. In this chapter, we touched in passing upon two notions that could at least in part play that critical possibility-focusing role: first, the residues, or non-empirical effects hiding in our more successful spatial models (Batty and March 1976), and second, ontologies, which provide structure and restrict meaning so as to help keep the semantics of data interpretations consistent. Combined with data-mining techniques in the broadest sense, a priori elements of order, reliability, and consistency such as these might shape the hybrid strategies that can do justice to our age's unprecedented data riches. If informatics is the science of information, we should look to it for answers to questions that go beyond big data and their role in ICTs.

And here ends the speculation. This book has a very concrete double objective, which is to provide a comprehensive overview of the methods that so far form the core of urban informatics, as well as a technical introduction to the research tools necessary for understanding and creating the smart city of tomorrow. This should help prepare the ground for answering two major questions that may be asked concerning the general subject of this book: (a) How can the new science of information lead to the new science of cities? and (b) How can big data lead to actionable wisdom under conditions of pervasive uncertainty and complexity? It is not within the scope of the present book to tackle these questions directly, though its original chapters contribute to the necessary discussion that has already begun.

References

Batty M (2018) Inventing future cities. The MIT Press, Cambridge, MA
Batty M, March L (1976) The method of residues in urban modeling. Environ Plann B 8:189–214
Brail RK, Klosterman RE (2001) Planning support systems: integrating geographic information systems, models, and visualization tools. ESRI Press, Redlands, CA
Brynjolfsson E, McAffee A (2014) The second machine age: work, progress and prosperity in a time of brilliant technologies. WW Norton & Co, New York, NY
Caragliu A, Del Bo C, Nijkamp P (2009) Smart cities in Europe. In: Proceedings of the third Central European conference in regional science—CERS 2009, Kosice, pp 45–50
Casti JL (1984) Simple models, catastrophes and cycles. Kybernetes 13:213–229
Couclelis H (1984) The notion of prior structure in urban modelling. Environ Plann A 16:319–338
Couclelis H (2009) Rethinking time geography in the information age. Environ Plann A 41:1556–1575
Couclelis H (2010) Ontologies of geographic information. Int J Geogr Inf Sci 24(12):1785–1809
Doxiadis C (1968) Ecumenopolis: tomorrow's city. Britannica Book of the Year 1968. Encyclopedia Britannica, Inc., London
European Commission (2020) https://ec.europa.eu/info/eu-regional-and-urban-development/topics/cities-and-urban-development/city-initiatives/smart-cities_en#what-are-smart-cities. Accessed 1 May 2020
GaWC (Global and World Cities) (2020) www.lboro.ac.uk/gawc/group.html. Accessed 1 Jan 2020
Geertman S, Stillwell J (eds) (2009) Planning support systems: best practice and new methods. Springer, New York, NY
Geertman S, Ferreira J, Goodspeed R, Stillwell J (2015) Ch. 1, Introduction to planning support systems and smart cities. In: Geertman S, Ferreira J, Goodspeed R, Stillwell J (eds) Planning support systems and smart cities. Springer, Berlin, pp 1–17
Goh K (2015) Who's smart? Whose city? The sociopolitics of urban intelligence. In: Geertman S, Ferreira J, Goodspeed R, Stillwell J (eds) Planning support systems and smart cities. Springer, Berlin, pp 169–187

Gottman J (1961) Megalopolis: the urbanized northeastern seaboard of the United States. The Twentieth Century Fund, New York, NY
Hoyler M, Parnreiter C, Watson A (eds) (2018) Global city makers: economic actors and practices in the world city network. Edward Elgar Publishing, Cheltenham, UK
Jankowski P, Nyerges T (2001) Geographic information systems for group decision making: towards a participatory geographic information science. Taylor and Francis, New York, NY
Koch R, Latham A (2017) Key thinkers on cities. Sage, London, UK
Latham R, Sassen S (2005) Digital formations: IT and new architectures in the global realm. Princeton University Press, Princeton, NJ
McBride E, Davis A, Goulias KG (2019) Fragmentation in daily schedules of activities using activity sequences. Transportation Research Record 1–11. National Academy of Sciences, Washington, DC
Reades J, De Souza J, Hubbard P (2018) Understanding urban gentrification through machine learning. Urban Stud 56(5):922–942

Helen Couclelis is Professor of Geography Emerita at the University of California, Santa Barbara. Her interests in GIS and other forms of representing space and process reflect her academic and professional background, bridging the analytical social sciences and the synthetic sciences of civil engineering, architecture and planning.

Part II
Urban Systems and Applications

Chapter 10
Introduction to Urban Systems and Applications

Mei-Po Kwan

Abstract As new information technologies and large amounts of data from a wide range of sources become available to government agencies and the public, urban researchers have started to investigate how these data can be used to enhance the planning and management of various urban systems. As a result, new methods for collecting and analyzing complex space–time data about urban systems have been developed to address various urban issues. These urban systems include transportation systems, energy systems, and health systems. In recent years, considerable new work has been conducted to examine how new information technologies and data can enhance our understanding of and ability to address urban issues. The eight chapters in this section present various applications of urban informatics to specific urban systems or phenomena, including human mobility and travel, urban freight systems, urban resilience and disaster response, urban crime, urban governance, the use of remote sensing for environmental monitoring, health and wellbeing, and urban energy systems. All of them emphasize how new, big, or open data are useful for helping us to better understand and manage specific urban systems. They also highlight significant challenges in such applications of urban informatics, which would be particularly helpful to urban researchers and planners.

Keywords Urban informatics · Urban systems · Transportation systems · Energy systems · Health systems

Urban mobility patterns have been examined for decades using travel-survey data, which are useful for the management and planning of urban infrastructures and facilities (e.g., transport systems) but are costly and time-consuming to collect. The sample sizes for travel surveys are often limited when compared to other sources of urban big data such as point-of-interest (POI) data. In Chap. 11, Pierre Melikov and colleagues illustrate how passively collected data can be used to examine human mobility patterns based on a case study of Mexico City. Using POIs registered on

M.-P. Kwan (✉)
Department of Geography and Resource Management and Institute of Space and Earth Information Science, The Chinese University of Hong Kong, Hong Kong, China
e-mail: mpk654@gmail.com

W. Shi et al. (eds.), *Urban Informatics*, The Urban Book Series,
https://doi.org/10.1007/978-981-15-8983-6_10

Google Places to approximate trip attraction in the city, the chapter compares the trip distribution patterns obtained with the POI data and those obtained using conventional datasets based on travel surveys. The study finds that the POI data provide good estimates of the trip flows in the study area when compared to the estimates obtained with the official origin–destination matrices.

As tracking and sensing technologies are increasingly used to collect a wide range of urban data, new sources of urban data have become widely available. This, in turn, allows for the development of highly detailed transportation models that facilitate the analysis of urban freight movement and the generation of policy recommendations. In Chap. 12, André Romano Alho and colleagues review the recent developments in data-collection methods in urban freight transportation and how the new data can be used in state-of-the-art transport modeling. The chapter describes two software platforms for enhancing freight movement research. The first platform is called Future Mobility Sensing (FMS), which is a data-collection platform that integrates tracking devices and mobile applications for collecting highly accurate mobility data. The second platform is called SimMobility, which is an open-source, agent-based urban simulation platform for modeling disaggregate urban passenger and freight movements. The authors discuss how the two platforms can be used jointly to advance behavioral modeling for passenger and goods movements in urban areas.

As populations continue to increase and migrate to cities, disaster risks from events like hurricanes, earthquakes, or wildfires are increasing and becoming more pronounced in urban areas. In a world that is rapidly urbanizing, the safety of rapidly increasing numbers of urban residents is at risk. In Chap. 13, Susan Cutter discusses how the resilience concept (as an outcome or as a process of building capacity) has become more central in the last decade as a means for understanding how cities prepare for and recover from disaster events. Using selected case studies of several cities as examples, she reviews research that attempts to develop urban informatics for facilitating intervention or mitigation strategies and fostering urban resilience. She suggests that shifting from passive to active sensor data and making low-cost, near-real-time data more accessible would greatly enhance research on and responses to urban risks.

Researchers have long been interested in the relationships between urban environments and crime. Environmental criminologists now commonly accept that environmental factors have considerable influence on criminal behavior, and understanding these influences would help to shed light on what measures are effective for crime prevention. Chapter 14 by Tao Cheng and Tongxin Chen provides a useful review of the development of crime research, including historic criminology and data-driven policing, and its implications for urban security and crime prevention in practice. It discusses various analytical tools for analyzing and preventing urban crime (e.g., crime hotspot mapping and police resource allocation). The chapter proposes a comprehensive data-driven policing system as a framework for urban crime prevention and security improvement.

Transparency is a critical element in urban governance. It encourages civic engagement, ensures that elected officials are accountable for their decisions, and limits the potential for corruption. To achieve transparency in urban governance, a wide range of data about cities have to be widely available to the public. Chapter 15 by Alex Singleton and Seth Spielman addresses the need for and challenges in providing adequate data to the public to enhance transparency and civic engagement. It discusses how open-source data platforms in urban governance may facilitate the realization of these goals and how the availability of the new data offers the potential to transform urban governance. The chapter, however, highlights the risks of reproducing or developing new social inequalities as a result of the proliferation of new data and their integration into software that automatically generates results based on certain algorithms.

Recent advances in sensing technologies and retrieval methodologies (e.g., the much finer spatial and temporal resolutions of modern sensors) have greatly increased the applicability of remote sensing in urban environmental applications. Chapter 16 by Janet Nichol and colleagues reviews the latest developments in the use of remote sensing in urban pollution monitoring, including assessment of urban air quality, urban heat islands, and water quality around urban coastlines. It discusses the main sensors used and the developments in retrieval algorithms for environmental monitoring in urban areas.

The technology and information available to urban residents may help increase their access to health and health-enhancing information and thus may help enhance their health and wellbeing. Chapter 17 by Clive Sabel and colleagues explores how information technology and everyday devices connected via the Internet (the Internet of Things) are shaping global research on the health and wellbeing of urban populations. It reviews various types of data used in health research in the context of smart cities. Using examples from the big data Centre for Environment and Health (BERTHA) Project at the Aarhus University of Denmark, innovative methods for collecting individual data for examining the health and wellbeing of urban residents, such as machine learning, mobile sensing, and tracking, are discussed. The chapter also reviews ethical, privacy, and confidentiality issues related to the use of sensitive personal data in health research.

The development and maintenance of urban infrastructures are highly energy-intensive. The complex interactions between human dynamics and critical infrastructures in urban areas have significant implications for traffic congestion, emissions, and energy consumption. Chapter 18 by Budhendra Bhaduri and colleagues highlights recent research at Oak Ridge National Laboratory (ORNL) in the USA on the integration of four distinct components (i.e., data, critical infrastructure models, scalable computation, and visualization) for understanding the complex interactions between physical and social systems in urban areas. It discusses four main themes in such research: population and land use, sustainable mobility, energy-water nexus, and urban resiliency. It describes how ORNL promotes innovative interdisciplinary research that integrates its expertise in critical infrastructures and their interactions with the human population using scalable computing, data visualization, and unique data sets from a variety of sources.

Chapter 11
Characterizing Urban Mobility Patterns: A Case Study of Mexico City

Pierre Melikov, Jeremy A. Kho, Vincent Fighiera, Fahad Alhasoun, Jorge Audiffred, José L. Mateos, and Marta C. González

Abstract Seamless access to destinations of value such as workplaces, schools, parks or hospitals, influences the quality of life of people all over the world. The first step to planning and improving proximity to services is to estimate the number of trips being made from different parts of a city. A challenge has been representative data available for that purpose. Relying on expensive and infrequently collected travel surveys for modeling trip distributions to facilities has slowed down the decision-making process. The growing abundance of data already collected, if analyzed with the right methods, can help us with planning and understanding cities. In this chapter, we examine human mobility patterns extracted from data passively collected. We present results on the use of points of interest (POIs) registered on Google Places to approximate trip attraction in a city. We compare the result of trip distribution models that utilize only POIs with those utilizing conventional data sets, based on surveys. We show that an extended radiation model provides very good estimates

P. Melikov · J. A. Kho · V. Fighiera · M. C. González (✉)
University of California, Santa Barbara, USA
e-mail: martag@berkeley.edu

P. Melikov
e-mail: pierre_melikov@berkeley.edu

J. A. Kho
e-mail: jerkho@berkeley.edu

V. Fighiera
e-mail: vincent.fighiera@berkeley.edu
URL: https://github.com/VincentFig/urban_computing_mexico

F. Alhasoun
Massachusetts Institute of Technology, Cambridge, USA
e-mail: fha@mit.edu

J. Audiffred
Data Lab MX, Mexico City, Mexico
e-mail: ja@digitalstate.mx

J. L. Mateos
Universidad Nacional Autónoma de México, Mexico City, Mexico
e-mail: mateos@fisica.unam.mx

W. Shi et al. (eds.), *Urban Informatics*, The Urban Book Series,
https://doi.org/10.1007/978-981-15-8983-6_11

when compared with the official origin–destination matrices from the latest census in Mexico City.

Keywords Trip distribution models · Transit use · Clustering methods · Mobility science

11.1 Introduction

As more people continue to migrate from rural to urban settings, the challenges of improving cities increase in pace and complexity. Planning for daily mobility within metropolitan areas is one important topic of the coming years. The estimates of the total daily trips specific to a metropolis are the first step to establish efficient strategies that inform the transportation-planning process. However, the lack of reliable and accessible data sources of individual mobility greatly slows down the planning progress. Data on human mobility have thus far been collected through individual surveys with small and potentially biased sample sizes because they require active participation and often rely on self-reporting (Cottrill et al. 2013). While conventional travel surveys provide a wealth of valuable information, they are very expensive and time-intensive. For most major cities, these surveys are conducted about once a decade; for smaller cities and towns, it is less frequent than that or not at all. Between the publication of these surveys, a lot can happen that could change the dynamic of the city: new attractions, redevelopment of entire city blocks, changing economic trends, the impact of a natural calamity, or just the gradual shift of a city's characteristics. These changes would not be captured until the next travel survey is issued, which could be anywhere from the following year to a decade. With the abundance of information and connectivity today, other sources of easily accessible data could prove to be useful as a proxy for the data obtained in conventional surveys. One example of this is the use of triangulated mobile phone data to form mobility networks and extract individual trip chains (Jiang et al. 2013). Another such potential is points of interest (POIs) registered on Google Places, a feature of the mapping service developed by Google LLC (Google), which are extensive, updated frequently, and relatively accessible for most people. Google Places lists various types of establishments, such as restaurants, schools, offices, and hospitals, allowing it to serve as a good indicator of trip attraction. For an overview of mining POI data for urban land-use classification and disaggregation, see the work of Jiang et al. (2015).

As a complement to the development of statistical methods to carefully treat travel diaries (Ben-Akiva and Lerman 1985; Hall 1999; de Dios Ortúzar and Willumsen 2011), alternative, cheaper, and larger data sources are necessary to push our understanding of human mobility efforts further. The evolution of technology over the past decade has given rise to ubiquitous mobile computing, a revolution that allows billions of individuals to access people, information, and services through information technologies such as their cellular or mobile phones. Using today's large-scale computing infrastructure and data gathered from sensing technologies, one can combine methods from computer science with urban planning, transportation, and environmental science, to tackle specific problems with fined-tuned methodologies in a data-centric computing framework.

Urban-science methods for characterizing human mobility should take into account the complexity of these dynamics. However, despite being a complex system, recent results have indicated some patterns or general features that can clarify these dynamics. These features are called universals in analogy with phenomena in the physical sciences. First, there is a set of models to analyze aggregated human mobility in cities or large-scale migrations. On the one hand, we have gravity-like models, and on the other radiation models (Simini et al. 2012). In 2008, González et al. (2008) used data from mobile phones to show that the step-length distribution can be described by a truncated power law. To understand the mechanism that gives rise to this distribution, the authors used the radius of gyration: a quantity that characterizes the radius enclosing the most visited locations of an individual over months of observation. Simulations suggest that the step-length distribution of the entire population is produced by the convolution of Lévy flight processes, each with a different characteristic jump size within the individual radius of gyration of each person. The observed power law is the result of the heterogeneity in the radius of gyration of the population. While the great majority of users have a radius of a few kilometers, there is a minority of users that cover thousands. Similar to the income and other variables following a power law, following the Pareto principle 80% of the distance covered comes from 20% of the subjects.

Another interesting pattern of human mobility is the interplay between randomness and predictability. There is a high rate of return to previously visited locations such as home or work. The nature of these returns follows a probability inversely proportional to the rank of the location, following then a Zipf law. Subsequent work by Song et al. (2010a, b) using data from mobile phones, revealed two important characteristics of human behavior. First, the number of distinct visited locations increases as a power of time with exponent less than 1, indicating a very slow rate of explorations. Second, the probability that an individual returns to a previously visited place scales with the inverse of the rank of that location, a phenomenon labeled as a preferential return. With a perspective from information theory, Song et al. (2010a, b) used different kinds of entropy measures to analyze the limits of predictability of human mobility.

Another approach to study human mobility is by mobility motifs, introduced by Schneider et al. (2013) as an abstract (semantic) way to define periodic trajectories in the daily movements of individuals. A daily mobility motif is a directed network (digraph) where unlabeled nodes represent locations and the edges are trips from one location to another. Counting motifs in data from mobile phones and traditional travel surveys, they amazingly found that despite over 1 million unique ways to travel between 6 or fewer locations, just 17 motifs are used by 90% of the population. For an overview of these works, see the papers by Jiang et al. (2013) and Toole et al. (2015), and the recent review of human mobility by Barbosa et al. (2018).

In this chapter, we focus on statistical methods of the type described above in the analysis and modeling of human mobility both in the aggregate and individually. We take advantage of novel data sources passively collected, to enrich the information on human mobility patterns. Namely we parse an alternative source of geospatial data, apply trip distribution models to estimate aggregated trips, and implement unsupervised machine learning to characterize different types of commuters by their mode of transportation and travel time.

As a sample case, we focus on Mexico City, one of the largest cities in the world with over 21 million people in the greater metropolitan area. It is also one of the most important cultural and historical centers in the Americas. With such a large number of people and a high level of vibrancy, mobility in the region can be quite a challenge. In 2017, a major household travel survey (Encuesta Origen-Destino en Hogares de la Zona Metropolitana del Valle de Mexico 2017) was completed for the Metropolitan Zone of the Valley of Mexico. Conducted from January–March 2017, the survey obtained information to facilitate a better understanding of the mobility of the inhabitants in the metropolitan region. This includes data on trip generation, trip attraction, mode choice, trip purpose, trip duration, socio-demographics, and more, which is representative of 34.56 million daily trips occurring in our study zone.

11.2 Data Collection of POIs

In order to obtain POIs (Jiang et al. 2015) from Google Places, programming scripts were written to utilize the application programming interface (API) that Google provides (Documentation of Google Maps API no date). However, Google sets limits on the number of POIs a single request can return and on the number of API requests an account is allowed to make in order to differentiate commercial and non-commercial applications. While the conduct of this undertaking is non-commercial, the data to be collected tend to exceed Google's limitations. Hence, an efficient algorithm needs to be implemented to collect the most information from a minimal number of API requests.

To achieve this, API requests were framed and constrained by geometries defined by the Hexagonal Hierarchical Geospatial Indexing System (H3) of Uber Technologies, Inc (Uber Engineering 2018). Uber's H3 system is an application of the concept of fractals. Maps are divided into large hexagonal tiles, with each tile further divided into seven smaller hexagons. With 16 supported resolutions, the system is flexible to most use cases. Figure 11.1a shows a sample resolution applied to a district in Mexico City.

Hexagons serve as good approximations of circles while minimizing the overlap between cells. This is useful as the Google Places API requires a radius parameter within which the search for POIs will be made.

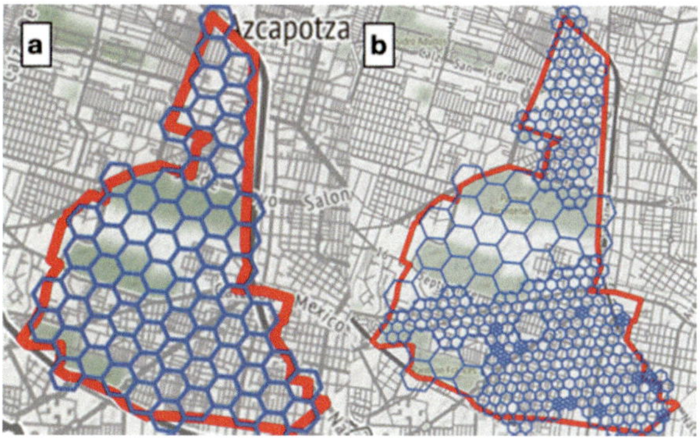

Fig. 11.1 Hierarchical sampling method to extract POIs. **a** Initial state and resolution of parsing algorithm, **b** Final state after recursively increasing ressolution in hexagons that reach the API request limit

11.2.1 Parsing Algorithm

An initial resolution for the size of the hexagons was determined. The coarser the initial resolution, the more efficiently the script is likely to run, as excessive requests are avoided in sparsely developed areas. On the other hand, coarse resolutions also increase the marginal areas near the borders of irregular shapes that are unaccounted by the algorithm. Before issuing any API request, the initial resolution was tuned and visualized to balance these tradeoffs.

For each hexagon, an API request was made at the centroid. If the request reaches the limit of POIs that it can return, the algorithm subdivides that hexagon into smaller hexagons. This process is repeated until each request is met without reaching the limit. In Fig. 11.1b, some areas, such as parks and nature reserves, do not need numerous API requests. Downtown city blocks and dense neighborhoods, on the other hand, are recursively splintered.

11.3 Spatial Distribution of POIs

In the use case for this chapter, the parsing algorithm returned a total of over 733,000 POIs from Google Places across the Metropolitan Zone of the Valley of Mexico. These points of interest provide new dimensions to analyze data from the travel survey that could generate insights on the characteristics of the megacity.

For instance, the API requests return tags for each POI, indicating the nature of the establishment. This may include broad categories, such as store, or more specific labels, such as electronic store. Clustering relevant tags together, POIs may be classified as either commercial or public-service establishments. Combining these data with the travel survey, Fig. 11.2a maps the relationship of the sociodemographic

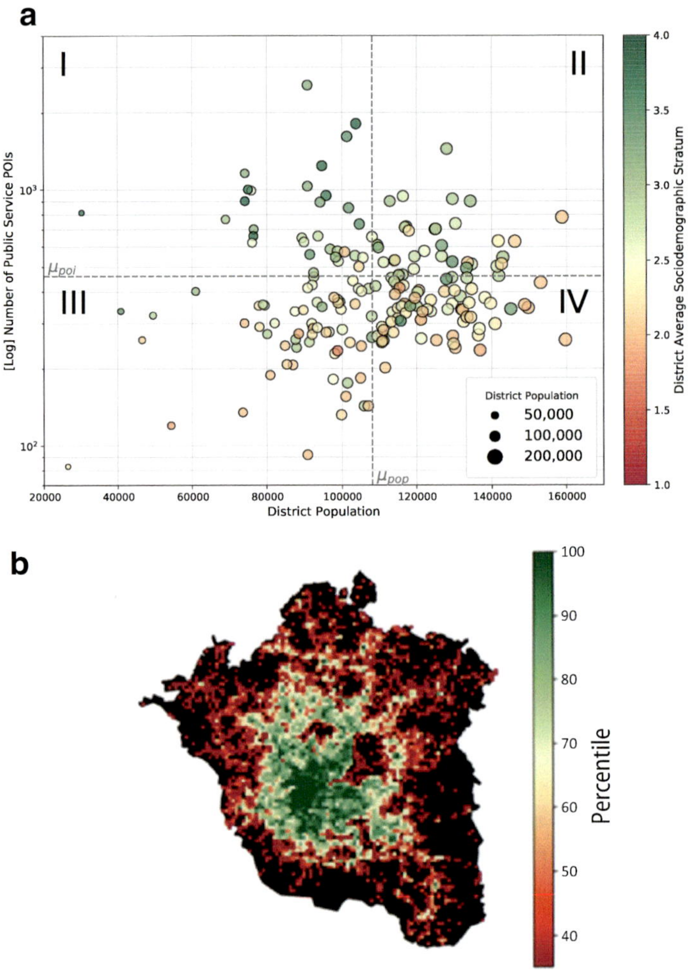

Fig. 11.2 Spatial distribution of population and services. **a** Relationship of the sociodemographic stratum of a district with the ratio of the number of public service establishments to the population, **b** Percentiles of the number of public service POIs for every 1 km^2 block

status of a district with the ratio of the number of public service establishments to the population.

In this case, sociodemographic strata are indices defined by the travel survey to characterize a respondent's social and economic conditions, with numbers from 1 to 4 denoting increasing economic well-being. In Quadrant I, the number of public-service establishments is above average and the population is below average: such districts tend to enjoy the highest sociodemographic stratum. Quadrant II has districts of intermediate sociodemographic status, still benefiting from an above-average number of POIs. Quadrant III has both less than the average population and number of facilities and a lower socio-economic stratum. Interestingly, Quadrant IV has districts on opposite ends of the sociodemographic spectrum, possibly due to the diversity of inner cities and the efficiencies of density that allow fewer establishments to serve more people in a small amount of space. These enrich the spatial information of the surveys and deserve further research.

Another advantage gained through the POIs is the spatial granularity of the collected data. Travel survey respondents are often organized by the district of residence, whereas establishments on Google Places are pinpointed to street address coordinates. Since cities and districts are not homogeneous, this level of detail provides a more realistic perspective on city dynamics, highlighting functional interaction over arbitrary political boundaries.

In Fig. 11.2b, the coordinates of public-service establishments are truncated to two decimal places, binning them to grids that are approximately a kilometer per side. Due to the orders of magnitude in the difference between the urban core and more rural areas, the number of public-service establishments is abstracted to intervals of 5 percentile points. As it is, mapping these establishments may have a strong dependency on population density. Nevertheless, a hidden structure to the city is revealed, with a strong urban core, some urban corridors expanding outwards from the city center, and regional centers further away from the center. Significantly, there are large regions on the outskirts of the study area where public services are sparse. Further insights may be gained when supplemented by population distribution data at a similar level of granularity.

11.3.1 Extended Radiation Model for Human Mobility

Counting the number of POIs per district is necessary for direct comparison with the 2017 travel survey data, which have the smallest granularity only at the level of districts. Mapping these per district in Fig. 11.3a, b, a direct comparison can be made with trip attraction reported in the 2017 travel survey.

While the correspondence is not perfect, the distribution of points of interest makes a good approximation to the distribution of trip attraction obtained from the travel survey. Most notably, the difference between the city center and the rest of the region is similarly stark.

Plotting the relationship between trip attraction and points of interest in Fig. 11.3c yields a quantitative plot, with the correlation coefficient of the two variables determined to be quite high at 0.81. This comparison will be of great relevance later, where the POIs are used to model mobility patterns in the city, in place of travel-survey data.

Many models have been developed in order to predict population movement at different scales. In the context of Greater Mexico City, we want to investigate how accurate such models are and how well they perform to reconstruct mobility patterns. The models of trip distribution can be divided into gravity-model types (Barthélemy 2010; Erlander and Stewart 1990; Jung et al. 2008; Lenormand et al. 2016), or

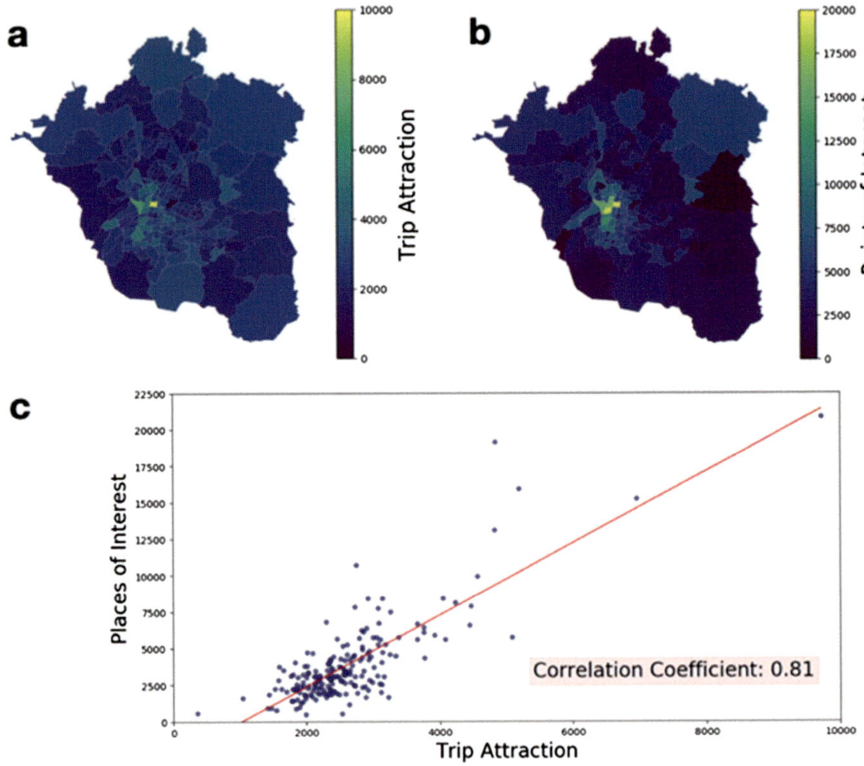

Fig. 11.3 Trip attraction versus POIs. **a** Values of trip attraction, **b** The number of points of interest, **c** Correlation plot of trip attraction and points of interest

intervening-opportunity types (Lenormand et al. 2016). In this chapter, we present an application of the latter, named the extended radiation model (Yang et al. 2014), to estimate trip distributions in Mexico City.

The radiation model (Simini et al. 2012, 2013) is based on a stochastic process that is parameter-free and enables, without previous mobility measurements, estimates of trip distributions in good agreement with mobility and transport patterns (Simini et al. 2013). The original radiation model only relies on population densities to estimate commuting patterns between US counties (Simini et al. 2013).

Here, we use the natural partition of the city in districts. The model states that a trip occurs based on the number of opportunities that can be found in each district if the two following steps are met: (1) an individual seeks opportunities from all districts, including his or her home district (the number of opportunities in each county is proportional to the resident population); (2) the individual goes to the closest district that offers more opportunities than his or her home district. To analytically predict the commuting fluxes with the radiation model, we consider locations i and j with population m_i and n_j, respectively, at distance r_{ij} from each other. We denote with s_{ij} the total population in the circle of radius r_{ij} centered at i (excluding the source and destination population). The average flux T_{ij} from i to j is:

$$\langle T_{ij} \rangle = T_i \frac{m_i n_j}{(m_i + s_{ij})(m_i + n_i + s_{ij})} \tag{11.1}$$

where $T_i = \sum_{i \neq j} T_{ij}$ is the total number of commuters that start their journey from location i, or the trip production of location i.

The extended radiation model aims at predicting flows without first calibrating the data. Thus, it introduces a scaling parameter $\propto$ by combining the derivation of the original radiation model with survival analysis and gives:

$$\langle T_{ij} \rangle = \gamma T_i \frac{[(a_{ij} + m_j)^{\propto} - a_{ij}^{\propto}](n_i^{\propto} = 1)}{(a_{ij}^{\propto} + 1)[(a_{ij} + m_j)^{\propto} + 1]} \tag{11.2}$$

where $a_{ij} = n_i + s_{ij}$, γ, is the percentage of trips between all places found between the origin and destination, and empirically set $\propto = (\frac{1}{36[\mathrm{km}]})^{1.33}$, where ***i*** is the characteristic length of the study area, and $\propto$ accounts for the fact that the trip distributions depend on the area of study.

The extended radiation model was meant to be used when we lack trip data for calibration. When there are actual trip data as in this case, one can evaluate them with the common part of commuters based on the Sørensen index (Lenormand et al. 2016):

$$\mathrm{CPC}(T, \tilde{T}) = \frac{2 \sum_{i=1}^{n} \sum_{j=1}^{n} \min(T_{ij}, \tilde{T}_{ij})}{\sum_{i=1}^{n} \sum_{j=1}^{n} T_{ij} + \sum_{i=1}^{n} \sum_{j=1}^{n} \tilde{T}_{ij}} \tag{11.3}$$

It gives a quantitative measure of the goodness of the flow estimation, 0 meaning no agreement found and 1 perfect estimation. CPC compares the model estimates T_{ij} versus the empirical observations $\tilde{T}_{ij}$, between all origin–destination pairs.

11.3.2 Results

From the survey data, we extracted the different variables to run the extended radiation model. First, we extracted the 194 districts that compose Greater Mexico City with their respective population, trip attraction (number of daily trips coming to the district), trip production (number of daily trips leaving from the district), points of interest, and characteristic length, given as the square root of the area of the district.

Then, we set i as the mean of the characteristic length of each district. We also constructed the distance matrix that gives for every row i and column j the distance between the centroids of the districts i and j. Finally, γ was set to the total number of trips as a proportion of the total population.

Table 11.1 Comparison of the goodness of fit depending on different input data in the model

Origin	Trip production	Trip production	Population	Population
Destination	Trip attraction	POI	Trip attraction	POI
CPC	0.69	0.67	0.64	0.63

Four different setups were then used to compare the performance of the model based on different approximations of the trip production from the origin districts and the trip attraction of the destination districts: (1) we used trip attraction and trip production as a baseline, (2) we used the number of POIs as a proxy for trip attraction, (3) we used population as a proxy for trip production, and (4) we combined (2) and (3). The resulting CPC values are shown in Table 11.1.

Table 11.1 shows that the CPC of the estimates of the extended radiation model was close to other recently proposed models (Lenormand et al. 2016). Moreover, we investigated the impact of different proxies for flow generation and attraction volumes as input in our model and found that the use of more easily acquired data sources such as population and POI density achieves nearly the same level of accuracy. POIs seem particularly interesting because they enable good estimates without travel surveys, but with data of much cheaper access. On the other hand, the use of population in place of trip production aims at predicting future mobility patterns given the knowledge of γ, the proportion of the total population of the system commuting, and assuming changes in this ratio. Here, we extracted γ from the 2017 survey and used it for the models. Consequently, we cannot validate the predictive power of the model; but nonetheless, when distorting the population data of each district by multiplying it by γ, we still observe encouraging results.

11.4 Analyzing Human Mobility by Mode of Transportation

This section is devoted to the analysis of individual travelers within Mexico City. One advantage of a broad user survey is to identify types of dominant behavior in the population, with respect to the modes of transportation used, their geographic distribution, and socio-demographic characteristics.

We analyzed the large database collected by the Mexico City survey, containing information on individual residents; it details information on more than half a million trips. For each trip identified, we have the mode of transportation, the districts of departure and arrival, the time of departure and arrival, the purpose of the trip, the gender of the traveler, and his or her age and socio-demographic stratum. As many as twenty different modes of transportation can be identified among the 196 districts of the survey.

We wanted to reduce the complexity of this information by grouping the trips based on transportation mode, without associating the other metrics. The latter would then be involved in the analysis of clusters formed. In doing so, we sought to distinguish the main mobility behaviors, which would, in turn, combine various proportions of the possible transport modes and trip purposes.

By simple inspection, it is clear that all the means of transport mentioned in the database were not significantly present in the main groups of behaviors. We expected to see certain modes of transport, such as cars or walking, as the majority in certain behaviors and others, such as the category "Other means of transport," very poorly represented or even absent. It is, therefore, not necessary for such a large number of variables, initially twenty, to describe the individual trip database. We applied principal component analysis (PCA) to determine the main variables. This allowed us to reduce computation time and complexity when using a clustering algorithm. Projecting into a lower dimensional base informs our understanding (Eagle and Pentland 2009; Ibes 2015).

The PCA method aims to capture as much of the total variance of the data as possible with a reduced number of variables, called principal components (PC). Since the objective was to set the size of the new projected database such that the first N PCs had to account for 85% of the total variance, we, therefore, chose to keep only the first five PCs for the rest of the study (Shlens 2005).

To group trips around main behaviors, we used k-means clustering (Jiang et al. 2012). Each journey of the database was initially represented as a vector composed of zeros and ones, depending on the mode of transportation used. We only considered its projection in the PCs database when applying the k-means algorithm. K-means works iteratively to ultimately minimize the sum of the distances between each projected journey and the centroids of the clusters determined by the algorithm, and thus allows patterns to be identified within the dataset. As a result, we obtained a list that reflected the membership of each trip in a particular cluster. We also calculated the proportions of the modes of transport for each cluster to determine their average behavior (Jiang et al. 2012). While the ideal number of clusters can be estimated via various metrics, such as the elbow method, the best number of clusters depends on the interpretability of the data available. In this case, we decided to keep six clusters.

11.4.1 Detected Mobility Groups

Figure 11.4a at the top shows the six clusters that characterize daily mobility in Mexico City and their percentages. They represent the main ways of moving around the city. Since the database reports journeys, several of which may have been made by the same person, and residents can have several trips. The analysis groups journeys and not individuals. Note that these journeys also have the purposes of these trips such as: going home, going to work, errands, shopping, etc. Their average percentage is shown at the bottom of Fig. 11.4a. In the top of Fig. 11.4a, only the three most reported modes of transportation in each cluster are shown. Each of these components is associated in the y-axis with its fraction within the cluster. The % in the x-axis

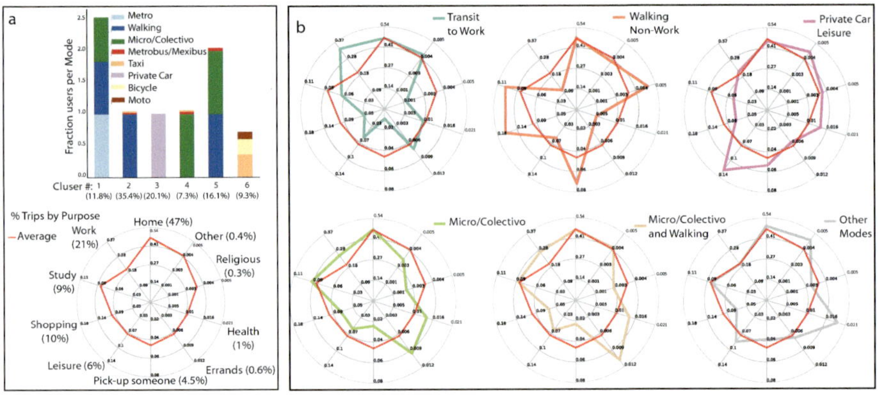

Fig. 11.4 Mobility groups in Mexico City. **a** The fraction of users per mode in each behavioral group or cluster. The lower part shows the legend displaying the percentage of trip purposes averaging the entire population. **b** Comparison of the percentage of trip purposes by cluster in contrast to the mean. The clusters are from 1 to 6 from left to right, starting at the top. We see that certain purposes are more present in each group. Cluster 1 uses combined transit modes with a higher percentage of work travel, Cluster 2 groups shopping, school, and social activities (picking someone up) by walking. Cluster 3 groups leisure trips via private car. Clusters 5 and 6 group errands done by Micro/Colectivo or combining Micro/Colectivo and Walking

shows the fraction of the total journeys in each cluster. We can see that the majority of journeys in Clusters 1 and 5 combines three or two modes respectively.

Cluster 2 contains 35% of all the trips in the Mexico City survey. The fraction of walking on the ordinate is equal to one, while that of the second most present mode of transportation in this cluster, Mexibus & Metrobus, has a fraction of 0.027. Thus, only about 2.7% of the trips attached to this cluster combined their walking with Mexibus or Metrobus. It can therefore be said that these trips are made almost exclusively by walking.

Figure 11.4b shows, for each of the six clusters, the proportion, per cluster, of each of the ten purposes of the trips considered in the survey: going to home, going to work, going to school, shopping, leisure, errands, picking someone up, religion, health purposes, or all other purposes.

We compared the average percentage of trip purposes with the average within each cluster. Cluster 1 represents 11.8% of all the trips and has 33% of them with work as its purpose, larger than the average of 21% among all trips. We see that when people walk (Cluster 2), the shopping purpose is twice the average. While about 16% of the trips associated with the second cluster are for shopping purposes, the average number for all trips is around 10% for this category. On the contrary, it seems that walking is not commonly used for commuting or going to the doctor.

In addition, since the average travel time of this cluster is about 20 min while the average travel time for the total population is about twice as long, this cluster can therefore be associated with local trips. This suggests that workplaces or healthcare centers are generally located further from family homes than shops, schools, or religious places.

Cluster 3 groups 20% of the daily trips made in Mexico City; it is exclusively composed of private cars as a mode of transportation. This case has leisure in higher proportion compared to other clusters. This can be a consequence of the lack of transit to cover distant journeys, or being inconvenient for such purpose.

Cluster 5 contains 16% of the trips and includes the routes that exclusively combine walking and micro/colectivo, while Cluster 4 with 7% of the trips does not include walking. These two clusters are similar in purpose to the average and their average travel time is the longest, about one hour per trip.

The use of walking, metro and micro/colectivo during the same journey is also observed in the first cluster. Indeed, metro obtains a proportion equal to 1, walking 0.83 and micro/colectivo 0.71. Not all the journeys in this cluster, therefore, systematically combine these three means of transport, but on average in the great majority of cases these three means of transport are combined. This group is over-represented in the heart of the capital's historic district, where more than 55% of the trips undertaken are associated with this cluster. On the other hand, it becomes absent as soon as one moves away from this geographical area. This is due to the high concentration of metro and micro/colectivo in this part of the city, making travel much faster and more convenient by linking these modes of transport, particularly to get to work.

Cluster 6 is not possible to interpret, because it does not represent any particular mode. However, it should be noted that it is mainly concentrated in the agricultural regions that make up some districts.

Koelbl and Helbing analyzed data from the UK National Travel Surveys during nearly three decades, in the years 1972–98, observing that the average journey times for different modes of transport are inversely proportional to the energy consumption rates measured for the respective human physical activities. In Figure 11.5a, we show the distribution of the travel times per mode divided by their mean, inspired by the

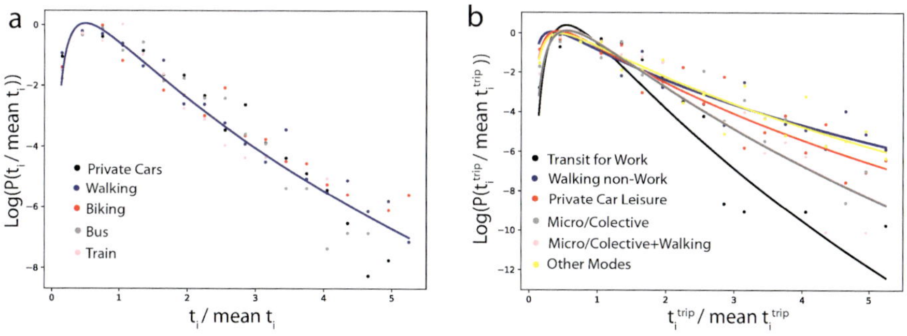

Fig. 11.5 Comparison of travel times by mode and by cluster group. **a** Lognormal fit for the scaled time-averaged travel-time distributions for different modes of transport on a logarithmic scale as reported by Schneider et al. (2013) based on UK surveys. **b** Lognormal fit for the scaled time-averaged travel-time distributions for the clusters found in the Mexico City travel survey

Table 11.2 Comparison of the fitted parameters for the clusters

Cluster	μ	var	mean trip time
Transit to work	−0.03	0.21	89
Walking non-work	−0.28	0.63	20
Private car + leisure	−0.19	0.69	40
Micro/collective	−0.07	0.41	49
Micro/collective + walking	−0.11	0.41	58
Other modes	−0.30	0.47	30
Results Kölbl et al. (2003)	−0.14	0.51	N/A

results reported by Kölbl and Helbing (2003). The authors presented five transport modes, and they all collapse well in one lognormal distribution with parameters reported in Table 11.2. To further investigate our clusters, we made the same analysis of the travel time of the individual trips divided by the mean travel time. We observed a lognormal with different parameters for each cluster; only Cluster 5 has closer parameters to the ones reported by Kölbl and Helbing (2003). Given the challenges of mobility in Mexico City, we observed larger variance among the members of each cluster, except for the trips of Cluster 1, which groups a higher fraction of the journeys to work. The differences between the results reported in the UK and Mexico City could be related to more strained transit service and longer commuting journeys in a vast metropolis. The universal scaling which is shown in different modes by Kölbl and Helbing (2003) could still serve as a guide to target improvements in the transit system. Note that the variance of private-car travel times is less than half that for transit. If the travel times were more similar, transit could be more attractive for those that can afford traveling by private car.

11.5 Conclusions

Data-informed analysis of complex socio-technical systems has become the interest of interdisciplinary groups around the world. These techniques can inform urban planning with an analytical angle in the complex task of amending current cities and their infrastructures. This increases its relevance to better accommodate the continued expansion of major cities and metropolises around the world. The purpose of this study was to summarize statistical methods to analyze human mobility in the urban context. We combined alternative data sources and methods in the topic that has mostly used travel diaries and econometric methods. The common aim of the data analysis presented is to reduce the complexity of the dataset at hand, while simultaneously extracting useful information. To this end, the recent growth of passively collected data lends important opportunities to the understanding and the implementation of these and other methods. In particular, we analyzed and modeled human mobility in Greater Mexico City, one of the largest cities in the world with over

21 million people. We explored a data set of a recent major travel survey conducted in 2017, using clustering methods, and compared the trip distributions with the one inferred from an extended radiation model that uses population and points of interest.

Future extensions should include the sociodemographic stratum, and possible interventions to plan for social equity and accessibility.

Acknowledgements We are grateful to Emmanuel Landa and Irving Morales of DataLabMX for collaborating with us in collecting data and in gaining better insights into the Metropolitan Zone of the Valley of Mexico. These contents of this chapter were initiated as a class project based on the content covered in CYPLAN 257: Data Science for Human Mobility and Sociotechnical Systems. The codes and data used in this chapter are available at https://github.com/VincentFig/urban_computing_mexico.

References

Barbosa H, Barthelemy M, Ghoshal G, James CR, Lenormand M, Louail T, Menezes R, Ramasco JJ, Simini F, Tomasini M (2018) Human mobility: models and applications. Phys Rep 734:1–74

Barthélemy M (2010) Spatial networks. Phys Rep 499:1–101

Ben-Akiva ME, Lerman SR (1985) Discrete choice analysis: theory and application to travel demand. MIT Press, Cambridge

Cottrill CDA, Pereira FCA, Zhao FA, Dias IF, Lim HB, Ben-Akiva ME, Zegras PC (2013) Future mobility survey. Transport Res Record 2354:59–67

de Dios OJ, Willumsen LG (2011) Modelling transport. Wiley, Chichester

Documentation of Google Maps API. https://developers.google.com/places/web-service/search

Eagle N, Pentland AS (2009) Eigenbehaviors: identifying structure in routine. Behav Ecol Sociobiol 63:1057–1066

Encuesta Origen-Destino en Hogares de la Zona Metropolitana del Valle de Mexico (2017) Instituto Nacional de Estadistica y Geografia, Mexico. https://en.www.inegi.org.mx/programas/eod/2017/. Accessed 11 Oct 2018

Erlander S, Stewart NF (1990) The gravity model in transportation analysis: theory and extensions, vol 3. Vsp

González MC, Hidalgo C, Barbási AL (2008) Understanding individual human mobility patterns. Nature 453(7196):779–782

Hall RW (ed) (1999) Handbook of transportation science. In: International series in operations research and management science, vol 23. Springer, Boston

Ibes DC (2015) A multi-dimensional classification and equity analysis of an urban park system: a novel methodology and case study application. Landscape Urban Plann 137:122–137

Jiang S, Ferreira J, González MC (2012) Clustering daily patterns of human activities in the city. Data Min Knowl Disc 25:478–510

Jiang S, Fiore GA, Yang Y, Ferreira Jr J, Frazzoli E, González MC (2013) A review of urban computing for mobile phone traces: current methods, challenges, and opportunities. Paper presented at the 2nd ACM SIGKDD international workshop on urban computing, Chicago, Illinois, August 2013.

Jiang S, Alves A, Rodrigues F, Ferreira J Jr, Pereira FC (2015) Mining point-of-interest data from social networks for urban land use classification and disaggregation. Comput Environ Urban Syst 53:36–46

Jung WS, Wang F, Stanley HE (2008) Gravity model in the Korean highway. Europhys Lett 81(4):48005

Kölbl R, Helbing D (2003) Energy laws in human travel behaviour. New J Phys 5(1):48

Lenormand M, Bassolas A, Ramasco JJ (2016) Systematic comparison of trip distribution laws and models. J Transp Geogr 51:158–169
Schneider C, Belik V, Couronné T, Smoreda Z, González MC (2013) Unravelling daily human mobility motifs. J R Soc Interface 10(84):20130246
Shlens JA (2005) Tutorial on principal component analysis (December 10, 2005; Version 2)
Simini F, González MC, Maritan A, Barabási AL (2012) A universal model for mobility and migration patterns. Nature 484(7392):96–100. https://doi.org/10.1038/nature10856
Simini F, Maritan A, Neda Z (2013) Human mobility in a continuum approach. PLoS ONE 8:e60069
Song C, Koren T, Wang P, Barabási AL (2010a) Modeling the scaling properties of human mobility. Nature Phys 6:818–823
Song C, Qu Z, Blumm N, Barabási AL (2010b) Limits of predictability in human mobility. Science 327(5968):1018–1021
Toole JL, de Montjoye YA, González MC, Pentland AS (2015) Modeling and understanding intrinsic characteristics of human mobility. In: Goncalves B and Perra N (eds) Social phenomena: from data analysis to models. Springer
Uber Engineering (2018). https://eng.uber.com/h3/
Yang Y, Herrera C, Eagle N, González MC (2014) Limits of predictability in commuting flows in the absence of data for calibration. Sci Rep. https://doi.org/10.1038/srep05662

Pierre Melikov holds a Master of Science in Systems Engineering of the Department of Civil and Environmental Engineering of the University of California, Santa Barbara, and also a Master's degree from CentraleSupélec.

Jeremy A. Kho is a Master of Science graduate in Civil Systems Engineering at the University of California, Santa Barbara, and a Bachelor of Science graduate in Civil Engineering at the University of the Philippines, Diliman. He is the Data Science Lead in GrowSari, a leading enterprise e-commerce startup in the Philippines.

Vincent Fighiera is a master's degree student in Urban Computing at UC Berkeley and earned an engineering master's degree from the École Nationale Supérieure d'Arts et Métiers, Paris, France. His research mainly deals with a game theory approach on the impact of app use on traffic patterns with the Institute of Transportation Studies at UC Berkeley.

Fahad Alhasoun is a Ph.D. candidate in the Computational Science and Engineering program at Massachusetts Institute of Technology, his Ph.D. work focuses on applications of machine learning in urban computing using street view imagery. His research interest is in machine learning and applications across domains.

Jorge Audiffred is a Mexican entrepreneur who applies Data Science to design informed strategies for, among other things, the improvement of mobility in urban areas. He is the Founder of Data Lab Mx, based in Mexico City.

José L. Mateos is Research Professor at the Institute of Physics and Research Director of the Center of Complexity Sciences at the National Autonomous University of Mexico UNAM. His areas of expertise include Statistical Physics, Non-linear Dynamics, Network Science and Urban Mobility.

Marta C. González is an Associate Professor in Engineering and City & Regional Planning at UC Berkeley, she leads the Human Mobility and Networks Laboratory. Her group applies Complex Network and Complex Systems Sciences to understanding and planning for the interactions of humans with the built and natural environments.

Chapter 12
Laboratories for Research on Freight Systems and Planning

André Romano Alho, Takanori Sakai, Fang Zhao, Linlin You, Peiyu Jing, Lynette Cheah, Christopher Zegras, and Moshe Ben-Akiva

Abstract Advancements in information and communication technologies (ICT) and the advent of novel mobility solutions have brought about drastic changes in the urban mobility environment. Pervasive ICT devices acquire new sources of data that can inform detailed transportation simulation models, and are useful in analyzing new policies and technologies. In this context, we developed software laboratories that leverage the latest technological developments and enhance freight research. Future mobility sensing (FMS) is a data-collection platform that integrates tracking devices and mobile apps, a backend with machine-learning technologies and user interfaces to deliver highly accurate and detailed mobility data. The second platform, SimMobility, is an open-source, agent-based urban simulation platform which replicates urban passenger and goods movements in a fully disaggregated manner. The two

A. R. Alho (✉) · T. Sakai · F. Zhao
Future Urban Mobility Interdisciplinary Research Group, Singapore-MIT Alliance for Research and Technology, Queenstown, Singapore
e-mail: andre.romano@smart.mit.edu

L. You
School of Intelligent Systems Engineering, Sun Yat-sen University, Guangzhou, China
e-mail: lyou@mail.sysu.edu.cn

P. Jing
Intelligent Transportation Systems Lab, Massachusetts Institute of Technology, Cambridge, USA
e-mail: peiyu@mit.edu

L. Cheah
Department of Engineering Systems and Design, Singapore University of Technology and Design, Tampines, Singapore
e-mail: lynette@sutd.edu.sg

C. Zegras
Department of Urban Studies and Planning, Massachusetts Institute of Technology, Cambridge, USA
e-mail: czegras@mit.edu

M. Ben-Akiva
Department of Civil and Environmental Engineering, Massachusetts Institute of Technology, Cambridge, USA
e-mail: mba@mit.edu

W. Shi et al. (eds.), *Urban Informatics*, The Urban Book Series,
https://doi.org/10.1007/978-981-15-8983-6_12

platforms have been used jointly to advance the state of the art in behavioral modeling for passenger and goods movements. In this chapter, we review recent developments in freight-transportation data-collection techniques, including contributions to transportation modeling, and state-of-the-art transportation models. We then introduce FMS and SimMobility and demonstrate a coordinated application using three examples. Lastly, we highlight potential innovations and future challenges in these research domains.

12.1 Introduction

The urban mobility system, including passenger and goods movements, is becoming more complex. Demand for mobility is growing and, at the same time, the roles to be played, modes available, and system-wide synergies are becoming more diverse. These changes have been stimulated by the evolution of information and communication technologies (ICT). For example, crowdsourcing initiatives allow individuals to become temporary freight carriers. These and other changes show a clear need for simulation tools that allow researchers, industry practitioners, and urban planners to better grasp the potential impacts of technologies and policies in the urban mobility system. Despite their predominantly passenger-centric development, state-of-the-art behavioral simulation models are now capable of replicating business-to-business transactions between agents that can play multiple roles (shipper, carrier, and receiver) in a disaggregate manner. The next generation of models is expected to extend its capabilities to cover business-to-consumer and consumer-to-consumer flows, which are becoming more important as e-commerce plays a larger role in urban goods movements. Moreover, as the boundaries between passenger and goods movements become dimmer, new challenges to the development of integrated models will arise. The increasing ability to comprehensively represent relevant agents' decisions and behaviors is associated with a need for fine-resolution data. Still, data collection for freight remains a challenge, plagued by low participation rates for surveys and hard-to-reach key respondents. Innovations in the methods for collecting freight-transportation data are sought, leading to expectations of relying on sensing technologies and Big Data sources to overcome the data limitations. At this point in time, these new sources of data are minimally incorporated into transportation models for testing a wide range of policies and technologies.

This chapter consists of four sections, presenting (1) future mobility sensing (FMS), a freight data-collection platform, (2) SimMobility, an urban land-use and transport-simulation platform, and (3) examples of their coordinated use to move forward the current domain knowledge. The first two sections start with self-contained literature reviews on relevant research, including basic techniques, methods and applications. They are followed by a detailed account of the laboratories, FMS and SimMobility, as well as past and current applications. In Sect. 12.4, we provide examples of the coordinated use of the laboratories, and finally, we conclude with a summary and future research directions in Sect. 12.5.

12.2 Future Mobility Sensing, a Behavioral Laboratory

12.2.1 Background

The practice of transportation modeling and planning relies on a variety of data for both passenger and goods movements. Particularly for freight-transportation, high-quality data is required for the development of simulation models for commodity flows and freight-vehicle operations. Data-collection efforts in the urban freight domain need to deal with a variety of agents (e.g., companies, establishments, and vehicle drivers) in terms of decision-making mechanisms and behaviors. The heterogeneity of agents and agent types makes it challenging, compared to passenger movements, to collect a comprehensive dataset that portrays their joint decisions. As a result, multiple data-collection approaches are used which, in broad terms, can be categorized into four main groups.

12.2.1.1 Static and Count Data

These are data collected through fixed location sensors such as inductive loop detectors, automatic vehicle classifier systems, weight-in-motion (WIM) systems, or video systems. Although road-based sensors, such as inductive loop detectors, are inherently limited to capture fine-resolution freight counts, Tok (2008) developed a high-fidelity inductive loop sensor to achieve commercial vehicle classification based on the inductive signatures of vehicle types, demonstrating their potential to provide information-rich commercial vehicle traffic-count data.

The installation of video cameras made traffic counts easier than in the past, particularly for congested settings or when attempting to disaggregate the data by vehicle types. Zhang et al. (2007) detailed a video-based vehicle detection and classification (VVDC) system for collecting vehicle count and classification data using uncalibrated video images. The proposed approach was demonstrated with high accuracy, although there are a series of enhancements suggested to deal with longitudinal vehicle occlusions, severe camera vibrations, and headlight reflection problems. Mammes and Klatsky (2017) presented a video-based system to assess freight loading-bay demand and availability. Sun et al. (2017) have used video cameras for monitoring local freight traffic movements with fine resolution by developing computer-vision algorithms.

12.2.1.2 Dynamic and Mobile Data

These are data collected through sensors that move with vehicles, using devices such as GNSS, on-board diagnostics (OBD), or similar telematics. GPS data are often collected by companies for monitoring their vehicles. One of the most widely known truck GPS datasets is published by the American Transportation Research Institute

(ATRI). This dataset considerably contributes to freight research in the USA and has been used for multiple purposes, including the development of truck route-choice data (Kamali 2015) and the generation of statewide freight-truck flows (Zanjani 2014). It is often fused with other datasets because, despite its large size, it lacks details on commodities carried or trip purposes (Eluru et al. 2018). An alternative to data fusion is to complement GPS tracking with surveys, which will be discussed later in this chapter.

12.2.1.3 Survey Data

Data can also be collected through surveys that target drivers, fleet managers, or warehouse employees, among others. There are various designs of freight surveys. Freight survey design and its applications are summarized by Allen et al. (2012), covering establishment surveys, vehicle observation surveys, parking surveys, driver surveys, commodity-flow surveys, roadside-interview surveys, and other surveys. Cheah et al. (2017) provided a literature review focused on commodity and establishment-based freight surveys.

12.2.1.4 Indirect Data

This refers to data from sources that are not designed to inform freight models or derive freight-related insights, but could be used for such purposes. Some sources of Big Data would fit this category.

A challenge for freight-transportation data collection is that a single method only allows for a partial view of the urban freight distribution system, as indicated by Holguín-Veras and Jaller (2013). The same authors also detailed the strengths and weaknesses of several of the data-collection methods. Some of the above-mentioned surveys have leveraged novel technologies, although not to a great extent. Despite a greater number of freight data-collection efforts taking place, several surveys are still paper-based, although Web-based surveys reduce the burden of data entry and associated errors and are becoming more common (e.g., the Lisbon Establishment-based Freight survey described by Alho and de Abreu e Silva 2015). A major challenge lies in the fact that user-reported data are prone to inaccuracies as respondents often need to recall past activities. Furthermore, the aforementioned high-resolution data needed for modeling and simulation purposes can easily lead to extensive surveys which respondents might not be willing to fill in. Jeong et al. (2016) highlighted the challenges of ensuring sufficient participation to achieve a meaningful sample size, based on the experience of pairing a Web-based fleet manager survey and a smartphone app-based driver survey to pilot a preliminary design for the California Vehicle Inventory and Use Survey (CAL-VIUS).

In summary, we found three main research thrusts in freight data collection that call for greater attention. First, the innovative use of technology, including sensing technologies, as a means to reduce user burden requires further advances. Second, as it is challenging to recruit participants for freight surveys, there is a need to design incentive methods that can effectively increase response rate and encourage long-term participation. Some of these efforts have been piloted in household travel surveys (Nahmias-Biran et al. 2018) and are related to informational incentives, which can complement or be alternatives to monetary incentives. Third, new and alternative data sources have to be explored. Ludlow and Sakhrani (2017), present a report (NCFRP 49—New Source of Freight Data for Urban and Metropolitan Mobility) that focuses on new data sources to address urban and metropolitan freight challenges. The highlighted novel and potentially useful data sources include crowdsourced data, road and vehicle sensors (Bluetooth, RFID, connected vehicles), vehicle data streams, or image data (such as satellite-based). The FMS platform aims to address these three research areas and is a flexible and comprehensive behavioral laboratory for freight data collection.

12.2.2 FMS Architecture

Future mobility sensing (FMS) is a data-collection and visualization platform that leverages mobile sensing technology, machine-learning algorithms, and user verification to provide details of mobility behavior of passengers or freight. It was first developed as a smartphone-based automated household travel survey system. In a second iteration, it was extended to support commodity-flow surveys and track freight and commercial vehicles (FMS-Freight). FMS-Freight collects and processes survey data from business establishments related to the role(s) they play in goods movements (shipping, receiving, and transporting), associated shipments, and vehicle operations, and it also collects trip information from the drivers. FMS consists of the three distinct but interconnected components illustrated in Fig. 12.1:

- A mobile app and tracking devices that leverage various sensing technologies;

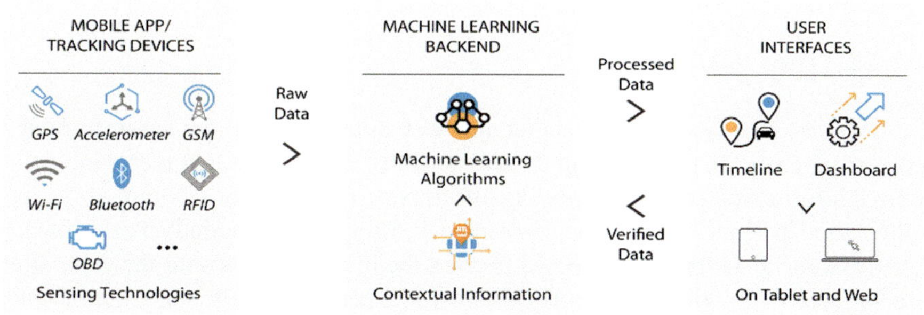

Fig. 12.1 Future mobility sensing (FMS) platform architecture

- A backend consisting of a server system with (a) a database and (b) custom algorithms to infer stops, trip purposes, and other trip details, to reduce user burden; and
- User interfaces, both mobile and Web-based, used for verification of activities by respondents and displaying summarized information (e.g., a dashboard as described by You et al. 2018).

When FMS is used to support freight data collection, the details of each component are as follows.

12.2.2.1 Mobile App/Tracking Devices

FMS-Freight supports the collection of raw data from various mobile sensing devices, such as tablets, GPS loggers, and OBD devices. GPS loggers and OBD devices are primary tools to collect data. Data are gathered from several sensors and uploaded to the backend for analysis. These devices can be easily installed and attached, respectively, to vehicles and shipments, and can collect location information with high accuracy. In the case of collecting vehicle trajectory data, the use of the vehicle battery to power the device allows for uninterrupted multi-day data collection.

12.2.2.2 Backend

Backend machine-learning algorithms process collected raw data together with the user-verified timeline (i.e., records of activities, verified through user interfaces detailed below) and contextual information (e.g., POI data) to infer stops and stop activities (Zhao et al. 2015). For shipment tracking, travel modes are also detected, which can be used to further reduce the user's verification burden. Verified data are fused and post-processed to support the identification of vehicle and shipment patterns.

12.2.2.3 User Interfaces

User-friendly interfaces on both tablet and Web applications allow a user to review and verify her or his timeline and activities. Daily verification includes confirming inferred information and filling missing information (i.e., activities, commodity type) as illustrated in Fig. 12.2. The data verified by the user are subsequently used to further train the algorithms for inferences. Moreover, the interface allows for the generation of a summary of activities in a dashboard for a user to review. An example of a shipment trace is presented in Fig. 12.3.

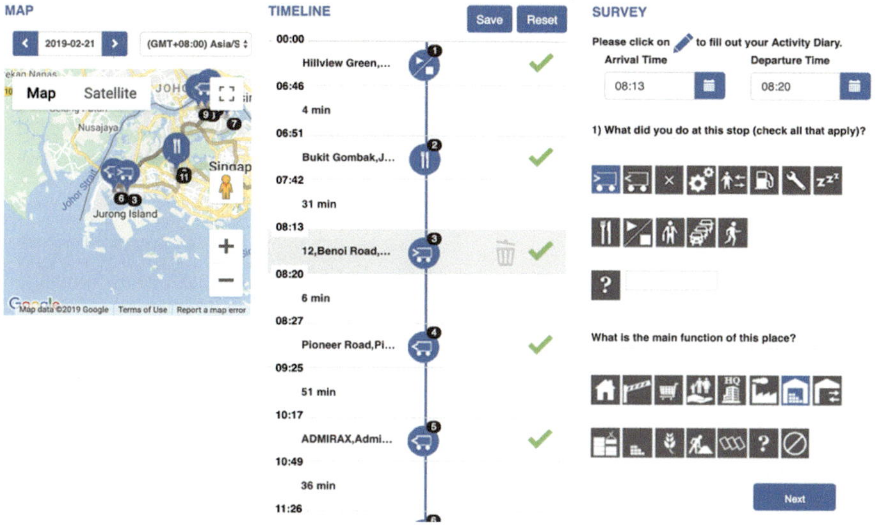

Fig. 12.2 FMS-freight stop verification interface for drivers

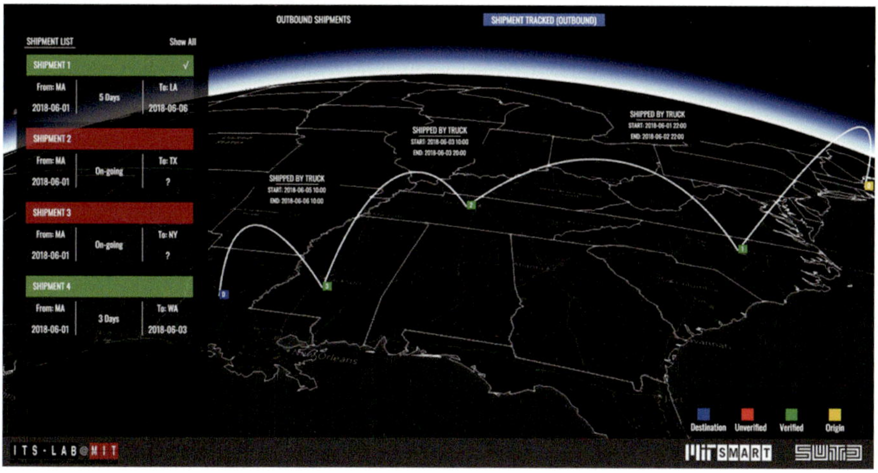

Fig. 12.3 Shipment dashboard, a form of informational incentive

12.2.3 Applications

FMS-Freight can be used to support applications ranging from truck-driver surveys, shipment-tracking surveys, or full-fledged integrated commodity-flow surveys (CFS). The survey process for integrated CFS is shown in Fig. 12.4, which consists of three steps: first, registration and pre-survey for establishment and driver information; second, shipment and freight-vehicle tracking; and lastly verification of

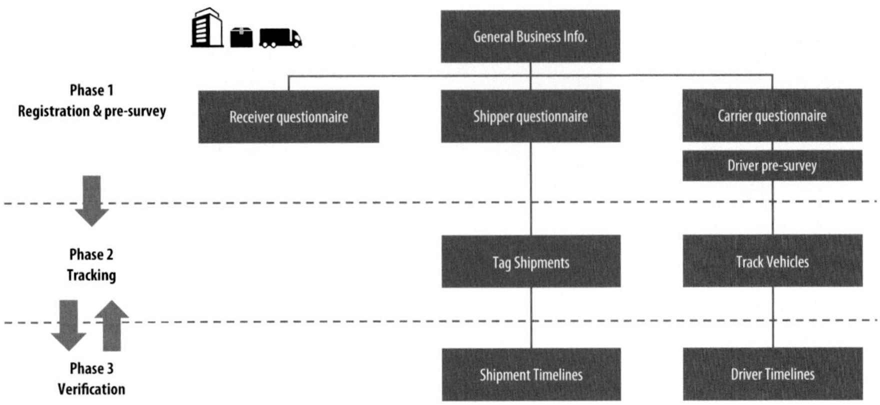

Fig. 12.4 Integrated commodity-flow survey process

inferred activities based on the tracking data. The tracking and verification steps are an iterative process that can span days or weeks depending on the survey needs.

While being continuously developed and enhanced, the FMS-Freight platform has so far been employed in the following pilots:

- A GPS-based inter-city truck-driver survey, which includes tracking, verification and stated preferences survey on driver routing behavior (Ben-Akiva et al. 2016).
- A large-scale GPS-based vehicle tracking and driver-activity survey to a sample of season parking ticket holders of the Urban Redevelopment Authority heavy vehicle parks to understand movements and parking patterns (Alho et al. 2018).
- A pilot of a commodity-flow survey in Singapore (Cheah et al. 2017), which is being followed by a larger deployment, to understand commodity flows and associated business characteristics.
- A shipment-tracking pilot in the USA and Singapore to gain additional understanding of the supply chain structures that shipments go through.

12.3 SimMobility, a Simulation Laboratory

12.3.1 Background

Simulation models have been developed and used to meet analytical and policy needs in city planning for decades. Regarding transportation, the models that simulate traffic flows are used to predict the future transportation environment and evaluate technology and the impacts of policy measures, providing the basis for policy decisions. With the increasing need for models that are able to handle a variety of technology and policy changes, the past few decades saw remarkable progress in the capability of transportation simulation tools. Classical aggregate models are being replaced with disaggregate, agent-based models. These novel simulation tools capture the complex mechanism of decisions associated with the movements of passenger and goods. As such, they enable the use of simulations to support the analysis of land-use and transportation systems changes, infrastructure management (e.g. dynamic road pricing), and emerging mobility services (e.g., shared and on-demand vehicles) among others.

The above-mentioned trend also applies to urban freight models for which advanced frameworks were proposed around 2000 and after. A number of agent-based urban freight models, which take into account behavioral mechanics in supply chain and logistics operations, have been proposed as alternatives to traditional aggregate commodity- or truck-based models (Chow et al. 2010). Those models simulate the decisions and behaviors of different agents, such as shippers, receivers, carriers (including drivers), and policymakers, and their interactions for commodity flows, logistics and transportation services, and transportation infrastructure usage (Boerkamps et al. 2000; Wisetjindawat et al. 2005; Fischer et al. 2005; Roorda et al. 2010). The resultant improvement of the granularity in decisions and behaviors allows a model to capture the inter-relations among them in a reasonable and reliable manner. The increase in data availability for specific regions and the advent of new data-science techniques further promote the development and application of disaggregate models, which, by their nature, require extensive data inputs. Thus, the potential for using them in real-world planning practices has been increasing. However, at a global level, a shortage of suitable data hampers the widespread applications of such models. In the USA, agent-based freight models were developed for some metropolitan regions, including the Chicago region (Outwater et al. 2013; RSG 2015) and the Arizona Sun Corridor Megaregion (Livshits et al. 2018). One example of this type of model is SimMobility (Adnan et al. 2016), an open-source urban simulation platform developed by the Singapore-MIT Alliance for Research and Technology (SMART) and the Intelligent Transportation Systems (ITS) Lab at Massachusetts Institute of Technology. Targeting urban freight modeling, a set of SimMobility components was estimated and calibrated for Singapore. This set of components adds the capability of simulating goods movements across supply chains, as well as agents' reactions to freight-focused policies. Examples of the latter are route restrictions, urban consolidation schemes, off-hour deliveries, and overnight, pickup, and delivery parking choices. We provide an overview of the simulation tool in this section. The details of the tool, including model specifications, are available in the paper by Sakai et al. (2019).

12.3.2 SimMobility Architecture

SimMobility is an agent-based simulation platform consisting of models for land-use changes and passenger and goods movements at the metropolitan scale. The simulations in SimMobility are fully disaggregated and maintain the consistency of agents. In SimMobility, three temporal layers are considered (Fig. 12.5): long-term (LT), mid-term (MT), and short-term (ST). The LT model covers the components of urban simulation, such as residential and firm locations, school and work locations, vehicle ownership, and parking locations, as well as business relationships among firms. The MT model, on the other hand, simulates activities of individuals, logistics operations, and vehicle and transportation-system operations at the daily level. The short-term (ST) model is a microscopic simulator for the movements of agents within a day. The different modules share a single database which maintains the data about agents, land use, transportation, and activities, enabling data exchange across the modules. The fine-resolution simulations also allow for keeping track of the behaviors of individual agents, or of knowing specifically on which vehicle a shipment was loaded.

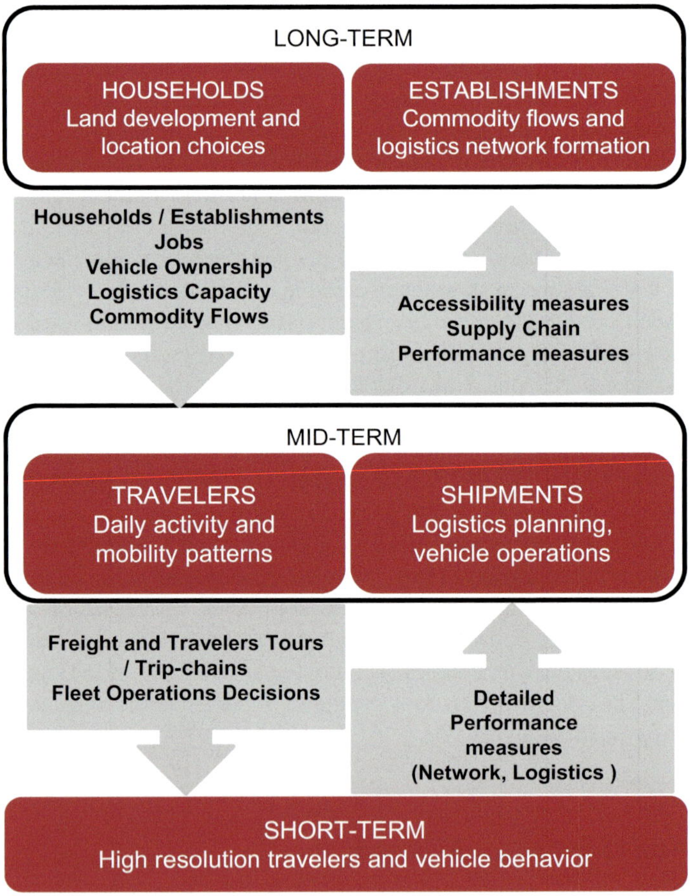

Fig. 12.5 SimMobility framework

To date, the platform has been deployed for the Greater Boston area, the Baltimore region, and Singapore as well as several prototypical cities. The freight models are currently estimated for Singapore. Further details of different components of SimMobility are available in the literature (Adnan et al. 2016; Zhu et al. 2018; Lu et al. 2015; Azevedo et al. 2017). The models incorporated in SimMobility were developed using a variety of datasets, including those obtained from FMS.

The set of components for freight simulation, termed the freight simulator hereafter, was designed for advancing the state of the art in urban freight modeling practices. It should be noted that the freight simulator is integrated with other components in SimMobility, sharing some modules, such as micro- and meso-scale traffic simulators, as well as taking inputs with passenger simulation. Figure 12.6 shows the main modules of the freight simulator, which follow the above-mentioned three temporal layers. The LT model simulates commodity contracts, which define commodity flows (i.e., selling and purchasing policies), and overnight parking choices for freight vehicles. The MT model simulates pre-day logistics planning and within-day vehicle operations, translating commodity flows to vehicle operations and behaviors, and subsequently to transport-network conditions. Lastly, the ST model simulates the behaviors of agents at an increased level of detail, particularly regarding driver behaviors, using car-following and lane-changing models. Each module, excluding the ST model, is briefly described below. A detailed description of the ST model, the microscopic traffic simulator, is available by Azevedo et al. (2017).

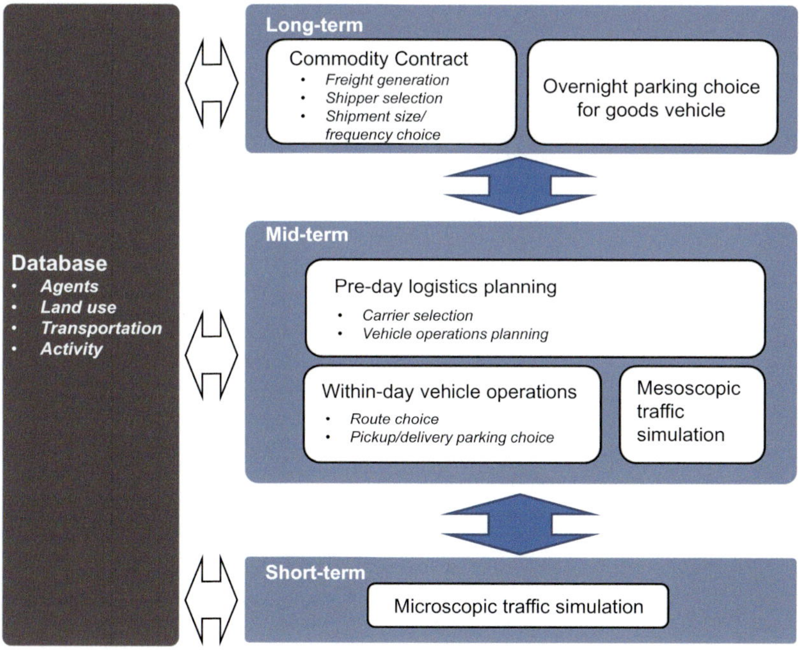

Fig. 12.6 Major components of the freight simulator

In freight simulations, business establishments play a key role. An establishment is characterized by location, employment and floor sizes, function, and industry. Establishments can play multiple roles, being able to behave as a receiver (or consumer), a shipper (or supplier), and a carrier (or a third-party logistics service provider). Commodity contracts and logistics planning are associated with establishment-level decisions. As for the application in Singapore, the synthetic population of establishments was developed based on various business statistics (Le et al. 2016).

12.3.2.1 Commodity Contract Estimation (LT Model)

Commodity contracts define selling and purchasing policies and are the basis of the commodity flows between establishments. Each commodity contract specifies shipper and receiver locations, commodity type, amount of goods, and shipment size and frequency. The commodity contract estimation is composed of three separate steps: (1) freight generation, (2) shipper selection, and (3) size and frequency choice (Fig. 12.7). Freight generation starts with identifying whether each establishment is a shipper or receiver, using a logit model. Then, multinomial logit models simulate the selection of commodity types for outbound and inbound shipments. Finally,

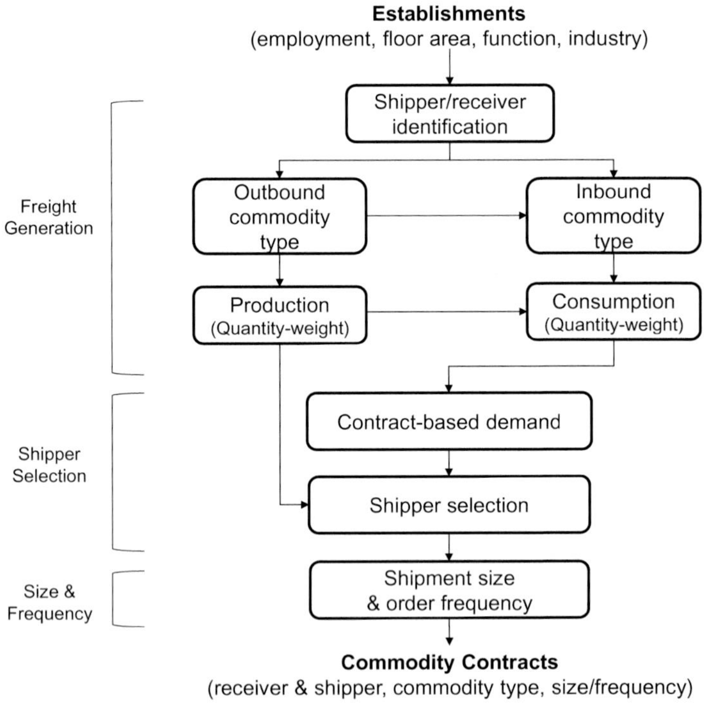

Fig. 12.7 Flow of the commodity contract estimation

the quantities of production and consumption, which are quantities shipped and received, respectively, for a certain time period, are determined using linear models. In the following step—shipper selection—the estimated consumptions are used to generate contract-based demands. Each contract-based demand requires a single shipper (supplier), and each contract is made for a single receiver–shipper pair. A receiver can make one or more contracts with shippers. Logit mixture models with error components simulate shipper selection, considering the correlations among the alternative shippers with the same distribution channel type (Sakai et al. 2018). In the third step, linear models estimate shipment size and order frequency based on factors associated with the volume of goods, and transportation and inventory costs.

12.3.2.2 Overnight Parking Choice (LT Model)

Overnight parking choice is considered a long-term decision. We simulate the decisions of vehicle owners to assign parking lots for freight vehicles using multinomial logit models, using freight-vehicle population and overnight parking supply for freight vehicles as inputs. This module enables the simulations to evaluate the impacts of parking supply policies and to define their starting and end point of daily trips.

12.3.2.3 Pre-day Logistics Planning (MT Model)

Logistics planning processes convert shipment demand into vehicle-operation plans (VOPs). The VOPs define trips or tours of vehicles to be performed in a given day, including details about stop locations and the purposes (e.g., delivery of a specific shipment) and duration of stops. The logistics planning process has sub-modules for carrier selection and vehicle-operation planning, both of which are rule-based. A carrier is assigned to each shipment based on the distances from the shipment origin to potential carriers (i.e., transportation service providers), subject to their transport capacities. Vehicle-operation planning simulates the process of assigning shipments to vehicles as well as determining the orders of pickups and deliveries. In this sub-module, a custom algorithm is applied to consolidate shipments and estimate stop duration for pickups and deliveries in a realistic manner.

12.3.2.4 Within-Day Vehicle Operations (MT Model)

VOPs are used as inputs for simulating vehicle operations and network traffic within a given day. Multinomial logit models simulate route choices for trips (i.e., movements from one location to another) based on route attributes, and driver and vehicle characteristics. Furthermore, another set of multinomial logit models simulates pickup and delivery parking choices considering cost, capacity, and congestion of parking facilities near the stop points (i.e., the activity locations), subject to parking-infrastructure data availability. A mesoscopic traffic simulation is run jointly with these simulations while updating network conditions.

12.3.2.5 Visualization of Outputs

The freight simulator runs at a metropolitan scale, which allows the measurement of the impacts of policies, technologies, or other system-related changes. Figures 12.8 and 12.9 show the examples of outputs from the LT and MT models, respectively. It should be noted that these figures are made only for illustrative purposes using a test data set and are not representative of the predicted flows. Figure 12.8 covers industry-to-industry and zone-to-zone commodity flows and overnight parking locations of freight vehicles. Figure 12.9 includes delivery locations by freight vehicles, durations of vehicle usage in VOPs, and network traffic volume.

12.3.3 Applications

SimMobility supports the evaluation of a wide range of policies, from long-term land-use development plans to short-term parking-infrastructure operations. A series of urban freight case studies have been conducted for policy analysis purposes, with others being designed, including:

- Land-use changes, specifically those related to new industrial development as well as regulatory policies designed to mitigate negative impacts;
- Overnight parking-infrastructure supply policies (Gopalakrishnan et al. 2019);
- Urban consolidation policies involving participation by shippers, carriers, and receivers;
- Regulations to promote off-hour deliveries; and
- Route restrictions for goods vehicles.

12.4 Demonstrations

The two laboratories have been used jointly to advance the state of the art in behavioral modeling and simulation. We provide three cases demonstrating such joint use, focusing on their complementarity rather than the applications of the tool for decision-making processes, which is the subject of other publications (e.g., Gopalakrishnan et al. 2019).

The first case is the estimation of freight route-choice models, the second is the quantification of the performance of freight models (applied to vehicle tour formation models), and the third is the replication of freight and non-freight-vehicle tours for specific vehicular operation patterns that are not captured by conventional

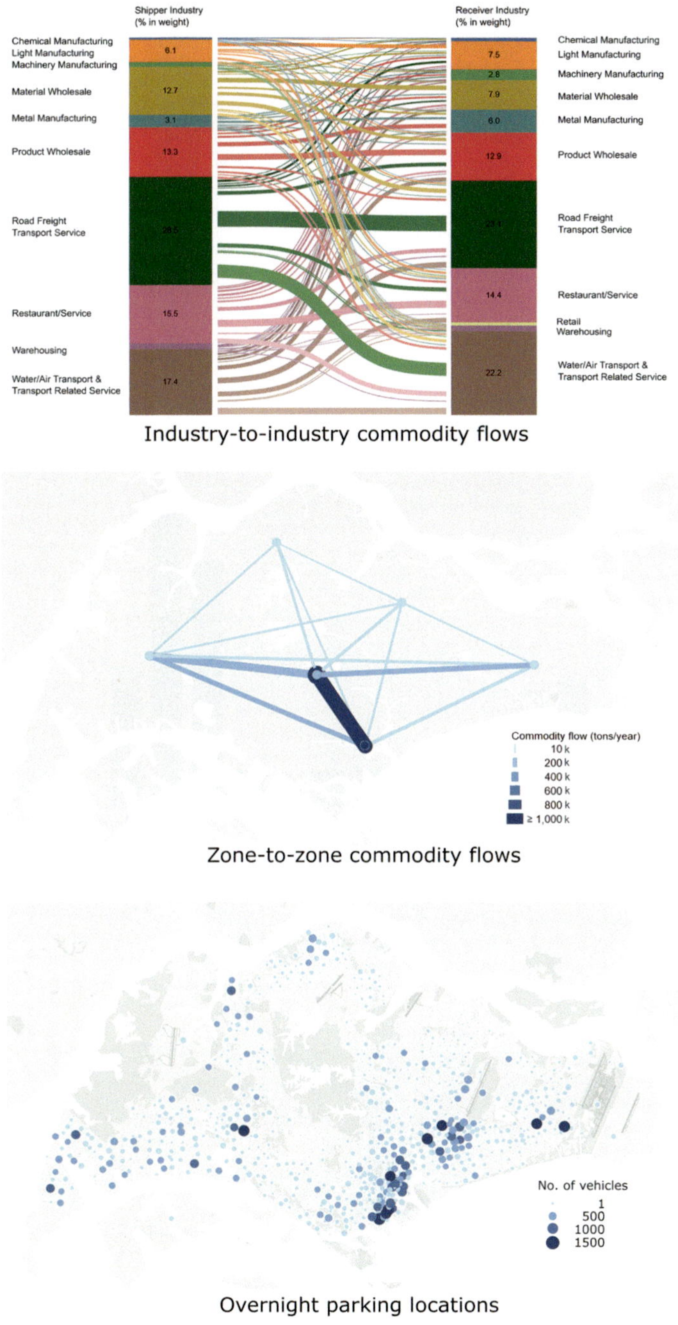

Fig. 12.8 Illustrative outputs from the long-term model

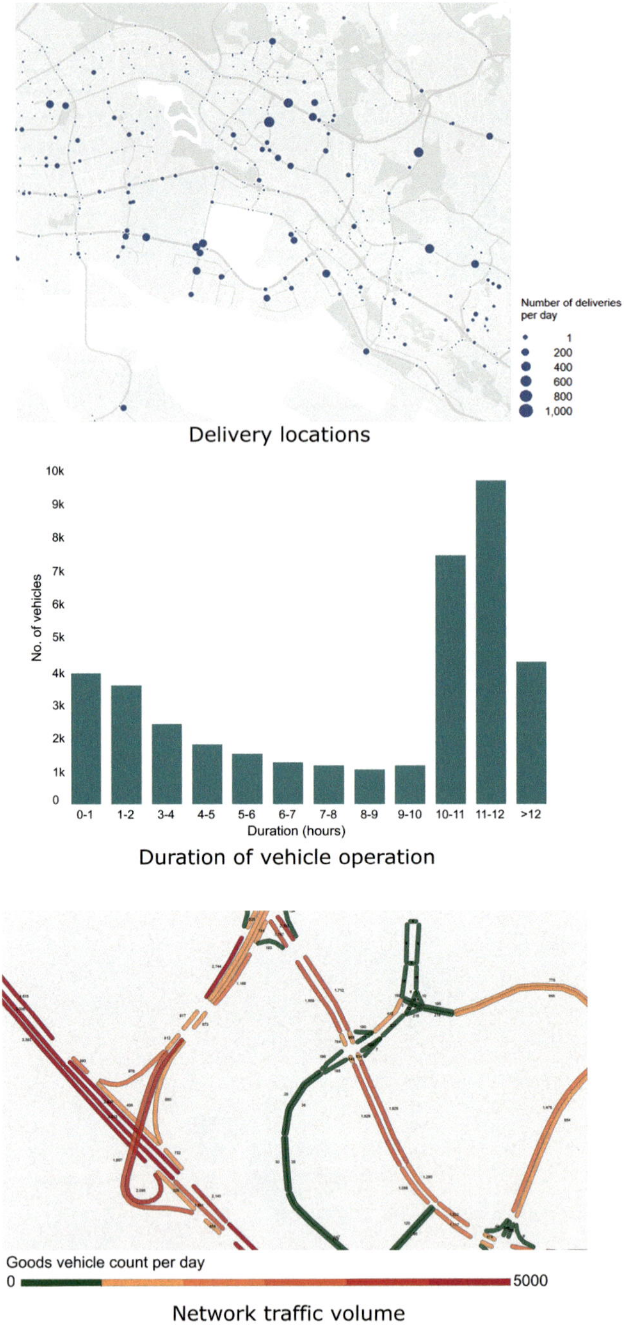

Fig. 12.9 Illustrative outputs from the mid-term model

demand models. More details about these applications can be found in the following references: Toledo et al. (2018), Alho et al. (2019b), and Gopalakrishnan et al. (2019).

12.4.1 Freight-Vehicle Route-Choice Model

The first application is the estimation of a freight-vehicle route-choice model. The route-choice decision of freight-vehicle drivers differs from that of passenger-vehicle drivers in terms of higher sensitivity to traffic conditions, and greater heterogeneity among driver types and associated commodity attributes, among other factors. The first step was to develop a truck-driver survey using FMS-Freight, which was conducted in the USA (Ben-Akiva et al. 2016). The survey collected user-annotated GPS data and characteristics of operational practices, vehicles, and drivers. A multinomial logit model was estimated using the dataset and applied to simulate the within-day route choice of drivers in SimMobility using the mid-term model. Explanatory variables include (1) traffic network attributes, which are generated by the supply simulation (e.g., travel time) or stored in the SimMobility database (e.g., road class, distance); and (2) characteristics of the driver and the vehicle, which are generated in the SimMobility long-term model. The model takes the value of explanatory variables as inputs and predicts the route between a given set of OD pairs with a Monte Carlo procedure. Figure 12.10 illustrates how the data collected using FMS-Freight are used to develop a freight route-choice model and how the model is applied in SimMobility.

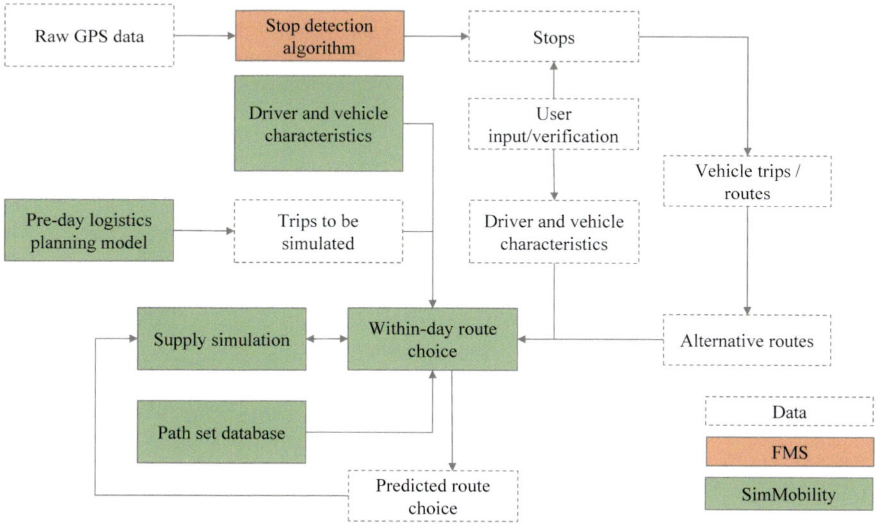

Fig. 12.10 Data and model flow for freight-vehicle route choice

12.4.2 Quantification of Model Performance

The second case is the application of the laboratories to explore the research question: What is the value of using additional data and more sophisticated model formulations? We targeted the research question specifically at vehicle-operation planning, which generates tours in the freight simulator, and used data collected using FMS-Freight to compare the model formulations' outputs against observed truck flows. We evaluate discrepancies in zone-to-zone flows, realizing that some of the proposed methods applied in SimMobility achieve superior performance against state-of-the-practice methods. The process of integrating the data between both laboratories is summarized in Fig. 12.11. In broad terms, verified vehicle stops are associated with specific vehicle tours. Further details on the algorithms that can be used for this purpose can be found in papers by Alho et al. (2019a, b). Once tours are identified, specific tour-types allow commodity flows to be estimated (Alho et al. 2018). These commodity flows are used as an input to the SimMobility mid-term logistics planning model. By varying the formulation of this model, different vehicle OD flows are generated, which can then be compared with the original OD flows revealed from the data to assess model performance in replicating such flows.

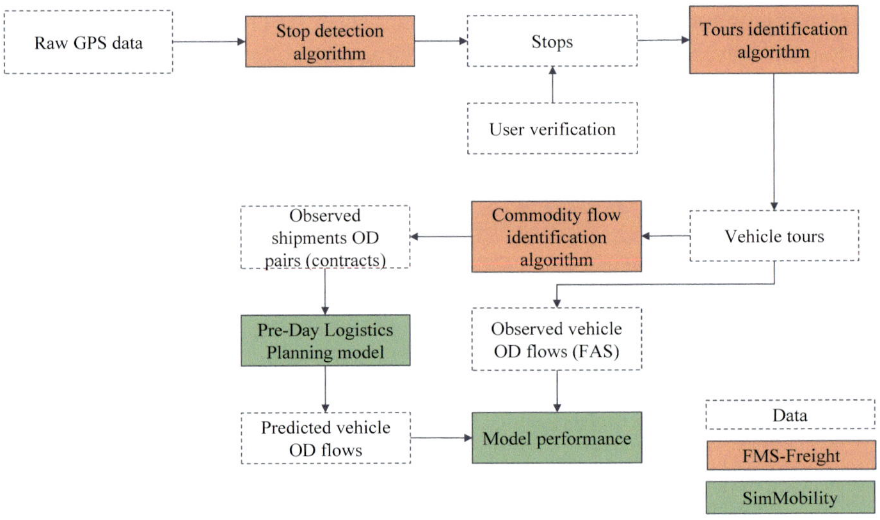

Fig. 12.11 Data and model flow for model performance quantification

12.4.3 *Replication of Specific Freight and Non-Freight-Vehicle Tours*

The final selected application is related to the replication of specific freight and non-freight-vehicle tours. The research team has performed a case-study in Singapore where simulation was used to assess a hypothetical scenario of overnight parking-infrastructure re-organization, and associated tours performance. If the overnight parking infrastructure and the assignment of vehicles to it are optimized, this can contribute to reducing empty travel, and reducing traffic congestion and air pollution. For this purpose, vehicle trips to and from overnight parking locations had to be replicated. Since the overnight parking lots are not only occupied by conventional freight vehicles, but also by private buses (on-demand, for use by companies, tourism, among other uses) and service vehicles (e.g., some construction vehicles such as cranes), there was a need to replicate the tours of both these vehicle and operation types. It should be noted that demand models for these vehicle and operation types are commonly estimated as OD matrices and not at a level of detail we required for our simulations. Thus, the approach illustrated in Fig. 12.12 was applied. This required expanding the sampled tours to the relevant vehicle populations of subscribers of the overnight parking lots.

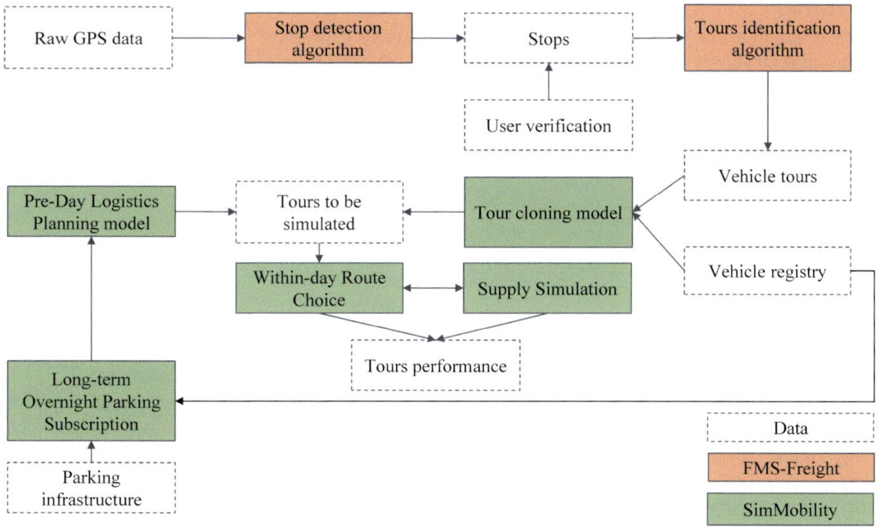

Fig. 12.12 Data and model flow for sample replication of tours of specific vehicle types

12.5 Concluding Remarks

Urban freight data-collection and modeling techniques are currently portrayed at a transition point. Meersman and Van de Voorde (2019) question whether past and current data-collection methods are suitable to inform current and future modeling needs. For all we know, the evolution of methods is predominantly incremental. We put forward that laboratories such as those demonstrated in this chapter are key to the assessment of new approaches to data collection and modeling, including a quantitative assessment of the alternative's performance against the prior. Furthermore, we demonstrate that the research progress in either data collection, or modeling and simulation, can be augmented by coordinated use of their capabilities.

The pace of change in urban freight transport appears to grow faster, and with critical implications to the relevance of freight models in assessing technological and policy impacts. This calls for further attention to the representation of relevant agents in the urban freight system in simulations, as well as their behaviors and interactions. For the latter cases, the role of sensing technologies is key to reducing survey fatigue and allowing for lengthier and deeper data-collection efforts.

Acknowledgements This research is supported in part by the Singapore Ministry of National Development and the National Research Foundation, Prime Minister's Office, under the Land and Liveability National Innovation Challenge (L2 NIC) Research Programme (L2 NIC Award No L2 NICTDF1-2016-1). Any opinions, findings, and conclusions or recommendations expressed in this material are those of the author(s) only and do not reflect the views of the Singapore Ministry of National Development and National Research Foundation, Prime Minister's Office, Singapore. We thank the Urban Redevelopment Authority of Singapore, JTC Corporation, Land Transport Authority of Singapore and Housing and Development Board of Singapore for their support.

References

Adnan M, Francisco CP, Carlos LA, Kakali B, Milan L, Sebastian R, Yi Z, Joseph F, Christopher Z, Moshe B (2016) SimMobility: a multi-scale integrated agent-based simulation platform. Paper presented at the 95th annual meeting of the transportation research board, Washington

Alho AR, de Abreu e Silva J (2015) Lisbon's establishment-based freight survey: revealing retail establishments' characteristics, goods ordering and delivery processes. Eur Transp Res Rev 7:16. https://doi.org/10.1007/s12544-015-0163-7

Alho AR, You L, Lu F, Lynette C, Fang Z, Moshe B (2018) Next-generation freight vehicle surveys by supplementing truck GPS tracking with a driver activity survey: account and insights. In: 21st international conference on intelligent transportation systems (ITSC), Maui, 4–7 Nov 2018, pp 2974–2979. https://doi.org/10.1109/ITSC.2018.8569747

Alho AR, Sakai T, Chua MH, Jeong K, Jing P, Ben-Akiva M (2019a) Exploring methods for revealing freight vehicle tours, tour-types and tour-chains from GPS vehicle tracking and driver survey data. J Big Data Anal Transp

Alho AR, Sakai T, Chua MH, Raven M, Hara Y, Ben-Akiva M (2019b) The influence of tour-formation assumptions on the reproducibility of freight vehicle flows. In: Proceedings of the eleventh city logistics conference, Dubrovnik

Allen J, Browne M, Cherrett T (2012) Survey techniques in urban freight transport studies. Transp Rev Transnatl Transdiscipl J 32(3):287–311. https://doi.org/10.1080/01441647.2012.665949

Azevedo CL, Deshmukh NM, Marimuthu B, Oh S, Marczuk K, Soh H, Basak K, Toledo T, Peh L, Ben-Akiva M (2017) SimMobility short-term: an integrated microscopic mobility simulator. Transp Res Rec 2622:13–23

Ben-Akiva ME, Toledo T, Santos J, Cox N, Zhao F, Lee YJ, Marzano V (2016) Freight data collection using GPS and web-based surveys: insights from US truck drivers' survey and perspectives for urban freight. Case Stud Transp Policy 4(1):38–44. https://doi.org/10.1016/j.cstp.2015.11.005

Boerkamps J, van Binsbergen A, Bovy P (2000) Modeling behavioral aspects of urban freight movement in supply chains. Transp Res Rec 1725:17–25

Cheah L, Zhao F, Stinson M, Lu FP, Dingmastera J, Marzano V, Ben-Akiva M (2017) Next-generation commodity flow survey: a pilot in Singapore. Paper presented at city logistics conference, Phuket

Chow JY, Yang CH, Regan AC (2010) State-of-the art of freight forecast modeling: lessons learned and the road ahead. Transportation 37(6):1011–1030

Eluru N, Li X, Pinjari A, Abdelaty M, Anowar S, Momtaz SU, Iraganaboina NC, Keya N, Dey B, Zhao D, Balusu SPK, Sheela PV (2018) Freight data fusion from multiple data sources for freight planning applications in Florida. Final report. University of Central Florida, Department of Civil, Environmental and Construction Engineering

Fischer M, Outwatern M, Cheng L, Ahanotu DN, Calix R (2005) Innovative framework for modeling freight transportation in Los Angeles County, California. Transp Res Rec 1906:105–112

Gopalakrishnan R, Alho AR, Sakai T, Cheah L, Ben-Akiva M (2019) Assessing overnight parking infrastructure policies for commercial vehicles in cities using agent-based simulation. In: Proceedings of the eleventh city logistics conference, Dubrovnik

Holguín-Veras J, Jaller M (2013) Comprehensive freight demand data collection framework for large urban areas. In: Gonzalez-Feliu J et al (eds) Sustainable urban logistics: concepts, methods and information systems. EcoProduction. Springer-Verlag, Berlin Heidelberg, pp 91–112. https://doi.org/10.1007/978-3-642-31788-0_6

Jeong K, Tok A, Ritchie SG, Park J (2016) California vehicle inventory and use survey: pilot study insights. Transp Res Rec 2547(1):32–40. https://doi.org/10.3141/2547-06

Kamali M (2015) Development of truck route choice data using truck GPS. M.Sc. thesis, University of South Florida. https://scholarcommons.usf.edu/cgi/viewcontent.cgi?referer=https://www.google.com/&httpsredir=1&article=7164&context=etd. Accessed 27 Feb 2019

Le DT, Cernicchiaro G, Zegras C et al (2016) Constructing a synthetic population of establishments for the simmobility microsimulation platform. Transp Res Procedia 19:81–93. https://doi.org/10.1016/j.trpro.2016.12.070

Livshits V, You D, Zhu H, Jeon K, Vallabhaneni L, Camargo P, Hong S, Nippani S, Sun A, Noh H, Tokishi J, Kuppam A, Selby B, Beagan D, Murray D, Smith D, Lemp J, Short J, Viswanathan K, Ehlen M, Stinson M, Khurana M, Bingham P, Bostrom R, Golub S, Carlin G, Ravulaparthy S, Pourabdollahi Z (2018) Mega-regional multi-modal agent-based behavioral freight model. Final report. Strategic highway research program 2 (SHRP2 C20 IAP grant). https://www.azmag.gov/Portals/0/Documents/MagContent/TRANS_2017-02-13_SHRP2-TRANS_2017-06-06-C20-MAG-Next-Generation-Freight-Demand-Model-Update.pdf

Lu Y, Adnan M, Basak K, Pereira FC (2015) SimMobility mid-term simulator: a state of the art integrated agent based demand and supply model. In: 94th annual meeting of the transportation research board, Washington

Ludlow D, Sakhrani V (2017) NCFRP 49—new source of freight data for urban and metropolitan mobility. In: 2017 international urban freight conference (I-NUF), Long Beach. https://www.metrans.org/sites/default/files/Special%20Session%201%20Ludlow%20.pdf. Accessed 27 Feb 2019

Mammes N, Klatsky M (2017) Approaches to monitor truck loading activity in New York City using video analytics. In: TRB innovations in freight data workshop, Irvine, 18 May 2017. https://onlinepubs.trb.org/onlinepubs/conferences/2017/FreightData/MammesPPT.pdf. Accessed 27 Feb 2019

Meersman H, Van de Voorde E (2019) Freight transport models: ready to support transport policy of the future? Transp Policy 83:97–101. https://doi.org/10.1016/j.tranpol.2019.01.014
Nahmias-Biran B, Han Y, Bekhor S, Zhao F, Zegras C, Ben-Akiva M (2018) Enriching activity-based models using smartphone-based travel surveys. Transp Res Rec 2672(42):280–291
Outwater M, Smith C, Wies K et al (2013) Tour based and supply chain modeling for freight: integrated model demonstration in Chicago. Transp Lett 2:55–66
Resource Systems Group, Inc (2015) User's guide and model documentation: agent-based supply chain modeling tool. The report for Chicago metropolitan agency for planning. https://www.cmap.illinois.gov/data/transportation/modeling
Roorda MJ, Cavalcante R, McCabe S, Kwan H (2010) A conceptual framework for agent-based modelling of logistics services. Transp Res Part E 46(1):18–31
Sakai T, Bhavathrathan BK, Alho AR, Hyodo T, Ben-Akiva M (2018) Commodity flow estimation for a metropolitan scale freight modeling system: supplier selection considering distribution channel using an error component logit mixture model. Transportation 1–29. https://doi.org/10.1007/s11116-018-9932-1
Sakai T, Alho AR, Bhavathrathan BK, Dalla Chiara G, Gopalakrishnan R, Jing P, Hyodo T, Cheah L, Ben-Akiva M (2019) SimMobility freight: an agent-based urban freight simulator for evaluating logistics solutions. Transp Res Part E Logist Transp Rev
Sun X, Ding J, Dalla Chiara G, Cheah L, Cheung NM (2017) A generic framework for monitoring local freight traffic movements using computer vision-based techniques In: 5th IEEE international conference on models and technologies for intelligent transportation systems (MT-ITS), Naples, pp 63–68. https://ieeexplore.ieee.org/abstract/document/8005592
Tok YC (2008) Commercial vehicle classification system using advanced inductive loop technology. UC Berkeley: University of California Transportation Center. Retrieved from https://escholarship.org/uc/item/92x23786
Toledo T, Jing P, Atasoy B, Ding Mastera J, Santos JO, Ben-Akiva M (2018) Intercity truck driver route choice incorporating drivers' heterogeneity in toll road usage: data collection, model estimation, and model application. In: Transportation research board 97th annual meeting, Washington
Wisetjindawat W, Sano K, Matsumoto S (2005) Supply chain simulation for modeling the interactions in freight movement. J East Asian Soc Transp Stud 6:2991–3004
You L, Zhao F, Cheah L, Jeong K, Zegras C, Ben-Akiva M (2018) Future mobility sensing: an intelligent mobility data collection and visualization platform. In: 21st IEEE international conference on intelligent transportation systems, Maui, 4–7 Nov
Zanjani AB (2014) Deriving statewide freight truck flows from global positioning system (GPS) data. M.Sc. dissertation, University of South Florida. https://scholarcommons.usf.edu/cgi/viewcontent.cgi?referer=https://www.google.com/&httpsredir=1&article=6373&context=etd. Accessed 17 Feb 2019
Zhang G, Avery RP, Wang Y (2007) A video-based vehicle detection and classification system for real-time traffic data collection using uncalibrated video cameras. Transp Res Rec 1993(1):138–147. https://doi.org/10.3141/1993-19
Zhao F, Ghorpade A, Pereira FC, Zegras C, Ben-Akiva M (2015) Stop detection in smartphone-based travel surveys. Transp Res Procedia 11:218–226
Zhu Y, Diao M, Ferreira J Jr, Zegras PC (2018) An integrated microsimulation approach to land-use and mobility modeling. J Transp Land Use 11(1):633–659

André Romano Alho is a Research Scientist in the Future Urban Mobility Interdisciplinary Research Group, at the Singapore-MIT Alliance for Research and Technology. He is passionate about developing and applying research methods to provide innovative solutions for Transportation Systems, particularly focusing on Urban Freight operations.

Takanori Sakai is a Senior Postdoctoral Associate in the Future Urban Mobility Interdisciplinary Research Group at the Singapore-MIT Alliance for Research and Technology. His research mainly focuses on urban freight transportation planning and modeling and transportation geography. He is a member of the Transportation Research Board—Freight Transportation Planning and Logistics Committee.

Fang Zhao is a Research Scientist at the Singapore-MIT Alliance for Research and Technology, Future Urban Mobility Interdisciplinary Research Group. She has an inter-disciplinary background with expertise in machine learning, communication networks, and travel surveys. She received her Ph.D. degree in Electrical Engineering from the Massachusetts Institute of Technology.

Linlin You is an Associate Professor at the School of Intelligent Systems Engineering, Sun Yat-sen University, and also a Research Affiliate at the Intelligent Transportation Systems Lab, Massachusetts Institute of Technology. He is a member of IEEE and ACM, and interested in Smart Service Orchestration, ITS and Multi-source data fusion.

Peiyu Jing is a Transportation Ph.D. candidate in the Department of Civil and Environmental Engineering at Massachusetts Institute of Technology. Her research interest is freight data collection, agent-based freight modeling and simulation, and freight congestion pricing.

Lynette Cheah is an Associate Professor of Engineering Systems at the Singapore University of Technology and Design. She leads the Sustainable Urban Mobility research group, which develops data-driven models and tools to reduce the environmental impacts of passenger and urban freight transport.

Christopher Zegras is Professor of Mobility and Urban Planning at the Massachusetts Institute of Technology where he is Head of the Department of Urban Studies and Planning. He is also a Principal Investigator of the Future of Urban Mobility Interdisciplinary Research Group, under the Singapore MIT Alliance for Research and Technology.

Moshe Ben-Akiva is the Edmund K. Turner Professor of Civil and Environmental Engineering at Massachusetts Institute of Technology (MIT), Director of MIT's Intelligent Transportation Systems Lab, and Principal Investigator at the Singapore-MIT Alliance for Research and Technology. His interests include Smart Mobility and discrete choice analysis with machine learning capabilities.

Chapter 13
Urban Risks and Resilience

Susan L. Cutter

Abstract The resilience concept has become more significant in the past decade as a means for understanding how cities prepare and plan for, absorb, recover from, and more successfully adapt to adverse events. Definitional differences—resilience as an outcome or end-point versus resilience as a process of building capacity—dominate the literature. Lagging behind are efforts to systematically measure resilience to produce a baseline and subsequent monitoring, in order to gauge what, where, and how intervention or mitigation strategies would strengthen or weaken urban resilience. The chapter reviews research and practitioner attempts to develop urban informatics for resilience and provides selected case studies of cities as exemplars.

13.1 Introduction

Disaster risks are increasing and becoming more pronounced in urban areas as populations increase and migrate to cities, turning them into megacities, and ultimately megaregions. Whether originating from natural forces such as hurricane-produced flooding (Houston), hurricanes (San Juan), wildfires (Los Angeles), earthquakes (Mexico City), or anthropogenic sources like unhealthy air pollution days (New Delhi), or the more insidious slow-onset events such as sea-level rise with increased "blue sky" coastal flooding (Jakarta), the health, safety, and welfare of urban residents is clearly at risk. In a world that is rapidly urbanizing, where more than 70% of the global population will live in cities by 2050, the nature and significance of urban disaster risk has garnered attention in research, policy, and practice. The looming question is how can urban informatics assist in the reduction of such disaster risks, and equally enhance resilience to them?

The need to reduce disaster risk in cities roared into public consciousness in 2010 when two violent earthquakes struck Port-au-Prince, Haiti (7.0Mw) and Concepcion, Chile (8.8Mw) within six weeks of each other. The impacts were catastrophic but unequal: more than 316,000 estimated lives lost in Haiti compared to 520 in Chile, and

S. L. Cutter (✉)
University of South Carolina, Columbia, USA
e-mail: scutter@sc.edu

W. Shi et al. (eds.), *Urban Informatics*, The Urban Book Series,
https://doi.org/10.1007/978-981-15-8983-6_13

$30 billion in damages in Chile compared to the $14 billion in Haiti (Table 13.1). Such disparities in earthquake impacts reflected the pre-existing vulnerabilities in both places and brought more attention and pressure to address disaster risk reduction in cities (International Federation of Red Cross and Red Crescent Societies 2010). In many urban areas where poor-quality, overcrowded housing, and basic infrastructure and services are insufficient to protect people from harm, health hazards such as cholera or an infectious disease outbreak, an extreme environmental condition like a heat wave or harmful or unhealthy air pollution episode becomes more deadly. Reducing disaster risk, especially in urban areas, has become the rallying call for civil

Table 13.1 Selected urban disasters 2010–2018

Date	Urban area	Event	Deaths	Damage[a]
2010	Japan cities	Heat wave	1718	
	Port-au-Prince, Haiti	Earthquake	316,000	~$14b
	Concepcion, Chile	Earthquake/tsunami	520	$30b
2011	Bangkok, Thailand	Flooding	815	$32b
	Christchurch, New Zealand	Earthquake	185	$24b
	Tohoku, Japan	Earthquake/tsunami	20,000	$211b
	Rio de Janeiro, Brazil	Flooding/landslides	900	$1.2b
	Mindanao Island, Philippines	Tropical Storm Washi (Sendong)	1300	<$1b
2012	New York City	Hurricane Sandy	44	$71.4b
	Ibadan and Lagos, Nigeria	Flooding	363	$7.2b
2013	Tacloban and Cebu City, Philippines	Typhoon Haiyan (Super Typhoon Yolanda)	7300	$10b
	Passau, Magdeburg, Halle, and Wittenberge, Germany	Flooding	9	$13b
2015	Southern India	Heat wave	2500	
	Southern Pakistan	Heat wave	2000	
	Katmandu, Nepal	Earthquake	9000	$10b
2016	Kunamoto, Japan	Earthquake	205	$32b
2017	Houston	Hurricane Harvey	103	$125b
	San Juan, Puerto Rico	Hurricane Maria	4475	$90b
	Puebla, Mexico	Earthquake	369	$6b
2018	Palu, Sulawesi, Indonesia	Earthquake/tsunami	4340	<$1b
	Southern California	Wildfires	3	$5.2b
	Denver, Dallas–Ft. Worth	Hail storms		$3.6b
	Osaka, Japan	Super Typhoon Jebi	11	$15b

[a]Estimates of deaths and economic damage (in US$ billions) vary widely depending on the source and when the estimation was done. They illustrate the magnitude of the events, but are not definitive of the real loss or damage. Information is compiled from a variety of Internet sources

society globally in the second decade of the twenty-first century. One of the avenues for reducing risk is to increase the resilience of cities to absorb and withstand the everyday stressors and occasional shocks that lead to disastrous outcomes (Rodin 2014). The foundation for increasing resilience is the creation and application of relevant information and data for assessment and monitoring.

The resilience concept is not new (Alexander 2013), but has gained currency in the past two decades as a means for understanding how communities prepare for, absorb, recover from, and successfully adapt to stressors or adverse events. There are multiple disciplines engaged in conceptualizing resilience and methods for operationalizing it that run the gamut from descriptive to normative to analytical approaches (Meerow et al. 2016). The units of analysis are equally variable ranging from individuals (person, building, bridge) to functional groups (households, economic sector) or social groups (elderly) to systems (ecosystem, infrastructure, community) (Cutter 2016a). A community or a city functions as a system of systems where resilience is measureable within individual systems (e.g., governance, environment, financial) and in the interactions and interdependencies between and among systems. In this respect, cities operate as complex adaptive systems. Given the multiple, and often conflicting meanings of resilience, the objects of study, and the types of resilience examined (social, economic, etc.), application tensions arise between policy discourses and local actions.

Ultimately, however, the development of strategies for enhancing resilience in urban places requires three sets of information: (1) the existing and potential vulnerabilities and exposures to risks and hazards; (2) the inherent resilience or capacity to cope with such risks; and (3) empirical measurements, in order to gauge what, where, and how intervention or mitigation strategies would strengthen or weaken resilience. The chapter reviews research and practitioner attempts to develop urban informatics for resilience during the past decade.

13.2 Risks, Exposure, and Vulnerability

There are a variety of social and environmental trends from local to global scales contributing to increasing disaster risk and vulnerability (Ismail-Zadeh et al. 2017; UN Office for Disaster Risk Reduction 2019). This is partly a function of the ongoing global patterns of urbanization not only in the world's megacities, but also in small to mid-sized cities. Infrastructure assets in hazard-prone coastal and riverine areas create more physical exposure with potentially catastrophic economic damage because of the changing frequency in weather extremes and sea-level rise due to climate change (Wong et al. 2014). Another process affecting increasing exposure is globalization and economic interdependencies, whereby production and consumption activities are no longer locally or regionally constrained, but occur within a larger global economic system. The juxtaposition of economic globalization with climate change produces the double exposure of impacts across regions, social groups, or sectors (Leichenko and O'Brien 2008).

Along with increasing risk exposure, there is also growing population vulnerability. As income and wealth gaps widen between and within urban areas, the most disadvantaged bear most of the risk burdens. These often relate to lack of locational choice, whereby formal and informal housing locates in high-risk areas such and floodplains, low-lying coastal areas subject to tidal inundation, or on steep slopes subject to failure. In many cases, the settlements lack basic municipal services such as potable water, sanitation, and power, which in turn generate additional public health risks such as diarrhea, cholera, typhoid, or asthma from indoor pollutants from open-fire cooking.

As the demographic profiles of urban areas change, many cities in Western Europe and the USA are seeing increased levels of dependent social groups, especially the elderly and immigrant populations. The elderly in western cities live on fixed retirement incomes, with fewer and fewer living in multi-generational homes. Elderly persons living alone become more socially isolated and suffer daily stressors related to medical disabilities, limited mobility, limited financial resources, and fear of crime. When a shock occurs such as a heat wave, mortality among this vulnerable cohort is especially high, leading to further inequalities in risk impacts (Fleming et al. 2018; Klinenberg 2002).

The escalation of risk exposure and vulnerability in urban areas is also a function of the variability in coping capacities and resilience, the latter of particular concern for small to mid-sized cities (Birkmann et al. 2016). Strong governance structures, political, and social engagement by stakeholders, and understanding of cities as interdependent systems of systems all influence coping capacities (the term used in hazards and disasters) or adaptive capacities (the term preferred in climate change research) in either negative or positive ways (Cutter et al. 2008). Equally influential are culture, institutions, infrastructure, technology, collective action, historical experience, environmental quality, and planning (e.g., growth management, climate change, hazard mitigation) (Carter et al. 2015).

The social transformations that are taking place globally occur within the context of hazard extremes from not only climate-sensitive hazards, but equally from geophysical events. Table 13.1 provides a sampling of these singular events (shocks) in terms of death tolls and economic damage associated with urban disasters in the past decade. While the periodicity of geophysical hazards is uncertain, it is clear that weather-related extremes are increasing globally, affecting many of the world's urban areas. Declining air quality, water scarcity, and food insecurity are everyday stressors, which compound the impacts of the shocks, but also serve to reduce the coping capacity when such shocks do occur.

13.3 Urban Resilience and Capacities

As complex adaptive systems with social, infrastructural, and ecological networks, cities are a particular focus for resilience research given their scale, spatial form, and overlapping governance structures. While definitions of urban resilience abound based on disciplinary and theoretical orientation, this chapter defines urban resilience in its simplest form as "… the ability of a city or urban system to withstand a wide array of shocks and stresses" (Leichenko 2011, p. 164). Definitions and the range of approaches to urban resilience are as varied as the interdisciplinary schools of thought involved, ranging from socio-ecological systems, to engineering, to ecology, to public health. Despite nuanced differences, there is consistency among the perspectives in terms of fostering positive social change, leading to longer-term sustainability, in other words moving forward to what could be, not bouncing back to what was.

13.3.1 The Definitional Quagmire

The exponential growth in urban resilience research began in earnest in the early twenty-first century. According to bibliometric analyses of the academic literature (Meerow and Newell 2019; Meerow et al. 2016; Moser et al. 2019; Nunes et al. 2019; Wang et al. 2018), studies were primarily focused on definitions, characterizations, unpacking of a number of conceptual tensions, and theoretical inconsistencies in the literature. Among these are resilience as an equilibrium or non-equilibrium state; resilience as a positive construct (e.g., return to normal); resilience as a system trait, outcome, or process; pathways for achieving a resilient state (persistence, transition, transformation); adaptation versus adaptability; and timescale (rapid or slow).

Resilience resonates among a wide array of disciplines and stakeholders precisely because it is a descriptively flexible term that enables different parties to adapt the term for their own usage, or what is often termed a boundary object (Brand and Jax 2007). It also projects a positive action (becoming resilient) rather than its affiliate (reducing vulnerability), recognizing that vulnerability and resilience are not the opposite of one another—just because an individual, group, or system is vulnerable does not mean that it lacks resilience (Cutter 2018). The definitional quagmire presents both opportunities and constraints. The opportunities are the flexible definitions, as well as a robust academic discourse on terminology and philosophy, which has permeated the literature in the past decade. The constraints include an inability to move beyond the semantics into measurement, let alone into policy and practice. As it now stands, there is little integration in the research literature within the social sciences on resilience (based on climate change adaptation versus disaster risk reduction fields), let alone integration among disciplinary perspectives (engineering, health, ecology, social sciences) even when working with the same unit of analysis (a city).

13.3.2 Objects of Analysis

During the past decade, much of the urban resilience literature focused on climate change, urban ecological systems, and disasters with specific threats (floods, earthquakes) as stressors. There were relatively few examples of integrated urban system resilience. Instead, the literature remained stove-piped by discipline into three main types (or schools of thought) of urban resilience: ecological resilience, engineering resilience, and socio-ecological resilience. Focusing on the dynamics of ecological processes and patterns within cities, ecological resilience narrowly focused on understanding ecosystem dynamics in specific cities, making broader comparisons and generalizations across cities difficult. For example, much has been learned from the program of long-term ecological research in urban areas (LTER sites in Baltimore and Phoenix) in the USA. This includes the role of urban ecosystem services in resilience (McPhearson et al. 2015), and the increasing prevalence of green infrastructure (integration of ecology and urban design) as a mechanism for increasing urban resilience (Childers et al. 2015). Particularly, in the urban realm, convergence of urban ecology and socio-ecological perspectives in recognizing cites as complex and dynamic systems subject to natural and anthropogenic agents of change from local to global scales (Grimm et al. 2008; McPhearson et al. 2016) has prompted new research approaches and measurements for analyzing the ecology of cities.

Engineering resilience, also termed equilibrium or functional resilience, conveys intrinsic value-neutral decision making, whereby the attributes of the systems in the resilient city are described in network performance terms: rapidity of systems restoration; robustness to withstand damage without losing form or function; and systems backup and redundancies (Borsekova et al. 2018; Bristow 2019; Heeks and Ospina 2016). There were some attempts to transcend boundaries through socio-technical studies but much of that research is either system-specific (e.g., transportation, ICT, power, or water), or asset-specific such as buildings or roads. Integration with socio-ecological perspectives is less common, but increasing in the disasters field.

Given the increasing normative interpretation of resilience, scholars began to question the apolitical nature of urban resilience by asking "Resilience for whom?" and "Resilience to what?" (Cutter 2016b) or what Meerow and Newell (2019) call the "five Ws of urban resilience"—whom, what, when, where, and why. Such concerns about equity fundamentally challenged the asset-based approaches in engineering resilience. Resilience actions within a city shaped by contested views and differing value sets, and further manipulated by unequal power and competing interests, necessitate negotiated implementation strategies and planning (Borie et al. 2019; Leitner et al. 2018; White and O'Hare 2014). Increasingly such evolutionary or transformative resilience is both dynamic and more sensitive to social conditions and change, but also highlights the value-laden nature of urban resilience embedded within the existing sociocultural structure of a city with its own historical identity and context that is as variable as the cities themselves. It also becomes more difficult to assess.

13.4 Measurement and Assessment Informatics

The definitional ambiguity of urban resilience is significant insofar as it influences its assessment and measurement. For example, the engineering perspective focuses on the efficiency of the built environment to resist or absorb shocks (robustness), redundancies in systems to maintain functioning, and the return time for such systems to return to normal operations—all static approaches. On the other hand, socio-ecological frameworks presume dynamic interactive processes that learn, transform, and adapt to new conditions in nonlinear and uncertain ways, thereby building capacity to withstand the next shock while simultaneously maintaining both social and ecosystem services. As many authors have recognized, resilience measurement is in its nascent state, whereby resilience policy is further ahead than the science of resilience assessment and measurement (The National Academies 2012).

A number of reviews of existing resilience measurement schemes appear in the recent literature (Asadzadeh et al. 2017; Beccari 2016; Brown et al. 2018; Cai et al. 2018; Ostadtaghizadeh et al. 2015; Rus et al. 2018; Sharifi 2016; The National Academies of Sciences, Engineering and Medicine 2019). Many of these are not specific to urban resilience, but instead focus more broadly on community resilience and resilience to climate change or natural hazards. Evaluation or assessments of resilience generally include one of the following: measuring baselines, measuring initiatives against accepted definitions or pre-determined indicators, or measuring resilience compared to achieving project or program goals (Brown et al. 2018).

As described in these reviews, many of the measurement efforts are mesoscale top-down quantitative efforts employing secondary data collected by governmental agencies, to produce an empirically-based view of resilience characteristics and drivers at metropolitan, county, or community scales. Many studies use indexing procedures with weighted or unweighted composite indices to derive a value for the entire enumeration unit, arguing that such a baseline or screening approach (pre-stressor or impact) is an important starting point for subsequent measurement and policy intervention (Cutter et al. 2014, 2016; Cutter and Derakhshan 2018; González et al. 2018; Harwell et al. 2019). A slightly different conceptual orientation by Kammouh et al. (2019) added additional interdependency matrices to their indicator-based approach and then tested it on a post-event case study of 1989s Loma Prieta earthquake. Many of the composite indices referenced above employ geospatial analytics in their construction and visualization of results.

The non-indexing methods incorporate fragility analyses (Barría et al. 2019), graph theory and network analytics (in spatial and non-spatial forms; Bristow 2019; Sharifi 2019), and agent-based modeling and simulations (Kanno et al. 2019; Moghadas et al. 2019). Locally based approaches such as those of Eisenman et al. (2014) and Plough et al. (2013) use pre-and post-testing of subjects to assess resilience-building programmatic activities to enhance resilience outcomes. Lastly, while relative few in number, the use of qualitative methods (narratives, focus groups; Borie et al. 2019; Huck and Monstadt 2019) are adding richness to the understanding of bottom-up (or locally based visions) of urban resilience.

What is surprising about the emerging field of resilience measurement is the lack of big data and more sophisticated and innovative geospatial methodologies. The development of crisis informatics (Liu and Palen 2010; Palen and Anderson 2016) is now well-established, but primarily used for emergency response such as during the 2010 Haiti earthquake or more recently in Hurricane Harvey in Houston and Hurricane Maria in Puerto Rico. A review of remote-sensing-based proxies for urban resilience (Ghaffarian et al. 2018) highlights the utility of reflectance of building materials and texture as proxy indicators for resilience (wood versus reinforced concrete structures in seismic areas, for example), or night-time lights as a proxy for economic resilience, as was illustrated with Hurricane Maria in Puerto Rico.

There are increasing numbers of analyses employing passive citizen-sensor data to support measurement of disaster resilience using mobile-phone or smart-card data. For example, Wilkin et al. (2019) suggest that the use of mobile-phone data for social-network analyses is one unexplored opportunity of big data. Another usage is to track population movements post-event, which is more focused on disaster recovery than on risks or resilience (Bengtsson et al. 2011). Experimentally, Wi-Fi signal data has been used to estimate the location of buried people in a hypothetical building collapse (Moon et al. 2016). The use of social-media data (with a geospatial digital trace) is more prevalent, but again primarily focused on emergency preparedness. Mainly used to show population movements out of mandated hurricane evacuation zones, Twitter data was used to gauge residential compliance with evacuation orders (Martín et al. 2017). Despite data access issues for mobile-phone data in near-real time, and biased demographics and lack of validation of social-media data such as Twitter, opportunities exist to use such data in better understanding urban resilience and its visualization (Li et al. 2015; Zou et al. 2018).

13.5 Science Informs Practice and Practice Informs Science

While research on urban resilience continues its previous bifurcations into the primary schools of thought, there is increasing convergence among them with integration between research and methods from socio-ecological and socio-technical systems approaches, largely led by the social sciences working in conjunction with urban ecologists and engineers. What is absent in much of the work to date is what is called the implementation gap, or turning the science into practice, mindful of urban governance, stakeholder engagement, and local value systems. Instead, cities have moved forward in the resilience space, implementing strategies and projects on their own, often devoid of any theoretical, conceptual, or methodological understanding of differences in the academic resilience concept or orthodoxies. At the same time, transdisciplinary science has been slow to engage practitioners in this arena as well.

One of the largest (and most well funded) of these efforts is the Rockefeller Foundation's 100 Resilient Cities project. The goal of the project was to embed resilience into city policies, programs, and practices using a comprehensive resilience strategy.

Recognizing that cities might be unable to do this alone, the Rockefeller Foundation provided the initial funding for a resilience officer for each of the 100 cities. The project developed standardized domains for measurement in order to eventually compare the global cities using locally generated and collected data based on a top-down matrix of attributes provided by Rockefeller through their City Resilience Index (Arup 2015). The identification of risks and the hazards they face, and the pathways to reduce such exposure, provided the basis for prioritizing implementation projects for enhancing resilience. The entire process was designed to build local capacity to withstand future shocks and stressors within the cities by the people and institutions that were located there.

The 100 Resilient Cities effort was not without critics (Fainstein 2018; Leitner et al. 2018). A mid-term evaluation (5 years into the program) of the experiment in urban transformation found generally positive results in building cooperation and adopting the prescriptive resilience strategy and in developing a peer-to-peer network (Martín et al. 2018). Yet in 2019, the Rockefeller Foundation decided to phase out the program, as it had grown too costly and no longer aligned with Foundation goals (Bliss 2019).

Other communities of practice continue to work toward making cities resilient and measuring progress toward that goal (Table 13.2). The UNDRR has more than 4200 cities participating in its Making Cities Resilient effort, starting with a list of the ten essentials for making a city resilient. The UNDRR also supports using the benchmark Disaster Resilient Scorecard for cities to use in resilience planning, and monitoring progress toward the implementation of the Sendai Framework for Disaster Risk Reduction. Similarly, the World Bank and the Global Facility for Disaster Reduction and Recovery (GFDRR) have an urban resilience initiative. They produced a rapid diagnostic tool to first identify sectoral resilience in cities, and then procedures for integrating the sectors and other cross-linkages for the entire city. The tool provides a locally based, bottom-up qualitative assessment for each city. UN Habitat, through their city resilience profiling tool, provides a framework for data collection and analysis to create a city profile complete with urban characteristics, crosscutting issues, internal stressors, and expected shocks and stresses for use in planning, what-if scenario development, and impact monitoring. Knowledge sharing is the primary purpose of the ICLEI and US National Academies efforts (Resilient America). Other efforts to develop specific metrics for resilient cities include the 100 Resilient Cities City Resilience Index (CRI), and ISO standardized indicators for measuring resilience in cities for benchmarking and comparisons with other cities.

Some of these efforts include remote and smart sensing and citizen science but none is as advanced as New York City's Climate Action Plan. The current plan includes an integrated science-stakeholder-community indicator and monitoring framework embodied in an operational New York City Climate Change Resilience Indicators and Monitoring (NYCLIM) system (Rosenzweig and Solecki 2019).

Table 13.2 Communities of practice focused on assessment and measurement of urban resilience

Group/entity	Tool/program	Metric URL
UN Office of Disaster Risk Reduction (UNDRR)	Making cities resilient campaign	https://www.unisdr.org/campaign/resilientcities/assets/toolkit/documents/UNDRR_Making%20Cities%20Resilient%20Report%202019_April2019.pdf
	Disaster resilient scorecard for cities	https://www.preparecenter.org/sites/default/files/unisdr_disaster_resilience_scorecard_for_cities_preliminary.pdf
Global Facility for Disaster Reduction and Recovery (GFDRR), World Bank	Urban resilience initiative, city strength diagnostic	https://www.worldbank.org/en/topic/urbandevelopment/brief/citystrength
UN Habitat	City resilience profiling tool	https://urbanresiliencehub.org/wp-content/uploads/2018/02/CRPT-Guide.pdf
European Union URBACT	Resilient Europe	https://urbact.eu/ready-future-urban-resilience-practice
Rockefeller Foundation	100 resilient cities	https://www.100resilientcities.org/about-us/ and their City Resilience Index developed by Arup https://www.cityresilienceindex.org/#/resources
International Standards Organization (ISO)	Indicators for resilient cities (ISO 37123)	https://www.iso.org/obp/ui#iso:std:iso:37123:dis:ed-1:v1:en
ICLEI	Resilient cities	https://iclei.org/en/publication/resilient-cities-report-2018
US National Academies	Resilient America	https://sites.nationalacademies.org/PGA/resilientamerica/
Urban Land Institute	Urban resilience program	https://americas.uli.org/research/centers-initiatives/urban-resilience-program/
Mississippi-Alabama Sea Grant Consortium	Climate and resilience community of practice	https://masgc.org/climate-resilience-community-of-practice/about1
Resilience Measurement Evidence and Learning	Community of practice	https://www.measuringresilience.org/
C40 Climate Leadership Group	C40 cities	https://www.c40.org/about
Urban Climate Change Research Network (UCCRN)		https://uccrn.org/what-we-do/goals-and-activities/

(continued)

Table 13.2 (continued)

Group/entity	Tool/program	Metric URL
Sustainable Development Solutions Network (SDSN)	Sustainable cities	https://unsdsn.org/what-we-do/thematic-networks/sustainable-cities-inclusive-resilient-and-connected/

13.6 Moving Forward

It is quite clear that the present state of knowledge is insufficient in understanding resilience with its many forms and constructs, especially when applied communities or more specifically cities. More attention is needed on the details of measuring and assessing resilience (informatics), but these methodologies must advance quickly to be of use to cities who want to enhance or build resilience. As stated earlier, the science of resilience measurement in general, and urban resilience metrics specifically, must mature rapidly to be of any practical use to cities who are eager to move to more resilient and sustainable pathways. Efforts to incorporate mixed methodological approaches that engage stakeholders and local knowledge (the so-called bottom-up perspective) with top-down and more quantitative approaches hold the most promise. Similarly, locally grounded input data that serve multiple purposes (resilience indicators, general plans, land-use plans, economic development, emergency plans, etc.) is a must. Aligning city data collection and syntheses with global frameworks such as the Sendai Framework for Disaster Risk Reduction, the Sustainable Development Goals, the Paris Agreement on Climate Change, the World Humanitarian Summit's Agenda for Humanity, and Habitat III's New Urban Agenda saves time and effort in reporting requirements to different entities. It also creates opportunities for enhanced data collection, as the routine parameters are already collected.

Smart cities should be able to make citizen-sensor and geospatial digital trace data more accessible for research purposes (while protecting individual privacy) in near-real time and at a lower cost than at present. Moving from passive to active sensor data, including the use of remote-sensing technologies and data, is another source of proxy data on urban risks and resilience that is underutilized.

Finally, it is incumbent upon researchers and practitioners who are interested in urban risks and resilience to engage more widely beyond their specific and often limited domains of interest. Not only is the urban system complex and multi-faceted, but so too is its resilience. Knowledge across the domains and schools of thought is important, but what is really needed given the complexity and urgency is a new way of thinking about how to achieve urban resilience. Convergence research, spanning beyond multi-, inter- or transdisciplinary framings, is one avenue, as long as it truly integrates societally relevant knowledge, methods, expertise, and values to not only solve problems, but also to advance scientific discovery and innovation and produce usable outcomes for cities in the process.

References

Alexander DE (2013) Resilience and disaster risk reduction: an etymological journey. Nat Hazards Earth Syst Sci 13(11):2707–2716. https://doi.org/10.5194/nhess-13-2707-2013

Arup (2015) City resilience index: understanding and measuring city resilience. Available from https://www.arup.com/perspectives/publications/research/section/city-resilience-index. Accessed 5 Sept 2019

Asadzadeh A, Kötter T, Salehi P, Birkmann J (2017) Operationalizing a concept: the systematic review of composite indicator building for measuring community disaster resilience. Int J Disaster Risk Reduct 25:47–162. https://doi.org/10.1016/j.ijdrr.2017.09.015

Barría P, Cruzat ML, Cienfuegos R et al (2019) From multi-risk evaluation to resilience planning: the case of central Chilean coastal cities. Water (Switzerland) 11(3):1–24. https://doi.org/10.3390/w11030572

Beccari B (2016) A comparative analysis of disaster risk, vulnerability and resilience composite indicators. PLoS Curr 1. https://doi.org/10.1371/currents.dis.453df025e34b682e9737f95070f9b970

Bengtsson L, Lu X, Thorson A, Garfield R, von Schreeb J (2011) Improved response to disasters and outbreaks by tracking population movements with mobile phone network data: a post-earthquake geospatial study in Haiti. PLoS Med 8(8):e1001083. https://doi.org/10.1371/journal.pmed.1001083

Birkmann J, Welle T, Solecki W, Lwasa S, Garschagen M (2016) Boost resilience of small and mid-sized cities. Nature 537:605–608

Bliss L (2019) The rise, fall, and possible rebirth of 100 resilient cities. City Lab. Available https://www.citylab.com/environment/2019/06/climate-change-resilience-cities-rockefeller-foundation/589861/

Borie M, Pelling M, Ziervogel G, Hyams K (2019) Mapping narratives of urban resilience in the global south. Glob Environ Change 54:203–213. https://doi.org/10.1016/j.gloenvcha.2019.01.001

Borsekova K, Nijkamp P, Guevara P (2018) Urban resilience patterns after an external shock: an exploratory study. Int J Disaster Risk Reduct 31:381–392. https://doi.org/10.1016/j.ijdrr.2018.05.012

Brand FS, Jax K (2007). Focusing the meaning(s) of resilience: resilience as a descriptive concept and a boundary object. Ecol Soc 12(1):23. Retrieved from https://www.ecologyandsociety.org/vol12/iss1/art23/

Bristow DN (2019) How spatial and functional dependencies between operations and infrastructure leads to resilient recovery. J Infrastruct Syst 25(2):04019011. https://doi.org/10.1061/(ASCE)IS.1943-555X.0000490

Brown C, Shaker RR, Das R (2018) A review of approaches for monitoring and evaluation of urban climate resilience initiatives. Environ Dev Sustain 20(1):23–40. https://doi.org/10.1007/s10668-016-9891-7

Cai H, Lam NSN, Qiang Y et al (2018) A synthesis of disaster resilience measurement methods and indices. Int J Disaster Risk Reduct 31:844–855. https://doi.org/10.1016/j.ijdrr.2018.07.015

Carter JG, Cavan G, Connelly A et al (2015) Climate change and the city: building capacity for urban adaptation. Progr Plann 95:1–66. https://doi.org/10.1016/j.progress.2013.08.001

Childers DL, Cadenasso ML, Morgan Grove J et al (2015) An ecology for cities: a transformational nexus of design and ecology to advance climate change resilience and urban sustainability. Sustainability 7(4):3774–3791. https://doi.org/10.3390/su7043774

Cutter SL (2016a) The landscape of disaster resilience indicators in the USA. Nat Hazards 80(2):741–758. https://doi.org/10.1007/s11069-015-1993-2

Cutter SL (2016b) Resilience to what? Resilience for whom? Geogr J 182(2):110–113. https://doi.org/10.1111/geoj.12174

Cutter SL (2018) Linkages between vulnerability and resilience. In: Fuchs S, Thaler T (eds) Vulnerability and resilience to natural hazards. Cambridge, pp 257–270. https://doi.org/10.1017/9781316651148
Cutter SL, Ash KD, Emrich CT (2014) The geographies of community disaster resilience. Glob Environ Change 29:65–77
Cutter SL, Ash KD, Emrich CT (2016) Urban–rural differences in disaster resilience. Ann Am Assoc Geogr 106(6):236–1252. https://doi.org/10.1080/24694452.2016.1194740
Cutter SL, Barnes L, Berry M et al (2008) A place-based model for understanding community resilience to natural disasters. Glob Environ Change 18(4):598–606. https://doi.org/10.1016/j.gloenvcha.2008.07.013
Cutter SL, Derakhshan S (2018) Temporal and spatial change in disaster resilience in US counties, 2010–2015. Environ Hazards. https://doi.org/10.1080/17477891.2018.1511405
Eisenman D, Chandra A, Fogleman S et al (2014) The Los Angeles County community disaster resilience project: a community-level, public health initiative to build community disaster resilience. Int J Environ Res Public Health 11(8):8475–8490. https://doi.org/10.3390/ijerph110808475
Fainstein SS (2018) Resilience and justice: planning for New York City. Urban Geogr 39(8):1268–1275. https://doi.org/10.1080/02723638.2018.1448571
Fleming A, Michaelson R, Youssef A et al (2018) Heat: the next big inequality issue. The Guardian. Available from https://www.theguardian.com/cities/2018/aug/13/heat-next-big-inequality-issue-heatwaves-world
Ghaffarian S, Kerle N, Filatova T (2018) Remote sensing-based proxies for urban disaster risk management and resilience: a review. Remote Sens 10(11):1760. https://doi.org/10.3390/rs10111760
González DP, Monsalve M, Moris R, Herrera C (2018) Risk and resilience monitor: development of multiscale and multilevel indicators for disaster risk management for the communes and urban areas of Chile. Appl Geogr 94:262–271. https://doi.org/10.1016/j.apgeog.2018.03.004
Grimm NB, Faeth SH, Golubiewski NE et al (2008) Global change and the ecology of cities. Science 319(5864):756–760. https://doi.org/10.1126/science.1150195
Harwell LC, Vivian DN, McLaughlin MD, Hafner SF (2019) Scientific data management in the age of big data: an approach supporting a resilience index development effort. Front Environ Sci 7:1–13. https://doi.org/10.3389/fenvs.2019.00072
Heeks R, Ospina AV (2016) Analysing urban community informatics from a resilience perspective. J Community Inform 11(1):1–14
Huck A, Monstadt J (2019) Urban and infrastructure resilience: diverging concepts and the need for cross-boundary learning. Environ Sci Policy. https://doi.org/10.1016/j.envsci.2019.05.008
International Federation of Red Cross and Red Crescent Societies (2010) World disasters report 2010: focus on urban risk. Available from https://www.ifrc.org/Global/Publications/disasters/WDR/WDR2010-full.pdf
Ismail-Zadeh AT, Cutter SL, Takeuchi K, Paton D (2017) Forging a paradigm shift in disaster science. Nat Hazards 86(2):969–988. https://doi.org/10.1007/s11069-016-2726-x
Kammouh O, Zamani Noori A, Cimellaro GP, Mahin SA (2019) Resilience assessment of urban communities. J Risk Uncertain 5(1):04019002. https://doi.org/10.1061/ajrua6.0001004
Kanno T, Koike S, Suzuki T, Furuta K (2019) Human-centered modeling framework of multiple interdependency in urban systems for simulation of post-disaster recovery processes. Cogn Technol Work 21(2):301–316. https://doi.org/10.1007/s10111-018-0510-2
Klinenberg E (2002) Heat wave: a social autopsy of disaster in Chicago. University of Chicago Press, Chicago
Leichenko R (2011) Climate change and urban resilience. Curr Opin Environ Sustain 3(3):164–168. https://doi.org/10.1016/j.cosust.2010.12.014
Leichenko RW, O'Brien KL (2008) Environmental change and globalization: double exposures. Oxford University Press, Oxford

Leitner H, Sheppard E, Webber S, Colven E (2018) Globalizing urban resilience. Urban Geogr 39(8):1276–1284. https://doi.org/10.1080/02723638.2018.1446870

Li K, Lam NSN, Qiang Y et al (2015) A cyberinfrastructure for community resilience assessment and visualization. Cartogr Geogr Inf Sci 42(sup1):34–39. https://doi.org/10.1080/15230406.2015.1060113

Liu SB, Palen L (2010) The new cartographers: crisis map mashups and the emergence of neogeographic practice. Cartography 37(1):69–90

Martín C, Mctarnaghan S, Malik A et al (2018) Institutionalizing urban resilience: a midterm monitoring and evaluation report of 100 resilient cities. Rockefeller Foundation. Available from www.rockefellerfoundation.org

Martín Y, Li Z, Cutter SL (2017) Leveraging Twitter to gauge evacuation compliance: spatiotemporal analysis of Hurricane Matthew. PLoS One 12(7):1–22. https://doi.org/10.1371/journal.pone.0181701

McPhearson T, Andersson E, Elmqvist T, Frantzeskaki N (2015) Resilience of and through urban ecosystem services. Ecosyst Serv 12:152–156. https://doi.org/10.1016/j.ecoser.2014.07.012

McPhearson T, Pickett STA, Grimm NB et al (2016) Advancing urban ecology toward a science of cities. Bioscience 66(3):198–212. https://doi.org/10.1093/biosci/biw002

Meerow S, Newell JP (2019) Urban resilience for whom, what, when, where, and why? Urban Geogr 40(3):309–329. https://doi.org/10.1080/02723638.2016.1206395

Meerow S, Newell JP, Stults M (2016) Defining urban resilience: a review. Landscape Urban Plan 147:38–49. https://doi.org/10.1016/j.landurbplan.2015.11.011

Moghadas M, Asadzadeh A, Vafeidis A et al (2019) A multi-criteria approach for assessing urban flood risk in Tehran, Iran. Int J Disaster Risk Reduct 35:1–14. https://doi.org/10.1016/j.ijdrr.2019.101069

Moon H, Kim C, Lee W (2016) A UAV based 3-D positioning framework for detecting locations of buried persons in collapsed disaster area. ISPRS Int Arch Photogramm Remote Sens Spatial Inf Sci XLI-B8:121–124. https://doi.org/10.5194/isprs-archives-XLI-B8-121-2016

Moser S, Meerow S, Arnott J, Jack-Scott E (2019) The turbulent world of resilience: interpretations and themes for transdisciplinary dialogue. Clim Change 153(1–2):21–40. https://doi.org/10.1007/s10584-018-2358-0

Nunes DM, Piinheiro MD, Tome A (2019) Does a review of urban resilience allow for the support of an evolutionary concept? J Environ Manage 244:422–430. https://doi.org/10.1016/j.jenvman.2019.05.027

Ostadtaghizadeh A, Ardalan A, Paton D et al (2015) Community disaster resilience: a systematic review on assessment models and tools. PLoS Curr 1:1–19. https://doi.org/10.1371/currents.dis.f224ef8efbdfcf1d508dd0de4d8210ed

Palen L, Anderson KM (2016) Crisis informatics—new data for extraordinary times. Science 353(6296):224–225. https://doi.org/10.1126/science.aag2579

Plough A, Fielding JE, Chandra A et al (2013) Building community disaster resilience: perspectives from a large urban county department of public health. Am J Public Health 103(7):1190–1197. https://doi.org/10.2105/AJPH.2013.301268

Rodin J (2014) The resilience dividend: being strong in a world where things go wrong. Public Affairs, New York

Rosenzweig C, Solecki W (eds) (2019) Special issue: advancing tools and methods for flexible adaptation pathways and science policy integration. Ann N Y Acad Sci 1439. Available from https://www.nyas.org/annals/special-issue-advancing-tools-and-methods-for-flexible-adaptation-pathways-and-science-policy-integration-new-york-city-panel-on-climate-change-2019-report-vol-1439/

Rus K, Kilar V, Koren D (2018) Resilience assessment of complex urban systems to natural disasters: a new literature review. Int J Disaster Risk Reduct 31:311–330. https://doi.org/10.1016/j.ijdrr.2018.05.015

Sharifi A (2016) A critical review of selected tools for assessing community resilience. Ecol Ind 69:629–647. https://doi.org/10.1016/j.ecolind.2016.05.023

Sharifi A (2019) Resilient urban forms: a review of literature on streets and street networks. Build Environ 147:171–187. https://doi.org/10.1016/j.buildenv.2018.09.040
The National Academies (2012) Disaster resilience: a national imperative. The National Academies Press, Washington
The National Academies of Sciences, Engineering and Medicine (2019) Building and measuring community resilience: actions for communities and the Gulf Research Program. The National Academies Press, Washington
United Nations Office for Disaster Risk Reduction (2019) Global assessment report (GAR) on disaster risk reduction 2019. Available from https://gar.unisdr.org
Wang L, Xue X, Zhang Y, Luo X (2018) Exploring the emerging evolution trends of urban resilience research by scientometric analysis. Int J Environ Res Public Health 15(10):2181. https://doi.org/10.3390/ijerph15102181
White I, O'Hare P (2014) From rhetoric to reality: which resilience, why resilience, and whose resilience in spatial planning? Environ Plann C 32(5):934–950. https://doi.org/10.1068/c12117
Wilkin J, Biggs E, Tatem AJ (2019) Measurement of social networks for innovation within community disaster resilience. Sustainability 11(7):1–16. https://doi.org/10.3390/su11071943
Wong PP, Losada IJ, Gattuso J-P et al (2014) Coastal systems and low-lying areas. In Field CB, Barros VR, Dokken DJ et al (eds) Climate change 2014: impacts adaptation, and vulnerability. Part A: global and sectoral aspects. Cambridge University Press, Cambridge, pp 361–409
Zou L, Lam NSN, Cai H, Qiang Y (2018) Mining Twitter data for improved understanding of disaster resilience. Ann Am Assoc Geogr 108(5):1422–1441. https://doi.org/10.1080/24694452.2017.1421897

Susan L. Cutter is Carolina Distinguished Professor of Geography at the University of South Carolina where she directs the Hazards and Vulnerability Research Institute. Her research focuses on vulnerability and resilience science, the development of metrics, and their application to evidentiary-based hazards and disaster policy and emergency management practice.

Chapter 14
Urban Crime and Security

Tao Cheng and Tongxin Chen

Abstract Scientists have an enduring interest in understanding urban crime and developing security strategies for mitigating this problem. This chapter reviews the progress made in this topic from historic criminology to data-driven policing. It first reviews the broad implications of urban security and its implementation in practice. Next, it focuses on the tools to prevent urban crime and improve security, from analytical crime hotspot mapping to police resource allocation. Finally, a manifesto of data-driven policing is proposed, with its practical demand for efficient security strategies and the development of big data technologies. It emphasizes that data-driven strategies could be applied in cities due to their promising effectiveness for crime prevention and security improvement.

Keywords Urban security · Crime mapping and analysis · Road network · Crime prediction · Data-driven policing

14.1 Introduction

Crime is largely an urban phenomenon (Baldwin et al. 1976). Globally, crime and violence are typically more serious in some urban areas than others and are exacerbated due to rapid urban growth. According to a UN report (UN Habitat 2007), though the crime rates have significantly decreased in some developed countries of North America and Western Europe over the past two decades, in other districts, such as Africa and Latin America, the total crime rate increased. Specifically, the report has shown that 60% of urban inhabitants in developing countries have been victims of crimes and the rate of victimization has reached 70 percent in some cities of Latin America and Africa over five years (UN Habitat 2007). On the other hand,

T. Cheng (✉) · T. Chen
SpaceTimeLab, Department of Civil, Environmental and Geomatic Engineering, University College London, London, UK
e-mail: tao.cheng@ucl.ac.uk

T. Chen
e-mail: tongxin.chen.18@ucl.ac.uk

W. Shi et al. (eds.), *Urban Informatics*, The Urban Book Series,
https://doi.org/10.1007/978-981-15-8983-6_14

security is usually considered as a concept (Baldwin 1997) that confronts the crime problem, by incorporating both the policing to implement crime prevention and the public's perception of crime and safety. Therefore, understanding urban crime and security would mitigate urban crime and violence, as well as enhancing the quality of inhabitants' life and improving urban sustainability (Cozens 2008).

Conventionally, crime pattern theory, routine-activity theory, and rational-choice theory—which extensively investigate criminal behaviors to explain how and why crimes occur—have been the main approaches for crime prevention. Environmental criminologists have a long and enduring interest in place and its effect on producing crime (Weisburd et al. 2009). They think that environmental factors have a substantial influence on criminal behaviors so that crime prevention should focus on solving the problems at the place of crime. Inspired by such perspective, crime prevention through environmental design (CPTED) and situational crime prevention (SCP) have been developed to tackle urban crime problems. Thus, the environmental perspective can bridge the gap between urban crime occurrence, crime understanding, and crime mitigation using scientific and effective crime prevention practices.

Recently, big data technology has gained much attention. Such technology enables a further understanding of the dynamics of crime, and it can lead to developments and improvements in crime and security analysis tools. These improvements range from retrospective to prospective approaches, from grid-based to network-based methods, and from isolated to integrated analysis. For example, network-based crime hotspot mapping or the online police patrolling deployment toolkit have been developed and applied in crime prevention. It is difficult to separately discuss urban crime and urban security due to their interdependence in complex urban environments. From the viewpoint of intelligent data-driven policing, the whole procedure, from data collection to policing outcomes, should be addressed when tackling the urban crime and security issues.

The rest of this chapter is organized as follows. Section 14.2 reviews the development of crime studies, including their historic roots in understanding urban crimes and the latest development of environmental criminology. Section 14.3 presents the concerns and theories in urban security which is devoted to reducing the urban crime problems and protecting citizenship. Section 14.4 introduces the improvement of crime analysis and security applications and the latest tools for tackling the challenges in security practices. Finally, Sect. 14.5 proposes a holistic and intelligent data-driven policing system that serves as a synthetical framework for urban crime prevention and security improvement.

14.2 Urban Crime

As an urban-related issue, crime has been extensively discussed in many research areas including ecology, sociology, geography, economics, and political science. For example, income inequality, wage structure, and labor market are considered as important contributors to the crime rate from the perspective of economics (Freeman 1999). Researches have also shown that there exists a strong relationship between crime, the criminal, and the urban environment, which provides an environmental perspective that can explore and analyze crime at different geographic levels (Wortley and Mazerolle 2008).

Nowadays, the environmental perspective in criminology has been popular among many urban and criminological research areas and has gradually shaped a multi-disciplinary approach: environmental criminology. In this section, we will first depict the historical roots of understanding urban crime from an environmental perspective. We then outline the key concepts and theories in environmental criminology.

14.2.1 Historical Roots in Understanding Urban Crime: An Environmental Perspective

Traditional criminological research focuses on the criminality of offenders and explores how biological factors, life-course experiences, and social forces influence and create criminals. Therefore, the crime is seen as the expression of the offender's deviance, influenced by events that occurred in his or her childhood. However, the concerns of the environmental perspective differ greatly from other criminological approaches. They argue that the criminal is just one portion of the crime event, and the concern is the dynamic of crime pattern, such the time, space, victim, and type.

In addition, there has been an enduring interest in place (environmental perspective) in criminology (Weisburd et al. 2012). Different crime theories explain crime at different spatial levels, ranging from the country level, province level, city level, and community level to the street segments level. Brantingham and Brantingham (2017) suggested three geographic levels of analysis—the macro-level, the meso-level, and the micro-level—within the domain of environmental criminology.

This classification matches the development of the unit of analysis in geographic analysis, which also reflects the historical roots of understanding urban crime from an environmental perspective. Briefly, studies started in the nineteenth century were mainly referred to as macro-level (e.g., countries, provinces) analysis (Guerry 1833).

Then, the early twentieth century witnessed the urban crime studies led by the Chicago School, which mainly focused on the meso-level of analysis, such as cities and big urban areas (e.g., Burgess 1928). Lately, micro-level (e.g., community and street segments) studies, starting from the late twentieth century, have attempted to achieve a fine-resolution analysis of urban crime (e.g., Sherman and Weisburd 1995), which makes crime more predictable than before.

14.2.1.1 Macro-Level Studies

Macro-level studies focus on analyzing crime distribution between countries, states, or provinces. The world's first crime map was made by Guerry and Balbi (1829).

Leveraging the geographic map, they demonstrated that crime in urban areas was more than that in the rural areas in some provinces in France.

Many interesting findings were obtained based on macro-level studies. For example, Quetelet (1831) explored the correlations between crime and many factors (e.g., levels of poverty, ethnicity, the attraction of city) in different cities of different countries. Especially, in terms of common sense, poverty may cause crime, even if violent crimes were more prevalent in poorer rural districts, and property-related crimes showed a higher level in wealthy districts than in rural areas. Such findings indicated that poverty was not highly associated with property crime, but the opportunities existed because wealthy provinces contained more valuable targets (Guerry 1833).

After that, similar studies have compared crime between different areas, such as countries. In the mid- and late nineteenth century, empirical studies in England showed distinctive differences in crime levels and rates across various counties. This study also reported higher crime rates in urban and industrialized areas than in rural areas (Mayhew 1851).

14.2.1.2 Meso-Level Studies

Meso-level studies involve the analysis of crime patterns within cities or metropolises. Studies at this level investigate crime concentrations based on a medium scale of geographic areas. For example, concentration tends to exhibit a difference between central urban areas and suburbs.

In the 1900s, a group of American sociologists known as the Chicago School took a leadership role in the development of environmental criminology at the meso-level. They treated crime as a social problem that is spatially distributed in urban areas. Park (1915) argued that urban life must be studied for crime analysis, such as "its physical organization, its occupations, and its culture" and especially the changes therein. Neighborhoods in his view were the elementary form of social cohesion in urban life. In addition Thomas and Znaniecki (1927), introduced an important concept of social disorganization, which means a decrease of the influence of existing social rules of behavior upon individual members of a group. This concept has drawn attention to communities and neighborhoods. Then, Burgess (1928) split the city into five concentric rings, and he also suggested that the urban functional zone strongly shaped the crime pattern. Inspired by the zone model developed by Burgess (1928), Shaw and Mckay firstly detected the spatial distribution of urban crime by an original method of crime mapping (Shaw and Mckay 1942). Shaw and Mckay (1942) also explored the spatial patterns of juvenile delinquency in Chicago City by comparing the spot maps of delinquency rate with the urban racial zone map and showed that crime rates varied over the urban area.

14.2.1.3 Micro-Level Studies

Micro-level studies examine crime patterns based on spatial areas at a fine resolution, such as the community level, the street level, and prime locations. In the 1980s, urban crime researchers still focused on using social disorganization theory to explain the dynamics of crime patterns at the community level. For example, Bursik Jr (1986) found that long-term crime stability was affected by community stability. More typically, Sampson et al. (1997) proposed a concept of collective efficacy which significantly influences crime in different communities. Since then, research attention has been shifted from macro- or meso-level analysis to micro-level crime study (Weisburd et al. 2009).

After the emergence of various sophisticated spatial analysis tools (e.g., GIS) in the late twentieth century, researchers could explore how various environmental factors influence specific crime locations in practice. These micro-level areas include buildings, addresses (Sherman et al. 1989), street segments (Johnson and Bowers 2010), or locations (Sherman and Weisburd 1995). Current studies confirm that street- or location-level analyses about crime sustainably enrich environmental criminology and make crime more readily forecasted (Cozens 2011).

14.2.2 Theoretical Concepts in Environmental Criminology

Environmental criminology (i.e., the environmental perspective in criminology) emphasizes the influence of the environment on crime patterns, considering that crime is the convergence of offenders, victims, and law enforcement at particular times and places (Wortley and Mazerolle 2008). Research in this area explores the spatiotemporal patterns of crime events and explains the patterns by referring to the features from the urban fundamentals—street networks, road segments, buildings, and so on. Consequently, the strategies of crime prevention derived from the explanations are becoming popular among both urban managers and inhabitants who want to manage and live in an environmentally friendly city.

Environmental criminology is mainly based on three hypotheses, which have their own implications for crime prevention (Scott et al. 2008). First, apart from the offender's ability or the accessibility of victim information, the instant environment where crime occurs could significantly affect the offender's behavior by affecting the criminal's person–situation interaction. In this principle, environmental criminology not only argues that crime is derived from criminogenic individuals but also aims to explore and explain how the environment affects the offender and why some places are criminogenic. Second, the spatiotemporal distribution of crime is not random. Crimes are spatially concentrated at places where the environmental features would promote crime opportunities. They are also concentrated around the intersection of routine activities between offenders and victims. Such crime patterns explain why crime hotspots are stable during extended periods in particular areas, a phenomenon known as the law of crime concentration (Weisburd 2015). Third, knowledge of the criminogenic environment and crime patterns could help law enforcement to allocate resources to mitigate crime in a particular location. Practically, environmental criminology could provide new insights into solutions for proactive crime prevention, such as crime prevention through environmental design, or situational crime prevention, which will be further discussed in the next section in the context of urban security implementation issues.

14.3 Urban Security

Security involves various concepts within a complex social system. As Zedner (2010) suggested, security is a strong emotion carrying multiple meanings simultaneously arising from individuals. Traditionally, security refers to the supply of private services to protect people or information from crime or violence, and properties for individual- or community-level safety (Smith and Brooks 2012). Security also relies on the public policing that is operated by the government or public services, including but not limited to crime prevention, security technology, and risk management (Brooks 2010). In the context of the urban environment and the aforementioned urban crime, urban security refers not only to crime prevention practices and implementations but also to the public perception of crime. In this section, we will review the literature about the fear of crime in urban areas and about the necessity of studying urban security, followed by a depiction of contemporary crime prevention.

14.3.1 Fear of Crime in Urban Areas

In the 1960s, a fear of crime emerged in the USA where national public opinion polls started to involve open-ended questions relating to the public perception of crime (Furstenberg 1971). The national survey reported by The President's Commission on Law Enforcement and Administration of Justice (1967) stated that the fear of crime could influence the basic life-quality of citizens. The report also found that fear of crime varied with race, income, gender, and the experience of victimization.

However, the results from public opinion polls showed that high levels of fear were found not only in areas with high crime rates but also in areas with low crime rates (McIntyre 1967). The mismatch between the fear of crime and crime rates has been evidenced in public polls in Australia (Borooah and Carcach 1997), New Zealand (Doeksen 1997), the UK (Smith 1987), and Switzerland (Killias and Clerici 2000) and has aroused the interest of researchers.

Though the fear of crime is possibly irrational and expressed in individual perceptions, it still attracts the attention of policymakers. The motivation to study the fear of crime stems from the belief that the results of these studies could be translated into practical policies for reducing fear (Box et al. 1988). Such claims are based upon the assertion that perceptions of crime are more essential than the actuality in terms of the influence on urban lives.

14.3.2 Implementation of Crime Prevention

Crime prevention from the perspective of environmental criminology differs from many other approaches. It focuses on the criminals or the reason for committing a crime and the places in which crime occurs. Here, we will review two crime prevention approaches: crime prevention through environmental design (CPTED) and situational crime prevention (SCP), both of which are highly practical and effective ways of mitigating urban crime.

14.3.2.1 Crime Prevention Through Environmental Design

CPTED, also known as designing out crime, aims at reducing crime through the design and handling of the built environment in urban areas. It focuses predominantly upon designing out crime opportunities before they occur (Armitage 2007). As a multi-disciplinary crime prevention method, CPTED derives strong theoretical support from environmental criminology, that is, the correlation between crime and environment. CPTED is concerned about the identification and modification of the social and physical conditions that potentially may generate criminal opportunities, in the hope of mitigating urban crime (Brantingham and Faust 1976).

The basis of CPTED is the concept of defensible space proposed by Newman (1972). Defensible space aims to depict the features by design that improves territorial behaviors, such as by utilizing such space among local residents. Then Poyner (1983), developed the principles of CPTED comprising surveillance, movement control, activity support, and motivational reinforcement. Cozens et al. (2005) extended to six principles: access control, territoriality, surveillance, target hardening, image, and activity support.

In practice, the US Department of Housing and Urban Development and the US Department of Justice both expressed interest in CPTED based on inspiration from the early research of Newman and Franck (1982). The concept of defensible space in CPTED is now commonly considered in many processes of urban planning, in Florida, British Columbia, the Netherlands (Saville and Cleveland 2008), the UK, South Africa, Australia, and New Zealand (Cozens et al. 2005). In this way, CPTED linked with urban sustainability is devoted to improving the quality of urban living.

14.3.2.2 Situational Crime Prevention

SCP is an efficient strategy for analyzing and reducing specific crime issues. Specifically, it aims to change the situational factors of crime so as to reduce crime opportunities. Similar to CPTED, situational prevention is grounded in theoretical perspectives in environmental criminology and environmental psychology.

In early literature, the situational prevention opportunity was used synonymously with the situation (Clarke 1980). Nevertheless, later studies concluded that situations provide not only opportunities for criminals but also temptations, inducements, and provocations (Wortley 2001). This argument emphasizes that crime is always a personal choice, which widens the scope of situational prevention. Specifically, the interaction between motivation obtained and the situation involved must be mediated in the process of an offender's decisions making (Cornish 1994).

For crime prevention Clarke (1997), offered a framework for evaluating security with 25 techniques for SCP under five main headings: increase the effort, increase the risks, reduce the rewards, reduce provocations, and remove excuses. This discussion of solutions argues that situational prevention could be easier to utilize than long-term social efforts to change the situation. The effectiveness of situational prevention is shown in its impact on most property crime, such as burglary, theft, or vandalism (Smith et al. 2002) and has recently been applied to child abuse (Wortley and Smallbone 2006) and terrorism (Clarke and Newman 2007).

However, like CPTED, situational prevention provides very simple strategies for crime prevention so that it simply displaces crime instead of preventing it; that is, it moves crime somewhere else or changes its form after such intervention. In contrast Clarke (2008), stated that crime is rarely a compulsion and the displacement is overstated. It may be credible for some types of crimes, but not for all. For example Hesseling (1994), found no evidence of crime displacement in 22 of the 55 areas he examined. In the remaining 33 areas, though some evidence of displacement was found, the crime displaced was less than what had been prevented in every examined case.

14.4 Latest Tools in Urban Crime Analysis and Security

Crime analysis is an investigative tool, defined as "the set of systematic, analytical processes that provide timely, pertinent information about crime patterns and crime-trend correlations" (Wortley and Mazerolle 2008). It utilizes crime and police data to examine crime problems, involving the features of crime scenes, offenders, victims, and crime patterns. Crime analysis aims to provide tactical suggestions to policing with respect to criminal investigations, deployment of resources, planning, assessment, and crime prevention strategies.

In this section, we will review the development of the tools that help the police deter crime and secure the city; in particular, the crime analysis tools of hotspot mapping and security approaches to online police patrolling.

14.4.1 Crime Hotspot Mapping: From Retrospective Analysis to Prediction

Crime hotspots are small geographic areas with high rates of criminal activity (Weisburd and Telep 2014). Various studies define the geographical features of hotspots differently, ranging from street segments to individual addresses. Weisburd (2015) proposed an essential attribute of a crime hotspot: stability, which suggests that crime concentrations tend to remain hot over space and time. This provides an important implication for effective policing: crime problems can be migitated by gathering appropriate data. Crime hotspot mapping is a spatial technique that concentrates on the detection of clusters of crime events across an urban area (Zhao and Tang 2018). There are several methods to producing crime hotspot maps for different purposes, such as the standard deviational ellipse, the Getis-Ord Gi* statistic, and kernel density estimation. Empirically, these analytical methods can evaluate the concentration effects across various crime types. For example, kernel density estimation (KDE) is a kind of nonparametric spatial statistical approach for calculating the probability density function of crime incidents. This method is quite popular for crime mapping owing to its fast parameter inference process. In addition, a reaction-diffusion-based technique has been proposed to explain the dissipation and displacement of hotspots (Short et al. 2010).

Traditional methods of crime hotspot mapping mainly aim to generate risk surfaces that suggest where the crime events have clustered previously. Due to fast and automatic data acquisition and computation, both the researchers and practitioners are trying to make the traditional methods suitable to predict the crime risk in customized space and time.

For example, Bowers et al. (2004) proposed a method of predictive crime mapping named ProMap. The risk at a location for a particular period could be calculated by the density function of crime that has occurred at or near that location. Continuously, empirical studies have shown that the prediction precision of ProMap is reliable (Johnson et al. 2007). Kennedy et al. (2011) advocated risk terrain modeling (RTM) to forecast monthly crime risk and focused more attention on exploring why criminogenic places generate crime rather than the crime itself. To predict crime within a short interval, Mohler et al. (2011) utilized a self-exciting point process (SEPP), which was initially used to model the propagation of earthquake aftershock or disease, to predict future crime risk based on grid cells. This approach is capable of forecasting the next day's crime risk, and it has been allied in some law enforcement in the USA. Lately, Rosser et al. (2017) proposed a network-based crime hotspot predictive mapping, and the authors showed that its predictive accuracy outperforms the state-of-the-art grid-based model. This prospective crime mapping technique based on the road network provides micro-level prediction results based on which police resources could be deployed precisely and effectively.

14.4.2 Advanced Police Patrolling Strategies

Police patrols aim to deliver police services to prevent crimes (Novak et al. 2016) and to make response to crime incidence more rapid. Police patrolling strategies are of significant importance to improving policing effectiveness and public security. Nowadays, various models have been developed for police patrolling area allocation and patrol route planning.

Allocating patrol areas aims to arrange management precincts derived from urban areas for police officers. Gholami et al. (2015) proposed a computational learning framework that leveraged a dynamic Bayesian network to connect police officers with crime events. Further, Mukhopadhyay et al. (2016) developed a bi-level optimization method, including a linear programming patrol response formulation and Bender's decomposition, to optimize police patrolling allocation so as to reduce the expected crime response time. However, offenders may commit new crimes in different locations and times. To solve this problem, Zhang and Brown (2012) used an iterative Bender's decomposition with a discrete-event simulation model to optimize patrolling area allocation, speed up response, and reduce work variation.

The goal of patrol route planning is to design routes to make patrols more effective, to deter crime or to make a quick response when crime incidents happen, which should be more impartial and effective than a random patrolling mode. For instance, Chen and Yum (2010) proposed an efficient algorithm leveraging cross-entropy for real-time police patrolling in dynamic environments. However, there exists a time lag between consecutive patrols and target visits. To solve this issue, a real-time cooperative routing strategy using online agent-based simulation was introduced to improve the effectiveness of police patrol (Chen et al. 2017). Furthermore, Chen et al. (2018) designed a street-network-based patrolling algorithm, which enables multiple police operators to patrol across different police districts on street networks and enhances effectiveness and workload balance.

In addition, the assessment of the effectiveness of police patrolling in crime deterrence has been studied for decades. It concerns where police officers visit and what they actually do during patrolling, which is useful to avoid diluting benefits and to enhance the effectiveness of resource allocation. Sherman and Weisburd (1995) compared the patrolling time in crime hotspots with associated crime reduction to assess police strategies. Lastly, Shen and Cheng (2016) proposed a framework to identify groups of police officers by clustering their GPS trajectories. This approach helps to synthetically understand police officers' patrolling behaviors across space and time, which is essential for the evaluation, planning, and optimization of police patrolling strategy.

14.5 Intelligent Data-Driven Policing

Recently, big data and AI technology have changed the traditional structure of industries such as finance and online retail industries and have been employed in a diverse range of domains. However, the application of big data technology in policing has been limited, in sharp contrast to other domains (Babuta et al. 2018).

The use of big data technology could tackle the current difficulties associated with time-consuming data analysis tasks. It could improve the effectiveness of policing by automatic or data-driven decision-making, rather than manual experience-based decision-making. Instead of simply responding to crime events when they occur, this advanced technology might allow police forces to develop proactive crime prevention strategies and targeting.

Intelligent data-driven policing is an approach that integrates such techniques as hotspot policing, intelligence-led policing, and predictive policing (Cheng et al. 2016). In particular, it emphasizes the interactions of crime, policing, and citizens in space–time. Measuring, modeling, and predicting these interactions may lead to an intelligent and holistic approach to policing in the big data age. Conceptually, it includes four inter-related issues that arise in the process from data collection to policing outcomes (Cheng et al. 2016).

First, data-driven tools must be easy to utilize and must transfer directly into policing practices. Nevertheless, the outputs of most existing tools are far from suitable on these criteria: the current large box or grid hotspots identified by predictive mapping methods, for instance, include many road sections and cannot suggest precisely where police officers should be deployed. To ensure their suitability, tools should be explicitly designed with police operation in mind. For this, network-based crime hotspot mapping tools developed by Rosser et al. (2017) and Zhang and Cheng (2020) should be deployed to enhance the chance of technology adoption, because these tools pin the crime hotspots to road segments, the fundamental structure supporting urban life and human activities, as well as police patrolling.

Second, predictive accuracy is paramount if police forces are to adopt the tools, and thereby to enhance policing efficiency. Accuracy evaluation is important to enhance the confidence of the application. For example, Adepeju et al. (2016) proposed a practical evaluation tool in different metrics for spatiotemporal crime prediction. This requires the refinement of analytical techniques for specific policing contexts, as well as the selection of appropriate units of analysis, so that police resources can be effectively deployed. In addition, given that police and offender activities are constrained by road networks in urban areas, the greater accuracy and precise methods on road networks will have a higher chance for deployment.

Third, police patrol strategies should be coordinated to enhance the efficiency and effectiveness of crime deterrence. Police need to deal with emergencies and routine patrolling, involving the movement and placement of police officers in large numbers and spatial diversity. It is vital to effectively allocate the tasks and design the routing (Chen et al. 2018). For this purpose, police resources should be first districted in a balanced way, and then a dynamic real-time online dispatch strategy could be adopted to deal with emergencies and patrolling implementation (Chen et al. 2017).

Finally, it is necessary to evaluate the implementation and refine policing strategies, as part of an intelligent policing system. To evaluate policing implementations, Davies and Bowers (2015), proposed to compare the supply of policing (i.e., police activities) and the demand for policing (i.e., call for services) in order to support the commanding officer's decision. Examining police patrolling patterns across space and time could help our understanding of patrolling behaviors (Shen and Cheng 2016). In addition, public confidence in policing is always a top priority of the government agenda (Skogan 2006). However, public views of data-driven policing are ambiguous with the advent of big data and artificial intelligence technologies due to worries about the use of machine decision-making in conducting policing activities.

To put all these principles together, an end-to-end solution with functions of prediction, online patrolling, and real-time feedback is needed for intelligent policing. For this purpose, a Web-based prototype has been developed and is shown in Fig. 14.1. This prototype integrates analysis and evaluation across crime events, policing strategies, and citizenship, and it establishes an entire framework to secure the public.

Fig. 14.1 Spatiotemporal patterns formed by crime, policing, and citizenship activity form dynamic, interdependent networks (Cheng et al. 2016)

14.6 Summary

Urban crime and security play a continuing and essential role in the sustainable development of urban cities and the quality of citizens' life. In this chapter, we gave an overview of urban crime and security from a historical and practical perspective. We first reviewed the theories of environmental criminology and the historical roots of understanding urban crime, and then the state-of-the-art crime and security applications; predictive crime hotspot mapping and police patrolling strategies. Finally, we proposed an intelligent data-driven policing associated with big data and AI, a comprehensive perspective that ranges from spatial units and accuracy of data analysis to police patrolling and effectiveness evaluation, leading to an intelligent and holistic policing system for urban crime prevention and security enforcement.

References

Adepeju M, Rosser G, Cheng T (2016) Novel evaluation metrics for sparse spatio-temporal point process hotspot predictions: a crime case study. Int J Geogr Inf Sci 30(11):2133–2154

Armitage R (2007) Sustainability vs. safety: confusion, conflict and contradiction in designing out crime. In: Imagination for crime prevention: essays in honour of Ken Pease. Lynne Rienner Publishers,

Babuta A, Oswald M, Rinik C (2018) Machine learning algorithms and police decision-making. https://rusi.org/publication/whitehall-reports/machine-learning-algorithms-and-police-decision-making-legal-ethical

Baldwin DA (1997) The concept of security. Rev Int Stud 23(1):5–26

Baldwin J, Bottoms AE, Walker MA (1976) The urban criminal: a study in Sheffield. Tavistock Publications

Borooah VK, Carcach CA (1997) Crime and fear: evidence from Australia. Br J Criminol 37(4):635–657

Bowers KJ, Johnson SD, Pease K (2004) Prospective hot-spotting: the future of crime mapping? Br J Criminol 44(5):641–658

Box S, Hale C, Andrews G (1988) Explaining fear of crime. Br J Criminol 28(3):340–356

Brantingham PL, Brantingham PJ (2017) Notes on the geometry of crime. In: Principles of geographical offender profiling. Routledge, pp 97–124

Brantingham PJ, Faust FL (1976) A conceptual model of crime prevention. Crime Delinquency 22(3):284–296

Brooks DJ (2010) What is security: definition through knowledge categorization. Secur J 23(3):225–239

Burgess EW (1928) Residential segregation in American cities. Annals Am Acad Polit Soc Sci 140(1):105–115

Bursik RJ Jr (1986) Ecological stability and the dynamics of delinquency. Crime Justice 8:35–66

Chen X, Yum T-SP (2010) Cross entropy approach for patrol route planning in dynamic environments. In: IEEE international conference on intelligence and security informatics, pp 114–119

Chen H, Cheng T, Wise S (2017) Developing an online cooperative police patrol routing strategy. Comput Environ Urban Syst 62:19–29

Chen H, Cheng T, Shawe-Taylor J (2018) A balanced route design for min-max multiple-depot rural postman problem (mmmdrpp): a police patrolling case. Int J Geogr Inf Sci 32(1):169–190

Cheng T, Bowers K, Longley P, Shawe-Taylor J, Adams T, Davies T, Rosser G, Adepeju M, Chen H, Shen J (2016) CPC: Crime, policing and citizenship: intelligent policing and Big Data. UCL SpaceTime Lab, London
Clarke RV (1980) Situational crime prevention: theory and practice. Br J Criminol 20:136
Clarke R (1997) A revised classification of situational crime prevention techniques. In: Crime prevention at a crossroads. Cincinnati
Clarke RV (2008) Situational crime prevention. In: Wortley R, Mazerolle L (eds) Environmental criminology and crime analysis. CRC Press, Boca Raton, FL, p 178
Clarke RV, Newman GR (2007) Police and the prevention of terrorism. Polic J Policy Pract 1(1):9–20
Cornish DB (1994) The procedural analysis of offending and its relevance for situational prevention. Crime Prev Stud 3:151–196
Cozens P (2008) Crime prevention through environmental design in Western Australia: planning for sustainable urban futures. Int J Sustain Dev Plann 3(3):272–292
Cozens PM (2011) Urban planning and environmental criminology: Towards a new perspective for safer cities. Plann Pract Res 26(4):481–508
Cozens PM, Saville G, Hillier D (2005) Crime prevention through environmental design (CPTED): a review and modern bibliography. Prop Manage 23(5):328–356
Davies T, Bowers K (2015) Quantifying the deterrent effect of police patrol via GPS analysis. In: Proceedings of GIS research UK (GISRUK), pp 1–4. DOI: https://doi.org/10.1007/s00033-005-0018-5.
Doeksen H (1997) Reducing crime and the fear of crime by reclaiming New Zealand's suburban street. Landscape Urban Plann 39(2–3):243–252
Freeman RB (1999) The economics of crime. Handb Labor Econ 3:3529–3571
Furstenberg FF (1971) Public reaction to crime in the streets. Am Sch 601–610
Gholami S, Zhang C, Sinha A, Tambe M (2015) An extensive study of Dynamic Bayesian Network for patrol allocation against adaptive opportunistic criminals
Guerry A-M (1833) Essai sur la statistique morale de la France. Clearwater
Guerry A-M, Balbi A (1829) Statistique comparée de l'état de l'instruction et du nombre des crimes dans les divers arrondissements des Académies et des Cours Royales de France. Renouard, Paris
Hesseling R (1994) Displacement: a review of the empirical literature. Crime Prev Stud 3(1):97–230
Johnson SD, Bowers KJ (2010) Permeability and burglary risk: are cul-de-sacs safer? J Quant Criminol 26(1):89–111
Johnson S, Birks D, McLaughlin L, Bowers K, Pease K (2007) Prospective mapping in operational context. Home Office Online Report 19/07. Home Office, London
Kennedy LW, Caplan JM, Piza E (2011) Risk clusters, hotspots, and spatial intelligence: risk terrain modeling as an algorithm for police resource allocation strategies. J Quant Criminol 27(3):339–362
Killias M, Clerici C (2000) Different measures of vulnerability in their relation to different dimensions of fear of crime. Br J Criminol 40(3):437–450
Mayhew H (1851) London labour and the London poor: a cyclopaedia of the condition and earnings of those that will work, those that cannot work and those that will not work, vol 1. Cass
McIntyre J (1967) Public attitudes toward crime and law enforcement. Annals Am Acad Polit Soc Sci 374(1):34–46
Mohler GO, Short MB, Brantingham PJ, Schoenberg FP, Tita GE (2011) Self-exciting point process modeling of crime. J Am Stat Assoc 106(493):100–108
Mukhopadhyay A, Zhang C, Vorobeychik Y, Tambe M, Pence K, Speer P (2016) Optimal allocation of police patrol resources using a continuous-time crime model. In: International conference on decision and game theory for security. Springer, pp 139–158
Newman O (1972) Defensible space. Macmillan, New York
Newman O, Franck KA (1982) The effects of building size on personal crime and fear of crime. Popul Environ 5(4):203–220
Novak KJ, Fox AM, Carr CM, Spade DA (2016) The efficacy of foot patrol in violent places. J Exp Criminol 12(3):465–475

Park RE (1915) The city: Suggestions for the investigation of human behavior in the city environment. Am J Sociol 20(5):577–612
Poyner B (1983) Design against crime: beyond defensible space. Butterworths, London
President's Commission on Law Enforcement and Administration of Justice (1967) The challenge of crime in a free society. US Government Printing Office, Washington, DC
Quetelet A (1831) Research on the propensity to crime of different ages. Hayez, Brussels (trans: Sawyer F, Test S). Anderson Publishing Co, Cincinnati, OH
Rosser G, Davies T, Bowers KJ, Johnson SD, Cheng T (2017) Predictive crime mapping: arbitrary grids or street networks? J Quant Criminol 33(3):569–594
Sampson RJ, Raudenbush SW, Earls F (1997) Neighborhoods and violent crime: a multilevel study of collective efficacy. Science 277(5328):918–924
Saville G, Cleveland G (2008) Second-generation CPTED: the rise and fall of opportunity Theory. In: 21st century security and CPTED. Auerbach Publications, pp 93–104
Scott M, Eck J, Knutsson J, Goldstein H (2008) Problem-oriented policing and environmental criminology. In: Wortley R, Mazerolle L (eds) Environmental criminology and crime analysis. CRC Press, Boca Raton, FL, p 221
Shaw CR, McKay HD (1942) Juvenile delinquency and urban areas.
Shen J, Cheng T (2016) A framework for identifying activity groups from individual space-time profiles. Int J Geogr Inf Sci 30(9):1785–1805
Sherman LW, Weisburd D (1995) General deterrent effects of police patrol in crime "hot spots": a randomized, controlled trial. Justice Q 12(4):625–648
Sherman LW, Gartin PR, Buerger ME (1989) Hot spots of predatory crime: routine activities and the criminology of place. Criminology 27(1):27–56
Short MB, Brantingham PJ, Bertozzi AL, Tita GE (2010) Dissipation and displacement of hotspots in reaction-diffusion models of crime. Proc Natl Acad Sci 107(9):3961–3965. https://doi.org/10.1073/pnas.0910921107
Skogan WG (2006) Police and community in Chicago: a tale of three cities. Oxford University Press, New York
Smith SJ (1987) Fear of crime: beyond a geography of deviance. Prog Hum Geogr 11(1):1–23
Smith C, Brooks DJ (2012) Security science: The theory and practice of security. Butterworth-Heinemann,
Smith MJ, Clarke RV, Pease K (2002) Anticipatory benefits in crime prevention. Crime Prev Stud 13:71–88
Thomas WI, Znaniecki F (1927) The Polish peasant in Europe and America: 1–2
UN Habitat (2007) Enhancing urban safety and security: global report on human settlements 2007. Routledge
Weisburd D (2015) The law of crime concentration and the criminology of place. Criminology 53(2):133–157
Weisburd D, Telep CW (2014) Hot spots policing: What we know and what we need to know. J Contemp Crim Justice 30(2):200–220
Weisburd D, Bruinsma GJ, Bernasco W (2009) Units of analysis in geographic criminology: historical development, critical issues, and open questions. In: Putting crime in its place. Springer, pp 3–31
Weisburd D, Groff ER, Yang S-M (2012) The criminology of place: street segments and our understanding of the crime problem. Oxford University Press, New York
Wortley R (2001) A classification of techniques for controlling situational precipitators of crime. Secur J 14(4):63–82
Wortley R, Mazerolle L (2008) Environmental criminology and crime analysis: situating the theory, analytic approach and application. In: Wortley R, Mazerolle L (eds) Environmental criminology and crime analysis. CRC Press, Boca Raton, pp 1–18
Wortley R, Smallbone S (2006) Applying situational principles to sexual offenses against children. Crime Prev Stud 19:7

Zedner L (2010) Security, the state, and the citizen: the changing architecture of crime control. New Crim Law Rev Int Interdiscip J 13(2):379–403
Zhang Y, Brown DE (2012) Police patrol district design using agent-based simulation and GIS. In: 2012 IEEE international conference on intelligence and security informatics, pp 54–59
Zhang Y, Cheng T (2020) Graph deep learning model for network-based predictive hotspot mapping of sparse spatio-temporal events. Comput Environ Urban Syst 79:101403
Zhao X, Tang J (2018) Crime in urban areas: A data mining perspective. ACM SIGKDD Explor Newsl 20(1):1–12

Tao Cheng is a Professor in Geo-Informatics at University College London (UCL). She is also the Founder and Director of UCL SpaceTimeLab for Big Data Analytics which aims to gain actionable insights from geo-tagged and time-stamped data for smart city applications including transport, policing and business. Her research interests span network complexity, GIScience, urban analytics, and applied AI and machine learning.

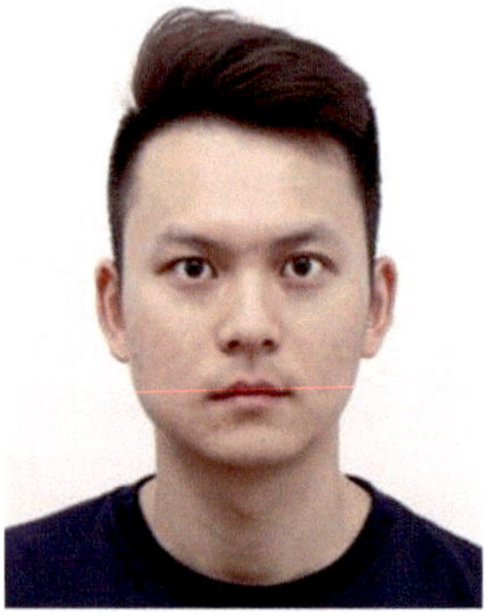

Tongxin Chen is a Ph.D. student at UCL SpaceTimeLab. He has the educational background of criminology and crime justice. His research interests include crime mapping, spatio-temporal crime pattern analysis and applied machine learning for crime analysis

Chapter 15
Urban Governance

Alex D. Singleton and Seth E. Spielman

Abstract In this chapter, we discuss how the availability of new urban data has the potential to transform the governance of cities. Such effects are realized in several ways: by increasing transparency; creating greater scope to appropriately set and measure municipal policy outcomes; and by design of well-planned and managed digital infrastructure, better empower citizens to hold decision-makers to account. However, such potential is not without risks, and without critical reflection, the proliferation of new data and their integration into software delivering algorithmic insight or automation may reproduce or develop new inequalities. We conclude that for digital urban governance to make a future that we want, it is important that we reflect upon how and where these technologies are implemented to ensure these are optimized in favor of the public good.

Keywords Governance · Dashboards · Open Data · Algorithms

15.1 Transparency and City Open Data

Transparency in the processes of city governance both limit the potential for corruption, while also ensuring that the citizens of urban areas can hold democratically elected officials to account for their use of public funding. UN Habitat (2004) argues that greater transparency can reduce urban poverty and enhance civic engagement; and by promoting engagement through a range of different policy instruments, can reduce citizen apathy, make service delivery better contribute to poverty reduction, increase ethical standards, and grow city revenues. Transparency within urban governance is an expansive topic. However, we focus here on the role of Open Data within this context.

A. D. Singleton (✉)
Department of Geography and Planning, University of Liverpool, Liverpool, UK
e-mail: alex.singleton@liverpool.ac.uk

S. E. Spielman
Department of Geography, University of Colorado Boulder, Boulder, USA
e-mail: seth.spielman@colorado.edu

W. Shi et al. (eds.), *Urban Informatics*, The Urban Book Series,
https://doi.org/10.1007/978-981-15-8983-6_15

Data about our cities are legion and include both traditional sources such as surveys or censuses, and those new forms of data related to other collection mechanisms such as sensors (e.g., noise, pollutants, etc.), social media, or as an operational by-product (e.g., meeting minutes, expenses, administrative records). The ownership and control of access to such data are a key facet of transparency, and much data about cities are held within the private realm. For example, geolocated Tweets posted by citizens of urban areas is owned by the private company Twitter, with public access restricted to either limited subsets of Tweets or commercially procured full access. The costs of accessing these data may, however, be prohibitively expensive to all but a few users. By contrast, Open Data are distributed under very different licensing conditions, typically enabling data to be supplied without cost, and to be reused and re-distributed without downstream licensing implications. Within some countries, an Open Data license has a more formal definition; for example, the UK adopts an Open Government License (https://www.nationalarchives.gov.uk/doc/open-government-license/version/3/) for officially defined Open Data.

There are several common rationales given for the release of Open Data. The first is to provide a resource that can enhance civic engagement in the processes of governance. For example, through the provision of data about the expenses of government employees, these are open to scrutiny and oversight. Secondly, Open Data can be integrated into platforms design to improve aspects of public service (e.g., school and healthcare comparison). Finally, Open Data can act as a driver for innovation and has the potential to create both direct and indirect economic benefits. Despite such diverse potential benefits, the release of Open Data is however not free, as the preparation, maintenance, and hosting of data assets have costs attributed (Spielman and Singleton 2015; Johnson et al. 2017). Furthermore, their release or availability is often governed by complex political data economies. For example, the permanence of Open Data can be somewhat illusionary, and there are examples of where Open-Data licenses have been revoked retrospectively and for future releases, or where guidance associated with such a license has been adapted so that this constrains future use. In the USA, the removal of the website open.whitehouse.gov followed the election of Donald Trump; and in the UK, the Land Registry switched its policies for data previously distributed with an Open Government License to terms that are more restricted.

15.1.1 Open Data Platforms

Within many municipalities, Open Data are disseminated through online portals, with two popular platforms including Socrata (https://www.tylertech.com/products/socrata) and CKAN (https://ckan.org/). An example of an Open Data platform running CKAN is shown in Fig. 15.1.

There are a number of reasons why such data portals provide better tools for transparency over simply sharing data through a static Web site. Most platforms provide access to search, highlighting the breadth of the available data; and results

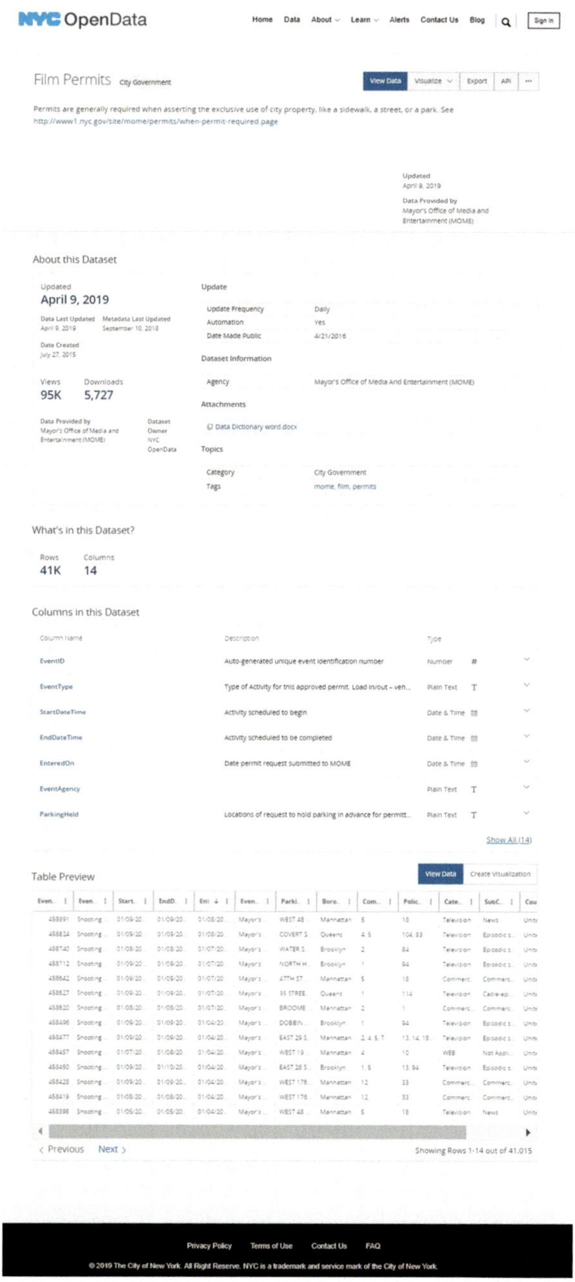

Fig. 15.1 Open Data portal for New York City showing a catalog entry for film permits

are typically returned alongside detailed metadata, sample extracts and some limited visualization capability. With many portals, data sit within a database that, in addition to being presented to the catalog's visual interface, are often also made available through publicly accessible application programming interfaces (API), enabling integration into a wide variety of software and tools. Such API endpoints and associated document object identifiers (DOIs) provide permanent and direct links to Open Data that enhance both usability and reproducibility.

However, the extent to which a community can benefit from engagement with sources of Open Data or those platforms designed to turn these assets into information can be variable; and differences may manifest between social, racial, ethnic, and economic groups. Mitigating access differentiation has to be a priority in urban governance if the implementation of Open-Data systems is to be maximized in the interests of the public good.

However, it is important to recognize that the creation of effective Open Data platforms requires significant investment. Organizationally, it is complex to initiate buy-in from stakeholder data owners, and additionally to facilitate the creation of effective management, storage, dissemination, outreach, and training associated with such new data infrastructure investments. Glasgow, which is the largest city in Scotland, was the recipient of £24 m of government funding to deliver a Future Cities demonstrator project (Sarf 2015). Around £7 m of this investment was allocated to build "Open Glasgow," which is a data platform providing access to numerous and previously siloed urban data. The project involved making 372 different datasets available through a CKAN-based Open Data portal alongside an online mapping platform provided by Esri. Around 21 different roles were associated with this project, and beyond the technical implementation, included additional support for Open Data development, engagement, and hackathons.

15.1.2 Open Data and Accountability

The growing adoption of Open Data platforms is a positive development, but in and of themselves these platforms have little impact on the lives of citizens. To have an impact, Open Data platforms have to be used by people and organizations. This means that the usability and accessibility of the platform itself are essential, but more importantly, it means within either the city agencies or the public at large, there must be constituencies who have the skills and time to transform the data assets into information.

The potential benefit of Open Data is only realized if certain conditions are met. We argue that Open Data repositories for urban governance should follow a set of principles that are accepted by scientific communities. These are sometimes referred to as the FAIR principles: findable, accessible, interoperable, and reusable.

- Findable: Data are published to stable and publicly accessible URLs. The URL is advertised and made known within the government and across agencies.

- Accessible: Data should be published in a usable format with stable and well-documented procedures for access. For example, pdfs are not a usable format for data. Access protocols should be well-documented and standardized; for example, APIs should remain stable over time. Individual data files should have a static URL. Data have to be documented and documentation must be maintained.
- Interoperable: Data should be organized such that linkages between data sets, and/or over time, are possible.
- Reusable: Data should have licensing provisions that allow flexible reuse of data.

In an effort to boost engagement some data-savvy communities sponsor events to encourage public consumption of the data published on open platforms. A consortium in New York City regularly organizes events around Open Data. For example, in Boulder, CO, USA, the city sponsored an "Art of Data" exhibition which encouraged local artists to create physical works of art from digital data. Some forms of digital data, such as text, can be difficult to work with in traditional forms of analysis—in the City of Boulder's Art of Data Exhibit, one artist built an installation based on individuals' test responses to survey questions about safety and other aspects of city life. Creative use of public data can be strikingly impactful. However, getting residents or the public, private, and not-for-profit sectors to use Open Data, and to communicate their findings to a broader audience, can be difficult yet is critical to closing the loop and allowing Open Data platforms to achieve their potential. Incentivizing creative use of data seems like a wonderful way to spur innovation; however, the lack of well-established norms of use and goals for Open Data platforms inhibits the impact of these resources.

We believe that the most impactful uses of public data focus on accountability; that is, using data to track progress toward institutional, individual, or collectively defined goals. However, there are not well-established models around how Open Data platforms might be integrated with participatory social and political processes to guide and track progress at the city-scale. Identifying and tracking progress toward goals can be non-trivial in the urban context.

Cities are large and complex systems bureaucratically, physically, and socially. Developing an understanding of the components and their interrelationships within systems is enormously difficult. For the average citizen, it can be hard to know where a city's responsibilities begin or end and observing the scope of a city's operations in a particular domain can be very difficult. Cities are a patchwork of public and private land, with city agencies often having overlapping jurisdictions and conflicting priorities. For example, a transportation department might want to increase the number of vehicles moving through an intersection and the planning department might want to improve pedestrian safety by reducing traffic volume. Given such organizational complexity, assessing accountability and progress toward goals can be complex. Goals may not be shared between various parts of the city's administrative structure. Moreover, the institutional goals may not be shared by the residents of the city, and in some communities, residents may have different priorities than others.

Open Data potentially simplifies some of this complexity by providing citizens and other interested groups with mechanisms to observe these large systems and to understand where cities are, and are not, investing resources. That is, if the right data are made available at the right level of aggregation, citizens can begin to observe the city not just as the space within which their daily activities take place but as an organizational unit.

Here we focus on the conditions required to realize the potential for Open Data to improve governance and in particular to drive accountability; in this context, using data systems to track progress toward measurable social and organizational goals. While ideally, these goals would emerge from participatory public processes, we omit discussion of these mechanisms here.

15.1.3 Why Are Goals Important?

Simply stated, the concept of accountability as applied to public data is that citizens (and municipal leaders) can hold public-sector agencies accountable for their work. However, large and complex projects that are undertaken without clear goals can be difficult to assess. For example, consider the partnership between Kansas City, Google, Sprint, and Cisco to develop a highly instrumented corridor with WiFi and advanced traffic control systems. In spite of millions of dollars in investment, it is difficult to say whether the project has been successful. The media report that the project reduced travel time an average of 37 s. Sprint, as a company, harvested data from thousands of citizens. But did the project achieve its goals? Was it a success? If so, for whom? Without clearly stated and measurable criteria, it is difficult to answer such questions.

A framework of accountability can, however, have powerful and positive social impacts. When police departments around the USA started to publish data about the racial characteristics of people they stop and question, glaring social inequalities were laid bare. In cities across the USA, data highlighted and confirmed the long-running perception that racial minorities in the USA are disproportionately targeted by the police. The use of Open Data to hold police departments accountable for seemingly biased patterns of enforcement is an excellent example of citizen empowerment in the challenge of existing doctrines. Our implicit goals in this example refer to widely held beliefs around how public institutions ought to function; for example, that enforcement of laws should be uniformly applied, not based on race or class.

15.1.4 Dashboards and Performance Indicators

Open Data dashboards simply make data or information available to municipal stakeholders. Data in their raw form are only consumable by people with those technical skills (and time) to both effectively frame questions and then investigate. Dashboard interfaces provide a more widely accessible visual interface to data. Often, a dashboard will display indicators that are derived from data. An indicator can be simple and direct, such as the number of traffic citations written in the preceding 30 days, or complex and derived such as the social vulnerability of the population. Kitchin et al. (2015) document the spread of the dashboard and its increasingly widespread use around the world. They critically argued that rather than simply "reflecting cities, [dashboards] actively frame and produce them." Whether they are mirrors reflecting data or instruments of power seems secondary to the fact that dashboards are widely used, and in governance, they can be used productively or unproductively.

In and of themselves, dashboards accomplish very little. They find their utility through linkage with implicit or explicit social goals and incorporation into some governmental process that links action (or incentives) to the indicators on the dashboard. A dashboard that simply displayed data, disconnected from meaningful administrative or social goals, would have little impact. For example, to provide insight into racial bias, the police department in Minneapolis, Minnesota, USA, publishes a dashboard breaking down police stops by race, location, gender, and age (https://www.insidempd.com/datadashboard/); while this dashboard is not linked to explicit goals and targets, it is squarely addressing implicit social goals. On the other end of the spectrum, the City of Boulder, Colorado, USA, uses a dashboard to track progress toward explicitly stated targets around safety, health, livability, sustainability, housing, and governance (Fig. 15.2). While rudimentary, the dashboard uses a simple system of green checks for targets that are met and red exclamation points for missed goals. A public process determined the indicators to be tracked on the dashboard; these were derived from the city's "Sustainability and Resilience Framework" which was a document designed to guide "budgeting and planning processes by providing consistent goals necessary to achieve Boulder's vision of a great community and the actions required to achieve them" (https://www-static.bouldercolorado.gov/docs/Sustainability_+_Resilience_Framework-1-201811061047.pdf).

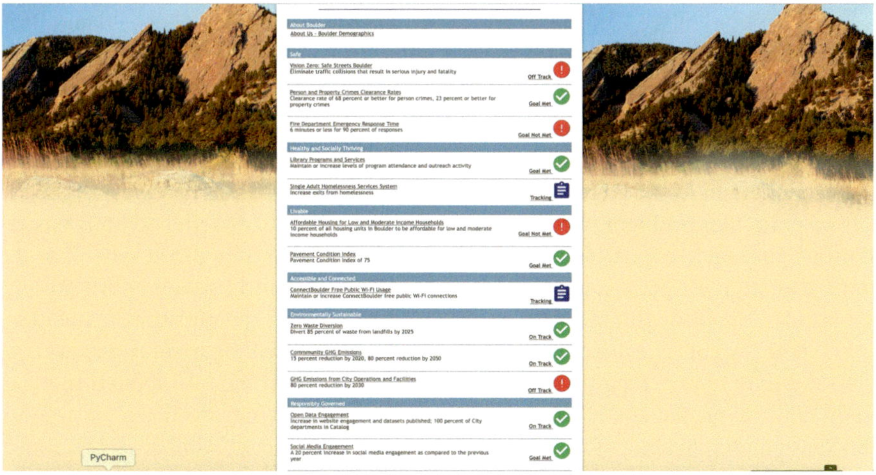

Fig. 15.2 A goal-based dashboard from the City of Boulder, Colorado, USA

The use of quantitative targets, such as those employed by Boulder, is a widespread practice in the private sector where such indicators are sometimes called key performance indicators (KPIs). Performance indicators are powerful tools in so far as several criteria are met:

- Urban KPIs must measure the right things. That is, they must quantify social, political, or economic processes of interest to the leaders and residents of the city.
- Urban KPIs must be actionable; measuring things that residents and leaders have no power to change is of no consequence. Dashboards should in some meaningful way drive action.
- Urban KPIs must be correctly measured: Data quality is a serious concern for public dashboards. Linking data to public goals creates incentives to manipulate or misreport data.
- Not all goals are quantifiable: It is important that KPIs and dashboards play an appropriate role. Critical social goals, such as well-being, may be unmeasurable but this does not mean that public institutions should not strive toward them.

There are, however, critiques of dashboards and urban data more broadly, notwithstanding that it seems to us that they are rooted in a genuine effort to provide transparency and accountability. While data may be imperfect and the social processes that produce them may be loaded and flawed, we strongly argue that providing access to information is better than not. Dashboards, when made public, reflect a kind of self-imposed, publicly stated accountability toward targets. While it is true that measuring what matters to the residents of a city is a non-trivial exercise, and that data systems are more likely to reflect things that can be measured than things of direct concern to residents, there is some meaningful overlap. It is within this space of overlap where data can help advance the governance of cities.

15.2 Algorithmic Decision-Making

There is a proliferation of increasingly granular measures or insights that can be extracted from urban data, which is necessitating new methods for both their management and their analysis. Algorithms are computational processes that are designed to solve a particular problem, which within an urban context can relate to both aspects of urban analytics (e.g., which communities are best served by green space), or the implementation of operational models (e.g., traffic light control systems). Algorithms can also have differing degrees of autonomy through their specification, estimation, or implementation. The use of computational algorithms within urban contexts is not new, and they have a lengthy history of application, from models applied to make predictions about the spatial organization of human activities, to those teasing out geodemographic structure from multidimensional spatial data (Webber 1975), alongside those which have been implemented operationally to guide decision-making (Foot 1982).

15.2.1 Positioning Algorithms

The argument is made that the successful implementation of algorithms can augment or supplant human expertise. For example, a fire inspector may have knowledge of the city in which he or she works and might choose buildings to inspect based on his or her expertise. Alternatively, an algorithm might rank buildings based upon the probability that they contain a building code violation. In one realization of an algorithmic process, an inspector could be dispatched to all buildings scored as risky by the algorithm. Alternatively, the algorithm could augment the inspector's expertise, providing him or her with a way to guide attention. In either case, the use of algorithms in law enforcement raises questions about the biases, fairness, and transparency in algorithms, especially when algorithms are trained or validated based on historically biased enforcement actions.

We believe that there are three broad use cases for models and algorithms in urban governance. By models, we mean tools that use learned or estimated parameters to produce classifications, probabilities, or scores. Algorithms are computational procedures that may or may not involve data and models. We use the two terms somewhat interchangeably, preferring the term "algorithmic decision-making" to refer to the use of computation to augment municipal operations. The use cases for algorithmic decision-making are:

- Augmentation: This refers to the use of models to guide or enhance human expertise. For example, using machine learning to augment the building inspector's expertise and to help focus efforts on buildings likely to contain a violation.
- Replacement: Using an algorithm in place of a human: for example, using combinations of cameras and radar to automate traffic enforcement. In this case, the machines determine if a violation occurred and take action. The computational enforcement system replaces a human system.
- Efficiency: Using models or algorithms to manage urban systems. Computation enables a kind of dynamic optimization that is difficult in the absence of sophisticated systems. For example, heating, ventilation, and air conditioning systems in buildings may take into account occupancy, outdoor temperatures, historical norms, and other factors. Transport systems may make small adjustments to signal timing system-wide in order to continuously adapt to variations in traffic and demand, thus optimizing flow.

At their best, across these use cases, algorithms potentially present an unbiased way to improve public welfare and the operation of cities. That is, well-designed systems can make people safer and urban systems more efficient. Machines potentially remove individual biases and capacities from urban management and enforcement. When algorithms and models are transparent and interpretable by humans, they move decisions out of the subjective and political domain into the public sphere. Open algorithms and models can also force conversations about principles, such as what kinds of actions or places should be targeted, or what publicly generated training or validation data should be used. Such models can then embed these collectively generated principles. Enforcement actions are then the result of a public process around the kinds of factors that contribute to risk or that the community wants to minimize or maximize.

At their worst, algorithms could become super enforcers of institutional biases and racism, and reinforce existing structural inequalities, or at the extreme create new ones. When algorithms replace humans (or are positioned at the extremes of augmentation), there are valid concerns that the system-automated surveillance that emerges violates basic human rights to privacy and equal (unbiased) enforcement of laws. For example, it is not possible to place surveillance cameras everywhere; from the perspective of a police department, placing cameras in high-crime areas might be an efficient use of limited resources. However, if algorithmic tools are used to augment enforcement or replace policing it means that people in high-crime areas have a higher probability of being found guilty of crimes than those in areas without cameras, even if algorithms are fair and unbiased.

15.2.2 Challenges for Operationalizing Algorithms

Unlike inferential models that have historically been applied within urban contexts, many contemporary and emerging methods from the cannon of data science, AI, and machine learning focus instead on prediction, which produces models with operational utility, but because the structural manifestations of causal effects are often hidden, their value can be argued as limited in terms of explaining how processes operate over time and space, and as such, we have weaker understanding of the dynamics of systems. Although we may be able to make very good forecasts from such new modeling paradigms, this is in tension with generalizable models of how the world functions, and the development of theory.

Additionally, many new algorithms that are used to create predictions rely on big data that are used to train models, which is the process by which an algorithm learns from the past to make new or future predictions. However, in doing so, an analyst has to be certain that there are no systematic biases in such data, and that any measures taken are likely to be stable over time. The non-compliance of such issues has been argued as integral to cases where previously successful models stop making effective predictions: for example, inaccuracies in magnitudes predicted by Google Flu Trends (Lazer et al. 2014).

Beyond issues of measurement, it has also been noted that most if not all big data are socially constructed, which also leads to potential bias, and should drive ethical considerations and framing. If such data are integral to the function of algorithms, and those decisions that they advise or take, the algorithms themselves can inherit such same bias; and as such may ensue real-world implications if adopted uncritically (Kitchin 2014). For example, the content of social-media data is only representative of those people who generate it, and so may under- or over-represent certain socioeconomic or demographic characteristics; or for georeferenced data, accuracy may be impacted by both where the social-media data were collected (e.g., the built environment impacting GPS signal reflection) or by people's prevailing attitudes to location sharing. More generally, crowdsourcing refers to the process of the public contributing attributes of observed phenomena for some particular purpose. Such data collection does not have an a-priori sample design, and as such the data's underlying collection is influenced by those who engage with a project. For example, the Street Bump (https://www.streetbump.org/) application was created for the city of Boston, USA, and collected data using the accelerometer in phones when a depression in a car was recorded as it passed over a pothole. These readings were pooled and analyzed to identify where remedial action may be required on a street. The representativeness of such data was, however, bound up in the collection process, with the application only being available to those with an iPhone, those who could afford one of these handsets, and additionally a subsection of this population who would be likely to install the application, and additionally volunteer geolocated information. Such a segment of the population may also have particular travel patterns, and there is additionally potential that only a partial survey of the city is conducted through such a tool. Understanding such bias and how this might impact algorithmic governance is a fundamental issue that should be considered by decision-makers.

15.3 Conclusion

In this chapter, we have outlined how the processes and operationalization of urban governance are being enhanced and challenged through the emergence of new digital technologies that relate to the instrumentation of cities, how those data being generated, and how the information derived can be used within urban contexts to enhance decision-making. For digital urban governance to be effective we posit that the inclusion of stakeholders by design, aligned to principles of transparency and openness, is essential in order to mitigate risks of associated negative dystopian consequences. The power of new digital frameworks has great potential to improve the health, prosperity, inclusivity, and sustainability of cities; yet it is essential that these technologies do not end up reinforcing past injustices, or at their most extreme create new inequalities. Future cities will be digitally augmented, and the challenge for us now is to critically reflect on the impacts that ensue from these new technologies, and to make sure we plan for a future that we want.

References

Foot D (1982) Operational urban models. Taylor and Francis, London

Johnson PA, Sieber R, Scassa T, Stephens M, Robinson P (2017) The cost(s) of geospatial open data. Trans GIS 21(3):434–445

Kitchin R (2014) The real-time city? Big data smart Urbanism. GeoJournal 79(1):1–14

Kitchin R, Lauriault TP, McArdle G (2015) Knowing and governing cities through urban indicators, city benchmarking and real-time dashboards. Reg Stud Reg Sci 2(1):6–28

Lazer D, Kennedy R, King G, Vespignani A (2014) The parable of Google Flu: traps in big data analysis. Science 343(6176):1203–1205

Sarf N (2015) Learning from future cities demonstrators. https://innovateuk.blog.gov.uk/2015/10/29/learning-from-future-cities-demonstrators/. Accessed 22 May 2015

Spielman SE, Singleton A (2015) Studying neighborhoods using uncertain data from the American Community Survey: a contextual approach. Annals Am Assoc Geogr 105(5):1003–1025

UN Habitat (2004) Transparency in local governance. United Nations Human Settlements Programme. Nairobi, Kenya. https://unhabitat.org/books/tools-to-support-transparency-in-local-governance/. Accessed 10 April 2019

Webber RJ (1975) Liverpool social area study, 1971 Data. PRAG Technical Paper 14. Centre for Environmental Studies, London

Alex D. Singleton is a Professor of Geographic Information Science at the University of Liverpool. He has been PI or Co-I to over £14m of research income and is Deputy Director of the ESRC Consumer Data Research Centre (CDRC) and Director of the ESRC Data Analytics & Society CDT. His research is concerned with how the complexities of individual behavior, attitudes, and contexts manifest spatially and can be represented and understood through the framework of Geographic Data Science.

Seth E. Spielman is an Associate Professor of Geography and Information Science at the University of Colorado where he also serves as the Assistant Vice Chancellor for Data Science and Strategy. His research focuses on how to create and reason with digital representations of social and economic space.

Chapter 16
Urban Pollution

Janet E. Nichol, Muhammad Bilal, Majid Nazeer, and Man Sing Wong

Abstract This chapter depicts the state of the art in remote sensing for urban pollution monitoring, including urban heat islands, urban air quality, and water quality around urban coastlines. Recent developments in spatial and temporal resolutions of modern sensors, and in retrieval methodologies and gap-filling routines, have increased the applicability of remote sensing for urban areas. However, capturing the spatial heterogeneity of urban areas is still challenging, given the spatial resolution limitations of aerosol retrieval algorithms for air-quality monitoring, and of modern thermal sensors for urban heat island analysis. For urban coastal applications, water-quality parameters can now be retrieved with adequate spatial and temporal detail even for localized phenomena such as algal blooms, pollution plumes, and point pollution sources. The chapter reviews the main sensors used, and developments in retrieval algorithms. For urban air quality the MODIS Dark Target (DT), Deep Blue (DB), and the merged DT/DB algorithms are evaluated. For urban heat island and urban climatic analysis using coarse- and medium- resolution thermal sensors, MODIS, Landsat, and ASTER are evaluated. For water-quality monitoring, medium spatial resolution sensors including Landsat, HJ1A/B, and Sentinel 2, are evaluated as potential replacements for expensive routine ship-borne monitoring.

J. E. Nichol (✉) · M. S. Wong
Department of Land Surveying and Geo-Informatics, The Hong Kong Polytechnic University, Hong Kong, China
e-mail: janet.nichol@connect.polyu.hk

J. E. Nichol
Department of Geography, University of Sussex, Brighton, UK

M. Bilal
School of Marine Sciences, Nanjing University of Information Science and Technology, Nanjing, China

M. Nazeer
Key Laboratory of Digital Land and Resources, East China University of Technology, Nanchang, China

Earth and Atmospheric Remote Sensing Lab (EARL), Islamabad, Pakistan

W. Shi et al. (eds.), *Urban Informatics*, The Urban Book Series,
https://doi.org/10.1007/978-981-15-8983-6_16

16.1 Monitoring Air Quality in Urban Areas

The gathering of air-quality data for urban areas and their source regions is a major challenge because the large areas involved cannot be represented by ground stations. Although satellite sensing systems and methodologies have recently been developed with an adequate spectral and temporal resolution for monitoring aerosols, it is difficult to obtain fine spatial resolution because the atmospheric signal being sensed is only a small proportion of the total image reflectance; thus large areas corresponding to large pixels, giving a higher measurable signal, are required.

The most accessible remotely sensed parameter of air quality is aerosol optical depth (AOD). This is a unit-less measure of the total amount of aerosol in the atmospheric column and is based on the opacity of the atmosphere in a particular waveband. There is no general algorithm which can retrieve aerosol properties over every kind of surface. Instead, different algorithms have been developed for (i) water, (ii) dark vegetation, (iii) bright surfaces, and (iv) heterogeneous land surfaces respectively, the latter two of which include urban surfaces. However, techniques for retrieving aerosol over low-reflecting surfaces of water and vegetation are better developed than those over land, because assumptions can be made that the surface reflectance is either zero or near zero. Based on this, Kaufman and Tanré (1988) developed an algorithm which first uses the NDVI (Normalized Difference Vegetation Index) to detect dense dark vegetation (DDV) pixels, then used the short-wave infrared (SWIR, 2.1 μm) band, which is not affected by aerosol, to obtain the surface reflectance for the DDV pixels. Then based on the relationship

$$\begin{aligned} Lsurf_{0.49} &= 0.25 * Lsurf_{2.1} \\ Lsurf_{0.66} &= 0.5 * Lsurf_{2.1} \quad \text{(Kaufman and Sendra 1988)}, \end{aligned}$$

the apparent surface reflectance in the blue (0.49 μm) and red (0.66 μm) bands can be obtained. The difference between the actual surface reflectance in these bands and the observed (top of the atmosphere, TOA) reflectance is assumed to be due to aerosol. This amount is then fitted to a best-fit aerosol model, with knowledge of the expected aerosol types in the study area—for example, continental, industrial/urban, biomass burning, and marine—to arrive at AOD from the image blue and red wavebands.

From this DDV concept, NASA developed the MODIS Dark Target (DT) AOD product (MOD04; Kaufman and Tanré 1998) covering the globe. Although the DT product at 10 km spatial resolution only provides meaningful depictions on a broad regional scale, it is capable of giving an overview of air-quality conditions prevailing over a city's region. The expected error (EE) of the DT algorithm is $\pm$ $(0.05 + 0.15 \times \text{AOD})$ (Levy et al. 2013), which represents about 66% of retrievals within the EE on a global scale (Levy et al. 2010). The most recent version of the DT algorithm is the MODIS Collection 6.1 (C6.1) AOD product (Bilal et al. 2018a; Gupta et al. 2016). The C6.1 product addresses uncertainties due to the heterogeneity of urban surfaces, and updates the surface reflectance ratios using NASA's MOD09 surface reflectance product, which newly incorporates information on land cover type for pixels with urban cover > 20% (Gupta et al. 2016). The Deep Blue (DB) AOD retrieval algorithm (Hsu et al. 2004) provides estimates of AOD over bright urban and desert, as well as dark surfaces, using the deep blue channels 412 and 470 μm in which these surfaces appear dark, as well as the red channel (0.65 μm) for dark surfaces. The EE of DB depends on geometry (Hsu et al. 2013; Sayer et al. 2013). The MODIS C6 product (including DT and DB algorithms) has been evaluated over urban areas with varying accuracies. For example, over Beijing, both the DT and DB C6 products (MOD04 and MYD04) were found to overestimate during highly polluted days due to a large error in the surface reflectance estimation (Bilal and Nichol 2015; Tao et al. 2015).

Within C6, a combined DT/DB algorithm has also been produced at 10 km, which combines both DT and DB algorithms in the same image, to retrieve AOD over both dark and bright surfaces including urban areas (Levy et al. 2013). However, accuracy over Asian cities was observed to be low, with only 57% of retrievals falling within the expected error. Bilal et al. (2017) introduced a customized algorithm which specifies the use of the DB algorithm when NDVI > 0.3, which cancels out the tendency of the DT and DB algorithms respectively, to under- and overestimate the surface reflectance, and which improved the percentage of retrievals within the expected error to 65%.

Although both DT and DB algorithms use MODIS 500 m resolution wavebands, their AOD products are produced at the spatial resolution of 10 km because the 500 m pixels are amalgamated into windows of 20 × 20 (400) pixels to increase the signal-to-noise ratio. Then, to eliminate clouds and water surfaces, dark and bright pixels, which are unsuitable for retrieval of AOD, are deselected, with at most 120 pixels remaining. Because the MODIS DT and DB products are unable to resolve city-level features, the MODIS aerosol team produced a global DT product at 3 km, the MOD04_3K/MYD04_3K, within the operational C6 aerosol product (Remer et al. 2013). Comparison with AERONET (AErosolROboticNETwork) ground stations suggests that the MOD_3K is less reliable than the 10 km products (Bilal et al. 2018b). This may be because only a maximum of 11 pixels remain in the deselection window, making the product noisier than that at 10 km.

Yang et al. (2018) conducted a preliminary investigation of an AOD product at 1 km resolution using the geostationary Advanced Himawari Imager (AHI) satellite, based on the DT algorithm, with results showing some overestimation compared to AERONET data, with a correlation coefficient of 0.83 and RMSE of 0.11. Due to the recent availability of AHI, the AOD retrievals could not be thoroughly evaluated but are considered promising. In view of the superior temporal resolution of geostationary satellites (10-minutes for AHI), along with future improvement in spatial resolution, semi-continuous monitoring of particulate concentrations at the city district scale will be possible.

Contributions of the DB and DT retrievals to future global aerosol monitoring projects such as ESA's EarthCARE mission (Illingworth et al. 2015), with 10 km radar and LiDAR, WMO's GALION project, a ground-based aerosol LiDAR system (Bösenberg et al. 2008), ESA's ADM-AEOLUS mission, a space-based wind profiler system launched in 2018 (Lolli et al. 2013), and NASA's on-going CALIPSO mission with satellite-based aerosol LiDAR (Winker et al. 2010), will be very important.

As with AOD retrieval, the estimation of other gaseous pollutants from satellite-image wavebands is constrained by the weakness of the signal relative to the total image reflectance, thus necessitating large pixel sizes. The MOPITT (Measurement of Pollution in the Troposphere) sensor, which measures CO emissions from the Earth's surface, has 22 km spatial resolution at nadir, and OMI (Ozone Monitoring Instrument) for ozone and NO_2 estimation with a spatial resolution of 13 km × 24 km, are not readily applicable for retrieval of urban-scale pollutant concentrations. Although Bechle (2013) found that the OMI sensor aboard NASA's Aura satellite was able to measure spatial variability in NO_2 exposure over a large urban area, detailed district-level concentrations were constrained by the coarse resolution of the sensor. These constraints have been lessened somewhat by the TROPOMI sensor onboard the European Space Agency's Sentinel 5P satellite launched in October 2017, which measures ozone, NO_2, SO_2, methane, and CO at 7 km × 3.5 km resolution. However, this is still too coarse for application at urban scales, and since algorithms developed for complex land areas are difficult to apply, the task of deriving accurate air-quality products for urban areas remains challenging.

16.2 Remote Sensing of the Urban Heat Island

Urban heat islands are caused by the replacement of natural evaporative and porous land surfaces with non-evaporative human-made surfaces (Chandler 1965). These disperse a much greater proportion of energy received into the surrounding atmosphere as sensible heat, compared with the predominantly latent heat loss of rural surfaces. Along with the generally lower albedo of urban surfaces, this results in significantly higher air temperatures in cities compared with their rural surroundings, and the difference ($\Delta T(u\text{-}r)$) reaches a maximum at night. As most cities have few air-monitoring stations, the level of detail of intra-city temperatures is inadequate, whereas satellite thermal data provide a dense grid of continuous and time-synchronized land surface temperatures (LSTs) over a whole city. Since cities are identifiable on thermal satellite images for their temperature contrasts, as much as for their optical differences with surrounding rural areas, many remote-sensing studies have taken place (Roth et al. 1989; Weng 2009; Zhou et al. 2019). However, there are numerous constraints to the use of the data in urban climatology, which are discussed below.

16.2.1 Spatial Resolution of Satellite Sensors Related to Scales of Urban Climate

Due to the inverse relationship between wavelength and signal strength, longer-wavelength thermal infrared sensors generally have a coarse resolution. Therefore the thermal waveband of MODIS, at 1 km resolution, has only been used for general temperature-trend analysis over city regions (Bonafoni 2016; Hulley et al. 2014). The 60 m and 90 m resolution sensors of Landsats 5–7/8 and 90 m of ASTER have also been used for urban climatic analysis at the district and even the street scale within cities (Nichol 1996a; Nichol et al. 2009; Feng and Myint 2016; Meng et al. 2018). To overcome the limitation of spatial resolution, various ways of disaggregating the thermal signal to provide more spatial detail have been presented (Nichol 2009; Rodriguez-Galliano et al. 2012; Zhou et al. 2019). Figure 16.1 shows the effects of emissivity modulation on an ASTER thermal image of a suburban area of Hong Kong. The original resolution of 90 m (Fig. 16.1c) is disaggregated to a 10 m pixel size (Fig. 16.1a), while correcting for surface emissivity differences (Nichol et al. 2009).

16.2.2 Relationship Between Surface Temperature and Air Temperature

The conception as well as the usefulness of the UHI concept derives from its representation of urban air temperatures which affect human comfort. More specifically these are air temperatures within the urban canopy layer comprising the space within streets between the surface and the top of the buildings (Oke 1976). However, satellite thermal sensors measure the surface radiometric temperature or land surface temperature (LST). Thus, the surface heat island (SUHI) represents the radiometric temperature difference between urban and non-urban surfaces (Zhou et al. 2019). Since the satellite-derived heat island is based on LST, the optimum usefulness of these data depends on defining their relationship to a more conventional view of the urban heat island, such as screen-level air temperature at the time of imaging (Nichol et al. 2009; Schwarz et al. 2012; Clay et al. 2016). Li et al. (2018) developed an air-temperature dataset at 1 km resolution covering the entire USA by combining daily air-temperature data from weather stations with gap-filled MODIS LST data and an elevation model. The method proved satisfactory, generating root mean square errors of 2.1 and 1.9 °C, and R^2 of 0.95 and 0.97 for daily minimum and maximum air temperature, respectively. Sun et al. (2015) estimated air temperatures over Beijing from MODIS LST data combined with vegetation indices, obtaining accuracies of approximately 2°K compared with weather station data.

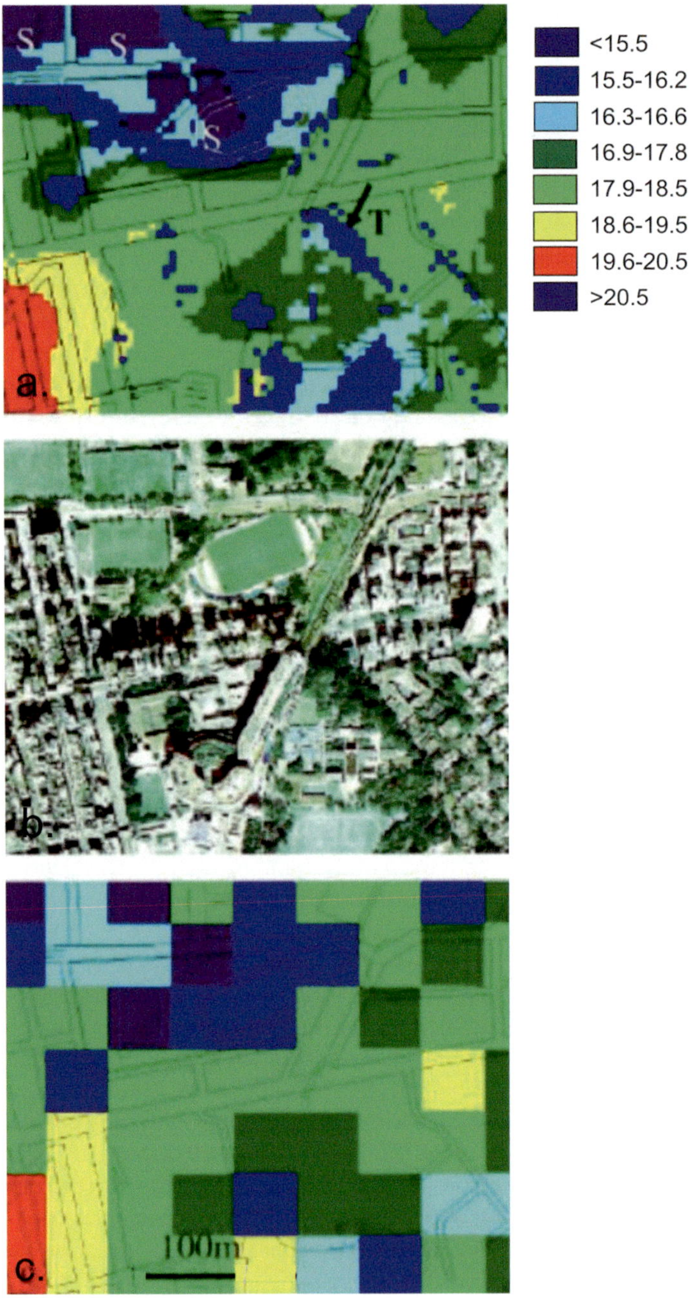

Fig. 16.1 Surface temperatures of a mixed urban/suburban district in Hong Kong from: **a** ASTER nighttime thermal image at 10.42 pm on 31.01.07 after emissivity modulation, **b** Aerial photograph showing land cover types, **c** Original ASTER thermal image with 90 m resolution

16.2.3 Time of Imaging in Relation to Heat Island Maximum

Most space-borne thermal sensors such as the Landsat series and ASTER record mainly during the daytime when densely built, high-rise areas may constitute a heat sink (Nichol 2005; Rasul et al. 2017). Tropical cities (Nichol 2003) or arid zones in summertime (Nassar et al. 2016; Rasul et al. 2017) may also exhibit heat sinks during the day. Furthermore, the timing of the satellite overpass may not be ideal for detecting temperature differences. Landsat for example at 9.30–10.30 am local time is near the morning thermal crossover time when minimal thermal contrasts would be expected. Differences in surface temperature are largest during the daytime, thus the surface heat island based on LST is more pronounced than that of the conventional UHI based on air temperature, for which the greatest differences are at night (Nichol 2005). Additionally, Sun et al. (2015) observed that LST was more similar to air temperatures within the urban canopy layer at night but considerably different during the day. The relationship may even be negative, as LST in urban districts increases due to early-morning warming, while high-rise urban districts in shadow when the sun angle is still low may constitute a heat sink (Nichol 2005).

In changing environmental conditions, satellite images taken at a single instant may be unrepresentative. However, Nichol and To (2012) found that in Hong Kong, due to a more stable boundary layer at night, nighttime ASTER thermal images were representative of commonly occurring climatic conditions for a 13-h period surrounding the image acquisition time, and were significantly correlated with ground air temperatures over the city, for 93% of hot summer nights.

16.2.4 Anisotropy of the Satellite View

Satellites record the temperature of horizontal surfaces, which may only represent the complete radiating surface in flat rural areas. The effective (active) surface area of a city, especially in high-rise areas, and using narrow field-of-view sensors, is much larger than the equivalent countryside of the same size (Voogt and Oke 1996). In high-rise housing estates in Singapore, for example, the active surface was found to be 1.7 times greater than the planimetric (satellite seen) surface (Nichol 1998). Thus nadir views would be warmer or cooler than off-nadir views depending on the sun position. Hu et al. (2016) quantified anisotropic effects for two high-rise cities—New York and Chicago—observing that daytime maximum temperature bias due to anisotropy was up to 9°K for the most urbanized areas. When averaged over the entire SUHI as measured by MODIS LST, the UHI magnitude was modified by 2.3°K, that is, 25–30%, due to surface anisotropy. Voogt and Oke (1996) recommended using ground-based observations to construction models for the weighting of temperatures according to area and sun position (see also Nichol et al. 2014).

16.2.5 The Need for Emissivity and Atmospheric Correction

Although satellite-derived radiance values can readily be converted to equivalent black-body temperature (or brightness temperature) using Planck's Law, this under-estimates the surface radiometric temperature if corrections for emissivity differences according to the type of land cover are not carried out. For example, a metal roof of emissivity 0.92, and tile roof of emissivity 0.98, both with a radiometric temperature of 27 °C, will have brightness temperature (image) values of 20.8 and 25.5 °C respectively. However for UHI studies, measurement of individual surface temperatures is both impossible and unnecessary, as emitted radiation from each pixel is an aggregated value of all surfaces within the pixel, and subject to anisotropic effects according to look angle and the pixel's horizontal/vertical surface ratio. To address this, Yang et al. (2015, 2016) developed an urban emissivity model based on the sky view factor (SVF), which accounted for surface material type and building geometry, and found that a decrease in SVF was accompanied by increased emissivity due to multiple scattering among buildings. Another potential source of error in thermal image values is that they can only be considered accurate in clear, dry atmospheres, and a further correction using atmospheric data in a radiative transfer model such as MODRAN (Berk et al. 2014) should be made, if absolute temperatures are desirable. In humid atmospheres, energy absorption by atmospheric water vapor may account for brightness temperatures up to 15 °C cooler than the surface radiometric temperature (Nichol 1996b).

16.3 Monitoring Water Quality Along Urban Coastlines

Coastal waters are spatially complex, as they comprise a mixture of both saline and brackish water, as well as containing different types of land runoff. Urban coastlines are especially complex due to additional anthropogenic inputs, from both point and non-point sources, with often severe impacts on water quality (WQ). For this reason, WQ along urban coasts is subject to greater spatial and temporal variability than other coastlines, and WQ monitoring from remote-sensing platforms requires sensors with fine spatial as well as temporal resolution. A further challenge is due to the wide range of organic and inorganic inputs to urban coastal waters making them optically complex for ocean color monitoring. A common problem in countries with unregulated drainage is high nutrient inputs from agricultural, industrial, and urban waste, resulting in eutrophication and algal bloom events. These may be toxic to humans as well as affecting a wide variety of marine organisms.

Due to these factors, sensors frequently used for marine applications such as the Sea-Viewing Wide-Field-of-View Sensor (SeaWiFS), the Moderate Resolution Imaging Spectroradiometer (MODIS), the Visible and Infrared Imager/Radiometer Suite (VIIRS), the Geostationary Ocean Color Imager (GOCI), and the Ocean and Land Color Imager (OLCI), with spatial resolutions of several hundred meters, are unable to resolve the necessary spatial detail, although they may have good temporal and spectral resolutions. Recent space-based sensors with moderate resolution used for retrieval of water-quality indicators (WQIs) include NASA's Landsat, the Chinese HJ1 A/B, and ESA's Sentinel series. The most recent Landsat 8 carries the Operational Land Imager (OLI), with 9 spectral wavebands, 5 in the optical spectrum from 430–880 nm, which are being used for ocean color monitoring (Franz et al. 2015; Vanhellemont and Ruddick 2015). OLI has 30 m spatial resolution and a repeat cycle of 16 days, which is increased to 8 days if combined with Landsat 7. The MultiSpectral Instrument (MSI) on ESA's Sentinel-2 platform carries 12 wavebands, including three ocean color bands, blue (490 nm), green (560 nm) and red (665 nm) at 10 m resolution, and three Near InfraRed (NIR) bands (705–783 nm) at 20 m resolution. OLI has a 16-day repeat cycle.

Clear water shows low reflectance in the visible spectrum and absorbs most energy in the NIR region, but the optical properties of water are affected by a range of substances. These have given rise to the concept of ocean color sensing (Morel and Prieur 1977), as dissolved organic matter (DOM) is strongly absorptive in the blue (490 nm) spectral region, chlorophyll-a (Chl-a) in phytoplankton and algal pigments mainly absorbs sunlight in the blue and red regions of the spectrum, and suspended solids (SS) mainly reflect in the red and NIR regions (600–800 nm). Due to the difficulty of retrieving an adequate reflected signal from the water column which absorbs most light energy, the atmospheric component may be dominant unless it is first removed, thus atmospheric correction is an essential pre-processing step (Pahlevan et al. 2017). Algorithms for retrieval of WQIs from the water column have undergone refinement as the spatial and spectral resolutions of space-borne sensors and computing power have improved. Improvements in temporal resolution with more satellite sensors and more frequent repeat cycles have released more data for testing and validation of retrievals, which require close synchronization with sea-station data (Pahlevan et al. 2019). Algorithms for retrieval of WQPs are usually based on obtaining a substantial number of synchronous image and station samples for regression against image wavebands, and a further substantial number for validating the results.

For example, a study of water quality around the heavily urbanized coastlines of Hong Kong and the Pearl River Delta (PRD; Nazeer and Nichol 2016a) was able to obtain 240 co-located samples of Chl-a and SS within two hours of image acquisition when combining images from Landsat TM/ETM + and HJ1 A/B sensors over a 13-year period (2000 to 2012). However due to the complexity of the coastal waters, with PRD river sediments to the west, urban runoff in the central section, and clear waters of the South China Sea to the east, retrieval algorithms developed across the whole region were less accurate than those applied to individual water-quality zones delineated by fuzzy *c*-means clustering. Thus, for Chl-a a low root mean square error (RMSE) of 1.61 μg/l was obtained for individual water-quality zones compared with 4.59 μg/l when applied to the whole spectrum of different water types across the region. For SS concentrations, a significant improvement was also observed, with the RMSE reducing from 2.72 mg/l to 1.19 mg/l when the models were applied to individual zones. These results are good, considering the wide range of concentrations obtained in the ship-sampled datasets, namely a Chl-a range of 0.30 to 13.0 μg/l and SS concentration range of 0.5 to 56.0 mg/l, and suggest that space-borne sensors are capable of providing spatially detailed, accurate, and cost-effective water-quality status around urban coastlines.

With urbanization of coastlines, an increasing incidence of red tide events caused by massive algal blooms from high nutrient inputs is being seen around the world, but especially in rapidly urbanizing parts of Asia such as China and the Philippines (Azanza et al. 2008; Nazeer et al. 2017). Such events are toxic to the marine ecosystem and pose dangers to human health; thus, environmental authorities need timely and detailed information on their occurrence. However, since the occurrence of a red tide does not usually correspond with routine ship-borne water sampling missions (monthly in Hong Kong), many go undetected. In Hong Kong, which is a thriving international port but still has diverse coastal ecosystems, a severe red tide event from December 2015 to February 2016 saw 220 tons of fish kills reported (SCMP 2016). A remote sensing study of chlorophyll-a concentrations around the complex coastal waters of Hong Kong using Landsat TM/ETM + (Fig. 16.2; Nazeer and Nichol 2016b) observed that a ratio of the red (630–690 nm) with the square of the blue (450–520 nm) bands were most capable of representing actual Chl-a concentrations due to the differential response of the red and blue wavebands to the Chl-a signal. A correlation coefficient of 0.89 and mean absolute error (MAE) of 1.02 μg/l obtained for the study indicated a good degree of confidence in remote sensing for routine monitoring of red tide events along urban coastlines.

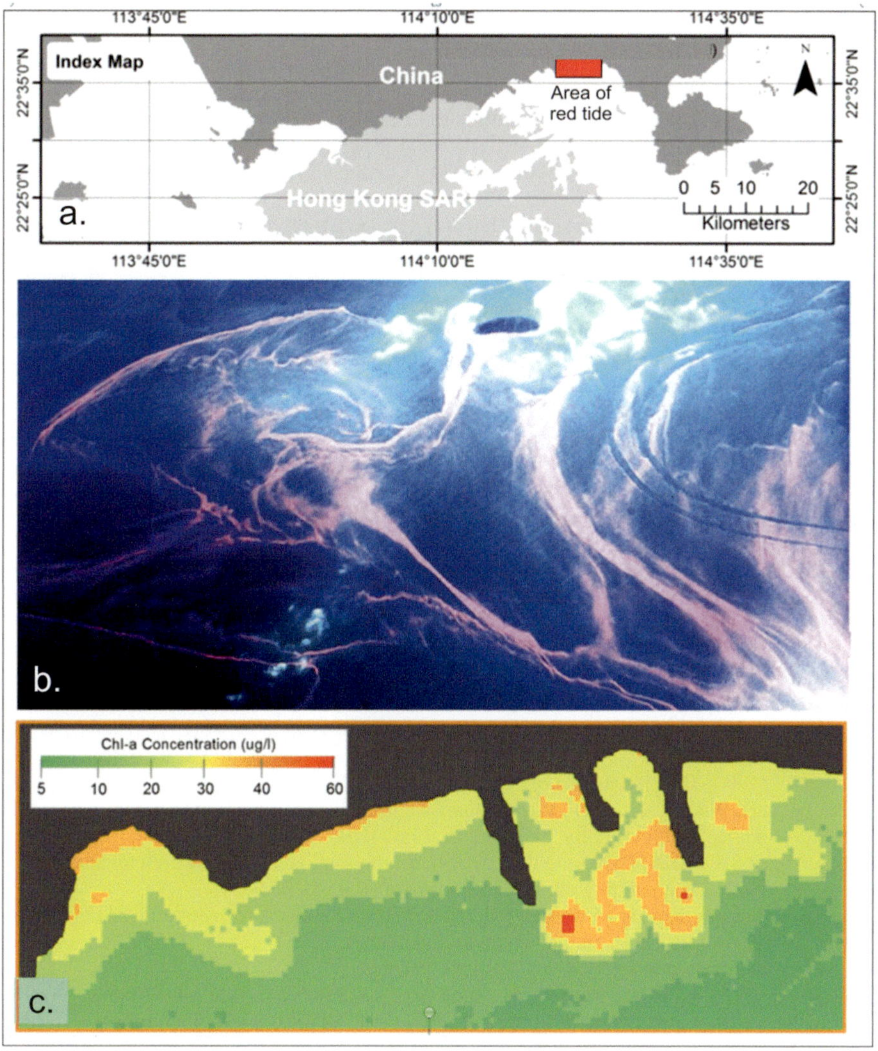

Fig. 16.2 Red tide along the Chinese coast adjacent to Hong Kong, on 25th November 2014. **a** Location of red tide, **b** Aerial photograph of red tide (photo credits Xinhua), **c** Chl-a concentration map in μg/l of red-tide-affected area using the ratio of Landsat/HJ1 blue (450–520 nm) and red bands(630–690 nm)

References

Azanza RV, David LT, Borja RT, Baula IU, Fukuyo Y (2008) An extensive Cochlodinium bloom along the western coast of Palawan, Philippines. Harmful Algae 7:324–330. https://doi.org/10.1016/J.HAL.2007.12.011

Bechle (2013) Remote sensing of exposure to NO_2: satellite versus ground-based measurement in a large urban area. Atmos Environ 69:345–353

Berk A, Conforti P, Kennett R, Perkins T, Hawes F, van den Bosch J (2014) MODTRAN6: a major upgrade of the MODTRAN radiative transfer code. In: Proceedings of SPIE 9088, Algorithms and Technologies for Multispectral, Hyperspectral, and Ultraspectral Imagery. p. 20: 90880H. https://doi.org/10.1117/12.2050433

Bilal M, Nichol JE (2015) Evaluation of MODIS aerosol retrieval algorithms over the Beijing-Tianjin-Hebei region during low to very high pollution events. J Geophy. Res. Atmos 120:7941–7957. https://doi.org/10.1002/2015JD023082

Bilal M, Nichol JE, Wang L (2017) New customized methods for improvement of the MODIS C6 dark target and deep blue merged aerosol product. Remote Sens 197:115–124. https://doi.org/10.1016/j.rse.2017.05.028

Bilal M, Nazeer M, Qiu Z, Ding X, Wei J (2018a) Global validation of MODIS C6 and C6.1 merged aerosol products over diverse vegetated surfaces. Remote Sens 10(3):475. https://doi.org/10.3390/rs10030475

Bilal M, Qiu Z, Campbell JR, Spak S, Shen X, Nazeer M (2018b) A new MODIS C6 dark target and deep blue merged aerosol product on a 3 km spatial grid. Remote Sens 10:463. https://doi.org/10.3390/rs10030463

Bonafoni S (2016) Downscaling of Landsat and MODIS land surface temperature over the heterogeneous urban area of Milan. IEEE J Sel Top. Appl Earth Obs Remote Sens 9:2019–2027

Bösenberg J, Hoff R, Ansmann A, Müller D, Antuña JC, Whiteman D, Sugimoto N, Apituley A, Hardesty M, Welton Jet al. (2008) GAW aerosol lidarobservation network (GALION). WMO, Geneva, Switzerland

Chandler TJ (1965) The climate of London. Hutchinson, London

Clay R, Guan H, Wild N, Bennett J, VinodkumarEwenz C (2016) Urban heat island traverses in the city of Adelaide, South Australia. Urban Clim. 17:89–101

Feng X, Myint SW (2016) Exploring the effect of neighboring land cover pattern on land surface temperature of central building objects. Build Environ 95:346–354

Franz BA, Bailey SW, Kuring N, Werdell PJ (2015) Ocean color measurements with the Operational Land Imager on Landsat-8: implementation and evaluation in SeaDAS. J Appl Remote Sens 9(1):096070. https://doi.org/10.1117/1.JRS.9.096070

Gupta P, Levy RC, Mattoo S, Remer LA, Munchak LA (2016) A surface reflectance scheme for retrieving aerosol optical depth over urban surfaces in MODIS Dark Target retrieval algorithm. Atmos Meas Tech 9:3293–3308. https://doi.org/10.5194/amt-9-3293-2016

Hsu NC, Tsay SC, King MD, Herman JR (2004) Aerosol properties over bright-reflecting source regions. IEEE Trans Geosci Remote Sens 42:557–569

Hsu NC Jeong MJ, Bettenhausen C, Sayer AM, Hansell R, Seftor CS, Huang J, Tsay SC (2013) Enhanced deep blue aerosol retrieval algorithm: the second generation. J Geophys Res Atmos 118:9296–3015

Hu L, Monaghan A, Voogt JA, Barlage M (2016) A first satellite-based observational assessment of urban thermal anisotropy. Remote Sens Environ 181:111–121

Hulley G, Veraverbeke S, Hook S (2014) Thermal-based techniques for land cover change detection using a new dynamic MODIS multispectral emissivity product (MOD21). Remote Sens Environ 140:755–765

Illingworth AJ, Barker HW, Beljaars A, Ceccaldi M, Chepfer H, Clerbaux N, Cole J, Delanoë J, Domenech C, Donovan DP et al (2015) The Earthcare satellite: The next step forward in global measurements of clouds, aerosols, precipitation, and radiation. Bull Am Meteor Soc 96:1311–1332

Kaufman YJ, Tanré D (1998) Algorithm for remote sensing of tropospheric aerosols from MODIS. NASA MODIS algorithm theoretical basis document. Goddard Space Flight Cent 85(1998):3–68

Levy RC, La Remer, Kleidman RG, Mattoo S, Ichoku C, Kahn R, Eck TF (2010) Global evaluation of the Collection 5 MODIS dark-target aerosol products over land. Atmos Chem Phys 10:10399–10420

Levy RC, Mattoo S, Munchak LA, Remer LA, Sayer AM, Patadia F, Hsu NC (2013) The Collection 6 MODIS aerosol products over land and ocean. Atmos Meas Tech 6:2989–3034

Li X, Zhou Y, Asrar GR, Zhu Z (2018) Developing a 1 km resolution daily air temperature dataset for urban and surrounding areas in the conterminous United States. Remote Sens Environ 215:74–84

Lolli S, Delaval A, Loth C, Garnier A, HFP (2013) 0.355-micrometer direct detection wind lidar under testing during a field campaign in consideration of ESA's ADM-Aeolus mission. Atmos Meas Tech 6:3349–3358

Meng Q, Zhang L, Sun Z, Meng F, Wang L, Sun Y (2018) Characterizing spatial and temporal trends of surface urban heat island effect in an urban main built-up area: A 12-year case study in Beijing, China. Remote Sens Environ 204:826–837

Morel A, Prieur L (1977) Analysis of variations in ocean color. Limnol Oceanogr 22:709–722. https://doi.org/10.4319/lo.1977.22.4.0709

Nassar AK, Blackburn GA, Whyatt JD (2016) Dynamics and controls of urban heat sink and island phenomena in a desert city: development of a local climate zone scheme using remotely-sensed inputs. Int J Appl Earth Obs Geoinf 51:76–90

Nazeer M, Nichol JE (2016a) Improved water quality retrieval by identifying optically unique water classes. J Hydrol 541:1119–1132. https://doi.org/10.1016/j.jhydrol.2016.08.020

Nazeer M, Nichol JE (2016b) Development and application of a remote sensing-based Chlorophyll-a concentration prediction model for complex coastal waters of Hong Kong. J Hydrol 532:80–89. https://doi.org/10.1016/j.jhydrol.2015.11.037

Nazeer M, Wong MS, Nichol JE (2017) A new approach for the estimation of phytoplankton cell counts associated with algal blooms. Sci Total Environ 590:125–138. https://doi.org/10.1016/j.scitotenv.2017.02.182

Nichol JE (1996a) High resolution surface temperature patterns related to urban morphology of in a tropical city: a satellite-based study. J Appl Meteorol 35(1):135–146

Nichol JE (1996b) Analysis of the urban thermal environment of Singapore using LANDSAT data. Environ Plan 23:733–747

Nichol JE (1998) Visualisation of urban surface temperatures derived from satellite images. Int J Remote Sens U.K. 19(9):1639–1649

Nichol JE (2003) Heat island studies in the third world cities using GIS and remote sensing. Chapter 13. In: MesevV (ed) Remotely sensed cities. Taylor and Francis

Nichol JE (2005) Remote sensing of urban heat islands by day and night. Photogram Eng Remote Sens 71(5):613–621

Nichol JE (2009) An emissivity modulation method for spatial enhancement of thermal satellite images in urban heat island analysis. Photogram Eng Remote Sens 75(5):547–556

Nichol JE, Fung WY, Lam KS, Wong MS (2009) Urban heat Island diagnosis using ASTER satellite images and 'in situ' air temperature. Atmos Res 94:276–284

Nichol JE, To PH (2012) Temporal characteristics of thermal satellite images for urban heat stress and heat island mapping. ISPRS J Photogram Remote Sens 74:152–162

Nichol J, Bilal M, Ashley WWS (2014) Retrieval of Aerosol Optical Thickness (AOT) from urban shadows using fine resolution WorldView-II images. In: Remote sensing and photogrammetry society annual conference, Aberystwyth, 02–05 Sept

Oke TR (1976) The distinction between canopy and boundary-layer heat islands. Atmosphere 14:268–277

Pahlevan N, Chittimalli SK, Balasubramanian SV, Vellucci V (2019) Sentinel-2/Landsat-8 product consistency and implications for monitoring aquatic systems. Remote Sens Environ 220:19–29. https://doi.org/10.1016/J.RSE.2018.10.027

Pahlevan N, Schott JR, Franz BA, Zibordi G, Markham B, Bailey S, Schaaf CB, Ondrusek M, Greb S, Strait CM (2017) Landsat 8 remote sensing reflectance (Rrs) products: Evaluations, intercomparisons, and enhancements. Remote Sens Environ 190:289–301. https://doi.org/10.1016/J.RSE.2016.12.030

Rasul A, Balzter H, Smith C, Remedios J, Adamu B, Sobrino JA, Srivanit M, Weng Q (2017) A review on remote sensing of urban heat and cool islands. Land 6:38. https://doi.org/10.3390/land6020038

Remer LA, Mattoo S, Levy RC, Munchak LA (2013) MODIS 3 km aerosol product: algorithm and global perspective. Atmos Meas Tech 6:1829–1844

Rodriguez-Galiano VF, Pardo-Igúzquiza E, Chica-Olmo M, Mateos J, Juan P, Rigol-Sánchez JP, Vega M (2012) A comparative assessment of different methods for Landsat 7/ETM + pansharpening. Int J Remote Sens 33(20):6574–6599. https://doi.org/10.1080/01431161.2012.691612

Roth M, Oke TR, Emery WJ (1989) Satellite derived urban heat islands from three coastal cities and the utilisation of such data in urban climatology. Int J Remote Sens 10(11):1699–1720

Sayer AM, Hsu NC, Bettenhausen C, Jeong MJ (2013) Validation and uncertainty estimates for MODIS collection 6 "deep blue" aerosol data. J Geophys Res Atmos 118:7864–7872

Schwarz N, Schlink U, Franck U, Grossmann K (2012) Relationship of land surface and air temperatures and its implications for quantifying urban heat island indicators—An application for the city of Leipzig (Germany). Ecol Ind 18:693–704

South China Morning Post (SCMP) 2016 Hong Kong fish farmers claim proliferation of red tides is worst 'UnnaturalDisaster' to hit industry in years, 30th April 2016

Sun H, Chen Y, Zhan W (2015) Comparing surface- and canopy-layer urban heat islands over Beijing using MODIS data. Int J Remote Sens 36:5448–5465

Tao M, Chen L, Wang Z, Tao J, Che H, Wang X, Wang Y (2015) Comparison and evaluation of the MODIS collection 6 aerosol data in China. J Geophys Res Atmos 120:6992–7005

Vanhellemont Q, Ruddick K (2015) Advantages of high quality SWIR bands for ocean colour processing: examples from Landsat-8. Remote Sens Environ 161:89–106. https://doi.org/10.1016/j.rse.2015.02.007

Voogt JA, Oke TR (1996) Complete urban surfacetemperatures. In: Proceedings of the 12th conference on biometeorology and aerobiology. American Meteorological Society, Atlanta, USA, pp 438–441

Weng Q (2009) Thermal infrared remote sensing for urban climate and environmental studies: Methods, applications, and trends. ISPRS J Photogram Remote Sens 64:335–344

Winker DM, Pelon JAC Jr, Ackerman SA, Charlson RJ, Colarco PR, Flamant P, Fu Q, Hoff RM, Kittaka C et al (2010) The calipso mission. Bull Am Meteor Soc 91:1211–1230

Yang F, Wang Y, Tao J, Wang Z, Fan M, de Leeuw G, Chen L (2018) Preliminary investigation of a new AHI Aerosol Optical Depth (AOD) retrieval algorithm and evaluation with multiple source AOD measurements in China. Remote Sens 10:848. https://doi.org/10.3390/rs10050748

Yang J, Wong MS, Menenti M, Nichol J (2015) Modeling the effective emissivity of the urban canopy using sky view factor. ISPRS J Photogram Remote Sens 105:211–219

Yang JX, Wong MS, Menenti M, Nichol JE, Voogt J, Krayenhoff ES (2016) Development of an improved urban emissivity model based on sky view factor for retrieving effective emissivity and surface temperature over urban areas. ISPRS J Photogram Remote Sens 122:30–40

Zhou D, Xiao J, Bonafoni S, Berger C, Deilami K, Zhou Y, Frolking S, Yao Y, Qiao Z, Sobrino JA (2019) Satellite remote sensing of surface urban heat islands: progress, challenges, and perspectives. Remote Sens 11:48. https://doi.org/10.3390/rs11010048

Janet E. Nichol is an Applied Geographer, specializing in Remote Sensing, Geo-Informatics, and Environmental Change. Her main research interests are in the application of remote sensing to urban areas and landscape change, especially in the context of global climatic change. She has previously worked in universities in Nigeria, the Republic of Ireland, Singapore, and Hong Kong, and is currently a Visiting Professor at the University of Sussex, UK.

Muhammad Bilal is a Professor of Remote Sensing at the Nanjing University of Information Science and Technology (NUIST), Nanjing, China. In October 2018, the Jiangsu Provincial Education Department conferred on him the special title of "Distinguished Professor" based on his outstanding research achievements.

Majid Nazeer is serving as Associate Professor at the Key Laboratory of Digital Land and Resources, East China University of Technology (ECUT), Nanchang, Jiangxi, China. His research interests include oceanic remote sensing, land cover and land use classification, spatial modeling, air pollution, and atmospheric correction of satellite imagery.

Man Sing Wong is an Associate Professor of the Department of Land Surveying and Geo-Informatics, The Hong Kong Polytechnic University, and also a NASA's AERONET Hong Kong station site manager. He is a chartered member of the Royal Institution of Chartered Surveyors, the Hong Kong Institute of Surveyors, and a Fulbright scholar supported by the United States Department of State. He has published over 100 SCI journals and received over HKD 58 million of research funding as PI in the last couple of years.

Chapter 17
Urban Health and Wellbeing

Clive E. Sabel, Prince M. Amegbor, Zhaoxi Zhang, Tzu-Hsin Karen Chen, Maria B. Poulsen, Ole Hertel, Torben Sigsgaard, Henriette T. Horsdal, Carsten B. Pedersen, and Jibran Khan

Abstract This chapter explores how the Internet of Things and the utilization of cutting-edge information technology are shaping global research and discourse on the health and wellbeing of urban populations. The chapter begins with a review of smart cities and health and then delves into the types of data available to researchers. The chapter then discusses innovative methods and techniques, such as machine learning, personalized sensing, and tracking, that researchers use to examine the health and wellbeing of urban populations. The applications of these data, methods, and techniques are then illustrated taking examples from BERTHA (Big Data Centre for Environment and Health) based at Aarhus University, Denmark. The chapter concludes with a discussion on issues of ethics, privacy, and confidentiality surrounding the use of sensitive and personalized data and tracking or sensing individuals across time and urban space.

17.1 Smart Cities and Health

Smart cities have become popular in urban discourse, research, and policy environments; yet the term remains ambiguous. Here, we conceptualize smart cities as enabled by the Internet of Things (IoT), where sensing citizens and authorities employ information and technology to better navigate their lives and manage resources more

C. E. Sabel (✉) · P. M. Amegbor · Z. Zhang · T.-H. K. Chen · M. B. Poulsen · O. Hertel · T. Sigsgaard · H. T. Horsdal · C. B. Pedersen · J. Khan
BERTHA Big Data Centre for Environment and Health, Aarhus University, Roskilde, Denmark
e-mail: cs@envs.au.dk

C. E. Sabel · P. M. Amegbor · Z. Zhang · T.-H. K. Chen · M. B. Poulsen · O. Hertel · J. Khan
Department of Environmental Science, Aarhus University, Roskilde, Denmark

T. Sigsgaard
Department of Public Health - Institute of Environmental and Occupational Medicine, Aarhus University, Roskilde, Denmark

H. T. Horsdal · C. B. Pedersen
Department of Economics and Business Economics, CIRRAU - Centre for Integrated Register-based Research, Aarhus University, Roskilde, Denmark

W. Shi et al. (eds.), *Urban Informatics*, The Urban Book Series,
https://doi.org/10.1007/978-981-15-8983-6_17

efficiently. The utilization of information technology presents unique opportunities for understanding individual behavior and interactions in the urban space and their implications for human health and wellbeing. Often the aim is to combine the use of digital technologies and green city planning to optimize wellbeing and at the same time improve the physical environment and mitigate climate change. Boulos and Al-Shorbaji (2014) assert that an important component of smart cities is that they contain the ingredients necessary for improving the quality of life and wellbeing of residents. The technology and information available to urban residents have the potential to affect their health positively or negatively.

On the one hand, technology and the interconnection of people via the Internet present the opportunity for increasing access to health and health-enhancing information while reducing the cost of health care, particularly for the socioeconomically vulnerable (Aborokbah et al. 2018; Solanas et al. 2014). Remote monitoring of individuals can help quantify individual-level risks and provide vital information for effective person-centered health care (Aborokbah et al. 2018). For instance, real-time individual physiological and environmental information could help healthcare providers understand contextual factors that expose an individual to adverse health outcomes or improve their health and psychosocial wellbeing (Bryant et al. 2017; Lomotey et al. 2017; Rocha et al. 2019).

Others talk about the use of technology and information to deliver services to vulnerable and disadvantaged persons in the urban context with the aim of increasing their independence and wellbeing (Gilart-Iglesias et al. 2015; Rodrigues et al. 2018; Turcu and Turcu 2013). Just as studies show the myriad advantages associated with using personal information and technology in advancing health and wellbeing, they also highlight their negative effect on health outcomes (Do et al. 2013). The use of the Internet has opened new health and wellbeing challenges, beyond the traditional methods of providing and sustaining health and wellbeing, including misinformation, cyberbullying, cyber-fraud, and victimization. Do et al. (2013) observed that excessive use of the Internet among adolescents contributes to a higher incidence or likelihood of reporting depressive symptoms, suicidal thoughts, overweight, and lower self-reported health status due to sleep deprivation. Likewise, studies also show that the Internet has given an impetus to anti-vaccination campaigns through misinformation, contributing to lower acceptance and hesitation in accepting vaccine (Dubé et al. 2014).

This chapter is structured into four main sections, all considering health and wellbeing in an urban context. We begin by discussing data in an informatics era, before considering existing and emerging analytical techniques and methods. Example applications are taken from our BERTHA center, before we round off the discussion with the important issues surrounding privacy and confidentiality.

BERTHA (Big Data Centre for Environment and Health) is our interdisciplinary research center, based at Aarhus University, Denmark, bringing together urban geographers, environmental modelers, data scientists, and medical practitioners. BERTHA aims to muster the huge potential opportunities from the big data revolution in medical, environmental and population registers, personalized sensors, and crowdsourced data mining to disentangle the complex interactions between whole-life-course environmental and social exposures, and human health. Key to this overarching aim is assembling, linking, and analyzing diverse, huge datasets, developing algorithms, and intelligent data analytics.

17.2 Data

17.2.1 Big Data

There has been a lot of hype and hyperbole in the past decade over the Big Data paradigm. Big Data from a variety of data sources from government and citizens can be applied to improve urban health and wellbeing (Fleming et al. 2014). Within BERTHA, we see Big Data as not just about using large datasets, but critically, the combination of (huge) datasets to reveal value greater than the sum of the individual parts. The Big Data term has also been used to encompass the use of predictive data analytics and the computational analysis of extremely large, multi-source datasets to reveal patterns, trends, and associations. Thus, we prefer Rich Data rather than Big Data.

17.2.2 Individual and Population Data

Decisions on the health and wellbeing of a population are often informed by data and knowledge available on individual citizens. Generally, there are two sources of data for this decision-making process: individual or population data, and environmental data. Traditionally, administrative records and censuses were the main sources of individual or population-level data. While these data sources have their flaws, the data from some countries, including the Scandinavian countries, contain rich information about individuals from the onset of their lives till their demise (Frank 2000). The data from these registers enable detailed analyses and research on each individual in the population. The information from the various registers can be linked to each member of the population through a unique personal identification number. Examples of such unique identification numbers are Denmark's Centrale Personregister (Central Person Register, CPR) number, Norway's Fødselsnummer (national identification number), and Sweden's personnummer. In Denmark, these unique identifiers enable researchers to link data and information from nearly 200 databases from information on places of residence, employment, to medical records and socioeconomic data on salaries and tax. The records of some databases extend as far back as 1924 (Pedersen 2011; Pedersen et al. 2006), but the critical ones have been digital since 1968. In other countries, the information about individuals from government registers and databases can be extracted or linked using social-security numbers; for example,

Canada's Social Insurance Number (SIN). Similar to the Scandinavian personal identification numbers, these unique social-security numbers are normally assigned at birth. Information from the registers and the databases, such as a residential address, workplace, and school, can also be geocoded, enabling researchers to identify environmental exposures over each individual's total life course (Pedersen 2011). Particularly in the case of the data from Scandinavian registers, it is possible to define location histories of each individual in the population, accurately georeferenced to 1 m (Pedersen 2011).

In the digital era, tracking and sensing of an individual's activities in urban environments has become commonplace (Lupton 2013, 2017; Swan 2009, 2012). Advances in technology and miniaturization have facilitated the ability to track time-activity patterns of individuals, via GPS-enabled smartphone apps, watches, or proprietary wearable devices. These digital devices and social-media platforms not only enable individuals to generate and analyze personalized health data, but also enable them to share this information directly or indirectly with others (Gimpe et al. 2013; Lupton 2013, 2017). Prior to this, the accepted practice was to use daily research diaries to record life events and activities. These diaries may be intimate journals with uncensored information about one's thoughts, opinions, or experiences; or memoirs often written with an audience in mind; or a log of events and activities that occurred in one's life (Elliott 1997).

17.2.3 Environmental Data

Records of air pollution, water quality, housing conditions, recreational space, and exposure to chemicals traditionally came from field surveys, household surveys, or stationary observations. However, these data are usually limited in sample size and are not often available for longitudinal studies. Increasingly, environmental data are obtained from modeling or simulation, informed from field monitoring.

Remote sensing is a valuable source of environmental data, which are complementary to survey data and help to capture the dynamics of urban environments. Time-series satellite images allow understanding of urban sprawl and shrinkage in many parts of the world. For instance, urban expansion has been investigated with Landsat time-series images over more than two decades in India (Sharma and Joshi 2013), the USA (Li et al. 2018; Sexton et al. 2013), Japan (Bagan and Yamagata 2012), and China (Shi et al. 2017). The variations of urban greenness across the years can also be monitored via remote-sensing data and used to predict the outbreaks of mosquito-borne diseases in cities (Chen et al. 2018). On the other hand, building damage and land-use changes due to environmental disturbances, such as the 2003 Bam earthquake in Iran (Chini et al. 2008) and the 2011 Fukushima nuclear disaster in Japan (Sekizawa et al. 2015), were traced by satellite. In complex human-environment systems, researchers also utilize satellite images to understand different pathways of agricultural damage (Chen and Lin 2018).

Many recent epidemiological studies have evaluated the health impacts of specific land-cover types and the configuration of urban land use, including commercial, residential, and recreational areas, green space, agricultural areas, and proximity to blue space. The literature shows that natural environments, such as green or blue space, can have health-enhancing (or salutogenic) properties that improve the physical and psychosocial wellbeing of urban residents (Bornioli et al. 2018; Duarte et al. 2010; Olsen et al. 2019; Stigsdotter et al. 2017); however, the associations between environmental measures and health remain uncertain (Briggs et al. 2009; Wheeler et al. 2015). Other studies have questioned the relationship between salutogenic spaces and health outcomes (Gren et al. 2018). For instance, while green space may mitigate pollution levels through removing pollutants from the air, it is also a source of pollens, aggravating allergies and increasing particulate-matter counts.

Researchers have also been critical of the proxies used in measuring environmental exposures. Determining exposure metrics of various land covers that potentially impact health is complex. Early work (Pearce et al. 2006) used distance as a proxy for exposure to green space, by defining either a radius around the residential home or using the road network distance. Nearly, all studies have focused on the residential home, or neighborhood, as the location of analysis, often ignoring places of work or education and the more complex daily-life trajectories (Sabel et al. 2000, 2009; Steinle et al. 2013). However, proximity does not equate to accessibility. The literature highlights the distinction between the two concepts and stresses that physical and socioeconomic barriers (including, highways, or gated communities) may impede the ability of individuals in proximity to these natural environments from fully benefitting from their health-enhancing properties (Markevych et al. 2017). More recently, research has moved on to consider the quality and configuration of urban space, since there is evidence that homogeneous spaces are less beneficial to health than heterogeneous, biodiverse ones (Wheeler et al. 2015).

Air pollution is traditionally measured by costly devices at fixed-site monitoring stations. It is absolutely crucial that such devices are advanced and accurate, since they are usually used in air-pollution monitoring programs legislated by governments to test compliance with air-quality guidelines. However, it is increasingly being questioned whether assessing personal exposure to air pollution using fixed-site monitoring data might provide an error in the individual exposure as the impact of the mobility pattern is ignored (Buonanno et al. 2014; Steinle et al. 2013). However, newly developed low-cost, portable sensor nodes provide new options for personal-exposure monitoring (PEM) by mobile measurements. The sensor nodes can easily be carried around during our daily life, where we constantly move in time and space through different environments both indoor and outdoor. We commute between home and work, spend time indoors with household activities and work, and maybe we play with our kids at the local playground. Thus, we are constantly exposed to highly variable concentrations of air pollution with documented evidence for negative health effects. However, these low-cost personal air-pollution sensors are not as robust scientifically as the fixed-site monitors, and it is still uncertain how measurements are affected when the sensor nodes are moving: how does it affect the performance of the sensors when one moves between different microenvironments, especially when one moves from indoors to outdoors, exposing the sensor to rapid changes in temperature and humidity.

17.3 Methods and Techniques

Recent advances in information technology have contributed new sources of individual data for researchers in their quest to understand human-environment interactions and their impact on health and wellbeing in urban space. Mobile digital devices, such as smartphones, smartwatches, tablets, and sensors, together with apps on the devices, can collect users' data on physical activity, sporting performance, and daily routines, as well as demographic and health data. These mobile devices also simultaneously provide spatiotemporal geolocational data of the user, using GPS or cellphone-network triangulation. The information from these devices has radically changed the opportunities for researchers and practitioners within the health and wellbeing arena. For researchers, it has extended the traditional boundaries and the methods, techniques, or approaches used in conducting our studies; and also makes us critical of existing models and concepts of health and wellbeing (Lupton 2013; Swan 2009). For medical practitioners, the data can provide additional information about patients, the inclusion of the individual in the healthcare process, and the ability to provide holistic care for patients (Dingler et al. 2014).

Compared with traditional methods, multi-source big data could be collected from many other aspects passively and unconsciously. Wang et al. (2019a, b) in their survey about sensor-based human activity recognition (HAR) catalog common-used sensors into four types: (1) Inertial sensors, including accelerometer, gyroscope, and magnetometer applied in detecting multiple motions; (2) Physical health sensors, such as electrocardiograms, skin temperature, heart rate, and force sensors, used to detect people's health conditions, while new technology products like sports watches and fitness tracking bracelets have a similar function; (3) Environmental sensors like temperature, light, and barometer sensors, delivering context information related to activities; (4) Others: other wearable devices like cameras, microphones, and GPS. GPS can track people's routes and record locations simultaneously and is useful in studies of urban space and people's behavior (Bohte and Maat 2009). The cell phone has been applied in public-health studies and can be combined with gyroscope (Shoaib et al. 2014) and barometer (Muralidharan et al. 2014) to identify physical activity and sleep quality. Image sensors like wearable cameras have been applied in recording people's daily exposure (Wang and Smeaton 2013), including dietary intake (Zhou et al. 2019), and environmental exposure (Chambers et al. 2017).

The emergence of social media and smartphone technologies more generally has opened new sources of data for understanding health and wellbeing in the urban context. However, the data from these sources are subject to potential biases since users are often not fully representative of society, under-representing persons of lower socioeconomic status, and older and non-tech savvy persons. It can be argued that socioeconomic factors are as important as the physical environment in determining health impacts on human populations, since a disproportionate share of the burden of environmental exposure falls on vulnerable groups of society, including low SES, ethnic minorities, women, and the elderly and young, due partly to issues of environmental (in)justice. In addition, SES can explain differences in external exposure because of the different prevalence of specific behaviors in some groups; for example, differences in diet between SES groups. Individual health and wellbeing are influenced by many factors including past and present behavior, healthcare provision, and wider determinants including social, cultural, and environmental factors. Traditional sources of data, such as government registers, and demographic and health surveys, offer information on these broader contextual factors that are often absent in individual data from smart technologies. The breadth of the traditional data means they are relatively less susceptible to selection bias compared to the new sources of data.

Additionally, traditional data also bring the ability to construct area-level exposures and their influence on health and wellbeing, such as to address the context versus composition debate (Macintyre et al. 2002), regarding the wider question of which is more important for shaping health: the area in which people live (context) or the people who make up the inhabitants of that area (composition). Area-level SES is often estimated by means of a weighted index of factors from published secondary data, such as the UK Index of Multiple Deprivation (IMD) and the Vancouver Area Neighborhood Deprivation Index (VANDIX) (Bell and Hayes 2012; Ellaway et al. 2012; Macintyre et al. 2008; Schuurman et al. 2007). Weighted factors might typically include measures of education, income, homeownership, and access to transport.

Another informatics area experiencing fast adoption is using citizens as sensors (Goodchild 2007) to obtain evidence of citizens' experiences in the urban landscape (Zook 2017). An emerging field in the health arena, supported by smartphone technology, is ecological momentary assessment. Here apps are utilized such as in the Mappiness project (MacKerron and Mourato 2013; Seresinhe et al. 2019) to ask people to describe their responses to the environment directly, with the advantage that input is related to the current location via GPS. This allows researchers to explore the more psychological aspects of how people are responding to their environments.

Modeling, as opposed to monitoring, of urban environments has been enabled by the digital era. As a branch of artificial intelligence, machine learning is a field of study growing in popularity in urban modeling that provides computers with the ability to automatically learn and improve their own algorithms from data. Machine-learning studies often investigate urban dynamics based on remotely sensed data. The approach of mapping the urban environment with machine-learning methods goes back to the 1990s. For instance, Gong et al. (1992) used a maximum-likelihood classifier and USGS Landsat imagery to automate urban land-use mapping. Such development, however, was slow until the 2000s, when satellite images at 30 m and finer resolution became affordable and publicly readable (Weng 2012).

Machine learning has the potential to automate the process of urban mapping, which traditionally relies on intensive labor. Automatic image recognition, from sources such as Google Streetview, encourages urban scientists to detect more nuanced features in cities. With the capability of increasing computation power, deep-learning methods, such as convolutional neural networks (CNNs), have increased the dimension of detectable urban attributes. Because of CNN's capabilities in recognizing the spatial patterns of image patches, recent studies have applied CNN to streetview images and aerial photographs for quantifying a sky view of street canyons (Gong et al. 2018), mapping local climate zones (Qin et al. 2017), and classifying specific types of urban facilities (e.g., church, park, and garage) (Kang et al. 2018). Remote sensing and machine learning are complements to urban simulation models (Batty 2013), which can forecast dynamics and growth, but not represent spatial details.

Similarly, researchers have also applied machine-learning methods to data from personalized sensors and streetview images to understand dynamism in the urban space and its effect on mental health as well as susceptibility to crime (Goin et al. 2018; Helbich 2018; Helbich et al. 2016; Mohr et al. 2017; Wang et al. 2019a, b). Machine learning can also be used to improve the prediction accuracy of models that seek to understand the effect of individual and community factors on health outcomes. Machine-learning approaches, such as least absolute shrinkage and selection operator (LASSO) and random forest, have been used to identify optimal individual-level and community-level factors that predict firearm violence in urban communities (Goin et al. 2018).

17.4 BERTHA Studies

17.4.1 AirGIS

Models are used in academic research to enhance our knowledge of reality by simplifying the complexity of the phenomena we study as researchers. For instance, GIS models are used to estimate and assess exposure to adverse environmental conditions. In Denmark, the Danish AirGIS (Jensen et al. 2001) and Operational Street Pollution Model (OSPM) (Berkowicz 2000) are routinely used to estimate street- or local-scale air pollution. In an effort to improve this model system and increase its accessibility, researchers in BERTHA developed an open-source GIS model for computing local-scale air-pollution estimates (Khan et al. 2019a, b). The new model is able to reproduce both temporal (correlation range: 0.45–0.96) and spatial (correlation range: 0.32–0.92) variations in observed air pollution, and subsequently to estimate both short- and long-term exposures to air pollution, which enables researchers to better understand its duration and effects on human health and wellbeing. The AirGIS system is currently being extended to estimate noise mainly originating from urban transport.

At present, the AirGIS is being further extended to estimate dynamic time-activity exposure to air pollution by tracking individuals in urban commuting environments, and making use of measured and modeled air-pollution data (Khan et al. 2019a,

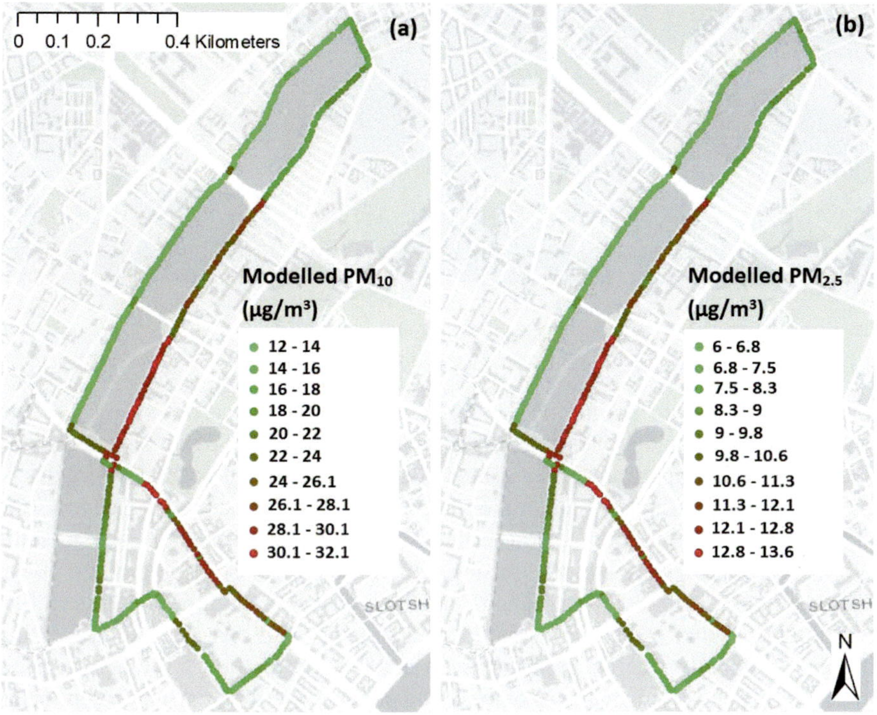

Fig. 17.1 **a** Modeled PM10 ($\mu g\ m^{-3}$) at GPS track points of the walking-based activity of the study participants in Copenhagen, Denmark. The modeled values are for Monday, February 4, 2019, during 7:00–10.00 am **b** the same for modeled PM2.5 ($\mu g\ m^{-3}$)

b). The focus is on developing a novel exposure assessment framework to facilitate health-related studies. As an example a walking-based activity was performed in Copenhagen, Denmark (Khan et al. 2019a, b). At GPS track points, air-pollution concentrations (NO_x, NO_2, PM10, and PM2.5 in $\mu g\ m^{-3}$) were calculated using the AirGIS system to analyze dynamic exposure to modeled air pollution (Fig. 17.1). Preliminary findings suggest that exposure estimates based on time-activity patterns of individuals depend on the level of one's mobility as well as on the location of one's workplace relative to home.

17.4.2 Personalized Tracking and Sensing

Wearable devices are practically ubiquitous in the informatics era. Among these devices, the wearable camera has attracted increasing attention, since it can capture details of daily life by images or videos, which can enhance researchers' understanding of people's movements, behaviors, and preferences. Zhang and Long (2019)

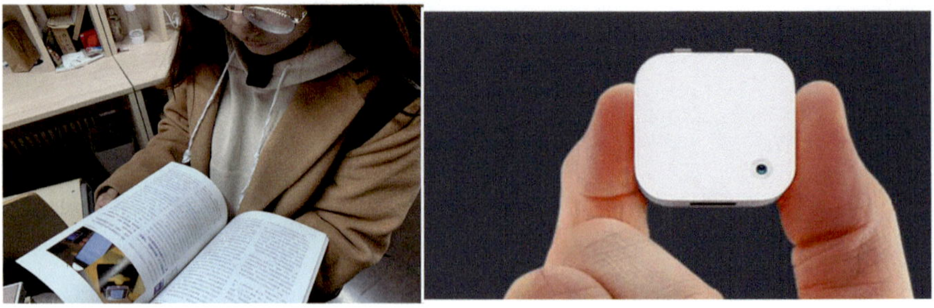

Fig. 17.2 Wearable camera (also appears in Zhang and Long 2019)

conducted research in Beijing, validating applying wearable cameras (Fig. 17.2) in built-environment studies. Through identifying and analyzing 8598 images collected from a one-week experiment, they summarized the spatiotemporal characteristics of the user while wearing the camera, and compared the frequency of greenery (the ratio of green) and outdoor exposure (the ratio of blue) by means of color identification. The images were classified using artificial intelligence, and common image elements (tags) were identified (Zhang and Long 2019), including building, traffic, figure, food, digital screen, and greenery. Results showed that as a kind of digital lifelogging, an individual image database is an effective support for future interdisciplinary studies involving the environment and personal wellbeing from a micro-scale perspective. In the future, as the popularization of IoT technology becomes real, an increasing number of wearable gadgets such as wristbands (pulse, blood pressure, and heartbeat), glasses (eyesight, eye pressure, distance to screen) and so on, can be utilized to build a more comprehensive profile of individual health and exposure.

17.4.3 Personalized Air-Pollution Sensors

Computer and sensor technologies have developed tremendously over the past ten years, and air-pollution sensors have been miniaturized, are reasonably accurate, cheap, and have a fine time resolution. This development enables personal-exposure monitoring, and deploying such measurements might improve our knowledge about how we are exposed to air pollution during our regular activities. However, personalized sensors require a user-friendly interface to ease their use by those who wish to monitor their daily exposures. This is often done by visualizing data via an app. However, the design of such apps demands that some decisions be made in advance. How much information should the user of the app be presented with and how are data visualized in the most useful way? Will the idea of using different color zones make air-pollution data more understandable or will it misinform; for example, if green, yellow, and red are used to indicate low, medium, and high concentration ranges, then there is a risk that the color red will scare the user and that the color green will

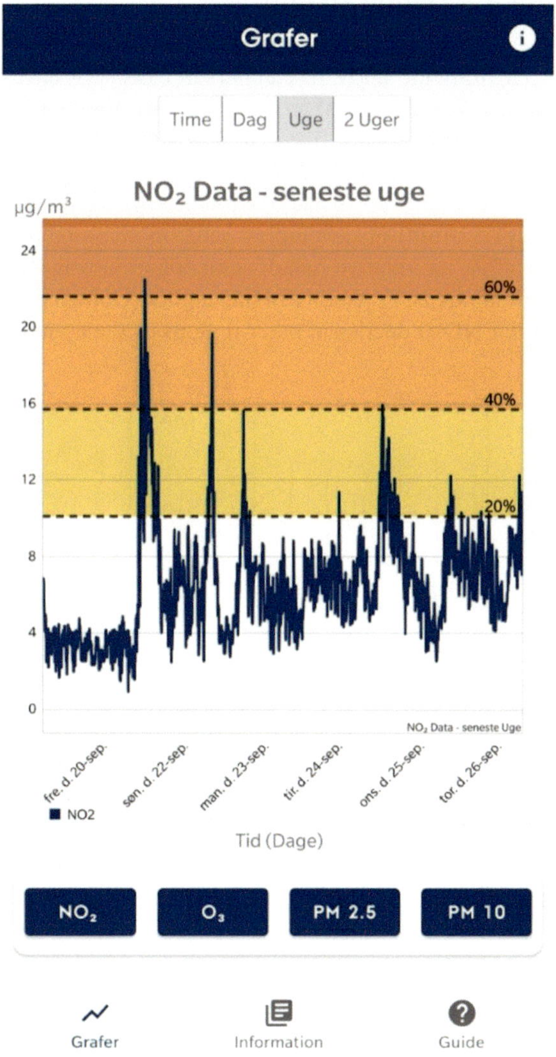

Fig. 17.3 User interface of personalized air-pollution monitoring app

misinform, as low concentrations do not necessarily mean a healthy environment. Another important thought is whether GPS positions are presented or not and how are these are secured in accordance with the EU's General Data Protection Regulation (GDPR). Our work with the personalized air-pollution sensors focuses on optimizing sensor performance in a mobile environment, along with app development to convey data to the users (Fig. 17.3).

17.4.4 Mental Health

In a nationwide study, researchers in BERTHA have combined data from the Danish Psychiatric register and green space, measured by NDVI from 30 m by 30 m Landsat imagery, in Denmark from 1985 to 2013 in order to understand the potential effect of green space exposure on schizophrenia. The study reveals that individuals with childhood exposure in places with the lowest amount of greens pace have an increased risk (1.52-fold) of developing schizophrenia (Engemann et al. 2018, 2019). From Fig. 17.4, the relative risk of schizophrenia was shown to be higher among persons in urban areas, especially in the capital (Copenhagen) compared to people living in similar NDVI deciles in other regions of the country.

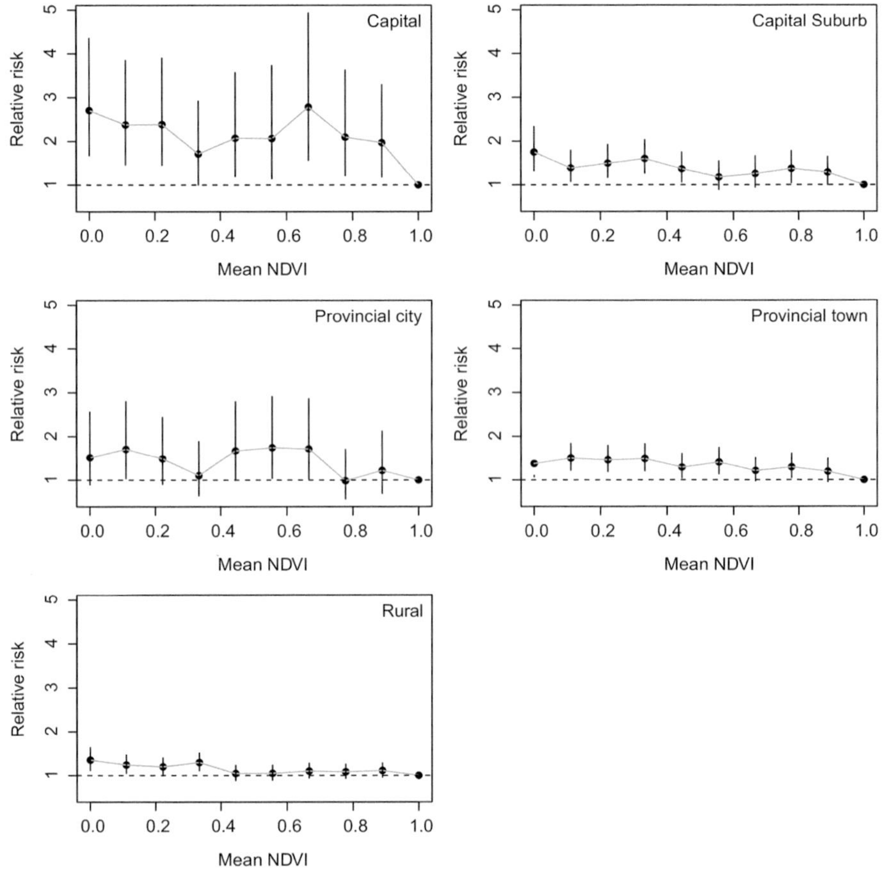

Fig. 17.4 From Engemann et al. (2018)

Further ongoing work is investigating a broader range of psychiatric disorders and natural environment exposure. Initial results suggest that growing up in natural environments is associated with lower levels of psychiatric disorders.

17.4.5 Physical Activity

BERTHA collaborates with RUNSAFE,[1] a non-commercial, multidisciplinary research group based at Aarhus University Hospital, Denmark. In collaboration with Garmin, RUNSAFE has launched a worldwide study recruiting runners willing to monitor their running habits with a Garmin device and report their injury and health status on a weekly basis over an 18-month period. With other big data, the relationship between running activity, personal characteristics, and risk of running-related injuries will be investigated (Nielsen et al. 2019). This data source is fundamental for BERTHA, as the fitness data will be combined with air pollution data to investigate if physical activity in polluted areas increases the risk of heart-rate variability as a sign of effects of air quality on the cardiovascular system.

17.4.6 Danish Blood-Donor Study

In combination with personal sensors, we are aiming at a study examining the obstacles and drivers of mobility in different age groups with a special interest in life periods—children, teenagers, adults, and seniors—as mobility has been shown to differ between these groups. The Danish blood-donor study is targeting susceptibility factors related to air pollution, taking advantage of the repetitive sampling of plasma. This enables the study of biomarkers of air pollution in the total population, or strata related to genetic markers of susceptibility, for example, atopy, gender, and age (Hansen et al. 2019).

17.5 Privacy

We live in an increasingly monitored world. People can be tracked as they navigate their urban lives, via cameras, monitoring of their smartphones, or their social media accounts. Norms and expectations are rapidly evolving. What might be considered ethically acceptable by young people might be viewed as intrusive for older generations. While this offers the urban researcher unparalleled data access, there are important ethical issues to be considered. Particularly in the health and wellbeing

[1] Garmin RunSafe: Running Health Study (n.d.) Retrieved October 7, 2019. https://garmin-runsafe.com/.

domain, there are multiple privacy issues to consider. Some of these have been covered in other chapters, notably Chap. 32, but there are specific issues to consider when handling personal health information.

Taking the example of Denmark, but similar procedures apply elsewhere, access to all individual-level data is regulated by Danish legislation. Research studies needing additional information directly from study participants also need approval from the relevant ethical committee, followed by informed consent from study participants. Updated individual-level information originating from national registers may only be accessed at secure research platforms, including Statistics Denmark or the Danish Health Data Authority. All data must comply with the recently introduced EU GDPR Regulation 2016/679 (General Data Protection Regulation).

Standard epidemiological protocols around ethics, privacy, and confidentiality also apply to data derived from personalized sensors and smartphone apps. Online consent is normally sought, for example, when users sign up to a new service, be it a wearable device or a social-media account. When users sign up, are the users aware of exactly what they are consenting to? Most apps or devices cannot be used without agreeing to the often long list of terms and conditions, and many users will not read the full terms. Once signed up, often the terms and conditions allow the service provider or sensor developer to store, analyze, make public, or sell for profit, an individual's data. Researchers can then legally access these data, often without the individual's knowledge. This is particularly challenging in a big data environment, when users might have given consent individually but may not be aware of the ability to link data across platforms to infer much more.

Lastly, the public debate around data privacy needs to balance the individual's right to privacy versus the opportunities to make new scientific discoveries from wider data availability. Globally, governments are leaning more toward the protection of citizen's rights over the exciting opportunities that wider data access could offer to make fundamental scientific breakthroughs.

17.6 Conclusions

This chapter started by sketching the relationship of smart cities and urban informatics to human health and wellbeing. We talked about the how advancement in information technology and mobile devices has enhanced health and wellbeing for urban residents through the provision of person-centered solutions to understand how the social and built environment impacts their lives. The technology and its associated platforms offer less costly ways for delivering vital health and wellbeing services to the wider population at a minimal cost. They have also encouraged individuals to be proactive participants in the healthcare delivering system, as well as offered them resources for engaging in healthy lifestyles via tracking their health behavior. Nevertheless, the emergence of these innovative and smart technologies is not without caveats. Within a rapidly changing technological world, researchers and policy-makers have to keep abreast of changing behavior and the preferences of the population, particularly the urban population who are often at the forefront of this technological drive. IoT has also exposed people to new forms of health risks, such as cyber victimization, misinformation, and addiction. As researchers, we need to develop new tools and techniques (beyond the traditional ones) to understand these risks and their implications on individuals and the wider population. Researchers and policymakers also have to maintain a delicate balance between the desire to improve health and wellbeing (using the newly available technology and data), and respecting individual privacy (and other ethical considerations). Considering the sociodemographic characteristics of users of these smart devices and technology, critical questions also remain about whether the research will perpetrate inequalities in the urban space through the policy and planning of health and wellbeing that emerge from the new IoT.

References

Aborokbah MM, Al-Mutairi S, Sangaiah AK, Samuel OW (2018) Adaptive context aware decision computing paradigm for intensive health care delivery in smart cities—a case analysis. Sustain Cities Soc 41:919–924

Bagan H, Yamagata Y (2012) Landsat analysis of urban growth: how Tokyo became the world's largest megacity during the last 40 years. Remote Sens Environ 127:210–222

Batty M (2013) The new science of cities. MIT Press, Cambridge, Massachusetts

Bell N, Hayes MV (2012) The vancouver area neighbourhood deprivation index (VANDIX): a census-based tool for assessing small-area variations in health status. Can J Public Health 103(2):S28–S32

Berkowicz R (2000) A simple model for urban background pollution. Environ Monit Assess 65(1–2):259–267

Bohte W, Maat K (2009) Deriving and validating trip purposes and travel modes for multi-day GPS-based travel surveys: a large-scale application in the Netherlands. Transp Res Part C: Emerg Technol 17(3):285–297

Bornioli A, Parkhurst G, Morgan PL (2018) The psychological wellbeing benefits of place engagement during walking in urban environments: a qualitative photo-elicitation study. Health Place 53:228–236

Briggs DJ, Sabel CE, Lee K (2009) Uncertainty in epidemiology and health risk and impact assessment. Environ Geochem Health 31(2):189–203

Bryant N, Spencer N, King A et al (2017) IoT and smart city services to support independence and wellbeing of older people. In: 25th international conference on software, telecommunications and computer networks. Split, Croatia, 21–23 Sept

Boulos MNK, Al-Shorbaji NM (2014) On the Internet of Things, smart cities and the WHO healthy cities. Int J Health Geographics 13:1–6

Buonanno G, Stabile L, Morawska L (2014) Personal exposure to ultra fine particles: the influence of time-activity patterns. Sci Total Environ 469:903–907

Chambers T, Pearson AL, Kawachi I, Rzotkiewicz Z, Stanley J, Smith M, Mhurchu CN, Signal L (2017) Kids in space: Measuring children's residential neighborhoods and other destinations using activity space GPS and wearable camera data. Soc Sci Med 193:41–50

Chen THK, Chen VYJ, Wen TH (2018) Revisiting the role of rainfall variability and its interactive effects with the built environment in urban dengue outbreaks. Appl Geogr 101:14–22

Chen TH, Lin KHE (2018) Distinguishing the windthrow and hydrogeological effects of typhoon impact on agricultural lands: an integrative OBIA and PPGIS approach. Int J Remote Sens 39(1):131–148

Chini M, Pierdicca N, Emery WJ (2008) Exploiting SAR and VHR optical images to quantify damage caused by the 2003 Bam earthquake. IEEE Trans Geosci Remote Sens 47(1):145–152

Dingler T, Sahami A, Henze N (2014) There is more to well-being than health data: Holistic lifelogging through memory capture. In: Meyer J, Simske S, Siek KA, Gurrin CG, Hermens H (ed) Proceedings of the CHI 2014 workshop on beyond quantified self: data for wellbeing. ACM, New York, NY

Do YK, Shin E, Bautista MA, Foo K (2013) The associations between self-reported sleep duration and adolescent health outcomes: what is the role of time spent on Internet use? Sleep Med 14(2):195–200

Duarte CS, Chambers EC, Rundle A, Must A (2010) Physical characteristics of the environment and BMI of young urban children and their mothers. Health Place 16(6):1182–1187

Dubé E, Vivion M, MacDonald NE (2014) Vaccine hesitancy, vaccine refusal and the anti-vaccine movement: influence, impact and implications. Expert Rev Vaccines 14(1):99–117

Ellaway A, Benzeval M, Green M, Leyland A, Macintyre S (2012) "Getting sicker quicker": does living in a more deprived neighbourhood mean your health deteriorates faster? Health Place 18(2):132–137

Elliott H (1997) The use of diaries in sociological research on health experience. Soc Res Online 2(2):38–48

Engemann K, Pedersen CB, Arge L, Tsirogiannis C, Mortensen PB, Svenning JC (2018) Childhood exposure to green space—a novel risk-decreasing mechanism for schizophrenia? Schizophr Res 199:142–148

Engemann K, Pedersen CB, Arge L (2019) Residential green space in childhood is associated with lower risk of psychiatric disorders from adolescence into adulthood. Proc Natl Acad Sci 116(11):5188–5193

Fleming LE, Haines A, Golding B, Kessel A, Cichowska A, Sabel CE, Depledge MH, Sarran C, Osborne NJ, Whitmore C, Cocksedge N (2014) Data mashups: potential contribution to decision support on climate change and health. Int J Environ Res Public Health 11(2):1725–1746

Frank L (2000) When an entire country is a cohort. Science 287(5462):2398–2399

Gilart-Iglesias V, Mora H, Pérez-delHoyo R, García-Mayor C (2015) A computational method based on radio frequency technologies for the analysis of accessibility of disabled people in sustainable cities. Sustainability 7(11):14935–14963

Gimpe H, Nißen M, Görlitz RA (2013) Quantifying the quantified self: A study on the motivation of patients to track their own health. Thirty fourth international conference on information systems: reshaping society through information systems design. Italy, Milan, pp 3286–3301

Goin DE, Rudolph KE, Ahern J (2018) Predictors of firearm violence in urban communities: a machine-learning approach. Health Place 51:61–67

Gong P, Marceau DJ, Howarth PJ (1992) A comparison of spatial feature extraction algorithms for land-use classification with SPOT HRV data. Remote Sens Environ 40(2):137–151

Gong FY, Zeng ZC, Zhang F, Li X, Ng E, Norford LK (2018) Mapping sky, tree, and building view factors of street canyons in a high-density urban environment. Build Environ 134:155–167

Goodchild MF (2007) Citizens as sensors: the world of volunteered geography. GeoJournal 69(4):211–221

Gren Å, Colding J, Berghauser-Pont M, Marcus L (2018) How smart is smart growth? Examining the environmental validation behind city compaction. Ambio 48(6):1–10

Hansen TF, Banasik K, Erikstrup C, Pedersen OB, Westergaard D, Chmura PJ, Nielsen K, Thørner L, Hjalgrim H, Paarup H, Larsen MAH (2019) DBDS Genomic Cohort, a prospective and comprehensive resource for integrative and temporal analysis of genetic, environmental and lifestyle factors affecting health of blood donors. BMJ Open 9(6):1–7

Helbich M (2018) Toward dynamic urban environmental exposure assessments in mental health research. Environ Res 161:129–135

Helbich M, Emmichoven MJZ, van Dijst MJ, Kwan MP, Pierik FH, de Vries SI (2016) Natural and built environmental exposures on children's active school travel: a dutch global positioning system-based cross-sectional study. Health Place 39:101–109

Jensen SS, Berkowicz R, Hansen HS, Hertel O (2001) A Danish decision-support GIS tool for management of urban air quality and human exposures. Transp Res Part D: Trans Environ 6(4):229–241

Kang J, Körner M, Wang Y, Taubenböck H, Zhu XX (2018) Building instance classification using street view images. ISPRS J Photogram Remote Sens 145:44–59

Khan J, Kakosimos K, Raaschou-Nielsen O, Brandt J, Jensen SS, Ellermann T, Ketzel M (2019a) Development and performance evaluation of new AirGIS—a GIS based air pollution and human exposure modelling system. Atmos Environ 198:102–121

Khan J, Poulsen MB, Ketzel M, Ørby PV, Christensen JH, Dalgaard R, Sabel CE, Hertel O (2019) Estimation of exposure to air pollution in Denmark–A step towards activity-based dynamic exposure assessment framework. In: 19th international conference on harmonisation within atmospheric dispersion modelling for regulatory purposes. Bruges, Belgium, 3–6 June 2019

Li X, Zhou Y, Zhu Z, Liang L, Yu B, Cao W (2018) Mapping annual urban dynamics (1985–2015) using time series of Landsat data. Remote Sens Environ 216:674–683

Lomotey RK, Pry J, Sriramoju S (2017) Wearable IoT data stream traceability in a distributed health information system. Pervasive Mob Comput 40:692–707

Lupton D (2013) Quantifying the body: monitoring and measuring health in the age of mHealth technologies. Crit Publ Health 23(4):393–403

Lupton D (2017) Self-tracking, health and medicine. Health Soc Rev 26(1):1–5

Macintyre S, Ellaway A, Cummins S (2002) Place effects on health: How can we conceptualise, operationalise and measure them? Soc Sci Med 55(1):125–139

Macintyre S, Macdonald L, Ellaway A (2008) Do poorer people have poorer access to local resources and facilities? The distribution of local resources by area deprivation in Glasgow, Scotland. Soc Sci Med 67(6):900–914

MacKerron G, Mourato S (2013) Happiness is greater in natural environments. Glob Environ Change 23(5):992–1000

Markevych I, Schoierer J, Hartig T, Chudnovsky A, Hystad P, Dzhambov AM, De Vries S, Triguero-Mas M, Brauer M, Nieuwenhuijsen MJ, Lupp G (2017) Exploring pathways linking greenspace to health: theoretical and methodological guidance. Environ Res 158:301–317

Mohr DC, Zhang M, Schueller SM (2017) Personal sensing: Understanding mental health using ubiquitous sensors and machine learning. Annu Rev Clin Psychol 13(1):23–47

Muralidharan K, Khan AJ, Misra A, Balan RK, Agarwal S (2014) Barometric phone sensors—More hype than hope! In: Proceedings of the 15th workshop on mobile computing systems and applications, HotMobile 2014. Santa Barbara, California, 26–27 Feb 2014

Nielsen RØ, Bertelsen ML, Ramskov D, Damsted C, Brund RK, Parner ET, Sørensen H, Rasmussen S, Kjærgaard S (2019) The Garmin-RUNSAFE running health study on the aetiology of running-related injuries: rationale and design of an 18-month prospective cohort study including runners worldwide. BMJ Open 9(9):e032627

Olsen JR, Nicholls N, Mitchell R (2019) Are urban landscapes associated with reported life satisfaction and inequalities in life satisfaction at the city level? A cross-sectional study of 66 European cities. Soc Sci Med 226:263–274

Pearce J, Witten K, Bartie P (2006) Neighbourhoods and health: a GIS approach to measuring community resource accessibility. J Epidemiol Community Health 60(5):389–395

Pedersen CB (2011) The Danish civil registration system. Scand J Publ Health 39(7):22–25

Pedersen CB, Gøtzsche H, Møller JØ, Mortensen PB (2006) The Danish civil registration system. A cohort of eight million persons. Dan Med Bull 53(4):441–449

Qin Y, Xiao X, Dong J, Chen B, Liu F, Zhang G, Zhang Y, Wang J, Wu X (2017) Quantifying annual changes in built-up area in complex urban-rural landscapes from analyses of PALSAR and Landsat images. ISPRS J Photogram Remote Sens 124:89–105

Rocha NP, Dias A, Santinha G, Rodrigues M, Queirós A, Rodrigues C (2019) Smart cities and healthcare: a systematic review. Technologies 7(3):58

Rodrigues M, Santos R, Queiros A, Silva AG, Amaral J, Gonçalves LJ, Pereira A, da Rocha NP (2018) Meet SmartWalk, smart cities for active seniors. In: 2nd international conference on technology and innovation in sports, health and wellbeing (TISHW), Thessaloniki, Greece, 20–22 June 2018

Sabel CE, Gatrell AC, Löytönen M, Maasilta P, Jokelainen M (2000) Modelling exposure opportunities: estimating relative risk for motor neurone disease in Finland. Soc Sci Med 50(7–8):1121–1137

Sabel CE, Boyle P, Raab G, Löytönen M, Maasilta P (2009) Modelling individual space-time exposure opportunities: a novel approach to unravelling the genetic or environment disease causation debate. Spat Spatio-Temporal Epidemiol 1(1):85–94

Schuurman N, Bell N, Dunn JR, Oliver L (2007) Deprivation indices, population health and geography: An evaluation of the spatial effectiveness of indices at multiple scales. J Urban Health 84(4):591–603

Sekizawa R, Ichii K, Kondo M (2015) Satellite-based detection of evacuation-induced land cover changes following the Fukushima Daiichi nuclear disaster. Remote Sens Lett 6(11):824–833

Seresinhe CI, Preis T, MacKerron G, Moat HS (2019) Happiness is greater in more scenic locations. Sci Rep 9(1):1–11

Sexton JO, Song XP, Huang C, Channan S, Baker ME, Townshend JR (2013) Urban growth of the Washington, DC-Baltimore, MD metropolitan region from 1984 to 2010 by annual, Landsat-based estimates of impervious cover. Remote Sens Environ 129:42–53

Sharma R, Joshi PK (2013) Monitoring urban landscape dynamics over Delhi (India) using remote sensing (1998–2011) inputs. J Indian Soc Remote Sens 41(3):641–650

Shi L, Ling F, Ge Y, Foody GM, Li X, Wang L, Zhang Y, Du Y (2017) Impervious surface change mapping with an uncertainty-based spatial-temporal consistency model: a case study in Wuhan city using Landsat time-series datasets from 1987 to 2016. Remote Sens 9(11):1148

Shoaib M, Bosch S, Incel OD, Scholten H Havinga PJ (2014) Fusion of smartphone motion sensors for physical activity recognition. Sensors 14(6):10146–10176

Solanas A, Patsakis C, Conti M, Vlachos IS, Ramos V, Falcone F, Postolache O, Pérez-Martínez PA, Di Pietro R, Perrea DN, Martinez-Balleste A (2014) Smart health: a context-aware health paradigm within smart cities. IEEE Commun Mag 52(8):74–81

Steinle S, Reis S, Sabel CE (2013) Quantifying human exposure to air pollution—Moving from static monitoring to spatio-temporally resolved personal exposure assessment. Sci Total Environ 443:184–193

Stigsdotter UK, Corazon SS, Sidenius U, Kristiansen J, Grahn P (2017) It is not all bad for the grey city—A crossover study on physiological and psychological restoration in a forest and an urban environment. Health Place 46:145–154

Swan M (2009) Emerging patient-driven health care models: An examination of health social networks, consumer personalized medicine and quantified self-tracking. Int J Environ Res Publ Health 6(2):492–525

Swan M (2012) Crowdsourced health research studies: An important emerging complement to clinical trials in the public health research ecosystem. J Med Internet Res 14(2):186–198

Turcu CE, Turcu CO (2013) Internet of Things as key enabler for sustainable healthcare delivery. Procedia Soc Behav Sci 73:251–256

Wang P, Smeaton AF (2013) Using visual lifelogs to automatically characterize everyday activities. Inf Sci 230:147–161

Wang R, Yuan Y, Liu Y, Zhang J, Liu P, Lu Y, Yao Y (2019a) Using street view data and machine learning to assess how perception of neighborhood safety influences urban residents' mental health. Health Place 59:102186

Wang Y, Cang S, Yu H (2019b) A survey on wearable sensor modality centred human activity recognition in health care. Expert Syst Appl 137:167–190

Weng Q (2012) Remote sensing of impervious surfaces in the urban areas: requirements, methods, and trends. Remote Sens Environ 117:34–49
Wheeler BW, Lovell R, Higgins SL (2015) Beyond greenspace: an ecological study of population general health and indicators of natural environment type and quality. Int J Health Geographics 14(1):17
Zhang Z, Long Y (2019) Application of wearable cameras in studying individual behaviors in built environments. Landscape Archit Frontiers 7(2):22
Zhou Q, Wang D, Mhurchu CN, Gurrin C, Zhou J, Cheng Y Wang H (2019) The use of wearable cameras in assessing children's dietary intake and behaviours in China. Appetite 139:1–7
Zook M (2017) Crowd-sourcing the smart city: Using big geosocial media metrics in urban governance. Big Data Soc 4(1):1–13

Clive Sabel is Professor of Health Geography and Director of BERTHA, Denmark's Big Data Centre for Environment and Health, based at Aarhus University. He is a spatial data scientist, using data mining, machine learning and personalized apps to understand how micro-scale mobility affects individual level environmental exposure.

Prince M. Amegbor is a Post-Doctoral research fellow in the Department of Environmental Science and BERTHA—Big Data Centre for Environment and Health (Aarhus University, Denmark). His research interests include environment and health, social determinants of health, public policy and health, victimization, and ageing.

Zhaoxi Zhang is a Ph.D. student In the Department of Environmental Science and BERTHA—Big Data Centre for Environment and Health. Her current research focuses on using urban multi-data and human sensors to understand how microenvironment quality influences personal exposure and health, in the context of smart and healthy cities.

Tzu-Hsin Karen Chen is a Ph.D. fellow of BERTHA, Denmark's Big Data Centre for Environment and Health at Aarhus University. She is a quantitative human geographer using remote sensing, machine learning, and spatiotemporal analysis to understand urban dynamics and the implications for human well-being.

Maria Bech Poulsen is a Ph.D. student in the Department of Environmental Science and BERTHA—Big Data Centre for Environment and Health, Aarhus University. Her research focuses on using portable sensors to measure personal exposures to air pollution.

Ole Hertel is deputy head of department, head of the Ph.D. program, and full professor in the Department of Environmental Science, Aarhus University. His research is within air pollution modelling and assessment of human exposure to air pollution. In recent years, he has worked with low-cost air pollution sensors for pollution mapping and personal exposure monitoring.

Torben Sigsgaard is Professor in Environmental and Occupational Medicine with more than 30 years of experience in studying human exposure to airborne contaminants and health. He is a co-PI in BERTHA, Denmark's Big Data Centre for Environment and Health at Aarhus University.

Henriette Thisted Horsdal is a senior researcher in the National Centre for Register-based Research at Aarhus University. She is in charge of managing the Danish national-level registries data and a data manager of BERTHA—Big Data Centre for Environment and Health, Aarhus University (Denmark).

Carsten B. Pedersen is a Professor in the Department of Economics and Business Economics (Aarhus University). He is also Head of the National Centre for Register-based Research and the Centre for Integrated Register-based Research at Aarhus University. Carsten is a co-PI in BERTHA, Denmark's Big Data Centre for Environment and Health, based at Aarhus University.

Jibran Khan a postdoctoral researcher in the Department of Environmental Science and BERTHA—Big Data Centre for Environment and Health, Aarhus University (Denmark). His research focuses on using GIS, Remote Sensing, and machine learning to study human exposure to environmental risk factors, including air and noise pollution.

Chapter 18
Urban Energy Systems: Research at Oak Ridge National Laboratory

Budhendra Bhaduri, Ryan McManamay, Olufemi Omitaomu, Jibo Sanyal, and Amy Rose

Abstract In the coming decades, our planet will witness unprecedented urban population growth in both established and emerging communities. The development and maintenance of urban infrastructures are highly energy-intensive. Urban areas are dictated by complex intersections among physical, engineered, and human dimensions that have significant implications for traffic congestion, emissions, and energy usage. In this chapter, we highlight recent research and development efforts at Oak Ridge National Laboratory (ORNL), the largest multipurpose science laboratory within the U.S. Department of Energy's (DOE) national laboratory system, that characterizes the interactions between the human dynamics and critical infrastructures in conjunction with the integration of four distinct components: data, critical infrastructure models, and scalable computation and visualization, all within the context of physical and social systems. Discussions focus on four key topical themes: population and land use, sustainable mobility, the energy-water nexus, and urban resiliency, that are mutually aligned with DOE's mission and ORNL's signature science and technology capabilities. Using scalable computing, data visualization, and unique datasets from a variety of sources, the institute fosters innovative interdisciplinary research that integrates ORNL expertise in critical infrastructures including energy, water, transportation, and cyber, and their interactions with the human population.

18.1 Introduction

The Earth is urbanizing rapidly, experiencing an unprecedented rate of population growth that is increasing demand for energy, food, water, and other natural resources, and raising concern about environmental impacts and matters of human security such as poverty, crime, and pandemics. Urban areas account for 67–76% of global final

B. Bhaduri (✉) · R. McManamay · O. Omitaomu · J. Sanyal · A. Rose
Oak Ridge National Laboratory, Oak Ridge, USA
e-mail: bhaduribl@ornl.gov

R. McManamay
Baylor University, Waco, USA

W. Shi et al. (eds.), *Urban Informatics*, The Urban Book Series,
https://doi.org/10.1007/978-981-15-8983-6_18

energy consumption, and 71–76% of fossil-fuel-related CO_2 emissions (Seto et al. 2014). Increases in urban energy use have mirrored the growing global population, increasing urbanization promoted by the migration of population from rural to urban areas for a better quality of life, and rapid evolution of housing, transportation, food, and water, and other associated infrastructures necessary to support urban lifestyle. According to a recent estimate by the World Health Organization (WHO 2019), the urban population in 2014 accounted for 54% of the total global population, up from 34% in 1960. Following this trend, it is widely anticipated that over 70% of the world's nine billion population will live in urban areas by 2050. Also, by 2050, there will be a nearly 50% increase, compared to 2018, in the consumption of energy, water, transportation, healthcare, urban infrastructure, and food (U.S. EIA 2019). Most of this growth comes from countries where strong economic growth is driving demand, particularly in Asia. While generation and consumption of electricity dominate urban energy use, it is a combined effect of the growing population and per capita electricity consumption which is higher for developed countries.

Urban areas are characterized by the complex interactions between the critical infrastructure components, such as buildings, utility networks, and mobility systems, and their users at multiple spatial and temporal scales. There are tremendous opportunities to design optimal, resilient urban systems by exploiting the inherent complexity of these interactions; for example, assessment of the impact of new technologies changing the dynamics between energy end-users and distribution and storage systems. Our ability to observe and measure through direct instrumentation of our environment and infrastructures from buildings to the planet scale, coupled with the explosion of data from citizen sensors, provides a unique opportunity to manage and increase efficiencies of existing built environments as well as design a more sustainable future. We can take advantage of both the enormous amounts of spatial and non-spatial data, in traditional and non-traditional forms, as well as new approaches in data science, particularly in geospatial applications, to answer questions for which data had previously not available.

With its mission to deliver scientific discoveries and technical breakthroughs that accelerate the development and deployment of solutions in clean energy and global security, coupled with leadership-class data and high-performance computing infrastructures, the Urban Dynamics Institute (UDI) at ORNL was established in 2014 to develop novel science and technology to observe, measure, analyze, and model urban dynamics from the city to the global scale. UDI's research themes focus on key urban energy issues that drive energy demand, consumption, and efficiency, and efforts to address questions such as: How does distribution and morphology of human settlements and associated population influence energy usage? How do we design mobility systems that make urban transportation energy efficient? How does water use for urban energy production impact our ecological systems? How do we design urban infrastructures that enable cities to reduce energy and environmental costs? To illustrate some of ORNL's contributions to the understanding of such complex urban systems, the following sections are organized into four key themes that reflect the primary dynamics of urban energy systems and have the potential for data-driven analysis:

1. Population and land use: Provide insights into the evolving spatial and sociodemographic patterns of human population distribution and activity that respond to and transform urban landscapes and systems at varying spatial and temporal scales.
2. Sustainable mobility: Improve transportation sustainability, safety, and accessibility through enhanced understanding of the energy and environmental implications of emerging transportation systems and their interdependencies with other critical infrastructures.
3. Energy-water nexus: Maximize the efficiency, sustainability, and resiliency of interconnected energy and water systems in the planning, development, and operation of urban infrastructures.
4. Urban resiliency: Enhance understanding of the physical and cyber-risks, challenges, and opportunities of the integrated framework of population, energy, water, transportation, and policy to improve reliability and resiliency of infrastructure services under changing and extreme climate conditions.

18.2 Population and Land Use

One of the biggest challenges in urban energy applications is the lack of data for population and land use that would be required to adequately investigate urban issues, particularly those tied to energy access and use. Further, even when data are available, the resolution of the analyses we would like to conduct is often much finer than the data available in support. In this section, recent innovative approaches developed at the UDI are discussed that address existing data gaps so that energy access and consumption patterns may be better modeled and evaluated both locally and globally.

18.2.1 Big Data and GeoAI to Create Population and Land-Use Data

Urban areas continue to grow both in expanse and magnitude of population, which heightens the need for increasing environmental awareness. Population distribution and dynamics data are foundational to assessing energy demand and usage patterns, which in turn guide energy generation and distribution scenarios. For the past two decades, ORNL has provided the community with LandScan Global fine-resolution (1 km) population distribution data for the world utilizing global-scale remotely sensed data through a smart interpolation technique (Bhaduri et al. 2002). This approach was further extended to LandScan USA, a 90 m population distribution and dynamics dataset for the USA, that used over sixty different geographic datasets to create both nighttime residential and daytime population (Bhaduri et al. 2007). Recently, Weber et al. (2018) have demonstrated a further refinement of this smart interpolation approach for census-poor regions by developing 90 m population distribution estimates for Nigeria, where human-settlement data from fine-resolution satellite images, categorization of settlements in different land-use classes, and population-density appraisals from census-independent sources were employed.

Understanding the existing structures of cities and their futures is an important component of urban sustainability and resiliency, particularly for assessing present and future energy usage. Up-to-date and highly resolved land-use maps allow researchers, policymakers, and other stakeholders to inform the better allocation of resources to communities. However, accurate and complete land-use data remain scarce for most of the developing world. Even in the developed world, this information is often geographically disjointed and incomplete. An important step in addressing this need is to develop robust, scalable, and automated methods to differentiate development patterns in fine-resolution satellite imagery by semantic segmentation. A recent collection of work by ORNL researchers (Arndt et al. 2019; Kurte et al. 2019; Lunga et al. 2018; Yang et al. 2018) have tackled various challenges associated with developing machine-learning models for urban-feature characterization and extraction. CNN-based deep-learning methods were used for automated land-use classification and to develop a typology for urban land-use data that captures the variation in structural patterns within cities. These development patterns, or more generally land use, can be used to spatialize variables within cities. These variables can include socioeconomic indicators such as electricity consumption patterns, as discussed later in this section, which are traditionally difficult to capture. Given that land use is shaped by human activities, researchers have utilized cellular phone-call data records (CDR) to infer land use. Using tower-based call data from Dakar, Mao et al. (2017) analyzed aggregated call volume and applied non-negative matrix factorization to identify fundamental behavioral classes of human activity patterns, and successfully inferred two fundamental land-use patterns: commercial/business/industrial (C/B/I) and residential (Fig. 18.1).

Evaluating energy consumption patterns, particularly in conjunction with highly resolved maps of settlement types, can be a useful first step in identifying areas that lack access to energy and other urban services. Many of these areas are considered slums, housing nearly 1 billion people worldwide (UN Habitat 2016). On a global scale, locating and monitoring the magnitude and composition of these areas is critical for making progress toward improving the lives of those who live there. This goal is the focus of the Millennium Development Goal 7 Target 7D (http://www.un.org/millenniumgoals/), "to have achieved by 2020 a significant improvement in the lives of at least 100 million slum dwellers", as well as a proposed measure of the Sustainable Development Goal 11 Target 11.1 (https://sustainabledevelopment.un.org/), "By 2030, ensure access for all to adequate, safe and affordable housing and basic services and upgrade slums".

Recent work by Brelsford et al. (2018; and see https://www.youtube.com/watch?v=YuRjeUkNf9o) shows how maps of these areas can be put into action. Once slums are identified, this study shows how we can address the problem of accessibility in these neighborhoods using topological analysis. Ultimately, the study revealed that urban slums showed a different topological structure than that of developed

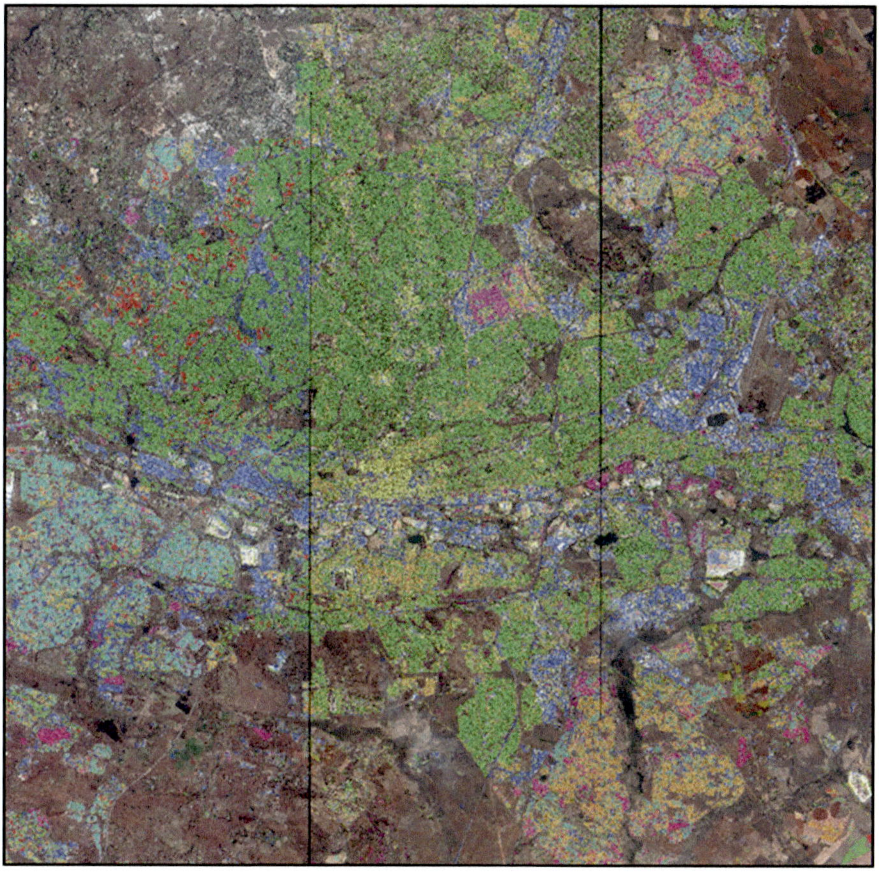

Fig. 18.1 Different land use in Johannesburg, South Africa, delineated from deep learning on fine-resolution satellite imagery. Various residential areas are shown based on different levels of formality of structures

cities—a critical piece of information to address the problem of accessibility to services. This work investigates the potential to increase that accessibility in these areas with minimal cost by growing road networks in existing slums and demonstrates its effectiveness through examples in Mumbai, India; Cape Town, South Africa; and Harare, Zimbabwe.

18.2.2 Estimating Urban Electricity Use in Data-Poor Regions

In many parts of the world, sustainable and universal energy access is a persistent challenge. This is particularly problematic considering that urban areas, which are the most rapidly growing areas of population, presently consume around three-quarters of the global energy supply. Understanding these urban-energy consumption patterns would be a strong first step toward addressing challenges as related to urban sustainability and energy security. Yet the required urban-energy datasets are virtually non-existent for the developing countries where this information is most critical. This creates an urgent need to develop new research methods for capturing and quantifying urban-energy use patterns. Without available urban-level energy statistics, capacity building and accessibility planning and assurance become prohibitive, particularly in data-poor regions of the world where future urban growth is expected to be the largest.

In a recent study conducted by Roy Chowdhury et al. (2020), a data-driven approach to characterize urban settlements based on their formality was conducted to assess intra-urban-energy consumption in three cities. Since electricity is the fastest-growing energy fuel, the premise of the study is to evaluate the relationship between urban settlement types and corresponding nighttime light emission, which is considered a proxy of electricity consumption. This study presents an approachable and scalable solution to fill the existing data gap to better understand differential electricity consumption patterns.

Three cities in the developing world—Ndola, Zambia; Sana'a, Yemen; and Johannesburg, South Africa—were used in this study as they collectively displayed considerable variation in population size and socioeconomic characteristics. These variations were useful in order to examine which characteristics may result in distinct electricity consumption profiles. Following an approach developed by Yuan et al. (2015), human settlement areas within these cities were classified into different functional types. Those distinct settlement types were then correlated with nighttime lights emission from VIIRS DNB (https://earthdata.nasa.gov/viirs-dnb) data following the assumption that lights are a reasonable socioeconomic indicator and can help us to understand electricity consumption. In all three study cities, a statistically significant correlation between human settlement types and nighttime lights emission (considered as a surrogate of electricity consumption) was discovered, which demonstrates the potential to develop and generalize this method to other geographic areas in order to understand energy consumption patterns within cities, specifically when no other data are available.

The data-driven approach captured in this study not only mitigates issues where no ground information is available, but the patterns of energy consumption that are uncovered can be used in myriad analyses, particularly when combined with other information such as land-use maps, to inform urban planning where energy resources may be limited (Fig. 18.2).

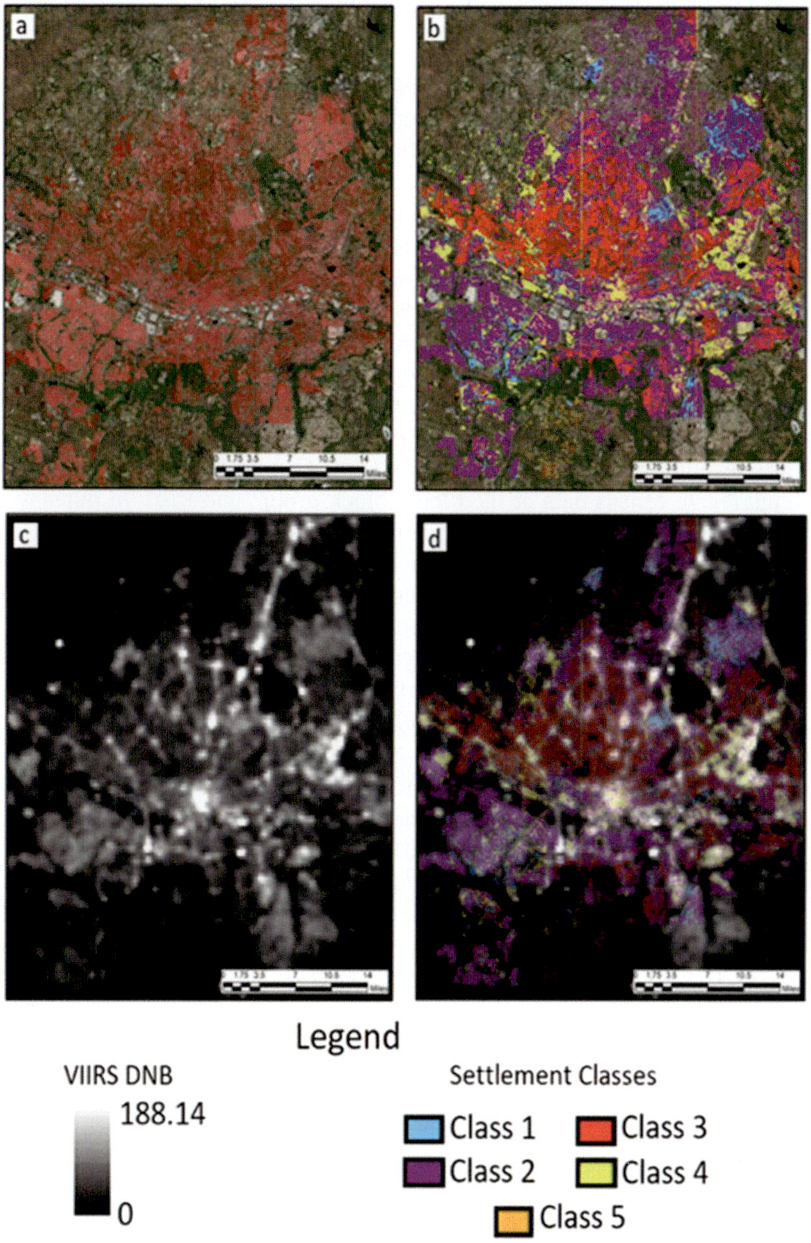

Fig. 18.2 Clockwise from top-left: Settlement map, settlement classes, settlement classes overlaid on VIIRS DNB image, and VIIRS (from Roy Chowdhury et al. 2020)

18.2.3 Estimating Household-Level Energy Consumption

Understanding residential energy consumption patterns is of critical importance since this sector alone accounts for nearly 30% of all energy consumption worldwide (IEA 2016). One limitation of current approaches to model energy consumption is that they are highly dependent on region-specific data sources requiring building-level detail, which are generally not openly available. Surveys that capture population and housing characteristics are commonly conducted for small segments of the population and provide household-level or individual-level samples from a single neighborhood, city, region, or country to provide very detailed information. Although these data contain considerable sociodemographic depth, they are not available for a full population. To address this disparity, synthetic spatial microdata—a high-performance, data-driven simulation of the American population—for modeling urban dynamics, termed UrbanPop, were developed at ORNL to simulate the American population with fine-resolution human demographics (e.g., Census block/block group) that match aggregate census data at the block, block group, and tract. In other words, given a set of demographic attributes of interest, the algorithm can recreate joint distributions of these attributes at the block or block-group level that when aggregated, return the census results within a certain margin of error. The algorithms in UrbanPop consider the full demographic profile of commuters and trace the movements of the profile from the nighttime (home) and daytime (work).

In a recent study by Morton et al. (2017a, b), a fine-resolution residential electricity consumption model was developed by merging a dasymetric model with a complementary machine-learning algorithm. The foundation of this approach is the use of publicly available data, supporting a model that is applicable to a wide range of regions. The authors used UrbanPop data to estimate residential energy consumption, combined with the 2008–2012 household-level Public Use Microdata Sample (PUMS; https://www.census.gov/programs-surveys/acs/data/pums.html) of the American Community Survey (ACS), to provide detailed demographic and household characteristics, as well as the average monthly electricity cost per household. The 2008–2012 ACS summary tables, which contain both tract- and block-group-level average totals, were used as constraints. The model was tested on three counties in Tennessee (Anderson, Knox, and Union) by using a dasymetric approach to disaggregate a weighted sample of surveyed households into smaller geographic areas and then using a learning algorithm to estimate electricity consumption for each of the households. These estimated values at the household-level were then aggregated to larger areas for analysis.

This approach demonstrated its utility by estimating and evaluating aggregate block-group-level residential consumption within a growing urban area. Further, it also provides a well-defined method for handling the uncertainty that enters into the model via input data sources. The ability to estimate the residential energy consumption while still capturing measures of uncertainty provides analysts with an improved set of data to evaluate spatio-demographic factors that may impact energy use. This deeper understanding can then translate into the implementation of effective energy-efficiency measures, particularly in urban areas that are experiencing rapid growth.

This study illustrates a practical path forward for estimating highly resolved energy consumption patterns while overcoming data limitations through the use of openly available data. Although this specific study does not include a formal validation process, both internal and external validations have been conducted on the algorithm used here (Rose and Nagle 2017).

18.3 Sustainable Mobility

In 2018, vehicles moved an estimated 11 billion tons of freight, more than $32 billion of goods per day, and traveled 3 trillion vehicle miles in the USA according to the U.S. Department of Energy's Vehicle Technologies Office. Transportation typically accounts for about a third of all energy used in the nation, and developing sustainable transportation solutions is imperative as the nation's economy expands and the global economy grows. In recent years, the word mobility is increasingly used to refer to various aspects of human interactions with transportation systems. Mobility encompasses the notion of being inclusive of multi-modal transportation options, smart connectivity, crowdsourced data-enabled transportation alternatives, ride-hailing and ride-sharing options, as well as system-scale efficiencies for transportation system design. Clearly, developing sustainable means of mobility has societal, economic, as well as environmental benefits (Bigazzi and Bertini 2009). Recent advances in ubiquitous sensing, big data, social media platforms, and the growth of app-based mobility options has heralded an unprecedented shift in not just mobility but also vehicle ownership. Increasingly, people are considering not owning vehicles and accessing their mobility needs as a provided service.

Significant changes are also here from an infrastructural standpoint. The variety and types of deployed sensors on and about roadways have gone up. Typically, cities these days have a number of fine-resolution cameras deployed with real-time video feeds, radar detector sensors every few hundred yards recording speed and volume every couple of seconds, induction loops coupled with spherical cameras detecting stationary queues and turning vehicles, control algorithms to coordinate signals, as well as Bluetooth sensors to detect flows through urban environments. Coupled with advances in connected and automated vehicles, opportunities are ripe for data-driven system-wide approaches for control and optimization.

18.3.1 Human Interactions with Transportation Systems

In complex urban environments, population, transportation, building energy, and urban climate are interdependent. The modeling of each individual component is fairly mature; however, the modeling and simulation of complex urban interactions pose significant challenges. By coupling the individual systems, one moves from studying different aspects in isolation toward studying a city as a whole. Active transportation can be defined as any self-propelled, human-powered mode of transportation, such as walking or bicycling that is often mixed with public transportation and helps to alleviate congestion, reduce energy consumption and greenhouse gas emissions, and fight against chronic health conditions such as obesity, diabetes, heart disease, and stroke. Promoting active transportation modes requires analysis of factors that substantially influence a transportation mode-choice process. Each transportation mode has a unique set of influencing factors for individuals, including sociodemographic attributes, transportation cost and network characteristics, and social interactions. This emphasizes the need to understand macro aspects of transportation-mode choices by modeling millions (or even billions) of commuters and their complex, simultaneous, and mutually dependent decision processes. Agent-based modeling and simulation (ABM) approaches offer a mechanism to represent such a complex system as a collection of autonomous agents and their environments, in which the agents interact with one another and with their environments.

Recent research by Aziz et al. (2018a, b) and Park et al. (2018), explored the effects of traffic safety, walk-bike network facilities, and land-use attributes on walk and bicycle mode-choice decision in New York City for the home-to-work commute. Applying the flexible econometric structure of random parameter models, they captured the heterogeneity in the decision-making process and simulated scenarios considering the improvements in the walk-bike infrastructure such as sidewalk width and length of bike lanes. They utilized fine-resolution sociodemographic data from UrbanPop to estimate likely night and day locations for individuals matching a demographic profile, and suggested appropriate origins and destinations (OD pairs) for synthetic commuters. The determination of OD pairs is a fundamental input for transportation and mobility applications. Using the UrbanPop simulated population, an agent-based model was implemented on ORNL's Titan supercomputer (Park et al. 2018) to simulate mode choices for commuters in New York City, and how these mode choices might be tipped in favor of bike or walking (Park et al. 2018; Morton et al. 2017a, b). Creating agent-based models from the simulator allows the exploration of how improvements in sidewalk conditions or having bike lanes may impact commuter choices to bike or walk instead of driving or using public transit. The results from the New York City case study indicate that infrastructure investments such as widening sidewalks and increasing bike lane networks can positively influence active transportation mode choices (Fig. 18.3). The impact varies with geographic locations. The ABM simulation results indicate that social promotions focusing on active transportation can positively reinforce the impacts of infrastructure changes.

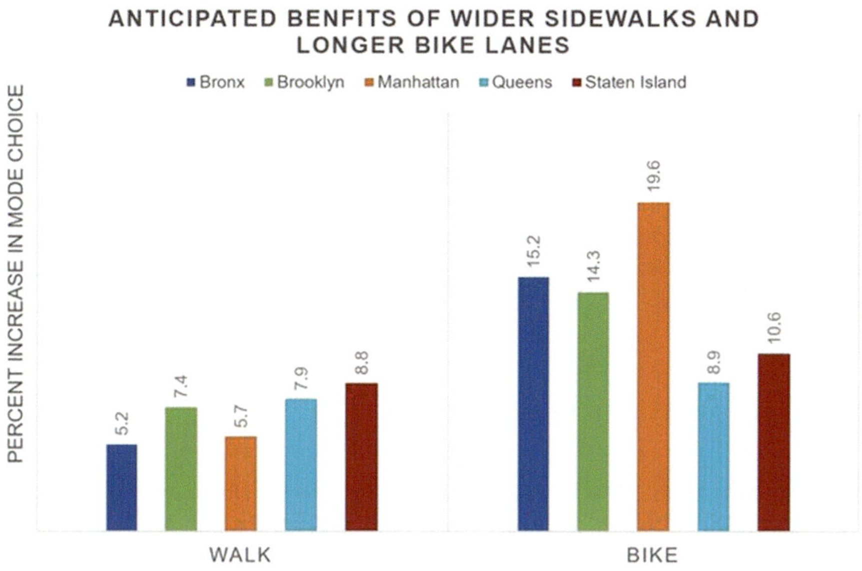

Fig. 18.3 Effect of wider sidewalk and building more bike lanes in the five boroughs of New York City

18.3.2 Emerging Options for Freight Delivery for Businesses

Freight, and particularly intra-city freight delivery, is a key aspect of the course of business activity that depends on mobility. Due to the relatively recent shift in consumer preferences to purchase items online rather than making purchases in brick-and-mortar stores, and the preference for next-day and same-day delivery, logisticians and parcel delivery companies have been prompted to search for new ways to move and deliver parcels to improve efficiency and reduce costs associated with energy usage. ORNL conducted a study to consider innovative modes of parcel delivery, and modal configurations involving multiple modes of freight transport, with a focus on the last mile (Moore 2019). The data for this study consisted of GPS traces of delivery-truck tours from a portion of the truck fleet at the UPS depot outside of Columbus, Ohio. Delivery locations were extracted from the dataset and used along with socioeconomic and land-use data obtained from the metropolitan planning organization for Columbus, to develop a delivery-demand model to estimate parcel deliveries in areas lacking GPS data.

Alternative scenarios were developed involving the use of electric Class-Six trucks, electric delivery vans, parcel delivery lockers, the use of drones, as well as electric passenger vehicles. Energy usage in kilowatt-hour per mile was estimated for the scenarios and compared with energy estimates for the baseline case involving the standard Class-Six delivery truck. The findings suggest that electric Class-Six delivery trucks paired with parcel delivery lockers reduce energy usage, especially in suburban neighborhoods. The findings also suggest the use of parcel lockers in suburban areas, which typically have less connectivity and more cul-de-sacs. Pairing both electric Class-Six delivery trucks with parcel lockers significantly reduced energy usage in outlying TAZs (traffic analysis zones) in suburban Columbus. The scenarios involving drones, on the other hand, were found to be energy-intensive, and suggest the need for more optimized drone scenarios which consider improved drone technology, such as increased battery range and payload, and more efficient use of the technology, possibly including the use of multiple drones, mid-air transfers, and improved flightpaths.

18.4 Energy–Water Nexus

To date, there is no widely accepted and consistent definition of the energy-water nexus, although the EWN is broadly conceptualized as the interdependencies between the energy and water, such as the water required to produce electricity, or the amount of electricity required to treat and distribute water. However, when applied to urban dynamics and informatics, defining the EWN becomes even more obscure. For instance, in the context of urban systems, the need to expand the EWN to consider linkages and dependencies among other sectors, such as agricultural development and natural and human-built environments, becomes quite apparent. Planning for urban growth or infrastructure expansions requires understanding complex relationships and feedbacks among multiple sectors, and the potential consequences of population growth and climate extremes on infrastructure resilience, operations, and resource availability and stress. Characterizing these relationships requires consideration of appropriate scales and overcoming challenges to data and analytical limitations. In this section, we expand upon research within the Urban Dynamics Institute that has used informatic-type approaches to explore the urban EWN through consideration of scale and removing obstacles to data challenges. First, however, we discuss the importance of scale, and data and analytical challenges, to linking the EWN to urban informatics.

Scale considerations As with all research that examines the hierarchical complexity of systems, the difficulty of developing a consistent working definition of the EWN is a matter of scale (Allen and Star 2017). For example, the broadest definition of the EWN includes research spanning multiple spatial and temporal scales, from developing efficient membrane technologies for desalination (micro-scale) to agent-based modeling of electricity and water use by water treatment systems (meso-scale), to the development of plausible socioeconomic scenarios of future global communities (macro-scale). In this respect, a focus on urban dynamics actually helps to constrain the scope of the EWN in the following ways. First, an urban focus imposes a requirement of scales that examine collective behaviors of more than one human, who might move substantial distances within short periods of time and utilize a range of resources that impact many sectors that are internal and external to urban boundaries. Second, dynamics suggests a need to understand the behavior of systems, which are composed of multiple interacting parts. Finally, a central construct for the ORNL Urban Dynamics Institute is that almost all research has a spatial or mappable component. Hence, when we apply these constraints to the field of multi-sector research, the scales would indeed be restricted to consider spatial units no smaller than neighborhood levels (possibly buildings), whereas the temporal scales remain unrestricted.

Challenges Accurately depicting and characterizing multi-sector relationships and interdependencies comes with many challenges, primarily related to data. These include limited data availability for both energy and water infrastructures and use, mismatches in spatial and temporal scales of data across different sectors, heterogeneity in data types, and lack of standards for data collection and availability (US DOE 2014; Zaidi et al. 2018). For example, Chini and Stillwell (2018) reported that data on urban water resources are highly limited, and data on energy requirements for water treatment and distribution are virtually absent. Obviously, this prohibits the accurate characterization of urban-energy–water dynamics to support infrastructure investments and predict resiliency under climate uncertainty. Even if data are available, practitioners and research communities may be unaware of the wealth of analytical approaches that are available for characterizing urban EWN dynamics (Allen et al. 2018). Possibly more troublesome is how to integrate the disparate modeling platforms that are used to characterize patterns and processes within different sectors (Brewer et al. 2018). Furthermore, the multi-dimensionality and sheer complexity of the EWN, in conjunction with limited data, may constrain which components and relationships are evaluated, leaving major gaps of knowledge in understanding the implications of urban growth for sustainability and resiliency.

EWN interface with Urban Dynamics Institute To address these challenges, ORNL, through support from the DOE Biological and Environmental Research Integrated Assessment Research Program, developed the Energy–Water Nexus Knowledge Discovery Framework (EWN-KDF) (https://climatemodeling.science.energy.gov/projects/energy-water-nexus-knowledge-discovery-framework). The KDF provides a data management and geovisual analytics platform to enable efficient characterization of energy-water relationships and decision making regarding present and future infrastructures (Bhaduri et al. 2018). As stated previously, obstacles to discovering complex relationships within the EWN relate to time expenditures associated with the acquisition and storage of data, but also the fusion of disparate data sources and data types from mismatched spatiotemporal scales. In part, the KDF platform expedites this process by harnessing Argonne National Laboratory's Globus cloud-data transfer service, which bypasses the need for the EWN community to download and manipulate data locally. The KDF also provides quick access to widely applicable climate, physical (or physiographic), and socioeconomic datasets. To address the challenge of accelerating knowledge discovery, the KDF provides real-time coupled analytic and visualization capabilities for users to explore anomalies or anomalous behavior in datasets as well as spatiotemporal clustering and trend analysis. As an example, suppose a user desires to understand complex spatial and temporal relationships (or tradeoffs) among land and water use in regions experiencing elevated population growth and water stress. A commonly used dataset available through the KDF is the US Geological Survey's Water Use in the United States (USGS 2018), which provides county-level estimates of surface and groundwater use among eight major economic sectors from 1985 to 2015. The KDF also assembled land-cover estimates within counties for the same period of record. To allow users to explore spatiotemporal patterns, the KDF provides dynamic time warping, which uses algorithms to measure the similarity between temporal sequences, such as water-use and land-cover changes over time. Similarity matrices are seamlessly incorporated into clustering algorithms to explore regions or counties that share similarities in temporal signatures or behaviors. These analytics and visualizations are rendered in real time, allowing users to quickly explore and understand dynamic patterns; it would take hours, if not days, to conduct analogous exploration on local machines. By increasing the rate at which users can observe new phenomena, the KDF creates a robust learning platform that changes the rate and nature of hypothesis generation for urban EWN dynamics.

Another application of EWN to urban dynamics is through examining dependencies between cities and their neighboring regions. To support the resource demands of dense populations, cities rely on expansive infrastructure that supplies numerous commodities, such as energy, water, food, and material goods and services (Ruddell et al. 2014). Therefore, city and utility governance must remain cognizant of these external supply chains, as well as how offsetting their resource burdens to outside regions induces stress on natural resources, particularly water availability (McManamay et al. 2017). These increasing stressors are important to quantify, as limited resource availability makes cities more vulnerable to climate extremes. However, a significant challenge to effective decision-making across sectoral boundaries is that of transcending disparate policies and jurisdictions, since each sector is governed by different entities, which operate on different scales and rely on different information. For instance, how does a city planning official translate population growth and land zoning at the parcel scale into estimates of stress on water intake and treatment infrastructures at the stream level (i.e., water policies), or stress to the electricity grid at the power-plant level (i.e., energy policies)? Creating spatially explicit maps of interconnected infrastructures and relationships between demand and regional sources of commodities provides transparency and interpolicy coordination to all parties involved in planning for future urban growth. Of course, for the reasons stated earlier, capturing these relationships is difficult due to limited data availability, heterogeneous data, or mismatched scales.

A couple of recent projects through ORNL's UDI use informatics to overcome these challenges by developing spatially explicit interconnections between cities and their regional infrastructures. One example is the development of city energy sheds, that is, a region outlying an urban center and comprised of the transmission infrastructure and electricity production at powerplants that are required to offset high electricity consumption occurring within urban areas (McManamay et al. 2017; DeRolph et al. 2019; Fig. 18.4). Over 100 US cities have established goals to transition to 100% renewable energy (Sierra Club 2018); however, detailed strategies for how to make these transitions effective vary immensely across cities. Furthermore, we surmise

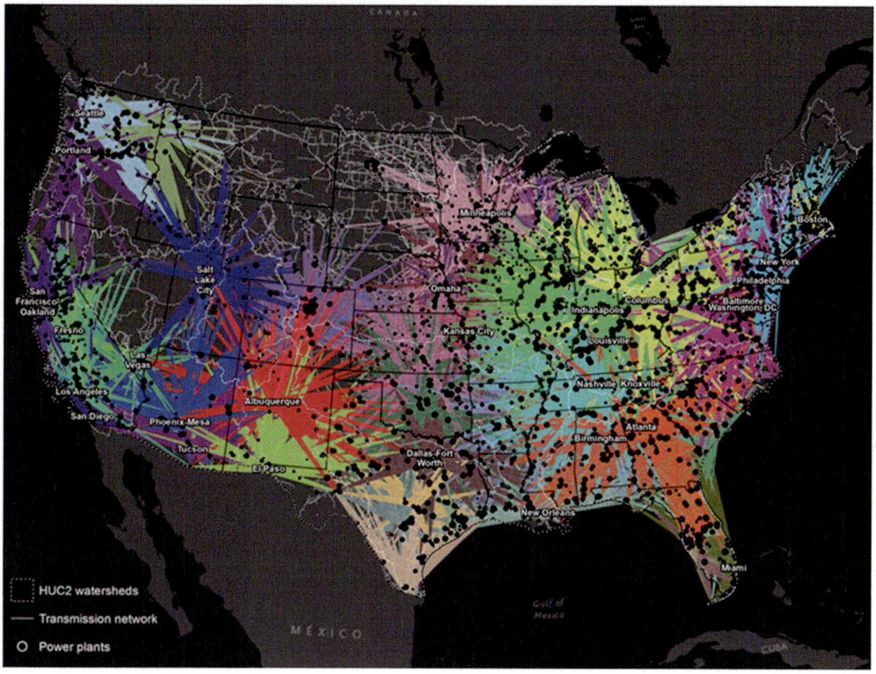

Fig. 18.4 City energy sheds depicting sources of electricity supplying urban epicenters. Taken and modified from DeRolph et al. (2019)

that most city governance and sustainability officials are unaware of the electricity footprint and the magnitude of infrastructural investments required to make these transitions. Using available information on transmission and substation infrastructure and electricity production at powerplants, DeRolph et al. (2019) used a market-share network allocation optimization in ArcMap (Esri, Redlands CA) to balance the electricity grid for the conterminous USA The grid was amended to include connections between substations and census block groups, and annual electricity demand was downscaled from state-level electricity consumption (from the Energy Information Administration). Such an exercise was computation-intensive: The grid considered that any one of the nation's 200,000 block groups could receive electricity from >5000 power plants weighted by transmission voltage, which creates over 1 billion unique combinations; however, electrical impedance increased with distance, and lower transmission voltages were used to constrain the optimization. By isolating only block groups within urban boundaries, DeRolph et al. (2019) identified the powerplants providing the majority of a city's electricity demand (Fig. 18.4). Additionally, this provides a template to quantify a city's indirect carbon and water footprints through electricity production. The analysis yielded very important insights: First, the majority of US cities, especially those with aggressive renewable-energy transition plans, have energy mixes that are far from attaining 100% renewable status.

Hence, the transition will require massive infrastructure investments. Secondly, those cities facing electricity congestion challenges from immense population growth and electricity demand do not consistently have public support or local and state policies to enable renewable energy transitions to meet growing demands.

Understanding the implications of city growth on regional water availability is also critical. Another UDI project examined the fine-resolution impacts of city land transformation, electricity production, and water supply infrastructure on hydrologic alteration and biodiversity loss in streams (McManamay et al. 2017). Such an analysis requires multiple steps to isolate the individual effects of city infrastructures on aquatic ecosystems experiencing cumulative anthropogenic stress from areas outside the influence of cities. Furthermore, each step has unique information challenges: (1) estimating commercial and residential energy and water demands at fine resolutions, (2) mapping detailed infrastructures required to meet those demands, (3) geospatially summarizing infrastructures in ways meaningful for stream-network analysis, (4) using statistical models to estimate hydrologic alteration, (5) statistically isolating roles of individual sectors in contributing to cumulative hydrologic alteration, and (6) assembling biodiversity occurrence information to estimate species losses due to urban drivers. We highlight a few ways in which informatics approaches provided opportunities to characterize these complex relationships. Landscape alterations induce up-to-downstream impacts on river systems; therefore, predicting how infrastructures may alter hydrology required accumulating geospatial information for dendritic stream networks. Additionally, translating these geospatial variables into measures of hydrologic alteration requires either calibrating mechanistic models (i.e., time consuming) or using novel statistical approaches, which are far less time consuming, but no less accurate. McManamay et al. (2017) summarized geospatial variables in NHDPlus stream reaches (Horizon Systems Corporation 2019) using the network analyst in ArcMap, and then assembled discharge information for streams from the US Geological Survey National Water Information System. After calculating metrics depicting hydrologic departures from natural or reference conditions, the authors then used machine-learning algorithms (random forests) to relate geospatial characterizations of city infrastructures to hydrologic alterations at the stream-reach level. Isolating the roles of individual sectors (e.g., electricity production, water supply) on hydrologic conditions in streams becomes very difficult in situations of compounded stress from upstream sources. Hence, McManamay et al. (2017) extracted partial dependency functions (PDFs) from random forests to estimate how individual variables (or combinations of variables) associated with a given sector influence hydrologic conditions. Once sector-specific hydrologic alterations were isolated in streams, millions of occurrences of aquatic species were organized by taxa and conservation concern and then overlain with those areas to characterize the city–aquatic biodiversity nexus.

A remaining challenge of supporting multi-sector decision-making for urban dynamics is creating user-centric Web-visualization and analytic platforms. As a brief example, ORNL developed a stream classification Web application to guide decision-making for stream restoration and mitigation (McManamay and Derolph 2019a, b). Such a tool is highly relevant to urban dynamics, as stream restoration in the USA is related to remediating the impacts of urban landscape transformation (Bernhardt et al. 2005, 2007). The premise of the stream classification is guiding users to appropriately select reference streams to guide restoration practice, through the selection of streams that share similar physical typologies (McManamay et al. 2018). The Stream Classification Web-App allows users to query any of the nation's 2.6 million stream reaches and find streams that share similar natural properties or anthropogenic disturbance regimes. Unfortunately, seeking more complex platforms that support urban EWN dynamics induces tradeoffs between flexibility, application breadth, and computational expense. For instance, one strategy might provide highly flexible applications seeking maximum relevance to a wide spectrum of user groups, but possibly only supporting superficial decision making. The opposite endpoint might consist of applications with far less flexibility but substantial depth to support decision making from a narrow user group or a narrow range of applications. This tradeoff becomes critical when designing platforms for EWN relationships to urban dynamics, as finding an optimal balance between flexibility and provision of meaningful outcomes becomes very difficult when considering multiple sectors and their complex (and uncertain relationships). Nonetheless, platforms that achieve this optimal balance are in increasing demand from all sectors of government and the economy.

18.5 Urban Resiliency

Urban resiliency indicates how a city recovers better and stronger after a shock. Such a shock could be due to natural or humman-made disasters, failure of engineered infrastructure, economic downturns, and so on. Long-term climatic trends and short-term extreme weather events (e.g., 2011 earthquake and tsunami in Japan, 2012 Superstorm Sandy in Northeast U.S., 2018 Hurricane Maria in Puerto Rico, 2018 wildfires in Northern California, etc.) have renewed interest in the concept of urban resiliency. The resiliency of urban water and energy infrastructures is of relevance in this context. For example, in the longer term, estimating renewable energy potential, assessing existing renewable energy infrastructures, managing urban flooding with green infrastructures to minimize energy cost for pumping water out of flooded areas, reducing energy usage for snow and ice removal, and water-quality impacts from urban de-icing are of key interest for cities. For near-term disruptions, having a distributed renewable (solar) energy infrastructure builds resiliency when the electric grid is disrupted by disasters; and also developing a situational awareness for the nation's energy infrastructures is critical during the emergency preparedness, response, and recovery phases of natural or technological disasters. Consequently, researchers at ORNL are developing new methods and approaches for building a more resilient urban infrastructure by utilizing scientific and open-source data resources. In this section, three approaches are discussed that focus on one of the most important agendas that decision-makers will be facing in the coming decades—integration of resilience thinking into urban planning to improve response to known and unknown risks.

18.5.1 Renewable Energy-Infrastructure Assessment

Solar photovoltaic (PV) is the fastest-growing source of distributed generation of renewable energy. In fact, renewable-energy capacity is projected to expand by 50% between 2019 and 2024, led by solar PV. This increase of 1200 GW is equivalent to the total installed power capacity of the USA today. Estimating solar potential in urban environments, namely on building rooftops utilizing LiDAR-derived 3D elevation models with solar radiation data, has shown to be an effective approach (Nguyen et al. 2012; Latif et al. 2012; Kodysh et al. 2013). However, data for the actual spatiotemporal distribution of installed solar panels greatly benefits applications related to energy policy-making, power systems, and solar PV market analysis but was not available on a large scale till recently (Yu et al. 2018; Hou et al. 2019). Recognizing this data challenge, as early as 2012 ORNL researchers were among the first to develop a machine-learning approach based on a convolutional neural network (CNN) that exploited large-scale, fine-resolution (0.3 m) aerial imagery to efficiently and accurately detect rooftop-installed solar panels covering large areas in two US cities (Bradbury et al. 2016; Yuan et al. 2016).

18.5.2 Optimizing Energy and Safety Through Precision De-icing

In the USA, more than $1.5 billion is spent every year for winter road maintenance programs. In addition to these direct costs, each state in the country incurred between $300 and $700 million per year in indirect costs (Transportation Research Board 1991). As the number and severity of snowfall events grow, the need for safer urban roads during snowfall events is also growing. In 2014, the Pennsylvania Department of Transportation dispensed 686,000 tons of salt for road treatment; that is, 200,000 more tons than was used in the average year (Black and Arking 2014). While overtreating roads with salt and brine has energy, environmental, and financial burdens, undertreatment can lead to decreased safety on the roadways as described in a study that has shown that snow depth correlates with the number of traffic accidents (Seeherman and Liu 2015).

Road-treatment chemicals, such as brine solutions and common road salt, together with plowing, are effective tools for snow and ice removal. However, there are two challenges that impact the resiliency of cities during snowfall events: (1) lack of enough resources to treat all roads in a city, thus limiting social and economic activities in the city; and (2) excessive use of road salt increases urban environmental impacts. The first challenge is addressed by preselecting roads to be treated based solely on traffic counts. Thus, streets with high traffic volumes are treated, while feeder streets, trouble spots, and neighborhood roads often go untreated. Consequently, many residents are unable to safely make it to the treated roads, lowering the overall utility gained from the treated roads. With enough resources, all the roads in a city can be treated, thus leading to the second stated challenge.

The impacts of excessive use of road salt are: (i) increase salinity of groundwater and surface water adjacent to roadways, potentially impacting human health and resulting in localized decreases in the biodiversity of organisms; (ii) creation of unfavorable changes in the physical properties of roadside soils leading to increased surface runoff, erosion, and sedimentation of rivers and streams; (iii) increased corrosion rates of automobiles, highway components, steel reinforcement bars, and concrete; (iv) increasing incidence of vehicle-animal accidents—birds and mammals are attracted to road salt; and (v) decreasing health and vigor of roadside plants due to water stress and soil nutrient imbalances (Kelting and Laxson 2010).

In order to make more urban roads safer without using excessive road salt, researchers at ORNL developed a new metric called the Road Vulnerability Index to snowfall accumulation (RVI). The premise of this index is that road segments should be classified based on their capacity to melt snowfall quickly and their elevation value. The behavior of snowmelt in a given situation depends on temperature, precipitation, humidity, wind, and cloudiness (NRCS 2004). The developed methodology divides the urban roads into road segments of 50 m length as suggested in the literature (e.g., Chapman and Thornes 2011). The rate of snowmelt (RoSM), based on the thermodynamics of snowmelt, is then calculated for each road segment using the U.S. Army Corps of Engineers formulation (USACE 1998) during non-rainy periods and rainy periods. The incident solar radiation data is obtained using the hemispherical viewshed algorithm and LiDAR (Light Detection and Ranging) data (Kodysh et al. 2013). Using the rate of snowmelt and slope data, the road segments are then classified into RVI categories (Chapin et al. 2017) using the classification rules shown in Table 18.1.

The RoSM data are grouped into five classes based on their solar insolation values. The RVI has four categories: Least Vulnerable (1), Less Vulnerable (2), More Vulnerable (3), and Most Vulnerable (4) as shown in Table 18.1. A map showing the RVI categories for the City of Knoxville, Tennessee is shown in Fig. 18.4. The city has 6555 lane miles, of which 722 miles are classified as Categories 1 and 2 roads, 4916 miles are classified as Category 3 roads, and 917 miles are classified as Category 4 roads. Using the RVI approach, Category 4 roads need more attention and

Table 18.1 Classification rules for RVI categories

Slope	RoSM				
	1-Sunny	2	3	4	5-Shaded
0—Flat ($\leq$10% grade)	1	2	2	3	3
1—Incline ($\geq$10% grade)	2	2	3	3	4

Table 18.2 Cost of treating all roads in the city of Knoxville using the current method and the RVI method

Approach	Treatment option	Lane miles	Road-treatment gauge	Treatment cost per lane mile	Total cost for treatment
Current approach	All roads	6555	Full treatment	$49.80	$326,415
Proposed approach	Vulnerable roads	722	25% of full treatment	$12.45	$8989
	More vulnerable roads	4916	50% of full treatment	$24.90	$122,402
	Most vulnerable roads	917	Full treatment	$49.80	$45,655
	All roads	6555			$177, 046

should be given full treatment according to the current practice; Category 3 roads should remain safer with one-half of the full treatment; and Categories 1 and 2 roads should not need more than one-quarter of the full treatment to remain motorable. The simple cost analysis of Table 18.2 shows that using the RVI approach will not only reduce cost to about 54% of the total cost of treating all roads in the city using the current approach, but also will significantly decrease the amount of road salt used to achieve complete treatment (Fig. 18.5).

18.6 Situational Awareness of National Energy Infrastructure

The ability of the USA to effectively respond to and facilitate the restoration of energy infrastructures during disaster preparedness, response, and recovery depends on the ability of local, state, and federal government agencies, and private-sector electricity and fuel providers, to have access to timely, accurate, and actionable information about the status and potential impacts of energy-sector disruptions. Among the many critical requirements for decision support, two important challenges arise in (i) effective spatiotemporal representation of dynamic data and (ii) efficient integration of such data from disparate and distributed sources. This capability is currently provided by the U.S. Department of Energy (DOE) via its Environment for Analysis of Geo-Located Energy Information (EAGLE-I™) system that is developed and maintained at ORNL. EAGLE-I™ and associated energy-infrastructure awareness capabilities provides an energy-sector-specific wide-area visualization and serves as the authoritative federal source for historical and real-time situational awareness for the nation's energy infrastructure through the National Outage Map (NOM), which shows the number of customers without electricity for every county in the USA.

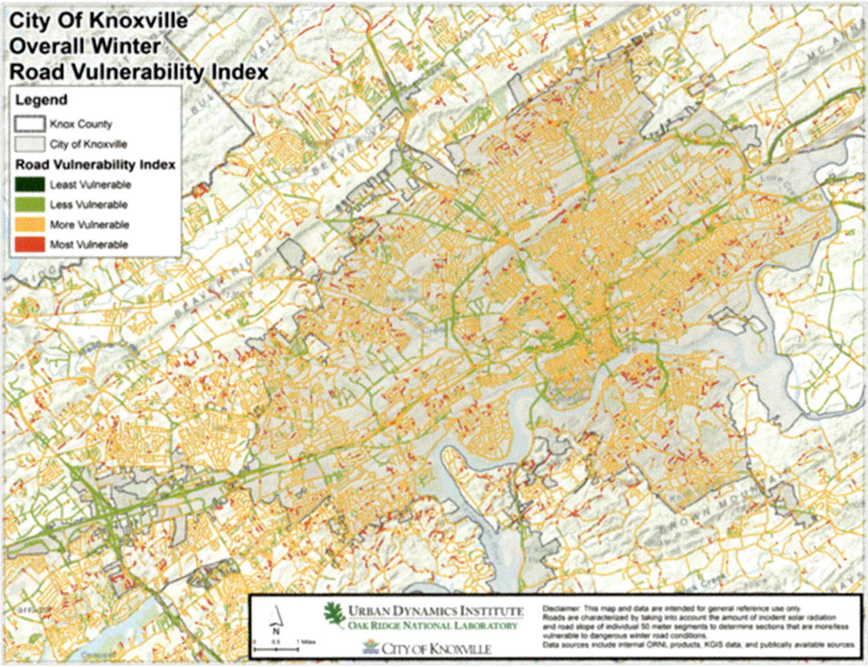

Fig. 18.5 RVI categories for each 50 m road segments in the city of Knoxville

Most utility companies provide customer outage status information covering their service regions via their websites. Having an integrated view of outage status across the nation is crucial for subject matter experts; but it is a challenging task because of data-source variations and changes, since utility companies may change the URLs of their outage information data sources and data formats. They may also support various data granularities such as latitude and longitude, county, zip code, city, census area, etc., they may change service areas, and they may need to handle too many utility companies. EAGLE-I^{TM} provides an integrated, NOM system that has been systematically designed and developed. It is composed of several Python scripts that scrape data from utility company websites, standardize and store collected information into database tables, and track erroneous scripts. This capability incorporates the most current and relevant data, to provide effective and comprehensive support for energy-infrastructure awareness and response capabilities (Fig. 18.6).

Timely detection of electricity outage and restoration is a critical component of situational awareness during disruptive events for utility companies and emergency responders. Restoration is often slow because of significant delays in gathering efficient power-outage information and problems in allocating limited power resources. Crowdsourced data from social-media platforms are an attractive source to assess electricity outage in near-real time. Recent research by Mao et al. (2018) provided a

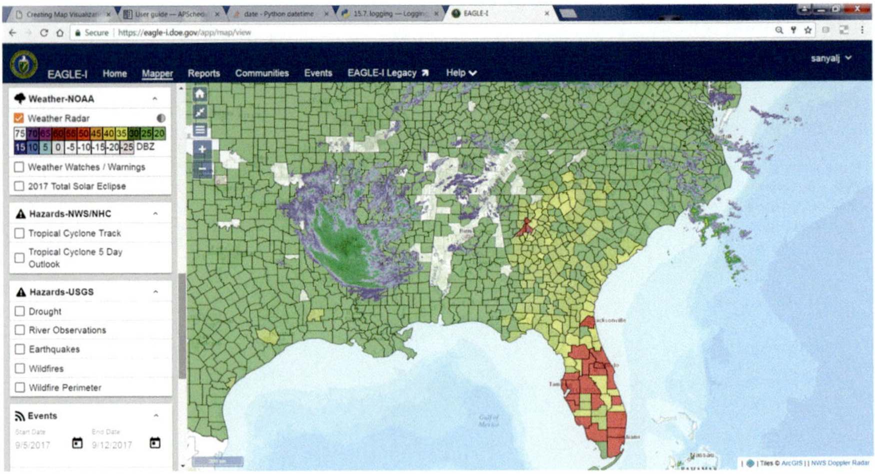

Fig. 18.6 Eagle-I™ displaying locations of over 7 million customers who lost electricity as an aftermath of Hurricane Irma in the southeastern USA during September 2017

novel two-stage framework based on machine learning and deep learning for power-outage detection from Twitter. First, a probabilistic classification model was applied to find true power-outage tweets. Subsequently, a new deep-learning method (bidirectional long short-term memory networks) was implemented to extract outage locations from text. Results showed a promising classification accuracy (86%) in identifying true power-outage tweets, and approximately 20 times more usable tweets can be located compared with simply relying on geotagged tweets.

18.7 Conclusion

As cities continue to grow and create more demand for resources, it is imperative that scientists and policymakers alike embrace and leverage the power of data science. This chapter discussed ways in which researchers at the U.S. Department of Energy's Oak Ridge National Laboratory are leveraging geographic data at scale to explore the population and land-use characteristics of cities in order to better inform urban issues such sustainability, particularly as it pertains to energy accessibility and consumption. The example of developing a synthetic population to estimate residential energy consumption at the household level demonstrates a generalizable method to fill existing data gaps in order to better understand and evaluate patterns of energy use. This is a useful approach for the USA and potentially other areas of the developed world where good-quality public-use microdata and complementary census summary tables exist. Where even those data are scarce, for example, in much of the developing world, other new approaches are needed. Using machine-learning algorithms to extract human-settlement areas from fine-resolution imagery and then correlating the results with nighttime lights data presents an example of this. The approach is scalable and provides an understanding of electricity consumption in urban areas where no ground data are available. Further, discerning types of human settlement can aid efforts to understand where underserved populations live and target these areas to improve access to basic services. Finally, it is important to make the connection between the science and how it can be used to make a positive impact for people and their environment. The example of a topological analysis of slums to increase the accessibility to services in urban areas is but one. An interdisciplinary approach to integrate foundational R&D, operational communities, and industry is critical for the future success of UDI. By collaborating with public- and private-sector partners, researchers can connect foundational research and development, the operational community, and industry. While urbanization magnifies our current challenges of energy sustainability, resilience, and efficiency, it also provides a unique science and technology opportunity to learn from the past, bend the present, and shape the future of urban systems where our energy, environment, and mobility goals are collectively achieved.

Acknowledgements This manuscript has been authored by UT-Battelle, LLC under Contract DE-AC05-00OR22725 with the U.S. Department of Energy. The US Government retains and the publisher, by accepting the article for publication, acknowledges that the US Government retains a non-exclusive, paid-up, irrevocable, worldwide license to publish or reproduce the published form of this manuscript, or allow others to do so, for US Government purposes. The Department of Energy will provide public access to these results of federally sponsored research in accordance with the DOE Public Access Plan. The authors would like to acknowledge the funding support from US Government agencies and the Bill and Melinda Gates Foundation for the research discussed here. Sincere thanks to Dalton Lunga, Pranab Roy Choudhury, Husain Aziz, and Chris DeRolph for their help with some of the figures used here. Tremendous assistance from Ava Ianni in the manuscript preparation process is greatly appreciated and acknowledged.

References

Allen TFH, Star TB (2017) Hierarchy: perspectives for ecological complexity, 2nd edn. The University of Chicago Press, Chicago

Allen MR, Zaidi SMA, Chandola V, Morton AM, Brelsford CM, McManamay RA, Binita KC, Sanyal J, Stewart RN, Bhaduri BL (2018) A survey of analytical methods for inclusion in a new energy-water nexus knowledge discovery framework. Big Earth Data 2(3):197–227

Arndt J, Lunga D, Weaver J, Ledoux, ST Tennille S (2019) Multiscale based characterization and classification of urban land-use. In: 2019 IEEE international geoscience and remote sensing symposium (IGARSS). Yokohama, Japan, pp 9470–9473

Aziz HM, Park B, Morton A, Stewart RN, Hilliard M, Maness M (2018a) A high resolution agent-based model to support walk-bicycle infrastructure investment decisions: a case study with New York City Transportation. Transp Res Part C: Emerg Technol 86:280–299

Aziz HM, Nagle N, Morton A, Hilliard MR, White DA, Stewart RN (2018b) Exploring the impact of walk-bike infrastructure, safety perception, and built-environment on active transportation mode choice: a random parameter model using New York City commuter data. Transportation 45:1207–1229

Bernhardt ES, Palmer MA, Allan JD, Alexander G, Barnas K, Brooks S, Carr J, Clayton S, Dahm C, Follstad-Smith J, Galat D, Glass S, Goodwin P, Hart D, Hassett B, Jenkinson R, Katz S, Kondolf GM, Lake PS, Lave R, Meyer JL, O'Donnell TK, Pagano L, Sudduth E (2005) Synthesizing U.S. river restoration efforts. Science 308:636–637

Bernhardt ES, Sudduth KB, Palmer MA, Allan JD, Meyer JL, Alexander G, Follastad-Shah J, Hassett B, Jenkinson R, Lave R, Rumps J, Pagano L (2007) Restoring rivers one reach at a time: results from a survey of U.S. river restoration practitioners. Restor Ecol 15:482–493

Bhaduri B, Bright E, Coleman P, Dobson J (2002) LandScan: locating people is what matters. Geoinformatics 5(2):34–37

Bhaduri B, Bright E, Coleman P, Urban ML (2007) LandScan USA: A high-resolution geospatial and temporal modeling approach for population distribution and dynamics. Geo J 69:103–117

Bhaduri B, Simon AJ, Allen MR, Sanyal J, Stewart RN, McManamay RA (2018) Energy-water nexus knowledge discovery framework, experts' meeting. Oak Ridge National Laboratory. ORNL/TM-2017/753. Jan 2018. https://doi.org/10.2172/1426570

Bigazzi AY, Bertini RL (2009) Adding green performance metrics to a transportation data archive. Transp Res Rec 2121(1):30–40

Black J, Arkin D (2014) Pinched: Salt shortage leads to dangerously slippery streets. NBC News [Online]. Available: http://www.nbcnews.com/storyline/deep-freeze/pinched-salt-shortage-leads-dangerously-slippery-streets-n22516

Bradbury K, Saboo R, Johnson TL, Malof JM, Devarajan A, Zhang W, Collins LM, Newell RG (2016) Distributed solar photovoltaic array location and extent dataset for remote sensing object identification. Sci Data 3:160106. https://doi.org/10.1038/sdata.2016.106

Brelsford C, Martin T, Hand J, Bettencourt LM (2018) Toward cities without slums: Topology and the spatial evolution of neighborhoods. Sci Adv 4(8):eaar4644. https://doi.org/10.1126/sciadv.aar4644

Brewer SK, Worthington TA, Mollenhauer R, Stewart DR, McManamay RA, Guertault L, Moore D (2018) Synthesizing models useful for ecohydrology and ecohydraulic approaches: An emphasis on integrating models to address complex research questions. Ecohydrology. https://doi.org/10.1002/eco.1966

Chapin F, Omitaomu OA, Bhaduri BL (2017) Making roads safer: Optimizing de-icing using snowmelt rates and slope data. In: Proceedings of the 2nd World Congress on Civil, Structural, and Environmental Engineering (CSEE'17), Barcelona, Spain, April 2–4

Chapman L, Thornes JE (2011) What spatial resolution do we need for a route-based road weather decision support system? Theoret Appl Climatol 104(3–4):551–559

Chini CM, Stillwell AS (2018) The state of US urban water: data and the energy-water nexus. Water Resour Res 54(3):1796–1811

Chowdhury PK, Weaver JE, Weber EM, Lunga D, LeDoux STM, Rose AN, Bhaduri BL (2020) Electricity consumption patterns within cities: application of a data-driven settlement characterization method. Int J Digital Earth 13(1):119–135. https://doi.org/10.1080/17538947.2018.1556355

DeRolph CR, McManamay RA, Morton AM, Nair SS (2019) City energysheds and their implications for renewable energy penetration. Nature Sustainability 2:412–420. https://doi.org/10.1038/s41893-019-0271-9

Horizon Systems Corporation (2019) NHDPlus Version 2. http://www.horizon-systems.com/nhdplus/. Accessed 8 Mar 2019

Hou X, Wang B, Hu W, Yin L, Wu H (2019) SolarNet: A deep learning framework to map solar power plants in China from satellite imagery. ArXiv, abs/1912.03685

IEA (2016) World Energy Outlook 2016, IEA, Paris https://www.iea.org/reports/world-energy-outlook-2016

Kelting DL, Laxson CL (2010) Review of effects and costs of road de-icing with recommendations for winter road management in the Adirondack Park. http://www.protectadks.org/wp-content/uploads/2010/12/Road_Deicing-1.pdf. Accessed 5 Mar 2019

Kodysh JB, Omitaomu OA, Bhaduri BL, Neish BS (2013) Methodology for estimating solar potential on multiple building rooftops for photovoltaic systems. Sustain Cities Soc 8:31–41

Kurte K, Sanyal J, Berres A, Lunga D, Coletti M, Yang HL, Graves D, Liebersohn B, Rose A (2019) Performance analysis and optimization for scalable deployment of deep learning models for country-scale settlement mapping on Titan supercomputer. Concurrency Computation: Pract Experience e5305. https://doi.org/10.1002/cpe.5305

Latif, ZA, Zaki, NA, Salleh, SA (2012) GIS-based estimation of rooftop solar photovoltaic potential using LiDAR. In: 2012 IEEE 8th international colloquium on signal processing and its applications. pp 388–392

Lunga D, Yang HL, Reith A, Weaver J, Yuan J, Bhaduri B (2018) Domain-adapted convolutional networks for satellite image classification: a large-scale interactive learning workflow. IEEE J Sel Topics Appl Earth Observations Remote Sens 11(3):962–977

Mao H, Thakur G, Sparks K, Sanyal J, Bhaduri B (2018) Mapping near-real-time power outages from social media. Int J Digital Earth 12(11):1285–1299

Mao H, Ahn Y, Bhaduri B, Thakur G (2017) Improving land use inference by factorizing mobile phone call activity matrix. J Land Sci 12(2–3):138–153. https://doi.org/10.1080/1747423x.2017.1303546

McManamay RA, Surendran Nair S, DeRolph CR, Ruddell BL, Morton AM, Stewart RN, Troia M, Tran L, Kim H, Bhaduri BL (2017) US cities can manage national hydrology and biodiversity using local infrastructure policy. Proc Natl Acad Sci 114:9581–9586. https://doi.org/10.1073/pnas.1706201114

McManamay RA, Smith JG, Jett RT, Mathews TJ, Peterson MJ (2018) Identifying non-reference sites to guide restoration and long-term monitoring. Sci Total Environ 621:1208–1223. https://doi.org/10.1016/j.scitotenv.2017.10.107

McManamay RA, DeRolph CR (2019a) A stream classification system for the conterminous United States. Sci Data 6:190017

McManamay RA, DeRolph CR (2019b) Stream classification Web app. https://hydrosource.ornl.gov/environmental-information/us-stream-classification-system

Moore A (2019) Optimization of intra-city freight movement with new delivery methods. In: 98th Annual Meeting of the Transportation Research Board, Washington DC

Morton A, Nagle N, Piburn J, Stewart RN, McManamay R (2017) A hybrid dasymetric and machine learning approach to high-resolution residential electricity consumption modeling, In: Advances in Geocomputation, Geocomputation 2015. Springer International Publishing, Berlin, pp 47–58

Morton A, Piburn J, Nagle N, Aziz HM, Duchscherer SE, Stewart RN (2017) A simulation approach for modeling high-resolution daytime commuter travel flows and distributions of worker subpopulations. Geocomputation 2017, Leeds UK, September

Nguyen HT, Pearce JM, Harrap R, Barber G (2012) The application of LiDAR to assessment of rooftop solar photovoltaic deployment potential in a municipal district unit. Sensors 12(4):4534–4558. https://doi.org/10.3390/s120404534

NRCS (2004) Part 630 Hydrology; Chapter 11—Snowmelt. National resources conservation service, United States Department of Agriculture

Park BH, Aziz HM, Morton A, Stewart R (2018) High performance data driven agent-based modeling framework for simulation of commute mode choices in metropolitan area. In: 21st IEEE international conference on intelligent transportation system, Hawaii

Rose AN, Nagle NN (2017) Validation of spatiodemographic estimates produced through data fusion of small area census records and household microdata. Comput Environ Urban Syst 63:38–49

Ruddell BL, Adams EA, Rushforth R, Tidwell VC (2014) Embedded resource accounting for coupled natural-human systems: An application to water resource impacts of the western U.S. electrical energy trade. Water Resour Res 50:7957–7972

Seeherman J, Liu Y (2015) Effects of extraordinary snowfall on traffic safety. Accid Anal Prev 81:194–203

Seto K, Dhakal S, Bigio A, Blanco H, Delgado GC, Dewar D, Huang L, Inaba A, Kansal A, Lwasa S, McMahon J, Müller D, Murakami J, Nagendra H, Ramaswami A (2014) Human settlements, infrastructure and spatial planning. In: Edenhofer O et al (eds) Climate change 2014: Mitigation of climate change. Contribution of Working Group III to the Fifth Assessment Report of the Intergovernmental Panel on Climate Change, IPCC, Geneva, pp 923–1000

Sierra Club (2018) https://www.sierraclub.org/sites/www.sierraclub.org/files/blog/1846%20RF100-CaseStudies2018_Report_05_web.pdf

Transportation Research Board (1991) Highway deicing: Comparing salt and calcium magnesium acetate. Special Report 235. http://onlinepubs.trb.org/onlinepubs/sr/sr235/00i-012.pdf. Accessed 5 Mar 2019

UN Habitat (2016) Slum almanac 2015–2016: Tracking improvements in the lives of slum dwellers. United Nations Human Settlements Programme, Nairobi, Kenya. http://unhabitat.org/slum-almanac-2015-2016/

USACE (1998) HEC-1 flood hydrograph package. Hydrological Engineering Center, Davis, CA

USDOE (2014) The water-energy nexus: Challenges and opportunities. US Department of Energy. Available at: https://www.energy.gov/downloads/water-energy-nexus-challenges-and-opportunities

US Energy Information Administration (2019) International Energy Outlook 2019 with projections to 2050. https://www.eia.gov/outlooks/ieo/pdf/ieo2019.pdf

USGS (2018) Water use in the United States. U.S. Geological Survey. https://water.usgs.gov/watuse/

Weber E, Seaman V, Stewart R, Bird TJ, Tatem AJ, McKee JJ, Bhaduri B, Moehl J, Reith AE (2018) Census-independent population mapping in Northern Nigeria. Remote Sens Environ 204:786–798. https://doi.org/10.1016/j.rse.2017.09.024

World Health Organization (2019) Global Health Observatory (GHO) data. https://www.who.int/gho/urban_health/situation_trends/urban_population_growth_text/en/

Yang HL, Yuan J, Lunga D, Laverdiere M, Rose A, Bhaduri B (2018) Building extraction at scale using convolutional neural network: Mapping of the United States. IEEE J Sel Topics Appl Earth Observations Remote Sens 11(8):2600–2614. https://doi.org/10.1109/jstars.2018.2835377

Yu J, Wang Z, Majumdar A, Rajagopal R (2018) DeepSolar: a machine learning framework to efficiently construct a solar deployment database in the United States. Joule 2:2605–2617. https://doi.org/10.1016/j.joule.2018.11.021

Yuan J, Wang D, Cheriyadat AM (2015) Factorization-based texture segmentation. IEEE Trans Image Process 24(11):3488–3497

Yuan J, Yang H-H, Omitaomu OA, Bhaduri BL (2016) Large-scale solar panel mapping from aerial images using deep convolutional networks. In: IEEE international conference on big data. Washington, DC

Zaidi SMA, Chandola V, Allen MR, Sanyal J, Stewart RN, Bhaduri BL, McManamay RA (2018) Machine learning for the energy-water nexus: challenges and opportunities. Big Earth Data 2(3):228–267. https://doi.org/10.1080/20964471.2018.1526057

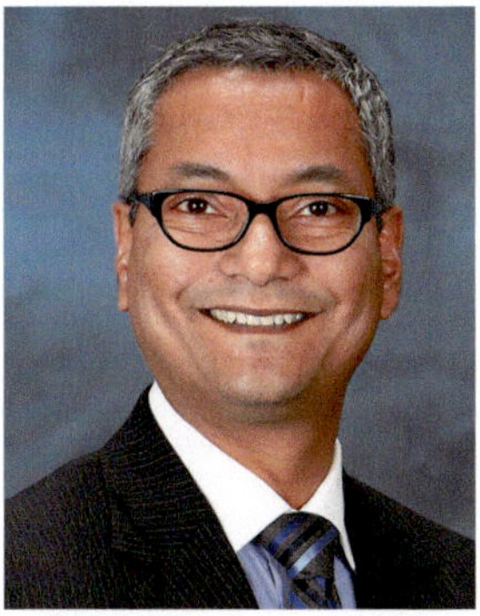

Budhendra Bhaduri is a Division Director and Corporate Research Fellow at Oak Ridge National Laboratory and holds joint professorial appointments at the University of Tennessee, Knoxville. He is a fellow of the American Association for the Advancement of Science and interested in GIScience and technology applications across energy, environment, and national security missions.

Ryan McManamay is an Assistant Professor within the Department of Environmental Science at Baylor University. His research examines human-environmental systems aimed at balancing ecosystem and societal needs, specifically the effects of macro-scale urban expansion with respect to changes in land cover, energy and water infrastructure, shifts in regional to global water budgets, and ultimate consequences to biodiversity.

Olufemi Omitaomu is a Senior R&D Staff member in Computational Systems Modeling at Oak Ridge National Laboratory (ORNL), and also a Joint Faculty Associate Professor in the Department of Industrial and Systems Engineering at the University of Tennessee, Knoxville. He is a Senior Member of IEEE and IISE.

Jibo Sanyal is currently the group leader for the Computational Urban Sciences Group at Oak Ridge National Laboratory. He holds a Ph.D. in Computer Science from Mississippi State University from 2011. He is a member of ACM, AGU, IBPSA, and a Senior Member of IEEE.

Amy Rose is a Senior R&D Staff in Human Geography at Oak Ridge National Laboratory and serves as Joint Faculty Assistant Professor in the Department of Industrial and Systems Engineering at The University of Tennessee. Her interests are in geocomputational methods to characterize the spatiotemporal and demographic patterns of human populations.

Part III
Urban Sensing

Chapter 19
Introduction to Urban Sensing

Wenzhong Shi

Urban sensing can be regarded as the collective of technologies to sense and obtain information about physical space and human activities in urban areas. The urban objects to be sensed include, for example, the overall city, its land cover and its land use, buildings, roads, cars, or individual persons. The properties that can be sensed include static ones like the existence of a building with its geometry and other relatively stable features, as well as dynamic ones like the moving trajectory and speed of a car, or the change of land uses which reflects the change of people's activity in the space. Urban sensing can result in spatial, temporal, and attribute data for an urban area, which will then be used for urban analytics and will finally provide urban service and urban governance.

The technologies for urban sensing have been developed for a long time and have progressed very fast in recent years with the advances of sensor technologies and computation power. Urban objects can be sensed from different perspectives, sensors, and platforms. These include optical or interferometric synthetic aperture radar (InSAR) images from satellites in space, light detection and ranging (LiDAR) or optical images and digital signals from aircraft or unmanned aerial or autonomous vehicles (UAVs), ground-based laser scanning data from a car with mobile mapping systems, ground-penetrating radar (GPR) on underground utility information from a trolley, or sonar signals mapping underwater terrain from a multi-beam sonar sensor on a boat. For individuals, their indoor or outdoor locations can be obtained based on information from the sensors in a mobile phone, and their properties like body temperature can be obtained from wearable devices.

The full set of urban sensing technologies covers a very wide range, especially with the latest technologies, such as edge computing, the Internet of Things (IoT), and sensor networks. Part III of this book introduces the urban sensing technologies

W. Shi (✉)
Department of Land Surveying and Geo-Informatics and Smart Cities Reserach Institute, The Hong Kong Polytechnic University, Hong Kong, China
e-mail: lswzshi@polyu.edu.hk

W. Shi et al. (eds.), *Urban Informatics*, The Urban Book Series,
https://doi.org/10.1007/978-981-15-8983-6_19

mainly from a geomatics perspective, and more sensing technologies can be further identified in a full and more comprehensive review.

In Chap. 20, Man Sing Wong, Xiaolin Zhu, Sawaid Abbas, Coco Yin Tung Kwok, and Meilian Wang present the history and latest developments in optical remote sensing, and introduce the representative optical satellite sensors. They elaborate on the processing of remotely sensed satellite images and update the applications of optical remote sensing in remotely analyzing the attributes of groups of objects.

Optical satellite images can provide rich attribute and geometric information, while data produced by synthetic aperture radar (SAR) can produce high-accuracy geometric data for monitoring deformation. Chapter 21 by Hongyu Liang, Wenbin Xu, Xiaoli Ding, Lei Zhang, and Songbo Wu introduces the working mechanisms of SAR and InSAR, as well as the implementation of multitemporal InSAR. InSAR applications in generating digital elevation models (DEMs) and monitoring subsidence and building deformation are illustrated with various examples, and the advantages of this technology in remote geometric analysis with millimeter-level accuracy are demonstrated.

LiDAR is another data acquisition method focusing on the geometry of objects. As one of the most advanced technologies for acquiring quasi-continuous urban geometric data, airborne laser ranging technology and a machine-learning-based application in detection and characterizing urban objects are discussed by Wei Yao and Jianwei Wu in Chap. 22. Multispectral images and airborne LiDAR data are co-registered to classify buildings, trees, and natural terrain, as well as moving artifacts along with estimates of their velocity.

Often compared with LiDAR, photogrammetry is one of the most time-honored surveying techniques. The presence of corresponding texture and common points is used to create binocular pairs to generate geometric information, while the texture can be used for prompt texture projection with no extra registration required. In Chap. 23, Bo Wu presents the history and principles of photogrammetry, its state-of-the-art developments with computer vision and 3D mapping, and its modern applications and potential in generating both geometric and texture data of urban environments.

Most of the surveying technologies are based on direct line-of-sight, while there is no such convenience in underground utility surveying. The objective of using GPR is to see the unseen underground world. In Chap. 24, Wallace W.L. Lai compares and discusses the sensors and working principles for detecting invisible underground objects using electromagnetic induction (EMI) and GPR, as well as the in-line technologies for direct checking of pipelines. The chapter also introduces future trends in developing imaging and diagnosis of underground utilities.

In contrast to most static mapping technologies that can only provide data captured at discrete positions, mobile mapping based on sensors embedded on moving platforms has become a highlight of research in recent decades. Conventional surveying techniques, including GNSS (global navigation satellite system) positioning, inertial measurement unit (IMU) dead reckoning, LiDAR data acquisition, and photogrammetry, are synergized to achieve mobile mapping. Chapter 25 by Kai Wei Chiang, Guang-Je Tsai, and Jhih Cing Zeng introduces the history of mobile mapping and elaborates on its recent developing progress. Also reviewed are the common implementations and applications of mobile systems in disaster response, indoor mapping, and autonomous driving, as well as future trends in mobile mapping technology.

With detailed seamless mapping, ubiquitous positioning becomes feasible and practical. Mobile phones are common platforms to realize ubiquitous positioning. In Chap. 26, Ruizhi Chen and Liang Chen review indoor positioning technologies based on radio frequency and built-in sensors, with discussions and comparisons of their pros and cons in the context of different applications. The difficulties and future trends of indoor positioning are also presented with a comparison of various mobile-phone-based indoor positioning technologies.

With the development of computer technology and the widespread installation of surveillance cameras, data processing and extraction from them also become research highlights. Deployed on urban facilities, cameras are organic components of urban sensor networks. Chapter 27 by Fábio Duarte and Carlo Ratti discusses the applications of computer vision and machine learning in analyzing urban landscape data to understand the characteristics of human mobility, moving patterns, and public spaces.

The technologies presented in Chaps. 20 to 27 mostly produce professionally generated content. As an important complement, Chaps. 28 and 29 focus on the emerging approach of urban sensing by user generated content (UGC). In Chap. 28 by Song Gao, Yu Liu, Yuhao Kang, and Fan Zhang, background, definition, and characteristics of UGC and processing frameworks are introduced systematically. Applications of UGC in extracting citizen demographics, mobility patterns, and place semantics, and uncovering urban spatial structures are also demonstrated.

Based on the UGC acquired, a number of new urban study areas have been explored, especially those related to individual citizens. In Chap. 29, Wei Tu, Qingquan Li, Yatao Zhang, and Yang Yue present UGC-driven urban studies within this general framework. These new urban studies have revealed invisible landscapes of urban dynamics and demonstrated how urban space is perceived by the public. Challenges and future directions of UGC-based urban studies are also discussed.

During recent decades, the development of information technology has changed the surveying and mapping of the real world and raised the urgent needs of urban informatics. While Part III of this book intends to cover the essential and trending urban sensing technologies, many technologies are beyond the coverage of this book due to their large variety, with a few key examples as follows.

Besides indoor positioning, satellite positioning with the Global Positioning System (GPS) by the US, Global Navigation Satellite System (GLONASS) by Russia, Galileo by the European Union, Beidou by China, and other regional satellite positioning systems is a more classical positioning technology and has been widely adopted in precise measurement in open-sky environments. With an appropriate differential positioning link established, the accuracy of satellite positioning can achieve centimeter level.

Wearable devices are also widely used for sensing the properties and movements of individual persons. These devices monitor the wearer's physical and emotional status through embedded sensors, such as IMU, optical sensors, electrodes, force and pressure sensors, thermometers, microphones, and GNSS modules. By collecting physical data like moving acceleration, pose changes, and heart beats, wearable devices can determine the movement, health, and safety status of the wearer. By collecting data from a significant number of wearers, implicit moving patterns, living habits, and urban traffic flows can be revealed and visualized.

Another key technology lies in the Internet of Things (IoT; Chap. 38), which is a collection of machines, objects, animals, or humans with embedded sensors, connected by a linked network and transferring data over a network. The embedded sensors can be connected directly as the components of the sensor network for fluent exchange and comprehensive management of the data. IoT has been widely applied to smart traffic, smart home, and public security. A typical example of IoT is the smart lamp post, where camera, Wi-Fi hotspot, thermometer, decibel meter, and pollutant sensors are integrated onto a normal lamp post alongside urban streets. It provides closer monitoring of the environment and better incident response for public safety, and acts as an effective data source for urban planning.

Chapter 20
Optical Remote Sensing

Man Sing Wong, Xiaolin Zhu, Sawaid Abbas, Coco Yin Tung Kwok, and Meilian Wang

Abstract Applications of Earth-observational remote sensing are rapidly increasing over urban areas. The latest regime shift from conventional urban development to smart-city development has triggered a rise in smart innovative technologies to complement spatial and temporal information in new urban design models. Remote sensing-based Earth-observations provide critical information to close the gaps between real and virtual models of urban developments. Remote sensing, itself, has rapidly evolved since the launch of the first Earth-observation satellite, Landsat, in 1972. Technological advancements over the years have gradually improved the ground resolution of satellite images, from 80 m in the 1970s to 0.3 m in the 2020s. Apart from the ground resolution, improvements have been made in many other aspects of satellite remote sensing. Also, the method and techniques of information extraction have advanced. However, to understand the latest developments and scope of information extraction, it is important to understand background information and major techniques of image processing. This chapter briefly describes the history of optical remote sensing, the basic operation of satellite image processing, advanced methods of object extraction for modern urban designs, various applications of remote sensing in urban or peri-urban settings, and future satellite missions and directions of urban remote sensing.

20.1 Introduction

A major part of the global population now lives in cities; consequently, cities are growing in complexity and dynamics. For example, a city's expansion is not restricted to horizontal expansion as most of the developed cities are now growing vertically as well. In addition, new urban designs with a variety of construction materials pose unique environmental challenges. Thus, innovative urban information technologies

M. S. Wong (✉) · X. Zhu · S. Abbas · C. Y. T. Kwok · M. Wang
Department of Land Surveying and Geo-Informatics, The Hong Kong Polytechnic University, Hong Kong, China
e-mail: ls.charles@polyu.edu.hk

W. Shi et al. (eds.), *Urban Informatics*, The Urban Book Series,
https://doi.org/10.1007/978-981-15-8983-6_20

are needed to provide a solution to problems associated with contemporary urban design and development models, especially in the era of smart cities.

Rapid development and the dynamic growth of urban areas require innovative technologies to provide a huge amount of increasing information about an urban landscape. Remote sensing (RS) is defined as the science of collecting, extracting, and analyzing information about objects, on images obtained without having physical contact with the objects. Wide spatial coverage from space or airborne remote sensors complements the information obtained from extensive field-based inventories of urban landscapes. Remote sensing has a strong potential to play a pivotal role in developing the urban informatics of evolving urban spaces.

Ever-increasing improvements in spatial (from coarse-resolution to fine-resolution image models) and spectral resolution (from a few spectral bands to more than a hundred spectral bands) of remote sensing images, along with development in cyberinfrastructure and algorithms to extract information from the images, have accelerated the urban applications of remote sensing. These applications focus on various domains of urban settings, such as urban geometric and morphological models, traffic modeling, 3D urban models, urban noise and pollution management, solid waste management, tourism, and rapid-response mapping for disaster-risk reduction, and several other environmental and socioeconomic dynamics.

Since the launch of the first Earth-observation satellite in the 1970s, a wide range of remote sensing satellites has been launched, acquiring Earth-observation data in the visible (VIS) and near-infrared (NIR) portions of the electromagnetic spectrum. All the acquired Earth-observation data require that rigorous processing and algorithms are ready for analysis, and then another set of techniques are applied to extract relevant information from images. Therefore, knowledge of the essential characteristics of remote sensing platforms and sensors, along with an understanding of the basic and advanced information extraction methods, are required to reconstruct urban models. To this aim, this chapter will focus on providing background information about the history and the latest developments in optical remote sensing, processing of remote sensing images to analyze and extract information, examples of remote sensing applications in urban or peri-urban settings, and a broad outlook on future directions and the latest developments of remote sensing-based operations in urban informatics.

20.2 History of Optical Remote Sensing

The term remote sensing (RS) first appeared in 1962, but its origin dates back to the employment of photography and the development of flight at the beginning of the nineteenth Century (Olsen 2016). The balloonist Gaspard Tournachon took photographs of Paris from a balloon in 1859, starting the era of RS. Then a wide range of scientists followed Tournachon's experiment and made many improvements. For example, Germans used aerial photographs to measure features and areas in forests. The Bavarian Pigeon Corps used pigeons to take aerial photos, and Albert Maul used a rocket to take an aerial photograph. Until the 1910s, systematic RS and aerial photography were rapidly developed with the purpose of military surveillance and photoreconnaissance during World War I. A series of related technologies were also developed and reached a climax during the war. The most significant development of RS technology took place in World War II. Several imaging systems, such as photography using near-infrared and thermal infrared, aiming to differentiate real vegetation from camouflage, and airborne imaging radar that was used for nighttime bombing, were also achieved (Blaschke et al. 2011).

After the war and in the 1950s, RS systems advanced to a global scale and substantial progress in radar development was achieved. The first Earth-observation satellite, Landsat launched in 1972, began a new RS era. Various Earth-observing and weather satellites, like AVHRR, Landsat, and SPOT, provided global measurements of various data for all kinds of purposes. Attention was also paid to the development of image processing of satellite imagery and fine-resolution imagery. The first hyperspectral sensor was developed in 1986 and the first fine-resolution satellite, IKONOS, was launched in 1999 (Blaschke et al. 2011). Currently, online platforms, such as Google Earth and Google Maps, collect and store massive satellite images and make them accessible to the general public, thus accelerating the development of RS technology.

20.3 Latest Developments in Optical Remote Sensing

Over the past decades, extensive research and development in sensor technology have been carried out, making it possible to collect fine-resolution and hyperspectral imagery. All of the sensors have different spatial, spectral, radiometric, and temporal resolutions. The major characteristics of the well-known optical RS satellite sensors are summarized in Tables 20.1, 20.2 and 20.3. As shown in Table 20.1 and Fig. 20.1, most satellites were launched by the USA. There was a total of 791 Earth-observation and Earth-science satellites in orbit by March 2019, among which 481 were optical/multispectral/hyperspectral imaging satellites (Fig. 20.1; UCS Satellite Database 2005).

20.3.1 Introduction to Representative Optical Satellite Sensors

A variety of optical RS satellites have been launched for Earth-observation applications. A brief description of representative sensors is given in this section.

Since 1972, there have been eight Landsat satellites launched, with Landsat 9 planned to be launched in 2021. Landsat 5 was the longest operating Earth-observation satellite, continually collecting data for 28 years from its launch in March 1984 until it was decommissioned in January 2013. Imagery from the series

Table 20.1 Representative satellites and launching date

Sensors (Country)	Launching date
TIRO (USA)	1960
NIMBUS (USA)	1964, 1966, 1969, 1970, 1972
Landsat (USA)	1972, 1975, 1978, 1982, 1984, 1993, 1999, 2013
METEOSAT (ESA)	1977, 1981, 1988, 1989, 1991, 1993, 1997
SIR (USA)	1981, 1984
SPOT (France)	1985, 1990, 1993, 1998, 2002, 2012, 2014
IRS (India)	1988, 1991, 1995, 1996, 1997, 1999
ERS (ESA)	1991, 1995
JERS (Japan)	1992
Orbview (USA)	1995, 1997, 2003, 2008
QuickSCAT (USA)	1999
IKONOS (USA)	1999
KOMPSAT (South Korea)	1999
MODIS (USA)	1999
Tsinghua (China)	2000
EORS (ESA)	2000
JASON (USA)	2000, 2008, 2016
EOS(USA)	2000, 2002
Quickbird (USA)	2001
ENVISAT (ESA)	2002
GRACE (USA)	2002, 2016
ALOS (Japan)	2003
Worldview (USA)	2007, 2009, 2014, 2016
Sentinel-2 (ESA)	2015, 2017
Hyperion	2000

of Landsat satellites has been archived in the US and at Landsat receiving stations around the world, providing unique resources for global-change research and applications in agriculture, cartography, geology, forestry, regional planning, surveillance, and education; and the data can be accessed through the United States Geological Survey (USGS) EarthExplorer website.

SPOT (Satellite Pour l'Observation de la Terre) is a part of the RS program set up in 1978 by France in collaboration with Belgium and Sweden. Each SPOT is comprised of two identical fine-resolution optical imaging instruments that can be operated in either panchromatic or multispectral mode. It has been designed to

Table 20.2 Characteristics of representative optical satellites

Satellites	Spatial resolution (meters)	Revisit time (days)	Spectral range (μm) and number of bands
ASTER	15–90	15	0.52–11.65 (15 bands)
Landsat	15–120	16	0.45–12.5 (11 bands)
SPOT	10–20	26	0.45–1.75 (5 bands)
IKONOS	1–4	1–4	0.45–0.90 (5 bands)
MODIS	250–1000	0.25	0.4–14.4 (36 bands)
Quickbird	0.61–0.72	1–6	0.45–0.9 (4 bands)
WorldView-2	0.46–2.4	1.1–3.7	0.4–1.05 (8 bands)
WorldView-3	0.31–30	<1.0–4.5	0.40–23.6 (26 bands)
WorldView-4	0.31–1.24	<1.0–4.5	0.65–0.92 (4 bands)
Pleiades	0.5–2	1	0.47–0.94 (5 bands)
IRS	5.8–70	5–24	0.52–1.7 (4 bands)
Sentinel-2	10–60	5	0.04–2.19 (12 bands)
Hyperion	30	16	0.35–2.58 (220 bands)
ALI	30	16	0.40–2.40 (7 bands)
CHRIS	18–36	7	0.40–1.05 (19 bands)
AVNIR2	10	46	0.42–0.89 (4 bands)
RapidEye	5	5	0.44–0.85 (5 bands)
Gaofen	0.8	2	0.45–0.89 (4 bands)
SkySat	0.8–1	1	0.45–0.90 (4 bands)
Jilin-optical	0.72–2.88	3.3	0.45–0.90 (4 bands)
Jilin-HypSpec	5–150	2–3	0.45–13.5 (28 bands)
TH	2–10	5	0.43–0.90 (4 bands)
Dove	2.7–3.2	1	0.42–0.90 (4 bands)
GeoEye	0.46–1.84	2.1–8.3	0.45–0.92 (4 bands)
SuperView	0.5–2.0	2	0.45–0.89 (4 bands)

explore the Earth's resources, detect and forecast phenomena involving climatology and oceanography, and monitor human activities and natural phenomena.

ASTER (the Advanced Spaceborne Thermal Emission and Reflectance Radiometer) consists of three subsystems: Visible and Near-Infrared (VNIR), Shortwave Infrared (SWIR), and Thermal Infrared (TIR). ASTER data are often used to derive maps of land surface temperature, reflectance, and elevation. It also has many applications, including monitoring vegetation, hazards, geology, land surface, hydrology, and land-cover change.

IKONOS is the first civilian fine-resolution sensor, providing images with a comparable resolution to aerial photos. It is useful for applications such as urban geography, land-use, agriculture, and natural-disaster management due to its fine-resolution. Quickbird was launched in 2001 and decommissioned in 2015. It has

Table 20.3 Primary applications of representative optical satellites

Sensor	Applications
ASTER	Vegetation and ecosystem dynamics, land surface temperature, geology, hazard monitoring, land-cover change, land surface climatology, hydrology
Landsat	Global-change research, agriculture, cartography, geology, forestry, regional planning, surveillance, education
SPOT	Exploring the Earth's resources, detecting and forecasting phenomena involving climatology and oceanography, and monitoring human activities and natural phenomena
IKONOS	Urban geography, land-use, agriculture, and natural-disaster management
Quickbird	Map publishing, land and asset management, and risk assessment
WorldView	Mapping clouds, ice, snow and correcting for aerosol and water vapor
Pleiades	Crisis monitoring
Hyperion	Hyperspectral land imaging for various applications
PROBA-CHRIS	Atmosphere, land, agriculture, and oceans and coasts
ALOS-AVNIR-2	Agriculture, forest, and natural disasters
RapidEye	Regional and global agricultural mapping
Sentinel	Land and maritime monitoring, emergency management, and surveillance
Gaofen	Urban monitoring and precision agriculture
SkySat	Defense, agriculture, and environmental monitoring
Jilin-optical	Development, urban monitoring, and agriculture
Jilin-HypSpec	Environment, agriculture, and forestry
TH	Terrain modeling, surveying, and mapping
Dove	Urbanization, deforestation, disasters, and agriculture

very-fine-resolution sensors that can acquire images in panchromatic and multispectral modes concurrently. It is designed to support applications such as map publishing, land and asset management, and risk assessment. WorldView consists of very-fine-resolution satellites with a short average revisit time. WorldView-1, launched in 2007 and still operating today, is only capable of collecting panchromatic imagery but having the finest resolution of 0.41 meters. WorldView-2, launched in 2009 and still in operation, has the capabilities to capture eight spectral bands. WorldView-3 was launched in 2014 with fine-resolution imagery captured in sixteen multispectral bands. WorldView-4, launched in 2016, is a multispectral, fine-resolution commercial satellite with four multispectral bands and a panchromatic band.

The Indian Remote Sensing (IRS) satellite series was launched to technically support the development of agriculture, water resources, forest and ecology, geology, water-conservancy facilities, fisheries, and coastline management in India. Gravity Recovery and Climate Experiment (GRACE), a collaboration between National Aeronautics and Space Administration (NASA) and the German Aerospace Center, is a satellite mission that monitors Earth's gravitational field. Scientists can infer

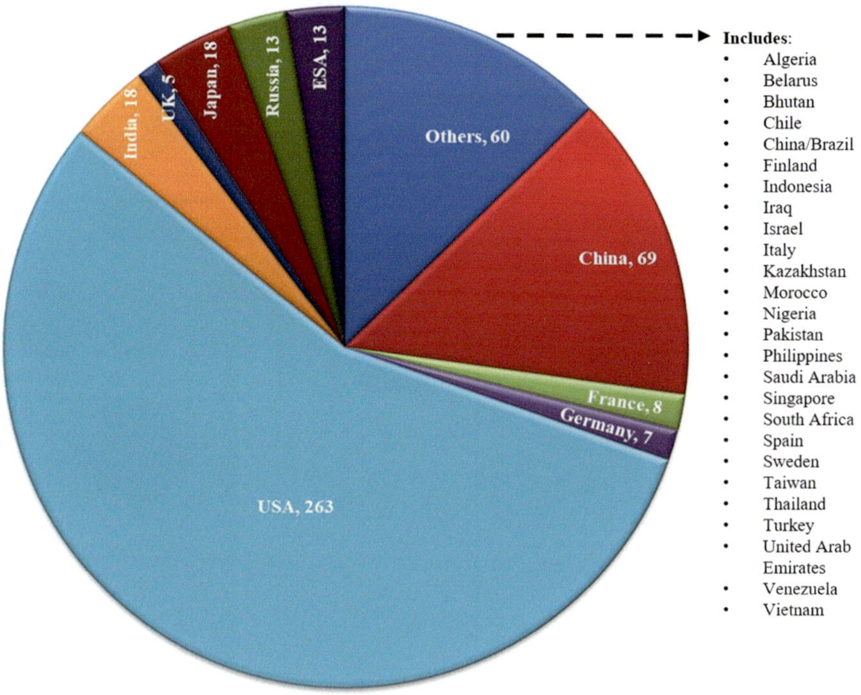

Fig. 20.1 Earth-observation satellites in orbit by March 2019

changes in groundwater by measuring the changes in the gravitational field. The summary of primary applications of different satellites is shown in Table 20.3.

In recent years, with the development of commercial images and the launch of satellite-based sensors, hyperspectral imaging is becoming the mainstream in the RS field. And the rapid development of artificial intelligence may provide a new era of applications for RS in the future.

20.4 Processing of Remote Sensing Satellite Images

Not all the acquired RS images are ready to use, because there are many distortions or deviations in raw images. The distortions can be divided into random distortions (Fig. 20.2) and systematic distortions. Random distortions can be caused by changes in altitude, attitude, and speed of the sensor platform, atmospheric refraction, or relief displacement, while systematic distortions are caused by panoramic distortion, skew distortion (Fig. 20.3), and the Earth's curvature. Before we use RS images, it is important to correct these errors.

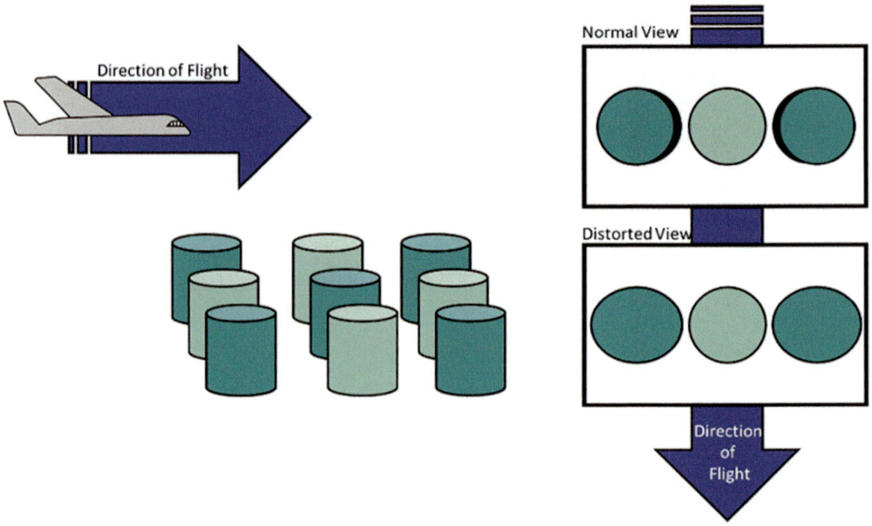

Fig. 20.2 A graphical illustration of random distortion

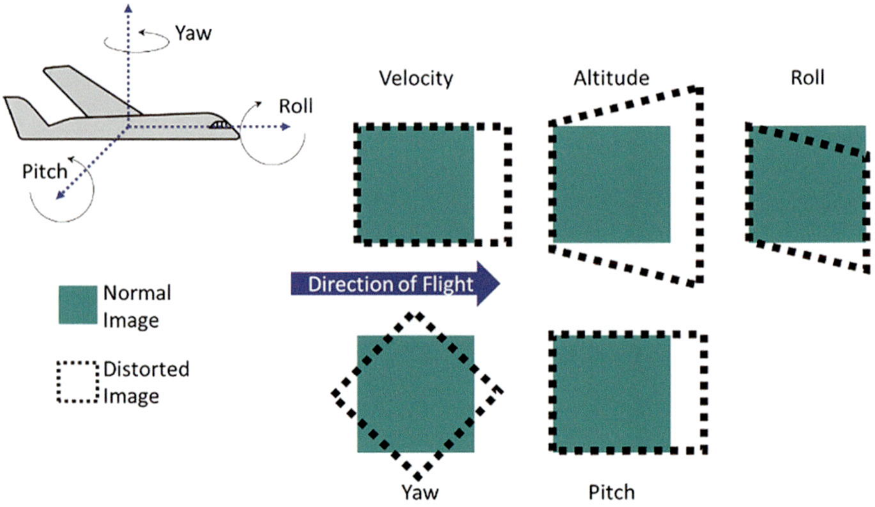

Fig. 20.3 A graphical illustration of skew distortion

Generally, satellite image processing operations can be divided into three stages: (i) image pre-processing, (ii) image processing, and (iii) image post-processing. Image pre-processing aims to correct distortion and to reduce noise in the data. The purpose of image processing is to understand the information stored in remotely sensed images and to optimize the appearance for the visual system by using or not using enhancement technology, so the operation involves filtering, and band ratio or contrast enhancement to enhance or mask image features or classify images. The objective of post-processing is to further reduce the errors of image processing based on expert knowledge and ancillary information.

20.4.1 Image Pre-processing

Primary image pre-processing procedures include image rectification, also known as a geometric correction, and radiometric correction which deals with atmospheric error correction and conversion from digital number (DN) to radiance. The process of rectification is to correct distortions, including image-to-image registration and image-to-map registration (Fig. 20.4). In this process, the coordinates in an image match the selected points in a map or an image to derive geometric transformation coefficients; then these coefficients may be used to rectify the image geometrically. The root-mean-square error (RMSE) is used to assess the correction accuracy. The closer the value is to zero, the smaller the residuals, representing a more accurate correction. The procedure of radiometric correction includes atmospheric correction and DN-to-radiance conversion. It is used to calibrate the system and reduce the systematic calibration effect and atmospheric effect. The particles in the atmosphere can cause scattering and absorption depending upon the physical and chemical characteristics of the atmospheric particles. Atmospheric correction can be conducted through an empirical method using empirical line calibration, which forces the RS image data to match the in situ spectral reflectance measurements, and through the dark pixel method, which finds the minimum pixel value from each band using histograms, and subtracts that value from all of the pixels in the band.

The pre-processing procedures produce consistent images with high scientific quality that can be directly used for scientific applications and subsequent analysis.

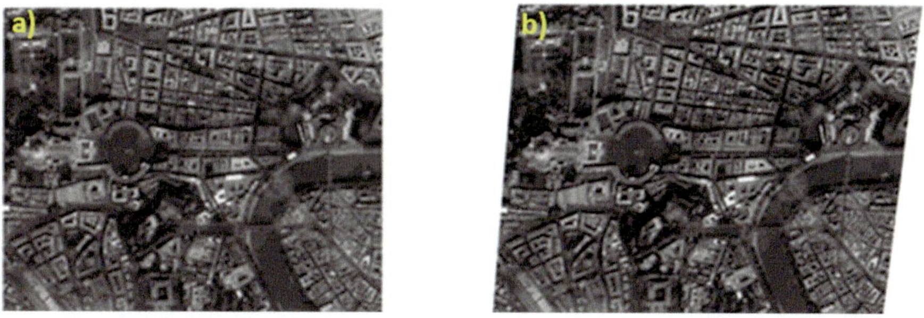

Fig. 20.4 A typical example of geometric correction of a satellite image; **a** raw image and **b** geometrically corrected image

20.4.2 Image Processing

Satellite image processing includes: (i) masking or clipping area of interest (AOI), (ii) contrast enhancement, (iii) spatial filtering, (iv) spectral enhancement, (v) image classification, and (vi) object recognition and extraction.

Masking of a study area or area of interest is the foremost processing step in which an image (or mosaic of images) is clipped over a region of interest. The clipping helps to reduce the size of the image and processing time as well as to focus on the desired study area or region of interest.

Contrast enhancement is used to transform satellite images for visual enhancement by stretching the input values to the maximum available range. The contrast enhancement procedures can be applied on the entire image for a better contrast among different land-cover or land-use types, or it can be used to enhance specific features in an image to emphasize a specific land-cover or land-use type (e.g., vegetation, soil, water, or snow) by diminishing others. Sometimes image displays may not clearly show all the features, especially when dealing with monochromes. This is where contrast enhancement comes in. Contrast enhancement is done through spectral feature manipulation. It can maximize the contrast between the features according to the image histogram. The most common method is a linear stretch (Fig. 20.5).

Spatial filtering is a process to emphasize or de-emphasize various spatial frequencies in the image data or tonal variations in an image. An example of spatial enhancement (filtering) is shown in Fig. 20.6. Filtering makes use of kernels, a square matrix that is moved pixel by pixel and is designed to increase the brightness of the central pixel, depicted as a single positive value surrounded by negative values. The larger the kernel, the more blurred the pixels. A low-pass filter emphasizes low-frequency changes in the brightness and de-emphasizes or smooths local details such as by taking the mean, while high pass filters de-emphasize more general low-frequency details and emphasize the high-frequency components by exaggerating local contrast.

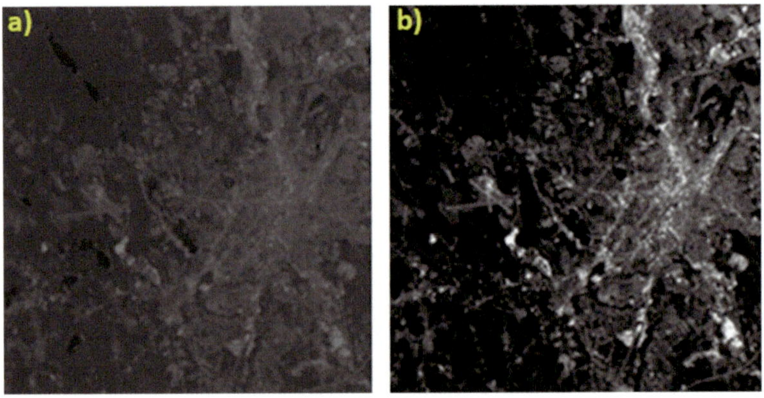

Fig. 20.5 Contrast enhancement of an image: **a** original image and **b** linearly stretched image

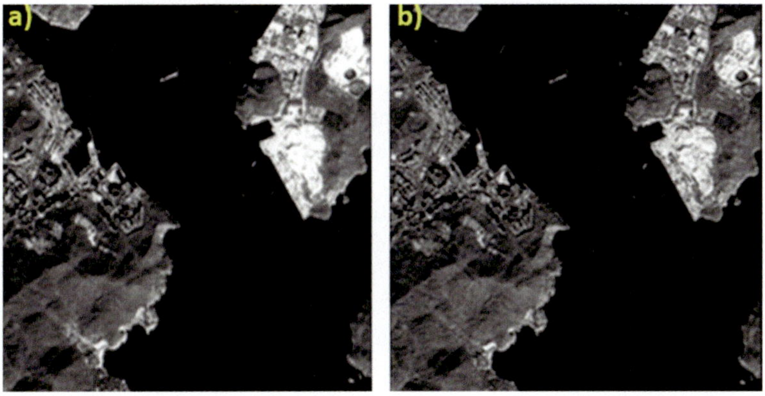

Fig. 20.6 An example of spatial enhancement (filtering): original image (**a**) and filtered image (**b**)

Filters can also be used for edge preservation and noise removal. For example, the median filter is better at preserving edges on an image, and a model smoothing filter can remove the "salt and pepper" effect on a classified image, leaving a more homogeneous output.

Spectral enhancement comprises image transformation processes used to extract unique spectral information, combine the information in different spectral bands, and compress information from multiple wavebands into fewer bands.

Once the data have been processed, it is then up to the operator to analyze what is captured in an image. In order to interpret an image, the operator first has to detect, identify, and classify the object. Normally, classification methods mainly follow two approaches: unsupervised classification and supervised classification. The unsupervised approach clusters pixels based on spectral statistics, without sampling and training, while the supervised approach employs classifiers based on the results of sampling and training land-cover classes, and users need to define useful information about categories and examine the spectral separability before classification.

The information in a satellite image can be extracted and classified at various processing units of the image; for example, pixel level, a unit defined by the image spatial resolution; sub-pixel level, a pixel is spectrally unmixed to identify a portion of a land-cover feature in the pixel; and object-based classification, which is based on the concept of grouping homogeneous pixels and primarily applied on a very-fine-resolution image where an object is divided and stored into many pixels. Generally, sub-pixel and object level (object-based) classification routines are implemented for information extraction over urban areas. For example, a linear spectral unmixing model was applied to an IKONOS (4 m spatial resolution) image to estimate the contribution of trees and grasses in the urban landscape of Hong Kong (Nichol and Wong 2007).

Supervised techniques rely on user-defined training sites describing the nature and number of possible land-cover classes (Mather 2011). The most significant and conventional decision rules of supervised classification include maximum likelihood decision rule, nearest neighbor decision rule, and parallelepiped decision rule.

The unsupervised approach is optimal when there is no enough prior ground truth information about the area of interest (Mather 2011). According to analyst-defined parameters, unknown image pixels are iteratively clustered until either the proportion of pixel class values remains unchanged or a maximum number of iterations is reached (Jensen 2009). The three most commonly used clustering algorithms are: *k*-means clustering, fuzzy *c*-means (or modified *k*-means), and ISODATA (iterative self-organizing data analysis technique).

In 1999, with the launch of IKONOS (Goetz et al. 2003), intra-class spectral variations and inter-class spectral confusion had increased in fine-resolution satellite imagery. Due to higher pixel-to-pixel variability and information contained in patch-based landscape structures, classical approaches of image analysis are becoming out of date. The recently developed object-based image analysis techniques of pattern recognition overcome these difficulties by first segmenting the image into multi-pixel image object primitives according to both spatial and spectral features of groups of pixels.

Over the past decade, there has been a noticeable shift in the analysis of Earth-observation (EO) data, from what has been predominantly 30 years of per-pixel multispectral-based approaches, towards the development and application of multi-scale object-based analysis. New concepts of object-based analysis, such as the fractal net evolution approach (FNEA), linear scale-space and blob-feature detection (SS), and multi-scale object-specific segmentation (MOSS) were developed for information extraction from RS data stored in the form of digital images (Mallinis et al. 2008).

In addition, a wide range of advanced classification approaches has been developed in recent years to solve a variety of problems arising with fine-resolution data sets and complex urban environments. The new methods and approaches from machine learning and pattern recognition include artificial neural networks (ANN), deep learning methods, decision trees, support vector machines, extreme learning machines, an artificial immune system, active learning, semi-supervised learning, binary tree support vector machine, and random forest. Other modern techniques also include ensemble learning based on multiple learners, spatial-spectral classification, multi-kernel support vector machine, wavelet analysis, phenology-based classification, kernel *k*-means, and expectation-maximization (Xue et al. 2015; Du et al. 2012; Fernandez-Delgado et al. 2014; Lu and Weng 2007; Mountrakis et al. 2011; Tan and Du 2011).

Combining multiple RS data sets, advanced urban feature extraction algorithms, and accurate classification algorithms, an urban information system has been developed to effectively monitor the rapidly evolving urban areas and their impact on the environment (Kadhim et al. 2016). Recent urban applications of RS comprise urban green spaces mapping, aerosol monitoring, urban heat island effect, automatic feature extraction (e.g., roads, buildings, and trees), relationships between land-use and surface temperature, 3-dimensional geometric models for urban heat island, urban energy-efficiency models, and mapping migrant housing in mega-urban centers (Blaschke et al. 2011; Hamdi 2010; Jin et al. 2011; Hofmann et al. 2011; Miyazaki et al. 2011; Hermosilla et al. 2011; Rinner and Hussain 2011; Hay et al. 2011; Geiß et al. 2011; Liu and Zhang 2011; d'Oleire-Oltmanns et al. 2011). Also, some modern urban RS methods are focusing on integrating multiple RS (night light imagery and multispectral indices) and geolocation datasets using machine learning approaches for urban informatics application of RS (Xia et al. 2019).

In the past couple of decades, with the advent of very-fine-resolution remote sensing images (1 m or less), there has been a major shift in information extraction from conventional pixel-based classification towards object-based classification and target-object extraction over urban areas. Modern techniques of machine learning focus on extracting typical urban features such as roads, buildings (more specific characteristics of buildings), cars, and urban trees, rather than classifying whole images or mapping urban sprawl.

20.4.3 Image Post-Processing

After determining the classes of image objects, image post-processing procedures usually include map production, raster to vector conversion, and image interpretation. The information on images needs to be converted to land-cover classes. Applying a majority filter to remove salt and pepper in pixel-based land-cover maps is the most commonly applied post-classification process. In urban areas, expert knowledge and ancillary information, such as population density, may be required to distinguish between spectrally similar high-density residential areas and commercial buildings. Current technologies have some automated procedures, enabling automated detection and identification, but ultimately it would be left up to the operator to interpret the results.

20.5 Applications of Optical Remote Sensing

Recent advanced technologies have improved what we can do in RS. Since 1995, RS is no longer restricted to military and government use. And rapidly developing technologies also allowed for the expansion of applications, such as urban and population growth, town planning, weather forecasting, crop prediction, and forecasting, forest and rangeland monitoring, air-quality monitoring and assessment, and surface-material detection, just to name a few. Infrared cameras become commercially available, which can be used to detect the health condition of vegetation, and hand-held devices can be carried on helicopters to record heat signatures and to monitor the urban heat island effect.

For coastal water-quality monitoring, RS data sets which combine a synoptic viewpoint with the ability to measure the reflected energy from the water surface in different spectral regions, are increasingly available for coastal water-quality applications. For example, improved estimation of chlorophyll-a concentrations for the coastal area of Hong Kong has helped in the detection of algal blooms, including their intensity and extent. For vegetation monitoring, aerial photographs and fine-resolution satellite images can be used for mapping secondary vegetation succession. When dealing with the mapping of deforestation and degradation, medium-resolution Landsat satellite images can provide satisfactory results, while coarse-resolution satellite images are required when monitoring the impact of drought on vegetation moisture conditions, using photos captured by MODIS. Research on atmospheric aerosols using satellite RS is popular. Aerosols are suspended particles in the atmosphere emitted from natural and anthropogenic sources. These particles are responsible for climate change, poor air quality, and atmospheric visibility, and also associated with public health. Satellite RS is an effective and unique technique for retrieving spatial aerosol optical thickness over the globe. Different satellite sensors such as MISR, MODIS, and Visible Infrared Imaging Radiometer Suite (VIIRS) can retrieve aerosol optical thickness.

20.5.1 Land-Use and Land-Cover Mapping

Land-cover refers to the features on the Earth's surface, and land-use indicates the human activities on the particular land parcel (Lillesand et al. 2008). Detailed land-cover mapping can be utilized in urban planning, land-use monitoring, change-detection analysis, and policymaking. With the development of RS technology, satellite images achieve a good visual performance and are brought into more practical applications at local or territory-wide scales, such as for urban land-use classification (Lu and Weng 2009; Pacifici et al. 2009), environmental monitoring (Knight et al. 2013), and land-cover change detection (Potapov et al. 2017).

20.5.1.1 Multi-scale Object-Oriented Segmentation and Classification Method (MOOSC)

In order to improve land-use land-cover (LULC) mapping effectively and efficiently, a study of the multi-scale object-oriented segmentation and classification method (MOOSC) was developed (Nichol and Wong 2008). This method was implemented for habitat mapping to study a mountainous and ecologically diverse area of Tai Mo Shan and Shing Mun Country Parks in Hong Kong using fine-resolution IKONOS satellite images. The method started with grouping homogeneous pixels into image objects or segments at their respective scales. Then a five-level decision tree classification was constructed to classify each feature or object. Apart from the four native multispectral bands of the IKONOS images, additional layers of NDVI (Normalized Difference Vegetation Index), chlorophyll index, digital elevation model (DEM), and three texture bands were used in segmentation and classification procedures. The minimum mapping unit (MMU) of the classification map was about 150 m^2.

This study provides appropriate and optimal results to substitute the traditional methods of mapping using aerial photographs. The major merits of this method are: (i) the potential to produce more accurate results than traditional classification due to its wide range of parameters such as spectral information, texture, shape, and size; (ii) object-based classifications use a segmentation process to identify and delineate meaningful targets on images (it is important that the segmentation process is an automated digitizing method for delineating the target boundaries; the availability of classification outcomes in vector format is considerable merit of an object-based approach as compared with raster-based maps using conventional classification methods); and (iii) the developed object-based classification method is cost-effective since it can achieve accuracy comparable to the manual interpretation of aerial photographs but at only one-third of the cost.

20.5.1.2 Hybrid Object and Pixel-Based Classification (HOPC)

The object-based classification works well in homogeneous areas with similar spectral signatures, while pixel-based classification works on heterogeneous or fuzzy areas. Neither of them can be applied alone on broad land-cover classification especially over vegetation areas. A new approach, hybrid-MOOSC, has been developed by integrating multi-scale object-based segmentation, decision tree classification, and pixel-based classification technologies to classify heterogeneous natural landscapes of Hong Kong from fine-resolution satellite images. The approach combines SPOT-6 multispectral images, a fine-resolution DEM, and a digital surface model (DSM). The rationale of this hybrid-MOOSC is to utilize an object-based approach over homogeneous areas and a pixel-based approach over fuzzy or uncertain areas. The individual accuracy of habitat classification of mixed classes such as isolated trees and shrubs in open grassland has been significantly improved using the approach. The classification results derived from hybrid-MOOSC, as shown in Fig. 20.7, can be fully utilized in urban planning, land-use monitoring, and change-detection analysis in local and territory-wide classification with a promising potential to classify urban areas from very-fine- and fine-resolution satellite images.

Multi-resolution segmentation was applied to create objects with coherent spectral characteristics. It is a process during which pixels with similar spectral characteristics are merged into an image object. Then, classification is conducted on the image objects by assigning them to specific land-cover types. Ideally, an image object comprises only one class, but any resolution of satellite image does not void the availability of similar spectral values from mixed-class objects. Therefore, this study used a rule-based separation of pure objects and fuzzy objects (decision rules for each class). The thresholds were defined by analyzing the sampling histograms of various features (such as NDVI, blue-red ratio, red ratio, and object height) of image objects corresponding to each land-cover class. Most of the image objects were correctly classified into corresponding classes, which correspond to the homogeneous classes.

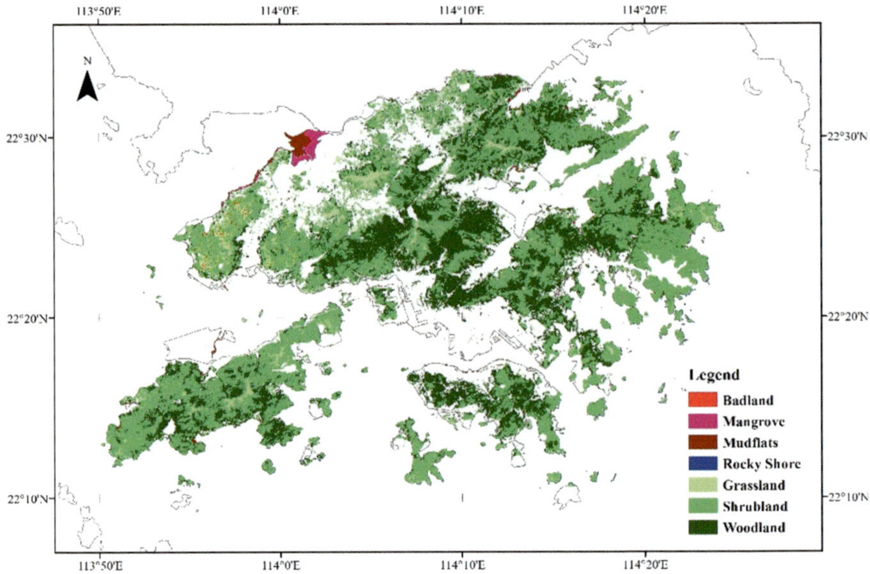

Fig. 20.7 Land-cover map of the entire territory of Hong Kong using hybrid-MOOSC

However, some image objects cannot be classified efficiently due to overlapping in their feature properties, such as spectral response, resulting in fuzzy areas.

A fuzzy object contains two or more classes at a certain spatial scale. For example, an object may contain both grassland and open shrubs which cannot be separated into two objects in the multi-resolution segmentation stage. In these fuzzy objects, their feature properties are averaged over classes that are not distinctive from pure classes, as their feature properties usually overlap in the sampling histograms. Therefore, for fuzzy objects, refinement is needed in order to achieve a more accurate classification result. For this purpose, a pixel-based segmentation was performed on the fuzzy objects, which is a method of dividing large objects into smaller pixels. When the objects are broken down into pixels, they will be reclassified into their corresponding classes. The advantage of the object-based approach is to alleviate the original noise, while the pixel-wise method is good at preserving the details of ground objects, especially in fuzzy areas which are transition stages of habitat classes in a landscape. The proposed HOPC is useful for improving the classification of a fine-resolution image by combining both approaches.

The high accuracy of the HOPC result may be mainly due to its hybrid approach which combines the advantages of object-based classification and pixel-based classification, with flexible expert judgment. The object-based fuzzy areas were further broken down into pixels and reclassified to the corresponding class. This advanced method helped to increase the overall accuracy significantly. However, if only pixel-based classification is adopted, for example, MLC, it does not consider in an object aspect, so that many homogeneous areas contain inconsistent classes after classification, such as the salt and pepper effect. For object-based classification, homogeneous objects can be segmented first and then classified, but this does not deal with the borders of the objects, which usually introduces fuzzy areas.

20.5.2 *Urban Vegetation Phenology*

Vegetation phenology is the timing of seasonal developmental stages in plant life cycles. It has been gaining considerable attention due to its implications for water, carbon, and energy cycles, and even human health. Vegetation phenology is sensitive to environmental conditions. As we know, urbanization can change environmental conditions (e.g., alter the local climate and bring more artificial light), and thus affect vegetation phenology. Studying urbanization-induced vegetation phenology shifts will provide insights on how vegetation responds to environmental changes. Considering that urbanization is accelerating around the world, addressing this question will further help to investigate future ecosystem scenarios under the pressure arising from global climate change and growing population.

Several studies have used RS data to investigate the urbanization effects on vegetation spring phenology in different cities (Li et al. 2017). These investigations have reached the same conclusion, that vegetation spring phenology in urban areas occurs earlier than in surrounding rural areas.

However, the magnitude of this rural-urban difference is quite different among these studies. Yao et al. applied 2001–2015 MODIS EVI data to study phenology change in all cities of northeast China and revealed that the spring phenology in urbanized areas advanced 0.79 days/year more than in rural areas in this period (Yao et al. 2017). Li et al. used 2003–2012 MODIS EVI data to study phenology change in more than 4500 urban clusters in the conterminous United States (Li et al. 2017). They found that phenology changes are related to urban area size. A tenfold increase in the size of a city could lead to earlier spring phenology of about 1.3 days. More studies are needed to explore the reasons for these diverse urban effects on vegetation phenology.

20.5.2.1 Urban Vegetation Phenology of Beijing

A study was conducted to implement phenology-based vegetation monitoring methods in Beijing city (i) to explore the spatial pattern of vegetation phenology along the urban–rural gradient; and (ii) to examine the relationship between vegetation phenology and urban environmental factors including both air temperature and artificial light (Yao et al. 2017). The data used in this study included MODIS EVI time series in 2012 (MOD13Q1 Version 6, 16-day composite, 250 meters), the hourly air temperature in 2012 from 232 meteorological stations in Beijing, and nighttime light data from the VIIRS in 2012.

The method proposed by Piao et al. was used to detect the start of the season (SOS) and end of the season (EOS) from the EVI time series (Piao et al. 2006). This method first computes a reference EVI curve by averaging multi-year EVI curves and then finds SOS (when 20% of the seasonal amplitude is reached during the green-up period) and EOS (when 60% of the seasonal amplitude is reached during the brown-down period) in the reference EVI curve. Next, the EVI values in the reference curve corresponding to SOS and EOS are selected as thresholds. Then, an EVI curve in each year is fitted by a polynomial function. Finally, the SOS and EOS of each year can be detected from the fitted curve and the thresholds.

The result for SOS (Fig. 20.8a) shows a spatial distribution of green-up onset in 2012, from which we can see the onset dates of vegetation green-up in the urban area occurred earlier than the surroundings. The spatial distribution of EOS (Fig. 20.8b) shows that the onset date of vegetation dormancy in urban areas is generally later than the surroundings, especially in the rural area. Besides, both SOS and EOS in the urban expansion area distribute intricately, indicating that the vegetation in the urbanization area is heterogeneous.

The correlation analysis between air temperature and phenology shows that SOS is negatively correlated to spring air temperature ($R = -0.23$, p-value <0.01) while EOS is positively correlated with autumn air temperature ($R = 0.16$, p-value <0.1). SOS is negatively correlated to nighttime light intensity ($R = -0.22$, p-value <0.01), while EOS has no significant correlation with nighttime lights. Above results suggest that both urban heat island and artificial lights may have impacts on the vegetation growth in the urban environment, and this effect is more significant in urban centers and decreases toward rural areas.

20.5.3 Urban Heat Island Mapping

Urban heat island (UHI) refers to the phenomenon that air and surface temperatures in an urban area are higher than those in rural areas. This temperature difference can range from 1.5 to 4 °C in summer daytime to 2–6.5 °C in winter daytime. However, a more significant UHI effect is expected at night and in the early morning. The main causes of UHI include (i) compact urban structure such as high-rise buildings with high-density; and (ii) anthropogenic heat released by human activities, for example from transportation and electricity. Then, heat will be released and trapped, resulting in a higher temperature in urban areas (for a discussion of the computational issues of UHI, see Chap. 41).

20.5.3.1 New Emissivity and Land Surface Temperature Retrieval Method

Hong Kong as a city suffers from the UHI effect due to high-rise buildings and high building density. Therefore, UHI monitoring is significantly required and studies have been conducted to improve UHI modeling by developing different sets of algorithms to enhance the retrieval of heat-relating parameters.

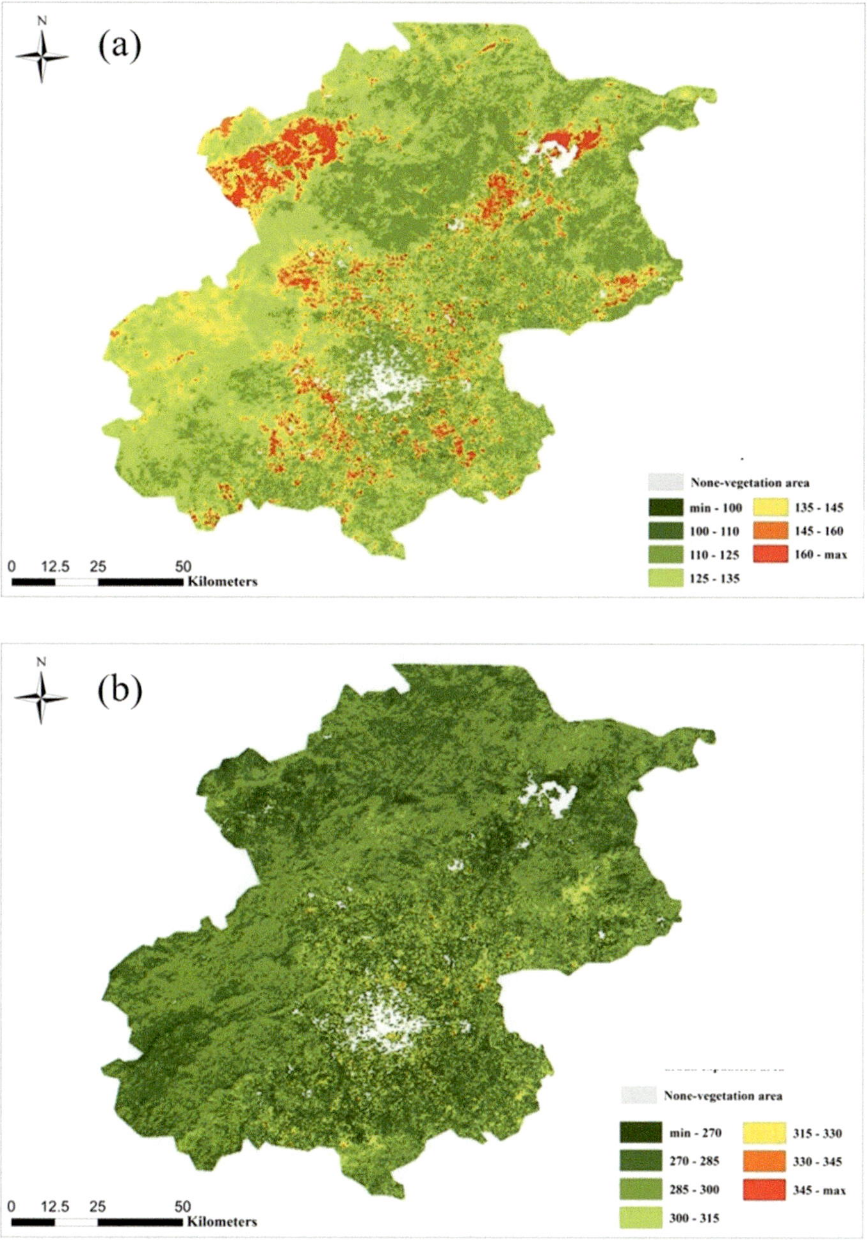

Fig. 20.8 SOS (**a**) and EOS (**b**) of Beijing detected from MODIS EVI time series in 2012

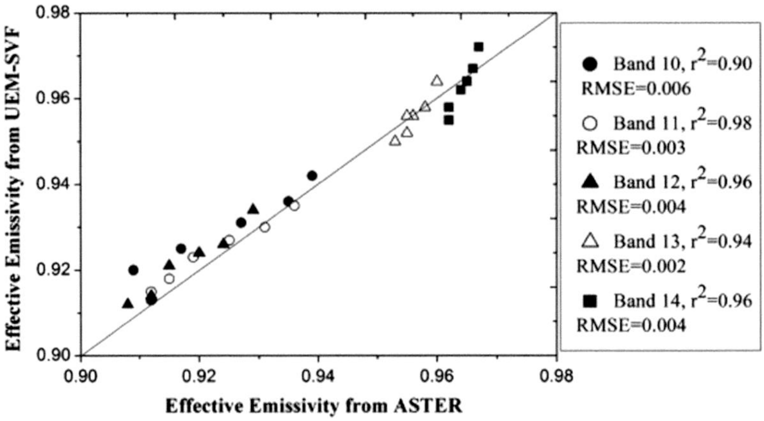

Fig. 20.9 Validation of effective emissivity derived from the UEM-SVF model

Emissivity, accounting for the percentage of radiation emitted from a surface, is a crucial parameter in retrieving land surface temperature (LST) and hence accurate retrieval of emissivity is needed. Yang et al. (2015) proposed a method estimating the effective emissivity using a sky-view factor. This factor represents the portion of the sky that can be seen from the ground and is derived from airborne LiDAR data, land-cover classification data, and building data. This study shows that there exists a high correlation between effective emissivity and the sky-view factor, attaining a correlation coefficient of more than 0.90. By additionally considering scattering, that is, the reflection effect of adjacent pixels, the refined model, named the urban emissivity model based on the sky-view factor (UEM-SVF), was developed to estimate effective emissivity in an accurate manner. Figure 20.9 shows the validation results of the emissivity derived from the UEM-SVF model and ASTER satellite images.

In addition to the sky-view factor, more urban geometry factors were included to improve emissivity retrieval, resulting in an improved urban emissivity model based on the sky-view factor (IUEM-SVF) (Yang et al. 2015). The new geometrical consideration factors include (i) facet emission within an instantaneous field of view (IFOV); (ii) reflection of facet emission due to adjacent facets; and (iii) scattering of emitted and reflected radiation in 3D space. Temperatures of urban facets in 3-D (TUF-3D), a microscale radiative transfer code using an energy-balance model, was employed to assess the accuracy of IUEM-SVF. Results suggested that the inclusion of geometrical considerations could improve the retrieval accuracy of effective emissivity by showing a good agreement between IUEM-SVF and TUF-3D. However, when there is more variance in emissivity, the retrieval accuracy of effective emissivity decreases.

With an accurate determination of effective emissivity, the results could then be used in several applications such as LST retrieval. Yang et al. (2016) applied the effective emissivity derived from IUEM-SVF to obtain LST for a nighttime ASTER satellite image.

20.5.3.2 Anthropogenic Heat Flux Modeling

Anthropogenic heat modeling is another important area in understanding, UHI since it is one of the major causes in a city. Wong et al. (2015) developed a novel algorithm retrieving anthropogenic heat using satellite images over Hong Kong with consideration of the complex land-cover in Hong Kong. The algorithm is based on the conventional energy-balance model with modification based on the heterogeneous characteristics of land-cover. The anthropogenic heat flux derived over Hong Kong on October 11, 2012, is illustrated in Fig. 20.10, and the anthropogenic heat was found to be correlated to building height and building density (Fig. 20.11 and 20.12). In urban areas, results showed that commercial areas emit the most anthropogenic heat flux, followed by industrial areas (Fig. 20.13).

With the modeling of anthropogenic heat flux over entire Hong Kong using satellite images, firstly, the general pattern of anthropogenic heat can be extracted; secondly, different relationships between anthropogenic heat and urban geometry and characteristics can be investigated. These findings can improve our understanding of the formation, distribution, and magnitude of UHI and can assist different experts in their decision-making about mitigating the UHI effect.

20.5.4 Rock Outcrops Identification

Rock outcrops are part of the bedrock that is completely exposed on the surface of terrain, and they are strongly related to geologic hazards, such as landslides and rockfall. The exposed rock surface is subject to chemical and physical weathering,

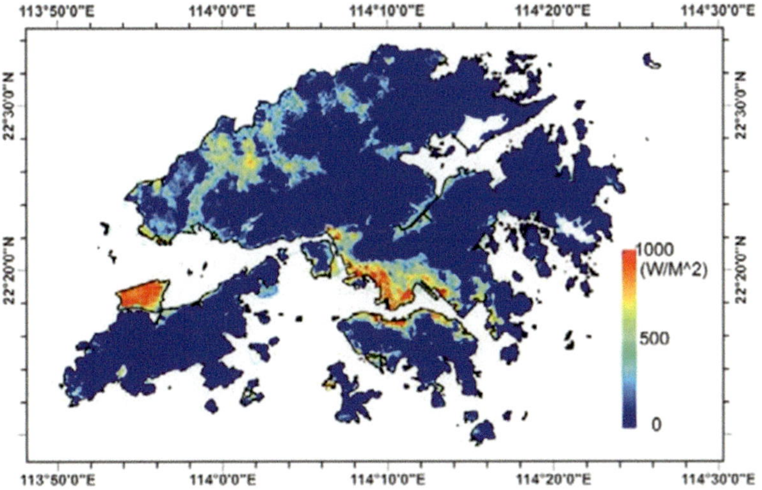

Fig. 20.10 Anthropogenic heat flux over Hong Kong on October 11, 2012

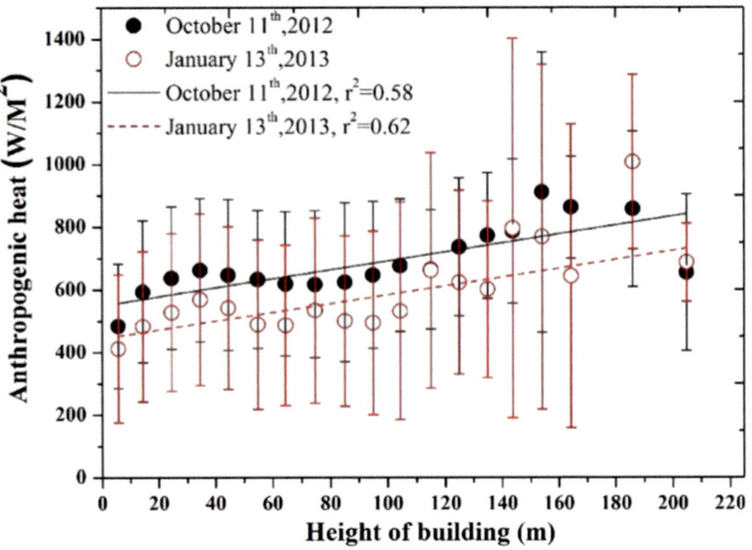

Fig. 20.11 Relationship between anthropogenic heat and building height

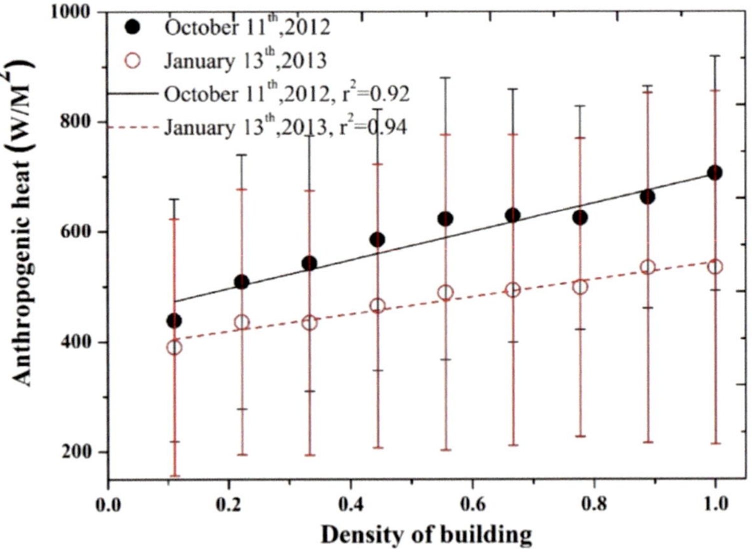

Fig. 20.12 Relationship between anthropogenic heat and building density

which increases the risk of landslides or rock falls. In high-density cities, a high density of buildings and infrastructure developed on steep slopes become a concern towards the stability of urban infrastructure and city development (Owen and Shaw 2007). The traditional ways to map the rock outcrops include field measurement and aerial photo interpretation (API). Field measurement can be conducted using

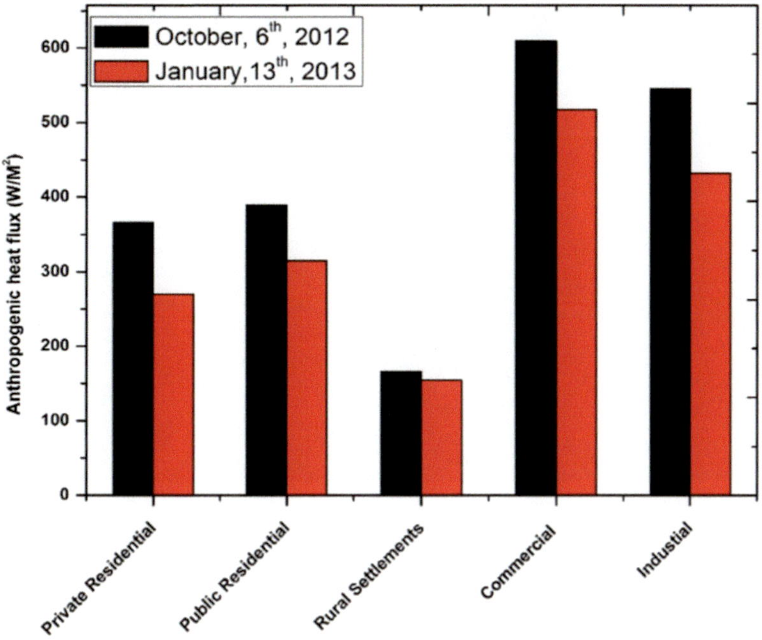

Fig. 20.13 Comparison of anthropogenic heat with different land-use types

the following approaches: (i) a structural geologist carrying a GPS tracker to locate the exposed segments; (ii) identification of angle and direction based on clinometer and geologic compass; (iii) identification of the geological faults and rock types of each exposed segment based on mineral characteristics, fossils, and geological ages. However, there are several limitations of the field measurement, including the accessibility of rock outcrops and the time-consuming work of mapping. To tackle these problems, API has been used for mapping rock outcrops. The advantage of using API is that it can locate the rock outcrops in areas that are inaccessible by fieldworkers. With the extensive coverage of a flight plan, it is able to cover a larger spatial extent that can be used for mapping rock outcrops of an entire city, such as Hong Kong. The major issue of using the API method is that it is time-consuming since rock outcrops are identified based on a knowledge-based process (Outcalt and Benedict 1965). It is essential in the process of identifying rock outcrops because the classification is mainly based on the differentiation of colors, tones, shape, and association (Outcalt and Benedict 1965). Based on human interpretation, there can be a high rate of misclassification.

20.5.4.1 Deep Learning Method to Identify Rock Outcrops in Hong Kong

In order to reduce the potential bias from a pixel-based RS application, object-based techniques have been developed. An innovative methodology combining the deep learning technique of convolutional neural networks and RS techniques was

Fig. 20.14 Examples of Rock outcrops

developed to leverage the balance between spatial resolution and spectral resolution for mapping rock outcrops in Hong Kong.

Five target land-cover types were selected as the training and testing samples in this study, including rock outcrops, grassland, tree, badland, and urban. The examples of rock outcrops are shown in Fig. 20.14. They were trained with a 16-layers VGGNet (Simonyan and Zisserman 2014) with a pre-trained model from ImageNet. Training accuracy increases significantly from the first epoch of around 50% to the third epoch of 80% and increases steadily until the end of the training. While the testing accuracy increases from the first epoch of 70% to the 20th epoch of 90% and then remains oscillating between 90 and 92% until the end of the training, it indicates that there is no more improvement in the testing accuracy after the 20th epochs. Therefore, the trained network can provide high accuracy for land-cover classification of over 90% accuracy on both training set and testing set.

After training the model, the trained network was applied to the whole selected digital orthophoto (DOP) of the whole of the Hong Kong territory. For each of the DOPs, a 20 $\times$ 20 m kernel was input into the CNN network for classification and the probability of that kernel belonging to rock outcrops was predicted. The land-cover classification map (Fig. 20.15) and rock outcrops probability map (Fig. 20.16) were then generated, and finally, the rock outcrops map of Hong Kong (Fig. 20.17) was produced.

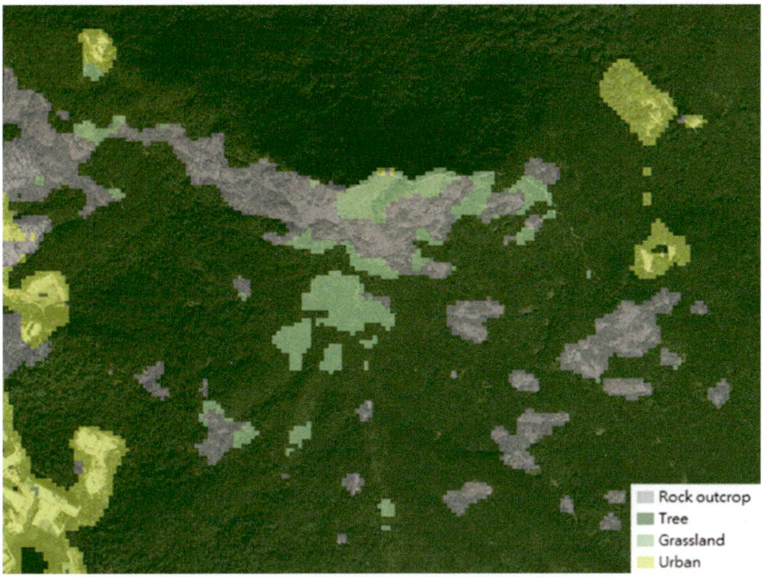

Fig. 20.15 Classification result of High West, Hong Kong Island with the year 2015 DOP

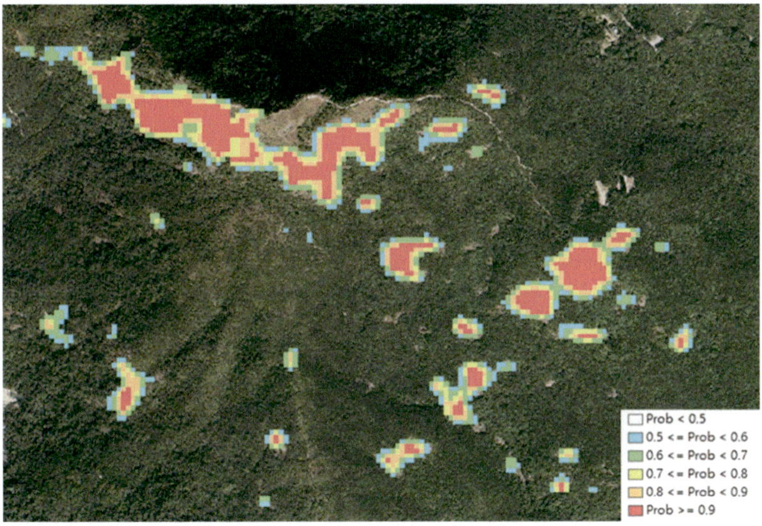

Fig. 20.16 Probability map of rock outcrops of High West, Hong Kong Island with the year 2015 DOP

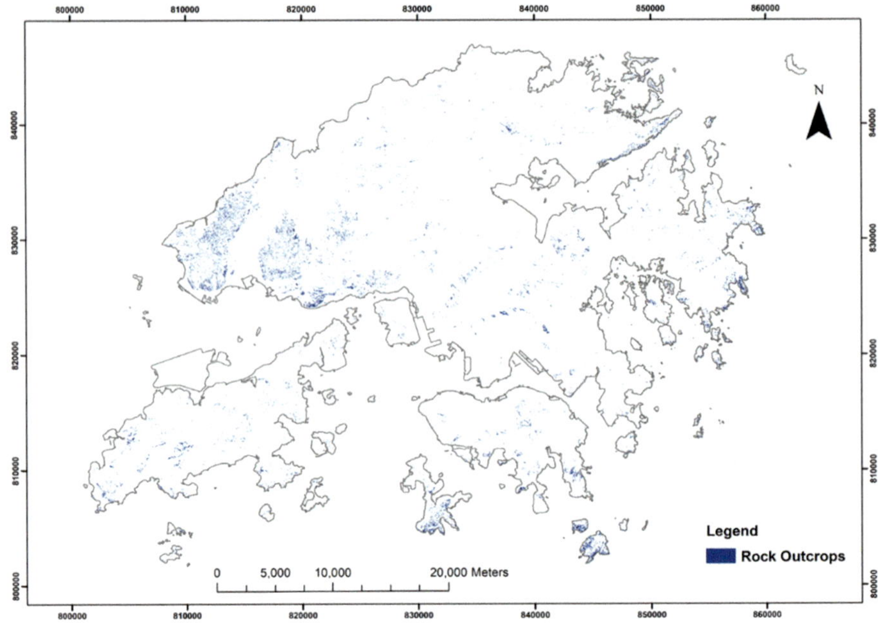

Fig. 20.17 A rock outcrops map of Hong Kong

20.6 Summary

Presently, the development of smart cities is highly dependent on spatial information derived from remote sensing technologies. However, prior to using modern tools and techniques, knowledge about the characteristics of remote sensing datasets, interpretation theories, automatic extraction of urban objects, and problems associated with these methods is essential. It has been thoroughly discussed in this chapter. With the advent of very-fine-resolution images, contemporary research is focused towards information extraction using big data analytics, due to the huge volume of data with finer and finer spatial, spectral, and temporal resolution. In addition, analysis paradigms are shifting towards a high precision of geometric details and vertical developments; a trade-off between spectral and spatial information of the remote sensing datasets; the automatic object-oriented feature extraction to update changes in urban space; the development of urban spectral libraries from image spectroscopy to detect and classify numerous urban surface materials; cutting-edge technologies for 3D building generation from LiDAR point clouds; land-use type classification along the vertical surfaces of skyscrapers; dynamics of urban sprawl and population migration as a result of economic developments; population estimation from satellite images; sustainable urban ecology in the context of future development; disaster-risk reduction in the context of extreme weather events and earthquakes, urban noise pollution and air-pollution monitoring; urban trees and biodiversity for environmental conservation; and smart transportation systems. Thus, the enormous amount of remote sensing data and big data analytics will be the backbone of mandatory geospatial cyberinfrastructure for the development of future smart cities.

References

Blaschke T, Hay GJ, Weng Q, Resch B (2011) Collective sensing: integrating geospatial technologies to understand urban systems-An overview. Remote Sens 3:1743–1776

d'Oleire-Oltmanns S, Coenradie B, Kleinschmit B (2011) An object-based classification approach for mapping migrant housing in the mega-urban area of the Pearl River Delta (China). Remote Sens 3:1710–1723

Du P, Xia J, Zhang W (2012) Multiple classifier system for remote sensing image classification: a review. Sensors 12:4764–4792

Fernandez-Delgado M, Cernadas E, Barro S, Amorim D (2014) Do we need hundreds of classifiers to solve real world classification problems? J Mach Learn Res 15:3133–3181

Geiß C, Taubenböck H, Wurm M, Esch T, Nast M, Schillings C, Blaschke T (2011) Remote sensing-based characterization of settlement structures for assessing local potential of district heat. Remote Sens 3:1447–1471

Goetz SJ, Wright RK, Smith AJ, Zinecker E, Schaub E (2003) IKONOS imagery for resource management: tree cover, impervious surfaces, and riparian buffer analyses in the mid-Atlantic region. Remote Sens Environ 88:195–208

Hamdi R (2010) Estimating urban heat island effects on the temperature series of Uccle (Brussels, Belgium) using remote sensing data and a land surface scheme. Remote Sens 2:2773–2784

Hay GJ, Kyle C, Hemachandran B, Chen G, Rahman MM, Fung TS, Arvai JL (2011) Geospatial technologies to improve urban energy efficiency. Remote Sens 3:1380–1405

Hermosilla T, Ruiz LA, Recio JA, Estornell J (2011) Evaluation of automatic building detection approaches combining high resolution images and LiDAR data. Remote Sens 3:1188–1210

Hofmann P, Strobl J, Nazarkulova A (2011) Mapping green spaces in Bishkek: how reliable can spatial analysis be? Remote Sens 3:1088–1103

Jensen JR (2009) Remote sensing of the environment: An earth resource perspective, 2nd edn. Pearson Education India, Delhi

Jin MS, Kessomkiat W, Pereira G (2011) Satellite-observed urbanization characters in Shanghai, China: aerosols, urban heat island effect, and land-atmosphere interactions. Remote Sens 3:83–99

Kadhim N, Mourshed M, Bray M (2016) Advances in remote sensing applications for urban sustainability. Euro-Mediterr J Environ Integr 1:1–22

Knight JF, Tolcser BP, Corcoran JM (2013) The effects of data selection and thematic detail on the accuracy of high spatial resolution wetland classifications. Photogram Eng Remote Sens 79:613–623

Li X, Zhou Y, Asrar GR (2017) Response of vegetation phenology to urbanization in the conterminous United States. Glob Change Biol 23:2818–2830

Liu L, Zhang Y (2011) Urban heat island analysis using the Landsat TM data and ASTER data: a case study in Hong Kong. Remote Sens 3:1535–1552

Lillesand TM, Kiefer RW, Chipman JW (2008) Remote sensing and image interpretation, 6th edn. Wiley, Hoboken

Lu D, Weng Q (2007) A survey of image classification methods and techniques for improving classification performance. Int J Remote Sens 28:823–870

Lu D, Weng Q (2009) Extraction of urban impervious surfaces from an IKONOS image. Int J Remote Sens 30:1297–1311

Mallinis G, Koutsias N, Tsakiri-Strati M (2008) Object-based classification using Quickbird imagery for delineating forest vegetation polygons in a Mediterranean test site. ISPRS J Photogrammetry Remote Sens 63:237–250

Mather PM (2011) Computer processing of remotely-sensed images: An introduction, 4th edn. Wiley-Blackwell, Chichester, West Sussex

Miyazaki H, Iwao K, Shibasaki R (2011) Development of a new ground truth database for global urban area mapping from a gazetteer. Remote Sens 3:1177–1187

Mountrakis G, Im J, Ogole C (2011) Support vector machines in remote sensing: a review. ISPRS J Photogrammetry Remote Sens 66:247–259
Nichol J, Wong MS (2007) Remote sensing of urban vegetation life form by spectral mixture analysis of high-resolution IKONOS satellite images. Int J Remote Sens 28:985–1000
Nichol J, Wong MS (2008) Habitat mapping in rugged terrain using multispectral Ikonos images. Photogram Eng Remote Sens 74:1325–1334
Olsen R (2016) Remote sensing from air and space, 2nd edn. SPIE Press, Washington
Outcalt SI, Benedict JB (1965) Photo-interpretation of two types of rock glacier in the Colorado Front Range, USA. J Glaciol 5:849–856
Owen B, Shaw R (2007) Hong Kong landscapes: shaping the barren rock, vol 1. Hong Kong University Press, Hong Kong
Pacifici F, Chini M, Emery WJ (2009) A neural network approach using multi-scale textural metrics from very high-resolution panchromatic imagery for urban land-use classification. Remote Sens Environ 113:1276–1292
Piao S, Mohammat A, Fang J (2006) NDVI-based increase in growth of temperate grasslands and its responses to climate changes in China. Glob Environ Change 16:340–348
Potapov P, Hansen M, Laestadius L, Turubanova S, Yaroshenko A, Thies C, Smith W, Zhuravleva I, Komarova A, Minnemeyer S et al (2017) The last frontiers of wilderness: Tracking loss of intact forest landscapes from 2000 to 2013. Sci Adv 3:e1600821. https://doi.org/10.1126/sciadv.1600821
Rinner C, Hussain M (2011) Toronto's urban heat island: exploring the relationship between land use and surface temperature. Remote Sens 3:1251–1265
Simonyan K, Zisserman A (2014) Very deep convolutional networks for large-scale image recognition. arXiv preprint arXiv1409:1556
Tan K, Du P (2011) Combined multi-kernel support vector machine and wavelet analysis for hyperspectral remote sensing image classif ication. Chin Opt Lett 9:11003
UCS Satellite Database (2005) UCS union of concerned scientists. https://www.ucsusa.org/nuclear-weapons/space-weapons/satellite-database. Accessed 20 Sep 2019
Wong MS, Yang Jinxin, Nichol J (2015) Modeling of anthropogenic heat flux using HJ-1B Chinese small satellite image: a study of heterogeneous urbanized areas in Hong Kong. IEEE Geosci Remote Sens Lett 12:1466–1470
Xia N, Cheng L, Li MC (2019) Mapping urban areas using a combination of remote sensing and geolocation data. Remote Sens 53(1):70–84
Xue Z, Li J, Cheng L (2015) Spectral-spatial classification of hyperspectral data via morphological component analysis-based image separation. IEEE Trans Geosci Remote Sens 53:70–84
Yang J, Wong MS, Menenti M, Nichol J (2015) Modeling the effective emissivity of the urban canopy using sky view factor. ISPRS J Photogrammetry Remote Sens 105:211–219
Yang J, Wong MS, Menenti M, Nichol J, Voogt J, Krayenhoff ES, Chan PW (2016) Development of an improved urban emissivity model based on sky view factor for retrieving effective emissivity and surface temperature over urban areas. ISPRS J Photogrammetry Remote Sens 122:30–40
Yao R, Wang L, Huang X, Guo X, Niu Z, Liu H (2017) Investigation of urbanization effects on land surface phenology in Northeast China during 2001–2015. Remote Sens 9:66

Man Sing Wong is an Associate Professor of the Department of Land Surveying and Geo-Informatics, The Hong Kong Polytechnic University, and also a NASA's AERONET Hong Kong station site manager. He is a chartered member of the Royal Institution of Chartered Surveyors, the Hong Kong Institute of Surveyors, and a Fulbright scholar supported by the United States Department of State. He has published over 100 SCI journal papers and received over HKD 58 million of research funding as PI in the last couple of years.

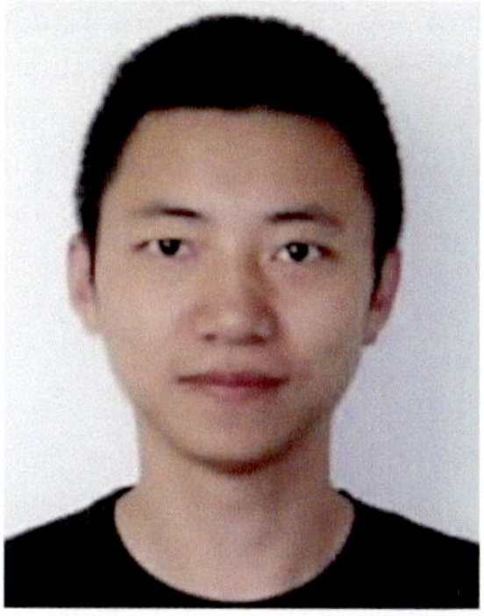

Xiaolin Zhu is an Assistant Professor of the Department of Land Surveying and Geo-Informatics, The Hong Kong Polytechnic University. He is interested in remote sensing and geospatial analysis. He received a Presidential Fellowship from Ohio State University and the Robert N. Colwell Memorial Fellowship from the American Society of Photogrammetry and Remote Sensing.

Sawaid Abbas is a Research Assistant Professor at the Department of Land Surveying and Geo-informatics, at The Hong Kong Polytechnic University, Hong Kong. He was awarded a Hong Kong Ph.D. fellowship by the Research Grants Council of Hong Kong for the period 2013–2016. He has worked on the 70-years Forest Succession project in Hong Kong's Country Parks.

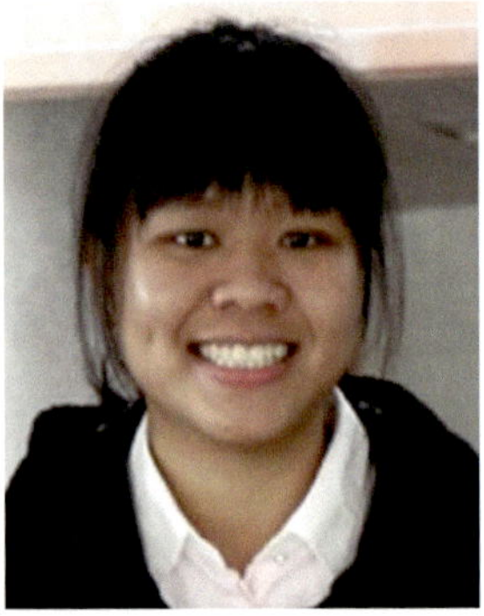

Coco Yin Tung Kwok received a BSc in Geomatics (Land Surveying) from The Hong Kong Polytechnic University in 2015 and is currently pursuing a Ph.D. degree in remote sensing at the same Institute. Her research interests include remote sensing, geographical information systems, and indoor positioning.

Meilian Wang is a Ph.D. student of the Department of Land Surveying and Geo-Informatics, The Hong Kong Polytechnic University. She is interested in remote sensing research on the identification of vegetation health, and the application of light detection and ranging (LiDAR) on urban vegetation.

Chapter 21
Urban Sensing with Spaceborne Interferometric Synthetic Aperture Radar

Hongyu Liang, Wenbin Xu, Xiaoli Ding, Lei Zhang, and Songbo Wu

Abstract Synthetic aperture radar (SAR) and interferometric SAR (InSAR) are state-of-the-art radar remote sensing technologies and are very useful for urban remote sensing. The technologies have some very special characteristics compared to optical remote sensing and are especially advantageous in cloudy regions due to the ability of the microwave radar signals used by the current SAR sensors to penetrate clouds. This chapter introduces the basic concepts of SAR, differential InSAR, and multi-temporal InSAR, and their typical applications in urban remote sensing. Examples of applying the various InSAR techniques in generating DEMs and monitoring ground and infrastructure deformation are given. The capabilities and limitations of InSAR techniques in urban remote sensing are briefly discussed.

21.1 Synthetic Aperture Radar

A radar (RAdio Detection and Ranging) system typically sends out electromagnetic pulses and receives the pulses scattered back by objects. By precisely determining the time delay and Doppler frequency shift between the emitted and received pulses, a radar system can measure the distance to, and the moving velocity of, an object with respect to the radar. Synthetic-aperture radar (SAR) is a commonly used radar remote sensing technique that achieves finer spatial resolution imaging (i.e., up to meter level or better), in comparison with the real aperture radar, by taking advantage of the movement of the radar antenna along a particular trajectory to mathematically create a virtual radar antenna that has a much larger size than that of the physical antenna. The radar system is usually mounted on an aircraft or a satellite with a side-looking imaging geometry (Fig. 21.1). Most spaceborne SAR antennas are 10–15 m long and result in a ground spatial resolution of 1–20 m by using the SAR principle. Since the first spaceborne SAR satellite was launched in 1978 by the U.S. National Aeronautics and Space Administration (NASA), many SAR satellites have

H. Liang · W. Xu · X. Ding (✉) · L. Zhang · S. Wu
Department of Land Surveying and Geo-Informatics, The Hong Kong Polytechnic University, Hong Kong, China
e-mail: xl.ding@polyu.edu.hk

W. Shi et al. (eds.), *Urban Informatics*, The Urban Book Series,
https://doi.org/10.1007/978-981-15-8983-6_21

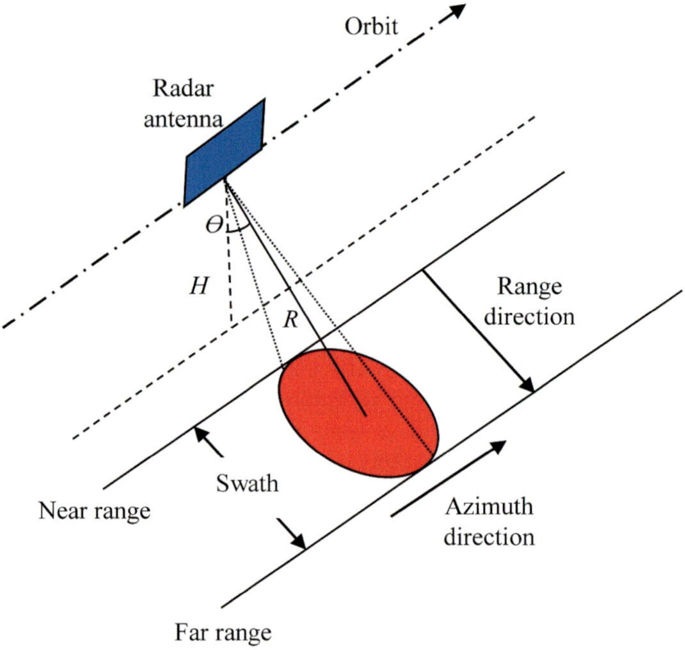

Fig. 21.1 Typical SAR imaging geometry. The antenna receives the backscattered signal from the illuminated area. The moving direction of the satellite is called the azimuth direction of the image, while the direction of radar illumination is referred to as the range direction of the image. H and R are the height of the satellite and the slant range between the satellite and a ground resolution cell, respectively. θ represents the look angle

been developed (Table 21.1). Over ten SAR satellites are currently in operation or to be launched in the near future.

A SAR system obtains information on both the intensity and the phase of the returned signal from each ground resolution cell, referred to as pixel. The intensity depends primarily on the roughness and dielectric property of the scattering surface while the phase is determined by the time delay between signal transmission and reception. The signal in a pixel can be represented by

$$y_1 = a_1 + b_1 i = A_1 \cdot e^{i\varnothing_1} \tag{21.1}$$

where a_1 and b_1 are the real and imaginary parts of the complex value; and A_1 and $\varnothing_1$ represent the amplitude and phase of the signal.

Table 21.1 SAR satellites launched to date

Satellite	Operator	Band/wavelength (cm)	Operational period
Seasat	U.S. National Aeronautics and Space Administration (NASA)	L/23.5	1978
ERS-1	European Space Agency (ESA)	C/5.66	1991–2000
JERS-1	Japan Aerospace Exploration Agency (JAXA)	L/23.5	1992–1998
ERS-2	European Space Agency (ESA)	C/5.66	1995–2011
Radarsat-1	Canadian Space Agency (CSA)	C/5.66	1995–2013
Envisat	European Space Agency (ESA)	C/5.66	2002–2012
ALOS	Japan Aerospace Exploration Agency (JAXA)	L/23.5	2006–2011
Radarsat-2	Canadian Space Agency (CSA)	C/5.66	2007–
TerraSAR-X	German Aerospace Center (DLR)	X/3.1	2007–
COSMO-SkyMed constellation	Italian Space Agency (ASI)	X/3.1	2007–
TanDEM-X	German Aerospace Center (DLR)	X/3.1	2010–
Sentinel-1A	European Space Agency (ESA)	C/5.66	2014–
ALOS-2	Japan Aerospace Exploration Agency (JAXA)	L/23.5	2014–
Sentinel-1B	European Space Agency (ESA)	C/5.66	2016–
Gaofen-3	China National Space Administration (CNSA)	C/5.66	2016–
PAZ	Hisdesat	X/3.1	2018–

21.2 Interferometric Synthetic Aperture Radar

Basic interferometric synthetic aperture radar (InSAR) involves a pair of focused complex SAR images of the same ground area and acquired with the same or similar imaging geometries, often referred to as single look complex (SLC) images. InSAR extracts very useful information from the interferometric combination of the two SAR images separated in space and time. The spatial separation between the two images is termed the spatial baseline, while the temporal separation forms the temporal baseline when the SAR images are acquired from repeat-pass orbits using the same antenna.

After alignment and resampling of the two SAR images into the same geometry, a complex interferogram is generated by coherent cross-multiplication of the two SAR images,

$$v = y_1 \cdot y_2^* = A_1 A_2 \cdot e^{(i\varnothing_1 - \varnothing_2)} \tag{21.2}$$

where v represents the signal in a pixel of the interferogram. The phase component of the signal $\varnothing_1 - \varnothing_2$ gives the phase difference between the SAR images. For a single SAR image, although the phase values appear quite random in space, the difference between the two images offers very useful information (see Fig. 21.2). The phase difference $\varnothing_1 - \varnothing_2$ can be decomposed into two components,

$$\varnothing = \varnothing_1 - \varnothing_2 = -\frac{4\pi}{\lambda}(R_1 - R_2) + \left(\psi_{\text{scat},1} - \psi_{\text{scat},2}\right) \tag{21.3}$$

where λ is the wavelength of the radar signal, R_1 and R_2 are the slant ranges from the antenna positions to the ground target for two SAR acquisitions, and $\psi_{\text{scat},1}$ and $\psi_{\text{scat},2}$ are related to the interactions between the radar signal and the ground scatterers.

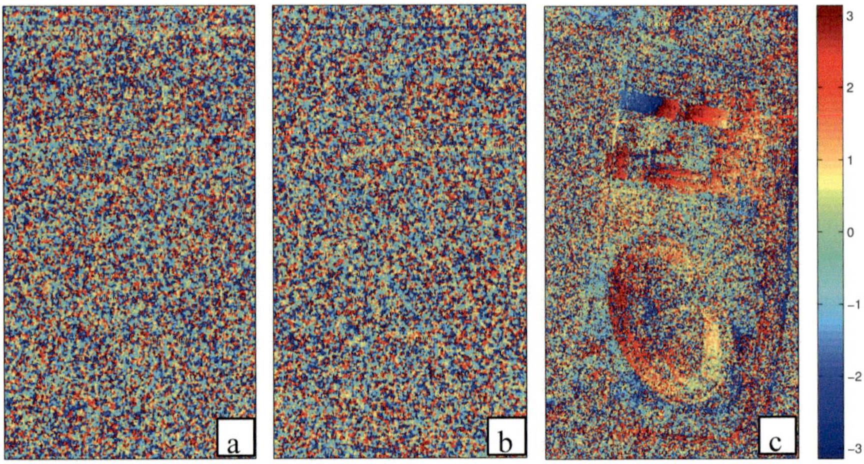

Fig. 21.2 **a** Phase image of a TerraSAR SLC image acquired on July 22, 2011, over East Asian Games Dome, Macau. **b** Phase image of a TerraSAR SLC image acquired on October 7, 2011, over the same area. **c** Interferometric phase generated by differencing image a and b; the interferometric phase values show some regular patterns which contain information about the ground surface topography, deformation, etc. The units are in π. The phase values of a, b, and c are modulo 2π, ranging from $-\pi$ to π

Although the interactions are unpredictable in real cases, the scattering will remain coherent if the spatial and temporal separations between the SAR acquisitions are small. As a consequence, the phase difference is mainly dependent on the range difference $R_1 - R_2$ as the interaction phase contributions mostly cancel out.

From a geometry perspective, the interferometric phase can also be defined as:

$$\varnothing = \varnothing_{\text{flat}} + \varnothing_{\text{topo}} + \varnothing_{\text{defo}} + \varnothing_{\text{atm}} + \varnothing_{\text{orb}} + \varnothing_{\text{noise}} \tag{21.4}$$

where $\varnothing_{\text{flat}}$ is the flattening phase and is due to the slant range variation with the elevation of the reference surface; $\varnothing_{\text{topo}}$ is the phase component resulted from the topography; $\varnothing_{\text{defo}}$ is the phase caused by ground surface displacement; $\varnothing_{\text{atm}}$ is due to the phase propagation delay when the radar signal travels through the atmosphere; $\varnothing_{\text{orb}}$ is related to the phase induced by inaccurate orbit data, and $\varnothing_{\text{noise}}$ is the phase caused by the noise. Since the wavelength of the radar signal is normally in the cm range (see Table 21.1), the phase contributions can be measured to an accuracy of mm, i.e., a fraction of the wavelength.

In early applications, radar interferometry was primarily used to map land surface topography, with a comparable accuracy (i.e., meter level) to photogrammetric methods and capability of working under all weather conditions. It was then soon demonstrated that repeat-pass interferometry could measure relative surface displacement, yielding a cm to mm accuracy. InSAR has been used extensively to retrieve ground surface deformation that is related to natural or anthropogenic activities, such as earthquakes (e.g., Fialko 2004), volcano eruption (e.g., Lu and Dzurisin 2014), glacier change (e.g., Goldstein et al. 1993), landslides (e.g., Sun et al. 2015), and land subsidence due to extraction of water or other resources (e.g., Qu et al. 2015). We will briefly introduce below how to use SAR interferometry to produce a ground surface deformation map.

The method of using two SAR images to perform interferometry for deformation mapping is called differential InSAR (DInSAR) (Massonnet and Feigl 1998). Disregarding atmospheric propagation delay and satellite orbit errors, before obtaining a deformation image, the flattening and topographic phase contributions need to be removed from the interferogram,

$$\begin{aligned} \varnothing_{\text{flat}} &= -\frac{4\pi}{\lambda}\frac{B_{\perp}s}{R\tan\theta} \\ \varnothing_{\text{topo}} &= -\frac{4\pi}{\lambda}\frac{B_{\perp}h}{R\sin\theta} \end{aligned} \tag{21.5}$$

where $B_{\perp}$ is the perpendicular baseline; R is the slant range from the antenna to the ground point; θ is the incidence angle of the radar signal; and s and h represents the differences of slant range and elevation with respect to a reference point, respectively. These parameters can be obtained from a SAR system configuration. The operation of removing the flattening phase is called interferogram flattening and the result is a flattened interferogram (see Fig. 21.3b, c). The removal of the topographic

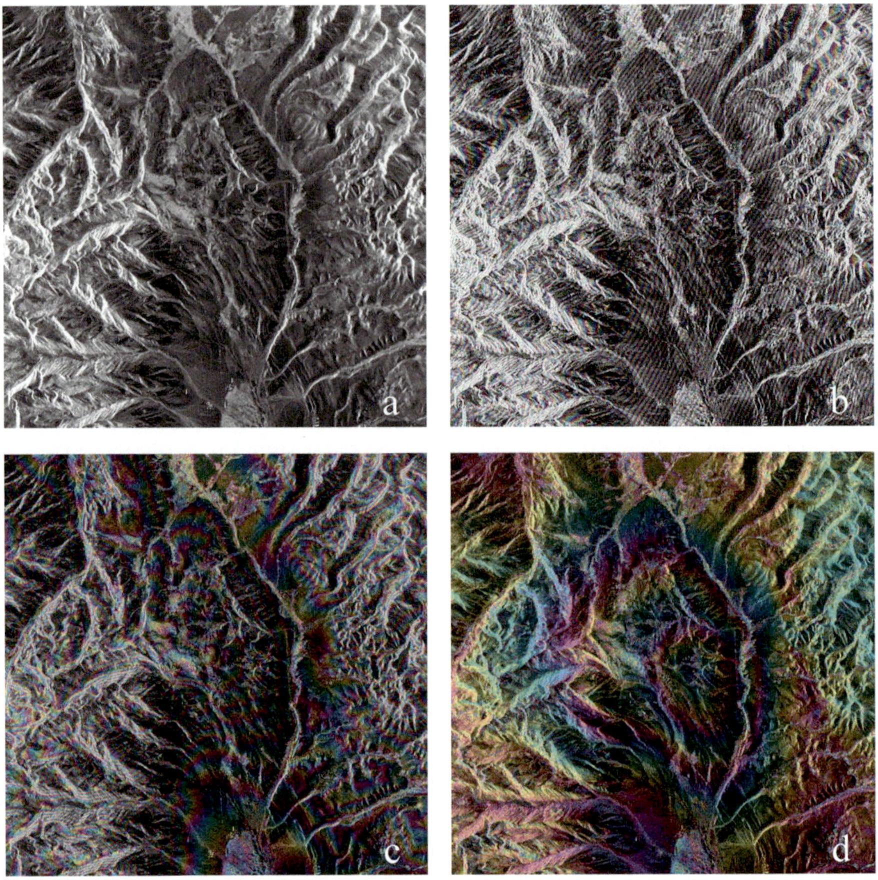

Fig. 21.3 **a** Amplitude image of a PALSAR image acquired on July 3, 2008, over Dangxiong, China. **b** Original interferogram formed by differencing two PALSAR images acquired from July 3, 2008, and February 18, 2009, respectively. The fringes with 2π phase-cycle reflect the contributions of the reference surface, topography, deformation, etc. **c** Interferogram after flattening. **d** Interferogram after flattening and removal of topographic phase. The resulting fringes mainly contain surface deformation produced by the Mw 6.3 earthquake that occurred on October 6, 2008

phase can be achieved by deploying an external digital elevation model (DEM) and the InSAR imaging geometry to simulate a synthetic interferogram and then subtracting the phase contribution from the flattened interferogram (Massonnet and Feigl 1998). Currently, there are several global DEM datasets generated based on this technique, including results from the Shuttle Radar Topographic Mission (SRTM; Farr et al. 2007) and ALOS Global Digital Surface Model "ALOS World 3D-30m" (AW3D30m; Tadono et al. 2016). Alternatively, the synthetic interferogram can be directly formed from other SAR acquisitions of the same area with short temporal separation and then can be scaled to the spatial baseline of the original interferogram. The combination of the original interferogram with a third or fourth SAR acquisition is called three-pass or four-pass InSAR (Zebker and Rosen 1994), as the approaches use additional SAR images to produce the DEM interferogram that is assumed to solely contain the topographic contribution.

Subtracting flattening and topographic phases from the original interferogram results in a differential interferogram (see Fig. 21.3d). Since atmospheric propagation delay and other systematic errors are neglected at this point, the resulting phase observations can be regarded as the sum of two contributions: (1) the relative ground displacement that occurs during the time interval between the SAR acquisitions, and (2) phase noise due to ground scattering characteristics that are related to the variation of spatial and temporal baselines. The phase noise propagates into the derived displacement map and degrades the quality of the results. To mitigate the noise effect, a low-pass filter can be applied to improve the signal-to-noise ratio (SNR) of the phase measurement, but at the cost of possible image resolution reduction (Goldstein and Werner 1998).

The filtered interferogram contains information mainly on the ground motion. However, it is impossible to directly convert the filtered differential interferogram into a displacement map as the interferometric phase values are modulo 2π, ranging from $-\pi$ to π. The wrapped phase values require adding the correct multiple of 2π to recover the absolute phase values. This procedure is referred to as phase unwrapping. Many different phase unwrapping methods have been proposed, such as the residue cut (Goldstein et al. 1988), least squares (Ghiglia and Romero 1994; Pritt and Shipman 1994), and minimal cost flow methods (Costantini 1998). Each of the methods has its own pros and cons and their performance depends on the noise level, the characteristics of terrain, and other conditions. Once the interferometric phases are unwrapped, the deformation map in the line-of-sight (LOS) direction can be obtained with respect to a reference point. As a summary, the workflow of DInSAR in extracting terrain deformation is shown in Fig. 21.4.

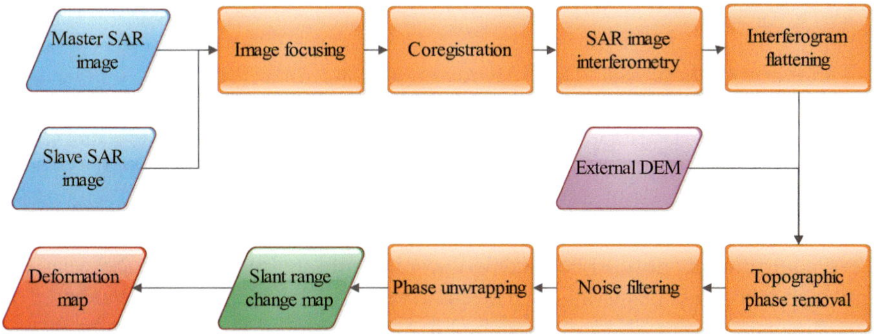

Fig. 21.4 Workflow of DInSAR processing for extracting a deformation map

21.3 Multi-temporal InSAR (MTInSAR)

The effectiveness of the DInSAR approach is limited by several factors including errors in the external DEM that are used to remove the topographic phase, atmospheric propagation delays, phase ramps induced by orbit errors, spatial and temporal decorrelation, and phase unwrapping errors. The limitations have motivated the development of the multi-temporal InSAR (MTInSAR) technique that attempts to tackle the aforementioned problems by deploying a time series of SAR images covering the same area and focusing on scatterers with strong phase stabilities (i.e., persistent scatterers or PS).

After about 20 years of development, three categories of MTInSAR techniques are currently in existence. The first category of methods exploits single master (SM) interferograms and the methods include, for example, persistent scatterers InSAR (PSInSAR; e.g., Ferretti et al. 2001), the Stanford method for persistent scatterers (StaMPS; e.g., Hooper et al. 2004, 2007), and the spatiotemporal unwrapping network method (STUN; e.g., Kampes 2006; Kampes and Hanssen 2004). The second category of methods attempts to extract deformation information from scatterers with moderate phase stabilities (i.e., distributed scatterer or DS), where an interferogram stack is formed from multiple master (MM) interferograms. Examples include the small baseline subset (SBAS) technique (e.g., Berardino et al. 2002; Lanari et al. 2004), coherent point target (CPT; e.g., Mora et al. 2003), and temporally coherent point InSAR (TCPInSAR; e.g., Zhang et al. 2011a, b, 2014; Liang et al. 2019). In the third category, some newly developed techniques make use of all possible interferometric combinations to enhance the phase quality of DS, and then use the PS and the enhanced phase measurements of the DS to estimate the deformation information under the SM interferogram framework. The methods include SqueeSAR (e.g., Ferretti et al. 2011), component extraction and selection SAR (CAESAR; e.g., Fornaro et al. 2015), phase-decomposition-based InSAR (PD-PSInSAR; e.g., Cao et al. 2016), and joint-scatterer InSAR (JSInSAR; e.g., Lv et al. 2014).

The innovations of the MTInSAR techniques are three-fold. First, high-quality coherent points form the foundation of MTInSAR. Methods for identifying such points have been developed based on different criteria, including the amplitude dispersion index (ADI; Ferretti et al. 2001), signal-to-clutter ratio (SCR; Adam et al. 2005), spatial phase stability (Hooper et al. 2004), coherence map (Jiang et al. 2015; Mora et al. 2003), and pixel offsets (Zhang et al. 2011a, b). Second, the various phase contributions need to be modeled according to the relationships between the signals and the phase observations. The contributions can be separated either based on InSAR observation itself (e.g., topographic error, orbital inaccuracy, height-related tropospheric delays; e.g., Zhang et al. 2014; Liang et al. 2019) or external data (e.g., atmospheric delays Jolivet et al. 2014). Finally, ground surface displacement history can be estimated from the function model. The estimation complexity depends on the existence of the phase ambiguities. On the one hand, the phase observations after the spatial unwrapping procedure can be easily solved by least squares. When it is challenging to carry out spatial phase unwrapping, temporal unwrapping can be performed. Typical methods include periodogram method (Ferretti et al. 2001), 3D phase unwrapping (Hooper et al. 2004), integer least squares (Kampes 2006), and least squares with outlier detection (Zhang et al. 2011a, b).

21.4 Applications in Urban Areas

It can be seen from the above discussion that the main applications of InSAR are in DEM generation and surface deformation mapping. It is often necessary to build 3D models of urban areas for purposes such as environmental modeling and urban planning. Monitoring ground and infrastructure deformation can provide essential information for better management of geohazards such as land subsidence, landslides, and sinkholes, and for ensuring the safety of urban infrastructures such as buildings, bridges, and road surfaces. We will discuss below applications of InSAR in DEM generation, land subsidence measurement, and infrastructure monitoring.

21.4.1 Construction of Fine Resolution DEM

Mapping urban topography is essential for a variety of scientific and practical applications, such as modeling urban heat island effects, urban landscape design, and urban planning. InSAR techniques can be used to generate DEM products of fine resolution in metropolitan areas. Especially data from the TanDEM-X mission has been used for generating accurate and detailed DEMs that cover the global area with an effective resolution of 6 m (Zhu et al. 2018). Based on the tandem SAR satellites TerraSAR-X and TanDEM-X, the mission performs single-pass SAR interferometry based on advanced algorithms for phase filtering and unwrapping. The single-pass bistatic interferogram has the advantage that the derived interferogram does not suffer from temporal decorrelation and atmospheric artefacts (Rossi and Gernhardt 2013), at the cost of spatial resolution due to phase filtering. Alternatively, by making use of repeat-pass acquisitions with full resolution, the MTInSAR technique can produce accurate urban DEMs with even finer spatial resolution (Perissin and Rocca 2006). Figure 21.5 presents the point cloud of a DEM product over Shenzhen, China. A total of 79 TerraSAR-X images spanning from May 2008 to May 2013 were used to generate the DEM product. The adopted methodology follows the MTInSAR processing framework (Wu et al. 2018), which has the characteristics of limiting atmospheric delays and mitigating decorrelation effects. It can be seen from Fig. 21.5 that the high-rise buildings (i.e., those higher than 100 m) are clearly identified with regular spatial patterns. Figure 21.6 presents more detail of the DEM product, in which the point clouds match well with the 3D model of the buildings in Google Earth, demonstrating the effectiveness of MTInSAR for mapping urban topography.

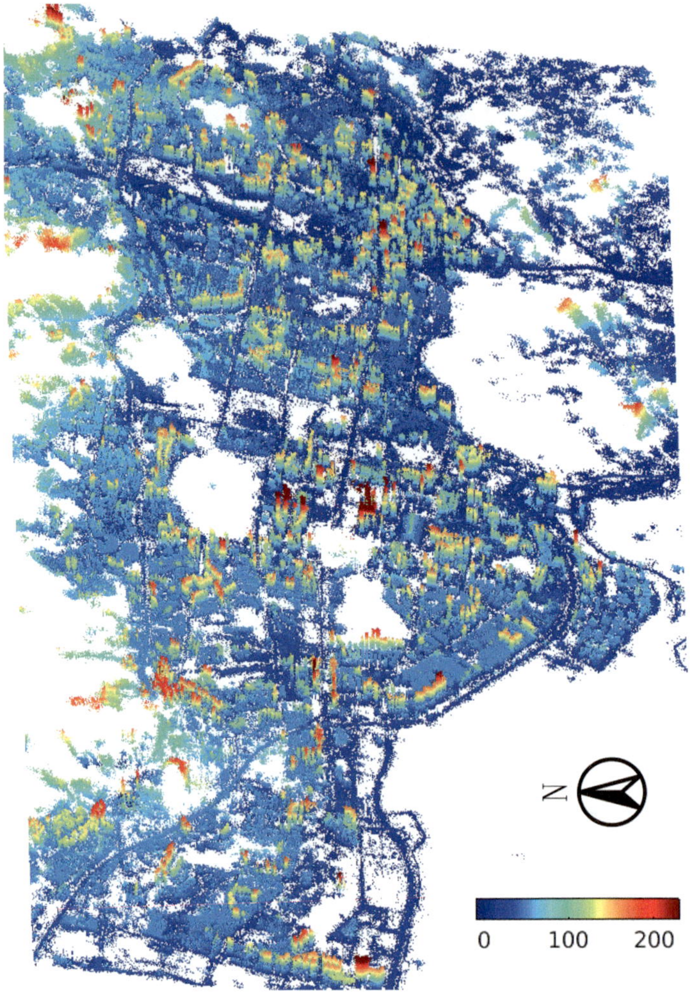

Fig. 21.5 Surface elevation model of part of Shenzhen from 79 TerraSAR-X images and MTInSAR processing

The InSAR technique is similar to stereophotogrammetry in that both use a pair of images to infer the target elevation. However, InSAR is also like the LiDAR technique as they both use range measurements. Compared to other topographic mapping techniques, the operation cost of InSAR is usually lower.

The weakness of InSAR in mapping urban topography includes specular reflection of signals, signal sidelobe, and geometric distortions of SAR images. Specular reflection of signals occurs when the ground surface is smooth, like a mirror. Little signal is backscattered in this case, leading to weak signal returns and loss of phase information. Sidelobe is caused by strong scatterers that contaminate the phase values of neighboring pixels. The geometric distortions, due to the oblique viewing geometry

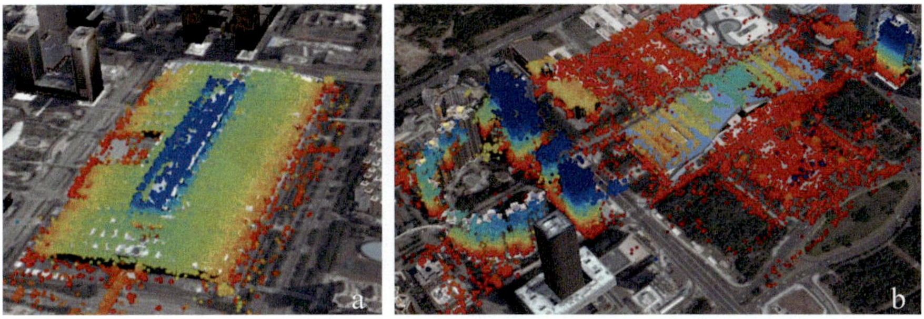

Fig. 21.6 Geocoded height maps of buildings over Shenzhen. **a** Shenzhen Convention & Exhibition Center, **b** Shenzhen Citizen Center. The maps are superimposed on a Google Earth image (© 2019 Google)

of SAR systems, have two main issues in urban environments, that is, shadowing and layover. Shadows occur when the radar signals are obscured by buildings or natural terrain, while layover is the result of superposition of multiple scatterers when terrain slope exceeds the radar incidence angle. With the development of advanced InSAR technologies, the effect of geometric distortions can be mitigated to a certain extent. SAR data from different viewing geometries (i.e., ascending and descending orbits) can be used complementarily to reduce the areas affected by shadows. For the layover problem, the elevation and deformation rate of the superimposed scatterers can be separately estimated by extending InSAR measurements into 4D (space–time) space. This operation is called differential SAR tomography (TomoSAR; (e.g., Lombardini 2005; Zhu and Bamler 2010).

21.4.2 Subsidence Measurement

The MTInSAR technique has enabled extraction of urban-area deformation with unprecedented spatial resolution. Due to the ample persistent scatterers in typical urban environments (e.g., buildings and other man-made structures), the temporal decorrelation effect is largely mitigated (Ferretti et al. 2001). The capability of InSAR in monitoring urban-area deformation has been extensively demonstrated in recent years.

Land subsidence caused by extracting groundwater is one of the emphases (Qu et al. 2015). Many areas in the world suffer from water shortages, especially in areas that are being rapidly urbanized. Figure 21.7 presents the area subsidence due to overuse of groundwater in Beijing. A total of 12 TerraSAR-X images were used to retrieve the subsidence field and its temporal evolution. The deformation results show that the largest deformation rate reaches 1.3 cm/year and the accumulative subsidence is 2.2 cm from 2010 to 2012. The InSAR derived deformation maps provide useful information on the amount and location of groundwater extraction.

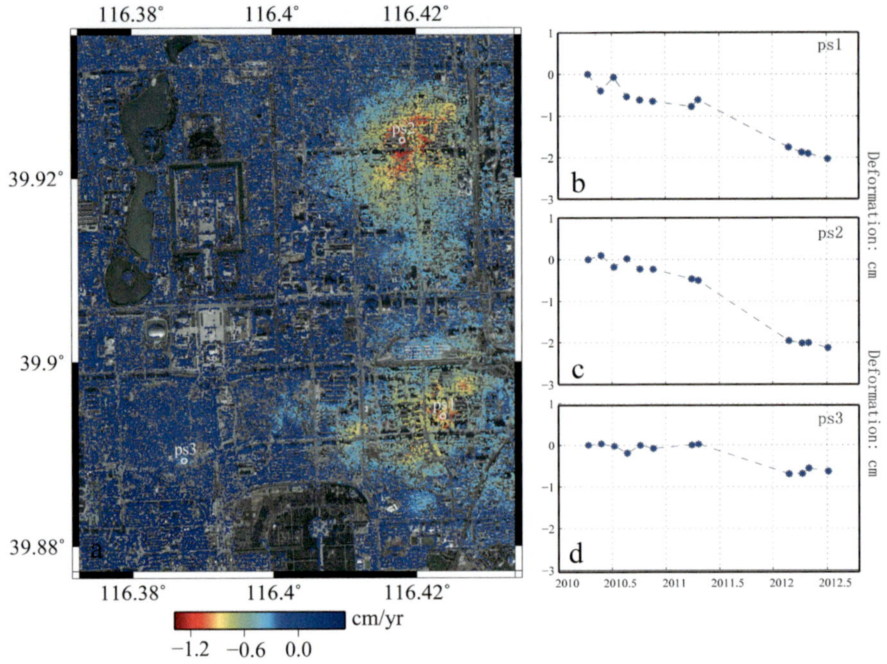

Fig. 21.7 **a** Deformation rate map over Beijing from 12 TerraSAR-X images and MTInSAR processing; **b** Deformation time series of ps1; **c** Deformation time series of ps2; **d** Deformation time series of ps3

Due to a shortage of useable land, many coastal cities reclaim land from the sea to support further urban development. A complex submarine geology can pose challenges for controlling the stability of reclaimed land (Shi et al. 2018). Figure 21.8 presents the rapid subsidence within only nine days over a man-made island. The InSAR technique has become a safe and efficient technique to extract terrain-motion information for analyzing geological stability and managing construction progress.

Subsidence caused by underground construction can be conveniently monitored by InSAR techniques (e.g., Serrano-Juan et al. 2017). Figure 21.9 shows the subsidence areas along subway lines revealed by processing 50 TerraSAR-X images from December 2013 to July 2016. Settlement due to the subway construction poses a potential threat to the surrounding areas. InSAR measurements can be used as input for analyzing the cause of subsidence.

Other land deformation such as that caused by sinkholes and landslides can also be monitored with InSAR. The feasibility of InSAR techniques for such applications depends on the rates of ground subsidence and surface features.

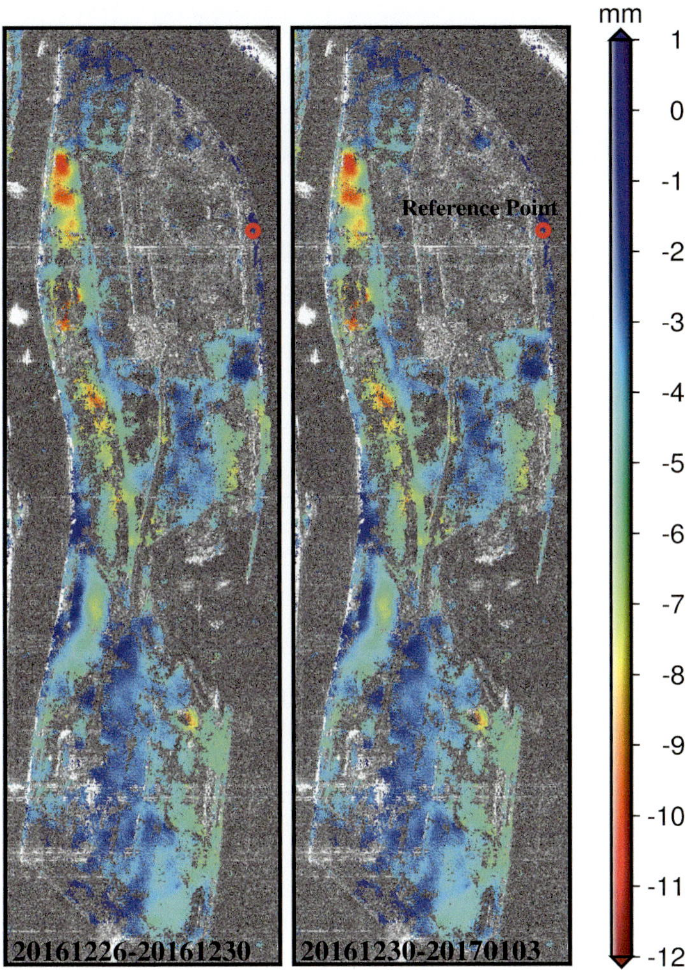

Fig. 21.8 Deformation map over a man-made island in Macau from three COSMO-SkyMed images and DInS AR processing

21.4.3 Monitoring Stability of Infrastructures

Urban infrastructures such as buildings and bridges are essential in supporting the daily lives of urban dwellers. It is important to check the stability of the infrastructures as any structural failure can lead to hazardous consequences. In-situ sensors such as accelerometers and traditional survey methods provide useful information on structural stabilities. It is however expensive to measure a large number of urban structures with these methods. InSAR, in particular MTInSAR, can be used to monitor both ground and structural deformation over a large area. It is therefore very efficient and provides very useful complementary information to the existing techniques (Ma and Lin 2016).

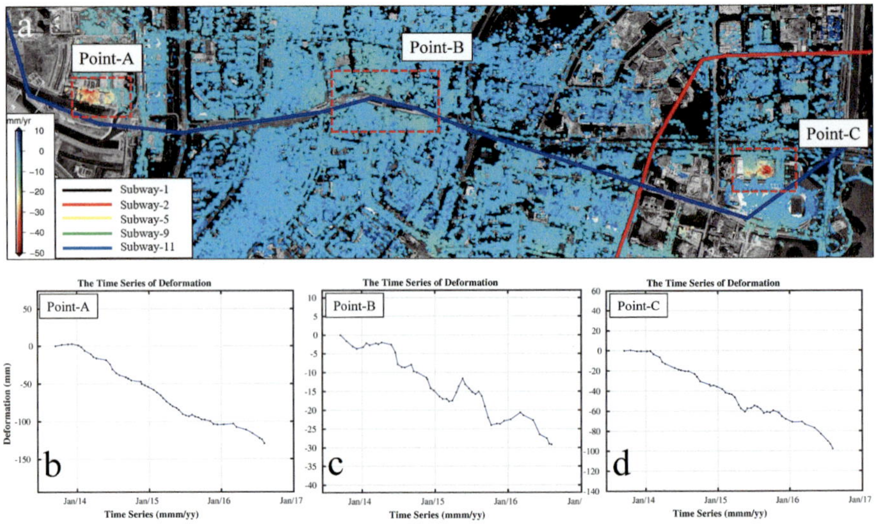

Fig. 21.9 **a** Deformation rate map along a metro line in Shenzhen from 50 TerraSAR-X images and MTInSAR processing; **b** Deformation time series of Point-A; **c** Deformation time series of Point-B; **d** Deformation time series of Point-C

In general, structural displacement observed with InSAR contains both thermal dilation and long-term deformation. Thermal dilation is caused by temperature variation of the measured structures (Crosetto et al. 2015; Qin et al. 2018). Figure 21.10 presents an example of the relationship between structural deformation and temperature for a high-rise building in Hong Kong. The thermal dilation coefficient of a structure depends on its materials.

Figure 21.11 shows two mean deformation velocity maps of road viaducts in Hong Kong, obtained by processing 29 TerraSAR-X images from 2013 to 2014. It can be seen that the deformation rates varied along the longitudinal direction of the roads. Figure 21.12 presents the deformation rate map of Stonecutter Bridge in Hong Kong after removing thermal expansion effects. The deformation rate map shows some clear deforming areas on the bridge deck.

Processing multiple SAR images from a single orbit provides information on the deformation along the line-of-sight only (Gernhardt and Bamler 2012; Schunert and Soergel 2012). By fusing multiple tracks of SAR data, infrastructures can be better observed and different deformation components, for example, the vertical and the horizontal components, can be resolved (e.g., Hu et al. 2014).

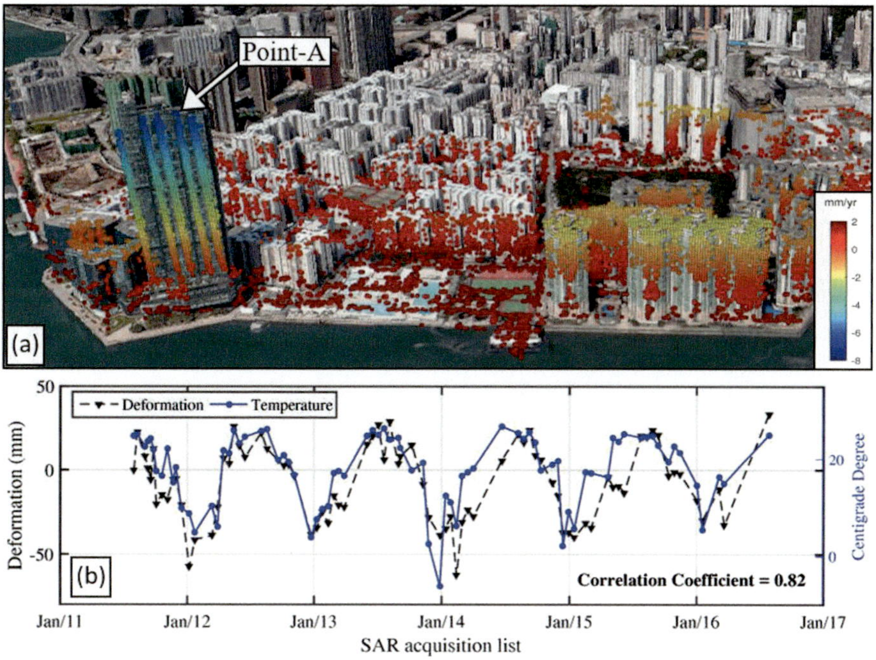

Fig. 21.10 **a** Geocoded deformation rate map of part of Kowloon Peninsula of Hong Kong from 80 COSMO-SkyMed images and MTInSAR processing. The map is superimposed on a Google Earth image (© 2019 Google). **b** Deformation time series and temperature variations of Point-A

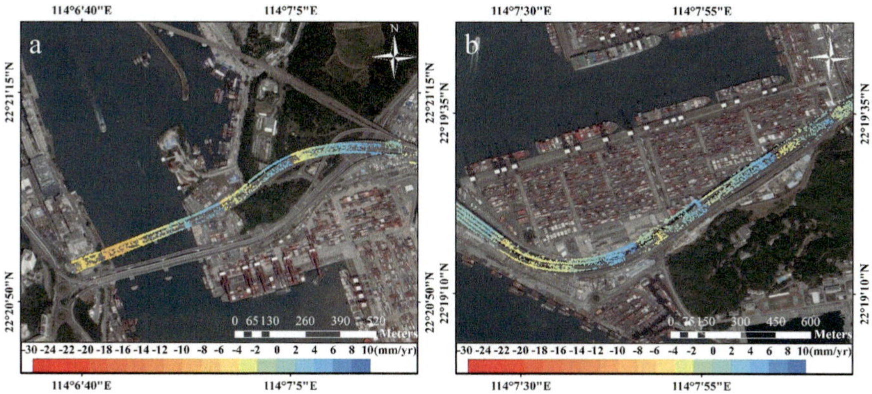

Fig. 21.11 Examples of road viaduct deformation in Hong Kong from 29 TerraSAR-X images and MTInSAR processing. **a** Tsing Kwai highway, **b** Tsing Sha highway

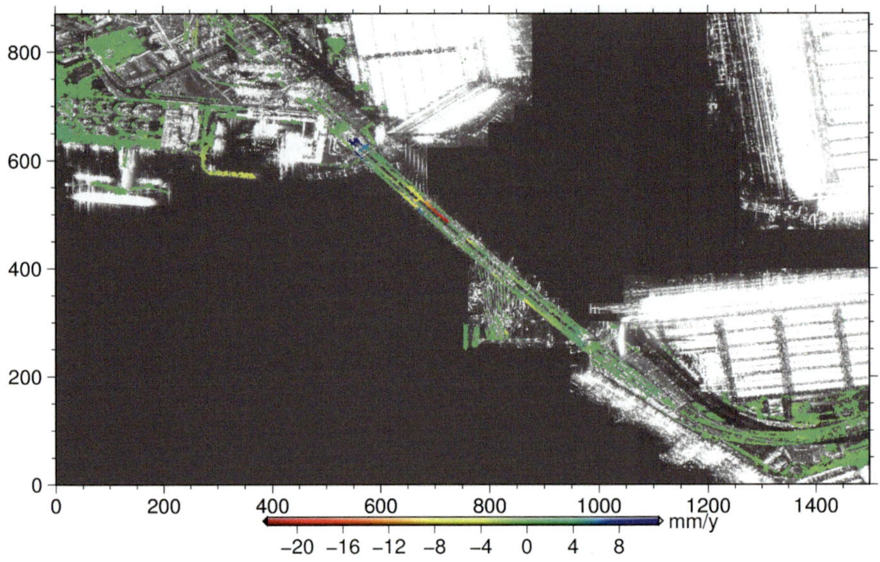

Fig. 21.12 Deformation rate map of Stonecutters Bridge in Hong Kong from 51 TerraSAR-X images and MTInSAR processing. The map was derived after removing the thermal dilation effect

21.5 Summary

We have reviewed the basic concepts of SAR, InSAR, and MTInSAR and their applications in urban environments. InSAR has benefited from the recent advances in spatial resolution and orbit control of spaceborne radar sensors and has become a vital technology in generating DEMs and in monitoring deformation phenomena related to, for example, ground subsidence and instability of infrastructures. InSAR techniques offer several advantages in such applications. For example, they can be applied in all weather conditions. This ability is especially useful in cloudy regions. Spaceborne InSAR technology can easily cover a large ground area with spatial and temporal resolutions hardly matched by any other technologies. InSAR however still has some shortcomings in these and other related applications. Further, research is still necessary to advance technology in terms of developing new SAR sensors, systems, and data processing algorithms. For example, geostationary satellite SAR constellations and P-band SAR sensor systems are currently being investigated. It can be expected that the capability of InSAR technology will be significantly enhanced in the near future.

References

Adam N, Kampes B, Eineder M (2005) Development of a scientific permanent scatterer system: modifications for mixed ERS/ENVISAT time series. Envisat and ERS symposium

Berardino P, Fornaro G, Lanari R, Sansosti E (2002) A new algorithm for surface deformation monitoring based on small baseline differential SAR interferograms. IEEE Trans Geosci Remote Sens 40(11):2375–2383

Cao N, Lee H, Jung (2016) A phase-decomposition-based PSInSAR processing method. IEEE Trans Geosci Remote Sens 54(2):1074–1090

Costantini M (1998) A novel phase unwrapping method based on network programming. IEEE Trans Geosci Remote Sens 36(3):813–821

Crosetto M, Monserrat O, Cuevas-González M, Devanthéry N, Luzi G (2015) Measuring thermal expansion using X-band persistent scatterer interferometry. ISPRS J Photogr Remote Sens 100:84–91

Farr TG, Rosen PA, Caro E, Crippen R, Duren R, Hensley S, Kobrick M, Paller M, Rodriguez E, Roth L, Seal D, Shaffer S, Shimada J, Umland J, Werner M, Oskin M, Burbank D, Alsdorf D (2007) The Shuttle Radar Topography Mission. Rev Geophys 45(2)

Ferretti A, Fumagalli A, Novali PC, Rocca F, Rucci A (2011) A new algorithm for processing interferometric data-stacks: SqueeSAR. IEEE Trans Geosci Remote Sens 49(9):3460–3470

Ferretti A, Prati C, Rocca F (2001) Permanent scatterers in SAR interferometry. IEEE Trans Geosci Remote Sens 39(1):8–20

Fialko Y (2004) Evidence of fluid-filled upper crust from observations of postseismic deformation due to the 1992 Mw7.3 Landers earthquake. J Geophy Res Solid Earth 109(B8):B08401

Fornaro G, Verde S, Reale D, Pauciullo A (2015) CAESAR: an approach based on covariance matrix decomposition to improve multibaseline-multitemporal interferometric SAR processing. IEEE Trans Geosci Remote Sens 53(4):2050–2065

Gernhardt S, Bamler R (2012) Deformation monitoring of single buildings using meter-resolution SAR data in PSI. ISPRS J Photogr Remote Sens 73:68–79

Ghiglia DC, Romero LA (1994) Robust two-dimensional weighted and unweighted phase unwrapping that uses fast transforms and iterative methods. J Opt Soc Am A 11(1):107–117

Goldstein RM, Engelhardt H, Kamb B, Frolich RM (1993) Satellite radar interferometry for monitoring ice sheet motion: application to an Antarctic ice stream. Science 262:1525

Goldstein RM, Werner CL (1998) Radar interferogram filtering for geophysical applications. Geophys Res Lett 25(21):4035–4038

Goldstein RM, Zebker HA, Werner CL (1988) Satellite radar interferometry: two-dimensional phase unwrapping. Radio Sci 23(4):713–720

Hooper A, Segall P, Zebker H (2007) Persistent scatterer interferometric synthetic aperture radar for crustal deformation analysis, with application to Volcán Alcedo, Galápagos. J Geophys Res 112(B7):B07407

Hooper A, Zebker H, Segall P, Kampes B (2004) A new method for measuring deformation on volcanoes and other natural terrains using InSAR persistent scatterers. Geophys Res Lett 31(23):L23611

Hu H, Li ZW, Ding XL, Zhu JJ, Zhang L, Sun Q (2014) Resolving three-dimensional surface displacements from InSAR measurements: a review. Earth-Sci Rev 133:1–17

Jiang M, Ding X, Hanssen RF, Malhotra R, Ll C (2015) Fast statistically homogeneous pixel selection for covariance matrix estimation for multitemporal InSAR. IEEE Trans Geosci Remote Sens 53(3):1213–1224

Jolivet R, Agram PS, Lin NY, Doin MP, Peltzer G, Li Z (2014) Improving InSAR geodesy using global atmospheric models. J Geophys Res Solid Earth 119(3):2324–2341

Kampes BM (2006) Radar interferometry: persistent scatterer technique. Springer, Dordrecht
Kampes BM, Hanssen RF (2004) Ambiguity resolution for permanent scatterer interferometry. IEEE Trans Geosci Remote Sens 42(11):2446–2453
Lanari R, Mora O, Manunta M, Mallorqui JJ, Berardino P, Sansosti E (2004) A small-baseline approach for investigating deformations on full-resolution differential SAR interferograms. IEEE Trans Geosci Remote Sens 42(7):1377–1386
Liang H, Zhang L, Ding X, Lu Z, Li X (2019) Toward mitigating stratified tropospheric delays in multitemporal InSAR: a quadtree aided joint model. IEEE Trans Geosci Remote Sens 57(1):291–303
Lombardini F (2005) Differential tomography: a new framework for SAR interferometry. IEEE Trans Geosci Remote Sens 43(1):37–44
Lu Z, Dzurisin D (2014) InSAR imaging of Aleutian volcanoes: monitoring a volcanic arc from space. Geophysical sciences. Springer-Verlag, Berlin Heidelberg
Lv X, Birsen Y, Mourad Z, Victoria B, Tarek A (2014) Joint-scatterer processing for time-series InSAR. IEEE Trans Geosci Remote Sens 52(11):7205–7221
Ma P, Lin H (2016) Robust detection of single and double persistent scatterers in urban built environments. IEEE Trans Geosci Remote Sens 54(4):2124–2139
Massonnet D, Feigl KL (1998) Radar interferometry and its application to changes in the Earth's surface. Rev Geophys 36(4):441–500
Mora O, Mallorqui JJ, Broquetas A (2003) Linear and nonlinear terrain deformation maps from a reduced set of interferometric SAR images. IEEE Trans Geosci Remote Sens 41(10):2243–2253
Perissin D, Rocca F (2006) High-accuracy urban DEM using permanent scatterers. IEEE Trans Geosci Remote Sens 44(11):3338–3347
Pritt MD, Shipman JS (1994) Least-squares two-dimensional phase unwrapping using FFT's. IEEE Trans Geosci Remote Sens 32(3):706–708
Qin X, Zhang L, Yang M, Luo H, Liao M, Ding X (2018) Mapping surface deformation and thermal dilation of arch bridges by structure-driven multi-temporal DInSAR analysis. Remote Sens Environ 216:71–90
Qu F, Lu Z, Zhang Q, Bawden GW, Kim JW, Zhao C, Qu W (2015) Mapping ground deformation over Houston-Galveston, Texas using multi-temporal InSAR. Remote Sens Environ 169:290–306
Rossi C, Gernhardt S (2013) Urban DEM generation, analysis and enhancements using TanDEM-X. ISPRS J Photogr Remote Sens 85:120–131
Schunert A, Soergel U (2012) Grouping of persistent scatterers in high-resolution SAR data of urban scenes. ISPRS J Photogr Remote Sens 73:80–88
Serrano-Juan A, Pujades E, Vázquez-Suñè E, Crosetto M, Cuevas-González M (2017) Leveling vs. InSAR in urban underground construction monitoring: Pros and cons. Case of La Sagrera railway station (Barcelona, Spain). Eng Geol 218:1–11
Shi G, Lin H, Ma P (2018) A hybrid method for stability monitoring in low-coherence urban regions using persistent and distributed scatterers. IEEE J Sel Top Appl Earth Obs Remote Sens 11(10):3811–3821
Sun Q, Zhang L, Ding X, Hu J, Li Z, Zhu J (2015) Slope deformation prior to Zhouqu, China landslide from InSAR time series analysis. Remote Sens Environ 156:45–57
Tadono T, Nagai H, Ishida H, Oda F, Naito S, Minakawa K, Iwamoto H (2016) Generation of the 30 M-Mesh Global Digital Surface Model by Alos Prism. ISPRS Int Arch Photogr Remote Sens Spatial Inf Sci XLI-B4:157–162
Wu S, Zhang L, Ding X, Perissin D (2018) Pixel-wise MTInSAR estimator for integration of coherent point selection and unwrapped phase vector recovery. IEEE Trans Geosci Remote Sens 57(5):2659–2668
Zebker HA, Rosen P (1994) On the derivation of coseismic displacement fields using differential radar interferometry: the Landers earthquake. In: Proceedings of IGARSS '94—1994 IEEE international geoscience and remote sensing symposium,vol 1, pp 286–288
Zhang L, Ding XL, Lu Z (2011a) Modeling PSInSAR time series without phase unwrapping. IEEE Trans Geosci Remote Sens 49(1):547–556

Zhang L, Ding XL, Lu Z (2011b) Ground settlement monitoring based on temporarily coherent points between two SAR acquisitions. ISPRS J Photogr Remote Sens 66:146–152
Zhang L, Ding X, Lu Z, Jung HS, Hu J, Feng G (2014) A novel multitemporal InSAR model for joint estimation of deformation rates and orbital errors. IEEE Trans Geosci Remote Sens 52(6):3529–3540
Zhu XX, Bamler R (2010) Tomographic SAR inversion by L1-norm regularization: the compressive sensing approach. IEEE Trans Geosci Remote Sens 48(10):3839–3846
Zhu XX, Baier G, Lachaise M, Shi Y, Adam F, Bamler R (2018) Potential and limits of non-local means InSAR filtering for TanDEM-X high-resolution DEM generation. Remote Sens Environ 218:148–161

Hongyu Liang received a B.S. from Southwest Jiaotong University, Chengdu, China in 2013, and an M.Sc. in Geomatics from The Hong Kong Polytechnic University, Hong Kong, China, in 2014. He is currently working towards his Ph.D. in the Department of Land Surveying and Geo-Informatics, The Hong Kong Polytechnic University. His research interests include multi-temporal InSAR modeling and parameter estimation, and large-scale geohazard monitoring.

Wenbin Xu is Professor of Satellite Tectonic Geodesy at Central South University, Changsha, China. He is interested in studying and modeling earthquakes, volcanoes and tectonics.

Xiaoli Ding is Chair Professor, the Department of Land Surveying and Geo-Informatics, The Hong Kong Polytechnic University. His research interests are in satellite positioning technologies (such as GPS), Synthetic Aperture Radar (SAR) and Interferometric Synthetic Aperture Radar (InSAR) technologies for ground deformation and structural health monitoring.

Lei Zhang is currently a Senior Research Fellow at The Hong Kong Polytechnic University. His research mainly focuses on developing advanced satellite radar data processing algorithms, and the application of interferometry technologies to image deformation associated with natural hazards and urban infrastructure instability.

Songbo Wu received a B.S. from Xinjiang University, Xinjiang, China in 2012, and an M.Sc. from South-West Jiaotong University, Sichuan, China in 2015. He is currently working towards his Ph.D. in the Department of Land Surveying and Geo-Informatics (LSGI), The Hong Kong Polytechnic University, Hong Kong. He is interested in InSAR modeling and the study of urban resilience using satellite remote sensing data.

Chapter 22
Airborne LiDAR for Detection and Characterization of Urban Objects and Traffic Dynamics

Wei Yao and Jianwei Wu

Abstract In this chapter, we present an advanced machine learning strategy to detect objects and characterize traffic dynamics in complex urban areas by airborne LiDAR. Both static and dynamical properties of large-scale urban areas can be characterized in a highly automatic way. First, LiDAR point clouds are colorized by co-registration with images if available. After that, all data points are grid-fitted into the raster format in order to facilitate acquiring spatial context information per-pixel or per-point. Then, various spatial-statistical and spectral features can be extracted using a cuboid volumetric neighborhood. The most important features highlighted by the feature-relevance assessment, such as LiDAR intensity, NDVI, and planarity or covariance-based features, are selected to span the feature space for the AdaBoost classifier. Classification results as labeled points or pixels are acquired based on pre-selected training data for the objects of building, tree, vehicle, and natural ground. Based on the urban classification results, traffic-related vehicle motion can further be indicated and determined by analyzing and inverting the motion artifact model pertinent to airborne LiDAR. The performance of the developed strategy towards detecting various urban objects is extensively evaluated using both public ISPRS benchmarks and peculiar experimental datasets, which were acquired across European and Canadian downtown areas. Both semantic and geometric criteria are used to assess the experimental results at both per-pixel and per-object levels. In the datasets of typical city areas requiring co-registration of imagery and LiDAR point clouds a priori, the AdaBoost classifier achieves a detection accuracy of up to 90% for buildings, up to 72% for trees, and up to 80% for natural ground, while a low and robust false-positive rate is observed for all the test sites regardless of object class to be evaluated. Both theoretical and simulated studies for performance analysis show that the velocity estimation of fast-moving vehicles is promising and accurate, whereas slow-moving ones are hard to distinguish and yet estimated with acceptable velocity accuracy.

W. Yao (✉)
Department of Land Surveying and Geo-Informatics, The Hong Kong Polytechnic University, Hong Kong, China
e-mail: wei.hn.yao@polyu.edu.hk

J. Wu
School of Remote Sensing and Information Engineering, Wuhan University, Wuhan, China
e-mail: jianwei_wu@whu.edu.cn

W. Shi et al. (eds.), *Urban Informatics*, The Urban Book Series,
https://doi.org/10.1007/978-981-15-8983-6_22

Moreover, the point density of ALS data tends to be related to system performance. The velocity can be estimated with high accuracy for nearly all possible observation geometries except for those vehicles moving in or (quasi-)along the track. By comparative performance analysis of the test sites, the performance and consistent reliability of the developed strategy for the detection and characterization of urban objects and traffic dynamics from airborne LiDAR data based on selected features was validated and achieved.

22.1 Introduction

Urban scene classification and object detection are important topics in the field of remote sensing. Recently, point cloud data generated by LiDAR sensors and multispectral aerial imagery have become two important data sources for urban scene analysis. While multispectral aerial imagery with fine resolution provides detailed spectral texture information about the surface, point cloud data is more capable of presenting the geometrical characteristics of objects.

LiDAR has become a common active surveying method to directly realize the digital 3D representation of targets through a laser ranging, positioning, and orientation system (POS). Based on different platforms, LiDAR technology can cover terrestrial, mobile, airborne, and spaceborne applications. This chapter focuses on airborne applications. Airborne LiDAR (ALS) has attracted plenty of research attention for more than two decades. The ALS technique has been widely applied in diverse fields such as forest mapping (Næsset and Gobakken 2008; Reitberger et al. 2008; Zhao et al. 2018), coast monitoring (Earlie et al. 2015; Bazzichetto et al. 2016), smart urban applications (Garnett and Adams 2018) and so on. As it can directly derive accurate and highly detailed 3D surface information, and because more than one half of the population resides in urban areas, ALS was able to achieve significant applications in urban areas such as urban modeling (Zhou and Neumann 2008; Lafarge and Mallet 2012; Chen et al. 2019), land cover and land use classification (Azadbakht et al. 2018; Balado et al. 2018; Wang et al. 2019), environment monitoring and tree mapping (Liu et al. 2017; Degerickx et al. 2018; Lafortezza and Giannico 2019), urban population estimation (Tomás et al. 2016), energy conservation (Jochem et al. 2009; Dawood et al. 2017) and so on. Urban modeling with ALS data includes the 3D reconstruction of buildings (Bonczak and Kontokosta 2019; Li et al. 2019), roads (Chen and Lo 2009), bridges (Cheng et al. 2014), powerlines (Wang et al. 2017) and so on. Very recently, ALS data are also helpful to improve accuracy for urban mapping and land cover classification. Degerickx et al. (2019) applied ALS data as an additional data source to enhance the performance of multiple endmember spectral mixture analysis for urban land-cover classification using hyperspectral and multispectral images, and found that implementing height distribution information from ALS data as a basis for additional fraction constraints at the pixel level could significantly reduce spectral confusion between spectrally similar, but structurally different land-cover classes. Accurate and highly detailed height information from

ALS data is also used to enhance urban mapping accuracy based on the 3D rational polynomial coefficient model (Rizeei and Pradhan 2019).

Besides the above-mentioned applications, ALS can also be used to detect and monitor dynamic objects. Compared to traditional optical imagery, airborne LiDAR data are characterized by involving not only rich spatial but also temporal information. It is theoretically possible to extract vehicles from single-pass airborne LiDAR data, to identify the vehicle motion, and to derive the vehicle's velocity and direction based on the motion artifacts effect. Thus, besides common applications of airborne LiDAR, it should also be regarded as a demonstrator for traffic monitoring from the air.

Urban scene analysis can be categorized by different object types, different data sources, and also algorithms. During the past decades, more work referring to urban scene analysis has concentrated on the classification or detection of specified objects. Much marvelous research (Clode et al. 2007; Fauvel 2007; Sohn and Dowman 2007; Yao and Stilla 2010; Guo et al. 2011; Xiao et al. 2012) has been done in extracting objects like buildings and roads, while trees and vehicles are also interesting objects for intelligent monitoring of natural resources and traffic in urban areas (Höfle and Hollaus 2010; Yao et al. 2011). However, detection and modeling of diverse urban objects may involve more complicated situations due to the various characteristics and appearances of the objects. As ALS data became widely available for the task of creating 3D city models, there was an increasing amount of research on developing automatic approaches to object detection from images and LiDAR data, which showed the great potential of 3D target modeling and surface characterization in urban areas (Schenk and Csatho 2007; Mastin et al. 2009). In this chapter, we focus on analyzing airborne LiDAR data by the adaptive boosting (AdaBoost) classification technique for urban object detection based on selected spatial and radiometric features. In this chapter, we will develop and validate a robust classification strategy for urban object detection through fusing LiDAR point clouds and imagery.

As mentioned above, ALS data have become an important source for object extraction and reconstruction for various applications such as urban and vegetation analysis. However, traffic monitoring remains one of the few fields which are still not intensively analyzed in the LiDAR community. There are several motivations driving us to perform traffic analysis using airborne LiDAR in urban areas:

- The penetration ability of laser rays towards volume-scattering objects (e.g., trees) can improve vehicle detection;
- The motion artifacts generated by the linear scanning mechanism of airborne LiDAR can determine object motion;
- The explicit extraction of vehicles can refine the results of operations such as DTM filtering and road detection where vehicles are regarded as stubborn disturbances.

The task of detecting moving vehicles with ALS has been addressed in several scientific publications. The research most relevant to our work came from Toth and Grejner-Brzezinska (2006). In this chapter, an airborne laser scanner coupled with a digital frame camera was adopted to analyze transportation corridors and acquire traffic flow information. However, the testing of this system was limited to a motorway; the same problem needs to be investigated in more challenging regions using the system equipped solely with LiDAR. In the contribution from Yao et al. (2010a), a context-guided approach based on gridded ALS data was used to delineate single instances of vehicle objects and results demonstrated the feasibility of extracting vehicles for motion analysis. A vehicle extraction method was presented, running directly on LiDAR point clouds that integrate height, edge, and point shape information in a segmentation step to improve the vehicle extraction through object-based classification (Yao et al. 2011). Based on the extracted vehicles, Yao et al. (2010b) proposed a complete procedure to distinguish vehicle motion states and to estimate the velocity of moving vehicles by parameterizing, classifying, and inverting shape deformation features. In contrast to applications monitoring military traffic, civilian applications include more constraints regarding the objects to be detected. We can assume that vehicles are bound to roads on a known road network, which might not be true in military applications. Such knowledge provides a priori information for motion estimation.

This chapter concerns the detection of selected urban objects and the characterization of traffic dynamics with ALS data. In Sect. 22.2, a robust and efficient supervised learning method for detecting urban objects is proposed, and the analysis of urban traffic dynamics is performed in Sect. 22.3. Section 22.4 presents the experiment and results of detecting urban objects and their dynamics. Finally, conclusions are drawn in Sect. 22.5.

22.2 Detection of Urban Objects with ALS and Co-registered Imagery

22.2.1 General Strategy

The workflow of the entire strategy for detecting three urban object classes (buildings, trees, and natural ground) with ALS data and co-registered images is depicted in Fig. 22.1.

22.2.2 Feature Derivation

In this chapter, we combine point clouds and image data, while multispectral and LiDAR intensity information is also available. In total 13 features are defined (Wei et al. 2012).

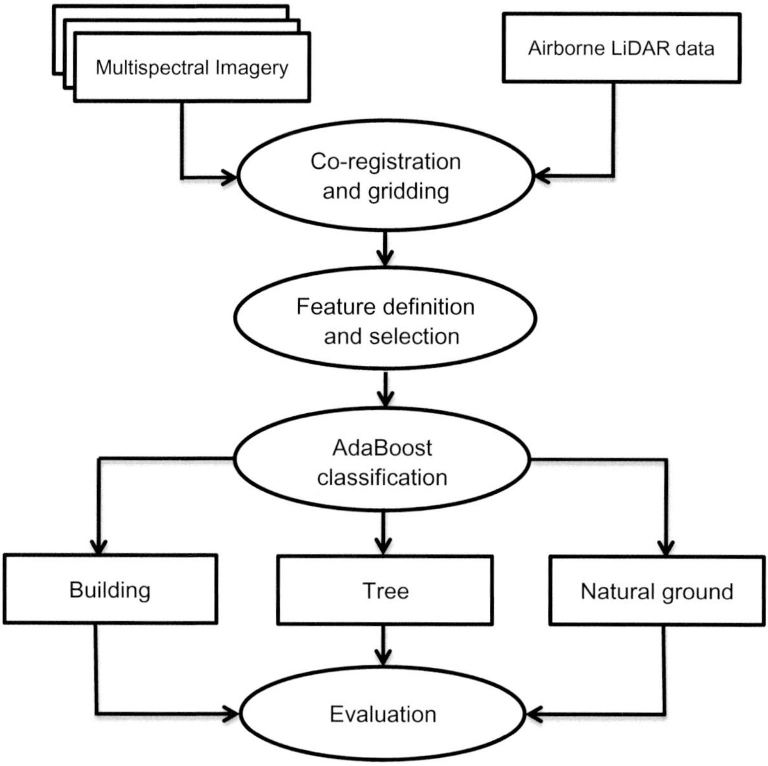

Fig. 22.1 Overview of the entire strategy

22.2.2.1 Basic Features

The so-called basic features contain the features that can be directly retrieved from the point cloud and image data, respectively:

- *R*, *G*, *B*: The three color channels of the digital image. As two data sets are used for experiments and one of them (named data set Vaihingen) provides color-infrared images, features R, G, B stand for infrared, red, and green spectra, But in the other data set (Toronto), the features R, G, and B are normal bands of Red, Green, and Blue. To avoid confusion, we always use the symbols R, G, B to indicate the three color channels of the image in order.
- *NDVI*: Normalized Difference Vegetation Index, defined as:

$$NDVI = \frac{(NIR - VIS)}{(NIR + VIS)} \tag{22.1}$$

NDVI can assess whether the target being observed contains green vegetation or not. This feature is specified for data set Vaihingen because it provides color-infrared imagery.

- Z: The vertical coordinate of each point in the LiDAR data, as the topography of datasets used here, is assumed to be flat.
- I: Pulse intensity, which is provided by the LiDAR system for each point.

22.2.2.2 Spatial Context Features

Based on the basic features, we intend to extract more features. Therefore, a 3D cuboid neighborhood is defined with the help of a 2D square with radius of 1.25 m in horizontal dimension as shown in Fig. 22.2. All points located within the cell volume will be counted as the neighbors; the value 1.25 m is chosen empirically.

- ΔZ: Height difference between the highest and lowest points within the cuboid neighborhood.
- σ_Z: standard deviation of height of points within the cuboid neighborhood.
- ΔI: Intensity difference between points having the highest and lowest intensities within the cuboid neighborhood.
- σ_I: Standard deviation of intensity of points within the cuboid neighborhood.
- E: Entropy, here being different from the normal entropy of images, we measure the entropy using LiDAR intensities I_k of the points within the cuboid neighborhood by Eq. 22.2 with K being the number of neighbors:

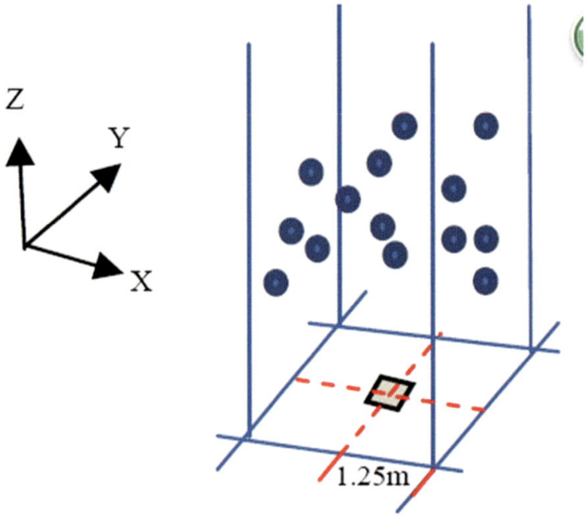

Fig. 22.2 The 3D cuboid neighborhood used to acquire spatial context features

$$E = \sum_{k=1}^{K} \left[(-I_k) \cdot \log_2^{I_k} \right] \tag{22.2}$$

The following two features O and P are based on the three eigenvalues of the covariance matrix from the xyz coordinates of points within the cuboid neighborhood. The three eigenvalues λ_1, λ_2, and λ_3 are arranged in descending order, and they can present the local tridimensional structure. This allows us to distinguish between a linear, a planar, or a volumetric distribution of the points.

- O: Omnivariance, which indicates the distribution of points in the cuboid neighborhood. It is defined as:

$$O = \sqrt[3]{\prod_{i=1}^{3} \lambda_i} \tag{22.3}$$

- P: Planarity, defined as:

$$P = (\lambda_2 - \lambda_3)/\lambda_1 \tag{22.4}$$

P has high value for roofs and ground, but low values for vegetation.

22.2.3 AdaBoost Classification

AdaBoost is an abbreviation for adaptive boosting (Freund and Schapire 1999), which is an improved version of boosting. AdaBoost is an attractive and powerful supervised learning algorithm of machine learning and it has been successfully applied in both classification and regression cases. For classification cases, AdaBoost is adapted to take full advantage of the weak learners and solves the problem of combining a bundle of weak classifiers to create a strong classifier which is arbitrarily well correlated with the true classification. It consists of iteratively learning weak classifiers with respect to a distribution and adding them to a final strong classifier. Once a weak learner is added, the data are reweighted according to the weak classifier's accuracy; misclassified samples gain weight and correctly classified samples reduce weight. No other requirement is essential for the weak learners used in the AdaBoost except that their classification accuracy is better than the random classification, which means that the weak learners only need to achieve a classification accuracy better than 50%. In this chapter, we use an open-source AdaBoost toolbox with one tree weak learner CART (classification and regression tree), more details of which can be found in the reference (Freund and Schapire 1999).

Like other supervised learning algorithms, AdaBoost contains two phases as well: training and prediction. In the training phase, it repeatedly trains T weak classifiers through T rounds. In this chapter we implemented the multiclass classification task through iterating corresponding binary classifiers, as shown in the following pseudocode for the binary classification:

$Input\text{-}Training\ Data\ with\ m\ samples$: $(x_i, y_i), y_i \in Y = \{-1, +1\}, i \in [1, m]$;

$Initialize$: $W_1^i = \frac{1}{m}, h_1^i = 0$;

$for\ t = 1{:}T$

$train\ the\ t^{th}\ weak\ classifier\ h^t\ with\ weight\ vector\ of\ sample\ distribution\ W_t$;

$choose\ \varepsilon_t = \sum_i^m W_t^i * I\left(h_t^i(x_i) \neq y_i\right)$;

$\alpha_t = \ln\left(\frac{1-\varepsilon_t}{\varepsilon_t}\right)/2$;

$Z_t = \sum_{i=1}^m W_t^i e^{(-\alpha_t h_t(x_i) y_i)}$;

$W_{t+1}^i = W_t^i * e^{(-\alpha_t h_t(x_i) y_i)}/Z_t$; $\quad for\ i = 1{:}m$

end

end

The T weak classifiers are combined and output-weighted as follows:

$$H(x) = \operatorname{sgn}\left(\sum_{t=1}^{T} \alpha_t h_t\right) \tag{22.5}$$

where the *sgn* function is defined as:

$$\operatorname{sgn}(x) = \begin{cases} -1, x < 0 \\ 0, x = 0 \\ 1, x > 0 \end{cases} \tag{22.6}$$

In the above, pseudocode (x_i, y_i) represents the ith training sample with x_i standing for its feature vector and y_i for its class type; m represents the amount of training data; W_t^i is a weight for the ith training sample being selected to train the tth classifier h^t and W_t is a vector of W_t^i; ε_t is the weighted prediction error of h^t; α_t is the weight coefficient for updating the sample distribution; the value of $I\left(h_t^i(x_i) \neq y_i\right)$ is 1 if $h_t^i(x_i) \neq y_i$, else it equals 0; Z_t is a normalization factor. At beginning, each sample is assigned an equal weight equal to $W_1^i = 1/m$, which means that each training sample is selected with the same probability to train h^1. In the tth training round, the AdaBoost algorithm updates W_{t+1}^i as follows: training samples correctly identified by classifier h_t are weighted less while those incorrectly identified are weighted more. Then when training h^{t+1}, the algorithm tends to select samples wrongly classified by previous classifiers with higher probability. After T rounds of training, T-weak classifiers are trained and finally combined into a weighted classifier $H(x)$ as the training phase's output, which has better prediction performance.

The prediction phase uses the combined classifier for classification. Compared to boosting, AdaBoost two advantages for learning a more accurate classifier. First, for each weak classifier's training, boosting randomly chooses training samples, while AdaBoost chooses samples misclassified in the previous training rounds with greater probability. Thus, AdaBoost can better train the classifier. Second, AdaBoost determines each sample's classification label through weighting each classifier's output, which makes an accurate classifier contribute more to the final classification result.

22.3 Detection of Urban Traffic Dynamics with ALS Data

In this section, we give a brief review of deriving the theory for detecting object dynamics in ALS. We refer to the dimension perpendicular to the sensor heading synonymously as across-track. The dimension along the sensor path will be denoted by a along-track.

22.3.1 *Artifacts Effect of Vehicle Motion in ALS Data*

In order to assess the feasibility of extracting information on traffic dynamics from airborne LiDAR sensors installed on the airborne platform, the main characteristics of the sensor, including the data formation method, should be considered first. In most airborne LiDAR scanning processes, exclusive of flash LiDAR which are predominantly based on mechanical scanning, a rotating laser pointer rapidly scans the Earth's surface with continuous scan angles during flight. While the sensor is moving it transmits laser pulses at constant intervals given by the pulse repetition frequency (PRF) and receives the echoes. With respect to moving objects, the fundamental difference between scanning and the frame camera model is the presence of motion artifacts in the scanner data. Due to short sampling time (camera exposure), the imagery preserves the shape of moving objects; if the relative speed between the sensor and the object is significant then increased motion blurring may occur. In contrast, scanning will always produce motion artifacts, since the distance between sensor and target is usually calculated based on the stationary-world assumption; fast-moving objects violate this assumption and therefore image the target incorrectly depending on the relative motion between the sensor and the object. The dependency can be seen by adding the temporal component into the range equation of the LiDAR sensor. Here, it is assumed that the sampling rate is consistent among all the vehicles independent of the scan angle. That is to say that all the vehicles are scanned with enough points to represent their shape artifacts.

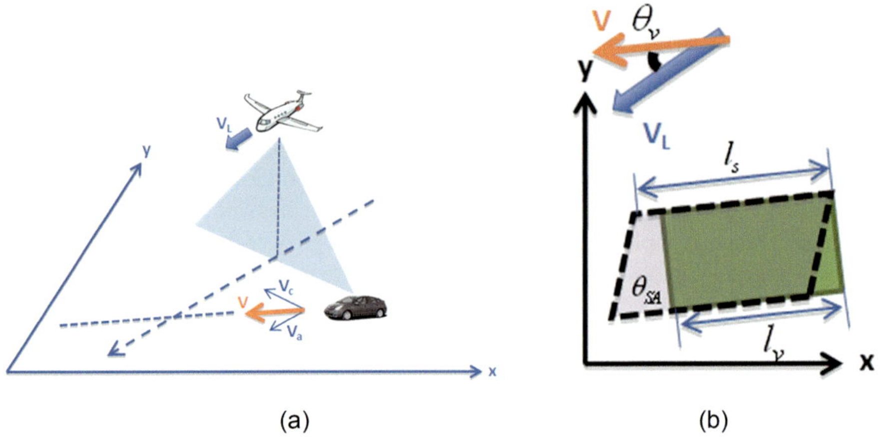

Fig. 22.3 Moving objects undergo the scanning of airborne LiDAR. Copyright © 2010 IEEE, reproduced by permission

In Fig. 22.3a the geometry of data acquisition is shown. The sensor is flying at a certain altitude along the dotted arrow. An example of shape artifacts generated by moving objects is also depicted in Fig. 22.3b, where the black dotted box indicates the vehicle shape obtained in the scanning process of airborne LiDAR while the original vehicle is depicted as a rectangle nearby. It can be perceived that the moving vehicle is imaged as a stretched parallelogram. Let θ_v be the intersection angle between the moving directions of sensor and vehicle where $\theta_v \in [0°, 360°]$, v_L and v the velocity of aircraft and vehicle respectively, l_s and l_v the sensed and original lengths of the vehicle, respectively; and θ_{SA} the shearing angle that accounts for the deformation of the vehicle as a parallelogram. The analytic relations between shape artifacts and object-movement parameters can be derived as:

$$l_s = \frac{l_v \cdot v_L}{v_L - v \cdot \cos(\theta_v)} = \frac{l_v}{1 - \frac{v}{v_L} \cdot \cos(\theta_v)} \tag{22.7}$$

$$\theta_{SA} = \arctan\left(\frac{v \cdot \sin(\theta_v)}{v_l - v \cdot \cos(\theta_v)} \right) + 90° \tag{22.8}$$

where $\theta_{SA} \in (0° \ 180°)$ and is found as the left-bottom angle of the observed vehicle.

For the sake of full understanding of the appearance of moving objects in the airborne LiDAR data, object motions are to be divided into the following different components and investigated for their respective influences on the data artifacts generated.

First, the target is assumed to move with constant velocity v_a following the along-track direction, which leads to the stretching effect of the object shape depending on the relative velocity between target and sensor as illustrated in Fig. 22.4.

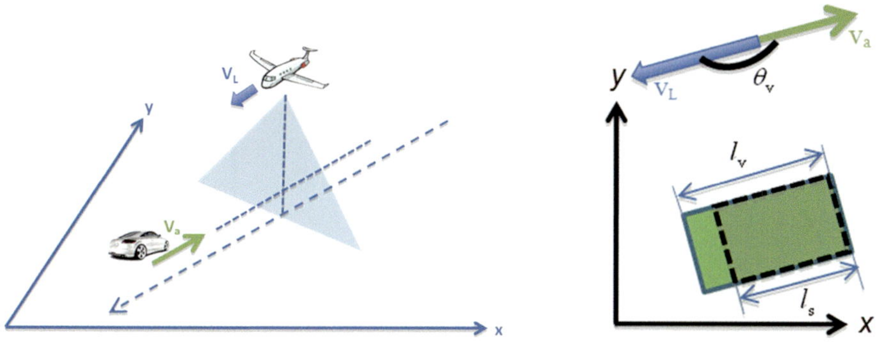

Fig. 22.4 Along-track object motion. Copyright © 2010 IEEE, reproduced by permission

The analytic relation between the object velocity in along-track direction v_a and the observed stretched length l_s thus can be summarized in Eq. 22.9. The relation in Eq. 22.9 is further modified to Eq. 22.10 which explicitly connects v_a with the variation in the aspect ratio of vehicle shape in a mathematical way, thereby making motion detection and velocity estimation more feasible and reliable:

$$l_s = \frac{l_v}{1 - \frac{v_a}{v_L}} \tag{22.9}$$

$$Ar_s = \frac{l_s}{w_v} = \frac{Ar}{1 - \frac{v_a}{v_L}} \tag{22.10}$$

where Ar_s is the sensed aspect ratio of the vehicle in ALS data while Ar is the original aspect ratio of the vehicle and w_v is the width of the vehicle.

Secondly, the target is assumed to move in the across-track direction with a constant velocity v_c. This results in a scanline-wise linear shift of laser footprints that hit upon the target in the direction of movement when the sensor is sweeping over so that the observed vehicle shape in ALS data is deformed (sheared) to a certain extent as illustrated in Fig. 22.5.

Let v_c be the across-track motion component of the object velocity. Since $v_c = v \cdot \sin(\theta_v)$, Eq. 22.8 can be rewritten as Eq. 22.11 for describing the analytic relation between the object velocity v_c and the observed shearing angle θ_{SA} through the sensor velocity v_L and the intersection angle θ_v:

$$\begin{aligned} \theta_{SA} &= \arctan\left(\frac{1}{v_L/v_c - \cot(\theta_v)}\right) + 90^\circ \quad \text{where}\ \theta_v \neq 0^\circ/180^\circ \wedge v_c \neq 0 \\ \theta_{SA} &= 90^\circ \quad \text{where}\ \theta_v = 0/180^\circ \vee v_c = 0 \end{aligned} \tag{22.11}$$

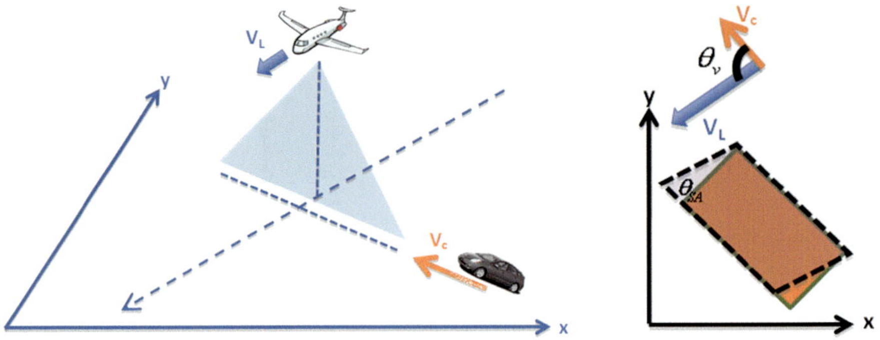

Fig. 22.5 Across-track object motion. Copyright © 2010 IEEE, reproduced by permission

22.3.2 *Detection of Moving Vehicles*

All of the effects of moving objects described above can be exploited to not only detect vehicles' movement but also measure their velocity. Our scheme for vehicle motion detection relies on a strategy consisting of two basic modules successively executed: (1) vehicle extraction; and (2) determination of the motion state.

For vehicle extraction, we used a hybrid strategy (Fig. 22.6) that integrates a 3D segmentation-based classification method with a context-guided approach. For a detailed analysis of vehicle detection, we refer the readers to Yao et al. (2010a, 2011).

To determine the motion state, a support vector machine (SVM) classification-based method is adopted. A set of vehicle points can be geometrically described as a spoke model with control parameters, whose configuration can be formulated as

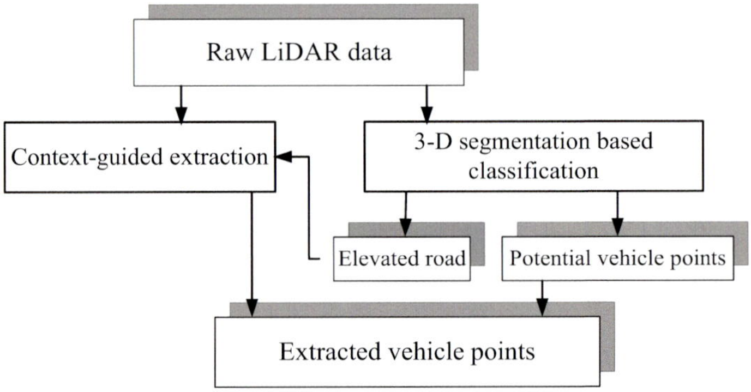

Fig. 22.6 Workflow for vehicle extraction

$$\mathbf{X} = \begin{pmatrix} \mathbf{U}_1 \\ \cdot \\ \cdot \\ \mathbf{U}_k \end{pmatrix}, \mathbf{U}_i = \begin{pmatrix} \theta^i_{SA} \\ Ar_i \end{pmatrix} \tag{22.12}$$

where k denotes the number of spokes in the model. It can be seen that the vehicle shape variability can be represented as a two-dimensional feature space (if the number of spokes $k = 1$). Thus, the similarity between vehicle instances of different motion states needs to be measured by a nonlinear metric. The SVM has advantages in nonlinear recognition problems and finds an optimal linear hyperplane in a higher dimensional feature space that is nonlinear in the original input space. The trick of using a kernel avoids direct evaluation in the feature space of higher dimension by computing it through the kernel function with feature vectors in the input space. The SVM classifier can be used here again to perform binary classification on those vehicles which still remain after excluding the ones of uncertain state obtained by the shape parameterization step. In addition, the classification framework for distinguishing 3D shape categories (Fletcher et al. 2003) can be adapted to the motion classification schema based on exploiting the vehicle shape features.

22.3.3 Concept for Vehicle Velocity Estimation with ALS Data

The estimation of the velocity of detected moving vehicles can be done based on all motion artifacts effects in a single pass of ALS data by inverting the motion artifacts model to relate the velocity with other observed and known parameters. Thus, different measurements and derivations might be used to estimate the velocity. The estimation scheme can be initially divided into two main categories, depending on whether the moving direction of vehicles is known or not:

First, given the intersection angle which can be further separated into the following three situations using respective observations to estimate the velocity:

(a) The measure for shearing angle of the detected moving vehicles from their original orthogonal shape of rectangles;
(b) The measure for the stretching effect of detected moving vehicles from their original size; and
(c) The combination of the along-track and across-track velocity components which are estimated based on the above-mentioned effects, respectively.

Second, if the intersection angle is not given:

(a) The solution to a system of bivariate equations constructed by uniting the two formulas.

The three methods in the first category assume that the moving directions of vehicles are given beforehand, whereas the last one from the second category does not. To estimate the velocity, the first three methods either utilize the shape stretching or shearing effect or combine them together when applicable. For the last case, the moving direction of vehicles can be estimated along with the velocity by uniting the variable of velocity with the variable of the intersection angle to build a system of bivariate equations and solving it, thereby giving the motion estimation great flexibility to deal with many arduous cases encountered in real-life scenarios. That means that not only the quantity but also the direction of vehicles' motion can be derived. All possible approaches have their advantages and disadvantages and differ in the accuracy of their results, which are to be analyzed and evaluated in the following subsections, respectively.

22.3.3.1 Velocity Estimation Based on the Across-Track Deformation Effect

The shearing angle of moving vehicles caused by the across-track deformation allows for direct access to the velocity only if the moving direction is known a priori and input as an observation. Still, information about the orientation of the road axis relative to the vehicle motion is needed to derive the real velocity of vehicles. The velocity estimate v of the vehicle based on the shearing effect of its shape is derived by inverting Eq. 22.8 as

$$v = \frac{v_L \cdot \tan(\theta_{SA} - 90^\circ)}{\cos\theta_v \cdot \tan(\theta_{SA} - 90^\circ) + \sin(\theta_v)} \tag{22.13}$$

The value of the intersection angle θ_v can be determined based on principal axis measurements of vehicle points as the flight direction of the airborne LiDAR sensor can always be assumed to be known thanks to sustained navigation systems. Given Eq. 22.13 which shows that the accuracy of the velocity estimate based on the across-track deformation effect σ_v^c is a function of the quality of the moving vehicle's heading angle relative to the sensor flight path θ_v and the accuracy of the shearing angle measurement θ_{SA}, the standard deviation of the velocity estimate is calculated using the error propagation law (Wolf and Ghilani 1997) and derived as

$$\begin{aligned}\sigma_v^c &= \sqrt{\left(\frac{\partial v}{\partial \theta_v}\right)^2 \sigma_{\theta_v}^2 + \left(\frac{\partial v}{\partial \theta_{SA}}\right)\sigma_{\theta_{SA}}^2} \\ &= \sqrt{\begin{matrix}\left(\frac{v_L \cdot \tan(\theta_{SA}-90^\circ)\cdot(\cos(\theta_v)-\tan(\theta_{SA}-90^\circ)\cdot\sin(\theta_v))}{(\sin(\theta_v)+\tan(\theta_{SA}-90^\circ)\cdot\cos(\theta_v))^2}\right)^2 \sigma_{\theta_v}^2 \\ +\left(\frac{2v_L\cdot\sin(\theta_v)\left(\tan(90^\circ-\theta_{SA})^2+1\right)}{\cos(2\theta_v)\cdot\tan(90^\circ-\theta_{SA})^2-2\sin(2\theta_v)\cdot\tan(90^\circ-\theta_{SA})-\cos(2\theta_v)+\tan(90^\circ-\theta_{SA})^2+1}\right)^2 \sigma_{\theta_{SA}}^2\end{matrix}}\end{aligned} \tag{22.14}$$

with v_L being the instantaneous flying velocity of the sensor system.

22.3.3.2 Velocity Estimation Based on Along-Track Stretching Effect

Besides the above mentioned approach, the velocity of a moving vehicle can be derived by measuring its along-track stretching effect from its original vehicle size. The functional relation is given by:

$$v = \frac{(1 - Ar/Ar_s) \cdot v_L}{\cos(\theta_v)} \tag{22.15}$$

where $Ar_s = l_s/w_v$ is the sensed aspect ratio of the moving vehicle, while Ar is the original aspect ratio and assumed to be constant. The accuracy of the velocity estimate based on the along-track stretching effect σ_v^a is a function of the quality of the aspect ratio measurement for detected moving vehicles and the accuracy of the vehicle's heading relative to the sensor flight path. σ_v^a can be calculated by the error propagation law as follows:

$$\begin{aligned}\sigma_v^a &= \sqrt{\left(\frac{\partial v}{\partial \theta_v}\right)^2 \sigma_{\theta_v}^2 + \left(\frac{\partial v}{\partial Ar_s}\right)^2 \sigma_{Ar_s}^2} \\ &= \sqrt{\left(-\frac{v_L \cdot \sin(\theta_v) \cdot (Ar/Ar_s - 1)}{\cos(\theta_v)^2}\right)^2 \sigma_{\theta_v}^2 + \left(\frac{Ar \cdot v_L}{Ar_s^2 \cdot \cos(\theta_v)}\right)\sigma_{Ar_s}^2}\end{aligned} \tag{22.16}$$

22.3.3.3 Velocity Estimation Based on Combining Two Velocity Components

Both estimation methods presented above might fail to give a reliable velocity estimate if vehicles are moving in such a direction that generated deformation effects for the vehicle shape are not dominated by either one of what the two moving components account for (e.g., a moving vehicle with intersection angle $\theta_v = 35°$ and velocity $v = 40$ km/h). To fill this gap and enable a velocity estimate in an arbitrary traffic environment, it is proposed to use both shape deformation effects for estimating velocities. The functional dependence of the velocity estimate can be given by the sum of squares of the two motion components, which are derived based on two the shape deformation parameters Ar_s and θ_{SA}, respectively:

$$v = \sqrt{(v_a)^2 + (v_c)^2} \tag{22.17}$$

$$\text{where}\begin{cases} v_a = v_L \cdot \left(1 - \frac{Ar}{Ar_s}\right) \\ v_c = \frac{v_L}{\cot(\theta_{SA} - 90°) + \cot(\theta_v)} \end{cases} \tag{22.18}$$

and where v_a and v_c are along and across-track motion components. The accuracy of the velocity estimate based on combining the two components σ_v^{a+c} is a function of the quality of the along-track and across-track motion measurements for the detected moving vehicle and σ_v^{a+c} can be first calculated with respect to these two motion components by the error propagation law as:

$$\sigma_v^{a+c} = \sqrt{\left(\frac{\partial v}{\partial v_a}\right)^2 \partial^2 v_a + \left(\frac{\partial v}{\partial v_c}\right)^2 \partial^2 v_c}$$
$$= \sqrt{\frac{v_a^2}{v_a^2 + v_c^2}\sigma_{v_a}^2 + \frac{v_c^2}{v_a^2 + v_c^2}\sigma_{v_c}^2} \quad (22.19)$$

where σ_{v_a} and σ_{v_c} are the standard deviations of along- and across-track motion derivations, respectively. They can be further decomposed into the accuracy with respect to the three observations concerning the vehicle shape and motion parameters based on Eq. 22.18. Using the error propagation law, σ_{v_a} and σ_{v_c} are inferred as:

$$\sigma_{v_a} = \frac{\partial v_a}{\partial Ar_s}\sigma_{Ar_s} = \frac{Ar \cdot v_L}{Ar_s^2}\sigma_{Ar_s} \quad (22.20)$$

$$\sigma_{vc} = \sqrt{\left(\frac{\partial v_c}{\partial \theta_v}\right)^2 \sigma_{\theta_v}^2 + \left(\frac{\partial v}{\partial \theta_{SA}}\right)\sigma_{\theta_{SA}}^2}$$
$$= \sqrt{\left(\frac{v_L \cdot \left(\cot(\theta_v)^2 + 1\right)}{\left(\cot\left(90^\circ - \theta_{SA}\right) - \cot(\theta_v)\right)^2}\right)^2 \sigma_{\theta_v}^2 + \left(\frac{v_L \cdot \left(\cot\left(90^\circ - \theta_{SA}\right)^2 + 1\right)}{\left(\cot\left(90^\circ - \theta_{SA}\right) - \cot(\theta_v)\right)^2}\right)^2 \sigma_{\theta_{SA}}^2} \quad (22.21)$$

Finally, after substituting Eqs. 22.20 and 22.21 into Eq. 22.19, the error propagation relation for the velocity estimate is based on combining the two velocity components with respect to the three variables Ar_s, θ_{SA}, and θ_v is derived.

22.3.3.4 Joint Estimation of Vehicle Velocity and Direction by Solving Simultaneous Equations

So far, all of the estimation methods are not able to give velocity estimates if they are moving in an unknown direction or their moving detections cannot be accurately determined in advance. To solve this problem, we propose to jointly consider velocities and the intersection angle θ_v as unknown parameters simultaneously, with the variables describing the deformation effects caused by the motion components as observations. Actually, two analytic formulas for the motion artifacts model can be directly viewed as an equation system to which the velocity and the intersection angle are formulated as a set of solutions. This system of bivariate equations relating unknown parameters to observations is given by:

$$\begin{cases} \theta_{SA} - 90^\circ = \arctan\left(\frac{v \cdot \sin(\theta_v)}{v_L - v \cdot \cos(\theta_v)}\right) \\ 1 - \frac{v}{v_L} \cdot \cos(\theta_v) = \frac{Ar}{Ar_s} \end{cases} \tag{22.22}$$

The system is to be solved using the substitution method. First, transform the second sub-equation of Eq. 22.22 into

$$v = \frac{v_L}{\cos(\theta_v)} \cdot \left(1 - \frac{Ar}{Ar_s}\right) \tag{22.23}$$

and substitute it into the first sub-equation of Eq. 22.22, which has been converted into a more solution-friendly expression in advance:

$$\tan(\theta_{SA} - 90^\circ) \cdot v_L = v \cdot (\tan(\theta_{SA} - 90^\circ) \cdot \cos(\theta_v) + \sin(\theta_v)) \tag{22.24}$$

After substitution, the expression of Eq. 22.24 can be rewritten as:

$$\begin{aligned} \tan(\theta_{SA} - 90^\circ) \cdot v_L = v_L\left(1 - \frac{Ar}{Ar_s}\right) \cdot \tan(\theta_{SA} - 90^\circ) \\ + \tan(\theta_v) \cdot v_L \cdot \left(1 - \frac{Ar}{Ar_s}\right) \end{aligned} \tag{22.25}$$

Further, we transform to facilitate the solution and get:

$$\begin{aligned} \tan(\theta_v) &= \frac{\tan\left(\theta_{SA} - 90^\circ\right) \cdot \left[\left(1 - \left(1 - \frac{Ar}{Ar_s}\right)\right)\right]}{1 - \frac{Ar}{Ar_s}} = \tan\left(\theta_{SA} - 90^\circ\right)\left(\frac{Ar_s}{Ar_s - Ar} - 1\right) \\ \Rightarrow \theta_v &= \arctan\left[\tan\left(\theta_{SA} - 90^\circ\right) \cdot \left(\frac{Ar_s}{Ar_s - Ar} - 1\right)\right] \end{aligned} \tag{22.26}$$

Finally, substitute the second sub-equation in Eq. 22.26 into Eq. 22.23 again and the velocity estimate of the moving vehicle v can be derived as follows:

$$v = v_L \cdot \left(1 - \frac{Ar}{Ar_s}\right) \cdot \sec\left\{\arctan\left[\tan(\theta_{SA} - 90^\circ) \cdot \left(\frac{Ar_s}{Ar_s - Ar} - 1\right)\right]\right\} \tag{22.27}$$

It can be seen that the velocity of a moving vehicle can be directly estimated based on the shape deformation parameters without the need to know the intersection angle θ_v a priori. θ_v can be estimated as an intermediate variable solely based on two shape deformation parameters Ar_s, and θ_{SA} and is independent of the sensor flight velocity v_L. For accuracy analysis, two accuracy measures can be estimated, namely the moving direction and the velocity. The accuracies of the intersection angle σ_{θ_v} and the velocity estimate σ_v can be derived as functions of the quality of the along-track stretching and across-track shearing measures. Equivalently, σ_{θ_v} and σ_v can be calculated with respect to the two deformation parameters by the error propagation law as:

$$\sigma_{\theta_v} = \sqrt{\left(\frac{\delta\theta_v}{\delta Ar_s}\right)^2 \sigma_{Ar_s}^2 + \left(\frac{\delta\theta_v}{\delta Ar_{\theta_{SA}}}\right)\sigma_{\theta_{SA}}^2}$$

$$= \sqrt{\left(\frac{Ar\cdot\tan(90^\circ-\theta_{SA})}{Ar^2\cdot\tan(90^\circ-\theta_{SA})^2\cdot(Ar-Ar_s)^2}\right)^2\sigma_{Ar_s}^2 + \frac{Ar\cdot\left(\tan(90^\circ-\theta_{SA})^2+1\right)+(Ar-Ar_s)}{Ar^2\cdot\tan(90^\circ-\theta_{SA})^2+(Ar-Ar_s)^2}\sigma_{\theta_{SA}}^2} \quad (22.28)$$

$$\sigma_v = \sqrt{\left(\frac{\delta v}{\delta Ar_s}\right)^2 \sigma_{Ar_s}^2 + \left(\frac{\delta v}{\delta\theta_{SA}}\right)^2 \sigma_{\theta_{SA}}^2}$$

$$= \sqrt{\left(\frac{Ar\cdot v_L\cdot\left(Ar\cdot\tan(90^\circ-\theta_{SA})^2+Ar-Ar_s\right)}{Ar_s^2(Ar-Ar_s)\cdot\sqrt{\frac{Ar^2\tan(90^\circ-\theta_{SA})^2+(Ar-Ar_s)^2}{(Ar-Ar_s)^2}}}\right)^2\sigma_{Ar_s}^2 + \left(\frac{Ar\cdot v_L\cdot\tan(90^\circ-\theta_{SA})\cdot\left(\tan(90^\circ-\theta_{SA})^2+1\right)}{Ar^2(Ar-Ar_s)\cdot\sqrt{\frac{Ar^2\tan(90^\circ-\theta_{SA})^2+(Ar-Ar_s)^2}{(Ar-Ar_s)^2}}}\right)^2\sigma_{\theta_{SA}}^2} \quad (22.29)$$

The empirical error values for two observations σ_{Ar_S} and $\sigma_{\theta_{SA}}$ was also assessed to the same values as used in the preceding methods. The accuracies of intersection angle σ_{θ_v} and velocity estimates σ_v based on the joint estimation of moving velocity and direction are derived by inserting the empirical errors for the observations into Eqs. 22.28 and 22.29. The error of intersection angle σ_{θ_v} is shown in Fig. 22.7a as a function of vehicle velocity and relative angle between vehicle heading and the sensor flying path; the relative error is indicated in Fig. 22.7b. The (relative) velocity errors σ_v and σ_v/v are shown in Fig. 22.8 as a function of vehicle velocity v and intersection angle θ_v. It can be seen from the plots that most of the vehicles on road sections of urban areas could not allow for high accuracy of moving direction

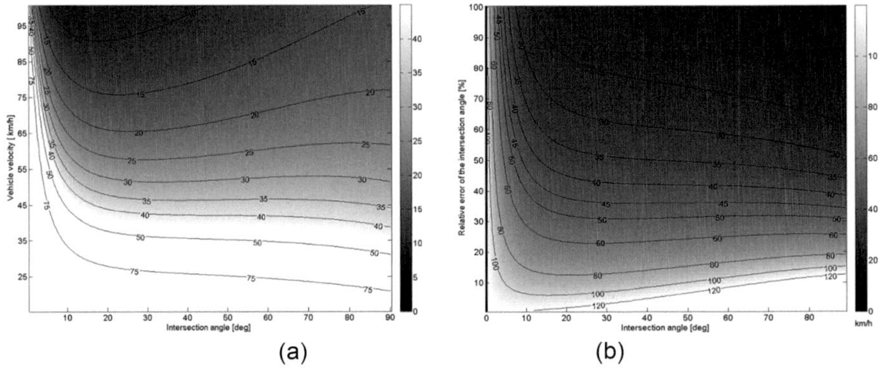

Fig. 22.7 **a** Relative error of the intersection angle $\sigma_{\theta v}/\theta_v$ of intersection angles obtained based on the joint estimation of velocity and heading as a function of target velocity v and the intersection angle θ_v, $\sigma_{\theta v}/\theta_v$ is given in %; **b** Vehicle velocity v (given in km/h) as a function of $\sigma_{\theta v}/\theta_v$ and θ_v

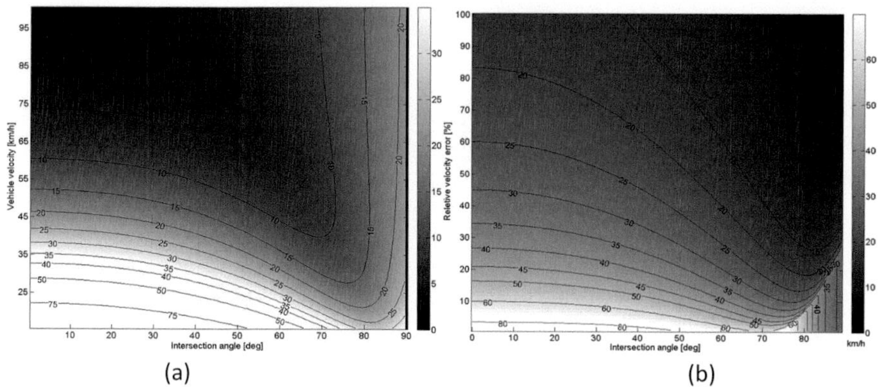

Fig. 22.8 **a** Relative velocity error σ_v/v of vehicle velocities obtained based on the joint estimation of velocity and heading as a function of target velocity v and the intersection angle θ_v, σ_v/v is given in %; **b** Vehicle velocity v (given in km/h) as a function of σ_v/v and θ_v.

estimation ($\sigma_{\theta_v}/\theta_v < 25\%$) unless they move a little bit faster (>70 km/h). The high accuracy of velocity estimates could be only guaranteed for vehicles that obviously don't travel in an across-track direction ($\theta_v < 75\%$). The overall accuracy of velocity estimation derived in this way is slightly degraded compared to other solutions where the moving direction is given beforehand.

22.4 Experiments and Results

22.4.1 Detection of Urban Objects with ALS Data Associated with Aerial Imagery

22.4.1.1 Experimental Data for Urban Objects Detection

Two datasets were used in this chapter for an urban scene object detection test, which both include aerial images and airborne LiDAR data. The first dataset (yellow areas in Fig. 22.9) was captured over Vaihingen in Germany and is a subset of the data used for the test of digital aerial cameras carried out by the German Association of Photogrammetry and Remote Sensing (DGPF; Cramer 2010). The other dataset covers an area of about 1.45 km^2 in the central area of the City of Toronto in Canada (red areas in Fig. 22.10).

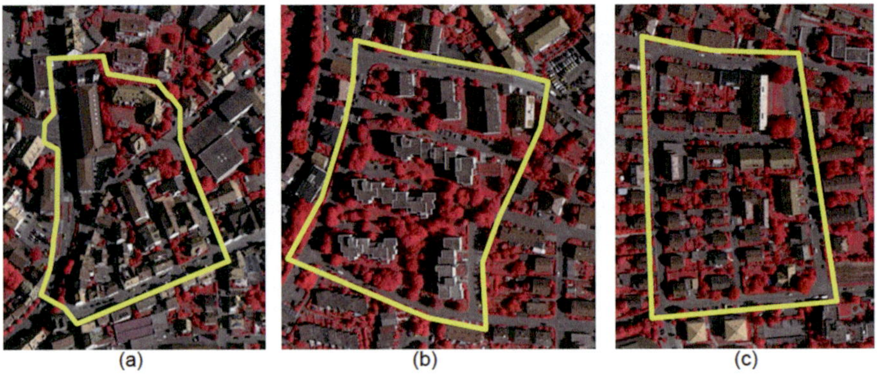

Fig. 22.9 Three test sites in Vaihingen: **a** Area 1; **b** Area 2; **c** Area 3

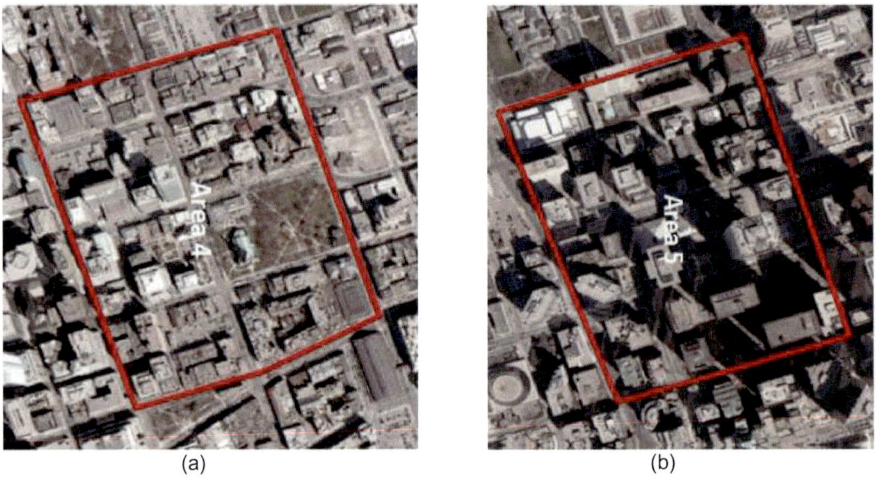

Fig. 22.10 Two test sites in Toronto: **a** Area 4; **b** Area 5

22.4.1.2 Experimental Design for Urban Objects Detection

The following steps are considered in this experiment:

Data preprocessing. For both datasets, the aerial images and airborne LiDAR data were acquired at different times. Thus, they are co-registered by geometrical back-projecting the point cloud into the image domain with available orientation parameters. After that, all data points are grid-fitted into the raster format in order to facilitate acquiring spatial context information per-pixel or point. We apply grid-fitting using an interval of 0.5 m on the ground, ensuring that each resampled pixel can be allocated at least with one LiDAR point.

Feature selection. For Dataset 1, as color-infrared images, point cloud data including intensity information are available. All 13 features (*R*, *G*, *B*, *NDVI*, *Z*, *I*,ΔZ, σ_Z, ΔI, σ_I, *E*, *O*, and *P*) introduced in Sect. 2.2 are extracted and used for the object detection test. For Dataset 2, there is no infrared band image and thus 12 features are used in the experiment only, without NDVI.

Training samples' selection. Since training samples are essential and important for supervised learning classification, it is necessary to adopt a suitable approach to derive valid samples considering the characteristics of the used classifier. In this chapter, AdaBoost using the one tree weak learner (CART) is adopted as the final strong classifier (Freund and Schapire 1999), which chooses training samples randomly to some extent. Therefore, for each test site, we first classify the whole test area manually and then randomly choose 10% of the whole test area's corresponding labeled samples as input training samples for the AdaBoost classifier.

Classifier control and classification procedure. This chapter uses the binary AdaBoost classifier to detect buildings, natural ground, and trees from the urban scene. To do so, the binary AdaBoost classifier is iteratively generated and applied: (1) the classifier for detecting building is generated by training the randomly chosen building samples and non-building samples corresponding to 10% of the whole data amount, and applied to classify the building from the urban scene; (2) 10% natural and non-natural ground samples are randomly selected to train and generate the classifier for natural ground detection, which is then used to separate the natural ground from the complex urban scene; (3) tree detection proceeds by using the binary AdaBoost classifier which is trained on the randomly selected 10% tree and non-tree samples. To test and validate the methods, several areas are chosen for the object detection test according to the actual urban scene. For the building detection, all the five test areas (three in Vaihingen and two in downtown Toronto) are used, whereas Areas 1–4 are used to test the detection of natural ground. And finally, Areas 1–3 in Dataset 1 are used for the detection of trees. The implementation code of the AdaBoost classifier used in this chapter was adapted from that published by Vezhnevets (2005).

Evaluation methods. The evaluation of object detection results is obtained from the ISPRS Test Project on Urban Classification and 3D Building Reconstruction, which conducts the evaluation based on the method described by Rutzinger et al. (2009) and Rottensteiner et al. (2005). The software used for evaluation reads in the reference and the object detection results, converts them into a label image, and then carries out the evaluation as described by Rottensteiner et al. (2013). Since the output of binary AdaBoost classifiers consists of samples labeled by class but not segmented objects, the topological clarification for detected objects described by Rutzinger et al. (2009) is applied to perform the object-based evaluation, which was automatically implemented by the evaluation software. The evaluation output consists of a text file containing the evaluation results and a few images that visualize these results, which include many accuracy indexes such as geometric accuracy, pixel-based completeness, and correctness, object-based completeness, and correctness, balanced completeness and correctness, etc., and the middle evaluation includes attributes like an evaluation on a per-object level as a function of the object area, etc.

This chapter applies the binary AdaBoost classifier by fusing the image and LiDAR features to detect buildings, natural ground, and trees in several different complex urban scenes. The detection accuracies of buildings, natural ground, and trees are presented in Tables 22.1, 22.2, and 22.3, respectively. In these tables pixel-based evaluation accuracy (Compl area [%], Corr area [%], Pix-Quality [%]), object-based evaluation accuracy (Compl obj [%], Corr obj [%], obj-Quality [%]), balanced evaluation accuracy (Compl obj 50 [%], Corr obj 50 [%], obj-Quality 50 [%]), and detected objects' geometric accuracy (RMS [m]) are listed for evaluating the detection result of buildings in Areas 1–5, natural ground in Areas 1–4, and trees in Areas 1–3, respectively.

22.4.1.3 Results of Urban Objects Detection

As stated in Sect. 22.2, this chapter applies the binary AdaBoost classifier by fusing the image and LiDAR features to detect buildings, natural ground, and trees in several different complex urban scenes. The detection accuracy of buildings, natural ground, and trees are presented in Table 22.1, Table 22.2, and Table 22.3 respectively. In Tables 22.1, 22.2 and 22.3, pixel-based evaluation accuracy (Compl area [%],Corr area [%], Pix-Quality [%]), object-based evaluation accuracy(Compl obj [%],Corr obj [%], obj-Quality [%]), balanced evaluation accuracy (Compl obj 50 [%], Corr obj 50 [%], obj-Quality 50 [%]) and detected objects' geometric accuracy (RMS [m]) are listed for evaluating the detection result of buildings in Areas 1–5, natural ground in Areas 1–4, and trees in Areas 1–3, respectively.

Building detection result. It can be noticed from Table 22.1 that all the five test sites obtain 85% or higher pixel-based completeness, while the object-based completeness is lower due to the area of overlap of objects, especially for Test Sites 2 and 3 with object-based completeness of less than 80%. With regard to correctness, the three test sites in Dataset 1 perform better than the two test sites in Dataset 2 with respect to all evaluation aspects: evaluation methods of pixel-based, object-based, and pixel-object balanced. Thus, it can conclude that the building detection of Dataset 1 is more robust than that of Dataset 2. Concerning the geometric aspect, Test Area 2 obtained the best geometric accuracy of RMS 0.9 m, followed by Area 3 with RMS 1.0 m, and Area 1 with RMS 1.2 m, while both test sites in Dataset 2 obtain the worst geometric accuracy with RMS 1.6 m. Among the five test sites, Area 2 achieved the best overall building detection accuracy completeness of 92.5%, correctness of 93.9%, detection quality of 87.2% using pixel-based evaluation, completeness of 100%, correctness of 100%, and detection quality of 100% based on evaluation balanced between pixels and objects, correctness of 100% based on object-based evaluation, and geometric accuracy of RMS 0.9 m. Due to the small number of buildings, three false negatives on detected objects gave Test Site 2 lower completeness than Test Sites 1, 4, and 5 based on object-based evaluation, even though there are more false negatives.

Table 22.1 Detection accuracy of buildings

Area	Compl area [%]	Corr area [%]	Pix-Quality [%]	Compl obj [%]	Corr obj [%]	Obj-Quality [%]	Compl obj 50 [%]	Corrobj 50 [%]	Obj-Quality 50[%]	RMS [m]
Area 1	89.8	92.2	81.7	89.2	97.1	82.5	100.0	100.0	100.0	1.2
Area 2	92.5	93.9	87.2	78.6	100.0	78.6	100.0	100.0	100.0	0.9
Area 3	86.8	92.5	81.0	75.0	100.0	75.0	97.4	100.0	97.4	1.0
Area 4	85.1	80.0	70.2	86.2	92.3	80.4	87.7	94.1	83.1	1.6
Area 5	85.0	81.1	70.9	81.6	88.2	73.6	88.6	90.9	81.4	1.6

Table 22.2 Detection accuracy of natural ground

Area	Compl area [%]	Corr area [%]	Pix-Quality [%]	Compl obj [%]	Corr obj [%]	Obj-Quality [%]	Complobj 50 [%]	Corrobj 50 [%]	Obj-Quality 50 [%]	RMS [m]
Area 1	60.4	82.8	53.6	57.9	75.0	48.5	75.0	85.7	66.7	1.3
Area 2	80.5	85.7	71.0	47.4	50.0	32.1	83.3	100.0	83.3	1.1
Area 3	78.9	84.2	68.7	68.0	76.3	56.2	90.5	87.0	79.7	1.1
Area 4	49.8	81.5	45.0	45.0	16.5	13.7	50.0	76.9	43.5	1.7

Table 22.3 Detection accuracy of trees

Area	Compl area [%]	Corr area [%]	Pix-Quality [%]	Compl obj [%]	Corr obj [%]	Obj-Quality [%]	Compl obj 50 [%]	Corrobj 50 [%]	Obj-quality 50 [%]	RMS [m]
Area 1	69.3	71.2	54.1	61.0	58.3	42.4	96.3	96.2	92.3	1.4
Area 2	72.0	78.5	60.1	63.0	82.4	55.5	89.3	98.6	88.2	1.4
Area 3	69.5	80.1	59.3	53.5	76.4	46.0	93.9	100.0	93.9	1.3

Natural Ground Detection Result. The results of Dataset 1 are better than those of Dataset 2 on all indexes. Concerning the pixel-based evaluation result, the detection completeness is lower than the correctness for all the test sites, while it is the same for the object-based evaluation result except for Test Site 4. For this test site, the object-based correctness is very low compared to the pixel-based correctness, which shows that the natural ground of Test Site 4 is fragmented and cannot be detected well at the object level. Regarding the geometric aspect, Areas 2 and 3 obtain the best geometric accuracy of RMS 1.1 m, followed by Area 1 with RMS 1.3 m, while test site 4 in Dataset 2 obtains the worst geometric accuracy with RMS 1.7 m. Among the four test sites, Site 2 achieves the best overall natural ground detection accuracy with completeness of 80.5%, correctness of 85.7%, detection quality of 71.0% based on pixel-based evaluation, completeness of 83.3%, correctness of 100%, detection quality of 83.3% based on a balanced evaluation of pixels and objects, and geometric accuracy of RMS 1.1 m. Due to the larger number of small-sized natural ground objects and fewer larger ones, Test Site 2 obtains lower detection accuracy using object-based evaluation.

Tree-detection result. Only Dataset 1 was tested. From Table 22.3, it can be noticed that the tree-detection accuracy is lower than 80%, being lower than that of building detection in the same test site. Although the accuracy indexes obtained based on both pixel-based and object-based evaluation are not so good, this is related to the definition of trees in the reference data since the balanced accuracy is good. On the geometric aspect, Area 3 obtains the best geometric accuracy of RMS 1.3 m, followed by Area 1 and 2 with RMS 1.4 m. The geometric accuracy for tree detection is worse than that of both buildings and natural ground, due to the more complex shape of trees in 2D and 3D. Among the three test sites, Area 2 achieves the best overall tree-detection accuracy with the completeness of 72.0%, correctness of 78.5% based on pixel-based evaluation, completeness of 63.0%, correctness of 82.4% based on object-based evaluation, completeness of 89.3%, and correctness of 98.6% using the balanced evaluation of pixels and objects, and geometric accuracy of RMS 1.4 m.

The detection results presented above show that the proposed AdaBoost-based strategy can detect objects very well in complex urban areas based on relevant spatial and spectral features that have been obtained by combining point clouds and image data. First, most detected objects only suffer from errors in boundary regions, especially with respect to buildings in Test Sites 1–3, which means that the proposed method can successfully separate desirable objects from the background using the combined spatial-spectral features. Second, the trees and natural ground can be discriminated efficiently in Dataset 1 in spite of similar spectral features, which demonstrates that the method can take full use of the advantages of fusing features and an ensemble classifier. Third, the detection achieves the best geometric accuracy for buildings, with RMS 0.9 m, partly biased by data co-registration error, which demonstrates the proposed high accuracy of the method. Fourth, larger-sized objects achieve better detection completeness and correctness; for example, all the buildings with area larger than 87.5 m^2 are detected correctly for Test Sites 1–3, while some smaller buildings are omitted due to being classified as false positives, which justifies the reliability of the AdaBoost-based strategy for urban objects detection.

22.4.2 *Accuracy Prediction for Vehicle Velocity Estimation Using ALS Aata*

To demonstrate the quality of the velocity estimation for real-life scenarios and to deliver quantitative guidance on the planning of LiDAR flight campaigns for traffic analysis, real road networks in urban areas will be used in an experiment to simulate the prediction of velocity and estimate its accuracy. This will be useful for exploiting boundary conditions in applying the proposed strategy in real airborne LiDAR campaigns for traffic analysis. Generally, it can be stated that this simulation has been designed by considering the following points:

- Validate the feasibility and repeatability of velocity estimation results;
- Verify the velocity estimation scheme, which provides rational results with sufficient accuracy in a wide range of datasets acquired over urban areas; and
- Demonstrate the potential of velocity-accuracy analysis to provide valuable guidance on optimizing flight planning for traffic monitoring.

The accuracy of the estimated velocity σ_v is simulated for two road network sections north of Munich which represent the most typical scenarios in urban areas. In this area, several main roads and large express roads are situated and are highly frequented during rush hours. For each test site, two general schemes are assumed to exist, where the four different velocity estimators presented above are applied: First, the moving direction of a vehicle relative to the sensor flight path is known (here the moving direction is derived based on the road orientation); and second, the moving direction of the vehicle relative to the sensor flight path is unknown.

As three methods within the first scheme complement each other concerning performance, we finally combined the estimators depending on the relative orientation between the vehicle heading and the sensor flight path to get optimal results. For every relative orientation the estimator that provides the best results is chosen. That means that the maximum of estimated velocity accuracies is assumed to be selected as the accuracy value for a velocity estimate at that road location. Parameters of real flying using the Riegl LMSQ560 sensor have been used in this simulation and an average speed of 120 km/h was assumed (concrete configurations can be found in Table 22.4). The average velocity of moving vehicles on the roads is set to 60 km/h.

Table 22.4 Parameters of typical airborne topographic LiDAR

Flight height	h	420 m
Pulse repetition rate	PRR	110 kHz
Sensor velocity	v_L	120 km/h
Scan angle	α_s	60°
Point density	PD	4 points/m^2
Swath	S_w	450 m
View mode		Nadir
Scan pattern		Parallel line

The error measures for the shearing angle and intersection angle of moving vehicles can be assessed empirically from shape parameterization: for our case, $\sigma_{Ars} = 0.4$, $\sigma_{\theta_{SA}} = 2°$, and $\sigma_{\theta_v} = 2°$. The orientation of the roads relative to the planned flying path and the resulting σ_v values obtained by combining the estimators in the first scheme are shown in Fig. 22.11a, c, while the resulting values of σ_v using second scheme for the same sites are shown in Fig. 22.11b, d. σ_v is given in % of the absolute velocity. With the algorithm described earlier, velocities can be estimated with an accuracy better than 10% for about 80% of the investigated road networks. Figure 22.12 indicates which estimator is chosen in which parts of the road network. It shows that the across-track shearing-based estimator (Method 1) provides the best

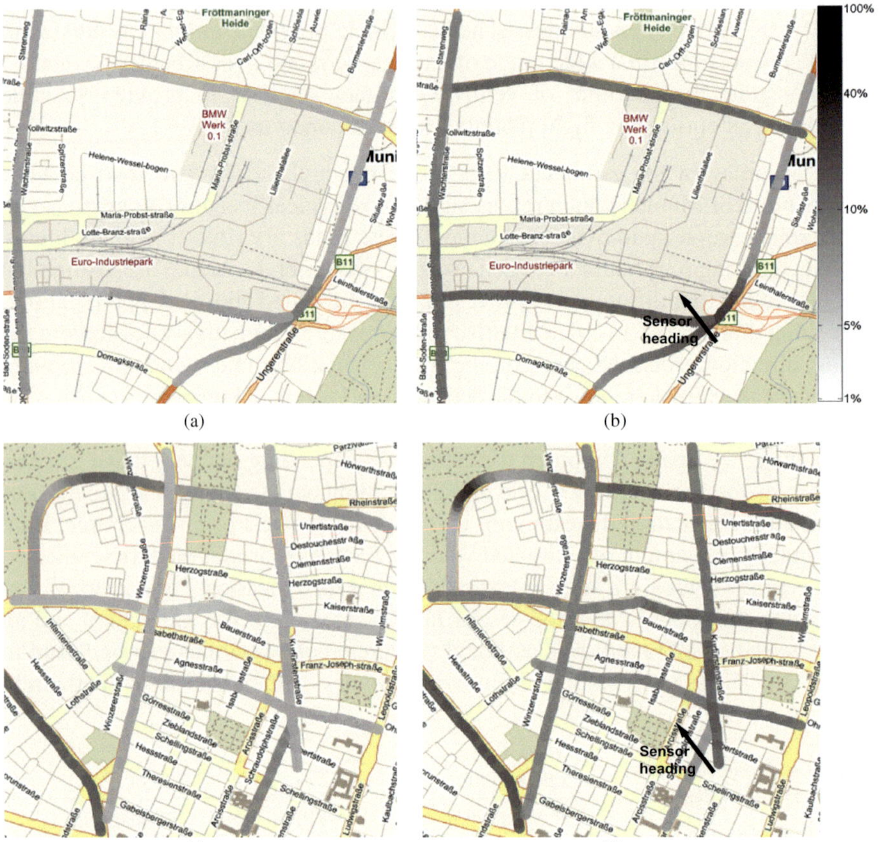

Fig. 22.11 Simulation of σ_v for two road networks north of Munich using the velocity estimation schemes: **a** The estimation accuracy for the first road network in % of the absolute velocity using the second scheme; **b** The estimation accuracy for the first road network in % of the absolute velocity using the first scheme; **c** The estimation accuracy for the second road network in % of the absolute velocity using the first scheme; **d** The estimation accuracy for the second road network in % of the absolute velocity using the second scheme

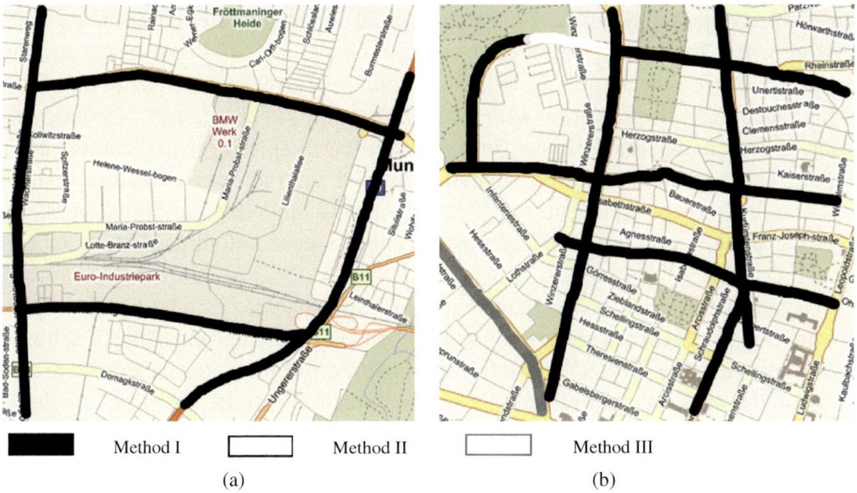

Fig. 22.12 Indication of velocity estimation methods used for the two road networks under the first scheme for velocity estimation (moving direction relative to sensor flight is known): **a** Indicating which estimation method is chosen in which parts of the first road network; **b** Indicating which estimation method is chosen in which parts of the second road network

results for large parts of the road network. The along-track stretching-based (Method 2) and combined (Method 3) estimators outperform the across-track shearing-based approach only in areas where the road is extended roughly in the along-track direction (i.e., $\forall\, \theta_v \leq 25°$). For example, in the second test site (Fig. 22.12b), Dachauer Street (in the bottom-left part) requires Method 3 to be used for velocity estimation, whereas one part of Ackermann Street (curved, in the top-left part) requires Method 2 to be used. Moreover, in most parts of the road network, the accuracy of velocity estimation using the first scheme is generally higher than that obtained using the second scheme, especially when vehicles move along a direction that is close to across-track. This is due to the fact that the joint estimation of velocity and moving direction angle can incorporate additional error sources caused by the unknown moving direction of vehicles relative to the sensor flight path, leading to an accumulative error for final velocity estimates.

22.5 Summary

This chapter is concerned with detecting urban objects and traffic dynamics from ALS data. Urban object detection in complex scenes is still a challenging problem for the communities of both photogrammetry and computer vision. Since LiDAR data and image data are complementary for information extraction, relevant spatial-spectral features extracted from ALS point clouds and image data can be jointly applied to detect urban objects like buildings, natural ground objects, and trees in complex urban environments. To obtain good object detection results, an AdaBoost-based strategy was presented in this chapter. It includes: First, co-registering LiDAR point clouds with images by back-projection with available orientation parameters; Second, grid-fitting of data points into the raster format to facilitate acquiring spatial context information; Third, extracting various spatial-statistical and radiometric features using a cuboid neighborhood; and Fourth, detecting objects including buildings, trees, and natural ground by the trained AdaBoost classifier whose output consists of labeled grids.

The performance of the developed strategy towards detecting buildings, natural ground, and trees in urban areas was comprehensively evaluated using the benchmark datasets provided by ISPRS WGIII/4. Both semantic and geometric criteria were used to assess the experimental results. From the detection results, it can be concluded that the AdaBoost-based classification strategy can detect urban objects reliably and accurately, achieving the best detection accuracy for buildings with completeness of 92.5% and correctness of 93.9%, for natural ground with completeness of 80.5% and correctness of 85.7%, and for tree detection with completeness of 72.5% and correctness of 78.5% based on per-pixel evaluation. The quality indexes for the detection of tree and natural ground, evaluated on per-object level, seem not to be as high as for buildings. Nevertheless, the overall accuracy is high for such complex urban scenes, as can be concluded from the balanced evaluation of pixels and objects. With further research, the detection results might be refined with graph-based optimization, which is expected to improve the detection accuracy by accounting for label smoothness both locally and globally. Moreover, in order to further ensure the reliability of object detection, we still need to refine the co-registration accuracy of multimodal data via hierarchical feature matching and optimize alterable parameters through sensitivity analysis.

For characterizing urban traffic dynamics, a method to identify vehicle movement from airborne LiDAR data and to estimate respective velocities has been developed. Besides a description of the developed methods, theoretical and simulation studies for performance analysis were shown in detail. The detection and velocity estimation of fast-moving vehicles seems to be promising and accurate, whereas slow-moving vehicles are harder to distinguish from non-moving ones and it is harder to obtain estimates with acceptable accuracy. Moreover, the point density of LiDAR datasets tends to be directly proportional to the performance of motion detection. The estimation of the velocity of detected vehicles can be done with high accuracy for nearly all possible observation geometries except for those ones which are moving in the (quasi-)along-track direction while sensors are sweeping over instantaneously.

Although the results shown in this chapter cannot directly be compared with those of induction loops or bridge sensors, they show nonetheless great potential to support traffic monitoring applications. The big advantages of ALS data are their large coverage and certain penetrability through trees, and thus, the possibility to derive traffic data throughout an extended road network that may be occluded by trees on the roadsides. Evidently, this complements the accurate but sparsely sampled measurements of fixed mounted sensors. A natural extension of the presented approach would be an integration of the accurate, sparsely sampled traffic information with the less accurate but area-wide data collected from space or air-borne sensors. Existing traffic flow models would provide a framework to do this.

Acknowledgements This was partially supported by The Hong Kong Polytechnic University grants 1- ZE8E and 1-YBZ9, by PhD research excellence grant of Elite Network of Bavaria, and partially supported by the National Key Research and Development Program of China (No. 2016YFF0103503) and NSFC (No. 41771485). The experimental data set over Vaihingen for urban objects detection was provided by the German Society for Photogrammetry, Remote Sensing, and Geoinformation (DGPF) (Cramer 2010): https://www.ifp.uni-stuttgart.de/dgpf/DKEP-Allg.html.

References

Azadbakht M, Fraser CS, Khoshelham K (2018) Synergy of sampling techniques and ensemble classifiers for classification of urban environments using full-waveform LiDAR data. Int J Appl Earth Obs Geoinf 73:277–291

Balado J, Díaz-Vilariño L, Arias P, González-Desantos LM (2018) Automatic LOD0 classification of airborne LiDAR data in urban and non-urban areas. Eur J Remote Sens 51(1):978–990

Bazzichetto M, Malavasi M, Acosta ATR, Carranza ML (2016) How does dune morphology shape coastal EC habitats occurrence? A remote sensing approach using airborne LiDAR on the Mediterranean coast. Ecol Ind 71:618–626

Bonczak B, Kontokosta CE (2019) Large-scale parameterization of 3D building morphology in complex urban landscapes using aerial LiDAR and city administrative data. Comput Environ Urban Syst 73:126–142

Chen LC, Lo CY (2009) 3D road modeling via the integration of large-scale topomaps and airborne LIDAR data. J Chin Inst Eng 32(6):811–823

Chen Z, Liu C, Wu H (2019) A higher-order tensor voting-based approach for road junction detection and delineation from airborne LiDAR data. ISPRS J Photogr Remote Sens 150:91–114

Cheng L, Wu Y, Wang Y, Zhong L, Chen Y, Li M (2014) Three-dimensional reconstruction of large multilayer interchange bridge using airborne LiDAR data. IEEE J Sel Topics Appl Earth Obs Remote Sens 8(2):691–708

Clode S, Rottensteiner F, Kootsookos P, Zelniker E (2007) Detection and vectorization of roads from LiDAR data. Photogr Eng Remote Sens 73(5):517–535

Cramer M (2010) The DGPF test on digital aerial camera evaluation—overview and test design. Photogrammetrie – Fernerkundung – Geoinformation 2(2010):73–82

Dawood N, Dawood H, Rodriguez-Trejo S, Crilly M (2017) Visualising urban energy use: the use of LiDAR and remote sensing data in urban energy planning. Vis Eng 5(1):22

Degerickx J, Roberts DA, McFadden JP, Hermy M, Somers B (2018) Urban tree health assessment using airborne hyperspectral and LiDAR imagery. Int J Appl Earth Obs Geoinf 73:26–38

Degerickx J, Roberts DA, Somers B (2019) Enhancing the performance of Multiple Endmember Spectral Mixture Analysis (MESMA) for urban land cover mapping using airborne LiDAR data and band selection. Remote Sens Environ 221:260–273

Earlie CS, Masselink G, Russell PE, Shail RK (2015) Application of airborne LiDAR to investigate rates of recession in rocky coast environments. J Coastal Conser 19(6):831–845

Fauvel M (2007) Spectral and spatial methods for the classification of urban remote sensing data. Ph.D. thesis, Grenoble Institute of Technology, France, and University of Iceland, Iceland

Fletcher PT, Lu C, Joshi S (2003) Statistics of shape via principal geodesic analysis on Lie groups. In: IEEE computer society conference on computer vision and pattern recognition, Madison, Wisconsin, 16–22 June 2003

Freund Y, Schapire R, Abe N (1999) A short introduction to boosting. J Jpn Soc Artif Intell 14(5):771–780
Garnett R, Adams M (2018) LiDAR—A technology to assist with smart cities and climate change resilience: a case study in an urban metropolis. ISPRS Int J Geo-Inf 7(5):161
Guo L, Chehata N, Mallet C, Boukir S (2011) Relevance of airborne LiDAR and multispectral image data for urban scene classification using Random Forests. ISPRS J Photogr Remote Sens 66(1):56–66
Höfle B, Hollaus M (2010) Urban vegetation detection using high density fullwaveform airborne LiDAR data—combination of object-based image and point cloud analysis. Int Arch Photogr Remote Sens Spatial Inf Sci 38(B7):281–286
Jochem A, Höfle B, Rutzinger M, Pfeifer N (2009) Automatic roof plane detection and analysis in airborne LiDAR point clouds for solar potential assessment. Sensors 9(7):5241–5262
Lafarge F, Mallet C (2012) Creating large-scale city models from 3D-point clouds: a robust approach with hybrid representation. Int J Comput Vision 99(1):69–85
Lafortezza R, Giannico V (2019) Combining high-resolution images and LiDAR data to model ecosystem services perception in compact urban systems. Ecol Ind 96:87–98
Li M, Rottensteiner F, Heipke C (2019) Modelling of buildings from aerial LiDAR point clouds using TINs and label maps. ISPRS J Photogr Remote Sens 154:127–138
Liu L, Coops NC, Aven NW, Pang Y (2017) Mapping urban tree species using integrated airborne hyperspectral and LiDAR remote sensing data. Remote Sens Environ 200:170–182
Mastin A, Kepner J, Fisher J (2009) Automatic registration of LIDAR and optical images of urban scenes. In: EEE conference on computer vision and pattern recognition, Miami Beach, FL, 22–24 June 2009
Næsset E, Gobakken T (2008) Estimation of above-and below-ground biomass across regions of the boreal forest zone using airborne laser. Remote Sens Environ 112(6):3079–3090
Reitberger J, Krzystek P, Stilla U (2008) Analysis of full waveform LiDAR data for the classification of deciduous and coniferous trees. Int J Remote Sens 29(5):1407–1431
Rizeei HM, Pradhan B (2019) Urban mapping accuracy enhancement in high-rise built-up areas deployed by 3D-orthorectification correction from WorldView-3 and LiDAR imageries. Remote Sens 11(6):692
Rottensteiner F, Trinder J, Clode S, Kubik K (2005) Using the Dempster-Shafer method for the fusion of LiDAR data and multi-spectral images for building detection. Inf Fusion 6(4):283–300
Rottensteiner F, Sohn G, Gerke M (2013) ISPRS test project on urban classification and 3D building reconstruction. https://www2.isprs.org/tl_files/isprs/wg34/docs/ComplexScenes_revision_may13.pdf. Accessed 12 June 2013
Rutzinger M, Rottensteiner F, Pfeifer N (2009) A comparison of evaluation techniques for building extraction from airborne laser scanning. IEEE J Sel Topics Appl Earth Observ Remote Sens 2(1):11–20
Schenk T, Csathó B (2007) Fusing imagery and 3D point clouds for reconstructing visible surfaces of urban scenes. In: Urban remote sensing joint event, Paris, France, 11–13 April
Sohn G, Dowman I (2007) Data fusion of high-resolution satellite imagery and LiDAR data for automatic building extraction. ISPRS J Photogr Remote Sens 62(1):43–46
Tomás L, Fonseca L, Almeida C, Leonardi F, Pereira M (2016) Urban population estimation based on residential buildings volume using IKONOS-2 images and LiDAR data. Int J Remote Sens 37(sup1):1–28
Toth CK, Grejner-Brzezinska D (2006) Extracting dynamic spatial data from airborne imaging sensors to support traffic flow estimation. ISPRS J Photogr Remote Sens 61(3–4):137–140
Vezhnevets A (2005) GML AdaBoost Matlab Toolbox. https://graphics.cs.msu.ru/en/science/research/machinelearning/adaboosttoolbox. Accessed 20 June 2013
Wang Y, Chen Q, Liu L, Zheng D, Li C, Li K (2017) Supervised classification of power lines from airborne LiDAR data in urban areas. Remote Sens 9(8):771
Wang C, Shu Q, Wang X et al (2019) A random forest classifier based on pixel comparison features for urban LiDAR data. ISPRS J Photogr Remote Sens 148:75–86

Wei Y, Yao W, Wu J, Schmitt M, Stilla U (2012) Adaboost-based feature relevance assessment in fusing LiDAR and image data for classification of trees and vehicles in urban scenes. ISPRS Ann Photogr Remote Sens Spatial Inf Sci 1(7):323–328
Wolf RP, Ghilani DC (1997) Adjustment computations: statistics and least squares in surveying and GIS, 3rd edn. Wiley , New York
Xiao J, Gerke M, Vosselman G (2012) Building extraction from oblique airborne imagery based on robust façade detection. ISPRS J Photogr Remote Sens 68:56–68
Yao W, Hinz S, Stilla U (2010a) Automatic vehicle extraction from airborne LiDAR data of urban areas using morphological reconstruction. Pattern Recogn Lett 31(10):1100–1108
Yao W, Hinz S, Stilla U (2010b) Airborne analysis and assessment of urban traffic scenes from LiDAR data—theory and experiments. In: Proceedings of workshops of IEEE conference on computer vision and pattern recognition, San Francisco, CA, USA, 13–18 June
Yao W, Hinz S, Stilla U (2011) Extraction and motion estimation of vehicles in single-pass airborne LiDAR data towards urban traffic analysis. ISPRS J Photogr Remote Sens 66(3):260–271
Yao W, Stilla U (2010) Mutual enhancement of weak laser pulses for point cloud enrichment based on full-waveform analysis. IEEE Trans Geosci Remote Sens 48(9):3571–3579
Zhao K, Suarez JC, Garcia M, Hu T, Wang C, Londo A (2018) Utility of multitemporal LiDAR for forest and carbon monitoring: tree growth, biomass dynamics, and carbon flux. Remote Sens Environ 204:883–897
Zhou QY, Neumann U (2008) Fast and extensible building modeling from airborne LiDAR data. In: Proceedings of the 16th ACM SIGSPATIAL international conference on advances in geographic information systems, Irvine, USA, Nov 5–7

Wei Yao received his Ph.D. degree in photogrammetry and remote sensing from the Technical University of Munich, Germany in 2010. Currently, he is an assistant professor with the Department of Land Surveying and Geoinformatics, The Hong Kong Polytechnic University. His research interests include 3D remote sensing techniques for urban and environmental informatics.

Jianwei Wu received his Ph.D. degree in photogrammetry and remote sensing from Wuhan University, Wuhan, China in 2008. Currently, he is an assistant professor with the School of Remote Sensing and Information Engineering, Wuhan University. His research interests include laser scanning, remote sensing in forest, and geospatial data registration.

Chapter 23
Photogrammetry for 3D Mapping in Urban Areas

Bo Wu

Abstract Photogrammetry is the technology for obtaining 3D geometric information from photographs or images. This chapter describes the fundamental knowledge and latest advances in photogrammetry for 3D mapping in urban areas. First, the key fundamental techniques in photogrammetry for deriving 3D information from imagery are presented. Then, the latest advances in photogrammetry for 3D mapping in urban areas, including structure-from-motion (SfM), multi-view stereo (MVS), and integrated 3D mapping from multiple-source data, are described and discussed. Examples of using photogrammetry for 3D mapping and modeling in urban applications are presented. Finally, concluding remarks and future outlooks are addressed.

23.1 Introduction

Photogrammetry is the science and technology for obtaining reliable 3D geometric and physical information about objects and the environment from photographic images (ASPRS 1998). Practically, photogrammetry allows 3D measurements of geometric information of objects (e.g., positions, orientations, shapes, and sizes) from photographs.

Photogrammetry has a long history and can be dated back to the 1850s (Konecny 1985). In its earlier stage, the main purpose of photogrammetry was map generation from aerial photographs. Since the 1960s, the emerging of satellite and close-range imaging and measurements has facilitated the application of photogrammetry to various areas, such as 3D mapping and modeling, industrial inspection, architecture, robotics, civil engineering, and hazard monitoring. Advances in photogrammetry had been insignificant over the past 50 years until the recent decade. The latest advances from the photogrammetry and computer vision communities, such as aerial oblique photogrammetry, structure-from-motion (SfM) and multi-view stereo (MVS), and

B. Wu (✉)
Department of Land Surveying and Geo-Informatics, The Hong Kong Polytechnic University, Hong Kong, China
e-mail: bo.wu@polyu.edu.hk

W. Shi et al. (eds.), *Urban Informatics*, The Urban Book Series,
https://doi.org/10.1007/978-981-15-8983-6_23

integrated 3D mapping, have facilitated the development of photogrammetry towards a more automatic solution for 3D mapping and modeling, with better quality, even for challenging cases such as in urban areas.

This chapter first describes the key fundamental knowledge for obtaining 3D information from images through photogrammetry. Then, the latest advances in photogrammetry for 3D mapping in urban areas, including SfM, MVS, and integrated 3D mapping from multiple-source data, are described and discussed. Examples of using photogrammetry for 3D mapping and modeling in Hong Kong and other typical urban areas are presented. Finally, summary remarks are given and future outlooks are discussed.

23.2 Fundamentals of Photogrammetry

The following describes the fundamental techniques for obtaining 3D information from images via photogrammetry, including image orientation, bundle adjustment, and image matching.

23.2.1 Image Orientation

Image orientation is the procedure of recovering the positional and orientation information of the optical ray when the image is collected. Image orientation includes two consecutive steps: interior orientation (IO) and exterior orientation (EO).

IO defines the transformation from the pixel coordinates measured on the image to the image-space coordinates referring to the focal plane. Taking a traditional aerial image as an example, typically, there are four to eight fiducial marks distributed in the corners and along the edges of the image. Their pixel coordinates can be directly measured on the image. Also, the coordinates of these fiducial marks in the image-space coordinate system are usually known. They can be used to determine the principal point (x_0, y_0) in the image-space coordinate system. They can also be used to derive a 2D transformation model between the image-space coordinates and the image measurements, and then the 2D transformation model can be used to transform any other pixel coordinates measured on the image to the image-space coordinates.

The coordinates of the principal point (x_0, y_0) and the principal distance (or focal length) f are the intrinsic parameters of the camera. The camera intrinsic parameters normally do not change. However, there are usually distortions existing on images, such as lens distortions, different pixel spacing, and stretching or shrinkage of the images. They have to be calibrated before using the images for 3D mapping. Errors in these parameters will lead to errors in the IO process and the subsequent 3D measurement. These parameters and distortions can be calibrated using a particular control field with calibration targets precisely measured by a total station or differential GPS. They can also be computed during the 3D mapping task through self-calibration approaches (Wu 2017).

EO defines the transformation from the image-space coordinates to the 3D object space coordinates, which can be formulated using the following co-linearity equations (Wang 1998):

$$
\begin{aligned}
x - x_0 &= -f\frac{m_{11}(X - X_S) + m_{12}(Y - Y_S) + m_{13}(Z - Z_S)}{m_{31}(X - X_S) + m_{32}(Y - Y_S) + m_{33}(Z - Z_S)} \\
y - y_0 &= -f\frac{m_{21}(X - X_S) + m_{22}(Y - Y_S) + m_{23}(Z - Z_S)}{m_{31}(X - X_S) + m_{32}(Y - Y_S) + m_{33}(Z - Z_S)}
\end{aligned}
\tag{23.1}
$$

The co-linearity equations connect a point (x, y) on the image and its corresponding position (X, Y, Z) in the 3D object space. (X_S, Y_S, Z_S) represent the coordinates of the camera perspective center in the object space when the image is taken. m_{ij} are the components of a rotation matrix, which is derived from three rotation angles (φ, ω, κ) of the camera frame referring to the object space. These six parameters—three positions (X_S, Y_S, Z_S) and three rotation angles $(\varphi, \omega, \kappa)$—are called EO parameters.

Each set of co-linearity equations represents a straight line that links an image point, the camera perspective center, and a 3D point in the object space. To determine the object point's 3D position, at least two straight lines are necessary to form an intersection. In other words, a pair of corresponding points measured on a stereo pair of images will be necessary to compute their corresponding 3D position in the object space. This process is called space intersection.

The EO parameters of each image can be measured by sensors (e.g., GPS and IMU) mounted on the same platform as the camera when it takes the image so that 3D measurements can be achieved by using at least two images together with their EO parameters. However, direct measurement of the EO parameters by the sensors will usually have errors and sometimes no direct measurement of the EO parameters will be provided. Therefore, in photogrammetry, the EO parameters are usually derived or improved in one of three ways: space resection, relative orientation (RO) followed by absolute orientation (AO), or simultaneous orientation through bundle adjustment.

Space resection is based on the above co-linearity equations. If three control points (their coordinates in the image-space and object space are known) are available, they offer six observations based on the co-linearity equations and provide a unique solution to the six EO parameters. Normally, more control points are used to calculate the EO parameters through the least-squares adjustment for improved accuracy. Usually, space resection is used to determine the EO parameters of a single image. For an image block, other methods are used as they require fewer control points.

RO is used to determine the internal relationship between two images. RO is able to generate a scale-free 3D model of the imaged scene within an arbitrary coordinate system. Before the 3D model obtained from RO can be used for actual measurement, it must be scaled, rotated, and translated to the actual coordinate system in object space. This is the procedure of AO. AO uses 3D transformations (e.g., 3D conformal transformation) to convert the model coordinates obtained by RO into real object coordinates. The RO and AO can be performed on a single stereo pair or on large image blocks.

23.2.2 Bundle Adjustment

Bundle adjustment (BA) is an alternative method to the above RO and AO procedures. Based on the principles of the co-linearity equations, an optical ray can be defined that starts from the image point, passes through the perspective center of the camera, and finally reaches the 3D point in the object space. This produces an observation based on the co-linearity equations. Giving some tie points matched on a stereo pair of images or multiple images, a bundle of optical rays determined by the tie points can link the images together, and subsequently link the image-space to the object space. In the ideal situation, the optical rays from the tie points on different images should exactly intersect at the same object point. However, this will usually not be true in the reality due to uncertainties and errors of different levels in the image orientation parameters. Therefore, BA is used to improve the image orientation parameters, from which the bundle of optical rays can intersect at the 3D point in the object space correctly.

BA is based on the least-squares principle. Usually, four types of observation equations can be formulated in a BA system, as listed in the following.

$$\begin{aligned} Av + B\Delta &= f \\ v_x - I\Delta &= f_x \\ A_c v_c + C\Delta_c &= f_c \\ A_{\mathrm{ap}} v_{\mathrm{ap}} + D\Delta_{\mathrm{ap}} &= f_{\mathrm{ap}} \end{aligned} \tag{23.2}$$

The first observation equation is for the image measurements (tie points matched on the images), which is based on the co-linearity equations that connect the image measurements with their 3D coordinates. Δ is the vector of the unknown EO parameters. A is the matrix of observation coefficients. B is the matrix of parameter coefficients. v is the vector of residuals. The second observation equation is for the unknown EO parameters and the 3D object coordinates of the tie points to be calculated. The third observation equation is for constraints of the parameters. For instance, a stereo camera system with a fixed camera base can provide a constraint that the distance between the three positional EO parameters of the left image and those of the right image should equal to the length of the camera base. The fourth observation equation is for self-calibration, of which the additional parameters (e.g., principal distance, lens distortions) can be solved simultaneously in the BA system.

Based on the observation equations and provided with a small number of 3D control points and a large number of tie points matched on the images, BA is able to compute the unknown parameters and the 3D object coordinates of tie points simultaneously. BA is actually the simultaneous process of space resection and intersection as described previously. In the BA system, different weights can be assigned to different types of observations based on their *a priori* precision or practical analysis, so that the contributions of different observations can be controlled. For example, observations with higher precision (less uncertainty) will be assigned with higher weights, so that they will contribute more and be adjusted less in the BA system. Observations with less knowledge (large uncertainties) will be assigned with lower weights so that they will contribute less and be adjusted more. BA is fully rigorous through corrections for systematic errors and provides abundant statistical information. The residuals of all parameters can be calculated and they can be used to evaluate the performance of BA.

23.2.3 Image Matching

Image matching is for identifying image correspondences in two or more images with overlapping coverages. The corresponding points on images represent the same point in the object space. They usually have similar appearances on different images. Generally, image matching is based on finding the similarities in grey levels of small local patches on images or matching an image patch with an image template. Image matching may be implemented on a pixel-by-pixel basis, known as dense matching, or by matching individual point or pattern features, which is called feature matching.

In the photogrammetry and computer vision communities, much research has been done regarding image matching. A straightforward image matching method is the normalized cross-correlation (NCC) matching (Lhuillier and Quan 2002). NCC directly examines the level of similarity between two small image patches or local windows by calculating their cross-correlation score in terms of the grey levels. A significant development about feature point matching is the scale-invariant feature transform (SIFT) method (Lowe 2004) in the computer vision community. SIFT first detects feature points based on the local extrema in the scale space that are invariant to scale changes and distortions, and then matches the feature points according to the descriptors constructed based on their gradients in local regions. However, SIFT only provides sparse feature matching results. Semiglobal matching (SGM; Hirschmuller 2008) is another important development in dense image matching. SGM combines global and local methods for pixel-wise matching through optimization of an energy function. SGM is able to produce dense matching results; however, the global optimization strategy used in SGM may lead to an over-smoothing problem in 3D surface reconstruction.

Wu et al. (2011, 2012) presented a hierarchical image matching method, named self-adaptive triangulation-constrained matching (SATM). SATM includes a feature matching step followed by a dense matching step. It uses triangulations to constrain the matching of feature points and edges, of which the triangulations are dynamically updated along with the matching process by inserting the newly matched points and edges into the triangulations. Dense matching is conducted during the densification of the triangulations. In the matching propagation process, the most distinctive features are always successfully matched first; therefore, the densification of triangulations self-adapts to the textural pattern on the image, and provides robust constraints for reliable feature matching and dense matching. Ye and Wu (2018) further extended the SATM algorithm by incorporating image segmentation into the image matching framework to solve the surface discontinuity problem for dense and reliable matching of images in urban areas. Figure 23.1 shows an example of the matching results using SATM and SGM for a stereo pair of aerial images for generating a digital surface model (DSM) in an urban area. As can be seen from the DSMs generated by SATM (Fig. 23.1b) and SGM (Fig. 23.1c), the former performs better than the latter in terms of feature preservation and recovery of building boundaries.

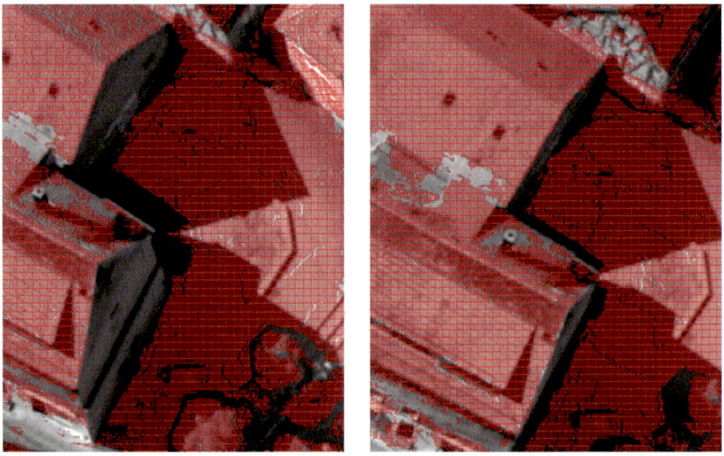

(a) A pair of aerial images with the matched results using SATM marked in red

(b) The generated DSM from SATM

(c) The generated DSM from SGM

Fig. 23.1 An example of the image matching algorithms SATM and SGM for DSM generation in urban areas

23.3 Advances in Photogrammetry for 3D Mapping in Urban Areas

Traditional photogrammetry has limited use for 3D mapping and modeling in urban areas (Qiao et al. 2010; Ye and Wu 2018). This is mainly due to the fact that traditional photogrammetry usually captures near-nadir images by cameras mounted on aircraft, and image matching in urban areas is particularly challenging. Most traditional photogrammetry systems require tremendous human labor to process images in urban areas, especially in metropolitan regions with tall buildings that are densely located. With the development of hardware and software in data acquisition and image processing in recent years, the image quality, automation degree, efficiency, and accuracy of photogrammetry have been boosted extensively in the past decade (Rupnik et al. 2015). The state-of-the-art oblique photogrammetry systems collect aerial oblique images in urban areas with high redundancy (e.g., with every ground point visible in over five or more images), which significantly improves the automatic image matching in urban areas and also provides information on building façades. Off-the-shelf solutions for 3D city modeling from aerial oblique images include two key steps: structure from motion (SfM) (Gerke et al. 2016) and multi-view stereo (MVS) (Galliani et al. 2015).

23.3.1 Structure from Motion and Multi-view Stereo

In the SfM method, feature points are used to obtain tie points between overlapped views of images automatically. For structured aerial images that are captured with designed flight plans, the connectivity between different images could be estimated accordingly. However, if the images are unordered, trying out all the possible image pairs is exhaustive for large datasets. Hence, image retrieval algorithms based on vocabulary trees (Gálvez-López and Tardos 2012) are used to find the putative image pairs that are similar and may have overlaps. After that, the initial orientation parameters are estimated and then refined by BA. BA approaches are typically divided into three categories in SfM, namely sequential, hierarchical, and global adjustment (Schonberger and Frahm 2016). Sequential adjustment methods start from a minimal image cluster (such as two or three well-connected images) and incrementally add new images to the existing clusters. The computation cost of this approach increases with each increment in reconstruction. Hence, a divide-and-conquer strategy can be adopted to reduce computation cost, which performs the BA hierarchically (Snavely et al. 2008). The scene graph is divided into several clusters first, and then these clusters are reconstructed individually. After that, these clusters are merged by a transformation with 7 degrees of freedom (DoF). Global methods normally estimate relative orientations of all the images at the same time, and estimate global rotation and translation separately (Toldo et al. 2015). However, it might be difficult for global optimization algorithms to achieve convergence, requiring good initial estimations and robust outlier detection and removal.

The resulting image orientation parameters and the scene graph of SfM serve as the foundation for the MVS (Schonberger and Frahm 2016). However, the sparse point clouds obtained by BA do not contain any solid geometry about the scene. Hence, MVS algorithms are employed to turn oriented 2D images into dense 3D point clouds using multiple images (Musialski et al. 2013). An example of the widely adopted MVS algorithm in the photogrammetry community is the patch-based multi-view stereo (PMVS) invented by Furukawa and Ponce (2010). In this method, corresponding points in multiple images are used to construct an initial set of patches to represent the scene, and the patches are repeatedly expanded to improve their density through enforcing photometric consistency and global visibility constraints to improve reconstruction accuracy. Based on the oriented images and the corresponding dense point clouds, a 3D mesh model of the surface can be reconstructed and textured using algorithms such as the Poisson reconstruction algorithm (Waechter et al. 2014), which produces watertight surfaces from oriented point clouds. Figure 23.2 is an example of automatically generated 3D models in Central Hong Kong using aerial oblique images based on SfM and MVS.

23.3.2 Integrated 3D Mapping from Multiple-Source Data

Apart from the above advances in oblique photogrammetry, there is a trend of integrating multiple-source images and laser-scanning data collected from different remote sensing platforms—for example, satellite, aircraft, unmanned aerial vehicle (UAV), and mobile mapping systems (MMS)—for better 3D mapping and modeling in urban areas (Wu et al. 2015, 2018).

Images and laser-scanning point clouds collected by different types of remote sensing platforms are widely used for 3D mapping and modeling. However, the 3D mapping results derived from different sensors and platforms usually show inconsistencies in the same area. Wu et al. (2015) presented an integrated 3D mapping model for the integrated processing of satellite imagery and airborne LiDAR data. In this model, the EO parameters of images, tie points matched in the overlapping images, and selected LiDAR points are used as inputs for a combined adjustment, and local constraints, including a vertical constraint and a horizontal constraint, are applied to ensure the consistency between these two types of data. After the integrated processing, the inconsistencies between the two types of data are reduced and the geometric accuracies of the mapping results are improved.

The integrated 3D mapping model was further extended for integrated processing of images and laser scanning point clouds collected from UAV and MMS platforms (Wu et al. 2018). Aerial oblique photogrammetry offers promising solutions for 3D mapping and modeling in urban areas. However, in metropolitan areas such as Hong Kong, where high-rise buildings are densely distributed, there are usually

(a) Aerial oblique images collected in Central, Hong Kong

(b) Automatically generated 3D models from the aerial oblique images

Fig. 23.2 SfM and MVS for automatic 3D modeling from aerial oblique images

geometric defects in the 3D models generated from aerial oblique imagery, and the textures on building façades are usually blurred. These problems are related to the common occlusion situations and large camera tilt angles of aerial oblique imagery. Meanwhile, MMS can collect ground images and laser scanning point clouds on the ground, which provides a dataset complementary to the aerial data. The integrated processing of images and laser scanning data collected from UAV and MMS platforms offers promising opportunities to optimize 3D modeling in urban areas. The integrated 3D mapping of aerial and ground datasets includes three main steps: (1) automatic feature matching between the aerial and ground images to link these two types of data; (2) combined adjustment of aerial and ground data to remove their geometric inconsistencies; and (3) optimal selection of aerial and ground data for the best textural quality and minimum occlusions. Figure 23.3 shows an example of the integrated 3D mapping from UAV and MMS images collected in Kowloon Bay, Hong Kong. Figure 23.3 indicates that the integration of aerial and ground data

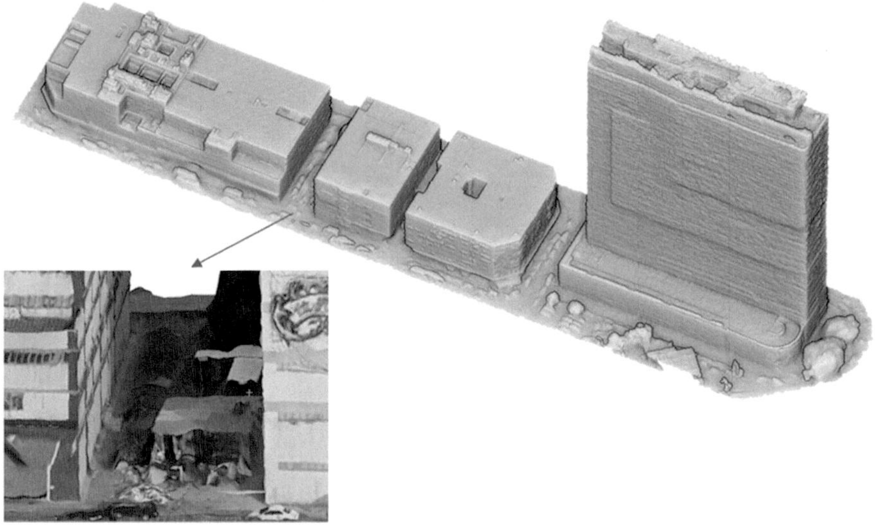

(a) 3D models from UAV images

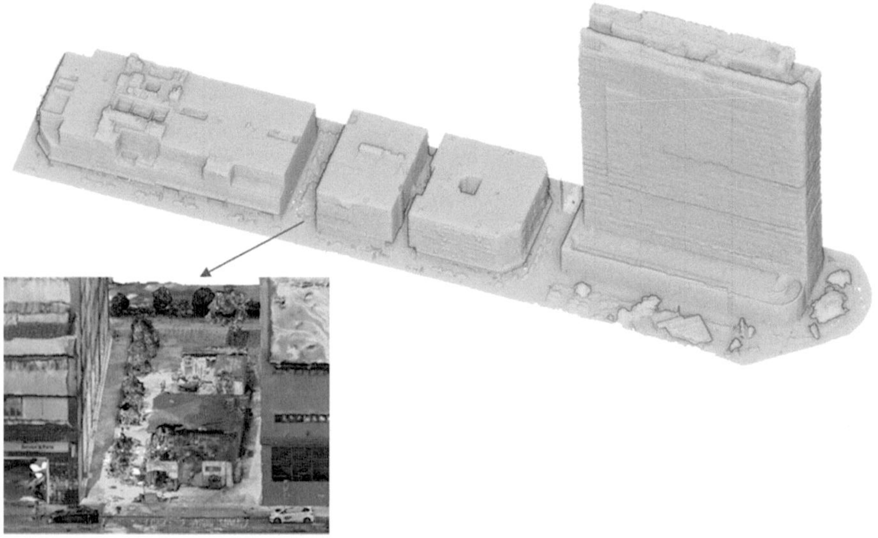

(b) 3D models from integrated processing of UAV and MMS images

Fig. 23.3 Integrated 3D mapping of UAV and MMS images in Kowloon Bay, Hong Kong shows a promising solution for generating 3D city models of the best geometry and quality. With the MMS data, the geometry and quality of the 3D mesh models at the street level are significantly improved, compared with those from aerial images only.

23.4 Summary

Photogrammetry is the most robust, efficient, economical, and flexible method for 3D mapping and modeling, regardless of the challenges ahead. Photogrammetry has been and will continue to be the representative and influential technology for obtaining 3D information. The latest advances in photogrammetry such as SfM, MVS, and integrated 3D mapping, offer great potential for optimized and enhanced 3D mapping and modeling in urban areas at both city scale and street level. Photogrammetry can be used as the primary technology to create the 3D spatial-data infrastructure for a digital city, which can be widely used to support applications in, for example, urban planning and design, urban management, urban environmental studies, and the development of smart cities.

References

ASPRS (1998) Definitions of photogrammetry, remote sensing, and GIS. American Society for Photogrammetry and Remote Sensing. https://www.asprs.org/a/society/about.html. Accessed 20 Oct 2019

Furukawa Y, Ponce J (2010) Accurate, dense, and robust multiview stereopsis. IEEE Trans Pattern Anal Mach Intell 32(8):1362–1376

Galliani S, Lasinger K, Schindler K (2015) Massively parallel multiview stereopsis by surface normal diffusion. In: Proceedings of the IEEE international conference on computer vision, pp 873–881

Gálvez-López D, Tardos JD (2012) Bags of binary words for fast place recognition in image sequences. IEEE Trans Rob 28(5):1188–1197

Gerke M, Nex F, Remondino F, Jacobsen K, Kremer J, Karel W, Huf H, Ostrowski W (2016) Orientation of oblique airborne image sets: experiences from the ISPRS/EUROSDR benchmark on multi-platform photogrammetry. Int Arch Photogramm Remote Sens Spat Inf Sci 41:185–191

Hirschmuller H (2008) Stereo processing by semiglobal matching and mutual information. IEEE Trans Pattern Anal Mach Intell 30(2):328–341

Konecny G (1985) The international society for photogrammetry and remote sensing—75 years old, or 75 years young. Photogramm Eng Remote Sens 51(7):919–933

Lhuillier M, Quan L (2002) Match propagation for image-based modeling and rendering. IEEE Trans Pattern Anal Mach Intell 24(8):1140–1146

Lowe DG (2004) Distinctive image features from scale-invariant key points. Int J Comput Vis 60(2):91–110

Musialski P, Wonka P, Aliaga DG, Wimmer M, Gool LV, Purgathofer W (2013) A survey of urban reconstruction. In: Proceedings of computer graphics forum. Wiley online library, pp 146–177
Qiao G, Wang W, Wu B, Liu C, Li R (2010) Assessment of geo-positioning capability of high-resolution satellite imagery for densely populated high buildings in metropolitan areas. Photogramm Eng Remote Sens 76(8):923–934
Rupnik E, Nex F, Toschi I, Remondino F (2015) Aerial multi-camera systems: accuracy and block triangulation issues. ISPRS J Photogramm Remote Sens 101:233–246
Schonberger JL, Frahm J (2016) Structure-from-motion revisited. In: Proceedings of 2016 IEEE conference on computer vision and pattern recognition (CVPR), Las Vegas, 26 June–1 July 2016
Snavely N, Seitz SM, Szeliski R (2008) Skeletal graphs for efficient structure from motion. In: Proceedings of 2008 IEEE conference on computer vision and pattern recognition (CVPR), Anchorage, 24–26 June 2008
Toldo R, Gherardi R, Farenzena M, Fusiello A (2015) Hierarchical structure-and-motion recovery from uncalibrated images. Comput Vis Image Underst 140:127–143
Waechter M, Moehrle N, Goesele M (2014) Let there be color! Large-scale texturing of 3D reconstructions. In: Proceedings of 2014 European conference on computer vision, pp 836–850
Wang Z (1998) The research program of the full digitized and automatic mapping system. J Wuhan Tech Univ Surv Map 23(4)
Wu B (2017) Photogrammetry—3D from imagery. In: Richardson D, Castree N, Goodchild MF, Kobayashi A, Liu W, Marston RA (eds) The international encyclopedia of geography. Wiley, New York, pp 1–13. https://doi.org/10.1002/9781118786352.wbieg0942
Wu B, Zhang Y, Zhu Q (2011) A triangulation-based hierarchical image matching method for wide-baseline images. Photogramm Eng Remote Sens 77(7):695–708
Wu B, Zhang Y, Zhu Q (2012) Integrated point and edge matching on poor textural images constrained by self-adaptive triangulations. ISPRS J Photogramm Remote Sens 68:40–55. https://doi.org/10.1016/j.isprsjprs.2011.12.005
Wu B, Tang S, Zhu Q, Tong K, Hu H, Li G (2015) Geometric integration of high-resolution satellite imagery and airborne LiDAR data for improved geopositioning accuracy in metropolitan areas. ISPRS J Photogramm Remote Sens 109:139–151
Wu B, Xie L, Hu H, Zhu Q, Yau E (2018) Integration of aerial oblique imagery and terrestrial imagery for optimized 3D modeling in urban areas. ISPRS J Photogramm Remote Sens 139:119–132
Ye L, Wu B (2018) Integrated image matching and segmentation for 3D surface reconstruction in urban areas. Photogramm Eng Remote Sens 84(3):35–48

Bo Wu is with the Department of Land Surveying and Geo-Informatics, The Hong Kong Polytechnic University. His research interests are mainly in the areas of Photogrammetry and Planetary Remote Sensing. He has worked on 3D city modeling projects, and the Mars and Moon exploration missions funded by NASA and China.

Chapter 24
Underground Utilities Imaging and Diagnosis

Wallace Wai-Lok Lai

Abstract The invisible and congested world of underground utilities (UU) is an indispensable mystery to the general public because their existence is invisible until problems happen. Their growth aligns with the continuous development of cities and the ever-increasing demand for energy and quality of life. To satisfy a variety of modern requirements like emergency or routine repair, safe dig and excavation, monitoring, maintenance, and upscaling of the network, two basic tasks are always required. They are mapping and imaging (where?), and diagnosis (how healthy?). This chapter gives a review of the current state of the art of these two core topics, and their levels of expected survey accuracy, and looks forward to future trends of research and development (Sects. 24.1 and 24.2). From the point of view of physics, a large range of survey technologies is central to imaging and diagnosis, having originated from electromagnetic- and acoustic-based near-surface geophysical and nondestructive testing methods. To date, survey technologies have been further extended by multi-disciplinary task forces in various disciplines (Sect. 24.3). First, it involves sending and retrieving mechanical robots to survey the internal confined spaces of utilities using careful system control and seamless communication electronics. Secondly, the captured data and signals of various kinds are positioned, processed, and in the future, pattern-recognized with a database to robustly trace the location and diagnose the conditions of any particular type of utilities. Thirdly, such a pattern-recognized database of various types of defects can be regarded as a learning process through repeated validation in the laboratory, simulation, and ground-truthing in the field. This chapter is concluded by briefly introducing the human-factor or psychological and cognitive biases, which are in most cases neglected in any imaging and diagnostic work (Sect. 24.4). In short, the very challenging nature and large demand for utility imaging and diagnostics have been gradually evolving from the traditional visual inspection to a new era of multi-disciplinary surveying and engineering professions and even towards the psychological part of human–machine interaction.

W. W.-L. Lai (✉)
Department of Land Surveying and Geo-Informatics, The Hong Kong Polytechnic University, Hong Kong, China
e-mail: wllai@polyu.edu.hk

W. Shi et al. (eds.), *Urban Informatics*, The Urban Book Series,
https://doi.org/10.1007/978-981-15-8983-6_24

24.1 Mapping and Imaging

One day, a patient visits a doctor describing a body pain. How does the doctor react? Will he or she immediately perform surgery or suggest a scan first to diagnose a serious health problem? Of course, the latter is the standard protocol when it comes to a doctor evaluating a patient. Unfortunately, the choice of surgery first dominates in construction work that can involve costly infrastructures such as bridges, buildings, heritage, foundations, road pavement, tunnel liners, and underground utilities. Even at home, it is not rare that someone might drill without a scan, and then inadvertently hit a gas pipe which may be damaged or even explode. An important difference between a patient and infrastructure is that a patient is more likely to take proper steps for taking care of themselves and seek out expert diagnosis, whereas the care of infrastructure which is shared by many (with most unaware of the risks and costs) is often neglected. Since the first X-ray image was captured in 1895, the diagnostic science of medicine has changed completely and become very advanced. No one would question the power of medical imaging for diagnosis and medication. But in the infrastructure world, modern scanning, mapping, and imaging methods are still not regularly practiced.

According to the Highways Department in Hong Kong, there is about 47 km of UU per kilometer of road. Such density is probably the greatest in the world. More than 20 utility companies are continually developing the underground utility network, but they occupy only the first few meters of urban underground space. In comparison with other cities, the density of underground pipelines in Hong Kong's utility network is 3.5 times greater than that of Singapore, 24 times denser than that of England, and 85 times denser than that of the United States (Wong 2014). Hong Kong and other compact cities and mega-cities probably have one of the most challenging environments for near-surface geophysical survey, mapping, imaging, and diagnosis. If the problems of UU detection in the dense environment can be solved for Hong Kong by new innovative solutions, the underground mapping problems for the rest of the world, which has the less dense underground infrastructure, will be much easier to solve.

UU accidents cause not only loss of money or valuable water resources, but also casualties such as the case of the Kaohsiung underground gas explosion in 2014 in Taiwan and the fatal Kwun Ling Lau landslide in 1994 in Hong Kong. The lack of visibility of UU and poor updating of records, in the long run, affect the design, construction, and maintenance stages of any building projects. Failures to identify the existence of UU at an early-stage can cause later design faults, leading to construction delays. The maintenance and rehabilitation of underground utilities have become difficult tasks due to the unknown location, complexity, aging, and negligence from a commonly-held mindset of "out of sight, out of mind." These factors are time bombs and increase the risk of UU damage during excavation.

In an urban area, utilities are mostly laid in a complicated manner under carriageways between buildings and pedestrian footpaths. Geophysical and non-destructive utilities surveys are always needed in the design, construction, or maintenance stages of urban development and redevelopment projects in order to avoid damage to existing UU. Several international specifications or standards are currently in use. In 2003, the American Society of Civil Engineers (ASCE) published a Standard Guideline for the collection and depiction of existing subsurface utility data in the United States (ASCE 2002). Four different quality levels for detection are stated (QL-D to QL-A), which indicate the different levels of effort required. For example, QL-D refers to a statutory record search, while QL-A refers to exposing a utility through trial holes or trenches. QL-B means geophysical surveys using equipment such as electromagnetic locators (EML) and ground-penetrating radar (GPR) (Anspach 2002).

Each of the four different quality levels represents different levels of required accuracy in defining the location of underground infrastructure. These different levels are also subdivided further into finer location requirements. There are two ways to express accuracy based on which error is greater, as shown in Tables 24.1, 24.2 and 24.3 (ICE 2014). The table indicates the reduction of location accuracy with increasing depth. Some of the higher quality levels require an absolute value of accuracy without any concern for depth. An example of the former is the British Standards Institution (BSI) which published the PAS 128:2014 standard for supplementing ASCE 38-02. Similarly, there are four quality levels for underground utility detection in the PAS 128:2014 standard. At a minimum, GPR and EML techniques are required for the quality level QL-B (ICE 2014) in Tables 24.1, 24.2 and 24.3. Another example of such expression can be found in the Competent Person Performance Monitoring Point System of the Electrical and Mechanical Services Department of the HKSAR government (EMSD). Horizontal accuracy of live power cable detection is required to be within 25% of depth. The second expression is an alternative that requires an absolute accuracy; for example, ±150 mm, ±250 mm, and ±500 mm in QL-B as shown in Tables 24.1, 24.2 and 24.3. This expression is designed for shallow utilities like telecommunication cables buried at a depth in the scale of tens of centimeters. In such cases, depth-dependent accuracy would be unnecessarily stringent, given the shallow buried depth. In terms of implementation, the quality levels used to express the accuracy of detection are somehow dependent on a clients' expectation. A recent initiative in Hong Kong established a specification with simplified accuracy levels for all types of utility detection, including pipes and cables, using only PCL/EML (LSGI 2019a). The specification also follows the rationale of both expressions of accuracy, that is, a utility survey is only declared reliable if it is within the range ±150 mm or ±15% of detected depth, whichever is greater. Uncertainties outside this range are declared unreliable. This accuracy level reflects a compromise after three rounds of consultation, and the need to balance technical constraints and expectations among different service providers, consultants, and clients of utility surveys.

Table 24.1 Quality level standard for underground utility detection in PAS 128:2014

Quality level on detection	Location accuracy		Supporting data
	Horizontal	Vertical	
Verification			
QL-A	±50 mm	±25 mm	Exposed the utility on verification
Detection			
QL-B1	±150 mm or ±15% of detected depth whichever is greater	±15% of detected depth	Horizontal and vertical location of the utility detected by multiple geophysical techniques
QL-B2	±250 mm or ±40% of detected depth whichever is greater	±40% of detected depth	Horizontal and vertical location of the utility detected by one of the geophysical techniques used
QL-B3	±500 mm	Undefined	Horizontal and vertical location of the utility detected by one of the geophysical techniques used
QL-B4	Undefined	Undefined	A utility segment which is suspected to exist but has not been detected and is therefore shown as an assumed route
Site reconnaissance			
QL-C	Undefined	Undefined	A segment of utility whose location is demonstrated by visual reference to street furniture, topographical features, or evidence of previous street work
Desktop utility records search			
QL-D	Undefined	Undefined	Desk study of the record drawings

Table 24.2 Recommended quality levels and accuracies of PCL/EML test/survey (LSGI 2019a)

Survey mode	Quality level	Location accuracy	
		Horizontal	Vertical
Active	Reliable	±150 mm or ±15% of detected depth whichever is greater	±15% of detected depth for utility buried shallower than or equal to 3 m
	Survey unreliable (SU)	Undefined	Undefined
	Survey not successful (SNS)		
Passive	Reliable	Undefined	Undefined
	Survey unreliable (SU)		
	Survey not successful (SNS)		

Table 24.3 Recommended quality levels and accuracies of GPR tests and surveys (LSGI 2019b)

Quality level	Horizontal location accuracy
Reliable	±150 mm or ±15% of detected depth whichever is greater. This accuracy level is only valid if the alignment of the utility is continuously observed in C-scan
Survey unreliable (SU)	Undefined
Survey not successful (SNS)	

24.1.1 EMI/PCL

Given the worldwide use of these specifications and standards, the quality levels and accuracies required in many projects are part of contract negotiations with the clients. The actual site constraints, such as overlaid materials and interference from neighboring utilities can impact the actual quality levels that can be achieved at a site and are not considered. For example, horizontal and vertical resolution limits of the survey are rarely studied, not to mention cases for instance, like steel bars in concrete masking the EM induction signal. Siu and Lai (2019) aims to assess such subsurface conditions as well as EM coupling effects as a major source of uncertainty in electromagnetic induction studies of UU positioning. The induced electromagnetic fields from neighboring current-carrying utilities crossing each other causes interference with the detected magnetic field, as shown in Fig. 24.1.

The results of this work can provide a reference for a better understanding of the complexity of UU mapping using EML. It provides information for UU design and survey, such as minimum clearance distances between live power-supply cables and nearby metallic utilities for the sake of later positioning.

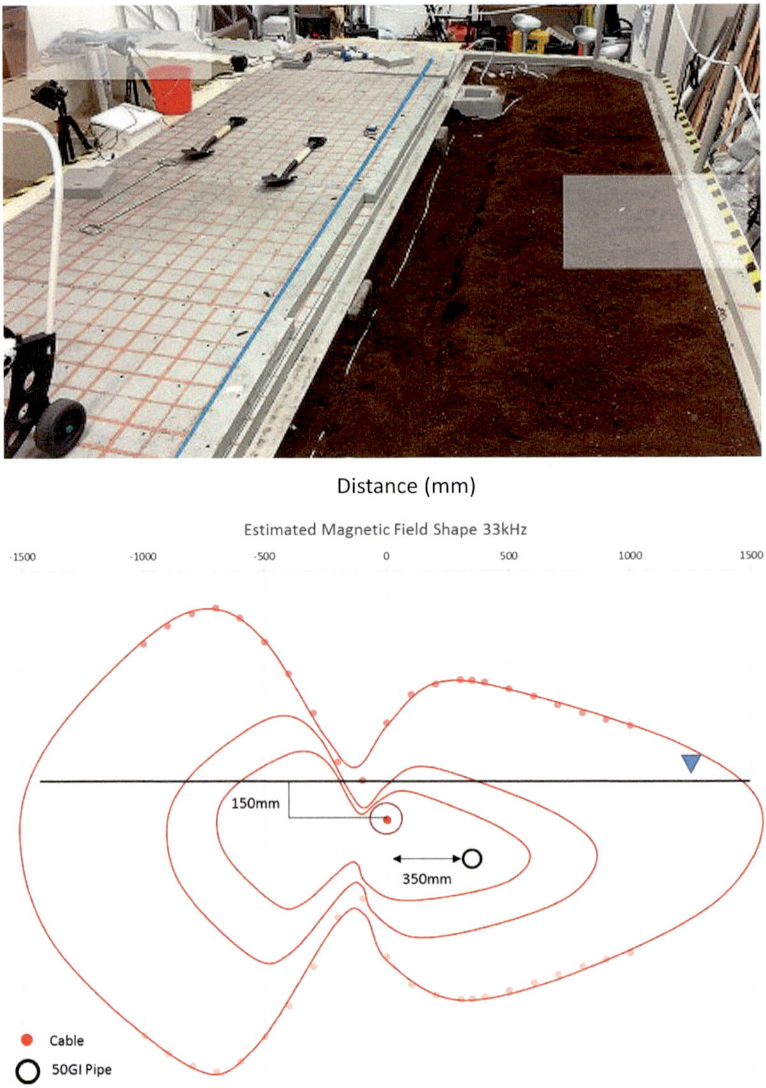

Fig. 24.1 **a** Experimental setup in HK PolyU's underground utility survey lab: *X*: horizontal separation (350, 550, 750, or 950 mm) *Y*: vertical position (150, 300, or 450 mm); **b** estimated magnetic field shape for a cable at 150 mm depth and separated horizontally from the pipe by 350 mm

24.1.2 GPR

The second means of detection is GPR, composed of a transmitter emitting and receiving radio waves in materials at a frequency of hundreds of MHz. The basic received signals are called an A-scan waveform, and B- and C-scans are used for

GPR data presentation in two and three dimensions, respectively. A C-scan images any horizontal plane at a specified depth below the ground surface. B-scan images are vertical depth sections, and both scans provide details of the reflected wave characteristics in the medium, such as phase changes, energy attenuation, and propagation velocity. These characteristics are controlled by the properties of the host medium. Through forward and inverse modeling, the subsurface world can then be reconstructed. Normally for data collection, a series of adjacent GPR profiles have to be collected in order to determine the positions and sizes of any subsurface target. 3D C-scans are increasingly useful as they provide a straightforward and easily understandable presentation. Furthermore, other forms of 3D GPR representations were developed recently, for instance, iso-surfaces, semantic images based on energy or similarity, and feature enhancements (Böniger and Tronicke 2010a, b; Leckebusch 2003). They are all derivative presentations of fully covered measurements in 3D. A sequence of high-quality C-scans with accurate geo-referencing is essential for correctly imaging underground. However, its first use was in the 1990s (Goodman et al. 1995; Lai et al. 2018a). The parameters used for the generation of slices are mainly determined by the experience of operators, leading to inevitable human bias (Millington and Cassidy 2010) because the choice of different parameter settings may result in completely different images. GPR 3D imaging has been widely applied in diverse fields of civil engineering: for example, in mapping underground utilities (Birken et al. 2002; Lai et al. 2016; Metwaly 2015); measuring change of physical properties in materials (Kowalsky et al. 2005; Léger et al. 2014; Leucci et al. 2003); and inspecting structural conditions (Alani et al. 2013; Baker et al. 1997; Lai et al. 2012, 2013). Goodman et al. (1995) summarized the processing flow of 3D time-slice reconstruction from a series of radargrams (B-scans) and focused on three major steps: setting up the survey grid, cutting slices, and interpolation, as shown in Fig. 24.2.

But a more rigorous workflow, likewise in 2D processing (Jol 2009), was developed empirically by Luo et al. (2019) after 25 sets of field and lab experiments with ground-truthing or known object arrangements. This work established a bridge connecting GPR theories and survey practice, and balance among physical principles and constraints, acceptable imaging quality, and survey workload based on the

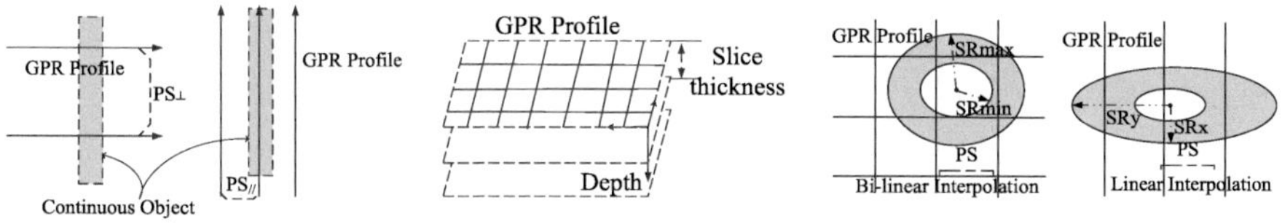

Fig. 24.2 **a** GPR profile spacing with a linear object: profile may be perpendicular or parallel to the object orientation; **b** illustration of slice thickness; **c** illustrations of profile spacing and radius of associated bilinear or linear interpolation, with SRmax and SRmin representing maximum and minimum acceptable search radii, respectively, while SRy and SRx denote the long axis and short axis of the elliptical search radius in linear interpolation, respectively (Luo et al. 2019)

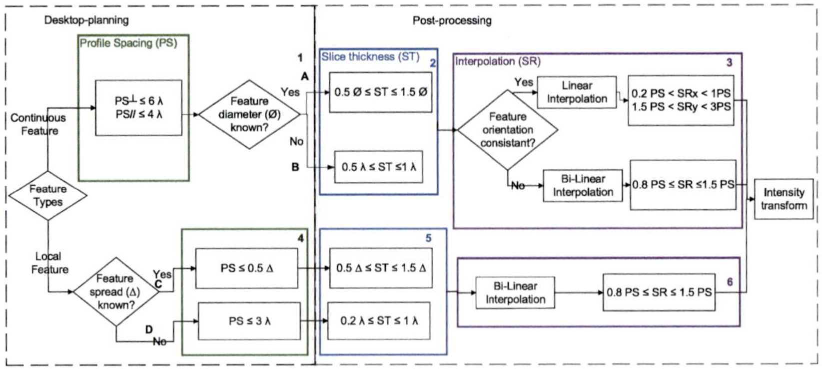

Fig. 24.3 3D GPR imaging workflow based on empirical experiments. Remarks: (1) based on Eq. (24.1), where v can be determined by common offset velocity analysis (Sham and Lai 2016), f can be determined by wavelet transform (Lai et al. 2013); (2) a feature spread (Δ) denotes feature's maximum spread along a traverse

work of Jol (2009). It is necessary because unlike remote sensing from satellite-based images, the features present in GPR responses are indeed a proxy of their true appearance. Post-processing and interpretation are needed in order to reconstruct an approximation of the real feature geometry. Basically, an underground feature can be categorized into two main groups: continuous features with linear shapes, or local features with round or irregular shapes, as shown in Fig. 24.3. Continuous reflections of linear features must appear at traverses across a series of parallel radargrams. Underground utilities and rebars in concrete are two examples of buried linear features. These linear features appear as continuous reflections in C-scan displays. Local features are non-continuous structures, such as small voids or cracks, which appear in GPR radargrams as discrete reflections. The most critical factor in identifying local features from GPR C-scans is the known or estimated feature size, and if not available, estimated GPR wavelength in the medium. A good slice imaging depends also on the adequate dielectric contrast between the two materials to record a reflection.

24.1.3 Comparison Between EMI/PCL and GPR

Two of the most important and useful EM technologies for underground mapping are EMI/PCL and GPR. Compared to the most often used mechanical waves methods such as impact echo and ultrasonic, EM-based EMI and GPR technologies are superior in terms of fast data acquisition in shallow (<6 m) underground characterization. The advantages of these methods are that they do not require physical contact with the surface during measurement, unlike mechanical wave methods, which also require much longer survey times. GPR and EMI are complementary to each other (Table 24.4).

Table 24.4 Comparison of horizontal and vertical accuracy requirement in different specifications

Quality Level	Sub-QL	Survey Method	Accuracy	
			Horizontal	Vertical
QL-D	D (ASCE38-02)	Review records, Interview	/	/
	D (AS 5488-)2013	Records, Cursory Site Inspection, Anecdotal Evidence	Indicative Location	/
	D (Malaysia)	Search/collect/analyze records	/	/
	QL-D (ICE)	Desktop utility records search	/	/
QL-C	C (ASCE38-02)	Survey and plot visible above-ground utility features	/	/
	C (AS 5488-2013)	Surface Feature Correlation and Interpretation, Site survey of visible evidence	Approximate Location	/
	C (Malaysia)	Survey surface appurtenances of utilities	/	/
	QL-C (ICE)	Site reconnaissance	/	/
QL-B	B (ASCE38-02)	Geophysical Methods	Tolerance defined by the project	/
	B (AS 5488-2013)	Survey and Trace	±300mm	±500mm
	B (Malaysia)	Geophysical Methods	/	/
	QL-B4 (ICE)	Detection	/	/
	QL-B3 (ICE) QL-B3P* (ICE)		±500mm	/
	QL-B2 (ICE) QL-B2P* (ICE)		±250 or ±40% of detected depth whichever is greater	±40% of detected depth
	QL-B1 (ICE) QL-B1P* (ICE)		±150mm or ±15% of detected depth whichever is greater	±15% of detected depth
QL-A	A (ASCE38-02)	Actual exposure and subsequent measurement of subsurface utilities	Applicable horizontal survey and mapping accuracy as defined or expected by the surveyor	±15mm
	A (AS 5488-2013)	Potholing	±50mm	±50mm
	A (Malaysia)	Excavate test holes	±100mm	±100mm
	QL-A (ICE)	Verification	±50mm	±25mm

Colour	Specification
(yellow)	(American)ASCE 38-02 Standard Guideline for the Collection and Depiction of Existing Subsurface Utility Data
(light orange)	(Australia) AS 5488-2013 Classification of Subsuface Utility Information (SUI)
(pink)	(Malaysia) Standard Guideline for Underground Utility Mapping
(green)	(UK) ICE PAS 128-2014 Specification for Underground Utility Detection, Verification and Location
* "P"	Means post-processing of signal which means GPR in the specification

24.2 Diagnosis

Utility service lives are limited due to deterioration; and proactive assessment and diagnosis are necessary before any accidents occur. However, accidents can occur without visible signs or warnings. For example, leakage from a sewage pipe or water pipe triggers soil erosion and causes a road to collapse (Hadjmeliani 2015), or the gas leak may cause an explosion (McKirdy 2014). Such problems disturb our daily life, such as the cutting off of services. Therefore, studies are necessary for developing different technologies for condition assessment and diagnosis of underground utilities. Condition assessment results help diagnosis, which is critical to maintenance schedules and rehabilitation work for underground utilities.

Thanks to the exponential growth of computation power, many technologies have been developed and used for condition assessment of underground utilities in the past decade. Some examples are (1) high-definition videos by closed-circuit television (CCTV); (2) an advanced visual method specifically for pipeline condition assessment: sewer scanning and evaluation technology; (3) acoustic methods such as sonar techniques; and more recently (4) laser-based scanning and (5) ground-penetrating radar; (6) in-line acoustic survey.

24.2.1 Ground-Based Technologies

24.2.1.1 Ground-Based Noise Logging for Leak Localization

Apart from imaging as reported in Sect. 24.1.2, GPR is also sensitive to changes in water content in the subsurface. It can detect early-stage water leakages in different pipe materials, not limited to PVC pipes and metallic pipes, as found in different lab-scale experiments (Ayala-Cabrera et al. 2011; Bimpas et al. 2010; Cataldo et al. 2014; Crocco et al. 2009; Demirci et al. 2012; Glaser et al. 2012; Goulet et al. 2013; Lai et al. 2016, 2017b; Ocaña-Levario et al. 2018). GPR is widely used as a non-destructive method for detection and mapping of buried, near-surface utilities (e.g., Metwaly 2015; Prego et al. 2017; Sagnard et al. 2016). The primary reason for GPR being used in the detection of pipe-water leakages is the mechanism of dielectric polarization, where water molecules in free form contained in a material are polarized by an incident GPR wave, thus reducing GPR wave velocity. In our present research, this mechanism is used to study underground water leakages. GPR also allows efficient and fine-resolution assessment of hazards like subsurface voids and washouts (e.g., Cassidy et al. 2011; Lai et al. 2017a; Nobes 2017). This is because the physical contact between the sensors and the objects is not required in GPR, in contrast to some acoustic methods such as leak-noise correlator or pipe cable detectors (Liu and Kleiner 2013). With the wide frequency ranges that are available, various GPR antennae allow applications addressing numerous physical properties and structures in the underground environment. GPR has been used on different pavement materials including asphalt, concrete pavements, and block pavements in road networks in most densely populated cities (e.g., Cassidy et al. 2011; Fernandes et al. 2017; Loizos and Plati 2007; Metwaly 2015; Shangguan et al. 2014; Tosti et al. 2016, 2018; Yehia et al. 2014).

The mapping of water leakage through scanning of GPR data in sliced horizontal planes is a tested approach. Because electromagnetic waves attenuate more with increasing free-water content, horizontal scans of GPR data have proven to be useful in locating leakages in water pipes in materials like sand and concrete (e.g., Lai et al. 2016, 2017b). However, the complex subsurface environment is usually densely packed with various utilities. This makes tracing the leakage or seepage of water pipes in such an environment a challenging task.

For GPR data, different velocity-estimation approaches have been proposed, including those utilizing the depth to a known reflector, velocity sounding, hyperbolic curve-fitting approaches, and estimation of GPR wave velocity assuming the value of the dielectric constant (ASTM D6432 2011). The approach of velocity analysis used in this research provides arguably a better diagnostic because it involves a comparison of wave velocities before and after the water leakage. The hyperbolic fitting method can be used to estimate GPR wave velocity from data acquired in a common offset transmitter–receiver configuration, as in ASTM D6432-11 (2011):

$$D = \frac{x}{\sqrt{\left(\frac{t_x}{t_0}\right)^2 - 1}}; \tag{24.1}$$

$$v = \left(\frac{2}{t_0}\right)\left[\frac{x}{\sqrt{\left(\frac{t_x}{t_0}\right)^2 - 1}}\right], \tag{24.2}$$

where t_x is the two-way travel time of the transmitted electromagnetic wave to the target and back to the antenna, t_0 is the two-way travel time of the transmitted electromagnetic wave to the target and back to the antenna, x is the distance between the two positions along the ground surface, and v is wave velocity (in m/ns).

Cheung and Lai (2019) compared the radargrams and velocity changes before and after the pressurized tests, to indicate if a leakage exists or not. A 10% reduction of wave velocity using a midfrequency GPR antenna (e.g., 600 MHz) is likely to be a sign of water leakage spreading upward, and a significant reverberation underneath the first arriving reflection from a buried pipe would be a sign of water leakage spreading downward. Second, for water pipes that are already in service but water leakage is suspected, if the measurements before water leakage are not available, then an examination of lateral changes in the pipeline reflections of GPR waves and changes in wave velocity would permit tracing the location of upward- or downward-spreading water leakages. This approach is based on the assumption that water leakage does not occur everywhere along the length of the pipe, and that the changes in GPR wave velocity are detectable using the equation (Sham and Lai 2016).

Noise loggers record the amplitude distribution of acoustic levels in dB. The graph record of a logger showing a sharp peak when compared with the background noise level can usually identify a point closest to the location of a possible leak. Comparing the results of multiple noise loggers at minimum flow at 2–4 am can localize the suspected leak area and extent but the exact pinpointing of the leak requires the following leak locating and pinpointing methods.

24.2.1.2 Ground-Based Leak Noise Correlation (LNC) for Leak Locating and Leak Pinpointing

A leak noise correlator is an electronic device used for pinpointing leak(s) in pressurized water or gas lines. Typically, two or more microphones or acoustic sound sensors are put in contact with the pipe at two or multiple points of access. The device records the sound emitted by a leak (e.g., a hissing noise) between the contact points by using the pipe as an acoustic waveguide. The sound data is processed to correlate the two recordings to determine the time difference that the noise takes to travel from one sensor to the others. Distance between the sensors is required to be known in advance for estimating the leak point. The cross-correlation signal of one continuous function with another is defined as

$$(f * g)(t) = \int_{-\infty}^{\infty} f * (\tau) g(t + \tau), \tag{24.3}$$

where f^* is the complex conjugate of f, and f and g are the two sound recordings of the noise produced by the leak, if any. The time delay can be found by estimating the time offset for which the cross-correlation product $(f*g)(t)$ reaches a maximum. When more than two sensors are used, the correlation process can be conducted at multiple sensor stations. This approach is accurate as long as the sound of the leak received at each sensor is adequately similar over a period of time, say a few minutes. After estimating the time delay of a leak, any leak correlators require (1) the sound travel velocity and (2) the prior measured length between the two access points, for identifying the exact distance of the leak from the sensors. For leak localization, the sound velocity depends on the size and material types of the pipe, which are standard inputs in most LNC devices. For leak pinpointing, it requires that the alignment of the pipe is determined by another method: pipe cable locating or electromagnetic locating. Leak detection is only accurate when these two methods provide a confident cross-correlation.

Leak noise correlator (LNC) and pipe pigging are widely adopted methods to detect water leakages by calculating the variances in time delay and predicting the speed of acoustic waves in the pressurized water pipe networks (Hao et al. 2012). LNC requires recording of leakage-induced noises under circumstances where sound and vibrational disturbances are negligible during the detection process. The LNC method, like all non-destructive testing methods, is limited due to a number of factors such as limited coupling of the pipe with the surroundings, inadequate pressure, and variation in pipe material and pipe size (Gao et al. 2005; Hao et al. 2012). In some cases, such as in the early-stage water leakages in gravity pipes, leaking pipes that have lost pressure or large-diameter trunk pipes, where acoustic wave transmission is considered to be unfavorable, the location of leakage points has not been possible (Gao et al. 2005; Hao et al. 2012; Liu and Kleiner 2013).

24.2.2 In-Line Technologies

In-line technologies mean putting sensors directly inside the utilities and letting the fluid (water and gas) drive the sensors automatically. These technologies avoid attenuation due to increasing depth and loss of resolution in the ground-based technologies. There are several methods for in-line condition assessment of pipelines available in the market. The following section will focus on those used by the largest group of agencies in the pipeline industry. The riskiest method is always that requiring human entry into the underground environment, and the application of the following methods reduces that need.

24.2.2.1 Closed-Circuit Television (CCTV)

Closed-circuit television (CCTV) is the commonest technique for pipeline condition assessment. The apparent advantage of CCTV is that it is a technically simple method that can directly capture illuminated images of defects on the pipe's interior wall. When necessary, the captured images can be examined in detail by further zooming the camera from different angles by controlling the tractor. CCTV was first introduced in the 1960s for the inspection of pipe interiors, and it consists of a small optical camera mounted on a tractor, which is a self-propelled platform with wheels. Nowadays, high-definition cameras permit the capture of better images for interpretation, and the system is remotely controlled by an operator on the ground surface. The natural limitation of CCTV is that it can only be applied above the water's surface, and the movements of CCTV tractor along the pipe may affect the quality of captured images (Kirkham et al. 2000). Besides, it can only determine defects that are already exposed on the surface of the pipe's interior wall. The interpretation of collected images is highly subjective, largely depending on the experience of the interpreter; any factors such as uneven and inadequate lighting may also affect the interpretation. About 2% of the main sewer network in the UK had been inspected by 2004 and at least 20% of those observations obtained by CCTV inspection were thought to be inaccurate (OFWAT 2004).

24.2.2.2 Sonar Techniques

Sonic techniques can be used to measure mass loss of exposed steel due to corrosion, and can also identify the deformation of pipes and the volume of debris inside a pipeline. The basic principle of sonar techniques is that a sound wave is excited from a transmitter, and the time for transmission and reflection is measured. The distance between the transmitter and the target can then be estimated by using the speed of sound traveling in the medium, for example, water; from this information, a sonar profiling image of a pipe's interior condition can be constructed and assessed (Hao et al. 2012). The advantage of sonar techniques is that they are not limited to pipelines that are free of fluids, which largely removes the cost of dewatering and reduces the possibility of uninspected pipelines (Schrock 1994). It is important to note that sonar images captured above and below the water surface should be constructed and interpreted separately because the traveling speeds of sound in air and water are different (Eiswirth et al. 2000).

24.2.2.3 Sewer Scanning and Evaluation Technology (SSET)

Optical scanner and gyroscope techniques were adapted for pipe-interior inspection in the late 1990s, integrated as sewer scanning and evaluation technology (SSET), and specially developed for pipe-interior condition assessment. Unlike CCTV, SSET allows defect interpretation after the device has finished running through the whole length of the pipe. There are studies in the literature on automating the assessment process in order to increase the efficiency and interpretation accuracy (Chae and Abraham 2001). Similar to CCTV, SSET also involves the interpretation of visual images collected by the device and only surface defects can be assessed. Therefore, SSET has recently been combined with other inspection techniques such as ground-penetrating radar (GPR; Koo and Ariaratnam 2006).

24.2.2.4 Laser-Based Scanning

Laser-based scanning started to be employed for pipeline inspection in the early twenty-first century. The basic principle of laser-based scanning is that it will continuously generate a laser beam, which is projected around the pipe-interior. It highlights and profiles the crown shape at each point along the pipe alignment (Read 2004). The limitation of laser-based scanning survey is that it can only be used reliably above the water surface. Recently, 3D laser scanning and modeling have been developed, which makes it possible to provide a 3D profile of the pipe (Garvey 2012).

24.2.2.5 Infrared Thermography

Sham et al. (2019) presented a first case study of customizing an in-pipe infrared thermographic system built in-house (IPITS). It makes use of thermo-images for imaging and diagnosis of pipe crown conditions in underground sewer pipelines. Active and passive infrared thermography (IRT) was attempted in two gravity sewer pipes in Singapore in July 2017. The results show that images captured with active IRT (with heating) can reveal the invisible lining defects not readily revealed by traditional visual inspection using CCTV. These defects include delamination and bubbles, water seepage, wrinkling, or construction details (like anchor knobs in the inspected HDPE material), for which sizes were estimated using an image-processing algorithm customized in an in-house program. The results are believed to pave the way for parallel inspections using a combination of CCTV and infrared cameras in composite-lined pipelines.

24.3 Future Trends of Research and Development

24.3.1 Multi-array and Fully Automated GPR

The single-channel GPR system discussed above is restricted by its limited underground footprint over a particular traverse; hence multiple traverses in x–y planes are required to generate an underground 3D image. With the advent of instrumentation and improved computer processing power, antenna arrays can be formed by aligning multiple antennae to cover a larger footprint. The advantage of this setup is that it allows the survey of a wide section in a single traverse, which can even be accomplished at highway speeds; thus it avoids tedious temporary traffic blockage and bureaucratic procedures as required in single-channel GPR imaging. Also, the configuration of the array is flexible, and spacing between antenna and number of channels can be user-defined to achieve the necessary resolution required for a survey. In addition, while the traditional pulse-GPR used a fixed center frequency and was limited to a certain bandwidth, new GPR arrays include step frequency continuous wave (SFCW) technology, which generates almost a flat response over a wide bandwidth (e.g., 10–1500 MHz). This newer setup can image satisfactorily at multiple depths and multiple resolutions in a single traverse.

24.3.2 In-Line Robotic Imaging with Micro-robots Carrying Small Sensors in Pressurized and Gravity Utilities

An increasingly popular type of in-line technology for condition diagnosis uses an installed inspection tool as an alternative for minor leaks and seepages that are not detectable by ground-based technologies in a pressurized utility, such as in-line acoustic emission (AE). When AE sensors are inserted in any pressurized water utility, leaks and defects can be detected following the same principle as the noise logger and LNC. This overcomes most of the ground-based AE's limitations and can reach the defective area directly. The in-line AE tool may consist of an acoustic hydrophone, magnetometer, gyroscope, accelerometer, and an internal power-supply, or in some cases, may employ free-swimming within the utility without power. The in-line AE tool, with appropriate water-proof and dust-proof housing, is conveyed through the utility and is driven by the flowing current without disrupting normal service. The quality of the in-line AE tool, the transport medium, and the current (water or gas) transmission velocity control the sensitivity. For an exact pinpointing of the leak or defect within the utility, the in-line AE tool is driven by the flow current, in which chainage is measured by an odometer wheel or regular time tag. The start (insertion), intermediate (tracking), and end (extraction) nodes (e.g., air valves) must be geo-referenced with GPS or topographic surveys.

24.3.3 Multi-disciplinary Research on Sensors, Robotics, Electronics, Pattern Recognition, and Change Detection

For any successful utility mapping, imaging, and diagnosis, there are three key technological elements:

Physics Sensors such as an antenna array, an induction coil, a piezoelectric device, CCD, a laser, or an echo sounder have to be designed to compromise (1) the survey purposes of imaging and diagnosis, and (2) its interaction with the utility material properties or media around the utility, such as attenuation, resolution, scattering, and environment. Results should be within reasonable ranges of uncertainty and acceptable levels of accuracy in the above-mentioned two modes of the survey: (1) ground-based technologies where the sensors and the utilities are remotely separated by materials like soil, and (2) in-line technologies where the sensors are directly driven by fluid flow along with the utilities.

Robots and electronics Current surveys have limited efficiency because of insufficient sampling of data resulting from the manual nature of the operation. For full-field utility imaging and diagnosis, the underground's confined space and its large volume of captured data require robots carrying sensors and electronics for seamless positionings, like an inertia motion unit (IMU), simultaneous localization and mapping (SLAM), and wire or wireless communication between the ground control station and the sensors.

Pattern recognition and change detection Comprehensive databases of signatures of subsurface defects are required to define defects as diagnostics for pattern recognition. Matching of physical methods and failure modes due to utilities are required: for example, GPR and void, IR and delamination, PCL/EML and pipe alignment, CCTV and surface defects, etc. With the matching defined in a database, operators will be liberated from the massive amount of data interpretation. Next, change detection enables establishment of a medical record of underground utilities with a series of time-lapse utility imaging and diagnosis, for extracting development of potential subsurface defects longitudinally rather than when failure happens. A successful pattern recognition system should be able to distinguish (1) true positive (TP; i.e., identified defects do exist) and false negative (FN; no defects identified and confirmed after ground-truthing); and (2) false alarm: true negative (TN; i.e., identified defects do not exist) or false positive (FP; defects exist but are not identified).

24.3.4 Utility Lab

An underground utility survey lab is very much in need for research on these topics. In the Department of Land Survey and GeoInformatics of the Hong Kong Polytechnic University, a lab was designed and built and has been in operation since July 2014. Scale-down networks and a matrix consisting of metallic and non-metallic fresh and saltwater supply pipes, drainage, and sewerage pipes connected with manholes, power cables, and gas cables, and valve chambers of various kinds are embedded in a big tank in the lab. These networks of underground utilities and back-filled soil serve as a scaled-down model comparable to actual field conditions. The lab provides an indoor and controllable environment where orientations, depths, sizes, material types, and coordinates of various utility networks are carefully designed and recorded. All these attributes are geo-referenced and integrated into a geographic information system.

Students and practitioners can operate various survey instruments to position and map the networks and the matrix of underground utilities and other objects, as well as to carry out condition surveys, and assessment and monitoring with advanced nondestructive instrumentation and software. The instrumentation includes ground-penetrating radar, electromagnetic induction, acoustic leak-noise correlation, noise logger, etc. The software consists of commercial and programs developed in-house, which support signal processing and multi-dimensional subsurface imaging of the collected electromagnetic, acoustic, and thermographic signals. In the lab, users can practice with the survey instruments, software, and standard survey procedures; understand what can and what cannot be done; and understand the relationships between accuracies and uncertainties of each survey method and any particular problem. Such an indoor and controlled environment enhances the confidence of students and practitioners who carry out underground utilities surveys, assessment, and monitoring in actual site environments, where most utilities are unseen and accuracies of records are not guaranteed.

The lab also serves as a hub to validate non-standardized survey methods and procedures for particular problems in two categories. The first is positioning and mapping, such as orientations, depths, sizes, and material types of utilities. The second is condition survey, assessment, and monitoring, including the effects of water leaks, subsurface voids, soil types, and moisture content, and coverage of concrete and asphalt structures for various types of survey signals. Each individual validation between any particular survey technology and any particular problem characterizes itself via the provision of the signal fingerprints. These validated fingerprints in the lab serve as a basis for pattern matching in actual field surveys. The lab is an essential step to substantiate any interpretation of imaging and diagnostic findings. The setting in the lab provides an ideal environment for such a validation process for better interpretation of positioning, mapping, condition survey, assessment, and monitoring of the very complicated and congested underground utilities in urban areas (Fig. 24.4).

24.4 Conclusion and the Way Forward

This chapter has reviewed the current state of the art's technologies of underground utility mapping, imaging, and diagnosis, and future trends of development, namely sensor physics, robotics and electronics, and pattern recognition and change detection. These are all still relatively new areas for practical imaging and diagnosis.

A literature review always tells of the successful rather than failed case studies of utility imaging and diagnostic applications in various underground problems. However, in reality, it is very normal for survey results to be less than satisfactory, especially when the introduced technologies are inappropriately carried out in commercial contracts. If one attempts to look beyond the successes, one finds that at least one or a combination of the following five factors (abbreviated to 4M1E) account for the outcome of those unsatisfactory results.

- Men (and women): qualified personnel who are trained and experienced
- Methods: procedures of data collection, processing, and interpretation
- Machines: functions, calibration, and verification in a fixed period of time
- Material: wave attenuation, resolution limits, wave scattering, etc.
- Environment: temperature, humidity, visibility, site constraints, etc.

These problems give rise to many opportunities for research and development, and can loosely be divided into the human (24.4.1) and technological (24.4.2) perspectives corresponding to 4M1E, leading to research and development opportunities.

24.4.1 Human-Factor Perspective

The first and the most important reason for the less than satisfactory cases is the first M, the staffing factor, which is more or less related to human factors and associated errors; for example, manipulation of an intensity scale for drawing favorable but not genuine conclusions. Urban geophysics for underground object imaging is becoming a regular technology, rather than one carried out by a small group of elite researchers. Its nature is similar to the function of radiographers assisting medical doctors in

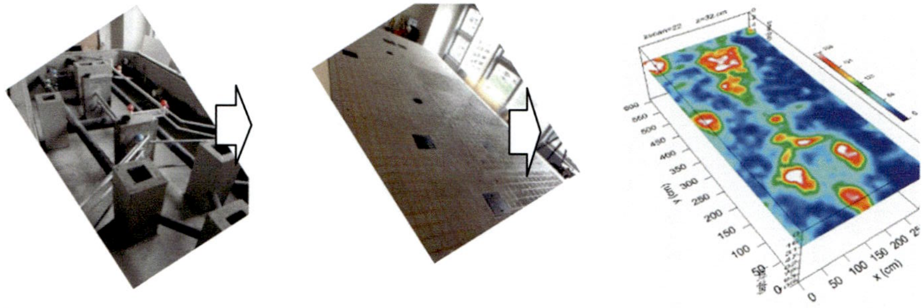

Before soil backfill After soil backfill and decking 3D subsurface imaging

Fig. 24.4 3D subsurface imaging of the buried utilities in the underground utility survey lab in The Hong Kong Polytechnic University making diagnoses for patients. But these crucial yet arbitrary functions always require indirect evidence and human judgement, which are heavily dependent on perception of the tasks and cognitive biases. They are often the least considered factor in the scientific community and in practice.

Yet they can be more important than, or at least as important as, the other uncertainties in other Ms and E. So, a blind test is the most efficient method for evaluating the capability of staff (Lai et al. 2018b). Research on the blind test's rationale aims (1) to identify and understand the common cognitive biases in the blind test systematically, (2) to investigate the effect of corresponding cognitive biases on the quality of decision-making, and (3) to establish a bias-alleviation model and guideline with de-biasing techniques specific for the blind test exercise on any tasks in utility imaging and diagnosis. In practice, regular certification and accreditation of service providers can also help to alleviate part of these problems.

24.4.2 Technological Perspective

Biases from human judgment or survey setting can be reduced but not completely, and therefore doubts arise about imaging and diagnostic purpose. Apart from multi-disciplinary hardware research (sensors, robotics, and electronics), systematic, bias-free, automatic, or semi-automatic workflow for urban underground diagnosis based on forward and inverse methods is surely the way forward. Development of methods integrated with image processing algorithms for extracting spatial and temporal features (i.e., hazards) from utility-surveying methods are of utmost importance because of the large amount of data and point clouds. The process imitates the decision-making process normally made by skilled professionals but in a semi-automatic and more robust fashion, especially when even the most skilled professionals would fall short in their ability to handling huge volumes of data.

This initiative contributes to the research, engineering, and surveying community in the following four aspects for each of the utility imaging and diagnostic methods described in the sections above. First, object- or hazard-oriented workflow for generalizing reliable images should be developed, with empirical, statistical, or learned thresholds and ranges of identified and crucial parameters. The workflow should be validated after comparing images and reality through ground truth. Secondly, the responses of underground hazards, for example, void, leak, pipe wall thinning, should be quantitatively analyzed with laboratory and fieldwork. Thirdly, a workflow integrating pattern recognition techniques should be developed to identify hazards automatically or semi-automatically and suggest rates of true positives. Last, development of a workflow is required to identify temporal changes from time-lapse datasets with change detection techniques commonly used in remote sensing, for example, k-means clustering to classify pixels into changed or unchanged. These four directions provide a gateway towards reliable and consistent imaging and diagnosis, and a basis of time-lapsed comparison with a well-established pattern recognition database. In short, this research and development direction, if implemented in practice, will establish a healthy diagnostic approach for the urban underground, so that human subjective interventions and other unfavorable factors in 4M1E are reduced as much as possible.

References

Alani AM, Aboutalebi M, Kilic G (2013) Applications of ground penetrating radar (GPR) in bridge deck monitoring and assessment. J Appl Geophys 97:45–54

Anspach JH (2002) Standard guideline for the collection and depiction of existing subsurface utility data. In: Pipelines 2002: beneath our feet: challenges and solutions, p 1

ASCE (2002) Standard guidelines for the collection and depiction of existing subsurface utility data. Reston

ASTM (2011) Standard guide for using the surface ground penetrating radar method for subsurface investigation. D6432

Ayala-Cabrera D, Herrera M, Izquierdo J, Pérez-García R (2011) Location of buried plastic pipes using multi-agent support based on GPR images. J Appl Geophys 75(4):679–686. https://doi.org/10.1016/j.jappgeo.2011.09.024

Baker JA, Anderson NL, Pilles PJ (1997) Ground-penetrating radar surveying in support of archeological site investigations. Comput Geosci 23(10):1093–1099

Bimpas M, Amditis A, Uzunoglu N (2010) Detection of water leaks in supply pipes using continuous wave sensor operating at 2.45 GHz. J Appl Geophys 70(3):226–236. https://doi.org/10.1016/j.jappgeo.2010.01.003

Birken R, Miller DE, Burns M, Albats P, Casadonte R, Deming R, Oristaglio M et al (2002) Efficient large-scale underground utility mapping in New York City using a multichannel ground-penetrating imaging radar system. Paper presented at the ninth international conference on ground penetrating radar (GPR 2002)

Böniger U, Tronicke J (2010a) On the potential of kinematic GPR surveying using a self-tracking total station: evaluating system crosstalk and latency. IEEE Trans Geosci Remote Sens 48(10):3792–3798

Böniger U, Tronicke J (2010b) Symmetry based 3D GPR feature enhancement and extraction. Paper presented at the 13th international conference on ground penetrating radar (GPR)

Cassidy NJ, Eddies R, Dods S (2011) Void detection beneath reinforced concrete sections: the practical application of ground-penetrating radar and ultrasonic techniques. J Appl Geophys 74(4):263–276. https://doi.org/10.1016/j.jappgeo.2011.06.003

Cataldo A, De Benedetto E, Cannazza G, Giaquinto N, Savino M, Adamo F (2014) Leak detection through microwave reflectometry: from laboratory to practical implementation. Measurement 47:963–970. https://doi.org/10.1016/j.measurement.2013.09.010

Chae MJ, Abraham DM (2001) Neuro-fuzzy approaches for sanitary sewer pipeline condition assessment. J Comput Civ Eng 15(1):4–14

Cheung BW-Y, Lai WW-L (2019) Field validation of water-pipe leakage detection through spatial and time-lapse analysis of GPR wave velocity. Near Surf Geophys 17(3):231–346

Crocco L, Prisco G, Soldovieri F, Cassidy NJ (2009) Early-stage leaking pipes GPR monitoring via microwave tomographic inversion. J Appl Geophys 67(4):270–277. https://doi.org/10.1016/j.jappgeo.2008.09.006

Demirci S, Yigit E, Eskidemir IH (2012) Ground penetrating radar imaging of water leaks from buried pipes based on back-projection method. Non-Destruct Test Eval Int 47:35–42. https://doi.org/10.1016/j.ndteint.2011.12.008

Eiswirth M, Heske C, Hötzl H, Schneider T, Burn L (2000) Pipe defect characterisation by multi-sensor systems. In: Proceedings of 18th international conference of no-dig, Oct 2000

Fernandes FM, Fernandes A, Pais J (2017) Assessment of the density and moisture content of asphalt mixtures of road pavements. Constr Build Mater 154:1216–1225
Gao Y, Brennan MJ, Joseph P, Muggleton J, Hunaidi O (2005) On the selection of acoustic/vibration sensors for leak detection in plastic water pipes. J Sound Vib 283(3–5):927–941
Garvey M (2012) 3D laser scanning technology benefits pipeline design: the world isn't flat. Pipeline Gas J 239(10):32–38
Glaser DR, Werkema DD, Versteeg RJ, Henderson RD, Rucker DF (2012) Temporal GPR imaging of an ethanol release within a laboratory-scaled sand tank. J Appl Geophys 86:133–145. https://doi.org/10.1016/j.jappgeo.2012.07.016
Goodman D, Nishimura Y, Rogers JD (1995) GPR (ground penetrating radar) time slices in archaeological prospection. Archaeol Prospect 2:85–89
Goulet J-A, Coutu S, Smith IFC (2013) Model falsification diagnosis and sensor placement for leak detection in pressurized pipe networks. Adv Eng Inform 27(2):261–269. https://doi.org/10.1016/j.aei.2013.01.001
Hadjmeliani M (2015) Degradation of sewage pipe caused sinkhole: a real case study in a main road. Paper presented at the Congrès Français de Mécanique
Hao T, Rogers CDF, Metje N, Chapman DN, Muggleton JM, Foo KY, Saul AJ et al (2012) Condition assessment of the buried utility service infrastructure. Tunn Undergr Space Technol 28:331–344. https://doi.org/10.1016/j.tust.2011.10.011
Institute of Civil Engineers (2014) PAS 128:2014 British standard: specification for underground utility detection, verification and location
Jol HM (2009) Ground penetrating radar theory and applications. Elsevier, Oxford
Kirkham R, Kearney PD, Rogers KJ, Mashford J (2000) PIRAT—a system for quantitative sewer pipe assessment. Int J Robot Res 19(11):1033–1053
Koo D-H, Ariaratnam ST (2006) Innovative method for assessment of underground sewer pipe condition. Autom Construct 15(4):479–488
Kowalsky MB, Finsterle S, Peterson J, Hubbard S, Rubin Y, Majer E, Gee G et al (2005) Estimation of field-scale soil hydraulic and dielectric parameters through joint inversion of GPR and hydrological data. Water Resour Res 41(11)
Lai WL, Kou SC, Poon CS (2012) Unsaturated zone characterization in soil through transient wetting and drying using GPR joint time–frequency analysis and grayscale images. J Hydrol 452–453:1–13. https://doi.org/10.1016/j.jhydrol.2012.03.044
Lai WL, Kind T, Stoppel M, Wiggenhauser H (2013) Measurement of accelerated steel corrosion in concrete using ground-penetrating radar and a modified half-cell potential method. J Infrastruct Syst 19(2):205–220. https://doi.org/10.1061/(asce)is.1943-555x.0000083
Lai WWL, Chang RKW, Sham JFC, Pang K (2016) Perturbation mapping of water leak in buried water pipes via laboratory validation experiments with high-frequency ground penetrating radar (GPR). Tunn Undergr Space Technol 52:157–167. https://doi.org/10.1016/j.tust.2015.10.017
Lai WW, Chang RK, Sham JF (2017a) Detection and imaging of city's underground void by GPR. Paper presented at 2017 9th international workshop on advanced ground penetrating radar (IWAGPR)
Lai WW, Ho ML-Y, Chang RK, Sham JF, Poon CS (2017b) Tracing and imaging minor water seepage of concealed PVC pipe in a reinforced concrete wall by high-frequency ground penetrating radar. Construct Build Mater 151:840–847
Lai WWL, Dérobert X, Annan P (2018a) A review of ground penetrating radar application in civil engineering: a 30-year journey from locating and testing to imaging and diagnosis. Non-Destruct Test Eval Int 96:58–78
Lai WLW, Chang KWR, Sham JFC (2018b) A blind test of nondestructive underground void detection by ground penetrating radar (GPR). J Appl Geophys 149:10–17
Leckebusch J (2003) Ground-penetrating radar: a modern three-dimensional prospection method. Archaeol Prospect 10(4):213–240

Léger E, Saintenoy A, Coquet Y (2014) Hydrodynamic parameters of a sandy soil determined by ground-penetrating radar inside a single ring infiltrometer. Water Resour Res 50(7):5459–5474. https://doi.org/10.1002/2013wr014226

Leucci G, Negri S, Carrozzo MT (2003) Ground penetrating radar (GPR): an application for evaluating the state of maintenance of the building coating. Ann Geophys 46(3)

Liu Z, Kleiner Y (2013) State of the art review of inspection technologies for condition assessment of water pipes. Measurement 46(1):1–15. https://doi.org/10.1016/j.measurement.2012.05.032

Loizos A, Plati C (2007) Accuracy of pavement thicknesses estimation using different ground penetrating radar analysis approaches. Non-Destruct Test Eval Int 40(2):147–157. https://doi.org/10.1016/j.ndteint.2006.09.001

Luo TX, Lai WW, Chang RK, Goodman D (2019) GPR imaging criteria. J Appl Geophys 165:37–48

McKirdy EE (2014) Dozens dead as Taiwan gas explosions tear up streets. CNN

Metwaly M (2015) Application of GPR technique for subsurface utility mapping: a case study from urban area of Holy Mecca, Saudi Arabia. Measurement 60:139–145. https://doi.org/10.1016/j.measurement.2014.09.064

Millington TM, Cassidy NJ (2010) Optimising GPR modelling: a practical, multi-threaded approach to 3D FDTD numerical modelling. Comput Geosci 36(9):1135–1144

Nobes DC (2017) Ground penetrating radar response from voids: a demonstration using a simple model. Non-Destruct Test Eval Int 91:47–53

Ocaña-Levario SJ, Carreño-Alvarado EP, Ayala-Cabrera D, Izquierdo J (2018) GPR image analysis to locate water leaks from buried pipes by applying variance filters. J Appl Geophys 152:236–247

OFWAT (2004) Out of sight—not out of mind. In: 13th report of session

Prego F, Solla M, Puente I, Arias P (2017) Efficient GPR data acquisition to detect underground pipes. Non-Destruct Test Eval Int 91:22–31

Read GF (2004) Sewers: replacement and new construction. Elsevier

Sagnard F, Norgeot C, Derobert X, Baltazart V, Merliot E, Derkx F, Lebental B (2016) Utility detection and positioning on the urban site sense-city using ground-penetrating radar systems. Measurement 88:318–330. https://doi.org/10.1016/j.measurement.2016.03.044

Schrock B (1994) Existing sewer evaluation and rehabilitation. In: ASCE manual and report on engineering practice 62

Sham JFC, Lai WWL (2016) Development of a new algorithm for accurate estimation of GPR's wave propagation velocity by common-offset survey method. Non-Destruct Test Eval Int 83:104–113. https://doi.org/10.1016/j.ndteint.2016.05.002

Sham JF, Lai WW, Chan W, Koh CL (2019) Imaging and condition diagnosis of underground sewer liners via active and passive infrared thermography: a case study in Singapore. Tunn Undergr Space Technol 84:440–450

Shangguan P, Al-Qadi IL, Lahouar S (2014) Pattern recognition algorithms for density estimation of asphalt pavement during compaction: a simulation study. J Appl Geophys 107:8–15. https://doi.org/10.1016/j.jappgeo.2014.05.001

Siu K, Lai WW (2019) A lab study of coupling effects of electromagnetic induction on underground utilities. J Appl Geophys 164:26–39

Tosti F, Benedetto A, Ciampoli LB, Lambot S, Patriarca C, Slob EC (2016) GPR analysis of clayey soil behaviour in unsaturated conditions for pavement engineering and geoscience applications. Near Surf Geophys 14(2):127–144

Tosti F, Ciampoli LB, Calvi A, Alani AM, Benedetto A (2018) An investigation into the railway ballast dielectric properties using different GPR antennas and frequency systems. Non-Destruct Test Eval Int 93:131–140

UUS-SPEC (2019a) Specification for nondestructive testing, surveying, imaging and diagnosis for underground utilities (NDTSID-UU) 1.1 pipe cable locating/eletromagnetic locating. Department of Land Surveying and Geo-informatics, The Hong Kong Polytechnic University

UUS-SPEC (2019b) Specification for nondestructive testing, surveying, imaging and diagnosis for underground utilities (NDTSID-UU) 1.2 ground penetrating radar. Department of Land Surveying and Geo-informatics, The Hong Kong Polytechnic University

Wong O (2014) A small army of specialists struggles to keep Hong Kongers safe from labyrinth of pipes and cables. South China Morning Post

Yehia S, Qaddoumi N, Farrag S, Hamzeh L (2014) Investigation of concrete mix variations and environmental conditions on defect detection ability using GPR. Non-Destruct Test Eval Int 65:35–46. https://doi.org/10.1016/j.ndteint.2014.03.006

Wallace W. L. Lai is an Associate Professor of the Department of Land Surveying and Geo-informatics at The Hong Kong Polytechnic University, and also a Visiting Scientist at the Federal Institute of Research and Testing of Materials in Berlin. His research and teaching interest is in the engineering and near-surface geophysics of the urban underground world.

Chapter 25
Mobile Mapping Technologies

Kai Wei Chiang, Guang-Je Tsai, and Jhih Cing Zeng

Abstract This chapter introduces the historic development as well as the latest progress of mobile mapping systems. First, mobile mapping technologies, including the introduction of positioning and mapping sensors, and how they can be integrated together, are briefly reviewed. Then the development of land-based, aerial, marine, and mobile portable mapping platforms is presented. The latest progress in mobile-mapping technologies is further discussed, along with sensor fusion schemes, seamless indoor and outdoor mapping strategies, and disaster response applications. In addition, this chapter explores future and potential applications, such as high-definition (HD) maps and autonomous mapping with autonomous systems.

25.1 Introduction

The recent growing market for geospatial data and its applications has increased the demand for collecting geospatial data efficiently and economically. Mobile mapping technologies, including multi-sensor integration and multi-platform mapping technology, have clearly established a modern framework moving towards efficient geospatial data acquisition for various applications such as conventional mapping scenarios, rapid disaster response, smart city, and autonomous vehicle applications. Among those applications, applying mobile mapping systems to build indoor maps for pedestrian navigation and high-definition (HD) maps for autonomous vehicles are the most popular topics driven by the booming business opportunities in geospatial communities.

Mobile mapping refers to a means of collecting geospatial data using mapping sensors mounted on a moving platform (El-Sheimy 1996). The original idea of adopting mobile mapping technologies was limited to applications that allowed the determination of exterior orientation parameters using existing ground control points. This procedure is known as georeferencing. In fact, the concept of mobile mapping has been rooted in the geomatics communities ever since photogrammetry

K. W. Chiang (✉) · G.-J. Tsai · J. C. Zeng
National Cheng Kung University, Tainan, Taiwan
e-mail: kwchiang@mail.ncku.edu.tw

W. Shi et al. (eds.), *Urban Informatics*, The Urban Book Series,
https://doi.org/10.1007/978-981-15-8983-6_25

was adopted. Research concerning mobile mapping was mainly driven by the need for highway infrastructure mapping and transportation corridor inventories in the late 1980s (El-Sheimy 1996).

Over the next decades, advances in satellite navigation and inertial sensing technology altered the development of mobile mapping in a different way. The trajectory and attitude of the mobile mapper are now determined directly, instead of using ground control points as references for positioning and orienting the images in space. The determination of time-variable position and orientation parameters for a mobile digital imager is known as direct geo-referencing (DG), which is the core ingredient of modern mobile mapping technology (El-Sheimy 1996). Figure 25.1 illustrates the evolution of georeferencing technology over the past decades.

Cameras and laser scanners or light detection and ranging (LiDAR), along with positioning and orientation sensors, are integrated and mounted on a moving platform for mapping purposes. Objects of interest can be directly measured and mapped from georeferenced images or point clouds. The most common technologies used for this purpose today are satellite positioning using global navigation satellite systems (GNSS) and inertial navigation using an inertial measuring unit (IMU). They are usually integrated to provide seamless time-variable position and orientation parameters for mobile mapping systems. Figure 25.2 illustrates the scope of mobile mapping technology, including components, platforms, and applications, respectively. Figure 25.3 illustrates the example of sensors applied by an image-based mobile mapping system and their functions, respectively.

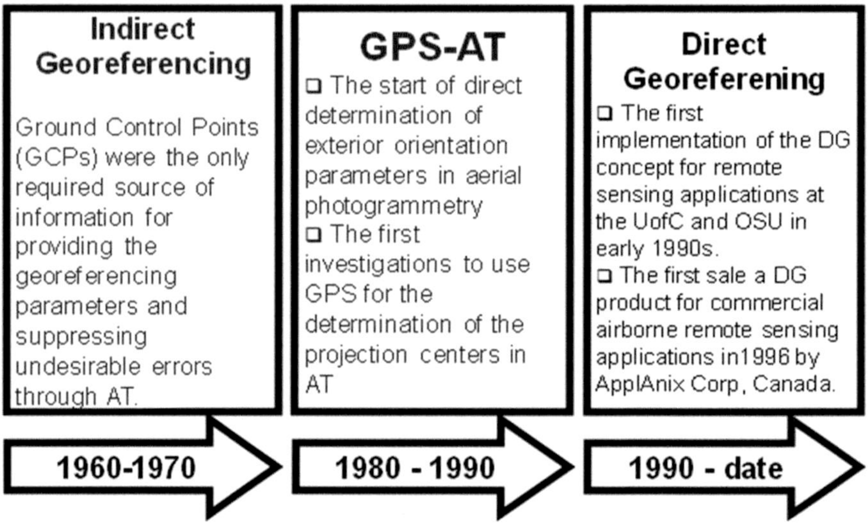

Fig. 25.1 The evolution of georeferencing technology

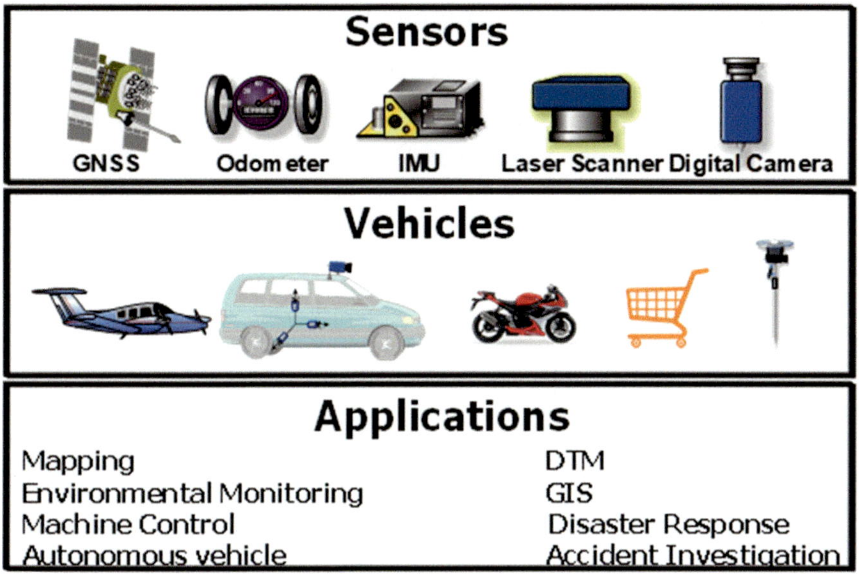

Fig. 25.2 The scope of mobile mapping technologies

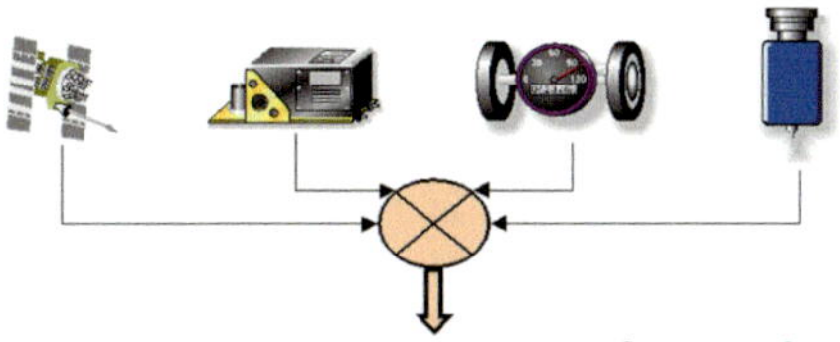

$$\mathbf{r}^m_{Point} = \mathbf{r}^m_{INS}(t) + \mathbf{R}^m_b(t)\,(S\,\mathbf{R}^b_c\,\mathbf{r}^c - \mathbf{a}^b)$$

Sensor	Primary Function	Secondary Function
GNSS	❖ Controls the INS error propagation ❖ Provides system synchronization ❖ Gives coordinates in WGS84	Provides the camera position in 3D space
INS	❖ Bridges GNSS outages and Corrects GNSS cycle slips ❖ Gives precise interpolation between GNSS fixes	Provides sesnor orientation in 3D space
Cameras	❖ Provides redundancy, i.e., more than two cameras/images of the same object ❖ Can be used to update navigation data through triangulation procedures	Two georeferenced cameras provide the position of objects in 3D space
Odometer	Updates the INS data during GNSS signal blockage	Triggers the cameras at constant distance intervals

Fig. 25.3 Sensor functions

25.2 Roadmap of Mobile Mapping Technologies

Pilot demonstrations of land-based mobile mapping technology date back to the demand for a mobile highway inventory system (MHIS) proposed by some Canadian provincial governments and US state governments in the early 1980s. Since the 1980s, at least 1000 land-based mobile mapping systems (including street-view cars) are currently practicing around the world to perform rapid geospatial information acquisition for various applications. The important milestones in this process can be divided into three stages: The first stage is the pre-INS period, from 1983 to 1993; the second stage is the post-INS period, from 1993 to 2000, and the last stage is the LiDAR period, from 2000 to the present. To meet the demands of different users, land-based mobile mapping technology has changed significantly in terms of its positioning and orientation systems over the past 30 years. The first representative system of the pre-INS era is the Alberta MHIS developed jointly by the Alberta Government of Canada and the University of Calgary (Schwarz and El-Sheimy 2008). Early land-based mobile mapping technology adopted dead-reckoning sensors such as gyroscopes, accelerometers, and odometers to derive positioning solutions using the principle of relative positioning, where in the 1980s, the imaging sensors utilized were mostly analog cameras. The images taken recorded the status of the road facilities and provide near-real-time road information for maintenance agencies. The second representative system during this period was a land-based mobile mapping system called GPSVan from the Center for Mapping at The Ohio State University. The system used the Global Positioning System (GPS) and odometers to provide navigation parameters, as illustrated in Fig. 25.4. The primary imaging sensors were two cameras that could continuously capture stereo pairs. The three-dimensional coordinates of the features were obtained by the principle of close-range photogrammetry. The positioning accuracy of GPSVan was 0.3–3 m (Grejner-Brzezinska 2001).

The representative system of the post-INS era was the VISAT series developed by the University of Calgary, Canada. The school has been developing land-based mobile mapping technology for nearly 40 years. First, the INS/GPS system was successfully integrated into the Alberta MHIS in 1994. The first generation of mobile mapping

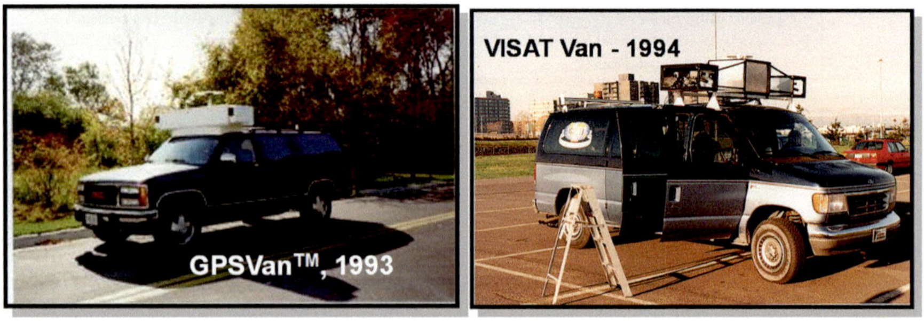

Fig. 25.4 The first land-based mobile mapping technology

technology architecture, called the first generation of VISAT Van (Shin 2005), is shown in Fig. 25.4.

The second generation of VISAT was a complete architecture, for example, INS/GPS integrated systems, odometers, and color charge-coupled device (CCD) cameras (El-Sheimy 1996). This system was the first in the world to introduce a navigation-grade INS (a gyro drift of less than 0.01°/h) using a ring laser gyroscope (RLG) with a positioning accuracy of 0.1–1 m. The system features an adjustable shooting interval at high moving speed (100 km/h). The LiDAR period began in the 2000s, and compared to the mobile mapping technology in the first two stages, the primary difference is the addition of LiDAR in the imaging-sensor component. Numerous geospatial information-associated companies around the world, such as Google, Apple, and their competitors, are adopting mobile mapping technology and building a solid digital foundation of countless exciting applications driven by geospatial information for the coming decades.

In addition to Google's sustained development of various applications based on Street View technology, Apple also implemented the development of independent mobile mapping technology in 2014 and developed the exclusive Apple Van to catch up with the progress of Google's geospatial information technology. At the same time, Finland's Nokia-funded world-class navigation map maker, Here, also developed its own mobile mapping technology, which was also acquired by Germany's three major automakers to produce accurate navigation maps to meet the demands of the automotive industry. Even Toyota exhibited a map-production technology for passenger cars at CES 2016. Therefore, mobile mapping technology plays an important role in the development of autonomous driving technology as it provides the digital world to meet the navigation safety requirements of future autonomous-vehicle applications.

The development of airborne mobile mapping technology dates back to the early 1990s, similar to the development of land-based mobile mapping technology. The important milestones can be divided into three stages as well: The first stage is the pre-INS period, from 1985 to 1995; the second stage is the post-INS period, from 1995 to 2000; and the last stage is the LiDAR period, from 2000 to the present. In the pre-INS period, many researchers in Europe and America proposed providing the orientation parameters for aircraft using a GPS multi-antenna array (Cohen and Parkinson 1992; El-Mowafy and Schwarz 1994), but the accuracy provided (0.1–0.03°) was limited by the baseline (2–10 m) of the multi-antenna array placed on the aerial survey aircraft and the solution of the GPS integer ambiguity values.

Since the early 1990s, many researchers in Europe and the United States have recognized the necessity of INS for the development of airborne mobile mapping technology (Cannon and Schwarz 1990). The earliest configuration of airborne mobile mapping technology with an INS was developed by the Department of Geomatics Engineering at the University of Calgary, Canada (Skaloud et al. 1996). Its DG accuracy without using ground control points was about 30–40 cm. The reason why the development of the airborne system lags behind the land-based system is the acquisition of high-precision INS. Most of the land-based systems developed in the early 1990s applied odometers and gyroscopes, while the demand for accurate orientation parameters using an INS for an airborne system is higher than that of a land-based system. The first land-based system using an INS was deployed in 1993. Therefore, it is not difficult to understand why the development of airborne mobile mapping technology was slightly behind that of land-based systems.

At the same time, the Center for Mapping at Ohio State University developed a similar Airborne Integrated Mapping System (AIMS) in 1998 with a DG accuracy of about 20–30 cm (Grejner-Brzezinska 2001). The operational flexibility of the DG mode was greatly enhanced, and its practical costs were considerably reduced, especially in applications where few or no ground control points are available for airborne applications. Ip et al. (2004) combined the traditional aerial triangulation using ground control points and DG to develop an integrated sensor orientation (ISO) procedure to improve the stability of airborne mobile mapping systems using limited ground control points. The last stage is the LiDAR period. Compared to the first two stages of the airborne mobile mapping technology, the main difference is the addition of a LiDAR system as an additional imaging sensor. The earliest experiment on airborne scanners dates back to the 1970s and 1980s, but only since the maturity of the data-processing and hardware technologies related to LiDAR- and INS/GPS-integrated positioning and orientation systems, have such airborne mobile mapping systems been widely applied in geomatics communities since 1996 (Axelsson 1999).

However, there are some limitations to conventional airborne mobile mapping systems. The expenses for practicing aerial photogrammetry are high, and there are strict regulations for the permits necessary to practice airborne surveys in most countries. Numerous studies have been conducted to adopt unmanned aerial vehicles (UAVs) for photogrammetry applications. For small and remote-area mapping, UAVs provide an appropriate and inexpensive platform, especially in developing countries. In recent years, more and more UAV-based photogrammetric platforms have been developed, and their performance has been proven in certain scenarios (Chiang et al. 2012).

Nagai et al. (2008) first proposed a UAV-borne mapping system using an unmanned helicopter as the platform equipped with an INS/GPS system to facilitate the DG capability, as shown in Fig. 25.5.

Chiang et al. (2012) developed a DG-based UAV photogrammetric platform where an INS/GPS integrated POS system was implemented to provide the DG capability of the platform. Rehak et al. (2013) developed a low-cost UAV for direct georeferencing. The advantage of such a system lies in its high maneuverability and operation flexibility as well as its ability to acquire image data without the need to establish GCPs.

Chiang et al. (2017) proposed a LiDAR-based unmanned aerial vehicle (UAV). The UAV integrates an IMU, a GNSS receiver, and low-cost LiDAR, as illustrated in Fig. 25.6. An unmanned helicopter was introduced, and a multi-sensor payload architecture for direct georeferencing was designed to improve the capabilities of the vehicle.

The development of shipborne mobile mapping technology dates back to 2005 (Zach et al. 2011). Its primary system architecture follows that of the land-based mobile mapping system and adds a stabilizer function to overcome the walrus' accuracy. Zach et al. (2011) applied a shipborne system using the RIGEL VMX-250 with

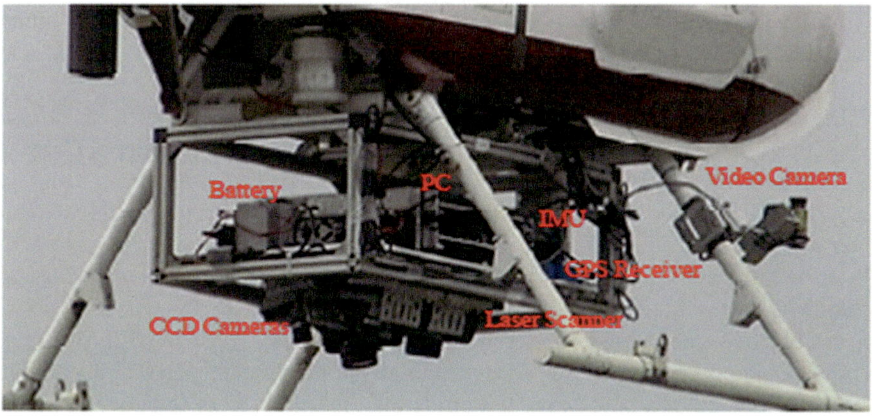

Fig. 25.5 An example of a DG-ready UAV helicopter-based photogrammetric platform. Adopted from Nagai et al. (2008, p. 1217)

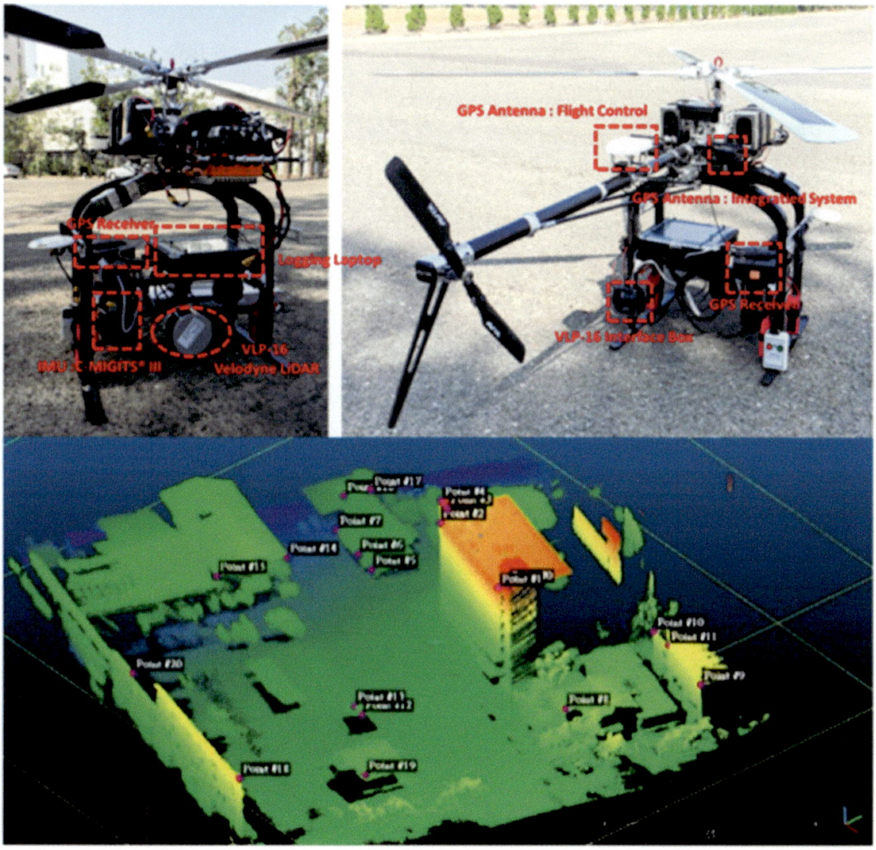

Fig. 25.6 An unmanned helicopter-based LiDAR mapping system

a GNSS receiver and a tactical-grade IMU to scan the relevant monuments along a canal in Venice, Italy. The objects on both sides of the river were scanned and recorded along the driving track.

The development of portable mobile mapping technology can be traced back to the early 2000s. The Department of Geomatics Engineering at the University of Calgary in Canada developed a prototype of a lightweight and low-cost personal mobile mapping system. The DG horizontal positioning accuracy of the system without control points was about 20 cm, and the vertical positioning accuracy was about 10 cm (Ellum 2001). This prototype utilized a digital magnetic compass instead of an IMU to provide attitude information; however, a digital magnetic compass is vulnerable to magnetic-field interference in urban areas and is unstable (Ellum 2001). A portable mapping system is especially beneficial for disaster response applications. The disadvantage of a land-based system is the discontinuity of image acquisition due to the limitations of road-network connections in some narrow lanes. Therefore, portable mobile mapping systems are designed to cope with such situations, as illustrated in Fig. 25.7.

Fig. 25.7 Example of portable mobile mapping systems

25.3 Recent Progress on Mobile Mapping Technology

A mobile mapping system comprises digital imaging systems, positioning and orientation systems, and various practicing platforms and application scenarios, as illustrated in Fig. 25.1. On the other hand, the development, hardware cost, and accuracy requirements of mobile mapping systems are highly correlated. In recent years, due to increasing demand for automation of mapping processes in the geospatial information industry, mobile mapping systems have gradually become commercially viable products, since the prototype development stage performed by professional research institutions before 2005 enabled an innovative solution in the geospatial information industry. Besides, the robotics industry also extensively applies similar concepts and sensors to develop perception technologies to navigate robots in unknown environments. Compared with the current mobile mapping systems developed by the geospatial information industry, the environmental perception technology developed by the robotic industry has the advantage of low prices, but its accuracy is not sufficient to meet the demands for geospatial applications. The development of mobile mapping technology in these two areas will definitely stimulate a lot of interest and further expand the penetration of geospatial information in other communities. Therefore, mobile mapping technology will continue to evolve based on the fundamental requirements of users, who are pursuing lower hardware costs, higher accuracy, and higher profits. Therefore, future development trends can be discussed according to the evolution of different levels of digital imaging systems, positioning and orientation systems, different operating platforms, and application scenarios.

25.3.1 *Digital Imaging Systems*

Current mobile mapping systems have fully adopted image sensors for digital electronic components. These image sensors include digital cameras using image frames, multi-spectral line scanners using line-scan technology, and optical and IFSAR/INSAR. The development of mobile mapping systems is closely related to the progress of digital imaging technology. Among imaging sensors, the evolution of image-based digital cameras has played the most important role. These cameras are in line with the development of LiDAR mobile mapping systems, but due to the limited resolution of CCD cameras used in the 1990s, these CCD digital cameras were used for land-based mapping systems because the distance of effective measurement in a land-based scenario is much smaller than the altitude requirements of airborne applications.

In recent years, the resolution and image size of CCD cameras have gradually improved. Numerous high-performance digital reflex cameras with single lenses have been developed and tested for airborne mobile mapping systems, and the results are quite encouraging. The advantages of using a digital camera are obvious. The user does not have to scan a film negative to improve mapping efficiency; the digital image processing technology improves the automation of feature extraction, and the updating and storage of digital images are easier.

In the evolution of these digital imaging systems, the IFSAR airborne mapping system has received more attention in the geospatial information community in recent years (see Chap. 21). It is characterized by rapid deployment, a nearly weather-free operation mode, and effective penetration of clouds. Another important development of digital imaging technology with an airborne mapping system is the airborne hyperspectral imaging system. Through a combination of different spectral images, many important features can be derived to provide environmental monitoring, mining exploration, vegetation inspection, disaster prevention, and land-resource management.

Recently, sensors adopted in low-cost mobile mapping systems have gradually been replaced by Kinect's depth cameras. For indoor scenes, such systems have the advantage of being low cost and offering mass production for the consumer market. Google and Apple are competing to develop inertial sensing, depth cameras, and CCD cameras to create indoor 3D models with mobile devices.

25.3.2 Positioning and Orientation Systems

GPS is a navigation satellite positioning system developed by the United States in the late 1970s. Currently, 32 satellites operate in orbits about 20,000 km from the Earth's surface. Since the design has been around for 30 years, the United States has implemented a GPS modernization plan, adding new, improved quality measurements to meet the demands of the coming years. More importantly, the GPS modernization plan upgrades the original dual-frequency system to a tri-frequency system.

In 2001, the Russian government decided to continue to maintain the operation of GLONASS and proposed a plan similar to GPS modernization. The program added 24 new satellites by the end of 2010 in order to provide accurate navigation services worldwide. Like the modernized GPS, the future GLONASS can provide tri-frequency civilian signals for accurate positioning, navigation, and time-related applications.

The Beidou Navigation Satellite System is the GNSS developed by China. It is committed to providing fine-precision positioning, navigation, and time services to users around the world, and can further provide services to authorized users with high accuracy requirements for both military and civilian users.

The Galileo system is the GNSS built by the European Union. After the US GPS, Russia's GLONASS, and China's Beidou system, it is the fourth system to provide civilian global satellite navigation services. The primary purpose of the Galileo system is to provide civilian navigation, which is different from the three systems mentioned earlier.

The GPS Block IIF satellites and the new generations of GPS III that are currently being launched are capable of transmitting tri-frequency signals, and the GLONASS-M and the GLONASS-K introduced after 2014 have also added the third frequency. After the completion of the Galileo and Beidou systems, the multi-frequency observation using multi-system GNSS is bound to bring higher satellite visibility and improved accuracy to mobile mappers around the world. In the future, whether it is real-time kinematic positioning for navigation purposes or post-processing kinematic or static-baseline solutions for geodetic requirements, users can use multi-system GNSS receivers to enjoy better positioning results. It is expected that after 2020, a general user will be able to use the multi-frequency measurements provided by GNSS to achieve improved positioning accuracy.

At present, e-GPS or e-RTK technology for kinematic positioning with virtual reference stations has been widely used in the geomatics community. For mobile mapping applications, the real-time information transfer for high-speed motion platforms required by RTK is a challenge; therefore, e-GPS or e-RTK is not a viable option for mobile mapping applications at the present time. Therefore, in the future, in terms of the multi-sensor positioning and orientation software used in the mobile mapping system, determining how to achieve differential kinematic positioning using GNSS virtual reference stations in the post-processing architecture is an important issue.

The development of the mobile mapping system was highly correlated with the development of strapdown inertial sensing technology. From a DG perspective, there would be no booming mobile-mapping-related industries without the advancement of inertial sensing technology. In principle, an IMU has three gyroscopes and accelerometers, and it provides compensated raw measurements, including velocity changes and orientation changes in three directions of its body frame. Those who require real-time navigation solutions with the use of an IMU require an external computer that has inertial navigation mechanization algorithms. On the other hand, an INS is an IMU combined with a navigation computer to provide navigation solutions in the chosen navigation frame directly in real-time. In addition, it also provides compensated raw measurements. Therefore, the main distinction between an IMU and INS is the ability to provide real-time navigation solutions. The former only provides compensated inertial measurements while the latter can provide real-time navigation solutions as well as compensated inertial measurements.

For mobile mapping system applications, the standard operating procedure in the calculation of the precise positioning and orientation solution through the post-processing procedure. Taking the same measurements as an example, in the same GNSS signal outage period, the positioning accuracy obtained by the post-processing software using smoothing algorithms is nearly 60% better than the real-time solution with filtering algorithms. Therefore, the IMU is suitable for mobile mapping applications.

In recent years, the rapid evolution of inertial sensing technology using micro-electro-mechanical systems (MEMS) has led to another advance in the sustainable development of mobile mapping technology. The MEMS IMU is low cost and provides acceptable performance compared to an IMU with a fiber optic gyroscope

(FOG) with the same specifications. The price is only one half of its counterpart with FOG, and the stability of the MEMS IMU will continue to improve over time. At present, MEMS IMUs with gyroscopes with a drift of 0.5°/h are available for mobile mapping applications.

25.3.3 Sensor Fusion Algorithms

The Kalman filter (KF) approach has been widely recognized as the standard optimal estimation tool for current sensor-fusion schemes. However, the major inadequacy related to the utilization of KF for sensor fusion is the necessity to have a predefined accurate stochastic model for each of the sensor errors. Furthermore, prior information about the covariance values of each sensor measurement as well as the statistical properties (i.e., the variance and the correlation time) of each sensor system must be accurately known (Schwarz and El-Sheimy 2008). Furthermore, for mobile mapping applications (where the process and measurement models are nonlinear), the extended Kalman filter (EKF) operates under the assumption that the state variables behave as Gaussian random variables. Naturally, the EKF may also work for nonlinear dynamic systems with non-Gaussian distributions, except in the case of heavily skewed nonlinear dynamic systems, where the EKF may experience problems (Chiang et al. 2009).

When compared to real-time filtering, post-processing has the advantage of utilizing an entire data set to estimate a trajectory. This is not possible when using filtering because only a fraction of the data is available at each sample instance. When filtering is used in the first step, an optimal smoothing method, such as a Rauch-Tung-Striebel (RTS) backward smoother, can be applied (Chiang et al. 2009). For most of the surveying applications that require superior accuracy, only data acquisition has to be implemented in real-time, and data processing and analysis are post-processed. The procedures for general mobile mapping applications include data acquisition, georeferencing, measurement, and GIS processing. Only real-time data acquisition is desired for acquiring IMU, GNSS, CCD image data, and LiDAR point clouds. For georeferencing processes that put position and orientation stamps on images, and measurement processes that obtain 3-D coordinates of all important features and store them in a GIS database, only post-mission processing can be implemented based on the accuracy requirements of these processes (El-Sheimy 1996).

According to Chiang et al. (2009), the development of the multi-sensor fusion algorithms for mobile mapping applications can be divided into the following categories:

- Sampling filter approach: The main feature is to establish an error dynamic model and sensor error model based on the statistical characteristics according to the concept of the traditional KF; the nonlinear INS/GNSS integration problem is linearized when the KF is used. On the contrary, most of these new sampling filter algorithms use nonlinear models to deal with navigation and positioning problems.

The traditional KF provides the best solution for the approximate model, and such sampling filters can provide approximate solutions for accurate models.

- Artificial intelligence approach: The main common feature of such algorithms is to establish nonlinearity by imitating human learning, where the dynamic models are approximated with artificial intelligence.
- Hybrid approach: Such fusion algorithms mainly combine the current KF smoother-based algorithms with AI to develop a hybrid algorithm.

25.3.4 Collaborative Mobile Mapping Schemes

The shortcomings of airborne mobile mapping technologies are similar to those of traditional aerial survey technologies such as weather dependence and limitations related to operating ranges. Compared to traditional surveying technologies, land-based mobile mapping technologies are less intrusive and provide better efficiency in geospatial information acquisition. While the land-based mobile mapping system can operate under poor weather conditions, it is sensitive to the quality of the GNSS signal, and its operating environment is also limited by the existing road network. The mobility of portable mobile mapping technology is much higher than the other two referred to above, and it has better operating flexibility.

Land-based mobile mapping systems can conduct control surveying, surface feature collection, rapid mapping, and image-database updating. The ability to directly georeferencing an image with an airborne mobile mapping system can provide the features of the surface entities under observation. Through the images provided by the vehicle, the user can quickly complete the mapping process and establish a large volume of attribute data required by the GIS for further analysis. At the same time, the portable system provides fast property updates to maintain the correctness of terrain features and database properties. In other words, mobile mapping technologies with collaborative mapping schemes are able to complete the mapping process rapidly, compared to a large amount of manpower and cost required to perform the same task using an aerial survey or geodetic survey. Therefore, the savings in manpower and operational costs are considerable with collaborative mobile mapping schemes. Figure 25.8 illustrates an example of collaborative mobile mapping with airborne and land-based mobile mapping technologies.

25.3.5 Mobile Mapping Technology for Rapid Disaster Response Applications

In recent years, numerous natural disasters have occurred due to drastic climate changes at the global level. It is very important to rapidly obtain geospatial information in disaster areas to provide subsequent analysis and decision-making. In this situation, collaborative mobile mapping technology can provide sufficient capacity to

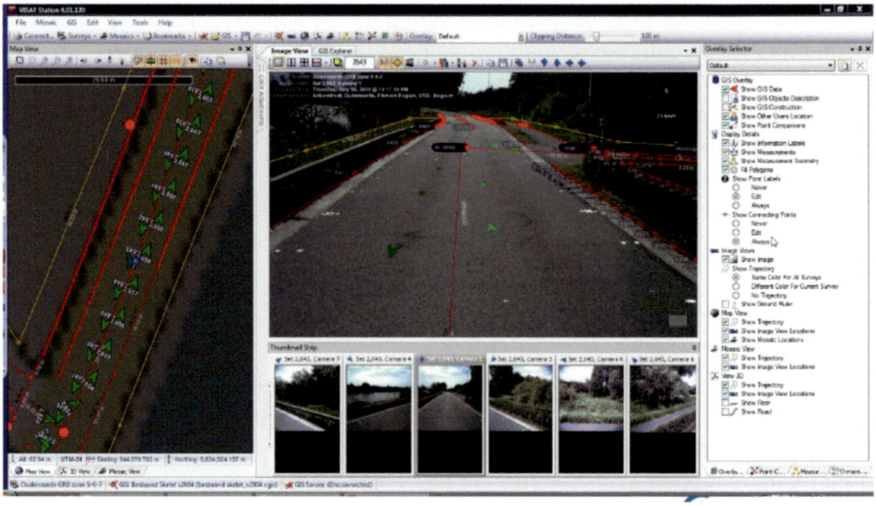

Fig. 25.8 An example of collaborative mobile mapping

solve this problem. Therefore, the development of low-cost, high-mobility mapping systems for timely intelligence acquisition and processing for disaster response is an attractive research theme among the geomatics community.

Satellite imagery has many limitations, such as weather conditions, overlap percentages, spatial and temporal resolution, and price. Aerial vehicles such as airplanes, helicopters, hot air balloons, and unmanned aircraft are relatively inexpensive options, especially with the recent development of airborne mobile mapping technology. Unmanned aerial mobile-mapping systems have high mobility in small areas. In the case of post-disaster rescue and assessment, they can be used to provide timely information that is necessary to cope with emergency situations. Today, high-resolution satellite imagery is still used to improve disaster response and relief. However, unmanned aerial vehicles are the best choice for small-area surveys, especially in developing countries.

On the one hand, mobile devices are popular, and their built-in sensors are quite suitable for certain mobile mapping applications. They usually include GNSS receivers, IMUs, and high-definition cameras. Mobile devices have the advantages of being low cost and popular compared to the classic mobile mapping systems, thus providing considerable convenience for rapid data acquisition missions, as shown in Fig. 25.9. The achievable 2D positioning accuracy of the smartphone mobile mapping system shown in Fig. 25.9 using commercial smartphones is around 1 m with object distances ranging from 10 to 15 m.

Such devices are suitable for disaster response applications with low accuracy requirements because their high penetration rate can efficiently accelerate disaster relief efforts. Therefore, future of mobile mapping technologies utilizing mobile devices will have considerable economic benefits and business potential.

Fig. 25.9 Smartphone mobile mapping technology

25.3.6 Mobile Mapping Technology for Indoor Mapping Applications

Geospatial information is becoming increasingly popular with the penetration of mobile devices into daily life. With the expanding demands of location-based services (LBS), the geospatial information industry's attention is shifting from outdoor to indoor environments. In buildings, more business opportunities can be discovered at the same time. Google, Microsoft, and their competitors around the world are showing high interest in indoor mapping and navigation applications. Google is currently implementing indoor business maps in the United States, Australia, Japan, and Taiwan, which has aroused high interest within the industry. However, the biggest technical challenge of indoor mapping systems lies in the lack of a unified source of maps, unlike an outdoor map, which can be obtained through the existing collaborative mobile mapping systems. Another major problem is the frequency of updating indoor maps. For example, counters in department stores change frequently, resulting in maintenance difficulties. The main methods of building indoor maps include the use of architectural blueprints or traditional surveying processes, but this method is time-consuming and laborious, and it is difficult to achieve the relevant standards. Therefore, the application of collaborative mobile mapping can be extended to the development of indoor mobile mapping technologies, such as the use of pedestrians and strollers as platforms for indoor mapping applications. Figure 25.10 illustrates a map of indoor parking lots produced with an indoor mapping cart that has electric power. The 3D positioning accuracy of this map is 30 cm.

In addition, LiDAR-based indoor mapping platforms can be applied for underground environmental exploration in the field of mining as well as underground facility inspections.

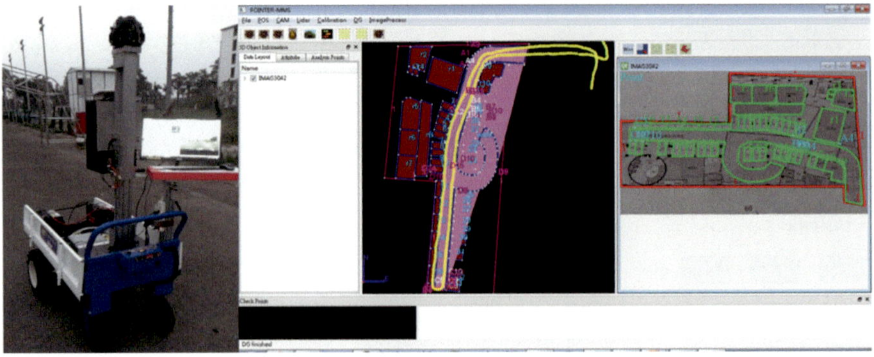

Fig. 25.10 Indoor mobile mapping technology

25.3.7 Mobile Mapping Technology for Autonomous Vehicle Applications

Autonomous driving vehicles, or self-driving cars, have made enormous progress in recent years. According to the classification method proposed by the Society of Automotive Engineers (SAE) International, the driving system can be divided into six levels. The first level (Level 0) is the most primitive system. The driver controls the mechanical and physical functions of the vehicle without any automatic driving intervention. In order to improve the overall driving feeling and driving safety, individual functions or devices, such as the electronic stability program (ESP) or anti-lock braking system (ABS), are added to improve driving safety. This system can be upgraded to Level 1; high-intermediate model vehicles are mainly controlled by the driver, but additional automation functions are added to reduce the user's operating burden. For example, the adaptive cruise control (ACC) system automatically adjusts a safe distance from vehicles ahead and warns about lane departures. The autonomous emergency braking (AEB) system combines blind-spot detection and the technologies of the collision avoidance system to reduce vehicle accidents caused by collisions. The system belongs to Level 2. Level 3 is conditional automation, that is, the driver must still be involved at any time in case of emergency; Level 4 or above is a fully automated driving category; and Level 5 has the best car communication system for communication between vehicles. However, in order to achieve a fully autonomous driving level, self-driving cars still face the following three major challenges:

- Autonomous vehicles must know their location and navigation information
- Overcoming the problem of in-vehicle sensors on autonomous vehicles that cannot be perceived due to obscuration or distance
- Connecting the autonomous vehicles with other vehicles to ensure road safety.

In order to achieve Level 4 or higher functional safety, obtaining the precise position information of the vehicle on the road is the most basic requirement for autonomous vehicles to be able to drive on the correct road in a known environment. In addition, according to advanced vehicle-safety research, if navigation equipment needs to be upgraded to the level of autonomous driving, it is necessary to improve the navigation accuracy of the vehicle to the sub-meter level or higher. Due to the limited shading or reflection of satellite reception in urban areas, autonomous vehicles cannot be accurately positioned in the right lane. With advances in computing and sensor technologies, onboard systems, the integrated system of cameras, LiDAR, GNSS, INS, and other perception sensors, can deal with a large amount of data and achieve real-time processes continuously and accurately. These systems also handle several specialized functional schemes such as positioning, mapping, perception, motion planning, and control. These key components are essential for the vehicle to achieve fully autonomous operation. On the other hand, taking the safety and hardware costs into considerations, the maps with navigation information for autonomous vehicles can provide reliable and robust prior information on the environment. The maps are called HD maps and are essential for the operation of autonomous driving technology.

Compared with the 2D digital navigation maps based on human visual viewpoints, autonomous vehicles need to make real-time decisions through map feedback during driving to allow passengers to reach their destinations safely. HD maps provide detailed map information for navigating autonomous vehicles to ensure navigation safety. The map itself serves as an additional pseudo-sensor in the car and significantly enhances the performance and accuracy of the perception and positioning algorithms necessary for the vehicle to drive autonomously. The difference between HD maps and current 2D digital navigation maps is that the use of the map is transferred from a person to a machine. The mapping accuracy and the road attributes on the map, and even the geometrical relationships of lanes, traffic signs, and roads, must be precisely defined to meet the safety requirements of autonomous vehicles. Thus, the current mapping specifications for producing navigation maps can no longer meet the needs of production, maintenance, and inspection in the case of HD maps. The conditions and definitions required for HD maps are given below:

- HD maps need to achieve sub-meter accuracy or better.
- All map information must be in 3D with sufficient accuracy.
- Features (including lanes, road boundaries, traffic signs, etc.) in the real world must be clearly defined on the map, and detailed attribute data should be attached.
- The scale of the HD maps must be consistent with the real world; that is, there can be no tolerance for scale problems.
- The maps must provide dynamic map information for the vehicle to make driving decisions.

Thus, the navigation system can accurately guide the vehicle and handle the situation, such as the non-planar places, viaducts, and underpasses. Figure 25.11 shows the difference and accuracy requirements of the digital map used by the land vehicle system, the ADAS map used by the advanced driver assistance system, HD maps for autonomous vehicles, and the requirements of accuracy.

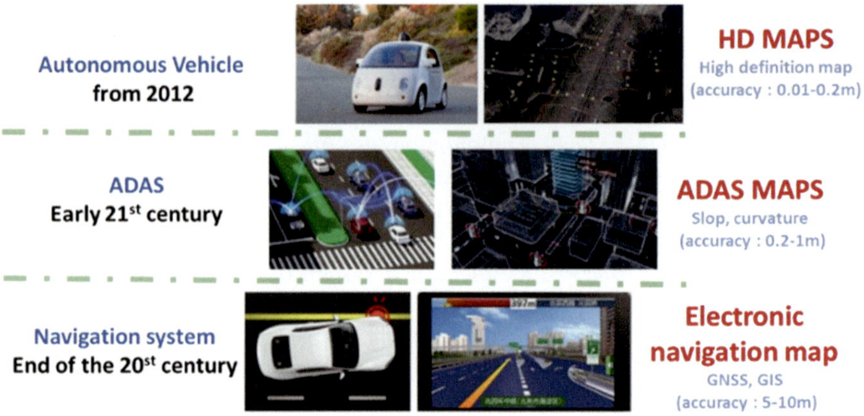

Fig. 25.11 Difference between existing navigation maps and HD maps

To produce HD maps, multi-sensor integration schemes are necessary to perceive the surrounding scenes, which can be divided into active and passive sensing components. Active components will actively emit laser waves to obtain the distance from the target. As in LiDAR and radar, it is more limited in terms of range but is less sensitive to the external environment. Passive sensors only need to receive external information, such as integrated navigation devices with GNSS, IMU, and visual odometers that use cameras to navigate. Multi-sensor integrated schemes are most commonly used in stationary terrestrial laser scanners (STLSs), mobile terrestrial laser scanners (MTLSs), and aerial laser scanners (ALSs). Their characteristics are illustrated in Table 25.1. Among them, the accuracy of STLS is consistent with HD map production, but the cost of practicing mapping and collecting road information in a large area with STLS is too high; the ALS can be free of road obstacles to complete the collection of urban HD maps, but it is still dangerous to fly in cities with a lot of high-rise buildings, and its resolution is not sufficient for producing HD maps; therefore, the most suitable option for an HD map production scheme is MTLS. Google, Apple, Here, and their competitors around the world are applying land-based mobile mapping technologies with MTLS to map the high definition digital world for autonomous vehicles (Fig. 25.12).

25.3.8 The Latest Developments of HD Maps for Autonomous Driving Applications in Taiwan

The 3D coordinates of lane markers, traffic signs, and other relevant parameters, such as curvature and slope, in HD maps, are essential for controlling driving behavior.

Table 25.1 Sensor matrices for building HD maps (after Farrell et al. 2016)

	Technology	Purpose for inclusion in sensor suite	ECEF accuracy	Feature detection capability
Individual sensor technologies	INS	Bandwidth, sample rate, continuity	N/A	No
	GNSS	ECEF accuracy	cm	No
	Camera	Feature detection and photolog	N/A	Yes
	LiDAR	Feature detection, accurate feature georectification	N/A	Yes
Sensor suites	STLS	GPS, camera, LiDAR	cm	Yes
	MTLS	INS, GPS, camera, LiDAR	cm	Yes
	ALS	INS, GPS, camera, LiDAR	Submeter	Yes

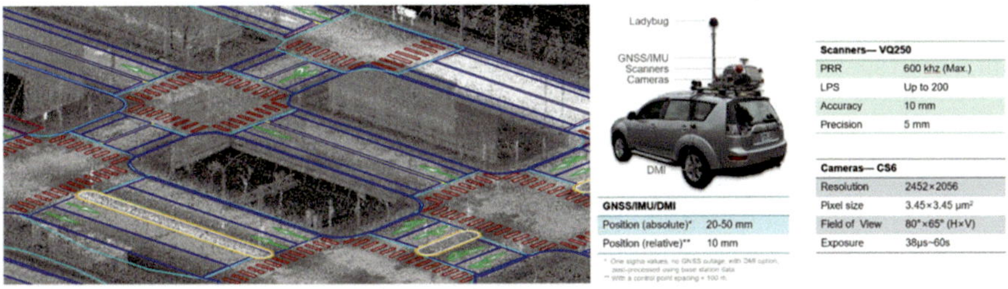

Fig. 25.12 HD map production with mobile mapping technology (Chiang et al. 2019)

They are the last reference information when the vision or radar-based vehicle environment sensing systems are failed. Moreover, they provide important multiple guarantees for the safe driving of vehicles. When machines surpass humans' ability to sense, reason, make decisions in real-time, and artificial intelligence technology guides vehicles safely and comfortably, then HD Maps may not be needed in the long run. However, it is necessary to be aware of the navigation, research, and development of autonomous vehicles through HD maps at the present time. Table 25.2 illustrates the list of autonomous driving classifications, required map types, and accuracy requirements according to Fig. 25.11 and the SAE classification of the driving system, respectively.

In terms of industry trends, since the huge business opportunities of autonomous driving and mapping technologies are promising in the future, international manufacturers have successively conducted preliminary arrangement competitions. In addition to Google's continued development of various applications based on Street

Table 25.2 Classification and map types requirement of autonomous driving

Grade	Title	Map	Accuracy of map	Typical conditions
Driver scenario				
1 (DA)	Driver assistance	ADAS map	Submeter level	Optional
2 (PA)	Partial automation	ADAS map	Submeter level	Optional
Automatic driving system ("system") scenario	ADAS map + HD map	Submeter level	Optional	
3 (CA)	Conditional automation	Centimeter level		
4 (HA)	High automation	ADAS map + HD map	Submeter level Centimeter level	Required
5 (FA)	Full automation	HD map	Centimeter level	Required (update automatically)

View technology, Apple also implemented its own development of mobile mapping technology in 2014 and developed an exclusive Apple Van to complement its disadvantages in spatial information compared to Google. The original mapping company HERE, owned by Nokia of Finland, supplies a chain of products and services that includes data collection, a map information office, and user map design. It has more than 300 surveying and mapping vehicles in the world to synchronously generate HD maps. It is the main map supplier to traditional car manufacturers such as BMW, Benz, Audi, for the development of autonomous driving technology. One of the map suppliers, TomTom, has more than 150 countries worldwide with vehicle graphics resources totaling more than 60 million kilometers, which includes existing business areas such as map authorization and cooperation with the automotive industry. In recent years, TomTom has focused on the production of HD maps based on the needs of autonomous driving navigation technology, and has proposed 3D mapping technology known as RoadDNA to construct and update HD maps. In Japan, with the support of the resources of the national government, a dynamic mapping platform (DMP) was established by the electronic information industry in partnership with domestic automakers to quickly achieve the demands of HD maps for the automotive industry in Japan. To sum up, at present, major international mapping companies and car manufacturers utilize MMS to generate HD maps based on their mapping technology and autonomous driving technology requirements.

The Department of Land Administration of the Ministry of the Interior in Taiwan proposes the Taiwan HD maps infrastructure that consists of three major pillars including qualified point clouds, qualified digital vector maps, and a Taiwan HD map format composed of the Opendrive format with local extension modules. In addition to the concept of an open base map, this architecture possesses interoperability between various HD map formats, as it is designed to provide map makers

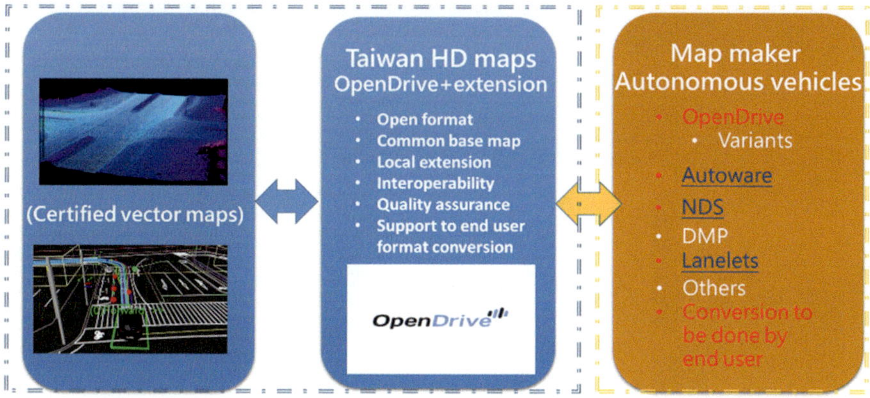

Fig. 25.13 The construction of Taiwan HD maps

and autonomous driving operators with an exchange format to facilitate added-value applications for the conversion to specific formats used by different autonomous vehicle platforms. In addition, it is also designed to support non-autonomous driving applications, such as disaster prevention, asset management, and the traditional surveying and mapping industry through verified fine-precision point clouds and diversified vector layer designs to achieve the concept of data sharing. Figure 25.13 illustrates the overall structure of Taiwan HD maps as well as certain formats used by different end-users (Chiang et al. 2019).

Currently, most of the Taiwanese autonomous driving platforms apply HD maps from Autoware, developed by the Tier 4 Company in Japan as well as the Open drive format. Therefore the Department of Land Administration of the Ministry of the Interior has been producing two HD map formats, Taiwan HD map format, and Autoware map format, for two primary autonomous vehicle test facilities in Taiwan in order to meet the growing demands for HD maps from various end-users. At the same time, the conversion tools between Taiwan HD Maps and certain end-user formats listed in Fig. 25.13 are also under development by the Land Department of the Ministry of the Interior (Chiang et al. 2019).

The scenarios for HD maps applications in Taiwan are proposed based on the concept of a local dynamic map (LDM; Shimada et al. 2015), as shown in Fig. 25.14. The exchange of time data (such as the signal transformation of traffic lights) and geospatial data (such as GNSS location information) of traffic participants can provide real-time information through communication sensors to improve the safety, efficiency, and comfort of the transportation system, and reduce the impact of traffic on the environment. This allows for the integration of static, temporary, and dynamic traffic information and the input of data with time-stamped and geo-referenced information into LDM as an integrated platform.

The LDM is a database that integrates real-time autonomous vehicles and traffic information into HD maps to achieve dynamic map data sharing. The meaning of local derives from the demand for geospatial information for the autonomous vehicle

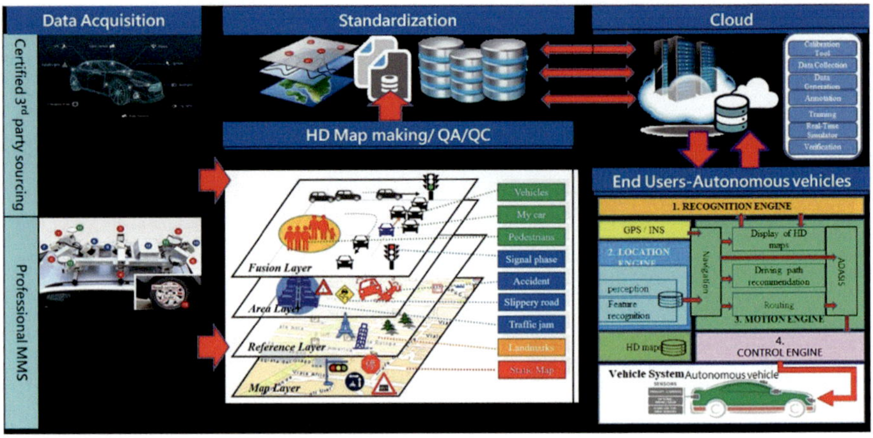

Fig. 25.14 The scenario of HD maps application

since it is close to the points of interest; the meaning of dynamic derives from the requirements of using dynamic traffic information to avoid collisions in a very short time. Therefore, the data requires the timestamp; the meaning of map depends on the association with a map. Local dynamic maps contain (Shimada et al. 2015):

- Static information and permanent static data: The first layer comes from geographic information system (GIS) map providers, including roads, lanes, intersections, road signs, traffic signs, road facilities, and points of interest (POI), phase data, and building location information, which are created by using a professional mobile mapping system. Update frequency is at least once a month.
- Semi-static information and transient static data: This layer mainly contains information about roadside infrastructure, including traffic regulations, traffic control schedules, road engineering traffic attributes, and area weather forecasts provided by the road-traffic control department. The information is obtained from outside the autonomous vehicle. Updating frequency of this information is at least once an hour.
- Semi-dynamic information and transient dynamic data: This mainly includes temporary regional traffic information, traffic control information, accident information, congestion information, phase conditions of traffic lights on roads or traffic signs, and local weather. The information is obtained from outside the autonomous vehicle. Updating frequency is at least once per minute.
- Dynamic information (highly dynamic data): This layer contains information detected by dynamic communication node V2X information, real-time status information such as traffic participants, surrounding vehicles, pedestrians, and the timing of traffic signals. The information is updated in real-time. Dynamic information is composed of the environmental information and the road-ahead information provided by an intelligent transportation system (ITS).

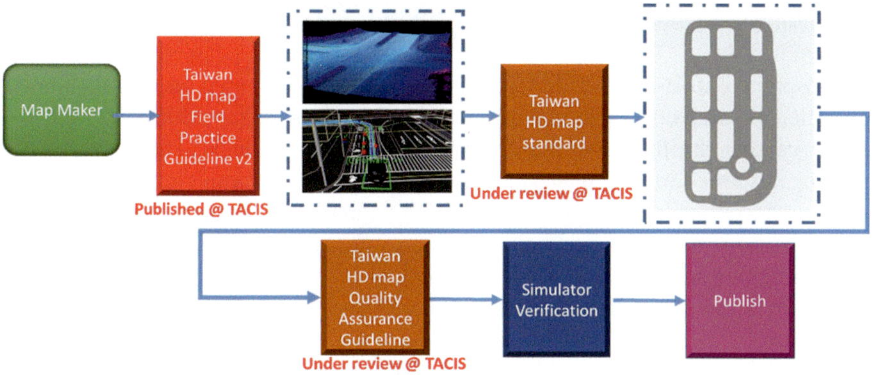

Fig. 25.15 Taiwan HD maps production procedure

In order to extend the spectrum of local development in the mapping and autonomous driving market, it is urgent to establish autonomous vehicle testing facilities and implement a unified HD maps format standard and regulation in Taiwan. The format standard for the static HD map layer is the primary task at the present time in Taiwan. The ultimate task is to build a static map to provide rich semantic information with sufficient accuracy to restrict and control vehicle behavior. This mainly includes the lane network, transportation facilities, the road network, and the positioning layer. Therefore, the Land Department of the Ministry of the Interior proposes to implement the production process of static layers of Taiwan HD maps using a professional mobile mapping system, as shown in Fig. 25.15, to meet the requirements for the production, maintenance, verification, and correctness according to "HD Maps Field Practice Guidelines v2," "Quality Verification Guidelines for HD Maps," and "HD Maps Data Contents and Formats Standard," to be published by the Taiwan Association of Information and Communication Standards soon.

Meanwhile, the applicability of HD maps is further evaluated by autonomous vehicle simulators and real vehicles to further ensure that the Taiwan HD maps format standards and services satisfy the requirements of autonomous vehicle applications in Taiwan and are in line with international standards (Chiang et al. 2019).

25.4 Future Trends in Mobile Mapping Technology

The recent big data market boom and deep-learning-related applications have been fueled by geospatial intelligence. Thus the importance of multi-platform mobile mapping technologies is being recognized by various communities. In fact, the widespread of mobile mapping technologies among various communities, such as the geospatial, robotics, computer vision, artificial intelligence, and navigation communities, is exceeding the expectation of the pioneers from the geospatial community who initially developed such technologies thirty years ago and continue to promote them even now.

Geospatial data are collected with mapping sensors mounted on various human-controlled or unmanned platforms, such as aircraft or helicopters, land vehicles, marine vessels, strollers, and those hand-carried by individuals. Therefore, mobile mapping systems certainly play a crucial role in urban informatics applications since timely and accurate geospatial data are the key ingredient in implementing the digital infrastructure serving the backbone of urban informatics. Figure 25.16 depicts an indoor mapping scenario to build a floorplan with a robot and indoor UAV, respectively, where the 3D positioning accuracy achieved was around 1–1.5 m based on the scenario.

Ultimately, the future technological trends in mobile mapping that will advance urban informatics applications can be characterized by (1) fulfilling seamless mapping scenarios; (2) increasing use of low-cost direct georeferencing devices; (3) increasing use with artificial intelligence; and (4) increasing use with unmanned multi-platforms for collaborative mapping.

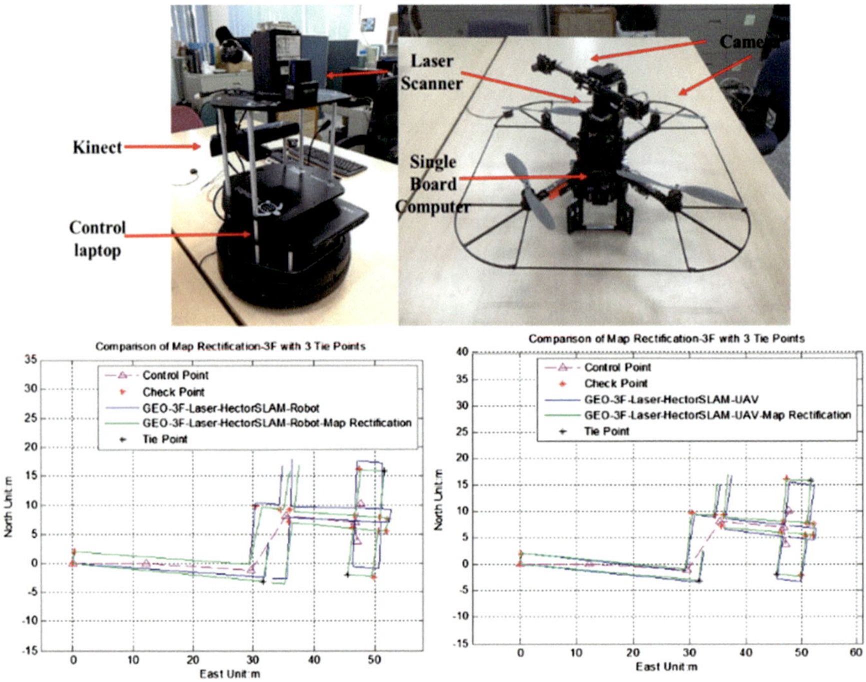

Fig. 25.16 An example of unmanned mobile mapping technology

25.5 Conclusion

This chapter has comprehensively discussed mobile mapping technologies. From the labor-consuming indirect georeferencing to the efficient DG, it is clear that evolution has been rapid and that researchers have contributed to the development of this technology. Nowadays, this technology also plays an important role in future applications, such as autonomous driving and rapid disaster response. In other words, accurate geospatial data become one of the game-changers in the future. It is worth mentioning that the individual components of mobile mapping technologies take part in every geospatial technology for data acquisition, such as computer vision, simultaneous localization and mapping (SLAM), and robotic mapping. In the foreseeable future, we are likely to see the ever-increasing importance of mobile mapping technologies.

References

Axelsson P (1999) Processing of laser scanner data—algorithms and applications. ISPRS J Photogramm Remote Sens 54(2–3):138–147

Cannon ME, Schwarz KP (1990) A discussion of GPS/INS integration for photogrammetric applications. In: Kinematic systems in geodesy, surveying and remote sensing, Banff, pp 443–452

Chiang KW, Chang HW, Li CY, Huang YW (2009) An artificial neural network embedded position and orientation determination algorithm for low cost MEMS INS/GPS integrated sensors. Sensors 9(4):2586–2610

Chiang KW, Tsai ML, Chu CH (2012) The development of an UAV borne direct georeferenced photogrammetric platform for ground control point free applications. Sensors 12:9161–9180

Chiang KW, Tsai GJ, Chan HW, Li YH, El-Sheimy N (2017) Development of LiDAR-based UAV system for environment reconstruction. IEEE Geosci Remote Sens Lett 14(10):1790–1794

Chiang KW, Tseng YH, Hong JH, Kuo CY, Wang CK, Lu HC (2019) The development of map standard format and mobile mapping technology report. Department of Land Administration, MOI

Cohen CE, Parkinson BW (1992) Aircraft applications of GPS-based attitude determination. Paper presented at the 5th international technical meeting of the satellite division of the institute of navigation (ION GPS 1992), Albuquerque, Sept 1992

Ellum CM (2001) The development of a backpack mobile mapping system. UCGE reports. Department of Geomatics Engineering, University of Calgary, Calgary

El-Mowafy A, Schwarz KP (1994) Epoch by epoch attitude determination using a multi-antenna system in kinematic mode. In: Proceedings of the international symposium on kinematic systems in geodesy, geomatics and navigation, KIS-94, Banff, pp 331–340

El-Sheimy N (1996) A mobile survey system for GIS applications. Dissertation, Department of Geomatics Engineering, The University of Calgary, Calgary

Farrell JA, Todd M, Barth M (2016) Best practices for surveying and mapping roadways and intersections for connected vehicle applications

Grejner-Brzezinska DA (2001) Direct sensor orientation in airborne and land-based mapping applications. Report no. 461. Department of Civil and Environmental Engineering and Geodetic Science, The Ohio State University

Ip AWL, El-Sheimy N, Mostafa MMR (2004) System performance analysis of IMU/DGPS integrated system for mobile mapping (MMS). In: 4th international symposium on mobile mapping technology (MMT 2004), Kunming, 29–31 Mar 2004

Nagai M, Chen T, Ahmed A, Shibasaki R (2008) UAV-borne mapping by multi sensor integration. Int Arch Photogramm Remote Sens Geospat Inf Sci 37(B1):1215–1222
Rehak M, Mabillard R, Skaloud J (2013) A micro-UAV with the capability of direct georeferencing. Int Arch Photogramm Remote Sens Geospat Inf Sci 40(1)/W2:317–323
Schwarz KP, El-Sheimy N (2008) Mobile mapping systems—state of the art and future trends. In: International society for photogrammetry and remote sensing TS SS 3, Beijing
Shimada H, Yamaguchi A, Takada H, Sato K (2015) Implementation and evaluation of local dynamic map in safety driving systems. J Transp Technol 5(02):102
Shin EH (2005) Estimation techniques for low cost inertial navigation. UCGE report 20219. Department of Geomatics Engineering, The University of Calgary, Calgary
Skaloud J, Cramer M, Schwarz KP (1996) Exterior orientation by direct measurement of camera position and attitude. Int Arch Photogramm Remote Sens 31(B3):125–130
Titterton DH, Weston JL (2004) Strapdown inertial navigation technology. In: Stewart N, Griffiths H (eds) IEE radar, sonar, navigation and avionics, series 17, 2nd edn. Institution of Electrical Engineers, Stevenage
Zach G, Studnicka N, Amon P, Pfennigbauer M (2011) Extraction of facade maps using a RIEGL VMX-250 mobile scanning system mounted on a boat. In: 7th international symposium on mobile mapping technology 2(1)

Kai Wei Chiang is Professor of Geomatics at the National Cheng Kung University, Tainan, and also a Director (Principal Investigator) at the High Definition Maps Research Center in Taiwan. He is a member of the US Institute of Navigation and interested in seamless multi-sensor mapping and navigation technologies.

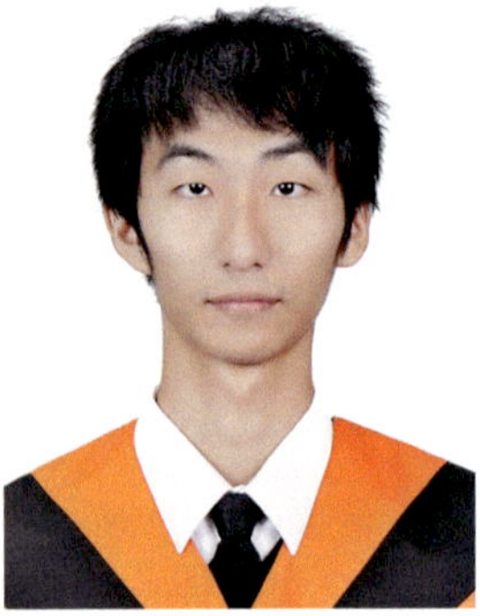

Guang-Je Tsai is currently a Ph.D. candidate of the Department of Geomatics at the National Cheng Kung University, Tainan. He is interested in multisensor fusion in terms of navigation and mapping technologies such as inertial navigation, simultaneous localization and mapping (SLAM), and mobile multisensor mapping systems.

Jhih Cing Zeng is a master degree student of the Department of Geomatics at the National Cheng Kung University, Tainan. She is interested in point cloud processing and HD Maps generation.

Chapter 26
Smartphone-Based Indoor Positioning Technologies

Ruizhi Chen and Liang Chen

Abstract Global Navigation Satellite Systems (GNSS) have achieved great success in providing localization information in outdoor open areas. However, due to the weakness of the signal, GNSS signals cannot be received well indoors. Currently, indoor positioning plays a significant role in many areas, such as the Internet of Things (IoT) and artificial intelligence (AI), but given the complexity of indoor spaces and topology, it is still challenging to achieve an accurate, effective, full coverage and real-time positioning solution indoors. With the development of information technology, the smartphone has become more and more popular. With a large number of sensors embedded in smartphones, it is thus possible to achieve low cost, continuity, and high usability for indoor positioning. In this chapter, we focus on indoor positioning technologies with smartphones, and in particular, emphasize the technologies based on radio frequency (RF) and built-in sensors. The pros and cons of the technologies are reviewed and discussed in the context of different applications. Moreover, the challenges of indoor positioning are pointed out and the directions for the future development of this area are discussed.

26.1 Introduction

Positioning is one of the core technologies of location-based services (LBS). It also plays a significant role in many applications of the Internet of Things (IoT) and artificial intelligence (AI). With the extensive urban development of recent years, indoor positioning is becoming more and more important. According to a report by the U.S. Environmental Protection Agency, people spend 70–90% of their time

R. Chen · L. Chen (✉)
State Key Laboratory of Information Engineering in Surveying, Mapping and Remote Sensing (LIESMARS), Wuhan University, Wuhan, China
e-mail: l.chen@whu.edu.cn

R. Chen
e-mail: ruizhi.chen@whu.edu.cn

Collaborative Innovation Center of Geospatial Technology, Wuhan University, Wuhan, China

W. Shi et al. (eds.), *Urban Informatics*, The Urban Book Series,
https://doi.org/10.1007/978-981-15-8983-6_26

indoors (Weiser 2002). A wide area of applications has emerged for indoor emergency rescue (Federal Communications Commission 2015), precision marketing in shopping malls, asset management and tracking in the smart factory, mobile health services, virtual reality games, and location-based social media (Sakpere et al. 2017; Davidson and Piché 2016; Ali et al. 2019). By 2025, the global indoor LBS market is expected to reach USD 18.74 billion (Globe Newswire 2019).

Global navigation satellite systems (GNSS) have achieved great success in positioning in outdoor open areas, and positioning accuracy is able to achieve a sub-meter level with various assisted technologies (Kaplan and Hegarty 2005). However, due to the weakness of signal power, GNSS signals cannot be received indoors sufficiently to provide continuous and reliable positioning. In many cases, especially in deep indoor areas, GNSS signals can even be totally blocked. Although various technologies have been developed for indoor positioning, which includes WiFi, Bluetooth, ultra-wideband (UWB), pseudolites, magnetic fields, sound and ultrasound, and pedestrian dead reckoning (PDR), it is still challenging to achieve an accurate, effective, full coverage and real-time positioning solution indoors (Maghdid et al. 2016). The main reasons are the constraints of spatial layout, topology, and the complex signal environment indoors (Zafari et al. 2019). To be more specific, the reasons are summarized as follows.

The indoor environment is complex and radio waves are often reflected, refracted, or scattered by obstacles indoors, which leads to non-line-of-sight (NLOS) propagation. NLOS propagation can cause a large deviation error in the positioning and seriously affect the localization accuracy.

Indoor space layout and topology are frequently changed and the number of people in the indoor space varies, for example, between peak and off-peak hours. Thus, signal propagation and the fields of sound, light, electricity, and magnetism can all be changed accordingly. Such changes will greatly affect the results when using the positioning methods with the feature or field matching.

The unpredictability of indoor pedestrian motions, such as frequent changes in speed and direction (Morrison et al. 2012), and motion without any predefined paths (Saeedi 2013) also increases the difficulty of continuous estimation of pedestrian position.

With the development of information technology, the smartphone has become more and more popular. As shown in Fig. 26.1, the smartphone has a large number of built-in sensors, such as accelerometers, gyroscopes, magnetometers, barometers, light sensors, microphones, speakers, and cameras, as well as Bluetooth chips and WiFi chips. Such sensors were not originally developed for the use of the positioning. Nevertheless, for applications in the mass market, it is promising to achieve low cost, continuity, and high usability mode for indoor positioning with the built-in sensors in a smartphone with appropriate technology (Davidson and Piché 2016).

In this chapter, we present a survey of indoor positioning with smartphone sensors. The state-of-the-art technologies will be reviewed. We will comprehensively compare the accuracy, complexity, robustness, scalability, and cost of different technologies, and comment on the pros and cons of the technologies in the context of different application scenarios. Moreover, from the perspective of developing the technology

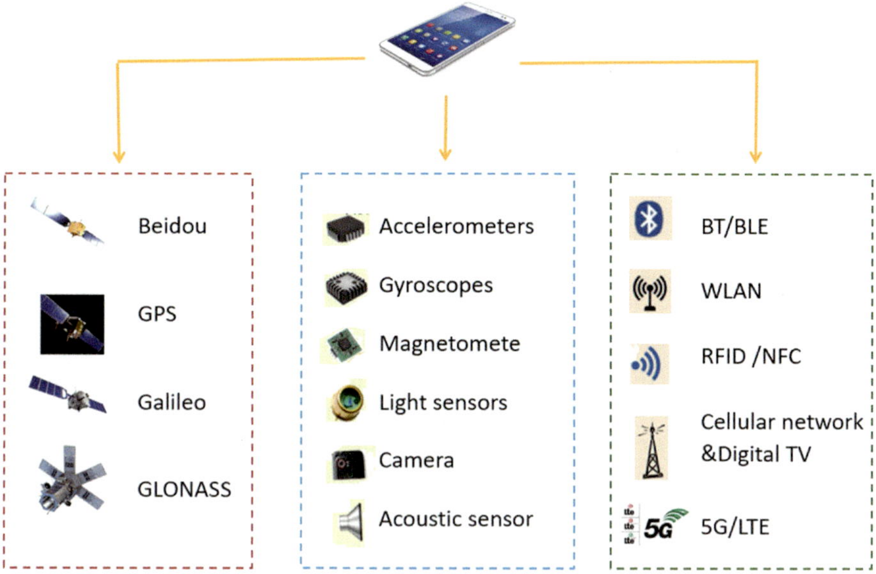

Fig. 26.1 Multiple sensors embedded in the smartphone

with high accuracy, high usability, high durability, and at low cost, we further discuss the directions of future development in this area.

The organization of the book chapter is as follows: in Sect. 26.2 we review the technologies of the smartphone for indoor positioning in detail. In Sect. 26.3 we summarize the difficulties in indoor positioning. In Sect. 26.4 potential future trends in smartphone indoor positioning are discussed. Conclusions are drawn in Sect. 26.5.

26.2 The State-of-the-Art Indoor Positioning with Smartphones

This section focuses on the state of the art of indoor positioning technology with smartphone sensors. The positioning technology can be classified into two categories: positioning with RF and positioning with built-in sensors.

26.2.1 Positioning Technology of RF Signals

Currently, WiFi, Bluetooth, and wireless cellular communication signals are the main radio-frequency signals that smartphones support for the purpose of data transmission. The methods of indoor positioning vary due to differences in carrier frequency, signal strength, and the effective transmission distance of the signals.

26.2.1.1 WiFi Positioning Technology

WiFi is a wireless local area network (WLAN) technology based on the IEEE 802.11 family of standards (IEEE Standard for Information Technology 2013). With the advantages of flexibility, convenience, rapid deployment, and low cost, WiFi technologies have now been widely deployed indoors and have been used for indoor positioning. There are basically two methods used for positioning with WiFi signals: triangulation and fingerprinting.

In the triangulation method, the smartphone measures the received signal strength index (RSSI) of each of multiple WiFi access points (APs), and then estimates the distances between the smartphone and each of the APs using a model of long-distance path loss (Liu et al. 2007). The model is a radio-propagation model that predicts the path loss a signal encounters inside a building or densely populated area. However, due to the strong reflections and scattering conditions indoors, RSSI measurements are seriously attenuated by multipath and NLOS signal propagation. Therefore, it is a challenging task to accurately estimate the position with RSSI measurements and the path loss model has given the various fading effects. In the method of triangulation, the other way to get the distance between the transceivers is to measure the time of flight (TOF; Schauer et al. 2013). Tests have shown that indoor multipath and the time-varying interruption service in WLAN have a great impact on the accuracy of TOF measurement. Ranging accuracy can be improved by proper design of filters and by smoothing of the raw measurements.

In the fingerprint positioning method (Bahl and Padmanabhan 2000), the basic idea is to match elements in a database to particular signal-strength fingerprints in the area at hand. The method operates in two phases: the training phase and the online positioning phase. In the training phase, a radio map is created based on the reference points within the area of interest. The radio map implicitly characterizes the RSSI position relationship through the training measurements at the reference points with known coordinates. In the online positioning phase, the smartphone measures RSSI observations and the positioning system uses the radio map to obtain a position estimate. The advantage of the method is that it does not need to know either the exact model of the channel attenuation between the transceivers or the coordinates of the WiFi APs. The disadvantage is that the signal is easily modified by the surroundings, the mismatch rate is relatively high in the open space indoors, and to build and update the fingerprint database is a time-consuming process. The fingerprinting method has been widely investigated in the literature. Recent surveys of the RSSI fingerprint method can be found by Khalajmehrabadi et al. (2017), He and Chan (2015), and Davidson and Piché (2016). In general, the methods can be divided into three types: deterministic approaches, probabilistic methods, and pattern-recognition methods. The main factors affecting the accuracy of WiFi positioning include inter-channel interference from different APs (Pei et al. 2012) and hardware differences in smartphones (Schmitt et al. 2014). Khalajmehrabadi et al. (2017), He and Chan (2015), and Davidson and Piché (2016) give a thorough summary of the factors that affect WiFi fingerprint positioning. Currently, WiFi positioning systems using RSSI fingerprints include RADAR (Bahl and Padmanabhan 2000), Ekahau (ekahau.com), and Horus (Youssef and Agrawala 2008), and the positioning accuracy is about 2–5 m.

Benefiting from the performance improvement of the WiFi receivers, commercial WiFi receiver modules are now able to provide channel state information (CSI; Wang et al. 2016). CSI gives more details on the multipath information of the channel attenuation than the RSSI measurements, which only provide the power measurement of a received radio signal. Research shows that using CSI information to build the fingerprint database can effectively improve the accuracy of indoor positioning (Wang et al. 2015b; Wu et al. 2012).

With the ratification of IEEE 802.11n standardization, the technology of multiple antennae has been introduced to WiFi transmission. Thus, angle of arrival (AOA) can be estimated in the WiFi positioning. The literature (Vasisht et al. 2016; Kotaru et al. 2015) simultaneously estimates the AOA and the time of arrival (TOA) to achieve positioning results with an accuracy of decimeter or centimeter, respectively. However, such methods are applied in the AP base station and are not applicable to a user-centric positioning with smartphones, in which only one antenna is embedded.

The main factor that limits WiFi fingerprint positioning in massive applications is the difficulty in effectively constructing and adaptively updating the radio map, which is both time and labor-consuming. The methods for reducing the costs of building and updating the radio map include crowdsourcing (Zhuang et al. 2015), LiDAR-based simultaneous localization and mapping (SLAM; Tang et al. 2015), and the use of interpolation (Zhao et al. 2016). In addition, with the increasing attention to the issues of information security and personal privacy (Chen et al. 2017), the scanning rate of WiFi signals have been adjusted to 1/30 Hz or even lower, which increases the latency for the positioning.

26.2.1.2 Bluetooth Positioning Technology

Bluetooth is a radio-frequency signal based on the IEEE 802.15.1 protocol, which is mainly developed for wireless personal area networks (WPAN). It operates in the 2400–2483.5 MHz range within the same ISM 2.4 GHz frequency band as WiFi IEEE 802.11 b/g. The transmission data is split into packets and exchanged through one of 79 designated Bluetooth channels, each of which has 1 MHz in bandwidth. Positioning with Bluetooth Classic (prespecification4.0) has used various techniques from proximity to trilateration to fingerprinting. The positioning accuracy is about 4 m (Chen et al. 2011a, 2013, 2015). However, in the specification, the scanning interval of a mobile handset to the nearby Bluetooth beacons can be more than 10 s, within which time the indoor pedestrian could travel 15 m or more. Due to the low scan rate, positioning using Bluetooth Classic has not proved popular (Faragher and Harle 2015).

In 2011, Bluetooth Low Energy (BLE), which was originally branded as Bluetooth 4.0, was created. Compared to classic Bluetooth, BLE provides an improved data rate of 24 Mbps and coverage range of 70–100 m with higher energy efficiency (Zafari et al. 2015). BLE also has a very short connection time (only a few milliseconds) and then goes into sleep mode until a connection is reestablished, which achieves low power consumption. With this property, BLE can be powered by a single battery which could last up to five years. Compared with WiFi, which is typically placed near power outlets, BLE, with its own batteries, is thus free to place beacons to provide good signal geometry with optimized signal coverage. In addition, with a much higher scan rate than WiFi, BLE can average out the occasional outliers caused by interference or multipath effects, and improve the tracking accuracy.

At the moment, the most popular BLE beacon ecosystems are Apple's iBeacon, Google's URI Beacon and Eddystone, and Radius Networks' Alt Beacon. Apple's iBeacon system (Apple 2014), based on RSSI ranging, has a positioning accuracy of 2–3 m in a typical office environment. A Bluetooth antenna array system, developed by Quuppa(2020), can achieve a sub-meters positioning accuracy. In January 2019, a new specification of Bluetooth 5.1 enhances location services with its new feature of direction-finding. With this new feature, it is possible that Bluetooth devices will be able to pinpoint physical location to centimeter accuracy indoors (How-To Geek 2019).

26.2.1.3 Cellular Positioning Technology

The cellular network is originally designed for dedicated mobile communication systems. Nevertheless, the large cellular communication infrastructure can still be reused for positioning purposes, providing an added value to network management and services (Del Peral-Rosado et al. 2017). In 2G/3G/4G mobile communication systems, cellular positioning is achieved by a localization module implemented in the base station, which is also known as the RAN (radio access network) positioning method. The most significant advantage of cellular positioning technology is to achieve seamless indoor and outdoor positioning, while the disadvantage is that the positioning accuracy is relatively low, generally in tens of meters to hundreds of meters (Zhao 2002; Lakmali and Dias 2008). Ericsson uses a long-term evolution (LTE) signal to adopt the OTDOA (observed time difference of arrival) method, and the positioning accuracy can reach 50 m, with a reliability of 97% (Ericsson Research Blog 2015). But the positioning results cannot meet the needs of most indoor positioning applications.

The upcoming fifth-generation (5G) of mobile communication systems are expected to improve positioning accuracy in cellular networks, which is a benefit of the key features of 5G, such as small cells, device-to-device (D2D) communication, heterogeneous networks (Het-Net), massive multi-input multi-output (MIMO), and millimeter-wave (mm-Wave) communication (Talvitie et al. 2017). In particular, through D2D communications, mobile stations or smartphones can determine their locations in a cooperative manner, which would not only increase the localization accuracy but also decrease the time delay. The massive MIMO technologies will offer more possibilities for accurate directional measurements. Dense networks with small cells will lead to a large number of line-of-sight (LOS) links, and higher signal bandwidths will improve the accuracy of range measurements, and increase the resolution of multipath.

26.2.2 *Positioning Technology Based on Embedded Sensors*

Built-in sensors for smartphones include accelerometers, gyroscopes, magnetometers, barometers, light intensity sensors, cameras, microphones, etc. These sensors are not designed for positioning, but measurements from such sensors can be used for indoor positioning with proprietary methods. The methods include PDR, geomagnetic matching, visual positioning, audio, and sound positioning.

26.2.2.1 Pedestrian Dead Reckoning

With the advances in micro-electro-mechanical system (MEMS) technology, more and more low-cost inertial measurement units (IMUs) are integrated into smartphones. Accelerometers, gyroscopes, and magnetometers are among the most popular sensors embedded; due to their low cost, their stability and measurement accuracies are relatively low. It is therefore difficult to use the strap-down inertial navigation method. As an alternative, PDR can be applied in indoor positioning using the measurements from low-cost MEMS sensors (Robert 2013). In more details, PDR uses an accelerometer to detect the number of steps, measures the walking speed, and determines the heading by magnetometer and gyroscope, and then calculates the relative position of the pedestrian by computing the speed and heading (Chen et al. 2011b; Deng et al. 2016).

The PDR algorithm (Fig. 26.2) is able to provide continuous positioning results. Without the process of integration, it is a relatively simple but effective method to use the raw measurements from the low-cost sensors. The difficulty of PDR lies in the heading estimation, which is affected by magnetic interference in the indoor environment. It is, therefore, necessary to integrate with other positioning algorithms, such as WiFi, BLE, or geomagnetic matching, which are able to provide absolute positioning results, to improve the heading estimate as well as to reduce the accumulating errors of relative positioning from PDR (Deng et al. 2016).

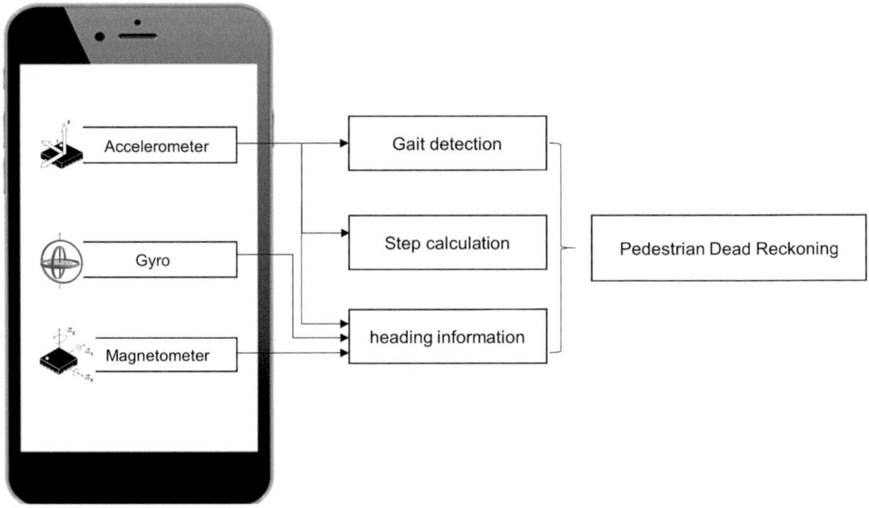

Fig. 26.2 PDR system block diagram

26.2.2.2 Magnetic Matching (MM) Positioning Technology

MM positioning technology takes the magnetic field as the signal for a fingerprint and fulfills the indoor positioning by matching characteristics of the magnetic field in the indoor environment. Similar to the process of WiFi fingerprinting, MM positioning is also divided into two steps: to set up a geomagnetic fingerprint database, and to match geomagnetic features for positioning. Because of the spatial correlation of the magnetic field, contour matching, for example, dynamic time warping, can be used in the MM to achieve more robust matching results. At present, most smartphones are integrated with magnetometers, and the magnetic field can be obtained when the phone is turned on. So, MM positioning technology is suitable for smartphone positioning. However, indoor magnetic field signals often change, so it is difficult to build an accurate fingerprint database of magnetic fields in practice. The University of Oulu in Finland proposed an indoor positioning system, named Indoor Atlas, which combines magnetic fields with built-in sensors (Thompson 2020), which is able to achieve a positioning accuracy of 0.1–2 m.

26.2.2.3 Visual Positioning Technology

The visual positioning for smartphones is mainly based on monocular vision since smartphones commonly use a monocular camera. One method is based on image matching, where the positioning is computed by matching the current photos with the photos stored in the image database. The methods of density matching and structure from motion (SFM) can be used to match the image features in the image feature database. Another method is based on visual gyroscopes and visual odometer technology (Ruotsalainen 2012; Ruotsalainen et al. 2013). The visual gyroscope uses a monocular camera to obtain a vanishing point of each image and uses a vanishing point change of two adjacent images to obtain the heading change rate. The visual odometer obtains the relative translation of pedestrians by matching photos taken in time series. The challenges of using the monocular camera as a visual gyroscope and visual odometer are in the sharp turns for the pedestrian where there are fewer feature points for matching in photos. The literature (Ruotsalainen et al. 2016) lists methods for merging visual gyroscopes and visual odometers with other IMUs.

Visual positioning technology can achieve decimeter-level or even centimeter-level accuracy in scenarios with sufficient light and image features. When an optical camera is combined with depth cameras (such as Google's Tango technology), the positioning accuracy can be further increased. But, in general, the algorithm of visual positioning is computationally complex and has high power consumption. With further improvement in the computation performance and storage capacity of smartphones, the method is promising in pedestrian navigation.

26.2.2.4 LED Visible Positioning Technology

Visible light positioning can be divided into two categories: the first is to locate a specific optical signal by modulating the light source. For example, an LED lamp emits a high-frequency flicker signal that is invisible to the naked eye, and the LED light signal is received by the smartphone sensors to calculate pedestrian position information. The byte light positioning system (Ganick and Ryan 2012) is based on such a principle, and the positioning accuracy can reach the one-meter level. The second is based on the pattern-matching method, which uses the time–frequency characteristics of ambient light to establish the environmental light fingerprint database in advance. In the real-time positioning phase, the measured light intensity is matched with the ambient light fingerprint database to achieve positioning (Liu et al. 2014). The built-in camera of the smartphone can sense light intensity and high-frequency light information, so the above optical positioning technology can be easily applied to indoor positioning of smartphones.

26.2.2.5 Ultrasonic Positioning Technology

Ultrasonic positioning technology uses the method of round-trip time ranging. The most popular ultrasonic positioning systems are the Active Bat system (Ward and Jones 1997) and the Cricket system (Priyantha et al. 2000). The positioning accuracy of the Active Bat system is within 9 cm with a 95% confidence interval. Although the ultrasonic positioning system has high positioning accuracy, the current smartphones have not been equipped with dedicated ultrasonic modules for transmitting or receiving ultrasound signals. However, the microphones in the current smartphones can monitor ultrasonic signals with the frequency ranging from 16 to 22 kHz. Determining the user's location with such ultrasonic signals has already attracted much attention in the area of smartphone positioning (Ijaz et al. 2013). In order to improve the accuracy of ultrasound indoor positioning, the main effort is to mitigate the echo signals, which have severe effects on the TOA detection of ultrasound.

26.2.3 Positioning Technology of Multi-source Fusion

As seen from the above, different positioning methods have their pros and cons in different scenarios of indoor positioning. For example, RF signals may have large coverage, however multipath interference, which is common indoors, will cause large positioning errors. Pedestrian-track estimation based on built-in sensors does not depend on the infrastructure indoors, but the errors from the IMUs accumulate over time. Currently, there has not yet been any method based on a single technology that suits all different scenarios of indoor positioning. Table 26.1 compares the performance of various technologies for the smartphone positioning in terms of positioning accuracy, complexity, robustness, scalability, and cost. Although there are many sources available for indoor positionings, such as sound, light, electrical signals, and magnetic fields, different positioning sources have their own limits and the usability depends on the actual environment in reality. For example, the method of WiFi fingerprinting requires a wide coverage of the signals with more APs and less radio interference, while the method of magnetic field matching requires significant magnetic features in the place of interest, where magnetic interference benefits positioning to some extent. As to the visual positioning, it works well in a bright environment, while it cannot work effectively in dark places.

With the improvement of computing performance and storage capacity on smartphones, the sensor fusion technology to integrate multiple positioning technologies has been a hot research topic in the field of indoor positioning with smartphones. The methods are broadly divided into loosely coupled and tightly coupled. The basic idea of the loosely coupled method is to fuse all the positioning results from different sensors and get the estimate of the position at a time epoch. This kind of fusion is easy to implement, but due to the heterogeneity of sensors in the smartphone positioning, it is difficult to analytically compute the weights on the position estimation from different sensors, which are sent to the sensor-fusion module. The tightly coupled method is to fuse different parameters estimated from different types of sensors and get the positioning estimate. At present, an effective way to implement tightly coupled fusion is based on Bayesian inference, which includes Kalman filtering (KF; Zhang et al. 2013), unscented Kalman filter (UKF; Chen et al. 2011c), and particle filter (PF; Quigley et al. 2010). In these methods, the state model and the measurement equations are first set up, and the moving states (position and velocity) of the pedestrian have been inferred in sequence based on the parameters estimated from different sensors, such as position, velocity, heading angle, and step size. The literature on sensor-fusion research includes: the hybrid positioning system with WiFi

Table 26.1 Comparison of different positioning technologies of smartphone sensors

Source	Precision	Robustness	Complexity	Scalability	Cost
WiFi	2–5 m with the fingerprint method, while the triangulation method is affected by different environments	Vulnerable to environmental, human body, and other interference	Time- and labor-consuming in building the fingerprint database	High	Using existed facilities with no additional cost
Bluetooth	Fingerprint method 2–5 m, iBeacon, antenna array mode <1 m	Vulnerable to environmental interference	Fingerprint matching is time-consuming and labor-intensive	High, iBeacon distance Less than 5 m	The cost of antenna is relatively high, but low cost with beacon technology
Infrared	One to several meters	Direct path required	Medium	High	Medium cost. It is necessary to set up an additional receiver Hair device
LED	1–5 m	Medium	Medium	High	Low
Ultrasonic	Centimeter	High	Low	Low	Medium, extra receiving module needed
Inertial navigation	Depending on the characteristics of the sensors, there are cumulative errors over time	High	Medium	High	Low
Geomagnetic	2–5 m	Vulnerable to environmental changes	High	High	Low
Computer vision	A few centimeters to several meters depending on the methods applied	Medium, affected by the strength of the ambient light	Very high	High	Medium

magnetic field and cellular signal (Kim et al. 2014); WiFi positioning fused with PDR results (Karlsson et al. 2015; Li et al. 2016); Bluetooth module, accelerometers, and barometers used for 3D indoor positioning (Jeon et al. 2015); and WiFi fingerprinting with PDR and magnetic field matching (Zhang et al. 2017). In addition, indoor maps are commonly used to assist indoor positioning. The positioning system can reliably achieve meter-level accuracy by integrating the map-constrained information with WiFi fingerprint and PDR positioning results (Wang et al. 2015a). Ruotsalainen et al. (2016) provide a solution to infrastructure-free indoor navigation by fusing the observations from IMUs, cameras, ultrasonic sensors, and barometer with the PF algorithm. The average positioning accuracy is about 3 m. Various sensor-fusion positioning methods are compared in Table 26.2. The test results have already shown that the accuracy and stability of the sensor-fusion systems are better than an indoor-positioning system with a single technology.

26.3 Difficulties in Indoor Positioning

Using the method of sensor fusion, the positioning accuracy of a smartphone is able to reach 2–5 m, and it is possible to achieve within 1 m in some specific environments. However, in general, it is still challenging to develop a technology with low cost, fine precision, and high usability for indoor and outdoor seamless positioning. The main difficulties of smartphone indoor positioning are summarized as follows.

26.3.1 Complex Channel Transmission and Spatial Topology in Indoor Environments

For the positioning with RF signals, multipath interference and NLOS transmission are the main errors for TOA-based measurements. However, due to the complex topology of the indoor environment, the multipath effect and the NLOS conditions are common and more severe indoors, which introduces large positioning errors when applying traditional RF positioning technologies developed for outdoor positioning. For example, the relocation of the appliances and furniture indoors, the increase or decrease of goods on shelves, and variations in the layout of the venue all affect the signal transmission and the magnetic field of the indoor environment. Such changes are the main difficulty for indoor positioning systems to maintain high accuracy. It is challenging to automatically sense and recognize the changes of the radio and magnetic fields incurred by the spatial and temporal changes of indoor topology, and thus improve the self-learning and self-adaptive ability of the positioning environment by updating the positioning database, including the WiFi fingerprint database, the geomagnetic fingerprint database, the image feature database, and the landmark

Table 26.2 Comparison of various available sensor-fusion positioning methods

Fusion methods	References	Advantages	Disadvantages
WiFi/PDR	Karlsson et al. (2015), Li et al. (2016), Wang et al. (2015b)	Solve the instability problems, overcome positioning error, and thus improve the reliability and robustness of indoor positioning	The workload of constructing the WiFi fingerprint database is large
WiFi/magnetic field/cellular signal	Kim et al. (2014)	Reduced cost and labor consumption because magnetic fields do not require any pre-installed infrastructure	Magnetic field is affected by the environment
Bluetooth module/accelerometers/barometers	Jeon et al. (2015)	It realizes 3D positioning and achieves a significant improvement of the positioning accuracy compared to the use of the Bluetooth RSSI alone	Errors accumulate over time
WiFi/PDR/magnetic field	Zhang et al. (2017)	The positioning accuracy and system robustness are greatly improved	The sampling time needs to be controlled
IMUs/camera/ultrasonic/barometer	Ruotsalainen et al. (2016)	The method provides beyond the state-of-the-art performance and is anticipated to result in a SLAM solution	It is affected by the light environment

information database. Automatic update for such metrics is still a problem that has not been solved in the field of indoor positioning.

26.3.2 *Heterogeneous Source of Positioning*

As shown in Fig. 26.1, there are over 12 types of sensors embedded in smartphones, including GNSS receiver modules, short-range RF transmitters, WiFi and Bluetooth modules, or receivers and other embedded sensors, such as accelerometers, magnetometers, gyroscopes, barometers, light-intensity sensors, microphones, speakers, and cameras. However, except for the GNSS receiver modules, other sensors and RF signal modules are not specifically designed for the purpose of positioning. Although many methods have been developed for these sensors to estimate the parameters of positioning, these measurements from different sensors are in essence heterogeneous, due to the fact that they observe different parameters of positioning (e.g., position, velocity, heading rate), different sampling rates, and different noise, which are in essence heterogeneous. As discussed in Sect. 26.3.1, it is possible to integrate different sensors that are embedded in the smartphone for indoor positioning. However, in order to achieve an optimal solution to sensor fusion for indoor positioning, the following problems have to be tackled.

26.3.2.1 Synchronization of Signal Measurements

Different smartphone sensors work independently and may have different sampling rates. For example, the scanning rate of the WiFi RSSI signal ranges from 1/3 to 1/30 Hz, while the sampling frequency of the accelerometer can reach 180 Hz. Even with the same sampling rate, the sampling time instant may be different too. Therefore, in order to compute position with the sensor-fusion algorithm, a synchronized measurement obtained from different sensors in different time instants has to be aligned to a specific time baseline. The baseline can be the main clock time of the smartphones in the user-centric positioning or the network time of the cloud server in a solution of network-centric positioning. To meet the requirement of most indoor location services, the update rate of indoor location should be greater than or equal to 1 Hz. The interpolation method works well on the time alignment of asynchronized measurements when the user is in the low-speed motion state (the motion speed is less than 2 m/s), which suits the scenarios of pedestrian indoor navigation.

26.3.2.2 Different Accuracy of Sensor Measurements

There are over 12 types of sensors embedded in smartphones. Different sensors have different measurement noise and quantification errors. Besides, there are different methods for different sensors to measure the positioning parameters, and thus, the measurement accuracy consequently varies. For example, MEMS sensors embedded in smartphones are low cost, and the measurement accuracy of such sensors is very poor, so they cannot be directly used in strap-down inertial navigation. But they can be used in step detection, and provide walking speed and length with acceptable accuracy. The indoor environment also has a different effect on different sensors. Some sensors or modules, such as a Bluetooth antenna array, visual positioning, or audio positioning, can provide fine-precision measurements of distances and angles in small-scale indoor spaces. In large-scale areas indoors, these sensors may have much larger measurement errors, which might lead to the failure of the positioning. It is therefore important to develop positioning algorithms that have enough flexibility to intelligently integrate different sensors with different observation accuracies.

26.3.2.3 Inconsistency in Different Smartphone Terminals

Different smartphone manufacturers may use different chipsets or components for the receiver modules or embedded sensors. Thus, the measurements from different smartphones may be biased due to the differences in the hardware of terminals. For example, different mobile phones have differences in the signal strength measurement of the same WiFi base station. Some deviations are actually quite large, which largely affects the positioning accuracy for fingerprinting-based positioning. Such inconsistencies also happen to cameras and MEMS sensors in different smartphones. A process of self-calibration can improve the consistency of the measurements from different smartphones to some extent. However, such difference or deviation is critical when considering fine-precision indoor positioning with accuracy within 1 m.

26.3.3 Limited Computing Resources on Mobile Terminals

As a handset, a smartphone is limited in its computing and storage capacity and power supply. Although the computing performance of smartphones has recently been increasing in accordance with Moore's Law, smartphones already perform multiple functions—phone calls, positioning, assistance with daily work, recreation, etc.—all of which demand a portion of computing and power resources. From the point of view of energy saving, it is therefore not suitable for the smartphone to keep running complicated positioning algorithms for a long time. Though some complex positioning algorithms such as visual positioning and particle filter are gradually implemented in smartphones, more complicated algorithms related to deep learning and AI are still inappropriate for the handset platform and will need continuing upgrade of the computation resources in smartphones in the future.

26.4 The Development Trends of Indoor Positioning Technology

Indoor positioning is one of the hot research topics in academia and industry. Google, as one of the leading IT companies, has promoted visual positioning service (VPS) as its core technology, which fully demonstrates the importance of indoor positioning in the future application of AI. Other internationally renowned IT companies, such as Apple, Baidu, Huawei, and Alibaba, have all listed indoor positioning as one of their strategic technologies. From the perspective of developing the technology with high accuracy, high utility, and low cost, the future directions of smartphone indoor positioning may include new positioning sources, effective fusion methods on heterogeneous positioning technologies, and cooperative positioning based on geographic information systems (GIS).

26.4.1 Explore New Positioning Sources for Fine-Precision, High-Utility Smartphone Indoor Positioning

More and more sensors are integrated into smartphones, providing the opportunity to develop new positioning technologies. Among them, audio positioning is one of the promising methods to achieve high-accuracy indoor positioning with smartphones. The position is determined by measuring the TDOA from the sound transmitter to the smartphone. The frequency for audio positioning can be set between 16 and 21 kHz, which is within the working frequency of the microphone, while above the frequency of audible sound. The advantage of sound positioning is that the requirement for time synchronization is not as strict as that for RF positioning. Because the speed of sound in the air is about 340 m/s, the time difference between acoustic transmitters is within 0.1 ms. At this time, the error of acoustic positioning is within 3.4 cm, although that is a quite large error for RF positioning.

Light-source coding and positioning is another candidate method for high-accuracy positioning with smartphones. The location of the smartphone is determined based on an LED light installed on the ceiling with on/off signals as the positioning source. By rotating the LED light, such a code has a unique pattern in each sector, which can be utilized by smartphone light sensors for positioning (Fig. 26.3). By measuring the relative position of the mobile phone in the sector, positioning accuracy of 5–10 cm can be achieved without changing the hardware of the mobile phone.

In terms of RF signal, Bluetooth 5.1 and 5G signals will play an important role in indoor positioning. Bluetooth technology has the characteristic of low power consumption, and BLE 5.1 has enhanced the indoor positioning with an angle-finding property, which will achieve sub-meter. 5G-based wireless positioning technology is likely to become one of the core technologies for future indoor positioning, as it has explicitly announced indoor and outdoor positioning accuracy to be better than 1 m (Koivisto et al. 2017; Laoudias et al. 2018). UWB signals have recently been integrated into Apple's smartphone. It is believed that UWB positioning in smartphones will attract more interest in applications.

Visual positioning based on cameras is still a promising method to achieve high accuracy with decimeter-level or even centimeter-level positioning errors, provided that the ambient lights and image features are sufficient. By integration with a depth camera, the visual positioning accuracy can be further improved, which has been verified in Google's Tango technology. However, the computation complexity is high, in particular in the processes of feature detection, image matching, and AI-related

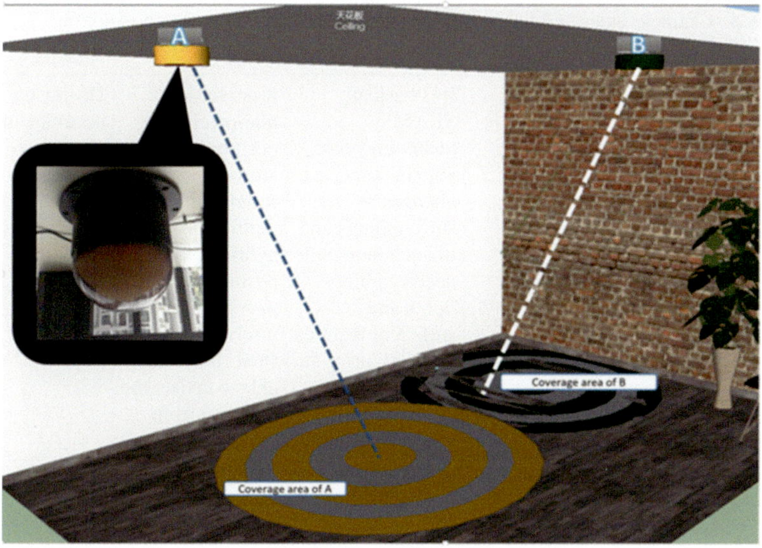

Fig. 26.3 Positioning with light coding

algorithms. With the 5G wireless communication systems coming into operation, their property of large bandwidth and low latency will allow smartphones to upload their photos to a cloud server, and get the positioning results from the server in real-time. It is, therefore, possible that all complicated algorithms will be computed in a high-performance cloud server.

Table 26.3 briefly analyzes the promising indoor positioning technologies mentioned above. Affected by the complex environment of indoor positioning, different positioning methods have their advantages and disadvantages in terms of positioning accuracy, reliability, availability, etc. In order to achieve continuous positioning estimates, fine-precision positioning technologies should intelligently fuse with each other.

26.4.2 Fusion of Heterogeneous Positioning Sources

At present, the technical development trend in the field of indoor positioning is to use a reliable estimation method to effectively integrate two or more positioning sources, to improve the accuracy and availability of the smartphone positioning system. In terms of the sensor fusion for indoor positioning, a complete solution needs to be developed, which should integrate the steps of heterogeneous hardware calibration, high-accuracy position estimation from a single technology, and the intelligent sensor-fusion method with the heterogeneous smartphone sensors. One possible way is to consider using the control points in the tightly coupled fusion

Table 26.3 Characteristics and function of future technologies for indoor positioning

Advanced technology	Characteristics	Function	Accuracy
Visual positioning	It is basically SLAM technology and able to perceive changes in surroundings and image features. It needs ambient lights and sufficient features	It provides fine-precision positioning and attitude estimation for multi-source hybrid positioning and provides initial information for PDR. It is also able to update the database from crowdsourcing	Decimeter accuracy in the scenarios of significant image features
RGB-D depth camera positioning	The depth information can be obtained and by using the method of angle and distance intersection, it is able to achieve decimeter level positioning The current price and power consumption are high, but in the future, it is likely to be more popular in mobile phones	The reliance on ambient light and image features is reduced. So, it can be used as a complement to visual positioning with optical cameras	Decimeter
Light-source code positioning	Fine precision and low power consumption on the smartphone side, and suitable for indoor open areas	As one of the main methods for smartphone positioning in open areas indoors	Decimeter
Fine-precision positioning based on new RF signals	Wide coverage and high availability, but suffers severely from multipath and NLOS errors	Provides fine-precision positioning results in a large space and the mainstream method of positioning with high availability	Sub-meter

(continued)

Table 26.3 (continued)

Advanced technology	Characteristics	Function	Accuracy
Sound indoor positioning	Independent of lighting and wireless base station; but needs special sound transmitters for positioning	Complement to the technology with visual and RF positioning. But not effective for car parking in the underground where the sound cannot penetrate into cars	Decimeter

method, where the control points are estimated from the high-accuracy positioning techniques mentioned in Sect. 26.2. To achieve a hybrid positioning solution with stability and reliability, it is also important to design appropriate filtering methods and cross-validation methods to identify the errors from heterogeneous measurements, in the case that the positioning sources are sufficient.

26.4.3 GIS-Based Semantic Constraint Location and Semantic Cognitive Collaboration Positioning

Currently, the research topics of GIS have gradually shifted from outdoors to indoors. Indoor GIS can on the one hand enhance the position estimates with indoor maps and indoor features, and on the other hand, fully utilize the potential value of indoor landmarks, providing semantic positioning capabilities with space constraints. However, all these supports are insufficient due to the lack of high-accuracy coordinates in current indoor GIS. Therefore, to establish a basic indoor GIS for a fine-precision intelligent indoor positioning system, the following key technologies need to be considered and properly addressed: (1) an indoor GIS model with a unified space–time reference system; (2) a simultaneous indoor modeling and positioning method with high-accuracy real-time coordinate computation; (3) an automatic update and instantaneous modeling method for maps using crowdsourcing; and (4) real-time visual positioning and 3D modeling with indoor semantics. At present, a new direction of indoor GIS research includes GIS-based semantic constraint positioning and semantic cognitive positioning.

26.5 Conclusions

Indoor positioning is one of the core technologies in the era of IoT, AI, and future super-AI (robots + human). Currently, smartphone-based indoor positioning technologies include RF positioning and sensor-based positioning. Many different methods have been developed for indoor positioning. However, all these technologies developed so far have their own shortcomings because they are affected by the complexity of space topologies, the heterogeneous data, and the limited computation capability from mobile terminals, and thus, are limited for developing a ubiquitous positioning solution. In order to meet the requirements of low cost, high accuracy, high usability, and high durability for mainstream applications, it is necessary to develop precise positioning solutions that are capable of adaptively fusing accurate observables, including visual images, light signals, acoustic signals, and RF signals. These precise locations can serve as the control points to prevent the propagation of positioning errors. To achieve full coverage, positioning solutions such as pedestrian dead reckoning and magnetic matching are needed to be integrated with the system.

References

Ali WH, Kareem AA, Jasim M (2019) Survey on wireless indoor positioning systems. Cihan Univ-Erbil Sci J 3(2):42–47

Apple (2014) Getting started with iBeacon Version 1.0

Bahl P, Padmanabhan VN (2000) RADAR: an in-building RF-based user location and tracking system. In: Proceedings of INFOCOM 2000, nineteenth annual joint conference of the IEEE computer and communications societies

Chen L, Kuusniemi H, Chen Y, Pei L, Kröger T (2011a) Information filter with speed detection for indoor Bluetooth positioning. In: Proceedings of the international conference on localization and GNSS (ICL-GNSS), Tampere, pp 47–52

Chen R, Chen W, Chen X, Zhang X, Chen Y (2011b) Sensing strides using EMG signal for pedestrian navigation. GPS Solut 15(2):161–170

Chen R, Pei L, Chen Y (2011c) A smart phone based PDR solution for indoor navigation. In: Proceedings of the 24th international technical meeting of the satellite division of the institute of navigation, Portland

Chen L, Pei L, Kuusniemi H, Kröger T, Chen R (2013) Bayesian fusion for indoor positioning using Bluetooth fingerprints. Wireless Pers Commun 70(4):1735–1745

Chen L, Kuusniemi H, Chen Y, Liu J (2015) Constrained Kalman filter for indoor Bluetooth localization. In: Proceedings of the 23rd European signal processing conference (EUSIPCO), Nice. IEEE, pp 1915–1919

Chen L, Thombre S, Jarvinen K, Simona E, Alen-Savikko A, Leppäkoski H, Bhuiyan M, Bu-Pasha S, Nunzia Ferrara G, Honkala S, Lindqvist J, Ruotsalainen L, Korpisaari P, Kuusniemi H (2017) Robustness, security and privacy in location-based services for future IoT: a survey. IEEE Access 5:8956–8977

Davidson P, Piché R (2016) A survey of selected indoor positioning methods for smartphones. IEEE Commun Surv Tutor 19(2):1347–1370

Del Peral-Rosado JA, Raulefs R, Lopez-Salcedo JA, Seco-Granados G (2017) Survey of cellular mobile radio localization methods: from 1G to 5G. IEEE Commun Surv Tutor 20(2):1124–1148

Deng Z, Wang G, Qin D, Na Z, Cui Y, Chen J (2016) Continuous indoor positioning fusing WiFi, smartphone sensors and landmarks. Sensors 16(9):1427–1447

Ericsson Research Blog (2015) Indoor positioning in LTE [EB/OL]. https://www.ericsson.com/research-blog/lte/indoor-positioning-in-lte/

Faragher R, Harle R (2015) Location fingerprinting with Bluetooth low energy beacons. IEEE J Sel Areas Commun 33(11):1–1

Federal Communications Commission (2015) Wireless E911 location accuracy requirements. Ps Docket

Ganick A, Ryan D (2012) Light positioning system using digital pulse recognition. US, 2012-08-21

Globe Newswire (2019) Indoor location-based services (LBS) market analysis report by product, by technology (RFID and NFC, Wi-Fi/WLAN and BT/BLE), by application (tracking, proximity, navigation), by end use, and segment forecasts, 2019–2025

He S, Chan SHG (2015) Wi-Fi fingerprint-based indoor positioning: recent advances and comparisons. IEEE Commun Surv Tutor 18(1):466–490

How-To Geek (2019) https://www.howtogeek.com/403606/bluetooth-5.1-whats-new-and-why-it-matters/. Accessed 17 Jan 2020

IEEE Standard for Information Technology (2013) Telecommunications and information exchange between systems—local and metropolitan area networks—specific requirements—part 11: wireless LAN medium access control (MAC) and physical layer (PHY) specifications; amendment 4: enhancements for very high throughput for operation in bands below 6 GHz. IEEE Std 802.11ac (TM)-2013 (Amendment to IEEE Std 802.11-2012, as amended by IEEE Std 802.11ae-2012, IEEE Std 802.11aa-2012, and IEEE Std 802.11ad-2012), pp 1–425

Ijaz F, Yang HK, Ahmad AW, Lee C (2013) Indoor positioning: a review of indoor ultrasonic positioning systems. In: Advanced communication technology. IEEE

Jeon JS, Kong Y, Nam Y, Yim K (2015) An indoor positioning system using Bluetooth RSSI with an accelerometer and a barometer on a smartphone. In: 10th international conference on broadband and wireless computing, communication and applications (BWCCA)

Kaplan E, Hegarty C (2005) Understanding GPS: principles and applications. Artech House

Karlsson F, Karlsson M, Bernhardsson B, Tufvesson F, Persson M (2015) Sensor fused indoor positioning using dual band WiFi signal measurements. In: Proceedings of the 2015 European control conference, Linz

Khalajmehrabadi A, Gatsis N, Akopian D (2017) Modern WLAN fingerprinting indoor positioning methods and deployment challenges. IEEE Commun Surv Tutor 19(3):1974–2002

Kim B, Kwak M, Lee J, Kwon T (2014) A multi-pronged approach for indoor positioning with WiFi, magnetic and cellular signals. In: International conference on indoor positioning and indoor navigation (IPIN)

Koivisto M, Hakkarainen A, Mário C, Kela P, Leppanen K, Valkama M (2017) High-efficiency device positioning and location-aware communications in dense 5G networks. IEEE Commun Mag 55(8):188–195

Kotaru M, Joshi K, Bharadia D, Katti D (2015) Spotfi: decimeter level localization using WiFi. In: Proceedings of the 2015 ACM conference on special interest group on data communication, London. ACM, pp 269–282

Lakmali BDS, Dias D (2008) Database correlation for GSM location in outdoor & indoor environments. In: Proceedings of the 4th international conference on information and automation for sustainability (ICIAFS), Colombo. IEEE, pp 42–47

Laoudias C, Moreira A, Kim S, Lee S, Wirola L, Fischione C (2018) A survey of enabling technologies for network localization, tracking, and navigation. IEEE Commun Surv Tutor 20(4):3607–3644

Li X, Wang J, Liu C, Zhang L (2016) Integrated WiFi/PDR/smartphone using an adaptive system noise extended Kalman filter algorithm for indoor localization. ISPRS Int J Geo-Inf 5(2):8–26

Liu H, Darabi H, Banerjee P, Liu J (2007) Survey of wireless indoor positioning techniques and systems. IEEE Trans Syst Man Cybern Part C (Appl Rev) 37(6):1067–1080

Liu J, Chen Y, Jaakkola A, Hakala T, Hyyppä J, Chen L, Tang J, Chen R, Hyyppä H (2014) The uses of ambient light for ubiquitous positioning. In: Position, location and navigation symposium, IEEE/ION

Maghdid HS, Lami IA, Ghafoor KZ, Lloret J (2016) Seamless outdoors-indoors localization solutions on smartphones: implementation and challenges. ACM Comput Surv 48(4):1–34. https://doi.org/10.1145/2871166

Morrison A, Renaudin V, Bancroft JB, Lachapelle G (2012) Design and testing of a multi-sensor pedestrian location and navigation platform. Sensors 12(3):3720–3738
Pei L, Liu J, Guinness R, Chen Y, Kröger T, Chen R, Chen L (2012) The evaluation of WiFi positioning in a Bluetooth and WiFi coexistence environment. In: Ubiquitous positioning, indoor navigation, and location based service (UPINLBS)
Priyantha NB, Chakraborty A, Balakrishnan H (2000) The cricket location-support system. In: Proceedings of the 6th annual international conference on mobile computing and networking. ACM, Boston
Quigley M, Stavens D, Coates A, Thrun S (2010) Sub-meter indoor localization in unmodified environments with inexpensive sensors. In: 2010 IEEE/RSJ international conference on intelligent robots and systems (IROS)
Quuppa (2020) https://quuppa.com/company/. Accessed 17 Jan 2020
Robert H (2013) A survey of indoor inertial positioning systems for pedestrians. IEEE Commun Surv Tutor 15(3):1281–1293
Ruotsalainen L (2012) Visual gyroscope and odometer for pedestrian indoor navigation with a smartphone. In: Proceedings ION GNSS
Ruotsalainen L, Kuusniemi H, Bhuiyan MZH, Chen L, Chen R (2013) A two-dimensional pedestrian navigation solution aided with a visual gyroscope and a visual odometer. GPS Solut 17(4):575–586
Ruotsalainen L, Kirkko-Jaakkola M, Chen L, Simo G, Robert G, Heidi K (2016) Multi-sensor SLAM for tactical situational awareness. In: 2016 international technical meeting of the institute of navigation, Monterey
Saeedi S (2013) Context-aware personal navigation services using multi-level sensor fusion algorithms. Doctoral dissertation, University of Calgary
Sakpere W, Adeyeye-Oshin M, Mlitwa NB (2017) A state-of-the-art survey of indoor positioning and navigation systems and technologies. S Afr Comput J 29(3):145–197
Schauer L, Dorfmeister F, Maier M (2013) Potentials and limitations of WiFi-positioning using time-of-flight. In: International conference on indoor positioning and indoor navigation. IEEE, pp 1–9
Schmitt S, Adler S, Kyas M (2014) The effects of human body shadowing in RF-based indoor localization. In: IEEE international conference on indoor positioning and indoor navigation (IPIN)
Talvitie J, Valkama M, Destino G, Wymeersch H (2017) Novel algorithms for high-accuracy joint position and orientation estimation in 5G mmWave systems. In: IEEE globecom workshops, pp 1–7
Tang J, Chen Y, Chen L, Liu J, Hyyppä J, Kukko A, Kaartinen H, Hyyppä H, Chen R (2015) Fast fingerprint database maintenance for indoor positioning based on UGV SLAM. Sensors 15(3):5311–5330
Thompson S (2020) IndoorAtlas. https://www.indooratlas.com
Vasisht D, Kumar S, Katabi D (2016) Decimeter-level localization with a single WiFi access point. In: Proceedings of the 13th usenix conference on networked systems design and implementation, Santa Clara. ACM, pp 165–178
Wang J, Hu A, Liu C, Li X (2015a) A floor-map-aided WiFi/pseudo-odometry integration algorithm for an indoor positioning system. Sensors 15(4):7096–7124
Wang X, Gao L, Mao S, Pandey S (2015) DeepFi: deep learning for indoor fingerprinting using channel state information. In: 2015 IEEE wireless communications and networking conference (WCNC). IEEE, pp 1666–1671
Wang X, Gao L, Mao S, Pandey S (2016) CSI-based fingerprinting for indoor localization: a deep learning approach. IEEE Trans Veh Technol 66(1):763–776
Ward A, Jones A (1997) A new location technique for the active office. IEEE Pers Commun 4(5):42–47
Weiser M (2002) The computer for the 21st century. IEEE Pervasive Comput 1(1):19–25
Wu K, Xiao J, Yi Y, Chen D, Luo X, Ni L (2012) CSI-based indoor localization. IEEE Trans Parallel Distrib Syst 24(7):1300–1309

Youssef M, Agrawala A (2008) The Horus location determination system. Wireless Netw 14(3):357–374
Zafari F, Papapanagiotou I, Christidis K (2015) Micro-location for internet of things equipped smart buildings. IEEE Internet Things J 3(1):96–112
Zafari F, Gkelias A, Leung KK (2019) A survey of indoor localization systems and technologies. IEEE Commun Surv Tutor
Zhang R, Bannoura A, Höflinger F, Reindl L, Schindelhauer C (2013) Indoor localization using a smart phone. In: Sensors applications symposium. IEEE
Zhao Y (2002) Standardization of mobile phone positioning for 3G systems. IEEE Commun Mag 40(7):108–116
Zhao H, Huang B, Jia B (2016) Applying kriging interpolation for Wi-Fi fingerprinting based indoor positioning systems. In: Proceedings of the 2016 IEEE wireless communications and networking conference, Doha, pp 1–6
Zhang M, Tingting Q, Jinhui Z, et al (2017) Indoor positioning tracking with magnetic field and improved particle filter [J]. Int J Distrib Sens Networks 13(11)
Zhuang Y, Syed Z, Li Y, El-Sheimy N (2015) Evaluation of two WiFi positioning systems based on autonomous crowd sourcing on handheld devices for indoor navigation. IEEE Trans Mob Comput 15(8):1982–1995

Ruizhi Chen is currently a professor and the director of the State Key Laboratory of Information Engineering in Surveying, Mapping and Remote Sensing (LIESMARS), Wuhan University. Before joining Wuhan University, he was as the Endowed Chair Professor in Texas A&M University Chorus Christi, U.S.A. Before that, he was the department head of Navigation of Positioning in Finnish Geodetic Institute. The main research of Prof. Ruizhi Chen is on the smart phone ubiquitous positioning and satellite navigation. He proposed a new concept of "mobile phone context thinking engine". He was president of the Global Chinese Navigation and Positioning Society, board member of the Nordic Institute of Navigation and he is the General Chair of the IEEE UPINLS conference.

Liang Chen is currently a professor in the State Key Laboratory of Information Engineering in Surveying, Mapping and Remote Sensing (LIESMARS), Wuhan University, China. He is currently the deputy director of the Institute of Spatial Intelligence, in LIESMARS of Wuhan University. Before he joined LIESMARS, he was a Senior Research Scientist in the Department of Navigation and Positioning at the Finnish Geospatial Research Institute (FGI), Finland. His research interests include seamless indoor/outdoor positioning, wireless positioning, signals of opportunity, and sensor fusion algorithm for indoor positioning. He has published more than 70 research papers. He serves as Associate Editor in NAVIGATION, The Journal of Institute of Navigation, and Journal of Navigation.

Chapter 27
What Urban Cameras Reveal About the City: The Work of the Senseable City Lab

Fábio Duarte and Carlo Ratti

Abstract Cameras are part of the urban landscape and a testimony to our social interactions with city. Deployed on buildings and street lights as surveillance tools, carried by billions of people daily, or as an assistive technology in vehicles, we rely on this abundance of images to interact with the city. Making sense of such large visual datasets is the key to understanding and managing contemporary cities. In this chapter, we focus on techniques such as computer vision and machine learning to understand different aspects of the city. Here, we discuss how these visual data can help us to measure legibility of space, quantify different aspects of urban life, and design responsive environments. The chapter is based on the work of the Senseable City Lab, including the use of Google Street View images to measure green canopy in urban areas, the use of thermal images to actively measure heat leaks in buildings, and the use of computer vision and machine learning techniques to analyze urban imagery in order to understand how people move in and use public spaces.

27.1 Introduction

Cameras have become part of the urban landscape and a testimony of our social interactions with the city. They are deployed on buildings and street lights as surveillance tools, carried by billions of people daily, or as an assistive technology in vehicles with different levels of self-driving capabilities. We rely on this abundance of images to interact with the city.

In fact, 2.5 quintillion bytes of data are created each day by billions of people using the Internet. Increasingly, social media are heavily based on visual data. Among the top social media channels, several are overwhelmingly and exclusively based on images: YouTube has 1.5 billion users and Instagram has 1 billion users—as a

F. Duarte (✉) · C. Ratti
Senseable City Lab and Department of Urban Studies and Planning, Massachusetts Institute of Technology, Massachusetts, USA
e-mail: fduarte@mit.edu

C. Ratti
e-mail: ratti@mit.edu

W. Shi et al. (eds.), *Urban Informatics*, The Urban Book Series,
https://doi.org/10.1007/978-981-15-8983-6_27

comparison, Facebook has 2.3 billion users. Such visually based social interactions are also extended to the interactions we have in our cities. In the USA, on average, a person is caught on camera 75 times per day, and over 300 times in London. Also, disruptive urban technologies such as autonomous vehicles use cameras. The challenge is to make sense of the amount of visual data generated daily in our cities in meaningful ways, beyond surveillance purposes.

In this chapter, we are not interested in the abundance of visual data available online collected by individuals and widely available on social media. The previous work used geotagged photographs available online to measure urban attractiveness (Paldino et al. 2016) or to assess the aesthetic appeal of the urban environment based on user-generated image (Saiz et al. 2018), and the visual discrepancy and heterogeneity of different cities around the world (Zhang et al. 2019). The focus of this chapter is not on the visual data produced by cameras carried by people for personal uses, but rather on the images collected by cameras specifically designed and deployed to gather visual data about the city—which we call here urban cameras.

Cameras deployed and controlled by a range of public and private organizations in urban areas are counted by the dozens of thousands in cities, from London and Beijing to New York and Rio de Janeiro. As an example, a Londoner is captured on camera more than 300 times every day; and during the same period, the UK captures over 30 million plate numbers (Kitchin 2016). Additionally, private companies, such as Google, collect and make available online hundreds of thousands of images of hundreds of cities worldwide.

Making sense of such large visual datasets is the key to understanding and managing contemporary cities. There are still many technical issues to be solved to make the use of such huge visual datasets actionable. Challenges include cloud versus local storage and processing; architecture integration, ontology building, semantic annotation, and search; and online real-time analysis and offline batch processing of large-scale video data (Shao et al. 2018; Xu et al. 2014; Zhang et al. 2015).

Besides the technical challenges, there are also ethical issues. The most prevalent among social scientists is the narrow understanding of cities when urban phenomena are equaled to available data, heading the operationalization of the urban (Luque-Ayala and Marvin 2015), mainly when "portions of the urban public space that are shadowed by the gaze of private cameras and security systems" (Firmino and Duarte 2015 p. 743) become subject to the datafication of the city, often leading to "social sorting and anticipatory governance" (Kitchin 2016 p. 4). Closed-circuit television (CCTV), deployed on public areas and aimed to assist police patrols with crime prevention, using video analytics to identify abnormal behaviors, fosters predictive policing by the profiling of subjects and places, and frequently triggers false alarms due to biases embedded in the algorithms (Vanolo 2016).

We are aware of these issues and have contributed ourselves to the literature on the risks of oversurveillance based on the abundance of data about people's behavior in public spaces. But, in this chapter, we would like to discuss the other side of this phenomenon: how novel computational techniques can be used to make sense of the huge amount of visual data generated about cities, and how such results reveal aspects of urban life that can contribute to better understanding and design of cities.

The projects discussed in this chapter are part of the extensive work using urban cameras done by the Senseable City Lab, at the Massachusetts Institute of Technology. These works can be divided into two types: the use of visual urban data available online, and the capture of visual data by the Lab with specifically designed devices.

In the first type, we take advantage of the visual urban data available online and develop machine learning techniques to make sense of these data. The datasets used in this research are Google Street View images, which we have been using to measure a critical aspect of cities with rapid urbanization: the quantification of green canopy in urban areas using a standard method that can be deployed cheaply, and that makes possible comparisons among hundreds of cities worldwide. And, at the same time, it provides a fine-grained analysis of greenery at the street level, allowing citizens and municipalities to assess tree coverage in different neighborhoods.

In the second type, we design specific devices to collect images and deploy them ourselves. In one example, we started by using thermal cameras mounted on vehicles to measure heat leaks in buildings. Using the same devices, we developed other techniques to use thermal data to quantify and track people's movements in indoor and outdoor areas. Besides the technical advantages of the method in terms of data transmission and processing, it also addresses an important concern about the use of cameras in public spaces: Thermal cameras allow us to have accurate data about people's behavior without revealing their identities, therefore avoiding privacy concerns. Also, as part of this type of research, we address the problem of indoor navigability in large public areas. It is a well-known problem that users often have difficulty in navigating areas such as shopping malls, university campuses, and train stations, due either to their labyrinthic design or to the repetitiveness of visual cues. Here, we collected thousands of images on the MIT campus and in train stations in Paris and trained a neural network to measure the easiness to navigate these spaces, comparing the results with a survey with users.

Visual data about cities will tend to increase in the coming years, with personal photographs and videos that people use to register their daily routines in cities posted on social media, the deployment of cameras for surveillance not only for policing purposes but also for traffic management and infrastructure monitoring, and the fact that visual data will be crucial in technologies such as self-driving cars. All work dealing with visual big data needs to overcome the hurdles of manually processing this massive amount of information and generating useful empirical metrics on visual structure and perception. In this chapter, we propose to discuss how the development of novel computation methods used to analyze the abundance of visual urban data can help us to better understand urban phenomena.

27.2 Computer Vision and the City: Google Street View Images

Some of the most prolific sources of spatial data are Google Maps, Earth, and Street View. These products offer Web mapping, rendering of satellite imagery onto a 3D representation of the Earth, terrain and street maps, and 360° panoramic views of hundreds of cities around the world. GSV in particular has several advantages that allow a quantitative study of the visual features of cities, including the availability of images in hundreds of cities in more than 80 countries, the use of similar photographic equipment everywhere, all images being georeferenced, and all images are available for download. As an example of the amount of visual urban data in GSV datasets, in New York City, there are approximately 100,000 sampling points: It sums up to approximately 600,000 images, since GSV captures six photographs at each sampling point. GSV and similar services have made available an unprecedented visual database of cities around the world with comparable characteristics.

Several researchers have been using GSV to analyze cities. Khosla, An, Lim et al. (2014) have analyzed 8 million GSV images from eight cities in different countries in order to compare how accurately humans and computers can predict crime rates and economic performance. Convolutional neural networks have been used by many researchers interested in measuring how physical features of cities affect different aspects of urban life, such as chronic diseases, the presence of crosswalks, building type, and vegetation coverage (Nguyen et al. 2018; Zhang al. 2019). GSV images have also used to quantify urban perception and safety (Dubey et al. 2016; Naik et al. 2014), to detect and count pedestrians (Yin et al. 2015), to infer landmarks in cities (Lander et al. 2017), and to quantify the connection between visual features and sense of place, based on perceptual indicators (Zhang et al. 2018).

Since 2015, the MIT Senseable City Lab has been using GSV to measure green canopy in cities. Xiaojiang Li pioneered this research with the Lab, using deep convolutional neural networks to quantify the amount of green areas at the street level. In this research initiative, called *Treepedia*, the focus is on the pedestrian exposure to trees and other green areas along the streets. Streets are the most active spaces in the city, where people see and feel the urban environment in their daily lives. Street-level images have a similar view angle with to pedestrians and can be used as proxies of physical appearance of streets as perceived by humans.

Li et al. (2015) and Seiferling et al. (2017) calculated the percentage of green vegetation in streets based on large GSV datasets. The process begins by creating sample sites, usually every 100 meters along the streets, and then collecting GSV metadata, static images, and panoramas. The basic technique involves the use of computer vision and DCNN to detect green pixels in each image. Once green pixels are detected, all the remaining part is subtracted, giving a general quantification of greenery. Thus, the percentage of the total green pixels from six images taken at each site to the total pixel numbers of the six images gives the Green View Index (Li et al. 2018).

Recent development in deep learning models allows us to improve the methodology to calculate the GVI. Initiated by Bill Cai (Cai et al. 2018), another researcher with the Senseable City Lab, the goal here is to quantify what is actually vegetation in GSV images, rather than using the ratio of green pixels as proxies to street-level greenery. The process begins by labeling images in a small-scale validation dataset. In this case, five cities with different climatic conditions were selected: Cambridge (Massachusetts, USA), Johannesburg (South Africa), Oslo (Norway), São Paulo (Brazil), and Singapore. One hundred images were randomly selected for each city, and vegetation was manually labeled. The DCNN model was then trained using the pixel-labeled Cityscapes dataset. Researchers also used a gradient-weighted class activation map (Grad-CAM) to interpret the features used by the model to identify vegetation. Results show that the DCNN models outperform the original Treepedia unsupervised segmentation model significantly, decreasing the mean absolute error from 10% to 4.7%.

The Treepedia Web site counts the Green View Index for 27 cities, and we have recently released an open-source Python library that allows anyone to calculate the GVI for a city where GSV images are available.

27.3 Thermals Images of the City

The richness of urban understanding that can be derived from video cameras is well known in urban studies. In groundbreaking research in the 1970s, William Whyte (2009) employed time-lapse cameras to understand people's behavior in public spaces and used this information to inform design. The negative reactions triggered by the deployment of cameras in public areas frequently happen due to a narrow understanding of their purposes (surveillance and policing) and poor analytical techniques, often based on officers watching footage (Luque-Ayala and Marvin 2015; Firmino and Duarte 2015).

In recent years, in research initiated by Amin Amjonshooa, the MIT Senseable City Lab has been addressing these three problems related to the deployment of cameras in urban areas. We do this by widening the spectrum of urban phenomena that we can understand using cameras, developing image processing techniques that are novel to urban studies, and employing cameras that by design do not capture people's identity features. Here, we discuss the quantification of traffic-related heat loss and people's trajectories in space using cameras mounted on street lights, and the assessment of building heat loss using cameras deployed on vehicles.

Human activities generate heat. Cooling and heating systems and transportation, to stay with examples that are part of our daily lives, generate anthropogenic heat and release it into the ambient environment. They are major sources of low-grade energy that have direct and indirect impacts on human health. Cars alone, either powered by gasoline or diesel, release 65% of the heat produced by engines into the urban environment. In order to assess vehicular heat emissions at the street level, and match such emissions to the number of pedestrians directly exposed, we have been using thermal cameras deployed in the existing infrastructures.

Thermal cameras capture wavelengths and measure the infrared radiation emitted from objects. They have a single channel, and thermal images have lower resolution, which makes thermal data much smaller in size, in comparison with RGB visual images. Smaller data size allows faster and better data transmission and processing, being less computational intensive. Thermal data only look like images when we apply the appropriate color maps.

The previous work has used thermal cameras to identify space occupancy and count people. Qi al. (2016) proposed the use of thermal images as a sparse representation for pedestrian detection. Gade et al. (2016) developed a system to automatically detect and quantify people in sport arenas, by counting pixel differences between two successive frames. Interestingly, they also showed that based on the movements captured by thermal cameras, they were able to differentiate the sport modality people are playing, based on the position, concentration, and trajectories of people in space.

We deployed FLIR Lepton micro thermal cameras on street lights next to MIT, in Cambridge, MA, with the goals of quantifying traffic-related heat loss and tracking pedestrian movements.

Internal combustion vehicles are one of the major sources of heat in cities. Based on the analysis of thermal images captured at this high-traffic intersection, we were able to quantify and visualize both heat intensity and traffic load. Thermal cameras showed another advantage in relation to RGB cameras: Besides the counting of vehicles and simple identification (motorcycles, cars, trucks, buses), thermal images also allowed us to measure whether the vehicle had been running for a short or long period before being scanned (Anjomshoaa et al. 2016). This analysis generated a thermal fingerprint of traffic flow at the intersection.

For the analysis of the thermal images, we propose a method based on accumulated Radon Transform, which computes the projection of images along various angles. The Radon Transform of thermal images reveals the warmer objects and at the same time preserves their locations. We used the same dataset to count pedestrians passing on the sidewalk near traffic. In order to optimize data transmission and processing, we limited the target area to a sidewalk segment next to the pedestrian crossing. It also helped us to eliminate the high thermal flux of cars, which would otherwise make detecting pedestrian thermal flux harder. With this research, we were able to study the exposure of pedestrians to various anthropogenic pollutants caused by internal combustion vehicles. Also, by detecting thermal peaks, we were able to differentiate between single individuals and groups of individuals; and by learning from many hours of image analysis and the varying amplitude of the peaks, we were able to estimate the number of people in the scene.

In the project called City Scanner, the Lab has been developing a drive-by solution in which we mount a modular sensing platform on ordinary urban vehicles—such as school buses and taxis—to scan the city. The advantage of this approach is that it does not require specially equipped vehicles, since our modular sensing platform can be deployed virtually on any vehicle. To prove this concept, in Cambridge, MA, we deployed the sensing platform on trash trucks (Anjomshoaa et al. 2018).

Among the sensors scanning the city for a period of eight months, we had two thermal cameras capturing data from the two sides of streets. These were non-radiometric thermal cameras, in which case the thermal output is not the scene temperature, but only a display of temperature fields. Scanning the thermal signature of all street segments of the city over different seasons, we created a thermal signature of the built environment in Cambridge. With these data and continuous scanning, any anomaly in the thermal difference between neighboring buildings might trigger a detailed analysis by city officials. In the case of Cambridge, a city that has programs to help residents to improve house insulation, this constant scanning can help the public authorities to be responsive when heat leaks are detected.

27.4 Navigating Urban Spaces Using Computer Vision

The explosion of big visual data is offering new sources of data that can overcome spatial and resource constraints that are common in studies of perception and legibility of urban spaces. At the Senseable City Lab, we have been using computer vision and deep convolutional neural networks to understand how people perceive, locate themselves, and navigate spaces.

As we have explained elsewhere (Wang et al. 2019), DCNN is based on probabilistic program induction, achieved by a bank of filters whose weights are adjusted during the training phase, with the goal of obtaining the key features of the images and, more importantly, the interplay of these features.

Here, we are particularly interested in addressing the problem of indoor navigability in large public areas. It is a well-known problem that users often have difficulty in navigating areas such as shopping malls, university campuses, and train stations, due to either their labyrinthic design or to the repetitiveness of visual cues.

In order to address this challenge, we have collected hundreds of thousands of images in two space types: university campuses and train stations. We trained a deep convolutional neural network to measure the easiness to navigate these spaces, and in the case of the train stations, we compared the results with a survey of users.

We first decided to test navigability on the MIT campus—in particular in a quite bland and disorienting space: the so-called infinite corridor, the interconnected indoors corridors and atriums that links several MIT buildings. The goal was to test DCNN to recognize different locations based on spatial features. Led by Fan Zhang (Zhang, Duarte, Ma et al. 2016), the study was based on 600,000 images extracted from video footage which we took using a GoPro camera for the training dataset, and 1,697 images taken with a smartphone for the test dataset. We compared our model with two commonly used in DCNN, and regarding the location in space, we achieved 96.90% top-1 accuracy on the validation dataset—higher than the other available models. We also proposed an evaluation method to assess how distinctive an indoor place is, when compared with all other spaces in the study area, and produced a distinctiveness map of buildings on the MIT campus, which might help to explain how people find their way (or get lost) in the infinite corridors of MIT.

Another indoor public space that might be disorienting is the train station (Wang, Liang, Duarte et al. 2019). In this research, we measured space legibility in two train stations in Paris: Gare de Lyon and Gare St. Lazare, each receiving more than 250,000 passengers daily. Legibility influences the ability of people to locate themselves and find their way—or navigate space (Herzog and Leverich 2003). We developed a device composed of a LiDAR sensor and a 360 camera. After the projection transformation, we cropped hundreds of thousands of images from panoramic images from each station to train our DCNN.

In our DCNN, we have removed the final labeling part of the neural network, because our goal was not to identify what objects are present in each image, but to understand how visual properties are used to navigate space based on visual similarities. For Gare de Lyon, we tested the model on 88,869 images and achieved 97.11% prediction accuracy of its top-1 choice, and 97.23% for Gare St. Lazare.

Although the model performed very well (more than 97% top-1 accuracy) overall, we noticed discrepancies in accuracy among different spaces in different floors and related to different uses, which could reflect different spatial legibility. Research using computer vision frequently employs surveys to test results. On one setting, in their study to compare how accurately humans and computers can predict the existence of nearby establishments, crime rates, and economic performance of urban areas, Khosla et al. (2014) used Amazon Mechanical Turk and asked participants to guess where are some establishments; on another setting, they trained the computer to recognize five visual features of the images. Their results show humans and computers with similar performance.

Thus, to prove the validity of our model, we deployed a survey on Amazon Mechanical Turk, collecting 4,015 samples. The human samples showed a similar behavior pattern and mechanism as the DCNN models. A 10-second video was shown to all participants on a Web-based survey. On the next page, we displayed one image snippet from the spatial segment shown in the video, in addition to three images (one from the same scene). From these three images, participants were asked to choose one that matched the same scene and were asked to point out three features that helped them to make the decision. We compared these results with the activation layer, which is the fully connected layer of the DCNN model. We created heatmaps of the main features used by the model and by humans to read spaces. Although in several situations both have focused on the same areas, discrepancies are also important: One example is that participants often used objects, such as TV screens or advertisement boards, to help recognize spaces and locate themselves—indicating that semantic values play an important role in spatial legibility, in addition to spatial features and visual cues. More importantly, the research showed that computer vision techniques can help us to understand space legibility even closer to how humans read space. Since the deployment of cameras is more easily reproducible than doing surveys, computer vision and DCNN are opening new avenues in the study of space legibility that can inform wayfinding and space design.

27.5 Conclusion

In this chapter, we discussed three initiatives by the Senseable City Lab, in which we proposed special devices, designed experiments, and developed machine learning methods to analyze visual urban data. Either by taking advantage of urban imagery available online or by collecting RGB and thermal images in urban areas, the goal is to demonstrate how these multiple images can help us to reveal different aspects of the city. It is only by creating novel approaches to understand the visual data generated in cities that we will be able to understand contemporary urban phenomena and inform design in innovative ways.

The abundance of images certainly raises several problems, mainly regarding individual privacy—and this topic must be taken seriously. However, we should raise other questions regarding ownership and proper use of images collected in urban areas. For example, plenty of breakthrough research has been done in the fields of urban design, computer science, and sociology, using the urban scenes available online in platforms such as Google Street View. This was done with the tacit understanding that a private company was taking pictures of public spaces and making them available for non-commercial use—including scientific research. It was almost a trade-off: We allow Google to put online images of the façades of our houses, our backyards, and our cars when parked on the streets, and, in exchange, we could use these images for the common good of deepening our understanding of cities. Recently, Google changed its rules and now forbids almost any use of Google Street View images, including for academic purposes. Thus, should we accept quietly that a private company can take millions of images of public spaces and make money out of it? And even of our private properties? The question of privacy is essential in an era of overabundance of images; but, likewise, is the question of allowing private companies to profit from common goods—and the cities are the essential common good of the modern age.

Another important aspect of the future of urban ambient sensing is that sensors will be increasingly embedded in our buildings and carried by people in different formats. In this chapter, we discussed research based on the collection of passive data from our cities: images. More and more, construction materials have sensors as their components, sensors that not only feel the environment, but also react to it. Fully transparent glass panels embedded with photovoltaic cells measure the amount of light, change the opacity to adjust to the luminosity set by the users, and, at the same time, generate energy. On the personal side, if we currently carry sensors in our cellphones, these sensors are also becoming the constituent material of our clothes, for instance. They measure the body temperature, the ambient temperature, and adjust the clothing to our optimal comfort. At the same time that glass panels or clothing are sensing and actuating at the individual level with building or user, they are also generating data that can help us to better understand the relations established between people, the built environment, and nature. Exploring new methods to understand these relations is the key to foster innovative urban design.

References

Anjomshoaa A, Duarte F, Rennings D, Matarazzo TJ, Desouza P, Ratti C (2018) City Scanner: building and scheduling a mobile sensing platform for smart city services. IEEE Internet of Things Journal 5:4567–4579. https://doi.org/10.1109/jiot.2018.2839058

Anjomshoaa A, Duarte F, Alvarez R, Britter R, Ratti C (2016) Quantifying the anthropogenic heat in urban areas using thermal images. 2016 International conference on computational science and computational intelligence (CSCI). https://doi.org/10.1109/csci.2016.0128

Cai BY, Li X, Seiferling I, Ratti C (2018) Treepedia 2.0: applying deep learning for large-scale quantification of urban tree cover. 2018 IEEE international congress on big data (BigData Congress). https://doi.org/10.1109/bigdatacongress.2018.00014

Dubey A, Naik N, Parikh D, Raskar R, Hidalgo CA (2016) Deep learning the city: quantifying urban perception at a global scale. Computer Vision—ECCV 2016 Lecture Notes in Computer Science 196–212. https://doi.org/10.1007/978-3-319-46448-0_12

Firmino R, Duarte F (2015) Private video monitoring of public spaces: The construction of new invisible territories. Urban Stud 53:741–754. https://doi.org/10.1177/0042098014567064

Gade R, Moeslund TB, Nielsen SZ, Petersen HS, Andersen HJ, Basselbjerg K, Dam HT, Jensen OB, Jørgensen A, Lahrmann H, Madsen TKO, Bala ES, Povey BØ (2016) Thermal imaging systems for real-time applications in smart cities. Int J Comput Applications in Technology 53:291. https://doi.org/10.1504/ijcat.2016.076790

Herzog TR, Leverich OL (2003) Searching for legibility. Environ Behav 35:459–477. https://doi.org/10.1177/0013916503035004001

Khosla A, An B, Lim JJ, Torralba A (2014) Looking beyond the visible scene. 2014 IEEE conference on computer vision and pattern recognition. https://doi.org/10.1109/cvpr.2014.474

Kitchin R (2016) The ethics of smart cities and urban science. Philosophical Transactions of the Royal Society A: Mathematical, Physical and Engineering Sciences 374:20160115. https://doi.org/10.1098/rsta.2016.0115

Lander C, Wiehr F, Herbig N, Krüger A, Löchtefeld M (2017) Inferring landmarks for pedestrian navigation from mobile eye-tracking data and Google Street View. In: Proceedings of the 2017 CHI conference extended abstracts on human factors in computing systems, CHI EA '17. https://doi.org/10.1145/3027063.3053201

Li X, Cai BY, Ratti C (2018) Using street-level images and deep learning for urban landscape studies. Landscape Architect Frontiers 6:20. https://doi.org/10.15302/j-laf-20180203

Li X, Zhang C, Li W, Ricard R, Meng Q, Zhang W (2015) Assessing street-level urban greenery using Google street view and a modified green view index. Urban For Urban Greening 14(3):675–685. https://doi.org/10.1016/j.ufug.2015.06.006

Luque-Ayala A, Marvin S (2015) The maintenance of urban circulation: an operational logic of infrastructural control. Environ Plan D: Soc Space 34:191–208. https://doi.org/10.1177/0263775815611422

Naik N, Philipoom J, Raskar R, Hidalgo C (2014) Streetscore–predicting the perceived safety of one million streetscapes. 2014 IEEE conference on computer vision and pattern recognition workshops. https://doi.org/10.1109/cvprw.2014.121

Nguyen QC, Sajjadi M, Mccullough M, Pham M, Nguyen TT, Yu W, Meng H-W, Wen M, Li F, Smith KR, Brunisholz K, Tasdizen T (2018) Neighbourhood looking glass: 360° automated characterisation of the built environment for neighbourhood effects research. J Epidemiol Commun Health 72:260–266. https://doi.org/10.1136/jech-2017-209456

Paldino S, Kondor D, Bojic I, Sobolevsky S, González MC, Ratti C (2016) Uncovering urban temporal patterns from geo-tagged photography. PLOS One. https://doi.org/10.1371/journal.pone.0165753

Qi B, John V, Liu Z, Mita S (2016) Pedestrian detection from thermal images: A sparse representation based approach. Infrared Phys Technol 76:157–167. https://doi.org/10.1016/j.infrared.2016.02.004

Saiz A, Salazar A, Bernard J (2018) Crowdsourcing architectural beauty: Online photo frequency predicts building aesthetic ratings. PLOS One. https://doi.org/10.1371/journal.pone.0194369
Seiferling I, Naik N, Ratti C, Proulx R (2017) Green streets – Quantifying and mapping urban trees with street-level imagery and computer vision. Landscape Urban Pla 165:93–101. https://doi.org/10.1016/j.landurbplan.2017.05.010
Shao Z, Cai J, Wang Z (2018) Smart monitoring cameras driven intelligent processing to big surveillance video data. IEEE Trans Big Data 4:105–116. https://doi.org/10.1109/tbdata.2017.2715815
Vanolo A (2016) Is there anybody out there? the place and role of citizens in tomorrow's smart cities. Futures 82:26–36. https://doi.org/10.1016/j.futures.2016.05.010
Wang Z, Liang Q, Khosla F, Zhang F, Charron L, Johnsen L, Cai B, Ratti C (2019) Quantifying legibility of indoor spaces using deep convolutional neural networks: case studies in train stations. Buil Environ 160:106099. https://doi.org/10.1016/j.buildenv.2019.04.035
Whyte WH, Underhill P (2009) City: rediscovering the center. University of Pennsylvania Press
Xu Z, Mei L, Liu Y, Hu C, Chen L (2014) Semantic enhanced cloud environment for surveillance data management using video structural description. Computing 98:35–54. https://doi.org/10.1007/s00607-014-0408-7
Yin L, Cheng Q, Wang Z, Shao Z (2015) 'Big data' for pedestrian volume: exploring the use of Google Street View images for pedestrian counts. Appl Geogr 63:337–345. https://doi.org/10.1016/j.apgeog.2015.07.010
Zhang F, Zhou B, Ratti C, Liu Y (2019) Discovering place-informative scenes and objects using social media photos. Roy Soc Open Sci 6:181375. https://doi.org/10.1098/rsos.181375
Zhang F, Zhou B, Liu L, Liu Y, Fung HH, Lin H, Ratti C (2018) Measuring human perceptions of a large-scale urban region using machine learning. Landscape Urban Plan 180:148–160. https://doi.org/10.1016/j.landurbplan.2018.08.020
Zhang F, Duarte F, Ma R, Milioris D, Lin H, Ratti C (2016) Indoor space recognition using deep convolutional neural network: a case study at MIT campus. arXiv:1610.02414
Zhang W, Xu L, Duan P, Gong W, Lu Q, Yang S (2015) A video cloud platform combining online and offline cloud computing technologies. Pers Ubiquit Comput 19:1099–1110. https://doi.org/10.1007/s00779-015-0879-3

Fábio Duarte is a Principal Research Scientist in the MIT Senseable City Lab and lecturer in the MIT Department of Urban Studies and Planning. Duarte is also professor at PUCPR, Brazil, and author of "Unplugging the city: the urban phenomenon and its sociotechnical controversies" (Routledge 2018)

Carlo Ratti is Director of the MIT Senseable City Lab and Professor of Practice in the MIT Department of Urban Studies and Planning. Ratti is the founder of Carlo Ratti Associati and author of "The city of tomorrow: sensors, networks, hackers, and the future of urban life" (Yale 2016).

Chapter 28
User-Generated Content: A Promising Data Source for Urban Informatics

Song Gao, Yu Liu, Yuhao Kang, and Fan Zhang

Abstract This chapter summarizes different types of user-generated content (UGC) in urban informatics and then gives a systematic review of their data sources, methodologies, and applications. Case studies in three genres are interpreted to demonstrate the effectiveness of UGC. First, we use geotagged social media data, a type of single-sourced UGC, to extract citizen demographics, mobility patterns, and place semantics associated with various urban functional regions. Second, we bridge UGC and professional-generated content (PGC), in order to take advantage of both sides. The third application links multi-sourced UGC to uncover urban spatial structures and human dynamics. We suggest that UGC data contain rich information in diverse aspects. In addition, analysis of sentiment from geotagged texts and photos, along with the state-of-the-art artificial intelligence methods, is discussed to help understand the linkage between human emotions and surrounding environments. Drawing on the analyses, we summarize a number of future research areas that call for attention in urban informatics.

28.1 Introduction

The urbanization process is accelerating in world cities and attracting large-scale job opportunities, human flows, business, and social activities. With the rapid development of information and communication technologies (ICT), location-aware devices, and sensor networks, the emergence of multi-source geospatial big data brings

S. Gao (✉) · Y. Kang
Geospatial Data Science Lab, University of Wisconsin, Madison, USA
e-mail: song.gao@wisc.edu

Y. Liu
Institute of Remote Sensing and Geographical Information Systems, Peking University, Beijing, China

F. Zhang
Senseable City Laboratory, Department of Urban Studies and Planning, Massachusetts Institute of Technology, Cambridge, MA, USA

W. Shi et al. (eds.), *Urban Informatics*, The Urban Book Series,
https://doi.org/10.1007/978-981-15-8983-6_28

new opportunities to understand the rich semantics of space and place and associated human activities in urban areas using large-scale user-generated content (UGC) and crowdsourcing data streams, such as geotagged social media posts, travel blogs, mobile phone data, smart card data from transportation, GPS-enabled ridesharing services, and so forth. In this chapter, we review state-of-the-art research in UGC-based urban informatics using crowdsourced geographic information.

28.1.1 Background and Definition

Following the development of Web technologies and mobile devices, people can easily produce large numbers of data and rich information irrespective of their expertise. This is known as user-generated content (UGC), which is a form of content created by users of a system or a service and made available publicly on that system. UGC ranges from social media data and crowdsourced GPS trajectory data, to smart card data and mobile location data from a variety of apps. UGC maximizes the opportunity to understand multiple facets of the cities that we inhabit. The uniqueness and potential of UGC are mainly demonstrated in two ways. On the one hand, UGC can be viewed as the complement of professional-generated content (PGC), as it is decentralized and can be collected from the bottom up and through citizen science (Goodchild 2007; See et al. 2016). Therefore, it can be utilized to capture public opinions and further be leveraged to understand place-based contexts and sociocultural perceptions. On the other hand, UGC can be produced in an economical yet effective manner, and individuals as sensors largely expand the data coverage within cities.

Generally speaking, UGC in geographic information applications can be categorized in two types. One is collaborative mapping platforms, such as Wikimapia and OpenStreetMap (OSM), in which volunteers create and contribute geographic features and detailed descriptions to the Web, where the entries are synthesized into databases and made available to both public and private sectors. This type of UGC is also known as volunteered geographic information (VGI; Goodchild 2007) and has lowered the barriers for the general public to not only consume geographic information but also to contribute to the platform. Different organizations can also produce, customize, and render the data sources based on their own preferences of map styles and application requirements, such as in natural disaster management and emergency routing (Longueville et al. 2010; De Albuquerque et al. 2015; Han et al. 2019). VGI demonstrates how geographic data, information, and knowledge are produced and circulated in practice among different communities and in society at large (Sui et al. 2012). In the past decade, there exist a couple of studies comparing the data quality of VGI to the authoritative mapping sources and proprietary geodata in different regions and countries (Haklay 2010; Girres and Touya 2010; Zielstra and Zipf 2010; Neis et al. 2012; Forghani and Delavar 2014; Yamashita et al. 2019; Tian et al. 2019), where developed countries generally had a better coverage and data quality compared to developing countries. And in some regions, OSM data had geographically imbalanced coverage and were missing various types of information such as roads, points of interest (POI), and land uses (Dorn et al. 2015; Kashian et al. 2019). The second type of UGC is socially constructed data streams from users, that is, data entries constructed from mobile phone apps including diverse social media sources, crowdsourcing, and location-based services (Facebook, Twitter, Weibo, Foursquare, Yelp, Flickr, Instagram, Waze, Uber, Lyft, Didi, etc.), where the general public use locations, place names, and geographic contexts to search for information, consume the service, describe their sense of place, and share diverse opinions and comments according to their experiences (Li et al. 2013; Liu et al.2015; Gao et al. 2017; Janowicz et al. 2019). Harvey (2013) argues that this would be more precisely labeled as user contributed data, since people may not consciously volunteer their data, but generate it in the process of using the platforms for their particular purposes.

In cities, as the most populated areas on the Earth, there have been increasing amounts of UGC data streams generated every day from social media platforms, location-based services, crowdsourcing, and sensor networks, which help in sensing and addressing the urban problems and challenges in the regional economy and in globalization (Martinez-Fernandez et al. 2012; Cheshire and Hay 2017), and also drive the new paradigm in urban analytics (Batty 2019) that combine big data, urban planning and design, and spatial information theory for future development of sustainable cities.

28.2 Characteristics of UGC

User-generated data have their own pros and cons (Martí et al. 2019). In urban studies, although researchers have successfully utilized this emerging source for assessing urban spatial structure and functional regions (Gao et al. 2017; Tu et al. 2017; Xu et al. 2019), analyzing human mobility patterns and transportation infrastructure (Cho et al. 2011; Noulas et al. 2012; Hawelka et al. 2014; Liu et al. 2014; Yue et al. 2014) and supporting the design of new urban development rules, a good understanding of the key characteristics of UGC data is a prerequisite for preventing the abuse of such data. Compared to traditional data sources (e.g. survey) used in urban studies, UGC data have the following advantages.

First, UGC has the five Vs (volume, velocity, variety, veracity, and value) characteristic of big data (Marr 2015; Yang et al. 2017). Millions of users from different countries and regions in the world are posting all kinds of information per second (Hu et al. 2015; Liu et al. 2015; Martí et al. 2019). For instance, on Twitter, as one of the most widely used social media platforms, there are more than 500 million tweets sent daily by 100 million active users from 160 countries (Aslam 2019). UGC covers all kinds of topics including news, sports, entertainment, education, economics, technology, travels, and lifestyle and provides various perspectives in sensing urban environments and human dynamics (Sagl et al. 2012). People share comments about their lives, surrounding environments, and nearby events. As social media records include the timestamps of users' contents and activities automatically, they provide valuable information for time-series data analytics and time-geography applications (Chen et al. 2016; Tirunillai and Tellis 2012; Kang et al. 2017; Li et al. 2016). Moreover, the UGC data-collection process for a large geographic area is faster, and the cost is reduced compared to traditional surveys (Li et al. 2013; Gao et al. 2014; Jiang, Li, and Ye 2019). Moreover, the resolution of UGC can be zoomed into the detailed individual level (Yue et al. 2014; Liu et al. 2015) rather than the aggregation level such as census data; and the data update period of UGC (i.e. seconds, minutes, hours, or days) is usually shorter than that of official surveys (i.e. months or years).

Second, UGC data are contributed by the users voluntarily or are collected from the users who use a service and agree to share their data. It is worth noting that some references may only use a strict definition of actively generated data or crowdsourcing. Citizens monitoring their surrounding urban environment can be considered as sensors (Goodchild 2007) in terms of expressions, perceptions, and behaviors, while producing streams of data on social media Web sites, which can help reveal different aspects of their own lives and their environment (Arribas-Bel 2014). Conventional data collection methods for urban studies usually require large community surveys, long-period observations, and high labor costs using questionnaires and fieldwork (Nawrath, Kowarik, and Fischer 2019; Oliveira and Campolargo 2015). In contrast, UGC is produced through the motivation of both the organizations and the individuals, for various purposes such as providing and using location-based services (Yap et al. 2012), and the desire to share with others to promote friendships and social connections (Ames and Naaman 2007; Hollenstein and Purves 2010). Through this procedure, massive data can be collected unobtrusively in which the response bias in traditional methods may be eliminated (Quercia et al. 2015).

While UGC offers promising opportunities, several internal challenges and limitations of the UGC should be addressed for urban studies as follows.

First, although large volumes of content are contributed by millions of users every second, we may get a very sparse data matrix (e.g. Lee et al. 2015) after slicing the UGC data into a fine spatiotemporal resolution (e.g. a city-block spatial unit with hourly temporal window), which is crucial in solving some urban problems such as transportation planning and traffic congestion control. The spatiotemporal data sparsity issue becomes more prominent in the regions with limited numbers of active users. Due to the reduced data volume, the uncertainty in each slice may increase when analyzing the data (Bao et al. 2012).

Second, a common concern about UGC refers to the lack of standardization for users in the data generation process, which causes poor data quality and low trustworthiness, as well as high uncertainty (Senaratne et al. 2017). Users produce geographic data based on their local knowledge and their perception of the place, which may vary across different users (Stephens 2013). And due to the vagueness and uncertainty in human conceptualization of location, space, and place, it is hard for users to express some geographic regions and spatial relations precisely (Montello et al. 2003; Goodchild and Li 2012). Thus, an approach driven by data synthesis (Gao et al. 2017b), combining UGC with an approach informed by fuzzy-set theory (Wu et al. 2019), and combining UGC with survey-based behavior approaches (Twaroch et al. 2019) has been proposed to address the abovementioned concerns. For instance, users may have different perceptions and cognitions for the same place, which can cause incorrect tagging behaviors for social media photos (Hollenstein and Purves 2010).

The third issue concerns the representativeness of UGC, which refers to the degree to which UGC observation samples can represent the actual population (Zhang and Zhu 2018). The results may be biased by data sampling. The existing studies have figured out that the information shared on social media platforms usually follows a power-law distribution, indicating that only a small proportion of users contribute most of the content online (Kwak et al. 2010; Longley and Adnan 2016; Gao et al. 2017a). Therefore, the content collected might be dominated by some specific features and can be another source of bias. Besides, the demographic bias in contributors also impedes the representativeness (Hecht and Stephens 2014). Not all people in the real world use social media frequently. People who have limited access to social media, such as the elderly and users in developing countries, may be less sampled by UGC. For example, the average age of users in Twitter is 28 (Longley and Adnan 2016), and most photos in the Yahoo Flickr Creative Commons (YFCC) dataset released by the Yahoo Labs are uploaded by users in USA (Thomee et al. 2015; Kang et al. 2018) and several other developed countries. It is worth noting that the users who send geotagged tweets are also not randomly distributed over the population but create bias in subtle ways (Malik et al. 2015).

Despite the existence of data bias, research driven by UGC data has achieved great success as a result of validation or through comparison with studies using traditional data sources (Al-ghamdi and Al-Harigi 2015; Blaschke et al. 2018; Gao et al. 2017b; Liu et al. 2016). Opportunities have arisen for urban studies using UGC data because of the abovementioned advantages: (1) big data with low collection cost; (2) fast data generation and update velocity; (3) high penetration rate among users. The next part of this chapter summarizes various examples of UGC-driven urban informatics research and applications and with a focus on the topics of urban spatial structure, urban functional regions, place semantics, and user sentiment analysis. We will first introduce an analytical and computational framework to process large-scale crowdsourced data, and followed this with various applications and case studies in the literature.

28.3 Analytical and Computational Framework to Process UGC Data

A general analytical and computational framework to process and analyze UGC data is shown in Fig. 28.1. It consists of three parts from the bottom up. First, researchers collect various sources of UGC datasets including Twitter, Weibo, Instagram, Facebook, Foursquare, Yelp, and Dianping and store the data (including structured table records and unstructured texts, images, and videos) in the computer server or a cloud data center with master server and data nodes. Second, the raw data must be cleaned, filtered, processed, and enriched to further extract the information about users, locations, and content (more details in Sect. 28.3). Lastly, spatiotemporal analyses, statistical methods, and machine learning models are employed to support urban analytics,

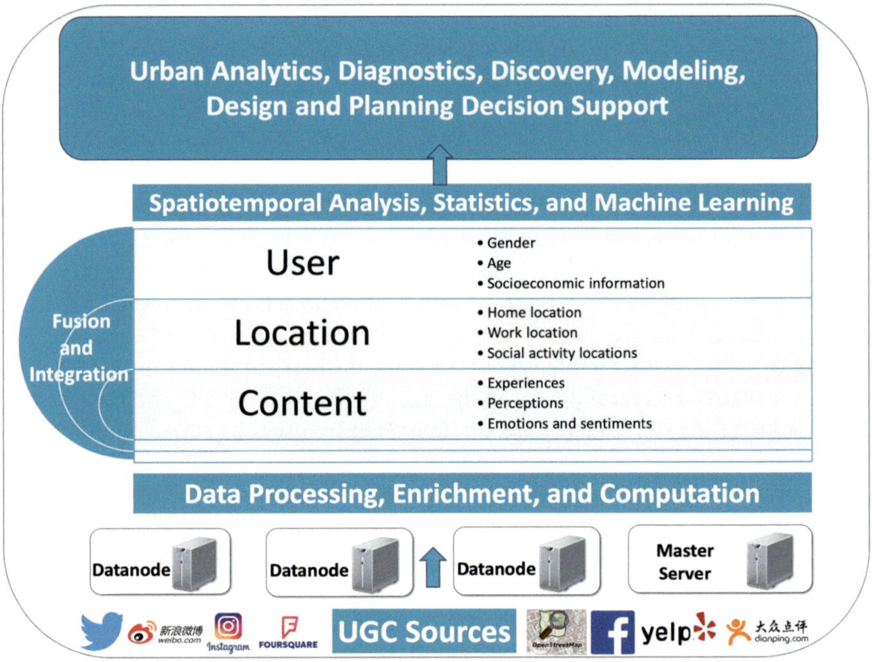

Fig. 28.1 A general analytical and computational framework to process and analyze UGC data

diagnostics, knowledge discovery, modeling, prediction, and decision-making applications. During this process, multi-source UGC and crowdsourced data can be integrated and fused. High-performance computing infrastructure (Cao et al. 2015; Gao et al. 2017; Yang et al. 2017) and open-source analysis toolkits as well as machine learning frameworks such as *scikit-learn, r-spatial, PySAL,* and *Tensorflow* can be utilized to facilitate the data processing and advanced analysis.

28.4 Single-Source UGC-Based Urban Studies

28.4.1 User Information and Citizen Demographics

User information in UGC refers to the metadata or the profile of a user, including the place of residence, name, gender, age, ethnicity, hobby, friends, and social connections, and so on. Users are the main entities who generate content. There are two ways to collect user information from UGC. On the one hand, some basic user information can be directly obtained from the public profile which users provide on social media Web sites. When they were registering and creating a new account, users were required to enter such information by filling out online forms. For example, some basic demographic information such as nationality, gender, and age can be directly extracted from the user profiles (Longley et al. 2015; Kang et al. 2018). Researchers can further utilize such demographic information about citizens to better understand the flow of people from different geo-demographic groups in cities (Longley and Adnan 2016; Huang and Wong 2016). In addition, the follower and friendship connections in social media platforms can also be obtained and have been used to examine theories in the social sciences (Sloan and Morgan 2015; Ugander et al. 2011; Hodas et al. 2013).

On the other hand, some missing user information may not be retrieved directly from the user profile but can be inferred by combining other data sources and further analyses. For instance, the gender, age, and ethnicity information can be inferred from the user identifiers with the forename–surname pairs (Chang et al. 2010; Mateos et al. 2011; Mislove et al. 2011; Longley et al. 2015; Luo et al. 2016). By tracking the location and time of user postings, residents and visitors can be identified and distinguished (García-Palomares et al. 2015; Liu et al. 2018; Su et al. 2016).

28.4.2 Human Mobility, Urban Spatial Structure, and Transportation

Understanding human mobility patterns is important for the planning and management of urban land use and transportation. The work location, the home location, and even social activity locations of UGC users can be identified through their geotagged posts and their activity patterns detected in social media platforms (Gao et al. 2014; Li et al. 2014; Yang et al. 2015; Wu et al. 2015; Liu, Huang, and Gao 2019). The home-to-job commuting trips and non-commuting trips can be extracted and aggregated for traffic analysis zones (TAZs) to support urban transportation analysis. For example, as shown in Fig. 28.2, researchers detected over 24,000 daily commuting trips with an estimated average commuting time of about 32 min and average commuting distance

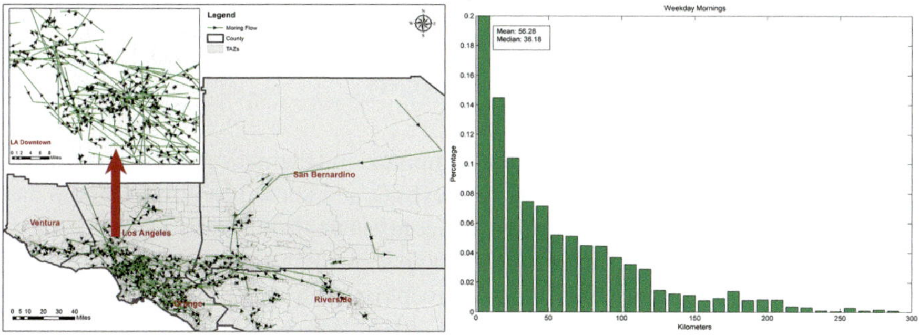

Fig. 28.2 Spatial and distance distributions of the detected commuting trips using geotagged Twitter data

of about 56 km in the Greater Los Angeles Area using millions of geotagged tweets (Gao et al. 2014). Moreover, when survey data and geotagged Twitter data were compared, the Pearson correlation coefficient of trips on weekdays was 0.91, and the correlation between detected trips using geotagged tweets and using a traditional travel demand model was 0.839 (Lee et al. 2015). While these correlations are far from perfect, the conclusions are nevertheless beneficial for urban transportation research.

Another benefit of using location-based check-in data from social networks is having access to information on place types (e.g. shops, offices, restaurants) for user activities, which is important to understand the spatial, temporal, and thematic distributions of human activities and activity-type transitions in cities (Noulas et al. 2011; Wu et al. 2014; McKenzie et al. 2015). For example, Wu et al. (2014) analyzed large-scale user check-in statistics in a location-based social-network platform in China and found different spatiotemporal activity transition probabilities among different types of places, including transportation facilities. Such activity-based transition patterns can also be extracted with pattern mining methods from call-detail-record data from mobile phones, allowing at-home, in-work, and social activity types to be annotated at each stay location (Cao et al. 2019). In addition, by combining information on user demographics, researchers found different movement patterns when comparing tourists and local residents (Chua et al. 2016; Liu et al. 2018), which could help transportation planning and management such as traffic congestion control and transportation regulations during events in cities. Moreover, the linkage between land use and urban dynamics can be identified through UGC and crowdsourcing data. For example, researchers found that human activities tended to decrease throughout the day for most land uses (e.g. offices, education, health) but remained constant in parks and increased in retail and residential zones (García-Palomares et al. 2018). Ren et al. (2019) examined the effect of land-use function complementarity on intra-urban spatial interactions using metro smart card records for different time periods and directions in the city of Shenzhen, China, which also demonstrates the trending use of individual-level big data in travel behavior studies in cities (Yue et al. 2014; Liu et al. 2015).

28.4.3 Place Semantics and Sentiments

Semantic signatures including the spatial, temporal, and thematic posed by McKenzie et al. (2015) and Janowicz et al. (2019) to extract and share high-dimensional data about types of places and neighborhoods. In contrast to spatial statistics, place-based analyses focus more on describing the topological and hierarchical relations between places and understanding various human perceptions and cognition at places (Li and Goodchild 2012; Gao et al. 2013; Zhu et al. 2016; Wu et al. 2019). Understanding the semantics of urban space and place could derive from the spatial, temporal, and thematic perspectives using geotagged texts, photos, and videos. These crowd-sourced geographic data could also help the identification of vibrant neighborhoods

(Cranshaw et al. 2012; Zhang et al. 2013) and urban areas of interest (AOI), which refers to the regions within an urban environment that attract people's attention (Hu et al. 2015). Urban AOIs often have high exposure to the general public and receive a large number of visits. UGC such as geotagged photos can reveal the visit popularity and scenery information for city planners, transportation analysts, and location-based service providers to plan new businesses. Besides, the existing studies have utilized POI information and user check-ins in location-based social networking platforms (such as Foursquare, Yelp, Jiepang, and Weibo) to investigate various urban informatics issues. For example, a location-distortion model was proposed to improve reverse geocoding (i.e. convert a latitude/longitude to a POI address) using behavior-driven temporal signatures (McKenzie and Janowicz 2015). Another Place2Vec model derives the reasoning about place type similarity and relatedness by learning embeddings from augmented spatial contexts and user check-in information (Yan et al. 2017). By combining the user check-in information in Foursquare with topic modeling approaches, researchers derived urban functional regions in the ten most populated US cities (Gao et al. 2017), which demonstrates a bottom-up data-driven perspective. In contrast, researchers also developed a top-down theory-informed approach to extracting urban functional regions. For example, a composition-pattern-based knowledge model was proposed to extract urban functional regions (Papadakis et al. 2019a). In this model, places are formalized as "patterns" which are defined as sets of components, composition rules, and functional implications. For example, a shopping plaza should consist of not only shopping stores but also restaurants, parking lots, and other facilities. Recently, an improved model was proposed using theoretical, empirical, and probabilistic patterns (Papadakis et al. 2019b) to enrich the knowledge-based model.

In addition, with advances in artificial intelligence (AI) technologies and open-source processing platforms as well as deep learning methods in the domains of natural language processing (NLP) and computer vision (CV), the extraction of human emotions (e.g. happiness, fear, anger, sadness, and surprise) and sentiments (i.e. positive, neutral, or negative) at different places and environments has become more accessible. For example, researchers applied advanced text mining techniques with spatial analysis to detect depressed Twitter users and their spatial clusters in US metropolitan areas. Socioeconomic variables from the Bureau of the Census and climate risk factors were found to have an impact on the prevalence of depression but may vary seasonally in different regions (Yang and Mu 2015; Yang et al. 2015). Human sentiment scores and their spatial distribution were extracted and explored in the city of Nanjing, China, using Weibo data (Zhen et al. 2018). High levels of air pollution were found to contribute to the urban population's reported low level of happiness in social media based on the analysis of over 210 million geotagged Weibo posts in China (Zheng et al. 2019). A semantic-specific sentiment analysis was conducted on Web-based neighborhood textual reviews in the city of New York for understanding the perceptions of citizens toward their living environments (Hu et al. 2019). As for image-based urban studies, researchers have used facial expression extraction techniques to explore human–environment interactions (as shown in Fig. 28.3) especially for the relationship between emotions and environments. A posi-

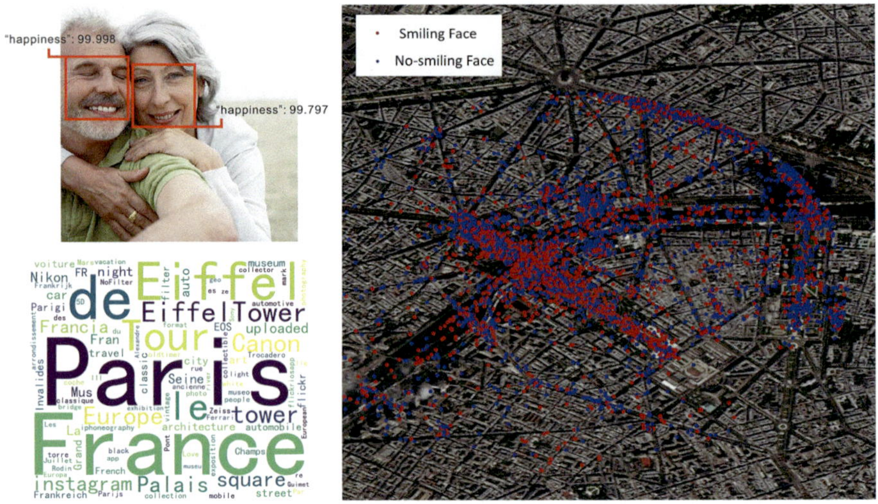

Fig. 28.3 Spatial distribution of smiling and no-smiling faces extracted from geotagged Flickr photos in Paris, France, and the associated word cloud of most frequent textual tags in these photos (Facial Expression subfigure was modified from the demo image of Face++ at https://www.faceplusplus.com/face-detection/)

tive correlation was found between the happiness score and the presence of natural environments such as water bodies and green vegetation in different types of place (Svoray et al. 2018; Kang et al. 2019). As another source of ambient sensing data, street view images can also be utilized to analyze human perceptions of places. For example, a data-driven machine learning approach with scene elements was proposed to measure how people perceive a place (including safe, lively, beautiful, wealthy, depressing, and boring) using street view images (Zhang et al. 2018a; Zhang et al. 2018b).

28.5 Multi-source Data-Driven Urban Studies

28.5.1 Fusion of Multiple UGC Sources

In traditional urban strategic planning or the classification results of remote sensing, many places in urban areas may be labeled as single land-use type; however, these areas may in reality contain multiple functions and land uses. In order to capture citywide dynamics of both human activities and urban functions at finer resolutions, multi-source UGC and crowdsourced information are combined to overcome their own limitations and to enrich the understanding of urban spatial structure and neighborhood demographics. Both mobile phone data and taxi trajectories usually cover large numbers of users and contain rich location information (and social network connections for mobile phone data) but lack place semantics (Liu et al. 2015). Social media data are sparsely distributed in space and time but contain rich content (Huang and Wong 2016; Martí et al. 2019). By combining both mobile phone data and social media, it is possible to extract citizen's home–job locations and social activity dynamics more effectively in space and time in cities (Tu et al. 2017). Also, by the integration of mobile-phone data and crowdsourced taxi trajectories, or the fusion of POI data and crowdsourced taxi trajectories, researchers have uncovered substantial differences between taxi trips and mobile-phone-based human movements in terms of spatial distribution and distance-decay effects (Kang et al. 2013) and explored the intensity of spatial interactions among different functional regions based on taxi origin–destination flows (Wang et al. 2018). In addition, researchers have used an online restaurant review platform with rich crowdsourced user-generated reviews and extracted machine learning features to further infer urban neighborhoods' population distribution and socioeconomic attributes in nine Chinese cities. They found a high predictability, in which the distributions of daytime and nighttime populations are estimated by mobile phone location data (Dong et al. 2019). UGC data can also be used to validate the urban spatial structure and place semantics extracted from ambient sensing and to reflect various urban environmental contexts. For example, as shown in Fig. 28.4, given only a certain number of street view images of a street, a deep learning model was trained to accurately estimate the hourly variation of human mobility patterns approximated by taxi trips along the streets (Zhang et al. 2019). In another study, researchers developed a mixed-use decomposition model based on temporal activity signatures extracted from social media check-in data, and taxi origin and destination (OD) trip data over one year were used to validate the land-use mixing results (Wu et al. 2019).

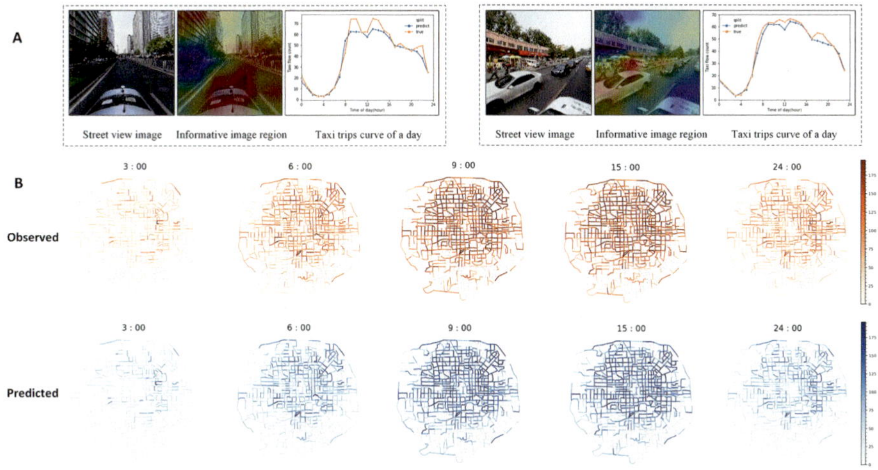

Fig. 28.4 **A** Predicting hourly variation of taxi trips using street view images; **B** Spatiotemporal variation of human mobility patterns approximated by taxi trips along the streets

28.5.2 Fusion of UGC and PGC

Compared to UGC, professional-generated content (PGC) mainly comes from domain experts and organizations who have the expertise and knowledge of study subjects, or the authority to collect and publish data, which is more trustworthy in social media platforms and in news media. The fusion of UGC and PGC can take advantages of both sides, to uncover urban spatial structures and dynamics, and to provide valuable information in the emergency management or disaster response scenarios. For example, crowdsourced geotagged photos and videos from social media users, volunteered geographic data, and authoritative storm surge data created by the U.S. Federal Emergency Management Agency (FEMA) were fused together to create a more accurate estimate of urban flood damage and updated road accessibility mapping in New York City during Hurricane Sandy (Schnebele et al. 2014). In urban planning and development, the integration of public participation from UGC big data sources together with the PGC-based expert design may provide a holistic approach through the process of idea generation, feedback, and evaluation for urban management and problem solving (Thakuriah et al. 2017).

In future, a number of multi-source data fusion research areas call for attention in urban informatics. First, the data sampling and fusing resolution requirements in space and time need to be investigated among different UGC sources to comprehensively understand human activities of different gender, age, and socioeconomic groups and place semantics for intra-urban and inter-city human mobility modeling. Second, combining UGC and PGC or combining data-driven and knowledge-driven approaches can solve urban problems such as traffic congestion and environmental pollution. Last but not least, there is a need to increase the engagement of citizen science in addressing urban changes in responsive cities through data-smart governance (Goldsmith and Crawford 2014).

28.6 Conclusion

UGC data contain rich information about human location, society, and human–environment interactions and have become a promising data source for urban informatics studies with unprecedented spatial, temporal, and thematic resolutions. This chapter summarized the key characteristics of UGC data with a focus on geographic information and urban studies. We discussed the analytical and computational framework to process UGC data and urban applications including citizen demographics, human mobility, urban spatial structure, place semantics, and sentiment analysis, to name a few. Considering the limitation of a single data source, various kinds of data fusion cases were discussed and suggested to advance future urban informatics studies. It is worth noting that we did not try to enumerate all possible fusion cases but just to list several scenarios with a focus on urban challenges. In sum, a combination of multi-source UGC-driven and theory-informed approaches provides a more holistic view for urban analytics, diagnostics, and human-centered sustainable urban planning and future development.

Acknowledgements Song Gao would like to thank the support of this research from the Office of the Vice Chancellor for Research and Graduate Education at the University of Wisconsin-Madison with funding from the Wisconsin Alumni Research Foundation; Yu Liu would like to thank the funding support from the National Natural Science Foundation of China (No. 41625003).

References

Al-ghamdi SA, Al-Harigi F (2015) Rethinking image of the city in the information age. Procedia Comput Sci 65:734–743

Ames M, Naaman M (2007) Why we tag: motivations for annotation in mobile and online media. In: Proceedings of the SIGCHI conference on human factors in computing systems, 971–980

Arribas-Bel D (2014) Accidental, open and everywhere: emerging data sources for the understanding of cities. Appl Geogr 49:45–53

Aslam S (2019) Twitter by the numbers: stats, demographics and fun facts. Retrieved from https://www.omnicoreagency.com/twitter-statistics/ on May 2019

Bao J, Zheng Y, Mokbel MF (2012) Location-based and preference-aware recommendation using sparse geo-social networking data. In: Proceedings of the 20th international conference on advances in geographic information systems, 199–208

Batty M (2019) Urban analytics defined. Environ Plan B: Urban Anal City Sci 46(3):403–405

Blaschke T, Merschdorf H, Cabrera-Barona P, Gao S, Papadakis E, Kovacs-Györi A (2018) Place versus space: from points, lines and polygons in GIS to place-based representations reflecting language and culture. ISPRS Int J Geo-Inf 7(11):452

Cao J, Li Q, Tu W, Wang F (2019) Characterizing preferred motif choices and distance impacts. PLoS One 14(4):e0215242

Cao G, Wang S, Hwang M, Padmanabhan A, Zhang Z, Soltani K (2015) A scalable framework for spatiotemporal analysis of location-based social media data. Comput Environ Urban Syst 51:70–82

Chang J, Rosenn I, Backstrom L, Marlow C (2010) epluribus: ethnicity on social networks. In: Fourth international AAAI conference on weblogs and social media

Chen BY, Yuan H, Li Q, Shaw SL, Lam WH, Chen X (2016) Spatiotemporal data model for network time geographic analysis in the era of Big Data. Int J Geogr Inf Sci 30(6):1041–1071

Cheshire PC, Hay DG (2017) Urban problems in Western Europe: an economic analysis. Routledge

Cho E, Myers SA, Leskovec J (2011) Friendship and mobility: user movement in location-based social networks. In: Proceedings of the 17th ACM SIGKDD international conference on knowledge discovery and data mining, 1082–1090

Chua A, Servillo L, Marcheggiani E, Moere AV (2016) Mapping Cilento: Using geotagged social media data to characterize tourist flows in southern Italy. Tourism Manag 57:295–310

Cranshaw J, Schwartz R, Hong J, Sadeh N (2012) The livehoods project: utilizing social media to understand the dynamics of a city. In: Sixth international AAAI conference on weblogs and social media

De Albuquerque JP, Herfort B, Brenning A, Zipf A (2015) A geographic approach for combining social media and authoritative data towards identifying useful information for disaster management. Int J Geogr Inf Sci 29(4):667–689

Dong L, Ratti C, Zheng S (2019) Predicting neighborhoods' socioeconomic attributes using restaurant data. In: Proceedings of the national academy of sciences 201903064

Dorn H, Törnros T, Zipf A (2015) Quality evaluation of VGI using authoritative data—a comparison with land use data in Southern Germany. ISPRS Int J Geo-Inf 4(3):1657–1671
Forghani M, Delavar M (2014) A quality study of the OpenStreetMap dataset for Tehran. ISPRS Int J Geo-Inf 3(2):750–763
Gao S, Janowicz K, McKenzie G, Li L (2013) Towards platial joins and buffers in place-based GIS. In: ACM SIGSPATIAL international workshop on computational models of place, pp 1–8
Gao S, Yang JA, Yan B, Hu Y, Janowicz K, McKenzie G (2014) Detecting origin-destination obility flows from geotagged Tweets in greater Los Angeles area. In: Short paper proceedings of the eighth international conference on geographic information science 1–4
Gao S, Janowicz K, Couclelis H (2017a) Extracting urban functional regions from points of interest and human activities on location-based social networks. Trans GIS 21(3):446–467
Gao S, Li L, Li W, Janowicz K, Zhang Y (2017b) Constructing gazetteers from volunteered big geo-data based on Hadoop. Comput Environ Urban Syst 61:172–186
Gao S, Janowicz K, Montello DR, Hu Y, Yang JA, McKenzie G, Ju Y, Adams B, Yan B (2017c) A data-synthesis-driven method for detecting and extracting vague cognitive regions. Int J Geogr Inf Sci 31(6):1245–1271
García-Palomares JC, Gutiérrez J, Mínguez C (2015) Identification of tourist hot spots based on social networks: A comparative analysis of European metropolises using photo-sharing services and GIS. Appl Geogr 63:408–417
García-Palomares JC, Salas-Olmedo MH, Moya-Gómez B, Condeco-Melhorado A, Gutierrez J (2018) City dynamics through twitter: relationships between land use and spatiotemporal demographics. Cities 72:310–319
Girres JF, Touya G (2010) Quality assessment of the French OpenStreetMap dataset. Trans GIS 14(4):435–459
Goodchild MF (2007) Citizens as sensors: the world of volunteered geography. GeoJournal 69(4):211–221
Goodchild MF, Li L (2012) Assuring the quality of volunteered geographic information. Spat Stat 1:110–120
Goldsmith S, Crawford S (2014) The responsive city: engaging communities through data-smart governance. Wiley
Haklay M (2010) How good is volunteered geographical information? A comparative study of OpenStreetMap and ordnance survey datasets. Environ Plan B: Plan Des 37(4):682–703
Han SY, Tsou MH, Knaap E, Rey S, Cao G (2019) How do cities flow in an emergency? Tracing human mobility patterns during a natural disaster with big data and geospatial data science. Urban Sci 3(2):51
Harvey F (2013) To volunteer or to contribute locational information? Towards truth in labeling for crowdsourced geographic information. In: Sui S, Elwood S, Goodchild MF (eds) Crowdsourcing geographic knowledge. Springer, Dordrecht, The Netherlands, pp 31–42
Hawelka B, Sitko I, Beinat E, Sobolevsky S, Kazakopoulos P, Ratti C (2014) Geo-located Twitter as proxy for global mobility patterns. Cartography Geogr Inf Sci 41(3):260–271
Hecht B, Stephens M (2014) A tale of cities: urban biases in volunteered geographic information. In: Eighth international AAAI conference on weblogs and social media
Hodas NO, Kooti F, Lerman K (2013) Friendship paradox redux: your friends are more interesting than you. In: Seventh international AAAI conference on weblogs and social media
Hollenstein L, Purves R (2010) Exploring place through user-generated content: using flickr tags to describe city cores. J Spat Inf Sci 1:21–48
Hu Y, Gao S, Janowicz K, Yu B, Li W, Prasad S (2015) Extracting and understanding urban areas of interest using geotagged photos. Comput Environ Urban Syst 54:240–254
Hu Y, Deng C, Zhou Z (2019) A semantic and sentiment analysis on online neighborhood reviews for understanding the perceptions of people toward their living environments. Ann Am Assoc Geogr 109(4):1052–1073
Huang Q, Wong DW (2016) Activity patterns, socioeconomic status and urban spatial structure: what can social media data tell us? Int J Geogr Inf Sci 30:1873–1898

Janowicz K, McKenzie G, Hu Y, Zhu R, Gao S (2019) Using semantic signatures for social sensing in urban environments. In: Mobility patterns, big data and transport analytics. Elsevier, 31–54

Jiang Y, Li Z, Ye X (2019) Understanding demographic and socioeconomic biases of geotagged twitter users at the county level. Cartography Geogr Inf Sci 46(3):228–242

Kang C, Sobolevsky S, Liu Y, Ratti C (2013) Exploring human movements in Singapore: a comparative analysis based on mobile phone and taxicab usages. In: Proceedings of the 2nd ACM SIGKDD international workshop on urban computing, 1–8

Kang Y, Wang J, Wang Y, Angsuesser S, Fei T (2017) Mapping the sensitivity of the public emotion to the movement of stock market value: a case study of Manhattan. Int Arch Photogrammetry Remote Sens Spat Inf Sci 42:1–8

Kang Y, Zeng X, Zhang Z, Wang Y, Fei T (2018) Who are happier? spatio-temporal analysis of worldwide human emotion based on geo-crowdsourcing faces. In: 2018 Ubiquitous positioning, indoor navigation and location-based services (UPINLBS), 1–8

Kang Y, Jia Q, Gao S, Zeng X, Wang Y, Angsuesser S, Fei T et al (2019) Extracting human emotions at different places based on facial expressions and spatial clustering analysis. Transactions in GIS 23(3)

Kashian A, Rajabifard A, Richter KF et al (2019) Automatic analysis of positional plausibility for points of interest in OpenStreetMap using coexistence patterns. Int J Geogr Inf Sci 33(7):1420–1443. https://doi.org/10.1080/13658816.2019.1584803

Kwak H, Lee C, Park H, Moon S (2010) What is twitter, a social network or a news media? In: Proceedings of the 19th international conference on world wide web, 591–600

Lee JH, Gao S, Goulias K (2015) Can Twitter data be used to validate travel demand models? In: IATBR 2015-WIND

Li L, Goodchild MF (2012) Constructing places from spatial footprints. In: Proceedings of the 1st ACM SIGSPATIAL international workshop on crowdsourced and volunteered geographic information, 15–21

Li L, Goodchild MF, Xu B (2013) Spatial, temporal, and socioeconomic patterns in the use of Twitter and Flickr. Cartography Geogr Inf Sci 40(2):61–77

Li G, Hu J, Feng J, Tan KL (2014) Effective location identification from microblogs. In: IEEE 30th international conference on data engineering, 880–891

Li J, Ye Q, Deng X, Liu Y, Liu Y (2016) Spatial-temporal analysis on Spring festival travel rush in China based on multisource big data. Sustainability 8(11):1184

Liu Y, Sui Z, Kang C, Gao Y (2014) Uncovering patterns of inter-urban trip and spatial interaction from social media check-in data. PLoS One 9(1):e86026

Liu Y, Liu X, Gao S, Gong L, Kang C, Zhi Y, Chi G, Shi L (2015) Social sensing: a new approach to understanding our socioeconomic environments. Ann Assoc Am Geogr 105(3):512–530

Liu L, Zhou B, Zhao J, Ryan BD (2016) C-IMAGE: city cognitive mapping through geo-tagged photos. GeoJournal 81(6):817–861

Liu Q, Wang Z, Ye X (2018) Comparing mobility patterns between residents and visitors using geo-tagged social media data. Trans GIS 22(6):1372–1389

Liu X, Huang Q, Gao S (2019) Exploring the uncertainty of activity zone detection using digital footprints with multi-scaled DBSCAN. Int J Geogr Inf Sci 33(6):1196–1223

Longley PA, Adnan M (2016) Geo-temporal Twitter demographics. Int J Geogr Inf Sci 30(2):369–389

Longley PA, Adnan M, Lansley G (2015) The geotemporal demographics of Twitter usage. Environ Plan A 47(2):465–484

Longueville BD, Luraschi G, Smits P, Peedell S, Groeve TD (2010) Citizens as sensors for natural hazards: a VGI integration workflow. Geomatica 64(1):41–59

Luo F, Cao G, Mulligan K, Li X (2016) Explore spatiotemporal and demographic characteristics of human mobility via Twitter: a case study of Chicago. Appl Geogr 70:11–25

Malik MM, Lamba H, Nakos C, Pfeffer J (2015) Population bias in geotagged tweets. In: Ninth international AAAI conference on web and social media

Marr B (2015) Big data: using SMART big data, analytics and metrics to make better decisions and improve performance. Wiley
Martí P, Serrano-Estrada L, Nolasco-Cirugeda A (2019) Social media data: challenges, opportunities and limitations in urban studies. Comput Environ Urban Syst 74:161–174
Martinez-Fernandez C, Audirac I, Fol S, Cunningham-Sabot E (2012) Shrinking cities: urban challenges of globalization. Int J Urban Reg Res 36(2):213–225
Mateos P, Longley PA, O'Sullivan D (2011) Ethnicity and population structure in personal naming networks. PLoS One 6(9):e22943
McKenzie G, Janowicz K (2015) Where is also about time: a location-distortion model to improve reverse geocoding using behavior-driven temporal semantic signatures. Comput Environ Urban Syst 54:1–13
McKenzie G, Janowicz K, Gao S, Yang JA, Hu Y (2015) POI pulse: a multi-granular, semantic signature-based information observatory for the interactive visualization of big geosocial data. Cartographica: The Int J Geogr Inf Geovisualization 50(2):71–85
Mislove A, Lehmann S, Ahn Y-Y, Onnela J-P, Rosenquist JN (2011) Understanding the demographics of twitter users. In: Fifth international AAAI conference on weblogs and social media
Montello DR, Goodchild MF, Gottsegen J, Fohl P (2003) Where's downtown? Behavioral methods for determining referents of vague spatial queries. Spat Cogn Comput 3(2–3):185–204
Nawrath M, Kowarik I, Fischer LK (2019) The influence of green streets on cycling behavior in European cities. Landscape Urban Plan 190:103958
Neis P, Zielstra D, Zipf A (2012) The street network evolution of crowdsourced maps: OpenStreetMap in Germany 2007–2011. Future Internet 4(1):1–21
Noulas A, Scellato S, Mascolo C, Pontil M (2011) An empirical study of geographic user activity patterns in foursquare. In: Fifth international AAAI conference on weblogs and social media
Noulas A, Scellato S, Lambiotte R, Pontil M, Mascolo C (2012) A tale of many cities: universal patterns in human urban mobility. PLoS ONE 7(5):e37027
Oliveira Á, Campolargo M (2015) From smart cities to human smart cities. In: 48th Hawaii international conference on system sciences, 2336–2344
Papadakis E, Resch B, Blaschke T (2019) Composition of place: towards a compositional view of functional space. Cartography Geogr Inf Sci 1–18
Papadakis E, Baryannis G, Petutschnig A, Blaschke T (2019b) Function-based search of place using theoretical, empirical and probabilistic patterns. ISPRS Int J Geo-Inf 8(2):92
Quercia D, Schifanella R, Aiello LM, McLean K (2015) Smelly maps: the digital life of urban smellscapes. AAAI Publications 327–336
Ren M, Lin Y, Jin M, Duan Z, Gong Y, Liu Y (2019) Examining the effect of land-use function complementarity on intra-urban spatial interactions using metro smart card records. Transportation 1–23
Sagl G, Resch B, Hawelka B, Beinat E (2012, July) From social sensor data to collective human behaviour patterns: Analysing and visualising spatio-temporal dynamics in urban environments. In: Proceedings of the GI-Forum. Herbert Wichmann Verlag, Berlin 54–63
Schnebele E, Cervone G, Waters N (2014) Road assessment after flood events using non-authoritative data. Nat Hazards Earth Syst Sci 14(4):1007–1015
See L, Mooney P, Foody G, Bastin L, Comber A, Estima J, Liu HY et al (2016) Crowdsourcing, citizen science or volunteered geographic information? The current state of crowdsourced geographic information. ISPRS Int J Geo-Inf 5(5):55
Senaratne H, Mobasheri A, Ali AL, Capineri C, Haklay M (2017) A review of volunteered geographic information quality assessment methods. Int J Geogr Inf Sci 31:139–167
Sloan L, Morgan J (2015) Who tweets with their location? Understanding the relationship between demographic characteristics and the use of geoservices and geotagging on twitter. PLoS One 10(11):e0142209
Stephens M (2013) Gender and the GeoWeb: divisions in the production of user-generated cartographic information. GeoJournal 78(6):981–996

Su S, Wan C, Hu Y, Cai Z (2016) Characterizing geographical preferences of international tourists and the local influential factors in China using geo-tagged photos on social media. Appl Geogr 73:26–37

Sui D, Elwood S, Goodchild M (eds) (2012) Crowdsourcing geographic knowledge: volunteered geographic information (VGI) in theory and practice. Springer Science and Business Media

Svoray T, Dorman M, Shahar G, Kloog I (2018) Demonstrating the effect of exposure to nature on happy facial expressions via Flickr data: advantages of non-intrusive social network data analyses and geoinformatics methodologies. J Environ Psychol 58:93–100

Thakuriah PV, Tilahun NY, Zellner M (2017) Big data and urban informatics: innovations and challenges to urban planning and knowledge discovery. In: Seeing cities through big data. Springer, Cham 11–45

Thomee B, Shamma DA, Friedland G, Elizalde B, Ni K, Poland D, Li LJ et al (2015) YFCC100M: The new data in multimedia research. ArXiv Preprint ArXiv:1503.01817

Tian Y, Zhou Q, Fu X (2019) An analysis of the evolution, completeness and spatial patterns of OpenStreetMap building data in China. ISPRS Int J Geo-Inf 8(1):35

Tirunillai S, Tellis GJ (2012) Does chatter really matter? Dynamics of user-generated content and stock performance. Mark Sci 31(2):198–215

Tu W, Cao J, Yue Y, Shaw SL, Zhou M, Wang Z, Li Q et al (2017) Coupling mobile phone and social media data: A new approach to understanding urban functions and diurnal patterns. Int J Geogr Inf Sci 31(12):2331–2358

Twaroch FA, Brindley P, Clough PD, Jones CB, Pasley RC, Mansbridge S (2019) Investigating behavioural and computational approaches for defining imprecise regions. Spat Cogn Comput 19(2):146–171

Ugander J, Karrer B, Backstrom L, Marlow C (2011) The anatomy of the facebook social graph. ArXiv Preprint ArXiv:1111.4503

Wang Y, Gu Y, Dou M, Qiao M (2018) Using spatial semantics and interactions to identify urban functional regions. ISPRS Int J Geo-Inf 7(4):130

Wu L, Zhi Y, Sui Z, Liu Y (2014) Intra-urban human mobility and activity transition: evidence from social media check-in data. PLoS One 9(5):e97010

Wu F, Li Z, Lee WC, Wang H, Huang Z (2015) Semantic annotation of mobility data using social media. In: Proceedings of the 24th international conference on world wide web 1253–1263

Wu L, Cheng X, Kang C, Zhu D, Huang Z, Liu Y (2019) A framework for mixed-use decomposition based on temporal activity signatures extracted from big geo-data. Int J Dig Earth 1–19. https://doi.org/10.1080/17538947.2018.1556353

Wu X, Wang J, Shi L, Gao Y, Liu Y (2019b) A fuzzy formal concept analysis-based approach to uncovering spatial hierarchies among vague places extracted from user-generated data. Int J Geogr Inf Sci 33(5):991–1016

Xu Y, Chen D, Zhang X, Tu W, Chen Y, Shen Y, Ratti C (2019) Unravel the landscape and pulses of cycling activities from a dockless bike-sharing system. Comput Environ Urban Syst 75:184–203

Yamashita J, Seto T, Nishimura Y, Iwasaki N (2019) VGI contributors' awareness of geographic information quality and its effect on data quality: a case study from Japan. Int J Cartography 1–11

Yan B, Janowicz K, Mai G, Gao S (2017) From itdl to place2vec: Reasoning about place type similarity and relatedness by learning embeddings from augmented spatial contexts. In: Proceedings of the 25th ACM SIGSPATIAL international conference on advances in geographic information systems 1–10

Yang W, Mu L (2015) GIS analysis of depression among Twitter users. Appl Geogr 60:217–223

Yang F, Jin PJ, Cheng Y, Zhang J, Ran B (2015a) Origin-destination estimation for non-commuting trips using location-based social networking data. Int J Sustain Transp 9(8):551–564

Yang W, Mu L, Shen Y (2015b) Effect of climate and seasonality on depressed mood among twitter users. Appl Geogr 63:184–191

Yang C, Huang Q, Li Z, Liu K, Hu F (2017) Big data and cloud computing: innovation opportunities and challenges. Int J Dig Earth 10(1):13–53

Yap LF, Bessho M, Koshizuka N, Sakamura K (2012) User-generated content for location-based services: a review. In: Lazakidou AA (ed) Virtual communities, social networks and collaboration 163–179

Yue Y, Lan T, Yeh AG, Li QQ (2014) Zooming into individuals to understand the collective: a review of trajectory-based travel behaviour studies. Travel Behav Soc 1(2):69–78

Zhang G, Zhu A-X (2018) The representativeness and spatial bias of volunteered geographic information: a review. Ann GIS 24(3):151–162

Zhang AX, Noulas A, Scellatos S, Mascolo C (2013) Hoodsquare: modeling and recommending neighborhoods in location-based social networks. In: 2013 international conference on social computing 69–74

Zhang F, Zhang D, Liu L, Lin H (2018a) Representing place locales using scene elements. Comput Environ Urban Syst 71:153–164

Zhang F, Zhou B, Liu L, Liu Y, Fung HH, Lin H, Ratti C (2018b) Measuring human perceptions of a large-scale urban region using machine learning. Landscape Urban Plan 180:148–160

Zhang F, Wu L, Zhu D, Liu Y (2019) Social sensing from street-level imagery: a case study in learning spatio-temporal urban mobility patterns. ISPRS J Photogrammetry Remote Sens 153:48–58

Zhen F, Tang J, Chen Y (2018) Spatial distribution characteristics of residents' emotions based on Sina Weibo big data: A case study of Nanjing. In: Shen Z, Li M (eds) Big data support of urban planning and management: the experience in China. Springer, Cham, Switzerland, pp 43–62

Zheng S, Wang J, Sun C, Zhang X, Kahn ME (2019) Air pollution lowers Chinese urbanites' expressed happiness on social media. Nat Hum Behav 3:237–243

Zhu R, Hu Y, Janowicz K, McKenzie G (2016) Spatial signatures for geographic feature types: examining gazetteer ontologies using spatial statistics. Trans GIS 20(3):333–355

Zielstra D, Zipf A (2010) A comparative study of proprietary geodata and volunteered geographic information for Germany. In: 13th AGILE international conference on geographic information science

Song Gao is an Assistant Professor in GIScience at the University of Wisconsin-Madison, where he leads the GeoDS Lab. His main research interests include place-based GIS, human mobility, and GeoAI. He is currently the associate editor of Annals of GIS.

Yu Liu is a Boya Professor of GIScience at the Institute of Remote Sensing and Geographic Information Systems, Peking University. His research interests mainly concentrate on the humanities and social sciences based on big geo-data. He is currently an associate editor of Computers, Environment and Urban Systems.

Yuhao Kang is a Ph.D. student at the GeoDS Lab, University of Wisconsin-Madison. He received his Bachelor's degree from Wuhan University. His research interests include place-based GIS, GeoAI and cartography.

Fan Zhang is a postdoctoral researcher at SENSEable City Lab, Massachusetts Institute of Technology. He received his Ph.D. from the Chinese University of Hong Kong. His research interests include place-based GIS, GeoAI and data-driven approaches for urban studies.

Chapter 29
User-Generated Content and Its Applications in Urban Studies

Wei Tu, Qingquan Li, Yatao Zhang, and Yang Yue

Abstract The emergence of Web 2.0 and mobile Internet produces massive user-generated content (UGC), including geo-tagged photos, social network posts, street view images, and crowdsourced GPS trajectories. UGC creates unprecedented opportunities to sense what was previously hidden in the physical surfaces of cities and to portray the interactions of infrastructures, geo-information, and people; therefore, it is not only a new lens for urban space but also leads to innovative applications. In this chapter, we will introduce several typical types of UGC, such as geo-tagged photos, social media data, crowdsourcing GPS trajectories, and videos. We showcase ways in which user-generated big data can be harvested and analyzed to generate invisible and impressionistic landscapes of urban dynamics and to stimulate innovative applications. We discuss typical UGC-driven applications to demonstrate the potential of UGC in revealing how urban spaces are perceived by the public, establishing links between tangible artifacts and physical-cyber-social spaces. This fosters alternative approaches to urban informatics that better capture the intricate nature of urban space and its dynamics.

W. Tu · Q. Li (✉) · Y. Yue
Guangdong Key Laboratory of Urban Informatics, Key Laboratory for Geo-Environmental Monitoring of Coastal Zone of Ministry of Natural Resource, and Shenzhen Key Laboratory of Spatial Smart Sensing and Service, Shenzhen University, Shenzhen, China
e-mail: liqq@szu.edu.cn

W. Tu
e-mail: tuwei@szu.edu.cn

Y. Yue
e-mail: yangyue@szu.edu.cn

Department of Urban Informatics, School of Architecture and Urban Planning, Shenzhen University, Shenzhen, China

Q. Li · Y. Zhang
State Key Lab of Information Engineering in Surveying, Mapping and Remote Sensing, Wuhan University, Wuhan, China
e-mail: yatau@foxmail.com

W. Shi et al. (eds.), *Urban Informatics*, The Urban Book Series,
https://doi.org/10.1007/978-981-15-8983-6_29

29.1 Introduction

Cities are the living spaces of more than 50% of the global population but occupy less than 2% of the Earth's land surface. Although the past decades have witnessed advances in the economy, the environment, and human health in urban areas, especially in developing countries, cities are still facing great challenges on the way toward a sustainable future. These challenges include traffic congestion, environmental pollution, waste management, vitality loss, and social inequality. Since 2000, the boom of information and communication technologies (ICT), Internet, and artificial intelligence (AI) has produced massive urban data. Therefore, urban studies are increasingly adopting an information-centric approach where they meet geographic information science (GIS), computer science, urban planning, etc. (Batty 2013; Li 2017).

When enabled with Web 2.0, mobile Internet, and smartphones, humans become sensors to perceive their immediate surroundings and thus produce multi-source and heterogeneous content, such as text, images, videos, and audio, that is, user-generated content (UGC) (Koskinen 2003; Wang et al. 2014). UGC denotes content that has been posted by users on online platforms, including Internet forums, blogs, wikis, Instagram, YouTube, Douyin, and social networks such as Weibo, Facebook, and Twitter (Cha et al. 2007; George and Scerri 2007; Goodchild 2007; Krumm et al. 2008; Lenders et al. 2008; Hollenstein and Purves 2010; Heipke 2010). The use of UGC has grown rapidly in recent years, because of its comparatively low cost, high penetration, and fast update. For instance, the popular Wikipedia (Fig. 29.1a), edited by worldwide volunteers, has become the largest encyclopedia in the world and continues to be updated following advances in science, technology, and society. Another example is OpenStreetMap (OSM; Haklay and Weber 2008; Fig. 29.1b) which attracts large numbers of volunteers who use GPS and fine-resolution imagery to produce a comprehensive base map covering 80% of all roads (Barrington-Leigh

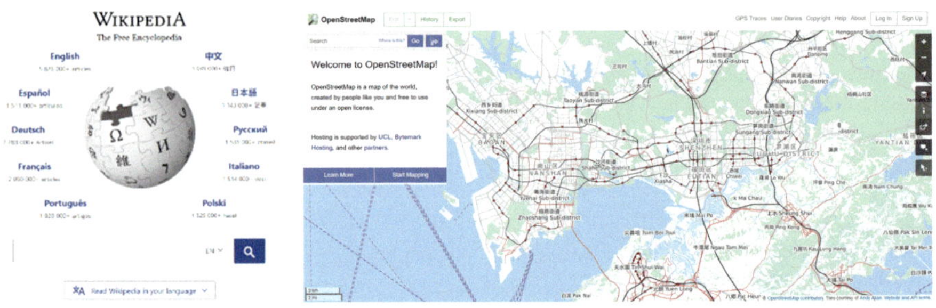

(a) (b)

Fig. 29.1 Representative user-generated content Web sites. **a** Wikipedia (https://www.wikipedia.org/); **b** OpenStreetMap in Shenzhen (https://www.openstreetmap.org/#map=11/22.5322/114.0912&layers=T)

and Millard-Ball 2017). Nowadays, OSM not only supports route planning and navigation services but also provides benefits to city planners with newly available urban data.

Classic urban studies generally rely on census data or field survey, which is expensive, labor-intensive, and of low temporal resolution. UGC enables urban studies to dive into the wave of big data (Aguilera et al. 2016). In general, UGC is produced by volunteers and thus contains volunteers' perceptions, preferences, or opinions about places, topics, and people. Accordingly, massive UGC provides unprecedented data sources for urban researchers to extra urban knowledge. On the other hand, UGC also motivates an alternative approach for conceptualizing and portraying the dynamics, structures, and characteristics of city. Consequently, UGC stimulates innovative urban applications which sense infrastructures, spaces, and people at all scales, reveals hidden urban knowledge, and makes real-time responses in support of urban emergency and long-term urban policies. Here, we sketch several types of UGC and their potential in urban sectors. The general framework of UGC-driven urban studies and insightful urban applications is reviewed. We discuss the challenges and future directions, including data quality and privacy, multi-source data fusion, integration of urban sensing, and urban governance.

The remainder of this chapter is organized as follows: Sect. 29.2 introduces four representative types of UGC, including geo-tagged photos, social media data, crowdsourcing GPS trajectories, and videos. Section 29.3 presents the general framework of UGC-driven urban studies and reviews typical urban applications. Section 29.4 discusses challenges and future directions. Section 29.5 concludes the chapter and discusses future work.

29.2 User-Generated Content

User-generated content has had a great impact on information-centric urban studies because of its appealing characteristics that crystallize the relationship between urban spaces and human activities with massive crowdsourcing data (Crooks et al. 2016; Jenkins et al. 2016; Thakuriah et al. 2016; Valdez et al. 2018). Accordingly, the sources and types of UGC are various (Heipke 2010; Mart et al. 2019; See et al. 2019). The focus here mainly concentrates on geo-tagged user-generated content as it provides opportunities to expose the hidden social, economic, and demographic information in urban spaces (Jenkins et al. 2016), which greatly benefits our understanding of the diversity of urban spaces and the complexity of urban dynamics. This section reviews several popular types of UGC and their characteristics, to provide a global overview of UGC, including geo-tagged photos, social media data, crowdsourcing GPS trajectories, and videos.

29.2.1 Geo-Tagged Photos

Geo-tagged photos are images uploaded to Internet forums and social networks by users. Usually, these photos are tagged with either explicit geographic coordinates or implicit forms of geo-information (e.g. point of interest or place name). There are two popular types of geo-tagged photos. One is sourced from the photo-sharing services, such as Flickr or Picasa, which allow users to share geo-tagged photos with text tags (Chen et al. 2018). Nowadays, there are many publicly available geo-tagged photos. For example, Yahoo Research Lab (Thomee et al. 2016) published one Flickr dataset YFCC100M containing 100 million images (https://webscope.sandbox.yahoo.com/catalog.php?datatype=i) for benchmarking purposes. MIT CSAIL (Zhou et al. 2018) published the dataset Place including 10 million photos of urban landmarks (http://places2.csail.mit.edu/). These photos, coordinates, and timestamps can be used to generate user footprints (Alivand and Hochmair 2017). Meanwhile, tagged texts provide auxiliary information with certain models, e.g. topic probability models. Through extracting the information hidden in these photos, researchers can effectively detect the temporal activities of photo takers and further analyze the behavior patterns of urban citizens.

Another type of geo-tagged photo is sourced from street view images collected by vehicles or volunteers, such as Google Street View (Hara et al. 2013; Li et al. 2015). Street view images usually contain one panoramic image and the corresponding location and therefore provide a sequence of images along a road. Different from remote sensing images monitoring geographic objects from above (aerial or space), the major advantage of street view images is the access they provide to urban landscapes from a pedestrian-like angle (Li et al. 2015; Cao et al. 2018). Consequently, street view images have had a significant impact on street level research, on such topics as urban greenery (Li et al. 2015), sidewalk accessibility (Hara et al. 2013), and the demographics of neighborhoods (Gebru et al. 2017).

Using innovative technologies such as computer vision and semantic annotations, geo-tagged photos have been used to extract massive knowledge about urban places and human beings. With regard to urban places, geo-tagged photos enable us to assess urban landscapes (Gebru et al. 2017; Li et al. 2015), including, for example, the distribution of urban infrastructure. In terms of human beings, they offer an opportunity to explore human social and mobility patterns at multiple geographic scales (Alivand and Hochmair 2017; Zhang et al. 2018). Furthermore, researchers can leverage them as a lens to articulate the relationship between urban spaces and human beings.

29.2.2 Social Media Data

Social media data contribute another valuable form of content to urban studies, especially location-based social networks (LBSN) (Kim et al. 2017; Shelton et al. 2015;

Thakuriah et al. 2016). In 2018, there were over 3 billion active social media users, and almost 3 billion active users of mobile social media (Mart et al. 2019). Generally, LBSN data provide various perspectives on social, economic, and demographic aspects in urban spaces. Through embedding social media data into urban spaces, the link to human beings is established, enabling the tangible and comprehensive understanding of human–environment interactions (Mart et al. 2019).

To date, there have been many substantial studies using LBSN data (e.g. Foursquare, Twitter, Airbnb, and Weibo) to portray urban dynamics. Table 29.1 lists the publicly available social media content. Foursquare data usually include place information, including check-ins, ratings, tips, and photos. Foursquare data have been used to identify users' perceptions and preferences in urban spaces through the identification of the most visited or checked-in places (Agryzkov et al. 2016; Mart et al. 2017). Twitter and Weibo are other commonly used social media datasets. The coordinates and timestamps associated with social media content of Twitter can be used to detect the spatiotemporal patterns in people's presence and activities (Crooks et al. 2015). Combined with natural language processing (NLP), Twitter is capable of detecting certain events, hot topics, culture distribution, urban functions, etc. (Yang et al. 2015; Tu et al. 2017; Tu et al. 2018a). Different from Twitter data, the content of Instagram is more visually related about the observed entity rather than text related, in the format of coordinates, photos, and corresponding descriptions (Giridhar et al. 2017). Thus, Instagram-based studies focus on the descriptions of a place through keywords and the activities happening in a place (Mart et al. 2019). Airbnb, one Web site offering information about temporal accommodation plays an important role in urban studies about rental homes. Meanwhile, Airbnb content provides an insight to observe tourism, especially in tourist cities.

Table 29.1 Publicly available social media data

Social media data	Description	Web link
Global Foursquare check-in dataset (Yang et al. 2015)	Contains 33,278,683 check-ins by 266,909 users on 3,680,126 venues (in 415 cities in 77 countries)	https://sites.google.com/site/yangdingqi/home/foursquare-dataset
Twitter dataset (Yang and Leskovec 2011)	Includes 467 million Twitter posts from 20 million users covering a 7-month period from June 1, 2009, to December 31, 2009	https://snap.stanford.edu/data/twitter7.html
Instagram dataset (Ferrara et al. 2014)	Contains information from 45,000 users of Instagram during the period from Jan 20 to Feb 17, 2014	http://www.emilio.ferrara.name/datasets/
Airbnb dataset	Contains reviews, listings, and neighborhood information in worldwide cities	http://insideairbnb.com/get-the-data.html

29.2.3 Crowdsourcing GPS Trajectories

The availability of crowdsourcing technologies facilitates the emergence and effective usage of geospatial data, which is of profound significance in the planning and management of urban spaces (Crooks et al. 2015; Jenkins et al. 2016). Crowdsourced GPS trajectories are usually collected by volunteers without professional services (Heipke 2010), implementing the concept of citizens as sensors proposed by Goodchild (See et al. 2019). So far, there have been many projects about crowdsourcing geospatial data (Heipke 2010), such as OpenStreetMap (OSM) (Planet 2019), Wikimapia, or HD Traffic™. OSM is probably the most prominent among all the crowdsourcing projects (Heipke 2010). The purpose of OSM is to establish a free, editable map across the world, supported by volunteers acting as sensors to collect geographic data (Barron et al. 2014). OSM has been widely used in a broad range of urban applications, from navigation to routing, from urban block division to urban function recognition (Crooks et al. 2015). In addition, digital footprints extracted from crowdsourced GPS trajectories are also important proxies. Digital footprints through time provide an insight to understand human mobility patterns and also offer access to the dynamic cognition of urban places.

29.2.4 Videos

Videos contain amounts of dynamic information about described phenomena and can greatly assist urban planning and management, such as urban scene understanding (Cordts et al. 2016), human activity analysis (Zhu et al. 2017), transportation surveillance (Chen et al. 2016), and emergency management (Schnebele et al. 2015). There are many ways to obtain video datasets, such as from YouTube videos (Douyin and Kuaishou), from social media platforms, urban surveillance videos, and street videos. Unlike the above three kinds of UGC data, although information in videos is wealthy and dynamic, it is relatively difficult to process videos quickly and efficiently due to their volume, noise, and diversity (Zhu et al. 2017). Lots of techniques for motion estimation, tracking, segmentation, and video filtering have been developed (Tekalp 2015). Nowadays, human activity and perception have become hot topics in urban studies. Videos from social media platforms, such as YouTube, can be utilized to perform spatiotemporal mapping of human activity, in the form of human activity recognition, sport mapping, weather impacts on human activities, crime detection, etc. (Zhu et al. 2017). Moreover, videos can reveal functions in urban scene understanding, such as those revealed by the Cityscapes dataset (https://www.cityscapes-dataset.com/; Cordts et al. 2016). This dataset provides a detailed annotated class list of urban stereo videos covering fifty cities, which can be used in semantic understanding of urban scenes (Cordts et al. 2016).

29.3 Urban Studies Driven by User-Generated Content

User-generated content contains massive hidden information, such as the users' socioeconomic status, preferences, opinions, and activity-mobility patterns (Jenkins et al. 2016; Mart et al. 2019; Thakuriah et al. 2016; Venerandi et al. 2015). Large volume UGC is stored, cleaned, and extracted to learn about phenomena in urban spaces and the interactions between urban functions and people. Consequently, UGC has been widely applied in urban studies, such as in urban planning, urban transportation, urban environment, and health. This section presents the general framework of UGC-driven urban studies and reviews representative urban applications.

29.3.1 Framework for UGC-Driven Urban Studies

Acquisition, integration, and analysis of UGC can be used to tackle the major issues that cities face, e.g. traffic congestion, urban growth, air pollution, public health, and urban safety. Generally, the framework of UGC-driven urban studies contains four layers from the bottom to the top as shown in Fig. 29.2: UGC harvesting, UGC management, UGC analytics, and smart urban applications.

In the UGC harvesting layer, single- or multi-source UGC is acquired from an online forum, vertical Web sites, and social networks. For example, posted Twitter messages about a city will be crawled for future data processing and analytics. In the second UGC management layer, the acquired UGC will be organized by locations, by users, or by associated topics. High-performance computing architectures and effective indexing structures that simultaneously incorporate spatiotemporal information, and texts will be built for efficient data manipulation. In the UGC analytics step, data mining (clustering and classification), and machine learning (e.g. logistics regression, decision tree, random forest, and support vector machine), deep learning (e.g. convolutional neural networks, deep residual networks, generative adversarial networks), and visualization will be used to recognize objects, patterns, and associations, and to speculate about causes and effects. In the smart urban application step, this extracted urban knowledge will be utilized by urban planners, transportation officials, environmentalists, and medical departments. In addition, the information will be disseminated to related people and organizations to improve urban living.

29.3.2 Urban Planning

Urban planning refers to social, economic, and political activities concerning the interconnectedness and complexity of urban spaces (Levy 2016). Urban planning is close related to many interactions of places and people, including urban form, land-use planning, locating transportation infrastructures, and designing urban interfaces.

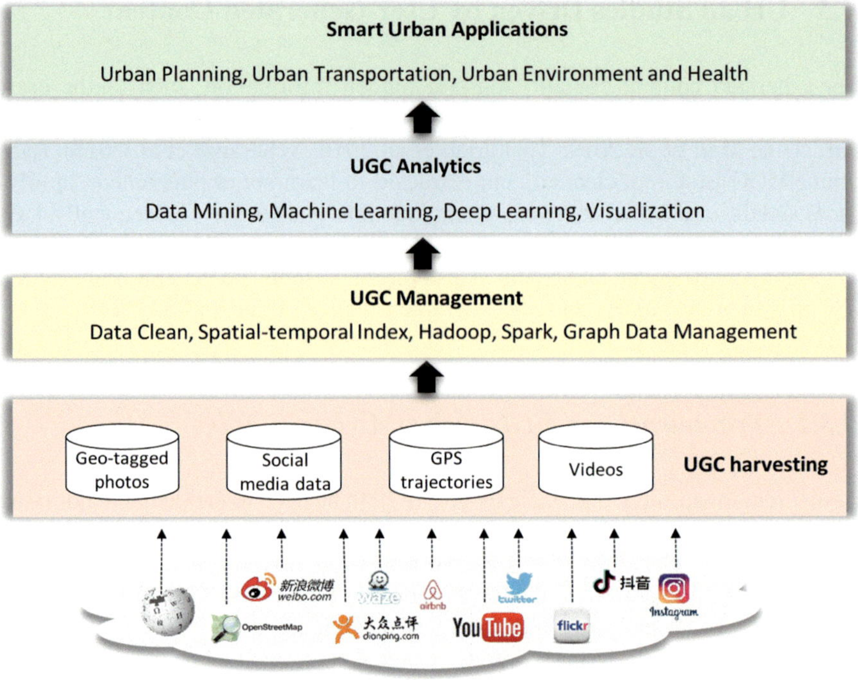

Fig. 29.2 General framework of urban studies using user-generated content

UGC not only provides rich representations about urban space, but also opens access to human activity research (Crooks et al. 2015; Li et al. 2017; Longley and Adnan 2016).

The focus here mainly lies on two parts, namely human activity, and urban form and function. Regarding human activity, social media data collected from a great number of users, such as Foursquare, Instagram, or Twitter, provide the detailed descriptions of human activities within urban spaces (Mart et al. 2019), with which researchers can recognize activity patterns at suitable spatiotemporal scales. Recently, Tu et al. (2018a) fused large volume social media check-in data and mobile phone positioning data to extract city-wide human activities and portray their diurnal patterns. Gebru et al. (2017) inferred demographic information at neighborhoods across the USA from massive street view images. Studies of urban form and function address the aggregation of the physical shapes of urban spaces and the human activities that happen in these spaces respectively (Crooks et al. 2016). UGC provides large amounts of information that can be used to understand urban form and function and highlights how they influence each other (Crooks et al. 2015). Street network maps of OSM give detailed insights into urban form and are of fundamental importance in a range of applications. Other types of UGC, such as geo-tagged photos and social-media data, can be used to understand urban function (Gebru et al. 2017; Li et al. 2015; Cao et al. 2018). For example, Zhong et al. (2018) presented a tweet-topic-function-structure framework to reveal spatial patterns from individual tweets. Their results demonstrated that when aggregating tweets by zones, the areas with the same topics form spatial clusters but of entangled urban functions. Using massive street view images, Zhang et al. (2018) developed a data-driven deep learning approach to map the distribution of city-wide human perception (e.g. safe, lively, beautiful, wealthy, depressing, or boring), which suggest the potential of massive UGC.

29.3.3 Urban Transportation

Transportation is essential to daily movements in the city. Quantities of urban-sensed data have been used to resolve problems in urban transportation and to build intelligent transportation systems (ITS; Wang et al. 2016). The social media platforms, mobile phones, and surveillance videos make it possible to generate rich social signals in a real-time manner and establish a data foundation for social transportation research (Zheng et al. 2016). UGC-based ITS can make use of various crowdsourced social signals to understand the social needs of transportation and combine needs and services to improve efficiency and effectiveness and make traffic conditions and citizen travel more convenient (Wang et al. 2016; Tu et al. 2019).

UGC can be used in a range of applications in urban transportation, for example, in mapping road networks, monitoring real-time traffic, or recommending travel routes. In terms of traffic monitoring, information obtained from social media platforms, such as Twitter, YouTube, and Flickr, encourages people to participate effectively in traffic tasks, such as identifying road hazards, and greatly cuts down on the related financial burden of government (Santani et al. 2015). In traffic management, social media data support shortest path computing, travel recommendation, etc., and can be improved by exploiting the content hidden in UGC (Wang et al. 2016). With respect to future green transportation, UGC that connects vehicles, people, and urban infrastructures can help to advance the efficiency of entire transportation systems and to promote reductions in fuel consumption and carbon emission (Wang et al. 2016).

29.3.4 Urban Environments and Health

The urban environment has a close relationship to the quality of human life and health, both of which should be emphasized in urban governance. The knowledge mined from social media data, mobile phones, and other UGC can provide opportunities to quantify aspects of the urban environment, such as urban green space (Li et al. 2015), air quality (Jiang et al. 2015), soundscapes (Aiello et al. 2016), and heat distribution (Overeem et al. 2013). Thus, fine-resolution maps of these environmental factors can help urban planners to improve residents' quality of life, surroundings, and health. For example, utilizing the green index of street view images, Li et al. (2015) assess street-level urban greenery and provide suggestions for urban planners to reasonably improve the distribution of urban green spaces. Jiang et al. (2015) analyzed the spatiotemporal tendency in social media data using Sina Weibo (Chinese Twitter) in an effort to monitor air quality dynamically in large cities. Also, maps can be drawn by establishing a relationship between human perceptions and soundscapes extracted from social media (Aiello et al. 2016). In addition, smartphone battery temperatures can be used to estimate urban daily mean air temperatures by utilizing a heat transfer model in real time (Overeem et al. 2013).

29.3.5 Urban Safety

Citizens residing in urban areas may face fires, storms, heavy rainfall, traffic jams, and other hazards, which affect urban safety and human life. Therefore, it is important to detect urban emergency events in real time (Xu et al. 2016). Lots of messages from UGC, such as social media, volunteered photos, and videos, contain information about urban events and are important data sources to derive emergency events, capture their physical and social features, and help urban management departments to react quickly (Schnebele et al. 2015; Xu et al. 2016). Thus, event detection becomes a crucial issue in urban emergency management. There have been many studies focused on urban event detection. For example, some studies proposed adaptive algorithms to detect urban events through geo-tagged data from photo-sharing services (Papadopoulos et al. 2010). Making use of crowdsourcing to build an emergency management system is another choice (Oliveira et al. 2017). In addition, in order to detect emergency events in real time, the 5 W (What, Where, When, Who, and Why) characteristics are proposed to depict the spatial and temporal information of social media and thus to achieve detection goals (Xu et al. 2016).

29.4 Challenges and Future Directions

Recent UGC research has made great advances in the domain of urban studies. Many innovative urban applications have stimulated thinking about better urban living. Because of the complexity of cities (Batty 2007), this research presents various challenges to information-centric cities.

29.4.1 Data Quality and Privacy

Recently, with the growing interest in artificial intelligence, it has become possible to produce UGC not only by people but also by machines. Several studies have reported that fake messages are posted on Twitter (Fourney et al. 2017). Many machine accounts have been created to disseminate special texts and photos with the objective of influencing specific groups of people. Consequently, UGC may be biased. When conducting UGC-driven urban studies, attention should be paid to the data quality issue to strengthen the reliability of the findings (Tu et al. 2018b; Jiang et al. 2019).

The privacy of UGC is another important issue. Scientific ethics should be highlighted for UGC research. Recently, a new General Data Protection Regulation (GDPR) was adopted in Europe and is likely to fundamentally reshape the way in which data are handled across every sector. The general public, Internet giants, and scientific communities should find an appropriate consensus on the collection, processing, and study of UGC.

29.4.2 Multi-source UGC Fusion

When thousands and even millions of users contribute to UGC, the results are often highly fragmented. For example, because most geo-tagged photos are shared by users with smartphones, the perceptions and preferences of people without smartphones cannot be captured. Tweets posted in tourist destinations and at landmarks tend to emphasize certain topics and opinions, resulting in bias with respect to the general population (Longley and Adnan 2016). Thus, careful selection of data sources is crucial if the reconstructed urban knowledge is to be complete and accurate. The results from a single source of UGC may be biased and contain only a part of urban knowledge. The misuse of UGC may consequently generate biased understanding. Fusion of multiple sources may be required to deepen our understanding of objects, people, and places in the city (Li et al. 2017). By integrating traditional urban data and alternative UGC, more and more comprehensive and wide-coverage urban solutions would be supported (Estima and Painho 2016).

29.4.3 Integrating Urban Sensing and Urban Governance

UGC can provide alternative data sources to sense the invisible city under the physical surface, for example, regarding urban deprivation (Venerandi et al. 2015), human mobility (Yang, Qu, Yang et al. 2019; Xu et al. 2019), urban areas of interest (Chen et al. 2018), urban vibrancy (Huang et al. 2019), and urban functions (Tu et al. 2017, 2018a; Zhong et al. 2018). UGC enables us to assess new dimensions of the city and to deepen our understanding of complex cities. However, these novel urban-sensing studies have not been well integrated with urban governance. How to take the sensed urban information into the workflow of urban governance is still an open question. UGC-driven urban policy-making will be necessary if we are to explore a new framework linking UGC to urban operation.

29.5 Conclusion

The prevalence of UGC provides an alternative data source for urban studies because of its characteristics of low cost, high penetration, and wide coverage. Massive UGC can not only sense invisible urban spaces but also provide fertile soil for breeding innovative applications. This chapter has summarized the four representative types of UGC: geo-tagged photos, social-media data, crowdsourced GPS trajectories, and videos. The general framework of UGC-driven urban studies has been presented, and smart UGC-driven applications in the city have been reviewed. The challenges and opportunities of UGC in urban studies have also been discussed, in order to provide insights for future urban informatics approaches. This will lead to the emergence of alternative urban informatics approaches that better capture the intricate nature of urban spaces and their dynamics.

Acknowledgements This research was jointly supported by the Natural Science Foundation of China (71961137003, 41401444), Shenzhen Scientific Research and Development Funding Program (#JCYJ20180305125113883), and China Scholarship Council (201708440434).

References

Agryzkov T, Martí P, Nolasco-Cirugeda A, Serrano-Estrada L, Tortosa L, Vicent JF (2016) Analysing successful public spaces in an urban street network using data from the social networks Foursquare and Twitter. Appl Netw Sci 1(1):12

Aguilera U, López-de-Ipiña D, Pérez J (2016) Collaboration-centred cities through urban apps based on open and user-generated data. Sensors 16(7):1022

Aiello LM, Schifanella R, Quercia D, Aletta F (2016) Chatty maps: constructing sound maps of urban areas from social media data. R Soc Open Sci 3(3):150690

Alivand M, Hochmair HH (2017) Spatiotemporal analysis of photo contribution patterns to Panoramio and Flickr. Cartography and Geographic Information Science 44(2):170–184

Barrington-Leigh C, Millard-Ball A (2017) The world's user-generated road map is more than 80% complete. PLoS ONE 12(8):e180698

Barron C, Neis P, Zipf A (2014) A comprehensive framework for intrinsic OpenStreetMap quality analysis. Trans GIS 18(6):877–895

Batty M (2007) Complexity in city systems: understanding, evolution, and design. In: de Roo G, Silva EA (eds) A planner's encounter with complexity, pp 99–122

Batty M (2013) Big Data, smart cities and city planning. Dialogues Hum Geogr 3(3):274–279

Cao R, Zhu J, Tu W, Li Q, Cao J, Liu B, Zhang Q, Qiu G (2018) Integrating aerial and street view images for urban land use classification. Remote Sens 10(10):1553

Cha M, Kwak H, Rodriguez P, Ahn YY, Moon S (2007) I tube, you tube, everybody tubes: analyzing the world's largest user generated content video system. In: Proceedings of the 7th ACM SIGCOMM conference on internet measurement, San Diego, California, USA, pp 1–14

Chen M, Arribas-Bel D, Singleton A (2018) Understanding the dynamics of urban areas of interest through volunteered geographic information. J Geogr Syst 21(1):89–109

Chen N, Chen Y, You Y, Ling H, Liang P, Zimmermann R (2016) Dynamic urban surveillance video stream processing using fog computing. In: IEEE second international conference on multimedia big data (BigMM) pp 105–112

Crooks A, Pfoser D, Jenkins A, Croitoru A, Stefanidis A, Smith D, Karagiorgou S, Efentakis A, Lamprianidis G (2015) Crowdsourcing urban form and function. Int J Geogr Inf Sci 29(5):720–741

Crooks AT, Croitoru A, Jenkins A, Mahabir R, Agouris P, Stefanidis A (2016) User-generated Big Data and urban morphology. Built Environ 42(3):396–414

Cordts M, Omran M, Ramos S, Rehfeld T, Enzweiler M, Benenson R, Franke U, Roth S, Schiele B (2016) The cityscapes dataset for semantic urban scene understanding. In: Proceedings of the IEEE conference on computer vision and pattern recognition (CVPR) pp 3213–3223

Estima J, Painho M (2016) User generated spatial content-integrator: Conceptual model to integrate data from diverse sources of user generated spatial content. ISPRS Int J Geo-Inf 5(10):183

Ferrara E, Interdonato E, Tagarelli A (2014) Online popularity and topical interests through the lens of Instagram. In: Proceedings of the 25th ACM conference on hypertext and social media, Santiago, Chile, 1–4 September 2014

Fourney A, Racz MZ, Ranade G, Mobius M, Horvitz E (2017) Geographic and temporal trends in fake news consumption during the 2016 US Presidential Election. In: Proceedings of the 2017 ACM conference on information and knowledge management, New York, pp 2071–2074

Gebru T, Krause J, Wang Y, Chen D, Deng J, Aiden EL, Fei-Fei L (2017) Using deep learning and Google Street View to estimate the demographic makeup of neighborhoods across the United States. Proc Natl Acad Sci 114(50):13108–13113

George C, Scerri J (2007) Web 2.0 and user-generated content: legal challenges in the new frontier. J Inf Law Technol 2. SSRN: https://ssrn.com/abstract=1290715

Giridhar P, Wang S, Abdelzaher T, Ganti R, Kaplan L, George J (2017) On localizing urban events with Instagram. In: IEEE INFOCOM, pp 1–9

Goodchild MF (2007) Citizens as sensors: Web 2.0 and the volunteering of geographic information. Int J Spat Data Infrastruct Res 7:24–32

Haklay M, Weber P (2008) Open street map: User-generated street maps. IEEE Pervasive Comput 7(4):12–18

Hara K, Le V, Froehlich J (2013) Combining crowdsourcing and Google street view to identify street-level accessibility problems. In: Proceedings of the SIGCHI conference on human factors in computing systems, pp 631–640

Heipke C (2010) Crowdsourcing geospatial data. ISPRS Journal of Photogrammetry and Remote Sensing 65(6):550–557

Hollenstein L (2010) Purves RJ (2010) Exploring place through user-generated content: Using Flickr tags to describe city cores. J Spat Inf Sci 1:21–48

Huang B, Yulun Z, Zhigang L, Song Y, Cai J, Tu W (2019) Evaluating and characterizing urban vibrancy using spatial Big Data: Shanghai as a case study. Environment and Planning B: Urban Analytics and City Science. https://doi.org/10.1177/2399808319828730

Jenkins A, Croitoru A, Crooks AT, Stefanidis A (2016) Crowdsourcing a collective sense of place. PLoS ONE 11(4):e152932

Jiang J, Li Q, Tu W, Shaw SL, Yue Y (2019) A simple and direct method to analyse the influences of sampling fractions on modelling intra-city human mobility. Int J Geogr Inf Sci 33(3):618–644

Jiang W, Wang Y, Tsou MH, Fu X (2015) Using social media to detect outdoor air pollution and monitor air quality index (AQI): a geo-targeted spatiotemporal analysis framework with Sina Weibo (Chinese Twitter). PLoS ONE 10(10):e141185

Kim HJ, Chae BK, Park SB (2017) Exploring public space through social media: an exploratory case study on the high line New York City. Urban Design International 23(2):69–85

Koskinen I (2003) User-generated content in mobile multimedia: empirical evidence from user studies. In: Proceedings ICME, pp 645–648

Krumm J, Davies N, Narayanaswami C (2008) User-generated content. IEEE Pervasive Comput 7:10–11

Lenders V, Koukoumidis E, Zhang P, Martonosi M (2008) Location-based trust for mobile user-generated content: applications, challenges and implementations. In: Proceedings of the 9th workshop on mobile computing systems and applications, pp 60–64

Levy JM (2016) Contemporary urban planning. Taylor and Francis

Li Q (2017) From geomatics to urban informatics. Geomatics Inf Sci Wuhan University 42(1):1–6

Li X, Tu W, Shen S, Yang Y, Luo NX, Li QQ. (2017) Revealing spatial variation and correlation of urban travels from big trajectory data. Int Arch Photogrammetry Remote Sens Spat Inf Sci 42:53–57

Li X, Zhang C, Li W, Ricard R, Meng Q, Zhang W (2015) Assessing street-level urban greenery using Google Street View and a modified green view index. Urban Forest Urban Greening 14(3):675–685

Longley PA, Adnan M (2016) Geo-temporal Twitter demographics. Int J Geogr Inf Sci 30:369–389

Martí P, Serrano-Estrada L, Nolasco-Cirugeda A (2017) Using locative social media and urban cartographies to identify and locate successful urban plazas. Cities 64:66–78

Martí P, Serrano-Estrada L, Nolasco-Cirugeda A (2019) Social media data: challenges, opportunities and limitations in urban studies. Comput Environ Urban Syst 74:161–174

Oliveira ACM, Botega LC, Saran JF, Silva JN, Neris VPA et al (2017) Crowdsourcing, data and information fusion and situation awareness for emergency management of forest fires: the project DF100Fogo (FDWithoutFire). Computers, Environment, and Urban Systems 77. https://doi.org/10.1016/j.compenvurbsys.2017.08.006

Overeem A, R Robinson JC, Leijnse H, Steeneveld GJ, Horn BK, Uijlenhoet R (2013) Crowdsourcing urban air temperatures from smartphone battery temperatures. Geophys Res Lett 40(15):4081–4085

Papadopoulos S, Zigkolis C, Kompatsiaris Y, Vakali A (2010) Cluster-based landmark and event detection for tagged photo collections. IEEE Multimedia 18(1):52–63

Planet (2019) Data volume of OSM. https://wiki.openstreetmap.org/wiki/ Planet.osm

Santani D, Njuguna J, Bills T, Bryant AW, Bryant R, Ledgard J, Gatica-Perez D (2015) Communisense: Crowdsourcing road hazards in Nairobi. In: Proceedings of the 17th international conference on human–computer interaction with mobile devices and services, pp 445–456

Schnebele E, Oxendine C, Cervone G, Ferreira CM, Waters N (2015) Using non-authoritative sources during emergencies in urban areas. Computational approaches for urban environments, Springer, Cham, pp 337–361

See L, Fonte CC, Antoniou V, Minghini M (2019) Volunteered geographic information: looking towards the next 10 years. J Geogr Syst 21:1–3

Shelton T, Poorthuis A, Zook M (2015) Social media and the city: Rethinking urban socio-spatial inequality using user-generated geographic information. Landscape Urban Plan 142:198–211

Tekalp AM (2015) Digital video processing. Prentice Hall Press, Upper Saddle River, NJ

Thakuriah P, Tilahun NY, Zellner M (2016) Introduction. Seeing cities through big data: research, methods and applications in urban informatics. Springer, Cham, pp 1–9

Thomee B, Shamma DA, Friedland G, Elizalde B, Ni K, Poland D, Borth D, Li LJ (2016) YFCC100M: The new data in multimedia research. Commun ACM 59(2):64–73

Tu W, Cao J, Yue Y, Shaw SL, Zhou M, Wang Z, Chang X, Xu Y, Li Q (2017) Coupling mobile phone and social media data: a new approach to understanding urban functions and diurnal patterns. Int J Geogr Inf Sci 31:2331–2358

Tu W, Hu Z, Li L, Cao J, Jiang J, Li Q, Li Q (2018a) Portraying urban functional zones by coupling remote sensing imagery and human sensing data. Remote Sens 10(1):141

Tu W, Cao R, Yue Y, Zhou B, Li Q, Li Q (2018b) Spatial variations in urban public ridership derived from GPS trajectories and smart card data. J Transp Geogr 69:45–57

Tu W, Santi P, Zhao T, He X, Li Q, Dong L, Wallington TJ, Ratti C (2019) Acceptability, energy consumption, and costs of electric vehicle for ridehailing. Appl Energy 250:147–160

Valdez AM, Cook M, Langendahl PA, Roby H, Potter S (2018) Prototyping sustainable mobility practices: user-generated data in the smart city. Technol Anal Strateg Manage 30(2):144–157

Venerandi A, Quattrone G, Capra L, Quercia D, Saez-Trumper D (2015) Measuring urban deprivation from user generated content. In: Proceedings of the 18th ACM conference on computer supported cooperative work and social computing, pp 254–264

Wang D, Amin MT, Li S, Abdelzaher T, Kaplan L, Gu S, Pan C, Liu H, Aggarwal CC, Ganti R, Wang X (2014) Using humans as sensors: an estimation-theoretic perspective. In: Proceedings of the 13th international symposium on information processing in sensor networks, Berlin, Germany, pp 35–46

Wang X, Zheng X, Zhang Q, Wang T, Shen D (2016) Crowdsourcing in ITS: The state of the work and the networking. IEEE Trans Intell Transp Syst 17(6):1596–1605

Xu Y, Chen D, Zhang X, Tu W, Chen Y, Shen Y, Ratti C (2019) Unravel the landscape and pulses of cycling activities from a dockless bike-sharing system. Comput Environ Urban Syst 74:184–203

Xu Y, Chen D, Zhang X, Mei L, Luo X, Wei X, Hu C (2016) Crowdsourcing based description of urban emergency events using social media Big Data. IEEE Trans Cloud Comput. https://doi.org/10.1016/j.compenvurbsys.2019.02.002

Yang D, Zhang D, Qu B (2015) Participatory cultural mapping based on collective behavior data in location based social networks. ACM Trans Intell Syst Technol (TIST) 7(3):1–23

Yang D, Qu B, Yang J, Cudre-Mauroux P (2019) Revisiting user mobility and social relationships in LBSNs: a hypergraph embedding approach. In: Proceedings of The Web Conference (WWW'19), San Francisco, USA

Yang J, Leskovec J (2011) Temporal variation in online media. In: ACM International Conference on Web Search and Data Mining (WSDM '11)

Zhang F, Zhou B, Liu L, Liu Y, Fung HH, Lin H, Ratti C (2018) Measuring human perceptions of a large-scale urban region using machine learning. Landscape Urban Plan 180:148–160

Zheng X, Chen W, Wang P, Shen D, Chen S, Wang X, Zhang Q, Yang L (2016) Big data for social transportation. IEEE Trans Intell Transp Syst 17(3):620–630

Zhong C, Zeng S, Tu W, Yoshida M (2018) Profiling the spatial structure of London: From individual tweets to aggregated functional zones. ISPRS Int J Geo-Inf 7(10):386

Zhou B, Lapedriza A, Khosla A, Oliva A, Torralba A (2018) Places: a 10 million image database for scene recognition. IEEE Trans Pattern Anal Mach Intell 40:1452–1464

Zhu Y, Liu S, Newsam S (2017) Large-scale mapping of human activity using geo-tagged videos. In: Proceedings of the 25th ACM SIGSPATIAL International Conference on Advances in Geographic Information Systems, p 68

Wei Tu is an associate professor of the department of Urban Informatics, Shenzhen University. He received his Ph.D. degree in photogrammetry and remote sensing from Wuhan University. He has been a visiting scholar at the Seneseable City Laboratory, MIT. His research interests include urban informatics, urban sensing, and trajectory analytics.

Qingquan Li is a Professor at Shenzhen University. He is the President of Shenzhen University, Shenzhen, China, and the Director of the Guangdong Key Laboratory of Urban Informatics. He is a member of the International Eurasian Academy of Sciences. He has received several national awards, including the State Scientific and Technological Progress Award, the state Technological Invention Award, and the Ho Leung Ho Lee Award.

Yatao Zhang received his B.S. degree from Sun Yat-sen University. He is currently working toward an M.S. degree in GIS in the State Key Laboratory of Information Engineering in Surveying, Mapping, and Remote Sensing, Wuhan University, Wuhan, China. His research interests include multi-source spatiotemporal data fusion and urban sensing.

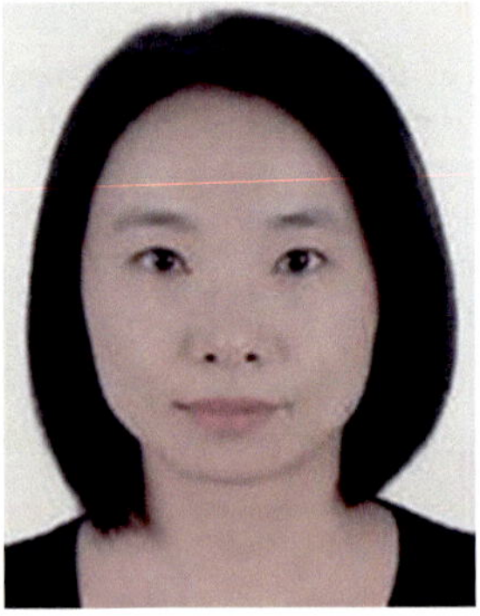

Dr. Yang Yue is a Professor of Urban Informatics, with an interdisciplinary background in Geomatics, GIS, and urban studies, and especially focuses on urban big data studies. She has published over 50 peer-reviewed research papers and serves as a committee member in local and international GIS, computer, transportation, and urban planning academic advising organizations.

Part IV
Urban Big Data Infrastructure

Chapter 30
Introduction to Urban Big Data Infrastructure

Michael F. Goodchild

Rapid progress is being made in the development of infrastructure for handling urban big data, as will be evident from even the most cursory examination of the eight chapters in this section. Big data require the ability to handle unprecedented volumes of data, often in near-real time, and to fuse and conflate data from multiple sources with different degrees of quality. But in addition, the nature of infrastructure should be interpreted broadly, as encompassing not only data, but also the software needed to handle the data, the people who possess the requisite skills, and the decision-makers and general public who make use of the products of urban big data and may also contribute data through crowdsourcing. Moreover, no discussion of urban big data can escape the ethical issues that are raised by the technology and its use, especially the thorny issue of privacy. Urban big data infrastructure is clearly a vast topic, and these eight chapters can do no more than scratch the surface. The following paragraphs give a brief introduction to each chapter and explain how the various contributions fit together. At the end, a short discussion suggests some of the topics that might be covered in a longer review, and gives an overall assessment of this part of the book.

In Chap. 31, Ningchuan Xiao and Harvey Miller expand on the definition of urban big data, explaining its role in concepts of smart mobility, the smart city, and enhanced digital infrastructure. They review many sources of urban big data, from sensors to crowdsourcing, and argue strongly for open access as a key to supporting many potential applications. Some well-chosen stories are used to identify use cases, and the example of access to real-time data on transit vehicles is used to demonstrate some of the technical challenges.

While ethical issues are often regrettably left till last, we have chosen to raise questions of privacy early in the section. Chapter 32 by Jerome Dobson and William Herbert discusses geoprivacy, the threat to individual privacy that originates with

M. F. Goodchild (✉)
University of California, Santa Barbara, Santa Barbara, USA
e-mail: good@geog.ucsb.edu

W. Shi et al. (eds.), *Urban Informatics*, The Urban Book Series,
https://doi.org/10.1007/978-981-15-8983-6_30

the widespread capturing of an individual's coordinates, often without that individual's knowledge and conscious consent. Regulation varies from country to country and even within countries, and while the European Union has recently adopted comprehensive protection of user privacy, there has been little progress in the USA.

Accurate surveying of property has existed for centuries, but it has generally been assumed that a point can lie in at most one property. Today, this may no longer be true: In condominiums, for example, properties can be stacked on top of each other, requiring a three-dimensional (3D) approach. In Chap. 33, Lin Li provides an extensive review of the complex ownership geometries that can now be dealt with using three-dimensional techniques and digital representations.

Chapter 34 follows directly from Chap. 33 by providing a comprehensive review of techniques for 3D digital modeling of city structures. Much of this interest comes from the construction industry, whose building information modeling (BIM) provides techniques for capturing not only architectural plans, but also as-built information on building infrastructure and use. The chapter compares BIM with City Geography Markup Language (CityGML), a product of the geospatial community that brings spatial database modeling indoors, allowing a full integration between outdoor applications that are largely 2D, and indoor functions in full 3D.

The sequence of chapters on 3D representations of cities ends with Chap. 35, based on Esri's CityEngine. City planning requires consideration of buildings in context and specifically with the ways in which planners regulate the development of neighborhoods. CityEngine was developed as a multipurpose planning tool that is capable of implementing regulations, providing perspective visualizations of plans, and supporting many of the functions of city government. The chapter provides ample illustration of the applications of the software and its implications for geodesign and the planning process.

Today's cities are complex and growing more so as a result of recent investments in digital infrastructure. The massive volumes of data that are now available, and the speed at which decisions are needed, argue in many cases for the use of high-performance computing (HPC). Cyber geographic information systems (CyberGIS), the topic of Chap. 36, use HPC to address many such applications, extending conventional GIS to take advantage of massive computational and communication technologies.

Chapter 37 focuses on spatial search, the process that allows users to find and assess big data resources and judge their fitness for a given application. Techniques of spatial search became necessary beginning in the early 1990s, as the availability of geospatial data began to outstrip any user's knowledge of where to look. Data warehouses, geolibraries, and geoportals are all responses to the need to be systematic about the storage of geospatial data. The chapter reviews the relevant techniques, including the concept of metadata, that is, data that allow a user to assess the fitness of a given data set.

Finally, Chap. 38 addresses the Internet of things (IoT), a term that describes sensors of various kinds that are connected to the Internet. Sensors might be fixed in space, such as closed-circuit television (CCTV) cameras, carried on vehicles, or carried by humans, often in the form of smartphone functions. IoT is clearly an important aspect of the smart city and of urban big data.

Big data infrastructure is a means to an end, rather than an end in itself. While Part IV has provided an overview of some of the foundational issues, the reader will have to look further for a complete view of the role of this infrastructure in enabling the functions of the modern city. Some of that can be found in other sections of this volume, and some is surely yet to emerge. While we can perhaps see and share some of the excitement over IoT or CityEngine, the eventual value of these tools is still difficult to predict. There is a "build it and they will come" sense to big data infrastructure, but also a sense that some of the eventual outcomes are unanticipated and may well have costs that exceed their benefits. Chapter 32 on privacy is perhaps a foretaste of what may arise as the technologies of surveillance proliferate.

Chapter 31
Cultivating Urban Big Data

Ningchuan Xiao and Harvey J. Miller

Abstract Urban big data often contain spatial and temporal elements that have increasingly become an integral part of various applications and projects such as smart mobility, smart city, and other digitally enhanced urban infrastructure. It is critical to develop an open and collaborative environment so that these data can be used by a wide range of users. This chapter first discusses some characteristics and sources of urban big data. Three hypothetical user stories are described to highlight the potential of these data. After describing the internal data structure of these data and techniques that can be used to retrieve the data, we discuss the difficulty in making the data useful for the general public and elaborate on a self-organizing agile approach to developing an urban big data infrastructure.

31.1 Introduction

Big data are one of the most popular topics of the past decade (Marr 2015). The concept of big data has evolved beyond the original context as a buzz word into the reality of daily life and has shown tangible values for businesses, governments, research communities, and the general public (Kim et al. 2014; Günther et al. 2017). Informally, big data refer to the vast amount of data that are generated, collected, or distributed at a high frequency or speed. More formal definitions of big data vary widely in the literature (Mergel et al. 2016), and researchers have generally agreed that big data all share certain characteristics, including volume, variety, veracity, velocity, and value (Chen and Zhang 2014).

Urban areas are a significant playground where multiple players are engaged in the generation, storage, and applications of big data (Kitchin 2014). For much of the urban population, big data have become an integral part of their daily lives. Many technological, economic, and demographic factors have contributed to this rapid

N. Xiao (✉) · H. J. Miller
Center for Urban and Regional Analysis, The Ohio State University, Columbus, USA
e-mail: xiao.37@osu.edu

Department of Geography, The Ohio State University, Columbus, USA

W. Shi et al. (eds.), *Urban Informatics*, The Urban Book Series,
https://doi.org/10.1007/978-981-15-8983-6_31

growth. Various sensor technologies used in domains such as environmental monitoring and shared transportation means are the data sources that provide continuous feeds (Cuff et al. 2008). These sensors have been connected through a network that forms what is dubbed the Internet of things or IoT (Atzori et al. 2010). In an urban area, the IoT plays an especially important role in everyday life because the so-called things in the IoT include both physical objects such as GPS devices and environmental sensors, and also people who are equipped with sensors that can provide information about the location and surrounding area of the person. In many cities around the world, public transportation systems have increasingly applied GPS to allow more accurate and accessible transit to their residents. For example, many public transit agencies instrument their vehicles with GPS receivers and share these data publicly to support real-time bus tracking and arrival applications. In the meantime, passengers of these transportation systems use new ticketing methods such as smart cards to pay the transit fare, which also allows the transportation authorities to record and track their movements. In addition, citizens in urban areas have become a special kind of sensor (Goodchild 2007). These "sensors" have multiple ways of generating data. For example, they may provide spatial and temporal data using technology developed by commercial companies, as in the case of Google Traffic, in exchange for services (Heipke 2010), or they collect data about gas prices or traffic and exchange them with companies such as GasBuddy or Waze for rewards or other types of membership benefits (Boulos et al. 2011). Telecommunication companies have established vast databases that contain user identities and spatiotemporal activities. Cell phones have been mostly replaced by smartphones where the original function of making phone calls has been reduced to merely one of a huge number of uses relying on the network provided by the telecommunication companies, where many of the other functions are enabled to track the user's location.

Urban big data generated through sensor technology have all the characteristics of big data in general, but more critically they have their own features. First, urban big data involve a wide range of users from the general public to those in private services. It is important to recognize that these groups of people are active in multiple roles in the entire ecosystem of urban big data, including the phases of data generation, maintenance, storage, and usage. The users of the data, for example, also contribute to the generation of the very data they are using, as in the case of GasBuddy[1] where members report gas prices at different stations and also use the information provided by the Web service. Second, urban big data always have a geographic footprint as the data must relate to an urban extent. This is different from other big data sources (e.g. Web search and tweets without geotags) where the geographic dimension is not salient. Along with the spatial dimension, urban big data also have an important and sensitive temporal dimension as many applications depend on the time stamp of the data (e.g., real-time bus information is important for users to schedule activities around bus operations). Third, urban big data as a whole are often ill-structured because many data sources often do not coordinate their data generation and collection efforts. Data tend to exist in a loosely managed environment where a particular

[1] www.gasbuddy.com.

data set may not be connected to other data sets and may not be known to other groups of people.

The purpose of this chapter is two twofold: We provide an overview of urban big data and discuss the technical aspects how data can be made useful for various purposes. We specifically focus on the part of big data within the urban context as described above. The remainder of this chapter starts with a discussion of data sources. We then discuss the elements of the data, followed by several hypothetical user stories. On the technical aspects of urban big data, we discuss several data-collecting techniques and then extend the discussion into the needs and requirements for developing an urban big data infrastructure.

31.2 Sources of Urban Big Data

Urban big data come from a wide range of sources, and it may not be straightforward to categorize these sources. For example, in a study of the characteristics of 26 data sets (Kitchin and McArdle 2016), seven types were used to categorize the data sets, including mobile communication, Web sites, social media/crowdsourcing, cameras/lasers, transactions of process-generated data, and administrative. Not all these data have the urban context. Here, we group big data sources by the type of data providers, which can be from private or public sectors. In addition, we also recognize the types of data that are generated voluntarily. Each data set can be open to the public to use or may be protected so that only authorized users can access it. The distinction between open and protected data is important, especially for the urban context, as many data sources may have limited uses because they are difficult to share among potential users of the data. Table 31.1 lists a number of example

Table 31.1 Example sources of urban big data

Provider	Open	Protected
Private	Bike sharing	Bike sharing Mobile phone calls Surveillance camera and CCTV Health data
Public	Real-time bus operation Census data LiDAR and remote sensing Traffic cameras and CCTV Air pollution sensors	Public transit usage Individual survey Public health data
Volunteers	Social media Community sensor network	Social media Health data on mobile devices

data sets for each category. The purpose of listing these examples is to give a brief overview of possible and practical data sources. We note that these are merely a small sample as different cities in different counties will certainly have more sources.

The private sector generates a huge amount of data on a daily basis. We only list a few examples that are more related to the urban context. Popular bike-sharing companies, for example, provide both open and protected data. The open slice of the data may include the number and locations of bike stations, and available bikes and docks at each location, while the protected part results from tracking the movement of each individual bike along with information about customers. Some companies (e.g. Waze) may choose to release an aggregated version of their individual data in the form of averages over space and time as the open part, while protecting the actual individual data. It is obvious that private companies have been collecting such data sets as phone calls, surveillance, and individual health information. These data are highly protected due to privacy laws and even the need to maintain good relationships with the public (Chap. 32).

Urban big data from sources in the public sector cover a variety of domains such as demography, transportation, environment, and public health. These data are not necessarily open to the general public due to privacy concerns. For example, while many municipal services provide public transit data (e.g. bus operations), individual usage of bus data that can be obtained through the records of bus passes is often protected. The duality also applies to census data, where the aggregated version of the demographic, housing, and economic data is open to the general public, but individual surveys are tightly guarded.

The third type of data source includes individuals or groups who volunteer their own data for various uses. These providers generate their own data as they are themselves sensors (Goodchild 2007; Chaps. 28 and 29), which is different from the other two provider types where data are passively collected. A significant source in this category is the social media data. Tweets, for example, can be harvested using different licensing policies granted by Twitter. While the users generate the data, they do not necessarily own their own data, and not all social media data are open to the public. Other important kinds of volunteered data are those generated by the general public using various sensors. One of the prominent examples is the use of affordable air quality sensors (Kumar et al. 2015), and the users of these sensors can share their data to form community sensor networks (Yi et al. 2015). Though the quality of such data may be questionable (Lewis and Edwards 2016), they have been used for mapping[2] or other analysis.[3]

---

[2] www.purpleair.com/map?#1/25/-30.

[3] www.citylab.com/environment/2018/07/cheap-sensors-are-democratizing-air-quality-data/563990/.

31.3 User Stories

Let us consider three user stories of urban big data. These stories are hypothetical, but they do represent some of the examples we have encountered in our previous applications. They are not limited just to the data but extend to the entire ecosystem of urban big data that includes, in addition to data, the software systems as deployed in a hardware or network setting. We assume the existence of the data, and we aim to demonstrate how such data can be used in meaningful ways to address real-life problems. These stories are based on examples from experiences in the USA, but we believe it is possible to find relevant examples in other countries. We note that we use the term user story instead of use case for a specific reason, as use cases are a software engineering term that requires more formal description of the system. However, in this chapter, as will be discussed later, the specific requirements of the data usages will be difficult to define, and we argue that an agile method is more suitable. More discussion about the agile method will be presented later in this chapter.

The first user story involves a resident, Jon, in an urban area. Jon plans to invite a few of his friends to a party over the weekend. He has a few requirements for the party venue. His friends like biking, and he wants to use the bike-sharing system so that his friends can rent bikes for some fun riding. The party location needs to have sufficient available bikes and be close enough to the trails. Not all of his friends have cars, so Jon must consider a place that can be accessed by public transit or only by biking. He also desires the place to be close to some respectable restaurants for a happy hour after the ride. There is no existing app that will help Jon plan the event. But Jon is data savvy and can use the openly available data and mapping tools to put together some candidate locations. He can also use historical data to tell roughly what will happen in the weekend. He then shares what he has found with his friends before he finalizes the party venue.

The second user story involves a group of individuals who are interested in the city's development direction. They are busy with their own daily work, and it is hard for them to find a good time to have face-to-face meetings. Most of their activities rely on the use of online communication tools. Recently, the county planning authority posted a statement that gives the overall environment of the county a low rating. But the group does not feel this rating fairly represents the progress the county has made over the past few years and would like to give the overall environment another look. Two group members, Rachie and Lieta, are especially critical of the county's rating. Rachie is interested in air quality, and he is able to collect official air quality data and unofficial, open-source data for the past year. These are daily average data. Leita works on water quality, and she acquires some environmental measures for the gauges in the major streams and lakes within the county. These are again daily averages. They make the data sets available on the group Web site where the members can see the maps and the dynamics of each of the environmental factors. In the discussion board, the group members eventually conclude that it is incorrect and unfair to use a single rating to represent the overall environment quality, and they will present their findings in a hearing.

A third user story involves, again, a group of citizens who are dissatisfied by the congressional redistricting plan put forward by the state commission. They believe the plan is biased toward a political party, even though the commission has clearly stated their anti-gerrymandering stance. The group collected population data at the census block level and voters' data to support their arguments that while the official plan has the overall population evenly divided into the congressional districts, the voters of one of the political parties are strongly concentrated in one district and diluted in others, which gives the other party the edge in the majority of districts. The group also wants to further their argument by establishing that there are multiple alternative plans that can be considered to be equally good. While there are software packages that can be used to generate different kinds of alternative aggregations, they also need to use different demographic and other social and economic data at various spatial resolutions. More importantly, the group uses the alternatives generated by the software and then each group member will start to modify those plans manually to create their own plans. The group members will then share their plans on an online platform that allows them to compare and even synthesize new plans.

Clearly, these user stories involve more than just data. For example, software tools and Web-based applications are essential, and developing those tools is a great challenge. However, it is also clear that data are the cornerstone of the entire ecosystem.

31.4 Elements of Urban Big Data

Urban big data exhibit different forms due to the standard chosen to suit the preferred application. For example, a public transit agency may tend to release data using the popular standard called the General Transit Feed Specification (GTFS, discussed later in this chapter). However, we can decompose the data into its smallest items where each can be formulated as a space–time–attribute (STA) tuple of three elements $d = (x, t, a)$, where x is the location or a representation of location of the data item, t is the time stamp to indicate when the observation of the data item occurs or is released, and a is a set of attributes that are associated with the data item.

The above encoding strategy is similar to that of a geo-atom (Goodchild et al. 2007). Here, we separate location and time and relax the way location and attributes can be represented. Location can be explicitly recorded using either a set of coordinates or a set of indicators such as identification numbers that can be used to uniquely refer to locations (see examples below). The attributes associated with the location and time together are a set that is considered as one item in the tuple. This can be done by formatting an attribute as an object formed by a pair of the name of the attribute and the actual value. For example, an attribute of a specific PM2.5 measure can be formed as {PM2.5: 65}. Multiple attributes can be put together in the same manner as {PM2.5: 65, Ozone: 35}, a format commonly used in many data encoding strategies such as JavaScript Object Notation (JSON) that is supported in many programming languages. Putting everything together, an example of ((−83, 40), Mon Jul 01 2019

23:52:00 GMT + 0800 (CST), {PM2.5: 65, Ozone: 35}) encodes two air quality measures at a location in Columbus, OH on Monday, July 1, 2019 at 11:52 PM. Another example is (101.1, 2010, {total: 1200}), indicating a total (population) of 1200 for census tract 101.1 in the year of 2010.

An STA tuple can be viewed as a special kind of observation that occurs at a certain time and location. The big data for an urban area is a set *d* for all available locations and time periods in the area for the kinds of attributes that can observed or collected. This data model can be used to represent different spatial and temporal phenomena. For example, air quality of an urban area can be represented by a sequence of measures at a number of air quality stations, where each station is marked by its coordinates. Air quality as a geographic phenomenon is a field where observations are possible at any point in space. However, as far as data are concerned, we often resort to discrete data points to represent the phenomenon. For areal data, locations can be represented by the identification numbers or other indicators. For example, different demographic data can be collected for census tracts for multiple years, where each tract is represented by an identification number. The actual geometry (shape and its corresponding coordinates) may not be crucial for the data collection purpose as each tract can be uniquely identified and referred to geographically through another data set containing the coordinates. Similar examples can be found for phenomena on linear features such as water quality measures along a stream, where discrete locations are used for observations.

An interesting case is social media data, which occur in huge volume and at high speed. Such data can still be captured using the STA tuple of three elements, where each social media event (such as a tweet, a Facebook post, and a weichat post) always has the time, location (though it may not be shared), and attribute (the content as in text or a mixture of multiple formats). Another example in the same manner is the vast volume of Web pages. While the location of a Web page may not seem to be essential, each Web page can be assigned a location since each will ultimately be either hosted by a Web site that has a physical and meaningful geographic location or created by a person at some location.

31.5 Data-Collecting and Processing Techniques

Urban big data can be obtained using various methods. Many data providers typically offer an application program interface (API) that allows users to collect the data through Internet connections. The APIs may have different constraints in terms of how data can be collected. In general, data providers have full control of how their data can be collected. For example, Twitter uses layers of data-streaming policies, where the free and public license only provides a tiny portion of the tweets, and the way those small numbers of tweets are sampled is not clear to users (Morstatter et al. 2013). Some other data providers, on the other hand, make their data more open. For example, many public transit systems use a particular data protocol to make their schedule and real-time vehicle positions available. In this section, we show how to stream urban big data using two examples. We focus on open data here, though similar techniques can be applied to more restricted data sources.

The first example is the public transit system. A commonly used format for public transit data (schedules and updates) is the General Transit Feed Specification or GTFS (Harrelson 2006). Since its invention in 2005, GTFS has become the standard for publishing public transit data by agencies such as TriMet in Portland, OR, and BART in San Francisco, CA, to bring data to the general public (McHugh 2013). GTFS data have also been incorporated into Google Maps, where users can find real-time transit information on a common platform. The actual data structure of GTFS consists of multiple text files in comma-separated values (CSV) format. Google also provides a Python package called `google.transit`,[4] where the `gtfs_realtime_pb2` module can be used to help extract information from GTFS without having to directly handle the text files.

The transit agency in Columbus, OH, Central Ohio Transit Authority (COTA), uses GTFS to publish the bus schedule and real-time information for bus trips and its vehicle positions. To retrieve data for vehicle positions, we first use the following four lines of code to import the necessary Python modules and request to open an online GTFS database. In the fourth line, the file called `VehiclePositions.pb` is not the database itself, but a Google Protocol Buffer that describes the structure of the data and the necessary encoding/decoding methods of the data.

```
>> from google.transit import gtfs_realtime_pb2
>>> import requests
>>> import datetime
>>> response = requests.get('http://realtime.cota.com/\
        TMGTFSRealTimeWeb Service/Vehicle/VehiclePositions.pb')
```

Now, we can establish the feed from the actual database and read the actual data using the following code:

```
>>> feed = gtfs_realtime_pb2.FeedMessage()
>>> feed.ParseFromString(response.read())
>>> print(len(feed.entity)) 182
```

There were 182 buses at the time of running the code, among which the first bus can be examined using the following code:

```
>>> bus = feed.entity[0]
>>> bus
id: "1001"
vehicle {
  trip {
    trip_id: "665028"
    start_date: "20190722"
    route_id: "001"
  }
  position {
    latitude: 39.944339752197266
```

[4]https://developers.google.com/transit/gtfs-realtime/examples/python-sample.

```
      longitude: -82.86833953857422
      bearing: 270.0
      speed: 7.93974322732538e-06
    }
    timestamp: 1563818766
    vehicle {
      id: "11001"
      label: "1001"
    }
}
>>>
d = datetime.datetime.fromtimestamp(bus.vehicle.timestamp)
>>>d.strftime("%h %d, %Y, %H:%M:%S")
'Jul 22, 2019, 14:06:06'
```

Along with the position of the vehicle, the data also include the trip ID on which the vehicle is currently running and the vehicle ID, and it will be straightforward to use an STA tuple to encode this information. The default timestamp uses the epoch time, and the last two lines of code show how to convert it into calendar date and time.

We can run the same code after a few seconds, and below is the result. The following example was obtained exactly 20 s after the previous result and the position has also changed, while the bus was running on the same trip.

```
id: "1001"
vehicle {
  trip {
    trip_id: "665028"
    start_date: "20190722"
    route_id: "001"
  }
  position {
    latitude: 39.94470977783203
    longitude: -82.87486267089844
    bearing: 270.0
    speed: 8.457552212348673e-06
  }
  timestamp: 1563818786
  vehicle {
    id: "11001"
    label: "1001"
  }
}
```

While the vehicle position feed provides real-time data about bus location, detailed information about bus stops must be obtained from another real-time feed. The following example uses a similar procedure to retrieve real-time stop information.

```
>>> response = requests.get('http://realtime.cota.com/\
...    TMGTFSRealTimeWebService/\
...    TripUpdate/TripUpdates.pb')
>>> feed = gtfs_realtime_pb2.FeedMessage()
>>> feed.ParseFromString(response.content)
```

Below we explore some information about the first trip. The following example reveals the information about the trip and the vehicle that was currently operating on this trip. This corresponds to the bus information from our previous example.

```
>>> feed.entity[0].trip_update.trip
trip_id: "665028"
start_date: "20190722"
route_id: "001"
>>> feed.entity[0].trip_update.vehicle
id: "11001"
label: "1001"
>>> len(feed.entity[0].trip_update.stop_time_update)
74
```

There are 74 stops made on this trip so far, and we look at the first two stops:

```
>>> feed.entity[0].trip_update.stop_time_update[0]
stop_sequence: 9
arrival {
  time: 1563818515
}
departure {
  time: 1563818515
}
stop_id: "LIVNOEW"
>>> ft.entity[0].trip_update.stop_time_update[1]
stop_sequence: 10
arrival {
  time: 1563818711
}
departure {
  time: 1563818711
}
stop_id: "LIVCOUNW"
```

Based on the difference in departure times between the two stops, the data show that the bus arrived at the second stop (coded "LIVCOUNW") after 156 s (3.3 min). Each stop has its unique code, and COTA maintains a master file for all the stops,[5] where each stop is associated with a set of attributes that include the address and coordinates.

With the above examples, it is clear that at a specific time and location, each bus is associated with certain attributes such as the trip information and speed, which can be encoded as an STA tuple. The same can be said about stops that are made by the busses. We can then write a program that automatically requests the real-time data for bus positions and stop updates at a desirable time interval (every second, for example). The information retrieved can then be recorded in a database where each record is an STA tuple (x, t, a). For the buses, for example, each record contains fields such as latitude, longitude, timestamp, vehicle ID, trip ID, bearing, along with any other information that is deemed to be useful. For each stop, we can do the same by

[5]https://github.com/joeshaw/cota-bus/blob/master/cota-gtfs/stops.txt.

recording fields such as the coordinates, arrival and departure times, trip ID, vehicle ID, and stop ID. The accuracy of the database is partly dependent on the time interval of data collection. A one-minute time interval may be sufficient for the purpose of information visualization and some analysis, and a smaller interval will be needed if we aim to provide real-time service to the general public for tasks such as trip planning that require higher accuracy.

The Environmental Protection Agency (EPA) of the USA maintains a network of air quality sensors across the country. EPA also provides an API to allow users to access air quality data.[6] This API provides a Web service based on a software architecture called REST (Richardson and Ruby 2008) that supports the use of a URL to query a database in order to retrieve data. For example, the following URL specifies the time frame, geography boundaries, and environment variable, along with other necessary parameters. The last parameter must be replaced by an actual API key that can be applied from the Web site.

```
https://airnowapi.org/aq/data/?
    parameters = pm25&
    bbox = -83.368244,39.586371,-82.269611,40.344184&
    startDate = 2019-05-19T03&endDate = 2019-05-19T04&
    DataType = B&format = application/json&verbose = 1&
    API_KEY = XXXX
```

This request will return the following data formatted in JSON. It shows that during the two-hour time frame specified, there are two PM2.5 sensors at two locations, and their data (e.g., locations, values, air quality index values) are provided. Again, we can write a program that automatically and repeatedly retrieves information like the above as STA tuples and store them into a database.


```
[
  {
    "Latitude": 40.11109, "Longitude": -83.065376,
    "UTC": "2019-05-19T03:00",
    "Parameter": "PM2.5",
    "Unit": "UG/M3", "Value": 14.8, "AQI": 57, "Category": 2,
    "SiteName": "Columbus NR - Smoky Row",
    "AgencyName": "Ohio EPA-DAPC",
    "FullAQSCode": "390490038", "IntlAQSCode":
"840390490038"
  },
  {
    "Latitude": 40.0845, "Longitude": -82.81552,
    "UTC": "2019-05-19T03:00",
    "Parameter": "PM2.5",
    "Unit": "UG/M3", "Value": 12.2, "AQI": 51, "Category": 2,
    "SiteName": "New Albany",
    "AgencyName": "Ohio EPA-DAPC",
    "FullAQSCode": "390490029", "IntlAQSCode":
"840390490029"
  },
```


[6] https://docs.airnowapi.org.

```
    {
      "Latitude": 40.11109, "Longitude": -83.065376,
      "UTC": "2019-05-19T04:00",
      "Parameter": "PM2.5",
      "Unit": "UG/M3", "Value": 14.7, "AQI": 56, "Category": 2,
      "SiteName": "Columbus NR - Smoky Row",
      "AgencyName": "Ohio EPA-DAPC",
      "FullAQSCode": "390490038", "IntlAQSCode":
  "840390490038"
    },
    {
      "Latitude": 40.0845, "Longitude": -82.81552,
      "UTC": "2019-05-19T04:00",
      "Parameter": "PM2.5",
      "Unit": "UG/M3", "Value": 12.1, "AQI": 51, "Category": 2,
      "SiteName": "New Albany",
      "AgencyName": "Ohio EPA-DAPC",
      "FullAQSCode": "390490029", "IntlAQSCode":
  "840390490029"
    }
  ]
```


The raw data collected in the above examples are merely STA tuples of the form (*x*, *t*, *a*) and must be processed to support purposes such as analyzing urban traffic status or mapping density of air pollution. In a bigger context, this is an area of data mining of big data (Vatsavai et al. 2012). In our example of using the GTFS feeds, two kinds of real-time raw data are acquired: vehicle positions and stop updates. Among all the GTFS text files, the file called `stop_times.txt` is used to store the bus schedule for all routes, containing detailed arrival and departure time as scheduled for each stop on each trip. By comparing the real-time trip updates of the actual arrival and departure time of each trip with the scheduled times, it is possible to compute the delay of each bus and conduct further analysis of how the delays propagate along the trip (Park et al. 2019). It is also possible to visualize the discrepancy in places that can be reached by the scheduled and actual buses (Fig. 31.1).

The above data collection examples show the general procedure of harvesting urban big data and the considerations of storing them in spatiotemporal databases. There are of course many other sources for urban big data that are designed for different purposes (e.g. Twitter data). Though these data sets differ in technical details such as data format and APIs, it can be argued that STA tuples can be used to capture most (if not all) of these data sets. To this extent, from a data perspective alone, it suffices to say that the data are "out there" for users to use. The real and more difficult challenge is how to make these data accessible to all.

31.6 Toward Urban Big Data Infrastructure

Urban big data as described above have the necessary elements to support the user stories described in the previous section of this paper. These data sets are also relatively straightforward to obtain. However, it should also be clear that the ecosystem

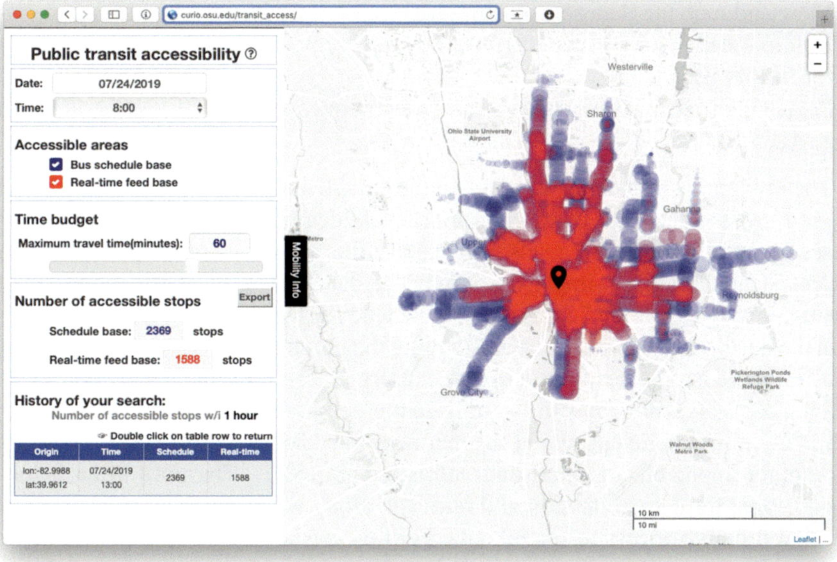

Fig. 31.1 Visualizing the difference between the scheduled stops (blue) and those that were actually reached (red) in a one-hour time frame from a given location (black pin icon). *Source* http://curio.osu.edu/transit_access/

of urban big data does not always suit regular users from the general public, who are often not trained to be as data savvy as the experts who generate the data. The difficulty these regular users may face can be as simple as where to find the data and as complicated as how to use them. These are the major limitations that make it difficult for the data to be accessible to a wide audience.

To address these problems, we advocate the idea of urban big data infrastructure under the spirit of data for all. The concept of infrastructure refers to the ubiquitous availability of resources such as electricity where a person, who does not need to be an electricity expert, can use it by simply plugging in. We would ponder if it is possible for a regular user to find a desired spatiotemporal data set by specifying it instead of by carrying out a process of searching and coding. For example, is it possible to ask a virtual assistant (e.g. Apple's Siri) on a smartphone to find the spatiotemporal data set by giving a description of the data? In the remainder of this section, we review some methods that may shed light in the future development of such an infrastructure.

There are a few existing methods that can be used to address *some* of the issues mentioned above. A geoportal (Tait 2005), for example, is designed as a gateway to serve geospatial data on a Web-based platform. More specifically, a geoportal can be used to allow users to do the following tasks:

- Discover geospatial data based on a catalog of the data maintained in the geoportal.
- Provide useful information about how to use each of the geospatial data sets.

- View and map the data sets discovered.
- Automatically harvest (collect) online data sources and store them in the geoportal for further uses.
- Provide data using various data query techniques such as REST, GeoRSS, and KML.

The implementation of a geoportal requires work on the server side and is suitable as a solution to data needs at the enterprise level. Ideally, by logging into a geoportal, a user can find relevant data sets and explore the properties of those data through mapping, tabulating, or simply describing the data. However, these geoportals are usually developed for data experts to use instead for the regular users, who may not have the necessary skill sets in understanding the portal and navigating the numerous data sets served. It is also difficult to expect users to develop their own geoportals or to develop data sets within existing portals. In this sense, the ultimate users (the general public in our case) are entirely at the mercy of the data experts or data enterprises.

Another approach is spatial data infrastructure (SDI). The term often involves technologies for data collection and retrieval, along with metadata, as well as policies that promote access to spatial data. For this reason, SDIs are not technological solutions to data problems but more of a social and political response to the data needs that emerge from communities at different scales. In an ideal situation, implementing an SDI requires the efforts of government agencies, the private sector, representatives of the general public, and even members of academia. In the past, SDIs have been effective in consolidating traditional data sets such as the cadastre, national base maps, large-scale topographic maps, and remotely sensed images. While it is well recognized that the success of SDIs is critically dependent on how the users, citizens, and institutions are engaged, their involvements have been a significant challenge (Erik de Man 2006; Elwood 2008). It should be noted that a major portion of the SDI literature is focused on the technological aspects, especially taking a GIS-centered perspective (Maguire and Longley 2005; Steiniger and Hunter 2012; Evangelidis et al. 2014; Helmi, Farhan and Nasr 2018). Through such a technological perspective, unfortunately, the concept of SDI tends to be reduced to merely a form of GIS or geoportal.

We argue that it is necessary to develop an urban big data infrastructure in order to address the issues discussed above and to fulfill the goals of using the data as mentioned in the user stories. The technical aspects of such an infrastructure, though still challenging, can be relatively straightforward, as much of the effort has already focused on how to utilize the technology in getting the data and making the data accessible. For example, the development of geoportals has already demonstrated that various data can be incorporated in commonly used formats and standards for users to discover and use. Many geospatial database management systems (e.g. GeoServer and Esri's geoportal) can be used to harvest data from different sources. More importantly, these systems typically also support data discovery. For example, Catalogue Services[7] is a specification standard proposed by the Open Geospatial Consortium

[7] https://www.opengeospatial.org/standards/cat.

(OGC) and has been supported by major software systems such as GeoServer[8] and Esri's geoportal.[9]

The fundamental challenge of developing urban big data infrastructures goes beyond the technological domain: It is the often ill-defined relationship among data, data providers, data users, and software developers and vendors that makes it difficult for such an infrastructure to be effective, as shown in the case of SDIs. From an engineering perspective, this challenge is due to the changing requirements as new user stories emerge whenever new data sources or new technology become available. There is no silver bullet that will solve all the problems. Instead, it is important to understand that a fully functional urban big data infrastructure (or SDIs at a lesser level of difficulty) takes time and must wait for collaborations to emerge.

We envision an agile process (Stellman and Greene 2014) where all parties involved in the use and production of urban big data will constantly engage with each other and revise any previous understandings about the data, even though the understandings may be preliminary and sometimes trivial at the early stages of development. A top-down approach to developing the infrastructure is bound to fail since such an approach is typically dependent on well-defined requirements, as shown repeatedly in the history and literature of software engineering (Sommerville 2016). The strong social and human aspects of urban big data infrastructure make it natural to consider an agile approach that stresses how the development process should actively engage with the system (data) users (Stellman and Greene, 2014). A typical agile development process starts from user stories that roughly but meaningfully describe the fundamental requirements of a system but often do not specify the details of how the system should be run and built. In order for the project to advance, the end user or client must constantly be involved in the process and provide feedbacks so that the requirements can become increasingly clear. Lack of user involvement will cause adverse consequences to both the team and the project (Hoda et al. 2011). User involvement in turn helps the developers understand the direction of the project and enables them to work together with the users, toward the end product.

Among the many agile methods, self-organizing agile methods are a promising recent development that have gained much recognition (Hoda et al. 2012) and can be especially suitable for the development of urban big data infrastructures. Researchers have studied the potential of such an approach from different perspectives, including organizational theory that focuses on how organizations may learn from past experience (Morgan 1998) and complex adaptive systems that show how feedback among individuals can help the system evolve (Lansing 2003). In addition to the customer/user, a regular agile team includes a product owner who maintains a close relationship with the customer and plays the role of a stakeholder, a coordinator (scrum master) who operates the daily routines of the team and keeps the team together, and team members who are dedicated to work on various parts of the project with a strong leadership from the coordinator and product owner. In the case of a self-organizing agile method, a team may still have those roles among team members,

---

[8]https://docs.geoserver.org/latest/en/user/services/csw/index.html.

[9]https://www.esri.com/en-us/arcgis/products/geoportal-server/overview.

but is a more autonomous group where the role of each member may change. A strong point of such an approach is that decisions about the project are made not by the product owner but more spontaneously from the collaborations among all team members, and more importantly with the customer (Hoda et al. 2011).

The key aspect of a self-organizing agile process is the collaborative leaders who play the most critical role. In the agile literature, these are team members who act as mentors and coordinators. Mentors are not bosses because they do not make decisions; instead, they are coaches who provide guidance and support the team's confidence. Coordinators are essential too because they work directly with users in order for the development to be on the right track as the users require.

Self-organizing agile methods are promising, and it should be noted that the development of an urban big data infrastructure will not emerge just because there are demands from users and data experts. Strong bonds between them are important, and leadership is required. We do not imagine that an infrastructure can be developed over just a few projects where big data are involved. Instead, given the fact that SDIs are still far from being functional despite the efforts of the past three decades (Erik de Man 2006; Grus et al. 2010), it is reasonable to believe that a fully functional urban big data infrastructure will also take a long time to materialize. However, with strong and collaborative leadership formed through the bond between the user (demand) and the developers (skills), it is possible to evolve the infrastructure through multiple projects where data and knowledge derived from the use of data will accumulate. An open and collaborative environment will be especially useful at the urban scale where similar tasks may repeat in different urban areas and therefore good practices can be adopted and improved through time.

31.7 Concluding Remarks

Urban big data have exhibited potential in helping us to better understand the city and make better and informed decisions. Such data have a wide range of sources, and the technology to retrieve the data is relatively straightforward. However, the social and human aspects have made the use of the data by the general public a real challenge. Cultivating urban big data requires long-term planning and sustainable collaboration between many parties. It is not reasonable to expect silver bullet solutions.

Technology aside, data have become the cornerstone of an ecosystem that is sustained by a chain of users, developers, companies, analysts, and investors. The roles of each player in this ecosystem are not the same as in the old economy. For example, while users are still using the services provided by companies such as Google and Facebook, they also contribute to data collection through using the Internet (e.g. conducting searches or posting on social media). To some extent, this era of urban big data is also an era where users act as products. Schneier (2015) describes the relationship between the (private) data provider and users as a feudalist system where the data "lords" have full and firm control on the properties (data) that are similar to the land in a feudal system, and the users receive benefits from the data "lords" through payment or other types of contribution (their own data, for example), similar to peasants in a feudal system who must trade their labor in order to have access to land and services. We do not believe such a feudalist world in the data domain is healthy for data to be used to its optimal extent. Through collaboration and policy, we can develop an open (though not necessarily free) urban big data infrastructure that will enable the data to be used by their true constituents: the general public.

References

Atzori L, Iera A, Morabito G (2010) The internet of things: a survey. Comput Networks 54(15):2787–2805

Boulos MNK, Resch B, Crowley DN, Breslin JG, Sohn G, Burtner R, Pike WA, Jezierski E, Chuang KYS (2011) Crowdsourcing, citizen sensing and sensor web technologies for public and environmental health surveillance and crisis management: trends, OGC standards and application examples. Int J Health Geogr 10(1):67

Chen C, Zhang C (2014) Data-intensive applications, challenges, techniques and technologies: A survey on Big Data. Inf Sci 275:314–347

Cuff D, Hansen M, Kang J (2008) Urban sensing: out of the woods. Commun ACM 51(3):24–33

Elwood S (2008) Grassroots groups as stakeholders in spatial data infrastructures: challenges and opportunities for local data development and sharing. Int J Geogr Inf Sci 22(1):71–90

Erik de Man WH (2006) Understanding SDI; complexity and institutionalization. Int J Geogr Inf Sci 20(3):329–343

Evangelidis K, Ntouros K, Makridis S, Papatheodorou C (2014) Geospatial services in the cloud. Comput Geosci 69:116–122

Goodchild MF (2007) Citizens as sensors: the world of volunteered geography. GeoJournal 69(4):211–221

Goodchild MF, Yuan M, Cova TJ (2007) Towards a general theory of geographic representation in GIS. Int J Geogr Inf Sci 21(3):239–260

Grus L, Crompvoets J, Bregt A (2010) Spatial data infrastructures as complex adaptive systems. Int J Geogr Inf Sci 24(3):439–463

Günther WA, Mehrizi MHR, Huysman M, Feldberg F (2017) Debating Big Data: a literature review on realizing value from Big Data. J Strategic Inf Syst 26(3):191–209

Harrelson C (2006) Happy trails with google transit. Google Official Blog (https://googleblog.blogspot.com/2006/09/happy-trails-with-google-transit.html). Accessed: July 25 2019

Heipke C (2010) Crowdsourcing geospatial data. ISPRS J Photogrammetry Remote Sens 65(6):550–557

Helmi AM, Farhan MS, Nasr MM (2018) A framework for integrating geospatial information systems and hybrid cloud computing. Comput Electrical Eng 67:145–158

Hoda R, Noble J, Marshall S (2011) The impact of inadequate customer collaboration on self-organizing agile teams. Inf Softw Technol 53(5):521–534

Hoda R, Noble J, Marshall S (2012) Self-organizing roles on agile software development teams. IEEE Trans Softw Eng 39(3):422–444

Kim GH, Trimi S, Chung JH (2014) Big-data applications in the government sector. Commun ACM 57(3):78–85

Kitchin R (2014) The real-time city? Big Data and Smart Urbanism. GeoJournal 79(1):1–14

Kitchin R, McArdle G (2016) What makes Big Data, Big Data? exploring the ontological characteristics of 26 datasets. Big Data and Society 3(1):1–10

Kumar P, Morawska L, Martani C, Biskos G, Neophytou M, Di Sabatino S, Bell M, Norford L, Britter R (2015) The rise of low-cost sensing for managing air pollution in cities. Environ Int 75:199–205

Lansing JS (2003) Complex adaptive systems. Ann Rev Anthropol 32(1):183–204

Lewis A, Edwards P (2016) Validate personal air-pollution sensors. Nature News 535(7610):29–31

Maguire DJ, Longley PA (2005) The emergence of geoportals and their role in spatial data infrastructures. Comput Environ Urban Syst 29(1):3–14

Marr B (2015) Big Data: using SMART Big Data, analytics and metrics to make better decisions and improve performance. Wiley, Hoboken

McHugh B (2013) Pioneering open data standards: THE GTFS story. In: Goldstein B (ed) Beyond transparency: Open Data and the future of civic innovation. Code for America Press, San Francisco, pp 125–135

Mergel I, Rethemeyer RK, Isett K (2016) Big Data in public affairs. Public Adm Rev 76(6):928–937

Morgan G (1998) Images of organization. SAGE, Thousand Oaks

Morstatter F, Pfeffer J, Liu H, Carley KM (2013) Is the sample good enough? comparing data from twitter's streaming API with twitter's firehose. In: Proceedings of the seventh international AAAI conference on weblogs and social media. Association for the Advancement of Artificial Intelligence, pp 400–408

Park Y, Mount J, Liu L, Xiao N, Miller HJ (2019) Assessing public transit performance using real-time data: spatiotemporal patterns of bus operation delays in Columbus, Ohio, USA. Int J Geogr Inf Sci 34(2):367–392. https://doi.org/10.1080/13658816.2019.1608997

Richardson L, Ruby S (2008) RESTful Web services. O'Reilly Media, Sebastopol

Schneier B (2015) Data and Goliath: the hidden battles to collect your data and control your world. WW Norton, New York

Sommerville I (2016) Software Engineering. Pearson, London

Steiniger S, Hunter AJ (2012) Free and open source GIS software for building a spatial data infrastructure. In: Bocher E, Neteler M (eds) Geospatial free and open source software in the 21st century. Springer, Dordrecht, pp 247–261

Stellman A, Greene J (2014) Learning agile: understanding scrum, XP, lean, and kanban. O'Reilly Media, Sebastopol

Tait MG (2005) Implementing geoportals: applications of distributed GIS. Comput Environ Urban Syst 29(1):33–47

Vatsavai RR, Ganguly A, Chandola V, Stefanidis A, Klasky S, Shekhar S (2012) Spatiotemporal data mining in the era of big spatial data: algorithms and applications. In: Proceedings of the 1st ACM SIGSPATIAL international workshop on analytics for big geospatial data. ACM. Washington, pp 1–10

Yi W, Lo K, Mak T, Leung K, Leung Y, Meng M (2015) A survey of wireless sensor network based air pollution monitoring systems. Sensors 15(12):31392–31427

Ningchuan Xiao is Professor of Geography and Associate Director of the Center for Urban and Regional Analysis (CURA) at The Ohio State University. He is interested in spatial data science and technology.

Harvey J. Miller is the Bob and Mary Reusche Chair in Geographic Information Science, Professor of Geography, and Director of the Center for Urban and Regional Analysis (CURA) at The Ohio State University. His research interests include mobility analytics, sustainable transportation, and time geography.

Chapter 32
Geoprivacy, Convenience, and the Pursuit of Anonymity in Digital Cities

Jerome E. Dobson and Willam A. Herbert

Abstract Cities demand spatial efficiencies that can be achieved only through sharing of information. Current technologies support collection, processing, and dissemination of unprecedented quantities of personal, public, and corporate information. Inherent in this milieu is an inevitable contest among societal efficiency, corporate profits, consumer convenience, personal privacy, and even freedom. The authors examine current trends in technology, data collection, legislation, and public acceptance. They find that without broad specific regulations limiting location data collection and use—including a universal protected right for individuals to pursue anonymity—governments, commercial enterprises, employers, and individuals increasingly will exploit tracking technologies at the expense of geoprivacy.

32.1 Introduction

Cities exist because of society's overriding need for spatial efficiency. Placing people close together, connected through systems that operate quickly and smoothly, can enhance productivity and leisure, resulting in the potential for relatively high standards of living for many, while also creating wide disparities in economic and social well-being. Information sharing is essential in commerce and marketing, which typically are concentrated in urban areas.

Here, we explain the range of urban information technologies and applications available now and likely to emerge soon. We discuss current policies, legislation, and court rulings governing geoprivacy—defined here as "individual rights to prevent [surveillance and] disclosure of the location of one's home, workplace, daily activities, or trips" (Kwan et al. 2004)—together with surveillance and control, including

J. E. Dobson (✉)
Department of Geography, University of Kansas, Lawrence, USA
e-mail: dobson@ku.edu

W. A. Herbert
Hunter College, City University of New York, New York, USA
e-mail: wh124@hunter.cuny.edu

W. Shi et al. (eds.), *Urban Informatics*, The Urban Book Series,
https://doi.org/10.1007/978-981-15-8983-6_32

the European Union's recent General Data Privacy Regulation (GDPR). We address the extent of government, corporate, and individual information gathering, and the risks involved in such data collection and use. We explore the processes and considerations by which corporations, groups, and individuals decide whether to accept or resist surveillance and control.

Delivering goods, managing traffic and mass transit, facilitating urban pleasures, and myriad other essential services such as crime prevention, depend on individuals merging their own activities with communal operations. Maximizing efficiency necessitates information sharing, which foments tension between societal demands and personal expectations of freedom and privacy. Tensions can rise to conflict when urban policymakers adopt "smart" technologies without studying and managing the impacts such technologies will have on privacy (Williams 2019).

How a society balances community needs with individual rights reflects collective values and priorities. The escalating growth of privatized urban spaces (Garrett 2015) impedes geoprivacy protections in the USA because, in general, private actors have more license to surveil and track than government agents who are subject to greater legal restrictions. More important, government regulations rarely reflect majoritarian views about geoprivacy, especially since Amazon, Apple, Facebook, Google, and Microsoft collectively spent $582 million over thirteen years to lobby the US Congress to promote their proprietary interests (Dellinger 2019).

In the USA, except for California, there is no comprehensive regulatory scheme (Swisher 2019). Instead, the burden of balancing convenience and privacy regarding data collection and accessibility is placed squarely on the individual. Hence, as Fowler (2018) warns, "Many of us will delete apps … disable as much tracking as we can on our phones … delete our Facebook accounts … delete our social media histories and old emails and text messages. But it won't be enough because most people will not care: The trade-off between privacy and convenience will be worth it to them, because the loss of their privacy will have little to no impact on their day-to-day lives. Most people will read (or perhaps ignore) the news stories about every new privacy scandal, and they will then go back to their phones." Even those who study and report on location privacy have a hard time retaining their location invisibility on the electronic surveillance grid (Swisher 2019).

Individuals routinely sacrifice some degree of privacy and personal choice for the common good or consumer convenience. These sacrifices are usually implicit tradeoffs without discernment or adequate information for informed consent. The extent of sacrifice is oftentimes mollified by extreme individual wealth, creating a non-egalitarian opt-out from shared sacrifice. In addition to economic inequality, a digital divide exists with respect to individual access to and sophistication with the use of technology (Slinn and Herbert 2011). Nevertheless, urban habits, design, customs, and laws frequently favor collective efficiency and commerce over individual self-determination with respect to privacy.

Traditionally, cities have provided individuals with a means of hiding in the crowd and maintaining relative anonymity. Many people crave the subjective perception of invisibility in crowded streets, parks, and trains. For centuries, they enjoyed an overarching sense of obscurity based on time, space, impermanence, and inherent limitations on human memory (Hartzog and Selinger 2019).

Collectively, however, people cannot have all they may want simultaneously. The more one seeks fame the less likely he or she can have anonymity or obscurity and so it goes for whole population segments within cities. Individuals and groups may choose open lifestyles—such as those of political and civic leaders, entertainers, entrepreneurs, and social media influencers. Others are forced into the public spotlight against their will or live a life in the shadows out of choice, necessity, or circumstances beyond their control.

New information technologies increase benefits and risks and make today's societal and individual choices ever more difficult. Some applications improve government, commercial, familial, and individual efficiencies and conveniences at the cost of privacy, but they are rarely designed to protect privacy. At the same time, emerging technologies enhance surveillance or control by government, employers, loved ones, or caregivers. Through the collection of location data by commercial enterprises, the most basic democratic rights of dissent and protest in the streets can be easily tracked (Warzel and Thompson 2019).

These technologies also can create a new form of slavery—geoslavery—based on location control, "a practice in which one entity, the master, coercively or surreptitiously monitors and exerts control over the physical location of another individual, the slave. Inherent in this concept is the potential for a master to routinely control time, location, speed, and direction for each and every movement of the slave or, indeed, of many slaves simultaneously. Enhanced surveillance and control may be attained through complementary monitoring of functional indicators such as body temperature, heart rate, and perspiration" (Dobson and Fisher 2003, pp. 47–48; 2007; Herbert 2006). Geoslavery violates a central component of personal liberty, namely freedom of locomotion, which includes the ability of a person to move from place to place without external restraint unless pursuant to law (see the works of Blackstone in Lemmings 2018).

Generalized fear of government or corporate electronic surveillance is common, even though the public barely knows the collective scope and magnitude of the data collection, sale, and use of such information. Moreover, the collection, use, and distribution of personal data by individuals—family, friends, and strangers—is routinely accepted without protest.

Health records, in particular, are considered sacrosanct in the USA. The Health Insurance Portability and Accountability Act of 1996 (HIPAA) contains a "Privacy Rule" so prominent that many people mistakenly dub the entire act the "Health Information Privacy Act." Its goals are to protect health insurance coverage when workers change or lose jobs and to protect health data confidentiality and availability. It guarantees a right of access to one's own health data on request (HIPAA Journal 2019). It was passed with the good intention of protecting individuals from any consequences that might result from divulging health information including workplace discrimination. Patients routinely are presented with a statement affirming their rights to privacy except for release to insurers, the one entity most likely to react detrimentally to a patient's interests if adverse health conditions are found. Concomitantly, HIPAA's disclosure rules restrict the release of health and geographic information on individuals so completely that the act itself stymies high-precision geographic research on factors, causes, and effects linking local health to local environments, thus fettering the complementary fields of medical geography and epidemiology.

Many people have acquiesced to the commodification of personal location data for advertising and consumer targeting, becoming willing subjects to what Shoshana Zuboff has labeled "surveillance capitalism" (Zuboff 2019). Some recognize a risk vs. benefit ratio; others do not. We explore the integration of location technology with social media platforms and deregulatory ideology in the age of social media. We discuss social and cultural changes arising from accelerated use of location technology, implications for precarious work (Uberization), and unwritten tradeoffs of "convenience" for loss of privacy. Here, we discuss such matters in the context of three illustrative applications that feature tracking technology.

32.1.1 Application #1: The Role of Cities in Slavery Prior to the Civil War

To contextualize the impact of twenty-first-century information technologies on urban geoprivacy, human rights, and property rights, consider an example from the nineteenth century based on analog technology rather than digital. From the earliest days of the American republic, surveillance and restraint were core components of the American slavery system. Freedom of movement was substantially restricted for those enslaved. Federal laws enabled slaveholders to track down, recapture, and return runaway slaves, then defined as human chattel with high monetary value. Self-emancipated slaves constituted a major economic loss for slaveholders, who spent substantial sums for location information to aid in the legal and frequently extra-legal capture by slave catchers (Foner 2016).

Fugitive slaves in the nineteenth century flocked to cities in search of anonymity, personal redefinition, and employment. Cities with large populations of free African-Americans were particularly attractive for escaped slaves. There they had a greater chance to attain obscurity and even mingle in crowds at public events (Franklin and Schweninger 1999). Black and white abolitionists assisted self-emancipated slaves in traveling to safer areas, creating new identities, and finding work and lodging. To personalize, W.C. Pennington arrived in New York City in 1828 after escaping slavery and stayed, establishing himself as a minister and educator. Another escaped slave, Frederick Bailey, traveled to New York a decade later. During his short stay, Bailey changed his name, married with Pennington officiating, and went off to become the famous abolitionist writer and orator Frederick Douglass (Foner 2016).

Even slaves emancipated by their former masters faced difficulties in avoiding discovery that could result in re-enslavement. Urban vigilance committees were formed to protect escaped slaves, free African-Americans kidnapped off city streets, and challenge legal proceedings intended to compel their enslavement in another state (Foner 2016). The importance of urban life to African-Americans explains, in part, the reluctance of many who received land grants from abolitionist and agrarian Gerrit Smith in the late 1840s to leave and start new lives in the remote Adirondack Mountains. Despite the continuing fear of slave catchers, the urban environment was more secure than attempting to create a safe community elsewhere (Stauffer 2002).

Imagine how modern tracking technologies, had they been available in antebellum times, would have maximized the efficiency of tracking down runaway slaves in cities and returning them to bondage. Indeed, such technologies might have negated the urban advantage in geoprivacy. The same principles apply to fugitive slaves in the nineteenth century or modern-day sex slaves seeking freedom and dignity or immigrants seeking refuge in the twenty-first century.

32.1.2 Application #2: Informed Delivery by the US Postal Service

How pervasive and vexing geoprivacy can be today. How integrally it is entangled with efficiency and convenience. In the first half of the twentieth century, it was generally assumed that a mailman could deliver a package by knocking on the door and handing it to a live person inside. Starting with World War II, however, changes in lifestyle rendered that premise untrue. More women were working, and fewer extended families lived together in the same house. Eventually, it became necessary to leave packages unattended at the door. That gave rise to "porch pirates"—scofflaws who steal unattended packages. Eventually, the problem became so rampant that critics objected to "porch pirate" as too frivolous a term for the damage done. An estimated 1.7 million packages are stolen every day across the USA (Hu and Haag 2019).

To counter theft, the US Postal Service (USPS) initiated a program called Informed Delivery. Any USPS customer could sign up for an electronic notice to inform him or her when a package would arrive so the customer could arrange to be home at or soon after its arrival. Unfortunately, USPS failed to install proper security procedures, and now it is fairly easy for crooks to sign up for someone else's account. Thus, some thieves receive convenient notices alerting them to deliveries at a time unknown to the resident. The problem could be solved by more stringent measures, such as holding the package for customer pickup at the Post Office, but that would incur unacceptable delays and additional travel on the part of the customer or mail carrier. It is a clear case of customers, bent on convenience, wanting a solution that turns out to be vulnerable itself.

Simultaneously, Amazon.com offered a program for customers to pre-approve delivery personnel to open the front door and place each package inside. Predictably, most customers recoiled at the thought. Next, Amazon offered to deliver inside the garage, but many urban dwellers do not have garages and acceptance among those who do is unclear.

Today, the most popular countermeasure to porch piracy is Amazon's Ring technology, which employs a video surveillance camera integrated into a doorbell (Wingfield 2018). Privacy concerns have been expressed because each installation surveils not only the owner's yard, but neighbors' yards, driveways, and streets as well, and formal agreements are being instituted for police departments (600 so far) to harvest and process data with the consent of owners but not the consent of neighbors, visitors, and other passersby (Harwell 2019a, Thorbecke 2019). Worse yet, hackers have frightened some residents (famously including an eight-year-old girl) by speaking to them through Ring security cameras inside the home (Chiu 2019).

32.1.3 Application #3: Geoslavery in the Middle East and China

In their initial article on geoslavery, Dobson and Fisher (2003) proposed "realistic scenarios of potential enslavement applications." Based on the real-life honor murder of Sevda Gok, "a teenage girl [in eastern Turkey] whose family held a council and voted to execute her in violation of their own country's laws," they envisioned the following hypothetical scenario, which would be anathema to Western societies, yet acceptable in some Middle Eastern countries: "Soon an enterprising businessman … may be able to purchase a central monitoring system … which can be locked onto the wrists of every member of the village (women, children, and men). Most likely, he will be able to offer a service to village parents at an affordable price that will cover his investment and a tidy profit."

At the time, some critics claimed the hypothetical scenario was futuristic and inflammatory. Yet in 2019, "U.S. Representative Jackie Speier and 13 colleagues wrote Apple CEO Tim Cook and Google CEO Sundar Pichai to call for the removal of a mobile app from the companies' app stores that allows Saudi men to track women and migrant workers…" The Congressional press release (Speier 2019) states, "The ingenuity of American technology companies should not be perverted to violate the human rights of Saudi women. Twenty-first century innovations should not perpetuate sixteenth century tyranny… Keeping this application in your [app] stores allows your companies and your American employees to be accomplices in the oppression of Saudi Arabian women and migrant workers… The app, Absher … allows a male "guardian" to take away permission for a woman or migrant laborer to exit the country and provides the man with notifications if there is an attempt to leave. Amnesty International has stated this app is another example of how the Saudi Arabian government has developed and employed tools to limit women's rights and freedoms."

When we first wrote about geoslavery (Dobson and Fisher 2003; Herbert 2006), the ultimate example we imagined was a nation tracking its entire population, and employers tracking their employees, surveilling with GPS, enhancing with government and corporate databases, and rewarding individuals for good behavior or punishing them for bad behavior.

In 2014, China announced plans to do exactly that. A year later, China's "omnipotent" Social Credit System was tested in pilot projects run by eight major companies for planned national implementation in 2020 (Hatton 2015). Today, the test involves more than twenty companies, where every individual is monitored through human tracking and surveillance to produce a social credit score used to rate each citizen's trustworthiness. The current concept is not a unified platform generating unique scores for 1.4 billion citizens. "Instead, the national program is envisioned as a web of individual systems run by cities, hospitals, businesses and agricultural-produce markets — all linked by data-sharing and using incentives and penalties to make people and businesses behave as the government wishes" (Mistreanu 2018). It is as if the US government were to explicitly appoint Google, Equifax, Sprint, and other corporations as guardians of every citizen's reputation, social success, job opportunities, and travel destinations.

The stated intention of China's original plan was to "allow the trustworthy to roam everywhere under heaven while making it hard for the discredited to take a single step" (Mistreanu 2018). By the end of 2018, "Citizens placed on black lists for social credit offences were prevented from buying train tickets 5.5 million times … [and] in 2017 … 6.15 million citizens had been barred from taking flights" (Kuo 2019). Data variables, held in vast national and corporate databases, include government information such as tax payments and traffic violations and corporate data such as consumer debt.

The program qualifies as geoslavery even with Dobson's original stipulation that geoslavery must be either "coercive or surreptitious." It is conspicuously not surreptitious, but surely it is coercive because the masters (currently the Chinese government and 26 large corporations) completely control every life that is being evaluated, including the decision to be watched. It cannot be consensual because the Chinese government and its corporate partners hold the ultimate power relationship over everyone submitting to it.

A Washington Post article (Song 2018) claims that the Chinese system is not as bad as it sounds, because, for instance, many of the worst offences (such as denying all travel requests for people who had traffic violations) happened in overzealous pilot projects and were then rejected from the national plan. We do not understand how that makes it better since the very same private companies running the tests are slated to continue running the program in a somewhat autonomous status, and private companies typically have more license to abuse than government itself does. Regardless, when citizens eagerly accept daily, continuous evaluation of any kind, as Chinese citizens are said to have done, there will be no turning back. Any future bureaucracy can add another and another at its whim, and no one can object without being down-scored.

China's Social Credit System is the ultimate digital-age version of the long-feared Panopticon. More than two centuries ago Samuel Bentham, an architect, designed a building that was actually a surveillance machine; his brother Jeremy Bentham fervently promoted the invention. Its optics were such that a single "inspector" could observe every occupant simultaneously. They called it the "Panopticon" (all seeing). It was, Jeremy said, "A new mode of obtaining power of mind over mind, in a quantity hitherto without example." Since its inception, surveillance technology has advanced in three major spurts, each of which triggered a new specter of surveillance and control. The first instance was the Benthams' building; the second was and is a tightly controlled closed-circuit television network (CCTV), and the third is today's electronic tracking services. Each had and has its own distinctive rationale: first the utopian perfection of society; second the enforcement of absolute tyranny; today safety, security, and convenience. Functionally, however, their root function is the same—total surveillance—and they are indeed three successive generations of Panopticons. Dobson and Fisher (2007) called them, respectively, Panopticon I, II, and III.

Clearly, China's Social Credit System qualifies as Panopticon III, a case of cultural acceptance that would not be acceptable in most western countries. But is western culture really that opposed? In 2019, the Trump Administration proposed a point-based plan to assign merit scores to immigrants applying for entry into the country (Shoichet 2019). US education officials are considering a new adversity score added to the SAT score that is so instrumental in determining social and financial opportunity (Jaschik 2019).

32.2 Tracking Technologies

New information technologies increase benefits and risks and make today's choices ever more crucial. Here, we explain the range of human tracking technologies and applications now available and how each is involved in tracking.

Human tracking technologies include Global Positioning System (GPS) receivers that are attachable or wearable with GPS chips embedded in cell phones, bracelets, or dedicated navigation devices, all of which may be connected to telecommunication networks that record coordinates and interact with geographic information systems (GIS) (Commonwealth v. Almonor 2019). A related form gets coordinates not from GPS but from less precise cell-site location information (CSLI) when a cell phone connects to a cell tower (Carpenter v. USA 2018).

Other ubiquitous sources of location data are the geosocial footprints extracted from social media activity and smartphones (Weidemann et al. 2018). A New York Times investigation described the extraordinary breadth of location information extracted from a million smartphones in New York City and stored in one database (Harris et al. 2018). Data from smartphones used in urban areas enables massive tracking of individuals regardless of their economic status, neighborhood, or worksite (Thompson and Warzel 2019).

The electronic exhibitionism inherent in social media is a major source of location data that are collected, analyzed, and sold. Until 2019, Facebook continuously collected location information on Android users even when the app was not in use (Gomez 2019). For close to a decade, Google has maintained a database called Sensorvault with detailed location information from millions of devices (Valentino-DeVries 2019).

Other tracking technologies include radio-frequency identification (RFID) and biometrics (Herbert and Tuminaro 2008). RFID chips can be imbedded in worn or carried objects such as urban transit cards and can be implanted in a person's body. Biometrics is an identification technology based on unique biological characteristics such as voice and facial recognition that is being utilized in immigration and even by landlords (Bellafante 2019). Wearable biometric devices are being used by professional sports teams to monitor the physical functions of athletes (Venook 2017). Location data from RFID are not spatially continuous and are limited to specific locations, but they are excellent for maintaining inventories of goods and people. Thus, a core use of RFID and biometrics is monitoring pedestrian traffic in buildings and transit systems. When integrated with surveillance cameras, these technologies can form the basis for a modern-day Panopticon II (Dobson and Fisher 2007).

Facial pattern recognition can be stationary, as when used to monitor crowds entering a stadium without necessarily following them home. However, frequent detection at ubiquitous geo-referenced sites or by mobile sensors creates a trail of geo-coordinates as effectively as GPS itself. Recently, Schuppe (2019) declared it a "routine policing tool in America." Yet, resistance is developing, and San Francisco has banned its use (Conger et al. 2019).

Increasingly, automobiles are equipped with surveillance devices capable of monitoring every aspect of engine performance but also direction, speed, and braking of the car itself, plus personal details such as eye movements to measure attentiveness.

Geoslavery is the most extreme application threatening privacy and personal freedom (Dobson and Fisher 2003; 2007; Fisher and Dobson 2003; Herbert 2006). The term was coined (Dobson 2002) soon after entrepreneurs started offering "kid-tracking" technology. Despite its kid name, then and now the devices can be used for tracking people of any age. Applications can be highly beneficial, and many are, but absolute control is a dangerous thing. The key to protecting the tracked is to establish applicable ethical standards, laws, and regulations.

Less extreme but still concerning is "nudging," a practice in which governments or corporations encourage mass behavior, and "big nudging," which uses big data to do it (Helbing et al. 2017; Dasgupta 2017). Insurance companies, for instance, reward customers for using location-based services (LBS) to enforce "safe" driving habits. State Farm Insurance offers a driving score that determines insurance rates, and they advertise it on TV making light of how it will dictate driving decisions such as workers being late for a meeting or a pregnant mother arriving late at the hospital for her baby's delivery (State Farm Insurance Company 2019).

Dasgupta views such nudging as "a modern form of paternalism. The new, caring government [or company] is … interested in what we do, but also … that we do [what] it considers to be right … To many this appears to be a sort of digital [prod] that allows one to govern the masses efficiently, without having to involve citizens in democratic processes." The technology used for nudging is ubiquitous computing and telecommunications systems, over which the individual consumer has little control. Laws and customs determine what is acceptable, but most collection and processing occurs in cloistered rooms. It is this separation of watcher and watched that frightens many people.

32.3 Informed Acceptance of Benefits and Adverse Acceptance of Risks

Society views geosurveillance—defined here as the practice, usually electronic, of monitoring and recording the geometries, topologies, and attributes of places and human and physical entities both stationary and moving—with two faces. When presented in the abstract, as CCTV was in George Orwell's 1984, geosurveillance is frightening in the extreme. When it is available commercially and used by many or even just a few, however, the specter subsides. This is particularly true when the technology is imbedded in smartphones, wearable devices, and apps. CCTV is now deployed routinely for surveillance in cities and sensitive rural sites, and the greatest fear for most people is merely a traffic fine. Likely, the key factor is individual perception of actual use. Prior to deployment, there is no such experience on which to judge. If then a device is widely deployed and seldom indicted for harm, the public is lulled into thinking the risk is small or nonexistent. We call this phenomenon an adverse acceptance.

The marketing of tracking technologies includes aggressive promotion of conveniences but reticence about dangers. Voluntary full disclosure of the scope, use, and sale of data collected would be self-defeating for proponents. The lack of understandable information renders it impossible for an urban dweller to make rational risk assessments connected to geoprivacy.

Excellent examples of this phenomenon are Hudson Yards—a new 28-acre "smart city" in Manhattan—and Waterfront Toronto, both owned by a subsidiary of Google's parent company Alphabet. In designing and promoting Hudson Yards, the developer emphasizes the conveniences of installed tracking technologies without disclosing what may be done with the data. As the developer's president proclaimed to a reporter: "The data is our data for the purposes of allowing us to make Hudson Yards function better" (Jeans 2019). Yet, privacy concerns ultimately forced Alphabet to scale back severely on certain onerous aspects of Waterfront Toronto (Bilefsky 2019).

Faced with such opaqueness, a resident, worker, visitor, or commercial customer at Hudson Yards has only three choices: accept the surveillance based on the developer's assurance of a positive or benign purpose; ignore the surveillance and accept an unknown risk concerning the use of the data by the developer or a third party; or refuse to enter the "smart city" to avoid surveillance and data collection. Buyers and renters must judge based on predominantly positive presentations. This situation is an example of what Attoh et al. (2019) have termed "idiocy in the smart city."

A similar dilemma is faced by Uber drivers and passengers because tracking technologies are imbedded in the labor relationship of the "gig" worker (Attoh et al. 2019). The driver can accept the cost of creating geodata for Uber as part of work or decline employment. Similarly, a potential customer can accept geosurveillance as a cost of the convenience of using the service or decline the ride (Smith and Leberstein 2015).

Consider the nature of this cost/risk versus benefit ratio at Hudson Yards, Uber, and anywhere else surveillance is installed. If the ratio is, say, 999 benefits to every

1 cost/risk, society may favor surveillance, but how can and should society protect itself from that one cost/risk? Consider this analogy: The benefits of white phosphorus matches are overwhelmingly positive, but we still have to devote some societal resources to match safety.

Some applications improve government and commercial efficiencies at the cost of privacy. Some yield control to government, loved ones, and caregivers. It is often said that the problem with privacy is not technology but rather misuse of technology. In turn, misuse is a function of societal norms and deviations from those norms. If a business offered a female tracking service in the USA similar to the one in Case Study #3, there would be wide public outrage including demands for government investigation, regulation, and prosecution. In Saudi Arabia, however, it fits within the norm of how women have been treated in the analog world. Still, some people in Saudi Arabia will object, and some Americans will try to do it anyway.

Already one tragedy complicated by geoslavery has been documented (Dobson 2007). When Stacy Peterson went missing in 2007, news reports claimed her husband Drew Peterson, a policeman in the Bolingbrook, Illinois Police Department, obsessively monitored her movements prior to her disappearance. She complained to family and friends that he was controlling her. She changed her cell phone number in a futile attempt to avoid his control. When confronted with the allegation that Drew was tracking Stacy's friends, his lawyer defended his actions in a frightening way. It was a common practice, the lawyer said, for local police officers to track their spouses, friends, and acquaintances. Stacy Peterson's body was never found. If she is dead, geoslavery is complicit in her murder. If she survived, geoslavery denied her the possibility of taking her children with her.

32.4 Legal and Regulatory Responses to Tracking Technologies

For decades, the European Union (EU) has been the international leader in regulating collection and use of personal electronic data, including location data (Herbert 2008). In May 2018, its General Data Privacy Regulation (GDPR) became effective, substantially broadening and improving protections for EU citizens. The regulation constitutes a significant step forward for protecting geoprivacy in European cities, particularly with its grant of the right to be forgotten.

The GDPR defines personal data to include location data as well as any other information related to a specific individual. The new regulations impose mandates that are relevant to geoprivacy, some particularly so: a requirement for informed and unambiguous individual consent; an insistence that data collection must be legitimate and necessary; a guarantee that individuals have rights to access and correct the information; and, most important, the provision of a right to be forgotten. The GDPR right to be forgotten, that is, to pursue anonymity, gives individuals a high degree of authority over their own location data. It is codified in GDPR, Article 17, which states, "The data subject shall have the right to obtain from the controller the erasure of personal data concerning him or her without undue delay and the controller shall have the obligation to erase personal data without undue delay." Erasure is enforceable under certain circumstances including when the data are "no longer necessary in relation to the purposes for which they were collected or otherwise processed."

The USA is far behind in developing such a comprehensive response to the privacy implications of electronic data. While American courts have grappled with some privacy disputes resulting from tracking technology, primarily involving criminal prosecutions, legislatures generally have been slow to respond. The delay in the USA is due, in part, to the fact that the rise of electronic tracking and social media occurred during the ascendancy and domination of neoliberal deregulation ideology.

The US Supreme Court and some state courts have ruled that the Fourth Amendment to the United States Constitution mandates that law enforcement obtain a judicial warrant before tracking with GPS or CSLI technologies. These rulings are interpretative of constitutional limitations on the use of tracking technologies by government actors. They are premised on concepts of property rights and reasonable expectations of privacy, rather than universal principles of human rights.

It is unlikely that federal legislation will be passed to grant strong privacy protections similar to GDPR in light of "the relationships between some members of Congress and Silicon Valley companies" (Fowler 2018). Therefore, the impetus for policy innovation concerning geoprivacy will more likely come from state legislatures and local governments unless a new national social movement arises to compel Congress to act with strong federal protections.

California has followed the EU's lead by adopting a right to be forgotten through passage of the California Privacy Act of 2018. Under the new state law, businesses that collect and/or sell personal consumer information, including geolocation data and biometric information, must notify the consumer, upon request, of the types or information being collected, used, and/or sold. More important, the law requires the deletion of such data, upon a consumer's request, except in certain specified situations. The City of Los Angeles sued an "IBM-owned app maker accused of sharing user location data with affiliates of its parent company and other advertisers, but also hiding the practice in a 10,000-word-long privacy policy" (Cimpanu 2019).

Other states have passed laws that seek to limit location tracking in narrower ways. The following examples highlight the lack of uniformity in such legislative measures. Montana and Utah statutes require law enforcement to seek a warrant before obtaining location data from a device under certain circumstances. It is a crime in Iowa and Wisconsin for a person to attach a GPS device to another person's vehicle without consent. Mandated or coerced RFID chip implants are prohibited by laws in California, Maryland, Utah, and New Hampshire. Some states have prohibited or regulated the collection of biometric information, particularly with respect to students.

Many people fear government or corporate surveillance, while ignoring the collection, use, and distribution of personal data by individuals, including family members, friends, and strangers. Some recognize a risk versus benefit ratio; most do not. Government and corporate surveillance and data collection are indiscriminate, applying to everyone for purposes of political control or corporate profit. In terms of everyday impact, however, the government might not care whether someone stops for a beer on the way home from work, while a spouse, parent, or caregiver may.

Surveys of public attitudes toward geosurveillance reveal a contradictory mixture of fear and acceptance. Rzeszewski and Luczys (2018) found, "The prevailing attitude that we identified [in Poznan, Poland and Edinburgh, UK] is neutral with a strong undertone of resignation—surrendering personal location is viewed as a form of digital currency. A smaller number of people had stronger, emotional views, either very positive or very negative, based on uncritical technological enthusiasm or fear of privacy violation. Such a wide spectrum of attitudes is not only produced by interaction with technology but can also be a result of different values associated with space and place itself."

Surveying public perception of privacy in the USA, Kar et al. (2013) found that respondents expect location data to be protected on the same level as health data and other personal information. However, respondents themselves are unaware of the legal implications of location privacy violations.

Indeed, public misunderstanding or outright ignorance of geoprivacy, geosurveillance, and geoslavery closely matches other manifestations of geographic ignorance and anti-intellectualism in the USA. The American purge of geography from all levels of education has left its mark on science and society (Kozak et al. 2015). In elementary school, geography has been misconstrued as "social studies," which deemphasize physical geography and spatial thinking. In high school, geography is required now by only 14 states. Geography is offered by most public universities but rarely by private universities. Only one geography department remains within the top twenty private US universities. To anyone who values education, it would seem remarkable if such neglect did not result in serious losses of public understanding. As one prominent example, a recent Pew Research Center (2018) report purporting to summarize "The State of Privacy in Post-Snowden America" missed its mark by failing to mention geoprivacy, spatial privacy, geosurveillance, geoslavery, or location (Pew Research Center 2018).

Citizens may fear government, but government agencies sometimes serve as their advocate and protector. The Federal Trade Commission (FTC 2014) has engaged in some limited efforts at challenging technology company misrepresentations concerning privacy. In 2014, the FTC issued a report entitled, "Data Brokers: A Call for Transparency and Accountability." In it, they named nine data brokers who amass and administer vast databases of personal information:

1. Acxiom: consumer data and analytics for marketing campaigns and fraud detection; information on about 700 million consumers worldwide.
2. CoreLogic: property, consumer, and financial information; more than 795 million historical property transactions, 93 million mortgage applications, and property-specific data covering over 99% of US residential properties; in total exceeding 147 million records.
3. Datalogix: businesses with marketing data on US households and more than a trillion dollars in consumer transactions; partnership with Facebook.

4. eBureau: predictive scoring and analytics services for marketers, financial services companies, online retailers; billions of consumer records.
5. ID Analytics: analytic services principally to verify identities or detect fraudulent transactions; 1.1 billion unique identity elements; 1.4 billion consumer transactions.
6. Intelius: background check and public record information; more than twenty billion records.
7. PeekYou: analyzes content from more than 60 social media sites, news sources, homepages, and blog platforms.
8. Rapleaf: data aggregator with at least one data point associated with more than 80% of all US consumer email addresses; supplements with the age, gender, marital status, and thirty other variables.
9. Recorded Future: historical data on consumers and companies; predicts future behavior.

Mirani and Nisen (2014) call them "The nine companies that know more about you than Google or Facebook." A representative list of what they know shows many variables that are spatial (address, address history, longitude and latitude); many reveal geographic identity (race, ethnicity, country of origin, religion, language); others relate to geographic habits (travel, vacation), not to mention dozens of variables that deeply probe finances, behavior, and lifestyle. The FTC report urged Congress to require the data broker industry to be more transparent and to give consumers greater control over their personal information.

32.5 Geoprivacy, the Inconscient Syndrome, and Control in the Academy

"We have entered a grand social experiment as momentous as any in our past and yet one so insidious that hardly anyone seems to have noticed" (Dobson 2009). For the first decade and more that we wrote about geoprivacy and geoslavery, there was precious little scholarly literature to cite. Today, there is a growing body based on empirical research, and we are especially thankful for those cited above. Still, technological and commercial advances are happening so fast that this chapter relies heavily on recent news media reports to augment the academic literature.

We encourage all applicable disciplines to join the quest for deeper understanding. Psychologists and sociologists, for instance, can study human motivations, responses, and behavioral issues. Technologists and legal scholars can develop alternative devices and regulations to thwart surveillance systems. Political scientists can explore better means for developing proactive and responsive public policies. Historians can search for antecedents to technologies, applications, and implications. Geographers and integrative teams of diverse disciplines can conduct interdisciplinary research.

Unfortunately, some academics have adopted tracking technologies with no more forethought than the general public. California physics professor Tom Bensky designed "a new mobile application and website … that tracks students' attendance using their cell phones," which is now used by "a couple hundred other professors and officials" (Bauer-Wolf 2019). He faced predictable complaints and answered in a typically naïve way, "But I can't convince them that I'm not going to do anything with the data I'm getting. It's just the app, server, and a database, but it is hard to convince people." Therein lies the ever-present question: Why should anyone trust anyone who holds the keys to his or her private world? One must ask, what happens if a student can't afford a smartphone or refuses to sign up? Is an accommodation (e.g., free phones, manual check-in) made, or does the student have to drop the class? Will only the compliant be educated?

At the very least, such impositions on students should be raised to a higher level, addressed in university policies to be developed through shared governance, and challenged in state and federal courts. Professor Bensky's app could form the basis for one of the first legal challenges under the new California Privacy Act. If Bensky were conducting a research experiment in precisely the same manner, federal law would require him to file an application and face an Institutional Review Board to ensure informed consent by those being tracked. A decade ago, privacy advocates were outraged when a research team published results from tracking 100,000 people without informed consent (González et al. 2008; Dobson 2009).

Bensky's quote above is a prime example of what we term the inconscient syndrome. In the course of our research, we have observed an inordinate number of inconscient actors who show no malice but also no forethought. Most simply do not think through the matter of surveillance deeply enough to perceive risks, and the geographic dimension makes the perception even more difficult. Manifestations include entrepreneurs who create and market new software and systems without realizing their potential dangers, consumers who persistently perceive benefits but not risks, workers and their unions acquiescing to geosurveillance, targeted individuals who naïvely trust their watchers, and commentators who trivialize risks in favor of benefits. Most seem genuinely convinced that no risk exists, but that perception often is influenced by sophisticated advertising aligned with commercial interests. Indeed, universities have become leading advocates and practitioners of geosurveillance to the concern of some faculty and others worried about intrusions into privacy (Vance 2019; Harwell 2019b).

32.6 Conclusions

Urbanization and the rapid rise of integrated location data technologies raise profound questions concerning societal values and priorities about privacy and control. The deregulated free market economy over the past four decades has empowered technology companies to develop products, platforms, and applications that maximize profits and data collection and effectively deliver individual conveniences while simultaneously eroding geoprivacy. Europe has responded with strong measures to protect privacy, freedom, and the pursuit of anonymity. Conversely, China's response is a perverse government assault on privacy. In the USA, use of tracking technologies against individuals is prohibited or regulated in certain areas, but true pro-active privacy regulation exists only in California.

The benefits of smartphones, GPS, social media, and other technologies are accepted for their conveniences with adverse acceptance of their risks and without a rigorous examination of potential means to balance benefits with risks. While such technologies help meet the need for urban spatial efficiencies, including infrastructure necessary for smart cities, they also feed massive corporate and government databases that can be used in urban areas to promote human control, manipulation, and even geoslavery. Developments in the Middle East and China, combined with memories of chattel slavery, demonstrate that the loss of geoprivacy is no longer a hypothetical proposition.

Regulation of geosurveillance to protect privacy is essential for cities to remain places where individuals can live and move about in relative obscurity. The EU's GDPR and the new California Privacy Act provide models for how societies can balance communal needs, consumer convenience, and individual autonomy. Central to such regulations are informed notice and consent; insistence on legitimacy and necessity in data collection; limitations of scope and duration of surveillance; rights of access and to correct the information; and a person's right to have the data destroyed. That last and crucial element would restore a vital aspect of urban living: the right to be forgotten—a guaranteed right to the pursuit of anonymity.

32.7 Epilogue

We submitted our final draft shortly before COVID-19 struck in earnest. The pandemic then hampered publication while dramatically changing the circumstances of our topic. Suddenly, geosurveillance was seen in a positive light as information technologies became essential for controlling the contagion country by country, enforcing social distancing, and tracing individuals exposed to the virus. When Apple and Google joined forces to support contract tracing, their offer was welcomed with fanfare. Simultaneously, the pandemic justified tracking workers, university students, and beachgoers. Some Americans envied China's apparent success without realizing how completely the country embraced geoslavery before the crisis. Conversely, some Americans resisted overhead drone surveillance while others objected even to preventive measures such as face masks.

We ourselves wrote an op-ed for the St. Louis Post Dispatch (May 6, 2020) condensing this whole chapter into a few points relevant to the pandemic. "For reopening," we said, "the goal must be to minimize deaths and illnesses while restoring essential goods and services, protecting fundamental rights, and maintaining acceptable life styles."

References

Attoh K, Wells K, Cullen D (2019) We're building their data: Labor, alienation, and idiocy in the smart city. Environ Plann D: Soc Space. https://doi.org/10.1177/0263775819856626

Bauer-Wolf J (2019) GPS to track student attendance. Inside Higher Ed, http://insidehighered.com/news/2019/06/20/professor-develops-new-app-gps-tracking-student-attendance. Accessed 25 June 2019

Bellafante G (2019) The landlord wants facial recognition in its rent-stabilized buildings. Why? New York Times, 28 Mar 2019, https://www.nytimes.com/2019/03/28/nyregion/rent-stabilized-buildings-facial-recognition.html?action=click&module=News&pgtype=Homepage. Accessed 25 June 2019

Bilefsky D (2019) Toronto's City of Tomorrow is scaled back amid privacy concerns. New York Times, October 31, 2019. https://www.nytimes.com/2019/10/31/world/canada/toronto-google-sidewalk.html. Accessed 6 Jan 2020

Carpenter v. United States (2018) 138 S. Ct. 2206

Chiu A (2019) She installed a Ring camera in her children's room for 'peace of mind.' A hacker accessed it and harassed her 8-year-old daughter. Washington Post, December 12, 2019. https://www.washingtonpost.com/nation/2019/12/12/she-installed-ring-camera-her-childrens-room-peace-mind-hacker-accessed-it-harassed-her-year-old-daughter/?arc404=true. Accessed 6 Jan 2020

Cimpanu C (2019) City of LA sues Weather Channel app for sharing location data with advertisers. https://www.zdnet.com/article/city-of-la-sues-weather-channel-app-for-sharing-location-data-with-advertisers/. Accessed 5 Aug 2019

Commonwealth v. Almonor (2019) 120 N.E. 3d. 35 (Mass.)

Conger K, Fausset R, Kovaleski SF (2019) San Francisco bans facial recognition technology, New York Times, May 14, 2019. https://www.nytimes.com/2019/05/14/us/facial-recognition-ban-san-francisco.html. Accessed 25 June 2019

Dasgupta A (2017) How geospatial justice can restrain misuse of data against humanity. Geospatial World, 10/16/2017. https://www.geospatialworld.net/blogs/need-to-develop-geospatial-justice/. Accessed 25 June 2019

Dellinger EJ (2019) How the biggest tech companies spent half a billion dollars lobbing Congress. Forbes, 30 Apr 2019. https://www.forbes.com/sites/ajdellinger/2019/04/30/how-the-biggest-tech-companies-spent-half-a-billion-dollars-lobbying-congress/#26329d3c57c9. Accessed 25 June 2019

Dobson JE (2002) Geoslavery. Paper presented at Annual Meeting of the American Association of Geographers, 19–22 Mar 2002, Los Angeles, CA

Dobson JE (2007) Under an electronic eye, 24/7. Ft. Worth Star-Telegram, December 27, 2007, also published as Geoslavery in the Stacy Peterson case, The Cleburne News, 3 Jan 2008

Dobson JE (2009) Big brother has evolved. Nature 458:968. https://doi.org/10.1038/458968a

Dobson JE, Fisher PF (2003) Geoslavery. IEEE Technol Soc Mag 22(1):47–52. https://doi.org/10.1109/mtas.2003.1188276

Dobson JE, Fisher PF (2007) The Panopticon's changing geography. Geogr Rev 97(3):307–323. https://doi.org/10.1111/j.1931-0846.2007.tb00508.x

Federal Trade Commission (2014) Data brokers: A call for transparency and accountability. https://www.ftc.gov/reports/data-brokers-call-transparency-accountability-report-federal-trade-commission-may-2014. Accessed 25 June 2019

Fisher PF, Dobson JE (2003) Who knows where you are, and who should, in the era of mobile geography? Geography 88(4):331–337

Foner E (2016) Gateway to freedom: the hidden history of the underground railroad. Norton & Company, New York. ISBN-1978-0393352191

Fowler S (2018) The 3 biggest challenges for tech in 2019. New York Times, 29 Dec 29 2018. https://www.nytimes.com/2018/12/29/opinion/tech-2018-trends-2019-predictions.html. Accessed 25 June 2019

Franklin JH, Schweninger L (1999) Runaway slaves: Rebels on the plantation. Oxford University Press, Oxford. ISBN-10: 0195084519, ISBN-13: 978-0195084511

Garrett BL (2015) The privatization of cities' public spaces is escalating. It is time to take a stand. The Guardian, 4 Aug 4 2015. https://www.theguardian.com/cities/2015/aug/04/pops-privately-owned-public-space-cities-direct-action. Accessed 25 June 2019

Gomez V (2019) Facebook adds new background location privacy controls to its Android app. RR-Magazine, 20 Feb 2019. https://rr-magazine.com/2019/02/20/facebook-adds-new-background-location-privacy-controls-to-its-android-app/. Accessed 25 June 2019

González MC, Hidalgo CA, Barabási AL (2008) Understanding individual human mobility patterns. Nature 453:779–782. https://doi.org/10.1038/nature07850

Harris R, Keller MH, Valentino-DeVries J (2018) 8 Places where smartphones tracked people's movements. New York Times, 15 Dec 2018. https://www.nytimes.com/2018/12/15/business/location-tracking-phones.html. Accessed 25 June 2019

Hartzog W, Selinger E (2019) Why you can no longer get lost in the crowd. New York Times, 17 Apr 2019. https://www.nytimes.com/2019/04/17/opinion/data-privacy.html. Accessed 25 June 2019

Harwell D (2019a) Police can keep Ring camera video forever and share with whomever they'd like, Amazon tells senator. Washington Post, 19 Nov 2019. https://www.washingtonpost.com/technology/2019/11/19/police-can-keep-ring-camera-video-forever-share-with-whomever-theyd-like-company-tells-senator/. Accessed 6 Jan 2020

Harwell D (2019b) Colleges are turning students' phones into surveillance machines, tracking the locations of hundreds of thousands. Washington Post, 24 Dec 2019. https://www.washingtonpost.com/technology/2019/12/24/colleges-are-turning-students-phones-into-surveillance-machines-tracking-locations-hundreds-thousands/. Accessed 6 Jan 2020

Hatton C (2015) China "social credit": Beijing sets up huge system. BBC 26 Oct 2015. https://www.bbc.com/news/world-asia-china-34592186. Accessed 25 June 2019

Helbing D, Frey BS, Gigerenzer G, Hafen E, Hagner M, Hofstetter Y, van den Hoven J, Zicari RV, Zwitter A (2017) Will democracy survive big data and artificial intelligence? Scientific American, 25 Feb 2017. https://doi.org/10.1007/978-3-319-90869-4_7

Herbert WA (2006) No direction home: will the law keep pace with human tracking technology to protect individual privacy and stop geoslavery? I/S: A J Law Policy Inf Soc 2(3):409–473. https://kb.osu.edu/bitstream/handle/1811/72738/1/ISJLP_V2N2_409.pdf. Accessed 26 June 2019

Herbert WA (2008) Workplace electronic privacy protections abroad: the whole wide world is watching. Univ Florida J Law Public Policy 19 J.L. & PUB POL'Y 379–420. https://works.bepress.com/william_herbert/12/. Accessed 26 June 2019

Herbert WA, Tuminaro AK (2008) The impact of emerging technologies in the workplace: who's watching the man (Who's watching me)? Hofstra Labor Employ J 25(2):355–393. https://scholarlycommons.law.hofstra.edu/cgi/viewcontent.cgi?article=1431&context=hlelj. Accessed 25 June 2019

HIPAA Journal (2019) What is the purpose of HIPAA? https://www.hipaajournal.com/purpose-of-hipaa/. Accessed 5 Aug 2019

Hu W, Haag M (2019) 90,000 packages disappear daily in N. Y. C. Is help on the way? New York Times, 3 Dec 2019. https://www.nytimes.com/2019/12/02/nyregion/online-shopping-package-theft.html. Accessed 5 Jan 2020

Jaschik S (2019) New SAT score: adversity. Inside Higher Ed. https://www.insidehighered.com/admissions/article/2019/05/20/college-board-will-add-adversity-score-everyone-taking-sat. Accessed 5 Aug 2019

Jeans D (2019) Related's Hudson Yards; Smart city or surveillance city. The Real Deal: New York Real Estate News, 15 Mar 2019. https://therealdeal.com/2019/03/15/hudson-yards-smart-city-or-surveillance-city/. Accessed 25 June 2019

Kar B, Crowsey RC, Zale JJ (2013) The myth of location privacy in the United States: surveyed attitude versus current practices. Prof Geogr 65(1):47–64. https://aquila.usm.edu/fac_pubs/7559. Accessed 25 June 2019

Kozak SL, Dobson JE, Wood JS (2015) Geography's American constituency: results from the AGS geographic knowledge and values survey. Int Res Geogr Environ Educ 24(3):1–22. https://doi.org/10.1080/10382046.2015.1034457
Kuo L (2019) China bans 23 m from buying travel tickets as part of 'social credit' system. The Guardian, Mar 2019. https://www.theguardian.com/world/2019/mar/01/china-bans-23m-discredited-citizens-from-buying-travel-tickets-social-credit-system?CMP=share_btn_fb. Accessed 26 June 2019
Kwan M-P, Casas I, Schmitz BC (2004) Protection of geoprivacy and accuracy of spatial information: how effective are geographical masks? Cartographica 39(2):15–28. https://doi.org/10.3138/x204-4223-57mk-8273
Lemmings D (ed) (2018) The Oxford Edition of Blackstone's commentaries on the laws of England: book I: of the rights of persons. Oxford University Press, Oxford. ISBN: 9780199600991
Mirani L, Nisen M (2014) The nine companies that know more about you than Google or Facebook. Quartz. https://qz.com/213900/the-nine-companies-that-know-more-about-you-than-google-or-facebook/. Accessed 25 June 2019
Mistreanu S (2018) Life inside China's social credit laboratory; The party's massive experiment in ranking and monitoring Chinese citizens has already started. https://foreignpolicy.com/2018/04/03/life-inside-chinas-social-credit-laboratory/. Accessed 6 Aug 2019
Pew Research Center (2018) The state of privacy in post-Snowden America. https://www.pewresearch.org/fact-tank/2016/09/21/the-state-of-privacy-in-america/. Accessed 25 June 2019
Rzeszewski M, Luczys P (2018) Care, indifference and anxiety—attitudes toward location data in everyday life. Int J Geo-Inf 7:383. https://doi.org/10.3390/ijgi7100383. Accessed 25 June 2019
Schuppe J (2019) How facial recognition became a routine policing tool in America. NBC News Online, 11 May 11 2019. https://www.nbcnews.com/news/us-news/how-facial-recognition-became-routine-policing-tool-america-n1004251. Accessed 25 June 2019
Shoichet CE (2019) What 'merit-based' immigration means, and why Trump keeps saying he wants it. https://www.cnn.com/2019/05/16/politics/merit-based-immigration-explainer/index.html. Accessed 5 Aug 2019
Slinn S, Herbert WA (2011) Some think of the future: Internet, electronic and telephonic labor representation elections. 56 St. Louis U. L.J. 171:190–192 (Fall 2011). https://digitalcommons.osgoode.yorku.ca/scholarly_works/461/. Accessed 26 June 2019
Smith R, Leberstein S (2015) Rights on demand: ensuring workplace standards and worker security in the in-demand economy. Natl Employ Law Project (Sept 2015) https://s27147.pcdn.co/wp-content/uploads/Rights-On-Demand-Report.pdf. Accessed 25 June 2019
Song B (2018) The West may be wrong about China's social credit system. https://www.washingtonpost.com/news/theworldpost/wp/2018/11/29/social-credit/?noredirect&utm_term=.2c7afc38be63. Accessed 25 June 2019
Speier J (2019) Rep Speier calls on Apple and Google to remove a Saudi Government app tracking women. Press Release, 22 Feb 2019. https://speier.house.gov/media-center/press-releases/rep-speier-calls-apple-and-google-remove-saudi-government-app-tracking. Accessed 26 June 2019
State Farm Insurance Company (2019) Kim's discount (drive safe & save). State Farm® commercial. https://www.youtube.com/watch?v=yMsmubcvY_k. Accessed 5 Aug 2019
Stauffer J (2002) The black hearts of men: radical abolitionists and the transformation of race. Harvard University Press, Cambridge. ISBN-10: 0674013670, ISBN-13: 978-0674013674
Swisher K (2019) Be paranoid about privacy. New York Times, 24 Dec 2019. https://www.nytimes.com/2019/12/24/opinion/location-privacy.html Accessed 6 Jan 2020
Thompson SA, Warzel C (2019) Twelve million phones, one dataset, zero privacy. New York Times, 19 Dec 2019 https://www.nytimes.com/interactive/2019/12/19/opinion/location-tracking-cell-phone.html. Accessed 6 Jan 2020
Thorbecke C (2019) Long Island police partner with Amazon's Ring to crack down on porch pirates. ABC News, 4 Dec 2019. https://abcnews.go.com/Technology/long-island-police-crack-porch-pirates-amazon-ring/story?id=67489715. Accessed 6 Jan 2020

Valentino-DeVries J (2019) Tracking phones, Google is a dragnet for the police. New York Times, 13 Apr 2019. https://www.nytimes.com/interactive/2019/04/13/us/google-location-tracking-police.html. Accessed 25 June 2019

Vance A (2019) Protecting student privacy. Opinion Letter. New York Times, 19 Dec 2019. https://www.nytimes.com/2018/12/19/opinion/letters/student-privacy-laws.html. Accessed 6 Jan 2020

Venook J (2017) The upcoming privacy battle over wearables in the NBA. The Atlantic, 10 Apr 2017. https://www.theatlantic.com/business/archive/2017/04/biometric-tracking-sports/522222/. Accessed 25 June 2019

Warzel C, Thompson SA (2019) How your phone betrays democracy. New York Times, 21 Dec 2019 https://www.nytimes.com/interactive/2019/12/21/opinion/location-data-democracy-protests.html. Accessed 6 Jan 2020

Weidemann C, Swift J, Kemp K (2018) Geosocial footprints and geoprivacy concerns. In: Thatcher J, Ekert J, Shears A (eds) Thinking big data in geography: New regimes, new research. University of Nebraska, Lincoln (doi:9781496205377)

Williams T (2019) In high-tech cities, no more potholes, but what about privacy? New York Times, 1 Jan 2019. https://www.nytimes.com/2019/01/01/us/kansas-city-smart-technology.html. Accessed 25 June 2019

Wingfield N (2018) Amazon buys Ring, maker of smart home products. New York Times, 27 Feb 2018. https://www.nytimes.com/2018/02/27/business/dealbook/amazon-buys-ring.html. Accessed 6 Jan 2020

Zuboff S (2019) The age of surveillance capitalism: the fight for a human future at the new frontier of power. Hachette Book Group, New York. ISBN-978-1-61039-659, ISBN-968-1-61039-570-0

Jerome E. Dobson is Emeritus Professor of Geography, University of Kansas; President Emeritus, American Geographical Society; and a Trustee of Reinhardt University. He is a Jefferson Science Fellow with the National Academies and the U.S Department of State.

William A. Herbert is a Distinguished Lecturer at Hunter College, City University of New York, and a Faculty Associate at the Roosevelt House Institute for Public Policy. He is also Executive Director of the National Center for the Study of Collective Bargaining in Higher Education and the Professions.

Chapter 33
3D Modeling of the Cadastre and the Spatial Representation of Property

Lin Li, Renzhong Guo, Shen Ying, Haizhong Zhu, Jindi Wu, and Chencheng Liu

Abstract An emerging technology, three-dimensional (3D) cadastres as extensions to the current parcel-based or two-dimensional (2D) cadastre, has been developed to meet the management of 3D urban land use and 3D properties. This chapter provides a brief review of the key issues of 3D cadastre and the spatial representation of ownership. In order to understand the importance of legislation for developing modeling technology for 3D property, the legislative context of ownership is addressed in specific reference to China. In light of spatial rights of land-use space, a 3D spatial model of property is presented in terms of polyhedra with four-layer structures. Being compatible with the existing 2D cadastre, this 3D spatial data structure is suitable as a hybrid cadastral system for 2D and 3D property and provides an available means to spatially represent 3D property with integrity. By analyzing the heterogeneity of the land space used for property, the ownership of condominiums with internal structure is addressed and spatial representation of ownership is presented by instantiation in a case study in China.

33.1 Introduction

A cadastre is generally regarded as a comprehensive land recording of the metes and bounds of a country's real property. According to the International Federation of Surveyors (FIG), a cadastre is normally a parcel-based and up-to-date land information system containing an official record of interests in land (i.e., rights, restrictions, and responsibilities or RRRs). In this record, the ownership, extent, and value of real property in a given area are explicitly and clearly registered and used for fiscal purposes (e.g., taxation), legal purposes, and to assist in the management of land and

---

L. Li · S. Ying · H. Zhu (✉) · J. Wu · C. Liu
School of Resource and Environmental Sciences, Wuhan University, Wuhan, China
e-mail: hhzhu@whu.edu.cn

R. Guo
Research Institute for Smart City, Shenzhen University, Shenzhen, China

W. Shi et al. (eds.), *Urban Informatics*, The Urban Book Series,
https://doi.org/10.1007/978-981-15-8983-6_33

land use (e.g., for planning and other administrative purposes). Registration of RRRs is the administrative core of cadastres and properties.

As ownership is defined as the lawful record of a property or a piece of land assigned to the people who own the property, the spatial extent and geographical location of the property are the critical elements for substantiating the ownership. Traditionally, a piece of land defined as a land parcel (or simply, a parcel) is a plane area with a clear boundary on the surface of the Earth. From the boundary on the ground, a spatial "cone" can be formed geometrically from the Earth's center to the sky, and ownership implicates the lawful record of all things within the spatial "cone." In this sense, the rights to land within the "cone" (space on, below, and above the ground) are hypothetically homogeneous and can be easily demarcated by the plane's extent. As such, a two-dimensional (2D) or parcel-based cadastre has so far dominated the administration of cadastres and has been adopted by various legal systems.

With the evolution of society and the economy, especially in urban areas, rapid urbanization presents a challenge to densely populated cities with limited urban land resources, and changes to land-use patterns in the form of urban sprawl have been increasing in recent years (Foley et al. 2005; Turner et al. 2007; Guo et al. 2013; Zulkifli et al. 2015; Li et al. 2016). Space on, below, and above the ground cannot be used merely for a single purpose. A piece of land must be shared by various parties for different contexts, and rights to it cannot be secured by its plane extent. The rights bounded to the space below or above ground are no longer fully consistent with that on the ground. Thus, the use of a land parcel in terms of cadastre inevitably evolves into the more general use of land space, which leads to a shift of focus from the surface of land parcels to the space above and below them in land use and development.

The emerging, spatially heterogeneous rights to land parcels break the spatial homogeneity of land rights within the cone, as long as required by the 2D parcel-based cadastre. The traditional concept of the 2D cadastre is augmented by dividing the utilization of land space vertically, in order to accommodate increased population density and intensive socioeconomic activities in urban areas. Three-dimensional (3D) cadastres have been developed to meet the management of 3D land-use space and 3D property (Guo et al. 2013; Stoter et al. 2013; Jazayeri et al. 2014; Karabin 2014). This emerging technology helps meet the increasing social demand for the precise management of immovable property (land and housing).

Here, a typical example quoted from the study by Guo et al. (2013) may present an intuitive understanding of the deficiency of a parcel-based cadastre. They cite a parcel with a complex building on it in Shenzhen, one of the fastest-growing and most economically advanced cities in China. This complex is made up of several plaza buildings containing many shops. Two main buildings are separated by a municipal road and connected by an arched structure. The buildings are registered on a parcel-based cadastral map (Fig. 33.1). The land space used for the over-ground arch is drawn on this map and labeled with H102-0037(B), which overlaps with the commercial shops and the underground parking lot. Two adjacent parcels, H102-0037 and H102-0038, contain the two main buildings, respectively. However, H102-0037(B) refers to the parcel above the surface, while H102-0037 and H102-0038 refer to parcels on the surface. The land space of the arch, a public pedestrian corridor (a kind of easement), belongs to the municipality, while the underground shops above the parking lot are

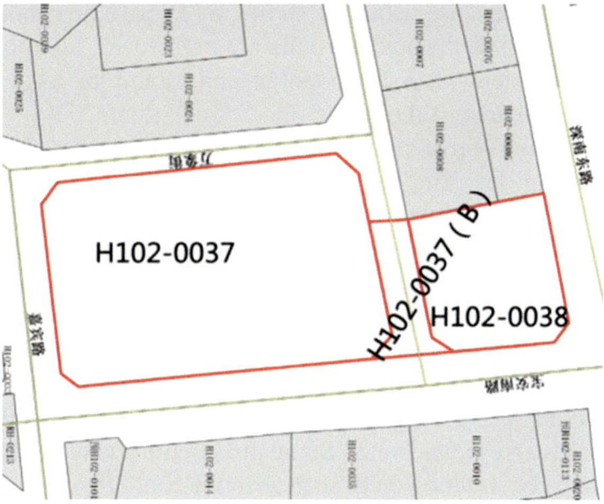

Fig. 33.1 Cadastral map (Guo et al. 2013)

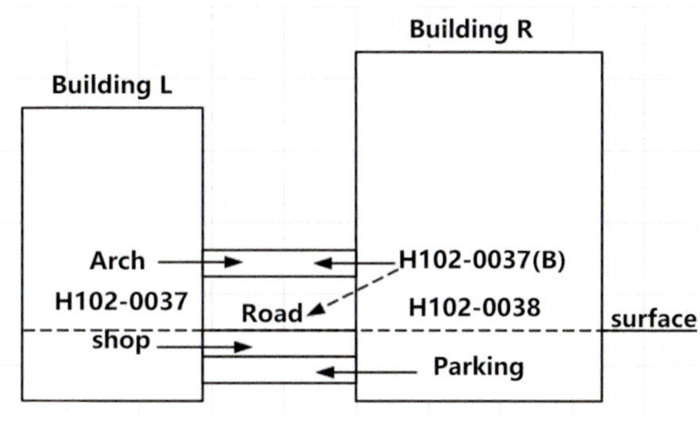

Fig. 33.2 The vertical profile of construction (modified from Guo et al. 2013)

owned by different individuals. The vertical configuration is illustrated in Fig. 33.2. However, it is found that this 2D cadastral map fails to record the spatial configuration of land space and may even confuse readers. The implications of a multi-purpose use of land in H102-0037(B) could not be geometrically clarified on the 2D cadastral map without adding a third dimension.

33.2 Spatial Rights to Real Property

33.2.1 Legal Context of a 3D Cadastre

When real property or a cadastre is registered on a 2D cadastral map, spatial rights to real property, or the spatial extents assigned by ownership, can only be directly presented in terms of 2D geometry, even though the rights are legally attributed in

3D. As the above example shows, a 2D cadastre cannot represent the 3D features of property. As the spatial rights are prescribed, interpreted, and implemented within legal systems, it is important to understand the legal context in order to model the spatial extent of the rights.

Ownership of land, or property in a wider sense, is set by legal systems and social conventions. The key issue in land administration is the management of various property or spatial rights on, in, and attached to the piece of land. These rights are embodied in the concept of property, which may have different meanings in different countries (Kalantari et al. 2008; Stubkjær 2004) that are largely dependent on legal systems (Paulsson and Paasch 2011). Some countries—such as the Netherlands, Germany, the UK, France, and Belgium—define ownership as the rights to the ground and of all space above and below it, including groundwater and fixtures (van der Molen 2003). Other countries understand ownership in a way that does not include mines and groundwater. Some jurisdictions may not allow separate rights to a parcel from construction on it, such as in the Netherlands and China. Other nations, such as Denmark, accept, through leasing, different ownerships for land and for buildings; in fact, the formation of a property "on top of another property" can be implemented under a special procedure (Sorensen 2011).

As most systems of land administrations in the world are set on the basis of 2D cadastres, the development of a 3D cadastre requires the amendment of property laws and regulations when land use extends spatially to a vertical from a horizontal plane. This is a big issue especially for those developed countries with comprehensive legal and administrative systems. It usually takes a quite long and arduous effort to finish an amendment. However, the laws in developing countries or regions are likely to be amended more easily than those of developed countries due to their imperfect legislation and administration.

China is a rapidly developing country and is currently perfecting her legislation and administration, which gives her room to adapt, update, or refine some items in her property laws where spatial rights of property have not been defined in great detail. It was in 2007 when the Real Right Law of the People's Republic of China was issued and took effect (October 1, 2007). The right to land is founded also on the principles of the parcel-based cadastre; however, Article 136 in this law states that "the right to use construction land may be created separately on the surface of or above or under the land. The newly established right may not injure the usufructuary right that has already been established." Article 138 further states that land space occupied by buildings, fixtures, and affiliated facilities shall be contained in a contract with the transfer of rights.

The separation of property rights for construction above and underground from those on the surface implies that uses of above and underground spaces may be different from those of the surface and that the parcel space may be multi-level, across boundaries, or without 2D geometric limitation. It indicates that the rights to land are always associated with some construction and no ownership will be created without construction (or buildings). This law provides a good legal basis for local governments to create their own rules and regulations for land use and makes it easier to develop a 3D cadastral system than in more developed countries or regions.

33.2.2 Geometry of 3D Property with Homogeneous Land Space

A property has both bona fide and legal aspects (Aien et al. 2013; Jazayeri et al. 2014; Ying et al. 2014), and it is considered a compound object that combines the physical object with the legal treatment of the object. The physical object (such as a usable unit of land space or an apartment) takes certain geometry and is the base of the ownership and other rights. The legal aspect of property is attached to the physical object and refers to or involves more space in various senses; for example, solar rights to an apartment involve a space beyond the space occupied just by the apartment and without a clearly defined boundary (Li et al. 2019). Thus, the spatial representation of the physical objects is the major task of modeling 3D property that is explicitly defined by spatial extent in the physical 3D space, that is, modeling ownership by spatial means.

As a building is always attached on a piece of land, a 3D property (containing both land and building or construction) consists spatially of two 3D geometries: a 3D model of the construction and a 3D container that is a derived spatial extent of land space used by the construction. Since a 3D model of construction is included in the container, the spatial relation of a property with others can be captured by the spatial relation among the containers. The architectural configuration of the construction may have some influence on rights to land space, such as the geometry of easement on neighboring spaces, and will be shaped by the access points of the architecture. However, this kind of influence is hardly depicted in an explicit geometry. Therefore, in terms of the cadastre, spatial modeling of a property in the form of land space is aimed at presenting an explicit 3D geometry of the containers, which simplifies the geometry of a property into a polyhedron. It comprises a prism or a combination of prisms that have vertical faces and flat tops or bottoms (Fig. 33.3).

Fig. 33.3 Geometry of 3D property in a cadastre

This simplicity results from the fact that land space for above or underground construction is plotted depending on a planar parcel. The faces of a polyhedron and the edges of the faces should satisfy the generalized Jordan curve theorem that refers to the orientability of these geometric elements. The interior of the container is hypothetically connected, which means that any container is simple, and no compound or multiple containers are allowed. If a container can be divided into two or more independent containers, each of the latter is treated as a simple one.

33.3 Integral Spatial Modeling of 3D Property

Spatial modeling of 3D objects long has been studied and is being addressed in the domain of geographical information systems (GIS) and related fields. Many 3D data models have been presented and are used to capture the spatial features of 3D objects in terms of geometry. 3D objects may be featured by simplexes (point, line, triangle, and tetrahedron; Carlson 1987), configured by a 3D formal data structure (FDS) (Molenaar 1990), represented by tetrahedronized irregular networks (Penninga et al. 2006), by polyhedra (Arens et al. 2005; Stoter 2004; Wenninger 1974; Zlatanova 2000), by polyhedral regular polytopes (Thompson 2007), or by a constructive solid geometry (CSG) and B-rep approach in computer graphics. Those data models have been commonly used for different fields and applications with certain semantic foci.

In spatially modeling of land administration and registration of property, an emphasis is placed on keeping these data consistent when developing a real 3D cadastre and extending its spatial dimension from 2D, since the semantics embedded in the data models are used to regulate and coordinate relationships among people and property under a given society, economy, and legal system. Therefore, the data model of 3D property should be compatible with the existing data model in 2D parcel-based cadastral systems so that the semantics recorded in the latter will not change.

The 2D data models with three-layer structure including topological features—faces, edges, and nodes (vertices)—are commonly adopted in 2D cadastre. A simple example is shown in Fig. 33.4 with Table 33.1, where an edge is terminated by its two nodes and a face is represented by its surrounding boundary as a series of edges. For example, in that figure f14 is composed of four edges {e25, e26, e27, and e28}.

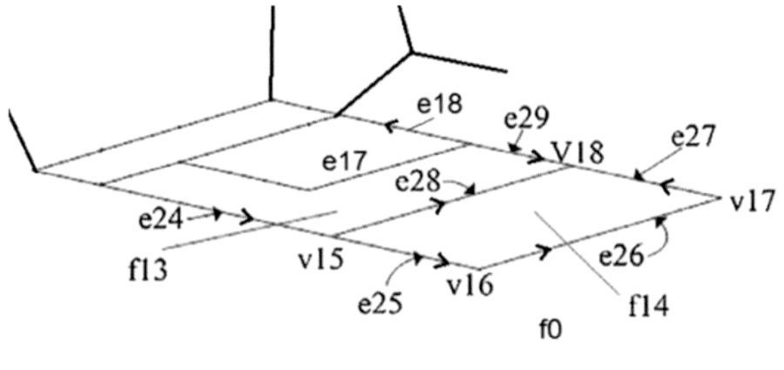

Fig. 33.4 2D data model for parcel-based property

Table 33.1 Table of the 2D data model shown in Fig. 33.4

Edge	From_Node	To_Node	Left_Face	Right_Face
…				
e25	v15	v16	f14	f0
e26	v16	v17	f14	f0
e27	v17	v18	f14	f0
e28	v15	v18	f13	f14
…				

Adding a 3D topological feature—a volume—to the 2D data model forms a 3D data model with four-layer structure for the 3D cadastral system. Consequently, a volume that is able to depict a container or polyhedron is represented by a set of faces that enclose a 3D space. Such a 3D data model may be operationally structured with a 3D piecewise linear complex (PLC), a commonly used geometric data structure in computer graphics (Cohen-Steiner et al. 2004; Miller et al. 1996; Si and Gartner 2005).

For example, two volumes (3D properties) in Fig. 33.5a are integrated with 2D parcels into a 3D spatial configuration of 3D space that accommodates both 2D properties and 3D properties shown in Fig. 33.5b. Volume Vol2 is represented by an

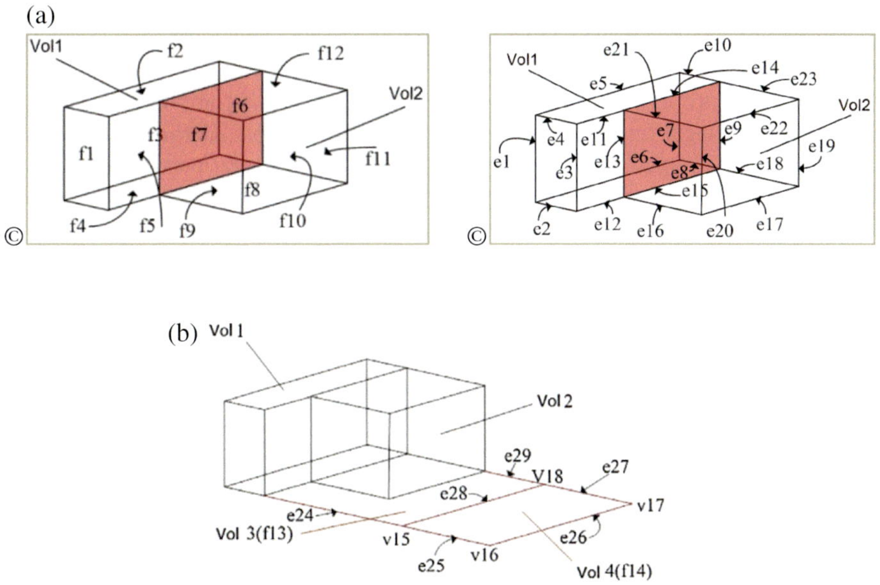

Fig. 33.5 A 3D data model of property compatible with a parcel-based 2D data model (modified from Guo and Ying 2010). **a** Two volumes (containers) with 3D geometry. **b** Compatible data model for 2D and 3D cadastre

enclosed face set {f7, f8, f9, f10, f11, f12}, and face f8 is demarcated by a set of edges {e15, e16, e17, e18}. Volume Vol4 is regarded as a special kind of 3D object, being degraded from 3D geometry into face f14 of the 2D geometry. This simple example shows that the 3D data model matches well with the commonly used 2D data model.

33.4 Heterogeneity of Land Space Used for Property

If an ownership includes a certain land space where all constructions lie within the space, a container mentioned above in the form of a polyhedron can be spatially modeled due to its homogeneous space with respect to ownership. However, in a densely populated urban area, many high-rise buildings are created to provide more housing and to accommodate more people. A unique owner of an apartment in a building is not an exclusive owner of a parcel of land that is undividable. Although an apartment uniquely occupies a chunk of land space and its ownership could be also spatially modeled by its polyhedral container geometry, different legal treatments associated with the ownership emerging from sharing integrity of land space break the homogeneity of the land space used by the apartment. In this case, the internal structure of the ownership should be clearly presented by its spatial representation. This poses a critical requirement for more precise management of property that includes not only land space and the vertical spatial extent of the property, but also the horizontal extent of the property and the ownership structure, which corresponds to the spatial components of the property.

In general, a property being viewed as a compound object combines the physical object with the legal treatment of the object in data models. However, a physical object (building or apartment) may be constructed with several parts with different functions or intentions, which lead to different legal treatments included in the ownership. An internal heterogeneity is then emerging in the ownership and reflects the disparity of the lawful recording of the different parts of an object and requires differentiating ownership in a property management system. A condominium unit is a typical property of this kind.

With a common or shared ground parcel, a building consisting of condominiums is divided into private and common parts. This co-ownership has been discussed by many studies (Çağdaş 2013; Pouliot et al. 2011, 2013; Rajabifard et al. 2013; Li et al. 2016). For this kind of ownership two types of ownership are found, exclusive ownership and shared ownership. Exclusive ownership means that an owner can dispose of his or her parts according to the corresponding laws. Shared (or common) ownership means that the common parts and the ground parcel cannot be disposed at someone's own will and must be disposed in common. It is also found that an ownership of a condominium is not the same as ownership of a piece of parcel or a chunk of land space. Its different spatial parts with certain rights should be represented in detail so that the internal structure of the ownership is expressed in a spatially explicit manner targeted toward more precise management of property.

Physical structural components associated with a condominium unit may have different rights to each part with internal homogeneity and those different rights come together to constitute the ownership of the condominium. For example, in China, an ownership of a condominium unit may include two physical objects: the exclusively owned apartment itself and some space (such as elevators and corridors) that is shared with others. The ownership includes at least two different internal rights to the parts. Even for exclusively owned objects (or spaces), the room space is physically recorded into the legal spatial extent, and a balcony (space) may be half-recorded into the legal spatial extent. Such subdivisions of ownership with legal space are critical in taxation, loans, and insurance.

As parts of land space corresponding to certain physical objects, each of these parts in general can be suitably modeled by an enclosed polyhedron in the four-layer structure. However, it becomes critical to clarify the semantics of those parts with ownership and spatial relations among them in spatial modeling of the ownership. As mentioned above, the meaning of ownership varies with different legal systems and social conventions; it would be much more helpful to discuss the spatial representation of the condominium ownership with a given legislative and institutional context. The following section uses China as an example.

33.5 A Case Study of Spatial Modeling of Ownership Structure in China

33.5.1 Ownership of Condominiums in China

According to the Land Administration Law in mainland China, urban land is administered differently from rural land. Any urban land is uniquely owned by the State and ownership cannot be altered. Ownership of the buildings or other constructions on urban land can be attributed to individuals or any legal parties. A property embodies the ownership of a house, a building, or buildings and the usufruct of land. In this legislative context as well as social conventions in China, condominiums are the predominant form of housing property in urban areas. Ownership is legislatively ensured by the Real Right Law of the People's Republic of China (People's Republic of China 2007), which offers provisions for the owners' co-ownership of building areas. Its Article 70 states that "as regards such exclusive parts within the buildings as the residential houses or the houses used for business purposes, an owner shall enjoy the ownership thereof, while as regards the common parts other than the exclusive parts, the owner shall have common ownership and the common management right thereof."

Ownership of a condominium unit refers to two types of objects, that is, exclusive objects and common or shared objects. In Specifications for Estate Surveying (People's Republic of China 2000), exclusive objects are further divided into two types of objects: the major body and annexes such as balconies, basements,

Table 33.2 Internal structure of ownership of a condominium unit

Physical parts	Physical objects	Sub-objects	Counted physical space	Remarks
Exclusive	*Major body*		Completely	Rooms
	Annexes	De facto	Totally	Balconies in the major structure
		Ratio	Partially	Balconies outside of the major structure
		Fiat	Non (height < 2.1 m) Partially (others)	Bay windows
Shared	*Apportionable*	De facto	Totally	Commonly owned indoor stairs
		Ratio	Partially	Commonly owned corridors
		Fiat	Non	Commonly owned roof gardens
	Non-apportionable		Non	Basements

and garages; common objects are further divided into apportionable and non-apportionable objects. Construction area is used to measure ownership in terms of magnitude. Apportionable means that the metric geometry of the objects is calculated in some approach to contribute the construction area of the corresponding condominium units, and non-apportionable means that the objects make no contribution to the construction area. That is, the legal construction area of a condominium unit consists of the construction area from its exclusive parts and from its shares of apportionable objects.

Since the spatial extent of physical objects from both types is the metric base for deriving the construction area and measures ownership in different ways, ownership of a condominium unit is structured by different parts in light of the physical configuration of the unit and buildings including the unit. The internal structure of ownership is tabulated in Table 33.2.

33.5.2 Implementation Tool for Spatial Modeling of Ownership

It is very clear from Table 33.2 that the structure of ownership can be presented by a 3D model of the physical building of a condominium unit. Although a condominium unit may be of complex physical structure, each part corresponds to a physical component of the building which can be modeled with the geometry of a 3D container as discussed above. It is known that CityGML models or building information models (BIMs) provide rich semantic and 3D information for the internal structure of a building (Li et al. 2019). A great effort has been made to adopt CityGML or BIMs in the field of land administration and property management (Amirebrahimi 2012; Çağdaş 2013; El-Mekawy et al. 2014; Góźdź et al. 2014). CityGML has shown its merits in exploring the internal heterogeneity of the ownership of condominiums and clarifying the spatial differences within the ownership.

The ISO19152 LADM is designed for offering a conceptual model that allows land administration objects and relationships to be described. Land administration is described as the process of determining, recording, and disseminating information on the relationship between people and land (or rather space). The LADM includes basic packages that are related to (1) parties; (2) basic administrative units and RRRs; and (3) spatial units (parcels, legal spaces of buildings, and utilities). The package, Spatial Unit, is composed of the surveying and spatial representation sub-packages, and has several different spatial profiles that describe geometrical and topological aspects. This package provides an available linkage to 3D models of building structures.

Although LADM and CityGML have different foci on spatial features, there is no obvious geometrical barrier between them because both LADM and CityGML are compatible with ISO19107. LADM provides a formal language to describe land administration in terms of its parties, administrative and spatial units, and sources and representations, while CityGML is a data encoding method that was created to exchange data. The representation of legal spaces from LADM can be mapped to and encoded as a CityGML ADE (application domain extension mechanism) (OGC 2012; Çağdaş 2013). That is, CityGML with LADM offers an effective way to develop a feasible 3D cadastral system which is able to model either homogeneous spatial rights of 3D property with integrity, or heterogeneous spatial rights with internal structure of ownership.

33.5.3 An Example of Spatial Representation of the Internal Structure of Ownership

A case study of a condominium in China (Li et al. 2016) is borrowed here as an example of the spatial modeling of the internal structure of ownership by CityGML with LADM. Modeling the ownership structure of a condominium unit is shown in Fig. 33.6. LADM packages (red color) are introduced and two separate hierarchies, a legal hierarchy (yellow color) and a physical hierarchy (light blue color), are modeled with CityGML independently, and an *n*:*n* relationship between these is established in the model. As a building unit might have a different legal spatial extent from its physical counterparts, an attribute "the numerical ratio" is designated as the ratio of the legal spatial extent to its physical spatial extent, such as 0.5, 1, or 0 for different types of building parts. Therefore, the legal spatial extent and relevant semantic information are attached to and combined with a corresponding physical object by extending the attributes and semantics in CityGML, which is implemented through the usage of the ADE mechanism. The legal object is described by its physical

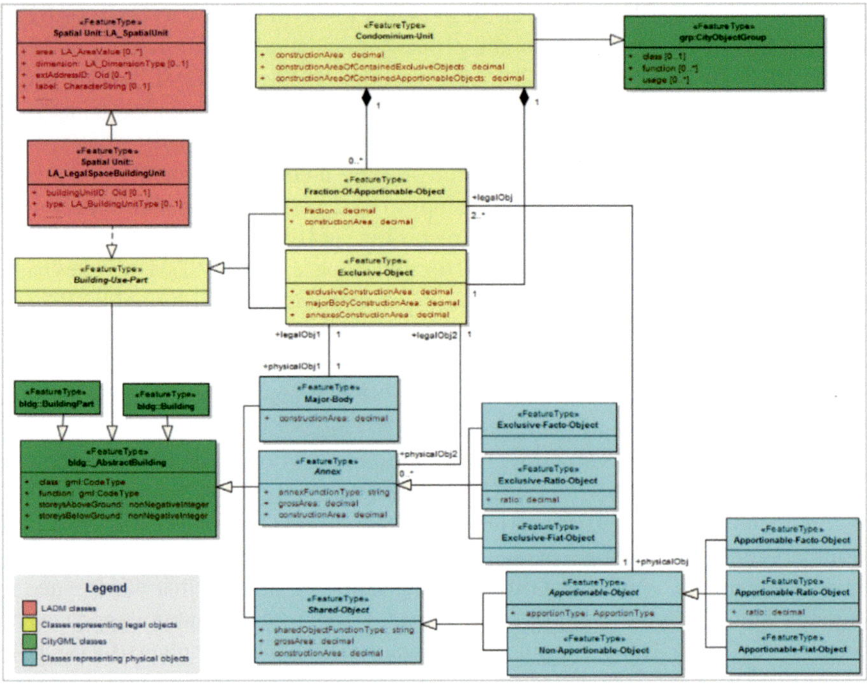

Fig. 33.6 UML diagram for modeling the ownership structure of a condominium unit (Li et al. 2016)

counterpart via semantic relations between them, which is also implemented by the use of the ADE mechanism.

A residential condominium with 28 stories is taken as an example of modeling. The internal structures of each story are similar to each other, so only the second story is viewed here. Three exclusive objects and seven shared objects are on this story. Each exclusive object is composed of one major body and some annexes, including de facto annexes, ratio annexes, and fiat annexes (Fig. 33.7). Apportionable de facto objects are also included, such as shared objects within a building (such as staircases) and shared objects in this story (such as corridors), apportionable ratio objects (such as a lanai), and apportionable fiat objects (such as a commonly used flowerbed).

Figure 33.8 shows the 3D representation of the interior structure of this second story. The semantic relations of the condominium units with their exclusive components and their physical counterparts in the second story, including the major bodies and annexes, are presented, for instance, in Fig. 33.9, which shows the semantic relations of Condominium Unit 1.

This example shows that although the ownership of a condominium unit is inherently complex, the internal structure can be subdivided into several sections in terms of homogeneity of rights, and the ownership structures can be modeled precisely by extending CityGML with the LADM. The spatial model here is mainly based on legal concepts specified by legislation in China. However, the modeling approach

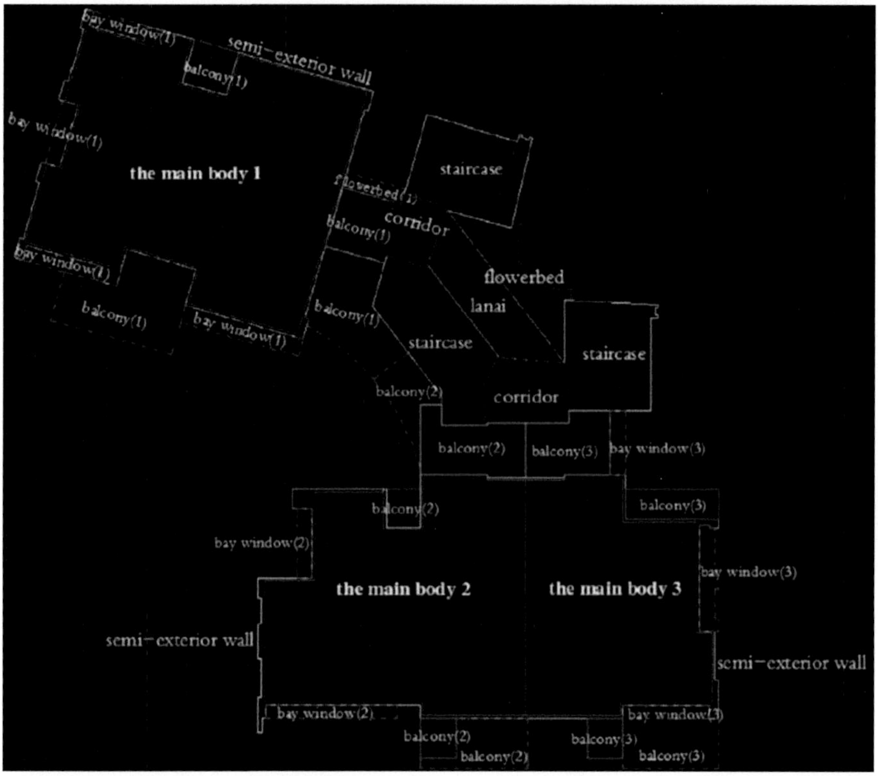

Fig. 33.7 Layout plan of the second story of the residential condominium building (Li et al. 2016). Red solid line: the major body; blue solid line: exclusive de facto object; green solid line: exclusive ratio object; blue dotted line: exclusive fiat object; yellow solid line: apportionable de facto object that is shared in the building; magenta solid line: apportionable de facto object that is shared in the story; cyan solid line: apportionable ratio object; magenta dotted line: apportionable fiat object; and number in brackets after the names of the annexes: the number of the major body to which the annexes are attached

may provide an available paradigm to model the ownership structure of a condominium unit, which could be adapted to other jurisdictions, especially in countries where similar legal concepts exist.

33.6 Summary

A transition in the administration of land or immovable property from land parcel (2D) to land space (3D) is a trend in urban areas, especially in populated cities, owing to both an increasing intensity of socioeconomic activities and a need to update to 3D technology. Although some rights to property may be completely or

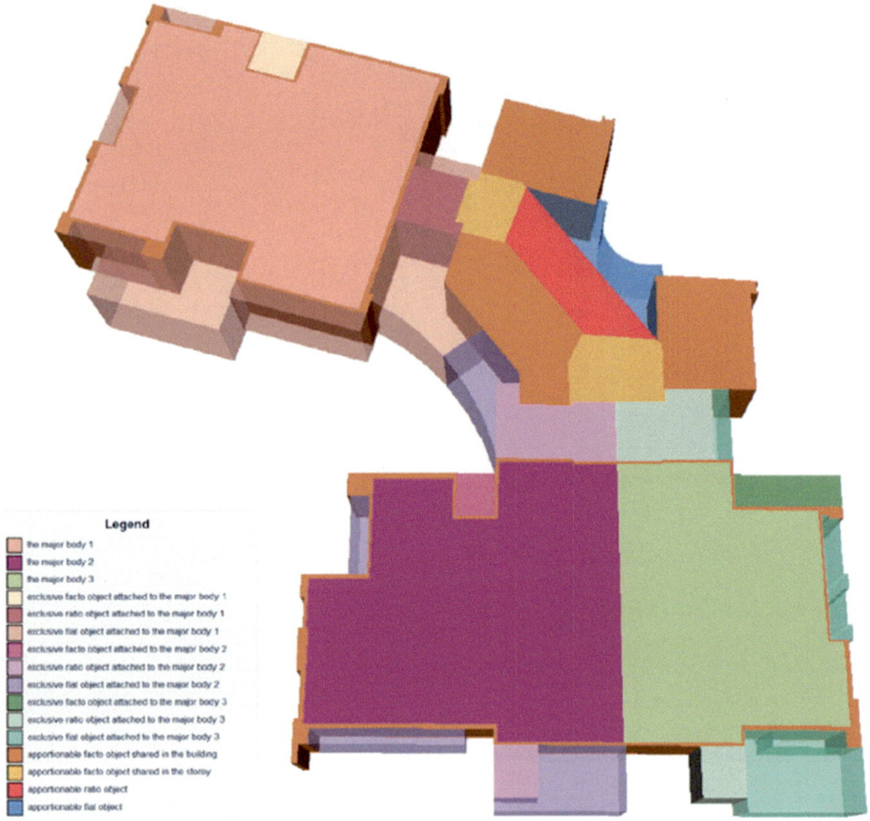

Fig. 33.8 3D representation of the interior structure of the second story (Li et al. 2016)

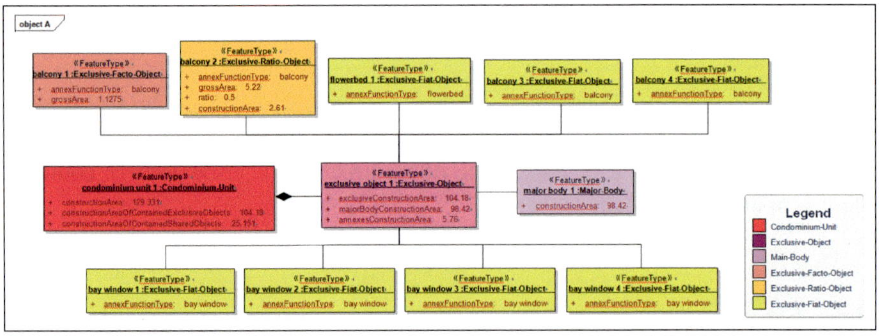

Fig. 33.9 Semantic relations between Condominium Unit 1 and its exclusive components (Li et al. 2016)

partially unclear with respect to space, the nature of the rights characterized by spatial features is crucial in managing and clarifying them. The use of the vertical space above and below ground, rather than horizontally defined surface parcels, is the key concept pushing property rights from a 2D to a 3D framework. Ownership, as the most important right to property, can be documented not only in text and in parcel-based 2D maps but also registered in terms of spatial extent, because it is determined and identified in the physical world. Spatial modeling of ownership can succeed in representing the spatial extent that is defined by the property's physical space.

For land management, a polyhedral container can be used for clarifying spatial rights to the use of land space. A PLC-based compatible 3D data model is an effective means to represent both 2D and 3D property, which is especially useful in the ongoing development of 3D cadastral systems, since 2D cadastres are the prevailing paradigm for the management of property. For housing property, the ownership may have a complex structure, so an individual polyhedral container may fail to capture the spatial extent of the ownership because of the heterogeneous rights to parts of property caused by sharing space. Therefore, explicitly demarcating the spatial extent of each part, clarifying the structure of ownership, and linking them with the legal spatial extent are the critical tasks for the precise management of properties.

It should be also noted that spatial modeling of property depends largely on its legal and institutional system. Here, cases in China are taken as an example, and the above-presented modeling details and data model are specific to the Chinese context. Nevertheless, it provides an available exemplar for applications in other legal systems, and its modeling paradigm may be very helpful for developing property management systems for various kinds of 3D property.

Acknowledgements This study is funded by the National Natural Science Foundation of China (No. 41871298)

References

Aien A, Kalantari M, Rajabifard A, Williamson I, Wallace J (2013) Towards integration of 3D legal and physical objects in cadastral data models. Land Use Policy 35:140–154

Amirebrahimi S (2012) BIM for facilitation of land administration systems in Australia. In: Rajabifard A, Williamson I, Kalantari M (eds) A national infrastructure for managing land information. Center for Spatial Data Infrastructures and Land Information, University of Melbourne, Melbourne, p 95

Arens C, Stoter J, van Oosterom P (2005) Modelling 3D spatial objects in a geo-DBMS using a 3D primitive. Comput Geosci 31(2):165–177

Çağdaş V (2013) An application domain extension to CityGML for immovable property taxation: a Turkish case study. Int J Appl Earth Obs Geoinf 21:545–555

Carlson E (1987) Three-dimensional conceptual modeling of subsurface structures. In: Proceedings of Auto-Carto 8. ACSM/ASPRS, Falls Church, Virginia, pp 336–345

Cohen-Steiner D, De Verdiere C, Yvinec M (2004) Conforming Delaunay triangulations in 3D. Comput Geom 28(2–3):217–233

El-Mekawy M, Paasch J, Paulsson J (2014) Integration of 3D cadastre, 3D property formation and BIM in Sweden. In: 4th international workshop on 3D Cadastres, Dubai, 9–11 Nov. International Federation of Surveyors (FIG)

Foley JA, DeFries R, Asner GP, Barford C, Bonan G, Carpenter SR (2005) Global consequences of land use. Science 309(5734):570–574

Góźdź K, Pachelski W, Van Oosterom P, Coors V (2014) The possibilities of using CityGML for 3D representation of buildings in the cadastre. In: Proceedings 4th international workshop on 3D cadastres, Dubai, 9–11 Nov 2014. International Federation of Surveyors (FIG)

Guo R, Ying S (2010) 3D cadastre analysis and data delivery. J China Land Sci 24(12):45–51 (in Chinese)

Guo R, Li L, Ying S, Luo P, He B, Jiang R (2013) Developing a 3D cadastre for the administration of urban land use: A case study of Shenzhen, China. Comput Environ Urban Syst 40:46–55

Jazayeri I, Rajabifard A, Kalantari M (2014) A geometric and semantic evaluation of 3D data sourcing methods for land and property information. Land Use Policy 36:219–230

Kalantari M, Rajabifard A, Wallace J, Williamson I (2008) Spatially referenced legal property objects. Land Use Policy 25(2):173–181

Karabin M (2014) A concept of a model approach to the 3D cadastre in Poland: technical and legal aspects. In: Proceedings 4th international workshop on 3D Cadastres, Dubai, 9–11 Nov 2014. International Federa-tion of Surveyors (FIG)

Li L, Wu J, Zhu H, Duan X, Luo F (2016) 3D modeling of the ownership structure of condominium units. Comput Environ Urban Syst 59:50–63

Li L, Lei Y, Tang L, Yan F, Luo F, Zhu H (2019) A 3D spatial data model of the solar rights associated with individual residential properties. Comput Environ Urban Syst 74:88–99

Miller GL, Talmor D, Teng SH, Walkington N, Wang H (1996) Control volume meshes using sphere packing: generation, refinement and coarsening. In: Proceedings of 5th international meshing roundtable, Pittsburgh, PA, pp 47–61

Molenaar M (1990) A formal data structure for 3D vector maps. In: Proceedings of EGIS'90, Amsterdam, pp 770–781

Open Geospatial Consortium (2012) Geography markup language (CityGML) encoding standard. OGC

Paulsson J, Paasch JM (2011) 3D property research—a survey of the occurrence of legal topics in publications. In: The 2nd international workshop on 3D Cadastres, Delft

Penninga F, Oosterom P, Kazar BM (2006) A tetrahedronized irregular network based DBMS approach for 3D topographic data modeling. In: Reidl A, Kainz W, Elmes GA (eds) Progress in spatial data handling. Springer, New York, pp 581–598

People's Republic of China (2000) Specifications for estate surveying. GB/17986–2000. TMOCATSBOSA Mapping

People's Republic of China (2007) The Real Right Law of People's Republic of China

Pouliot J, Vasseur M, Boubehrezh A (2011) Spatial representation of condomini-um/co-ownership: comparison of Quebec and French cadastral system based on LADM specifications. In: Proceedings of the 2nd international workshop on 3D ca-dasters, organized by FIGURE EuroSDR and TU Delft, Delft: The Netherlands

Pouliot J, Vasseur M, Boubehrezh A (2013) How the ISO 19152 land administration domain model performs in the comparison of cadastral systems: a case study of condominium/co-ownership in Quebec (Canada) and Alsace Moselle (France). Comput Environ Urban Syst 40:68–78

Rajabifard A, Williamson I, Marwick B, Kalantari M, Ho S, Shojaei D (2013) 3D-Cadastre, a multifaceted challenge. In: Proceedings of the XXV FIG International Congress 2014, Kuala Lumpur. International Federation of Surveyors (FIG)

Si H, Gartner K (2005) Meshing piecewise linear complexes by constrained Delaunay tetrahedralizations. In: Proceedings of the 14th international meshing roundtable. Springer, New York, pp 147–163

Sorensen EM (2011) 3 Dimensional property rights in Denmark, 3D property design is working—visualization not. In: 2nd international workshop on 3D Cadastres, Delft, pp 521–530

Stoter JE (2004) 3D cadastre. Ph.D. thesis, Delft University of Technology
Stoter JE, Ploeger H, van Oosterom P (2013) 3D cadastre in the Netherlands: developments and international applicability. Comput Environ Urban Syst 40:56–67
Stubkjær E (2004) Cadastral modeling: grasping the objectives. In: Proceedings of Joint FIG Commission 7 and COST Action G9 Workshop on Standardization in the Cadastral Domain, Bamberg, Germany
Thompson RJ (2007) Towards a rigorous logic for spatial data representation. Ph.D.'s Thesis, Delft University of Technology, The Netherlands
Turner BL, Lambin EF, Reenberg A (2007) The emergence of land change science for global environmental change and sustainability. Proc Natl Acad Sci 104(52):20666–20671
van der Molen P (2003) Institutional aspects of 3D cadastres. Comput Environ Urban Syst 27(4):383–394
Wenninger MJ (1974) Polyhedron models. Cambridge University Press, Cambridge
Ying S, Guo R, Li L, Van Oosterom P, Stoter J (2014) Construction of 3D volumetric objects for a 3D cadastral system. Trans GIS 19(5):758–779
Zlatanova S (2000) 3D GIS for urban development. Ph.D. thesis, Graz University of Technology
Zulkifli NA, Rahman AA, Van Oosterom P (2015). An overview of 3D topology for LADM-based objects. In: ISPRS Joint International Geoinformation Conference, Kuala Lumpur, Malaysia

Lin Li is a Professor and Luojia Outstanding scholar of Wuhan University, with a Ph.D. degree in Photogrammetry and Remote Sensing. He has been working at the School of Resource and Environmental Sciences, Wuhan University, and is currently interested in spatial modeling for 3D cadastre, indoor modeling from point cloud data, and the integration of semantic location.

Renzhong Guo is a Professor and Director of the Institute for Smart Cities at Shenzhen University. He is a member of the Chinese Academy of Engineering, and vice president of both the Chinese Society of Urban Studies and the China Land Science Society, and currently interested in the smart city.

Dr. Shen Ying is a Professor at the School of Resource and Environmental Sciences, Wuhan University, and a member of the "China National Special Support Program for High-Level Personnel." His interests include 3DGIS, cartography, and spatial analysis.

Haihong Zhu is a Professor in the School of Resource and Environmental Science, Wuhan University. She received her Ph.D. degree from Wuhan University. Her current research interests focus on 3D modeling and visualization, geographical ontology, map design, and navigation digital mapping.

Jindi Wu received a master's degree in Cartography and Geographical Information Systems from Wuhan University in 2016, focusing on 3D modeling of the ownership structure of condominium units, and is currently working at Tencent Tongtu Data Technology Company Limited, engaged in digital map production management.

Chengcheng Liu is a Ph.D. candidate in the School of Resource and Environmental Sciences, Wuhan University, majoring in Cartography and Geographic Information Engineering, and is currently interested in spatial modeling for 3D cadastre and Cartography.

Chapter 34
Semantic 3D City Modeling and BIM

Thomas H. Kolbe and Andreas Donaubauer

Abstract Semantic 3D city modeling and building information modeling (BIM) are methods for modeling, creating, and analyzing three-dimensional representations of physical objects of the environment. Digital modeling of the built environment has been approached from at least four different domains: computer graphics and gaming, planning and construction, urban simulation, and geomatics. This chapter introduces the similarities and differences of 3D models from these disciplines with regard to aspects like scale, level of detail, representation of spatial and semantic characteristics, and appearance. Exemplified by the international standards CityGML and Industry Foundation Classes (IFC), information models from semantic 3D city modeling and BIM and their corresponding modeling approaches are explored, and the relationships between them are discussed. Based on use cases from infrastructure planning, approaches for integrating information from semantic 3D city modeling and BIM, such as semantic transformation between CityGML and IFC, are described. Furthermore, the role of semantic 3D city modeling and BIM for recent developments in urban informatics, such as smart cities and digital twins, is investigated and illustrated by real-world examples.

34.1 Digital Models of the Built Environment

Many applications in the context of urban informatics require detailed information about the physical urban environment. For example, for the planning, design, and construction of buildings, detailed information about the location, the components, their materials and costs, and the construction schedule is required. For all kinds of urban simulations like noise propagation, air quality and pollution assessment, energy demand, and production estimation, but also for driving simulations and autonomous driving, comprehensive data are required on the urban topography.

Digital models of the built environment are computer representations of the objects, their characteristics, and their interrelationships within a specific urban

T. H. Kolbe (✉) · A. Donaubauer
Technical University of Munich, Munich, Germany
e-mail: thomas.kolbe@tum.de

W. Shi et al. (eds.), *Urban Informatics*, The Urban Book Series,
https://doi.org/10.1007/978-981-15-8983-6_34

terrain. This includes both the natural and man-made features like the digital terrain model (DTM), digital surface model (DSM), vegetation, water bodies, as well as man-made constructions like buildings, bridges, tunnels, and infrastructure. Key properties of the digital representations are spatial, temporal, graphical, and thematic information about the entities in and around cities, providing information on the location, shape, extent, visual appearance, classification, thematic attributes, functional aspects, and their interrelationships.

Different applications and use cases have different requirements regarding the resolution and level of detail of the objects of an urban model and their modeled aspects. For example, for the visual inspection of the urban topography by a human operator, it will be sufficient to represent the geometry and graphical appearance of the urban terrain. If thematic or spatio-thematic queries and analyses are to be carried out, like "list all windows of all buildings which have a line-of-sight to a specific place or route" or "find all buildings having a heating energy demand higher than 100 kWh/m^2/year", then thematic information has also to be represented, because the computer has to know which objects are buildings, their energy demand, which parts of them are windows, and what are their locations and orientations. For simulation applications like blast analysis or the propagation of radio waves, information about the materials of the different objects will also be required.

Urban models that only represent the 3D geometry and appearance information (visual models) will be referred to as virtual reality (VR) models in the following. Typical real-world examples of VR models are the 3D models of major cities in Google Earth or Apple Maps. They are just geometrical representations of the urban surface (3D meshes with graphical textures). A human viewer can easily recognize the different features, but for the computer, these data are not structured into separate meaningful objects. Models of real-world entities that also include the meaning of the objects, their thematic properties, and their logical relationships are generally referred to as semantic models or information models. Thus, urban models containing both the spatial and thematic aspects are called urban information models (UIM).

Now, urban modeling can be carried out in various ways and using different formal modeling techniques and data representations. This diversity results from the fact that 3D urban modeling has been approached from at least four different disciplines: computer graphics and gaming; geomatics (including the disciplines of geoinformatics, geodesy, photogrammetry, and remote sensing); planning and construction (including the disciplines of civil engineering and architecture, urban and landscape planning); and urban and environmental simulation. This is illustrated in Fig. 34.1.

It is important to understand that each discipline has its own scope and thus puts a different focus on the things that are modeled and on the way they are modeled. This has resulted in the development and usage of distinct modeling paradigms, conceptual data models, and data exchange formats, which frequently causes problems in discussions about urban models between people coming from different disciplines. On the other hand, system interoperability issues arise and have to be addressed when data from one discipline are to be brought into another discipline or if data from the different disciplines are to be used in an integrated way.

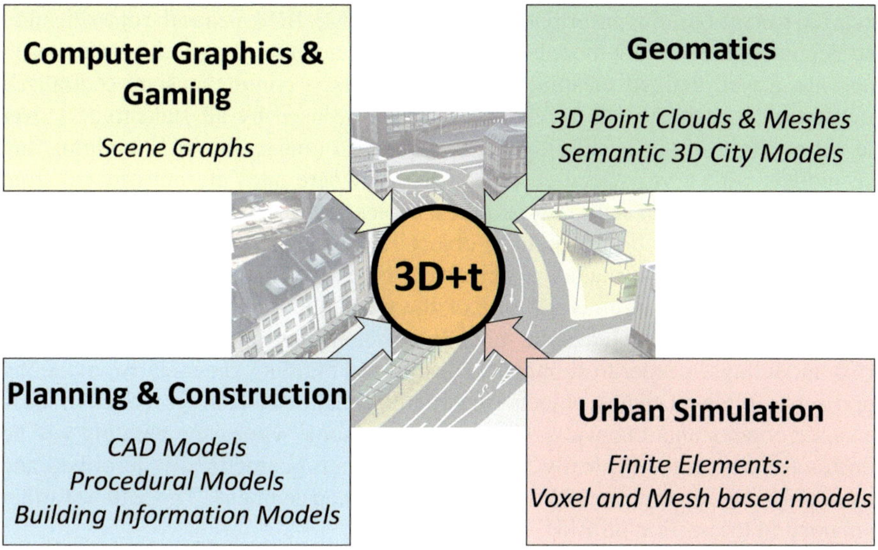

Fig. 34.1 Different disciplines and their approaches to the definition, generation, and usage of urban 3D/4D models

Data models and methods developed in the field of computer graphics (CG) and gaming aim at the efficient and high-quality 3D visualization of the cityscape and the elements in it. Thus, VR models are in the main focus of CG, containing information on geometry and (graphical) appearance. 3D objects are typically structured in so-called scene graphs, which allow for the definition and multiple instantiation of prototypical shapes and realize a hierarchical aggregation. Scene graphs may also contain light sources, virtual cameras, and information about the environment like fog density, and may provide the means for object animation, describing the dynamic behavior of objects, and user interaction (see, e.g., Foley et al. 1995). In CG, objects are typically modeled in a way that best supports rendering and visualization, which may suggest the aggregation of objects which might not be considered as a unit from a semantic point of view. The representation of semantic information is not a focus of CG and is often neglected.

Models and methods from the field of training simulation and computer games are quite similar to CG with respect to the representation of 3D objects. In addition, these models support the description of object physics (like weight, elasticity, mechanical connections, etc.), kinematic modeling, and complex object behaviors, in order to describe the functions and interactions to be considered by the simulator. Like in CG, object semantics are often not considered, apart from simulator control data.

The planning and construction domain focuses on the representation of man-made objects in fine detail in order to support the design and construction processes. While in the past computer-aided architectural design (CAAD) was mainly used to represent the geometry of the objects, in the past decade a strong transition has occurred toward building information modeling (BIM). BIM means the classification and decomposition of 3D models according to a semantic data model, where each class has a well-defined meaning. By these means, a comprehensive, centralized information repository will be created that can be used by all stakeholders over the entire life cycle of a building. BIM is focused (and tailored) to building and site models with a very detailed object model, where sites are constructed from components like walls, slabs, stairs, pipes, cables, power plugs, etc. BIM does not address the representation of natural objects like vegetation or water bodies and only recently started to include other object types like bridges, roads, or terrain. Nevertheless, since buildings are one of the most important entities in the urban terrain, and BIM also includes the modeling of their interiors, it is quite relevant to urban modeling. In order to support the design of a building, a generative modeling approach is followed, that is, objects are virtually constructed from a set of volumetric semantic components like walls, slabs, etc. in the same way as the building will be constructed in reality. Typically, the components are geometrically described and combined using constructive solid geometry and sweep geometry. This will be further explained in the section on BIM.

In geomatics, emphasis is given to the representation of the urban topography including natural objects, man-made objects, and the Earth's relief. While in the past 2D maps and 2D digital landscape models (DLM) have been used at different scales to visualize and represent the topographic structure of a region with respect to planimetric (horizontal/flat) shapes and extents, virtual 3D city and landscape models nowadays capture and visualize the 3D geometry, 3D topology, and appearance of the urban entities in different levels of detail (LoD). If the objects are structured according to a semantic model and have thematic attributes and logical interrelationships, these models are referred to as semantic 3D city models. They can be seen as a realization of the concept of urban information modeling. The modeling paradigm in geomatics is oriented toward the representation and mapping of observable features and thus is very close to the results that are obtained from data acquisition methods from photogrammetry, remote sensing, and surveying (see, e.g., Kolbe et al. 2009). Semantic 3D city models are explained in more detail in the next section.

More details about the similarities and differences of models from the planning and construction as well as geomatics domains are given in the fourth section of this chapter and by Kolbe and Plümer (2004) and Nagel et al. (2009).

The models used in the field of urban simulation often are based on regular or irregular decompositions of the urban space into finite elements. Both the air space and the space occupied by physical objects are represented by voxels, meshes of 3D tetrahedra, or 3D volumes bound by triangle meshes. Since all urban features use the same representation, they can be treated by the simulation tools in a similar way. The cells or elements of such a representation are parameterized by properties that are relevant for the respective simulation. For example, in pollution dispersion simulation, all voxels representing the urban air space have a parameter vector for wind direction, wind speed, air temperature, and concentrations of specific pollutants. Other kinds of simulations require the explicit spatio-semantic representation of urban objects. For example, in traffic simulations, the roads have to be represented together with traffic-related information such as speed limits, traffic lights, turning restrictions, and parking lots. For simulation of building heat-energy demand, 3D building models are required with information about usage type (e.g., residential, office, manufacturing) and about building physics like the wall, roof, and window insulation.

While digital models of the urban environment were often static in the past, that is, they just represented a snapshot of a specific timepoint, nowadays the time dimension plays an increasing role due to new application fields like smart cities and digital twins. In these application fields, sensors and their highly dynamic observations are related to the objects of the digital urban models. In the field of computer gaming, including training simulations, as well as in the field of urban simulations, the representation of dynamic behavior and changes over time has been addressed for long time. However, in the approaches of geomatics as well as of planning and construction to digital urban modeling, the time dimension has not yet been considered to a full extent (see, e.g., Chaturvedi and Kolbe 2019b).

In the remainder of this chapter, we will concentrate on the spatio-semantic modeling of the urban environment, namely semantic 3D city modeling and building information modeling.

34.2 Semantic 3D City Modeling

Semantic 3D city models are virtual models of the urban environment, that is, datasets representing the entities of the physical reality like buildings, streets, trees, bridges, and the terrain. In contrast to virtual reality (VR) models, they are structured (e.g., subdivided and attributed) according to thematic and logical criteria and not according to graphical or rendering considerations. The objects of a semantic 3D city model represent the respective real-world things with their thematic, geometrical, topological, and appearance properties. Furthermore, logical and spatial interrelationships between different objects are expressed. Objects belong to a set of predefined classes like Building, Road, CityFurniture, or WaterBody with spatial and thematic attributes whose semantics—that is, the meaning of the model components and properties—are explicitly defined in a specification. Complex objects are typically further decomposed into meaningful parts, for example, a building can be decomposed into building parts and these again are structured into roof, wall, and ground surfaces. Wall surfaces can further contain windows and doors. Objects can have thematic attributes on all aggregation levels. Their spatial properties are represented using geometric and topologic objects.

34.2.1 Purpose and Key Applications

3D city models are mostly used topographically, to describe the physical environment as it is with respect to the spatial, thematic, and appearance characteristics of the urban entities. They are used to create 3D maps for applications ranging from topographic mapping, cadastres, disaster management, visual exploration, navigation and autonomous driving, and urban simulations. Semantic 3D city models comprise all objects within larger geographical areas, typically starting from city blocks up to entire countries. They can be seen as the 3D successor of traditional 2D digital landscape models as created and maintained by mapping agencies. In fact, most semantic 3D city models today are being created and maintained by mapping departments on municipal, state, or country level. However, 3D city models are also produced by commercial companies as well as by initiatives like the Open Street Map project.

A semantic 3D city model could be seen (and is used) as an inventory of the relevant urban objects. As such, it is useful for applications related to property and asset management, as well as for life cycle management of the man-made and natural urban features. When it comes to urban data integration, semantic 3D city models play a key role, because data from different domains like urban planning, mobility, energy, and ecology are most often related to specific spatial urban objects. Since these objects are represented in a 3D city model, the domain-specific data can be linked with the respective city model objects. Alternatively, the urban objects could be enriched with the domain-specific data. The objects of a 3D city model then play the role of a common denominator, because data from different domains can be linked and interrelated via the urban objects. This is further illustrated below.

In their overview paper, Biljecki et al. (2015) enumerate and describe more than 100 applications of 3D city models. The authors distinguish mainly between use cases that are based on visualization and those where 3D models are being used for computations, queries, and more sophisticated analyses including simulations. While semantic 3D city models can also be used for visualization-based use cases, they are especially relevant for the second category and for many use cases are even required. Willenborg et al. (2018) explain in more detail how semantic 3D city models are being employed in three very different use cases: (1) solar irradiation analysis, (2) detonation simulation, and (3) building energy demand estimation.

34.2.2 Modeling Paradigm

Semantic 3D city models are typically being used to represent the existing physical objects of the urban environment. Hence, a descriptive modeling paradigm is being followed, which best supports the modeling of urban entities by observation methods from surveying, photogrammetry, remote sensing, and laser scanning. Direct results from these methods are typically 2D images and videos from different viewpoints (nadir and oblique views from airborne and space sensing, terrestrial views from mobile mapping) and 3D point clouds as resulting from laser scanning or stereo-photogrammetric dense image matching. 3D point clouds then can be triangulated, producing 3D meshes that describe the observed surface structures. In order to represent the 3D geometric extent and shape of separated objects, boundary representations (B-Reps) are being used, where volumetric geometries are specified by the accumulation of their bounding surfaces (see, e.g., Foley et al. 1995). In contrast to most other disciplines, geometries in the geomatics domain are always georeferenced with respect to a regional or global coordinate reference system (CRS). The exclusive usage of absolute coordinate values allows GIS and spatial databases to create and maintain spatial index structures, which facilitate efficient processing of spatial queries and analyses on very large datasets. This is not supported in comparable efficiency and completeness by the modeling paradigms which are followed in other disciplines.

Based on the reconstruction of 3D geometry, the semantic objects are then generated. Since only observable parts can be registered from surveying and remote sensing, the object decompositions are typically aligned with the visible surface parts. For example, buildings are decomposed into wall, roof, and ground surfaces as only the surfaces can be reliably detected, whereas in general the entire volumetric wall objects or other constructive elements like beams or slabs are not detectable. As a rule, each (relevant) real-world thing is represented by one classified object. Each object can have multiple representations, such as geometries of different types in multiple levels of detail, as well as multiple visual appearances. It is recommended that all objects should have globally unique identifiers and that these identifiers should also be kept stable over the lifetime of the real-world object. The reason is that this allows keeping track of the object in different applications and for linking information from different sources to it in a sustainable way.

Of course, 3D city models can also be used to represent future development states of cities, but the employed accumulative modeling principle (B-Rep geometries with absolute world coordinates) is not especially supportive regarding manual, interactive changes of object locations, extents, and shapes. This is in contrast to generative and parametric modeling principles that are typically used in building information modeling.

34.2.3 The International Standard CityGML

The City Geography Markup Language (CityGML), issued by the Open Geospatial Consortium (OGC), is the international standard for the representation and exchange of semantic 3D city and landscape models. CityGML defines a common information model and data exchange format for 3D urban and rural objects. It specifies the classes and relations for the most relevant topographic objects in cities and regional models with respect to their geometrical, topological, semantic, and appearance properties. Included are generalization hierarchies between thematic classes and aggregation and thematic relations between objects. CityGML is implemented as an application schema of the Geography Markup Language 3.1.1 (GML3; see Cox et al. 2004), the extensible international standard for geodata exchange and encoding issued by the OGC and the ISO TC211. It is further based on a number of standards from the ISO 191xx family, the OGC, the W3C Consortium, the Web 3D Consortium, and OASIS (Kolbe 2009; Gröger and Plümer 2012).

The data model consists of class definitions for the most important objects within virtual 3D city and landscape models. CityGML consists of a core module and several extension modules. Whereas the core module comprises the basic concepts and components of a virtual city, each extension module covers a specific thematic field like buildings, bridges, tunnels, digital terrain model, water bodies, vegetation, transportation, city furniture objects, etc. Implementations are not required to support the entire data model but may employ only a subset of modules according to their specific needs. Figure 34.2 shows an excerpt from the top-level class hierarchy of CityGML.

CityGML defines five consecutive levels of detail (LoD), where objects become more detailed with increasing LoD regarding both their spatial and thematic differentiation. Each object may have attached a separate representation for each LoD simultaneously. The five LoDs as defined by CityGML are illustrated in Fig. 34.3.

CityGML comprises class definitions for the representation of complex digital terrain models (DTMs) in various forms from point clouds over raster data or TINs, including break lines. All these DTM data types can be used to build composite or hybrid terrain representations. The LoD concept even allows for the maintenance of

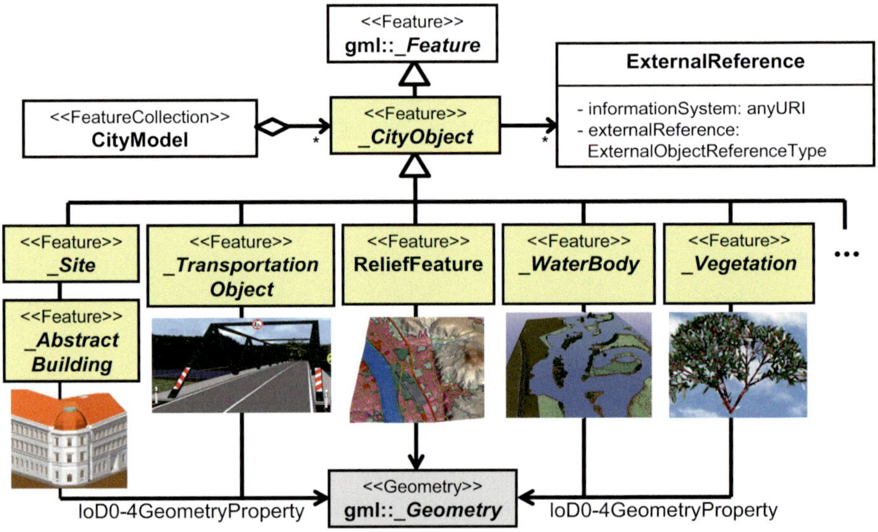

Fig. 34.2 UML diagram of the top-level class hierarchy of CityGML. All thematic objects are considered geographic features (according to ISO 19109), and their classes are derived from the abstract superclass CityObject. Attributes and subclasses are omitted here for the sake of readability

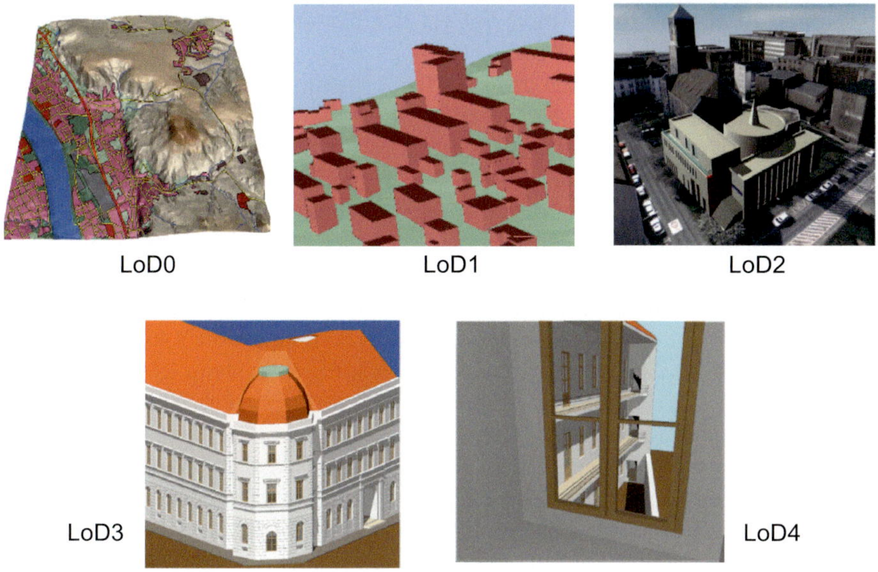

Fig. 34.3 Illustration of the five levels of detail defined by CityGML

several terrain variants in different resolutions. DTMs can be restricted by validity-extent polygons. Holes within these polygons allow for embedding of other DTM components, for example, a fine-resolution TIN embedded into a gridded DTM of a large area.

In CityGML, the coherent modeling of semantic and geometric/topological properties is supported. At the semantic level, real-world entities are represented by features such as buildings, walls, windows, or rooms. The description also includes attributes, relations, and aggregation hierarchies between them. At the geometric level, geometry is assigned to thematic features representing their spatial location and extent. Complex geometry objects are decomposed into geometric primitives. Thus, the model can consist of two aggregation hierarchies in which the corresponding objects are linked by relationships, but also simpler representations are supported (see, e.g., Stadler and Kolbe 2007).

Spatial properties of CityGML features are modeled according to the GML3 geometry model (see ISO 19107:2003; Cox et al. 2004) representing 3D geometry according to the boundary representation (B-Rep, see Foley et al. (1995), typically using a 3D coordinate reference system (CRS) with absolute world coordinates. Spatial database management systems, like Oracle Spatial and PostGIS, as well as many (3D) GIS, provide native support for GML3's geometry model enabling lossless storage, efficient management, and spatial indexing of CityGML data. Besides geographic and projected coordinates, also compound 3D CRS, that is, different CRS for planimetry and height, are supported.

In order to provide for a simple but yet flexible way of topological modeling, CityGML does not make use of GML's topology classes. Instead, topological neighborhood relations are expressed using GML's capability to establish XLinks from composite geometries to the shared geometry (parts). For example, a surface that is bounding both a house and a garage can be referenced by the two respective solid geometries assigned to each object. If a geometry object should be shared by different composite geometries or different thematic features, it only has to be assigned a unique identifier, which is then referenced by the corresponding GML geometry aggregate objects (see Gröger and Plümer 2012, for examples).

In addition to semantics and spatial properties, CityGML features can be assigned appearance information, that is, observable properties of a feature's surface. In most cases, these surface data are recorded by sensors, for example, a RGB or infrared camera. CityGML appearances are represented by textures, georeferenced textures, and material representations (the latter adopted from the CG standards X3D and COLLADA) of object surfaces, but are not limited to visual data. In contrast, appearance relates to any surface-based theme, such as infrared radiation, noise immission, radio-frequency absorption, and earthquake- or blast-induced structural stress. Consequently, appearance information can serve as input for both visualization and analysis tasks. CityGML supports feature appearances for each LOD and an arbitrary number of themes.

3D objects are often derived from or have relations to objects in external databases or datasets. In order to express these links, each object in the city model may have external references to its corresponding objects in external data sources, given as Uniform Resource Identifiers (URIs). Furthermore, explicit information which facilitates the integration of different 3D datasets/object types can be represented. The concept of the Terrain Intersection Curve (TIC) is introduced to integrate 3D objects with the digital terrain model at their correct height in order to prevent, for example, buildings from floating over or sinking into the terrain.

To allow for the aggregation of arbitrary city objects according to user-defined criteria, CityGML employs a generic grouping concept. Groups may be further classified by additional attributes and may contain other groups as members, allowing for nested grouping of arbitrary depth.

Attributes for classifying objects, such as roof types, often are restricted to a set of discrete values. To facilitate interoperability, in CityGML, these sets are specified as external codelists and implemented as GML simple dictionaries. External codelists can be (re)defined by the user.

Further objects which are not explicitly covered by the specification document can be represented using the concept of generic objects and attributes. In addition, the CityGML data model may be extended for specific applications through so-called Application Domain Extensions (ADEs). All datasets containing ADE can still be interpreted by applications that rely on the basic CityGML data model. By these means, the data model of CityGML balances between strictness and generality. This is realized by the three main parts: (1) the core thematic model with well-defined LoDs, classes, spatial and thematic attributes, and relations; (2) GenericCityObjects and generic attributes allow the extension of CityGML data on the fly; and (3) ADEs facilitate the systematic extension of the CityGML data model by new classes, attributes, and relations for specific application domains. Many ADEs have already been developed by different communities; for example, the Energy ADE (Nouvel et al. 2015) to support energetic analyses of buildings or the Utility Network ADE (Kutzner et al. 2018) supporting the simultaneous representation and analysis of multiple supply and disposal networks. A comprehensive discussion of existing CityGML ADEs is provided by Biljecki et al. (2018).

34.3 Building Information Modeling

34.3.1 Purpose and Key Applications

In the context of digital urban models, the acronym BIM stands for either building information modeling or building information model, two terms that were coined by the architecture, engineering, and construction (AEC) industry. Following Eastman et al. (2011), BIM is used as a verb in this contribution. This is to express that building information modeling (BIM) describes a modeling activity rather than just a collection of static object. According to Borrmann et al. (2015a), BIM is based on the idea of continuous usage of the digital representation of a building from its design, planning, and construction to operation and deconstruction. A basic premise of BIM is collaboration by different stakeholders in the different phases of the life cycle of a facility (National Institute of Building Sciences 2012). Therefore, BIM goes hand in hand with the idea of an improved exchange of data between all stakeholders involved and an increase in efficiency over the whole life cycle of a building. In contrast to computer-aided architectural design (CAAD) which mainly focuses on representing the geometry and appearance of man-made objects, BIM is focused (and tailored) to building and site models with a very detailed information model representing sites, buildings, and their components like walls, slabs, stairs, pipes, cables, power plugs as semantic objects, and the relations between them. The information model also allows representation of aspects like time (e.g., for scheduling tasks in the building project) and costs often referred to as 4D or 5D BIM.

Eastman et al. (2011) group the key applications of building information modeling according to the stakeholders involved in the BIM process as follows:

- Owners: assess design options from cost, time, sustainability and facility operation perspectives (requires quantity takeoff and computation, energy simulation, 3D visualization already in an early design phase); cost and schedule control; commissioning and asset management based on the as-built/as-maintained model
- Architects and engineers: space planning and program compliance, energy analysis, design communication/review (3D visualization), quantity takeoff and cost estimation, design and analysis/simulation of building systems (structure, mechanical and air handling systems, emergency systems, lighting, acoustics, etc.), design coordination (clash detection)

- Contractors: construction planning and scheduling (4D simulation), cost and schedule control, procurement purchasing and tracking, and safety management (4D simulation)
- Subcontractors and fabricators: automated manufacturing, preassembly, and prefabrication.

Common to all applications listed above is that they usually consider a single construction project or facility, not a whole district, a city, or even a larger geographical area.

While BIM in its early days was mainly applied in building construction, it is increasingly getting adopted in infrastructure construction today. An overview of BIM for infrastructure applications like planning, building and maintaining roads and railways, utility networks, etc. was provided by Bradley et al. (2016).

34.3.2 Modeling Paradigm

Although BIM can be applied for managing existing buildings (see applications for owners above), the majority of BIM applications is focused around the design and construction phase of a building. BIM models are therefore used as templates to create originals according to the model. This means that BIM adheres to a prescriptive modeling paradigm, as in most cases, the model already exists before the original (Brüggemann and von Both 2015). In addition, BIM follows a generative modeling approach since the model reflects the construction process (Kolbe and Plümer 2004). This requires highly detailed models with representations of all the constructive elements as components. However, the geometric representation of the constructive elements may vary in granularity depending on the state of planning (draft planning, execution planning, etc.). In order to provide the user of a model with information on the geometric granularity, BIM defines so-called levels of development (LoD). To support the dynamic nature of the planning process, the generative modeling approach followed in BIM must also enable changes to models of planned objects to be carried out quickly and efficiently. Therefore, mostly parametric and generative geometry models such as constructive solid geometry (CSG) and sweep representations are applied. Use of parametric representations and local transformations is making the interactive design of BIM models intuitive, as the characteristics of components can be changed easily by adjusting their parameters. For example, the thickness of a wall component can simply be changed by adjusting the width parameter; the change of geometry follows implicitly. Also the placement of a window within a wall could easily be modified by just moving the window object to some other place in the wall, that is, by changing the relative translation of the window object with respect to the wall object. The space taken by the window object then becomes subtracted from the wall in order to generate the hole in the wall. The same is true for the design and construction of a road, where the centerline describes the road alignment and a cross-section together with some parameters provide information about the width of the lanes and shoulders. If the road needs to be moved by 10 m to the left, for example, just the centerline has to be adjusted accordingly; the rest follows implicitly.

34.3.3 The International Standard IFC

The Industry Foundation Classes (IFC) (International Organization for Standardization 2018) defines a software-vendor-neutral product model and data exchange format for BIM that has been developed by buildingSMART, an international organization from the AEC domain. IFC is widely adopted: According to Borrmann et al. (2015a), IFC is supported by all major software vendors in the AEC domain and serves for realizing Open BIM, that is, for implementing a software-vendor-neutral BIM process which relies on exchanging data between the stakeholders in a standardized format and information model. IFC has been made mandatory for government projects in several countries such as Singapore, Finland, and Great Britain. The US National BIM Standard (National Institute of Building Sciences 2012) is specified based on IFC, and also the German national BIM strategy regards "Open BIM" realized using IFC as an important component for implementing BIM processes in public construction projects.

IFC provides a very detailed and rich information model (see Fig. 34.4) for 3D building representations using constructive elements like beams (class ifcBeam), walls (class ifcWall), etc., and also non-physical spatial objects like stories (class

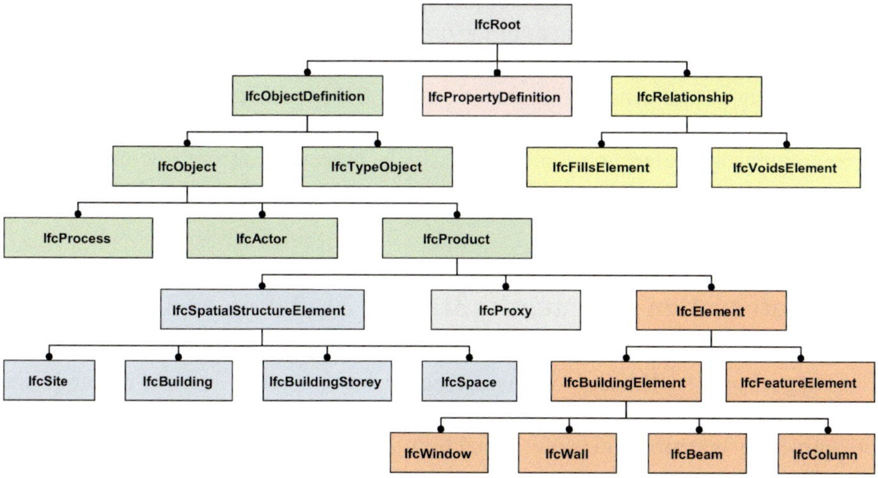

Fig. 34.4 Excerpt from the IFC information model showing the inheritance hierarchy of the most important top-level entities in EXPRESS-G notation. *Source* Borrmann et al. (2015a)

ifcBuildingStorey) and spaces (class ifcSpace). Diverse specializations are included for different crafts like steelworks, dry works, plumbing, electrical wirings, and air conditioning (HVAC). The information model includes material properties and costs, allowing, for example, cost calculations, planning of construction phases, and structural analyses to be carried out. Reflecting the scope and key applications of BIM, IFC not only allows buildings and their components to be modeled, but also processes that occur during a construction project and actors and non-physical objects that control other objects like legal directives and building regulations. Since IFC Version 4, the topic of BIM for infrastructure has been taken into account by defining objects for road and rail alignment. IFC data models for bridges and tunnels are in preparation.

The information model of IFC can be customized both by restriction and by extension. Model view definitions (MVD) can be created in order to restrict the data model to a specific purpose, for example, to define data exchange requirements for specific application domains. A range of predefined MVD documents can be found in the MVD database of buildingSMART International. They include an MVD for coordination between architectural, structural, and building services domains, for quantity takeoff, and an MVD for energy analyses. The standardized exchange format for MVD is mvdXML (Chipman et al. 2016). The concepts of property sets and quantity sets allow for a flexible extension of the semantic model by user-defined attributes. This may be done at runtime or can be defined using an MVD. The extension of IFC by new feature classes or the further refinement of existing feature classes by new subclasses is not supported.

IFC has a very comprehensive 2D and 3D geometry model. In line with the modeling paradigm suitable for BIM, IFC offers parametric geometry models like constructive solid geometry (CSG) and sweep, but also B-Rep geometries.

From Version 2.3, simple georeferencing has been included which allows one to specify the real-world coordinates of the origin of an entire site model in geographic coordinates (lat/long according to the WGS84 datum) plus ellipsoidal heights in meters. Along with the increasing importance of BIM for infrastructure and the need to handle objects with larger geographic extents, the current version of IFC 4 supports more complex georeferencing methods, which, however, are not yet sufficient for certain practical cases in large infrastructure projects (see Markič et al. 2018).

34.4 Integration of Semantic 3D City Modeling and BIM

The integration of BIM and GIS is currently the subject of intense research and development efforts in academia as well as in industry, and it has also found its way into university teaching and professional training courses (Hijazi et al. 2018; Noardo et al. 2019).

As a research area, BIM-GIS integration has developed over the past decade and is meanwhile described by several overview articles (e.g., Liu et al. 2017). The following classification of integration approaches builds upon Liu et al. (2017):

(a) Approaches transforming data between BIM and semantic 3D city modeling based on existing information models from the AEC and geospatial domains with IFC and CityGML being the most prominent information models of the respective domain (Stouffs et al. 2018).
(b) Approaches defining new information models (e.g., El Mekawy et al. 2012) or extensions of existing information models from the AEC and the geospatial domains (e.g., de Laat and van Berlo 2011). The aim of these approaches is to enable a data transformation between BIM and semantic 3D city modeling that is as lossless as possible. One of the most recent works in this field is described by Stouffs et al. (2018). Based on the use cases they identified with government agencies in Singapore, they extend the CityGML information model using the Application Domain Extension (ADE) mechanism in order to represent semantic information beyond what the CityGML information model provides. The transformation rules between IFC and CityGML are then defined using a triple graph grammar approach (Stouffs et al. 2018).
(c) Approaches integrating BIM and GIS at process level. According to Liu et al. (2017), this type of integration is characterized by the fact that BIM and GIS data reside in their original data formats and information models. Linking data from both information models can then be achieved, for example, by using semantic Web technologies or by encapsulating the data using Web services. However, it should be noted that although standardized Web-service interfaces exist in the geospatial domain (e.g., OGC WFS), comparable standardized interfaces currently do not exist for accessing BIM models. Researchers have also investigated querying BIM and GIS data residing in their original structures simultaneously. An example of such an approach is given by Daum et al. (2017). They define a spatio-semantic query language for the integrated analysis of 3D city models and building information models.
(d) Furthermore, application, vendor system, or project-specific approaches for BIM and GIS integration exist. These approaches do not necessarily rely on standardized information models on both sides. For example, GIS software vendors provide functionality to import the native format of a specific BIM authoring tool into their system. In case the geometry in the BIM data is parametric, it is transformed into explicit (mesh) geometry in the GIS software. Semantic transformations are not applied during import but could be applied by the GIS user.

The effort that researchers and software companies put into BIM-GIS integration indicates on the one hand the complexity of the topic, but on the other hand, it is also an indication of the need and benefit of such integration, as described in the following section.

34.4.1 Applications/Use Cases

Figure 34.5 names a selection of use cases for BIM-GIS integration related to the life cycle of a building or an infrastructure object. In the concept phase, an integration of a planned building with the virtual representation of its environment allows variant and feasibility studies and can facilitate stakeholder involvement and participatory planning by 3D visualization. In summary, it can be stated that BIM-GIS integration in the early design phase supports geodesign, according to Flaxman (2010) a "planning method which tightly couples the creation of design proposals with impact simulations informed by geographic contexts".

Simulations in the geographic context of a building can also be applied during the detailed design phase. This might include energetic simulations involving shadowing effects by adjacent buildings, vegetation, or topography. In infrastructure construction, simulations in the geographic context can also be helpful: When planning motorway junctions, for example, the glare effect is determined using virtual models of the surrounding topography. In the next section, we describe an overall approach to planning integration that enables many more applications based on a consistent virtual representation of existing and planned man-made and natural objects.

Also in the construction phase, a range of applications benefit from an integration. In construction-site logistics, for example, the locations of cranes and storage areas can be planned taking into account the surroundings. The planning and scheduling

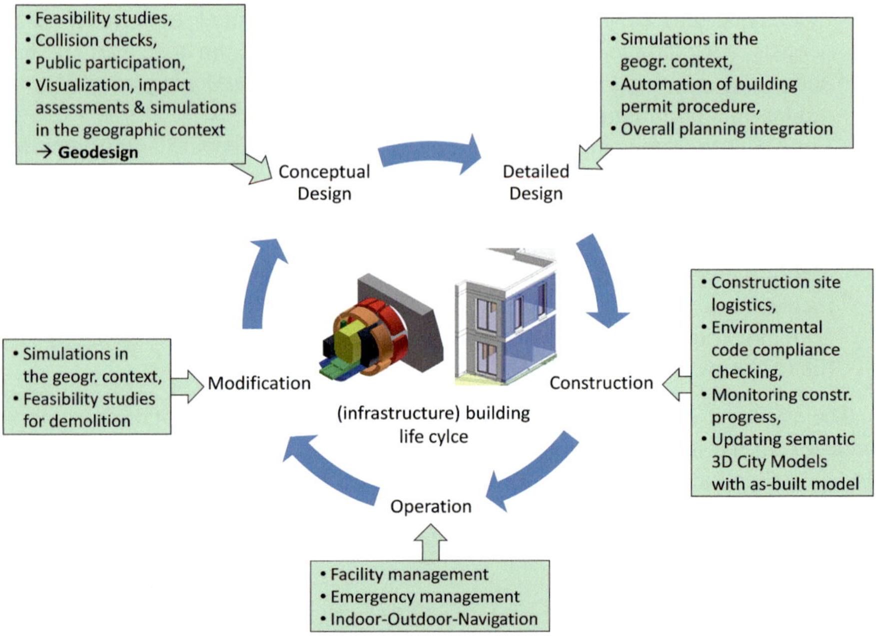

Fig. 34.5 BIM-GIS-integration use cases. Modified from Borrmann et al. (2015a)

of (heavy) transports can also be performed using geospatial data from semantic 3D city and landscape models. Environmental regulations must be observed during the construction phase. Schaller et al. (2017) describe, for example, how the construction sequence plan from BIM is compared with regulations for the clearing of woody plants in order to comply with species protection regulations. The species protection mapping is available in the form of geodata. At the end of the construction process, an as-built model of the structure is created. This can be used to update a semantic 3D city model.

Facility management, emergency management, and seamless indoor-outdoor transitions are examples for applications requiring the integration of BIM and semantic 3D city models from the maintenance phase of a building. Hijazi et al. (2011) show, for example, how indoor and outdoor utility networks can jointly be analyzed for building maintenance purposes.

Finally, in the modification phase an integration of BIM models into their geographic context supports feasibility studies for demolition works. Willenborg et al. (2018) show, for example, an approach to couple semantic 3D city models with a blast simulator in order to determine the safety zone around the detonation.

All the applications mentioned above can be classified into one of the following categories:

- Bringing BIM models into 3D city models for joint visualization, analyses, and simulation
- Bringing semantic 3D city models into BIM systems to import the surrounding environment for planned buildings or renovations
- Applications that make simultaneous use of indoor and outdoor representations.

It depends on the use case whether only the geometry, the geometry and the appearance, or whether also the semantics of the objects must be considered with the integration. Furthermore, the application determines whether the main focus is on BIM or semantic 3D city modeling, as the scope of both methods is complementary, with an overlap on the level of managing existing buildings, as explained in the following section.

34.4.2 Relationship of Semantic 3D City Modeling and BIM

Semantic 3D city modeling and BIM have in common that both methods deal with semantic modeling of the built environment. However, as we can see from the description of purpose and key applications of semantic 3D city modeling on the one hand and BIM on the other hand, there are different views on the same real-world objects which are manifested in the scope and scale as well as the different geometry modeling paradigms of the methods.

Figure 34.6 shows the differences in scope and scale. BIM's scale range includes a detailed view of a specific building, from the basic structure to the individual components. The scope is on the construction process (prescriptive modeling approach, see

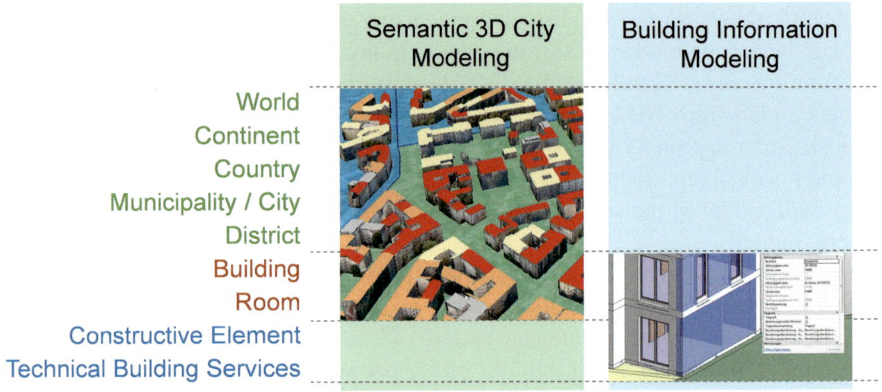

Fig. 34.6 Relation of semantic 3D city modeling and building information modeling with respect to scope and scale

section on purpose and key applications of BIM above). In contrast, semantic 3D city modeling includes the scale range of an entire region down to an individual room of a building, including further thematic areas like transportation, vegetation, and water bodies. Semantic 3D city modeling primarily describes the current state of the built environment. Semantic 3D city models can thus be seen as an inventory list of the physical objects of the built environment in a specific region and can therefore serve as a hub for linking information from various information systems (descriptive modeling approach, see section on purpose and key applications of semantic 3D city modeling above).

The different scopes and scale ranges of the two methods result in different geometry modeling paradigms, as shown in Fig. 34.7.

In semantic 3D city modeling, diverse sensors like airborne cameras and laser scanners, and terrestrial surveying instruments like tachymeters and terrestrial laser scanners, are applied to observe the surfaces of physical urban objects. Thus, objects are described by their observable surfaces like wall and floor surfaces, which can be accumulated to higher-level objects like rooms or buildings. The resulting geometry modeling paradigm is boundary representation (B-Rep), which means that geometric objects are recursively described by their boundaries (a solid by its bounding surfaces, a surface by its bounding rings, and so on). B-Rep has its strengths, for example, in its ability to be used with spatial indexing, which allows the storage and query of very large datasets. In contrast, BIM models reflect how a 3D object is constructed. Therefore, a generative modeling approach is applied, allowing the representation of constructive elements by volumetric and parametric primitives. The geometry modeling paradigm is often constructive solid geometry (CSG), where complex volumes are created from combinations of volumetric primitives; operators are union, intersection, and difference (set minus). CSG and other parametric geometry paradigms have their strength in the fact that changes can be carried out very efficiently. For example, to change the thickness of a wall in a CSG model means to

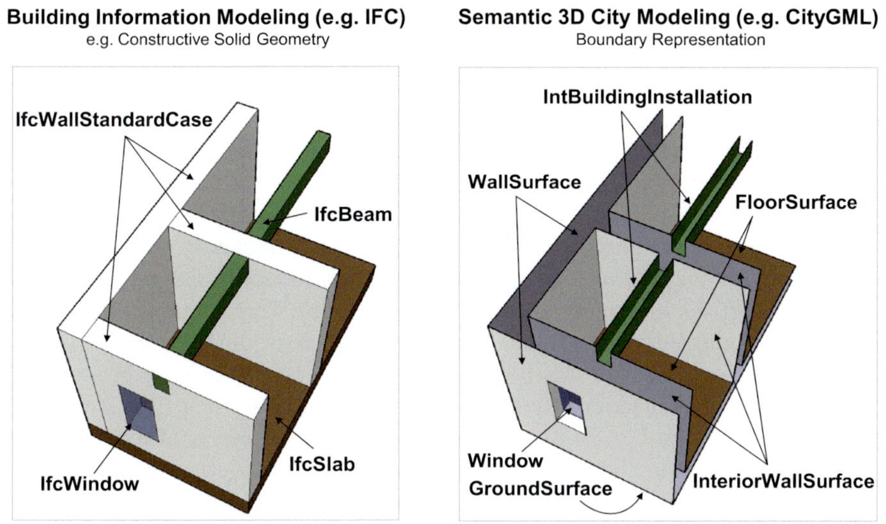

Fig. 34.7 Geometry modeling paradigms predominantly applied in BIM and semantic 3D city modeling (Nagel et al. 2009)

just alter one parameter, whereas in a B-Rep model many points would have to be moved individually, whereby inconsistencies could be introduced in the model.

While a CSG model can be uniquely mapped to exactly one B-Rep, the other way around is ambiguous: One B-Rep model can be created by an infinite number of different CSG models (see Kolbe and Plümer 2004; Nagel et al. 2009).

34.5 Recent Developments in Urban Informatics Involving Digital Models of the Built Environment

The following examples from the authors' project environment illustrate recent developments in urban informatics that involve semantic 3D city modeling, BIM, or a combination of the two methods.

34.5.1 Integrated Planning Models

As described in the previous section, the integration of semantic 3D city modeling and BIM can be employed for joint visualization and analysis of planned objects and their geographic environment. The authors of this chapter contributed to several research projects in the field of integrating BIM and semantic 3D city modeling for improving the planning process in infrastructure construction.

The project 3D Tracks (Breunig et al. 2017) developed new methods for collaborative subway track planning. A major research topic was the multi-scale nature of large infrastructure construction projects, with scale ranges from kilometer down to centimeter. Multi-scale representation is well established in the geospatial domain in general and in particular in semantic 3D city modeling (see the LoD concept of CityGML described above). However, as semantic 3D city models are rather static in nature (at least as far as the geometry of buildings is concerned), the LoD concept had to be adapted to the requirements of the highly dynamic planning process. Dependencies between the different levels of detail were introduced in a semantic model for representing shield tunnels (Borrmann et al. 2015b). This allows for the typical top-down planning approach from a coarser level, such as alignment (LoD 1), to a finer level. A key aspect of the model is that a refinement hierarchy between the representations of a tunnel in different LoDs is created with the help of space objects (see LoD 2–LoD 4 in Fig. 34.8), while the constructive elements of the tunnel are only represented in the highest LoD (LoD 5 in Fig. 34.8).

Figure 34.9 gives an example of the construction history of a shield tunnel in several levels of detail. Construction operations provided by parametric 3D CAD systems like sweeping, extrusion, etc. have been performed in a sequence, resulting in a graph structure which allows cross-LoD dependencies to be defined. Therefore, changes in a lower LoD will automatically take effect on objects in higher levels of detail. Although this modeling approach differs significantly from the way objects are represented in semantic 3D city modeling, Borrmann et al. (2015b) demonstrated that a geometric and semantic mapping, and geometric transformation of their tunnel objects to objects according to the CityGML representation of tunnels, is possible in an automated transformation workflow.

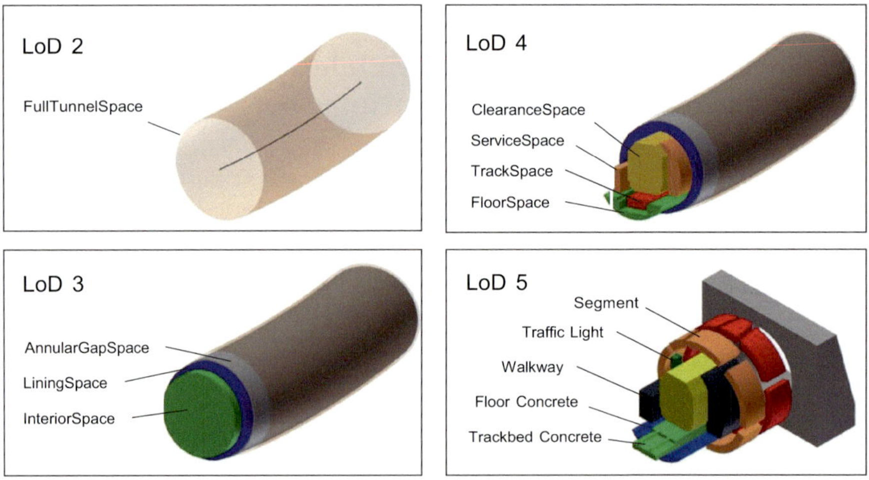

Fig. 34.8 A shield tunnel in different, dependent levels of detail (Borrmann et al. 2015b)

1.	CREATE SPLINE	LoD1
2.	CREATE SKETCH	LoD1
3.	CREATE SWEEP	LoD2
4.	CREATE SKETCH	LoD2
5.	CREATE SWEEP	LoD1
6.	BOOLEAN UNION	LoD1
7.	CREATE SKETCH	LoD2
8.	DEFINE CONSTRAINT	LoD2
9.	CREATE SWEEP	LoD2
10.	CREATE SWEEP	LoD3
11.	BOOLEAN DIFFERENCE	LoD3
12.	CREATE EXTRUSION	LoD3
13.	...	

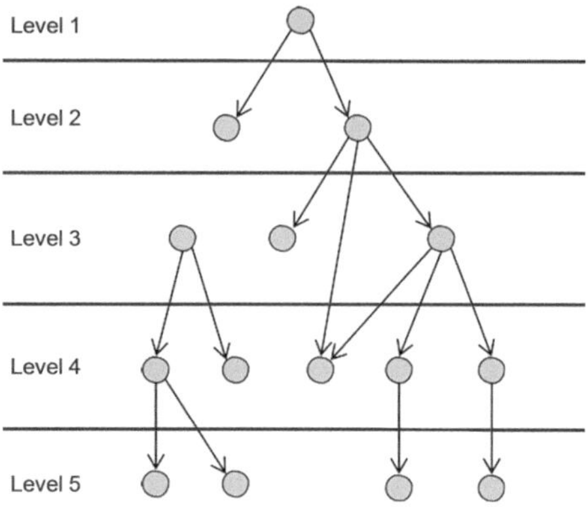

Fig. 34.9 Construction history and resulting cross-LoD dependency graph of a shield tunnel (Borrmann et al. 2015b)

Furthermore, in order to integrate parametric BIM authoring tools and analyses based on semantic 3D city models, the project team chose to encapsulate the geoprocessing workflows that had to be carried out for tasks like evaluating the planned rescue shafts of a subway track by standardized Web services provided in a distributed system. This allowed the team to keep the digital representations of the planned objects and the objects representing the geographic context in their own data structures, following integration approach (c) discussed earlier.

Schönhut (2018) describes a different approach of supporting subway planning by the integration of BIM and semantic 3D city modeling. Instead of keeping semantic 3D city models and BIM data in their original structures and bringing them together only encapsulated by processing services for specific analyses, she integrates data from both domains into a common information model (see Fig. 34.10). Her approach uses an integrated planning model and the CityGML schema as common information model. Since CityGML is not representing hydrogeological objects, which is critical for subway track planning, CityGML was extended using the Application Domain Extension (ADE) mechanism by classes of dedicated information models from the geology domain, namely the Geoscience Markup Language and the Groundwater Markup Language. An advantage of such an integration approach—besides a visualization of the BIM models in their environment—is that analysis and simulation methods developed on the basis of the CityGML standard for existing urban objects can now also be applied to the planned objects. Thus, what-if scenarios can be evaluated on different planning alternatives. This is useful not only in infrastructure planning but also in the context of smart cities.

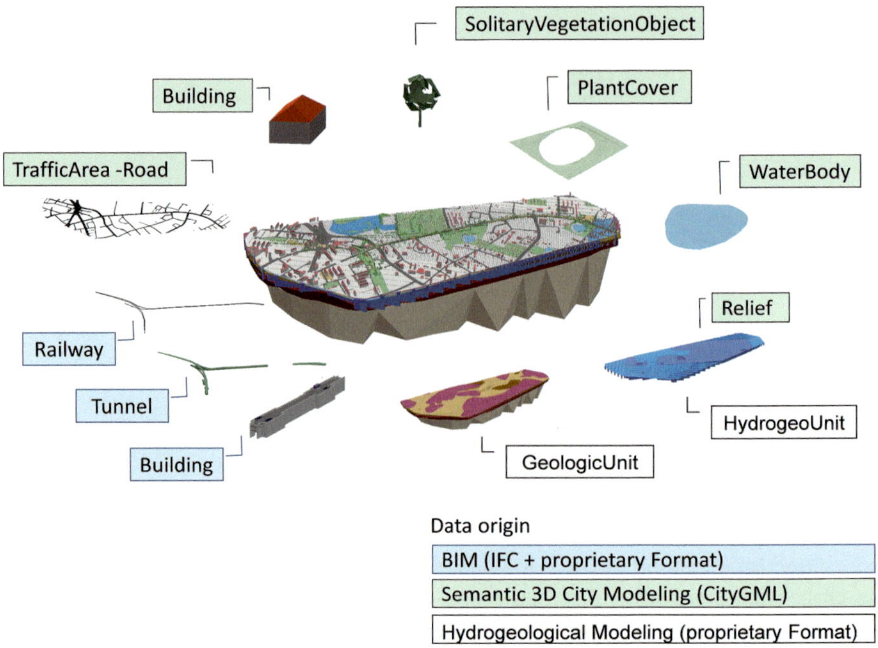

Fig. 34.10 An integrated planning model for subway planning based on a CityGML Underground Environment Application Domain Extension

34.5.2 Digital Models of the Built Environment, Smart Cities, and Digital Urban Twins

The notion of the digital twin (DT) was originally defined in product life cycle management for industrial machines (Datta 2017). The DT is a digital representation of the available information on a specific physical thing including its origin, state, history, as well as recorded performance data. It is used for documentation and predictive maintenance. Only very recently colleagues from geospatial information science and urban planning have started to discuss using DTs in the urban context, see Batty (2018). In contrast to industry, where all the information about a specific product is bundled by the manufacturer, the information about real-world objects of cities like buildings, streets, bridges, and so on is distributed across several organizations and stakeholders. Information about one and the same building is, for example, stored and managed by different departments of the city administration, by energy supply companies, and by the owners and users of the building. Creating and maintaining a digital twin therefore first of all means information integration. Due to the distributed and heterogeneous nature of the information about the built environment, creating the digital twin of a city is challenging, both technically and organizationally. In order to link and use such heterogeneous data, spatial data infrastructures for smart cities can play an important role in establishing interoperability between systems and platforms.

Moshrefzadeh et al. (2017) describe a concept for information integration in this context. Their smart district data infrastructure (SDDI) defines an organizational and technical framework for creating the digital twin of a city district. Their concept consists of actors, applications, sensors, urban analytics tools, a central resource registry of all the distributed information resources, and a 3D virtual district model as a central component (see Fig. 34.11). Based on the SDDI concept, Chaturvedi et al. (2019) present an approach for securing distributed applications and services which facilitates privacy, security, and controlled access to all stakeholders and the respective components and allows single-sign-on (SSO) authentication. Chaturvedi and Kolbe (2019a) describe an approach for interoperable access to sensor observations and time-series data from distributed, heterogeneous IoT and sensor platforms in the SDDI context.

A unique feature of SDDI is the fact that all the information, sensors, and applications coming from different domains are linked with the virtual 3D district model represented in CityGML. As shown in Fig. 34.12, digital representations of physical objects such as buildings and streets in semantic 3D city models can be used as anchor points for linking information from different domains and different stakeholders.

Thus, impacts of changes in the city can be simulated from different perspectives in the digital twin before they are implemented in the real city. Most smart city approaches today do not fully exploit this kind of information integration and therefore limit their view of the city to specific sectors, for example, smart mobility and smart energy, neglecting the interdependencies between those sectors.

A number of applications with real data from cities such as Berlin, London, and New York already show today that the concept of information integration based

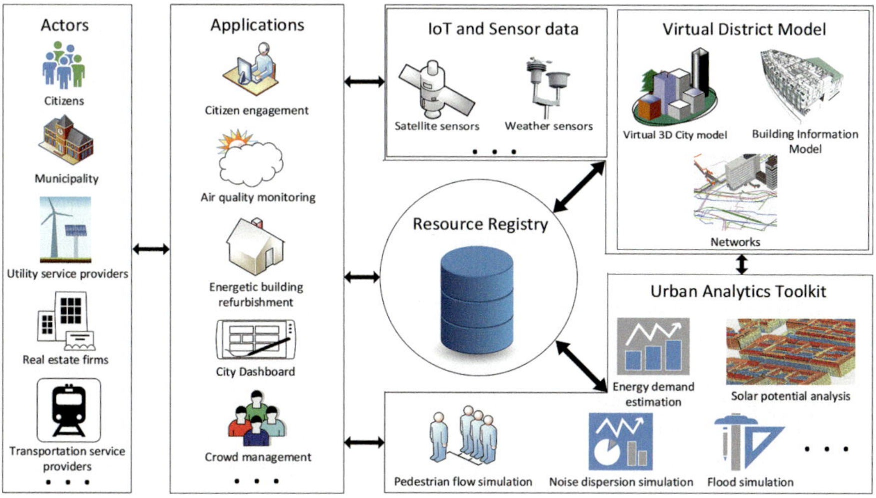

Fig. 34.11 Overview of the SDDI components

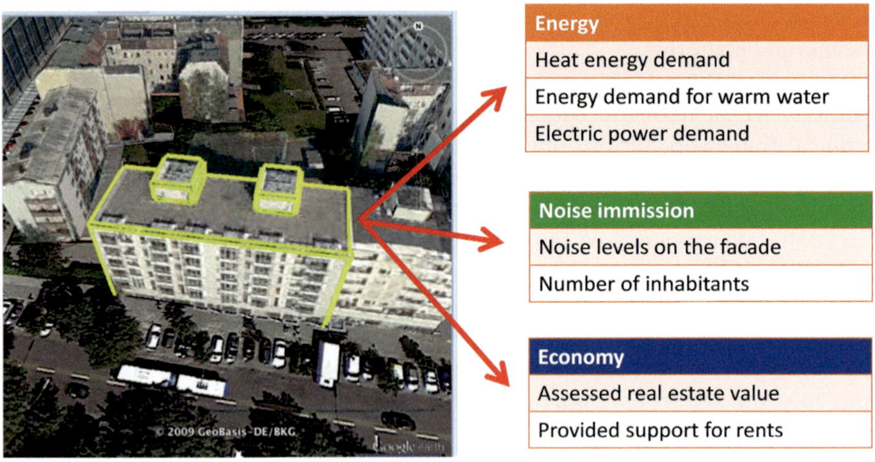

Fig. 34.12 Digital representations of physical objects in semantic 3D city models as anchor points for integrating information from different domains

on digital models of the built environment, especially semantic 3D city models, can make a valuable contribution to the planning and operation of cities. Examples of application domains are strategic energy planning (Kaden and Kolbe 2014) and solar potential analysis, as well as detonation simulations (Willenborg et al. 2018), traffic simulation (Beil and Kolbe 2017; Ruhdorfer et al. 2018), and flood-inundation simulation (Chaturvedi and Kolbe 2017).

34.6 Summary and Conclusions

Digital models of the built environment provide detailed information on the physical urban reality. Semantic 3D city modeling as well as building information modeling both address not only the representation of spatial and graphical aspects of urban entities, but especially focus on their thematic structuring and decomposition into meaningful objects. However, semantic 3D city modeling and BIM are following different modeling paradigms to achieve that goal. While the former is especially tailored to create descriptive models of the existing urban reality, BIM is tailored to create prescriptive models telling how reality should become. The different approaches are originating from different disciplines, that is, geomatics and AEC, and are supporting the typical applications within their disciplines very well. There is an increasing demand to combine the two representations, though, and a number of different approaches were explained in the chapter. Also, examples for use cases that require combinations of semantic 3D city models and BIM were given. In general, semantic urban models are key for a wide range of urban applications in a multitude of domains, including all kinds of simulations.

It is, however, important that urban models are structured and exchanged according to open standards. Standards play an important role in the acquisition and use of urban models, because data are typically captured, refined, visualized, and used by different parties and systems. Standards specify the exchange of information from the level of object definition and semantics down to the level of the physical file layout. The use of open standards ensures platform- and manufacturer-independent management and processing of data. Platform independence is also important to protect investments on collected datasets against arbitrariness, the risk of failure of a manufacturer, or abandoning of a specific software system.

In conclusion, it is important to point out that the achievable and manageable data quality of urban models is not only limited by the data collection processes (and thus by sensors and the subsequent interpretation of sensed data), but also from the employed standards concerning the data modeling frameworks and data exchange capabilities. Data loss may occur between two parties or systems, if the data exchange standard is not capable of preserving the original content, structure, and logic of a dataset.

CityGML and IFC are the most important open standards for semantic 3D modeling of the built environment.

References

Batty M (2018) Digital twins. Environ Plan B: Urban Anal City Sci 45(5):817–820. https://doi.org/10.1177/2399808318796416

Beil C, Kolbe TH (2017) CityGML and the streets of New York—a proposal for detailed street space modelling. In: Proceedings of the 12th international 3D GeoInfo conference 2017, ISPRS annals of the photogrammetry, remote sensing and spatial information sciences, vol IV-4/W5, pp 9–16

Biljecki F, Stoter J, Ledoux H, Zlatanova S, Çöltekin A (2015) Applications of 3D city models: state of the art review. ISPRS Int J Geo-Inf 4:2842–2889. https://doi.org/10.3390/ijgi4042842

Biljecki F, Kumar K, Nagel C (2018) CityGML Application Domain Extension (ADE): overview of developments. Open Geospatial Data, Softw Stand 3:13. https://doi.org/10.1186/s40965-018-0055-6

Borrmann A, König M, Koch C, Beetz J (2015a) Building information modeling. Springer, Wiesbaden

Borrmann A, Kolbe TH, Donaubauer A, Steuer H, Jubierre JR, Flurl M (2015b) Multi-scale geometric-semantic modeling of shield tunnels for GIS and BIM applications. Comput-Aided Civil Infrastruct Eng 3(4):263–281. https://doi.org/10.1111/mice.12090

Bradley A, Li H, Lark R, Dunn S (2016) BIM for infrastructure: an overall review and constructor perspective. Autom Constr 71(2):139–152. https://doi.org/10.1016/j.autcon.2016.08.019

Breunig M, Borrmann A, Rank E, Hinz S, Kolbe TH, Schilcher M, Mundani RP, Jubierre JR, Flurl M, Thomsen A, Donaubauer A, Ji Y, Urban S, Laun S, Vilgertshofer S, Willenborg B, Menninghaus M, Steuer H, Wursthorn S, Leitloff J, Al-Doori M, Mazroobsemnani N (2017) Collaborative multi-scale 3D city and infrastructure modeling and simulation. Tehran's Joint ISPRS Conferences of GI Research, SMPR and EOEC 2017 [The International Archives of the Photogrammetry, Remote Sensing and Spatial Information Sciences (ISPRS Archives)], ISPRS, pp 341–352. https://doi.org/10.5194/isprs-archives-XLII-4-W4-341-2017

Brüggemann T, von Both P (2015) 3D-Stadtmodellierung: CityGML. In: Borrmann A, König M, Koch C, Beetz J (eds) Building information modeling. Springer, Wiesbaden

Chaturvedi K, Kolbe TH (2017) Future city pilot 1 engineering report. Open Geospatial Consortium, OGC Doc. No. 16-098. http://docs.opengeospatial.org/per/16-098.html

Chaturvedi K, Kolbe TH (2019a) Towards establishing cross-platform interoperability for sensors in smart cities. Sensors 19(3). http://doi.org/10.3390/s19030562

Chaturvedi K, Kolbe TH (2019b) A requirement analysis on extending semantic 3D city models for supporting time-dependent properties. In: 4th international conference on smart data and smart cities, 1–3 October 2019, Kuala Lumpur, Malaysia. ISPRS annals of the photogrammetry, remote sensing and spatial information sciences, vol IV-4/W9

Chaturvedi K, Matheus A, Nguyen S, Kolbe TH (2019) Securing spatial data infrastructures for distributed smart city applications and services. Future Gener Comput Syst 101:723–736. https://doi.org/10.1016/j.future.2019.07.002

Chipman T, Liebich T, Weise M (2016) mvdXML—specification of a standardized format to define and exchange Model View Definitions with exchange requirements and validation rules. https://standards.buildingsmart.org/MVD/RELEASE/mvdXML/v1-1/mvdXML_V1-1-Final.pdf. Accessed 28 June 2019

Cox S, Daisy P, Lake R, Portele C, Whiteside A (2004) OpenGIS Geography Markup Language (GML3), Implementation Specification Version 3.1.0. OGC Doc. No. 03-105r1. Open Geospatial Consortium

Datta, SPA (2017) Emergence of digital twins—is this the march of reason? J Innov Manage 5(3):14–33. https://doi.org/10.24840/2183-0606_005.003_0003

Daum S, Borrmann A, Kolbe TH (2017) A spatio-semantic query language for the integrated analysis of city models and building information models. In: Abdul-Rahman A (ed) Advances in 3D geoinformation. Lecture notes in geoinformation and cartography. Springer, Cham. https://doi.org/10.1007/978-3-319-25691-7_5

de Laat R, van Berlo L (2011) Integration of BIM and GIS: the development of the CityGMLGeoBIM extension. In: Kolbe T, König G, Nagel C (eds) Advances in 3D geo-information sciences. Lecture notes in geoinformation and cartography. Springer, Berlin, Heidelberg. https://doi.org/10.1007/978-3-642-12670-3_13

Eastman C, Teicholz P, Sacks R, Liston K (2011) BIM handbook: a guide to building information modeling for owners, managers, designers, engineers, and contractors. Wiley, Hoboken

El Mekawy M, Östman A, Hijazi I (2012) A unified building model for 3D urban GIS. Int J Geo-Inf 1:120–145. https://doi.org/10.3390/ijgi1020120

Flaxman M (2010) Fundamentals of geodesign. In: Proceedings of digital landscape architecture 2010. Anhalt University of Applied Science

Foley J, van Dam A, Feiner S, Hughes J (1995) Computer graphics: principles and practice, 2nd edn. Addison Wesley

Gröger G, Plümer P (2012) CityGML—interoperable semantic 3D city models. ISPRS J Photogrammetry Remote Sens 71:12–33

Hijazi I, Donaubauer A, Kolbe TH (2018) BIM-GIS integration as dedicated and independent course for geoinformatics students: merits, challenges, and ways orward. ISPRS Int J Geo-Inf 7(8):319. https://doi.org/10.3390/ijgi7080319

Hijazi I, Zlatanova S, Ehlers M (2011) NIBU: an integrated framework for representing the relation among building structure and interior utilities in micro-scale environment. Geo-Spatial Inf Sci 14(2):98–108. https://doi.org/10.1007/s11806-011-0461-8

ISO (2018) Industry Foundation Classes (IFC) for data sharing in the construction and facility management industries—Part 1: data schema. ISO 16739-1:2018. International Organization for Standardization

Kaden R, Kolbe TH (2014) Simulation-based total energy demand estimation of buildings using semantic 3D city models. Int J 3D Inf Model 3(2):35–53. http://doi.org/10.4018/ij3dim.2014040103

Kolbe TH (2009) Representing and exchanging 3D city models with CityGML. In: Lee J, Zlatanova S (eds) 3D Geo-Inf Sci. Springer, New York, pp 15–31
Kolbe TH, Plümer L (2004) Bridging the gap between GIS and CAAD: geometry, referencing, representations, standards and semantic modelling. GIM Int 18:12–15
Kolbe TH, Nagel C, Stadler A (2009) CityGML—OGC standard for photogrammetry? In: Fritsch D (ed) Photogrammetric Week'09. Wichmann, Heidelberg
Kutzner T, Hijazi I, Kolbe TH (2018) Semantic modelling of 3D multi-utility networks for urban analyses and simulations—the CityGML utility network ADE. Int J 3D Inf Model (IJ3DIM) 7(2):1–34. http://dx.doi.org/10.4018/IJ3DIM.2018040101
Liu X, Wang X, Wright G, Cheng J, Li X, Liu R (2017) A state-of-the-art review on the integration of building information modeling (BIM) and geographic information systems (GIS). ISPRS Int J Geo-Inf 6:53. https://doi.org/10.3390/ijgi6020053
Markič Š, Donaubauer A, Borrmann A (2018) Enabling geodetic coordinate reference systems in building information modeling for infrastructure. Paper presented at the 17th international conference on computing in civil and building engineering ICCCBE 2018, Tampere, 5–7 June
Moshrefzadeh M, Chaturvedi K, Hijazi I, Donaubauer A, Kolbe TH (2017) Integrating and managing the information for smart sustainable districts—the smart district data infrastructure (SDDI). In: Kolbe M, Bill R, Donaubauer A (eds) Geoinformationssysteme 2017—Beiträge zur 4. Münchner GI-Runde. Wichmann, Heidelberg
Nagel C, Stadler A, Kolbe TH (2009) Conceptual requirements for the automatic reconstruction of building information models from uninterpreted 3D models. In: Proceedings of the academic track of the Geoweb 2009 conference in Vancouver, Canada, 27–31 July 2009, International archives of photogrammetry, remote sensing, and spatial information sciences, vol XXXVIII-3-4/C, ISSN 1682-1777. ISPRS
National Institute of Building Sciences (2012) National BIM Standard United States Version 2. http://www.nationalbimstandard.org. Accessed 28 June 2019
Noardo F, Arroyo Ohori K, Biljecki F, Krijnen T, Ellul C, Harrie L, Stoter J (2019) GeoBIM benchmark 2019: design and initial results. ISPRS International Archives of the Photogrammetry, Remote Sensing and Spatial Information Sciences, XLII-2/W13. https://doi.org/10.5194/isprs-archives-XLII-2-W13-1339-2019
Nouvel R, Bahu JM, Kaden R, Kaempf J, Cipriano P, Lauster M, Häfele KH, Munoz E, Tournaire O, Casper E (2015) Development of the CityGML application domain extension energy for urban energy simulation. In: Proceedings of the 14th conference of international building performance simulation association (BS2015) in Hyderabad, India, 7–9 Dec
Ruhdorfer R, Sindram M, Willenborg B (2018) Coupling of traffic simulations and semantic 3D city models. gis. Science (3) https://gispoint.de/artikelarchiv/gis/2018/gisscience-ausgabe-32018/4611-kopplung-von-verkehrssimulationen-und-semantischen-3d-stadt-modellen.html
Schaller J, Gnaedinger J, Reith L, Freller S, Mattos C (2017) GeoDesign: concept for integration of BIM and GIS in landscape planning. J Digital Landscape Archit 2:102–112
Schönhut S (2018) An underground environment data model for BIM and GIS integration. Master's Thesis, Chair of Geoinformatics, Technical University of Munich
Stadler A, Kolbe TH (2007) Spatio-semantic coherence in the integration of 3D city models. In: Proceedings of the 5th international symposium on spatial data quality, Enschede
Stouffs R, Tauscher H, Biljecki F (2018) Achieving complete and near-lossless conversion from IFC to CityGML. ISPRS Int J Geo-Inf 7(9):355. https://doi.org/10.3390/ijgi7090355
Willenborg B, Sindram M, Kolbe TH (2018) Applications of 3D city models for a better understanding of the built environment. In: Behnisch M, Meinel G (eds) Trends in spatial analysis and modelling. Geotechnologies and the environment, vol 19. Springer, Cham. https://doi.org/10.1007/978-3-319-52522-8_9

Thomas H. Kolbe is Full Professor and Chair of Geoinformatics at the Technical University of Munich, Germany. His interests are in GIScience, specifically in the fields of virtual 3D cities, landscape, and building information modeling. He is co-author of the OGC Standards CityGML and IndoorGML.

Andreas Donaubauer is a Senior Scientist in Thomas H. Kolbe's group at the Technical University of Munich, Germany. His interests are in spatial data infrastructures and interoperability, geodesign as well as semantic modeling, and transformation of geospatial data.

Chapter 35
CityEngine: An Introduction to Rule-Based Modeling

Tom Kelly

Abstract CityEngine is a rule-based urban modeling software package. It offers a flexible pipeline to transform 2D data into 3D urban models. Typical applications include processing 2D urban cartographic geographic information system (GIS) data to create a detailed 3D city model, creating a detailed visualization of a proposed development, or exploring the design space of a potential project. The rule-based core of Esri's CityEngine has some unique advantages: Huge cities can be created as easily as small ones, while the quality of the models is consistent throughout. Additionally, this rule-based approach means that large design spaces can be explored quickly, interactively, and analytically compared. Such advantages must be carefully balanced against the increased time to create and parameterize the rules and the sometimes stylistic or approximate models created; coming from more traditional workflows, CityEngine's pipeline can be initially overwhelming. We introduce the principal workflows and the flexibility they afford, sketch the procedural programming language used, and discuss the export pathways available.

35.1 3D: One Better than 2D

3D technologies are revolutionizing the way we plan, understand, communicate, and document our urban environments. Revolutions are, however, rarely easy; there are numerous issues and challenges around this transition from 2D to 3D toolchains.

Reading 2D plans and maps is often challenging because they are one dimension short of the 3D world we live in. The 3D data must be encoded using various tricks and conventions, such as contour lines, elevation diagrams, symbols, and shading. This is because there is more information in the 3D world than 2D plans contain. Technology now enables us to efficiently record, model, and plot in 3D. Collecting and sharing this 3D information has been, until recently, difficult and prohibitively

T. Kelly (✉)
University of Leeds, Leeds, UK
e-mail: twakelly@gmail.com

W. Shi et al. (eds.), *Urban Informatics*, The Urban Book Series,
https://doi.org/10.1007/978-981-15-8983-6_35

expensive. As various technologies such as commodity 3D CAD and photogrammetric reconstruction have matured, we are able to accurately construct virtual 3D models of our 3D world.

At the same time as making our data more accurate, 3D models make our data more accessible. While it has always been possible to create physical scale models of our environments, these are expensive, difficult to transport or share, and bulky to store. Technologies such as immersive virtual and augmented realities (VR, AR, often summarized as XR) allow anyone from children to city planners to understand complex designs by exploring them at real-world scales. 3D tools such as physical simulation (solar potential, window modeling) and viewpoint rendering help engineers design empirically better environments; because we are able to explore our design spaces more quickly, we understand them faster, produce better designs, and better comprehend any issues.

However, 3D modeling is difficult. The de facto 3D representation is the mesh. This is a set of corners (vertices) placed in 3D space, between which we create triangles. By creating many thousands of such triangles, we can build representations of complex 3D environments. We may even choose to apply colors or texture to each triangle.

There are many tools available for creating these polygonal meshes. Traditional manual 3D modeling tools offer a way to create multiple triangles at a time by creating more complex primitives (spheres, cubes, curves, surfaces, extrusions, etc.). Such manual tools include Autodesk Maya (2019), Trimble SketchUp (2019), or Blender (2019). Even though these manual tools have become incredibly sophisticated and general, they still require users to spend a lot of time positioning and editing triangles and primitives. For our use cases, we might imagine our long-suffering artist being employed to position a spherical doorknob on every rectangular front door, of every building, in the urban area we are modeling.

What we would rather do is to create a rule which encodes "attach a sphere to every front door". Luckily, computers are rather good at these repetitive tasks—if we can find a way to explain to them what to do. In this chapter, we introduce one way to instruct them: rule-based modeling. In particular, we will dive deeply into a particular modeling system: Esri's CityEngine. Such modeling systems offer tools to procedurally generate 3D meshes from systems of rules—they are able to create models with millions of vertices in seconds.

It is here that we see another advantage of working with virtual, rather than physical, 3D models. Computer programs can follow rules to create and manipulate virtual polygonal mesh models superhumanly quickly and accurately. We can repeatedly change the rules and view and explore the resulting environments on screen, in virtual reality, or physically produce them using a 3D printer. To perform the same changes in a physical 3D model would take many lifetimes.

35.2 2D Shapes + Rules = 3D Models

Because of the hierarchical, systematic, and often repetitive nature of urban environments, rule-based city modeling has been a driving force for general procedural modeling in general. We note in passing that other rule-based systems have been wildly successful in other domains. Of note are commercial systems such as SpeedTree (2019) for the rapid generation of trees and forests and Grome (Wikipedia 2019) for creating terrains and landscapes. For each different domain, different techniques and rules are appropriate. In CityEngine, as we will see, the rules and the operations they use have been carefully curated to allow rapid and accurate modeling of buildings and streets.

Before deciding to use a rule-based modeling pipeline, it is important to weigh the advantages and disadvantages against more traditional manual modeling pipelines. For smaller or more complex models, manual modeling may be faster and cheaper; the time to create the rules may be larger than the time that would be taken to perform the manual modeling. Rule-based modeling is particularly difficult for complex geometries where many decisions are involved in placement and evaluation. Translating each decision into a rule and ensuring that the decisions interact appropriately in all circumstances can be time-consuming. We note that many of the explanatory examples in this chapter would be more quickly created using manual modeling tools—only when scaling up to larger areas does rule-based modeling reward the time invested in creating the rules.

Writing rule files is a new skill that must be taught, studied, and maintained like any other. Because it is a newer technology, finding qualified personnel can be more difficult, especially because they may need a background in urban design, a basic knowledge of linear algebra, as well as the ability to (en)code our rules in a programming language.

These caveats aside, rule-based modeling is able to offer a flexible, quick, and responsive toolchain for quickly developing urban scenarios ranging from single building modeling, campus-scale designs, up to neighborhood and city-scale simulation. Once the rules are available, a large quantity of geometry can be created easily and quickly. Changes and modifications to scenarios can be made in real time. Both the level of detail ("do we draw chimneys on the buildings?", "do we draw roofs?"), the presentation format (Webviewer, VR), and the rule attributes ("how high is this building?") can be updated over an entire city at once, all thanks to rule-based modeling.

Esri's CityEngine is a software system for rule-based modeling in the urban domain. It provides a visual environment to apply rules, create new rules, and inspect the results. The historical context of CityEngine was that it was acquired by Esri during their transition from a 2D cartography company to a provider of 3D solutions. As witnessed by ArcGIS Pro, this transition has created a massively powerful pipeline with support for all the major industry formats. This business context underpins the CityEngine workflow—2D shapes are imported into the system, where rules are used to convert them to 3D models. These models are the 3D output which we

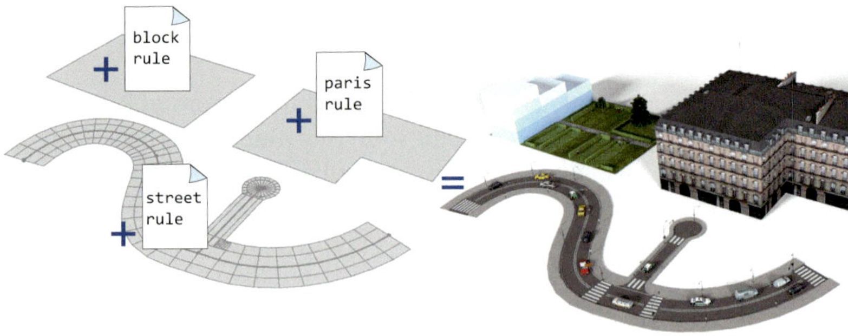

Fig. 35.1 The central paradigm of CityEngine is to apply rules to shapes (gray, left) to create 3D models (right). This approach is able to create a large variety of rule-driven models

may view in CityEngine or export to the Web or VR. Thus, the central process for modeling in CityEngine is to apply rules to shapes to create models (Fig. 35.1).

A CGA rule is a text file containing a list of instructions. In Fig. 35.2, we introduce a simple rule which extrudes a shape into a model of a 3D prism. While this rule only contains five lines of code, complex rule files can be thousands of lines long.

This chapter aims to be a broad introductory tour of the system with a deep dive into various implementation topics. We continue to describe shapes, the rules, analysis tools, and export paths from CityEngine. After reading this chapter, the kinesthetic learner is encouraged to spend a few days working through the CityEngine tutorials provided by Esri (2019a). Similarly, Esri's online documentation is an invaluable source of technical details (Esri 2019b).

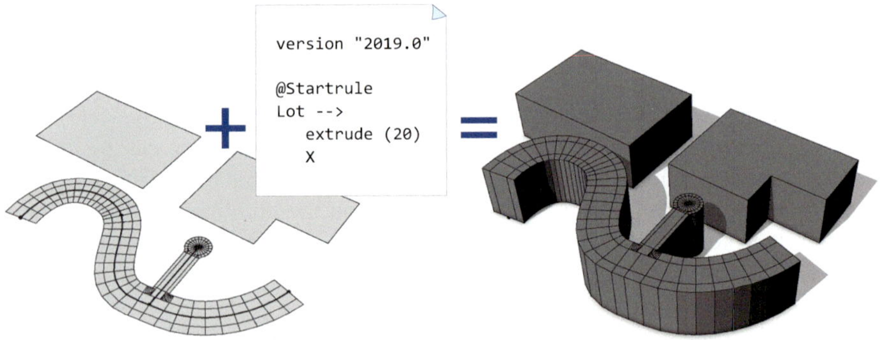

Fig. 35.2 A simple CGA rule file (center) is applied to several different shapes (left) to create the associated 3D models (right). This rule creates a prism of height 20 m over the shape

35.3 On the (Many) Origins of Shapes

CityEngine provides two workflows to instantly create entire cities with very little user input. The City Wizard (File → New... → CityEngine → CityWizard) uses an entirely procedural workflow to create an impressive quantity of shapes with complex rules in a few clicks. Of course, the resulting city is entirely fictional; if we wish instead to use an entirely data-driven set of shapes, we may use the Map Import (File → get Map Data...). This tool downloads satellite images, height maps, lot footprints, and street networks, to create shapes and terrain for a real-world area (Fig. 35.3). However, because there is no common data source for building rules, only simple rules are provided. Both the City Wizard and Map Import use shapes to model entire cities quickly but leave us with limited control over the shapes and rules. We continue to examine more controlled ways to create shapes.

Shapes are usually 2D polygons lying on the ground. Much of CityEngine's utility and complexity is driven by the different ways to create shapes. The various sources for shapes provide an overview of the different modeling workflows available in CityEngine:

- To create a 3D model of an existing area, we may use a collection of building lots from a geospatial data source (including FileGDB, DXF, Shapefile, or OBJ) as shapes.
- To plan a new urban area, we may draw our own shapes, for example by adding each corner of each lot at a time. The simplest way to create a shape is to use the Rectangular Shape Creation tool, which allows clicking and dragging to position two corners of a rectangle on the floor plane. To increase the accuracy, we may trace the outline of these shapes from images imported into CityEngine.

Fig. 35.3 A city created in 30 s using the Map Import functionality

- If we wish to use rules to add windows to the blank facades of a building, we could draw the building using the manual 3D modeling tools provided by CityEngine. This is an uncommon workflow because the shapes may not be horizontal. Such a workflow allows us to manually model a building and then apply rules only to specific façades. CityEngine has a range of tools for manual shape modeling, including rectangular, polygonal, and circle generation. Markus Lipp created this modeling system to use intelligent extrusions to quickly and manually model urban forms (Lipp et al. 2014).
- When modeling a street network, we may import a street graph (formats supported include DXF, FileGDB, and OpenStreetMap) and use CityEngine's dynamic shape system to automatically create street shapes, blocks, and lot shapes between the streets. We continue to explore the dynamic shape system in greater depth.

35.3.1 Dynamic Shapes: Streets, Blocks, and Lots

Dynamic shapes use algorithms to approximate the forms that we see in our urban environments. Because of this, they are only simulated designs that match general characteristics (the range of building lot widths) but not specific measurements (the width of a particular lot). We describe them as dynamic because they are generated dynamically from the street graph; if you move a street intersection, the adjoining roads and blocks are automatically recalculated. The flexibility of CityEngine allows for combinations of these shape generation approaches—manual, data-driven, and dynamic—to be used together. For example, streets can be imported from a GIS data source and the blocks between the streets can be dynamically subdivided to lots, or an area of the city where GIS data exist for streets and lots can be augmented by adjacent dynamically generated streets and lots.

A street graph describes the streets in a street network. Over this graph, dynamic street shapes are created for sidewalks, junctions, and the street themselves, as shown in Fig. 35.4. The graph edges describe the center lines, and the nodes (where the edges meet) describe the street junctions.

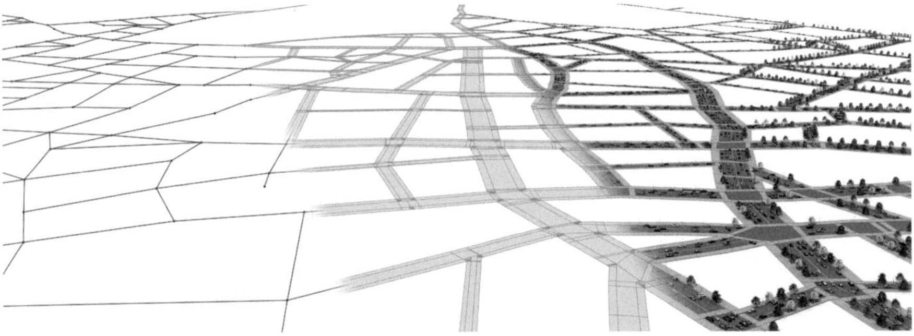

Fig. 35.4 Left: a blue street centerline graph; middle: the generated street shapes; right: 3D models generated by applying rules to the shapes

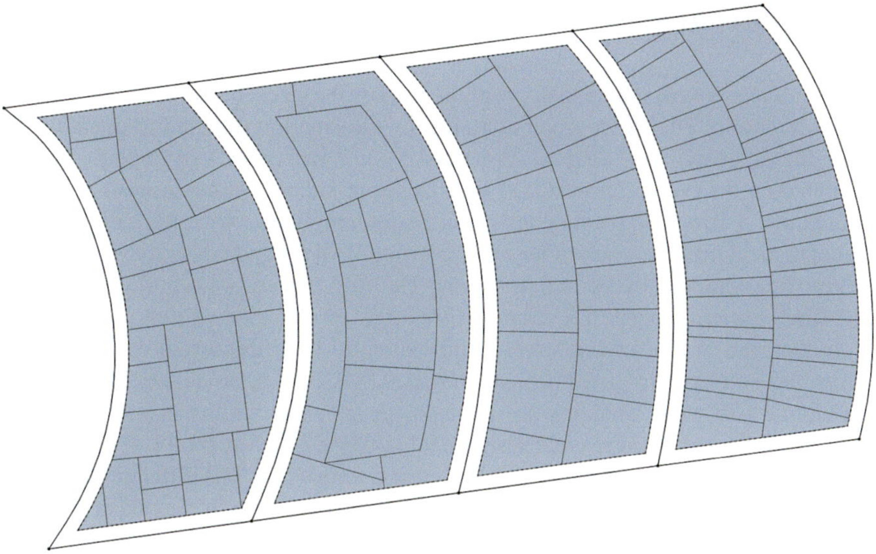

Fig. 35.5 Block subdivision algorithms used to create building lots. From left to right: *recursive*, *offset*, and *skeleton*. Far right: skeleton modified for a high irregularity and narrower lot width

Between streets, CityEngine dynamically generates blocks and from the blocks, lots. Generally, every loop of streets generates a block in its interior. The block contains a further selection of attributes which define its subdivision into lot shapes. The lot shape represents a parcel of land on which we will use rules to generate individual building models. When a block (or a street) is selected in CityEngine, the Inspector shows details about the object which drive the generation of the dynamic shapes. Block to lot subdivision algorithms are discussed by Vanegas et al. (2012) and are subdivided into two major categories: recursive subdivision and offsets. Each of these can be further controlled with attributes controlling on lot area, width, and variation, as in Fig. 35.5.

The generation sequence is an important part of the modeling paradigm used by CityEngine for dynamic shapes: Streets are created, between which blocks are found, and finally inside each block, lots are created. It is important to note this order when creating cityscapes and start with street creation before moving on to block and lot generation. This is because small changes in the street network will affect many blocks, whereas changing a block's subdivision settings will affect only the lots in the block. Similarly, changing a lot's rule or attributes will only affect the single lot's (building) model.

Remembering that our shapes will be the starting point for rules, it is also important to note the default starting rule names for each dynamic shape type. This name is used to automatically assign a start (initial) rule to the shape. For example, dragging a rule file onto a street's sidewalk shape will attempt to use the rule named Sidewalk (and taking no parameters), while the same file dragged onto a lot shape will use the rule Lot.

35.3.2 Graphs and Cities

The astute reader will notice that the street graphs (the street centerlines themselves) are not dynamic. The street graph contains the information required to dynamically create the other dynamic shapes. As we have come to expect, CityEngine provides manual, data-driven, and procedural approaches to creating street graphs.

Creating a street graph manually can be accomplished with the polygonal or freehand street creation tools. These allow graph vertices and edges to be created by clicking at corners or by sketching streets. The Edit Street tool can then be used to reposition vertices, curve streets, and adjust street or sidewalk widths.

An alternative to drawing street graphs directly is to import an existing graph from a GIS source. Supported formats include DXF, FileGDB, and OpenStreetMap. CityEngine can parse and map attributes such as street widths in some of these formats, which can avoid manual assignment with the Edit Street tool. Working with various data sources can take some experience because each has different properties such as distance between nodes or the presence of curved graph segments. To assist with working with these graphs, various tools are available to simplify a graph (*Graph → Simplify Graph...*), align the graph to the terrain (*Graph → Align Graph to Terrain*), or resolve crossing graph edges into bridges and underpasses (*Graph → Generate Bridges...*)

To create large street networks where there is no available GIS source, CityEngine provides the Grow Streets tool which creates a procedurally generated set of streets, as well as blocks and lots as described above. The origins of the street growth algorithms used are described in the paper by Parish and Müller (2001), although these have now advanced beyond the published details somewhat. In summary, self-sensitive L-Systems (Prusinkiewicz and Lindenmayer 2012) are employed to grow major and minor streets. Newly grown edges are snapped to attach to parts of the existing networks. By combining different patterns of growth for both the major and minor streets, a wide variety of different networks can be grown, illustrated in Fig. 35.6. The Grow Streets tool also allows the type of dynamic block subdivision to be specified.

Once a real street graph has been imported or synthetic graph has been grown, the Edit Street and Street Creation tools can be used to amend or fine-tune the data.

There are several use cases for graphs beyond their typical use of creating street models. Appropriate rules can be used to create various graph-like structures including walls, railroads, and power-lines as in Fig. 35.7.

We have seen an overview of the multitude of ways that CityEngine can be used to create different shapes; we continue to examine how we can obtain rules to transform our shapes into 3D models.

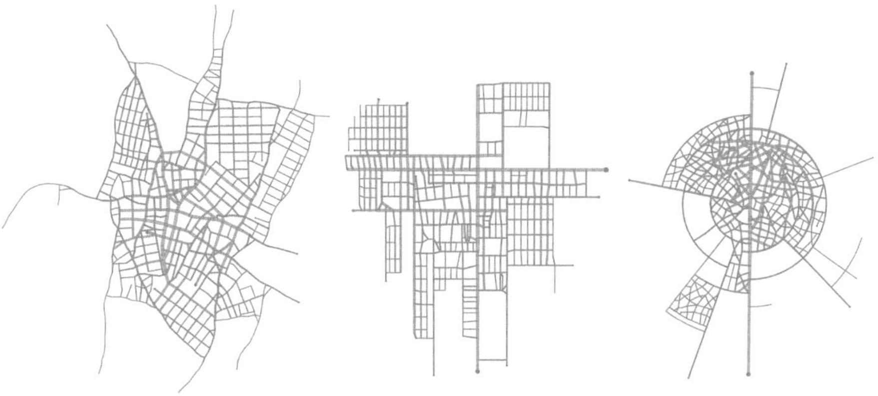

Fig. 35.6 A wide variety of street patterns can be generated by selecting the major and minor street patterns. Left: organic major and raster minor; Middle: raster major and raster minor; Right: radial major and organic minor

Fig. 35.7 Walls, streets, fences, and power-lines generated from rules executed on dynamic graph shapes

35.4 Writing CGA Rules for Fun and Profit

CityEngine rules are written in the Computer Generated Architecture (CGA) programming language. Writing a simple CGA rule can be quick and effortless; however, writing a realistic or flexible rule is an involved process. A library of existing rules is provided, and further rules can be found online. The fastest route to creating a 3D scene from a 2D map is by combining and parameterizing these existing rules, without ever writing CGA code ourselves.

Pre-installed rules can be found in the *ESRI.lib* project. A further selection of well-written rules for a variety of circumstances can also be found in the tutorials and

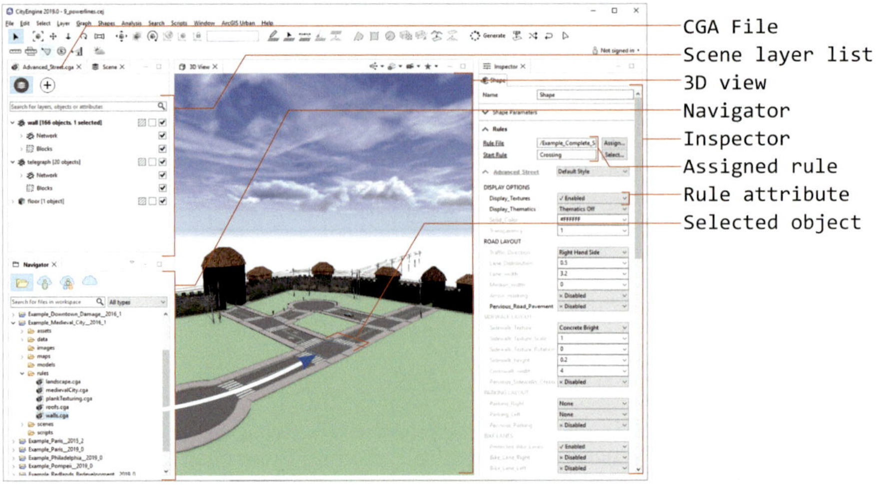

Fig. 35.8 CityEngine user interface elements. Orange: important elements of the interface. Blue: dragging a rule onto the selected shape to generate a 3D model

downloads dialog (*Help → Download Tutorials and Examples*). Finally, many user-generated rule packages (single .RPK files containing rules and resources) of varying quality can be found online ("ArcGIS content search" with keyword CityEngine; Esri 2019c). Exploring existing rules is a powerful way to understand how models can be generated using the CGA language. As rules can take a lot of time to write, reusing existing rules is advisable wherever possible; libraries should be used before writing CGA code ourselves.

To apply a rule or rule package, we may drag the rule package or file from the navigator onto a shape as shown in Fig. 35.8. By selecting a group of shapes before dragging, we may assign the rule to a number of shapes at once. The Inspector panel allows us to customize rules in a variety of ways. Various options exist for selecting shapes by layer or start rule can be found by right-clicking on a shape. After assigning a rule, there is a short delay while the rule is compiled and evaluated to create a model. If we desire more control, the Inspector contains more detailed options for the shape, including the CGA rule file, Start rule, and the previously mentioned rule attributes.

35.4.1 Writing Rules

While the mythos of "coders" and "software engineers" may have elevated programming to the status of a divine art, the reality is much more down to earth. CGA is a simpler language than the likes of Python, relying on a few basic operations which are repeatedly applied to write a rule. We find that undergraduate students are able to create their own rules after a few sessions with CityEngine. Those with experience of complex languages such as C or C++ must learn the CGA way of doing things which is more *functional* than they are used to. The dialect of CGA used in CityEngine has evolved from the version presented in the initial academic publication (Müller et al. 2006); care must be taken when comparing rules from different versions.

We take the opportunity here to untangle the term "shape" in CityEngine. This has been overused to describe both the input shapes (described in the previous sections) and the shapes which are passed between rules in CGA. CityEngine refers to these intermediate shapes as "CGA shapes"; here, we will use the term geometry. This regrettable confusion is somewhat caused by the academic origin of CityEngine, where our input shapes did not exist.

A CGA rule file is a text document containing a collection of rules. A rule is analogous to a *function* or *method* in other programming languages. Each rule is identified by its name and set of parameters: X(1) is a different rule to X(1,2). As the rule is executed, it can call various operations, as well as other rules. Operations are analogous to *library functions* in other programming languages. As parent rules use operations to create new geometries, they label each with a child rule. If this rule exists, it will then be executed on the child geometry. Unlike the academic description of CGA (Müller et al. 2006), there is no concept of priority; rules are evaluated purely according to their parent rule.

Each rule transforms a piece of geometry into new geometries (or nothing); the result is a 3D mesh model consisting of all the geometry that cannot be further transformed. The initial geometry is the input shape to which the initial rule (sometimes designated with the @Startrule annotation) is applied. The rule also has access to attributes, which allows the rule behavior to be customized by the user or a data source. Attributes and parameters are used in the same way other programming languages use *variables* to customize behavior. Most of the attributes' values can be set and read by various operations. Attributes are sometimes taken as additional context for operations to define and refine behavior. For example, predominant orientation and origin information are encoded in the scope and pivot attributes. When the split operation is used in the y-direction, this direction is relative to this orientation given by the scope and pivot locations stored in attributes.

The typical pattern of programming in CGA is to repeatedly expand-then-divide geometry. The rule to create a building model may start with a lot shape, expand with an extrude operation to create prism geometry as high as the building, and then use a comp operation to divide the prism into various faces. The face pointing upward expands to create a roof with a roofGable operation, while side faces are divided using the split operation to become floors and then windows. Another extrude operation finally recesses the windows into the façade. We continue to study such operations in more detail.

35.4.1.1 Operations

Learning to write CGA rules is predominantly the process of learning the various operations and their effects on geometry and attributes. While the complexity of existing rules can be overwhelming to the new user, the compact set of CGA operations presents a shallow learning curve.

CGA is a programming language designed to do one thing—model urban environments—and not much else. For this reason, we would describe it as a domain-specific (programming) language (DSL). For other domains, there are other programming languages: We may use L-Systems (Prusinkiewicz 1986) to generate flora or URDF (2019) to create robots. Because CGA is a DSL, its operations are carefully curated for the urban domain. A lot of theoretical effort was expended in finding a compact yet expressive set of operations. In contrast, general-purpose procedural modeling languages, such as Houdini (2019) and Rhino (2019), are not specialized in a single domain and have many complex operations to learn. Figure 35.9 introduces a handful of key CityEngine operations.

By repeatedly applying these operations, we can create a large variety of urban geometries. For example, the setback, extrude, comp, and roofGable operations can be used to create a house with a recessed top story and a gabled roof, as in the following Fig. 35.10.

An important observation is that CGA does not contain loop or repeat operations. To achieve repeating geometry (such as windows on a building façade or trees along a street), we can use the split operation with the asterisk (*) modifier to split a parent

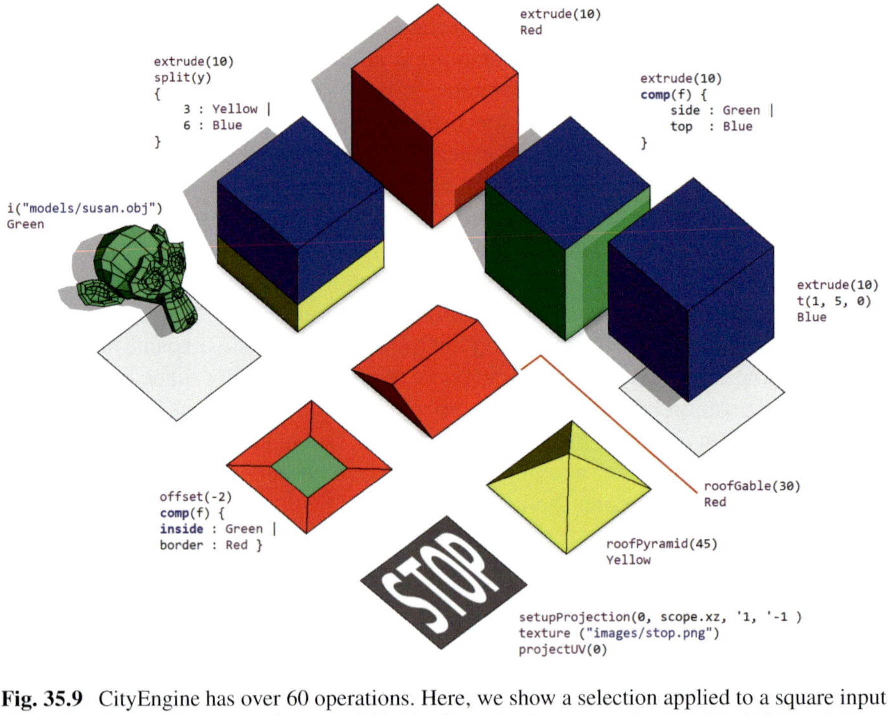

Fig. 35.9 CityEngine has over 60 operations. Here, we show a selection applied to a square input shape (gray), as well as example usage. Trivial rules with the names of colors (Red, Blue, etc.) are not shown, but would be included in the rule file

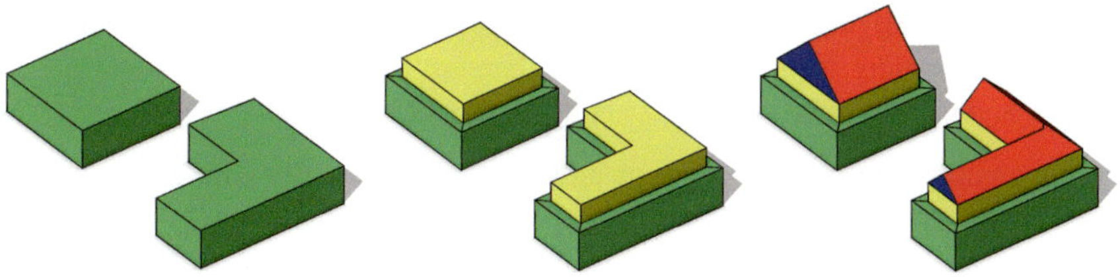

```
attr green="#00ff00" # this is the 'color hex string' for green.

@Startrule           # this is the rule to run first.
Lot -->              # define the rule 'Lot' which...
  Lower              # ...simply runs the rule 'Lower'.

Lower -->            # rule 'Lower':
  extrude (6)        # create a prism of height 6.
  color(green)       # color the faces green.
  X.                 # run a rule which doesn't exist (X)...
                     # ...to show all the green faces and finish(.)
```

```
attr green  = "#00ff00"
attr yellow = "#ffff00"    # let's create some yellow geometry.

@Startrule
Lot -->
  Lower

Lower -->                         # rule 'Lower':
  extrude (6)                     # create the same prism of height 6.
  comp (f) {                      # separate the prism faces...
    top : Upper |                 # ...the top face runs rule 'Upper'...
    side : color(green) X.        # ...while the side faces become green.
  }

Upper -->                         # rule 'Upper':
  offset (-1.5)                   # shrink the top-shape by 1.5 meters.
  comp(f) {                       # separate the shrunk shape from border.
    inside :                      # use the inside shrunk shape to...
      extrude (3)                 # ...extrude a prism of height 3...
      color(yellow) X.
    | border :                    # ...and for the remainder...
      color(green)  X.            # ...the side faces become green.
  }
```

```
attr green  = "#00ff00"
attr yellow = "#ffff00"
attr blue   = "#0000ff"    # we'll also use blue...
attr red    = "#ff0000"    # ...and red this time.

@Startrule
Lot -->
  Lower

Lower -->                     # rule 'Lower':
  extrude (6)
  comp (f) {
    top : Upper |
    side : color(green) X.
  }

Upper -->                     # rule 'Upper':
  offset (-1.5)
  comp(f) {
    inside :
      extrude (3)             # extrude a prism of height 3...
      comp (f) {              # separate the prism faces...
        top :
          Roof    |           # ...the top face runs rule 'Roof'...
        side :
          color(yellow) X.    # ...the sides remain yellow.
      }
    | border :                # as previously the border is green.
      color(green)  X.
  }

Roof -->                      # rule 'Roof':
  roofGable(45)               # create a gabled roof.
  comp (f) {                  # separate the roof faces...
    vertical :
      color(blue) X. |        # ...vertical faces become blue...
    all :
      color(red) X.           # ...and other faces become red.
  }
```

Fig. 35.10 A progression of three CGA rule files using operations including extrude, comp, and roofGable, accompanying models shown above. Note how we start with a simple rule and gradually extend it to create more complex geometries following the expand-then-divide paradigm. The green text highlights comments which are ignored by CityEngine, but help humans to understand the code

shape into a repeating number of child shapes with the same rules. This is illustrated in Fig. 35.11.

In our final example, we create geometry for streets. To create highway lanes, we wish to split down the long axis of the streets, which may be curved. The UV variant of the split operation achieves this. Finally, we may wish to add texture maps (bitmap images) over our geometry instead of simple colors using the texture operations, as in Fig. 35.12.

35.4.2 *Modeling Workflow*

Creating larger rule files can be a daunting task for those new to writing code. This is a skill that requires time to practice and learn, but when a little knowledge is gained is often intoxicating:

> The programmer, like the poet, works only slightly removed from pure thought-stuff. He builds his castles in the air, from air, creating by exertion of the imagination. (Brooks 1995)

This initial excitement often causes problems with inexperienced programmers; overconfidence causes a failure to understand the characteristics of a growing code base. As many small problems in the code ("bugs") become entrenched, it can become very time-consuming to make even small changes. We can provide some general guidance and tools which can help us build large CGA programs:

- Write small pieces of code at a time and test them frequently. This makes it much quicker to track down and isolate issues. If you cannot understand some behavior, it is frequently the case that too much code was written before trying to run it.
- Create reusable rules. A small rule that you have created which generates an "Acme brand window" may be reused if kept in separate file. CGA provides the import functionality to facilitate using this window rule in other rule files.
- Read the provided CGA documentation (*Help menu* → *CGA reference*).

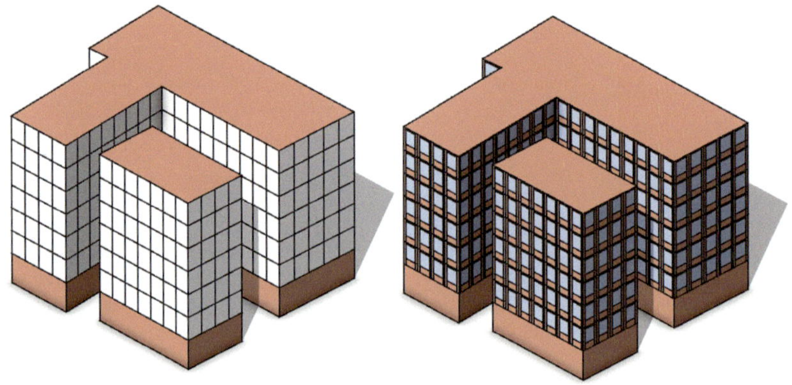

```
@Startrule
Lot -->                     # rule 'Lot'
    extrude (20)|
    comp(f) {
       side: Facade |
       top : Brick
    }

Brick -->                   # rule 'Brick'
    color("#d29a78") X.

Facade -->                  # rule 'Facade':
    split(y) {
      3.5: Brick |
      ~1 : TopFloors        # the remainder runs TopFloors.
    }

TopFloors -->               # rule 'TopFloors':
    split(y) {              # splits the face in the y direction...
      ~3: Floor             # aprox. 3m slices run the Floor rule.
    }*                      # the * causes the split to repeat.

Floor -->                   # rule 'Floor':
    split(x) {              # split along x (sideways)...
      ~2: X.                # ...creating geometry aprox. 2m wide.
    }*                      # the * causes the split to repeat.
```

```
@Startrule
Lot -->                            # rule 'Lot':
    extrude (20)
    comp(f) {
       side: Facade |
       top : Brick
    }

Brick -->                          # rule 'Brick':
    color("#d29a78") X.

Facade -->                         # rule 'Facade':
    split(y) {
        3.5: Brick |
        ~1 : TopFloors
    }

TopFloors -->                      # rule 'TopFloors':
    split(y) {
     ~3: Floor
    }*

Floor -->                          # rule 'Floor':
    split(x) {
        ~2: Tile                   # slices of the floor run Tile
    }*|

Tile -->                           # rule 'Tile':
   split (x) {                     # split tile sideways (x)
      0.3 : Brick |                # left 30cm becomes brick
      ~1  : split(y) {             # split the remainder vertically
        0.8 : Brick |              # bottom 80cm becomes brick
        ~1 : color ("#afc6e9")     # remainder is a blue 'window'
             X. |
        0.2 : Brick                # top 20cm becomes brick
      } |                          # finish off the first (x) split
      0.3 : Brick                  # right 30cm becomes brick
   }
```

Fig. 35.11 Example of using the split rule to subdivide a façade to create windows

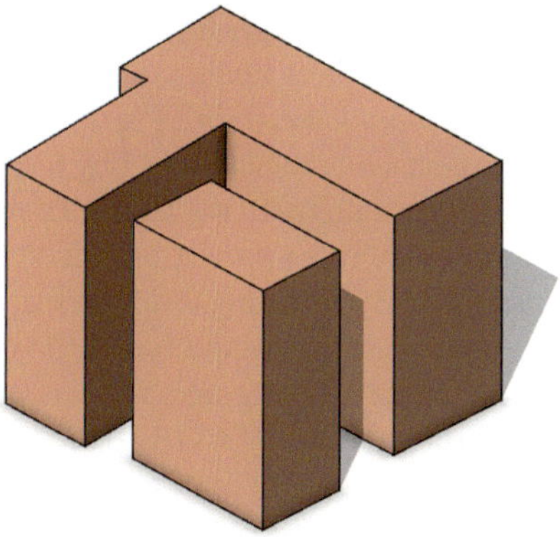

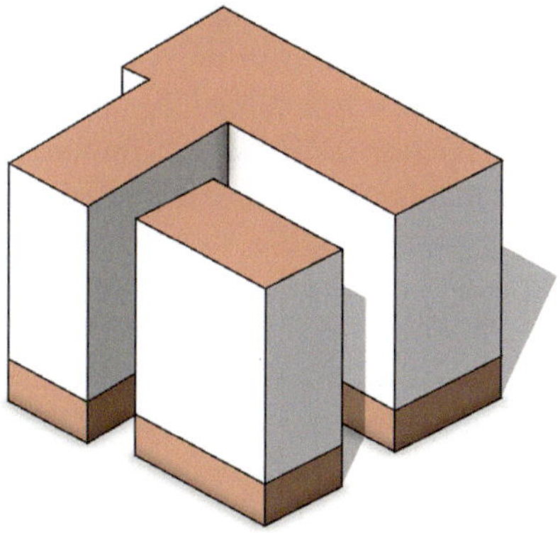

```
@Startrule
Lot -->                  # rule 'Lot':
    extrude (20)         # create a prism of height 20 meters.
    comp(f) {            # sepearate the faces of the prism...
       side: Brick |     # ...sides run the rule "Brick"...
       top : Brick       # ...as does the top.
    }

Brick -->                # rule 'Brick':
  color("#d29a78") X.    # color the shape, and display.
```

```
@Startrule
Lot -->                  # rule 'Lot':
    extrude (20)
    comp(f) {
       side: Facade |    # run the 'Facade' rule on the sides.
       top : Brick
    }

Brick -->                # rule 'Brick'
    color("#d29a78") X.

Facade -->               # rule 'Facade':
    split(y) {           # split the facade in the y (up)...
      3.5: Brick |       # ...the bottom 3.5 meters is brick...
      ~1 : X.            # ...the remainder is blank.
    }
```

Fig. 35.11 (continued)

- It is easy to get lost in the details of programming and write code that is easy to understand today but difficult to understand in a week's time when you have forgotten the details. Use code comments (sections of code which the computer does not see) to keep notes for yourself and inform future readers. CityEngine comments can be created in two ways:

  //everything on this line is a comment
  /* everything between the two asterisks is a comment */

- Collections of rule files can be large, written by multiple people, have multiple versions, or can even evolve different branches as they are developed. For these reasons, programmers will typically use a version control system (such as the insensitively named git (git 2019)) to manage their code.
- Be aware of the keyboard shortcuts and context (right-click) menus available in CityEngine. For example, if you have a shape selected with a rule and are editing the rule in the text editor, *Ctrl* + *S* followed by *Ctrl* + *G* (on Windows or Linux; use the command key instead of Ctrl on OS X) will save and show the updated 3D shape. In the 3D view, the *F* key will move the view to show the selected object, or *F9–F12* will show and hide various classes of objects.

Beyond general programming etiquette, CityEngine provides several bespoke mechanisms to help writing CGA rules. The Model Hierarchy panel shows a graph of the different rule applications (*Window* → *Show Model Hierarchy,* Fig. 35.13). This shows the *Inspect Model* tool button, which can be used to select a building to analyze (Note that *Inspect Model* is a different piece of functionality to the Inspector panel.). The resulting graph is shown in the panel, with every rule application illustrated by a gray arrow. Lines connect parent/child rule pairs. By selecting a rule in the graph, the 3D view will highlight the resulting geometry and show the scope, pivot, and trim planes valid for the application of the rule. Right-clicking on a rule node in the graph gives the option to jump to the corresponding portion of CGA. A single CGA rule will typically be applied in different locations and so will appear multiple times in the graph.

Another tool provided by CityEngine is the Façade Wizard (*Window* → *Show Façade Wizard*). For a single 2D façade, this aids in generating the split and extrude operations required for a well-parameterized façade.

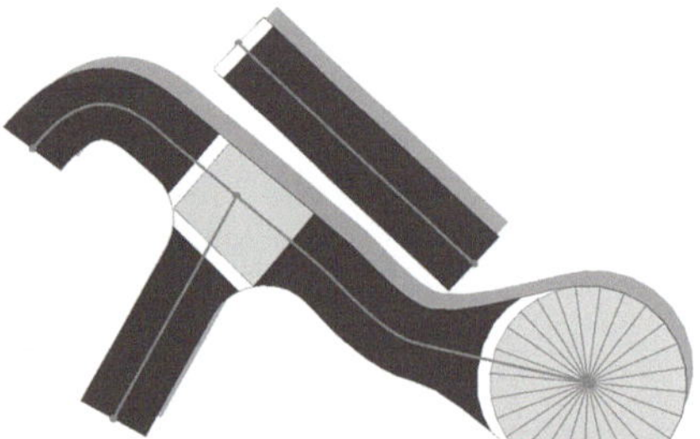

```
attr white  = "#ffffff"   # let's define white and gray...
attr gray   = "#555555"   # ...hex string colors.

@Startrule
Street -->                # for all Street (dynamic) shapes.
  split(u,uvSpace,1) {    # split the start of the street...
    0.7 : Create(white) | # ...and make it white.
      ~1 : Create(gray)   # the remainder is gray tarmac.
  }

Create(c) -->
    color (c)             # a utility rule to color...
    X.                    # ...and create geometry.
```

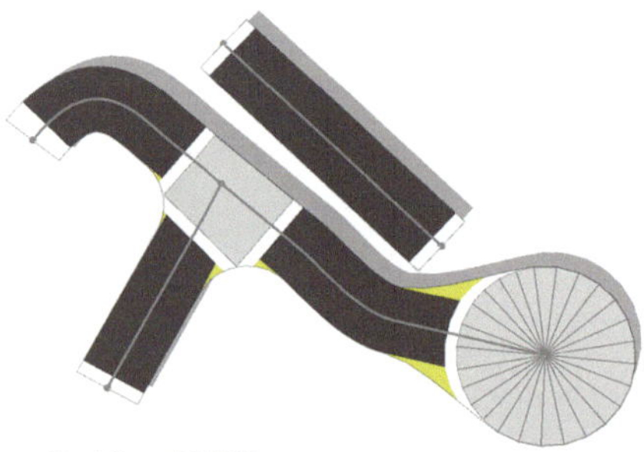

```
attr white  = "#ffffff"
attr gray   = "#555555"
attr yellow = "#ffff00"             # let's use yellow geometry too.

attr laneWidth = 3.5                # from the dynamic shape attr.
attr streetWidth = 7                # from the dynamic shape attr.

@Startrule
Street -->
  split(u,uvSpace,1) {
    0.7 : Create(white) |
      ~1 :
        split(u,uvSpace,2) {        # repeat the trick with UV set 2...
          0.7 : Create(white) |     # ...for white line at end of street.
          ~1 : WithGutters          # continue to process gutters.
      }
  }

/* gutters are wider parts of street shapes near junctions. */
WithGutters -->
  split(v,uvSpace,0) {              # split in v-direction on uv set 0 .
    -geometry.vMin :                # this value...
      Create(yellow)  |             # ...splits off left yellow gutter.
    rint ( streetWidth              # this equation gives the width of...
          / laneWidth) :
      Create(gray) |                # ...the central street.
      ~1 : Create(yellow) }         # split off right yellow gutter.

Create(c) -->
    color (c)
    X.
```

Fig. 35.12 Example of creating models for street shapes. The split rule is used with the UV parameter to split curved areas. The three different street UV sets split from different sides of the shapes. Finally, the normalize UV and texture commands create "stop" markings

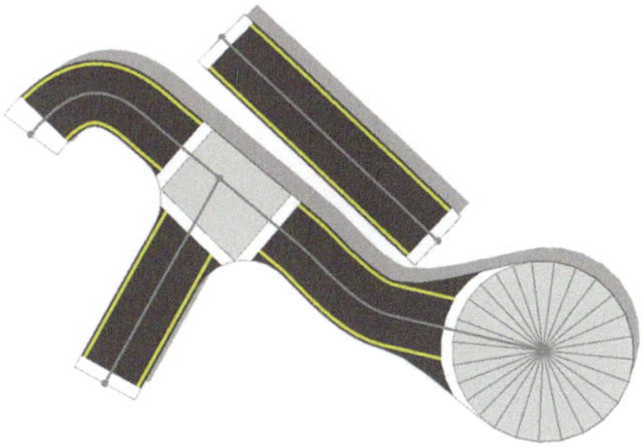

```
attr white  = "#ffffff"
attr gray   = "#555555"
attr yellow = "#ffff00"

attr laneWidth = 3.5
attr streetWidth = 7

@Startrule
Street -->
  split(u,uvSpace,1) {
    0.7 : Create(white) |
      ~1 :
        split(u,uvSpace,2) {
          0.7 : Create(white) |
          ~1 : WithGutters
      }
  }

WithGutters -->
  split(v,uvSpace,0) {
    -geometry.vMin :            # on second thought, let's...
      Create(gray)  |           # ...make the gutters gray.
    rint ( streetWidth
           / laneWidth) :
      WithoutGutter |           # create non-gutter gometry.
      ~1 : Create(gray) }

WithoutGutter -->
  split(v,unitSpace,0) {      # split along street edges (v).
    0.4 : YellowLine |         # to create left yellow line...
    ~1 : Create(gray) |        # ...central area fills remainder.
    0.4 : YellowLine           # and right yellow line.
    }

YellowLine -->
  split(v,unitSpace,0) {      # split along street edge (v).
    0.1 : Create(gray) |       # 5cm of gray...
    ~1 : Create(yellow) |      # ...then a 10cm yellow strip...
    0.1 : Create(gray)         # ...and a final 5cm of gray.
}

Create(c) -->
    color (c)
    X.
```

```
attr white  = "#ffffff"
attr gray   = "#555555"
attr yellow = "#ffff00"

attr laneWidth = 3.5
attr streetWidth = 7

@Startrule
Street -->
  split(u,uvSpace,1) {
    0.7 : Create(white) |
      ~1 :
        split(u,uvSpace,2) {
          0.7 : Create(white) |
          ~1 : WithGutters
      }
  }

WithGutters -->
  split(v,uvSpace,0) {
    -geometry.vMin :
      Create(gray)  |
    rint ( streetWidth
           / laneWidth) :
      WithoutGutter |
      ~1 : Create(gray) }

WithoutGutter -->
  split(v,unitSpace,0) {
    0.4 : YellowLine |
    ~1 : WithoutStop |         # create stop markings.
    0.4 : YellowLine
    }

YellowLine -->
  split(v,unitSpace,0) {
    0.1 : Create(gray) |
    ~1  : Create(yellow) |
    0.1 : Create(gray)
}

WithoutStop -->
  split(u,uvSpace,0) {         # split from the start of the street.
    -geometry.uMin :           # any junction areas...
      Create(gray) |           # ...become gray.
    ~1 :                       # everything else...
     split(u,unitSpace,0) {    # ...is split again in units (meters)...
       3 : Stop |              # ...to create a stop sign.
       ~1 : Create(gray) }     # everything else is a regular street.
   }

Stop -->
  normalizeUV (0, uv,          # stretch the image over all..
    collectiveAllFaces )       # ...the current geometry.
  texture("images/stop.png")   # texture with the stop.png image.

Create(c) -->
    color (c)
    X.
```

Fig. 35.12 (continued)

To deliver a CityEngine rule to an end user in a convenient format, use a rule package. This can be built by selecting the CGA file to export in the navigator, right-clicking, and selecting *Share As*.... Additional resources and metadata are specified in the dialog box. In this way, the resulting .RPK file may include many individual CGA files and other resources such as data in text files and texture images. Such a package is easily distributed as a single file, and Esri provides a cloud system to distribute rules.

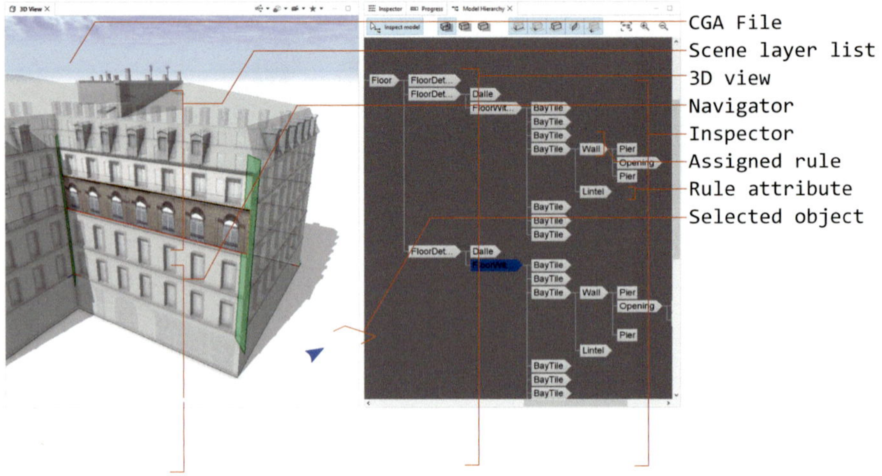

Fig. 35.13 The Model Hierarchy is a very useful tool for visualizing geometry. Left: a 3D view of a model from the first figure. The selected rule is highlighted and rendered with a solid color; the scope, pivot, and trim planes are also visualized. Right: the rule hierarchy identifies the rule which created the selected geometry. Clicking on another rule will show that rule's associated geometry. Note the Inspect Model button (top center) which is used to enable the Model Hierarchy functionality

35.4.3 Attributes

Having built our rules and assigned them to our shapes, we are often interested in further customizing the rule's expression using attributes.

Attributes are used to refine the evaluation of models within a rule application. They allow a rule to be generalized. For example, consider a number of otherwise identical buildings constructed from different materials; instead of a separate rule for each material, we may use a single rule with an attribute for the building material. Attributes can control any behavior of a rule, but typically, control features such as building height, age, or the number of pedestrians created on the sidewalks. CityEngine shows many of the available attributes for the selected shape and rule in the Inspector panel (Fig. 35.14); some rules have a great many attributes. The default attribute values are set by the rule. However, users can override the source of attributes to allow the rule to respond to different inputs.

The attributes in CityEngine have a multitude of different sources, and the interdependencies between them can be complex. Attribute sources include:

- Rule-sourced (Rule default), the default attribute behavior
- User-sourced
- Shape-sourced (Object attributes)
- Image- or shape-driven (Layer attributes).

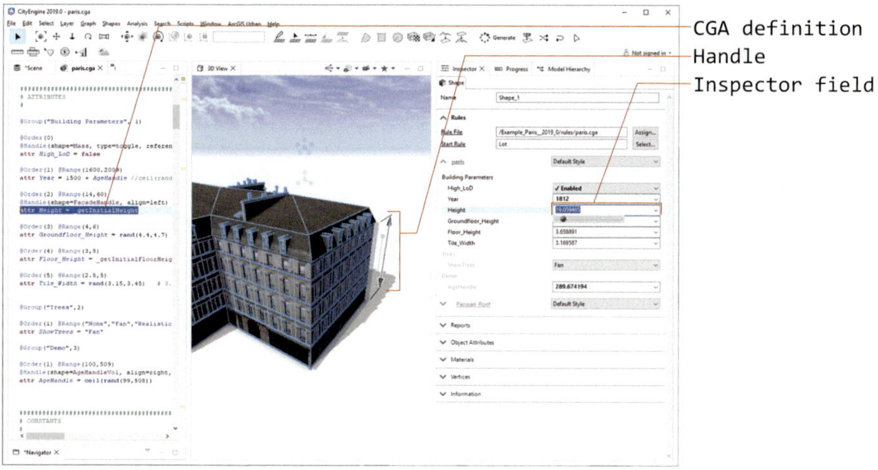

Fig. 35.14 Attributes are defined in the CGA file (left) and are edited either with handles (center) or using the Inspector (right)

These can be selected by clicking the down arrow next to an attribute in the Inspector panel and selecting *Connect Attribute*.... Rule-sourced attribute values are given in the CGA rule file. These attributes can be random; this feature can be used to add variation to a rule applied many times; for example, every building may be generated with the same rule, but given a height that is randomly selected between 10 and 20 m [attr height = rand (10,20)].

To allow users to change an attribute without editing the CGA file, attributes edited in the Inspector become user-sourced attributes. However, we may wish our attributes to come from other sources which may be driven by data. Object attributes are visible in the Inspector (under the *Object Attributes* heading) when a shape is selected. Object attributes can come from input data sources (e.g., OpenStreetMap data often gives every lot shape a building height attribute) or are created by dynamic shapes (e.g., the connectionStart and End attributes are added automatically to street shapes to specify the adjacent junction types).

Layer attributes sample their values from other shapes or a bitmap, as illustrated in Fig. 35.15. For example, we can drive the height-of-building attribute by using a georeferenced heightmap that has been captured by aerial LiDAR. In this way, we can control a rule using several different data sources. This approach significantly improves the accuracy of resulting geometry over a purely rule-driven procedural pipeline.

Finally, it is useful to know that the attributes for multiple shapes can be edited at once by selecting several shapes. Multiple shapes can be selected by shift-clicking or by dragging a selection box around them. Alternately, by right-clicking on shapes in the 3D view, various automatic selection options allow selection of many shapes within a layer. The Inspector shows the available attributes for the entire selection, and editing an attribute or source applies that attribute change to all the selected shapes.

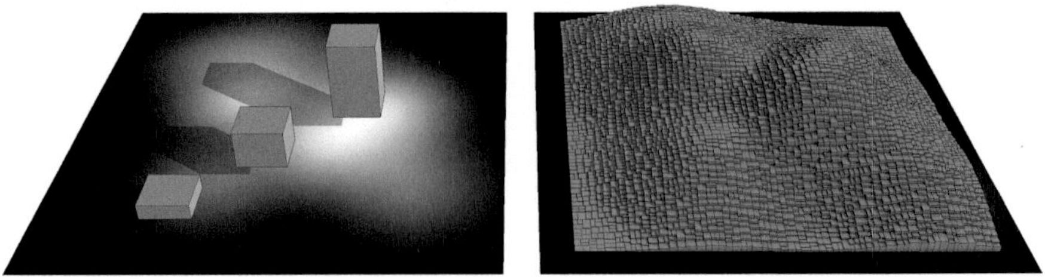

Fig. 35.15 Left: a black and white image imported as a texture is used to drive the height attribute of three rectangular suspended shapes, each with the same simple extrude rule. The white parts of the texture are sampled to large values, which are expressed as tall cuboids; black areas are small values which become short cuboids. Right: in this way, we may sample attributes from the same texture to vary building height (or any other attribute) across a city according to an image

35.4.4 Exploring Design Space

As a designer using CityEngine, the number of decisions that must be made can be very high. Complex rules present hundreds of attributes, and these must be aligned to user requirements, artistic visions, and practical considerations. Because every additional attribute adds a dimension to the design space, it can take a lot of time to explore large, heavily parameterized rules. Further, we may wish to design multiple scenarios: different rules, attributes, and shapes solving the same problem that we wish to compare side by side. CityEngine provides a Python interface for advanced programmers to control attributes (and many other scene elements) using custom code; typical uses are to create video animations of attributes or run custom design-space search algorithms. Most users, however, will want to avoid such complexities.

CityEngine presents a number of tools to help explore this design space of attributes visually. As we have seen, the simplest of these is the Inspector panel which arranges the attributes in groups specified by the rule file and allows the different attribute sources to be selected in a 2D interface. Given the large number of attributes in a rule such as the Paris example, it is often useful to see a visual representation of those attributes next to the 3D model. Handles present this functionality by showing the attributes (such as height) as controls in the 3D view. The handle system was inspired by the dimension lines of engineering diagrams, as introduced by Kelly et al. (2015). When a model with handle functionality is selected in the 3D view, the handles are shown at the edges of the model depending on the viewpoint. Various handles control different types of values: Boolean toggles, multiple-choice dials, distance-as-value dimension lines, and color selector triangular handles are available. The handle locations, behavior as the viewpoint moves, and appearance are defined by the @Handle annotation in the CGA rule file. They are designed by the rule creator and are only available if the rule author chooses to use them. Often the rule author will choose to expose only the most-used attributes using handles to avoid overcrowding the screen.

Handles change the value of an attribute throughout an entire rule evaluation for a single shape. There are situations where we wish to edit an attribute within a rule evaluation, for example, to make one story of a building taller than the others or to move the location of a single window in a large façade. In this situation, we can use local edits. These allow us to edit attributes with handles. Local edits are created by selecting the *Local Edits Tool*; depending on how the rule is structured, this tool may allow us to edit all local attributes in a row, column, or more complex patterns at once. Local edits are discussed further by Lipp et al. (2019).

As we modify rule attributes, we may be trying to achieve an objective target such as a target floor area for a building or group of buildings. CityEngine's reporting mechanism allows rules to collate such information and then prepare a summary report for each model. The report operation accumulates values whenever it is invoked, returning a sum total for the entire model [we may use the operation report ("area", 200)]. Multiple values (floor area, room volume, etc.) can be accumulated for each rule and displayed in the Inspector as a table. If CityEngine's dashboard functionality is used, these tables can be presented as a range of graphs which update automatically. They can show results over all models in the scene or only those selected.

By taking the time to add reports to your models and using the dashboard functionality, it becomes possible to explore the design space interactively with a wide range of users. For example, clients may appreciate being able to use the handles to edit building heights and receive instant feedback on the effects of available floor area and construction costs.

Beyond raw reported analytics, we may be interested in the visual consequences of our designs. CityEngine provides a range of tools for measuring distance and area in the 3D scene (Fig. 35.16), but most interestingly provides visibility calculations; this highlights the areas of models which are visible or not from a certain location under a given field of view.

Finally, scenarios allow us to compare different events. Each scenario can contain different layers of content on top of a shared background. For example, three different developments proposed for a city block with different height can be shown, while the surrounding city remains constant. A scenario can be duplicated and edited to explore a new design space.

Fig. 35.16 Analysis tools. Left: viewshed calculations showing visible (green) and occluded (red) areas. Middle: path length measuring tool. Right: area measuring tool

35.5 Beyond CityEngine: Export Pathways

After we have painstakingly created shapes, written rules, and adjusted parameters to generate our 3D reconstruction, we will want to view, export, and share our CityEngine scenes.

It should be noted that CityEngine's 3D view can create images with a reasonable-quality lighting model. There are options in the viewport panel (*View Settings*) to enable shadows (as cast by the sun), ambient occlusion (more accurate shadows in geometry creases), and field of view (the angle of the scene we see). Images can be saved from the 3D (*Bookmarks* → *Save Snapshot...*).

CityEngine's 3D view renderer is a real-time OpenGL renderer similar to those used for video games. If we would like more accurate physically based rendering (PBR) and are prepared to wait for each image to render, we can use a third-party renderer (such as POV-Ray, LuxRenderer, Unity game engine, Autodesk 3ds Max, or Blender) to create accurate images. These renderers are complex pieces of software in themselves, and the mechanics and artistry of setting up lighting and materials to create beautiful photorealistic images are beyond this chapter. However, in Fig. 35.17, we compare the default CityEngine rendering to the physically based Cycles renderer in Blender. We note the high quality of light simulation (reflections, shadows, and color bleeding) and material appearance.

To use an external renderer, we must export our models as 3D meshes from CityEngine to another package. CityEngine offers a variety of different formats to export models (*File* → *Export Models*): Wavefront's OBJ is a commonly used interchange format, but other more exotic formats include Collada, Autodesk FBX, and Alembic. Then a typical pipeline in a 3D modeling application such as Blender is to import the 3D meshes, set up textures, and position the camera and lights. Finally, a render operation is performed that might take minutes or even days to produce a large high-quality image.

To share our finished 3D meshes online with others as 3D objects, rather than 2D images, there are several options. There is a rapidly growing selection of Web-based 3D hosts (Sketchfab, SketchUp 3D Warehouse, or Google's Poly) who will host OBJ meshes online so that they may be viewed in a browser. Links to the resulting Web pages can be shared with clients and colleagues. However, these general 3D sites lack support for many details from a CityEngine scene. Esri provides two solutions to this problem: the CityEngine Web scene exporter (*File* → *Export Models...*) and the separate application ArcGIS Urban (*ArcGIS Urban* → *Synchronize all scenarios*). This ensures that details such as lighting information, different scenarios, and shape information remain visible and interactive for viewers, although editing attributes is not supported. Esri provides a convenient pipeline from CityEngine to host Web scenes on their online platform; this includes support for a "split-screen" to show two scenarios side by side in the browser.

Immersive technologies are a recent and popular trend in 3D visualization. Virtual reality (VR) is the most popular medium: Users wear a headset (such as the Oculus Rift or HTC Vive) which tracks head motions and shows different images to each

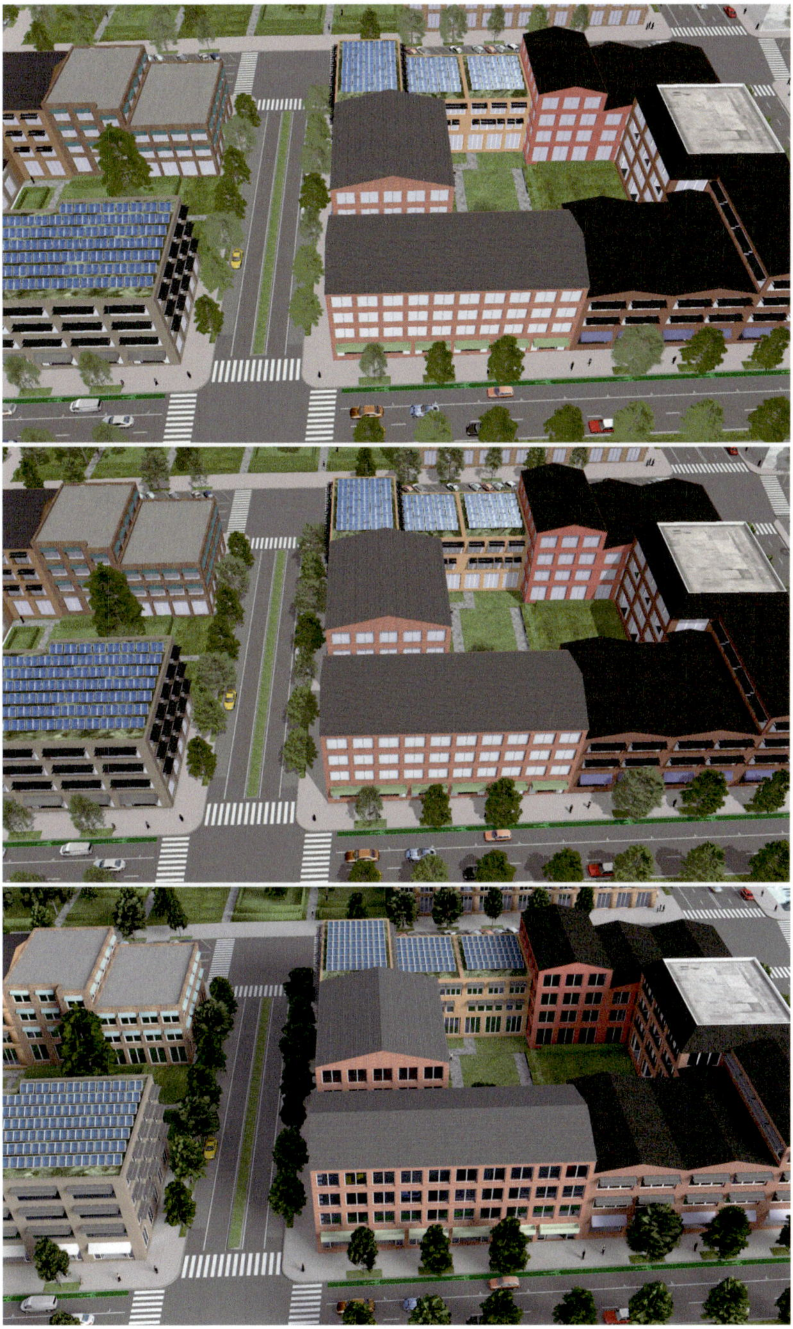

Fig. 35.17 Top: CityEngine's default OpenGL real-time renderer without ambient occlusion or shadows. Middle: with ambient occlusion and shadows. Bottom: Blender's Cycles renderer takes 12 min to render this image with soft shadows and reflective glass. The mesh was exported to Blender in the OBJ format

eye to create a realistic and immersive 3D experience. Creating these experiences is still a technical process and requires the use of a video game engine; the most developed CityEngine pipeline uses the Unreal Engine. CityEngine 2019.0 includes a beta Unreal Engine model exporter, the output of which can be imported into Unreal via the Datasmith toolkit. The technical details are documented online and are likely to change in the near future (Esri 2019d).

The CityEngine VR experience presents a tabletop containing the models (Fig. 35.18). This presents the exported models on a tabletop in a virtual office. Users are able to explore the models by dragging the model on the tabletop. Optionally, the user can teleport to pre-designated sites in the 3D world to get a street-level view of the model. These design decisions avoid some of the discomfort of moving users through VR at high speeds. The tabletop interface eliminates motion sickness by allowing users to stand over the scene and explore it from a "virtually static" location.

There are downsides to VR as a presentation format. A minority of people still experience motion sickness or discomfort, the headsets are not suitable to be worn for long periods of time, and they are still low resolution when compared to desktop monitors. These limitations are rapidly diminishing as improved hardware and software interfaces become available. However, for applications where immediate impact or immersion is important, they can be very powerful tools for stimulating discussion and gauging impact.

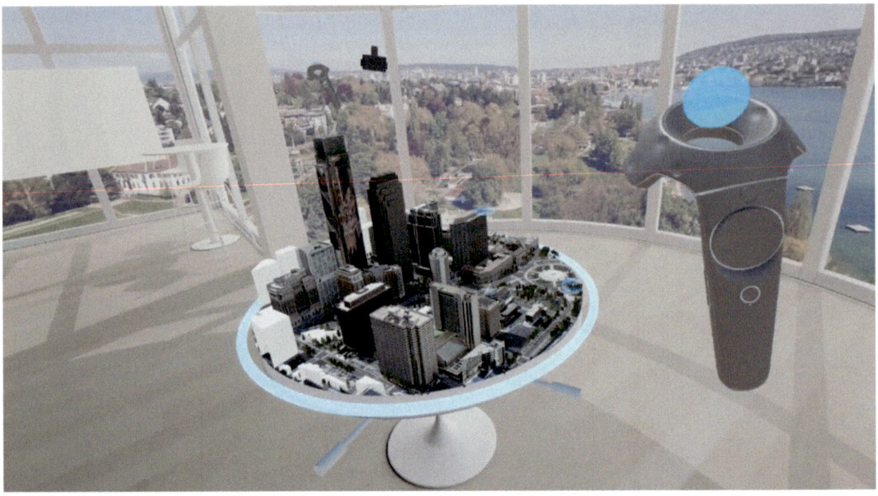

Fig. 35.18 CityEngine virtual reality presents a tabletop model to navigate using the controllers (right). Multiple users are supported (second user's headset shown top center)

35.6 Conclusion

CityEngine provides several pieces of unique functionality to the urban designer's toolkit. The ability to work with rules, rather than concrete manual models, can massively reduce the time, increase the scale, and lead to a multitude of new workflows for designing urban spaces. These new workflows allow us to quickly iterate solutions in a "client's office" situation; the solutions can be visualized and quantitatively analyzed on-the-fly. Such innovations allow faster user feedback as well as a better understanding of the problem and solution spaces.

All new workflows come with caveats and CityEngine is no exception. When a non-programmer (who does not write rules) uses CityEngine, he or she faces a limited selection of rule files. A programmer will usually have to invest substantial time learning CGA and creating rule files appropriate to the problem. However, there are substantial resources available to aid both groups of users: Large libraries of rules are available online, and comprehensive API documentation is provided for the programmer.

CityEngine originally grew out of Pascal Müller's academic work at ETH Zürich (Müller 2010). The continuing development of the CityEngine software product has been quietly shadowed by academic works detailing the future innovations in the system (Schwarz and Müller 2015); such technologies and features often flow between other Esri products and CityEngine itself. Recent innovations in dashboard data presentation and pipelines for virtual realities reflect the exciting ongoing development of the system at Esri R&D Center Zürich.

References

Blender (2019) https://www.blender.org/. Accessed 30 July 2019

Brooks FP (1995) The mythical man-month: essays on software engineering. Anniversary Edition, 2/E. Pearson Education India

Esri (2019a) https://doc.arcgis.com/en/cityengine/latest/tutorials/introduction-to-the-cityengine-tutorials.htm. Accessed 30 July 2019

Esri (2019b) https://doc.arcgis.com/en/cityengine/. Accessed 30 July 2019

Esri (2019c) http://www.arcgis.com/home/search.html. Accessed 30 July 2019

Esri (2019d) https://community.esri.com/docs/DOC-11563-cityengine-vr-experience-for-unreal-studio. Accessed 30 July 2019

git (2019) https://git-scm.com/. Accessed 30 July 2019

Houdini (2019) https://www.sidefx.com/products/houdini/. Accessed 30 July 2019

Kelly T, Wonka P, Müller P (2015) Interactive dimensioning of parametric models. Comput Graphics Forum 34(2):117–129

Lipp M, Wonka P, Müller P (2014) PushPull++. ACM Trans Graphics (TOG) 33(4):130

Lipp M, Specht M, Lau C, Wonka P, Müller P (2019) Local editing of procedural models. Comput Graphics Forum 38(2):13–25

Maya (2019) https://www.autodesk.co.uk/products/maya/overview. Accessed 30 July 2019

Müller P (2010) Procedural modeling of buildings. Thesis, University of Zürich. https://doi.org/10.3929/ethz-a-006397747

Müller P, Wonka P, Haegler S, Ulmer A, Van Gool L (2006) Procedural modeling of buildings. ACM Trans Graph 25(3):614–623

Parish YI, Müller P (2001) Procedural modeling of cities. In: Proceedings of the 28th annual conference on computer graphics and interactive techniques. ACM, pp 301–308

Prusinkiewicz P (1986) Graphical applications of L-systems. In: Proceedings of graphics interface and vision interface '86, pp 247–253

Prusinkiewicz P, Lindenmayer A (2012) The algorithmic beauty of plants. Springer Science & Business Media

Rhino (2019) https://www.rhino3d.com/. Accessed 30 July 2019

Schwarz M, Müller P (2015) Advanced procedural modeling of architecture. ACM Trans Graph 34(4):107

SketchUp (2019) https://www.sketchup.com/. Accessed 30 July 2019

SpeedTree (2019) https://store.speedtree.com/. Accessed 30 July 2019

URDF (2019) http://wiki.ros.org/urdf. Accessed 30 July 2019

Vanegas CA, Kelly T, Weber B, Halatsch J, Aliaga DG, Müller P (2012) Procedural generation of parcels in urban modeling. Comput Graphics Forum 31(2pt3):681–690

Wikipedia (2019) https://en.wikipedia.org/wiki/Grome. Accessed 30 July 2019

Tom Kelly is a member of faculty at the University of Leeds, where he conducts computer graphics research and teaches user interfaces. Previously, he worked as a software engineer at Esri and as a video game developer.

Chapter 36
Integrating CyberGIS and Urban Sensing for Reproducible Streaming Analytics

Shaowen Wang, Fangzheng Lyu, Shaohua Wang, Charles E. Catlett, Anand Padmanabhan, and Kiumars Soltani

Abstract Increasingly pervasive location-aware sensors interconnected with rapidly advancing wireless network services are motivating the development of near-real-time urban analytics. This development has revealed both tremendous challenges and opportunities for scientific innovation and discovery. However, state-of-the-art urban discovery and innovation are not well equipped to resolve the challenges of such analytics, which in turn limits new research questions from being asked and answered. Specifically, commonly used urban analytics capabilities are typically designed to handle, process, and analyze static datasets that can be treated as map layers and are consequently ill-equipped in (a) resolving the volume and velocity of urban big data; (b) meeting the computing requirements for processing, analyzing, and visualizing these datasets; and (c) providing concurrent online access to such analytics. To tackle these challenges, we have developed a novel cyberGIS framework that includes computationally reproducible approaches to streaming urban analytics. This framework is based on CyberGIS-Jupyter, through integration of cyberGIS and real-time urban sensing, for achieving capabilities that have previously been unavailable toward helping cities solve challenging urban informatics problems.

S. Wang (✉) · F. Lyu · S. Wang · A. Padmanabhan
Department of Geography and Geographic Information Science, and CyberGIS Center for Advanced Digital and Spatial Studies, University of Illinois at Urbana-Champaign, Urbana, USA
e-mail: shaowen@illinois.edu

S. Wang
e-mail: shaohua@illinois.edu

C. E. Catlett
Argonne National Laboratory, and the University of Chicago, Chicago, USA
e-mail: catlett@anl.gov

K. Soltani
Zillow Inc, Seattle, USA
e-mail: soltani2@illinois.edu

W. Shi et al. (eds.), *Urban Informatics*, The Urban Book Series,
https://doi.org/10.1007/978-981-15-8983-6_36

36.1 Introduction and Background

Harnessing urban big data to support scientific investigations into the impacts, challenges, and opportunities associated with increasing urbanization promises to enable the combination of analysis, observation, and modeling capabilities and to set and evaluate urban development policies and goals. Urban areas account for 70% of greenhouse gas emissions and energy use while contributing nearly 80% of total gross national product (GNP) (UN-Habitat 2011). They are consequently important levers to address environmental sustainability. For example, in the Chicago urban area, over 120 cities, towns, and villages have formally adopted a joint sustainability plan called the "Greenest Region Compact" (Marka 2019), understanding that challenges such as the reduction of greenhouse gas emissions or the improvement of air quality are regional in nature, requiring holistic approaches. Setting and tracking progress toward meeting those goals requires harnessing urban big data from not only traditional sources but from new sensor networks, high-bandwidth instruments such as light detection and ranging (LiDAR) and camera systems, and new sources such as those related to remote imaging or mobility. This will require a new approach to urban spatial analytics to support scientific investigations into the impacts, challenges, and opportunities associated with increasing urbanization. These investigations will require applying analysis, observation, and modeling capabilities to set and evaluate urban development policies and goals.

In this context, complex and massive urban data are increasingly collected for understanding and tackling such grand challenges, motivating many urban observatories that could play essential roles in resolving these challenges through science, engineering, and policy innovations (Miller et al. 2019). However, such observatories require innovative approaches to integrating dynamic and voluminous urban data with associated analytics for a variety of scientific problem-solving and decision-making purposes. Therefore, the overarching objective of this research is to develop an innovative cyberGIS (i.e. geographic information science and systems or GIS, based on advanced cyberinfrastructure: Wang 2010) framework for integrating urban sensing and analytics in a computationally reproducible way.

36.1.1 Urban Sensing Data

With recent rapid advances in and widespread adoption of location-aware devices and sensors, researchers in many fields now have an overwhelming wealth of dynamic urban data to investigate pressing scientific questions (Armstrong et al. 2019). These data streams from fixed as well as mobile platforms pose significant challenges to urban analytics. The past decade of open-data initiatives has similarly resulted in diverse new datasets related to urban infrastructure, operations, and activities (Huijboom and Van den Broek 2011). Anonymized open data is also available for many US cities such as the City of Chicago, with detailed records of over a decade of crimes, 311 service calls, permits, inspections, traffic flow, and other operational data. Integrating and analyzing these varied data sources will not only enable new questions about, and insights into the interdependencies of urban phenomena, but also new approaches to understanding complex environmental and urban systems (Xu et al. 2017). For example, a science question may be posed to explore the relationships between social factors such as crime or school performance and the environmental characteristics of urban neighborhoods (e.g. with or without green spaces, weak or strong local economy, etc.).

For many questions, data such as those related to air quality or urban heat lack the spatial and temporal resolutions that are needed to better understand neighborhoods. The National Science Foundation (NSF)-funded Array of Things (AoT), a partnership of the University of Chicago, Argonne National Laboratory, and the City of Chicago, set out to use new sensor technologies and embedded (or "edge") computation to create an experimental "instrument" comprising hundreds of intelligent sensing devices. The "nodes" were designed to measure Chicago's urban environment, air quality, and activity such as traffic or pedestrian flow at neighborhood resolution. The project integrates established and emerging sensor technologies to measure several dozen urban environmental conditions, with remotely programmable machine learning capabilities to measure factors for which no sensors are available, such as the flow of pedestrians through a park or of bicycles through an intersection (Catlett et al. 2017). AoT has deployed more than 130 nodes in Chicago. Test deployments are under way in over a dozen cities around the globe.

To illustrate the nature of data from such measurement instruments, a single month of AoT data is in the range of 2 GB compressed, or about 10 GB uncompressed. This is several times larger than the entire Chicago crimes database from 2001 to present (18 years) comprising 7 million rows of crime records.

36.1.2 CyberGIS

During the past decade, cyberGIS has emerged as a new generation of GIS, comprising a seamless integration of advanced cyberinfrastructure, GIS, and spatial analysis and modeling capabilities while leading to widespread research advances and broad societal impacts (Anselin and Rey 2012; Wang and Goodchild 2019). CyberGIS has provided a solid foundation for breakthroughs in diverse science, technology, and application domains, and contributed to the innovation of cyberinfrastructure overall (Wright and Wang 2011). During the past several years, cyberGIS has grown as a vibrant interdisciplinary field while the cyberGIS community has achieved significant advances in tackling challenging environmental and geospatial problems (e.g. Hu et al. 2017, Liu et al. 2018).

36.1.3 Spatial Data Synthesis

Substantial progress has been made through a data science project funded by NSF to establish core spatial data synthesis capabilities (e.g. integrating geotagged data streams from social media, census data, and urban infrastructure registry data; Wang 2016). The core capabilities were developed and deployed using cyberGIS supercomputing and cloud architecture to support spatial big data analytics. These capabilities include: (a) vector-data processing; (b) raster processing; (c) integration of heterogeneous spatial data streams; (d) spatial data visualization; and (e) spatial data retrieval and storage.

Developing synthesis capabilities for varied data from a multitude of sources poses new challenges due to the dynamic nature of the data sources and the user-driven nature of data synthesis, which requires the process to be always-on and highly available, demanding innovative computational capabilities. The NSF project has demonstrated powerful synthesis capabilities for spatial data that were developed to overcome the challenge of handling urban big data by researchers who may not be fully trained to employ advanced cyberinfrastructure (Soliman et al. 2017). The developed capabilities benefit from integrated high-performance and cloud computing to overcome some key challenges such as providing on-demand access to virtual distributed processing clusters with elastic resource provision. The cyberGIS framework described in this chapter integrates these capabilities to enable urban discovery and innovation based on streaming data and related urban analytics (Fig. 36.1).

36.1.4 Cyberinfrastructure

The varied types of urban data and associated analytics introduce critical requirements for innovating cyberinfrastructure and cyberGIS. The varied types, sizes, and formats of data pose a need for varied modalities of computing. For example, fast-streaming data from numerous AoT nodes will need an elastic and integrated high-performance computing (HPC) and cloud infrastructure to manage and process the data in near-real time, while historical datasets like census and topographic datasets can be processed in an HPC batch environment.

Resourcing Open Geospatial Education and Research (ROGER) has been established using experiences gained from an NSF Major Research Instrumentation project for computation- and data-intensive processing and analysis of geospatial data. It provides hybrid computing modalities, including high-performance computing (HPC) batch, data-intensive computing based on Hadoop and Spark, and cloud computing, backed by a petascale common data store (Wang 2017). Moreover, ROGER offers a wide variety of geospatial software packages, forming the core computational environment of the cyberGIS framework.

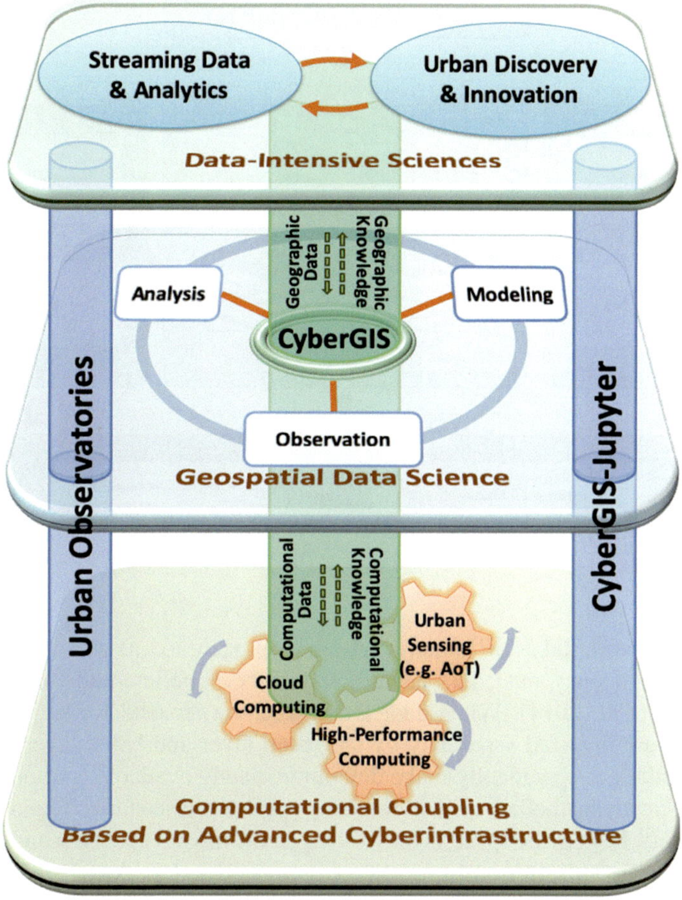

Fig. 36.1 A cyberGIS framework for streaming analytics to enable urban discovery and innovation

36.2 Framework

36.2.1 Architecture

The framework is designed to integrate cyberGIS with urban sensing data for (1) facilitating user interactions with streaming urban analytics through an online environment; (2) providing cyberGIS capabilities to achieve scalable urban analytics; and (3) managing the execution of analytics and their interactions with measurements. These functions are accomplished by: (a) the speed layer; (b) the batch layer; and (c) the serving layer, which are coupled with scalable computing capabilities including a workload-aware data and computation management capability (Fig. 36.2).

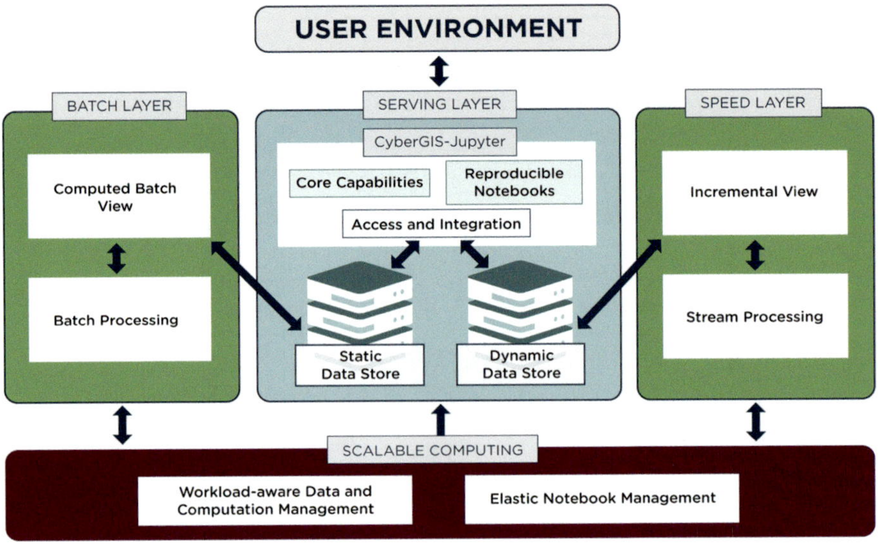

Fig. 36.2 Architecture

The framework takes a holistic system approach to: a) varying workloads including low-latency read, fast update, and ad-hoc queries; and b) linear scalability (Yang et al. 2014). When data arrive (e.g. via Apache Kafka; Kreps et al. 2011), they are ingested separately by the speed layer and batch layer. The speed layer is required to specifically make data immediately available for both real-time queries and analysis that are critical for some application scenarios (e.g. emergency management). Hence, the speed layer focuses on the most recent data and streaming analytics and is built on event-processing frameworks (e.g. Apache Storm 2020). On the other hand, the batch layer is designed to handle the integration with large historical datasets, with computationally intensive tasks performed on it. Therefore, the speed layer is designed to sustain high-frequency writes and provide a real-time view into the data while the batch layer is developed for read intensive and analytical workloads. Both batch and speed layers are connected to end users by the serving layer, which accesses the results of previous operations through a diverse range of data stores, including in-memory databases (e.g. REDIS 2020), NoSQL databases (e.g. Cassandra; Apache Cassandra 2020) and big data storage systems (e.g. HDFS; Shvachko et al. 2010). The serving layer provides the interactive user interfaces described in the following section.

36.2.2 User Environment

The user environment is established by enhancing CyberGIS-Jupyter to achieve reproducible and scalable computational tasks (Yin et al. 2019). Through this online environment, a user may invoke a CyberGIS-Jupyter notebook with a suite of analysis tasks, perform the tasks that can be executed on cyberinfrastructure resources, and customize the notebook for specific reproducible investigations that can be shared with other users. The user may also be interested to access automated workflows using cyberGIS visual analytics with a particular focus on specifying workflow parameters, interpreting workflow results, assessing visualizations, and sharing results and visualizations with pertinent collaborators and communities. The user environment is designed for a large number of users to simultaneously conduct streaming analytics.

36.2.3 Analytics

Spatial references and spatiotemporal resolutions are fundamental characteristics of urban data. Conflating urban data for both analytics and visualization purposes necessitates transforming the data into common projection systems and spatiotemporal units. For example, map reprojection achieves this transformation by applying common map operations such as coordinate translation, framing, forward- and inverse-mapping, and interpolation or resampling. Our earlier work has developed techniques to do reprojection using HPC resources (Finn et al. 2019). Another core capability aims to provide friendly interfaces through which users can interact with urban sensing data and related analyses based on map layers, charts, and tables. We have developed a Web-based and Open Geospatial Consortium (OGC) compliant solution capable of providing interoperable access to heterogeneous spatiotemporal data through the support of several Web services such as WMS, WFS, WCS, and WPS, and state-of-the-art mapping libraries (e.g. leaflet, d3.js) to enhance the visual representation of urban data.

36.3 Case Study

36.3.1 Study Area

The Chicago Metropolitan Area (CMA) provides an ideal test case for the framework. The CMA covers approximately 28,000 km^2 with a population of over 10 million people and is the third largest economy in the USA. It is at the crossroads of the rail, road, and air transportation infrastructures in North America. Extreme heat has already had detrimental effects on the Chicago urban population and by extension on the regional and US economy (Karl and Knight 1997). Elevated night temperatures over multiple days, exacerbated by urban heat-island (UHI) effects, are implicated in human health impacts (Semenza et al. 1996) as is neighborhood economic vitality (Browning et al. 2012). Given that the average summer time temperatures in the midwest are expected to increase by 3–6°F in the next 25–50 years (Wuebbles and Hayhoe 2004), the framework is crucially important for examining the urban microclimate at finer spatial and temporal granularities and directly coupling data with urban heat-related analytics to enhance our understanding of related issues in urban environments.

36.3.2 AoT Data

This case study uses data from AoT, coupled with computationally intensive spatial analyses, to explore a "smart city" vision that can make urban planning and policy adjustment possible on time scales of days or weeks rather than more traditional multi-year time windows. AoT nodes include both sensors (including cameras and a microphone) and embedded ("edge") computing resources, enabling remotely programmed machine learning to analyze data in situ. Currently, AoT nodes measure temperature, relative humidity, barometric pressure, light, vibration, carbon monoxide, nitrogen dioxide, sulfur dioxide, ozone, ambient sound pressure, and particulate matter. Nodes analyze images at 30 s intervals to count pedestrians and vehicles, transmitting these numbers along with readings from the sensors every 30 s to a central data repository. A map for the locations and types of sensors of AoT in Chicago is available at the project website (Catlett 2020).

Data are open and free, available for bulk download and through a real-time API. With respect to climate, AoT data have been used as part of a project funded by the Department of Energy's Exascale Computing Program for calibration and parameterization of fine-resolution weather models (Jain et al. 2018). Figure 36.3 shows a general workflow of how AoT measurement data can be translated into useful smart city applications.

Initiated with experimental nodes deployed in 2016, the project is implemented using Argonne's Waggle hardware/software platform (Beckman et al. 2016). As of late 2019, the 130 nodes in Chicago and over 60 nodes being deployed in partner cities represent the fourth generation of the platform (Fig. 36.4). Recent funding from NSF for the SAGE (Beckman et al. 2019) project aims to move to the fifth generation with substantially increased edge computing power, new sensors, and with experimental deployments in multiple observatories including the NSF's National Ecological Observation Network (NEON; Keller et al. 2008) and High-Performance Wireless Research and Education Network (HPWREN; Hansen et al. 2002).

The spatial distribution of nodes is illustrated in Fig. 36.5 showing the municipality of Chicago (589 km^2). The density of deployment varies from every block along several streets in the downtown area to more sparse distribution in residential areas. Locations are selected in cooperation with science teams, city officials, and community groups. An analysis by the University of Chicago's Center for Spatial Data

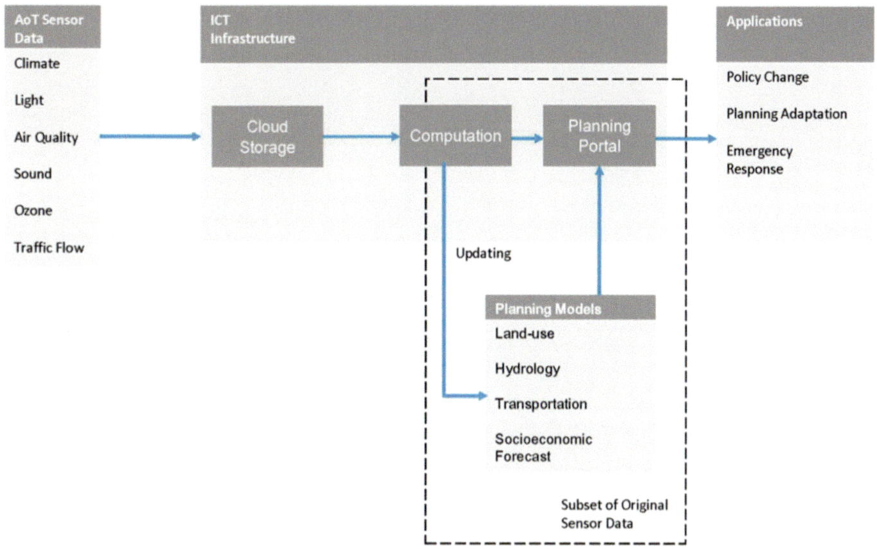

Fig. 36.3 A workflow from AoT sensor data to smart city applications

Science showed that 80% of Chicago's population lives within 2 km of an AoT node and 42% live within 1 km. While traditional sources for measurements such as air quality are available, for instance, there are fewer than 10 Environmental Protection Agency sites in the Chicago municipality, and most only measure 1 or 2 pollutants. AoT is an experimental instrument with respect to the technologies, and similarly, the density of nodes is aimed at optimal placement for various research or policy questions and their associated measurement requirements.

Another issue worth noting is that different generations (or "models") of AoT nodes (three models are in operation as of late 2019) vary with respect to sensors and capabilities. Only a few of the early nodes measured particulate matter, but all of the fourth-generation nodes are equipped with particulate-matter sensors. Similarly, the microphone in early nodes measured aggregate sound pressure, while new nodes provide measurements for ten octaves. As shown in Fig. 36.5, several nodes may be not working at a specific time, and during software updates and experimental software deployments, many nodes may be unavailable for periods of time. Figure 36.5 indicates that orange nodes are active while blue nodes denote inactive nodes. In reality, the number of nodes that are available may not equal the total number of AoT nodes deployed. The Waggle platform provides resiliency to communication outages, caching all measurements until the data have been transmitted to the central servers and acknowledged as received. Thus, in periods where nodes appear unavailable, the data for that period of time may become available later. Such factors are less visible in the bulk downloads than in using the real-time API.

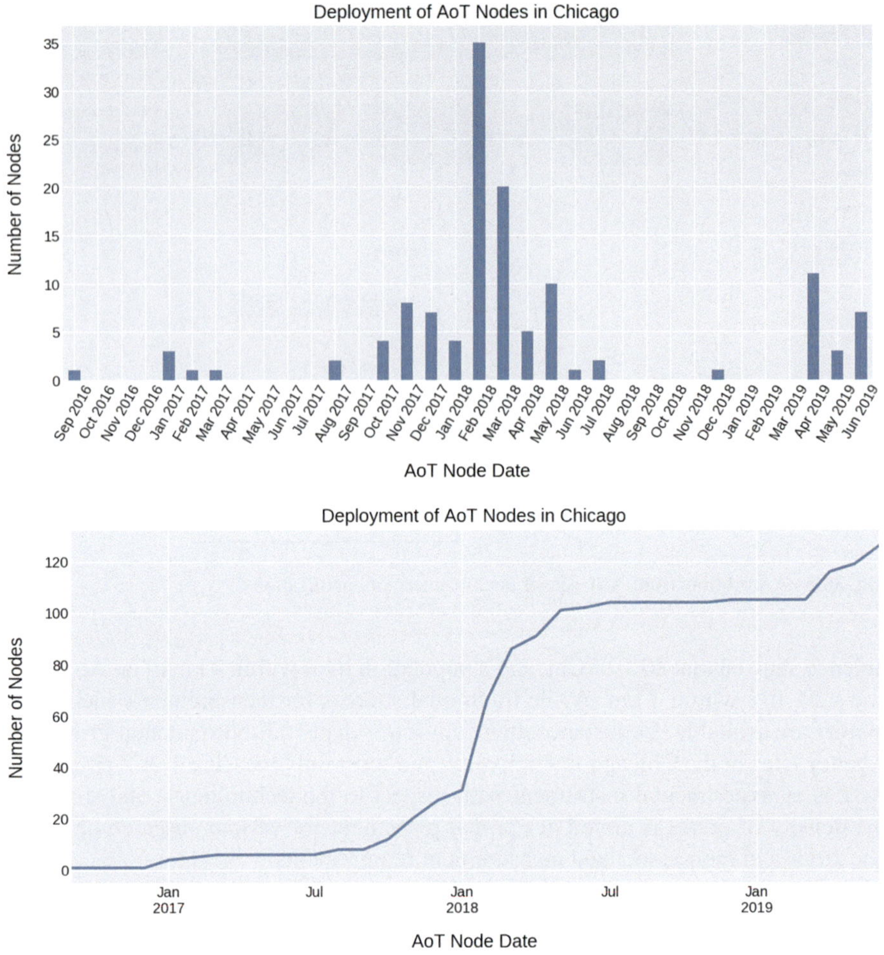

Fig. 36.4 The deployment information for AoT nodes in Chicago

36.3.3 CyberGIS-Jupyter

CyberGIS-Jupyter serves as the foundational engine for capturing and analyzing real-time streaming AoT data. CyberGIS-Jupyter is equipped with cyberGIS libraries scaling to both high-performance computing and cloud resources (Padmanabhan et al. 2019) and hence can support computationally intensive spatial analysis for users not only to capture the real-time, high-frequency data, but also to conduct urban analytics with AoT data. In this case study, real-time location-based AoT data can be used for understanding Chicago's heat environment. For example, temperature patterns can be derived based on AoT data as shown in Fig. 36.6. For all AoT nodes with temperature sensors, the temporal trend on September 30, 2019, is visualized

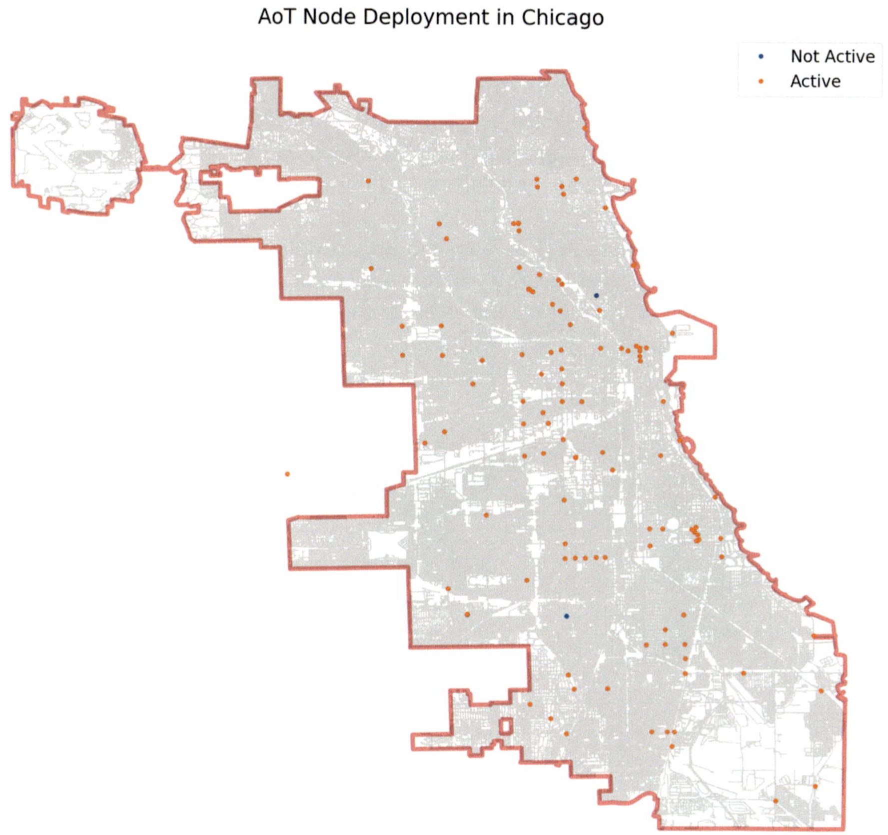

Fig. 36.5 The spatial distribution of AoT nodes in Chicago

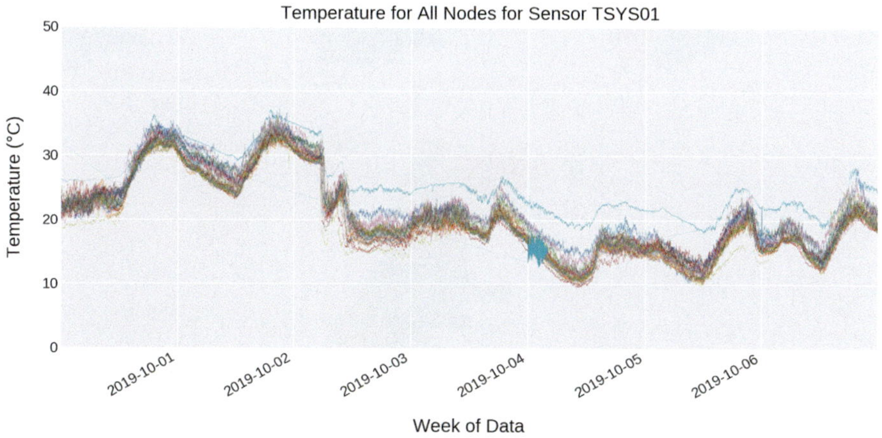

Fig. 36.6 The temperature curve derived from AoT data

on CyberGIS-Jupyter with different colors indicating different nodes. The AoT data have high frequency with the temperature data recorded every 26 s on average.

Due to the huge amount of data stored, with 2–3 GB of data captured from AoT every week, AoT's API cache only keeps 3–4 weeks of fresh data. In order to get the data back in 2017, for example, we need to download the whole dataset from the AoT bulk download website (or a subset of months of interest) and start our data processing from there.

Using the AoT streaming API as our data access option, spatial analysis of the temperature data and the geolocation of the AoT nodes can be conducted based on CyberGIS-Jupyter. Considering the need for identifying dense concentrations of high-temperature areas, Fig. 36.7 shows temperature patterns within one week

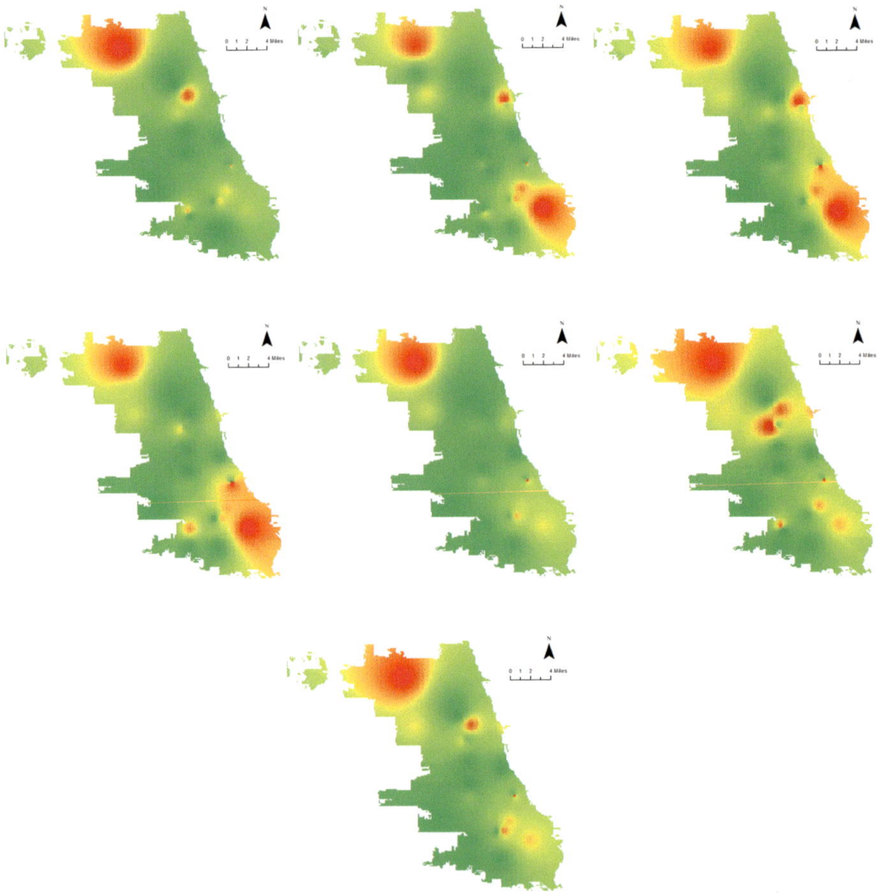

Fig. 36.7 Temperature maps in Chicago based on AoT sensors using a spatial interpolation algorithm. Temperature measurements are in degrees Celsius. From the top left to last map in the last row, each map represents the temperature distribution captured at 6am on September 30th, October 1st, October 2nd, October 3rd, October 4th, October 5th, and October 6th, respectively

in 2019. One can distinguish some hot spots from these heat maps. A workflow has been developed to capture the temperature data of the Chicago area based on CyberGIS-Jupyter from all of the available temperature sensors in Chicago using AoT's API at 6am in the morning from September 30th to October 6th on a daily basis. Combining with the geolocation of the sensors, the dynamic maps shown in Fig. 36.7 were generated using an inverse-distance weighted algorithm for spatial interpolation (Wang and Armstrong 2003). As shown in Fig. 36.7, throughout the week, the temperature in northwest Chicago, near Jefferson Park and North Park, and oftentimes in southeast and downtown Chicago, was higher than the average temperature in other areas. It is straightforward to understand that the temperature in downtown and southeast Chicago was higher due to human activities, since those areas have high population density. We investigated the sensor located in northwest Chicago (latitude 41.97 N, longitude 87.76 W, Fig. 36.8) and found it is installed near an underground transformer and some external air conditioners, which seem to be the heat sources. In addition, the density of sensors in northwest Chicago is lower than in other urban areas as shown in Fig. 36.5, leading to the skewed spatial interpolation result near Jefferson Park. The workflow for this analysis and associated data is represented as a CyberGIS-Jupyter notebook that can be shared with other users for reproducing the same results. The notebook can be adapted to accommodate data from different AoT nodes and time ranges and support different parameter values of the analysis (e.g. the number of the nearest neighbors in the spatial interpolation algorithm).

Fig. 36.8 A Google Steetview image of the AoT node located at latitude 41.97 N, longitude 87.76 W near Jefferson Park on Chicago's Northwest side

Similar to the example for analyzing temperature patterns demonstrated above, CyberGIS-Jupyter allows users to select other measurements from specific AoT nodes and specify temporal ranges to retrieve corresponding data streams for conducting computationally intensive analytics based on advanced cyberinfrastructure. Each workflow for combining AoT and other related data with specific analytics can be represented as CyberGIS-Jupyter notebooks that can record the provenance of computational steps in the workflow. Many users can simultaneously compose and run their notebooks on CyberGIS-Jupyter without noticing that their notebooks are executed on advanced cyberinfrastructure. While it is often challenging to "freeze" dynamic data streams to experiment with various analytical scenarios, CyberGIS-Jupyter notebooks can be shared among users to enable collaborative development and computational reproducibility of urban analytics with dynamic data (https://go.illinois.edu/CyberGIS-UrbanInformatics).

36.4 Concluding Discussion

Large cities like Chicago increasingly engage data-driven methods for urban planning and management, including for example land-use and transportation modeling, economic forecasts, and environmental monitoring. However, the ability to continuously monitor and alter policies of urban planning and management in a responsive manner is hampered by the difficulty of harnessing high-quality, spatially explicit, and temporally continuous data. In the USA, for example, large-scale land-use planning requires fine-resolution land cover data that is only available every five years from the National Land Cover Database. Similarly, socioeconomic models depend heavily on a census that is conducted on a ten-year interval. Due to these difficulties, though cities incorporate data-driven approaches in their planning processes, it is still challenging to implement the "smart city" vision based on fast data streams. A key barrier is the inability to make timely interventions and management decisions when environmental, social, or economic processes take place dynamically.

To address these challenges, this research has demonstrated that users can conduct computationally intensive streaming analytics using CyberGIS-Jupyter and AoT data without having to possess in-depth technical knowledge of cyberGIS or cyberinfrastructure. AoT data can be harnessed through CyberGIS-Jupyter to help users to monitor urban heat and other key indicators of urban dynamics. The cyberGIS framework described in this chapter is able to resolve the volume and velocity of urban big data through the support of advanced cyberinfrastructure; meet the computing requirements for processing, analyzing, and visualizing these datasets; and support concurrent online access to CyberGIS-Jupyter notebooks for collaborative development and computational reproducibility of urban analytics.

Regarding future research in urban informatics involving fast data streams, it is both important and challenging to achieve reproducible urban analytics. Without computationally reproducible urban analytics, it would be difficult, if not entirely impossible, to convince decision makers and practitioners to adopt such analytics in any real-world settings. Fast data streams produce data continuously and pose significant challenges that must be addressed through novel algorithms that treat spatial and temporal characteristics synergistically. Furthermore, exciting and important cyberGIS research is urgently needed to better understand and support computational reproducibility of urban analytics, which requires holistic approaches to optimizing access and management of cyberinfrastructure resources, trading off performance and uncertainty of spatial and spatiotemporal algorithms, and generalizing standards and specifications for the building blocks of urban analytics.

Acknowledgements This chapter and associated materials are based in part upon work supported by the National Science Foundation (NSF) under grant numbers 1443080, 1532133, 1743184, 1833225 and 1935984. The work used the ROGER supercomputer, which was supported by NSF under grant number 1429699. Any opinions, findings, and conclusions or recommendations expressed in this material are those of the authors and do not necessarily reflect the views of the funding agencies.

References

Anselin L, Rey SJ (2012) Spatial econometrics in an age of CyberGIScience. Int J Geogr Inform Sci 26(12):2211–2226

Apache Cassandra (2020) https://cassandra.apache.org/. Accessed 30 Jan 2020

Apache Storm (2020) https://storm.apache.org/. Accessed 30 Jan 2020

Armstrong MP, Wang S, Zhang Z (2019) The Internet of Things and fast data streams: prospects for geospatial data science in emerging information ecosystems. Cartography Geogr Inform Sci 46(1):39–56

Beckman P, Catlett C, Altintas I, Collis S, Kelly E (2019) Mid-Scale RI-1: SAGE: a software-defined sensor network. Award 1935984, national science foundation

Beckman P, Sankaran R, Catlett C, Ferrier N, Jacob R, Papka M (2016) Waggle: an open sensor platform for edge computing. IEEE SENSORS

Browning CR, Cagney KA, Iveniuk J (2012) Neighborhood stressors and cardiovascular health: crime and C-reactive protein in Dallas, USA. Soc Sci Med 75(7):1271–1279

Catlett CE (2020) https://arrayofthings.github.io/. Accessed January 30, 2020

Catlett CE, Beckman PH, Sankaran R, Galvin KK (2017) Array of things: a scientific research instrument in the public way: platform design and early lessons learned. In: Proceedings of the 2nd international workshop on science of smart city operations and platforms engineering. ACM

Finn MP, Liu Y, Mattli DM, Behzad B, Yamamoto KH, Shook E, Padmanabhan A, Stramel M, Wang S (2019) High-performance small-scale raster map projection empowered by cyberinfrastructure. In: Wang S, Goodchild MF (eds) CyberGIS for geospatial discovery and innovation. Springer, Dordrecht, pp 171–188

Hansen T, Yalamanchili P, Braun HW (2002) Wireless measurement and analysis on HPWREN. In: Proceedings of passive and active measurement workshop, fort collins, Co
Hu H, Lin T, Wang S, Rodriguez LF (2017) A cyberGIS approach to uncertainty and sensitivity analysis in biomass supply chain optimization. Appl Energy 203:26–40
Huijboom N, Van den Broek T (2011) Open data: an international comparison of strategies. Eur J EPractice 12(1):4–16
Jain R, Luo X, Sever G, Hong T, Catlett C (2018) Representation and evolution of urban weather boundary conditions in downtown Chicago. J Build Perform Simul 1–14
Karl TR, Knight RW (1997) The 1995 Chicago heat wave: how likely is a recurrence? Bull Am Meteor Soc 78(6):1107–1120
Keller M, Schimel DS, Hargrove WW, Hoffman FM (2008) A continental strategy for the national ecological observatory network. Front Ecol Environ 6(5):282–284
Kreps J, Narkhede N, Rao J (2011) Kafka: a distributed messaging system for log processing. In: Proceedings of NetDB
Liu YY, Maidment DR, Tarboton DG, Zheng X, Wang S (2018) A CyberGIS integration and computation framework for high-resolution continental-scale flood inundation mapping. J Am Water Resour Assoc 54(4):770–784
Marka E (2019) Greenest region compact. Metropolitan mayors caucus. https://mayorscaucus.org/initiatives/environment/rec/. Accessed 29 Jan 2020
Miller H, Akar G, Clifton K, Gopalakrishnan S, Tufte K (2019) Exploring a research network of urban sustainability observatories via data-enabled university-community partnership. NSF workshop report
Padmanabhan A, Yin D, Lyu F, Wang S (2019) Bridging local cyberinfrastructure and XSEDE with CyberGIS-Jupyter. In: Proceedings of the practice and experience in advanced research computing on rise of the machines (learning). ACM
REDIS (2020) https://redis.io. Accessed 30 Jan 2020
Semenza JC, Rubin CH, Falter KH, Selanikio JD, Flanders WD, Howe HL, Wilhelm JL (1996) Heat-related deaths during the July 1995 heat wave in Chicago. N Engl J Med 335(2):84–90
Shvachko K, Kuang H, Radia S, Chansler R (2010) The hadoop distributed file system. In: 26th IEEE symposium on mass storage systems and technologies (MSST). IEEE
Soliman A, Soltani K, Yin J, Padmanabhan A, Wang S (2017) Social sensing of urban land use based on analysis of Twitter users' mobility patterns. PLoS ONE 12(7):e0181657
UN-Habitat (2011) Cities and climate change: global report on human settlements 2011. https://unhabitat.org/?mbt_book=cities-and-climate-change-global-report-on-human-settlements-2011. Accessed 29 Jan 2020
Wang S (2010) A CyberGIS framework for the synthesis of cyberinfrastructure, GIS, and spatial analysis. Ann Assoc Am Geogr 100(3):535–557
Wang S (2016) CyberGIS and spatial data science. GeoJournal 81(6):965–968
Wang S (2017) CyberGIS. In: Richardson D, Castree N, Goodchild MF et al (eds) The international encyclopedia of geography. Wiley
Wang S, Armstrong MP (2003) A quadtree approach to domain decomposition for spatial interpolation in grid computing environments. Parallel Comput 29(10):1481–1504
Wang S, Goodchild MF (eds) (2019) CyberGIS for geospatial discovery and innovation. Springer
Wright DJ, Wang S (2011) The emergence of spatial cyberinfrastructure. Proc Natl Acad Sci 108(14):5488–5491
Wuebbles DJ, Hayhoe K (2004) Climate change projections for the United States Midwest. Mitig Adapt Strat Glob Change 9(4):335–363
Xu L, Kwan MP, McLafferty S, Wang S (2017) Predicting demand for 311 non-emergency municipal services: an adaptive space-time kernel approach. Appl Geogr 89:133–141

Yang F, Tschetter E, Léauté X, Ray N, Merlino G, Ganguli D (2014) Druid: a real-time analytical data store. In: Proceedings of the 2014 ACM SIGMOD international conference on management of data. ACM

Yin D, Liu Y, Hu H, Terstriep J, Hong X, Padmanabhan A, Wang S (2019) CyberGIS-Jupyter for reproducible and scalable geospatial analytics. Concurrency Comput: Pract Exp 31(11):e5040

Shaowen Wang is a Professor and Head of the Department of Geography and Geographic Information Science, and founding director of the CyberGIS Center for Advanced Digital and Spatial Studies at the University of Illinois at Urbana-Champaign. His research interests primarily include geographic information science and systems (GIS) and cyberGIS.

Fangzheng Lyu is a Ph.D. student in the Department of Geography and Geographic Information Science at the University of Illinois at Urbana-Champaign. He received his BE from the University of Hong Kong in computing engineering. His research interests focus on GIS, urban multi-sensing, remote sensing, cyberGIS, and big data analysis.

Shaohua Wang is a postdoctoral fellow in the Department of Geography and Geographic Information Science, and the CyberGIS Center for Advanced Digital and Spatial Studies at the University of Illinois at Urbana-Champaign. His major research interests include GIS, cyberGIS, spatial optimization, and spatiotemporal big data analytics.

Charlie E. Catlett is a Senior Research Scientist at the University of Illinois Discovery Partners Institute. He is also a Senior Computer Scientist at Argonne National Laboratory and Visiting Senior Scientist at the University of Chicago. His research focuses on intelligent sensing systems using edge computation to understand urban dynamics.

Anand Padmanabhan is a Research Associate Professor in the Department of Geography and Geographic Information Science at the University of Illinois at Urbana-Champaign. His research interests focus on middleware for advanced cyberinfrastructure; cyberGIS and geospatial problem solving; geospatial big data analysis; and interdisciplinary education.

Kiumars Soltani is a Software Engineer in Zillow Group. He received his Ph.D. from the University of Illinois at Urbana-Champaign. His research is focused on spatially aware distributed systems and algorithms for ingesting, processing, and visualizing large and high-throughput geosensor data.

Chapter 37
Spatial Search

Liping Di and Eugene G. Yu

Abstract Urban studies concern the evolution of spatial structure in cities, where information is often tied to location. The discovery of information is in a high-dimensional space based on spatial and temporal dimensions, where the spatial relationships of components play roles in studying urban evolution. Spatial search in urban studies has to deal with diverse aspects of data structures (structured versus unstructured), data spatial context (implicit versus explicit), data spatial relationships (containment versus intersection), data volume (large volume versus large variety), spatial search speed (speed against different requirements), and spatial search accuracy (exactness versus relevance). This chapter reviews the technology in mining and extracting spatial information into urban geographic information systems, spatially indexing the urban information for effective spatially aware search, spatial relationships and their search algorithms, improving spatial relevance with different spatial similarity measures and algorithms, and open standards and interoperability in spatial search in the Web environment. Emerging technologies for spatial search in urban studies are also reviewed. Applications of spatial search in urban studies are exemplified and evaluated.

37.1 Spatial Search in the Context of Urban Studies

Urban studies is a transdisciplinary field that encompasses different academic fields, including urban geography, urban sociology, urban economics, urban housing and neighborhood development, urban environmental studies, urban governance, politics and administration, urban planning, design, and architecture (Bowen et al. 2010; Harris and Smith 2011). Search is ubiquitous in these focused research areas (Ballatore et al. 2016). In its most general form, spatial search is the search for information in a spatial and temporal context (Miller 1992). The introduction of the spatial dimension in the search problem can be viewed from two perspectives: one is as part of the information sought (i.e. the search for a place) and the other is as

L. Di (✉) · E. G. Yu
Center for Spatial Information Science and Systems, George Mason University, Fairfax, USA
e-mail: ldi@gmu.edu

W. Shi et al. (eds.), *Urban Informatics*, The Urban Book Series,
https://doi.org/10.1007/978-981-15-8983-6_37

the context in which the search is carried out (e.g. the network of roads to be routed through with an optimal route; Miller 1992).

Spatial search in urban studies carries different connotations depending on the root subject and the application. In the context of technology and geoinformatics, spatial search includes spaceless point search, range search, *k*-nearest neighbor search, and aggregated spatial search (e.g. total area or total count). In economics and sociology, spatial search can be seen as a decision problem and behavior. The spatial search problem is formatted as a connected graph with physical dimensions (e.g. two-dimensional space). The spatial search problem can vary with options (e.g. perfect knowledge with fixed sample set, online without recall, online with recall, with imperfect information). In the environment of linked open data (LOD), spatial search can be described as a process of identifying the place (converting into geographic information), modeling the spatial dimensions, indexing spatially for improved performance or heuristic results, formulating the search problem, and searching for results in constrained cases.

Spatial search in urban studies involves the following components to manage and maintain a spatial information system:

- Geocoding: a process to parse and extract spatial references from a query request.
- Spatial indexing: a process to improve the performance of spatial information retrieval.
- Spatial search algorithms: a set of algorithms to achieve the efficient and effective discovery of spatial information for different applications.
- Catalog and federated catalog: a system to manage spatial metadata.

The chapter is organized as follows. The next section reviews the geocoding process. Information about popular geocoding approaches and tools is introduced in this section. This is followed by a review of the approaches and data structures used in indexing the spatial information. The third section describes the spatial search problem as expressed in computer algorithms, while the fourth section reviews the cataloging strategies of spatial data and their approaches in distributed environments. The final section briefly touches on some of the recent advances and research directions in spatial search.

37.2 Geocoding

In urban studies, place names and street addresses are commonly used in referencing data geospatially (Dueker 1974). Geocoding is the step to relate location to descriptive text or place names. In early literature, it was termed place naming (Dueker 1974; Tobler 1972). In urban areas, geocoding can be efficiently referenced using different approaches for different datasets. Street geocoding, parcel geocoding, and address-point geocoding are three of the commonly used approaches in geocoding to associate an address with spatial coordinates (Zandbergen 2008; Owusu et al. 2017). As more and more types of geocode have emerged, the levels of detail can be associated with geocodes at different granularities. Table 37.1 shows the major generations of geocoding technologies along with major software or services for the corresponding generation. Geocoding has evolved along with the development of geographic information systems (GIS). At the beginning of GIS development, in the 1960s, the simplest geocoding schemes and systems became available. Geocoded area units could be matched to a representative point. Because these geocodes (e.g. demographic information, economic metrics) can associate with many attributes, they can be used effectively as base areal units for analyzing spatial differentiation in urban areas.

In the Web environment or connected applications, the approach is to use the API provided by geocoding services. All these services support both geocoding and reverse geocoding. The responses of these APIs are mostly in JSON, which can be easily incorporated and used by JavaScript in the Web environment (Table 37.2).

A place name may evolve over time, and sometimes, a place may carry multiple alternative names. In such cases, a gazetteer (a searchable database of toponyms) is useful and may be adapted to provide specific geocoding assistance. A gazetteer also contains basic information about the place in addition to geographic coordinates. This basic information may include demographic statistics, physical features, literacy, and economic conditions. The NGA GEOnet Names Server (GNS) is one of the sources

Table 37.1 Brief history of geocoding development

Generation	Geocoding technologies	Representative system or service
1960s	City block codes; street segments; representative point; Address Coding Guides (ACG) (Dueker 1974)	Automatic Location Table (AULT; Dueker 1974) Street Address Conversion System (SACS; Dueker 1974)
1970s	Dual Independent Map Encoding (DIME) (Farnsworth and Curry 1970)	Address Matching System (ADMATCH), Geographic Base File System (DIME), Computer Mapping System (GRIDS) (Farnsworth and Curry 1970)
1980s	Geographic Base File (GBF) (Davis et al. 1992)	GBF/DIME (Davis et al. 1970)
1990s	Topologically Integrated Geographic Encoding and Referencing (TIGER) (Broome and Meixler 1990)	
2000s	Commercial geocoding scheme Multilevel geocoding (Zandbergen 2008; Goldberg 2017) ADDRESS-POINT™ (Mesev 2005) Geocoded National Address File (G-NAF) (Paull 2003) Open Street Map (OSM)	Commercial software and services (Goldberg et al. 2007)
2010s	Master Address File (MAF) (Trainor 2003)	MAF/TIGER (Galdi 2005; Trainor 2005) Commercial geocoding Application Programming Interface (API) (Panasyuk et al. 2019)

Table 37.2 List of selected geocoding web services

Name	Limitation for service	Reference data	Reference or endpoint
Google geocoding API	50 requests per second; free credit $200 each month	Google maps	https://maps.googleapis.com/maps/api/geocode/json
Bing Locations API	Maximum 2 jobs at the same time. 50 jobs/24 h. 5 data sources and 2500 entities per source	Bing maps (NAVTEQ)	https://docs.microsoft.com/en-us/bingmaps/rest-services/locations
Yahoo geocoding API	5000 queries per IP address per day	Yahoo maps (NAVTEQ)	https://local.yahooapis.com/MapsService/V1/geocode
Baidu geocoding/reverse geocoding API	One million times/day	Baidu maps	https://api.map.baidu.com/telematics/v3/geocoding https://api.map.baidu.com/telematics/v3/reverseGeocoding
Yandex geocoder API	25,000 total requests per day to the geocoder, router, and panorama service combined	Yandex maps (NAVTEQ)	https://tech.yandex.com/maps/geocoder
Gaode geocoder API	250 query requests per day (API calls)	Gaode map	https://lbs.amap.com/api/javascript-api/guide/services/geocoder
Nominatim	1 request per second	OpenStreetMap (OSM)	https://nominatim.openstreetmap.org/search https://nominatim.openstreetmap.org/reverse
Texas A&M Geoservices Geocoder	2500 queries	Combined resources	https://geoservices.tamu.edu/Services/Geocode

used in these services. These services from gazetteers have been found very useful in urban studies (Janowicz et al. 2019; Dimou and Schaffar 2009). Table 37.3 lists a few of the most widely used gazetteers for retrieving geographic dimensions or coordinates of a place name and basic information about the place. The capabilities of gazetteers in disambiguating place names and putting place in context have led to many applications in the semantic analytics of urban studies (Janowicz et al. 2019).

Table 37.3 List of selected gazetteers

Name	Coverage/reference data	Usage	Limits	URL
Getty thesaurus of geographic names	Worldwide	Find the modern analog	Not all records have geographic coordinates	https://www.getty.edu/research/tools/vocabularies/tgn/index.html
Geonames.org	Worldwide	Geocoding/reverse-geocoding	2000/hour and 30,000/day	https://geonames.org/
GEOnet names server (GNS)	Worldwide	Official spelling, map, info, OGC services	Maximum number of features 8500	https://geonames.nga.mil/gns/html/gns_services.html
JRC Fuzzy Gazetteer	Worldwide, 7 million place names (from GNS)	Spelling correction, old names	Browser-based	https://dma.jrc.it/services/fuzzyg
Falling Rain Global Gazetteer	Worldwide, directories	Easy to follow through country by country	Browser-based	https://www.fallingrain.com/world/index.html

37.3 Spatial Indexing

Spatial indexing is the process of creating an effective and efficient data structure to help in speeding up spatial queries. Spatial indexing differs from common database indexing in having spatial properties: the object is not just one value but has two or more dimensions, and the size of an object may be non-zero (that is, a line, area, or volume; Kriegel and Seeger 1988). These properties lead to spatial relationships that are more complex than simple linear relationships. Many spatial indexing schemes have been developed along with the development of computer technologies (Kriegel and Seeger 1988; Lu and Ooi 1993). The basic goal of such spatial indexing is to reduce the computation required to retrieve matched spatial objects, given a set of geometrical criteria.

To create a spatial index, it is first necessary to identify the features to be indexed. For example, in a 2D spatial world, geographic features are commonly expressed as points, lines, or areas. Points can be represented as a pair of coordinates, which can be treated as fields to be indexed in a spatial database. Most spatial indexing approaches are specially designed for points (Lu and Ooi 1993). Lines and areas cannot be represented accurately as fields fit for indexing in a spatial database without losing information. Representative features need to be either selected or extracted for complex geographic objects. The processes are analogous to feature selection and feature extraction in machine learning, statistics, and information theory. In other words, the selection of features does not change the values which can be interpreted as dimensions. For example, the minimum bounding rectangle (MBR), the two-dimensional case of the minimum bounding box, can be treated as a selected feature, since its value can be found in the array of coordinates representing the geographic object. Any selected coordinate from the represented arrays (e.g. start point, end point, or middle point) can also be selected as the basis of indexing. The process can be generalized as one of transforming a k-dimensional space to a 2 k-dimensional space as described by Kriegel and Seeger (1988). For example, a rectangle aligned with the axes in 2D space can be defined by four coordinates. One encoding can be the corner coordinates (either upper left coordinate plus lower right coordinate or lower left coordinate plus upper right coordinate) or the center coordinates plus extent distances to each side (Kriegel and Seeger 1988). The grid file could be a four-dimensional grid, with the rectangle snapped to the closest cell in the grid file. On the other hand, the extraction of features goes through a computerized process to compute a set of values from the objects. For example, a hashing value is computed from the object using a hashing function. A centroid can also be computed from the object. The object can be represented as the first n principal components using principal-component extraction algorithms. These derived features can be used as indexed fields in a spatial database.

The next question for spatial indexing is how to handle the overlapping of spatial objects defined by the indexing spatial feature. Two schemes are available to deal with the partition: a clipping scheme (C-scheme) and a bounding scheme (OR-scheme) (Kriegel and Seeger 1988). For example, when an MBR is used as the spatial feature, the coverage defined by one MBR may overlap with that of another MBR. One example is shown in Fig. 37.1. With the clipping scheme, the object is duplicated with both partitions when the partition line crosses the region. For example, Object R3 is duplicated in both partitions (Fig. 37.1a). With the OR-scheme, Object R3 is only included in one partition S1 (Fig. 37.1b). The advantages and disadvantages of the two schemes are described in Table 37.4.

The computerized data structures for spatial indexing are as follows:

- **Fixed grid index**: The simplest example is *uniform grid* scheme where the space is partitioned uniformly into regular grids by value ranges along each axis. The grid system can be predefined with specified intervals or units. Retrieval time for the closest spatial rectangle would be O(1), and on average for any spatial

Fig. 37.1 Partition scheme for overlapping regions

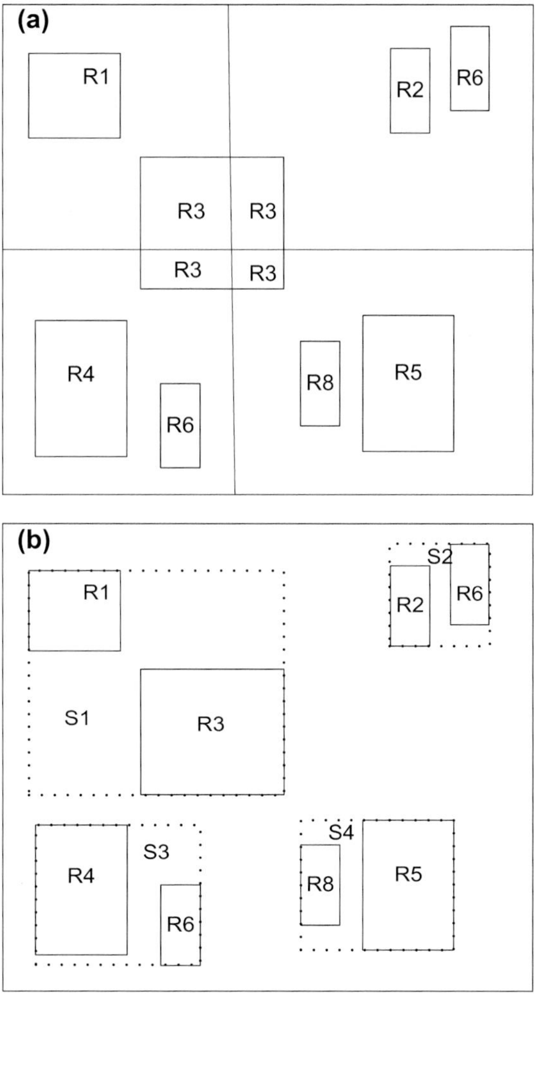

Table 37.4 Schemes for overlapping regions in a partition

Scheme	Pros	Cons
OR-scheme	Efficient storage utilization One file hosts both points and rectangles	Increased time to search, insert, or delete due to high overlap
C-scheme	Efficient inheritance of underlying point access methods One file hosts both points and rectangles	Duplications of MBRs Information redundancy

rectangle would be O($nCells + n$), where $nCells$ is the number of grid cells and n is the number of spatial objects, that is, the rectangles in the example. The memory requirement is O($nCells + n$).

- **Spatial hashing**: Because the distribution of spatial objects is often sparse, a uniform grid would result in many empty cells. A hash table can be used to store the index, and multi-level multi-key grid files can be used to index the multi-dimensional spatial data (Bentley and Friedman 1979).
- Spatial data partitioning trees
  - **Binary space partitioning (BSP) tree**: This is a general partition approach to partition space recursively into two convex sets using a hyperplane. It was developed as a general method in 3D video image processing (Schumacher et al. 1969). The *k*-dimensional binary search tree (*k-d* tree) is constructed by using one axis to split data at the median of the points along the axis (Bentley 1975). The Local Split Decision tree (LSD tree) is designed to handle both points and intervals (Henrich et al. 1989). The K-D-B tree is a derived tree structure that combines properties from the *k-d* tree and the B-tree (balanced tree) (Robinson 1981).
  - **Quad tree**: A quad tree builds a hierarchical representation of spatial data by dividing recursively into four quadrants (Finkel and Bentley 1974).
  - **Octree**: An octree is a hierarchical data structure that extends the quadtree to 3D, with all internal nodes having eight children (Meagher 1980).
  - **Balltree**: A balltree is "a complete binary tree in which a ball is associated with each node in such a way that an interior node's ball is the smallest which contains the balls of its children" (Omohundro 1989).
  - **R-tree**: An R-tree uses a minimum bounding rectangle (MBR) to determine its children (Guttman 1984). It is a balanced tree. Its variant trees include the Hilbert R (Kamel and Faloutsos 1984), R + (Sellis et al. 1984), Priority R (Arge et al. 2008), R* (Beckmann et al. 1990), GiST (Hellerstein et al. 1995), and G-tree (Zhong et al. 2015).
  - **Metric tree**: The vantage-point tree (vp-tree) is a space-partitioning algorithm to construct a tree with a sphere-like bounding area to partition the metric space (Yianilos 1993). Each part is defined within a threshold to each vantage point. A multi-vantage-point tree (MVP tree) is a variant of vp-tree which uses more than one point to partition at each level (Bozkaya and Ozsoyoglou 1999). The cover tree algorithms construct a leveled tree where each parent covers the

extent of all children (Begelzimer et al. 2006). The Bukhard-and-Keller tree (BK-tree) is adapted to discrete space by arranging points that are close to each other (Burkhard and Keller 1973).

37.4 Search Algorithms

A spatial search in urban studies can be viewed from different perspectives and formulated differently for different subject domains. In this section, two perspectives are examined. First, from the perspective of geography, spatial search is treated as a technology and method, and typical spatial queries and corresponding search algorithms are reviewed. Second, from the perspective of urban economics and urban sociology, spatial search is treated as a form of decision-making, generalized spatial search is formulated with graph theory, and related search algorithms are reviewed.

37.4.1 Spatial Queries

The following are the common types of spatial search used in urban studies:

- **Nearest neighbor search**: This is termed the *k*-nearest neighbor (*k*-NN) search. Typical questions can be "Find the *k* stores that are closest to a given point or current location" or "Find the closest restaurant."
- **Range search**: Range search is also common in urban studies. Example queries: "Find all the restaurants with 5 miles range" and "Find all the zones that can be reached between a half hour and a hour."
- **Aggregate search**: Questions can be often asked in urban studies that involve spatial aggregation. Examples are: "Get the number of hospitals for travel distance zones of under 10, 10–50, 50–100, and above 100 miles" or "Find the total area of green space in an urban district."

The *k*-NN search is well studied in computer science and geographic information systems (Knuth 1997). There are a suite of algorithms designed to solve the problem. There are two major categories of algorithms: exact search and approximate search. The simplest approach to find the *k*-nearest neighbors is sequential search that does not require any preprocessing of the spatial data (Bentley and Friedman 1979). The search time is O(*kn*), where *k* is the dimension and *n* is the total number of features. The storage requirement is also O(*kn*).

Spatial indexing can be used in preprocessing the data, creating a data structure that can be easily retrieved. BSP-trees, metric trees, and R-trees are three types of commonly used tree data structures in indexing spatial data. The *kd*-tree, one of the BSP-trees, uses axial rays to partition (ending up as rectangles), while the vp-tree, one of the metric trees, uses equidistance circles to partition data. The R-tree structure uses rectangles but has a focus on keeping the geographic object in a hierarchical structure. Most of these data structures lead to improvements by reducing the time to search to approximately O(log *n*) on average.

Different geographic information systems may support different spatial indexing algorithms. The R-tree and its variants are the most popularly implemented spatial indexing algorithms in geographic information systems, including PostGIS, MySQL, and Oracle. A grid-based spatial indexing scheme is popularly implemented in many geospatial databases, including Esri geodatabase, Oracle, and Microsoft SQL, due to its data-driven spatial indexing scheme.

Spatial search (*k*-NN, range search, or aggregate search) has been applied in many urban studies. Alternative site selections, such as the "spatial search" of Massam (1980), analyze spatial interactions and require range searches to assess the effect of selecting one alternative over another. For example, a firm searching for a location may consider the labor force that is available within a certain distance of each alternative location. In choosing a location for a retail store location, the analyst may need to conduct spatial queries on household purchasing power within a certain distance of each of the location alternatives. The results of such spatial queries would help in evaluating alternatives and making better plans.

37.4.2 Spatial Search with Graph Theory

Spatial search can be seen as a decision problem in urban studies, especially those studies with roots in economics. Economic Search Theory is well studied and has been used in studies of urban migration, urban markets, and urban agglomeration effects (Meier 2009,2010). Adding the spatial context, a generalized spatial search model can be formulated (Meier, 1995,2010). The spatial search problem is effectively defined within a connected graph. The vertices of the connected graph are alternatives at discrete locations in two-dimensional space. The edge connecting two vertices represents the cost, which may be a function of distance. The goal is to maximize the expected utility when the decision is to move from one vertex to another. Each alternative may be visited once.

The model of spatial search results from the tight bounding and integration of spatial context with a domain-specific model. In economics, this spatial model is tightly integrated with a model of economic search. This approach of integrating the spatial context with models in urban studies effectively converts the spatial search problem into an optimization problem on a graph.

The traveling salesperson problem is NP-hard. However, most problems in urban studies have a limited size, making them soluble. There are also heuristics to help in solving the optimization problem efficiently.

With the conversion of the spatial search problem to an optimization problem in a graph, the commonly used graph search algorithms become applicable to the spatial search model. These algorithms include breadth-first search, depth-first search, greedy best-first search, heuristic A*, and Dijkstra's shortest path algorithm.

The spatial search model has found applications in market area analytics, firm location, urban effect analysis, and urban modeling (Meier 1995). The simple distance or fuel cost-based spatial search model may be used in urban transportation planning and commercial truck routing (Zarezadeh et al. 2018; Moreno-Monroy and Posada 2018; Monte et al. 2018).

37.5 Distributed Search and Interoperability in the Web Environment

The abundance of geospatial information has grown beyond anyone's ability to manage it be properly. The introduction of live sensors and fast updating of information also suggests that the monolithic geographic information system cannot satisfy the requirements of spatial search in urban studies. Yet the data resources available for urban studies continue to grow.

There are several approaches to enable spatial search and geoprocessing to leverage the growing volume of information for urban studies. First, the information can be harvested and ingested into a local spatial catalog system through the harvesting of spatial metadata and data from different sources. The local spatial catalog system has to manage all the information. Each harvester may be updated or re-started (if incremental harvest is not supported by remote services). After each harvest, spatial indexing needs to be updated or re-built. The advantage for such a system is that the existing spatial indexing techniques are already supported. The major drawbacks are that the data can grow out of control and are not always current.

Second, the information is harvested, integrated, and indexed in a distributed manner. In this case, the local catalog system is replaced with a distributed catalog that clusters multiple cloud-computing instances. Each cloud-computing instance may handle a strip of information. A distributed spatial indexing scheme needs to be adopted to support the spatial search in such a distributed system (Priya and Kalpana 2018). The advantage for such a system lies in its capability to handle large datasets in a scalable cloud-computing environment. The major limitations are: (1) the freshness of the metadata and data cannot be warranted, (2) the remote services may not allow the duplication of their metadata and data for various reasons, and (3) the maintenance of a large distributed spatial catalog system can still be a challenge, and the distributed spatial search capability is still in development.

Third, a federated spatial catalog system can be adopted to support the on-the-fly integration of distributed search (Shao et al. 2013; Bai et al. 2007). The development of a federated spatial catalog depends on the adoption of open geospatial standards. The standard interface and response from catalogs make it possible to do translation on the fly. The idea of federated catalog is to set up a series of plug-in translators that handle the translation of request to and response from the remote catalog services. When a user sends in a spatial query, the query request is first translated into a format that matches the remote server and the translated request is sent out. The response from the remote service is then translated and integrated in the mediator to be sent back to the user. The advantages of such a federated catalog are: (1) it does not need extensive resources in manage the metadata and data since most of the resources are still maintained by the original provider; (2) the contents are in complete synchronization with remote services; and (3) spatial search is completed in a distributed environment. The drawbacks are: (1) the spatial search function and responses are tied to what the remote services offer, and (2) duplicates may not be removed properly if two remote services offer the same content.

37.6 Trends

The spatial search problem is a hard problem to solve. The performance of current solutions is acceptable only because either one of the following assumptions stands: (1) the size of data is limited, (2) optimal heuristics exist for the dataset, or (3) the best option executes in an acceptable time. This section reviews two frontiers in solving the spatial search problem: a quantum spatial search algorithm and semantic spatial search.

Quantum algorithms have emerged in solving the spatial search problem with improvements. Quantum computing is seen as the future of computing, to improve non-deterministic algorithms that consider multiple superpositions of states (Venegas-Andraca 2008; Chakraborty et al. 2016; Ambainis 2008). The spatial search problem is seen as one of the hard problems to be solved with classic computers (Meier 1995,2010), or as a decision problem to find the target vertex in a connected graph (Meier 1995). In a fully connected lattice graph of n vertices, the worst time to find the marked target is $O(n \log n)$ using a random walk in a classic computer. New algorithms in quantum computing have shown that the search can be improved many fold with quantum random walks (Portugal 2018). A discrete-time quantum walk (DTQW) algorithm improved the time to $O(\sqrt{n} \log n)$ (Ambainis et al. 2005). A controlled quantum walk (CQW) algorithm on a lattice using an ancilla qubit improved the time complexity to $O((n \log n)^{1/2})$ (Tulsi 2008). An improved version of DTQW also achieved the same time complexity (Ambainis et al. 2015). Portugal described an approach to the design of quantum algorithms for the spatial search problem that explains how Grover's algorithm (Grover 1996), the quantum algorithm for searching a database, "can be seen as a spatial search problem on the complete graph with loops using the coined model and on the complete graph without loops using the staggered model" (Portugal 2018).

The application of semantic technology improves the accuracy of spatial search with more explicit spatial semantics. Most current spatial search solutions treat spatial objects as a spaceless point. Spatial extents and spatial relationships are not taken into full consideration with current solutions. The augmentation of linked geodata (Stadler et al. 2012) with spatiotemporal semantics enables a semantic spatial search (Neumaier and Polleres 2019). A Transportation ontology domain can be added to a semantic-based public transportation geoportal to support semantic spatial search on concepts, relationships, and individuals (Gunay et al. 2014). Ontology provides additional semantic constraints in semantic spatial search (Jones et al. 2004,2001). A spatial entity can be described by its sub-components, and the search for a spatial entity can be modeled as a multi-component spatial search problem (MCSSP) (Menon and Smith 1989, Menon 1990). This effectively formulates the spatial search problem as a constraint satisfaction problem (CSP) in computer science. The suite of heuristic CSP algorithms can be applied to help in finding the best match, including backtracking, graph-based backjumping, arc consistency, and forward checking (Frost 1997).

37.7 Conclusion

Spatial search has been one of the most intensively researched topics in urban studies, and can be traced back to a pre-computer era. The classic spatial search in dealing with connectivity between spatial objects or entities has been thoroughly researched and supported by most geographic information systems. The spatial search problem can be integrated with models in urban studies to put the research in spatial context. Extending studies with spatial dimensions increases the complexity of problem solving. In a fully connected graph depicting the relationships among entities in a spatial context, the problem is NP-complete and is therefore difficult to solve. However, in actual applications in urban studies, the data size is often manageable and heuristics can be applied to solve the spatial search problem within a reasonable time interval.

New developments in alternative computing environments shed light on solving the spatial problem more efficiently. One of the most researched alternatives is to leverage random walk with quantum computing. Several algorithms have been proposed to solve the spatial search problem efficiently with quantum walks. Another frontier is the use of semantic Web technology in dealing with big data and heterogeneous data in the spatial context.

References

Ambainis A (2008) Quantum random walks—new method for designing quantum algorithms. In: Geffert V, Karhumäki J, Bertoni A, Preneel B, Návrat P, Bieliková M (eds) SOFSEM 2008: theory and practice of computer science. Springer, Berlin, pp 1–4. https://doi.org/10.1007/978-3-540-77566-9_1

Ambainis A, Bačkurs A, Nahimovs N, Ozols R, Rivosh A (2013) Search by quantum walks on two-dimensional grid without amplitude amplification. In: Iwama K, Kawano Y, Murao M (eds) Theory of quantum computation, communication, and cryptography. Springer, Berlin, pp 87–97. https://doi.org/10.1007/978-3-642-35656-8_7

Ambainis A, Kempe J, Rivosh A (2005) Coins make quantum walks faster. In: Proceedings of the Sixteenth Annual ACM-SIAM symposium on discrete algorithms. Society for industrial and applied mathematics, pp 1099–1108

Arge L, Berg MD, Haverkort H, Yi K (2008) The priority R-tree: a practically efficient and worst-case optimal R-tree. ACM Trans Algorithms 4:1–30

Bai Y, Di L, Chen A, Liu Y, Yaxing W (2007) Towards a geospatial catalogue federation service. Photogrammetric Eng Remote Sens 73:699–708

Ballatore A, Kuhn W, Hegarty M, Parsons E (2016) Special issue introduction: spatial approaches to information search. Spatial Cognit Comput 16:245–254. https://doi.org/10.1080/13875868.2016.1243693

Beckmann N, Kriegel HP, Schneider R, Seeger B (1990) The R*-tree: an efficient and robust access method for points and rectangles. In: Proceedings of the 1990 ACM SIGMOD international conference on management of data. ACM Press, Atlantic City, pp 322–331. https://doi.org/10.1145/93597.98741

Bentley JL (1975) Multidimensional binary search trees used for associative searching. Commun ACM 18:509–517. https://doi.org/10.1145/361002.361007

Bentley JL, Friedman JH (1979) Data structures for range searching. ACM Comput Surv 11:97–409. https://doi.org/10.1145/356789.356797

Beygelzimer A, Kakade S, Langford J (2006) Cover trees for nearest neighbor. In: Proceedings of the 23rd international conference on machine learning. ACM, Pittsburgh, pp 97–104

Bowen WM, Dunn RA, Kasdan DO (2010) What is "urban studies"? Context, internal structure, and content. J Urban Affairs 32:199–227. https://doi.org/10.1111/j.1467-9906.2009.00474.x

Bozkaya T, Ozsoyoglu M (1999) Indexing large metric spaces for similarity search queries. ACM Trans Database Syst 24:361–404. https://doi.org/10.1145/328939.328959

Broome FR, Meixler DB (1990) The TIGER data base structure. Cartography Geogr Inform Syst 17:39–47. https://doi.org/10.1559/152304090784005859

Burkhard WA, Keller RM (1973) Some approaches to best-match file searching. Commun ACM 16:230–236

Chakraborty S, Novo L, Ambainis A, Omar Y (2016) Spatial search by quantum walk is optimal for almost all graphs. Phys Rev Lett 116:100501. https://doi.org/10.1103/PhysRevLett.116.100501

Davis BA, George JR, Marx RW (1992) TIGER/SDTS: standardizing an innovation. Cartography Geogr Inform Syst 19:321–327. https://doi.org/10.1559/152304092783762164

Dimou M, Schaffar A (2009) Urban hierarchies and city growth in the Balkans. Urban Studies 46:2891–2906

Dueker KJ (1974) Urban geocoding. Ann Assoc Am Geogr 64(2):318–325

Farnsworth G, Curry JP (1992) The DIME geocoding system. U.S, Bureau of the Census

Finkel RA, Bentley JL (1974) Quad trees: a data structure for retrieval on composite keys. Acta Informatica 4:1–9. https://doi.org/10.1007/BF00288933

Frost DH (1997) Algorithms and heuristics for constraint satisfaction problems. Ph.D. dissertation, University of California, Irvine. https://www.ics.uci.edu/~dechter/publications/R69.pdf

Galdi D (2005) Spatial data storage and topology in the redesigned MAF/TIGER system. U.S, Census Bureau Geography Division

Goldberg DW (2017) Geocoding. In: Richardson D, Castree N, Goodchild MF, Kobayashi A, Liu W, and Marston RA (eds) International encyclopedia of geography: people, the Earth, environment and technology. Wiley, Oxford, pp 1–12. https://doi.org/10.1002/9781118786352.wbieg1051

Goldberg DW, Wilson JP, Knoblock CA (2007) From text to geographic coordinates: the current state of geocoding. URISA J 19:33–47

Grover LK (1996) A fast quantum mechanical algorithm for database search. In: Proceedings of the twenty-eighth annual ACM symposium on theory of computing, pp 212–219

Gunay A, Akcay O, Altan MO (2014) Building a semantic based public transportation geoportal compliant with the INSPIRE transport network data theme. Earth Sci Inf 7:25–37. https://doi.org/10.1007/s12145-013-0129-z

Guttman A (1984) R-trees: a dynamic index structure for spatial searching. In: Proceedings of the 1984 ACM SIGMOD international conference on management of data—SIGMOD '84. ACM Press, Boston, p 47. https://doi.org/10.1145/602259.602266
Harris R, Smith ME (2011) The history in urban studies: a comment. J Urban Affairs 33:99–105. https://doi.org/10.1111/j.1467-9906.2010.00547.x
Hellerstein JM, Naughton JF, Pfeffer A (1995) Generalized search trees for database systems. In: Proceedings of the 21th International conference on very large data bases. Morgan Kaufmann, pp 562–573
Henrich A, Six HW, Hagen F, Widmayer P (1989) The LSD tree: spatial access to multidimensional point and non-saint objects. In: 5th very large databases conference. pp 45–54
Janowicz K, McKenzie G, Hu Y, Zhu R, Gao S (2019) Using semantic signatures for social seing in urban environments. In: Antoniou C, Dimitriou L, Pereira F (eds) Mobility patterns, big data and transport analytics. Elsevier, New York, pp 31–54
Jones CB, Abdelmoty AI, Finch D, Fu G, Vaid S (2004) The SPIRIT spatial search engine: architecture, ontologies and spatial indexing. In: Egenhofer MJ, Freksa C, and Miller HJ (eds) Geographic information science. Springer, Berlin. pp 125–139. https://doi.org/10.1007/978-3-540-30231-5_9
Jones CB, Alani H, Tudhope D (2001) Geographical information retrieval with ontologies of lace. In: Proceedings of the international conference on spatial information theory: foundations of geographic information science. Springer. pp 322–335
Kamel I, Faloutsos C (1994) Hilbert R-tree: an improved R-tree using fractals. In: Proceedings of the 20th international conference on very large data bases. Morgan Kaufmann. pp 500–509
Kriegel H, Seeger B (1988) Techniques for design and implementation of efficient spatial access methods. In: Proceedings of the 14th VLDB Conference. pp 360–370
Knuth DE (1997) The art of computer programming. Addison-Wesley, Reading
Lu H, Ooi BC (1993) Spatial indexing: past and future. IEEE Data Eng Bull 16:16–21
Maier G (1995) Spatial search: structure, complexity, and implications. Physica-Verl, Heidelberg
Maier G (2009) Spatial prices, search behavior and location. Ann Reg Sci 43:27–48. https://doi.org/10.1007/s00168-007-0203-x
Maier G (2010) The spatial search problem: structure and complexity. Geogr Anal 25:242–251. https://doi.org/10.1111/j.1538-4632.1993.tb00295.x
Massam BH (1980) Spatial search: application to planning problems in the public sector. Pergamon Press, Oxford
Meagher D (1980) Octree encoding: a new technique for the representation, manipulation and display of arbitrary 3-D objects by computer. Rensselaer polytechnic instiute
Menon S (1989) Spatial search for multicomponent objects in a geographic information system using symbolic models and hierarchical data structures. PhD dissertation, University of California, Santa Barbara
Menon S, Smith TR (1989) Search for multi-component objects in large spatial databases. In: Litwin W, Schek HJ (eds) Foundations of Data Organization and Algorithms. Springer, Berlin. pp 246–251. https://doi.org/10.1007/3-540-51295-0_134
Mesev V (2005) Identification and characterisation of urban building patterns using IKONOS imagery and point-based postal data. Comput Environ Urban Syst 29:541–557. https://doi.org/10.1016/j.compenvurbsys.2005.01.003
Miller HJ (1992) Consumer spatial search behavior: a general modeling framework and operational models. Paper presented to the Association of American Geographers, San Diego, April
Monte F, Redding SJ, Rossi-Hansberg E (2018) Commuting, migration, and local employment elasticities. Am Econ Rev 108:3855–3890. https://doi.org/10.1257/aer.20151507
Moreno-Monroy AI, Posada HM (2018) The effect of commuting costs and transport subsidies on informality rates. J Dev Econ 130:99–112. https://doi.org/10.1016/j.jdeveco.2017.09.004
Neumaier S, Polleres A (2019) Enabling spatio-temporal search in open data. J Web Semant 55:21–36. https://doi.org/10.1016/j.websem.2018.12.007
Omohundro SM (1989) Five balltree construction algorithms. International Computer Science Institute, Berkeley

Owusu C, Lan Y, Zheng M, Tang W, Delmelle E (2017) Geocoding fundamentals and associated challenges. In: Karimi HA, Karimi B (eds) Geospatial data science techniques and applications. CRC Press, Boca Raton. https://doi.org/10.1201/b22052
Panasyuk A, Yu ESL, Mehrotra KG (2019) Improving geocoding for city-level locations. In: Proceedings, 2019 IEEE 13th international conference on semantic computing (ICSC). pp. 416–421. https://doi.org/10.1109/ICOSC.2019.8665524
Paull D (2003) A geocoded national address file for Australia: the G-NAF what, why, who and when. PSMA Australia Limited, Griffith
Portugal R (2018) Spatial search algorithms. In: Quantum Walks and Search Algorithms. Springer, Cham. pp. 175–200. https://doi.org/10.1007/978-3-319-97813-0_9
Priya M, Kalpana R (2018) Distributed processing of location based spatial query through vantage point transformation. Future Comput Informatics J 3:296–303. https://doi.org/10.1016/j.fcij.2018.09.002
Robinson JT (1981) The K-D-B-tree: a search structure for large multidimensional dynamic indexes. In: Proceedings of the 1981 ACM SIGMOD international conference on management of data. pp. 10–18
Schumacher R, Brand B, Gilliland M, Sharp W (1969) Study for applying computer-generated images to visual simulation. U.S, Air Force Human Resources Laboratory
Sellis T, Roussopoulos N, Faloutsos C (1987) The R+-tree: a dynamic index for multi-dimensional objects. In: Proceedings of the 13th international conference on very large data bases (VLDB)
Shao Y, Di L, Bai Y, Wang H, Yang C (2013) Federated catalogue for discovering Earth observation data. Photogrammetrie—Fernerkundung—Geoinformation 1:43–52. https://doi.org/10.1127/1432-8364/2013/0157
Stadler C, Lehmann J, Höffner K, Auer S (2012) LinkedGeoData: a core for a web of spatial open data. Semantic Web 3(4):333–354. https://doi.org/10.3233/SW-2011-0052
Tobler W (1972) Geocoding theory. In: Proceedings of the national geocoding conference
Trainor T (2003) U.S. Census Bureau geographic support: a response to changing technology and improved data. Cartography Geogr Inform Sci 30:217–223. https://doi.org/10.1559/152304003100011054
Trainor TF (2005) The MAF/TIGER enhancement program: the mechanics and maintenance of a large-scale national spatial database. Govern Publ Rev 13:181–201
Tulsi A (2008) Faster quantum-walk algorithm for the two-dimensional spatial search. Phys Rev A 78:012310. https://doi.org/10.1103/PhysRevA.78.012310
Venegas-Andraca SE (2008) Quantum walks for computer scientists. Synth Lect Quantum Comput 1:1–119. https://doi.org/10.2200/S00144ED1V01Y200808QMC001
Yianilos PN (1993) Data structures and algorithms for nearest neighbor search in general metric spaces. In: Proceedings of the fourth annual ACM-SIAM symposium on discrete algorithms. Society for industrial and applied mathematics, Austin. pp. 311–321
Zandbergen PA (2008) A comparison of address point, parcel and street geocoding techniques. Comput Environ Urban Syst 32:214–232. https://doi.org/10.1016/j.compenvurbsys.2007.11.006
Zarezadeh Z, Benckendorff P, Gretzel U (2018) Lack of progress in tourist information search research: a critique of citation behaviour and knowledge development. Current Issues Tourism 22(19):1–15. https://doi.org/10.1080/13683500.2018.1459511
Zhong R, Li G, Tan KL, Zhou L, Gong Z (2015) G-Tree: an efficient and scalable index for spatial search on road networks. IEEE Trans Knowl Data Eng 27:2175–2189. https://doi.org/10.1109/TKDE.2015.2399306

Liping Di is the professor and founding director of the Center for Spatial Information Science and Systems (CSISS), George Mason University. He served as the Chair for the US National Committee of the Geographic Information Standards from 2010–2016 and is currently the convener of the ISO TC 211 Working Group 9. Dr. Di has been engaged in geoinformatics research for over thirty years. He has achieved over 450 publications and has received more than $56 million in research grants.

Eugene Genong Yu is a Research Associate Professor in geographic information sciences and remote sensing at the Center for Spatial Information Science and Systems at George Mason University, USA. His research interests include geographic information systems, remote sensing, intelligent image understanding, sensor Web, semantic Web, computational vision, and robotics. He is a member of the American Geophysical Union, the IEEE Geoscience and Remote Sensing Society, and the IEEE Computer Society.

Chapter 38
Urban IoT: Advances, Challenges, and Opportunities for Mass Data Collection, Analysis, and Visualization

Andrew Hudson-Smith, Duncan Wilson, Steven Gray, and Oliver Dawkins

Abstract Urban Internet of Things (IoT) is in an early speculative phase. Often linked to the smart city movement, it provides a way of sensing and collecting data—environmental, societal, and transitional—both automatically, remotely, and with increasing levels of spatial and temporal detail. From city-wide data collection down to the scale of individual buildings and rooms, this chapter details the technology behind the rise of IoT in urban areas and explores the challenges (societal and technical) behind city-wide deployments. Drawing from a series of deployments at the Queen Elizabeth Olympic Park, London, it details the challenges and opportunities for mass data collection. Widening out the view, it looks at what is becoming known as "the humble lamp post" in Urban IoT fields to detail the potential of Urban IoT with the objects that already form part of the urban fabric. Finally, it examines the potential of Urban IoT for input into urban modeling and how we are on the edge of a shift in the collection, analysis, and communication of urban data.

38.1 The Urban Internet of Things

As Cellary (2013) notes, there is no common consensus about what "smart" really means in the context of information and communications technology (ICT). Although this term has become fashionable, it is also broadly used as a synonym of almost anything considered to be modern and intelligent (Anthopoulos 2017). In an urban context, Batty and others note that the term smart cities correspond with the rapid spread of computation into the kinds of public and open environments that others, from Hardin (1968) to McCullough (2013), have called the commons, meaning the spaces in the city that are notionally set aside for collective use and exploitation by the community. While the term smart has many competing definitions and public perceptions associated with it, we consider a focus on sensing and computation in public spaces to be its defining characteristic. In this way, the aspirations behind smart technologies we relate to self-monitoring, analysis, and reporting technology

A. Hudson-Smith (✉) · D. Wilson · S. Gray · O. Dawkins
The Bartlett Centre for Advanced Spatial Analysis, University College London, London, UK
e-mail: a.hudson-smith@ucl.ac.uk

W. Shi et al. (eds.), *Urban Informatics*, The Urban Book Series,
https://doi.org/10.1007/978-981-15-8983-6_38

(SMART) adapted from the association with computer hard disks as a way to internally monitor their own health and performance. SMART, in terms of disk drives, allows users to perform self-tests on the disk and to monitor a number of performance and reliability attributes and seems a useful close analogy. The ability to self-monitor, analyze, and report performance and reliability measures is, we argue, a closer definition of the smart city, especially when focusing on aspects of sensing the environment, communication, modeling, and analyzing based on data feeds from the urban context.

Covering urban areas in general and at multiple scales from the city as a whole down to the microscale of footfall at a given point in place and time, the potential for urban data collection is almost infinite and certainly satisfies accepted criteria for data to become big. In 2013, Ebbers, Abdel-Gayed, Budhi et al. stated that there are four main aspects of big data, these being data generated at a fast rate (velocity), very large and potentially unknown data quantities (volume), accuracy of the data (veracity), and different forms of data such as text, structured data, etc. (variety). Tennant et al. (2017) build upon this, noting that other aspects of big data have been added over the years, for example, volatility, referring to the length of validity of the data, which is particularly relevant when referring to real-time data streams; and value, referring to potential insights that can be derived by analyzing the data. Velocity, variety, volume, and veracity of data, interlinked with volatility and value, are central to the use of data within an urban context. This cuts across a broad spectrum of applications, but more especially applications consuming, analyzing, and visualizing data in an urban context from Internet of Things devices—or an Urban Internet of Things. Coulton et al. (2019) state that the term Internet of Things (IoT) was coined by Kevin Ashton in the late 1990s. Ashton explained how by using sensors to gather data that could be shared across the company's computer network, they could streamline their supply chain. He called these data-enabled parts of the supply chain the Internet of Things, and the phrase caught on.

The potential of the concept is immense, as it is linked to the automation of data collection en masse. As Ashton (2009) notes, if we had computers that knew everything there was to know about things—using data they gathered without any help from us—we would be able to track and count everything, and greatly reduce waste, loss, and cost. We would know when things needed replacing, repairing, or recalling, and whether they were fresh or past their best. Linking this to cities, Batty and Hudson-Smith (2007), in their often-cited paper, called this the computable city, stating that by the year 2050, everything around us will be some form of computer. In essence, they were predicting an Urban Internet of Things.

Building on this, the Mayor of London published a document entitled "The Smarter London Together" roadmap, in 2016. The roadmap, which is a non-statutory document, builds on the first Smart London Plan from the Greater London Authority (GLA) in 2013. It provides a new approach based on collaborative missions and calls for the city's 33 local authorities and various public services to work and collaborate better with the aid of data and digital technologies (GLA 2019). As part of this work, the city has developed a number of test beds, allowing the exploration of research-led deployments. One such location is the Queen Elizabeth Olympic Park (QEOP), to the east of the City of London and an area we will focus on to explore actual examples of Urban IoT. As the GLA note, the park's development is managed by the London Legacy Development Corporation (LLDC). Its ambition is to use the park as a test bed for new international standards in smart data, sustainability, and community building, sharing its successes across the city and beyond. This initiative has allowed the authors of this chapter to deploy a number of IoT-led initiatives within the park. Over the following sections, we explore these deployments while also focusing on the wider picture and also the current realities of Urban IoT in the context of our definition of smart—self-monitoring, analysis, and reporting technologies—and also within the view of what we define as the essential six Vs of Urban IoT: velocity, volume, veracity, variety, volatility, and value.

The Internet of Things is central to the collection of potentially all the types of data that are required to understand and manage an urban system. Link this further to knowing the location of each device and you have the potential of a real-time view of a city, or a representation of the city in software that is also known as a digital twin. As such, the development of digital twins has been used as one of the deployments for examination in QEOP.

38.2 The Digital Twin

Originally developed in the context of industrial design and manufacture during the early 2000s, the term digital twin was proposed as a means of monitoring the performance of industrial products with the aid of digital replicas. The digital twin would be connected to its physical counterpart, an aircraft engine for example, in such a way that any relevant changes in the state of the latter would be automatically sensed and registered (Grieves and Vickers 2017). In this way, the performance of complex and dynamic objects like aircraft engines, or even entire aircraft, could be modeled, monitored, and optimized throughout the entire industrial lifecycle, from design, through daily operation, and on to their eventual decommissioning and disposal. Each component could have its own digital twin, effectively giving us a nested hierarchy of digital twins all the way down to the most fundamental components.

New applications for digital twins are now being sought in other fields. At the urban scale, the digital twin is finding more immediate application in the convergence of IoT and building information modeling (BIM) (Deutsch 2017). A BIM model is a digital model of a building that has had the 3D geometric properties of the structure enhanced with quantitative values and semantic descriptions of the particular building components being represented (see Chap. 34). In principle, all of its components can be modeled, down to the smallest nut or bolt, in the same way as the original aircraft concept, to include information about their manufacture, appearance, physical properties, date of purchase, or installation and cost. The last two facilitate the additional time (4D) and cost (5D) dimensions used for scheduling BIM-based construction. Using open standards like the Industry Foundation Classes

(IFC), BIM models can be federated to enable multiple stakeholders to collaborate by reviewing and updating a BIM during the building's design and construction. At the same time, BIM is perhaps not an "obligatory point of passage" for a digital twin as some in the BIM industry might wish to suggest (cf. Law and Callon 1994).

While BIM provides an efficient means of constructing the 3D representations required for a digital twin of new builds, the models can quickly become static and outdated once they have been handed over to building owners. However, with the addition of embedded sensors and Internet-based connectivity, it is possible to continue monitoring aspects of the building's physical and environmental conditions in real time. In this way, IoT provides the potential for sensing, connectivity, and feedback through actuation that serve to animate and bring the building's digital twin to life by establishing its link to the physical counterpart. Figure 38.1 illustrates Here East (a building in QEOP), which was modeled in three dimensions and deployed with environmental IoT sensors to create a simple twin model. The model updates in real time, providing the twin aspect linked to the three-dimensional representation of the built form.

Even social aspects of the building's everyday life can be incorporated for a more holistic, responsive, and participatory approach to building management and operation (Dawkins et al. 2018). It is this broad spectrum of connectivity through multiple aspects of IoT, from environmental sensor data through to information occupation and across to social network information, that provides the real key to a digital twin.

Here, we find ourselves in the realm of connected environments. As Hudson-Smith et al. (2019) define them, a connected environment is any place—a home, a building, a street, a park—where sensors have been deployed and connected via the

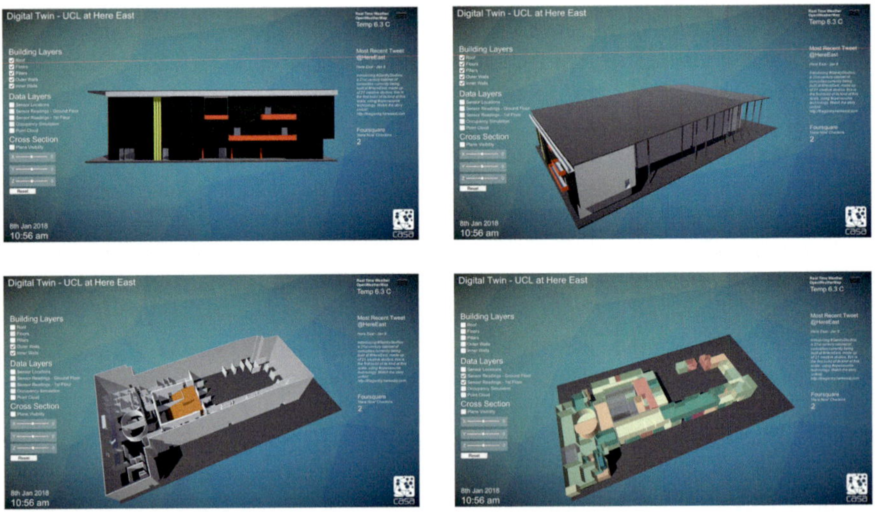

Fig. 38.1 A digital twin with IoT sensors of the here east building at the queen Elizabeth Park, London

Internet. Collecting data through these sensors allows them to be analyzed, checked for quality control, joined up with other data sets, and used to enhance the area, be it for management, social, environmental, or economic reasons. It is through the capture, processing, and analysis of longitudinal real-time operational data, increasingly performed in the Cloud, that the further possibilities for simulation and more exploratory and predictive use of a digital twin can be achieved. In this way, digital twins bestow on their users some of the powers of more enchanted objects like the crystal ball, insofar as they provide a digital means to see distant places and look into the past and future (Rose 2014). More prosaically, by representing the digital twin as a 3D model, and moving away from the use of abstract plots and graphs, the digital twin becomes more accessible to the public, and more relatable to a specific place. The digital twin is a new kind of enchanted object: a digital representation of the physical world that, with the addition of data collected from anything from building systems through to social and environmental feeds, gives each individual a kind of omniscience that can help one understand and act on one's environment.

Just as digital twins are the sum of their components, we can also aggregate them to create connected assemblages at coarser scales. The digital twin at the urban scale is still an emerging concept. Some imagine an urban digital twin as a swarm of connected systems collaborating autonomously to intelligently manage energy, traffic, utility, roads, and communication networks (Datta 2016). The digital twin can be viewed as a mirror held up to this world, one that not only reflects the environment as we ordinarily see it, but also the unseen or invisible patterns of phenomena that find themselves encoded in flows of sensor data. With mirror worlds, as conceived by computer scientist David Gelernter in the early nineties, "the whole city shows up on your screen, in a single dense, live, pulsing, swarming, moving, changing picture." This vision is currently being realized through the development of interactive virtual city models like Cityzenith, VU.CITY, Virtual Singapore, and CASA's own Virtual London (ViLo).

Commonly viewed on the computer screen, tablet, or mobile phone, new opportunities of interacting with these tools and the data they orchestrate are being opened up by increasingly immersive virtual, augmented, and mixed-reality devices. While virtual-reality systems enable us to visit other places and times and immerse ourselves within those environments, augmented and mixed realities can bring that informational content to us by overlaying it on the everyday environment (from room to building to street, neighborhood, and city). At different scales, data and reality can be mixed, viewed, and shared. Such mirror worlds then often engage new contexts and audiences while also providing new opportunities for learning and the exercise of personal and collective agency in the urban environment (Dawkins 2017). Digital twins can be used to view a variety of information in a multitude of ways. The ViLO model (Figure 38.2) allows viewing via a traditional computer desktop as well as via virtual reality, augmented reality, and mixed reality, all with real-time, geo-located data. Given the pace of technology, the creation of digital twins is inevitable, allowing the digitalization of our world and thus opening up the opportunity for new insights into physical worlds. Indeed, in the recent report "Data for the public good," the UK's National Infrastructure Commission (NIC 2018) proposes the creation of a digital

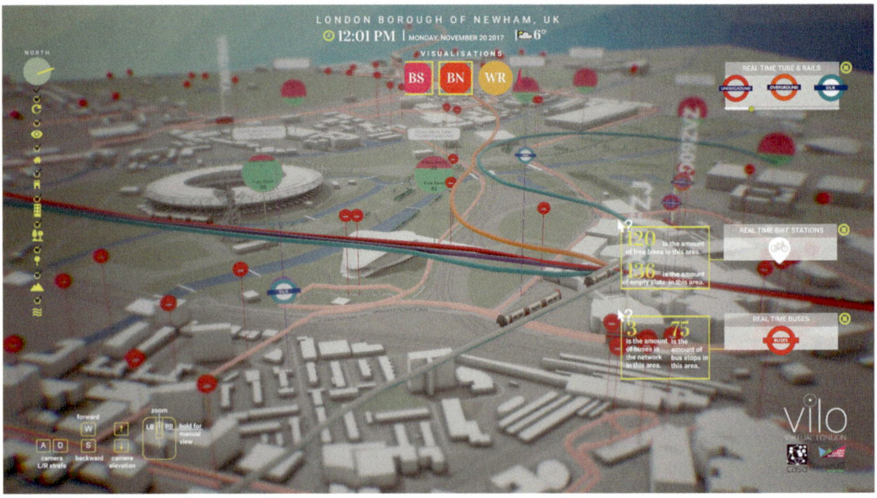

Fig. 38.2 QEOP in the "ViLO" model providing real-time IoT data within a 3D environment

twin to unify the management of data concerning transport, rail, power, water, and communications infrastructures alongside meteorology and demographics across the whole of the UK.

38.3 Potential Versus Reality

The potential of the Urban Internet of Things is such that it could be viewed as new data revolution, moving forward our understanding of the logistics of cities. There are already an estimated 26.6 billion IoT things in existence with a predicted 75 billion connected things by 2025 (Statista 2018).

Such numbers do not necessarily, however, mean that there are 26.6 billion operational devices. We would estimate that less than a tenth of these devices are currently live, transmitting data; a tenth of those probably have quality control on their data feeds; and a tenth of those have a known location, indeed probably even less. The potential is of course there, and all technological developments take time to become embedded into methodologies and systems, which are often developed on a wave of hype, expectations, and disillusionment, and then finally enter production. The Gartner Hype curve is a useful way to understand such adoption of technology; the most recent (Gartner 2018) has digital twins approaching the peak of inflated expectations.

The first realizations of cities inside a computer in iconic, rather than in more abstracted mathematical form, were mooted in the 1960s with the Skidmore, Owings, and Merrill wireframe model of Chicago, an early exhibit of these possibilities (Batty and Hudson-Smith 2007). The intervening years have seen the development of 3D models beyond the wireframe and into photorealism on a global scale. Indeed, as Goodchild (2018) notes, the technical ability to create and visualize 3D renderings of the Earth was unavailable in the mid 1960s at the birth of GIS, but it was achieved in the early 1990s, and led directly to Google Earth and its many competitors.

The technology continues to develop and the more recent introduction of the Google Earth Engine essentially now provides public access to a multi-petabyte curated collection of widely used geospatial datasets (Gorelick et al. 2017). Beyond this level of detail is the current domain of systems such as ViLO, linking in building information systems, with geographical information systems (GIS) providing the linkage between buildings, data, and geography. However, these merely provide the skeleton to the twin and arguably can be compared to the wireframe model of Chicago from the 1960s in terms of where we are in creating a true digital twin.

If the model is the skeleton of the city, then the Internet of Things can be compared to the neurons in the brain, communicating via wireless protocols rather than neurotransmitters. At the moment, however, the city does not have a brain, and the devices communicate to diverse systems, sometimes joined up, such as is the case in terms of public transport networks and deployed sensors, but often as part of local initiatives using devices deployed by hobbyists, or as part of small research trials. The data are, however, starting to flow, and developments in networking and computing technology are enabling small, low-power devices to be deployed in the field and communicate over long distances. This is the revolution on the horizon and it is just starting to become a reality, allowing data-collection devices to go from a small number to a number that has the potential to be compared to the number of neurons in the brain, collecting data about the city.

Data created en masse at a hyper-local level opens up the prospect of a data-driven view of the city that was unimaginable when the first computer models were created. It is the ability to sense and collect data at a range of time scales, now becoming dependent on need rather than technical ability, that opens up the potential of IoT within an urban environment. IoT data cover a wide range of themes, from data relating to transport flows through to the density of crowds, environmental data about air pollution and temperature, through to economic transactions and foot fall and data relating to buildings. It covers all scales, from the hyper-local presence sensor under a desk that infers occupation, through sensors of room temperature and use of energy, up to city-wide transport data and urban heat islands, with the integration of GIS and smart-cities systems.

The use of such devices for input into a smart-city system can be broken down into the following aspects as highlighted in Figure 38.3:

Although the diagram in Figure 38.3 appears complex, it can be broken down into its components, each allowing the data to be collected, processed, analyzed, and finally visualized. Sensing and actuating are ubiquitous in our modern cities, buildings, and consumer products. Sensors refer to the technology that "converts a physical measure into a signal that is read by an observer or by an instrument" (McGrath and Ní Scanaill 2014). The emergence of the first thermostat in 1883 (US Patent No. 281884) is considered by some to be the first modern sensor and is still common-place in most monitoring systems. The 1990s witnessed the large-scale use

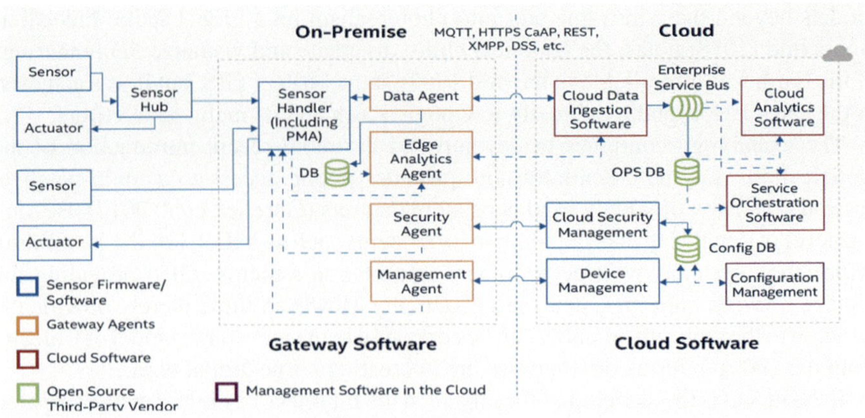

Fig. 38.3 Intel IoT reference architecture (Intel 2018).

of microelectromechanical (MEMS) sensors in automotive systems such as airbags and antilock braking, which introduced cheaper and more reliable sensing. The first consumer MEMS device, the Nintendo Wii controller of 2006, introduced a three-axis accelerometer which determined the motion and position of the controller. Economies of scale mean that similar technologies are now embedded in many consumer devices from phones to watches. From analog to digital, low cost to high, sensors cover a broad spectrum of operational parameters; for example, not all temperatures are equal and careful consideration needs to be given to the type of temperature sensor to be used (contact, non-contact, etc.).

Actuators on the other hand are the components of a machine that move or control some mechanism, by converting energy into motion. It is the mechanism by which a control system acts upon an environment. From the brute-force application in the construction site using hydraulics and pneumatics to the highly automated and controlled environment of the factory floor, all the applications have an ongoing operational cost—they are not fit and forget devices. The physical Internet has different maintenance requirements to those of the digital Internet.

Data generated by sensors or pushed to actuators are processed through gateways. These computational nodes can be on the same functional device (e.g. a mobile phone) or a separate compute module which gathers data from multiple sensing and actuating nodes (e.g. wireless sensor networks). The purpose of these data-collection devices is to capture, filter, and process data efficiently and to connect using wired or wireless communication technologies to legacy or Cloud infrastructure. This aggregation layer is often used to provide security, management, and data-preprocessing functions.

Data from gateways (or things) can be processed through any number of Cloud services, such as processing streams of data, implementing policies to make data available to different end consumers, or sending for storage. Data are typically stored for real-time analysis and presentation or archived to support offline analysis.

Smart technology can also use Cloud or edge architectures. These essentially describe where computing, storage, and analysis take place in the network. At the Cloud scale, data are typically sent to a centralized location where they are hosted on high-performance computing infrastructure and enjoy the benefit of compute power for complex analytic tasks. As an example, the meteorological network of weather stations maintained by the Meteorological Office around the UK all upload sensor data to servers where supercomputer facilities can be used to analyze and update rolling weather forecasts.

At the other end of the spectrum, there are many applications where it may be too expensive to send data via a data network to the Cloud, or where the latency in doing so means that useful analysis cannot be delivered in a timely manner. For example, autonomous vehicles need to operate at very low latency so that they can respond immediately to their surroundings; hence, many tasks are run locally in the vehicle, with non-time-critical information being sent to and from roadside infrastructure.

The final building block of IoT systems is the business intelligence layer, which both presents interfaces into the information being generated and provides the means to manage the system. IoT platforms provide the support software that facilitates communication, data flow, device management, and the functionality of applications. Outputs are typically screen-based and are increasingly accessed through virtual, augmented, or mixed-reality interfaces. As IoT systems mature, platforms are continually evolving to support the monitoring and management of connected devices at scale, since much of the value in the IoT supply chain is lost or made in the operational cost of those systems.

38.4 Putting It into Practice: Bats and Creatures

With the ability to visualize in three dimensions and collect data on the edge or within the Cloud via sensors and actuators linking into the digital twins of Urban IoT, the natural environment is often overlooked, especially by those focusing on city systems. Arguably, too many IoT test beds concentrate on smart transport systems, city logistics, or more traditional sensor-based devices. The opportunity of the Urban Internet of Things is the ability to look beyond the current normal and explore new possibilities. In terms of the health of an environment, bats are considered to be a good indicator species; a healthy bat population suggests a healthy biodiversity in the local area. As part of the QEOP test bed, Intel, in association with both University College London and Imperial College London, designed and deployed a "Shazam for Bats" project. Shazam is known for the ability to identify music through short audio clips, thus the aim to track and identify bats via IoT audio recording. A network of 15 smart bat monitors was developed and installed across the park in different habitats, creating a connected environment for monitoring wildlife.

The monitors (as pictured in Fig. 38.4) recorded the urban sound scapes via an ultrasonic microphone, with data processed by converting the sound into image files for data analysis. Each device processed the information locally using edge

Fig. 38.4 Echo box installed in the QEOP (https://naturesmartcities.com)

computing. As Premsankar et al. (2018) note, in edge architecture computing resources are made available at the edge of the network, close to (or even co-located with) end devices. Placing computing resources in close proximity to the devices generating the data reduces communication. Processing the data on the device has multiple benefits, firstly through reduced energy consumption and secondly through a dramatic decrease in the amount of data that has to be transmitted and processed on researchers' computers. During the first year of the trial (which is ongoing), the implementation of edge computing allowed a data reduction from 180Gb per day down to 2.2 Mb per day, a factor of 80000. Without the ability to process the data locally and instead relying on WiFi or such-like local infrastructure, neither the data collection nor the analysis would have been possible.

The use of the Internet of Things for longitudinal monitoring was carried out alongside more traditional survey techniques. The continuous data collection and analysis did however open up researcher time to focus on other aspects of the data and to note other shifts in bat activity. The use of IoT is notable as it provides an ongoing data stream without going into the field, allowing a background level of activity to be established and thus a series of interventions such as street lighting strategies to be implemented, with data accessible and therefore available for expert analysis on a daily basis. The trial is of interest in terms of the six Vs of Urban IoT: the velocity and volume led to the implementation of edge computing, while the veracity was tested as the identification of bat species was uncertain at the start of the trail. The data remained volatile, with hardware and power supply issues allowing approximately 70% uptime during the first year of testing. The sense of value is ongoing, but the ability to monitor remotely with data arriving in a preprocessed form creates intellectual, logistical, and economic values in terms of access to new data and analysis methodologies, the ability to carry on logistical trails in the park, and the saving of researchers' time.

Soft artificial intelligence (AI) is defined as non-sentient AI designed to perform at close to a human level in one specific domain. Soft AI is a reality now in the new generation of smart Internet of Things devices like Amazon's Alexa, Apple's Siri or Microsoft's Cortana (Milton et al. 2018). With over 100 million Alexa devices sold worldwide (The Verge 2019), the public at large are becoming used to talking to devices in their own home. As another part of the QEOP Urban IoT deployment, a series of 15 devices were placed in the park to allow the public to talk to them about the environment. The deployment was part of the project known as "Tales of The Park," looking at the wider issue of cybersecurity, trust, and risk within the Internet of Things. Using technology embedded into a series of 3D printed creatures (from bees through to otters and even garden gnomes), these geo-located devices used low-energy Bluetooth beacons to broadcast a URL to nearby users. A chatbot system then allowed users to converse with the devices via text-based messages using natural language. The IoT devices were aimed at communicating information about the local environment and the area's flora and fauna to the public at large, displayed on plinths at eye level, and spread across the park during the summer of 2018. We illustrate one such installation in Figure 38.5.

The majority of Urban IoT devices are small computers, often unseen, taking samples and communicating data out of sight. The aim of this part of the QEOP deployment was to make IoT visible, and to move beyond either the small hidden devices or devices in anonymous boxes, often found attached to lamp posts (more on lamp posts later in the chapter).

The creatures formed their own network of awareness, retaining information about the user as each device acted as a waypoint in the park. They opened up awareness of IoT devices being deployed with local environmental information, as well as moving the devices into a sense of awareness of the user as they learned more about the user at every interaction. In this sense, they open up the possibility of Urban IoT being more than invisible data collecting devices, and instead devices that chat and converse with users, allowing data to be both collected and communicated. Of course this opens up a whole issue around security and trust: how do you know which devices in the city to talk to? In the future, it may be necessary to address the possibility of a rogue Urban IoT, where devices are deployed to obtain information from the user without them either knowing or being aware. It is however an intriguing future to see Urban IoT as not only collectors but providers of information, and to have those devices be situated already within the environment, from trees to park benches and bus stops. All have the potential to be data collectors, and conversely, what could be more natural than talking to your bus stop for data on the bus times, weather, or air pollution, in the way you currently ask Alexa for information at home?

Fig. 38.5 One of the installations in the QEOP, in this case a gnome with embedded IoT technologies on a plinth

38.5 The Humble Lamp Post

The lighting of streets by electricity has brought a sense of security and wellbeing to our cities, towns, and villages for over 125 years. The first-ever electric streetlights in Britain were brought into operation in the 1870s in Holborn Viaduct and the Thames Embankment, London, and today, there are over 7.5 million streetlights in the UK (HTMA 2019). Lamp posts are part of the city; they are ubiquitous and almost unseen. As such, they make almost the perfect place for widespread, dense, and geo-located IoT sensors for the city. The process of transforming the lamp post into an IoT network is still in a conceptual stage, but test beds are in place at various locations around the world.

One such example is a trial to deploy customized multi-purpose lamp posts (MPLPs) in Kowloon East, Hong Kong's smart-city pilot area. The MPLPs will be interconnected with a telecommunication network to form an IoT backbone. Leveraging IoT sensors fixed on the lamp posts, the MPLP aims to enable real-time collection of city data, such as weather, air quality, temperature, and flows of people and vehicles, for city management and the support of various applications of smart-city initiatives (SCW 2019). Another example is the Humble Lamp Post, a cross-European initiative to upgrade and standardize the 90 million street lights across Europe with IoT services. Such envisioned services include: offering a (potentially free) public WiFi network; providing the powered foundations for a mesh network of (IoT) sensors across the city; helping drivers find a parking place; improving public safety; and supporting environmental monitoring (air quality, waste, flooding). Figure 38.6 illustrates the range of sensors and services envisaged. They can be a place for electronic street signage, public information, and advertising (revenue); be the home of sensors that help direct visually impaired people; a powered Web of electric vehicle (car, bike) charging points; or even pedestrian-flow monitors that can help keep the high street a vibrant place (BSI 2017).

A cross-technology and arts project, known as Hello Lamp Post, is an early example of using the lamp post as a social network. Using mobile-phone technology, the project started as an experimental urban-design intervention that operated in Bristol in July to September 2013. It used pre-existing identifier codes on street infrastructure to enable people to send text messages to objects such as lamp posts, post boxes, bins, telegraph poles, and so on. As Nansen et al. (2014) note, the project aimed to challenge ideas of efficiency tied up with the smart city by thinking about the city as a platform for social play. It allowed users to communicate with street furniture using SMS messages. Their exchanges with the objects were stored and used in exchanges to other people (Nijholt 2015), allowing a conversation to build, while the system was not directly automated (in comparison to the case of the chatbot creatures in QEOP). The project has been adapted for use in 12 cities around the world (Hello Lamp Post 2019) and was installed in the Queen Elizabeth Olympic Park during the summer of 2018 as part of the ongoing test bed for Smart London. Hello Lamp Post and the creatures in QEOP show that urban design and street furniture in cities can not only be conduits for more traditional digital data (data in binary form), but also for social data, collected from Urban IoT devices.

38.6 Urban Modeling

It is a little beyond this chapter to delve deeply into urban modeling, but it is worth noting that the first generation of urban models was designed and implemented in North America mainly during the years 1959–68, years which coincided with the launching of large-scale land-use transportation studies in major metropolitan areas (Batty 1979). In the intervening years, urban models and a variety of modeling techniques have been used to predict and forecast everything from the first transport

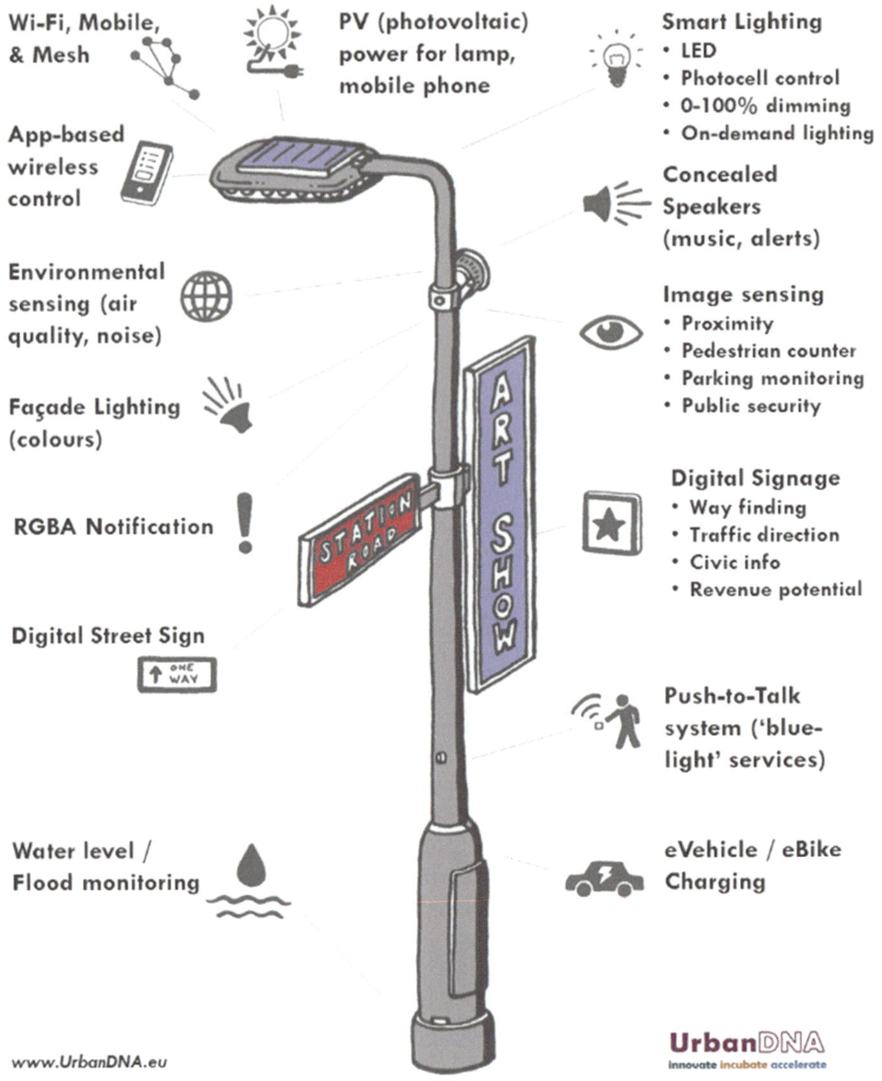

Fig. 38.6 Sensors on the humble lamp post, UrbanDNA (2018).

models to population growth, housing supply and demand, air pollution, the behavior of crowds, retailing, urban economics, and everything in between.

A number of techniques such as agent-based modeling are expanded upon within this book. All of them, however, rely on data and are arguably only as good as the data input to the model, and then also only as good as the methodology behind them. So while an increase in data may be seen as positive in terms of allowing a wider understanding of our cities, a focus needs to be made on understanding the veracity of the data. In terms of urban modeling, even small changes to an input's veracity can lead to a biased data set. As Harris et al. (2017) note, simulations that are based on biased data have the potential to increase biases by presenting results that are then used to influence policy. That said, the input from Urban IoT devices into urban modeling opens a new era in simulating and predicting our environment, but it requires standards and a joined-up approach to data analysis.

38.7 Talking to the Neighbors

As Summerson (2019) notes, the rapid rise of IoT devices within an urban context presents its own challenges. Summerson, leader of a UK government-funded organization known as the Future Cities Catapult (as of April 2019 renamed the Connected Places Catapult), notes that one problem is that much of IoT is still held in silos and separate systems that cannot communicate with each other. At the other end of the spectrum, however, irresponsible information usage raises serious—and arguably even dangerous—privacy and security concerns. Perera et al. (2018) highlight the issues by stating that IoT solutions often act as independent systems; the data collected by each of these solutions are used by them and stored in access-controlled silos. After primary usage, data are either thrown away or locked down in independent data silos.

A significant amount of knowledge and insight is hidden in these data silos that could be used to improve our lives; such data include our behaviors, habits, preferences, life patterns, and resource consumption. In short, at the current time, IoT devices often do not talk to each other; the data may be of high velocity and high volume and with a high level of veracity, but they are often isolated within a closed system. The system is often closed not only due to varying standards for sensing, communicating, and sharing data but also on a social-technical level, since IoT data is often private. As such the view of a self-monitoring, analysis, and reporting technology (SMART) city is complex and although often in close proximity, IoT devices are predominantly not aware of or communicating with their neighbors, making data collection and analysis within the IoT context an emerging challenge. As Summerson (2019) concludes, while IoT interoperability might be the key to accelerating improvements in traffic management, air quality and health, city planning, housing, and much more, the need to define and ensure the use of common languages and mechanisms—agreed IoT standards—has never been more urgent.

38.8 Conclusion

Digital twins are, according to Gartner (2018), at the peak of inflated expectations, while this arguably means the trough of disillusionment looms, before the arrival of wider use and a plateau of productivity. Their widespread use, and with it data collection, analysis, and use via Urban IoT devices, is on the horizon. To revisit the six Vs (velocity, volume, veracity, variety, volatility and value), without question, the volume and velocity are critical aspects of data in relation to Urban IoT devices. We are on the boundary of a change in the availability, use, and communication of data relating to cities. A majority of the estimated 75 billion IoT devices by 2025 will be in urban areas with a majority of them being able to provide data readings at a sub-minute and moving toward a sub-second frequency. In a similar change, the variety of data is increasing, from the ability to track foot-fall in real time, to pollutants at a hyper-local level, or levels of noise, through to the location of people and transport.

Advances in sensor technologies and networking are increasing the variety of information we are able to collect. Urban data, via the Internet of Things, are still in an early speculative phase and the veracity of the data is questionable. This is not only due to the quality of sensors but also to human factors. The volume of data can of course help with this; if you have enough devices deployed, then it is possible to identify rogue readings and delete them from any input or analysis. The value in terms of inputs into urban policy or urban modeling is long term, whereas the data collection is increasingly short term and high volume, raising issues around storage; and indeed, if data are simply used for the moment and then discarded due to excessive volume.

The opportunities for mass data collection via Urban IoT devices are immense, as are its potential inputs into urban modeling and policy. There are challenges, as we have noted, perhaps most notably in the veracity and volatility of data; but the value, volume, velocity, and variety of data collected from devices make the opportunities for Urban IoT almost limitless.

References

Anthopoulos LG (2017) The rise of the smart city. In: Understanding smart cities: a tool for smart government or an industrial trick? public administration and information technology 22. Springer

Ashton K (2009) "That 'internet of things' thing." The RFID journal. https://www.rfidjournal.com/articles/view?4986

Batty M (1979) Urban modeling: algorithms, calibrations, predictions, Cambridge University Press

Batty M, Hudson-Smith A (2007) Imagining the recursive city: explorations in urban simulacra. In: Miller HJ (ed) Societies and cities in the age of instant access. Springer, New York, pp 39–55. https://doi.org/10.1007/1-4020-5427-0_3.

BSI (2017) Exploiting the 'humble lamp post': a kick start to smart city. A guide for city leaders, Version 0,1. British Standard Institute.

Cellary W (2013) Smart governance for smart industries. In: Proceedings of the 7th international conference on theory and practice of eletronic governance (ICEGOV '13), October 22–25 2013. Seoul. pp 91–93

Coulton P, Lindley J, Cooper R (2019) The little book of design fiction for the Internet of Things. Lancaster University, ImaginationLancaster

Datta S (2016) Emergence of digital twins: is this the march of reason? arXiv preprint arXiv:1610.06467. https://arxiv.org/abs/1610.06467

Dawkins O (2017) Opening urban mirror worlds: possibilities for participation in digital urban dataspaces. In: Whose right to the smart city conference, plymouth. https://www.researchgate.net/publication/328497812_Opening_Urban_Mirror_Worlds_Possibilities_for_Participation_in_Digital_Urban_Dataspaces

Dawkins O, Dennett A, Hudson-Smith A (2018) Living with a digital twin: operational management and engagement using IoT and mixed realities at UCL's here east campus on the queen Elizabeth Olympic Park. Paper presented at GISRUK 2018, Leicester. https://www.researchgate.net/publication/324702983_Living_with_a_Digital_Twin_Operational_management_and_engagement_using_IoT_and_Mixed_Realities_at_UCL's_Here_East_Campus_on_the_Queen_Elizabeth_Olympic_Park

Deutsch R (2017) Convergence: the redesign of design. Wiley

Ebbers M, Abdel-Gayed A, Budhi V, Dolot F (2013) Addressing data volume, velocity, and variety with IBM. InfoSphere Streams V3.0

Gartner (2018) The gartner hype cycle. https://www.gartner.com

GLA (2013) Smart London plan. Greater London authority. https://www.london.gov.uk/sites/default/files/smart_london_plan.pdf

GLA (2019) Smarter London together. Greater London authority. https://www.london.gov.uk/what-we-do/business-and-economy/supporting-londons-sectors/smart-london/smarter-london-together

Goodchild MF (2018) Reimagining the history of GIS. Annals of GIS 24(1):1–8. https://doi.org/10.1080/19475683.2018.1424737

Gorelick N, Hancher M, Dixon M, Ilyushchenko S, Thau D, Moore R (2017) Google earth engine: planetary-scale geospatial analysis for everyone. Remote Sens Environ 202:18–27. https://doi.org/10.1016/j.rse.2017.06.031

Grieves M, Vickers J (2017) Digital twin: mitigating unpredictable, undesirable emergent behavior in complex systems. In: Kahlen FJ, Flumerfelt S, Alves A (eds) Transdisciplinary perspectives on complex systems. Springer, pp 85–113

Hardin G (1968) The tragedy of the commons. Science 162:1243–1248

Harris R, O'Sullivan D, Gahegan M, Charlton M, Comber L, Longley P, Brunsdon C, Malleson N, Heppenstall A, Singleton A, Arribas-Bel D, Evans A (2017) More bark than bytes? Reflections on 21+ years of geocomputation. Environ Plann B: Urban Analytics City Sci 44(4):598–617. https://doi.org/10.1177/2399808317710132

Hello Lamp Post (2019) Locations. https://www.hellolamppost.co.uk/#location

HTMA (2019) Street lighting. Highway terms maintenance association. https://www.htma.info/industry-topics/street-lighting.html

Hudson-Smith A, Wison G, Gray S (2019) The little book of connected environments. Lancaster University, ImagainationLancaster

Intel (2018) Intel® IoT Platform reference architecture. https://www.intel.com/content/www/us/en/internet-of-things/white-papers/iot-platform-reference-architecture-paper.html

Law J, Callon M (1994) The life and death of an aircraft: a network analysis of technological change. In: Bijker WE, Law J (eds) Shaping technology/building society. Studies in sociotechnical change, MIT, Cambridge, pp 21–52

McCullough M (2013) Ambient commons: attention in the age of embodied information. MIT Press, Cambridge

McGrath MJ, Ní Scanaill C (2014) Sensor technologies: healthcare, wellness, and environmental applications. Springer, New York

Milton R, Hay D, Gray S, Buyuklieva B, Hudson-Smith A (2018) Smart IoT and soft AI. IET conference publications

Nansen B, van Ryn L, Vetere F, Robertson T, Brereton M, Dourish P (2014) An internet of social things. In: Proceedings of the 26th Australian computer-human interaction conference on designing futures: the future of design. ACM, pp 87–96

National Infrastructure Commission (2018) Data for the public good. https://www.nic.org.uk/wp-content/uploads/Data-for-the-Public-Good-NIC-Report.pdf

Nijholt A (2015) Designing humor for playable cities. Procedia Manufacturing 3: 2175–2182. ISSN 2351-9789

Perera C, Vasilakos AV, Calikli G, Sheng QZ, Li KC (2018) Guest editorial: special section on engineering industrial big data analytics latforms for Internet of Things. IEEE Trans Industr Inf 14(2):744–747

Premsankar G, Di Francesco M, Taleb T (2018) Edge computing for the internet of things: a case study. IEEE Int Things J 5(2):1275–1284

Rose D (2014) Enchanted objects: design, human desire, and the Internet of Things. Simon and Schuster

SCW (2019) Hong Kong turns lamp posts into smart cities infrastructure Smart Cities World. https://www.smartcitiesworld.net/news/news/hong-kong-turns-lamp-posts-into-smart-infrastructure-3286

Statista (2018) Internet of Things (IoT), connected devices installed base worldwide from 2015 to 2025 (in billions). https://www.statista.com/statistics/471264/iot-number-of-connected-devices-worldwide/

Summerson G (2019) The emergence of interoperability standards in IoT. https://www.smartcitiesworld.net/opinions/opinions/the-emergence-of-interoperability-standards-in-iot-

Tennant M, Stahl F, Rana O, Gomes JB (2017) Scalable realtime classification of data streams with concept drift. Future Gener Comput Syst 75: 187–199. ISSN 0167739X https://doi.org/10.1016/j.future.2017.03.026

The Verge (2019) Amazon says 100 million Alexa devices have been sold –what's next? https://www.theverge.com/2019/1/4/18168565/amazon-alexa-devices-how-many-sold-number-100-million-dave-limp

UrbanDNA (2018) https://urbandnamedia.com

Andrew Hudson-Smith is Professor of Digital Urban Systems at The Centre for Advanced Spatial Analysis, University College London, and Visiting Professor at the University of Plymouth School of Art, Design and Architecture. He is a member of the Mayor of London's Smart London Board and has a current focus on IoT.

Duncan Wilson is Professor of Connected Environments in Centre for Advanced Spatial Analysis, University College London. His research focuses on how emerging technologies such as connected sensors and cognitive computing can augment our understanding of the built and natural environment.

Steven Gray is Principal Teaching Fellow in Centre for Advanced Spatial Analysis, University College London. His current research interests include large-scale data collection and data analysis, human computer interaction, mobile development, accessible web development with a focus on social media, web-based mapping and ubiquitous computing.

Oliver Dawkins is Data & Training Coordinator for the SFI funded Building City Dashboards project in the National Centre for Geocomputation at Maynooth University. He is also completing an EPSRC and Ordnance Survey funded PhD in mixed realities and urban IoT at the Bartlett Centre for Advanced Spatial Analysis, UCL.

Part V
Urban Computing

Chapter 39
Introduction to Urban Computing

Wenzhong Shi and Anshu Zhang

Abstract This chapter overviews Part V of this book themed urban computing. This part of the book covers the topics of visual analytics, cloud, edge, and mobile computing, data mining and knowledge discovery, AI and deep learning for urban computing, and a range of mainstream urban models and simulation methods. It provides a systematic review of computing technologies for urban governance and urban services, together with the examples of their usage, in the context of urban computing.

Within the context of urban informatics, urban computing is the processing of acquired urban data to serve urban applications. Urban computing can be regarded as the use of computing technologies to address urban issues, including those for urban governance and providing services to urban people. The computing technologies include those that are relevant to urban-related data communications, governance, analyses, mining, and visualization.

The basis of urban computing is the capability to perform highly scalable, fast, reliable, and flexible computation. The advances in cloud, mobile, and edge computing have greatly enhanced the computation capability for urban applications. Urban governance aims to improve the effectiveness and efficiency of urban management and decision making by addressing urban issues like traffic congestion, environmental pollution, disaster mitigation, aging population, large infrastructure maintenance, and housing. Urban services aim to provide a better experience for citizens in daily life. To achieve the goals of urban governance and urban services, urban computing needs to help people understand the data and extract actionable knowledge or other analytical results for alleviating urban issues and providing services. This leads to more dimensions of urban computing: urban data mining, analytics, modeling, and simulation.

W. Shi (✉) · A. Zhang
Department of Land Surveying and Geo-Informatics and Smart Cities Research Institute, The Hong Kong Polytechnic University, Hong Kong, China
e-mail: lswzshi@polyu.edu.hk

W. Shi et al. (eds.), *Urban Informatics*, The Urban Book Series,
https://doi.org/10.1007/978-981-15-8983-6_39

The chapters in Part V of this book describe urban computing from the perspectives of principles, models, and technologies in computing science and urban modeling. Emphases are put on the development and use of these principles, models, and technologies for urban contexts and urban applications.

While computations are carried out by machines, humans are the ones utilizing the computations to make decisions. Thus, Chap. 40 by Gennady Andrienko and 14 colleagues first introduces visual analytics, the study of the principles and methods for human-computer collaboration in solving complex problems, with a focus on visual analytics for urban mobility data. The chapter describes various visual and interactive analytical techniques and exemplifies the use of these techniques by analyzing Europe-wide data on the movement of passenger cars. By doing so, it shows how visual analytics greatly improve the ability of humans to see, interpret, link, and reason with data and their computation results, and then make decisions in urban contexts.

Chapter 41 by Chaowei Yang and his team introduces three backbone technologies for urban computing: cloud, mobile, and edge computing. Cloud computing provides scalability and on-demand availability of urban data computation. Mobile computing shifts the computation to mobile devices to reduce the load on central computation and enable more social interactions of citizens. Edge computing moves the computation to sensor networks to dramatically reduce the data communication load, speed up the response of sensors, and alleviate data-safety issues. The chapter systematically reviews the principles and characteristics of the three computing technologies and their applications in smart cities, and further illustrates their uses and integration by using the example of the urban heat island.

Chapter 42 by Chao Zhang and Jiawei Han moves to extracting succinct and easily interpretable knowledge from massive urban data. The review concentrates on discovering knowledge about urban activities from a type of crowdsourced and less-structured urban big data, that is, social sensing data contributed by users who share their experiences in the physical world online. The chapter first describes conventional and recently developed statistical and pattern-discovery methods for urban activity modeling, then presents the latest multimodal embedding techniques for learning urban activities, and concludes with future directions of urban knowledge discovery.

In the data-intensive era, approaches of mining knowledge from urban big data inevitably progress to leveraging the latest developments of artificial intelligence (AI), especially deep learning. In Chap. 43, Senzhang Wang and Jiannong Cao provide an overview of the challenges, methodologies, and applications of AI for urban computing. The chapter introduces the principles of mainstream AI techniques for urban computing, including popular deep-learning models that are commonly used in urban computing tasks. Then, the authors review the wide applications of urban computing based on AI and deep learning in urban planning, urban transportation, social networks, urban safety and security, and urban environment monitoring.

People use various urban models to understand cities and carry out urban governance and urban service tasks. The models run on real-world data with realistic complexity, as well as on simulation data that can overcome the sparsity of real-world data and be obtained with much lower cost and risk (e.g., for a disaster evacuation scenario). The remaining chapters in Part V introduce a number of mainstream urban models and simulation methods.

Chapter 44 by Mark Birkin presents microsimulation, the technique for generating synthetic population data of humans, households, or other entities at the individual level by using aggregate census data and individual-level sample data. Then, such synthetic data can support more analysis functions and result in deeper insight into the investigated problem than the original aggregate census tables. The chapter describes the principles of microsimulation, followed by the properties of microsimulation in computation, uncertainty, data assimilation, dynamics, and interdependence.

Chapter 45 by Anthony G. O. Yeh, Xia Li, and Chang Xia discusses cellular automata (CA) modeling for urban issues. With its unique strength in simulating complex nonlinear problems, CA has become a major analytical approach for creating what-if scenarios to facilitate urban policy making. The chapter covers the basics of CA models, the approaches to using CA models for urban modeling, different types of specialized urban CA models, applications of CA in urban studies and planning practices, and finally an outlook on further research for solving the remaining problems in urban CA modeling.

Chapter 46 by Andrew Crooks, Alison Heppenstall, Nick Malleson, and Ed Manley reviews agent-based modeling, the simulation technique that can create artificial worlds populated with individual agents, and investigate macroscopic processes in cities formed by interactions between the agents. A distinct advantage of agent-based modeling is its ability to assign diverse behaviors and rules to individual agents or groups of agents, which makes it a powerful way to simulate complex urban problems. The chapter presents the fundamentals of agent-based models and the applications of these models for solving urban problems. It further discusses how to capture decision-making processes in agent-based models, and new advances in agent-based modeling by utilizing big data, data mining, and machine-learning techniques.

Traveling and transportation have always been core topics in urban modeling. Chapter 47 by Eric J. Miller discusses the all-around evolution of transportation modeling driven by informatics. The chapter probes into this evolution from the changes in travel behavior due to real-time travel information and new mobility services and technologies; changes in transportation-system performance; new survey and tracking data available for transportation modeling; and the progress of modeling methods in response to new transportation phenomena and the latest computing and AI technologies. Finally, the chapter foresees new research problems where the theories and big data collide, that may fundamentally change transportation modeling in the future.

Due to space limitations, Part V only addresses a selection of core topics of urban computing. Many other important topics could be elaborated, for instance, urban data communication which is crucial for cloud, mobile, and edge computing. Urban data communication technologies include those for data transmission, wired and wireless data communication networks, devices, protocols, and security issues.

Also, the theories of modeling cities as complex systems have been discussed in Part I of this book, but much more discussion is needed on the computational aspect of complex system modeling for cities, particularly complex network modeling. Complex network models have been used not only on the topics traditionally employing network models, such as vehicle movements or road networks, but also on all kinds of dynamics and interactions in cities.

People will not stop pursuing higher computation capacity. Quantum computing, the computation based on principles of quantum mechanics such as superposition and entanglement, is a prominent example of the technologies in the experimental stage that aim to exponentially accelerate computation. Once some of these technologies become widely available, they are also likely to be applied to urban issues and to stimulate revolutions in urban computing.

Chapter 40
Visual Analytics for Characterizing Mobility Aspects of Urban Context

Gennady Andrienko, Natalia Andrienko, Fabian Patterson, Siming Chen, Robert Weibel, Haosheng Huang, Christos Doulkeridis, Harris Georgiou, Nikos Pelekis, Yannis Theodoridis, Mirco Nanni, Leonardo Longhi, Athanasios Koumparos, Ansar Yasar, and Ibad Kureshi

Abstract Visual analytics science develops principles and methods for efficient human–computer collaboration in solving complex problems. Visual and interactive techniques are used to create conditions in which human analysts can effectively utilize their unique capabilities: the power of seeing, interpreting, linking, and reasoning. Visual analytics research deals with various types of data and analysis tasks from numerous application domains. A prominent research topic is analysis of spatiotemporal data, which may describe events occurring at different spatial locations, changes of attribute values associated with places or spatial objects, or movements of people, vehicles, or other objects. Such kinds of data are abundant in urban applications. Movement data are a quintessential type of spatiotemporal data because they can be considered from multiple perspectives as trajectories, as spatial events, and as changes of space-related attribute values. By example of movement data, we demonstrate the utilization of visual analytics techniques and approaches in data exploration and analysis.

G. Andrienko (✉) · N. Andrienko · F. Patterson · S. Chen
Fraunhofer Institute IAIS, Sankt Augustin, Germany
e-mail: Gennady.andrienko@iais.fraunhofer.de

R. Weibel · H. Huang
University of Zürich, Zürich, Switzerland

C. Doulkeridis · H. Georgiou · N. Pelekis · Y. Theodoridis
University of Piraeus, Piraeus, Greece

M. Nanni
CNR, Pisa, Italy

L. Longhi
Sistematica, Terni, Italy

A. Koumparos
Vodaphone, Athens, Greece

A. Yasar
University of Hasselt, Hasselt, Belgium

I. Kureshi
Inlecom Group BVBA, Brussels, Belgium

W. Shi et al. (eds.), *Urban Informatics*, The Urban Book Series,
https://doi.org/10.1007/978-981-15-8983-6_40

40.1 Introduction

The science of visual analytics (Thomas and Cook 2005) develops principles, methods, and tools to enable synergistic work between humans and computers through interactive visual interfaces. Such interfaces support the unique capabilities of humans (such as the flexible application of prior knowledge and experiences, creative thinking, and insight) and couple these abilities with machines' computational strengths, enabling the generation of new knowledge from large and complex data.

In this chapter, we describe visual analytics approaches that are related to the study of urban mobility data and discuss how visual analytics can support analysis of such data and informed, justifiable decision making. We address different stages of the urban data science process, including data quality assessment, data transformation, exploration, and analysis, and indicate possibilities for model building, evaluation, and refinement. We conclude this chapter with a summary of achievements, unsolved problems, and future research directions.

We demonstrate the utilization of visual analytics techniques in a process of exploration and analytical reasoning using a real-world data set. In the EU-funded Track&Know project,[1] one of industrial partners collects Europe-wide tracks of passenger cars. The data are collected for insurance purposes under vehicle owners' informed consent, aiming at enabling transparent pricing and facilitating analysis of accidents. For these purposes, it is necessary to have an understanding of the context in which the vehicles move, which includes the surrounding traffic. There are several questions that require answers for understanding traffic: What are the major flows and their properties? How do they vary over time? What is the composition of the types of the cars appearing on streets? What are regular and irregular trips and how are they distributed in space and time? etc. Answers to these questions can be valuable for a variety of practical applications such as assessing which part of traffic can be potentially served by publicly shared vehicles or by electric cars, evaluating applicability of various car sharing schemes, identifying and assessing different driving styles, and investigating events, such as traffic accidents, in their context.

40.2 State of the Art

Batty (2013) considers a city as a system composed of flows (between locations and between activities) and networks of relationships and interactions among various entities. For understanding these factors of the urban context, a variety of different data sources is considered. There are studies (e.g., Kesting and Treiber 2013) based on stationary sensors such as traffic counters that record aggregated characteristics (how many cars passed a given street segment during some time interval and what was

[1] Track&Know, grant agreement 780,754: https://trackandknowproject.eu/.

their speed). Such sensors record aggregates but do not allow the tracking of vehicles. Another kind of stationary sensors is docking stations for rental bicycles (or, potentially, other kinds of shared vehicles). Usually, these sensors provide only general characteristics (overall capacity, numbers of docked bicycles, and empty slots) and their aggregates over time intervals. However, sometimes more detailed data are released, enabling analysis of the moves of the vehicles between the docking stations (Beecham and Wood 2014). Some researchers approximate mobility from space- and time-referenced social media records. A prominent example is provided by Lansley and Longley (2016) who studied in detail the distribution of the message topics in space and their variation over time. Itoh et al. (2016) studied data of smart-card usage in local trains together with social media records for reconstructing temporal characteristics of major flows and understanding abnormal situations.

Several review papers discussed visual analytics approaches to analyzing mobility and transportation. A review by Andrienko and Andrienko (2013a) considered approaches from the data processing perspective: looking at trajectories, clustering trajectories, transforming times in trajectories, and studying attributes, events, and patterns in trajectories, followed by generalization and aggregation of trajectories and tracing derived flows. In a more recent review on visual analytics of mobility and transportation, Andrienko et al. (2017) outline approaches used for the following problems: understanding details of individual movement, studying the variety of routes taken, assessing movement dynamics along a route, linking origins and destinations, characterizing collective movement over a territory, detecting events and studying their distributions, contextualizing movement, and studying impacts and risks.

Markovic et al. (2019) present a viewpoint of a road transportation agency, mentioning the following problems of interest: demand estimation, modeling human behavior, designing public transit, measuring and predicting traffic performance, assessing impact on the environment, and improving road safety.

The reviews indicate the need to consider movement data from multiple perspectives. We follow this approach in our work.

40.3 Mobility Data: Properties and Problems

To demonstrate the data analysis workflow, we use trajectories of 4521 passenger cars within the Greater London area that were recorded during two regular weeks in winter 2017; 4,284,493 position records in total. Each position record consists of an anonymized identifier of a vehicle, time-stamped geographic coordinates, and attributes such as momentary speed and heading, GPS signal quality. Transport for London estimates the number of all cars registered in London as about 2.6 million.[2] Respectively, our data set covers about 0.2% of the active "population" of the passenger cars. Figures 40.1 and 40.2 show the spatial and temporal distributions

[2]https://content.tfl.gov.uk/technical-note-12-how-many-cars-are-there-in-london.pdf.

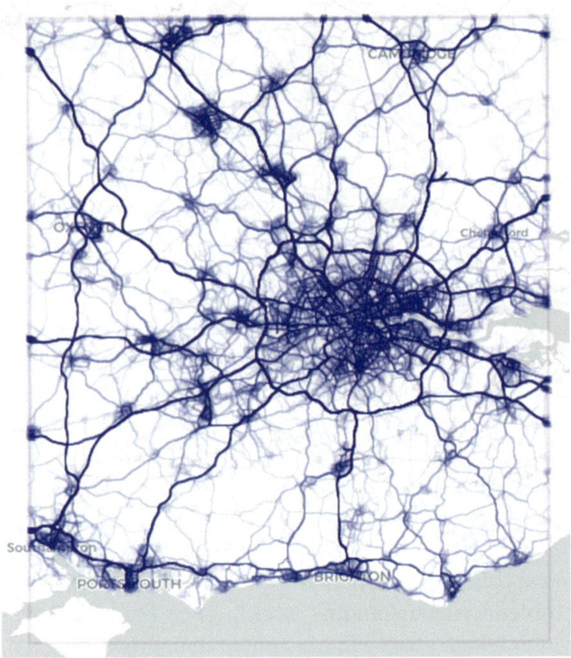

Fig. 40.1 Spatial footprint of all trajectories in the data set

of the recorded trajectories. From the map (Fig. 40.1), we can recognize the major roads and populated areas.

The time histogram (Fig. 40.2) reflects the distribution of the counts of distinct cars per hour, starting from Sunday midnight: 2 weeks $\times$ 7 days $\times$ 24 h $=$ 336 h in total. The time histogram clearly shows the weekly cycle and distinct profiles of weekdays and weekends.

For assessing the quality of the data set, we follow the approach proposed by Andrienko et al. (2016a). Possible problems in movement data include problems of coverage and accuracy that may occur in all components of the data, namely space, time, identifiers, and attributes. Respectively, we assess properties of all data components and their combinations.

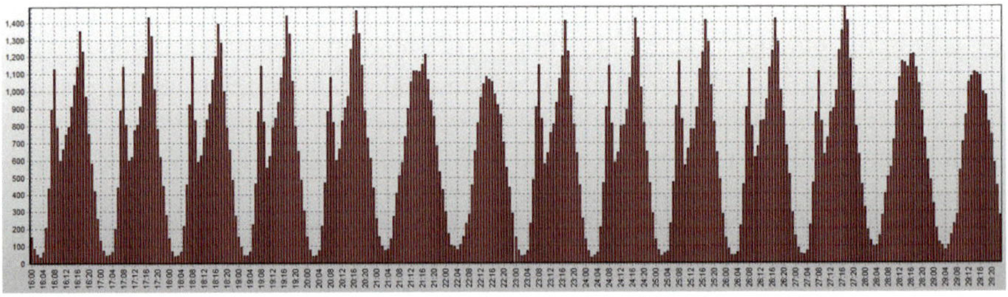

Fig. 40.2 Temporal profile of the data: the bars represent the car counts per hour

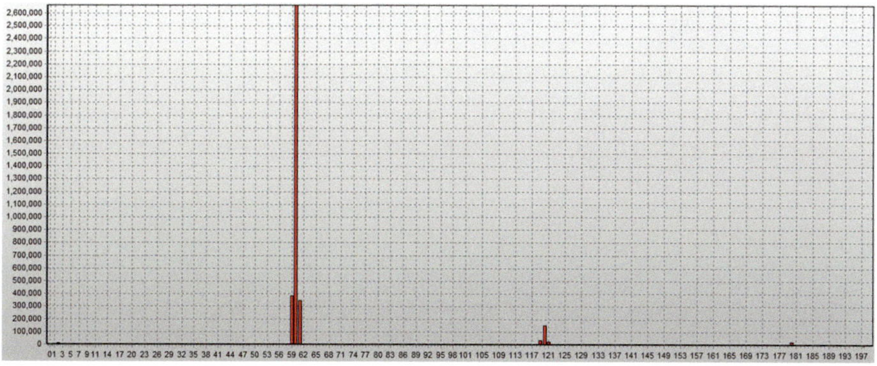

Fig. 40.3 Sampling rates

For the temporal component, we start with examining the sampling rates, i.e., the time intervals between consecutive position recordings for the same car. The statistics (Fig. 40.3) demonstrate that the most frequent sampling rate is around 1 min (59–61 s). A much smaller subset of points is characterized by the sampling rate of about 2 min, and only a few points have 3 min intervals to the next points. All other intervals appear in the data infrequently. Next, we checked if the sampling rate of 1 min is typical for all cars. For this purpose, we calculated the median sampling rate for each car. The results demonstrate that more than 98% of the cars have the median sampling rate of 1 min $\pm$ 1 s. However, we have identified a few outliers: about 100 cars that had only a few positions recorded and, correspondingly, rather arbitrarily sampling rates; 9 cars with many recorded positions but the median sampling rates of 3–5 min; and 2 cars with very high sampling rates (13 s). Such outliers need to be separated in further analysis. We have also identified several thousands of duplicate pairs of an identifier and a time stamp and excluded the duplicates.

Figure 40.4 shows the frequency distribution of the distances between consecutive position records, with the bins corresponding to 10 m intervals. We can observe major peaks at 420 and 1760 m. Since the typical sampling rate is 1 min, these peaks correspond to displacement speeds 25.2 and 105.6 km/h. We also observe narrower peaks at 100 m (6 km/h) and 2000 m (120 km/h). The former may correspond to small displacements caused by waiting at street intersections. We inspected the second peak separately. Such distances between points appear either at highways and may mean that some points were not recorded (e.g., due to bad satellite connection), or at the borders of the studied area (Fig. 40.4 bottom). These large displacements at the area boundaries are artifacts of data selection by a bounding rectangle.

Figure 40.5 presents the frequency distribution of the instant speed values in the positional data after excluding numerous (about 778,000) stationary points and a few outliers with speeds higher than 180 km/h. The clearly visible peaks roughly correspond to the speed limits on different categories of the UK roads.

Figure 40.6 shows the frequency distribution of the measured vehicle headings in the non-stationary points. There are two strange pits around the values 90° and

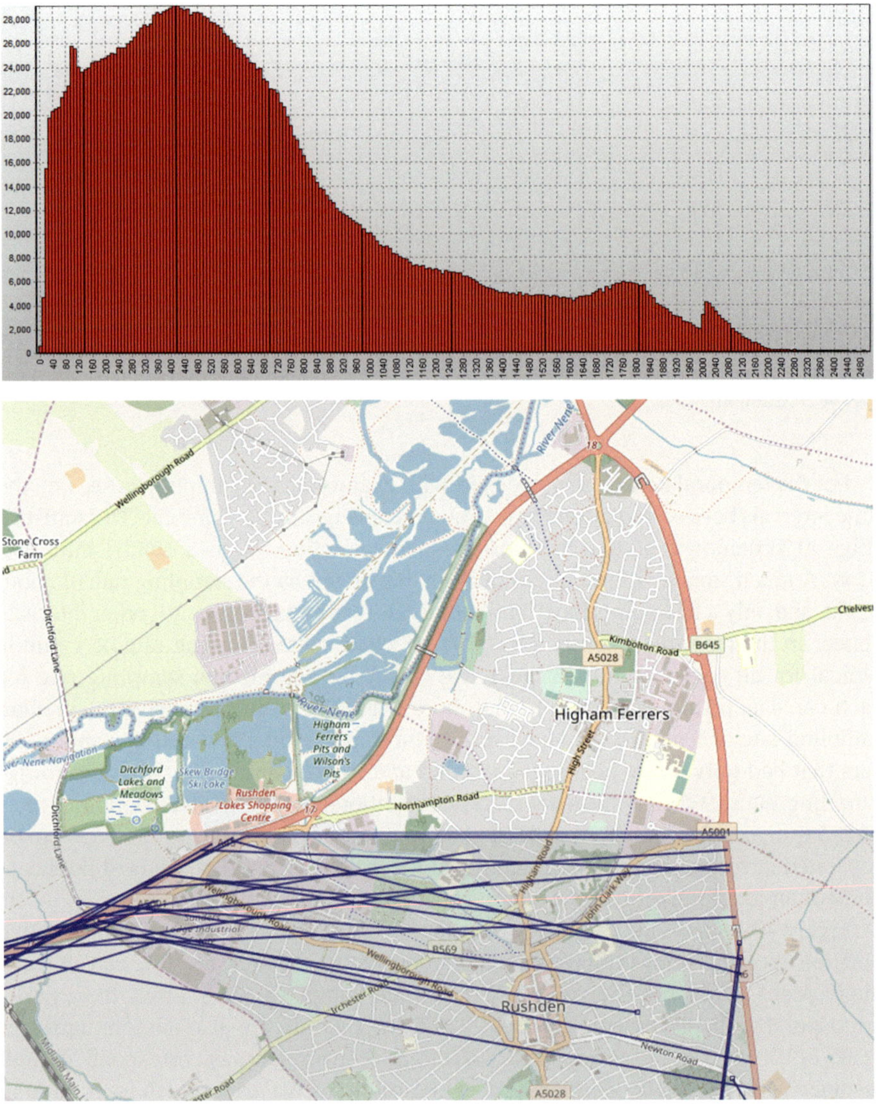

Fig. 40.4 Top: frequency distribution of the distances between consecutive points of trajectories. Bottom: long distances between consecutive points are caused by selecting data that fit in a chosen bounding rectangle (border effects)

270°. It is quite unlikely that these directions were really much less frequent than the others. The pits may be due to the method that is used by the tracking devices for determining the vehicle heading. The method may calculate the angle based on the ratio of the x- and y-differences between two consecutively measured positions (of

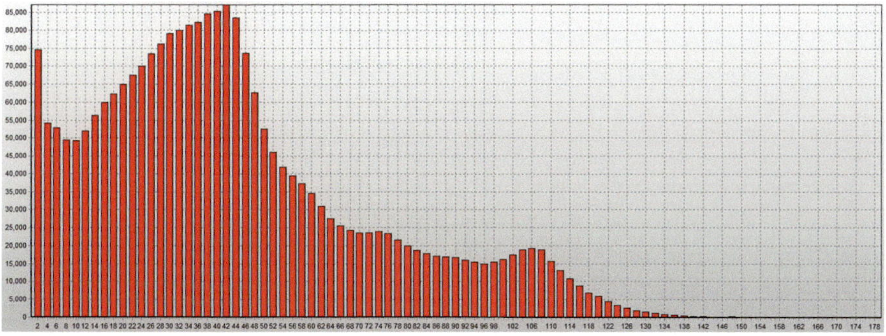

Fig. 40.5 Frequency distribution of the speeds after removing stationary positions and outliers

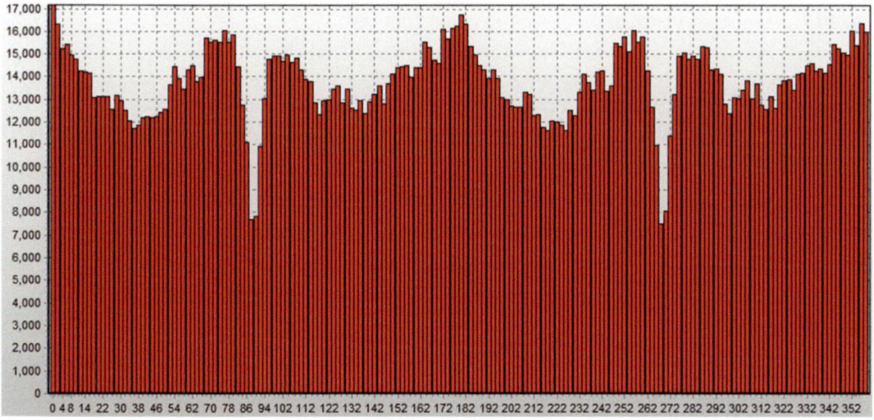

Fig. 40.6 Frequency distribution of the measured vehicle headings

which the second position is not recorded) and fail in cases when the *y*-difference equals zero. Whatever the reason, the measured heading values cannot be trusted.

For human mobility studies, it is important to divide trajectories into trips, e.g., between places of significant stops (Andrienko and Andrienko 2013a). There exist different criteria for separating trips: by positional attributes (e.g., taximeter is switched on or off), by temporal cycles (e.g., daily trips), by substantial displacement (e.g., if the next point is at least 5 km away) and by temporal gaps between points (no movement for at least 15 min). We used the latter criterion. For tolerating position measurement errors, the periods when positions remained within a small area during a time interval of a chosen length (15 min) were also treated as stops. In this way, we acquired 164,644 sub-trajectories, from which 3943 consisted of single points and were excluded from further consideration. The remaining sub-trajectories were treated as representing trips. Figure 40.7 presents the frequency distribution of the trip counts per car. About 300 cars had only 1 or 2 trips during the two weeks. Many cars performed from 30 to 50 trips, and only a few cars had more than 80 trips.

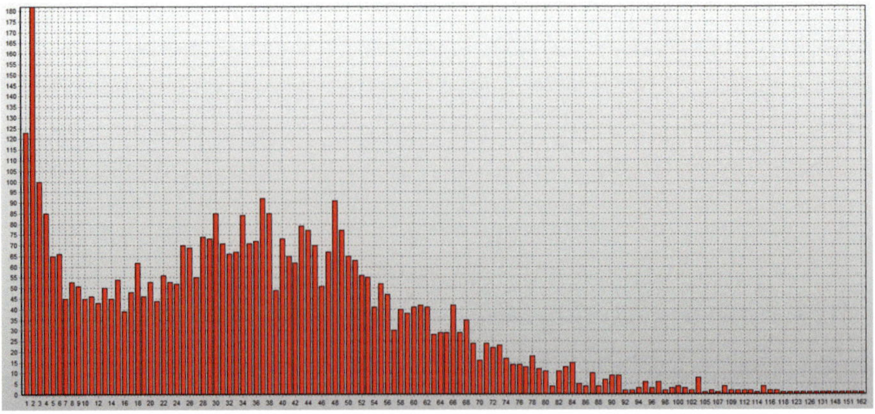

Fig. 40.7 Frequency distribution of the trip counts per car

Figure 40.8 presents an example of all trips of a single car during two weeks. The map on the left shows the spatial footprint. A space–time cube (Hägerstrand 1970; Kraak 2003) shows the same trips in space and time simultaneously. The vertical axis represents the time of the day. The colors encode the weekdays (green) and weekends (red). Generally, such a visualization may enable identifying the person whose track is shown; therefore, we have masked the locations on the map and will avoid disclosing any further potentially privacy-sensitive details in the text or illustrations.

After performing the investigation of the data properties and cleaning the data by excluding incomplete tracks and incorrect values, we can proceed with analysis.

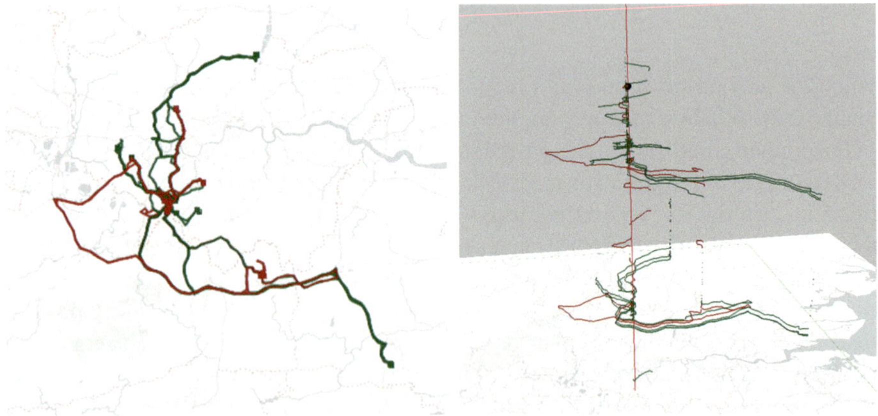

Fig. 40.8 Trips of a single car are represented on a map (left) and in space–time cube (right), in which the trips have been temporally aligned within the daily time cycle. The colors denote whether the trips took place on weekdays (green) or weekends (red)

40.4 Data Types: Events, Trajectories, Spatial Time Series, and Situations

There exists a range of transformations that can be applied to movement data for analyzing them in various ways and extracting different kinds of information. First of all, each recorded position is a *spatial event*, which is specified by a reference to the moving object *id*, time stamp *t*, and coordinates *x* (longitude) and *y* (latitude). An event may also have attributes: *id, t, x, y, attributes.*

The events of moving objects being at specific spatial positions at particular times can be called *position events* to distinguish them from other kinds of spatial events. Integration of chronologically arranged position events of the same moving object produces a *trajectory* of this object (Fig. 40.9). Such integration allows computation of derived attributes based on the positions of consecutive points: displacement distance and direction, time difference, speed estimate, etc. These derived attributes can be used for extracting secondary events from trajectories (e.g., stops) and dividing trajectories into smaller subsets (e.g., trips between stops). We applied these transformations when investigating the data properties.

Both trajectories and events can be spatially aggregated by a set of places. As a result, the places are characterized based on the visits by moving objects (e.g., counts of the objects and the visits, statistics of the duration of object presence in the area, etc.) or the events that occurred in them (e.g., counts of events of different kinds). The aggregation can be performed by time intervals producing *place-based time series* of the visits and presence. Additionally, trajectories can be aggregated according to the moves (transitions) between areas. The transitions link the areas, and these links can be characterized based on the number and properties of the transitions, such as the number of distinct objects that moved and the statistics of the speeds and durations. Aggregated transitions between places are usually called *flows*. The aggregation can also be made by time intervals resulting in *link-based time series* of flow characteristics.

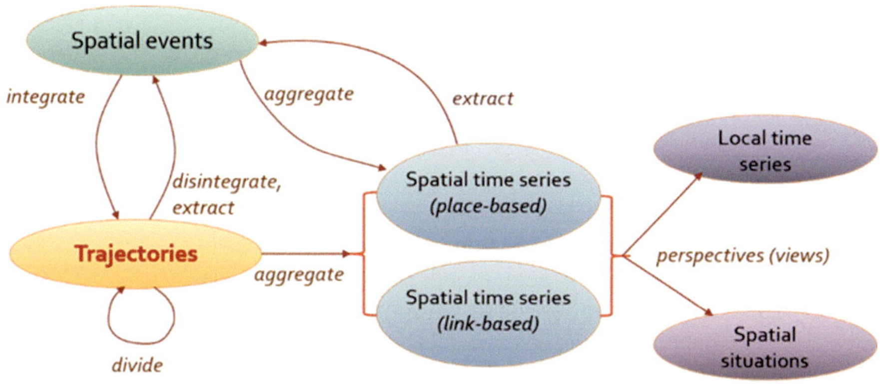

Fig. 40.9 A general scheme of movement data transformations

Spatial time series can be viewed in two complementary ways. On the one hand, they consist of sequences of values associated with individual places or links, which can be called *local time series*. Respectively, the places or links can be characterized and compared based on the temporal variation of the respective values. On the other hand, for each time step, there exists a particular distribution of the values over the set of places or links. This distribution can be called a *spatial situation*. The whole spatial time series can be seen as a sequence of such spatial situations. Respectively, the temporal variation of the spatial situations can be studied and characterized.

Further events (e.g., occurrences of extreme values) can be extracted from place- or link-based spatial time series.

Data transformations support investigation of different aspects of mobility phenomena. As our goal is characterization of urban context, we expect that transformations will allow us to enrich the context by different kinds of relevant information.

40.4.1 Context Acquisition from Movement Data

Traffic and mobility are important parts of the overall urban context. Information concerning movements of vehicles and people in an urban area may be relevant in studying various phenomena, such as air quality, noise, or disease spread, and events, such as traffic accidents, crimes, or disruptions in the work of public transport. Movement-related context information that can be extracted from trajectory data includes place visiting context, flow context, time context, trip context, and personalized semantic context. We consider a selection of the listed aspects in detail in the following sections.

40.4.1.1 Place Visiting Context

For describing the context in terms of place visits, it is necessary to have a suitable set of places. When there are no predefined places suiting the goals of an intended study, the places need to be appropriately defined. One possible way to do this is taking the neighborhoods of some positions of interest, e.g., circles of a chosen radius around the positions of studied events. Places relevant to transportation studies can be defined based on the street segments and intersections. However, the resulting level of detail and amount of data can be excessive for the envisaged spatial scale of the intended study. For studies of human mobility behaviors, places can be defined based on identifying areas of different kinds of human activities.

A set of places can also be derived by partitioning the territory into compartments based on the spatial distribution of some data, such as positions of stationary objects, events, or points from vehicle trajectories. Andrienko and Andrienko (2011) proposed to divide a territory based on the distribution of characteristic points of trajectories, which include the positions of stops and turns as well as trip starts and ends. The points are extracted from the trajectories and grouped according to their spatial locations. A special method for space-bounded point clustering produces spatial clusters whose radii do not exceed a given threshold. The medoids of the clusters (i.e., the points with the smallest mean distances to the other cluster members) are taken as generating seeds for Voronoi tessellation. When the points are not evenly spread throughout the territory but form dense clusters, the seeds tend to be taken from these clusters, which make the resulting places meaningful and interpretable. Depending on the chosen maximal radius of a point cluster, the territory is divided into larger or smaller compartments. Hence, an analyst can adjust the partitioning to the spatial scale of the intended analysis and the desired level of detail.

An example of territory partitioning based on trajectory data is shown in Fig. 40.10. The characteristic points have been grouped in clusters with the maximal radius 2.5 km. As a result, we have obtained 3535 places (compartments). It can be observed that the geometries and the spatial layout of the places reflect the topology of the

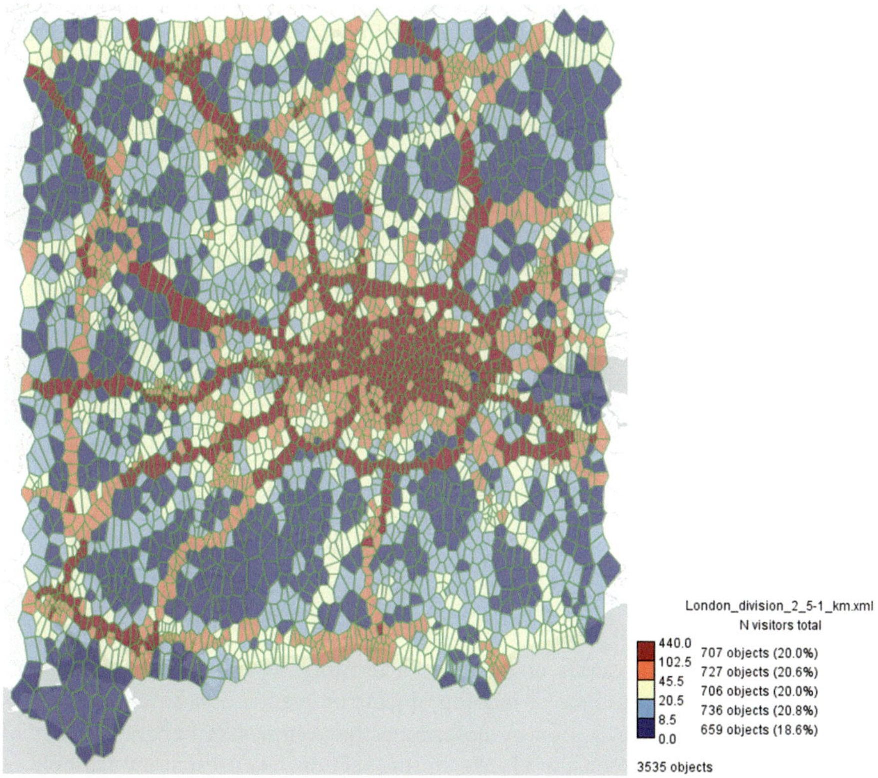

Fig. 40.10 Tessellation of the region into 3535 polygons based on point clustering bounded by a maximal cluster radius of about 2.5 km. Colors represent counts of distinct cars observed in each region, from blue (less than 8) to red (more than 102), using equal class size division

major roads. This is the effect of taking seeds for the tessellation from dense concentrations of trajectory points, which mainly occurred along these roads. The places in Fig. 40.10 are colored according to the numbers of distinct cars that visited them. As we mentioned earlier, other characteristics of places that can be derived from movement data are time series of place visits and their durations, and aggregate characteristics of the objects that visited the places.

Thus, our data allow us to characterize the places based on the "population structure" of the cars that visited them. The data set includes car manufacturer information for each anonymized car identifier. Respectively, it is possible to obtain separate car counts for different manufacturers. Using this information, we would like to cluster the places by the similarity of the car population structures. However, a straightforward application of clustering to the absolute counts just separates areas by total car counts, replicating the major patterns visible in Fig. 40.9. Therefore, it is necessary to normalize the counts by the total numbers of different cars recorded in each compartment, thus obtaining proportional values.

We have clustered the normalized counts using the partition-based clustering method *k*-means in combination with a projection of the cluster centroids onto a plane, as suggested by Andrienko and Andrienko (2013b). The results are presented in Fig. 40.11. The positions of the cluster centroids on the projection plane (top left) are used for selecting appropriate clustering parameters and then for assigning colors to clusters reflecting their similarities and differences. The cluster profiles in terms of the proportions of the cars from different manufacturers are shown in a bar chart (top right) and on a map (bottom left).

The clustering results show that the main motorways are dominated by Vauxhall, Ford, and VW, while central London and Brighton are characterized by a mix of everything, with some prevalence of Vauxhalls and Fords. One can find compact "villages" in rural areas populated mostly by Fiat, Ford, SEAT, Peugeot, or VW.

Places can also be grouped according to the place-based time series of visits or counts of distinct cars, either in absolute or normalized form. We omit such analysis here due to space restrictions. However, we shall consider link-based time series in the next section.

40.4.2 Flow Context

While place-based time series characterize a territory in terms of the spatiotemporal variation of the presence of moving objects or events, link-based time series complement the characterization by describing the volumes and characteristics of movements (flows) between the places. In this section, we present an example of analyzing the flows between the same places as in Figs. 40.10 and 40.11. For the set of 3,535 places, we obtain 13,153 directed links when we use the original trajectories and 12,654 links when we use the trajectories corresponding to the trips (resulting from dividing the original trajectories based on stops for 15 min or more). The divided trajectories are more appropriate for characterization of movement speeds.

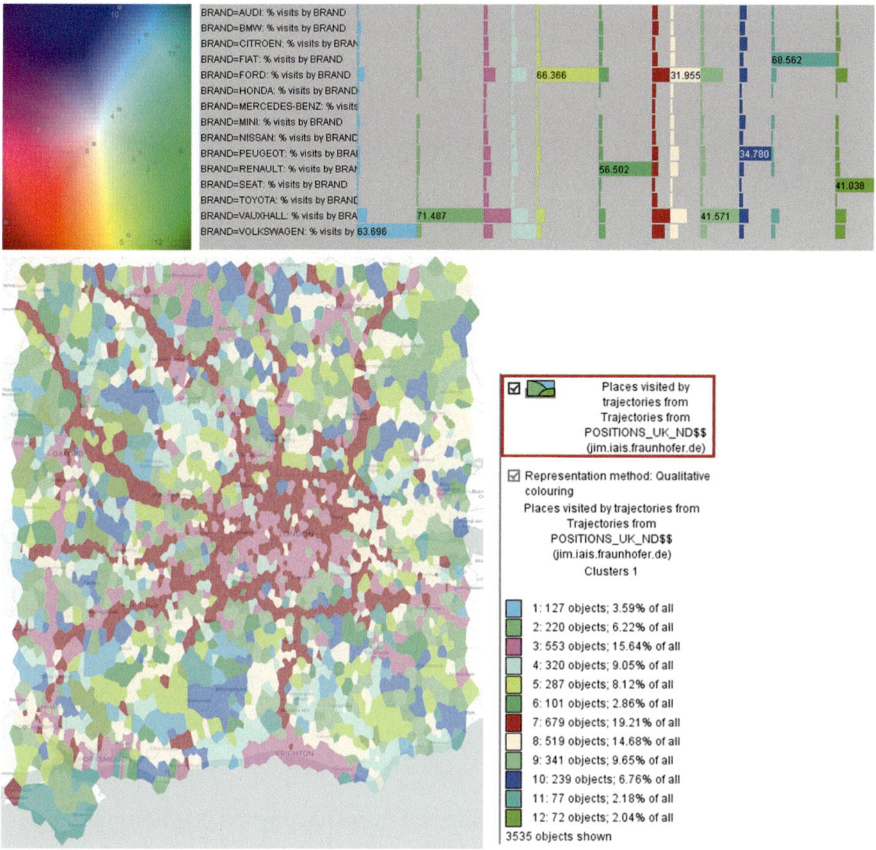

Fig. 40.11 Clustering of places by similarity of the car population structure. Top: a 2D projection of the cluster centers (left) and the profiles of the clusters in terms of the attributes involved in the clustering (right). Bottom: a map of the spatial distribution of the clusters (left) and the corresponding legend showing the cluster sizes (right)

Figure 40.12 presents a map where the links are represented by curved lines colored according to the average speeds during the transitions between the places. Similarly to Fig. 40.10, this map reflects the properties of the road network and the spatial distribution of the urban areas. Each pair of places is connected by two lines reflecting movements in opposite directions. We can notice that for the majority of the location pairs there is no substantial difference between the average speeds in the opposite directions. However, aggregates that reflect the temporal variation, such as the hourly flow volumes over the two weeks, may reveal asymmetry between the flows in opposite directions.

In Fig. 40.13, we have applied k-means clustering to the flow volumes normalized by the each link's mean value after exclusion of the links with very low flows (less than 50 moves in total during the 2 weeks period). As in the previous section (Fig. 40.11),

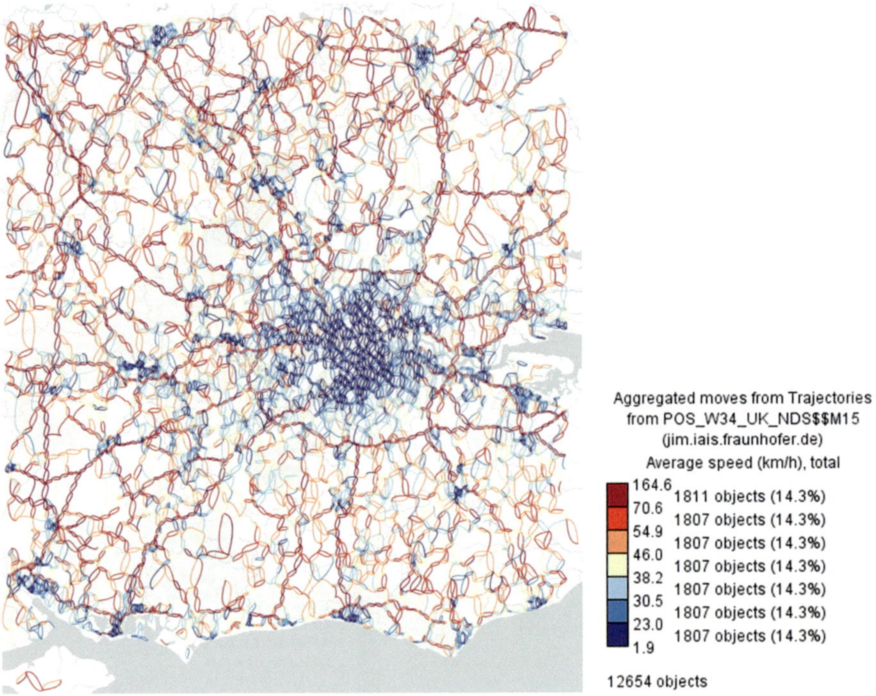

Fig. 40.12 Average speeds of the flows between the places

the parameters for the clustering were selected by inspecting the positions of the clusters centroids in the projection space, and the projection was also used for assigning colors to the clusters. Clusters whose centroids are close in the projection space due to the similarity of the respective attribute values receive similar colors. In the map in Fig. 40.13, we can observe the consistency of cluster affiliation along chains of links following the major roads; hence, the traffic has common patterns along the major transportation corridors formed by the most important motorways. We can also notice pairs of opposite links that were put in distinct clusters, which means that the temporal patterns of the respective flows differ.

40.4.3 Time Context

Mobility is essentially a temporal phenomenon; thus, the distribution of people and vehicles over a territory and their movements from place to place vary over time. As human activities are cyclic in general, we can expect temporal cycles to appear in aggregated representations of mobility, and we have observed them in the 2D histograms of the aggregated flows in Fig. 40.13.

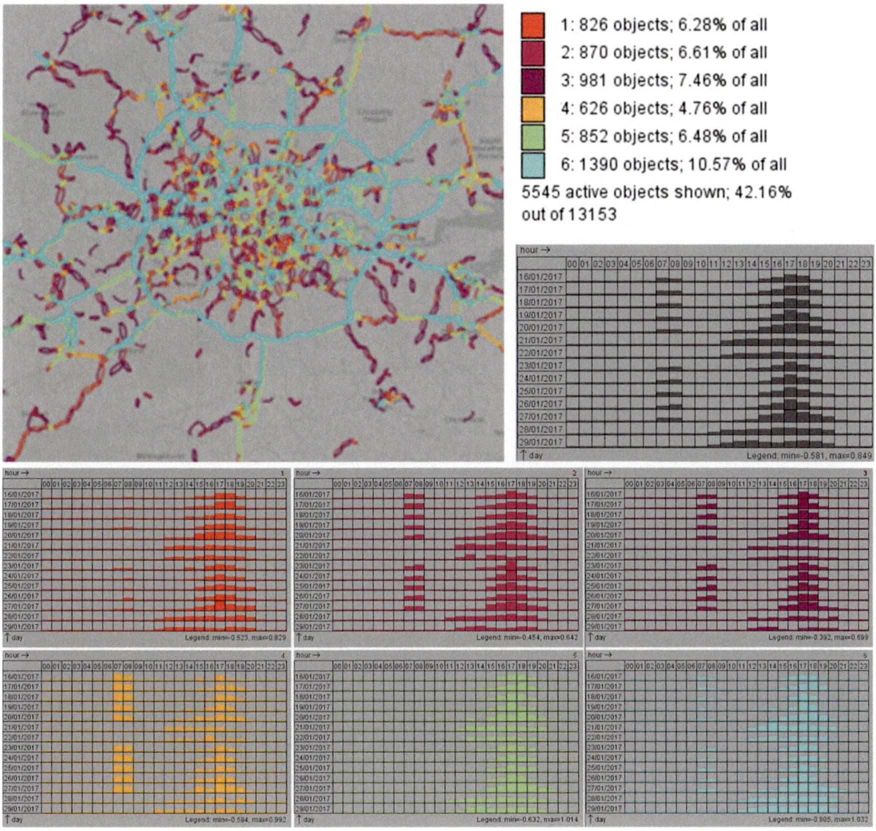

Fig. 40.13 Links clustered according to the similarity of the normalized time series of flow volumes. Top: a map with the links colored according to their cluster affiliation; the legend shows the cluster sizes. Bottom: the cluster profiles are represented in an aggregated form in two-dimensional histograms with the rows corresponding to days and columns to hours. The heights of the colored bars in the cells are proportional to the mean normalized hourly values for the clusters. The 2D histogram with the dark gray bar shows the average temporal variation for all links

As shown in Fig. 40.9, spatial time series can be viewed from two complementary perspectives: as spatially distributed local time series and as temporally varying spatial situations. Figure 40.13 corresponds to the former perspective: we applied cluster analysis to the local time series associated with the links. Now we are going to take the other perspective and apply clustering to the time steps of the time series. We cluster the time steps according to the similarity of the spatial distributions of the car presence (Figs. 40.14 and 40.15) and flow volumes (Figs. 40.16 and 40.17). The aggregates representing the presence have been obtained from the original (undivided) trajectories, to take stationary vehicles into account, and the link-based aggregates have been obtained from the divided trajectories representing the trips.

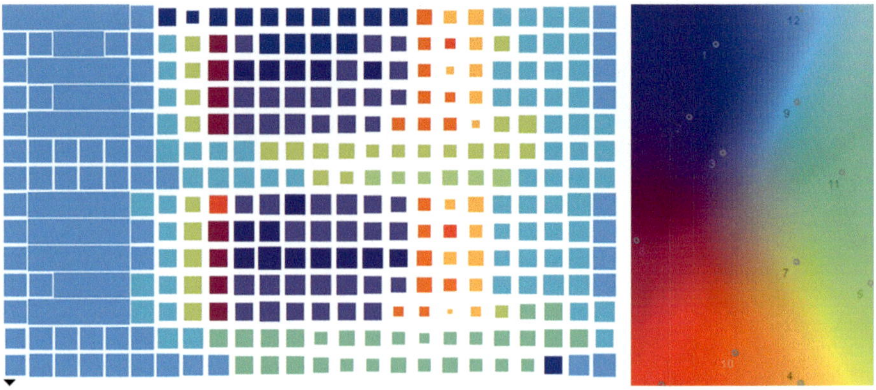

Fig. 40.14 Left: a calendar display of the clusters of the hourly time steps according to the distribution of the car presence over the set of places. The columns correspond to 24 h of the day and the rows to the 14 days from Monday (top) to Sunday of the next week (bottom). The colors correspond to different clusters, and the sizes of the colored rectangles represent the closeness of the cluster members to the cluster centroids (the closer, the bigger). Right: the colors for the clusters have been chosen by projecting the cluster centroids onto a continuously colored plane

The calendar view in Fig. 40.14, left, shows the daily and weekly patterns of the spatial distribution of the car presence, where the night hours are similar across the days; the morning and evening rush hours of the weekdays appear quite different from the midday times, and the weekend patterns are distinct from the weekday ones. The patterns on Friday evenings differ from the other weekdays by later beginnings of the evening- and night-specific distributions.

The small multiple maps in Fig. 40.15 demonstrate the spatial distribution of the mean volumes of the presence for each cluster. The clusters are arranged according to the succession of their numeric labels (from 1 to 12) in rows from left to right and from top to bottom. We can observe extremely prominent road network patterns, especially during the mass commuting times (e.g., Clusters 6 and 10). These patterns do not appear in late evenings and nights (Clusters 9 and 12).

Figures 40.16 and 40.17 present the results of applying clustering to the time steps of the link-based time series. The times have been clustered according to the similarity of the spatial distributions of the flow volumes. Figure 40.16 is analogous to Figs. 40.14 and 40.17 corresponds to Fig. 40.15, but the maps here show the spatial distributions of the mean flow volumes corresponding to the clusters. The volumes are represented by proportional widths of the flow lines.

The afternoon Clusters 1, 4, and 9 are characterized by intensive traffic on highways while the morning Clusters 6, 7, and 8 show higher traffic on local roads and in populated areas. Interestingly, the flow distribution patterns in Hours 9–14 on the weekdays are similar to those in the nights. Several clusters consist of only a few or even a single time moment with extraordinary traffic distributions. For example, Cluster 5 has a very high traffic on the inner ring of London.

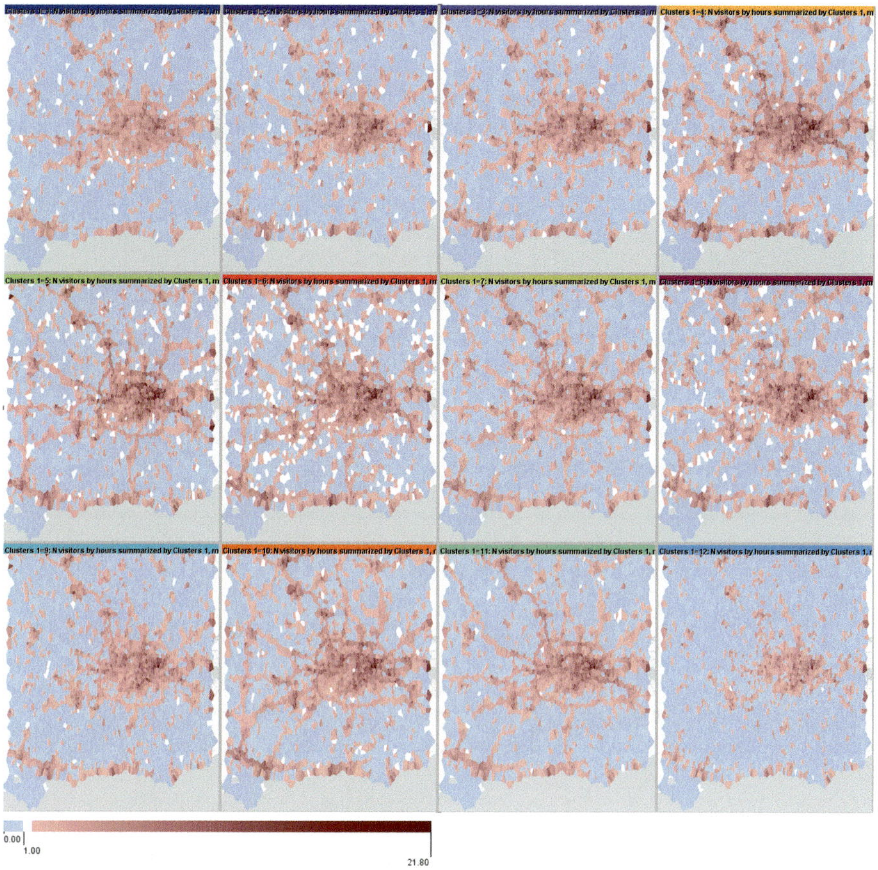

Fig. 40.15 Average spatial distributions of the car presence for the time clusters presented in Fig. 40.14. The mean car counts are represented by the darkness of the shades of red while light blue corresponds to zero values

40.5 Specifics of Episodic Movement Data

Depending on the temporal resolution and sampling regularity, movement data can be categorized as quasi-continuous or episodic (Andrienko and Andrienko 2013a). The example data used in this chapter can be ascribed to the former category, because the time intervals between the records are quite small and mostly of the same length. In episodic movement data, position measurements may be separated by large time gaps, in which the positions of the moving objects are unknown and cannot be reliably reconstructed. Such data require special approaches to analysis. Thus, like with quasi-continuous data, it is possible to aggregate episodic trajectories to flows between places. However, consecutive positions of a trajectory may fit in non-neighboring places. Flow maps constructed from episodic trajectories are typically extremely

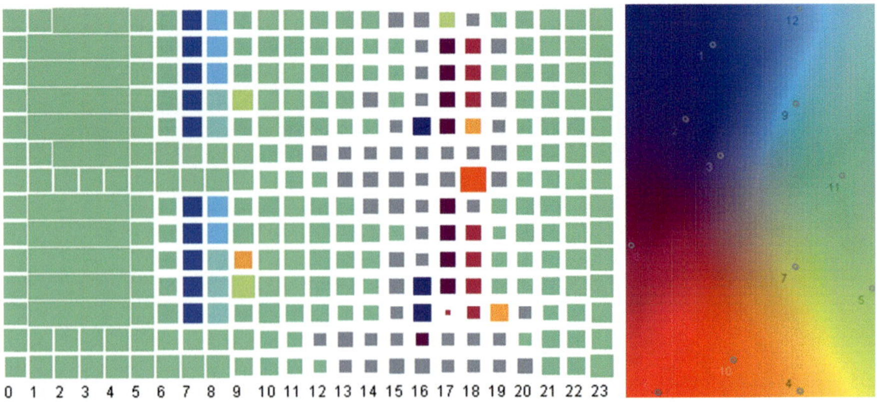

Fig. 40.16 Clusters of the hourly time steps according to the spatial distributions of the flow volumes. The representation is analogous to Fig. 40.14

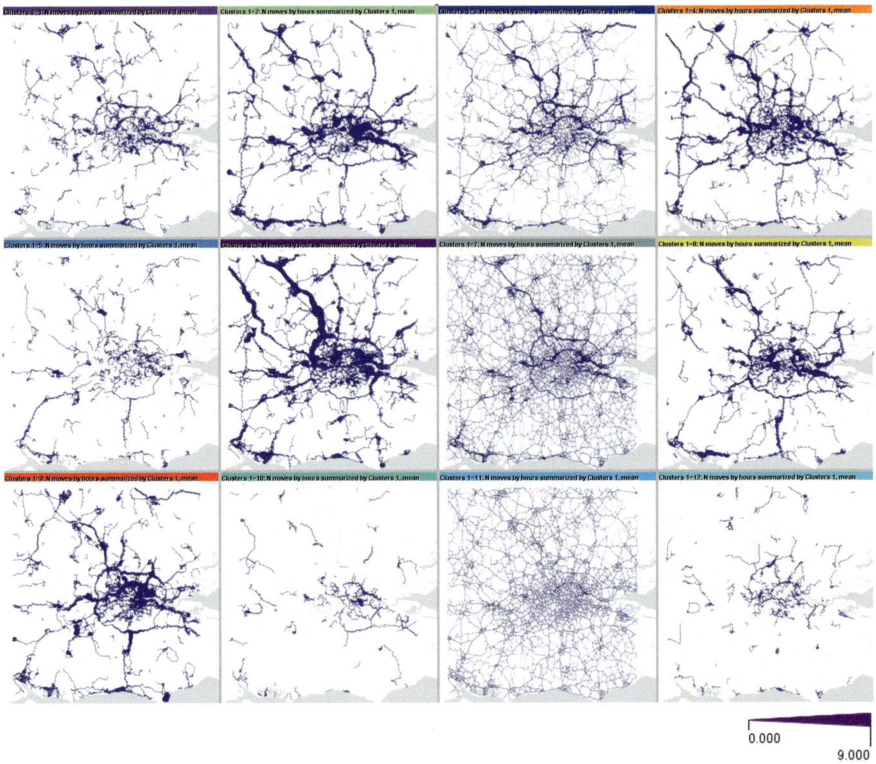

Fig. 40.17 Maps show the spatial distributions of the flow volumes, represented by proportional line widths, for the clusters shown in Fig. 40.16

cluttered due to a large number of intersecting flow lines connecting distant places. Moreover, time intervals between consecutive positions may be longer than the time intervals chosen for aggregation. Such trajectory segments must be ignored. It is also not possible to estimate the number of moving objects that were present in a place during a time interval because the exact times of coming to a place and leaving it is unknown.

In interpreting flow maps built from episodic movement data, analysts should keep in mind that they do not represent all movements that really happened. Nevertheless, such flow maps can be useful since there is a chance that mass movements or sufficiently frequent movement patterns can be adequately reflected.

As an example of episodic movement data, Fig. 40.18 demonstrates 11,671 trajectories reconstructed from georeferenced posts of social media (Twitter) users. Each trajectory consists of a chronological sequence of posts of one user. Similar trajectories can be constructed from data about mobile phone activities, including making calls, sending messages, and accessing Internet.

In Fig. 40.18, the locations of the social media posts are connected by lines, which are drawn with 97% transparency. Long lines mean unknown users' paths between the locations of their consecutive posts. In this data set, which spans a 28-days period in September, the median time interval between records of the same user is 14 min,

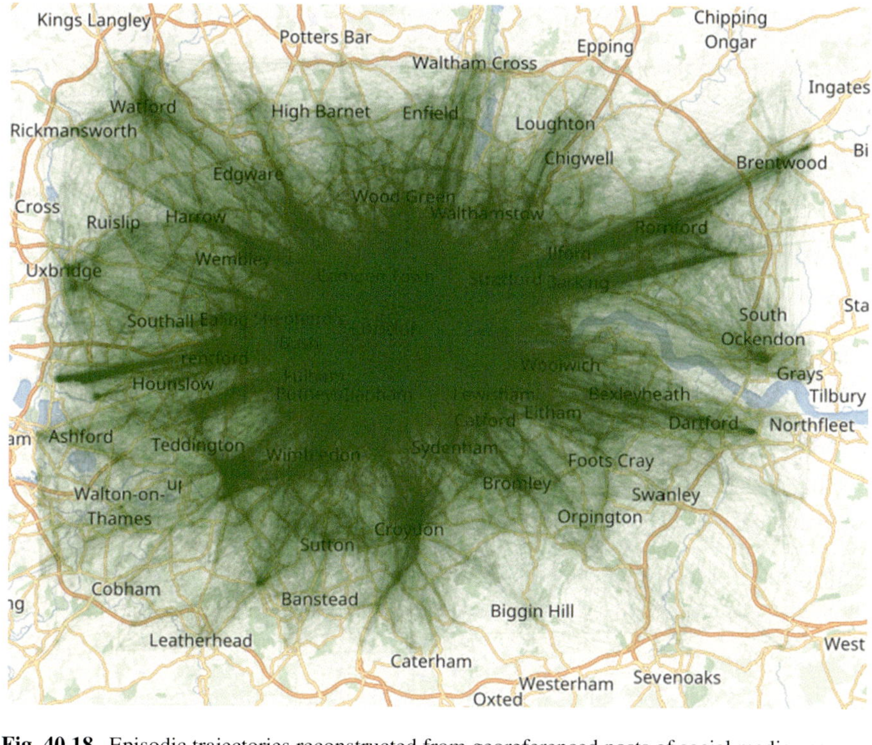

Fig. 40.18 Episodic trajectories reconstructed from georeferenced posts of social media users

the third quartile is about three hours, and the maximum is over 24 days. However, in most cases, the distances between the points are small, the third quartile being only 0.26 km. This means that people tend to make repeated posts from the same or nearly the same locations, which are, possibly, repeatedly visited.

Despite all uncertainties, episodic trajectories reconstructed from social media posts or mobile phone use registers can provide valuable information about mobility behaviors of people. Unlike trajectories of personal cars, taxis, or any particular kind of vehicles, these trajectories can reflect movements made with the use of diverse transportation modes. However, because of the uncertainties and inherent biases, such data need to be used cautiously as a complement to other mobility data rather than alone.

As we mentioned, special care needs to be taken in aggregation of episodic movement data. In our example, we partition the territory into spatial compartments using the method described earlier, that is, the same as we used for the vehicle trajectories. We want to aggregate the data by hourly time intervals; therefore, we split the trajectories into trips by time gaps longer than one hour. This means that, when the time interval between two points exceeds one hour, the later point is treated as the beginning of a new trip. Hence, the transition between the points is not used in the aggregation. Additionally, we split the trajectories by spatial gaps of more than 5 km, which is the average radius of a spatial compartment used for the aggregation. The flow map resulting from the aggregation is shown in Fig. 40.19. It reveals the importance of the central area of London for people's mobility: not only the major flows occurred in the center, but also there were relatively many radial movements to and from the central area. Besides, we can see "hubs," such as Camden Town and Wimbledon, with star-like patterns of flows around them.

Figure 40.20, left, demonstrates the temporal distribution of the aggregated movements of the social media users. In this two-dimensional temporal histogram, the rows correspond to the days, columns to the hours of a day, and the sizes of the squares are proportional to the numbers of moves made in the corresponding hourly intervals. Prominent patterns of more intensive movements in morning hours of the weekdays, with peaks at Hour 9, are clearly visible. Many movements also happen in the late afternoons and evenings of the weekdays, while on the weekends the movements are more uniformly distributed over a day starting from late morning. Interestingly, this temporal distribution differs from the temporal distribution of the counts of the posted messages shown on the right of Fig. 40.20.

This example shows that the approaches presented in this chapter are not specific to GPS tracks of vehicles but can be applied to other kinds of spatiotemporal data collected in various ways. However, the ways of data collection and the properties of the data need to be carefully taken into account in data transformation, analysis, and interpretation of visual displays and computation results.

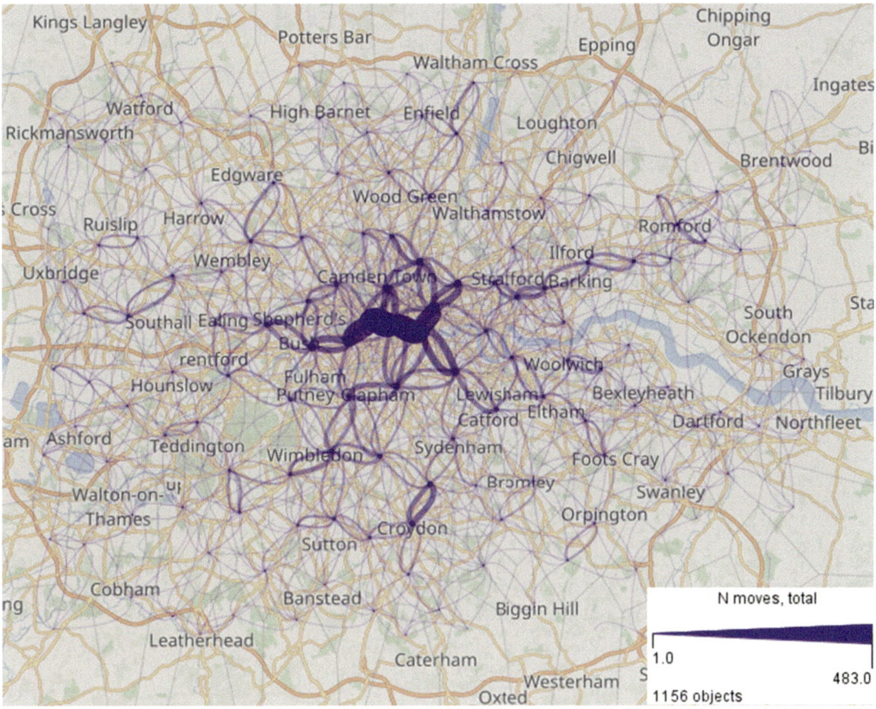

Fig. 40.19 Aggregated movements of social media users

40.6 Discussion and Conclusions

Our examples demonstrate how three major aspects of the urban context—places, flows, and times—can be characterized using trajectory data. We proposed methods to define a suitable set of places, aggregate trajectories into place- and link-based time series, and characterize the places, flows, and times taking two complementary perspectives in analyzing the time series. We demonstrated the use of methods of cluster analysis as a means of abstraction and as an aid in coping with large data volumes. Particularly, we showed that clustering by similarity can be applied to local time series, for characterizing places and links, and to spatial distributions, for characterizing times.

Due to the page limit, we shall only briefly outline the potential directions for extraction of further context information from trajectory data. One possibility is to consider attributes along trajectories, such as Andrienko et al. (2013b) have done:

- measured values, e.g., instant speed and direction, acceleration, turn, fuel consumption, CO_2 emission, etc.;
- spatial context, e.g., road type, land use, distances to stationary objects such as gas stations or other places of interest;

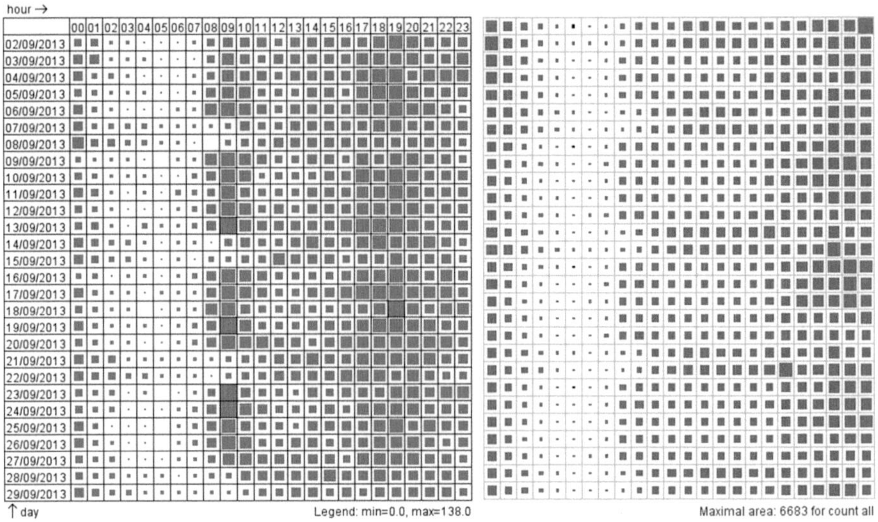

Fig. 40.20 Temporal patterns of the aggregated moves of the social media users (left) are compared with the temporal patterns of the number of posted messages (right). The rows correspond to the days, columns to the hours of a day, and the sizes of the squares are proportional to the numbers of moves or messages, respectively

- derived from sequences of positions of the same trajectory, e.g., computed speed and direction, curvature of the travelled path in a sliding time window; and
- computed based on trajectories of co-moving objects, e.g., count of trajectories in given space- and time-windows or distance to *n*th closest neighbor.

Acquired attributes can be aggregated by places, flows, or along trajectories, enabling selection of locations, connections, or vehicles with particular features. Such vehicles can be visualized on a trajectory wall (Tominsky et al. 2012).

Trajectory attributes can be used for identifying locations that are characterized by particular properties. Thus, density-based clustering of trajectory segments characterized by slow movement can be used for identifying locations of traffic jams and revealing their dynamics (Andrienko and Andrienko 2013b). Scalable methods are developed for identifying hotspots from big data (Nikitopoulos et al. 2018). Considering the parts of trajectories preceding traffic jams, one can study the traffic jam propagation over the street network (Wang et al. 2013).

Methods for time series analysis and modeling can be applied to place- or link-based local time series that have been clustered by similarity. The resulting models can be used for predicting traffic characteristics depending on time. Besides, link-based time series of flow volumes and average movement speeds not only can be modeled in separation but also used for representing and modeling the speed–volume dependencies as proposed by Andrienko and Andrienko (2013b). Such models can be utilized for simulation of regular and extraordinary traffic (Andrienko et al. 2016c) or for billboard pricing and informed decision making (Liu et al. 2017).

Division of trajectories into trips allows extraction of routine movement behaviors (Rinzivillo et al. 2014) and semantic interpretation of locations (Andrienko et al. 2016b). Analysis of semantically-annotated trajectory data (e.g. by state transition graphs, Andrienko and Andrienko 2018) allows finding important behavior patterns without compromising personal privacy.

Our study demonstrates that visual analytics approaches and techniques can support sophisticated analyses for gaining understanding of complex phenomena, such as urban mobility, which is necessary for building explainable models and making informed substantiated decisions. However, we see a need for further advances in visual analytics research and technical developments in the following major directions:

- Stronger support of joint analysis of multiple data sets of diverse structure and quality;
- Dealing with streaming data that are constantly generated and updated; and
- More specific approaches for supporting decision making, including development, evaluation, and comparison of decision options and performing what-if scenarios.

References

Andrienko N, Andrienko G (2011) Spatial generalization and aggregation of massive movement data. IEEE Trans Vis Comput Graphics 17(2):205–219. https://doi.org/10.1109/TVCG.2010.44

Andrienko N, Andrienko G (2013a) Visual analytics of movement: a rich palette of techniques to enable understanding. In: Renso C, Spaccapietra S, Zimanyi E (eds) Mobility data: modeling, management and understanding. Cambridge University Press, pp 149–173

Andrienko N, Andrienko G (2013b) A visual analytics framework for spatio-temporal analysis and modelling. Data Min Knowl Disc 27(1):55–83. https://doi.org/10.1007/s10618-012-0285-7

Andrienko N, Andrienko G (2018) State transition graphs for semantic analysis of movement behaviours. Inf Vis 17(1):41–65. https://doi.org/10.1177/1473871617692841

Andrienko G, Andrienko N, Bak P, Keim D, Wrobel S (2013a) Visual analytics of movement. Springer, Berlin

Andrienko G, Andrienko N, Hurter C, Rinzivillo S, Wrobel S (2013b) Scalable analysis of movement data for extracting and exploring significant places. IEEE Trans Vis Comput Graphics 19(7):1078–1094. https://doi.org/10.1109/TVCG.2012.311

Andrienko G, Andrienko N, Fuchs G (2016a) Understanding movement data quality. J Location Based Serv 10(1):31–46. https://doi.org/10.1080/17489725.2016.1169322

Andrienko N, Andrienko G, Fuchs G, Jankowski P (2016b) Scalable and privacy-respectful interactive discovery of place semantics from human mobility traces. Inf Vis 15(2):117–153. https://doi.org/10.1177/1473871615581216

Andrienko N, Andrienko G, Rinzivillo S (2016c) Leveraging spatial abstraction in traffic analysis and forecasting with visual analytics. Inf Syst 57(1):172–194. https://doi.org/10.1016/j.is.2015.08.007

Andrienko G, Andrienko N, Chen W, Maciejewski R, Zhao Y (2017) Visual analytics of mobility and transportation: state of the art and further research directions. IEEE Trans Intell Transp Syst 18(8):2232–2249. https://doi.org/10.1109/TITS.2017.2683539

Batty M (2013) The new science of cities. MIT Press, Cambridge

Beecham R, Wood J (2014) Exploring gendered cycling behaviours within a large-scale behavioural dataset. Transport Plann Technol 37(1):83–97. https://doi.org/10.1080/03081060.2013.844903

Hägerstrand T (1970) What about people in regional science? Papers Reg Sci Assoc 24:7–21

Itoh M, Yokoyama D, Toyoda M, Tomita Y, Kawamura S, Kitsuregawa M (2016) Visual exploration of changes in passenger flows and tweets on mega-city metro network. IEEE Transac Big Data 2(1):85–99. https://doi.org/10.1109/TBDATA.2016.2546301

Kesting A, Treiber M (2013) Traffic flow dynamics: data, models and simulation. Springer, Berlin

Kraak MJ (2003) The space-time cube revisited from a geovisualization perspective. In: Proceedings of the 21st international cartographic conference, Durban, 10–16 August 2003. International Cartographic Association, pp 1988–1995

Lansley G, Longley P (2016) The geography of twitter topics in London. Comput Environ Urban Syst 58:85–96. https://doi.org/10.1016/j.compenvurbsys.2016.04.002

Liu D, Weng D, Li Y, Bao J, Zheng Y, Qu H, Wu Y (2017) SmartAdP: visual analytics of large-scale taxi trajectories for selecting billboard locations. IEEE Trans Visual Comput Graphics 23(1):1–10. https://doi.org/10.1109/TVCG.2016.2598432

Markovic N, Sekula P, Vander Laan Z, Andrienko G, Andrienko N (2019) Applications of trajectory data from the perspective of a road transportation agency: literature review and Maryland case study. IEEE Trans Intell Transport Syst (accepted). https://doi.org/10.1109/TITS.2018.2843298

Nikitopoulos P, Paraskevopoulos AI, Doulkeridis C, Pelekis N, Theodoridis Y (2018) Hot spot analysis over big trajectory data. In: IEEE international conference on big data. https://doi.org/10.1109/BigData.2018.8622376

Rinzivillo S, Gabrielli L, Nanni M, Pappalardo L, Pedreschi D, Giannotti F (2014) The purpose of motion: learning activities from individual mobility networks. In: International conference on data science and advanced analytics. https://doi.org/10.1109/DSAA.2014.7058090

Thomas J, Cook K (2005) Illuminating the path: the research and development agenda for visual analytics. IEEE.

Tominski C, Schumann H, Andrienko G, Andrienko N (2012) Stacking-based visualization of trajectory attribute data. IEEE Trans Visual Comput Graphics 18(12):2565–2574. https://doi.org/10.1109/TVCG.2012.265

Wang Z, Lu M, Yuan X, Zhang J, van de Wetering H (2013) Visual traffic jam analysis based on trajectory data. IEEE Trans Visual Comput Graphics 19(12):2159–2168. https://doi.org/10.1109/TVCG.2013.228

Gennady Andrienko is a lead scientist responsible for visual analytics research at the Fraunhofer Institute for Intelligent Analysis and Information Systems and part-time professor at City University of London. His research interests are visual analytics and data science.

Natalia Andrienko is a lead scientist at the Fraunhofer Institute for Intelligent Analysis and Information Systems and part-time professor at City University of London. Results of her research have been published by Springer Verlag in two monographs, "Exploratory Analysis of Spatial and Temporal Data: a Systematic Approach" (2006) and "Visual Analytics of Movement" (2013).

Fabian Patterson is a Data Scientist with the Fraunhofer Institute for Intelligent Analysis and Information Systems. He is currently involved in software tools for the analysis and the visualization of data streams, such as those associated with moving entities in the context of smart cities and maritime security.

Siming Chen is a research scientist in Fraunhofer Intelligent Analysis and Information Systems. He is also a post-doc researcher at the University of Bonn. His main research focus is visualization and visual analytics. More information can be found at https://simingchen.me.

Robert Weibel is a Professor of Geographic Information Science at the University of Zurich, Switzerland. He is interested in mobility analytics with applications in transportation and health, spatial analysis for linguistic applications, and computational cartography.

Haosheng Huang is Professor of Cartography and GIS at Ghent University, Belgium. His research interests lie in location-based services, computational mobility analytics, and urban informatics. He is the Chair of the ICA Commission on Location-Based Services (2015–2023).

Christos Doulkeridis is Assistant Professor at the University of Piraeus in Greece, working on parallel and distributed query processing, large-scale data management, distributed knowledge discovery, and spatiotemporal data management.

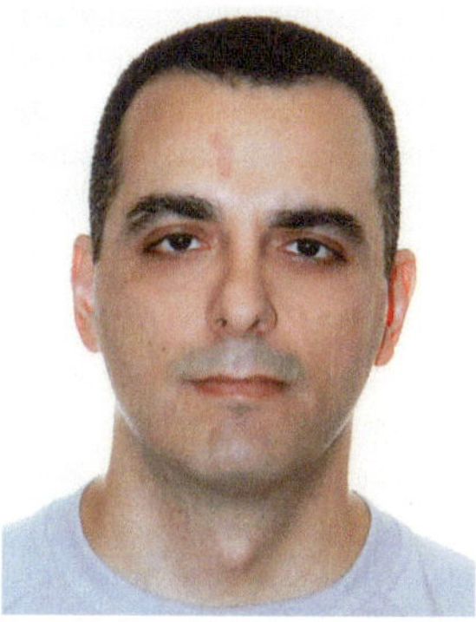

Harris Georgiou has been an independent R&D advisor in Artificial Intelligence, Machine Learning, Big data analytics, Signal Processing, and Medical Imaging for 20 years. He also worked as a post-doctorate researcher and associate professor with the National Kapodistrian University of Athens (NKUA) and currently with the University of Piraeus (UniPi). Since 2016 he is the active LEAR, team coordinator and scientific advisor with the Hellenic Rescue Team of Attica (HRTA) in R&D projects related to next-generation advanced technologies for first responders.

Nikos Pelekis is Associate Professor at the Department of Statistics and Insurance Science, University of Piraeus, Greece. His research interests include all topics in data science. He has been particularly working, for almost twenty years, in the field of Mobility Data Management and Mining. For more information: https://www.unipi.gr/faculty/npelekis/.

Yannis Theodoridis is Professor of Data Science at the University of Piraeus, Greece. His research interests include big data management & analytics for human mobility-related information. He has co-authored three monographs and over 100 refereed articles in scientific journals and conferences, with over 10,000 citations so far, according to Google Scholar. He holds a Dipl. Eng. (1990) and Ph.D. (1996) in Computer Engineering, both from the National Technical University of Athens (NTUA).

Mirco Nanni is a researcher at the ISTI institute of the National Research Council (CNR), Italy. His main research areas include data mining and machine learning, in particular mobility data analysis and applications to smart cities and sustainable transport.

Leonardo Longhi is a consultant for Sistematica SpA, one of the most important telematics companies in Europe. He specializes in image processing and computer-aided medical diagnostics, in particular ultrasound imaging, microscopic oncology imaging, and cardiovascular signal processing. Latterly, he has turned his attention to applied informatics in insurance and telematics products.

Athanasios Koumparos is a Solution Architect at Vodafone Innovus and is the principal engineer designing the company's IoT platform that manages thousands of live fleet devices. He is currently managing H2020 EU and other innovation projects.

Ansar Yasar is a Professor at the Transportation Research Institute (IMOB)—Hasselt University Belgium. He is currently responsible for the H202-Track&Know project, and his research interests include smart cities and communities, connected and intelligent mobility, drones system management, road safety using smart transport solutions and mobility management.

Ibad Kureshi is a Senior Research Scientist at Inlecom Group, Belgium. He currently holds multiple EU H2020 grants in the space of Big Data as it relates to Mobility, Transport & Logisitics, Cyber Security, and Urban Planning.

Chapter 41
Cloud, Edge, and Mobile Computing for Smart Cities

Qian Liu, Juan Gu, Jingchao Yang, Yun Li, Dexuan Sha, Mengchao Xu, Ishan Shams, Manzhu Yu, and Chaowei Yang

Abstract Smart cities evolve rapidly along with the technical advances in wireless and sensor networks, information science, and human–computer interactions. Urban computing provides the processing power to enable the integration of such technologies to improve the living quality of urban citizens, including health care, urban planning, energy, and other aspects. This chapter uses different computing capabilities, such as cloud computing, mobile computing, and edge computing, to support smart cities using the urban heat island of the greater Washington DC area as an example. We discuss the benefits of leveraging cloud, mobile, and edge computing to address the challenges brought by the spatiotemporal dynamics of the urban heat island, including elevated emissions of air pollutants and greenhouse gases, compromised human health and comfort, and impaired water quality. Cloud computing

Q. Liu · J. Yang · Y. Li · D. Sha · M. Xu · I. Shams · M. Yu · C. Yang (✉)
NSF Spatiotemporal Innovation Center & Department of Geography and GeoInformation Science, George Mason University, Fairfax, USA
e-mail: cyang3@gmu.edu

Q. Liu
e-mail: qliu6@gmu.edu

J. Yang
e-mail: jyang43@gmu.edu

Y. Li
e-mail: yli38@gmu.edu

D. Sha
e-mail: dsha@gmu.edu

M. Xu
e-mail: mxu6@gmu.edu

I. Shams
e-mail: ishams@gmu.edu

M. Yu
e-mail: myu7@gmu.edu

J. Gu
Beijing Institute of Surveying and Mapping, Beijing, China
e-mail: gujuan@bism.cn

W. Shi et al. (eds.), *Urban Informatics*, The Urban Book Series,
https://doi.org/10.1007/978-981-15-8983-6_41

brings scalability and on-demand computing capacity to urban system simulations for timely prediction. Mobile computing brings portability and social interactivity for citizens to report instantaneous information for better knowledge integration. Edge computing allows data produced by in-situ devices to be processed and analyzed at the edge of the network, reducing the data traffic to the central repository and processing engine (data center or cloud). Challenges and future directions are discussed for integrating the three computing technologies to achieve an overall better computing infrastructure supporting smart cities. The integration is discussed in aspects of bandwidth issue, network access optimization, service quality and convergence, and data integrity and security.

41.1 Introduction

41.1.1 Why Computing is Important in Smart Cities

Increasing global urbanization generates many problems, such as traffic congestion, energy consumption, industrial waste, and heat islands (Rao and Rao 2012; González-Gil et al. 2014; Li et al. 2012; Zhong et al. 2017; Rizwan et al. 2008). These problems produce serious negative impacts on urban residents. For example, an urban heat island (UHI) in an urban area or metropolitan area is significantly warmer than its surrounding rural areas due to human activities. UHI contributes directly to environmental warming, industrial waste, air pollution, and heat-related mortality (Petkova et al. 2016). In order to alleviate urban problems and achieve sustainable development, a number of smart-city solutions have been the subject of experiments in cities over the past two decades. Copenhagen Municipality uses monitor sensors installed in different trash containers and information systems to optimize waste handling (State of Green Denmark 2018). Seoul of South Korea has smart meters installed in residential houses, office areas, and industrial facilities to report in real time the consumption of electricity, water, and gas (Hwang and Choe 2013). Smart cities are supported by key information and communications technologies (ICT) including the Internet of things (IoT), computing platforms, big data, artificial intelligence (AI), geographical information, and others (Graham and Marvin 2002; Morán et al. 2016; Mitchell et al. 2013) (Fig. 41.1). Among them, diverse sensors, stable communication networks, and sophisticated computing platforms are three fundamental technologies for smart cities. Sensors are the smart-city's sensory organs, to capture and integrate data continuously in real time. Smart sensors, such as monitoring cameras, smart meters, and wearable devices, are widely employed to improve urban transportation, utility planning, parking-lot management, pollution monitoring, and health care. The number of connected devices on the Internet will exceed 50 billion by 2020 according to Cisco (2017). The communication network is the smart-city's transmission system, transmitting data from sensors to computing platforms. Reliable, scalable, and high-speed networks, including wired and wireless

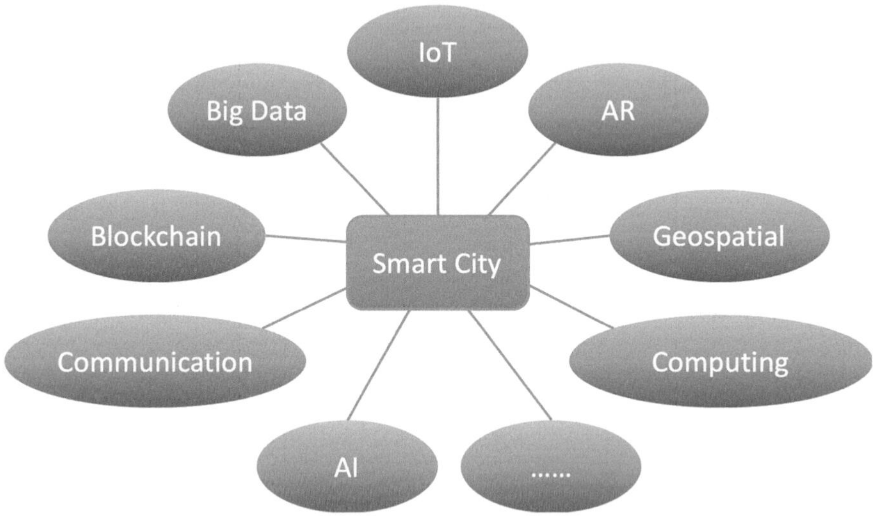

Fig. 41.1 Key technologies of smart cities

networks, are fundamental infrastructure for such transmission. Computing platforms support the management and analyses of relevant city data in a broader context, to identify city-relevant events that require processing and action. A large quantity of data is generated continuously from countless smart-city sensors. To store, process, and analyze the massive heterogeneous data, a stable, scalable, fast computing platform is required. For example, car drivers need a smart navigation system to provide them with the optimal driving route in real time, updated dynamically with traffic pattern and congestion changes. Different systems and devices using ICT have been developed to monitor and forecast UHI in the past years. For example, France developed a Heat Health Watch Warning System to monitor heat waves that may result in a large increase of mortality (Casanueva et al. 2019). Greece developed a UHI modeling system to simulate and forecast heat islands in Athens (Giannaros et al. 2014). Richmond has handmade devices equipped in cars and bikes to map UHI (Hoffiman 2018).

41.1.2 Major Computing Techniques in Smart City Studies

Washburn et al. (2009) described the smart city as using a collection of smart computing technologies to manage critical infrastructure components and services. A centralized cloud-computing architecture has been widely deployed in smart cities to extend the storage capability and improve the processing velocity with characteristics of elastically, on-demand, and pay-as-you-go computing resources (Yang and Huang 2013). Cloud computing maximizes the utilization rate of physical resources by adopting a series of technologies including virtualization and network security. Virtualization is a core technology supporting cloud computing, and abstracts actual hardware as virtual computer systems. Virtualization enables multiple operating systems to run on a computer system simultaneously and optimizes the use of computing and storage resources. Practically, cloud computing virtualizes computer resources and manages them in a resource pool to provide computing services over the network, reducing the idle time of resources including CPU, RAM, network, and storage. Public clouds (e.g., Amazon AWS, Microsoft Azure) are open to the public, who pay to use them. On the other hand, a private cloud is delivered via a secure private network and usually shared among people in a single organization. Cloud computing provides the smart city with the computing capability to store and access data and applications outside local computing environment through computer networks (Kakderi et al. 2016).

The proliferation of IoT enables smart cities to collect a large number of data and deploy a lot of applications at the edge to utilize these data (Shi et al. 2016). The data and applications also produce challenges of near-real-time response, privacy, and massive numbers of data for network transmission. Cloud computing alone is not sufficient to address such challenges. A new computing paradigm, edge computing, which shifts the data storage, processing and analyses to the end of the network, as close as possible to the devices, is deployed (Shi et al. 2016). With the aid of edge computing, the edges of network become data producers as well as data processors, addressing the challenge of response time, bandwidth, data safety, and privacy (Shi et al. 2016). Edge computing offers a number of benefits, including allowing services to continue to operate when there is no connection to the Internet, and processing data locally. This significantly reduces the network load with only processing results (which are normally smaller in volume than raw data) being transmitted across the network.

The past two decades have witnessed the increasingly use of mobile devices (such as mobile phones, portable computers, wearable devices, and smart vehicles) and rapid growth of wireless communication technology (Hashim Raza Bukhari et al. 2018). Data processing is shifted away from centralized computing centers to the mobile devices of end user. With battery volume and network bandwidth limitations, computing resources offered by mobile computing are not as reliable as the other two computing frameworks. Nevertheless, they are portable and able to collect and process data where cloud computing and edge computing are unavailable.

The three computing paradigms collaboratively provide a comprehensive and reliable data store and processing framework to overcome the disadvantages of a single device and enable a suite of applications of smart cities (Table 41.1) including: transport and traffic management, utilities and energy management, environmental protection and sustainability, public safety, and smart-city security.

Figure 41.2 illustrates the sensors and computing devices of a smart city and places them into three types: different sensors collecting different information for different purposes. The sensors also have embedded computing capabilities; for example, moving sensors can be used to provide flexible data collection to dynamically cover different regions with fast situation-aware processing capabilities such as navigation.

Table 41.1 Application examples of cloud, edge and mobile computing in smart cities

Application examples	Computing paradigm		
	Cloud computing	Edge computing	Mobile computing
Transport and traffic management	Using cloud computing for smart-city logistics (Nowicka 2014)	Connected parking meters (David 2018)	Location-aware mobile applications (Altman et al. 2015)
Utilities and energy management	Using cloud computing for smart grid energy management (Bera et al. 2015)	Street lighting (David 2018)	The use of GPRS technology for electricity network telecontrol (Souza et al. 2016)
Environmental protection and sustainability	Using cloud computing for climate analysis and simulation (Yang et al. 2017a, b)	Vehicular pollution system based on IoT (Rushikesh and Sivappagari 2015)	Location-aware weather report applications (Altman et al. 2015)
Public safety and smart-city security	Cloud computing services in medical heath care solutions (Kaushal and Khan 2014)	Smart home (Shi et al. 2016)	Healthcare applications (Hameed 2003) Lost child application (Satyanarayanan 2010)

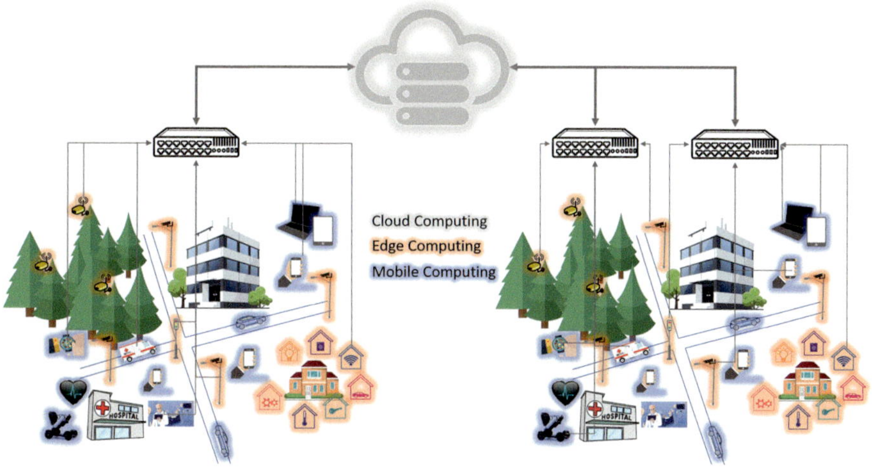

Fig. 41.2 Urban computing for smart cities include cloud computing (gray), edge computing (orange), and mobile computing (blue) devices and capabilities

Edge-computing sensors act as fixed data collectors with various computing powers depending on tasks assigned; for example, a higher edge-computing capacity enables handling analytics for a larger area, like a neighborhood. All data and processes can also be uploaded to the cloud's centralized computing, for extensive data processing and knowledge extraction or mining.

Computing serves as an indivisible capability to support effective and efficient smart-city applications and research, through which massive smart-city data can be processed in parallel and in a real-time manner. This chapter introduces the three computing paradigms' engagement in a smart city using UHI as a case study. A workflow was proposed to integrate three computing techniques as a seamless integration for handling UHI problem (one of the severe urban challenges facing us today especially with climate and global change).

This chapter starts with an introduction to urban computing in 41.1, followed by the current status and challenges of computing in different smart-city scenarios. Sections. 41.3, 41.4 and 41.5 introduce, respectively, cloud computing, edge computing, and mobile computing using UHI as a use case. The last section uses UHI as an example to integrate the three computing paradigms through collaborative workflow.

41.2 Computing for Smart Cities

41.2.1 Data and Model in Smart Cities

Smart cities require multiple data sources and reliable models to produce decision-supporting information. It becomes especially challenging when a massive number of smart devices and sensors are engaged. This section introduces five typical smart-city applications, the data engaged, corresponding models, and their requirements for computing.

41.2.1.1 Transport and Traffic Management

Transportation is one of the most important aspects for urban-living activities. Various sources of transportation data are related to people's travel and commuting, which is a complicated and indispensable part of smart cities. For example, traffic data are generated and collected by sensors in traffic vehicles (e.g., taxis, buses, metros, trains, vessels, and planes) or monitors installed along the roads (e.g., loop sensors and surveillance cameras). Commuting data refer to data that record people's regular movement in cities. Geo-tagged social network data collect posts (e.g., blogs, tweets) through social networks which are tagged with geoinformation. Road network data represent road segments and intersections, respectively. The transportation network is modeled as a directed graph which includes transit routes and stop facilities of buses and metro networks. Point of interest (POI) data depict related information for facilities, such as restaurants, shopping malls, parks, airports, schools, and hospitals in the city, which helps guiding people to find their destinations.

To handle and integrate the complex data from different sources efficiently and to satisfy various user groups, different models are used for intelligent transportation systems, such as agent-based traffic management models (Sciences et al. 2011), cognitive rationality-based decision-making models (Cascetta et al. 2015) and mixed-ranked logit models (Liu et al. 2017).

41.2.1.2 Utilities and Energy Management

The large volume of data for utilities and energy management is increasingly adding burden to urban computing systems, especially with the wide adoption of sensors, wireless transmission, and network communication (Zhou et al. 2017). The input data of smart-city energy systems include numeric data, text-based data, and audio-visual data. Numeric data refer to the observations and collections from sensors and meters, such as power quality, customer usage, and electrical production. Text-based data sources are mainly internal and external communications, regulatory documents, legal documents, and linguistic social media records. Audio-visual data are records and social media data in the form of sound and video (Schuelke-Leech et al. 2015).

The utilities and energy management systems should be green, sustainable, and with high operational speed and efficiency. Schuelke-Leech et al. (2015) demonstrate how future sustainable energy systems will be smart and integrated with smart grids, renewable sources, storage, and energy management and monitoring systems. The energy and utility systems of cities are complicated because they have to satisfy a huge number of requirements with comparably limited supply. The computational systems need not only to integrate intermittent power sources efficiently and effectively, but also to predict equipment failures and power outages, allowing utilities to optimize their maintenance budgets. For example, Sheikhi et al. (2015) presented an Energy Hub Model in a future vision of energy systems, which supported real-time and two-way computational communication between utility companies and smart energy hubs. Such models also allowed intelligent infrastructures at both ends, since to manage power consumption necessitates large-scale real-time computing capabilities to handle the communication and the storage of big data. These systems help managers, employees, and consumers to make informed decisions based on data and empirical investigation, rather than on intuition or past practice.

41.2.1.3 Environmental Protection and Sustainability

Environmental protection and sustainability also play important roles in smart cities. The environmental resources refer to minerals, forests and grasslands, wetlands, rivers, lakes, and the ocean. These natural resources have been exploited unduly, and the inappropriate management of natural resources has caused severe environmental degradation (Song et al. 2017). The data that urban environmental protection and sustainability management systems are dealing with include hydrogeological data, environmental surveillance data, ecological statistics, and meteorological data. The data quantity and dimensions are big according to the characteristics of big data. The functions of these data are not only to accurately present the current situation of the environment but also to effectively predict the future and sustainability. Therefore, powerful computational ability is needed to help governments and individual users to prevent and settle environmental challenges.

As environmental protection and sustainability are important factors for the development of smart cities, data collection and computational models have flourished in this domain. Take the IoT and its associated computing model as an example: the informational landscape of smart sustainable cities and big data applications is augmented to achieve the required level of environmental sustainability (Bibri 2018). For governments, the combination of 3D GIS and cloud computing is also offering effective services in the environmental management of smart cities (Lv et al. 2018).

41.2.1.4 Public Safety and Security

Public safety and security are directly related to citizens' wellbeing and their lives. With the growth of different kinds of monitoring devices and systems, data from the IoT, unmanned aerial vehicles (UAV) (Menouar et al. 2017), and social media are leveraged to make our cities more and more safe and stable. Usually, the safety and security issues are directly related to people's life and property, and needs immediate and accurate response from relevant personnel. Therefore, extremely high performance in efficiency and accuracy is needed for safety and security models and systems. Edge and mobile computing, which can share the burden of the central cloud and improve processing speed, are ideal for the applications such as finding a lost child (Shi et al. 2016). Wearable devices and medical sensors can measure users' health conditions and send health data to the processing unit for doctors' further diagnosis.

To address these challenges, safety systems should include the following data sources and model features: health care and monitoring systems; smart safety systems for surveillance; smart systems of crisis management to support decision making, early warning, monitoring and forecasting emergencies; centrally operated units of police and integrated rescue systems (IRS); safe Internet connection and data protection; and centers of data processing (Lacinák and Ristvej 2017).

41.2.1.5 Urban Heat Island and Urban Computing

Urban computing utilizes the three computing paradigms to store, process, integrate, model, and analyze various big data and phenomena, such as real-time data generated by diverse smart sensors and devices, fundamental urban geographical data, social media data, data on transportation on flooding, and on UHI. UHI is considered one of the major urban challenges and is caused by a set of complex factors, including urban land use changes, solar radiation, anthropogenic heat sources, climate change, urban development, and wind speed and direction (Memon et al. 2009). The negative effects of UHI include: (1) increasing temperature in cities (Voogt and Oke 2003); (2) contribution to global warming (Van Weverberg et al. 2008; EPA 2016); (3) air pollution (Sarrat et al 2006; Davies et al. 2008); (4) increasing energy demand (Santamouris et al. 2001; Santamouris 2015); and (5) heat-related mortality (Guest et al. 1999; Conti et al. 2005; Haines et al. 2006; Filleul et al. 2006; Hondula, et al. 2014).

To reduce the negative impact of UHI, remotely sensed data, stationary meteorological monitoring data, building data, digital elevation data and other data were integrated to model, monitor, simulate, and evaluate UHI in more than 100 cities in the past 50 years. However, UHI studies involve big data storage, processing, and modeling, which need complicated computing. There is no single efficient computing architecture for large-scale or long-term UHI studies. This chapter takes UHI as an example to introduce how the combination of cloud, edge, and mobile computing can help addressing the smart city challenges in sequence of: (1) what are the computing challenges of smart cities; (2) how the three computing paradigms can help address the challenges; and (3) how to integrate the three computing paradigms to address these challenges using UHI as an example.

41.2.2 Computing Challenges in Smart Cities

41.2.2.1 Big Data Handling

Urban data have been harvested from various sources including (1) remote sensing, (2) in-situ sensing, (3) social sensing, (4) IoT sensing, and (5) simulation. The collected data together provide a comprehensive view of the urban system: for example, the underground water distribution network for water usage management (Karwot et al. 2016), real-time parking prediction (Vlahogianni et al. 2016), and 3D city modeling for urban disaster management (Amirebrahimi et al. 2016). However, the sensing and simulation produce large numbers of data that far exceed the storage capacity of an individual computer. Taking remote sensing as an example, fine spatiotemporal resolution imagery grows exponentially with spatial resolution. For example, the volume of the Earth Observing System and Data Information System (EOSDIS) data archive was more than 27.5 petabytes (PB) at the end of fiscal year 2018 (NASA Earth Science Data Systems Program Highlights 2018). Efficiently storing such a large volume of data is a challenging task. Meanwhile, data are produced in high velocity in a continuous manner with the development of advanced techniques, such as water meters, which collect water usage data in a fixed interval (e.g., every 30 s). The velocity of data requires streaming data collection and analysis methods for near-real-time applications. In addition, the heterogeneous data are stored in various file formats, such as image, video, text, or audio, and pose grand challenges to data management.

41.2.2.2 Compute-Intensive Modeling and Processing

The smart city is becoming a sophisticated ecosystem where massive data are being collected and innovative solutions are being proposed to deliver smart services (Anthopoulos 2015). Generally, those solutions rely on complicated data models and analytics with the aid of the computer. Data models often represent objects or situations in the real world, and a digital model makes mathematical analysis possible. For example, a trend in smart cities is to build three-dimensional (3D) models for visualization and analytics such as skyline analysis, underground utility management, and route selection (Yao et al. 2017; and see Sects. 41.5 and 41.6). Although a 3D model can represent cities as virtual reality to support real 3D analysis, more computing resources are needed for effective 3D rendering and analysis. Data analytics is an important component of the big data paradigm. However, it comes after data collection, deduplication, completion, aggregation, harmonization, contextualization, and filtering. These components of the process are essential to enable analytics to derive useful insights. Different types of computing resources are required for different components in the data process workflow. For example, moving partial computing resources to the data collection sites for data cleaning can reduce the volume of data transferred to the core computing platform, result in a lower bandwidth cost and a higher analysis speed.

41.2.2.3 Data Security and Privacy

Security and privacy issues are two of the major challenges in smart-city computing due to the identification information within the data and the security issues located in the multiple computing layers. Generally, some of the raw data may contain confidential or sensitive information related to people or governments; such data processing should be protected against unauthorized usage. Taking cellular data for example, a phone number in each record represents a real person and makes an individual's daily activities traceable, which may divulge the private affairs of people. In the water distribution management system, a methodology for synthetic household water consumption was proposed to reproduce water consumption data due to privacy constraints (Kofinas et al. 2018). Simultaneously, in smart-city applications, data move over various computing layers through networks, some of which may be insecure. In an application, data may be processed with more than one computing technique including edge computing, mobile computing, and cloud computing. In most cases, mobile devices and edge computing nodes need to connect via Wi-Fi to upload data to the cloud-computing platform. Connection to unauthorized Wi-Fi may bring security risks to the system. Besides network connection, distributed open-source big data platforms like Hadoop and Elastic search are becoming increasingly popular for distributed data storage and analytics, However, compared to commercial solutions, these platforms lack sufficient security guarantees (Sharma and Navdeti 2014).

41.2.2.4 Efficiency

A trend in smart-city's applications is to extract information from big data, and thus, lack of efficiency becomes a bottleneck of most data-analytical applications. Different applications vary in levels of complexity and require different response times. Navigation needs immediate optimal route suggestion (e.g., fastest route option) based on real-time traffic data (Liebig et al. 2017). Predictions of hurricane intensity help people prepare for severe weather, saving properties, and human lives (Li et al. 2017). Applications like environmental sustainability are less sensitive to the response time. Meanwhile, although a series of open-source big data platforms, such as Apache Hadoop, Spark, HDFS, and MapReduce, have been developed and adopted in various domains, these platforms are not specifically designed to support spatiotemporal data. Performance issues are unavoidable when using these platforms to process spatiotemporal data without any modification. Some research has been done to customize these tools for domain adoption. Taking array-based raster data for example, a hierarchical index was proposed to speed up the query process of grid data stored in the HDFS file system (Hu et al. 2018). The development of an efficient spatiotemporal computing platform is still in an initial stage; how to utilize and optimize big data computing platforms to implement efficient smart-city applications remains a challenge.

41.2.3 Generic Computing Architecture for Smart Cities

Cloud, edge, and mobile computing support different functions and applications in the development of smart cities. To optimize the computation capability and further overcome the challenges discussed in Sect. 41.3, different types of computing paradigms should be utilized. Based on the characteristics and advantages of each type of computing, a computing architecture for a smart-city system is proposed (Fig. 41.3).

41.2.3.1 General Computing Modules in Smart Cities

The proposed architecture of computing system in smart cities contains the following five parts:

(1) Application acquisition: The function of the application layer is to collect requirements from users, then organize, and analyze them into the four aspects as mentioned in Sect. 41.2.1: Transportation and traffic management, utilities and energy management, environmental protection and sustainability, and public and smart-city security.
(2) Visualization: The visualization layer is designed to visualize the applications in the form of 2D and 3D maps, trajectories, images, charts, histograms, and others

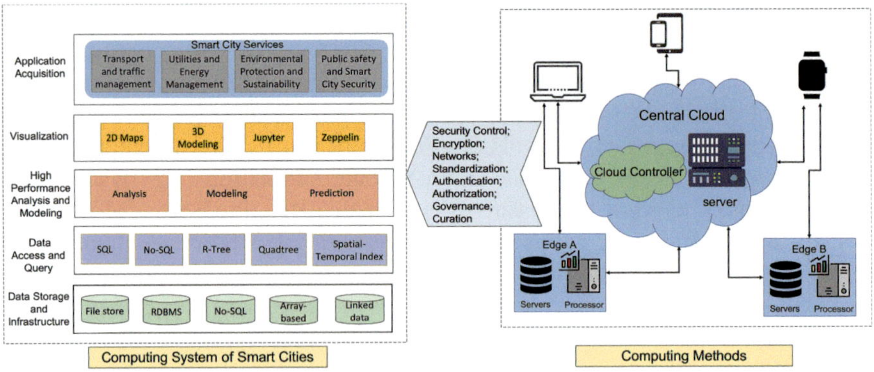

Fig. 41.3 Generic computing architecture for smart cities

using technologies and software such as 2D mapping, 3D modeling, Jupyter, and Zeppelin.

(3) High-performance analysis and modeling: As discussed in the former sections, computing for smart cities is usually encountered with big data issues, and high-performance computing techniques are essential to maintain a stable and efficient computation system. This layer implements data analysis, modeling, and prediction according to the applications.

(4) Data access and query: The system utilizes a data access and query layer to retrieve and select data sources that satisfy the needs and orders from users. Methods and techniques such as SQL, No-SQL, R-Tree, Quadtree, and spatial–temporal indexing will be adopted according to the category of data.

(5) Data storage and infrastructure: This layer provides the hardware and physical devices, including data storage facilities, as well as the servers and networks. The smart-cities-related data sources will be stored in different categories according to the requirements from uses, using database systems such as file storage, Relational Database Management System (RDMS), No-SQL, array-based, and linked-data databases.

41.2.3.2 Computing Methods Integration

Computing procedures are embedded in all the layers of the proposed computing architecture for smart cities, through a series of security controls, encryption, standardization, authentication, authorization, governance, curation, and network techniques. The core computing methods of smart cities contain central cloud computing, edge computing, and mobile computing. In the central cloud platform, data centers provide complex analysis and visualization capabilities, as well as hardware facilities and infrastructure for the cloud. The servers are linked with high-speed networks to provide services for clients. Normally, data centers are built and located in less populated places, with a high power-supply stability and a low risk of disaster (Dinh et al. 2013). The edge-computing platform is connected with the central cloud by the Internet. They have dual communication with each other to enable data interactions. The edge servers can share and reduce the burden of central servers, and as a result increase the speed of processing and delivering data. The mobile-computing platform is the mobile devices of the end users, which has a certain capability to process data along with mobility. Mobile devices can also be connected to central clouds by wireless networks for data transmission. Edge- and mobile-computing platforms are connected with each other in applications where interactions are needed.

In the architecture, the three computing paradigms are connected and assist each other, where there are distinctions between them in the collaboration of processing smart cities' services and applications. Different from cloud computing requiring all parts to be connected to the central cloud, where large volumes of data are processed to find optimization solutions or support decisions, edge computing relocates crucial data processing to the edge of the network, rather than constantly delivering data back to a central server. Therefore, edge-enabled devices can gather and process data in real time, allowing them to respond faster and more effectively, while mobile computing relates to the emergence of new devices and interfaces and has the data processing capability on the mobile devices. Moreover, the centralized cloud could perform extremely complex data processing, storing, and analytics. Edge computing usually performs less intricate data processing than central clouds, storing and forwarding. However, some mobile devices can only implement simple and limited data processing. By integrating the three computing paradigms, the efficiency challenges of intensive big data processing and computing can be remitted. Direct connection between edges, mobile devices, and the central cloud with a stable and secure network will guarantee the safety and security of the whole system.

41.3 Cloud Computing for Smart Cities

41.3.1 Methodology

Cloud computing is developed and improved based on the evolution of parallel computing, distributed computing, and grid computing (Jadeja and Modi 2012; Yang and Raskin 2009). Parallel computing allows many computation processes to run simultaneously, which achieves high performance in a divide-and-conquer fashion (Fu et al. 2015). Distributed computing contains components located on different networked computers which communicate and cooperate with each other to achieve a common computing objective (Yang et al. 2008). The inexpensive computer nodes and high-speed networks make possible the function of distributed computing systems (Jonas et al. 2017). Grid computing organizes a network of heterogeneous computer resources to work together and achieves high performance for processing and executing resource-hungry tasks like those normally allocated to supercomputers (Wang et al. 2018). Different from the above-mentioned computing modes, cloud computing is a model for enabling convenient, on-demand network access to a shared pool of configurable computing resources (NASA 2010), instead of a local machine or remote server handling applications.

Cloud computing is capable of scheduling and balancing the distribution of resources according to real utilization demand, and billing according to the usage. Using different techniques and according to different budgets, cloud computing extends subscription-based access to data, platforms, infrastructure, and software, approaches that are referred to as data as a service (DaaS), platform as a service (PaaS), infrastructure as a service (IaaS), and software as a service (SaaS) (Subashini and Kavitha 2011; Yang et al. 2011).

41.3.2 Challenges, Motivations and Opportunities

Past research (Gong et al. 2010; Zhang et al. 2010; Yang and Huang 2013; Mahmood 2011) identified the features and advantages of cloud computing as:

(1) Hyperscale. Some Internet companies have developed large-scale cloud-computing platforms for business applications, and the practical clouds have a considerable scale. For example, Google cloud computing (Xiong et al. 2017) has millions of servers; Amazon, IBM, Microsoft, Saleforce, Ali, and Tencent (Hashem et al 2015; Rittinghouse and Ransome 2016), and other agencies have hundreds of thousands of servers in their clouds. Conceptually, a cloud can provide users with unprecedented computing power.

(2) Virtualization. Cloud-computing supports users to access services at any location using a variety of terminals and devices. The requested resources come from the cloud, which uses virtualization techniques to separate computer resources and services from underlying fixed physical entities (Gong et al. 2010). The application runs above in the cloud without specifying a server. Simple network connection enables users to benefit from super-powerful services via multiple devices, such as a computer, a PAD, or a mobile phone.

(3) Reliability. Cloud computing uses the capability of fault tolerance and isomorphic interchangeability of computing nodes and other strategies to ensure high reliability and availability (Dai et al. 2009). Compared with traditional in-house computing infrastructures, cloud computing is more reliable and consistent.

(4) Universality. Cloud computing is not specific to any particular applications. It can support a variety of applications under the support of a single cloud. The same cloud infrastructure can be shared by different applications at the same time (Yang et al. 2016).

(5) Scalability. The capabilities and scales of the cloud can be modified and extended dynamically to meet the needs of applications and growth (Lehrig et al. 2015). Scalability allows cost-effective running of workloads that make a very high demand on servers but only for short periods of time or occasionally.

(6) On-demand. Users could request and receive access to cloud service offerings, like the traditional infrastructure utilities of water, electricity, and gas. Based on a pool of physical and virtual resources in the cloud, operations such as creating, stopping, and terminating could be conducted at any time without waiting for delivery and purchasing processes (Etro 2015). Usage monitoring tools of the cloud can record usage details for billing.
(7) Cost savings. The "pay-as-you-go" characteristic of cloud service enables personal and business clients to access the cloud from extremely cheap and price-flexible computing nodes. The automatic system of the cloud reduces the cost of data center management by deleting the basic maintenance budget. The lack of physical infrastructure removes the operational expenses of power, storage, administration and even labor costs.

Considering the advantages listed above, cloud computing can help to address the following computing challenges of smart cities:

(1) Unity and efficiency. Through the architecture of the IaaS model, cloud computing integrates various frameworks, hardware brands, and computing models of servers to the traditional data centers and provides a unified platform of application based on the cloud operating systems (Mitton et al. 2012). Meanwhile, with the virtualization techniques, cloud computing can be flexibly and effectively partitioned, allocated, and integrated over a potentially infinite number of storage and computing resources, and optimize the efficiency ratio according to application and requirements.
(2) Large-scale infrastructure. Infrastructure management of hardware and software is mainly responsible for the monitoring and management of large-scale foundational computing resources (Jin et al. 2014). Fundamental software resources include stand-alone operating systems, middleware, databases, and so on. Fundamental hardware resources include three main devices in the network environment: computing (server), storage (storage device), and network (switch, router, and other devices). The advantages of infrastructure management center are: (1) to manage the assets of the basic software and hardware resources; (2) to support the status and performance monitoring of the basic hardware; (3) to trigger alarms for abnormal situations, and remind users to maintain the abnormal equipment; (4) to carry out long-term statistical analysis of the basic software and hardware resources; and (5) to provide a decision-making basis for high-level resource scheduling.
(3) Sustainable and green energy. Facing the burden of large-scale fundamental software and hardware resources, green and energy-saving operation and maintenance management of this basic infrastructure is an inevitable demand for the supplier of cloud computing (Wibowo et al. 2018).
Presently, users often purchase large amounts of equipment to guarantee peak business operation demands. But for actual operation processes, the load of the equipment is generally low (Mastelic and Brandic 2015), especially in the low-loading period. A long-term low utilization rate will lead to a large waste of resources and energy.

A cloud-computing data center supports multi-tenant applications of resources. The utilization rate of resources can be effectively improved through the historical statistical information of business, and the coordination of business/resource scheduling management. In typical applications, a cloud-computing data center using energy-saving technology can increase the load of resources to a significantly higher level (Rong et al. 2016), remove the loss in the process of resources' scheduling, and double the resources' payload. During night operations, when the overall load of the data center decreases, the unused resources can be transferred to the idle mode, to maximize the green, low-carbon and energy-saving operation of the data center (Hao et al. 2012).

(4) Privacy and security. In the cloud-computing environment, the centralized and large-scale management of basic resources shifts the security problems to the server side in the data center. From the specialization perspective, end users can achieve business security through the security mechanism of the cloud data center, without consuming too much resources and power (Jin et al. 2014; Sen 2015). At the same time, cloud-computing centers will be directly responsible for the security of all users and specifically focus on the main security risks including data access risk, data storage risk, information management risk, data isolation risk, legal investigation support risk, as well as sustainable development and migration risk.

The security control of cloud computing can be integrated by the basic hardware and software security design. The architecture, strategy, authentication, encryption, and other aspects of a cloud-computing system ensure the information security of cloud-computing servers.

Cloud computing reduces the risk of data loss or leakage from individuals by storing data in a centralized database (Chang and Ramachandran 2015). At the same time, a cloud-computing center also uses a variety of backup methods in security and disaster recovery to guarantee that data will not be lost or illegally tampered with.

41.3.3 Urban Heat Island Use Case

Remote-sensing data analysis of a large area is a traditional approach to extract temperature information of cities for UHI modeling and prediction. Google Earth Engine (GGE) is a cloud-based platform sharing large numbers of satellite data online and allowing data analysis and processing on the fly (Gorelick et al. 2017).

Chakraborty and Lee (2019) implemented the SUE algorithm on the Google Earth Engine platform using MODIS images to calculate the UHI intensity for over 9500 urban clusters using over 15 years of data, making this one of the most comprehensive characterizations of the surface UHI to date. They designed an interactive, public-facing Web application to query UHI intensities of almost all urban clusters based on GGE. Ravanelli et al. (2018a,b) took advantage of GGE and the Climate Engine (CE) tool to process the huge amount of satellite Earth observation data (6000 Landsat images) over the period of 1992–2011 and realized wide spatiotemporal monitoring of surface UHI and its connection with land cover changes. Yu et al. (2019) utilized cloud-based computing of spatial and landscape analysis to identify the multi-scale spatiotemporal patterns and characteristics of regional heat islands.

Cloud-computing techniques enable researchers to calculate geophysical parameters from large numbers of remote-sensing data with high and efficient performance. The cloud-computing platform, like Google Earth Engine, assists users to store and manage original raw datasets and provides interactive SaaS for customized algorithms deployment and running for specific UHI-related use cases. These functions are successful in addressing the computing challenges of big data handling, efficiency, computing-intensive modeling and processing, and data security.

41.4 Edge Computing for Smart Cities

41.4.1 Methodology

With the development of computation technology and hardware, a large number of smart devices are integrated with sensors, enabling them to acquire real-time data and information from the environment. This phenomenon has culminated in the captivating concept of the IoT in which all smart things, such as smart cars (Morabito et al. 2018), wearable devices (Chen et al. 2017), sensors and industrial and utility components (Mehta et al. 2018) are connected via networks and empowered with data analytics that are significantly changing the way we work, live, and play. In the past few years, many scientific and industrial organizations have introduced and implemented the concept of IoT in various fields such as smart homes, smart cities, smart traffic, and smart environments. Edge computing is a new paradigm in which extensive computing and storage resources are placed to provide cloud-computing capabilities at the edge (variously referred to as cloudlets or micro data centers) of the Internet (Satyanarayanan 2010). Edge computing is a mesh network of micro-data centers that process or store data locally and push all received data to a centralized data center or cloud-storage repository (Butler 2017). By implementing computation closer to the edge of the network, analytics of complex data can be realized in near-real time. In applications, the forms of edge are various; for example, a gateway at a smart home is the edge between home devices and the central cloud; a micro-data center and a cloudlet are the edge between a smartphone and the central cloud.

The main function of edge computing is to ingest, store, filter, and send data to the central cloud systems ("What Is Edge Computing?|GE Digital" n.d.). At the heart of a smart city, there is widespread deployment of IoT sensor networks, which provide a regular flow of data that allows for effective and efficient management of services and assets. Typical deployment scenarios include a large scope of content: from bus tracking to traffic light management, street lighting control, air quality, and pollution monitoring. We envision that edge-computing could have similar impact on our society as that of cloud computing. Edge computing provides new possibilities in IoT applications, particularly for those tasks relying on AI techniques such as object detection (Ananthanarayanan et al. 2017), face recognition (Hu et al. 2016), language processing (Lewis et al. 2014), and obstacle avoidance (Zhang and Ye 2016).

41.4.2 Challenges, Motivations, and Opportunities

Nowadays, a smart city relies on the infrastructure of edge computing to leverage most of the up-to-date data-driven technologies. With edge computing, services can be ensured to flow continuously through local data processing even when the Web connection is interrupted (Abbas et al. 2017). For example, driverless cars and other modern IoT devices are designed to be built with enough processing capability, so that they can perform some of the computation themselves at the edge, without sending it to the central cloud. Edge-computing technology provides an attractive and resilient platform for cities, while at the same time reducing backhaul costs (Tran et al. 2017a, b), both in terms of the amount of data required and the sharing of connections by creating a mesh network.

There are challenges both in the big data generated and in creating the necessary network infrastructure to support an increasing number of end devices. Edge computing offers a solution to many of the challenges described in Sect. 41.2.2, which opens up many possibilities for smart cities. According to the advantages discussed above, edge computing can contribute to the following computing challenges of smart cities:

(1) Latency and efficiency. In a high-efficiency computing system, any device connected to the Internet has to be responsive in a short period of milliseconds. Any lag in the communication between network and devices is termed latency. Edge computing can eliminate the latency issue as it works on the principle of a more distributed network. This kind of system has the capability to guarantee real-time information processing and maintains a more reliable network (Hu et al. 2015). On the other hand, edge-computing processes the massive data generated by different types of IoT devices at the edge of network, instead of transmitting them to the centralized cloud infrastructure. Therefore, edge computing can provide services with faster response and greater quality in comparison with cloud computing, which greatly improves the efficiency of collecting, transferring, processing, and analyzing data generated by arrays of IoT devices.
(2) Privacy and security. Security concerns are more related to the transfer of data over a network to the central cloud. In an edge architecture, any outage would be limited to the edge devices and local applications. Therefore, edge computing will improve privacy and security by omitting the transmission since the data are stored and processed in or closer to the edge devices (He et al. 2018). With the improvement of authentication technology, the privacy and security

of edge computing can be further guaranteed by the emergence of biometric authentication such as fingerprint authentication, face authentication, touch-based, or keystroke-based authentication (Yi et al. 2015; Zhou et al. 2017).

(3) Internet load reduction. According to the Cisco Global Cloud Index ("Cisco Global Cloud" n.d.), the amount of traffic running through cloud-computing networks will increase to 14.1 zettabytes per year in 2020. This immersive amount of traffic can be removed from the central cloud by processing some of the data closer to the edge. Additionally, moving the processing of data away from the central cloud can minimize the network burden where the Internet bandwidth is limited (Lyu et al. 2018).

(4) Sustainability. Edge-computing systems provide the capability of decentralizing computation power, which support fault tolerance in that when one of the edge devices fails, other nodes and associated IT assets will still remain operational (Ning et al. 2019). This concept is similar to the cloud disaster recovery strategy ("Disaster Recovery Planning Guide|Architectures," n.d.) by using multiple available zones and regions to ensure that the data and applications are not lost in a catastrophic event.

Edge computing introduces a new concept that computing should happen as close as possible to the data sources. With this architecture, a request could be generated from the top of the computing paradigm and processed at the edge. By deploying edge computing, software engineers can create additional applications that utilize edge-computing platforms to leverage existing technology and benefit the smart cities in the following ways ("Smarter Cities with Edge Computing" n.d.):

(1) Streetlighting. A number of cities are in the process of upgrading their streetlights to lower-power LEDs. With the major cost of these upgrades being the physical fitting, edge appliances can be added to provide lighting controls (Xing et al. 2018).

(2) Security cameras. Nowadays, CCTV cameras have been a critical tool in modern policing systems. Edge computing can allow low-cost wireless IP cameras to be deployed in these systems, which will offer considerably less cost (Yi et al. 2017).

(3) Health emergency and public safety management. For applications that require real-time prediction and low latency such as health emergencies (Wang et al. 2017) and public safety (Zhang and Ye 2016) management, edge computing is also an appropriate paradigm since it could save the data transmission time as well as simplify the network structure. Decisions and diagnosis could be made and distributed from the edge of the network, which is more efficient compared with collecting information and making decisions at a central cloud.

(4) Location awareness. For geoinformatics-based applications such as transportation and utility management, edge computing exceeds cloud computing due to location awareness (Shi et al. 2016). In edge computing, data could be collected and processed based on geographic location without being transferred to the central cloud.

41.4.3 Urban Heat Island Use Case

Unlike cloud computing, edge devices are commonly decentralized. In order to monitor UHI from distributed sensors, edge computing offers closer contacts to each individual sensor, thus reducing energy consumption and response time during the transfer of observation data (Ngoko et al. 2018). Edge devices are those mounted directly on the edge for urban sensing of properties such as microclimate, having better durability compared to wireless devices. Densely distributed buildings in urban areas work as an ideal candidate for the deployment of edge devices, providing close proximity to the UHI impact factors such as temperature, humidity, and wind speed. Due to climate change, heating and cooling consume significant energy in buildings. These sectors contribute greatly to UHI and can be monitored by smart building sensors (Seitz et al. 2017). Lightweight tasks like data cleaning and basic decision support can be performed, and therefore contributes to UHI mitigation. Applications that support edge computing can benefit the field of UHI in: (1) allowing users to browse and query the UHI of cities around the world from a gateway; (2) providing a means to access real-time datasets from the edge without any latency; and (3) allowing users to search for a city of interest, query cities to generate charts of seasonal and long-term surface UHI, and download the UHI data.

41.5 Mobile Computing for Smart Cities

41.5.1 Methodology

Mobile computing could be described as a form of human–computer interaction where the computer is portable and transported during normal usage (Qi and Gani 2012; Akherfi et al. 2018). The fundamental concepts of mobile computing include: (1) communication, (2) hardware, and (3) software. Specifically, the communication concept refers to the wireless networks, data traffic, and protocols. The hardware could be any type of mobile device, which includes: (1) laptops, (2) tablets, (3) smartphones, (4) carputer, and others. The category boundaries of such devices are blurry, as more and more portable devices are installed with microchips and wireless modules, all of which have some computing power and the ability to transfer data through networks as a part of the mobile-computing hardware (Tong et al. 2016). The software in mobile computing consists of the applications in mobile device hardware, such as customized industry software, data collection applications, and Web browsers.

In the past decade, mobile computing has developed in two ways (Kumar et al. 2013): (1) deployment of sensors, and (2) growth in smartphones. It was also challenged by the explosion of big data (Laurila et al. 2012). Different from purpose-oriented IoT, mobile devices are integrated with multi-purpose sensors, such as GPS receivers, accelerometers, gyroscopes, and microphones. With the growth in both smartphone technologies and number of users, mobile devices are transitioning from specialized and customized platforms to powerful computing interfaces (Al-Turjman 2018). Mobile computing itself is also becoming a computing offloading contributor. The application layer of mobile computing faces various challenges due to its features. However, with the fast growth in communication technologies, including 4G and 5G networks and high-speed city Wi-Fi (Tran et al. 2017a, b), and mobile technologies in general, the number of applications running on mobile devices is growing at an exponential rate.

41.5.2 Challenges, Motivations, and Opportunities

In addition to most computing architectures in a wired network, mobile computing is different in the following aspects (Qi and Gani 2012): (1) Mobility: mobile-computing nodes or devices are expected to be portable and transportable; the computing power is not physically limited to a certain location and follows the principle of bringing computing to the data instead of transferring the data to computing resources. (2) The diversity of network conditions: the networks that mobile devices use are often not fixed; communication could be achieved through high-bandwidth or low-bandwidth networks; and the mobile device may even operate offline. (3) Inconsistency: as mobile devices are limited by their battery power and wireless network conditions, the inconsistency of communication and change of working status are expected and requires the mobile devices to switch modes to adapt to specific situations. (4) Asymmetric communication: wireless networks are often set with different bandwidths for downlink and uplink, which causes asymmetric communications between backend servers and local devices. (5) Low reliability: wireless communications are susceptible to interference; the security issues are enlarged in such networks and affect the reliability of mobile computing (Qi and Gani 2012).

The rapid development of mobile computing and smartphone applications is enabling integrated growth of smart-city applications. As stated in Sect. 41.2.2, mobile computing can help to improve the following challenges of smart-cities computing:

(1) Satisfy the need of users from different areas. Mobile computing supports smart-city computing in the forms of mobility and flexibility, which could help both end users and policy makers to meet different computing demands in different scenarios. Application use cases include services in higher education (Gikas and Grant 2013), and location-based services in general, which all utilize the mobility side of smart devices and allow them to act as both a data collector and data user (Raja et al. 2018). Another application of mobile computing is to utilize and integrate smart devices in smart spaces (Zheng and Ni 2010). The concept of the smart city is a big domain with enough space for the expansion and adaptability of mobile computing. Research topics including dynamic offloading for

mobile devices (Huang et al. 2012) and mobile cloud computing are all interactive examples of smart devices in smart spaces. Mobile cloud computing has been envisioned since 2009 as a combination of cloud computing and mobile computing, which leverages the mobility side of mobile computing and integrates with the elastic computing power from cloud computing (Tong et al. 2016; Dinh et al. 2013; Fernando et al. 2013). When integrated with cloud-computing power, it could also serve as an edge-computing device in the cloud-computing network.

(2) Computing efficiency and near-real-time analysis and feedback. Smart device holders are often fed with various information or data through sensors on the smart devices; with mobile-based computing power, stream-like data flow could be analyzed locally and uploaded to the centralized databases at the same time. End users with smart devices on hand could get feedback or results immediately; routing and mapping services, language translation services, and instant weather services are all good examples of this (Talukdar 2010). At the same time, public-security services and danger-awareness services could also be provided through mobile computing and locally based services (Aubry et al. 2014), such as the lost child and healthcare applications discussed in Sect. 41.1.2. The challenges in smart-city implementations bring new motivations and opportunities for the development of mobile computing and vice versa.

As one of its important components, mobile computing is enhancing the smart-city experience in the following aspects: (1) Transport and traffic management for both personal end users and policy makers; (2) Utilities and energy monitoring across the network, and (3) Improving public safety and smart-city security awareness.

41.5.3 Urban Heat Island Use Case

Mobile computing and mobile-based technologies are integrating innovative concepts and ideas to increase UHI awareness and aid city design to reduce the UHI effect. As Wong et al. (2014) mentioned in their reviews, tools have been developed and implemented to allow users to gather instantaneous energy performance feedbacks on their decisions and plans of building designing, such as the building orientation and thermal performance, through mobile-based applications (i.e., iPad/smartphone application). At the same time, mobile devices provide volunteered geographic information (VGI) to enhance the near-real-time estimation of UHI. For example, Koukoutsidis (2018) utilized mobile crowdsensing to estimate the mean area temperature in a linear region that exhibits the UHI effect.

41.6 Case Study

41.6.1 Urban Heat Island (UHI)

The direct cause of UHI is urbanization, which leads to the loss of more vegetation and causes more surfaces to be paved or covered with impervious materials such as cement, asphalt, buildings, and walls. Challenges are revealed due to the complexity of the composition of UHI impact factors. Major ones are stated by Oke (1982) in his previous studies and include: (1) the inherent complexity of the city-atmosphere system; (2) the lack of clear conceptual and theoretical frameworks; and (3) the expense and difficulty of observation in cities. UHI is a very common challenge to all urban areas in the world, although in megacities it is serious and less so in small towns.

UHI is usually measured in three scales: boundary UHI, canopy UHI, and surface UHI. Boundary UHI is measured from the altitude of the rooftop to the atmosphere. It is generally used to investigate the UHI effect at mesoscale and is acquired by using, for example, radiosondes. Canopy UHI is measured at the altitude that ranges from the ground surface to the rooftop. An assessment of canopy UHI is most suitable for a microscale study and is generally derived based on weather station data. Surface UHI is measured at the Earth surface level. Researchers have often used satellite images (e.g., thermal bands of Landsat TM/ETM/OLI, MODIS, AVHRR) to obtain the effect of surface UHI (Zhang et al. 2009). Researchers used remotely sensed data and stationary meteorological monitoring data to analyze the UHI changes and effects in the long or short term (Earl et al. 2016), as well as the relationship between UHI and land cover changes (Chen et al. 2006; Charkraborty and Lee 2019). A lot of research has simulated and evaluated UHI and its effect on the future by using numerical modeling based on real-time meteorological data (Morris et al. 2015).

41.6.2 UHI Challenges and Opportunities

From the aforementioned scientific challenges, UHI introduces its own computing challenges, mostly concentrated on handling the aspects of the expense and difficulty of observation in cities. These challenges include: (1) management of heterogeneous data sources; (2) integration of a huge volume of remotely sensed data and real-time meteorological data; and (3) a large amount of computation in modeling, visualizing, simulating, and predicting. Cloud computing has existed in the long term for allocating computing resources to enable the auto-scalable modeling and detecting in many study fields and has proved to be an efficient and economical solution (Yang et al. 2017a). Google Earth Engine is a cloud-computing platform, offering intrinsically parallel computational resources, and enabling monitoring and measurement of changes in the Earth's environment, at planetary scale, on a large catalog of Earth observation data (Moore and Hansen 2011). An implementation of large-scale correlation between land surface temperature and land cover alteration research is conducted upon this platform and has illustrated the capability of using cloud computing for efficient UHI monitoring (Ravanelli et al. 2018a, b).

The emergence of 5G and IoT technologies in the current era is bringing opportunities to facilitate advances in urban microclimate study with finer spatiotemporal resolution beyond just satellite imagery analysis (Li et al. 2018). Voogt and Oke (2003) argued that thermal remote sensors have a credible ability to observe the surface UHI and require consideration of the intervening atmosphere and surface radiative properties, leading to extra conversions and corrections. With implementing sensor device networks directly into the environment, urban environmental factors like air temperature are more accurately measured. These sensor networks can be designed and implemented for advanced urban microclimate and environment modeling (Jha et al. 2015). Challenges follow when considering the real-time streaming nature of IoT, as it requires the capacity of ingesting the large number of data and producing results with higher speed that is beyond the capability of conventional architectures (Rathore et al. 2018). Santamouris (2015) analyzed heat island magnitude and characteristics in one hundred cities and regions and indicated that analysis of 43% station measurements are only based on one station from urban and one from rural. According to the Gartner, up to 20.4 billion IoT devices will be connected machine-to-machine by 2020 (Meulen 2017), offering great potential to increase the number of sensors utilized for UHI research.

Since the first time it was introduced by Howard (1818), in the past 200 years, numerous studies have been developed to model UHI intensity, simulate, and predict UHI effects. However, it was proved from analyzing one hundred Asian and Australian cities and regions, that a systematic analysis like a workflow is still needed (Santamouris 2015). Coupling with aforementioned computing techniques (cloud computing, edge computing, and mobile computing), the following introduces a theoretical integrated workflow to enable the efficient data storage and processing for handling urban informatics challenges and using UHI as an example. This workflow targets the last two scientific challenges of UHI, and the overall architecture is illustrated in Fig. 41.4, starting from collecting urban observation data with mobile devices to the centralized cloud-based data analysis, and finishing with generating intelligent supportive materials for UHI monitoring and managing.

41.6.3 Integrated Workflow

41.6.3.1 Mobile Computing for Local Fast Response

Data in Fig. 41.4 are directly collected by sensors within a large sensor network deployed in the urban environment. Data streams into the workflow by entering the first gate: mobile computing. In general, the capacity of mobile devices is low, and due to the limitations like battery life, only lightweight preprocessing like data cleaning and reorganizing can be performed at the mobile computing stage. However, in situ

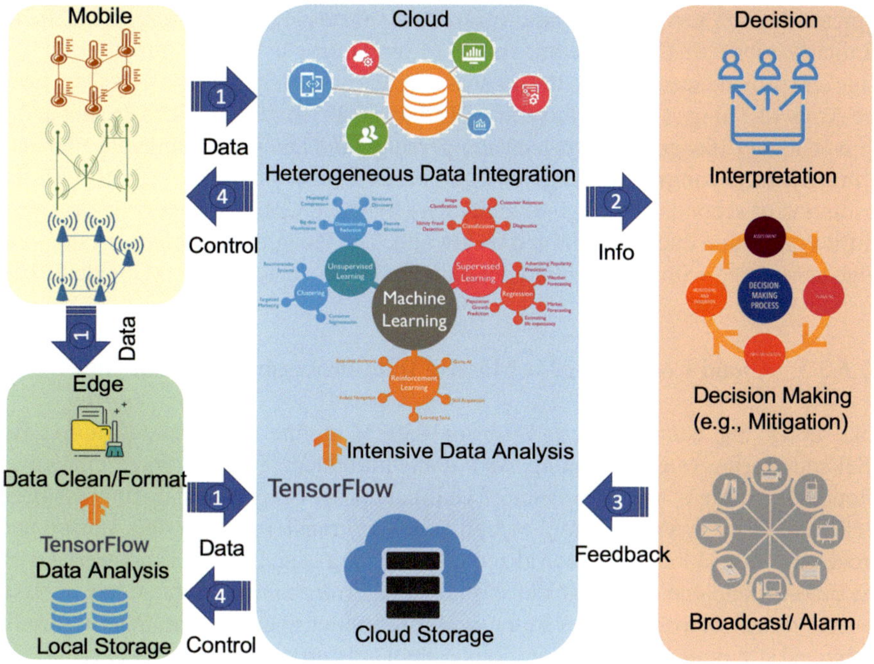

Fig. 41.4 Overall architecture of computing for UHI

monitoring coupled with light data understanding can reduce time latency for jobs that do not require extensive computation but only the ability to make simple judgments. For instance, alarms setup on a mobile device with constrained temperature threshold can be triggered responsively when unexpected heat is detected. Though the computing capabilities of mobile devices are low, with hundreds and thousands of contributions from them, appreciable computational resources are preserved for more intensive works like microscale UHI modeling (Mirzaei 2015).

41.6.3.2 Edge Computing for Data Preprocessing and Direct Microcontrol

Besides collecting data on the edge and passing the raw data to the cloud like mobile computing, edge computing offers more capacities for better data preprocessing. With the increasing data volume, uploading everything raw to the cloud can take a significant amount of time, and the heavy duty that is loaded to the center cluster can exceed the limit of the computing resources. To fill the gap between mobile computing and cloud computing, enhancing the performances regarding response time, data transform, data safety, and privacy, edge computing is integrated to the workflow to allow downstream data representing cloud services and upstream data representing IoT services (Sun and Ansari 2016; Shi et al 2016; Yannuzzi et al. 2014). Similarly, works that do not require much computation can be done directly from the edge and provide feedbacks to the sensors to reduce time lag (Gerla 2012). Data from the Array of Things (AoT) (University of Chicago 2019; see Sect. 4.7) project at the University of Chicago monitors local temperatures and other environmental elements from networks composed of hundreds of sensors, providing observations with the resolution of seconds. The high-velocity data transfers within the network can cause traffic congestion due to the limited bandwidth. The Google cloud platform supports edge computing with AI, enabling potential real-time data analytics (Google 2019).

41.6.3.3 Cloud Computing for Massive Data Processing and Analytics

Like every big data problem, a sensor dataset at fine temporal resolution for UHI monitoring (e.g., streaming AoT data) introduces a data storage challenge. Cloud computing as the final layer of UHI data processing and analyzing has been well studied for enabling heavy computations by transferring big data storing and processing from a local to a centralized cluster (Yang et al. 2017b). Empowered with the auto-expandable nature of the virtual storage mechanism, data streamed from sensors transfer through edges to the center for better management. With the well-resourced computing capacity, the cloud cannot only process the data that mobile and edge devices cannot, but also accelerate the processing beyond a standalone server.

IoT networks are massive and can be distributed with different protocols established by different management departments. Therefore, UHI-related attributes like temperature, humidity and wind speed from different networks are potentially captured with sensors powered by different standards. Data heterogeneity is one of the major concerns and the massive data cleaning workload requires significant computational capability. The cloud as a centralized computing resource pool offers sufficient capacity for such workload (Botta et al. 2014). As mentioned, there are many factors contributing to UHI study. Changing the composition leads to requirements for model parameter adjustments. SaaS as introduced in Sect. 41.3.1 and provided with cloud computing allows users to duplicate a model directly from a current version and customize the new one to fit the new environment. Advantages include reduced model-building time and decreased human error when transferring the experimental environment.

41.6.3.4 Mobile-Edge-Cloud Integrated Computing for UHI

A weather forecast example provided by a previous study indicated the basic workflow when the simulation is decomposed into a process-oriented pipeline (Tsahalis et al. 2013). Weather research shares conceptual similarities to UHI, and thus, their example is applied here as a base version of the conventional workflow. Heusinkveld et al. (2010) carried out an assessment of UHI intensity in Rotterdam using an innovative mobile bio-meteorological measuring platform mounted on a cargo bicycle.

Physiologically equivalent temperatures were calculated directly from the measurements and the intensity of UHI was evaluated in real time. Coupling with the IoT and mobile devices empowered a real-time urban microclimate analysis framework that integrated with the sensor network and cloud computing (Rathore et al. 2018); our workflow gains the experience from both. This enhanced framework composed of cloud computing, edge computing, and mobile computing is able to successfully address the previously introduced UHI challenges. Starting from measuring the geographic environmental of ground, air, and water, mobile computing can directly sense these parameters and give a quick response (e.g., a UHI detection alarming system) with minor data manipulation before entering the major processing and modeling procedures. Edge computing offers a higher computational capacity, mitigating the heavy workload that is initially carried by the centralized module. Building-scale UHI (i.e., building energy model) is limited to the study of an isolated building, requiring less computational resources as it considers less neighborhood environmental impacts (Mirzaei 2015). Therefore, UHI modeling, visualizing, simulating, and predicting for a smaller UHI study scale (i.e., building scale) can be directly computed on the edge for more efficiency. There are many tasks that cannot be satisfied with the limited resources from mobile computing or edge computing, such as heterogeneous data integration, and larger scale (e.g., microclimate) UHI modeling. The cloud as a big centralized resources pool is powered with enormous computing capabilities. UHI-related observation data like temperature, humidity, and windspeed are transferred from sensors to the cloud after a certain effort made by mobile computing and edge computing for data cleaning and preprocessing. Heterogeneous data integration on the cloud will be triggered for the massive data coupled with mixed data types and data standards. Large-scale UHI modeling, simulating, etc., are performed within the cloud. Elasticity that is offered as one of the key features of the cloud dispatches computing resources on demand and surpasses the traditional method of using a single computer for analysis, saving resources while providing enough capacity for the heavy tasks. All three computing paradigms work seamlessly from getting the sensor data to processing, analyzing, and decision support, enabling an efficient and effective workflow as a whole to handle the UHI challenges.

These three computing components should be leveraged and kept in balance when applied to UHI monitoring, data analysis, and problem solving. For instance, deploying edge nodes with higher computing capacity may increase the operational cost for processing the IoT data streams compared to processing them in the centralized cloud (Sun and Ansari 2016). Understanding the tradeoffs among the different interfacings of the three is crucial for maximizing the workflow efficiency and optimizing the computing architecture design. Many other smart-city applications are encountering similar problems, and the demonstrated UHI analytical workflow can be broadly applied when integrating computing components.

41.7 Summary

This chapter introduced the contribution and recent advances of computing for smart cities. The general challenges of computing in smart cities were introduced and include heterogeneous sources of big data, resulting from the unprecedented number of smart sensors and devices, various needs from users in multiple domains, data security, sustainability, and efficiency. To address the challenges, cloud computing, edge computing, and mobile computing were discussed for their advantages and limitations in smart-city applications. Cloud computing provides a unified and efficient platform, large-scale base infrastructure, sustainable and green software and hardware development and addresses system security and recovery issues. Edge computing helps reduce observation latency and increase the efficiency of data collection, improve data privacy and security, reduce data transmission load on computer network, and provide a sustainable decentralization of computing needs. Mobile computing contributes to the smart city with computational mobility and flexibility, and computing efficiency and near-real-time analysis. The characteristics of different computing paradigms were exemplified in the case study of urban heat island. With multiple computing paradigms leveraged, smart-city applications and services can be provided in a more efficient and effective fashion.

41.7.1 The Future of Urban Computing for Smart Cities

Big data and IoT are labeled as the primary drivers for the cloud, edge, and mobile computing. The development of mobile computing is increasing at an accelerating speed. With the fast implementation of 5G networks and closer integration with cloud computing, the mobile-computing system is merging with the cloud-computing network and serving as the network edge. The phrase mobile cloud computing has been frequently referenced in the mobile-computing field (Fernando et al. 2013; Akherfi et al. 2018). When the mobility of mobile computing interacts with the elastic computing power from cloud computing, it will push the whole computing network to a new decentralized computing stage and accelerate the smart-city process. Smarter devices, faster networks, and longer battery lives are the foreseeable future; the transformation of mobile computing and interaction with other computing fields will be the norm.

With the increasing number of mobile devices (phones, drones, cars, etc.), the need for interaction with nearby edge resources will become apparent. Coupled with better processing, computing, and power capacity, as well as the decentralized characteristic of mobile computing, edge computing is expected to provide significantly improved throughput, better performance, and real-time responses, moving both computing and data closer to the user and customizing the processing requirements from each user. Edge computing and mobile computing are both capable of handling localized data for fast action for a certain range of area size. However, the increasing urban data volume and cross-city geo-analysis are also driving centralized cloud computing.

Ever since the infrastructure was developed for cloud computing, the combined use of private and public clouds is engaged for many more individual and business purposes. As a mature platform to integrate powerful computing capabilities, large data storage and on-demand data analysis, cloud computing will lead cities toward a smart age—an age based on fully connected, interactive decision-supporting environment. Within the smart city, a variety of devices (e.g., domestic appliances and semiautomatic vehicles) will connect to the cloud-based Internet for sensing, recording, sharing, and analyzing numerous human-related activities. Coupled with the help from artificial intelligence algorithms, cloud computing will serve companies, governments, and individual residents with smarter solutions.

References

Abbas N, Zhang Y, Taherkordi A, Skeie T (2017) Mobile edge computing: a survey. IEEE Internet of Things J 5(1):450–465

Akherfi K, Gerndt M, Harroud H (2018) Mobile cloud computing for computation offloading: issues and challenges. Appl Comput Inform 14(1):1–16

Altman SH, Sivo NT, Jacobstein M, Knapp BR, Potter JF, Pernikoff TB (2015) Secure and private location sharing for location-aware mobile communication devices. U.S. Patent 8,989,778

Al-Turjman F (2018) Mobile couriers' selection for the smart-grid in smart-cities' pervasive sensing. Future Gener Comput Syst 82:327–341

Amirebrahimi S, Rajabifard A, Mendis P, Ngo T (2016) A framework for a microscale flood damage assessment and visualization for a building using BIM–GIS integration. Int J Digital Earth 9(4):363–386

Ananthanarayanan G, Bahl P, Bodík P, Chintalapudi K, Philipose M, Ravindranath L, Sinha S (2017) Real-time video analytics: the killer app for edge computing. Computer 50(10):58–67

Anon nd (2019) Array of things file browser. [Online]. https://aot-file-browser.plenar.io. Accessed on 10 May 2019

Anthopoulos LG (2015) Understanding the smart city domain: a literature review. Trans-forming city governments for successful smart cities. Springer, Cham, pp 9–21

Aubry E, Silverston T, Lahmadi A, Festor O (2014) CrowdOut: a mobile crowdsourcing service for road safety in digital cities. In: 2014 IEEE international conference on pervasive computing and communication workshops (PERCOM WORKSHOPS), pp 86–91

Bera S, Misra S, Rodrigues JJ (2015) Cloud computing applications for smart grid: a survey. IEEE Trans Parallel Distrib Syst 26(5):1477–1494

Bibri SE (2018) The IoT for smart sustainable cities of the future: an analytical framework for sensor-based Big Data applications for environmental sustainability. Sustain Cities Soc 38:230–253

Botta A, De Donato W, Persico V, Pescapé A (2014) On the integration of cloud computing and internet of things. In: 2014 international conference on future internet of things and cloud, pp 23–30

Butler B (2017) What is edge computing and how it's changing the network. Netw World 21(09)

Casanueva A, Burgstall A, Kotlarski S, Messeri A, Morabito M, Flouris AD, Nybo L, Spirig C, Schwierz C (2019) Overview of existing heat-health warning systems in Europe. Int J Environ Res Public Health 16(15):2657

Cascetta E, Carteni A, Pagliara F, Montanino M (2015) A new look at planning and designing transportation systems: A decision-making model based on cognitive rationality, stake-holder engagement and quantitative methods. Transp Policy 38:27–39

Chakraborty T, Lee X (2019) A simplified urban-extent algorithm to characterize surface urban heat islands on a global scale and examine vegetation control on their spatiotemporal variability. Int J Appl Earth Obs Geoinf 74:269–280

Chang V, Ramachandran M (2015) Towards achieving data security with the cloud computing adoption framework. IEEE Trans Serv Comput 9(1):138–151

Chen XL, Zhao HM, Li PX, Yin ZY (2006) Remote sensing image-based analysis of the relationship between urban heat island and land use/cover changes. Remote Sens Environ 104(2):133–146

Chen Z, Hu W, Wang J, Zhao S, Amos B, Wu G, Ha K, Elgazzar K, Pillai P, Klatzky R, Siewiorek D (2017) An empirical study of latency in an emerging class of edge computing applications for wearable cognitive assistance. In: Proceedings of the second ACM/IEEE symposium on edge computing, p 14

Cisco Global Cloud Index Projects Cloud Traffic to Nearly Quadruple Representing 92 Per-cent of Total Data Center Traffic by 2020 [WWW Document], n.d. URL https://newsroom.cisco.com/press-release-content?type=webcontent&articleId=1804748. Accessed on 13 May 2019

Conti S, Meli P, Minelli G, Solimini R, Toccaceli V, Vichi M, Beltrano C, Perini L (2005) Epidemiologic study of mortality during the Summer 2003 heat wave in Italy. Environ Res 98(3):390–399

Dai YS, Yang B, Dongarra J, Zhang G (2009) Cloud service reliability: Modeling and analysis. In: 15th IEEE pacific rim international symposium on dependable computing, pp 1–17

David R (2018) Smarter cities with edge computing. https://www.smartcitiesworld.net/opinions/opinions/smarter-cities-with-edge-computing

Davies M, Steadman P, Oreszczyn T (2008) Strategies for the modification of the urban climate and the consequent impact on building energy use. Energy Policy 36(12):4548–4551

Dinh HT, Lee C, Niyato D, Wang P (2013) A survey of mobile cloud computing: architecture, applications, and approaches. Wirel Commun Mobile Comput 13(18):1587–1611

Disaster Recovery Planning Guide|Architectures [WWW Document], n.d. Google Cloud 5.13.2019. https://cloud.google.com/solutions/dr-scenarios-planning-guide

Earl N, Simmonds I, Tapper N (2016) Weekly cycles in peak time temperatures and urban heat island intensity. Environ Res Lett 11(7):074003

EPA (2016) Climate Change Indicators in the United States, 2016. Available online: https://www.epa.gov/climate-indicators. Accessed on 8 April 2019

Etro F (2015) The economics of cloud computing. In: Cloud technology: concepts, methodologies, tools, and applications, pp 2135–2148. IGI Global

Fernando N, Loke SW, Rahayu W (2013) Mobile cloud computing: a survey. Future Gener Comput Syst 29(1):84–106

Filleul L, Cassadou S, Médina S, Fabres P, Lefranc A, Eilstein D, Le Tertre A, Pascal L, Chardon B, Blanchard M, Declercq C (2006) The relation between temperature, ozone, and mortality in nine French cities during the heat wave of 2003. Environ Health Perspect 114(9):1344–1347

Fu Z, Sun X, Liu Q, Zhou L, Shu J (2015) Achieving efficient cloud search services: multi-keyword ranked search over encrypted cloud data supporting parallel computing. IEICE Transac Commun 98(1):190–200

Gerla M (2012, June) Vehicular cloud computing. In: 2012 The 11th annual mediterranean ad hoc networking workshop (Med-Hoc-Net), pp 152–155. IEEE

Giannaros TM, Melas D, Daglis IA (2014) Development of an operational modeling system for urban heat islands: an application to Athens, Greece. Natural Hazards Earth Syst Sci 14(2):347–358

Gikas J, Grant MM (2013) Mobile computing devices in higher education: Student perspectives on learning with cellphones, smartphones and social media. Internet Higher Educ 19:18–26

Gong C, Liu J, Zhang Q, Chen H, Gong Z (2010) The characteristics of cloud computing. In: 39th international conference on parallel processing workshops, pp 275–279

González-Gil A, Palacin R, Batty P, Powell JP (2014) A systems approach to reduce urban rail energy consumption. Energy Convers Manage 80:509–524
Google (2019) AI at the edge. [Online] Accessed on 10 May 2019. https://cloud.google.com/edge-tpu/
Gorelick N, Hancher M, Dixon M (2017) Google Earth Engine: planetary-scale geospatial analysis for everyone. Remote Sens Environ 202:18–27
Graham S, Marvin S (2002) Telecommunications and the city: electronic spaces, urban places. Routledge.
Guest CS, Willson K, Woodward AJ, Hennessy K, Kalkstein LS, Skinner C, McMichael AJ (1999) Climate and mortality in Australia: retrospective study, 1979–1990, and predicted impacts in five major cities in 2030. Climate Res 13(1):1–15
Haines A, Kovats RS, Campbell-Lendrum D, Corvalán C (2006) Climate change and human health: impacts, vulnerability, and mitigation. Lancet 367(9528):2101–2109
Hameed K (2003) The application of mobile computing and technology to health care services. Telematics Inform 20(2):99–106
Hao L, Lei X, Yan Z, ChunLi Y (2012) The application and implementation research of smart city in China. In: 2012 international conference on system science and engineering (ICSSE), pp 288–292
Hashem IAT, Yaqoob I, Anuar NB, Mokhtar S, Gani A, Khan SU (2015) The rise of "Big Data" on cloud computing: review and open research issues. Inf Syst 47:98–115
Hashim Raza Bukhari S, Siraj S, Husain Rehmani M (2018) Wireless sensor networks in smart cities: applications of channel bonding to meet data communication requirements. In: Transportation and power grid in smart cities: communication networks and services, pp 247–268
He D, Chan S, Guizani M (2018) Security in the internet of things supported by mobile edge computing. IEEE Commun Mag 56(8):56–61
Heusinkveld BG, Van Hove LWA, Jacobs CMJ, Steeneveld GJ, Elbers JA, Moors EJ, Holtslag AAM (2010) Use of a mobile platform for assessing urban heat stress in Rotterdam. In: Proceedings of the 7th conference on biometeorology, vol 20, pp 433–438
Hoffiman J (2018) Where do we need shade? Mapping urban heat islands in Richmond, Virginia. https://www.climate.gov/news-features/climate-case-studies/where-do-we-need-shade-mapping-urban-heat-islands-richmond
Hondula DM, Georgescu M, Balling RC Jr (2014) Challenges associated with projecting urbanization-induced heat-related mortality. Sci Total Environ 490:538–544
Howard L (1818) The climate of London, vol 1. W. Phillips, sold also by J. and A. Arch
Hu YC, Patel M, Sabella D, Sprecher N, Young V (2015) Mobile edge computing—a key technology towards 5G. ETSI White Paper 11(11):1–16
Hu W, Gao Y, Ha K, Wang J, Amos B, Chen Z, Pillai P, Satyanarayanan M (2016) Quantifying the impact of edge computing on mobile applications. In: Proceedings of the 7th ACM SIGOPS Asia-Pacific workshop on systems, p 5
Hu F, Yang C, Jiang Y, Li Y, Song W, Duffy DQ, Schnase JL, Lee T (2018) A hierarchical indexing strategy for optimizing Apache Spark with HDFS to efficiently query big geospatial raster data. Int J Digital Earth 1–19
Huang D, Wang P, Niyato D (2012) A dynamic offloading algorithm for mobile computing. IEEE Trans Wirel Commun 11(6):1991–1995
Hwang JS, Choe YH (2013) Smart Cities Seoul: a case study. ITU-T technology watch report. pp 1–20
Jadeja Y, Modi K (2012) Cloud computing concepts, architecture and challenges. In: 2012 international conference on computing, electronics and electrical technologies (ICCEET), pp 877–880
Jha M, Marpu PR, Chau CK, Armstrong P (2015) Design of sensor network for urban micro-climate monitoring. In: IEEE first international smart cities conference (ISC2), pp 1–4
Jin J, Gubbi J, Marusic S, Palaniswami M (2014) An information framework for creating a smart city through internet of things. IEEE Internet Things J 1(2):112–121

Jonas E, Pu Q, Venkataraman S, Stoica I, Recht B (2017) Occupy the cloud: Distributed computing for the 99%. In: Proceedings of the 2017 symposium on cloud computing, pp 445–451
Kakderi C, Komninos N, Tsarchopoulos P (2016) Smart cities and cloud computing: lessons from the STORM CLOUDS experiment. J Smart Cities 2(1):4–13. https://doi.org/10.18063/JSC.2016.01.002
Karwot J, Kaźmierczak J, Wyczółkowski R (2016) Smart water in smart city: A case study. In: Proceedings of SGEM 16th international scientific conference on earth and geosciences 3:851–858
Kaushal DS, Khan Y (2014) Cloud computing services in medical healthcare solutions. Int J Res 1(4):312–324
Kofinas DT, Spyropoulou A, Laspidou CS (2018) A methodology for synthetic household water consumption data generation. Environ Modell Softw 100:48–66
Koukoutsidis I (2018) Estimating spatial averages of environmental parameters based on mobile crowdsensing. ACM Transac Sens Netw (TOSN) 14(1):2
Kumar K, Liu J, Lu YH, Bhargava B (2013) A survey of computation offloading for mobile systems. Mobile Netw Appl 18(1):129–140
Lacinák M, Ristvej J (2017) Smart city, safety and security. Procedia Eng 192:522–527
Laurila JK, Gatica-Perez D, Aad I, Bornet O, Do TMT, Dousse O, Eberle J, Miettinen M (2012) The mobile data challenge: Big Data for mobile computing research. In: Proceedings of the international conference on mobile computing
Lehrig S, Eikerling H, Becker S (2015) Scalability, elasticity, and efficiency in cloud computing: a systematic literature review of definitions and metrics. In: Proceedings of the 11th international ACM SIGSOFT conference on quality of software architectures, pp 83–92
Lewis G, Echeverría S, Simanta S, Bradshaw B, Root J (2014) Tactical cloudlets: Moving cloud computing to the edge. In: 2014 IEEE military communications conference, pp 1440–1446
Li Y, Zhou Y, Shi Y, Zhu X (2012) Investigation of a coupling model of coordination be-tween urbanization and the environment. J Environ Manage 98:127–133
Li Y, Yang R, Yang C, Yu M, Hu F, Jiang Y (2017) Leveraging LSTM for rapid intensifications prediction of tropical cyclones. In: ISPRS annals of photogrammetry, remote sensing and spatial information sciences, vol 4
Li S, Da Xu L, Zhao S (2018) 5G Internet of Things: a survey. J Indus Inf Integr 10:1–9
Liebig T, Piatkowski N, Bockermann C, Morik K (2017) Dynamic route planning with real-time traffic predictions. Inf Syst 64:258–265
Liu Y, Shi J, Jian M (2017) Understanding visitors' responses to intelligent transportation system in a tourist city with a mixed ranked logit model. J Adv Transport
Lv Z, Li X, Wang W, Zhang B, Hu J, Feng S (2018) Government affairs service platform for smart city. Future Gener Comput Syst 81:443–451
Lyu X, Tian H, Jiang L (2018) Selective offloading in mobile edge computing for the green Internet of Things. IEEE Netw 32(1):54–60
Mahmood Z (2011) Cloud computing: Characteristics and deployment approaches. In: IEEE 11th international conference on computer and information technology, pp 121–126
Mastelic T, Brandic I (2015) Recent trends in energy-efficient cloud computing. IEEE Cloud Comput 2(1):40–47
Mehta A, Bayuh Lakew E, Tordsson J, Elmroth E (2018) Utility-based allocation of industrial IoT applications in mobile edge clouds
Memon RA, Leung DY, Liu CH (2009) An investigation of urban heat island intensity (UHII) as an indicator of urban heating. Atmos Res 94(3):491–500
Menouar H, Guvenc I, Akkaya K, Uluagac AS, Kadri A, Tuncer A (2017) UAV-enabled intelligent transportation systems for the smart city: applications and challenges. IEEE Commun Mag 55(3):22–28
Meulen Rvd (2017) Newsroom. [Online] Accessed on 10 May 2019. https://www.gartner.com/en/newsroom/press-releases/2017-02-07-gartner-says-8-billion-connected-things-will-be-in-use-in-2017-up-31-percent-from-2016

Mirzaei PA (2015) Recent challenges in modeling of urban heat island. Sustain Cities Soc 19:200–206

Mitchell S, Villa N, Stewart-Weeks M, Lange A (2013) The Internet of everything for cities: connecting people, process, data and things to improve the livability of cities and com-munities. Cisco, San Jose

Mitton N, Papavassiliou S, Puliafito A, Trivedi KS (2012) Combining Cloud and sensors in a smart city environment

Moore RT, Hansen MC (2011) Google Earth Engine

Morabito R, Cozzolino V, Ding AY, Beijar N, Ott J (2018) Consolidate IoT edge computing with lightweight virtualization. IEEE Netw 32(1):102–111

Morán AJ, Profaizer P, Zapater MH, Valdavida MA, Bribián IZ (2016) Information and communications technologies (ICTs) for energy efficiency in buildings: review and analysis of results from EU pilot projects. Energy Build 127:128–137

Morris KI, Salleh SA, Chan A, Ooi MCG, Abakr YA, Oozeer MY, Duda M (2015) Computational study of urban heat island of Putrajaya, Malaysia. Sustain Cities Soc 19:359–372

NASA Cloud Computing Platform. https://www.nasa.gov/offices/ocio/ittalk/06-2010_cloud_computing.html. Accessed March 2010

NASA Earth Science Data Systems Program Highlights 2018, April 26, 2019, https://cdn.earthdata.nasa.gov/conduit/upload/11232/ESDISHighlights.pdf

Ngoko Y, Saintherant N, Cerin C, Trystram D (2018) How future buildings could redefine distributed computing. In: 2018 IEEE international parallel and distributed processing symposium workshops (IPDPSW), pp 1232–1240

Ning Z, Kong X, Xia F, Hou W, Wang X (2019) Green and sustainable cloud of things: Ena-bling collaborative edge computing. IEEE Commun Mag 57(1):72–78

Nowicka K (2014) Smart city logistics on cloud computing model. Proc Soc Behav Sci 151:266–281

Oke TR (1982) The energetic basis of the urban heat island. Q J R Meteorol Soc 108(455):1–24

Petkova EP, Vink JK, Horton RM, Gasparrini A, Bader DA, Francis JD, Kinney PL (2016) Towards more comprehensive projections of urban heat-related mortality: estimates for New York City under multiple population, adaptation, and climate scenarios. Environ Health Perspect 125(1):47–55

Qi H, Gani A (2012) Research on mobile cloud computing: review, trend and perspectives. In: 2012 second international conference on digital information and communication technology and its applications (DICTAP), pp 195–202

Raja CV, Chitra K, Jonafark M (2018) A survey on mobile cloud computing. Int J Sci Res Comput Sci Eng Inf Technol 3(3)

Rao AM, Rao KR (2012) Measuring urban traffic congestion: a review. Int J Traffic Transp Eng 2(4)

Rathore P, Rao AS, Rajasegarar S, Vanz E, Gubbi J, Palaniswami M (2018) Real-time urban microclimate analysis using internet of things. IEEE Internet of Things J 5(2):500–511

Ravanelli R, Nascetti A, Cirigliano R, Di Rico C, Leuzzi G, Monti P, Crespi M (2018) Monitoring the impact of land cover change on surface urban heat island through Google Earth Engine: proposal of a global methodology, first applications and problems. Remote Sens 10(9):1488

Ravanelli R, Nascetti A, Cirigliano RV, Di Rico C, Monti P, Crespi M (2018) Monitoring urban heat island through Google Earth Engine: Potentialities and difficulties in different cities of the United States. Int Arch Photogram Remote Sens Spatial Inf Sci 42(3)

Rittinghouse JW, Ransome JF (2016) Cloud computing: implementation, management, and security. CRC Press

Rizwan AM, Dennis LY, Chunho LIU (2008) A review on the generation, determination and mitigation of Urban Heat Island. J Environ Sci 20(1):120–128

Rong H, Zhang H, Xiao S, Li C, Hu C (2016) Optimizing energy consumption for data centers. Renew Sustain Energy Rev 58:674–691

Rushikesh R, Sivappagari CMR (2015) Development of IoT based vehicular pollution monitoring system. In: 2015 international conference on green computing and Internet of Things (ICGCIoT), pp 779–783

Santamouris M (2015) Analyzing the heat island magnitude and characteristics in one hundred Asian and Australian cities and regions. Sci Total Environ 512:582–598

Santamouris M, Papanikolaou N, Livada I, Koronakis I, Georgakis C, Argiriou A, Assima-kopoulos DN (2001) On the impact of urban climate on the energy consumption of buildings. Sol Energy 70(3):201–216

Sarrat C, Lemonsu A, Masson V, Guedalia D (2006) Impact of urban heat island on regional atmospheric pollution. Atmos Environ 40(10):1743–1758

Satyanarayanan M (2010) Mobile computing: the next decade. In: Proceedings of the 1st ACM workshop on mobile cloud computing and services: social networks and beyond (MCS)

Schuelke-Leech BA, Barry B, Muratori M, Yurkovich BJ (2015) Big Data issues and opportunities for electric utilities. Renew Sustain Energy Rev 52:937–947

Seitz A, Johanssen JO, Bruegge B, Loftness V, Hartkopf V, Sturm M (2017) A fog architecture for decentralized decision making in smart buildings. In: Proceedings of the 2nd international workshop on science of smart city operations and platforms engineering, pp 34–39

Sen J (2015) Security and privacy issues in cloud computing. In: Cloud technology: concepts, methodologies, tools, and applications. IGI Global, pp 1585–1630

Sharma PP, Navdeti CP (2014) Securing Big Data hadoop: a review of security issues, threats and solution. Int J Comput Sci Inf Technol 5(2):2126–2131

Sheikhi A, Rayati M, Bahrami S, Ranjbar AM, Sattari S (2015) A cloud computing framework on demand side management game in smart energy hubs. Int J Electr Power Energy Syst 64:1007–1016

Shi W, Cao J, Zhang Q, Li Y, Xu L (2016) Edge computing: vision and challenges. IEEE Internet of Things J 3(5):637–646

Smarter Cities with Edge Computing. n.d. Smart Cities World. Accessed on 26 Mar 2019. https://www.smartcitiesworld.net/opinions/opinions/smarter-cities-with-edge-computing-

Song M, Cen L, Zheng Z, Fisher R, Liang X, Wang Y, Huisingh D (2017) How would Big Data support societal development and environmental sustainability? Insights and practices. J Cleaner Product 142:489–500

Souza GBDC, Vieira FHT, Lima CR, Deus GADJ, De Castro MS, De Araujo SG, Vasques TL (2016) Developing smart grids based on GPRS and ZigBee technologies using queueing modeling–based optimization algorithm. Etri J 38(1):41–51

State of Green Denmark, 10 examples of smart city solution, Nov 01 2018. https://stateofgreen.com/en/partners/state-of-green/news/10-examples-of-smart-city-solutions/

Subashini S, Kavitha V (2011) A survey on security issues in service delivery models of cloud computing. J Netw Comput Appl 34(1):1–11

Sun X, Ansari N (2016) EdgeIoT: mobile edge computing for the Internet of Things. IEEE Commun Mag 54(12):22–29

Talukdar AK (2010) Mobile computing. 2nd edn. Tata McGraw-Hill Education

Tong L, Li Y, Gao W (2016) A hierarchical edge cloud architecture for mobile computing. In: IEEE INFOCOM 2016-The 35th annual IEEE international conference on computer communications, pp 1–9

Tran TX, Pandey P, Hajisami A, Pompili D (2017) Collaborative multi-bitrate video caching and processing in mobile-edge computing networks. In: 2017 13th annual conference on wireless on-demand network systems and services (WONS), pp 165–172

Tran TX, Hajisami A, Pandey P, Pompili D (2017) Collaborative mobile edge computing in 5G networks: new paradigms, scenarios, and challenges. IEEE Commun Mag 55(4):54–61

Tsahalis J, Tsahalis HT, Moussas VC (2013) Optimization of a heterogeneous simulations workflow. In: 5th international conference on experiments/process / system modelling/simulation / optimization (5th IC-EpsMsO)

Van Weverberg K, De Ridder K, Van Rompaey A (2008) Modeling the contribution of the Brussels heat island to a long temperature time series. J Appl Meteorol Climatol 47(4):976–990
Vlahogianni EI, Kepaptsoglou K, Tsetsos V, Karlaftis MG (2016) A real-time parking prediction system for smart cities. J Intell Transport Syst 20(2):192–204
Voogt JA, Oke TR (2003) Thermal remote sensing of urban climates. Remote Sens Environ 86(3):370–384
Wang H, Gong J, Zhuang Y, Shen H, Lach J (2017) Healthedge: Task scheduling for edge computing with health emergency and human behavior consideration in smart homes. In: 2017 IEEE international conference on Big Data, pp 1213–1222
Wang L, Jie W, Chen J (2018) Grid computing: infrastructure, service, and applications. CRC Press
Washburn D, Sindhu U, Balaouras S, Dines RA, Hayes N, Nelson LE (2009) Helping CIOs understand "smart city" initiatives. Growth 17(2):1–17
What Is Edge Computing?|GE Digital. n.d. Accessed on 26 Mar 2019. https://www.ge.com/digital/blog/what-edge-computing
Wibowo S, Wells M (2018) Green cloud computing and economics of the cloud: moving to-wards sustainable future. GSTF J Comput (JoC) 5(1)
Wong J, Wang X, Li H, Chan G (2014) A review of cloud-based BIM technology in the construction sector. J Inf Technol Constr 19:281–291
Xing J, Dai H, Yu Z (2018) A distributed multi-level model with dynamic replacement for the storage of smart edge computing. J Syst Architect 83:1–11
Xiong J, Thenkabail PS, Gumma MK, Teluguntla P, Poehnelt J, Congalton RG, Yadav K, Thau D (2017) Automated cropland mapping of continental Africa using Google Earth Engine cloud computing. ISPRS J Photogram Remote Sens 126:225–244
Yang C, Huang Q (2013) Spatial cloud computing: a practical approach. CRC Press, Baton Rouge, FL
Yang C, Raskin R (2009) Introduction to distributed geographic information processing re-search. Int J Geogr Inf Sci 23(5):553–560
Yang C, Li W, Xie J, Zhou B (2008) Distributed geospatial information processing: sharing distributed geospatial resources to support Digital Earth. Int J Digital Earth 1(3):259–278
Yang C, Goodchild M, Huang Q, Nebert D, Raskin R, Xu Y, Bambacus M, Fay D (2011) Spatial cloud computing: how can the geospatial sciences use and help shape cloud computing? Int J Digital Earth 4(4):305–329
Yang X, Huang X, Liu JK (2016) Efficient handover authentication with user anonymity and untraceability for mobile cloud computing. Future Gener Comput Syst 62:190–195
Yang C, Huang Q, Li Z, Liu K, Hu F (2017a) Big Data and cloud computing: innovation opportunities and challenges. Int J Digital Earth 10(1):13–53
Yang C, Yu M, Hu F, Jiang Y, Li Y (2017b) Utilizing cloud computing to address big geospatial data challenges. Comput Environ Urban Syst 61:120–128
Yannuzzi M, Milito R, Serral-Gracià R, Montero D, Nemirovsky M (2014) Key ingredients in an IoT recipe: fog computing, cloud computing, and more fog computing. In: 2014 IEEE 19th international workshop on Computer Aided Modeling and Design of Communication Links and Networks (CAMAD), pp 325–329
Yao Y, Peng Z, Xiao B, Guan J (2017) An efficient learning-based approach to multi-objective route planning in a smart city. In: 2017 IEEE International Conference on Communications (ICC), pp 1–6
Yi S, Qin Z, Li Q (2015) Security and privacy issues of fog computing: a survey. International conference on wireless algorithms, systems, and applications. Springer, Cham, pp 685–695
Yi S, Hao Z, Zhang Q, Zhang Q, Shi W, Li Q (2017) Lavea: latency-aware video analytics on edge computing platform. In: Proceedings of the second ACM/IEEE symposium on edge computing, p 15
Yu Z, Yao Y, Yang G, Wang X, Vejre H (2019) Spatiotemporal patterns and characteristics of remotely sensed region heat islands during the rapid urbanization (1995–2015) of Southern China. Sci Total Environ 674:242–254

Zhang H, Ye C (2016) An indoor navigation aid for the visually impaired. In: 2016 IEEE international conference on robotics and biomimetics (ROBIO), pp 467–472
Zhang X, Zhong T, Feng X, Wang K (2009) Estimation of the relationship between vegetation patches and urban land surface temperature with remote sensing. Int J Remote Sens 30(8):2105–2118
Zhang Q, Cheng L, Boutaba R (2010) Cloud computing: state-of-the-art and research challenges. J Internet Serv Appl 1(1):7–18
Zheng P, Ni L (2010) Smart phone and next generation mobile computing. Elsevier
Zhong S, Qian Y, Zhao C, Leung R, Wang H, Yang B, Fan J, Yan H, Yang XQ, Liu D (2017) Urbanization-induced urban heat island and aerosol effects on climate extremes in the Yangtze River Delta region of China. Atmos Chem Phys 17(8):5439–5457
Zhou J, Cao Z, Dong X, Vasilakos AV (2017) Security and privacy for cloud-based IoT: challenges. IEEE Commun Mag 55(1):26–33

Qian Liu is a Ph.D. candidate majoring in Geography and Geoinformation Science (GGS) at George Mason University (GMU). She serves as a graduate research assistant in the National Science Foundation (NSF) Spatiotemporal Innovation Center. Her research mainly focuses on geographical events detection and segmentation, machine learning applications in natural phenomena, climate data downscaling, global precipitation climatology analysis, remote sensing, and geographical data fusion.

Juan Gu is Senior Engineer of Geographical Information Science at Beijing Institute of Surveying and Mapping. She is interested in building smart cities using cut-edge geospatial technologies.

Jingchao Yang is a Ph.D. candidate majoring in Geoinformation Sciences at GMU. He has worked on several NSF and NASA funded projects as a Research Assistant for the NSF Spatiotemporal Innovation Center. He is currently applying the IoT dataset and machine learning algorithms to build a temperature forecasting model in urban areas.

Yun Li is a Ph.D. candidate majoring in Earth Systems and Geoinformation Sciences (ESGS) at GMU. Her research mainly focuses on improving geospatial data discovery using machine learning-based methods, high-performance computing, and outreaches to spatiotemporal analytics for environmental and climate data.

Dexuan Sha is a Ph.D. candidate majoring in ESGS at GMU. He serves as a graduate research assistant in the NSF Spatiotemporal Innovation Center. His research mainly focuses on cyberinfrastructure and big data platforms, high spatial resolution remote sensing, spatiotemporal computing, and knowledge graphing.

Mengchao Xu obtained his Ph.D. from the GGS Department at GMU. His research mainly focuses on cloud computing, high-performance computing networks, spatial database systems, and precipitation downscaling. He is a GIS data engineer for autonomous driving systems.

Ishan Shams is a Ph.D. student majoring in ESGS at GMU. His research mainly focuses on high-performance computing and spatiotemporal platform visualization.

Manzhu Yu is Assistant Professor of GIScience at the Department of Geography, Pennsylvania State University. She received her bachelor's degree in Remote Sensing from Wuhan University in 2012 and doctoral degree in Earth System and Geoinformation Science from George Mason University in 2017. Her research focuses on spatiotemporal theories and applications, atmospheric modeling, environmental analytics, big data and cloud computing, and the capability to use the above to solve pressing issues in natural hazards and sustainability.

Chaowei Yang is Professor of Geographical Information Science at George Mason University, where he founded and directs the Center for Intelligent Spatial Computing and the NSF Spatiotemporal Innovation Center. He is interested in analyzing, learning, mining, and identifying spatiotemporal patterns and principles to enable scientific discovery and engineering development.

Chapter 42
Data Mining and Knowledge Discovery

Chao Zhang and Jiawei Han

Abstract Our physical world is being projected into online cyberspace at an unprecedented rate. People nowadays visit different places and leave behind them million-scale digital traces such as tweets, check-ins, Yelp reviews, and Uber trajectories. Such digital data are a result of social sensing: namely people act as human sensors that probe different places in the physical world and share their activities online. The availability of massive social-sensing data provides a unique opportunity for understanding urban space in a data-driven manner and improving many urban computing applications, ranging from urban planning and traffic scheduling to disaster control and trip planning. In this chapter, we present recent developments in data-mining techniques for urban activity modeling, a fundamental task for extracting useful urban knowledge from social-sensing data. We first describe traditional approaches to urban activity modeling, including pattern discovery methods and statistical models. Then, we present the latest developments in multimodal embedding techniques for this task, which learns vector representations for different modalities to model people's spatiotemporal activities. We study the empirical performance of these methods and demonstrate how data-mining techniques can be successfully applied to social-sensing data to extract actionable knowledge and facilitate downstream applications.

42.1 Overview

Our physical world is being projected into cyberspace at an unprecedented rate. People nowadays visit different places and leave behind them million-scale digital traces such as tweets, check-ins, Yelp reviews, and Uber trajectories. The malls they

C. Zhang (✉)
School of Computational Science and Engineering, Georgia Institute of Technology, Atlanta, USA
e-mail: chaozhang@gatech.edu

J. Han
University of Illinois at Urbana-Champaign, Urbana, USA
e-mail: hanj@illinois.edu

W. Shi et al. (eds.), *Urban Informatics*, The Urban Book Series,
https://doi.org/10.1007/978-981-15-8983-6_42

go to, the restaurants they visit, the movies they watch, the concerts they attend—almost everything people do during a day can now result in rich cybertraces. For example, Foursquare has collected more than 8 billion check-ins as of today, Twitter has more than 10 million geo-tagged tweets published every day, and Instagram witnesses more than 20 million geo-tagged photos being shared every day. Such digital data represent a result of social sensing: people act as human sensors to probe different places in the physical world and leave online traces of their spatiotemporal activities.

The availability of massive online social-sensing data provides an unprecedented opportunity for modeling people's offline spatiotemporal activities. While traditional approaches to urban activity modeling often require costly surveys and field studies, the understanding is often coarse-grained and limited. In contrast, social-sensing data provide a fine-grained coverage of our physical world (Leetaru et al. 2013) and serve as a unique proxy for human activities (Cheng et al. 2011; Jurdak et al. 2015; Noulas et al. 2011). For the first time, it becomes possible to develop data-driven techniques for modeling people's spatiotemporal activities, which can potentially revolutionize many applications, including urban planning, traffic scheduling, disaster control, and trip planning.

Social-sensing data often comprise modalities (e.g., location, time, and text) that can have totally different representations and distributions. When using massive social-sensing data for spatiotemporal activity modeling, the key is to capture the correlations of these data modalities and make predictions across them. For a subset of the modalities (Fig. 42.1), the model is expected to predict the remaining ones. For example: (1) Given a location and time, what are the typical activities around that location and time? (2) Given an activity and time, where does this activity usually occur? and (3) Given an activity and a location, when does the activity usually occur?

In the remainder of this chapter, we first summarize key data-mining methods for urban analysis tasks (Sect. 42.2). Generally, these methods fall into four broad categories: (1) urban pattern discovery; (2) urban activity models; (3) urban mobility models; and (2) urban event detection. We will describe techniques in each category.

In addition to overviewing how data-mining techniques can address urban-analysis tasks, we introduce the latest development of urban activity modeling techniques based on multimodal embedding (Sect. 42.3). At a high level, multi-modal embedding directly captures cross-modal correlations by mapping items from

Fig. 42.1 An illustration of spatiotemporal activity modeling using social-sensing data

different modalities into the same latent space. If two elements are correlated (e.g., the JFK airport region and the keyword 'flight'), their latent representations are encouraged to be close to each other. Compared with existing generative models, multimodal embedding does not impose any distributional assumptions and incurs much lower computational cost in the learning process. We show the performance of the multimodal embedding method and demonstrate its superiority for urban activity modeling.

42.2 Data Mining for Urban Analysis

Generally, data-mining techniques for urban analysis tasks can be categorized into four classes: (1) urban pattern discovery; (2) urban activity modeling; (3) urban mobility modeling; and (4) urban event detection. In the following, we overview these tasks and describe key techniques for each task.

42.2.1 Urban Pattern Discovery

Urban pattern discovery aims to discover various forms of spatiotemporal patterns from social-sensing data. Sequential pattern is an important type of spatiotemporal pattern which captures sequential transition regularities of people's activities. Giannotti et al. (2007) defined a T-pattern as a region-of-interest sequence that appears frequently in the input trajectories. By partitioning the space, they used sequential pattern-mining techniques to extract the T-patterns. Zhang et al. (2014) extracted frequent movement patterns from semantic trajectory data. With a top-down approach, they first discovered coarse-grained sequential patterns, and then partitioned them into fine-grained sequential patterns by clustering pattern-matching snippets. Several studies have investigated how to find objects that frequently move together. Examples in this line include mining flock (Laube and Imfeld 2002), swarm (Li et al. 2010a), and gathering (Zheng et al. 2013) patterns.

Periodic patterns represent user behaviors that regularly occur with one or multiple time periods. To extract periodic patterns, Li et al. (2010b) first extracted reference spots by using density-based clustering, and then detected periodic patterns at those spots. They have also studied how to find periodic patterns from sequences with incomplete observations (Li 2012b). The idea is to partition the time series into small chunks and then overlay them for each candidate period. Cho et al. (2011) found that the mobility of each user usually centers around several regions. Based on this observation, they proposed a periodic mobility model that predicts a user's location by estimating the regions where a user most likely stays. Following this paper, Tarasov et al. (2013) modeled a region based on radiation models (Simini et al. 2012).

42.2.2 Urban Activity Modeling

Urban activity modeling aims to use statistical models to describe people's activity regularities and learn such models from data. There are two subcategories along this line: global activity models and personalized activity models.

Global activity models aim at characterizing people's activities over space and time at the global level without distinguishing personal preferences. Most existing techniques (Hong et al. 2012; Kling et al. 2014; Mei et al. 2006; Sizov 2010; Wang et al. 2007; Yin et al. 2011; Yuan et al. 2013) are latent variable models, which extend the classic topic models (Blei et al. 2003; Hofmann 1999) to handle spatiotemporal contexts. For example, Sizov (2010) extended LDA (Blei et al. 2003) by assuming that each latent topic was characterized by a multinomial distribution over text as well as two Gaussian distributions over latitudes and longitudes. Later, they further extended the model to discover topics that have non-Gaussian distributions (Kling et al. 2014). Yin et al. (2011) extended the PLSA model (Hofmann 1999) by modeling each region with a Gaussian distribution for location generation and a multinomial distribution for text generation.

In contrast, personalized activity models aim at describing spatiotemporal activities at an individual level. Hong et al. (2012) and Yuan et al. (2013) proposed to model the user factor in geographic topic models. In this way, users' individual-level preferences can be inferred. Yuan et al. (2017) later proposed a Bayesian non-parametric model, which can automatically discover the regions a user visits periodically.

42.2.3 Urban Mobility Modeling

The task of human mobility modeling is a corner-stone task for various applications, including urban planning, traffic scheduling, location prediction, and personalized recommendation. In the past years, this task has attracted much research attention from the data-mining community.

The first line of human mobility modeling is law-based methods. Such methods study the physical laws that govern human mobility. Brockmann et al. (2006) discovered that human mobility can be approximated by a continuous random-walk model with long-tail distributions. Gonzalez et al. (2008) used mobile phone data for human mobility modeling. They found that people return to a few locations periodically, and such mobility can be modeled by a stochastic process centered on a fixed point. Song et al. (2010) found that more than 93% of human movements are predictable, because of the high regularity of human mobility. They thus proposed a self-consistent microscopic model for individual mobility prediction.

Along another line, many model-based approaches have been explored to learn statistical models from human movement data. For example, Cho et al. (2011) found that a user usually moves around a few center locations (e.g., home, work) in fixed time periods. Based on this observation, they proposed to model user movement as a mixture of Gaussian distributions. Their model can be further extended by incorporating social influence, as a user is more likely to visit a location that is close to the locations of friends. Wang et al. (2015) proposed a hybrid mobility model, which improved location prediction by using heterogeneous mobility data.

One important area along the line of mode-based approaches is the hidden Markov model (HMM), which is a powerful statistical model for sequential data. In early work, Mathew et al. (2012) first partitioned the space into equally sized triangles using a hierarchical triangular mesh. Based on the assumption that each latent state imposes a multinomial distribution over the triangles, they trained an HMM for the input trajectories. Deb and Basu (2015) proposed a probabilistic latent semantic model. This model uses HMM to extract latent semantic locations from cell-tower and Bluetooth data. Ye et al. (2013) have explored how to use HMM to model user check-in data generated from location-based social networks (LBSNs). Their HMM model can incorporate the category information of places and thereby is capable of predicting the category for the user's next location. Zhang et al. (2016a) have applied HMMs to model people's sequential behaviors. The key idea of their model is that there are a few latent states underlying people's daily activities and that people typically move among these states with strong regularity. Instead of using one model for all the users, they proposed to group users based on their sequential patterns and learn a set of HMMs to characterize group-level activities.

42.2.4 Urban Event Detection

An urban event, such as a protest or a disaster, is an unusual activity occurring in a local area and having a specific time duration, while engaging a considerable number of participants. Detecting urban events in real time was nearly impossible years ago because of the lack of timely and reliable data. However, the recent availability of social-sensing data sheds light on this problem.

Many studies have explored how to detect urban events, which are also termed spatiotemporal events, from social-sensing data (Abdelhaq et al. 2013; Chen and Roy 2009; Feng et al. 2015; Lee et al. 2011; Sakaki et al. 2010; Zhang et al. 2016b). Existing techniques for identifying abnormal events can be categorized into document-based approaches and feature-based approaches. Document-based approaches consider documents as basic units and group similar documents to detect abnormal events. For example, Allan et al. (1998) performed single-pass clustering of the document stream and used a similarity threshold to determine whether a new document is a new topic or should be merged into an existing topic. Aggarwal and Subbian (2012) also proposed to detect events by clustering the tweet stream. However, their similarity measure jointly considers tweet content relevance and user social proximity. Zhang et al. (2016b) first detected geo-topic clusters as candidate events and then employed a z-score to identify abnormal clusters as true events.

The second line of event detection has adopted feature-based approaches (Fung et al. 2005; He et al. 2007; Li et al. 2012a; Mathioudakis and Koudas 2010; Weng and Lee 2011). The idea is to identify a set of bursty features (e.g., keywords or phrases) from the text stream and then cluster them into events. Specifically, Fung et al. (2005) modeled feature occurrences using a binomial distribution to extract bursty features. He et al. (2007) constructed the stream for each feature and then performed a Fourier transform to identify bursty events. Krumm and Horvitz (2015) monitored the spatiotemporal distributions of tweets and identified spikes in the spatiotemporal signal as abnormal events. There has also been work on detecting specific types of events. Sakaki et al. (2010) investigated real-time earthquake detection. They trained a classifier to judge whether a tweet was earthquake-related or not and then proposed to release an alarm whenever the number of earthquake-related tweets was large. Li et al. (2012a) detected crime and disaster events using a self-adaptive crawler, which can dynamically retrieve crime and disaster-related tweets. Abdelhaq et al. (2013) proposed the EvenTweet model, which could detect local events with the following steps: (1) examine several previous windows to identify bursty words; (2) compute the spatial entropy of each bursty word and discover localized words; (3) group localized words into clusters based on their spatial distributions; and (4) rank the resultant clusters based on event-indicative features such as burstiness and spatial coverage.

42.3 Multimodal Embedding for Urban Activity Modeling

We now describe the latest development of multimodal embedding techniques for urban activity modeling. Different from existing latent variable models that rely on latent states to bridge different modalities indirectly, such embedding-based methods can capture the cross-modal correlations directly. This is achieved by mapping all the modalities into a common vector space. In the following, we first describe the high-level idea (Sect. 42.3.1), then detail the multimodal embedding method for activity modeling (Sect. 42.3.2), and finally present the optimization process (Sect. 42.3.3).

42.3.1 Method Overview

At a high level, our embedding-based method, named CrossMap (Zhang et al. 2017a), maps items from different modalities into the same latent space with their correlations preserved, as shown in Fig. 42.2. Formally, it aims to learn the embeddings L, T, and W where: (1) L is the embeddings for regions; (2) T is the embeddings for hours; and (3) W is the embeddings for keywords. Take L as an example. Each element is a D-dimensional ($D > 0$) vector, which represents the embedding for region l. Once the embeddings are learned, cross-modal predictions can be made by simply searching for items nearest to the given query in the latent space.

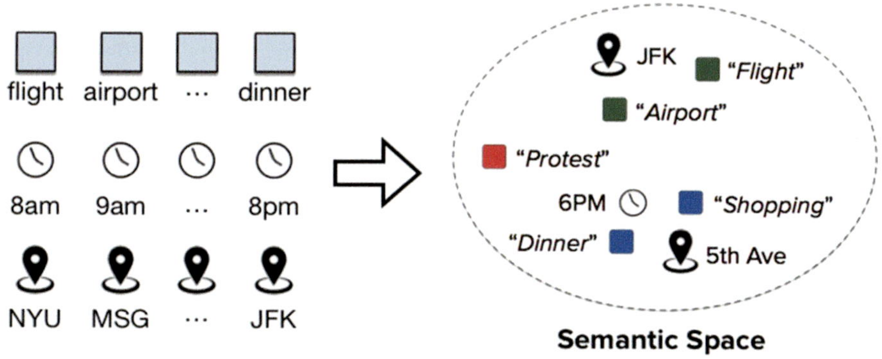

Fig. 42.2 An illustration of multimodal embedding for urban activity modeling. The idea is to map items from different modalities (e.g., location, time, text) into the same latent vector space to preserve their correlations. Their latent representations are then used for cross-modal prediction

42.3.2 Multimodal Embedding via Attribute Reconstruction

The key principle for multimodal embedding is to optimize the embeddings L, T, W such that the observed relationships among location, time, and text can be reconstructed. We thus define an unsupervised attribute reconstruction task. The goal is to learn the embeddings L, T, W such that the attributes of a record r can be reconstructed by assuming that the other attributes are observed.

Let r be a record. Given any attribute $i \in r$ with type X (could be location, time, or keyword), we compute the likelihood of observing attribute i as follows:

$$p(i|r_{-i}) = \exp(s(i, r_{-i})/\sum_{j \in X} \exp(s(j, r_{-i}))$$

where r_{-i} represents the set of all the attributes in r except for i, and $s(i, r_{-i})$ denotes the similarity between i and r_{-i}.

The key question for the above is how to define $s(i, r_{-i})$. A straightforward idea is to average the embeddings of all the attributes in r_{-i} and then compute $s(i, r_{-i})$ as $s(i, r_{-i}) = \mathbf{v}_i T \sum_{j \in r_{-i}} \mathbf{v}_j / |r_{-i}|$, where $\mathbf{v}_i$ denotes the embedding for attribute i. However, this simple definition fails to consider spatial and temporal continuities. Consider the spatial continuity as an example. According to the first law of geography, "everything is related to everything else, but near things are more related than distant things." To achieve spatial smoothness, two spatial items that are close to each other should be considered correlated instead of independent. We thus introduce spatial smoothing and temporal smoothing to capture the spatiotemporal continuities. With the smoothing technique, the method can not only maintain local consistency of neighboring regions and periods, but also alleviate data sparsity. One can refer to Zhang et al. (2017b) for more details about the smoothing techniques.

In addition to the above pseudo-region and period embeddings, we also introduce pseudo-keyword embeddings for notational ease. Given r_{-i}, its pseudo-keyword embedding is defined as:

$$\mathbf{v}_{\hat{\mathbf{W}}} = \sum_{w \in N_w} \mathbf{v}_w / |N_w|$$

where N_w is the set of keywords in r_{-i}. With these pseudo-embeddings, we define a smoothed version of $s(i, r_{-i})$ as $s(i, r_{-i}) = \mathbf{v}_i T \mathbf{h}_i$ where if i is a keyword then:

$$\mathbf{h}_i = (\mathbf{v}_l + \mathbf{v}_t + \mathbf{v}_{\hat{\mathbf{W}}})/3$$

If i is a region then:

$$\mathbf{h}_i = (\mathbf{v}_t + \mathbf{v}_{\hat{\mathbf{W}}})/2$$

If i is a period, then:

$$\mathbf{h}_i = (\mathbf{v}_l + \mathbf{v}_{\hat{\mathbf{W}}})/2$$

Let R_{U} be a collection of records for learning the urban activity model. The final loss function for the attribute reconstruction task is simply the negative log-likelihood of observing all the attributes of the records in R_{U}:

$$J_{R_{\mathrm{U}}} = -\sum_{r \in R_{\mathrm{U}}} \sum_{i \in r} \log p(i | r_{-i}) \tag{42.1}$$

42.3.3 The Optimization Procedure

To efficiently learn the embeddings, we can use stochastic gradient descent (SGD) and negative sampling (Mikolov et al. 2013) for optimizing the objective function shown in Eq. (42.1). At each step, we can use SGD to sample a record r and an attribute $i \in r$. Based on negative sampling, we then randomly select K negative attributes that have the same type as i but do not appear in r. Then the loss function for the selected samples becomes:

$$J_r = \log \sigma(s(i, r_{-i})) - \sum_{k=1}^{K} \log \sigma(-s(k, r_{-i}))$$

In the above, $\sigma(\cdot)$ is the sigmoid function. The updating rules for $\mathbf{v}_i$,$\mathbf{v}_k$, and $\mathbf{h}_i$ can be obtained by taking the derivatives of J_r. We omit the details because of the space limit.

42.4 Experiments

We now demonstrate the empirical performance of different algorithms on three real-life datasets:

- The first dataset, called LA, contains ~1.10 million geo-tagged tweets published in Los Angeles. We crawled the LA dataset by monitoring the Twitter Streaming API during 2014.08.01–2014.11.30 and continuously gathering the geo-tagged tweets in the bounding box of LA. We preprocessed the raw data as follows. For the text part, user mentions, URLs, stopwords, and the words that appear less than 100 times were removed. For space and time, we partitioned the LA area into small grids with size 300 m * 300 m and broke the one-day period into 24 one-hour windows.
- The second dataset, called NY, was also collected from Twitter. It consisted of ~1.20 million geo-tagged tweets published in New York City during the time period 2014.08.01–2014.11.30.
- The third dataset was called 4SQ. It was collected from Foursquare. It consisted of about 0.7 million Foursquare check-ins posted in New York City, during the time period 2010.08–2011.10. This dataset was mainly used to evaluate the performance of the multi-modal embedding method for the downstream task of activity classification. Similarly, user mentions, URLs, stopwords, and the words that appeared less than 100 times were removed.

We study the following methods for urban activity modeling: (1) the geographic topic model LGTA (Yin et al. 2011); (2) the non-Gaussian geographic topic model MGTM (Kling et al. 2014); (3) the tensor factorization method Tensor (Harshman 1970); (4) the SVD method, which first constructs the co-occurrence matrices between each pair of location, time, text, and category, and then performs singular-value decomposition on the matrices; (5) the TF-IDF method, which constructs the co-occurrence matrices between each pair of location, time, text, and category and then computes the TF-IDF weight for each entry in the matrix; (6) the multimodal embedding method CrossMap (Zhang et al. 2017a) as discussed in the previous section.

We investigated two types of urban activity prediction tasks. The first was to predict locations for a given textual query. Specifically, recall that each record reflects a user's activity with the following three attributes: a location, a timestamp, and a bag of keywords. In the location-prediction task, the input was the timestamp and the keywords, and the goal was to accurately pinpoint the ground-truth location from a pool of candidates. We predicted the location at two different granularities: (1) coarse-grained region prediction of the ground-truth region that r falls in; and (2) fine-grained POI prediction of the ground-truth POI that r corresponds to. Note that fine-grained POI prediction was only evaluated on the tweets that had been linked with Foursquare. The second task was to predict activities for a given location query. In this task, the input was the timestamp and the location, and the goal was to pinpoint the ground-truth activities at two different granularities: (1) coarse-grained category prediction of the ground-truth activity category of r (again, such a coarse-grained activity prediction was performed only on the tweets that had been linked with Foursquare); and (2) fine-grained keyword prediction of the ground-truth message from a candidate pool of messages.

To summarize, we studied four urban activity prediction subtasks in total: (1) region prediction; (2) POI prediction; (3) category prediction; and (4) keyword prediction. For each prediction subtask, we first generated a candidate pool by mixing the ground truth with a set of M random negative samples. Take region prediction as a concrete example. For the ground-truth region, we mixed with M randomly chosen regions. Then, we tried to pinpoint the ground truth from the size-($M + 1$) candidate pool by ranking all the candidates. Generally, the better a model captures the patterns underlying people's activities, the more likely it can rank the ground truth for top positions. We thus used mean reciprocal rank (MRR) to quantify the effectiveness of a model.

Tables 42.1 and 42.2 report the quantitative results of different methods for location and activity predictions, respectively. As shown, on all of the four subtasks, CrossMap and its variants achieved much higher MRRs than the baseline methods. Compared with the two geographic topic models (LGTA and MGTM), CrossMap showed as much as 62% performance improvement for location prediction, and 83% for activity prediction. Tensor, SVD, and TF-IDF had better performance than LGTA and MGTM by modeling time and category, yet CrossMap outperformed them by large margins. Interestingly, TF-IDF turned out to be a strong baseline, demonstrating the effectiveness of the tf-idf similarity for the prediction tasks. SVD and Tensor can effectively recover the co-occurrence matrices and tensor, but the raw co-occurrence seems a less effective measure for location and activity prediction.

Table 42.1 MRRs of various methods for location prediction. For each test tweet, we assume its timestamp and keywords are observed, and perform location prediction at two granularities: (1) region prediction retrieves the ground-truth region; and (2) POI prediction retrieves the ground-truth POI (for Foursquare-linked tweets)

Method	Region prediction		POI prediction	
	LA	NY	LA	NY
LGTA	0.3583	0.3544	0.5889	0.5674
MGTM	0.4007	0.391	0.5811	0.553
Tensor	0.3592	0.3641	0.6672	0.7399
SVD	0.3699	0.3604	0.6705	0.7443
TF-IDF	0.4114	0.4605	0.719	0.776
CrossMap	0.5373	0.5597	0.7845	0.8508

Table 42.2 MRRs of different methods for activity prediction. For each test tweet, we assume its location and timestamp are observed, and predict activities at two granularities: (1) category prediction of ground-truth category (for Foursquare-linked tweets); and (2) keyword prediction retrieves the ground-truth message

Method	Category prediction		Keyword prediction	
	LA	NY	LA	NY
LGTA	0.4409	0.4527	0.3392	0.3425
MGTM	0.4587	0.4640	0.3501	0.3430
Tensor	0.8635	0.7988	0.4004	0.3744
SVD	0.8556	0.7826	0.4098	0.3728
TF-IDF	0.9137	0.8259	0.5236	0.4864
CrossMap	0.6225	0.5874	0.5693	0.5538

We now performed a set of case studies to examine how well CrossMap predicted across modalities. Specifically, we performed one-pass training of CrossMap for LA and NY, and launched a bunch of queries at different stages. For each query, we retrieved the top-ten most similar items with different types from the entire search space.

Figure 42.3a shows the results when we queried with the keyword 'beach'. As shown, the retrieved items in each type are very meaningful: the top locations mostly fall around famous beaches in the Los Angeles area; the top keywords can well reflect people's activities on the beach, including 'sand' and 'boardwalk.' Fig. 42.3b shows the results for an example spatial query, at the GPS location of the centroid of LAX airport. One can see that the retrieved top spatial, temporal, and textual elements are closely related to the airport. Given the query at the airport, the top keywords are all concepts that reflect flight-related activities, such as 'airport,' 'tsa,' and 'airline.'

Figures 42.4a–c further show temporal-textual queries which can demonstrate the temporal dynamics of people's urban activities. When we fix the query keyword as 'restaurant' and vary the time point in the query, the retrieved top items vary obviously. By examining the top keywords, we can see the query '10am' results in many breakfast-related keywords, such as 'bfast' and 'brunch.' In contrast, when the query is changed to '2 pm,' many lunch-related keywords are retrieved. When '8 pm' is specified as the query, many dinner-related ones are retrieved. Another interesting observation is that the top locations for the queries '10am' and '2 pm' fall in working areas, while the results for '8 pm' distribute mostly in residential areas. Such results show that the time factor plays an important role in determining people's activities, and CrossMap captures such fine-grained temporal dynamics.

We proceeded to examine the performance of multimodal embedding models for downstream applications. For this purpose, we chose activity classification as an application. In the 4SQ dataset, every check-in belongs to one of nine categories: Food, College & University, Nightlife Spot, Shop & Service, Travel & Transport, Residence, Arts & Entertainment, Outdoors & Recreation, Professional & Other Places. We used those categories as the labels for people's urban activities and aimed

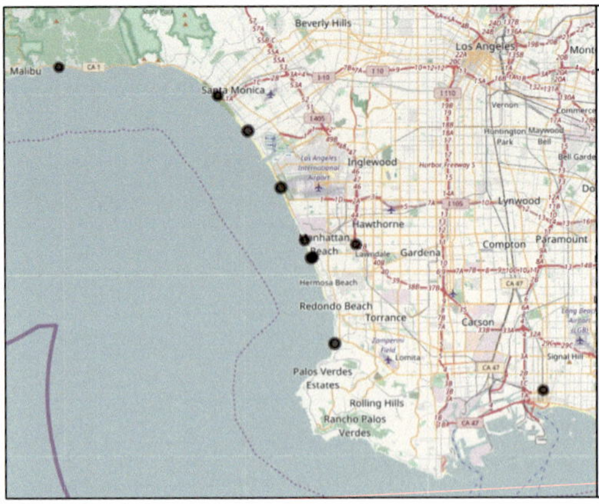

Text	Time
beach	19
beachday	18
beachlife	17
surfing	16
sand	20
boardwalk	14
pacificocean	15
longbeach	13
redondobeach	11
dockweiler	12

(a) Query = 'beach'

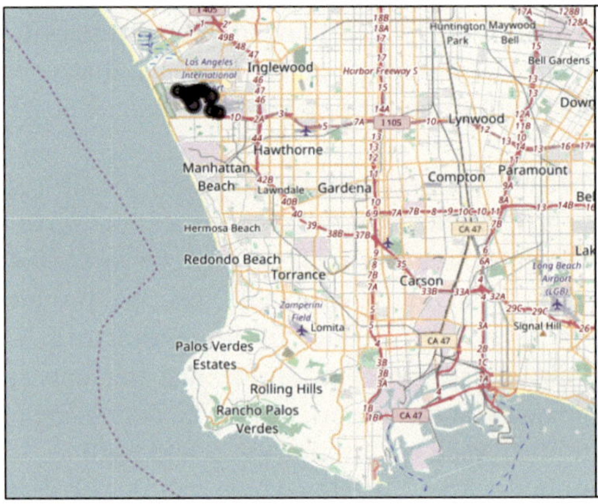

Text	Time
airport	7
tsa	10
airline	8
lax	6
southwester	11
americanair	9
delay	5
terminal	12
jfk	16
sfo	14

(b) Query = '(33.9424, -118.4137)' (LAX airport centroid)

Fig. 42.3 Two example queries and the top-ten results returned by CrossMap

to learn classifiers that can predict those labels for any given check-in. We performed a random shuffling of the dataset, and then randomly chose 80% for training and 20% for testing. For any check-in r, all the studied methods can obtain vector representations for the location, time, and text; we concatenated the vectors as the feature representation of a check-in.

With the above feature transformation, we then trained a multiclass logistic regression for activity classification. Figure 42.5 reports the performance of different methods for the activity classification task. As shown, CrossMap outperformed the other methods significantly. Using the simple linear classification model, the F1 score of the method can reach as high as 0.843. Such results show that the embeddings obtained by multimodal embedding can well distinguish the semantics of different categories. We further verified this fact using data visualization. As shown

Text	Time
restaurant	10
bfast	7
pastry	6
brunching	8
deli	9
brunch	5
yummm	11
bakery	12
thai	14
foodporn	16

(a) Query = ‘restaurant’ + ‘10am’

Text	Time
restaurant	14
lunch	15
seafood	13
deli	16
foodporn	17
vietnamese	12
lunchfood	7
instafood	6
dimsum	10
thai	8

(b) Query = ‘restaurant’ + ‘2pm’

Text	Time
restaurant	20
dinner	18
happyhour	19
seafood	17
bartender	16
thai	7
server	5
yummy	14
dating	20
mexican	15

(c) Query = ‘restaurant’ + ‘8pm’.

Fig. 42.4 Three temporal-textual queries and the top ten results returned by CrossMap

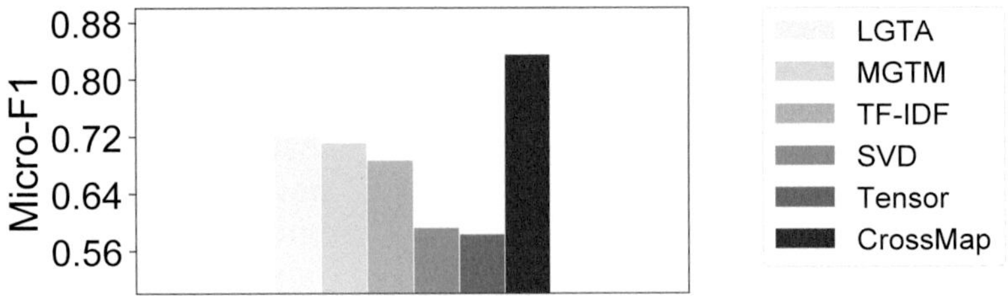

Fig. 42.5 Activity classification performance on 4SQ

in Fig. 42.6, we chose three categories and used the t-SNE method (Maaten and Hinton 2008) to visualize the feature vectors. One can observe that the learnt representations of the multimodal embedding method resulted in much clearer inter-class boundaries compared to the baselines such as geographic topic models.

42.5 Summary

We have presented data mining techniques for modeling people's urban activities from massive social-sensing data. We first overviewed data mining techniques for four important urban analysis tasks: (1) urban pattern discovery; (2) urban activity modeling; (3) urban mobility modeling; (4) urban event detection. Then, we presented the latest development of multimodal embedding techniques for urban activity modeling, which maps items from different data modalities into a common latent space with their correlation preserved. Compared with previous latent variable models, multimodal embedding techniques do not impose distribution assumptions of people's spatiotemporal activities, and scale well with the data size. We have studied the empirical performance of these methods on real datasets, and demonstrated that these techniques can enable the building of predictive urban activity models and can benefit downstream tasks like activity classification.

42.6 Future Directions

In the future, social-sensing data will continue to serve as an invaluable source for urban analysis. Data-mining techniques have already shown promising results when acquiring insights from social-sensing data for various tasks. However, there are still challenges that need to be addressed to fully unleash the power of social-sensing data. Below, we list several key challenges in this direction.

Integrating diverse data modalities. Modern social-sensing data often involve multiple modalities, such as text, image, location, and time. Considering the totally different representations of those data modalities and the complicated correlations

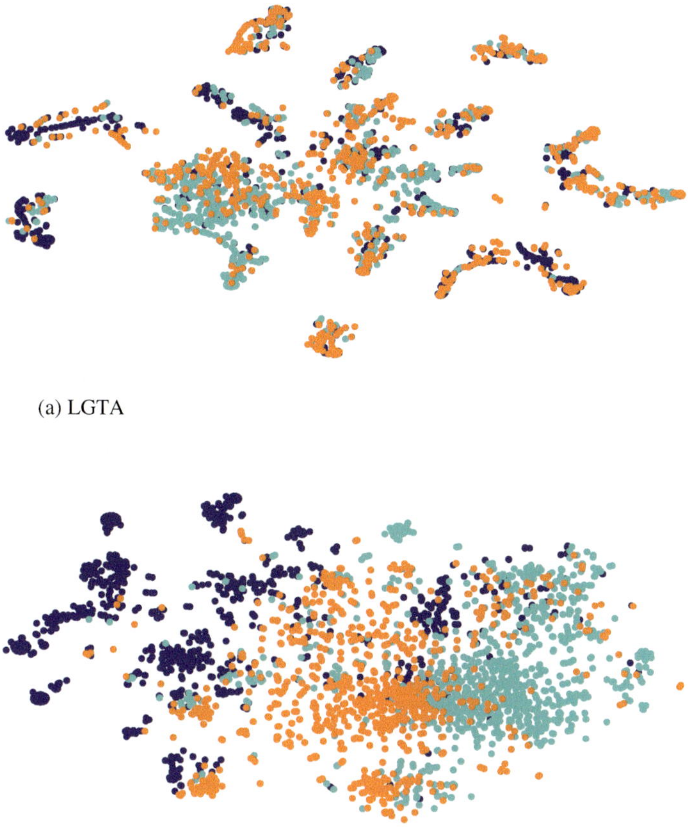

(a) LGTA

(b) CrossMap

Fig. 42.6 Visualizing the feature vectors generated by LGTA and CrossMap for three activity categories: 'Food' (cyan), 'Travel & Transport' (blue), and 'Residence' (orange). The feature vector of each 4SQ is mapped to a 2D point with t-SNE (Maaten and Hinton 2008)

among them, how to effectively integrate them for urban activity modeling and prediction remains a challenging problem.

Extracting insights from noisy data. Studies have shown that about 40% social-sensing data are pointless babbles. Even among those informative posts, most are rather short and noisy. It is nontrivial to analyze such noisy and short text messages and distill the information for end tasks.

Real-time data analysis. Many urban-analysis tasks require real-time performance. For instance, when an emergent event happens, it is important to report the event as soon as possible to allow for timely actions. As massive social-sensing data stream in, it is an important yet challenging problem to design on-line learning algorithms that can handle large-scale streaming data efficiently.

References

Abdelhaq H, Sengstock C, Gertz M (2013) Eventweet: online localized event detection from twitter. Proc VLDB Endowment 6(12):1326–1329

Aggarwal CC, Subbian K (2012) Event detection in social streams. In: SDM, SIAM international conference on data mining, pp 624–635

Allan J, Papka R, Lavrenko V (1998) On-line new event detection and tracking. In: SIGIR, international conference on information retrieval, pp 37–45

Blei DM, Ng AY, Jordan MI (2003) Latent Dirichlet allocation. J Mach Learn Res 3:993–1022

Brockmann D, Hufnagel L, Geisel T (2006) The scaling laws of human travel. Nature 439(7075):462–465

Chen L, Roy A (2009) Event detection from flickr data through wavelet-based spatial analysis. In: CIKM, international conference on knowledge management, pp 523–532

Cheng Z, Caverlee J, Lee K, Sui DZ (2011) Exploring millions of footprints in location sharing services. In: ICWSM, international conference on weblogs and social media, pp 81–88

Cho E, Myers SA, Leskovec J (2011) Friendship and mobility: user movement in location-based social networks. In: KDD, conference on knowledge discovery and data mining, pp 1082–1090

Deb B, Basu P (2015) Discovering latent semantic structure in human mobility traces. In: Wireless sensor networks. Springer, pp 84–103

Feng W, Zhang C, Zhang W, Han J, Wang J, Aggarwal C, Huang J (2015) Streamcube: hierarchical spatio-temporal hashtag clustering for event exploration over the twitter stream. IEEE international confeerence on data engineering

Fung GPC, Yu JX, Yu PS, Lu H (2005) Parameter free bursty events detection in text streams. In: VLDB, international conference on very large DataBases, pp 181–192

Giannotti F, Nanni M, Pinelli F, Pedreschi D (2007) Trajectory pattern mining. In: KDD, conference on knowledge discovery and data mining, pp 330–339

Gonzalez MC, Hidalgo CA, Barabasi AL (2008) Understanding individual human mobility patterns. Nature 453(7196):779–782

Harshman RA (1970) Foundations of the PARAFAC procedure: models and conditions for an "explanatory" multi-modal factor analysis. UCLA Working Papers Phonetics 16 (1):84

He Q, Chang K, Lim EP (2007) Analyzing feature trajectories for event detection. In: SIGIR, international conference on information retrieval, pp 207–214

Hofmann T (1999) Probabilistic latent semantic indexing. In SIGIR, international conference on information retrieval, pp 50–57

Hong L, Ahmed A, Gurumurthy S, Smola AJ, Tsioutsiouliklis K (2012) Discovering geographical topics in the twitter stream. In: WWW, international World Wide Web conference, pp 769–778

Jurdak R, Zhao K, Liu J, AbouJaoude M, Cameron M, Newth D (2015) Understanding human mobility from twitter. PLoS ONE 10(7):e0131469

Kling C, Kunegis J, Sizov S, Staab S (2014) Detecting non-gaussian geographical topics in tagged photo collections. In: WSDM, conference on web search and data mining, pp 603–612

Krumm J, Horvitz E (2015) Eyewitness: Identifying local events via space-time signals in twitter feeds. In: SIGSPATIAL, conference on advances in geographic information systems

Laube P, Imfeld S (2002) Analyzing relative motion within groups of trackable moving point objects. In: GIScience, international symposium on geographic information science, pp 132–144

Lee R, Wakamiya S, Sumiya K (2011) Discovery of unusual regional social activities using geo-tagged microblogs. World Wide Web 14(4):321–349

Leetaru K, Wang S, Cao G, Padmanabhan A, Shook E (2013) Mapping the global twitter heartbeat: the geography of twitter. First Monday 18(5)

Li C, Sun A, Datta A (2012a) Twevent: segment-based event detection from tweets. In: CIKM, international conference on knowledge management, pp 155–164

Li R, Lei KH, Khadiwala R, Chang KC (2012c) Tedas: a twitter-based event detection and analysis system. In: ICDE, IEEE international conference on data engineering, pp 1273–1276

Li Z, Ding B, Han J, Kays R, Nye P (2010b) Mining periodic behaviors for moving objects. In: KDD, conference on knowledge discovery and data mining, pp 1099–1108
Li Z, Ding B, Han J, Kays R (2010a) Swarm: Mining relaxed temporal moving object clusters. Proc VLDB Endowment 3(1):723–734
Li Z, Wang J, Han J (2012b) Mining event periodicity from incomplete observations. In: KDD, conference on knowledge discovery and data mining, pp 444–452
Maaten L, Hinton G (2008) Visualizing data using t-SNE. J Mach Learn Res 9(85):2579–2605
Mathew W, Raposo R, Martins B (2012) Predicting future locations with hidden Markov models. In: ACM international joint conference on pervasive and ubiquitous computing, pp 911–918
Mathioudakis M, Koudas N (2010) Twittermonitor: trend detection over the twitter stream. In: SIGMOD, international conference on management of data, pp 1155–1158
Mei Q, Liu C, Su H, Zhai C (2006) A probabilistic approach to spatiotemporal theme pattern mining on weblogs. In: WWW, international World Wide Web conference, pp 533–542
Mikolov T, Sutskever I, Chen K, Corrado GS, Dean J (2013) Distributed representations of words and phrases and their compositionality. In: NIPS, conference on neural information processing systems, pp 3111–3119
Noulas A, Scellato S, Mascolo C, Pontil M (2011) An empirical study of geographic user activity patterns in foursquare. In ICWSM, international conference on weblogs and social media, pp 570–573
Sakaki T, Okazaki M, Matsuo Y (2010) Earthquake shakes twitter users: real-time event detection by social sensors. In WWW, International World Wide Web Conference, pp. 851–860
Simini F, González MC, Maritan A, Barabási AL (2012) A universal model for mobility and migration patterns. Nature 484(7392):96–100
Sizov S (2010) Geofolk: latent spatial semantics in web 2.0 social media. In WSDM, conference on web search and data mining, pp. 281–290
Song C, Koren T, Wang P, Barabási A-L (2010) Modelling the scaling properties of human mobility. Nat Phys 6(10):818–823
Tarasov A, Kling F, Pozdnoukhov A (2013) Prediction of user location using the radiation model and social check-ins. In KDD workshop on urban computing, p 8
Wang C, Wang J, Xie X, Ma WY (2007) Mining geographic knowledge using location aware topic model. In GIR, AAMC Group on Information Resources, pp 65–70
Wang Y, Yuan NJ, Lian D, Xu L, Xie X, Chen E, Rui Y (2015) Regularity and conformity: Location prediction using heterogeneous mobility data. In KDD, conference on knowledge discovery and data mining, pp 1275–1284
Weng J, Lee BS (2011) Event detection in twitter. In ICWSM, international conference on weblogs and social media, pp 401–408
Ye J, Zhu Z, Cheng H (2013) What's your next move: User activity prediction in location-based social networks. In: SDM SIAM international conference on data mining
Yin Z, Cao L, Han J, Zhai C, Huang TS (2011) Geographical topic discovery and comparison. In: WWW, international World Wide Web conference, pp 247–256
Yuan Q, Cong G, Ma Z, Sun A, Thalmann NM (2013) Who, where, when and what: discover spatio-temporal topics for twitter users. In: KDD, conference on knowledge discovery and data mining, pp 605–613
Yuan Q, Zhang W, Zhang C, Geng X, Cong G, Han J (2017) Pred: Periodic region detection for mobility modeling of social media users. In: WSDM, conference on web search and data mining
Zhang C, Han J, Shou L, Lu J, La Porta T (2014) Splitter: mining fine-grained sequential patterns in semantic trajectories. Proc VLDB Endowment 7(9):769–780
Zhang C, Zhang K, Yuan Q, Peng H, Zheng Y, Hanratty T, Wang S, Han J (2017a) Regions, periods, activities: uncovering urban dynamics via cross-modal representation learning. In: WWW, international World Wide Web conference
Zhang C, Zhang K, Yuan Q, Tao K, Zhang L, Hanratty T, Han J (2017b) React: Online multimodal embedding for recency-aware spatiotemporal activity modeling. In: SIGIR, international conference on information retrieval

Zhang C, Zhang K, Yuan Q, Zhang L, Hanratty T, Han J (2016a) Gmove: Group-level mobility modeling using geo-tagged social media. In: KDD, conference on knowledge discovery and data mining

Zhang C, Zhou G, Yuan Q, Zhuang H, Zheng Y, Kaplan L, Wang S, Han J (2016b) Geoburst: Real-time local event detection in geo-tagged tweet streams. In: SIGIR, international conference on information retrieval

Zheng K, Zheng Y, Yuan NJ, Shang S (2013) On discovery of gathering patterns from trajectories. In: ICDE, IEEE international conference on data engineering, pp 242–253

Chao Zhang is an Assistant Professor at the School of Computational Science and Engineering, Georgia Institute of Technology. His research focuses on data mining and machine learning. He is a recipient of the Google Faculty Research Award (2020), the ACM SIGKDD Dissertation Award Runner-up (2019), the IMWUT Distinguished Paper Award (2018), the ECML/PKDD Best Student Paper Runner-up Award (2015), and the Chiang Chen Overseas Graduate Fellowship (2013).

Jiawei Han is Michael Aiken Chair Professor, University of Illinois at Urbana-Champaign. He is Fellow of ACM, Fellow of IEEE, and has received ACM SIGKDD Innovation Award (2004), IEEE Computer Society Technical Achievement Award (2005), IEEE Computer Society W. Wallace McDowell Award (2009), and Japan's Funai Achievement Award (2018).

Chapter 43
AI and Deep Learning for Urban Computing

Senzhang Wang and Jiannong Cao

Abstract In the big data era, with the large volume of available data collected by various sensors deployed in urban areas and the recent advances in AI techniques, urban computing has become increasingly important to facilitate the improvement of people's lives, city operation systems, and the environment. In this chapter, we introduce the challenges, methodologies, and applications of AI techniques for urban computing. We first introduce the background, followed by listing key challenges from the perspective of computer science when AI techniques are applied. Then we briefly introduce the AI techniques that are widely used in urban computing, including supervised learning, semi-supervised learning, unsupervised learning, matrix factorization, graphic models, deep learning, and reinforcement learning. With the recent advances of deep-learning techniques, models such as CNN and RNN have shown significant performance gains in many applications. Thus, we briefly introduce the deep-learning models that are widely used in various urban-computing tasks. Finally, we discuss the applications of urban computing including urban planning, urban transportation, location-based social networks (LBSNs), urban safety and security, and urban-environment monitoring. For each application, we summarize major research challenges and review previous work that uses AI techniques to address them.

S. Wang (✉)
College of Computer Science and Technology, Nanjing University of Aeronautics and Astronautics, Nanjing, China
e-mail: szwang@nuaa.edu.cn

J. Cao
Department of Computing, The Hong Kong Polytechnic University, Hong Kong, China
e-mail: csjcao@polyu.edu.hk

W. Shi et al. (eds.), *Urban Informatics*, The Urban Book Series,
https://doi.org/10.1007/978-981-15-8983-6_43

43.1 Background

In the big data era, sensing technologies (e.g., GPS and environment sensors) and large-scale computing infrastructures (e.g., distributed storage and computing) have produced and stored a variety of big data generated in urban space in real time, such as human-mobility data, air-quality data, transportation data, urban noise data, and urban crime data. Generally, big data can be defined as a field that studies the methodologies of effectively and efficiently storing, processing, extracting information from, discovering valuable knowledge from, and visualizing the datasets that are too large in data volume or too complex in data formats to be handled by traditional data storage, processing, and analytic paradigms. Usually, big data can be characterized by five Versus: volume, variety, velocity, veracity, and value (Ishwarappa and Anuradha 2015). The first primary characteristic of big data is its sheer volume. Variety means that the data can be unstructured, and the data types are much richer, including images, texts, videos, graphs, etc. As the data are usually generated in real-time and new data keep on coming, the characteristic of velocity requires that the new streaming data can be processed in near real time. Veracity refers to the trustworthiness of the data. Big data usually also mean big noise, such as in social-media data. The value hidden in the data can be low and may require carefully designed machine-learning or data-mining methods to discover useful knowledge from the massive data.

Mining knowledge hidden in the big data generated in urban areas is critically important to facilitate many real applications for smart cities, including relieving traffic congestion, urban crime prediction, real-time air pollution monitoring, urban planning, etc. To this aim, artificial intelligence (AI) techniques are urgently needed for knowledge discovery from the large-volume, noisy, heterogeneous, and ever-growing urban data (Zheng et al. 2014a, b). Recently, AI techniques driven by big data, such as the popular deep-learning models, have been widely used to solve diverse urban-computing tasks and have achieved success (Wang et al. 2019, 2020). For example, urban-traffic prediction and navigation driven by AI have been widely explored and applied in many applications such as the Gaode map for navigating and the City Brain system developed by Alibaba (Zhang et al. 2019a, b). As an interdisciplinary research field, knowledge discovery from urban big data is an indispensable part of urban computing, and AI techniques play a critically important role in mining correlations and patterns and predicting trends from the data.

Figure 43.1 shows a general framework to illustrate how AI techniques, especially machine learning, are used for various applications in urban computing. As shown in Fig. 43.1, there are three phases in general. The first phase is data acquisition. Diverse types of data generated from various sensors deployed in different locations in a city are collected, including GPS position data, air-quality data, weather data, data on social relations, points of interest (POIs), transportation networks, and social events. The collected raw data usually need to be preprocessed for further analysis.

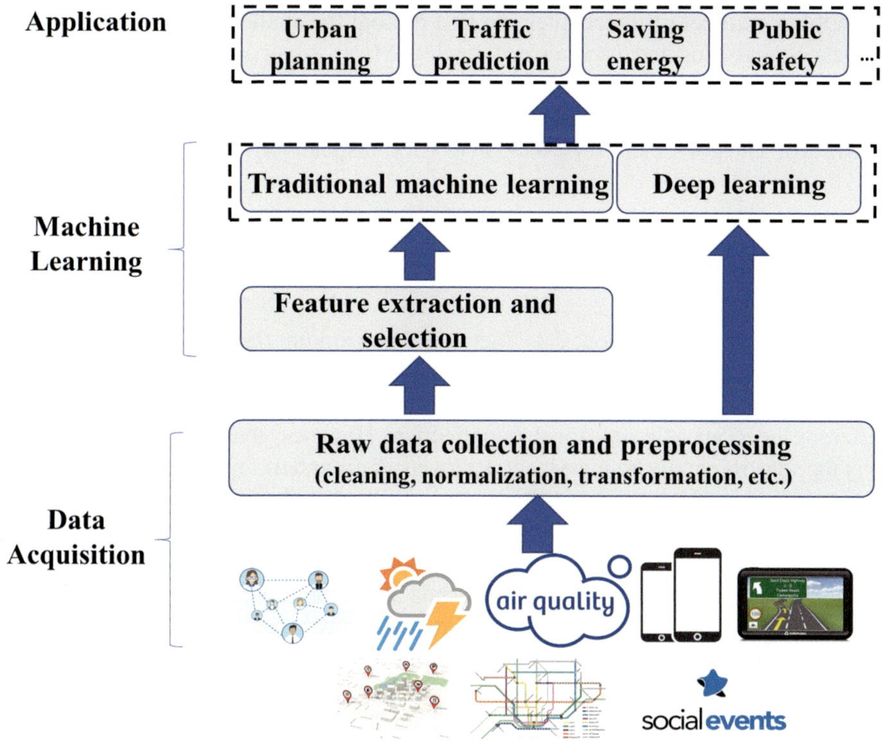

Fig. 43.1 Framework of applying AI techniques for urban computing

The data preprocessing operations include data cleaning, normalization, transformation, and instance selection. Next, the machine-learning phase performs pattern learning or knowledge discovery from the data. For traditional machine-learning methods, features need to be first extracted and selected from the data manually through feature engineering. In machine learning, features refer to a set of measurable properties or characteristics of the objects under study. They are used as the input of the machine-learning algorithms to be mapped to the output. Discriminating features can be extracted and selected from the raw data based on domain knowledge, and then fed into a machine-learning model such as the SVM classifier or logistic regression for training. Note that for the deep-learning models that are extremely popular nowadays, they do not need handcrafted features. Deep-learning models can automatically learn features from the raw data and integrate the feature learning and model learning in an end-to-end way, which is a significant advantage. The third phase is using the trained machine-learning models to support various urban-computing applications, such as urban planning, traffic prediction, public safety, and energy saving. The results of machine-learning models can provide us with knowledge, predictions, and guidance to help us make decisions on how to build a smarter city.

In the remainder of the chapter, we first present the challenges of using AI techniques for analyzing and discovering knowledge from urban data. Then, we introduce both traditional AI models and recent deep-learning models that are widely used in various tasks of urban computing. Next, we classify urban computing into several application categories and review-related work, respectively.

43.2 Challenges

Compared with other types of data, there are some unique challenges for conducting machine learning using the big data generated from various urban sensors.

Data acquisition: Usually, a large number of sensors should be deployed in different locations of a city for data collection. However, there are several reasons why the sensors cannot be massively deployed all around the city. First, some sensors are expensive, such as cameras and sensors in air-quality monitoring stations. Second, due to the energy consumption constraint, the number of sensors is usually limited. Sometimes it is difficult to select suitable locations to deploy sensors for data acquisition. It is also nontrivial to estimate the data at a location where there are no sensor readings, based on the observed sensor data from other locations.

Large volume and streaming data: The volume of the data generated from an urban area is usually very large considering the large number of sensors deployed in a city; and the data volume grows quickly, considering that the sensors generate data continuously in real time. Traditional machine-learning or data-mining techniques usually need a large number of labeled training samples and thus are time consuming. Many urban-computing tasks need real-time data analysis, such as traffic prediction and air-quality monitoring. Therefore, it is challenging for existing AI techniques to process this large volume of data continuously and almost instantly.

Heterogeneous data: Solving a specific task in urban computing usually involves multiple datasets rather than only one dataset. For example, city-wide air-pollution prediction involves the simultaneous study of multiple types of data, including traffic flow, weather, and land uses. Different datasets usually present diverse data formats or types. Traditional data-mining and machine-learning techniques are usually designed to handle one type of data, such as image, text, and graphics. How to fuse the heterogeneous data with different formats and structures involved in one learning task to serve the urban-computing application of interest is difficult, and also a hot research topic currently.

Complex dependencies among the data: Different types of urban data can be highly correlated, such as traffic data, air-quality data, and weather data. Traffic congestion is usually highly correlated with POI distribution, time of day, and social events. It is difficult for traditional statistics-based methods to capture the correlations and dependencies among the data without the help of domain expertise. Mining the dependencies among the data may be especially important to help improve various urban-computing applications such as urban planning, policy making, and intelligent transportation systems.

Noisy and incomplete data: Most data in urban computing are generated by urban sensors which are deployed in an open environment (e.g., the air-quality sensors deployed on the field). The sensors may fail to work normally and produce wrong or noisy data from time to time. In addition, some sensors are expensive, and only a limited number of sensors are deployed due to this cost limitation. For example, the road cameras for traffic monitoring are usually only installed in some intersections of a road network due to the high cost. Performing a task such as city-wide air-quality and traffic monitoring with such noisy and incomplete data is challenging.

Distributed data storage and processing: As the urban sensors are deployed at different locations, and the data volume increases rapidly, a distributed data-storage and processing infrastructure is usually required for more efficient computation of various machine-learning and data-mining algorithms. Considering the heterogeneity of the urban data, the complex dependencies among the data, and the nonuniform distributions of the data sensors, it is very challenging to design such a distributed data-storage and processing infrastructure.

Data privacy: Urban data are mostly collected from users. For example, users' mobility data can be collected from users' smartphones, and the urban-traffic data can be collected from the GPS module installed in private vehicles. How to protect the data privacy of the users and at the same time use the data to facilitate various applications such as navigation and travel route recommendation is a non-trivial problem. There needs a tradeoff between data privacy and data utility (see Chap. 32).

To address the above-mentioned challenges, various AI techniques are being explored in different application scenarios of urban computing, such as supervised learning, semi-supervised learning, unsupervised learning, matrix factorization, graphic models, deep learning, and reinforcement learning. Next, we briefly introduce the concept and preliminary knowledge of the methods and then discuss how these models can be used in different tasks of urban computing in detail.

43.3 Traditional AI Techniques

43.3.1 Supervised Learning

Supervised learning, such as classification and regression, is a type of machine learning that learns a function mapping the input features to an output label or variable, based on a set of training input–output pairs (Caruana and Niculescu Mizil 2006). Note that in supervised learning, a training dataset that contains both the input data and the corresponding output labels or variables is needed, and the goal is to learn a mapping function from the training dataset.

Supervised learning is widely used in many urban-computing tasks when a large number of labeled training data samples are available, such as traffic prediction

(Castro-Neto et al. 2009), region classification (Toole et al. 2012), and POI recommendation (Daniel and Sebastian 2000). For example, Toole et al. (2012) studied the problem of inferring the types of urban land-use from users' mobile-phone activity data. A supervised classification algorithm was used to identify four types of land uses with similar zoned uses and mobile-phone activity patterns. The training data of the algorithm contained three weeks of call records for about 600,000 users in the Boston region. Castro-Neto et al. (2009) proposed a supervised regression algorithm called online support vector machine to predict short-term freeway traffic flow under both typical and atypical conditions.

43.3.2 Unsupervised Learning

Significantly different from supervised learning, unsupervised learning does not need any labeled data for training. Unsupervised learning aims to capture the underlying structures, patterns, or distributions from the input data without the guidance of output labels or variables. Unsupervised learning can be generally grouped into clustering and association. Clustering is the task of grouping a set of objects so that objects in the same group are more similar to each other than to those in other groups. Each object group is called a cluster. Association-rule learning is a rule-based machine-learning method for discovering interesting relations between variables or patterns in large databases. Association-rule learning algorithms intend to identify such strong rules or patterns in the given dataset using measures of interestingness.

In many real application scenarios, there are no labeled data at all. In such a case, unsupervised learning techniques can be used for mining knowledge from the massive data. For example, mining patterns from the trajectories of moving objects is an important research topic in spatial–temporal data mining (Giannotti et al. 2007). There are no labeled training data for discovering new patterns in trajectories, and thus the unsupervised pattern-mining methods are applied. Another example is city-boundary detection driven by big data. This task aims to discover the real borders of a city according to the interactions between people, using GPS tracks or phone-call records, and there are no ground-truth labels for the boundary of a city. To solve this problem, Rinzivillo et al. (2012) proposed to first build a location network based on human interaction and then partition the network using an unsupervised community-detection method. The boundaries of regions can be thus characterized by the discovered location clusters, with denser interaction between locations in the cluster.

43.3.3 Semi-supervised Learning

Semi-supervised learning falls between unsupervised learning, which does not have labeled training data at all, and supervised learning which has complete labeled

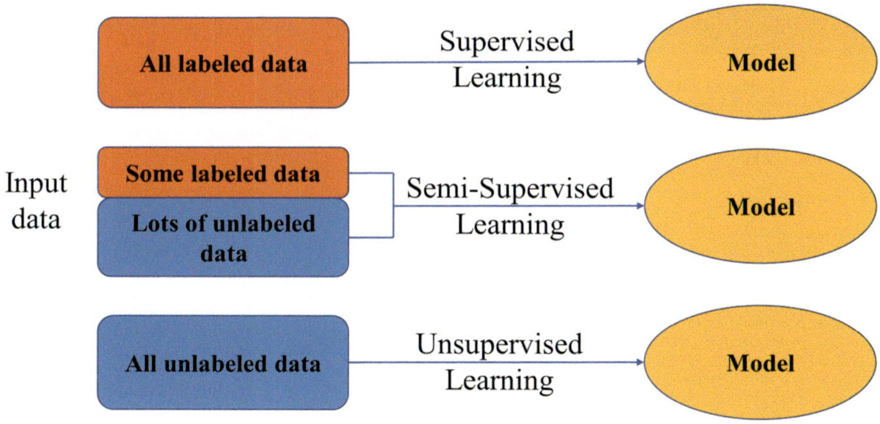

Fig. 43.2 Three types of machine-learning methods

training data. Semi-supervised learning makes use of both a small amount of available labeled data and a large amount of unlabeled data for training (Zhu 2005). As it is usually expensive and time-consuming to label a large number of training data for supervised learning, semi-supervised learning is widely used based on the observation that unlabeled data, when used in conjunction with a small amount of labeled data, can achieve considerable performance improvement over unsupervised learning. Semi-supervised learning also has broad applications in urban computing. For example, Zheng et al. (2013) proposed a semi-supervised learning approach based on a co-training framework to predict the air quality of a location where there is no air-quality monitoring station already. The used co-training framework consisted of two separated classifiers, with one using spatially related features and the other using temporally related features. Figure 43.2 compares three types of machine-learning methods.

43.3.4 Matrix Factorization

Matrix factorization, which is also called matrix decomposition, decomposes a matrix into a product of two or three smaller matrices. It is an approach that can simplify some complex matrix operations, since these can be performed on the decomposed smaller matrices rather than on the original large matrix (Daniel and Sebastian 2000). Popular matrix factorization methods include LU decomposition, QR decomposition, Jordan decomposition, and SVD. From an application point of view, matrix factorization can be used to discover the latent features underlying the interactions between two types of entities, such as users and items in recommendation systems. For example, SVD is widely used in collaborative filtering (Zhou et al. 2015), which factorizes the product-rating matrix A into the product of three smaller matrices, the left singular

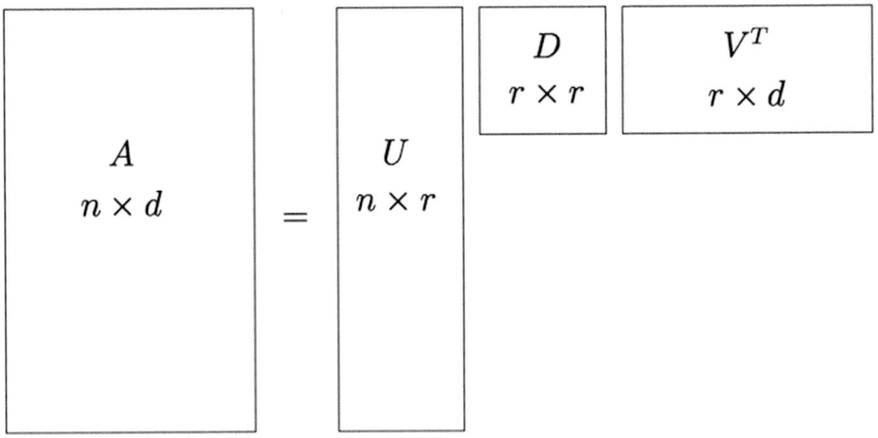

Fig. 43.3 Illustration of SVD

vectors U, the singular values D, and the right singular vectors V^T as shown in Fig. 43.3. Matrix factorization has very broad applications in machine learning, such as image processing, data compression, spectral clustering, recommendation, and matrix completion. For example, when the original matrix A is incomplete, with many unknown entry values, we can approximate it with three factorized low-rank matrices and estimate the missing entries in A to complete it.

Matrix factorization is widely used in many estimation or inference-related urban-computing tasks such as location recommendation, urban noise estimation, and urban-traffic estimation. For example, Zheng et al. (2010) proposed to collaboratively recommend location and activity to users through factorizing the location-activity matrix constructed from users' GPS historical trajectory data. Zheng et al. (2014a, b) integrated tensor composition and matrix composition to infer the fine-grained noise distribution at different times of day for each region of NYC. The noise distribution of NYC was modeled with a three-dimension tensor, whose three dimensions are regions, noise categories, and time slots. Supplementing the missing entries of the noise distribution tensor using the proposed tensor-matrix co-factorization approach, the noise distribution throughout the entire NYC can be inferred. Wang et al. (2019, 2020) proposed a locally balanced inductive matrix factorization model to infer the bike usage of a city at different hours of the day for dockless bike-sharing systems. The bike usage demand was modeled as a matrix whose two dimensions are region ID and time slot, and the entries are the needed number of bikes. The unknown entries of the bike-demand matrix are inferred through a proposed inductive matrix factorization method.

43.3.5 Graphical Model

A graphical model uses a graph to express the conditional dependency relationships among different random variables and is also called the probabilistic graphical model (PGM; Koller and Friedman 2009). It is widely used in probability theory, Bayesian statistics, and machine learning. Generally, graphical models use a graph-based representation to encode the variable distributions over a multi-dimensional space, which provides a general framework for modeling large collections of random variables with complex interactions. There are two types of commonly used graphical representations of variable distributions: Bayesian networks and Markov random fields. Figure 43.4 shows an example of a simple graphical model. Each node in the graph denotes a variable, and each arrow indicates a dependency relationship between two variables. In this example, D depends on A, B, and C; and C depends on B and D; whereas A and B are independent to each other.

In many urban-computing tasks, the data can be heterogeneous and collected from different sources, and the interactions and correlations among the data are usually complex. Graphical models can be used to model the dependencies among the data and make accurate estimates or inference. For example, in urban-traffic estimation and prediction, the traffic conditions of a road segment can be affected by both the neighboring road segments and the external factors such as weather, holidays, and rush hours. Wang et al. (2016a, b) proposed to use a coupled hidden Markov model for road-network-level traffic-congestion estimation. In this model, the traffic condition of a road segment at time t depends on its previous traffic condition at $t - 1$ and the traffic conditions of its neighboring road segments at $t - 1$. To model the complex dependencies among them, a graphical model that uses multiple coupled Markov chains was proposed. Shang et al. (2014) studied the problem of instantly inferring the gas consumption and pollution emission of the vehicles traveling on a road network of a city, based on the GPS trajectory data collected from a sample

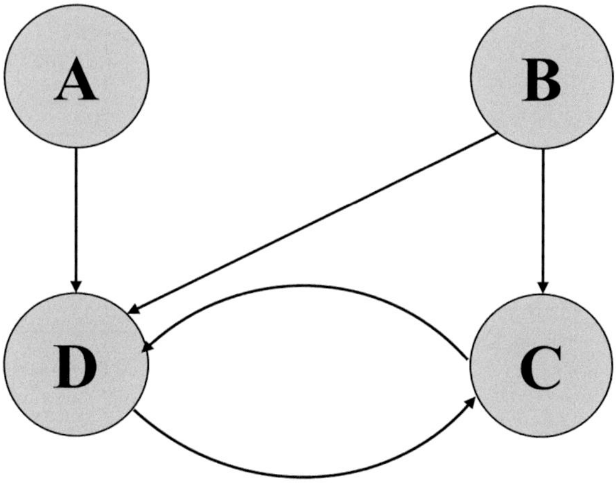

Fig. 43.4 A toy example of a graphical model

of vehicles. To address this task, they proposed an unsupervised dynamic Bayesian network model called the traffic volume inference model (TVI) to infer the number of vehicles passing each road segment per minute. TVI can model the effect of multiple external and internal factors on the traffic volume, including the travel speed, weather conditions, and the geographic features of a road.

43.4 Deep Learning

Deep learning is a type of machine-learning method whose structure, called an artificial neural network (ANN), is inspired by the structure and function of the human brain. The initial form of an artificial neural network is the perceptron, which was proposed in the 1950s (Rosenblatt 1957). Although ANNs have been proposed and studied for many years, early ANN models were not that successful compared with other machine-learning models, such as the Bayesian model and SVM, due to their shallow structures with only two or three layers of neurons. In recent years, ANN models with much deeper model structures containing tens of or even hundreds of neural layers are gaining popularity due to their supremacy in terms of prediction accuracy when trained with huge amounts of data (LeCun et al. 2015). Figure 43.5 shows the performance curves of deep-learning methods and most other traditional machine-learning methods with increasing amounts of training data. One can see that the learning performance of traditional methods first increases with an increase in the data amount and then reaches a performance bottleneck. More data will not lead to better performance due to the limited learning ability of traditional methods. For deep learning; however, the performance keeps on increasing with more and more training data, which is mainly due to its deep structure and powerful hierarchical feature-learning ability.

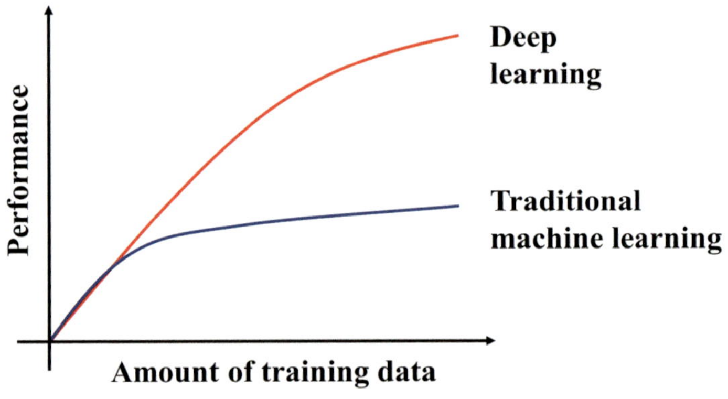

Fig. 43.5 Performance curves of deep learning and traditional machine learning with increasing amounts of training data

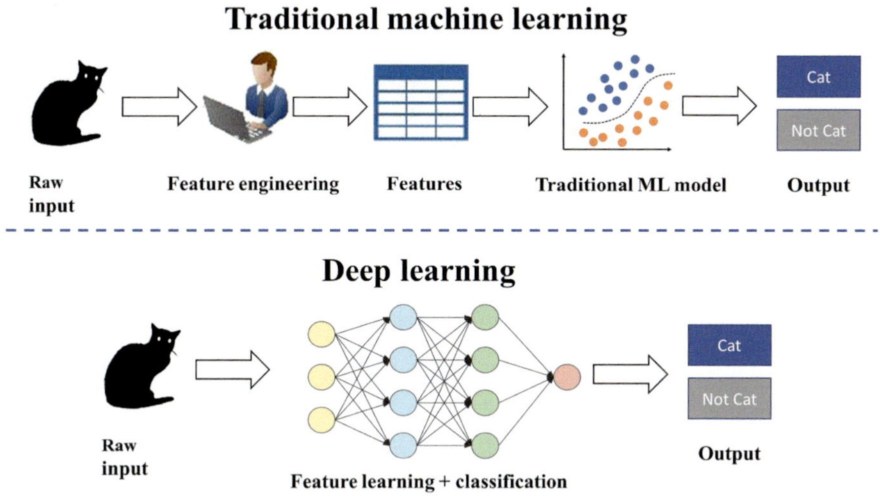

Fig. 43.6 Traditional machine learning *vs* deep learning

Besides the powerful learning ability from big data, another significant difference and advantage of deep learning compared with traditional machine learning is that deep learning does not need handcrafted features and can learn features from the input raw data automatically. Figure 43.6 shows a pipeline comparison between traditional machine learning and deep learning. We can see that for traditional machine-learning models, given the raw input data, feature engineering is first conducted to manually extract the features, and then, the features are input into the machine-learning model for classification. For deep-learning models, feature engineering is not needed any more. Feature learning and model learning are performed in an end-to-end learning way for deep-learning models.

Deep-learning architectures such as deep neural networks (DNN), deep belief networks (DBN), recurrent neural networks (RNN), and convolutional neural networks (CNN) have been widely applied in the fields of computer vision, speech recognition, natural-language processing, audio recognition, social-network analysis, machine translation, bioinformatics, medical-image analysis, and urban computing, where they have produced results comparable to and in some cases superior to humans. Next, we will briefly introduce some deep-learning models that are widely used in the tasks of urban computing.

43.4.1 Restricted Boltzmann Machines (RBM)

A restricted Boltzmann machine is a two-layer stochastic neural network (LeCun et al. 2015), which is broadly used for dimensionality reduction, classification, feature learning, and collaborative filtering. As shown in Fig. 43.7, RBM generally contains

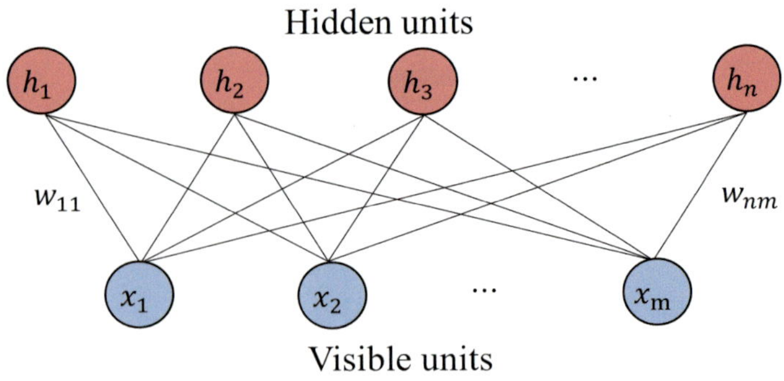

Fig. 43.7 Structure of RBM

two layers. The first layer of RBM is called the visible layer with the neuron nodes $\{x_1, x_2, \ldots, x_m\}$, and the second layer is the hidden layer with the neuron nodes $\{h_1, h_2, \ldots, h_n\}$. The structure of RBM can be considered as a fully connected bipartite undirected graph. All nodes in RBM are connected to each other across layers by undirected weight edges $\{w_{11}, w_{22}, \ldots, w_{nm}\}$, but no two nodes of the same layer are linked. The standard type of RBM has binary-valued neuron nodes and also bias weights. Depending on the particular task, RBM can be trained in either supervised or unsupervised ways.

43.4.2 CNN

A convolutional neural network (CNN) is initially designed to analyze visual imagery. Typically, CNN contains the following layers as shown in Fig. 43.8: the input layer, the convolutional layer, the pooling layer, the fully connected layer, and the output layer. Some CNN structures also have the normalization layer after the pooling layer. When it is used for image processing, the raw images are first input into the convolutional layer to learn the high-level and more abstract features. The convolutional layer captures the high-level latent features through multiple filters called kernels. A kernel is usually a $k \times k$ square matrix, which moves in the input image matrix from left to right and from top to bottom. A filtering operation is performed with the kernels on the corresponding positions of the input image matrix for generating high-level features. Then, the pooling layer performs a down-sampling operation on the high-level features based on the spatial dimensionality, to reduce the number of parameters. Finally, several fully connected layers are stacked to perform nonlinear transformation of the output high-level features from the pooling layers. Compared with a traditional multi-layer perceptron neural network, CNN has the following distinguishing characteristics that make it generalize well on vision problems: 3D volumes of neurons, local connectivity, and shared weights.

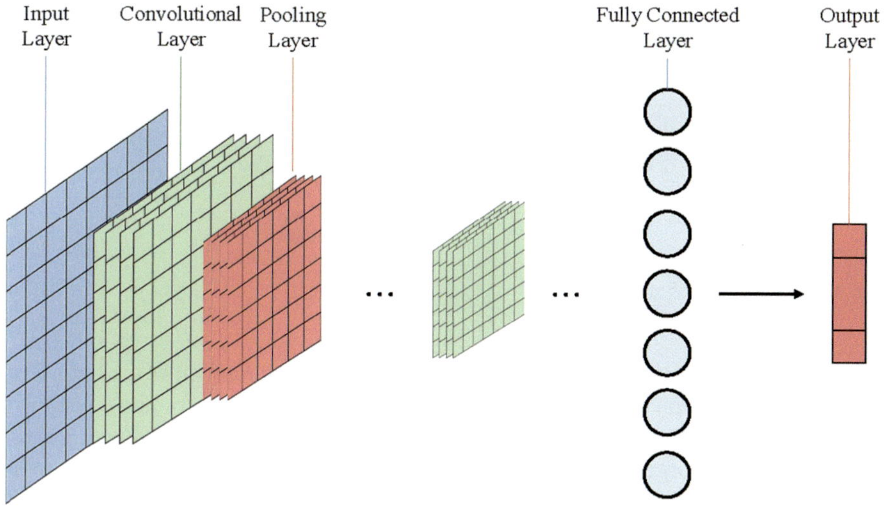

Fig. 43.8 Structure of CNN

43.4.3 RNN and LSTM

A Recurrent neural network (RNN) is designed to recognize the sequential characteristics of the input data and use the previous patterns to predict the future output. It is widely used in many areas such as speech recognition, natural-language processing, and time series data analysis. Figure 43.9 shows the general structure of an RNN network, where x_t is the input data, A are the parameters of the RNN network, and h_t is the learned hidden state. As shown in Fig. 43.9, the output of the previous time step $t - 1$ is input into the neurons of the next time step t. In this way, the historical information in the past time steps can be stored and conveyed to the future. A major shortcoming of the standard RNN is that it only has a short-term memory due to the issue of vanishing gradients. To solve this problem, the LSTM network was invented, which is capable of capturing the dependencies of the input data in a

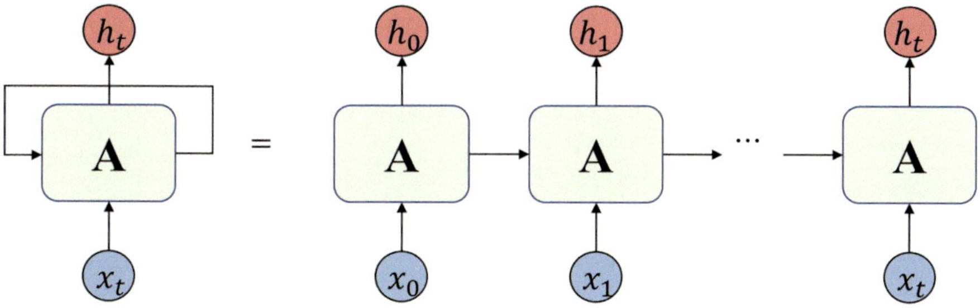

Fig. 43.9 Structure of an RNN

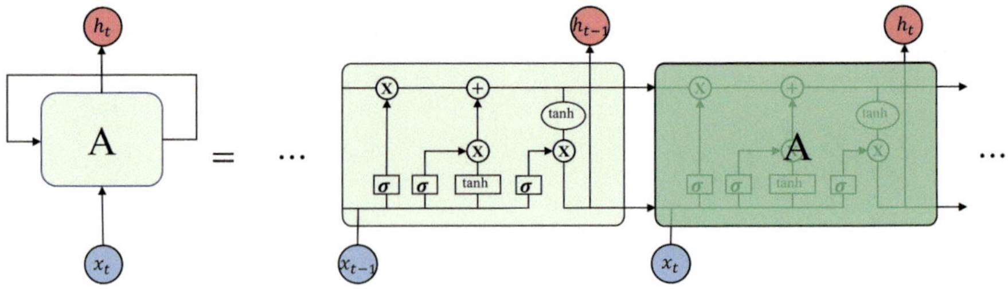

Fig. 43.10 Structure of an LSTM

much longer time period. Compared with RNN, LSTM can remember the long-term historical information of input due to its specially designed memory unit. As shown in the middle part of Fig. 43.10, an LSTM unit is composed of the following three gates: input gate, forget gate, and output gate. The input gate controls whether to let new input in, the forget gate controls whether to ignore some unimportant historical information, and the output controls whether to let the historical information impact the current output.

43.4.4 Autoencoder (AE)

An autoencoder is a type of artificial neural network that aims to learn compact data coding in an unsupervised manner (Hinton and Salakhutdinov 2013). As shown in Fig. 43.11, AE generally contains three types of layers: the input layer, the hidden layers, and the output layer. The raw data are first fed into the input layer, and then, one or multiple hidden layers are stacked to form an encoder for coding the input as compact latent representation vectors. Then, a decoder which is also composed of one or several hidden layers is used to reconstruct the raw input from the compact latent vector learned by the encoder. AE learns a compact representation of the input data in an unsupervised manner, which can be considered as a way of dimensionality reduction. As an effective learning technique for unsupervised feature representation, AE facilitates various downstream data-mining and machine-learning tasks such as classification and clustering. A stacked autoencoder (SAE) is a neural network consisting of multiple stacked AEs in which the outputs of the current AE are wired to the inputs of the successive AE (Bengio et al. 2006).

43.5 Reinforcement Learning

Reinforcement learning is more general than supervised/unsupervised learning (Richard and Andrew 1998). It learns from the interactions with the environment to get as much reward as it can over the long term. Intuitively, reinforcement learning

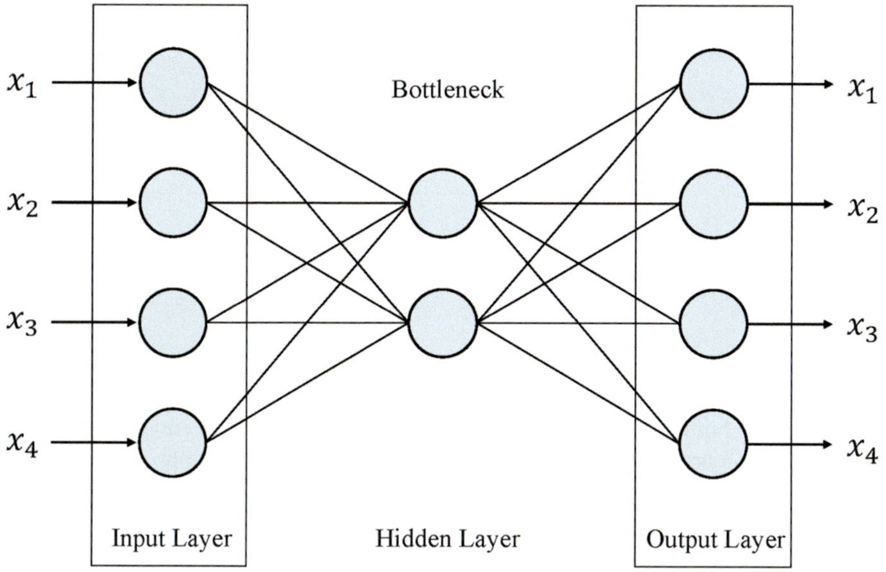

Fig. 43.11 Structure of an autoencoder

tries to imitate the human stress reaction. As shown in Fig. 43.12, imagine that you are a child in a living room with a stove in it, assume that you feel cold and are far from the stove, and then you try to approach it. You feel good and understand that the stove is a positive thing. But if you stay too close to the stove, your hand will be burned. From the interaction with the stove, you will learn that the stove is positive when you are a sufficient distance away because it produces warmth. But if you get too close to it, you will be burned. So too close to the stove will produce negative reward.

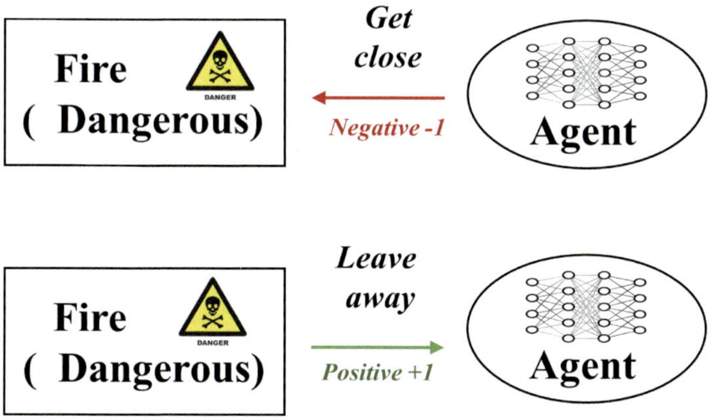

Fig. 43.12 A toy example to illustrate how humans learn through interaction with the environment

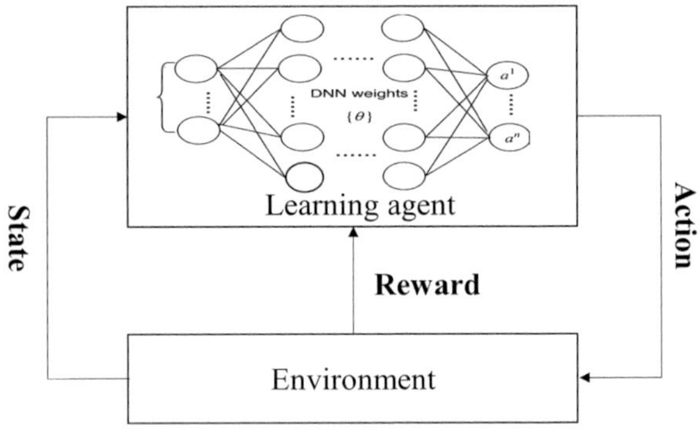

Fig. 43.13 General view of reinforcement-learning algorithms

Similar to humans learning through interaction with the environment, the reinforcement-learning algorithms learn to choose the most appropriate action through trail-and-error. The general idea of reinforcement-learning algorithms is illustrated in Fig. 43.13, which mainly consists of the four key elements: environment, reward, action, and state. A reinforcemsent-learning agent tries to learn how to best match states and actions in order to get the maximum long-term accumulated return (reward). As a result, the strategy will more frequently perform the actions that obtain positive rewards, while the actions that lead to negative punishment are less frequently performed.

Reinforcement-learning algorithms have broad application in the fields of robotics, optimal control, chess games, strategic games, flight control, missile guidance, predictive decision making, financial investment, and urban-traffic control, as they try to solve the general issues about how to best match the states and actions (Haldorai et al. 2019). Taking urban transportation as an example, where the city transportation network needs to control the traffic lights of multiple intersections and roads. Even without domain knowledge about how to control, by specifying the rule of reward, the reinforcement-learning algorithms can autonomously learn an optimal traffic light control strategy, such that all vehicles can pass the intersection in the shortest time (Rizzo et al. 2019). Even today, due to the complexity of urban-computing problems, learning control strategies through reinforcement-learning algorithms still face challenges of consuming a huge amount of computational time. However, with the development of computing power, reinforcement learning will enable an evolution from computational intelligence to artificial intelligence (Li et al. 2019).

43.6 Applications of AI Techniques in Urban Computing

The AI techniques described above have been widely applied in various urban-computing application scenarios, including urban planning, intelligent transportation systems, location-based social networks LBSNs, urban safety and security, and urban environmental monitoring. Next, we discuss these applications in detail. For additional discussion of the use of urban-mobility data, see Chaps. 28 and 29.

43.6.1 Urban Planning

Urban planning refers to the technical and political process concerned with the design and development of land use, and especially the spaces that the public share in urban areas. The goal of urban planning is to make cities safe, healthy, and enjoyable places to live. Urban planning is a very challenging task because a lot of complex factors should be considered, such as urban-traffic flow, human mobility, POI distribution, and urban functional regions. Traditionally, urban planners need to conduct surveys to guide them in making decisions on urban planning, which is less accurate, time consuming, and labor intensive. In the big data era, a lot of data generated in the urban area are increasingly available, and such data can be used to facilitate more effective and rational urban planning. Recently, research has tried to use big data and AI techniques in various urban planning tasks such as road-network planning (Zheng et al. 2011; Berlingerio et al. 2013), functional-regions discovery; (Zheng et al. 2014a, b; Yuan et al. 2012; Manley 2014), and city-boundary detection (Ratti et al. 2010; Rinzivillo et al. 2012).

Zheng et al. (2011) used the GPS trajectories of taxicabs traveling in urban areas to detect flawed urban planning in a city. They focused on detecting the pairs of regions with salient traffic problems and discovering the linking structure as well as correlations among them. The proposed model contains two steps: city-wide traffic modeling and flawed planning detection. In citywide traffic modeling, the urban area is first partitioned into disjoint regions based on major roads, and thus each region stands for a community containing some neighborhoods. Then, the origin–destination locations of the GPS trajectories of taxicabs are mapped to the partitioned regions, so that in each hour of a day the region transition matrices can be constructed. In flawed planning detection, the skyline of each region transition matrix is first detected, and then, a graph pattern-mining method is used to identify flawed planning from the skylines. Berlingerio et al. (2013) studied how to use large-scale cellphone mobility data of users to help transit operators better perform urban transportation planning. A system called AllAboard was developed for optimizing public transport with the guidance of people's cellphone data. AllAboard first infers the origin–destination (OD) flows in the city through a large volume of people's mobile phone location data. The OD flows are then converted to ridership on the existing transit network. Next, the sequential travel patterns are extracted from the flow data over the transit network, which can be used to propose new candidate transit routes. Finally, an optimization model is proposed to evaluate which new routes would best improve the existing transit network to increase ridership.

A functional region refers to a geographic area centered around a specific focal point with a specific function such as education, business, or transportation. Automatic functional-regions discovery and identification are particularly helpful to many urban-computing applications such as urban planning and city management. Yuan et al. (2012) proposed a data-driven approach called DRoF to discover different functional regions of a city by using both the human-mobility data among regions and the POI distributions in the regions. DRoF first segments a city into disjointed regions based on the major roads such as arterial roads, highways, and urban expressways. Then, the functions of each region are inferred by a proposed graphic-based probabilistic inference model. By borrowing the idea from topic model in natural-language processing, DRoF regards a region as a document, a region function as a topic, and the human-mobility trips (when people reach or leave which region) as words. The POI distribution in each region is also incorporated as the side information to help the model achieve more accurate inference accuracy. Evaluations are conducted on the three-month taxi GPS trajectory data generated by over 12,000 taxicabs in Beijing. Nine types of different functional regions labeled by humans are identified by DRoF. Manley (2014) applied the community-detection algorithm over the traffic network of a city to identify functional urban regions. The traffic network was constructed from the travel routes of about 1.5 million minicab trips. The region communities discovered from the large volume of traffic flow data can help identify areas of the road network that are used together, and thus help city planners to have a better understanding of the functional structure of the city. People's mobile phone data of a city can be also used to understand the spatio-temporal distribution of people in different regions of the city. For example, call detail records (CDR), which provide information on the locations of mobile phones where a call is made or a text message is sent, can be used to infer the dynamics of urban land use (Toole et al. 2012). A supervised classification algorithm is used to identify clusters of functional zones that present similar mobile phone activity patterns.

As the city expands rapidly and people move among different regions of the city, the boundaries of a city and its regions change quickly. It is very challenging for traditional methods to capture the dynamics of city boundaries. To tackle this issue, recently there have been studies using human-mobility data or activity data (e.g., GPS trajectories and CDR data) to better discover the real borders of city regions with data-driven approaches. Ratti et al. (2010) proposed a novel approach for regional delineation by analyzing networks of billions of individual human transactions. Given a geographic area and some measure of the strength of links between its inhabitants, Ratti et al. (2010) partitioned the area into disjoint smaller regions based on the rule that the disruption to each person's links in different regions should be minimized. The proposed method was tested on a large human interaction network containing 20.8 million nodes, which is inferred from a large telecommunications database in Great Britain. The human interaction network can be also inferred from other types of data such as the vehicle GPS tracks. Rinzivillo et al. (2012) first extracted region clusters from the human-interaction network constructed from the vehicle GPS data. Then, the region clusters were mapped back onto the territory of a city and were shown to match well with the existing administrative city borders.

43.6.2 Urban Transportation

Currently, most vehicles are installed with GPS devices for real-time positioning and navigation. The large-scale vehicle GPS data reflect the urban-traffic conditions in real time and thus are crucially important for intelligent transportation systems. Both deep-learning models and traditional machine-learning models are used to address various issues in urban transportation such as traffic flow prediction (Zhang et al. 2019a, b; Du et al. 2019) and traffic-congestion prediction (Wang et al. 2015; Wang et al. 2016a, b).

To address the issue that traditional traffic flow-prediction methods cannot effectively capture the nonlinear, stochastic, and time-varying characteristics of the traffic data, Zhang et al. (2019a, b) proposed a network-scale deep traffic-prediction model GCGAN. The framework of the GAGAN model is shown in Fig. 43.14, which combines adversarial training and graph CNN. GCGAN is a prediction framework

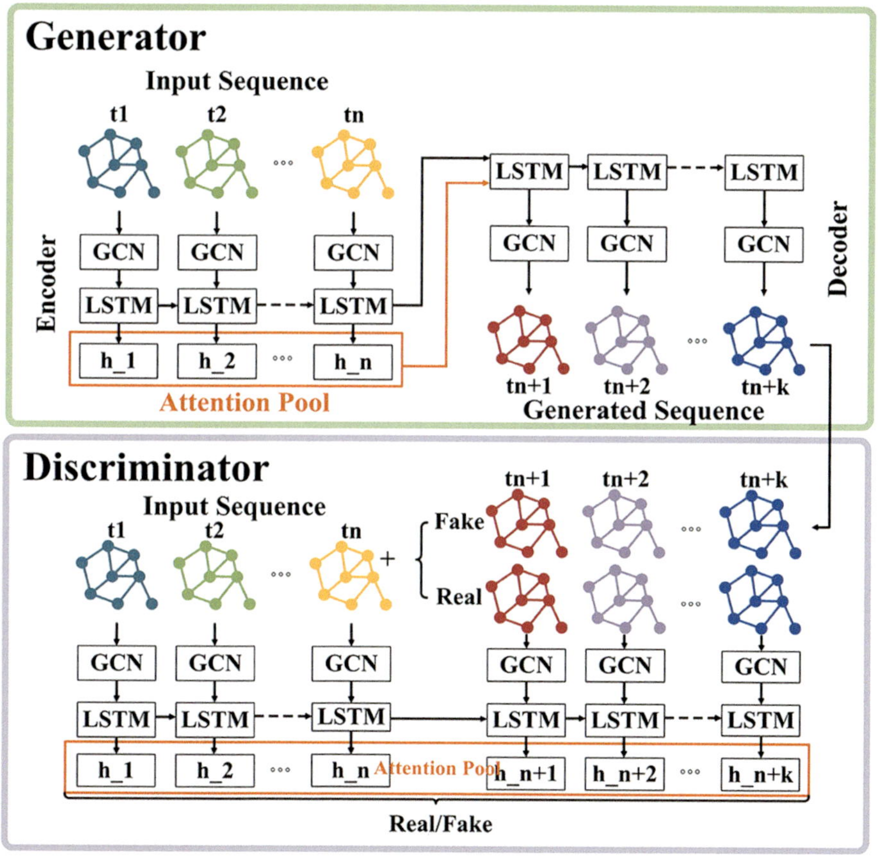

Fig. 43.14 Framework of the GCGAN model (Zhang et al. 2019a, b)

based on a Generative Adversarial Net, and thus can make more robust predictions by introducing adversarial training loss. As shown in the upper part of Fig. 43.14, GCGAN uses an encoder–decoder framework that is sequence-to-sequence based to encode the traffic conditions of a road network in previous time intervals and to decode the traffic conditions in future time intervals as the prediction. To model the spatial correlations among the road links of a transportation network, a graph convolution network (GCN) is used in both the generator and the discriminator for feature learning. LSTM is also used to capture the temporal dependencies. Du et al. (2019) studied the problem of predicting urban-traffic passenger flows with various types of traffic passenger flow data, including subway, taxi, and bus flows. Considering the complex factors such as hybrid transportation lines, mixed traffic models, transfer stations, and some extreme weather, a deep irregular convolutional residual LSTM network model called DST-ICRL was proposed by Du et al. (2019). The passenger flows among different traffic lines in a transportation network are first modeled as multi-channel matrices analogous to the RGB pixel matrices of an image. Then, a deep-learning framework that integrates an irregular convolutional residential network and LSTM units is proposed to learn the spatial–temporal feature representations from the passenger flow matrices. DST-ICRL samples both the short-term and long-term historical traffic data for model training to capture both the periodicity and the long-term trend of the traffic passenger flows.

Although deep-learning models are popular nowadays, some traditional machine-learning models such as matrix factorization and Markov models may perform better when there are multiple types of heterogeneous traffic data that need to be fused for traffic analysis. Wang et al. (2015) used a coupled matrix and tensor factorization model to infer city-wide traffic-congestion conditions by fusing multiple types of data including social-media data, social-event data, road physical features, and traffic-congestion patterns. As shown in Fig. 43.15, the proposed model used a coupled matrix and tensor factorization scheme to collaboratively factorize the traffic-congestion matrix X with the congestion correlation matrix Z, event tensor A, and the road feature matrix Y. By assuming that these matrices and tensor share the common latent factor matrix U in the road-segment dimension, these data are jointly factorized in order to fuse all the information. The traffic-congestion matrix of an entire city is then completed by multiplying the low-rank latent factor matrices U and V. Wang et al. (2016a, b) further extended the model of Wang et al. (2015) by incorporating GPS probe data. Wang et al. (2016a, b) constructed two traffic-congestion matrices: one was inferred from social-media data and the other from GPS probe data. The final estimation result is the weighted combination of the two matrices. Wang et al. (2016a, b) proposed an extended coupled hidden Markov model (E_CHMM) to combine GPS probe data and social-media data for traffic-congestion prediction. Figure 43.16 shows the framework of E_CHMM, which contains a data collection and processing part and the model part. Besides the vehicle GPS probe data, the tweets that report traffic events are also collected and used in this model. From each traffic-related tweet, the traffic event type, location, and time information are extracted. For each road link, Wang et al. (2016a, b) assumed that the occurrence

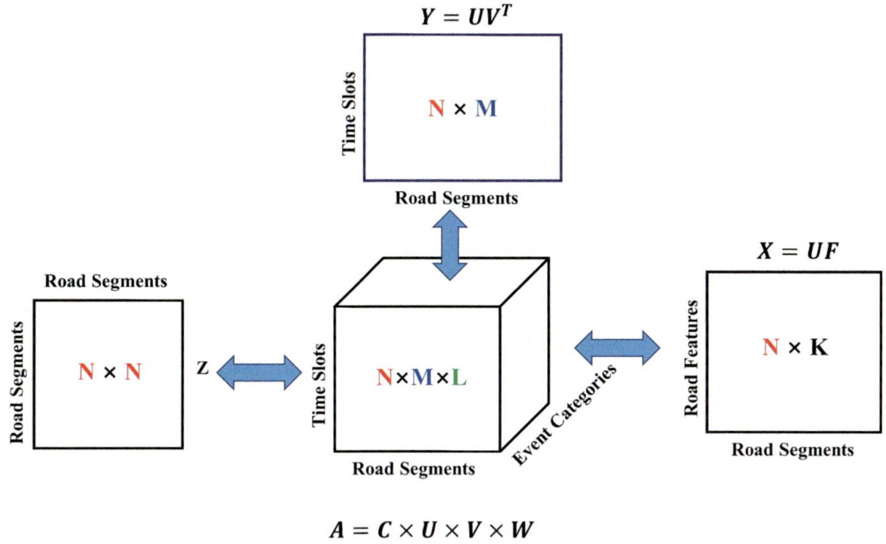

Fig. 43.15 Coupled matrix and tensor factorization model for traffic-congestion estimation (Wang et al. 2019, 2020)

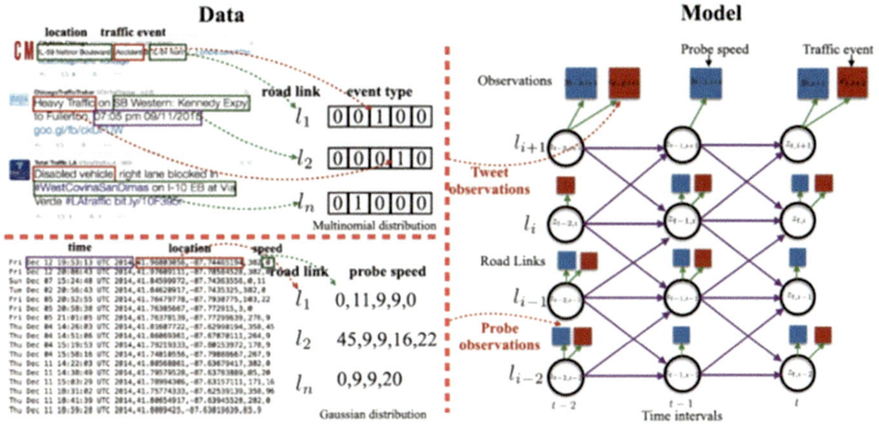

Fig. 43.16 Extended coupled hidden Markov model (E_CHMM) for traffic-congestion prediction (Wang et al. 2016a, b)

of traffic events follows a multinomial distribution, and the traveling speed of vehicles in a particular time interval follows a Gaussian distribution. In the model part, the traffic-congestion states of the road links in a road network are hidden and need to be inferred, while the GPS probe readings and traffic events extracted from tweets are observations. The goal of E_CHMM is to accurately infer the hidden traffic-congestion states of a road network based on the fusion of two types of observations: GPS probe readings and traffic event-related tweets.

43.6.3 Location-Based Social Networks (LBSNs)

LBSNs such as Foursquare and Flickr are social networks that use GPS features to locate users and enable users to share their locations and contents to their friends through mobile devices. They are more and more popular as they can connect users in both physical and virtual worlds. When users come to favorable restaurants, new POIs, or tourist attractions, they can check-in through their mobile phones immediately, so that their friends nearby can know their locations and join. AI techniques can be used to support many applications in LBSNs, including next check-in location prediction or recommendation (Ye et al. 2010; Gao et al. 2013; Bao et al. 2012), potential friends recommendation (Scellato et al. 2011; Bao et al. 2015), and check-in time prediction (Yang et al. 2018).

In LBSNs, there usually exist strong social and geospatial ties among users and their favorite locations. To take this into consideration for better check-in location recommendation, Ye et al. (2010) proposed a novel friendly collaborative filtering (FCF) approach for location recommendation based on the collaborative ratings on the places made by social friends. Motivated by the fact that a user's preferences for the check-in locations may change continuously over time, Gao et al. (2013) considered the temporal effects in location recommendation in LBSNs. Two types of temporal properties of a user's daily check-in preferences were considered: (1) non-uniformness, which means that a user has different check-in preferences at different hours of a day; and (2) consecutiveness, which means that a user's check-in preference in consecutive hours is more similar than that in non-consecutive hours. The two properties demonstrate that a user's check-in time and the corresponding preferred check-in locations can be highly correlated. Therefore, Gao et al. (2013) proposed a new check-in location recommendation framework by considering the temporal effects based on the observed two temporal properties. Besides a user's preference, other factors such as a user's current location and the opinions about a location given by the others may also be helpful for location recommendation. Bao et al. (2012) proposed a location-based and preference-aware recommender system that recommended POIs such as restaurants and shopping malls to a user by considering the user preferences, the current location of the user, and the opinions of the POIs given by other users.

Friend recommendation is a critically important service in social networks to help users find new friends and expand their social circles. In LBSNs, the location information can help to improve the effectiveness of social-friend recommendation. The basic intuition is that a user's preference can be revealed by his or her visited locations in LBSNs. Similar location histories imply similar preferences, thus such users are more likely to become friends (Bao et al. 2015). For example, Scellato et al.

(2011) analyzed the LBSN data from Gowalla, from which they found that the link-prediction space can be largely reduced by considering the similarity of the visited locations of the users. Based on this observation, a supervised link-predication model that considers the users' visited locations was proposed by Scellato et al. (2011) to predict which users will become friends in the future. Check-in time prediction aims to predict the time when a user will check-in to a given location. Generally, check-in time prediction can be formulated as a regression problem by considering time as a continuous variable. However, directly applying a regression model may not achieve desirable performance due to the check-in data scarcity issue. To deal with this, Yang et al. (2018) formulated check-in time prediction as a survival analysis problem and proposed a recurrent-censored regression (RCR) model to address it. RCR first uses the gated recurrent units (GRUs) to learn the latent representations of historical check-ins of a user and then inputs the latent representations into a censored regression model to predict the check-in time at a given location.

43.6.4 On-Demand Service

On-demand services (e.g., Uber, Mobike, DiDi, GoGoVan, etc.) are becoming increasingly popular nowadays due to the wide use of mobile phones and the prevalence of the sharing economy. A large volume of on-demand service data is generated continuously and needs to be analyzed in real time to help the service providers meet customer needs and improve the user experience. Many challenging tasks in on-demand services, such as demand–supply prediction (Wang et al. 2019, 2020) and user behavior prediction (Wang et al. 2017a, b), require effective AI techniques.

Wang et al. (2017a, b) studied the order response-time prediction problem in on-demand logistics services. In on-demand logistics services, users can make goods delivery orders via a mobile application, and registered van drivers would respond to take these orders in a very short period of time (usually less than several minutes). Making and taking orders through such an online app installed in mobile phones is much faster than the traditional way through van calling centers, and thus makes the logistics service much more efficient. An important task to help the service providers improve their services is the accurate prediction of the response time of the van drivers to the posted delivery orders, because the response time can largely reflect the preference of the drivers for the order. Wang et al. (2017a, b) formulated the response-time prediction task as a matrix factorization problem, and proposed a coupled sparse matrix factorization model to fuse the heterogeneous and sparse data from different domains, including historical order data, personalized requirements of the user, and location-relevant features, for more accurate prediction. Currently, dockless bike-sharing systems have emerged as a new type of on-demand service in China. Users can check-out and check-in a bike conveniently at any location through scanning the QR-code on the bike with an app installed in their mobile phones. The demand–supply analysis of the bikes in dockless bike-sharing systems is a very important yet challenging problem for efficient and effective system management.

Wang et al. (2019, 2020) proposed a data-driven approach for bike usage demand–supply inference in dockless bike-sharing systems. The idea is that before massively deploying a large number of bikes in an entire city, the system operator will first pre-deploy a relatively small number of bikes in certain regions of the city for data collection. The demands in some regions are first estimated from a small number of observed bike check-out/in data directly, and then, they are used as seeds to infer bike usage demands in other regions of the city. Wang et al. (2019, 2020) formulated the problem as a matrix completion task by considering the regions and time intervals as the two dimensions of the bike usage demand and supply matrices. As the two matrices are sparse and only partial entries are known due to the bike trip data in limited regions, a matrix factorization model was designed to complete the demand and supply matrices.

Deep-learning models such as CNN and LSTM are also widely used for demand–supply prediction in on-demand services. Lin et al. (2018) proposed a graph CNN model to predict the station-level hourly demand in a large-scale bike-sharing network. The model proposed by Lin et al. (2018) combined convolutional neural networks and LSTM to learn the underlying correlations of bike usage between the bike stations. Wang et al. (2017a, b) studied the supply–demand prediction problem for online car-hailing services with deep-learning methods. An end-to-end learning framework called DeepSD was proposed by Wang et al. (2017a, b) which used a novel deep neural network structure to automatically discover complicated supply–demand patterns from the car-hailing service data.

43.6.5 Urban Safety and Security

Crimes, traffic accidents, and environmental disasters can seriously threaten urban safety and security. In the big data era, urban safety- and security-related data such as crimes and traffic accidents can be recorded and stored in a database. Recently, there has been increasing research interest in studying whether and how AI techniques can be applied to analyzing these data, and to help address various urban safety- and security-related issues such as disaster detection (Lee and Sumiya 2010; Song et al. 2013) and crime prediction (Duan et al. 2017; Huang et al. 2018).

Lee and Sumiya (2010) developed a nation-wide geo-social event detection and monitoring system by collecting a large number of messages from Twitter. The proposed geo-social event detection model contains the following main steps: (1) collecting geo-tagged tweets using a Twitter monitoring system; (2) identifying regions of interest of Twitter users and measuring geographic regularities of crowd behaviors, and (3) detecting geo-social events through a comparison of the regularities. Song et al. (2013) analyzed and modeled the evacuation behaviors of people during the Great East Japan Earthquake and Fukushima nuclear accident based on a large volume of people's real mobility data in daily life. A population mobility database was constructed to store and manage people's mobility data of GPS records from approximately 1.6 million individuals throughout Japan over one year. A probabilistic inference model was developed to effectively represent people's mobility patterns. The proposed model can help researchers toward a better understanding of human evacuation behaviors during a disaster, and how those behaviors can be impacted by various cities during disasters. The system developed by Song et al. (2013) can be used to simulate and predict population mobility when disasters happen in cities so as to improve future disaster relief and management.

Many governments and law-enforcement agencies make city crime data (e.g., crime type, location, and time information) publicly available, so that researchers can use AI techniques for crime-data analysis. An important application of AI for crime-data analysis is crime prediction. Huang et al. (2018) developed a crime-prediction framework based on a deep neural network, called DeepCrime. DeepCrime can capture the dynamic crime patterns and explore the evolving inter-dependencies between different types of crimes to predict how many crime incidents will occur in the future in different regions of a city. A region-category interaction encoder is used to learn the complex interactions between regions and occurred crime categories. Then a hierarchical recurrent framework was proposed to jointly encode the temporal dynamics of crime patterns and capture the inherent interrelations between crimes and other ubiquitous data such as POIs. Finally, an attention mechanism was used to capture the unknown temporal relevance and automatically assign importance weights to the learned hidden states in different time frames. Duan et al. (2017) applied deep convolutional neural networks (CNNs) for automatic crime-referenced feature extraction and crime prediction. The urban area under study was first divided into grid regions. Then, the crimes in all the grid regions can be considered as an image, where each grid region is a pixel and the crime number is the gray value of the pixel. CNNs are applied on the image-like crime data of all the grid regions for feature learning.

43.6.6 Urban Environment Monitoring

Currently, a large number of diverse sensors are deployed all around a city to monitor environmental variables, weather conditions, and air-quality indexes (AQI) in real time. With a large amount of data collected from these sensors, AI techniques are required to process and analyze the data for smart environment monitoring.

Some air-quality monitoring stations have been built in different locations to collect a city's real-time air-quality indexes (AQI) such as $PM_{2.5}$, NO_2, and CO. However, due to the high cost of building and maintaining such stations, only a very limited number of stations can be built in a city; it is then a challenge to accurately obtain the AQI data of the entire city. Zheng et al. (2013) inferred the fine-grained AQI throughout a city by fusing the AQI data of limited locations with other types of data, including the meteorology, traffic flow, human mobility, structure of road networks, and POIs. A semi-supervised learning approach based on the co-training framework was proposed. This approach contains an artificial neural network to model the spatial correlation between the AQI of different locations, and a temporal classifier to model the temporal dependency of AQI in a location. Cheng et al. (2018) proposed a deep-learning model named ADAIN for urban air-quality inference. ADAIN combines feedforward and recurrent neural networks for modeling static and sequential features as well as capturing deep feature interactions effectively. An attention mechanism was also applied in a pooling layer of ADAIN to automatically learn the different weights of features from different monitoring stations.

Due to population expansion in big cities, urban noise pollution currently is becoming a more and more serious issue that threatens public health. AI techniques can also be used to help monitor, estimate, and analyze urban noise. Rana et al. (2010) designed an end-to-end participatory urban noise mapping system called Ear-Phone. Ear-Phone leverages compressive sensing to address the issue of recovering the noise map from the incomplete and random samples obtained by crowdsourcing noise-pollution data. The noise data are collected by the sound sensors installed in mobile phones. Zheng et al. (2014a, b) studied how to infer the fine-grained noise situation, including a noise-pollution indicator and the composition of noises at different times of a day in New York City, by using multi-sourced data including citizens' complaint data about city noise, social media, road-network data, and POIs. The noise situation of New York City was modeled as a three-dimensional tensor, where the three dimensions stand for regions, noise categories, and time slots. By filling in the missing entries of the tensor through a context-aware tensor decomposition approach, the noise situation throughout New York City can be recovered.

43.7 Conclusion

Recently, mining knowledge from the data generated in urban spaces for supporting urban-computing tasks to help build smart cities is a critically important and substantially challenging research topic. The large volume of heterogeneous data that are continuously generated in urban spaces, and recent advances in AI techniques, especially deep learning, have provided us with unprecedented opportunities to tackle the big challenges in urban computing. In this chapter, we conducted a comprehensive review of the challenges, methodologies, and frameworks that arise when AI techniques are applied in urban computing, and categorized the application domains of urban computing. To address the unique challenges for learning knowledge from urban data, we introduced both the traditional AI techniques and recently popular deep-learning models that are widely used for urban computing, including supervised learning, semi-supervised learning, unsupervised learning, matrix factorization, graphic models, deep learning, and reinforcement learning. We also categorized the utilization of AI techniques in different urban-computing applications including urban planning, urban transportation, location-based social networks (LBSNs), urban safety and security, and urban environmental monitoring.

References

Bao J, Zheng Y, Mokbel MF (2012) Location-based and preference-aware recommendation using sparse geo-social networking data. In: 20th international conference on advances in geographic information systems, Redondo Beach, California, 06–09 Nov

Bao J, Zheng Y, Wilkie D, Mokbel MF (2015) Recommendations in location-based social networks: a survey. GeoInformatica 19(3):525–565

Bengio Y, Lamblin P, Popovici D, Larochelle H (2006) Greedy layerwise training of deep networks. In: 19th international conference on neural information processing systems, Canada, 04–07 Dec

Berlingerio CM, Giusy F, Lorenzo GD, Nair R, Pinelli F, Sbodio ML (2013) All aboard: a system for exploring urban mobility and optimizing public transport using cellphone data. In: 12th European conference on machine learning and principles and practice of knowledge discovery in databases. Washington DC, USA, 12–16 Jan

Caruana R, Niculescu-Mizil A (2006) An empirical comparison of supervised learning algorithms. In: 23rd international conference on machine learning, Pittsburgh, Pennsylvania, USA, 25–29 June

Castro-Neto M, Jeong YS, Jeong MK, Han LD (2009) Online-SVR for short-term traffic prediction under typical and atypical traffic conditions. Expert Syst Appl 36(3):6164–6173

Cheng WY, Shen YY, Zhu YM, Huang LP (2018) A neural attention model for urban air quality inference: Learning the weights of monitoring stations. In: Thirty-second AAAI conference on artificial intelligence, New Orleans, Louisiana, USA, 02–07 Feb

Daniel DL, Sebastian SH (2000) Algorithms for non-negative matrix factorization. In: 13th international conference on neural information processing systems, Denver, CO, USA, 27–30 Nov

Du BW, Peng H, Wang SZ, Bhuiyan MZA, Wang LH, Gong Q, Liu L, Li J (2019) Deep irregular convolutional residual LSTM for urban traffic passenger flows prediction. IEEE Trans Intell Transp Syst. https://doi.org/10.1109/TITS.2019.2900481

Duan L, Hu T, Cheng E, Zhu JF, Gao C (2017) Deep convolutional neural networks for spatiotemporal crime prediction. In: International conference on Information and Knowledge Engineering (IKE), Las Vegas, Nevada, USA, 17–20 July

Gao HJ, Tang JL, Hu X, Liu H (2013) Exploring temporal effects for location recommendation on location-based social networks. In: 7th ACM conference on recommender systems, Hong Kong, China, 12–16 Oct

Giannotti F, Nanni M, Pedreschi D, Pinelli F (2007) Trajectory pattern mining. In: 13th ACM SIGKDD conference on knowledge discovery and data mining, San Jose, California, USA, 12–15 Aug

Haldorai A, Ramu A, Murugan S (2019) Artificial intelligence and machine learning for future urban development. Springer International Publishing, New York

Hinton GE, Salakhutdinov RR (2013) Reducing the dimensionality of data with neural networks. Science 313(5786):504–507

Huang C, Zhang J, Zheng Y, Chawla NV (2018) Deepcrime: attentive hierarchical recurrent networks for crime prediction. In: 27th ACM international conference on information and knowledge management, Torino, Italy, 22–26 Oct

Ishwarappa Anuradha A (2015) A brief introduction on big data 5Vs characteristics and Hadoop technology. Proc Comput Sci 48:318–324

Koller D, Friedman N (2009) Probabilistic graphical models: principles and techniques. The MIT Press, Cambridge, Massachusetts and London, England

LeCun Y, Bengio Y, Hinton G (2015) Deep learning. Nature 521:436–444

Lee R, Sumiya K (2010) Measuring geographical regularities of crowd behaviors for Twitter-based geo-social event detection. In: ACM SIGSPATIAL GIS workshop on location based social networks, San Jose, California, 02 Nov

Li GL, Randy G, Keisuke N, He B (2019) Human-centered reinforcement learning: A survey. IEEE Transac Hum Mach Syst 49(4):337–349

Lin L, He Z, Peeta S (2018) Predicting station-level hourly demand in a large-scale bike-sharing network: a graph convolutional neural network approach. Transport Res Part C: Emerg Technol 97:258–276

Manley E (2014) Identifying functional urban regions within traffic flow. Reg Stud Reg Sci 1(1):40–42

Rana RK, Chou CT, Kanhere SS, Bulusu N, Hu W (2010) Ear-phone: An end-to-end participatory urban noise mapping system. In: 9th ACM/IEEE international conference on information processing in sensor networks, Stockholm, Sweden, 12–16 April

Ratti C, Sobolevsky S, Calabrese F, Andris C, Reades J, Martino M, Claxton R, Strogatz SH (2010) Redrawing the map of Great Britain from a network of human interactions. PLoS ONE 5(12):e14248. https://doi.org/10.1371/journal.pone.0014248

Richard SS, Andrew G (1998) Barto. In: Reinforcement learning: a introduction. MIT press, Cambridge

Rinzivillo S, Mainardi S, Pezzoni F, Coscia M, Pedreschi D, Giannotti F (2012) Discovering the geographical borders of human mobility. KI - Künstliche Intelligenz 26(3):253–260

Rizzo GS, Vantini G, Chawla S (2019) Reinforcement learning with explainability for traffic signal control. In: 22nd IEEE Intelligent Transportation Systems Conference (ITSC), Auckland, New Zealand, 27–30 October

Rosenblatt F (1957) The perceptron, a perceiving and recognizing automaton project. Cornell Aeronautical Laboratory, Buffalo, New York

Scellato S, Noulas A, Mascolo C (2011) Exploiting place features in link prediction on location-based social networks. In: 17th ACM SIGKDD international conference on knowledge discovery and data mining, San Diego, California, USA, 21–24 Aug

Shang JB, Zheng Y, Tong WZ, Chang E, Yu Y (2014) Inferring gas consumption and pollution emission of vehicles throughout a city. In: 20th SIGKDD conference on knowledge discovery and data mining, New York, New York, USA, 24–27 Aug

Song X, Zhang Q, Sekimoto Y, Horanont T, Ueyama S, Shibasaki R (2013) Modeling and probabilistic reasoning of population evacuation during large-scale disaster. In: 19th SIGKDD conference on knowledge discovery and data mining, Chicago, Illinois, USA, 11–14 Aug

Toole JL, Ulm M, Bauer D, González MC (2012) Inferring land use from mobile phone activity. In: ACM SIGKDD international workshop on urban computing, Beijing, China, 12 Aug

Wang SZ, He LF, Stenneth L, Yu PS, Li ZJ (2015) Citywide traffic congestion estimation with social media. In: 23rd ACM SIGSPATIAL international conference on advances in geographic information systems, Seattle, Washington, 03–06 Nov

Wang SZ, He LF, Stenneth L, Yu PS, Li ZJ, Huang ZQ (2016a) Estimating urban traffic congestions with multi-sourced data. In: 17th IEEE international conference on mobile data management, Porto, Portugal, 13–16 June

Wang SZ, Li FX, Stenneth L, Yu PS (2016b) Enhancing traffic congestion estimation with social media by coupled hidden Markov model. In: Joint European conference on machine learning and knowledge discovery in databases, Riva del Garda, Italy, 19–23 Sept

Wang D, Cao W, Li J, Ye JP (2017a) Deepsd: supply-demand prediction for online car-hailing services using deep neural networks. In: IEEE 33rd International Conference on Data Engineering (ICDE), San Diego, CA, USA, 19–22 April

Wang YQ, Cao JN, He LF, Li WG, Sun LC, Yu PS (2017b) Coupled sparse matrix factorization for response time prediction in logistics services. In: ACM on conference on information and knowledge management, Singapore, Singapore, 06–10 Nov

Wang SZ, Chen H, Cao JN, Zhang JW, Yu PS (2019) Locally balanced inductive matrix completion for demand-supply inference in stationless bike-sharing systems. IEEE Transac Knowl Data Eng. https://doi.org/10.1109/TKDE.2019.2922636

Wang SZ, Cao JN, Yu PS (2020) Deep learning for spatio-temporal data mining: a survey. In: IEEE transactions on knowledge and data engineering. https://doi.org/10.1109/TKDE.2020.3025580

Yang GL, Cai Y, Reddy CK (2018) Spatio-temporal check-in time prediction with recurrent neural network based survival analysis. In: Twenty-Seventh international joint conference on artificial intelligence, Stockholm, Sweden, 13–19 July

Ye M, Yin PF, Lee WC (2010) Location recommendation for location-based social networks. In: 18th SIGSPATIAL international conference on advances in geographic information systems, San Jose, California, 02–05 Nov

Yuan J, Zheng Y, Xie X (2012) Discovering regions of different functions in a city using human mobility and POIs. In: 18th SIGKDD conference on knowledge discovery and data mining, Beijing, China, 12–16 Aug

Zhang JF, Hua XS, Huang JQ, Shen X, Chen JY, Zhou Q, Fu ZH, Zhao YR (2019a) City brain: practice of large-scale artificial intelligence in the real world. IET Smart Cities 1(1):28–37

Zhang YX, Wang SZ, Chen B, Cao JN (2019) GCGAN: generative adversarial nets with graph CNN for network-scale traffic prediction. In: International joint conference on neural networks, Budapest, Hungary, 14–19 July

Zheng WC, Zheng Y, Xie X, Yang Q (2010) Collaborative location and activity recommendations with GPS history data. In: 19th international conference on World Wide Web, Raleigh, North Carolina, USA, 26–30 April

Zheng Y, Liu YC, Yuan J, Xie X (2011) Urban computing with taxicabs. In: 13th international conference on ubiquitous computing, Beijing, China, 17–21 Sep

Zheng Y, Liu F, Hsieh HP (2013) U-Air: When urban air quality inference meets Big Data. In: 19th SIGKDD conference on knowledge discovery and data mining, Chicago, Illinois, USA, 11–14 Aug

Zheng Y, Capra L, Wolfson O, Yang H (2014a) Urban computing: concepts, methodologies, and applications. ACM Trans Intell Syst Technol 5(3):1–55

Zheng Y, Liu T, Wang YL, Zhu YM, Liu YC, Chang E (2014b) Diagnosing New York City's noises with ubiquitous data. In: 16th international conference on ubiquitous computing, Seattle, Washington, 13–17 Sept

Zhou X, He J, Huang GY, Zhang YC (2015) SVD-based incremental approaches for recommender systems. J Comput Syst Sci 81(4):717–733

Zhu XJ (2005) Semi-supervised learning literature review. Technical report 1530. University of Wisconsin-Madison

Senzhang Wang is an Associate Professor of the College of Computer Science and Technology at the Nanjing University of Aeronautics and Astronautics. He is a member of ACM, AAAI, and CCF. His research interests include spatio-temporal data mining and graph data mining.

Jiannong Cao is a Chair Professor of the Department of Computing at The Hong Kong Polytechnic University, and leads the Internet and Mobile Computing Laboratory. He is an IEEE Fellow, a distinguished member of ACM, and a senior member of CCF. His research interests include distributed compu-ting, wireless sensing, mobile computing, and big data.

Chapter 44
Microsimulation

Mark Birkin

Abstract From origins in economics and financial analysis, microsimulation has become an important technique for spatial analysis. The method relies on conversion of aggregate census tables, sometimes complemented by sample data at the individual level, to synthetic lists of people and households. The individual records generated by the microsimulation can be aggregated flexibly to small areas, linked to create new attributes, and projected forward in time under stable conditions, or in the context of 'what-if' policy scenarios. The chapter outlines the basic building blocks of microsimulation and shows how these are combined within a representative practical application. It is argued that further progress can be expected through advances in computation, assimilation of data into models, and greater capacity to handle uncertainty and dynamics. We also expect the creation of more sophisticated architectures to reflect the interdependence between population structures at the micro-scale, and the supply-side infrastructures and urban environments in which they evolve.

44.1 Background to Microsimulation

Microsimulation models were introduced to the literature by Guy Orcutt in the 1950s. The approach was initially conceived as a powerful way to evaluate the distributional impact of economic and financial policies. The essence and distinctive feature of the method is that it proceeds through the specification and analysis of discrete entities which typically represent persons or households, in contrast to array-based representations which count the number of occurrences of a particular type. Consider for example an appraisal of the consequences of a series of changes in taxation which depend on the age, marital status, and income of the subject. A microsimulation approach would specify the population as a list of individuals, including age, marital status, and income as characteristics, to which an updated set of taxation rules can easily be applied. The notion of applying one or more discrete rules to a list of elements in order to determine an outcome ("list processing," see below) is a

M. Birkin (✉)
School of Geography, University of Leeds, Leeds, UK
e-mail: m.h.birkin@leeds.ac.uk

W. Shi et al. (eds.), *Urban Informatics*, The Urban Book Series,
https://doi.org/10.1007/978-981-15-8983-6_44

central feature of the microsimulation modeling approach. The individual elements may then be combined into groups for cross-sectional analysis as required ("flexible aggregation," see below).

The addition of a spatial label to the list of population characteristics provides a straightforward means to introduce a geographical element. Spatial microsimulation approaches have been popular in the analysis of health-care systems, education, transport and mobility, labor markets, retailing, and demographic analysis. Often the spatial disaggregation of the model rules (or parameters) can add further value, for example by specifying place-based variations in migration rates within a demographic model, but this need not necessarily be a fundamental element of the approach. Just as economic microsimulation models were originally established to investigate the effect of changing rules, spatial microsimulation models (MSM) are equally well suited to the assessment of scenarios involving changing parameters (e.g. future demographic change) or in the provision of infrastructure or services. Hence, the models can be powerful components within spatial decision-support systems for city planning.

Another important feature of spatial MSM is that they can be used to determine the impacts of policy or scenarios across a population even when detailed profiles for individuals or households are not available. The relevant methods usually involve synthetic estimation of individual records, typically using iterative proportional fitting from aggregate data or equivalent methods. Aggregate data are often easily accessible from sources such as neighborhood-level census tables, and MSM can prove to be a very efficient means to leverage these data. However, the methods can also be adapted to exploit real individual records which are increasingly available in the age of big data, for example through government departments, service operators, and consumer-facing organizations. Since individual databases of this type are rarely comprehensive or completely representative, in this case a major interest is in reweighting samples in order to maximize their value.

In this chapter, we will provide an introduction to fundamental issues and concepts in microsimulation modeling. Through an idealized but meaningful example, the major features and techniques will be described. Against this background, a more practical and powerful implementation will be outlined, concentrating on a specific but wide-ranging program of MSM for infrastructure assessment. We will discuss—in relation to both the main case study, and other relevant applications—some of the major areas of interest and further development potential for MSM at the present time. Conclusions and reflections on the evidence will be presented.

44.2 Overview of Methods and Concepts

44.2.1 Population Synthesis

When dealing with spatial data, it is typically the case that a range of counts will be known for various attributes across an array of small areas. Consider the example in Table 44.1, where distributions are presented across four typical areas in a region. These are the kinds of data which have been available to researchers from population censuses and surveys for many years. The five dimensions of variation displayed are lifestage, household size, tenure, car ownership, and socio-economic status, and these vary in a natural way across area types. For example, there are more people living in flats (apartments) in urban areas, a heavy concentration of young adults in student areas, and the highest rates of car ownership in the countryside.

The essence of the microsimulation is to substitute synthetic individuals for the cell counts in each area. So for example, in Area 1, we will move to a list showing 1000 people, each with five attributes, rather than counts for every possible attribute of each state summing to 1000. In early applications (e.g. Birkin and Clarke 1988, 1989), a straightforward sequential estimation process is adopted. Let us suppose that the first attribute to be estimated is lifestage, and then, we would proceed immediately by creating 500 individuals in Area 1 who are young adults, 300 as family members, 100 as empty nesters and 100 as retired. In Area 2 there are 100 young adults, and so forth.

Next, we add car ownership as an attribute, and since the rate of car ownership in Area 1 is 40%, then 200 young adults become owners of a car, and 300 are not. We continue this process for tenure, household size, and socio-economic status. The number of simulated individuals adhering to each attribute combination can be

Table 44.1 Population distributions in four idealized urban areas

		1: City	2: Country	3: Students	4: Suburbs
Lifestage	Young	500	100	400	100
	Family	100	200	300	500
	Empty-nest	100	300	200	300
	Retired	300	400	100	100
Household	Single	600	200	750	200
	Multi-person	400	800	250	800
Tenure	House	400	800	200	800
	Apartment	600	200	800	200
Car-owners	Car	400	800	200	600
	No car	600	200	800	400
Socio-economic status	Managerial	250	600	200	800
	Manual	750	400	800	200

expressed as:

$$X_i^{km} = \prod_k \left(p_i^{km}\right) X_i^{**}$$

for characteristics m relating to attribute k in area i, where X is a count and p is a probability.

For example, the most numerous group in Area 1 (City) within the simulation will have a profile reflecting the most numerous characteristics for each attribute, that is, young non-car-owners, living alone in apartments, with manual occupations. Members of this group will appear 81 times (= 0.5 × 0.6 × 0.6 × 0.6 × 0.75 × 1000). A natural way to represent members of this group is simply as a list (11222)—lifestage is 1 (young), household is 1 (single), tenure is 2 (apartment), car ownership is 2 (does not have a car), and occupation is 2 (manual worker; see Table 44.1). The reader should be easily satisfied that the most numerous grouping in Area 2 is (42111); in Area 3, it would be (11222); and in Area 4 (22111).

Among many objections to this excessively simplified, presentation of the method is that the value in converting a small number of counts ($N = 12$) for each area into a list of 1000 people with 5 attributes ($N = 5000$) is not immediately apparent—but this should be more obvious by the end of this short exposition. Another problem is that it is unlikely a simple integer value will result from the product of a number of residents in an area (rarely likely to be as convenient a number as 1000 in practice) multiplied by a number of probabilities. This issue is usually addressed in MSM using Monte Carlo sampling—if there is a 60% chance that an individual lives alone then we draw lots, or random numbers, to assign household size. If that number is less than 0.6, then a single person household is the result (Lovelace and Ballas 2013 is one instance of a more sophisticated presentation and discussion of using integer weights to avoid any problems which might result from the assignment of fractions of individuals or households in spatial MSM).

44.2.2 *Iterative Proportional Fitting*

A third obvious objection to the simplified example in 2.1 is that independence between characteristics will rarely be a useful assumption. Thus, affluent white-collar workers are much more likely to be car owners than the unemployed, regardless of geographical location. Young people are more likely to be apartment dwellers, and so on.

This problem is usually handled using iterative proportional fitting (IPF). In the example above, it has in effect been assumed that compound probabilities for five attributes can be created as a linear combination of five independent constraint vectors, that is:

$$p\left(x_i^{k1}, x_i^{k2}, x_i^{k3}, x_i^{k4}, x_i^{k5}\right) = p\left(x_i^{k1}\right) p\left(x_i^{k2}\right) p\left(x_i^{k3}\right) p\left(x_i^{k4}\right) p\left(x_i^{k5}\right)$$

In practice, more complex tables will allow much better estimates to be generated. For example, in the UK Census 2011, it is possible to utilize tables of car ownership by age (V1, V4), socio-economic status by age (V1, V5), household size by age and tenure (V1, V2, V3), and household size by age and socioeconomic status (V1, V2, V5). IPF provides the means to assemble such multidimensional constraints into a single set of estimates of the combined probability distribution:

$$p\left(x_i^{k1}, x_i^{k2}, x_i^{k3}, x_i^{k4}, x_i^{k5}\right) = f^{IPF}\left[p\left(x_i^{k123}\right)p\left(x_i^{k125}\right)p\left(x_i^{k14}\right)p\left(x_i^{k15}\right)\right]$$

As the name implies, the mechanics of this procedure involve successive adjustment of the combined probability distribution for consistency with each probability subset. This iterative procedure is known to be robust and convergent for the great majority of relevant problems (Fienberg 1970; Lomax and Norman 2016). Furthermore, IPF can be extended to accommodate large numbers of constraints with complex interactions.

44.2.3 Reweighting

Thus, IPF provides a robust and effective way for creating combined probability distributions across attribute sets. Ultimately, however, the method relies on the statistical estimation of individual data from aggregate totals. An alternative approach is to use data which are directly generated at the individual level. For example, suppose that a local authority holds data on claimants of housing benefits, then it may be possible to make a direct estimate of the impact of changing benefits rules on that population. Even in this situation, however, a common situation would be that changing brings a new target population into view—hence, to identify those affected, some more comprehensive simulation of the population will be required. MSM provides the means for extensive assessment of this kind.

A more typical situation is that some sample of individual data may be accessible (e.g. a Sample of Anonymized Records in the UK Census, or the Public Use Micro-Sample or PUMS in its U.S. equivalent). Provided that the sampling is robust, then data of this kind can be relied on to preserve cross-attribute relationships in the underlying population. The task for microsimulation is now to reweight the sample data in order to represent the nature of small areas: So in our example above, one would wish to apply higher weights to young people still in education when reconstructing the population of a student area; in the countryside, one oversamples for car-owners; and so on. Now, the procedure must ensure that weights are generated in such a way that when the data are aggregated all known constraints are still observed. In practice, the common approach to this problem is to select at random from a sample population and then switch individual records in order to improve the fit to known constraints. Simulated-annealing algorithms which allow backward steps have been found to be particularly effective (Harland et al. 2012), although genetic algorithms and other heuristics such as tabu search have also been applied (Williamson et al. 1998; Zhu et al. 2015; Lidbe et al. 2017).

44.2.4 Data Linkage

An essential characteristic, and strength, of the MSM approach is an ability to thicken data sets, that is, to extend from a limited set of attributes into a much more extensive range of characteristics. In the simple example at Sect. 44.2.1, this is achieved by adding new characteristics from a different census table with independence. Once IPF is introduced, then the new attribute is related to the existing ones through a complex set of interrelationships. A more general approach to this problem, which is especially useful when data are reweighted from an individual sample, is to link between data sets.

Suppose we continue our example in which a population is characterized by age, socio-economic status, car ownership, etc. A lifestyle data set is made available in which respondents have declared their income based on age, car ownership, and occupation. The linkage problem is simply to add an income attribute by connecting the lifestyle data to the core demographics of the MSM. For straightforward problems, this can be achieved by creating a set of conditional probabilities for different income states in relation to the various independent variables and then using Monte Carlo sampling as above. A more general approach would be to create similarities between the individual records in each data set and then to combine the records. Where the number of records in the data is large relative to the attribute combinations, then this might result in multiple matching records in the target database. Again, this situation could be resolved by Monte Carlo sampling, that is, by selecting any of the matching records at random. Where the number of attribute combinations is very rich, or perhaps the linkage is to quite a small sample, then a perfect match may not be achievable. An alternative would be to create probabilistic linkages between the data sets, and so the linkage problem is to find a record in the target data set which has a high level of similarity to the origin record. This is tricky problem to resolve in view of the difficulty in equating (say) a situation in which two individuals are similar in every respect except they have different genders, as against two individuals who are identical except that one is a car owner and the other is not. Methods to resolve this difficulty, including a general application across ordinal, nominal, and categorical data sets, have been proposed and implemented by Burns et al. (2017). Of course, this method extends easily and naturally to the linkage of multiple attributes, either sequentially or simultaneously (e.g. if the lifestyle data set also includes expenditure, hobbies, or attitudes).

44.2.5 Efficient Representation and Flexible Aggregation

In Sect. 44.2.1 above, a question was raised as to why it might be advantageous to represent a city with a modest population as a list, rather than an array. Regardless of the other benefits described elsewhere, the value of this approach can quickly be seen as soon as the number of attributes and classes becomes more substantial. Van Imhoff and Post (1998) describe such an example in pure demographic terms, with a focus on a sub-model of reproduction. The likelihood of becoming pregnant might reasonably be supposed to vary substantially by single years of age in the mother, let us say in the range 15–44, but also according to marital status (married, single, widowed, or divorced), size of family (0,1,2,3,4 +), socio-economic group (6 classes), educational attainment (4 classes), employment status (3 classes), ethnicity (6 classes), and tenure (4 classes). In this situation, the number of potential unique states is evidently $30 \times 4 \times 5 \times 6 \times 4 \times 3 \times 6 \times 4 = 1.08$ million. So in any city or region with less than a million women of child-bearing age, it makes more sense to represent this population in the form of a list of individuals, rather than as a huge array with even more cells. Introduce some additional attributes (health status, socio-economic group, and educational attainment of the partner, perhaps), and the same consideration would apply across quite a large country.

This issue is doubly significant when considering small areas, especially when there are interactions, as for example in the consideration of migration, commuting, or retail flows. For example, the city of Leeds is frequently examined at a geography of more than 1000 census output areas, for example, when considering new housing developments, investments in transport infrastructure, or retail provision. Between these areas, there are evidently more than one million origin–destination pairs—many more than the number of workers, shoppers, or movers in the city. Hence, spatial MSM provides a powerful basis for efficient representation of both the structure and interaction patterns of population groups at a variety of geographical scales.

The representation of populations at the atomic level of individuals or households also permits flexible aggregation to any desired level of spatial or sectoral detail, provided only that the attributes of concern are appropriately embedded in the underlying data model. Of course, the census itself uses a complete (or almost complete) register of individual and household returns, and then aggregates these across specific topic areas for neighborhoods and regions—as we saw above, for example, in the case of car ownership or household composition by age of head. If car ownership, household composition, and age of head are included in the MSM along with a spatial identifier, then it is a straightforward matter to reproduce this logic, with the potential to cross-tabulate all three variables simultaneously if that is desirable. Should the MSM be extended to include twenty, thirty, or forty plus variables, then the potential attribute combinations become explosive, and the scope for diverse perspectives on a wide range of problems becomes very rich indeed.

44.2.6 List Processing

Another essential strength of MSM is the ability to apply rules for individual units of the population. A straightforward and common example of this would be in applying changing regimes for taxation: The impact of a new budget might be a change of income tax according to the earnings and marital status of a householder; the effect of changing fuel duty would depend on vehicle ownership and utilization; the impact of duties on cigarettes and alcohol would vary in relation to specific behaviors and habits. Each of these elements can quite easily be computed through a MSM, provided only that the determinants (i.e. income, car ownership, alcohol consumption, and so on) have already been represented in the base population. This means that not only is it possible to estimate potential benefits to the tax authorities, but also to evaluate distributional impacts on demographic sub-groups or small area populations in a city.

The concept of list processing can be applied in a different form, but with similar power and impact, to problems involving projection or forecasting of the population over time. For example, in relation to the attribute of age (in years), if we wish to project a population in time at single-year intervals, then age also increments by one at each interval. Other demographic processes, such as marriage, migration, or transitions within the labor market, may be subject to transition rates between classes. In this situation, changing states may be handled by Monte Carlo sampling of conditional probabilities (e.g. likelihood of marriage according to age, gender, and economic activity) as before.

44.3 An Example: Models of National Infrastructure

44.3.1 Overview

In 2010, partners from seven UK universities began working together on a Research Council program to explore future infrastructure options, requirements, and future scenarios. The Infrastructure Transitions Research Consortium (ITRC) considers the five sectors of transport, energy, water, wastewater, and IT, working in partnership with utilities, engineers, and regional and local providers, and acts as a trusted adviser to government through the National Infrastructure Commission. A second phase of funding with a focus on multi-scale infrastructure systems analytics (MISTRAL), including the translation of experience to international contexts, will continue until 2020.

Infrastructure projects are expensive and return on investment takes place over long-term horizons, regardless of whether these returns are measured in financial, social, or environmental terms. ITRC has a temporal framework which looks forward as far as possible toward the end of the twenty-first century. In order to create a more detailed understanding of the demand for infrastructure and its spatial and

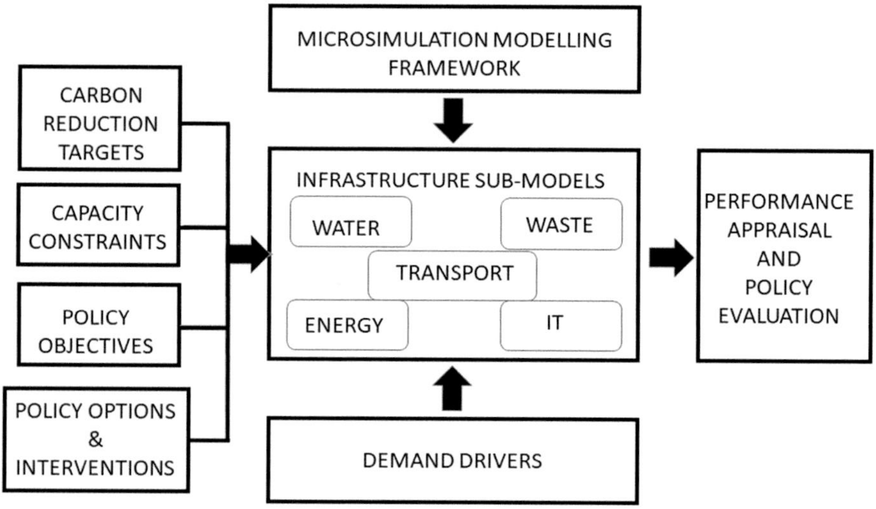

Fig. 44.1 Model structure for infrastructure assessment

sectoral composition, ITRC requires highly disaggregate estimates of future population in relation to individual attributes, household groupings, and the character of neighborhoods and small areas.

The overall structure of the ITRC assessment process is shown in Fig. 44.1 below. ITRC uses a spatial microsimulation model to provide demographic inputs to the demand-estimation process for each of the five infrastructure sectors. The MSM is specified to the level of individuals with rich attributes, including demographics, social and economic profiles, housing, health, and labor market characteristics. Working with domain specialists in the research team, a consensus is established on the attributes representing the most important direct or proxy measures for the major drivers of infrastructure demand. Linking to consumption data from market-research surveys or direct measures of service use, for example from smart meters, sensors, or utility bills, makes it easy to translate population estimates into demand for infrastructure. Each of the demand sub-models which are driven from the MSM is linked to supply-side representations and policy options in order to drive a rich decision-support structure for infrastructure assessment. In the next sub-section, we explore the detail and a specific example.

44.3.2 An Application of Spatial MSM to Energy Modeling

44.3.2.1 Population Reconstruction

In the first phase of development of the ITRC, the UK population was recreated from the Sample of Anonymized Records (SAR; Thoung et al. 2016). Each element of the

SAR represents a real individual or household from the 2011 census from which small area labels and other potential identifiers have been removed in order to maintain the privacy of the subjects. The SAR therefore contains all of the demographic and socio-economic identifiers of the census including age, marital status, ethnicity, general health, education, occupation, car ownership, household composition, tenure, dwelling type, and a number of others.

The SARs are reweighted to reflect the composition of each census output area (a neighborhood with a typical size of no more than 200 households) using a simulated-annealing algorithm developed at Leeds (Harland 2013).

An approach to creating demand estimates for an indicative sector (energy) is described by Zuo and Birkin (2014). The English Housing Survey (EHS) contains in-depth household interviews and physical surveys for 17,000 households. EHS facilitates profiles of energy consumption and expenditure by fuel type and purpose for a rich selection of population and housing characteristics. The MSM used a CHAID (chi-square automatic interaction detection) approach to cluster households in both the MSM and the EHS into 41 categories based on a combination of dwelling type, household size, age and occupation of the household head, lifestage, and household composition. A simple probabilistic match was applied to link records from the MSM and the EHS (i.e. records from the EHS were selected at random from the relevant cluster). Some contrasting energy-consumption profiles for different household types are shown in Fig. 44.2.

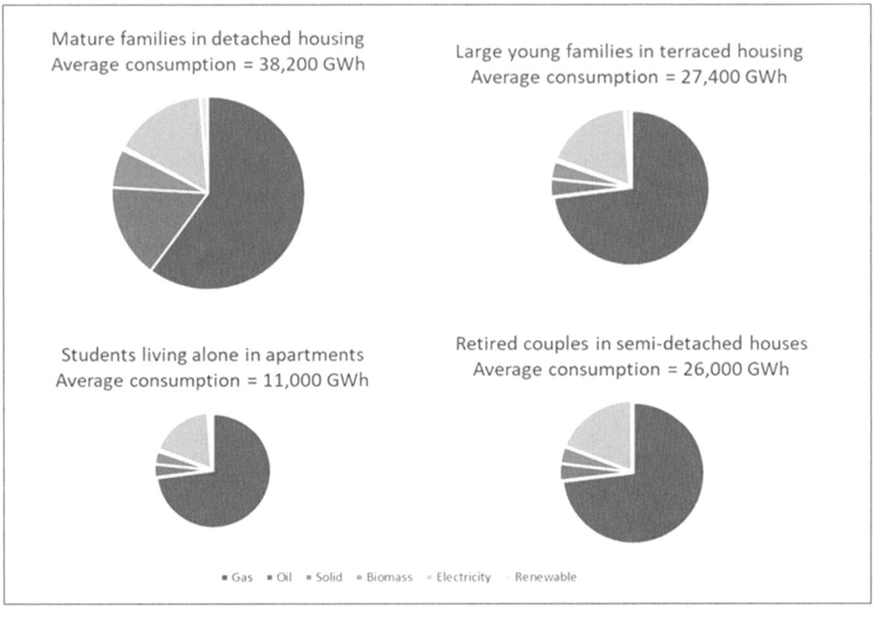

Fig. 44.2 Outputs from a microsimulation of energy consumption by household

44.3.2.2 Population Projection

The base populations within the ITRC MSM are projected forward in time using inputs from both the Office for National Statistics (ONS) National and Sub-National Population Projections (SNPP). The national projections provide the basis for estimation of aging, fertility, and mortality ("natural change") within the population, whereas the SNPP allows the introduction of migration and the calibration of the natural change parameters to local areas. The essence of this process is therefore to list-process the base populations using a combination of demographic change rates (for fertility, mortality, and migration). The parameter estimates are managed in order to ensure consistency of the simulation outputs with the ONS regional and population profiles. For more detail, see Zuo and Birkin (2014) and Thoung et al. (2016).

This simulation process adds considerable richness to the ONS estimates by permitting detailed spatial disaggregation on the sub-national projections—which are only available over a 25 year planning horizon—and by their extrapolation alongside the national medium (50 year) and long-term projections (75 years). The flexibility of MSM is also fully exploited in ITRC through the use of variant population projections. For much of the work which has been presented to policy-makers, eight scenarios are presented which illustrate the impact of future changes in technology, affluence, and political circumstances on the population (Thoung et al. 2016).

44.3.2.3 Scenarios

The spatial detail of the MSM is particularly important when considering future infrastructure investments which have strong local dependencies, including renewable energy, personal mobility, and the supply of water. In the outline above, it has been seen that energy consumption is expected to grow in relation to expansion of the population, and be subject to compositional shifts in relation to changes in supply. One of the major motivations of ITRC is to consider the potential impacts of climate change on infrastructure (Jenkins et al. 2014). In one published application from the ITRC, climate-change projections from the Met Office Hadley Center were combined with the spatial MSM, with modified energy consumption rules relating variations in energy use to regional and seasonal variations in the climate within the EHS. This scenario was extended to 2100. A significant reduction in household energy use was expected due to global warming (see Fig. 44.3). The authors note that the potential to counterbalance due to increased use of air conditioning was not examined because of limitations in the base data. However, a variety of other behavioral shifts were also considered, with evidence drawn from extant published studies. These included adoption of solar power, insulation, double glazing, adoption of low energy lighting, and shifts to more efficient central heating systems. Behavioral change was not expected to affect cooking or the use of electrical appliances (Zuo and Birkin 2014).

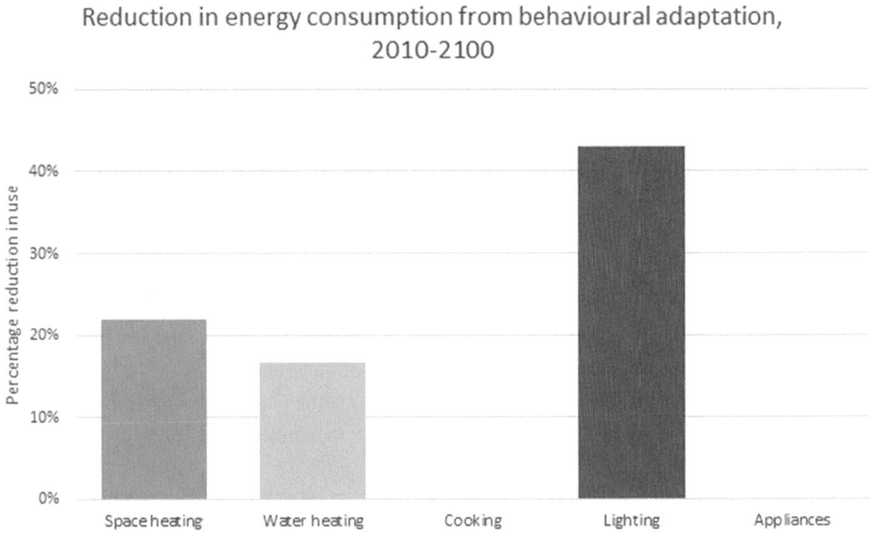

Fig. 44.3 Reductions in energy consumption from a behavioral simulation

44.3.3 Extensions

The architecture of spatial microsimulation which underpins the ITRC project has recently been completely overhauled. A technology platform for Synthetic Population Estimation and Scenario Projection (SPENSER) now services the infrastructure sub-models. It is also designed to support extensions to sectors such as education and health. The capability of the new system to represent diverse behavioral components has already been demonstrated through a flexible application to consumer spending across a full range of expenditure categories (James et al. 2019). This implementation is specifically aligned to the study of future meat consumption under various alternative scenarios for production, sustainability, affluence, and lifestyle preferences.

SPENSER has a more modular design than the previous deployment within ITRC, with separate routines for data mobilization, population recreation, forecasting, and scenario building. It is hoped that a more robust design will make SPENSER amenable to a wider range of substantive improvements in the underlying scientific approach. In the next section, some key elements of the agenda for future development are discussed.

44.4 Priorities for Spatial Microsimulation

44.4.1 Computation

The computational burden attached to spatial microsimulation models is often quite considerable. This need arises from a desire to represent the population with significant variety (i.e. many attributes) at a fine level of spatial resolution (i.e. a lot of zones), and potentially with complex spatial or behavioral interactions to model or represent. Significant computation is needed in both the generation of the initial population, including both reconstruction and linkage, and in projections of the model forward in time.

Simple approaches to reweighting baseline populations, or using conditional probabilities from iterative proportional fitting, are not especially expensive in computational terms when they are based on one-shot estimates of the parameters. Iterative approaches including genetic algorithms (GA) and especially simulated annealing (SA) have persistently yielded better results, but are often slow to converge. These techniques depend on complex evaluations of the fitness of a model: in principle a single step of either GA or SA involves exchanging the position of two elements in the simulation (e.g. moving and replacement of an individual from one zone to another), then reaggregating the population at zone level, calculating the fit to multiple constraining totals, and then applying an evaluation function to assess the utility of the switch. This activity can be repeated multiple times for each member of a population of millions, within a loop which could itself be executed hundreds of times within the algorithm. The dynamics of the modeling also involve complex processing across a large population size, often with small time steps and multiple scenario combinations. The impacts could become explosive if adopting methods such as ensemble modeling as a means for exploring sensitivities or robustness in the model outcomes. There is no doubt that the difficulty in accessing adequate computational resources has been an impediment to exploration of some potentially fertile approaches, such as the use of ensembles.

More intense applications of spatial MSM are being permitted to some degree by the availability of high-performance computing. For example, SPENSER has access to the Data Analytics Facility for National Infrastructure (DAFNI) as a platform for executing complex model runs. Similar capability exists within the Integrated Research Campus at the Leeds Institute for Data Analytics. Nevertheless, data-services infrastructures remain scarce, difficult, and expensive to access.

Rather than the provision of enhanced computational power, simplification of the models themselves is clearly an alternative to consider. A natural strategy would be to reduce the population size, for example by sampling, or the representation of subsets rather than individuals (Parker and Epstein 2011). This approach seems more feasible for national applications than those involving small spatial zones in which the full variety of the population must be retained. A more promising method which has been adopted in dynamic microsimulation is to lengthen the time interval between processing steps. When considering discrete events such as birth, migration or death, the usual method is to apply transition probabilities (or hazards; Clark and Rees 2017) to a population at risk at regular intervals, generally annualized. If the occurrence of such events is on average significantly less than once a year, then an option would instead be to process the time to next event and save the trouble of repeated assessments for change of state in the intervening period. This technique has been successfully introduced within the Canadian MSM DynaCan (Morrison 2007), and adopted elsewhere.

44.4.2 Uncertainty

The potential for error, and consequent uncertainty in model estimates and projections, is widespread in the microsimulation framework. While MSM are usually created from high-quality sources, including censuses and national statistics, these data are by no means free of bias and inaccuracy. For example, censuses are never completely enumerated, giving rise to errors in the imputation of missing records. Students, transient populations, and the homeless all have significant potential for misrepresentation. When these data are combined, then sophisticated models have the capability to reproduce aggregate constraints with minimal variations. However, the individual estimates are subject to unknown errors which are by definition unobservable to the extent that the purpose of the model is to simulate individual distributions which are not directly measured.

These issues become more challenging for more ambitious applications, for example if a demographic microsimulation is linked to big data for mobility, consumer spending, health, and behavior (Birkin 2018), because such data sets are themselves more variable in data quality and in view of distortions in the linkage process itself.

When the purpose of microsimulation modeling is to assess the effect of changing financial regulations, taxation, or benefits then modeling scenarios can be expected to be relatively robust. When the what-if models are reliant on changing infrastructure, uncertain behaviors, policy environments, and economic circumstances, then any attempts at projection and impact analysis are hugely uncertain. The MSM community has largely sidestepped the problems associated with uncertainty by offering single model estimates, occasionally flexed through defined scenarios with variant input assumptions. This may change if microsimulation chooses to align itself more closely with emerging disciplines in data science. A particular instance of this could be through the adoption of probabilistic programming (Improbable Research 2019). In this new style of model implementation, state variables are assigned distributions rather than discrete values, and operators may be treated in the same way. Hence, this approach lends itself naturally to the expression of outcomes in terms of likelihoods, confidence intervals, or other dimensions incorporating variability and uncertainty. A drawback of this style of research is that tools are still relatively inaccessible and in early stage of development, and experience of complex applications is limited.

44.4.3 Data Assimilation

The origins of spatial microsimulation are as a means to estimate unknown individual-level variations from aggregate data about neighborhoods and small areas. Later, applications incorporate more information by the addition of sample data, in which case the essence of the problem may be more about reweighting. In either of these cases, the ambition is to create simulations in detail from relatively restricted data, and in all circumstances, evaluation of the success of the models is a challenge, because by definition we are estimating things which are unobserved. In the age of Big Data, where increasingly more is known about the world at ever finer scales, the nature of the challenge is beginning to shift toward a view of the world in which it is possible to steer models toward more effective representations through the absorption of evidence. This could be facilitated by data assimilation.

It has been recognized for some time in the complex domain of weather forecasting that methods are needed to update models as new information becomes available. This process of data assimilation has been adopted into agent-based simulation, for example through the adaptation of pedestrian movement models to absorb movement data from street sensors (Ward et al. 2016). There seems no reason in principle that the philosophy and techniques of data assimilation might not be used to calibrate longer-term effects such as spatial diffusion or policy impacts in a microsimulation.

44.4.4 Dynamics

MSM is typically used in one of three modes, which can be characterized as static, comparative static, and dynamic. Static MSM may refer to population reconstruction processes in which aggregate data are decomposed to generate refined distributions at household or individual levels. These outputs may be valuable in their own right, for example to understand the prevalence of at risk groups, or provide inputs to agent-based models (ABM) or other policy models.

Linkage to other data sets is also a static or baseline process, for example using MSM to estimate expenditures or market potential in a retail model (James et al. 2019). As noted above, comparative static is a core mode for tax and benefits assessment (Sutherland and Figari 2013). Comparative-static applications are perhaps the most common in which some variation in the initial conditions allows the MSM to be applied in what-if mode. In SPENSER, many of the scenarios look to the future but are essentially comparative static since they start from the premise that higher level forecasts (such as ONS estimates of the future population) can be disaggregated, and then input to secondary models of demand for infrastructure or consumption of other services.

Truly dynamic models are not entirely absent (Morrison 2007; Li and O'Donoghue 2013; Rutter et al. 2011) but challenging in that they require the incorporation of longitudinal processes in relation to core demographics (e.g. fertility, mortality, and migration) or more specific elements such as morbidity or energy consumption. Backward propagation of MSM as a basis for validating both the structure and logic of dynamic MSM is another concept that might usefully be borrowed from climate-modeling literature, but is as yet relatively unexplored.

Fast and slow dynamics are also a consideration for MSM. Much more attention has been focused on long-term or slow dynamics, and these kinds of models are important for decision making in relation to major infrastructure investment and policy making. However, fast dynamics are becoming more relevant in relation to real-time observation. This makes a connection to data assimilation, and opportunities for real-time evaluation and model enhancement. We will see increasing use of machine learning techniques like reinforcement learning for traffic lights or store promotions, and blurring of boundaries between data science, MSM, ABM, and other forms of individual-based modeling. It is surprising that these approaches are relatively unexplored in commercial applications, where personalization and precision targeting are a priority with the growing availability and fidelity of individual data.

44.4.5 Interdependence

Applications of MSM are well-suited to the problems of demand estimation, which are typified by the uses of SPENSER as a tool within the ITRC framework for future infrastructure assessments. Similar applications can be seen in the estimation of retail expenditure (James et al. 2019), educational attainment (Kavroudakis et al. 2013), health care (Clark and Rees 2017) and even the incidence of crime (Kongmuang 2006) and the need for jobs (Ballas and Clarke 2000). The beauties of the technique in this regard are multiple (as we have seen), providing a powerful means to connect aggregate data to individual-level modeling, introducing rich and multiple simultaneous representations of individual attributes, and a sophisticated understanding in changing drivers of consumption over time.

Nevertheless, conceptual architectures which view microsimulation purely as a foundational layer in the modeling process are often in danger of simplifying away many of the subtle and vitally important interactions which underpin real-world problems. The importance of interaction and interdependence between individuals has always been fundamental to ABM, in which the capacity for complex structures to emerge—often in unexpected ways—is a cornerstone of the method (Schelling 1969). However, while conceptually rich in this sense, ABM is typically less strongly grounded in the empirical realities of everyday life.

The benefits of linking microsimulation to meso-scale representations of land-use and service provision have been recognized in early applications to a retail market (Birkin and Clarke 1987; Nakaya et al. 2007). In this framework, a microsimulation is used to create a rich population, which in turn forms the basis for expenditure assessments across a tapestry of small areas. These expenditure estimates are then combined with networks of service provision through a spatial interaction model (SIM), hence creating revenue flows from neighborhoods to shopping centers. These flows can then be sampled in order to create assignments of retail preferences for individual consumers, thus closing the loop from demand to supply. A similar process underlies a module within SPENSER which connects the microsimulation to migration flows through a spatial interaction model of internal migration (SIMIM; Lomax and Smith 2019). In order to fully embed microsimulation within land-use transport interaction models, however, it might be argued that the reciprocal dynamics of infrastructure systems including housing and transport must be fully incorporated within the model system.

The resulting applications would be somewhat analogous to the network planning models developed in Leeds by Geographical Modeling and Planning (GMAP) Limited the 1990s, in which service delivery was co-designed with retail demand. George et al. (1997) provide a good description of a representative problem. The broader significance, perhaps, of the GMAP experience (Birkin et al. 1996; Birkin et al. 2002, 2017) is in seeing spatial analysis approaches including MSM as elements of spatial decision-support systems (Geertman and Stillwell 2009). Robust translation of such ideas into the urban planning domain, for example through the integration of SPENSER with other models such as UCL's Quantitative Urban Analytics (QUANT) model of land-use and transport interactions, could provide stronger foundations for spatial decision support than hitherto.

While MSM is almost exclusively used to represent both individuals and households as the entities within a modeling system, there is no reason why other elements such as vehicles, houses, schools, hospitals, firms, or retail outlets might not equally be represented in a similar way, with rich characteristics and complex behavioral drivers. Indeed, one might argue whether cellular automata, in which the building blocks are land-use parcels changing in character through time, are so different to microsimulation. Hybrid models which combine MSM with SIM, land-use and transport interaction models, or even cellular automata are likely to become increasingly popular, but the absorption of more complex actors representing complementary sectors might be seen as a fully viable alternative strategy.

44.5 Conclusions

Spatial MSM has been developed as an important variant from the introduction of similar individual-based models in economics and financial policy. The technology of spatial microsimulation has progressed steadily over a period of more than thirty years, allowing population distributions in very small areas to be faithfully represented. The models benefit from increasingly detailed and diverse sources of data. This also provides underpinning for applications to a diverse range of problems.

The scope for further enrichment of spatial MSM is substantial, for example drawing on computational advances and progression of techniques in data science, machine learning, and artificial intelligence. This could help to increase the robustness of models, especially when their dynamic qualities are considered as a basis for projection and forecasting.

References

Ballas D, Clarke G (2000) GIS and microsimulation for local labour market policy analysis. Comput Environ Urban Syst 24:305–330
Birkin M (2018) Big data for social science research. Ubiquity, Jan 2018. 1–7. https://doi.org/10.1145/3158339
Birkin M, Clarke M (1987) Comprehensive models and efficient accounting frameworks for urban and regional systems. In: Griffith D, Haining R (eds) Transformations through space and time. Martinus Nijhoff, The Hague, pp 169–195
Birkin M, Clarke M (1988) SYNTHESIS: a SYNTHetic spatial information system for urban modeling and spatial planning. Environ Plan A 20:1645–1671
Birkin M, Clarke M (1989) The generation of individual and household incomes at the small area level using synthesis. Reg Stud 23:535–548
Birkin M, Clarke G, Clarke M, Wilson A (1996) Intelligent GIS: location decisions and strategic planning. Geoinformation International, Cambridge
Birkin M, Clarke G, Clarke M (2002) Retail geography and intelligent network planning. Wiley, Chichester
Birkin M, Clarke G, Clarke M (2017) Retail location planning in an era of multi-channel growth. Routledge, London
Burns L, Heppenstall A, See L, Birkin M (2017) Developing an individual-level geodemographic classification. Appl Spat Anal Policy 11:417–437. https://doi.org/10.1007/s12061-017-9233-7
Clark SD, Rees PH (2017) The drivers of health trends: a decomposition of projected health for local areas in England. In: Swanson DA (ed) Frontiers in applied demography. Applied demography series 9. Springer, Berlin, pp 21–40
Fienberg SE (1970) An iterative procedure for estimation in contingency tables. Ann Math Stat 41:907–917
Geertman S, Stillwell J (eds) (2009) Planning support systems: best practice and new methods. Springer, New York
George F, Radcliffe N, Smith M, Birkin M, Clarke M (1997) Spatial interaction model optimisation on parallel computers. Concurrency: Pract Exp 9(8):753–780
Harland K (2013) Microsimulation model user guide (Flexible modeling Framework). NCRM working paper. National centre for research methods
Harland K, Heppenstall A, Smith D, Birkin M (2012) Creating realistic synthetic populations at varying spatial scales: a comparative critique of population synthesis techniques. JASSS 15(1):1
Improbable Research (2019) Keanu: a probabilistic approach. https://github.com/improbable-research/keanu. Accessed 27 Oct 2019
James W, Lomax N, Birkin M (2019) Local level estimates of food, drink and tobacco expenditure for Great Britain. Sci Data. https://doi.org/10.1038/s41597-019-0064-z
Jenkins K, Hall J, Glenis V et al (2014) Probabilistic spatial risk assessment of heat impacts and adaptations for London. Climatic Change 124:105–117
Kavroudakis D, Ballas D, Birkin M (2013) Using spatial microsimulation to model social and spatial inequalities in educational attainment. Appl Spat Anal Policy 6(1):1–23
Kongmuang C (2006). Modeling crime: a spatial microsimulation approach. Ph.D. thesis. University of Leeds
Li J, O'Donoghue C (2013) A survey of dynamic microsimulation models: uses, model structure and methodology. Int J Microsimulation 6(2):3–55
Lidbe A, Hainen A, Jones S (2017) Comparative study of simulated annealing, tabu search and the genetic algorithm for calibration of the microsimulation model. Trans Soc Model Simul Int 93:21–33
Lomax N, Norman P (2016) Estimating population attribute values in a table: "get me started" in iterative proportional fitting. Prof Geogr 68(3):451–461
Lomax N, Smith A (2019) Spatial interaction models of internal migration. https://github.com/nismod/simim. Accessed 26 Oct 2019

Lovelace R, Ballas D (2013) 'Truncate, replicate, sample': a method for creating integer weights for spatial microsimulation. Comput Environ Urban Syst 41:1–11

Morrison R (2007) Model 6: DYNACAN (Longitudinal Dynamic Microsimulation Model). In: Gupta A, Harding A (eds) Modeling our future: population ageing, health and aged care. International symposia in economic theory and econometrics, vol 16. Emerald group publishing limited, Bingley, pp. 461–465

Nakaya T, Fotheringham AS, Hanaoka K, Clarke GP, Ballas D, Yano K (2007) Combining microsimulation and spatial interaction models for retail location analysis. J Geogr Syst 4:345–369

Parker J, Epstein J (2011) A distributed platform for global-scale agent-based models of disease transmission. ACM transactions on modeling and computer simulation 22(1): Article 2

Rutter C, Zaslavsky A, Feuer E (2011) Dynamic microsimulation models for health outcomes: A review. Med Decis Making 31:10–18

Schelling T (1969) Models of segregation. Am Econ Rev, Pap Proc 59:488–493

Sutherland H, Figari F (2013) EUROMOD: the European union tax-benefit microsimulation model. Int J Microsimulation 6(1):4–26

Thoung C, Beaven R, Zuo C et al (2016) Future demand for infrastructure services. In: Hall J, Tran M, Hickford A, Nicholls R (eds) The future of national infrastructure: a system-of-systems methodology. Cambridge University Press, Cambridge, pp 31–53

Van Imhoff E, Post W (1998) Microsimulation methods for population projection. Population 10(1):97–136

Ward JA, Evans AJ, Malleson NS (2016) Dynamic calibration of agent-based models using data assimilation. Roy Soc Open Sci 3:150703

Williamson P, Birkin M, Rees P (1998) The estimation of population microdata using data from small area statistics and samples of anonymised records. Environ Plan a 30:785–816

Zhu S, Tey L, Ferreira L (2015) Genetic algorithm based microscale vehicle emissions modeling. Mathematical problems in engineering, article ID 178490. https://doi.org/10.1155/2015/178490

Zuo C, Birkin M (2014) Spatial microsimulation modeling for residential energy demand of England in an uncertain future. GeoSpatial Inf Sci 17(3):157–169

Mark Birkin is Professor of Spatial Analysis and Policy at the University of Leeds. He is also Programme Director and Turing Fellow at the Alan Turing Institute and Director of the Leeds Institute for Data Analytics. He is a Fellow of the Academy of Social Sciences and of the Royal Geographical Society.

Chapter 45
Cellular Automata Modeling for Urban and Regional Planning

Anthony G. O. Yeh, Xia Li, and Chang Xia

Abstract In recent decades, cellular automata (CA) have become popular for evaluating and forecasting urban transformation over time and space, especially in rapidly developing countries. These models enhance the understanding of urban dynamics and the complex interplay between land-use changes and urban sustainability. CA help governments, planners, and stakeholders to predict and evaluate the potential outcomes of future policy alternatives before making decisions. Thus, CA are frequently used to create what-if scenarios for policy implementation. This chapter includes an overview of the basic and state-of-the-art concepts and methods in urban CA modeling, as well as the latest studies, applications, and current problems. First, we conduct a systematic review of urban CA modeling to provide critical comments on previous and recent studies. The basic techniques, including the components of a basic CA model, modifications for urban modeling, and collection of data sources, are then provided, along with a classification of different types of urban CA. Finally, the applications of CA in urban studies and planning practices are presented, as well as discussions of further research. We also point out the major problems in recent studies and applications for further research.

A. G. O. Yeh (✉) · C. Xia
Department of Urban Planning and Design and Centre of Urban Studies and Urban Planning, The University of Hong Kong, Hong Kong, China
e-mail: hdxugoy@hku.hk

C. Xia
e-mail: xia2016@whu.edu.cn

X. Li
School of Geographic Sciences, East China Normal University, Shanghai, China
e-mail: lixia@geo.ecnu.edu.cn

W. Shi et al. (eds.), *Urban Informatics*, The Urban Book Series,
https://doi.org/10.1007/978-981-15-8983-6_45

45.1 Introduction

Urbanization is a global issue characterized by continuous urban land expansion and rural–urban migration (Alcock et al. 2017; Seto et al. 2012). Urban development has brought social, economic, and technological changes, particularly, in developing countries, where cities are sprawling at high rates and metropolitan areas are emerging (Bai et al. 2012; Shahbaz et al. 2016; Zhou et al. 2004). However, large-scale population growth often leads to urban development beyond the carrying capacity of cities. Most of the urban development in developing countries is in the form of sprawl in urban fringes, causing many negative consequences to urban development and the eco-environment at unparalleled scales (Burak et al. 2017; Weeberb 2015). Thus, research into the mechanisms of urban expansion is of great significance for planners and governments to enhance their understanding of urban sustainability.

For understanding the complexity of urban systems, cellular automata (CA), that can provide a powerful simulation tool to predict and understand urban transformation over space and time, is one of the most prevalent urban modeling methods in recent years (Aburas et al. 2016; Santé et al. 2010; Musa et al. 2017). CA offer governments, planners, and stakeholders a tool to forecast and evaluate potential social benefits and environmental outcomes of urban development before implementation. CA also advance our fundamental understanding of urban dynamics and the complex relationships among urban changes, socio-economic development, and sustainable systems.

CA are a kind of discrete dynamic model with unique advantages for simulating complex nonlinear problems. CA originated in the 1940s, when S. Ulan and J. von Neumann considered the possibility of a self-replicating machine. Subsequently, many scholars undertook further studies of CA and helped with its advancement (Codd 1968; Gardner 1971). Wolfram (1984) demonstrated the capacities of CA for modeling complicated natural processes and generating spatio-temporal global changes through local interactions among components. The application of cellular-space models in geographic research was first proposed by Tobler in 1979. Then, the first theoretical approaches of urban CA modeling emerged in the 1980s (Batty and Xie 1994; Couclelis 1985; White and Engelen 1994). The integration of CA and geographic information systems (GIS) led to the simulation of real-world urban development. After the initial wave of urban CA modeling led by Batty, Couclelis, Clarke, and Tobler, research on urban CA moved to China quickly (Li et al. 2017; Zhuang et al. 2017). Since the end of the 1990s, Yeh and Li have developed a series of CA techniques, mainly combining CA with other models and extending cellular states, neighborhood definitions, and transition rules (Yeh and Li 2001; Li and Yeh 2002a). These models have been successfully applied to solving the environmental and ecological problems of rapid urban development in China.

The increasing popularity of CA in urban modeling could be largely attributed to their simplicity, flexibility, controllability, and ability to incorporate the spatial and temporal dimensions of urban development processes. CA can simulate complex dynamic urban systems through simple rules that can work with remotely sensed data and GIS (Santé et al. 2010; Musa et al. 2017). CA are more convenient than other models, such as agent-based models, because of methodologies developed in the past two decades. Another reason why CA have been widely applied in urban modeling is because CA can be easily integrated with GIS. The integration of CA with GIS provides a tool for performing complicated computations based on local information, thus producing better results than differential equations (Musa et al. 2017). However, despite the popular use of CA in urban modeling, errors in input spatial data sources and uncertainty in policies (Yeh and Li 2006) pose challenges in using CA to solve real planning problems (Poelmans and Rompaey 2010).

CA are increasingly being used to simulate spatio-temporal urban expansion and to address many environmental problems. However, defining the most suitable model structures for a specific application problem is difficult. To help users who are not familiar with CA, this chapter provides an overview of the basic and state-of-the-art concepts and methods in urban CA modeling, as well as the latest studies, applications, and current problems. The aim of this chapter is to provide an overview of defining, modifying, and applying CA for urban studies and planning from the perspectives of cell, cell space, neighborhood, time step, and transition rule, along with the collection of required data sources. The different types of CA and their characteristics are described, and the applications and urban issues involved in CA modeling are presented. These discussions attempt to answer the question, "what can and cannot CA provide for the modeler?" In addition, the strengths and weaknesses of CA are identified and common problems of current studies are discussed.

45.2 Methodology and Data Collection

45.2.1 Urban CA for Formulating Urban and Regional Planning Scenarios

The basic components of CA include cell space, cell, neighborhood, time steps, and transition rules. In an urban CA model, each component has geographic implications (Triantakonstantis and Mountrakis 2012). The cell space represents the two-dimensional geographic space composed of regular cells, and the states of cells represent different land uses. The core of a CA model is formed by transition rules. Each cell changes constantly in accordance with its states and the transition rules as time goes on, which represents the systemic deduction and change from an overall perspective.

A formal cell can be a regular grid consisting of square cells, which is particularly suitable for computer processing and compatible with remotely sensed data. Scholars have defined a hexagonal cell space such that the neighborhood could be homogeneous (Iovine et al. 2005). Besides, a cell space can be three-dimensional to represent the vertical growth of urban areas. To make the simulation process closer to the real world, relaxations to the two components are needed. The modified cell space can be based on irregular spatial units, such as Voronoi polygons (Shi and Pang 2000) or graphs (O'Sullivan 2001). Irregular cell space is sometimes presented as a patch-based space (Chen et al. 2014; Wang and Marceau 2013). The irregular spatial unit, such as a cadastral parcel or a census block, is usually represented as a polygon, to reflect land use, population, and economic conditions. Compared with regular cells, parcels or blocks provide a good representation of reality, but lead to complicated definitions of neighborhood. Cell space is normally assumed homogeneous in standard CA, indicating identical and exclusive cells characterized by their states. Nevertheless, the great influence of land attributes on land-use changes, such as transport accessibility or physical conditions, varies the suitability of different cells for certain land uses. Subsequently, the requirements for a non-uniform cell space emerge.

As for neighborhood, there are often two kinds of relaxations. In standard CA, neighborhood is isotropic and homogeneous for each cell (Wu 2002; Xie 1996) and consists of a fixed set of geometrically closest cells (i.e. Moore neighborhood). In urban applications, an extended neighborhood is adopted to consider the neighboring effect of geographic entities (White and Engelen 2000). Neighborhood size can be extended to a specified distance and a weight can be introduced according to the distance, to consider the effect of distance decay. If it is based on irregular units, adjacent units within a certain distance or degree of proximity are used to represent a neighborhood (Shi and Pang 2000). Another widely acknowledged modification is to a non-stationary neighborhood, which defines different neighborhood spaces for different cells (Couclelis 1985). However, this relaxation has been seldom applied due to the difficulty of implementation and vague geographic meanings.

As the core of CA model, transition rules usually entail substantial modifications, considering the particularities and complexity of specific applications. Original transition rules only depend on the states of a cell and its neighborhoods. Given that urban processes are influenced by numerous factors, such as transport accessibility and physical conditions, urban CA models are modified to consider external effects. As CA are flexible, transition rules can be defined in different ways according to the preferences of modelers. Randomness and uncertainty of urban growth, as well as many urban theories, can be reflected in the model structure. Besides, in standard CA, transition rules are static and the same at every time step. However, urban processes and determinants change over time and space, which leads to the necessity of calibrating transition rules based on the specific characteristics of different periods and areas (Clarke et al. 1997; Geertman et al. 2007; Li et al. 2008). For example, Clarke et al. (1997) proposed a self-modifying CA in which transition rules vary over time. The time steps in a formal CA are discrete, which assumes that urban growth occurs at the same time. Many urban CA models apply time steps of different lengths or various time steps for different cells to reflect the influence of specific events with different duration. However, compared with other components of CA, less relaxations have been implemented for time steps.

The future state of a cell depends on the transition rules and its state in the previous moment. A standard CA can be mathematically expressed as follows (Ahmed and Ahmed 2012):

$$S^{t+1} = f(S^t, N) \tag{45.1}$$

where t and $t + 1$ represent discrete time points, S^t and S^{t+1} represent the states of the cell at time t and $t + 1$, respectively, N represents the set of states of neighborhood cells, and f is a transition rule.

The straightforward nature of standard CA limits the ability to represent real-world geographic phenomena (Couclelis 1985). To adapt standard CA in urban applications, the particularities of geographic processes should be included for representing geographic heterogeneity, which leads to the relaxation of original CA components (Couclelis 1997). For example, geographic features in the neighborhood can be embodied in a simplified CA using rule-based structures (Batty 1997; Fig. 45.1):

By integrating CA with GIS databases, a constrained urban CA can be further developed for formulating planning scenarios. It is assumed that the evolution of real cities is influenced by a series of complicated factors which can be defined at various local, regional, and global levels. Some kinds of constraints should be used to regulate the simulation to improve modeling performance. Without constraints, urban

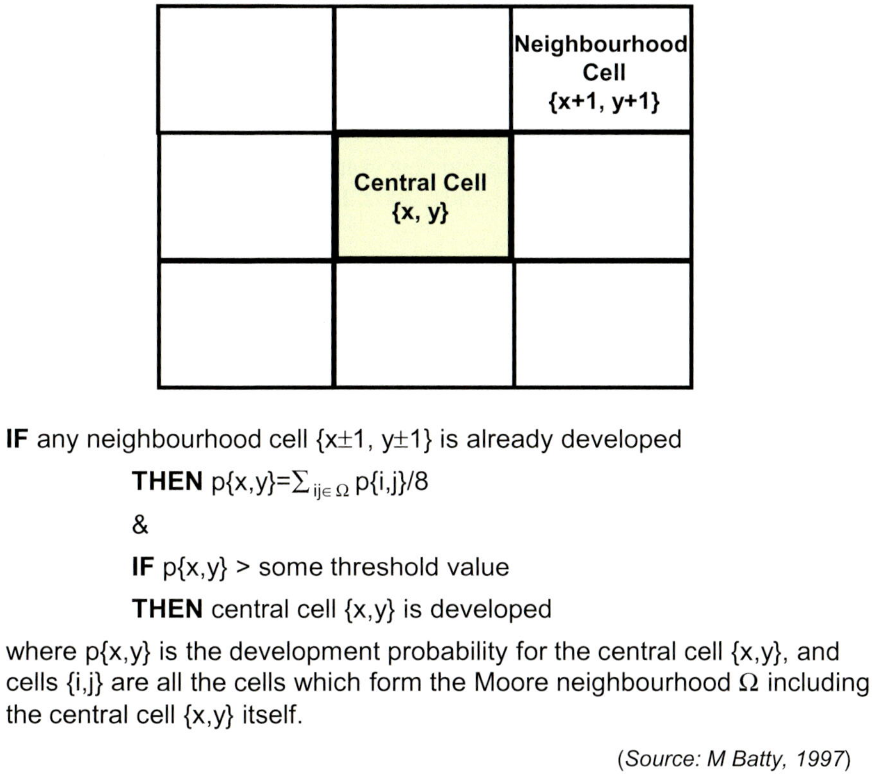

Fig. 45.1 Neighborhood and basic transition rules of cellular automata

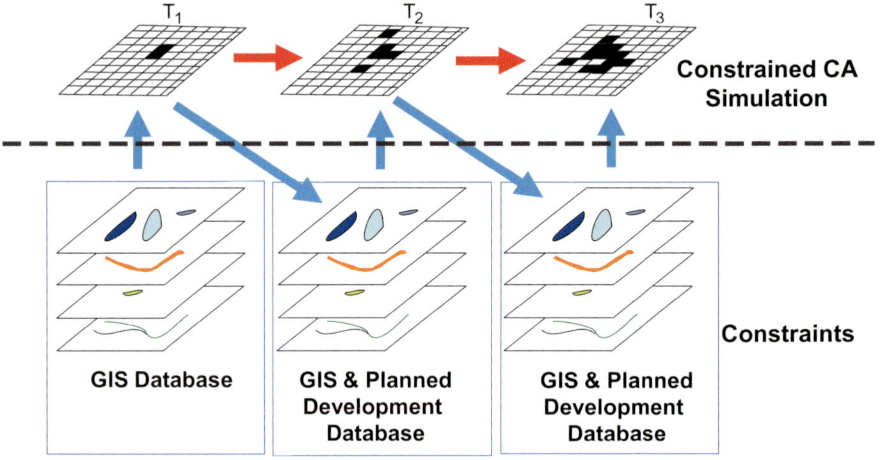

Fig. 45.2 Constrained CA with GIS and planned development database

simulation will generate patterns as usual based on historical trends. Constraints can be added into urban CA models to reflect environmental and sustainable development considerations. They are the important factors for the formation of idealized patterns. The generic constrained CA model takes into account not only the influences of neighboring states, but also a series of economic and environmental constraints. These constraints may include environmental suitability, urban forms, and development density (Yeh and Li 2001, 2002; Li and Yeh 2000; Fig. 45.2).

45.2.2 Data Collection and Model Calibration

As a bottom-up model, urban CA models are data hungry and usually require a large set of data input for real-world simulation. Remotely sensed data are often used for monitoring and measuring alterations and characteristics of land-use changes on the Earth's surface. Time series of historical remotely sensed images or land-use maps with different time phases in the same area can be used for model calibration and validation. In addition, traffic networks, natural attributes (i.e. elevation), and other physical factors are commonly used to evaluate the suitability of land for development. Land-use plans can provide land-development information, for example, a planned regional development center, which is crucial for considering the effects of urban planning on future development. Many studies have used fine socio-economic data, such as population density, to produce more realistic simulation results.

The data quality of these input data sources is a concern in urban CA applications (Aburas et al. 2016). Supervised classification is adopted to classify remote-sensing images into different land-use types: for example, urban and non-urban. Moreover, GIS software tools are used to create maps with different spatial resolutions for

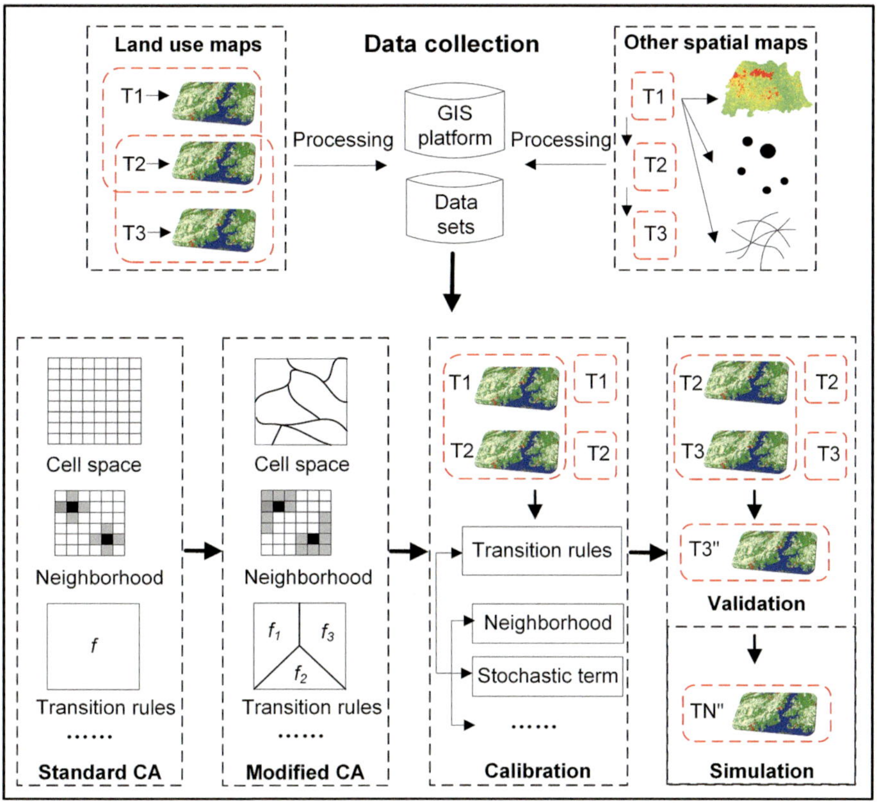

Fig. 45.3 Flow chart of urban CA modeling

comparative analysis. Errors and uncertainty can be produced by these common operations and the input data sources themselves, thus, influencing the results of urban simulation (Yeh and Li 2006). There are debates on whether urban CA models can provide meaningful results, especially for urban planning, due to inherent errors and uncertainty. Overall, considering the above two aspects, modelers can follow the flow chart in Fig. 45.3 to create an urban CA model.

45.3 Types of Urban CA Models

The model developed by Batty and Xie (1994) in Amherst, New York was one of the first applications of urban CA in real-world simulation. However, the first widespread empirical applications of urban CA were carried out by White et al. (1997) and Clarke et al. (1997). The application of White and Engelen was based on the previous work of White and Engelen (1993, 1997). In the model of White et al., the transition potential of conversion into different land uses is calculated for each cell, which can be regarded as a function of various factors, including suitability for different land uses, neighborhood and inertia effects, and stochastic disturbance. Several models of this functional type were applied to Cincinnati (White et al. 1997), the Netherlands (Engelen et al. 1999), Tokyo (Arai and Akiyama 2004), Dublin (Barredo et al. 2003), Lagos (Barredo et al. 2004), and San Diego (Kocabas and Dragicevic 2006). These applications confirmed the capacity of urban CA models in highly realistic simulation of urban transformation. Several improvements have been proposed to reinforce the methodological and theoretical basis of this type of model (Arai and Akiyama 2004; Caruso et al. 2005). Another application is the SLEUTH model, which is an acronym of the input maps: slope, land use, exclusion, urban extent, transportation, and hill shade (Clarke et al. 1997). SLEUTH considers four types of growth behaviors, which are spontaneous, diffusive, organic, and road-influenced. This model is designed to learn from the feedback of its local settings over time through self-modification, and its calibration is based on combining different metrics of the goodness-of-fit between observed and simulated results. SLEUTH has been applied to many cities, initially in North America (Berling-Wolff and Wu 2004; Clarke and Gaydos 1998; Dietzel and Clarke 2006; Herold et al. 2003; Yang and Lo 2003), and later in Europe (Silva and Clarke 2002), South America (Leao et al. 2004), and Asia (Feng et al. 2012; Mahiny and Gholamalifard 2007). Efforts have been made to improve SLEUTH, such as introducing new metrics and functionality (Guan and Clarke 2010; Jantz et al. 2010; Liu et al. 2012).

Other early urban CA models include those developed by Wu (2002, 1998), Wu and Webster (1998), and Wu and Martin (2002), in which the probability of urban development for each cell was calculated based on a group of factors, such as neighborhood. The first urban planning CA models proposed by Li and Yeh (2002b) and Yeh and Li (2001, 2002) adopted gray cells to represent continuous cell states and cumulative degrees of development. They developed a family of constrained CA urban planning models that can be used to generate different planning options according to different environmental considerations, urban forms, and densities, for the evaluation of urban development and planning for sustainable development. They added some constraint functions in CA modeling that incorporate environmental and urban-form data obtained from GIS.

The methods of multi-criteria evaluation and logistic regression were first introduced by Wu and Webster (1998) and Wu (2002) to allocate weights to different factors, which are simpler and require lesser computation compared with Monte Carlo (Chen et al. 2002). As urban development is a complicated and nonlinear process, Yeh and Li (2003) proposed to define transition rules using a neural network as a black box. Instead of mathematical transition rules, Li and Yeh (2004) defined explicit transition rules using IF–THEN statements, which are straightforward and intuitive. Several statistical, probabilistic, and artificial-intelligence algorithms were used to calibrate these types of urban CA models (Wu and Martin 2002; Almeida et al. 2008; Li and Liu 2006; Feng and Liu 2013).

Other popular urban CA models were derived from other research fields, such as DINAMICA, which is a CA-based model originally designed for deforestation simulation (Soares-Filho et al. 2002; Almeida et al. 2003,2005). As a bottom-up dynamic model, urban CA can be integrated with top-down models to gain complexity and power. The integration with the Markov approach compensates for its growth constraints and thus has received much attention recently (Al-Shalabi et al. 2013; Araya and Cabral 2010; Arsanjani et al. 2011; Li et al. 2014; Memarian et al. 2012; Samat et al. 2011; Deep and Saklani 2014; Olusina et al. 2014).

45.4 Applications of Urban CA in Urban Planning

The development of CA for urban and regional applications is considerably influenced by the intended use and functionality of models. Urban CA models are applied for exploring spatial complexity, testing urban theories and ideas, and as planning support tools (Fig. 45.4).

For exploring spatial complexity, urban CA models are used to advance the understanding of cities as complex adaptive and dynamic systems. Limited adjustments in the CA formalism are required for the models applied in exploring the principles governing urban spatial development. CA are the combination of a spatial structure and a set of states and transition rules. The idea behind CA is to find simple elements of complexity in cities and to compare these elements with similar models in other fields. The original work by Tobler and Couclelis in the 1970s and 1980s emphasized the conceptual and theoretical aspects of CA and related them to the theory of complex systems (Tobler 1979; Couclelis 1985). CA were taken as an epistemological tool to show how spatial development can be produced out of simple rules. CA for exploring spatial complexity were further developed along with fractal theory, chaos, nonlinearity, computer graphics, and complexity (Batty 2007; Torrens and O'Sullivan 2001).

CA can be used to test theories and ideas of urban development, examining the roles of complexity in the driving dynamics of urban processes, such as urban sprawl,

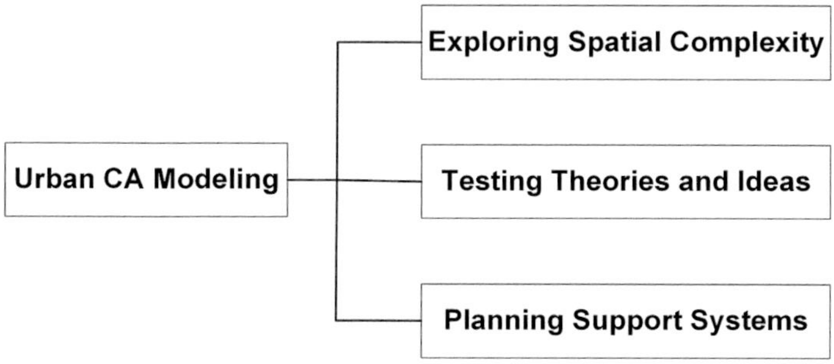

Fig. 45.4 Potential applications of urban CA modeling

diffusion and coalescence, and polycentricism. CA models are used as laboratories to test theories and ideas in urban economics, geography, and sociology. The formulation of transition rules is the key to developing close and direct links between urban CA models and urban theories. The transition rules derived from urban theories can help to explore various hypothetical ideas about cities. The complex relationships between physical and socio-economic processes and urban environments have been explored (Alberti 1999; Dietzel et al. 2005). Efforts have been extended to embrace other urban theories, including urban ecology, design, and sociology (Batty 1998; Benati 1997; Portugali et al. 1997). These studies have advanced the theoretical basis of urban CA models. However, CA models of urban theories are often concerned with details on how to build the model, but fail to explain the theories that they intended to explore (Torrens and O'Sullivan 2001). Thus, they are interesting but not well explored in urban CA modeling.

The use of urban CA models as planning support systems requires modifications of the above two applications of CA models to produce more realistic results relevant to urban planning, management, and policies. These CA models serve as planning support tools that can assist governments, planners, and stakeholders in evaluating the social benefits and environmental and ecological consequences of different urban planning goals, options, and policies. Various urban issues have been addressed in these types of urban CA models, including the delineation of urban growth boundaries, assessment of urban planning options, and prevention of illegal development (Jantz et al. 2010; Xia et al. 2020a). Despite the fact that urban CA models are increasingly developed in applied research, a gap exists in supporting practical planning of urban spaces and land uses (Santé et al. 2010).

In addition to using CA as a planning support system to (1) construct baseline growth simulation and prediction; (2) evaluate existing development as compared with optimal development; and (3) simulate development alternatives according to different planning objectives for assisting the urban planning process (Yeh and Li 2009), another example of using CA in urban planning is to delineate urban growth boundaries (UGBs). UGBs have become an important part of territorial planning in China. The objective is to ensure smart urban growth, which can increase the density of urban services and protect surrounding natural ecosystems (Jun 2004). UGBs have been regarded as an important element in designing land-use plans in China, although the concept can be traced to Great Britain's green belts in the 1930s (Nelson and Moore 1993). China needs to restrain its chaotic urban expansion via the delineation of UGBs to sustain its shrinking farmland stock.

The designers of UGBs should understand the mechanism of urban dynamics and consider various geographic factors. These models can assist planners in delimiting optimal UGBs for directing the future urban expansion from a spatial optimization perspective. Traditionally, evaluation models for land-use suitability provide a simple way for delimiting UGBs (Bhatta 2009). A major problem is that cities are dynamic systems influenced by anthropogenic activities and natural processes. These suitability-based methods ignore landscape characteristics during the delineation of UGBs (Santé et al. 2008). This approach requires efficient and feasible techniques to delimit those boundaries. CA can satisfy multiple objectives in delineation of

UGBs, including maximum urban suitability, high-quality farmland preservation to the greatest extent, and the most compact landscape pattern (Ma et al. 2017; Liang et al. 2018).

An example is to use the software GeoSOS-FLUS (https://www.geosimulation.cn), which is available on the Internet, to serve as an effective tool to delineate UGB. The implementation of UGB using GeoSOS-FLUS involves several procedures. First, we retrieved various spatial variables and historical land-use data for estimating the transition probability of each land-use type. Second, we defined the simulation subject to different planning visions according to a number of scenarios, such as baseline, economic zoning development, and excessive urban growth scenarios. Third, we carried out the simulation of UGBs on the basis of the above urban development probability and multi-scenarios constraints, as well other constraint factors. Fourth, the simulated UGBs can be further modified by using two common morphology operators, namely, dilation and erosion.

Figure 45.5 shows the example of using GeoSOS-FLUS to simulate UGBs in the study area of Guangdong-Hongkong-Macau Bay Area (GHMBA), which is one of the fastest-developing urban agglomerations in China, projected to 2030. This GeoSOS-FLUS has also been applied to the delineation of UGBs in other fast-growing cities of China, such as Foshan, Zhengzhou, and Chongqing. The simulated UGBs can be used to guide future urban master plans, which can prevent wastage of land resources.

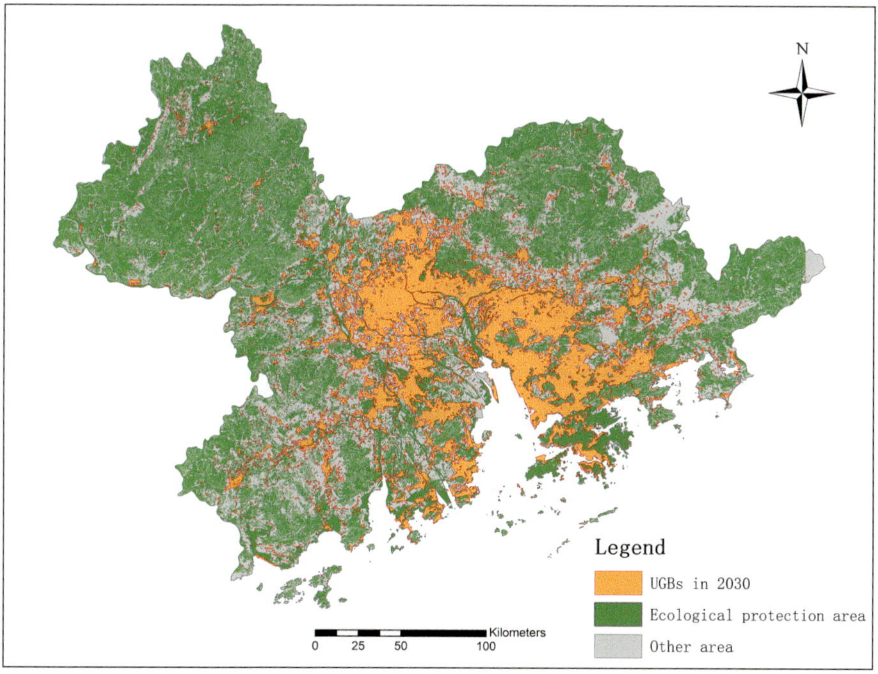

Fig. 45.5 Simulation of UGBs in the study area of GHMBA in 2030

45.5 Discussion and Conclusion

45.5.1 Current Issues in Urban CA Modeling

Urban CA models have strengths and weaknesses. The fast development of urban CA models is mainly due to their simplicity. However, simplicity often limits the CA capacity to represent realistic urban phenomena, leading to extensive modifications and introduction of complexity into the model. Questions are raised over whether these elaborated models actually constitute CA at all, if the relaxations are too much. Another strength of urban CA models is flexibility, which allows them to be adopted to different applications. However, flexibility may cause confusion and difficulties for users if there is no standard definition of transition rules. Although difficult, finding the balance between simplicity and realism, as well as between flexibility and standardization, is needed. As descriptive models, urban CA models have the ability to examine hypothetical ideas related to cities. In terms of data requirements, input data collected for different models can vary greatly. In the past, the software available for implementing general urban CA models has been very limited and inconvenient to use; users are usually required to modify or re-design their models for specific purposes (Xia et al. 2018, 2020b).

In recent years, more user-friendly CA packages have been developed to solve various simulation and planning problems, such as the CA_MARKOV module in IDRISI, and GeoSOS. The CA_MARKOV module in IDRISI adopts a hybrid Markov-CA model to allocate land use until the areas that are predicted by a Markov chain are achieved (Yang et al. 2014). GeoSOS also provides a variety of CA models (e.g. neural network CA, logistic regression CA, decision tree CA), which can be freely downloaded at https://www.geosimulation.cn. Moreover, GeoSOS for ArcGIS (a software add-in that runs in ArcGIS Desktop) has been developed to provide the full functions of simulating, predicting, optimizing, and displaying a variety of geographic patterns and dynamic processes, such as land-use changes, urban evolution, zoning of natural areas for protection, and facilities sitting. As the only software integrating spatial simulation and optimization capability together, GeoSOS for ArcGIS comprises a geographic simulator and optimizer, which use multiple CA models and ACO-based model, respectively, by coupling their results to solve complex spatial simulation and optimization problems. GeoSOS for ArcGIS is a free and open-source software and is also available for freely downloading at the GeoSOS Web site (https://www.geosimulation.cn). So far, this ArcGIS Desktop added-in component has been downloaded by users in 46 countries all round the world.

The current literature on CA applications reflects problems that have arisen from researchers who just applied CA, but were not familiar with the CA models themselves. First, many users have claimed that their simulation results can support urban planning and management without offering good examples of real-world applications. Successful applications should demonstrate that governments or planners can make better decisions due to the use of CA models. Second, many users have difficulty in obtaining details of the input data, especially the dates in acquiring them. In some cases, the present road network that was built after the simulated period was used in the simulation, making the simulation somewhat questionable. Third, they evaluated their simulation results by comparing the simulated map to the reference map of the entire study area, but failed to compare the percentage of errors to the percentage of converted areas (Liu et al. 2014; Pontius and Millones 2011). Therefore, they used flawed metrics for assessing model performance such as the goodness-of-fit (Pontius and Millones 2011). Finally, they just separated calibration information from validation information through space (by selecting pixels randomly), rather than through time (by using an urban map in another year), leading to overestimation of the accuracy of the model.

45.5.2 Summary and Future Research Directions

This chapter has summarized the basic concepts and techniques of CA modeling for urban and regional planning from the perspectives of basic CA components, formulation of urban CA, and data collection. Urban CA were classified into different types, and systematic and critical reviews on previous and recent studies and applications were provided. Finally, the strengths and weaknesses of urban CA models were pointed out for new modelers, along with current problems in the literature.

Further studies are needed to provide new insights into the uses of CA in geographic and urban theories, which would advance the theoretical basis of urban CA. The integration of urban CA models and other models may overcome the weaknesses of CA, such as with economic models, thus improving model performance. More effort should be made on improving CA by incorporating microlevel interactions and multiple processes. So far, the calibration is often based on two years of land-use maps. There is an issue of over-calibration because of bifurcation effects inherited from complex systems. Bifurcation refers to the fact that a small smooth change in the parameter values may cause a sudden change in the model's behavior. Finally, elaboration is also required to demonstrate how urban CA models can support planning and management in practice. Urban CA models should not be used to provide exact predictions of urban systems, but to simulate interactively different what-if scenarios for policy implementation through the modification of transition rules.

Concern for global changes has grown tremendously in recent years. CA should incorporate factors of climate change in urban planning, such as the effects of urban heat islands, changes in agricultural production, and changes in land-use patterns. CA simulation could be integrated with climate and hydrological models in future studies (Chen et al. 2020). For example, urban simulation could incorporate the universal climate scenarios developed by the Intergovernmental Panel on Climate Change, such that future land use can meet the demand required by economic and social development. This integration can facilitate the simulation of future changes in global and regional land covers. For example, the simulation of urban evolution with finer urban land categories should be attractive for actual planning practice. This requires the integration of current CA with big data or social media data.

References

Aburas MM, Ho YM, Ramli MF, Ash'aari ZH (2016) The simulation and prediction of spatio-temporal urban growth trends using cellular automata models: a review. Int J Appl Earth Obs Geoinf 52:380–389

Ahmed B, Ahmed R (2012) Modelling urban land cover growth dynamics using multi-temporal satellite images: a case study of Dhaka, Bangladesh. ISPRS Int J Geo-Inform 1(3):3–31

Alberti M (1999) Modelling the urban ecosystem: a conceptual framework. Environ Plann B: Plann Des 26(4):623–646

Alcock I, White M, Cherrie M, Wheeler B, Taylor J, McInnes R, Kampe EO, Vardoulakis S, Sarran C, Soyiri I, Fleming L (2017) Land cover and air pollution are associated with asthma hospitalisations: a cross-sectional study. Environ Int 109:29–41

Almeida CM, Batty M, Monteiro AMV, Camara G, Soares-Filho BS, Cerqueira GC, Pennachin CL (2003) Stochastic cellular automata modelling of urban land use dynamics: empirical development and estimation. Comput Environ Urban Syst 27(5):481–509

Almeida CM, Gleriani JM, Castejon EF, Soares-Filho BS (2008) Using neural networks and cellular automata for modelling intra-urban land-use dynamics. Int J Geogr Inform Sci 22(9):943–963

Almeida CM, Monteiro AMV, Camara G, Soares-Filho BS, Cerqueira GC, Pennachin CL, Batty M (2005) GIS and remote sensing as tools for the simulation of urban land-use change. Int J Remote Sens 26(4):759–774

Al-Shalabi M, Billa L, Pradhan B, Mansor S, Al-Sharif AA (2013) Modelling urban growth evolution and land-use changes using GIS based cellular automata and SLEUTH models: the case of Sana'a metropolitan city, Yemen. Environ Earth Sci 70(1):425–437

Arai T, Akiyama T (2004) Empirical analysis for estimating land use transition potential functions—case in the Tokyo metropolitan region. Comput Environ Urban Syst 28(1–2):65–84

Araya YH, Cabral P (2010) Analysis and modelling of urban land cover change in Setubal and Sesimbra, Portugal. Remote Sens 2(6):1549–1563

Arsanjani JJ, Kainz W, Mousivand AJ (2011) Tracking dynamic land-use change using spatially explicit Markov Chain based on cellular automata: the case of Tehran. Int J Image Data Fusion 2(4):329–345

Bai X, Chen J, Shi P (2012) Landscape urbanization and economic growth in China: positive feedbacks and sustainability dilemmas. Environ Sci Technol 46(1):132–139

Barredo JI, Demicheli L, Lavalle C, Kasanko M, McCormick N (2004) Modelling future urban scenarios in developing countries: an application case study in Lagos, Nigeria. Environ Plann B: Plann Des 31(1):65–84

Barredo JI, Kasanko M, Mccormick N, Lavalle C (2003) Modelling dynamic spatial processes: simulation of urban future scenarios through cellular automata. Landscape Urban Plann 64(3):145–160

Batty M (1998) Urban evolution on the desktop: simulation with the use of extended cellular automata. Environ Plann A 30(11):1943–1967

Batty M (2007) Cities and complexity: understanding cities with cellular automata, agent-based models, and fractals. The MIT Press, Cambridge, MA

Batty M (1997) Cellular automata and urban form: a primer. J Am Plann Assoc 63(2):266–274

Batty M, Xie Y (1994) From cells to cities. Environ Plann B: Plann Des 21(7):s31–s48

Benati S (1997) A cellular automaton for the simulation of competitive location. Environ Plann B: Plann Des 24(2):205–218

Berling-Wolff S, Wu J (2004) Modelling urban landscape dynamics: a review. Urban Ecosyst 19(1):119–129

Bhatta B (2009) Modelling of urban growth boundary using geoinformatics. Int J Digit Earth 2(4):359–381

Burak G, Zhou Y, Diana U, Güneralp B, Zhou Y, Ürge-Vorsatz D, Gupta M, Yu S, Patel PL, Fragkias M, Li X, Seto KC (2017) Global scenarios of urban density and its impacts on building energy use through 2050. Proc Natl Acad Sci 114(34):8945–8950

Caruso G, Rounsevell M, Cojocaru G (2005) Exploring a spatio-dynamic neighborhood-based model of residential behaviour in the Brussels periurban area. Int J Geogr Inform Sci 19(2):103–123

Chen G, Li X, Liu X, Chen Y, Liang X, Leng J, Xu X, Liao W, Qiu Y, Wu Q, Huang K (2020) Global projections of future urban land expansion under shared socioeconomic pathways. Nat Commun 11:537

Chen J, Gong P, He C, Luo W, Tamura M, Shi P (2002) Assessment of the urban development plan of Beijing by using a CA-based urban growth model. Photogram Eng Remote Sens 68(10):1063–1072

Chen Y, Li X, Liu X, Ai B (2014) Modelling urban land-use dynamics in a fast developing city using the modified logistic cellular automaton with a patch-based simulation strategy. Int J Geogr Inform Sci 28(2):234–255

Clarke KC, Gaydos LJ (1998) Loose-coupling a cellular automaton model and GIS: long-term urban growth prediction for San Francisco and Washington/Baltimore. Int J Geogr Inform Syst 12(7):699–714

Clarke KC, Hoppen S, Gaydos LJ (1997) A self-modifying cellular automaton model of historical urbanization in the San Francisco Bay area. Environ Plann B: Plann Des 24(2):247–261

Codd EF (1968) Cellular automata. Academic Press, New York

Couclelis H (1985) Cellular worlds: a framework for modelling micro–macro dynamics. Environ Plann A 17(5):585–596

Couclelis H (1997) From cellular automata to urban models: new principles for model development and implementation. Environ Plann B: Plann Des 24(2):165–174

Deep S, Saklani A (2014) Urban sprawl modelling using cellular automata. Egypt J Remote Sens Space Sci 17(2):179–187

Dietzel C, Clarke K (2006) The effect of disaggregating land use categories in cellular automata during model calibration and forecasting. Comput Environ Urban Syst 30(1):78–101

Dietzel C, Herold M, Hemphill JJ, Clarke KC (2005) Spatio-temporal dynamics in California's central valley: empirical links to urban theory. Int J Geogr Inform Sci 19:175–195

Engelen G, Geertman S, Smits P, Wessels C (1999) Dynamic GIS and strategic physical planning support: a practical application to the IJmond/Zuid-Kennemerland region. In: Stillwell S, Geertman S, Openshaw S (eds) Geographical information and planning. Springer, Berlin, pp 87–111

Feng HH, Liu HP, Lü Y (2012) Scenario prediction and analysis of urban growth using SLEUTH model. Pedosphere 22(2):206–216

Feng Y, Liu Y (2013) A cellular automata model based on nonlinear kernel principal component analysis for urban growth simulation. Environ Plann B: Plann Des 40(1):117–134

Gardner M (1971) On cellular automata self-reproduction, the garden of Eden and the game of 'Life.' Sci Am 224(2):112–118

Geertman S, Hagoort M, Ottens H (2007) Spatial-temporal specific neighborhood rules for cellular automata land-use modelling. Int J Geogr Inform Sci 21(5):547–568

Guan Q, Clarke KC (2010) A general-purpose parallel raster processing programming library test application using a geographic cellular automata model. Int J Geogr Inform Sci 24(5):695–722

Herold M, Goldstein NC, Clarke KC (2003) The spatiotemporal form of urban growth: measurement, analysis and modelling. Remote Sens Environ 86(3):286–302

Iovine G, D'Ambrosio D, Gregorio SD (2005) Applying genetic algorithms for calibrating a hexagonal cellular automata model for the simulation of debris flows characterised by strong inertial effects. Geomorphology 66(1–4):287–303

Jantz CA, Goetz SJ, Donato D, Claggett P (2010) Designing and implementing a regional urban modelling system using the SLEUTH cellular urban model. Comput Environ Urban Syst 34(1):1–16

Jun MJ (2004) The effects of Portland's urban growth boundary on urban development patterns and commuting. Urban Stud 41(7):1333–1348

Kocabas V, Dragicevic S (2006) Assessing cellular automata model behaviour using a sensitivity analysis approach. Comput Environ Urban Syst 30(6):921–953

Leao S, Bishop I, Evans D (2004) Simulating urban growth in a developing nation's region using a cellular automata-based model. J Urban Plann Dev 130(3):145–158

Li W, Wu C, Zang S (2014) Modelling urban land use conversion of Daqing city, China: a comparative analysis of 'top-down' and 'bottom-up' approaches. Stoch Env Res Risk Assess 28(4):817–828

Li X, Chen Y, Liu X, Xu X, Chen G (2017) Experiences and issues of using cellular automata for assisting urban and regional planning in China. Int J Geogr Inform Sci 31(8):1606–1629

Li X, Liu XP (2006) An extended cellular automation using case-based reasoning for simulating urban development in a large complex region. Int J Geogr Inform Sci 20(10):1109–1136

Li X, Yang Q, Liu X (2008) Discovering and evaluating urban signatures for simulating compact development using cellular automata. Landscape Urban Plann 86(2):177–186

Li X, Yeh AGO (2000) Modelling sustainable urban development by the integration of constrained cellular automata and GIS. Int J Geogr Inform Sci 14(2):131–152

Li X, Yeh AGO (2002a) Urban simulation using principal components analysis and cellular automata for land-use planning. Photogram Eng Remote Sens 68(4):341–352

Li X, Yeh AGO (2002b) Neural-network-based cellular automata for simulating multiple land use changes using GIS. Int J Geogr Inform Sci 16(4):323–343

Li X, Yeh AGO (2004) Data mining of cellular automata's transition rules. Int J Geogr Inform Sci 18(8):723–744

Liang X, Liu X, Li X, Chen Y, Tian H, Yao Y (2018) Delineating multi-scenario urban growth boundaries with a CA-based FLUS model and morphological method. Landscape Urban Plann 177:47–63

Liu X, Sun R, Yang Q, Su G, Qi W (2012) Simulating urban expansion using an improved SLEUTH model. J Appl Remote Sens 6(1):061709

Liu Y, Feng Y, Pontius R (2014) Spatially-explicit simulation of urban growth through self-adaptive genetic algorithm and cellular automata modelling. Land 3(3):719–738

Ma S, Li X, Cai Y (2017) Delimiting the urban growth boundaries with a modified ant colony optimization model. Comput Environ Urban Syst 62:146–155

Mahiny AS, Gholamalifard M (2007) Dynamic spatial modelling of urban growth through cellular automata in a GIS environment. Int J Environ Res 1(3):272–279

Memarian H, Balasundram SK, Talib JB, Sung CTB, Sood AM, Abbaspour K (2012) Validation of CA-Markov for simulation of land use and cover change in the Langat Basin, Malaysia. J Geogr Inform Syst 4:542–554

Musa SI, Hashim M, Reba MN (2017) A review of geospatial-based urban growth models and modelling initiatives. Geocarto Int 32(8):813–833

Nelson AC, Moore T (1993) Assessing urban growth management: the case of Portland, Oregon, the USA's largest urban growth boundary. Land Use Policy 10(4):293–302

Olusina JO, Abiodun EO, Oseke JI (2014) Urban sprawl analysis and transportation using cellular automata and Markov Chain. Phys Sci Int J 4(8):1191–1210

O'Sullivan D (2001) Graph-cellular automata: A generalised discrete urban and regional model. Environ Plann B: Plann Des 28(5):687–705

Poelmans L, Rompaey AV (2010) Complexity and performance of urban expansion models. Comput Environ Urban Syst 34(1):17–27

Pontius RG, Millones M (2011) Death to Kappa: birth of quantity disagreement and allocation disagreement for accuracy assessment. Int J Remote Sens 32(15):4407–4429
Portugali J, Benenson I, Omer I (1997) Spatial cognitive dissonance and sociospatial emergence in a self-organizing city. Environ Plann B: Plann Des 24(2):263–285
Samat N, Hasni R, Elhadary YAE (2011) Modelling land use changes at the peri-urban areas using geographic information systems and cellular automata model. J Sustain Dev 4(6):72–84
Santé I, Andrés MG, Miranda D, Crecente R (2010) Cellular automata models for the simulation of real-world urban processes: a review and analysis. Landscape Urban Plann 96(2):108–122
Santé RI, Crecente MR, Miranda BD (2008) GIS-based planning support system for rural land-use allocation. Comput Electron Agric 63:257–273
Seto KC, Guneralp B, Hutyra LR (2012) Global forecasts of urban expansion to 2030 and direct impacts on biodiversity and carbon pools. Proc Natl Acad Sci 109(40):16083–16088
Shahbaz M, Chaudhary AR, Ozturk I (2016) Does urbanization cause increasing energy demand in Pakistan? empirical evidence from STIRPAT model. Energy 122:83–93
Shi W, Pang MYC (2000) Development of Voronoi-based cellular automata—an integrated dynamic model for geographical information systems. Int J Geogr Inform Syst 14(5):455–474
Silva EA, Clarke KC (2002) Calibration of the SLEUTH urban growth model for Lisbon and Porto, Portugal. Comput Environ Urban Syst 26(6):525–552
Soares-Filho BS, Cerqueira GC, Pennachin CL (2002) DINAMICA—a stochastic cellular automata model designed to simulate the landscape dynamics in an Amazonian colonization frontier. Ecol Model 154(3):217–235
Tobler WR (1979) Cellular geography. In: Gale S, Olsson G (eds) Philosophy in geography. Reidel, Dordrecht, pp 379–386
Torrens PM, O'Sullivan D (2001) Cellular automata and urban simulation: where do we go from here? Environm Plann B: Plann Des 28(2):163–168
Triantakonstantis D, Mountrakis G (2012) Urban growth prediction: a review of computational models and human perceptions. J Geogr Inform Sys 4(6):555–587
Wang F, Marceau DJ (2013) A patch-based cellular automaton for simulating land-use changes at fine spatial resolution. Trans GIS 17(6):828–846
Weeberb JR, Henrique LR, Koutrakis P (2015) Novel land use approach for assessment of human health: the relationship between urban structure types and cardiorespiratory disease risk. Environ Int 85:334–342
White R, Engelen G (1993) Cellular automata and fractal urban form: a cellular modelling approach to the evolution of urban land-use patterns. Environ Plann A 25(8):1175–1199
White R, Engelen G (1994) Cellular dynamics and GIS: modelling spatial complexity. J Geogr Syst 1(3):237–253
White R, Engelen G (1997) Cellular automata as the basis of integrated dynamic regional modelling. Environ Plann B: Plann Des 24(2):235–246
White R, Engelen G (2000) High-resolution integrated modelling of the spatial dynamics of urban and regional systems. Comput Environ Urban Syst 24(5):383–400
White R, Engelen G, Uljee I (1997) The use of constrained cellular automata for high-resolution modelling of urban land-use dynamics. Environ Plann B: Plann Des 24(3):323–343
Wolfram S (1984) Universality and complexity in cellular automata. Physica D 10(1–2):1–35
Wu F (1998) SimLand: A prototype to simulate land conversion through the integrated GIS and CA with AHP-derived transition rules. Int J Geogr Inform Syst 12(1):63–82
Wu F (2002) Calibration of stochastic cellular automata: the application to rural-urban land conversions. Int J Geogr Inform Sci 16(8):795–818
Wu F, Martin D (2002) Urban expansion simulation of Southeast England using population surface modelling and cellular automata. Environ Plann A 34(10):1855–1876
Wu F, Webster CJ (1998) Simulation of land development through the integration of cellular automata and multicriteria evaluation. Environ Plann B: Plann Des 25(1):103–126

Xia C, Wang H, Zhang A, Zhang W (2018) A high-performance cellular automata model for urban simulation based on vectorization and parallel computing technology. Int J Geogr Inform Sci 32(2):399–424

Xia C, Zhang A, Wang H, Liu J (2020a) Delineating early warning zones in rapidly growing metropolitan areas by integrating a multiscale urban growth model with biogeography-based optimization. Land Use Policy 90:104332

Xia C, Zhang B, Wang H, Qiao S, Zhang A (2020b) A minimum-volume oriented bounding box strategy for improving the performance of urban cellular automata based on vectorization and parallel computing technology. GISci Remote Sens 57(1):91–106

Xie Y (1996) A generalized model for cellular urban dynamics. Geogr Anal 28(4):350–373

Yang X, Lo CP (2003) Modelling urban growth and landscape changes in the Atlanta metropolitan area. Int J Geogr Inform Sci 17(5):463–488

Yang X, Zheng XQ, Chen R (2014) A land use change model: integrating landscape pattern indexes and Markov-CA. Ecol Model 283:1–7

Yeh AGO, Li X (2001) A constrained CA model for the simulation and planning of sustainable urban forms by using GIS. Environ Plann B: Plann Des 28(5):733–753

Yeh AGO, Li X (2002) A cellular automata model to simulate development density for urban planning. Environ Plann B: Plann Des 29(3):431–450

Yeh AGO, Li X (2003) Simulation of development alternatives using neural networks, cellular automata, and GIS for urban planning. Photogram Eng Remote Sens 69(9):1043–1052

Yeh AGO, Li X (2006) Errors and uncertainties in urban cellular automata. Comput Environ Urban Syst 30(1):10–28

Yeh AGO, Li X (2009) Cellular automata and GIS for urban planning. In: Madden M (ed) Manual of geographic information systems. American Society for Photogrammetry and Remote Sensing, Bethesda, pp 591–619

Zhou L, Dickinson RE, Tian Y, Fang J, Li Q, Kaufmann RK, Tucker CJ, Myneni RB (2004) Evidence for a significant urbanization effect on climate in China. Proc Natl Acad Sci 101(26):9540–9544

Zhuang Y, Li W, Wang H, Hong S, Wang H (2017) A bibliographic review of cellular automaton publications in the last 50 years. J Cell Automata 12:475–492

Anthony G. O. Yeh is Chair Professor and Chan To-Haan Professor of the Department of Urban Planning and Design at the University of Hong Kong. He is a member of the Chinese Academy of Sciences and Hong Kong Academy of Sciences and a Fellow of TWAS. His interest is Planning Support Systems and GIScience.

Xia Li is a professor at School of Geographic Sciences, East China Normal University, Shanghai, P.R. China. His major research interests include land-use change analysis, urban simulation and spatial optimization, and global land-use modeling and analysis. He has developed GeoSOS and FLUS models, which have been widely used for urban and land-use simulation.

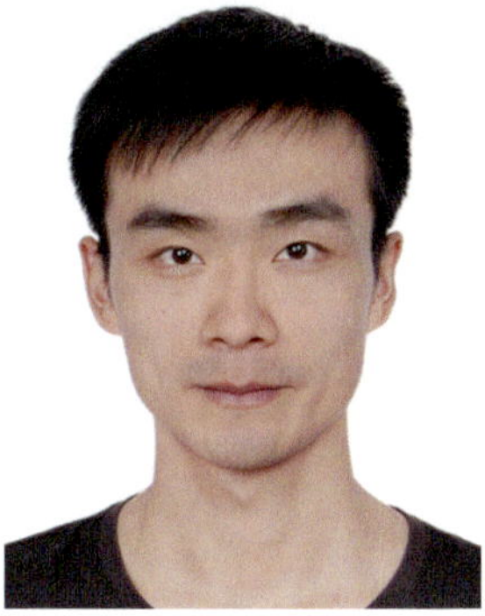

Chang Xia is a Ph.D. candidate in the Department of Urban Planning and Design at the University of Hong Kong. He received his Master's and Bachelor's degree in Land Resource Management from Wuhan University. His research interests include geo-informatics, urban management and planning, and environmental psychology.

Chapter 46
Agent-Based Modeling and the City: A Gallery of Applications

Andrew Crooks, Alison Heppenstall, Nick Malleson, and Ed Manley

Abstract Agent-based modeling is a powerful simulation technique that allows one to build artificial worlds and populate these worlds with individual agents. Each agent or actor has unique behaviors and rules which govern their interactions with each other and their environment. It is through these interactions that more macro-phenomena emerge: for example, how individual pedestrians lead to the emergence of crowds. Over the past two decades, with the growth of computational power and data, agent-based models have evolved into one of the main paradigms for urban modeling and for understanding the various processes which shape our cities. Agent-based models have been developed to explore a vast range of urban phenomena from that of micro-movement of pedestrians over seconds to that of urban growth over decades and many other issues in between. In this chapter, we introduce readers to agent-based modeling from simple abstract applications to those representing space utilizing geographical data not only for the creation of the artificial worlds but also for the validation and calibration of such models through a series of example applications. We will then discuss how big data, data mining, and machine learning techniques are advancing the field of agent-based modeling and demonstrate how such data and techniques can be leveraged into these models, giving us a new way to explore cities.

46.1 Introduction

The start of the twenty-first century marked a milestone in human history: for the first time more than half of the world's population, approximately 3.9 billion people, lived in urban areas. This trend is expected to continue in the foreseeable future,

A. Crooks (✉)
Department of Geography, RENEW Institute, University at Buffalo, Buffalo, USA
e-mail: atcrooks@buffalo.edu

A. Heppenstall · N. Malleson · E. Manley
School of Geography, University of Leeds, Leeds, UK

The Alan Turing Institute, London, UK

W. Shi et al. (eds.), *Urban Informatics*, The Urban Book Series,
https://doi.org/10.1007/978-981-15-8983-6_46

with 6.3 billion people living in cities by 2050 (United Nations 2014). Population growth will cause more urban land to be developed during the first 30 years of the twenty-first century than in all of human history (Angel et al. 2011). Less than five percent of the earth's surface is urbanized and with the urban population predicted to grow to 5 billion by 2030, the urban footprint will still be less than 10% (Seto et al. 2011). Combine this with the unprecedented urban expansion, especially in the form of megacities—cities with more than 10 million in population—which have grown from eight in the 1970s to 36 in 2016 and are expected to rise to 41 by 2030 as shown in Fig. 46.1, and society as a whole will be faced with unprecedented challenges and questions to be asked with respect to all aspects of city life. Will cities be sprawling or compact? How will cities adapt to climate change? How will new technologies such as autonomous cars, for example, affect our lives? These are challenging questions made more complicated by the fact that cities are excellent examples of complex systems, composed of people, places, flows, and activities (Batty 2013), all of which interact in a variety of different ways.

An exact definition of a complex system is difficult to pin down, as it has a different meaning to different people (Thrift 1999). A simple definition is one whereby a small number of rules or laws, applied at a local level and among many entities, are capable of generating complex global phenomena such as collective behaviors, extensive spatial patterns, and hierarchies, in such a way that the actions of their parts do not simply sum to the activity of the whole, due to self-organization, nonlinearities, feedbacks (both positive and negative), and path dependencies.[1] Cities are complex systems, composed of many parts, dynamic, and containing large numbers of discrete actors interacting within space and with other systems from nature and technology, and have a wide-ranging impact on the economy, public policy, national defense, social trends, public health, climate change, etc. As Wilson (2000) writes, understanding cities is "*…one of the major scientific challenges of our time.*" Human behavior cannot be understood or predicted in the same way as in the physical sciences such as physics or chemistry. The actions and interactions of the inhabitants of a city, for example, cannot be easily described in a physical-science theory such as that of Newton's Laws of Motion. This notion is captured quite aptly by a quote by Nobel laureate Murray Gell-Mann: "*Think how hard physics would be if particles could think.*" In the remainder of this chapter, we will introduce agent-based modeling (Sect. 46.2) as it offers a way to explore the processes that lead to the patterns we see in cities from the bottom up, but also allows us to incorporate ideas from complex systems (e.g. feedbacks, path dependency, emergence) along with providing a gallery of applications of geographically explicit agent-based models. Next, we discuss how we can incorporate various decision-making processes within such models, and also how we can integrate this style of modeling with data, with a specific emphasis on geographical and social information (Sect. 46.3). This section also discusses how

[1]Readers wishing to know more about cities and complexity are referred to the works of Allen (1997), Wilson (2000), and Batty (2007).

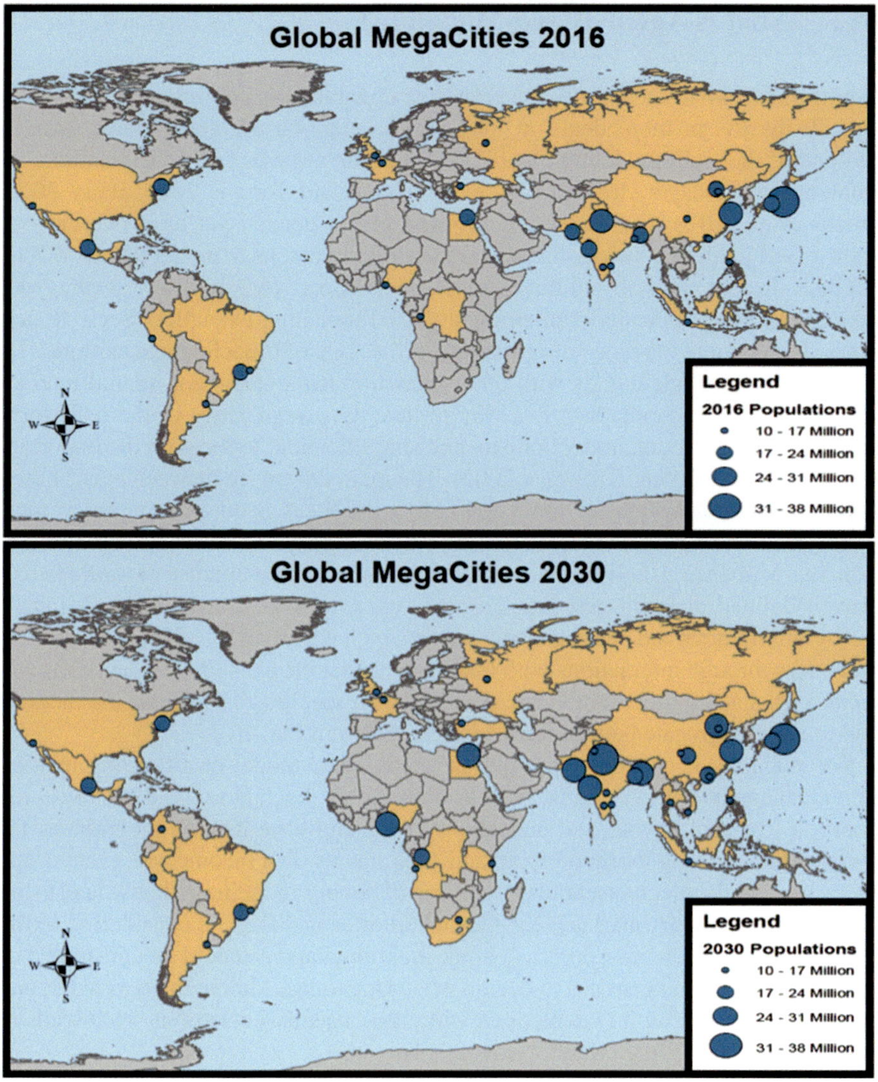

Fig. 46.1 Global megacities in 2016 and estimated megacities by 2030 (data source: United Nations 2016)

agent-based modelers are utilizing machine learning within their models. Finally, in Sect. 46.4, we will provide a summary and discuss new opportunities with respect to agent-based modeling and the city.

46.2 What is Agent-Based Modeling?

Over the past two decades, with the growth of computational power and data (which we will discuss in more detail in Sect. 46.3), agent-based models have evolved into one of the main modeling paradigms for urban systems and understanding the problems that today's cities face (see: Benenson and Torrens 2004; Batty 2005; Crooks et al. 2019). In this section, we first give a general yet brief overview of agent-based modeling before discussing the various reasons to model (Sect. 46.2.1). We then discuss steps in building such models (Sect. 46.2.2) before turning our attention to geographically explicit agent-based modeling examples (Sect. 46.2.3) which demonstrate the types of problems such a style of modeling can explore.

Agent-based modeling, as with other modeling techniques (e.g. spatial interaction models, microsimulation) is a way to take the complexities of the real world and, through abstraction, reductionism, and simplification, to focus on the important task at hand (Gilbert and Troitzsch 2005). The main difference between agent-based modeling and other styles of modeling is that the focus is on interactions of individual entities and their behaviors, and how more aggregate patterns emerge through such interactions (e.g. how individual cars can lead to the emergence of traffic jams). Broadly defined, an agent-based model can be considered as an artificial world inhabited by autonomous and heterogeneous agents, each with their set of goals and preferences. It is through interactions with other agents that the agent makes decisions and decides what actions are to be carried out based on specific goals. These interactions lead to more aggregate patterns emerging as shown in Fig. 46.2.

For example, if one were to build an agent-based model of a housing market, individual agents could be considered as households. Each household has to decide where to live and as with real households, each can have its own preferences for hosing style and neighborhood type, and each has its own income constraints. The interactions with other households in the form of buying and selling a house lead to the emergence of property markets (e.g. Geanakoplos et al. 2012). Or considering traffic congestion during the morning rush hour, individual agents could be considered as drivers of cars: each agent has to decide what time to leave home to go to work, and by driving on the road its interactions with other agents (i.e. cars) is what leads to traffic jams forming (e.g. Manley et al. 2014).

46.2.1 Examples of Why to Model

As with other modeling styles, within agent-based modeling, there are multiple reasons for why one should model, from understanding a certain phenomenon to predicting and forecasting (see Epstein 2008 for a discussion on the various reasons to model) and therefore agent-based models range from abstract thought experiments to more empirically applied applications. For example, Schelling's (1971) model of segregation is not only a classic example of an abstract model, but it also

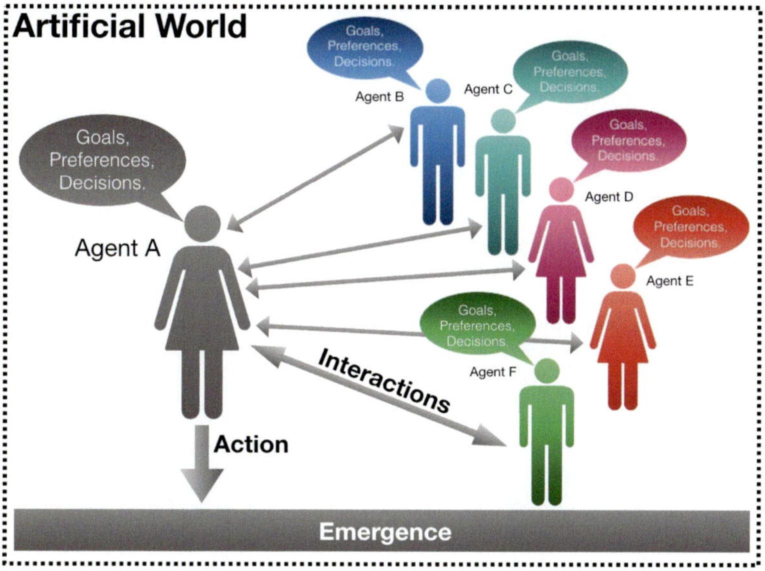

Fig. 46.2 Schematic of an agent-based model, showing how interactions between agents lead to emergent phenomena within an artificial world

demonstrates how emergent phenomena (in this case segregation) can occur through individual preferences. Moreover, it demonstrates how macro-level segregation does not necessarily reflect micro-level preferences. For example, in Fig. 46.3, we show two types of agents, those who prefer football versus those who prefer baseball. In this simple example, based on notions from Schelling's (1971) model, agents (i.e. individuals) want to be in locations (a cell on a 11 by 11 grid which acts as our artificial world) where a certain percentage of their neighbors are similar to themselves (in this example 30%).

Over time (T), agents move if their preference for their neighborhood composition is not met. As one can see, from an initial randomly distributed population, segregated neighborhoods emerge due to agents interacting with other agents and taking actions (in this case moving) and to the resulting feedbacks and past locational choices of others. Also, the model demonstrates how the actions of one agent might affect others. For example, an agent may be satisfied in a certain location but another agent moving into the neighborhood might cause this agent to become dissatisfied and therefore cause it to move. By altering the agent's preferences for certain neighborhood compositions (e.g. from 30 to 70% of similar neighbors), we can also see how individual preferences and interactions at the micro-level lead to more macro-level phenomena emerging as we show in Fig. 46.4; specifically in this example, we see how more segregated communities emerge as preferences are increased.

What is interesting about this phenomenon is that often when we see segregated neighborhoods, the process and actions that led to this pattern have already occurred. However, through agent-based modeling, we can explore what processes or actions

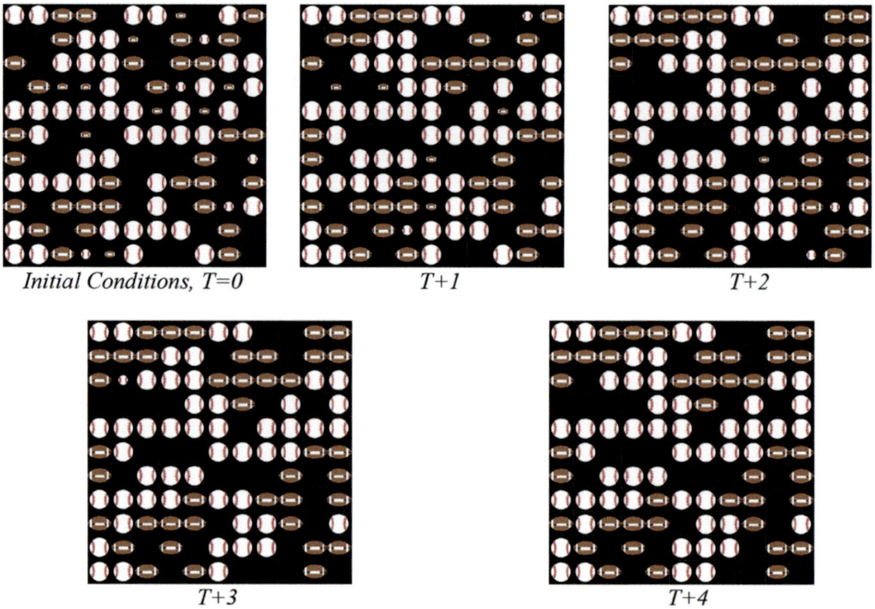

Fig. 46.3 Example of segregation emerging over time as agents move to locations where their preferences are met (note smaller balls are dissatisfied agents)

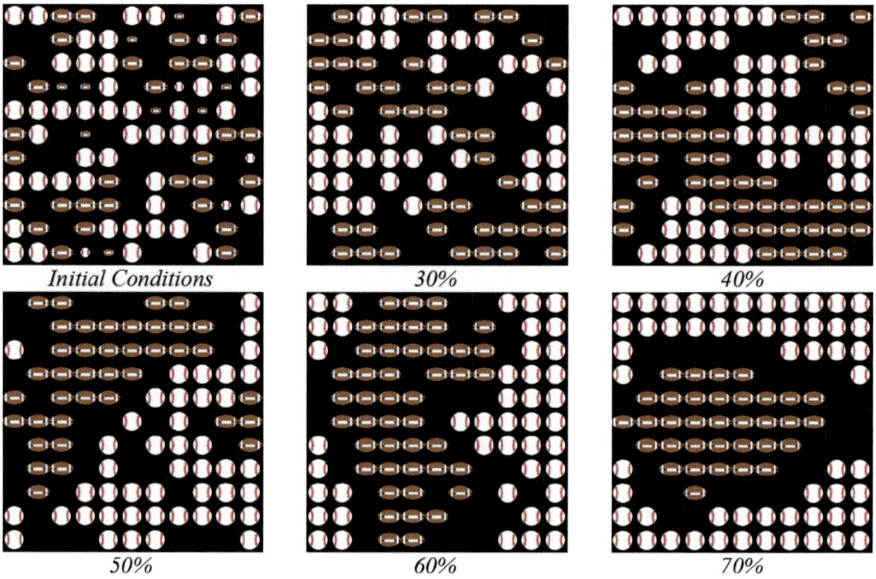

Fig. 46.4 Examples of how different preferences lead to different patterns of segregation

might have led to such patterns emerging in the first place, and thus devise potential interventions before it is too late. However, as noted above, agent-based models can also be empirically grounded. Take for example the work of Benenson et al. (2002), which explored how people's preferences for certain neighborhoods and building types lead to distinct residential patterns emerging in Tel Aviv, Israel. While both have their own purpose, Schelling's (1971) to explore basic behavior and that of Benenson et al. (2002) to explain residential choice based on empirical data and test various scenarios, both show that individual preferences for certain types of neighborhoods lead to distinct residential patterns emerging, which would be difficult to explain from just looking at aggregate data alone. It should however be noted that agent-based modeling is not just an academic exercise, but has been used by companies and organizations for a variety of decision-making purposes. These range from the potential impact of decimalization of the NASDAQ Stock Market (Darley and Outkin 2007), to that of understanding store design, consumer markets, or hiring strategies for companies (see Bonabeau 2003). Readers of this chapter might also be surprised to know that they have probably seen agent-based models while at the cinema or watching TV as they are often used for massive crowd scenes in movies, replacing the need for a large cast of extras (see Massive 2019). Companies, especially engineering ones, are also utilizing agent-based models to study pedestrian (e.g. products such as Legion 2019 and STEPS 2019) or traffic dynamics (e.g. PTV Visum 2019 and Paramics 2019) in order to assess new designs for buildings or traffic measures before they are built or implemented.

46.2.2 Steps in Building an Agent-Based Model

When it comes to building an agent-based model, the process can be broadly viewed as having three steps. First, before we can get to the model itself, we need to identify the research question we are trying to solve with the model (e.g. reasons for traffic patterns), define the target of the model, know specifically what we are we trying to solve (e.g. traffic dynamics), and consider if there are any observations of the target we wish to include to provide parameters and initial conditions for the model (e.g. origin–destination data). We then need to make assumptions and design the model. Once the model has been designed and implemented (often in computer code), the second step is to run (execute) the model, which creates an artificial world. This is then populated with agents (e.g. cars) that are assigned attributes and rules (depending on the application or phenomena of interest). We then run the model until a certain condition is met or a specific time epoch is reached, and report and observe the results which are shown in Fig. 46.5a (while Fig. 46.5b shows a simple worked example of the segregation model discussed in Sect. 46.2.1). While this figure and the description given above are highly generalized and simple, in essence, one could make the argument that agent-based models are just rule-based systems, in the sense that they could be considered as just a series of *if-then-else* statements. For example, *if* the fire alarm goes off, *then* exit the building, *else* stay in

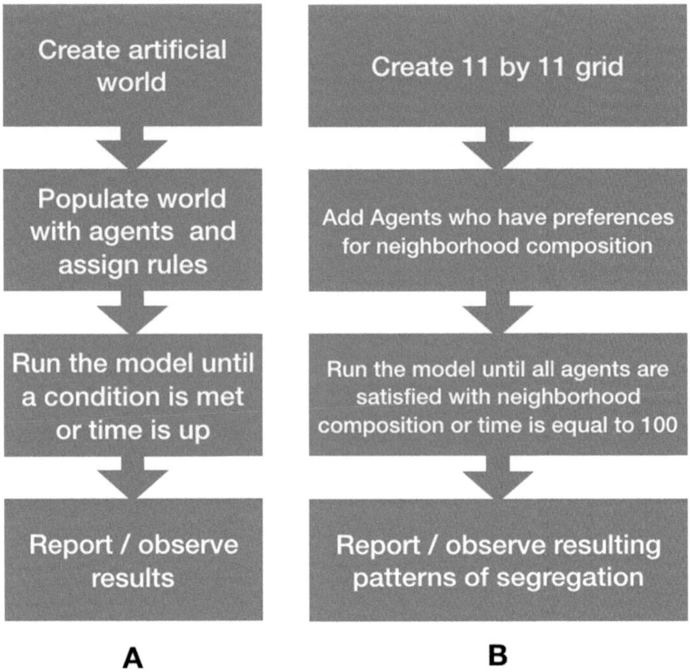

Fig. 46.5 Highly generalized flow of an agent-based model **a** and the corresponding flow of the basic segregation model **b**

the building. However, the richness of agent-based modeling is that while the agents themselves might be highly specified and their rules of interactions are well-known, and it is not until the model is run that we can know the outcome, due to the variety of possible interactions of autonomous heterogeneous decision-making agents. In essence, like complex systems themselves, agent-based models are more than the sum of their parts. Once the model is run, the third step is to evaluate the model (e.g. verification, calibration, validation, sensitivity analysis). For further guidelines on designing, implementing, and evaluating agent-based models, readers are referred to Gilbert and Troitzsch (2005) and Crooks et al. (2019).

46.2.3 Application Areas for Geographically Explicit Agent-Based Models

Geographically explicit agent-based models (i.e., those utilizing geographical information which we will go into more detail about in Sect. 46.3) have been developed to explore a range of problems which society faces over a variety of spatial and temporal scales from the micro-movement of pedestrians over seconds (e.g. Torrens 2012) to that of the macro-evolution of city systems over centuries (Pumain and Sanders 2013). The flexibility that the agent-based modeling approach provides has allowed such models to be used in a diverse set of applications. These range from archeology (Axtell et al. 2002), agriculture (Hailegiorgis et al. 2018), basketball (Oldham and Crooks 2019), crime (Malleson et al. 2013), diseases (Perez and Dragicevic 2009), disasters (Jumadi et al. 2018), invasive species (Anderson and Dragićević 2018), to urban growth (Xie and Yang 2011), housing markets (Geanakoplos et al. 2012), gentrification (Jackson et al. 2008), slum formation (Patel et al. 2018), and traffic (Manley and Cheng 2018). So, while agent-based modelers have been utilizing geographical data in their models, what has changed is the growth of data and ways of integrating such data within models (which will be discussed more in Sect. 46.3.2).

Open-source agent-based modeling toolkits such as GAMA (Taillandier et al. 2019), MASON (Luke et al. 2018), Repast (North et al. 2013), and NetLogo (Wilensky 1999) have evolved substantially over the past 20 years and many have built-in functionality to directly integrate data into models (e.g. raster and vector data structures), thus lowering the bar for creating geographically explicit models (for a review of these platforms and their applications readers are referred to Crooks et al. 2019). For example, in Fig. 46.6, we show a selection of models created utilizing the MASON toolkit and its GeoMason extension for GIS integration that span both spatial and temporal scales. These include such things as the micro-movement of pedestrians over seconds to that of the macro-movement of migrants over years, and many things in between such as modeling traffic, responses to disasters, disease outbreaks, and urban growth (for access to these models see MASON 2019, and for equivalent geographically explicit models in NetLogo see https://www.abmgis.org/).

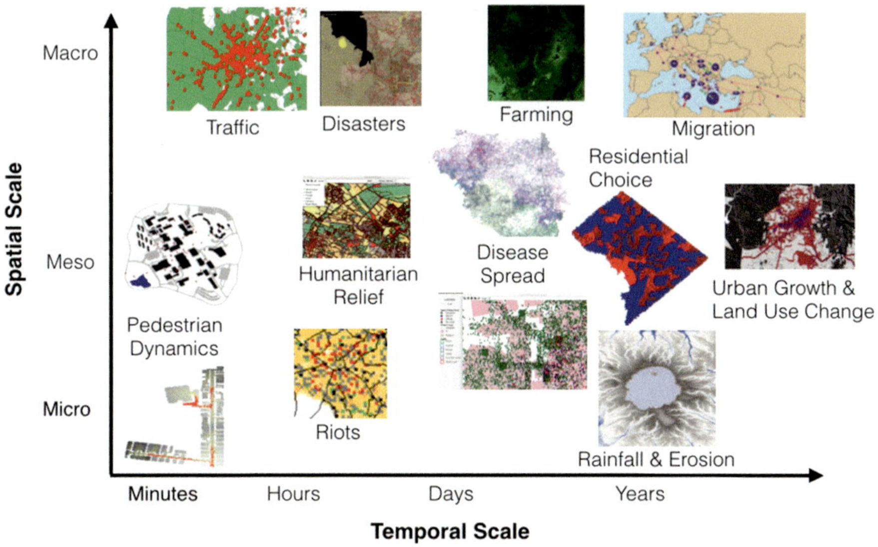

Fig. 46.6 Selection of GeoMason models across various spatial and temporal scales

In addition to these general-purpose open-source toolkits which allow for a range of urban phenomena to be simulated, where one could argue that the only constraint is that of the modeler's imagination, there are others that are dedicated to specific domains such as the open-source transportation simulations (e.g. MATSim of Horni, Nagel, Axhausen 2016, POLARIS of Auld et al. 2016, or TRANSIMS 2019), which are being used to study a wide range of transportation issues (e.g. daily trips, route planning, evaluation of intelligent transportation systems) in multiple cities around the world.

46.3 Integrating Data and Decision-Making into Agent-Based Models

Apart from the individual entities within agent-based models interacting with each other, these entities are also interacting and are affected by the artificial world (or environment) which they inhabit; similar to how the world around us affects our lives. For example, take land-use change. Developers may buy agricultural land, convert the land to residential use, and then sell it to residents who then move into it (e.g. Magliocca et al. 2011). Agents can also perceive their environment and respond to it (e.g. changing climatic conditions may alter farming practices as discussed in Hailegiorgis, Crooks, Cioff-Revilla 2018). Initially, many agent-based models represented space rather abstractly as we showed with the Schelling (1971) model in Sect. 46.2.1. However, perhaps with the demonstration of the Sugarscape model by Epstein and Axtell (1996), which showed how the environment can affect agents' wealth and survival, modelers started to realize that the artificial world that the agents inhabited could be stylized on geographical data. From earlier works such as those by Gimblett (2002) or Benenson and Torrens (2004) to current day work (e.g. Crooks et al. 2019), researchers have utilized data not only to represent the physical aspects of the artificial world (e.g. land cover, road networks) but also to help inform the social aspects (e.g. census data to help with knowing how many agents live in an area). Such data take the abstract representations of space and make it more grounded in real-world locations as we show in Fig. 46.7.

Different data layers in the form of rasters (e.g. land-use and land-cover, elevation) and vector formats (e.g. census areas, road networks) can act as the environment for the artificial world in which our agents interact. For example, vector data about roads can be used for a traffic simulation in the sense of allowing agents to navigate from one location to another. Or census data can be used to create a specified number of agents for a given location with associated socio-economic characteristics (e.g. Burger et al. 2017). Raster data such as those from the national land-cover dataset (Wickham et al. 2014) can be used for initialization of an urban growth simulation as they provide details on urban and non-urban land extents which affect where cities can and cannot grow (see Crooks et al. 2019 for further details and examples of how one can use such data in models). Such social and physical data layers in

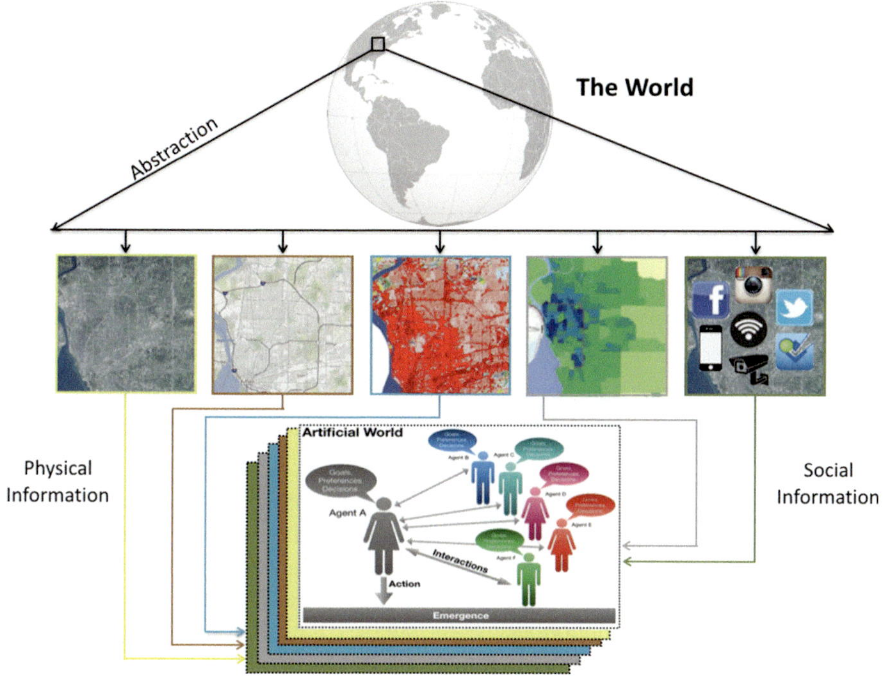

Fig. 46.7 Using geographic information as a foundation for artificial worlds

Fig. 46.7 replace the abstract artificial world presented in Fig. 46.2 and ground the model to actual real-world locations, which can have an impact on individual agents' interactions. Compare, for example, the abstract room in Fig. 46.8a which is used to test basic pedestrian movement to that of Fig. 46.8b which is based on actual

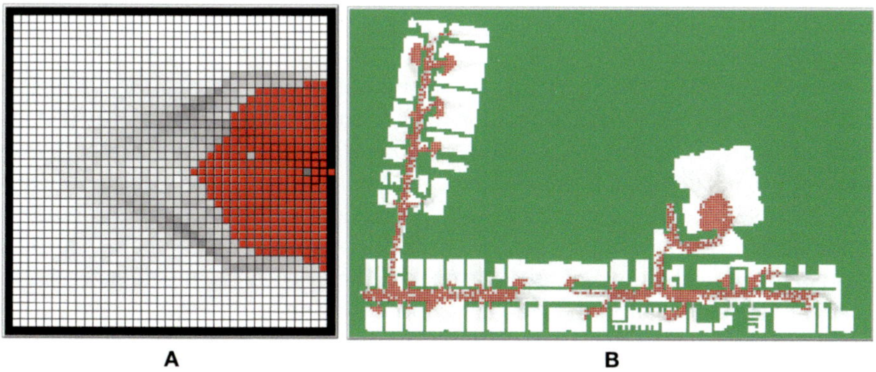

Fig. 46.8 Moving from an abstract room **a** to one where the artificial world is based on a real-world building floor plan **b**

CAD data of a real-world building. Here, actual walls, corridors, and exits constrain the agent's movement. While we already have discussed in Sect. 46.2.3 application areas, where researchers have created geographically explicit agent-based models to explore a wide range of phenomena, in the remainder of this section, we first discuss how one can incorporate decision-making into agent-based models (Sect. 46.3.1), before turning our attention to how new forms of data are being used in such models, to help inform decision-making (Sect. 46.3.2) and how with such data researchers are utilizing machine learning methods for various phases (steps) within the agent-based modeling (Sect. 46.3.3).

46.3.1 Incorporating Decision-Making into Agent-Based Models

As noted in Sect. 46.2.2, agent-based models are essentially rule-based systems in the sense that an agent's actions are programmed directly into them. Therefore, it is important to consider how we go about choosing these rules. However, as discussed in Sect. 46.1, modeling human behavior is not as simple as it sounds. This is because humans do not just make random decisions, but base their actions upon their knowledge and their abilities. In addition, it might be nice to think that human behavior is rational, but this is not always the case. Decisions can be based on emotions, such as self-interest, happiness, anger, or fear (see Izard 2007). In addition, emotions can influence one's decision-making by altering perceptions about the environment and future evaluations (Loewenstein and Lerner 2003). The question therefore is: how do we model human behavior? This is where agent-based models excel over other modeling approaches (as discussed in Sect. 46.2). Agent-based modeling allows us to focus on individuals or groups of individuals and give them diverse knowledge and abilities, which is not possible in other modeling methodologies. As such, agent-based models act as a testing ground for a variety of theoretical assumptions and concepts about human behavior (Stanilov 2012) within the safe environment of a computer simulation.

Broadly speaking, there are three main approaches to capturing such decision-making processes within agent-based models (Kennedy 2012). The first is a mathematical approach such as the use of ad hoc direct and custom coding of behaviors within the simulation, such as using random number generators to select a predefined possible choice (e.g. to buy or sell; Gode and Sunder 1993). But, people are not random, which has led researchers to develop other methods such as directly incorporating threshold-based rules; that is, when an environment parameter passes a certain threshold a specific agent behavior will result (e.g. move to a new location when the neighborhood composition reaches a certain percentage) as in the Schelling (1971) example introduced in Sect. 46.2.1. One could argue that these modeling approaches are appropriate when behavior can be well-specified. The second approach to modeling human behavior within agent-based models uses conceptual cognitive frameworks. Within such models, instead of using thresholds, more abstract concepts such as beliefs, desires, and intentions (BDI; Rao and Georgeff 1991) or physical, emotional, cognitive, and social factors (PECS; Schmidt 2002) are given to individual agents. Both the BDI and PECS frameworks have been successively applied to modeling human behavior in a number of applications, such as what drives people to crime (see Brantingham et al. 2005 and Malleson et al. 2010, respectively).

These conceptual cognitive frameworks and mathematical approaches for representing behavior, like agent-based models more generally, can both be considered as rule-based systems and are often applied to tens to millions of agents. The third approach, that of cognitive architectures, (e.g. Soar (Laird 2012) and ACT-R (Anderson and Lebiere 1998)) focuses on abstract or theoretical cognition of one agent at a time with a strong emphasis on artificial intelligence. This approach is rarely used to model more than a small number of agents, which makes their utility for modeling challenges faced by cities rather limited. However, while there are multiple ways of representing decision-making within agent-based models, why a modeler chooses one over the other is rarely discussed (Schlüter et al. 2017) or why a certain theory was chosen (if at all) to build upon (Groeneveld et al. 2017). Readers wishing to know more about decision-making within agent-based models are referred to Balke and Gilbert (2014) and to learn how such models can be used in a policy context see Calder et al. (2018).

46.3.2 The Growth of Data and Its Utilization Within Agent-Based Models

Coinciding with the ease of incorporating data into agent-based models (as discussed in Sects. 46.2.3) is the growth and availability of digital data (i.e. big data) for urban areas, many of which have an explicit or implicit geographic component (Stefanidis et al. 2013). Such data range from more traditional types such as census data, or remotely sensed imagery or in situ sensing devices (e.g. weather stations and air-pollution monitoring systems) to data from mobile sensors such as smartphones, GPS devices attached to taxis, or social media. This rise in data in a variety of shapes and forms coupled with increased computational resources has led to the rise of urban analytics. There are several definitions for urban analytics: for example, Singleton et al. (2017) defines it as a "*multidisciplinary area of research concerned with using new and emerging forms of data, alongside computational and statistical techniques to study cities*," while Batty (2019) places urban analytics in the wider scope of analytics more generally, stating the "*term analytics implies a set of methods that can be used to explore, understand and predict properties and features of any system, in our case of cities.*" What is common between the definitions is utilizing data and computational techniques to explore cities. If we first turn to data, we are not only referring to traditional datasets such as census and infrastructure (e.g. roads) traditionally collected and distributed by governmental organizations and industry but also to volunteered geographic information (e.g. OpenStreetMap) and social media, Internet of things (IoT), and cell phones, which are giving us new ways to explore the urban environment (Batty et al. 2012; Crooks et al. 2015b).

By bringing and analyzing these data together, we can begin to understand the wider patterns of cities. For example, smart-city data are founded at the individual level and through the analysis of travel cards can tell us how many people commute into a city every day (e.g. Zhong et al. 2015) and hint at the purpose of trips when combined with land-use information and social-media check-ins (Yang et al. 2019b). Dockless-bike data can provide information on urban flows and impacts of new infrastructure (e.g. Yang et al. 2019a) Similarly, cell-phone data can show origin–destination pairs for urban mobility (e.g. Louail et al. 2015) or patterns of movement and interactions (e.g. Malleson et al. 2018; Manley and Dennett 2019). What such data cannot tell us explicitly is the purpose of one's trip or their experience of the city while one is there. Bringing in data about the individual (social data) from multiple sources (e.g. Twitter, Facebook) might help complete the picture but still gives us only patterns and not necessarily the processes and the underlying motivations that led to the patterns emerging.

Identifying how and when these patterns will emerge is extremely difficult. Take for example congestion: it arises as a result of individual mobility decisions based on factors such as life stage, accessibility to workplace, shops, or other facilities which are constantly changing. Congestion can build locally at pinch points, placing sections of the city's transportation networks under severe strain. There is some irony that while we inhabit a data-rich world, without modeling it is extremely challenging to understand how the combination of physical environment and social dynamics contributes to how our cities function and grow. Data alone will not solve all the problems cities face, especially when using data from the past to look at the future. For example, with respect to financial or housing markets, we might have data on the stock market from 2010 to 2019 but this does not capture the 2007–2008 financial crisis. What happens if there is a structural change or some sort of evolution of the system or something happens outside of these bounds? Data capture only what they see, not necessarily extreme market events. Or to quote Heraclitus: "*No man ever steps in the same river twice, for it's not the same river and he's not the same man.*" This is one of the motivations for modeling, specifically agent-based models. We can explore such issues and pose *what-if* scenarios based on individuals making their own decisions. For example, what would be the implications of imposing congestion charging, in terms of improvements to both congestion and people's activities (e.g. Zheng et al. 2012)?

If we refer back to Fig. 46.7, we can utilize such data to inform our models, act as inputs to a model, or validate model outcomes. For example, there are numerous applications that are utilizing OpenStreetMap data to act as the foundation of their artificial worlds. These range from assessing route choice for humanitarian support after an earthquake (Crooks and Wise 2013), or utilizing building and infrastructure information during disease outbreaks (Crooks and Hailegiorgis 2014) to vehicle routing over a network (Horni et al. 2016) or as a basis for evacuation-route choice

(Goetz and Zipf 2012). If we turn our attention to pedestrian movement, which is of paramount importance if we wish to design more walkable cities, new sensor technology such as GPS has been used to test walking behaviors (Torrens et al. 2012), while others have utilized CCTV to calibrate how people move through small areas (Crooks et al. 2015a) or calibrate crowd densities (Batty et al. 2003). Crols and Malleson (2019), on the other hand, used footfall data collected via sensors to validate their pedestrian model of daily mobility in the town center of Otley, West Yorkshire in order to better understand how the town center is being used by its inhabitants. Similarly, Grübel et al. (2019) used footfall data to validate their model of pedestrian flows through Westminster in London.

New sources of data are also shedding light into how people navigate around the city; for example, Manley et al. (2015) found in analyzing GPS data from London minicabs that the shortest path models often used in transportation studies poorly predicted the actual behavior of minicab drivers; but through an agent-based model they showed how drivers used specific urban features (i.e., "anchor points") with respect to navigating around the city. Moving beyond just geographic data, others are using natural language processing (NLP) to mine textual data to inform agent decision-making (Runck et al. 2019). In another example, Wise (2014) developed an agent-based model to explore a wildfire event and subsequent evacuation in Colorado Springs over the space of a week in 2012. Specifically, Wise mined social media, in this case, Twitter, to derive the moods of people in the area and fed this into an evacuation model. For example, if one of the agents (i.e. a Colorado Springs resident) knew that the fire was nearby, and this information was passed along his or her social network to other agents who then decided whether to evacuate or not. This decision to evacuate or not also led to congestion, which was validated based on data that were harvested from the crowd and news outlets. What the above examples show is that new sources of data can be utilized in many aspects of agent-based modeling, especially those related to urban applications over a variety of spatial and temporal scales.

46.3.3 The Potential of Machine Learning and Agent-Based Modeling

While there has been a tremendous growth over the past decade in machine learning, a subfield of artificial intelligence, which is partly due to increases in computational power and the availability of data and is leading to new areas of research within urban analytics, and terms such as geographic data science are appearing (see Singleton and Arribas-Bel 2019). By using machine learning techniques (such as genetic algorithms, artificial neural networks, Bayesian classifiers, decision trees, or reinforcement learning) and data mining (i.e. finding patterns in the data), researchers have been exploring many aspects of city life such as the identification of slums via decision trees (Mahabir et al. 2018) and using natural language processing to find meanings of place (Jenkins et al. 2016).

However, while machine learning and data mining have seen a large growth in urban analytics, there has only been limited uptake of these methods in agent-based models, even though as Rand (2006) notes they are similar in the sense that both can be considered as rule-based systems (as we discussed in Sect. 46.2.2), and as both need to be initialized with a specific set of parameters. Both need to be run, and while in agent-based models, we observe the dynamics, in machine learning, we observe the outputs of the machine learning process (such as numbers, rules, or categories), and conclude when the stopping conditions are met (Rand 2006).[2] For example, in an agent-based model, this might be when all agents are happy, while in machine learning, it could be when the algorithm completes its processing (e.g. the value of the objective function cannot be further improved).

As noted in Sect. 46.2.2, agent-based modeling has broadly three major steps: the design of the model, the execution of the model, and evaluation of the model. Machine learning techniques have been applied to all three of these phases (see Abdulkareem et al. 2019). For example, in the first phase, the designing of the model, machine learning has been used to derive parameter values for agent-based models such as in cases of human mobility and obesity (e.g. Kavak 2007; Padilla et al. 2016). Machine learning has also been used during the running of the model, often for agents to learn from past experiences and make more informed decisions via reinforcement learning or genetic algorithms or random forests (e.g. Ramchandani et al. 2017; Rand 2006; Wolpert et al. 1999). Zhang et al. (2018) used neural networks for traffic prediction under various traffic configurations. In another example, Abdulkareem et al. (2019) used Bayesian networks and survey data to explore the spread of cholera in Kumasi, Ghana. Specifically, they used Bayesian networks with respect to improving risk perception and decision-making about where to get water during a cholera outbreak. Others have used reinforcement learning with respect to retirement planning (Ramchandani et al. 2017) or Bayesian networks to infer agents' locational choice and how this affects land-use change (Kocabas and Dragicevic 2013). Bone and Dragicevic (2010) used reinforcement learning to achieve optimal forest harvesting strategies. With respect to using machine learning algorithms to analyze model outputs (i.e. Step 3), Heppenstall et al. (2007) used a genetic algorithm to validate model outcomes of an agent-based model which simulates the retail gasoline market.

The examples above are just a few agent-based models utilizing machine learning and are intended to show the reader that researchers are exploring the use of such techniques in various aspects of the agent-based modeling process. However, unlike in the data science community, the use of machine learning is rather limited. Perhaps, this is because in the data science community packages exist (such as those implemented in Python or R) for machine leaning, but this is not the case for agent-based modeling. While agent-based toolkits exist, modelers still need to design and implement their

---

[2]For a greater discussion on the similarities between agent-based modeling and machine learning, readers are referred to Rand (2006).

own models, which in itself is a time-consuming task. Also, agent-based models focus on individual behavior, and to fully utilize machine learning one needs training data which are often not available (due to ethical implications, privacy concerns, etc.) at the level of detail for agent-based models (e.g. Runck et al. 2019; Weinberger 2011). We do not have space to delve deeper into why there has only been limited uptake of machine learning within agent-based models, but we envisage that with the growth of data, more agent-based modelers will utilize machine learning, especially as there are increasing calls to incorporate empirical data into models (e.g. Janssen and Ostrom 2006; Robinson et al. 2007) along with efforts to validate such models. For example, there might be abundant fine-resolution trajectory data about people's movement in cities which can be used to validate movement models and thus test ideas and theories of what motivates such patterns to emerge.

46.4 Summary and Outlook

As the world is increasingly becoming more densely urbanized, it is becoming more important to understand each city as a complex system whose whole is more than the sum of its parts. Without such understanding, it will be difficult to grapple with future societal challenges such as climate change. Cities are composed of many individuals whose interactions and behaviors lead to many issues emerging (Sect. 46.1). In this chapter, we have introduced agent-based modeling (Sect. 46.2) which allows one to model social systems from the bottom up. The focus of such models is the creation of artificial worlds in which individuals are given unique behaviors and rules and interact with each other and their environment. It is through such interactions that more macro-patterns emerge: for example, how individuals form crowds, or people going to and from work result in traffic jams, or people buying and selling homes lead to property markets emerging. By integrating geographic information into such models, we can turn abstract artificial worlds to those that mimic real-world locations (Sect. 46.3).

We also discussed how agent-based modeling has seen a large uptake over the past 20 years, spurred by the growth and availability of data (Sect. 46.3.2), which is providing many application domains for study. Such data when mined not only provide new ways to explore how people perceive and use the space around them, but also through machine learning methods can be integrated into the various aspects of agent-based modeling, from model parameterization to validation and calibration (Sect. 46.3.3). However, this is still an area which is evolving and there is still a significant amount of research to be done. New sources of data can potentially be mined to provide information pertaining to who, what, when, where, and why people do what they do. However, as Robert Axtell notes "*…there is a large research program to be done over the next 20 years, or even 100 years, for building good high-fidelity models of human behavior and interactions*" (cited by Weinberger 2011). Potentially, machine learning methods could help with, this especially with respect to improving decision-making within agent-based models.

Moreover, readers might have noted that a gallery of applications was discussed in this chapter, but there were very few attempts to integrate or couple various urban processes together, which was often the case with more traditional styles of land-use transportation interaction (LUTI) models (see Wise et al. 2017 for such a discussion). Perhaps, this is because agent-based models are being applied on a variety of spatial and temporal scales depending on the question at hand. For example, rush-hour traffic or various longer-term processes such as urban growth make it difficult to resolve temporal clocks or computational issues when scaling models to larger areas or greater numbers of agents, etc. However, the argument could be made that we are still in the initial stages of understanding cities from the bottom up, and the focus until now has been on specific problems but not on the city as a whole system. There is some justification for this based upon Simon's (1996) concept of the near-decomposability of systems, in which parts of a system interact among themselves in clusters or subgraphs, with interactions among subsystems being relatively weaker or fewer but not negligible, and therefore in the short term, one can study such systems (or problems) in isolation.

Looking ahead, as we noted above, today we are in a data-rich world and we discussed how one can utilize such data for model initialization, the parameterization of agents' attributes, or for the validation of model outcomes. However, as agent-based models are often used to simulate the behavior of complex systems, these systems often diverge rapidly from initial starting conditions. One way to prevent a simulation from diverging from reality would be to occasionally incorporate more up-to-date data and adjust the model accordingly. Data, especially streaming data produced through near-real-time observational datasets (e.g. social media or vehicle routing counters) could be utilized in such a case as shown in Fig. 46.9.

This process is known as dynamic data assimilation. There is a range of techniques that come under the banner of data assimilation that are designed for exactly this purpose. However, they have largely evolved from fields such as meteorology (i.e. to incorporate up-to-date environmental data into weather forecasts) and only recently have they started to be applied to agent-based modeling (e.g. Malleson et al. 2017; Rai and Hu 2013; Ward et al. 2016). The marriage of data assimilation methods and agent-based models could be transformative for the ways that some systems, for example, smart cities, are modeled. In addition to this, with new sources of big data and methods from machine learning and the growth of computational resources, we are perhaps nearing a point where we can explore and model cities from the bottom up at resolutions and scales that have not yet been possible.

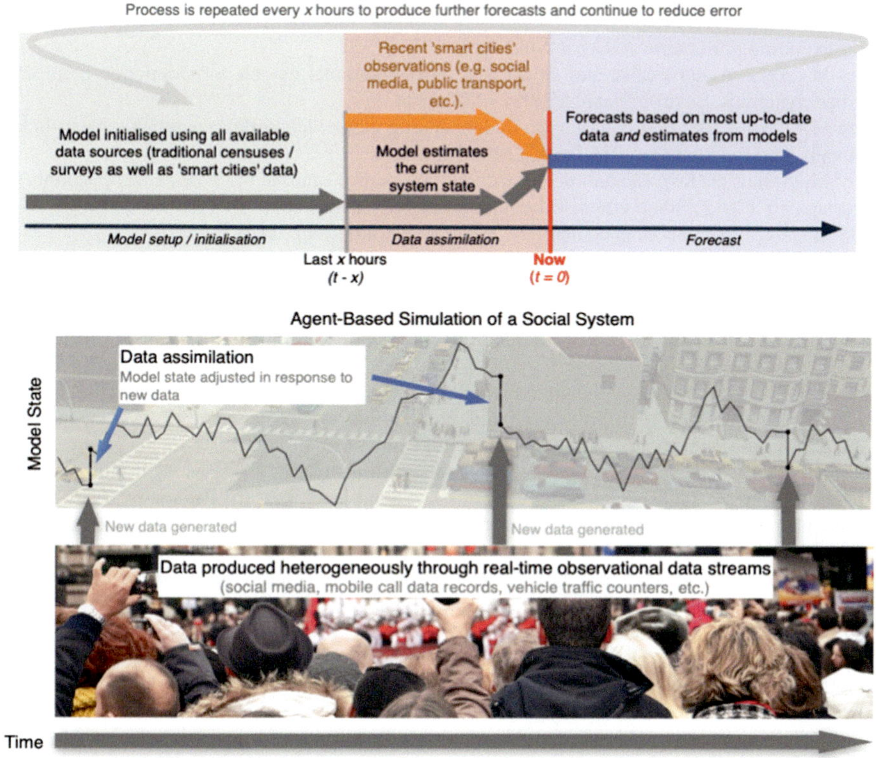

Fig. 46.9 Dynamic data assimilation and agent-based modeling

References

Abdulkareem SH, Mustafa YT, Augustijn E-W, Filatova T (2019) Bayesian networks for spatial learning: a workflow on using limited survey data for intelligent learning in spatial agent-based models. Geoinformatica 23(2):243–268

Allen PM (1997) Cities and regions as self-organizing systems: models of complexity. Gordon and Breach Science Publishers, Amsterdam, Netherlands

Anderson JR, Lebiere C (1998) The atomic components of thought. Erlbaum, Mahwah, NJ

Anderson T, Dragićević S (2018) A geographic network automata approach for modeling dynamic ecological systems. Geographical Analysis. https://doi.org/10.1111/gean.12183

Angel S, Parent J, Civco DL, Blei A, Potere D (2011) The dimensions of global urban expansion: estimates and projections for all countries, 2000–2050. Progress in Planning 75(2):53–107

Auld J, Hope M, Ley H, Sokolov V, Xu B, Zhang K (2016) POLARIS: agent-based modeling framework development and implementation for integrated travel demand and network and operations simulations. Transp Res Part C: Emerg Technol 64:101–116

Axtell R, Epstein JM, Dean JS, Gumerman GJ, Swedlund AC, Harburger J, Chakravarty S, Hammond R, Parker J, Parker M (2002) Population growth and collapse in a multiagent model of the kayenta anasazi in long house valley. Proc Natl Acad Sci 99(3):7275–7279

Balke T, Gilbert N (2014) How do agents make decisions? a survey. J Artif Soc Soc Simul 17(4):13

Batty M, Desyllas J, Duxbury E (2003) Safety in numbers? modelling crowds and designing control for the notting hill carnival. Urban Stud 40(8):1573–1590
Batty M (2005) Agents, cells, and cities: new representational models for simulating multiscale urban dynamics. Environ Plann A 37(8):1373–1394
Batty M (2007) Cities and complexity: understanding cities with cellular automata, agent-based models, and fractals. The MIT Press, Cambridge, MA
Batty M, Axhausen KW, Giannotti F, Pozdnoukhov A, Bazzani A, Wachowicz M, Ouzounis G, Portugali Y (2012) Smart cities of the future. Eur Phys J Spec Top 214(1):481–518
Batty M (2013) The new science of cities. MIT Press, Cambridge, MA
Batty M (2019) Urban analytics defined. Environ Plann B 46(3):403–405
Benenson I, Omer I, Hatna E (2002) Entity-based modelling of urban residential dynamics: the case of Yaffo. Tel Aviv. Environ Plann B 29(4):491–512
Benenson I, Torrens PM (2004) Geosimulation: automata-based modelling of urban phenomena. Wiley, London, UK
Bonabeau E (2003) Predicting the unpredictable. Harvard Bus Rev 80(3):109–116
Bone C, Dragicevic S (2010) Simulation and validation of a reinforcement learning agent-based model for multi-stakeholder forest management. Comput Environ Urban Syst 34(2):162–174
Brantingham PL, Glasser U, Kinney B, Singh K, Vajihollahi M (2005) Modeling urban crime patterns: viewing multi-agent systems as abstract state machines. In: Beauquier D, Börger E, Slissenko A (eds) Proceedings of the 12th international workshop on abstract state machines, Paris, France, pp 101–117
Burger A, Oz T, Crooks AT, Kennedy WG (2017) Generation of realistic mega-city populations and social networks for agent-based modeling. In: Proceedings of the computational social science society of America conference, Santa Fe, NM
Calder M, Craig C, Culley D, de Cani R, Donnelly CA, Douglas R, Edmonds B, Gascoigne J, Gilbert N, Hargrove C, Hinds D et al (2018) Computational modelling for decision-making: where, why, what, who and how. Royal Soc Open Sci 5(6):172096
Crooks AT, Wise S (2013) GIS and agent-based models for humanitarian assistance. Comput Environ Urban Syst 41:100–111
Crooks AT, Hailegiorgis AB (2014) An agent-based modeling approach applied to the spread of cholera. Environ Model Softw 62:164–177
Crooks AT, Croitoru A, Lu X, Wise S, Irvine JM, Stefanidis A (2015) Walk this way: improving pedestrian agent-based models through scene activity analysis. ISPRS Int J Geo-Inform 4(3):1627–1656
Crooks AT, Pfoser D, Jenkins A, Croitoru A, Stefanidis A, Smith DA, Karagiorgou S, Efentakis A, Lamprianidis G (2015) Crowdsourcing urban form and function. Int J Geogr Inform Sci 29(5):720–741
Crooks AT, Malleson N, Manley E, Heppenstall AJ (2019) Agent-based modelling and geographical information systems: a practical primer. Sage, London, UK
Crols T, Malleson N (2019) Quantifying the ambient population using hourly population footfall data and an agent-based model of daily mobility. Geoinformatica 23(2):201–220
Darley V, Outkin AV (2007) NASDAQ market simulation: insights on a major market from the science of complex adaptive systems. World Scientific Publishing, River Edge, NJ
Epstein JM, Axtell R (1996) Growing artificial societies: social science from the bottom up. MIT Press, Cambridge, MA
Epstein JM (2008) Why model? J Artif Soc Soc Simul 11(4):12
Geanakoplos J, Axtell R, Farmer D, Howitt P, Conlee B, Goldstein J, Hendrey M, Palmer N, Yang C (2012) Getting at systemic risk via an agent-based model of the housing market. Am Econ Rev 102(3):53–58
Gilbert N, Troitzsch KG (2005) Simulation for the social scientist, 2nd edn. Open University Press, Milton Keynes, UK
Gimblett HR (ed) (2002) Integrating geographic information systems and agent-based modelling techniques for simulating social and ecological processes. Oxford University Press, Oxford, UK

Gode DK, Sunder S (1993) Allocative efficiency of markets with zero-intelligence traders: market as a partial substitute for individual rationality. J Polit Econ 101:119–137

Goetz M, Zipf A (2012) Using crowdsourced geodata for agent-based indoor evacuation simulations. ISPRS Int J Geo-Information 1(2):186–208

Groeneveld J, Müller B, Buchmann CM, Dressler G, Guo C, Hase N, Hoffmann F, John F, Klassert C, Lauf T, Liebelt V et al (2017) Theoretical foundations of human decision-making in agent-based land use models—a review. Environ Model Softw 87:39–48

Grübel J, Wise S, Thrash T, Hölscher C (2019) A cognitive model for routing in agent-based modelling. AIP Conf Proc 2116(1):250005

Hailegiorgis AB, Crooks AT, Cioff-Revilla C (2018) An agent-based model of rural households' adaptation to climate change. J Artif Soc Soc Simul 21(4):4

Heppenstall AJ, Evans AJ, Birkin MH (2007) Genetic algorithm optimization of a multi-agent system for simulating a retail market. Environ Plann B 34(6):1051–1070

Horni A, Nagel K, Axhausen KW (eds) (2016) The multi-agent transport simulation MATSim. Ubiquity, London, UK

Izard CE (2007) Basic emotions, natural kinds, emotion schemas, and a new paradigm. Perspect Psychol Sci 2(3):260–280

Jackson J, Forest B, Sengupta R (2008) Agent-based simulation of urban residential dynamics and land rent change in a gentrifying area of Boston. Trans GIS 12(4):475–491

Janssen M, Ostrom E (2006) Empirically based, agent-based models. Ecol Soc 11(2):37

Jenkins A, Croitoru A, Crooks AT, Stefanidis A (2016) Crowdsourcing a collective sense of place. PLoS ONE 11(4):e0152932

Jumadi J, Heppenstall A, Malleson N, Carver S, Quincey D, Manville V (2018) Modelling individual evacuation decisions during natural disasters: a case study of volcanic crisis in Merapi Indonesia. Geosci 8(6):196

Kavak H (2007) A data-driven approach for modeling agents. PhD dissertation, Old Dominion University. https://doi.org/10.25777/6b7c-9a95, Norfolk, VA

Kennedy W (2012) Modelling human behavior in agent-based models. In: Heppenstall A, Crooks AT, See LM, Batty M (eds) Agent-based models of geographical systems. Springer, New York, NY, pp 167–180

Kocabas V, Dragicevic S (2013) Bayesian networks and agent-based modeling approach for urban land-use and population density change: a BNAS model. J Geogr Syst 15(4):403–426

Laird JE (2012) The Soar cognitive architecture. The MIT Press, Cambridge, MA

Legion (2019) Legion: pedestrian simulation software. https://www.bentley.com/en/products/brands/legion. Accessed 22nd August 2019

Loewenstein GF, Lerner JS (2003) The role of affect in decision making. In: Davidson RJ, Scherer KR, Goldsmith HH (eds) Handbook of affective science. Oxford University Press, Oxford, UK, pp 619–642

Louail T, Lenormand M, Picornell M, Cantú OG, Herranz R, Frias-Martinez E, Ramasco JJ, Barthelemy M (2015) Uncovering the spatial structure of mobility networks. Nat Commun 6:6007

Luke S, Simon R, Crooks AT, Wang H, Wei E, Freelan D, Spagnuolo C, Scarano V, Cordasco G, Cioffi-Revilla C (2018) The MASON simulation toolkit: past, present, and future. In: P. D, H. V (eds) Proceedings of the 19th international workshop on multi-agent-based simulation, stockholm, Sweden

Magliocca N, Safirova E, McConnell V, Walls M (2011) An economic agent-based model of coupled housing and land markets (CHALMS). Comput Environ Urban Syst 35(3):183–191

Mahabir R, Agouris P, Stefanidis A, Croitoru A, Crooks AT (2018) Detecting and mapping slums using open data: a case study in Kenya. Int J Digital Earth. https://doi.org/10.1080/17538947.17532018.11554010

Malleson N, Heppenstall A, See L (2010) Crime reduction through simulation: an agent-based model of burglary. Comput Environ Urban Syst 34(3):236–250

Malleson N, Heppenstall A, See L, Evans A (2013) Using an agent-based crime simulation to predict the effects of urban regeneration on individual household burglary risk. Environ Plann B 40(3):405–426
Malleson N, Tapper A, Ward J, Evans A (2017) Forecasting short-term urban dynamics: data assimilation for agent-based modelling. In: Proceedings of the 13th annual conference of the European social simulation association, Dublin, Ireland, pp 25–29
Malleson N, Vanky A, Hashemian B, Santi P, Verma SK, Courtney TK, Ratti C (2018) The characteristics of asymmetric pedestrian behavior: a preliminary study using passive smartphone location data. Trans GIS 22(2):616–634
Manley E, Cheng T, Penn A, Emmonds A (2014) A framework for simulating large-scale complex urban traffic dynamics through hybrid agent-based modelling. Comput Environ Urban Syst 44:27–36
Manley EJ, Addison JD, Cheng T (2015) Shortest path or anchor-based route choice: a large-scale empirical analysis of minicab routing in London. J Transp Geogr 43:123–139
Manley E, Cheng T (2018) Exploring the role of spatial cognition in predicting urban traffic flow through agent-based modelling. Trans Res Part A: Policy Pract 109:14–23
Manley E, Dennett A (2019) New forms of data for understanding urban activity in developing countries. Appl Spat Anal Policy 12(1):45–70
Massive (2019) Film gallery. https://www.massivesoftware.com/gallery.html. Accessed 17th June 2019
MASON (2019) Multi agent simulation of neighborhoods. https://cs.gmu.edu/~eclab/projects/mason/. Accessed 17th June 2019
North MJ, Collier NT, Ozik J, Tatara ER, Macal CM, Bragen M, Sydelko P (2013) Complex adaptive systems modeling with repast simphony. Complex Adapt Syst Model 1(1):3
Oldham M, Crooks AT (2019) Drafting agent-based modeling into basketball analytics. In: Proceedings of 2019 Spring simulation conference (SpringSim'19), Tucson, AZ
Padilla JJ, Diallo SY, Kavak H, Sahin O, Sokolowski JA, Gore RJ (2016) Semi-automated initialization of simulations: an application to healthcare. J Defense Model Simul 13(2):171–182
Paramics (2019) Traffic and pedestrian simulation software. https://www.paramics-online.com/. Accessed 17th June 2019
Patel A, Crooks AT, Koizumi N (2018) Spatial agent-based modeling to explore slum formation dynamics in Ahmedabad, India. In: Thill JC, Drajicavic S (eds) Geocomputational analysis and modeling of regional systems. Springer, New York, NY, pp 121–141
Perez L, Dragicevic S (2009) An agent-based approach for modeling dynamics of contagious disease spread. Int J Health Geogr 8. https://doi.org/10.1186/1476-072X-8-50
PTV Visum (2019) Traffic analyses software. https://vision-traffic.ptvgroup.com/. Accessed 17th June 2019
Pumain D, Sanders L (2013) Theoretical principles in interurban simulation models: a comparison. Environ Plann A 45(9):2243–2260
Rai S, Hu X (2013) Behavior pattern detection for data assimilation in agent-based simulation of smart environments. 2013 IEEE/WIC/ACM international joint conferences on web intelligence (WI) and intelligent agent technologies (IAT). Atlanta, GA. IEEE, pp 171–178
Ramchandani P, Paich M, Rao (2017) Incorporating learning into decision making in agent based models. In: Oliveira E, Gama J, Vale Z, Cardoso HL (eds) Progress in artificial intelligence: Proceedings of the 18th EPIA conference on artificial intelligence, Porto, Portugal. Springer, pp 789–800
Rand W (2006) Machine learning meets agent-based modeling: when not to go to a bar. In: Sallach D, Macal CM, North MJ (eds) Proceedings of the agent 2006 conference on social agents: results and prospects, university of Chicago and argonne national laboratory, Chicago, IL. pp 51–59
Rao AS, Georgeff MP (1991) Modeling rational agents within a BDI-architecture. In: Allen J, Fikes R, Sandewall E (eds) Proceedings of the second international conference on principles of knowledge representation and reasoning, San Mateo, CA, pp 473–484

Robinson DT, Brown D, Parker DC, Schreinemachers P, Janssen MA, Huigen M, Wittmer H, Gotts N, Promburom P, Irwin E, Berger T, Gatzweiler F, Barnaud C (2007) Comparison of empirical methods for building agent-based models in land use science. J Land Use Sci 2(1):31–55

Runck BC, Manson S, Shook E, Gini M, Jordan NR (2019) Using word embeddings to generate data-driven human agent decision-making from natural language. Geoinformatica 23(2):243–268

Schelling TC (1971) Dynamic models of segregation. J Math Soc 1(1):143–186

Schlüter M, Baeza A, Dressler G, Frank K, Groeneveld J, Jager W, Janssen MA, McAllister RR, Müller B, Orach K, Schwarz N (2017) A framework for mapping and comparing behavioral theories in models of social-ecological systems. Ecol Econ 131:21–35

Schmidt B (2002) The modelling of human behavior: the PECS reference model. In: Proceedings 14th European simulation symposium, Dresden, Germany

Seto KC, Fragkias M, Güneralp B, Reilly MK (2011) A meta-analysis of global urban land expansion. PLoS ONE 6(8):e23777

Simon HA (1996) The sciences of the artificial, 3rd edn. MIT Press, Cambridge, MA

Singleton AD, Spielman S, Folch D (2017) Urban analytics. Sage, London, UK

Singleton A, Arribas-Bel D (2019) Geographic data science. Geogr Anal. https://doi.org/10.1111/gean.12194

Stanilov K (2012) Space in agent-based models. In: Heppenstall A, Crooks AT, See LM, Batty M (eds) Agent-based models of geographical systems. Springer, New York, NY, pp 253–271

Stefanidis T, Crooks AT, Radzikowski J (2013) Harvesting ambient geospatial information from social media feeds. GeoJournal 78(2):319–338

STEPS (2019) Pedestrian movement software. https://www.steps.mottmac.com/. Accessed 17th June 2019

Taillandier P, Gaudou B, Grignard A, Huynh QN, Marilleau N, Caillou P, Philippon D, Drogoul A (2019) Building, composing and experimenting complex spatial models with the GAMA platform. GeoInformatica 23(2):299–322

Thrift N (1999) The place of complexity. Theor, Cult Soc 16(3):31–69

Torrens PM (2012) Moving agent-pedestrians through space and time. Ann Assoc Am Geogr 102(1):35–66

Torrens PM, Nara A, Li X, Zhu H, Griffin WA, Brown SB (2012) An extensible simulation environment and movement metrics for testing walking behavior in agent-based models. Comput Environ Urban Syst 36(1):1–17

TRANSIMS (2019) TRANSIMS: transportation analysis and simulation system. https://code.google.com/archive/p/transims/. Accessed 17th June 2019

United Nations (2014) World urbanization prospects: the 2014 revision. Department of Economic and Social Affairs, New York, NY

United Nations (2016) The world's cities in 2016. Department of Economic and Social Affairs, New York, NY

Ward JA, Evans AJ, Malleson N (2016) Dynamic calibration of agent-based models using data assimilation. Open Sci 3(4):150703

Weinberger S (2011) Web of war: can computational social science help to prevent or win wars? Nature 471:566–568

Wickham J, Homer C, Vogelmann J, McKerrow A, Mueller R, Herold N, Coulston J (2014) The multi-resolution land characteristics (mrlc) consortium—20 years of development and integration of USA national land cover data. Remote Sens 6(8):7424–7441

Wilensky U (1999) NetLogo. https://ccl.northwestern.edu/netlogo. Center for connected learning and computer-based modeling, Northwestern University, Evanston, IL

Wilson AG (2000) Complex spatial systems: the modelling foundations of urban and regional analysis. Pearson Education, Harlow, UK

Wise S (2014) Using social media content to inform agent-based models for humanitarian crisis response. PhD dissertation, George Mason University, Fairfax, VA

Wise S, Crooks AT, Batty M (2017) Transportation in agent-based urban modelling. In: Namazi-Rad M, Padgham L, Perez P, Nagel K, Bazzan A (eds) Agent based modelling of urban systems. Springer, New York, NY, pp 129–148

Wolpert DH, Wheeler KR, Tumer K (1999) General principles of learning-based multi-agent systems. In: Etzioni O, Müller JP, Bradshaw JM (eds) Proceedings of the third annual conference on autonomous agents, Seattle, WA. ACM. pp 77–83

Xie Y, Yang X (2011) Agent-based urban modeling: simulating urban growth and subsequent landscape change in Suzhou, China. In: Yang X (ed) Urban remote sensing: monitoring, synthesis and modeling in the urban environment. Wiley, Hoboken, NJ, pp 347–357

Yang Y, Heppenstall A, Turner A, Comber A (2019a) A spatiotemporal and graph-based analysis of dockless bike sharing patterns to understand urban flows over the last mile. Comput Environ Urban Syst 77:101361

Yang Y, Heppenstall A, Turner A, Comber A (2019b) Who, where, why and when? Using smart card and social media data to understand urban mobility. ISPRS Int J Geo-Inform 8(6):271

Zhang Y, Grignard A, Lyons K, Aubuchon A, Larson K (2018) Real-time machine learning prediction of an agent-based model for urban decision-making. In: Dastani M, Sukthankar G, André E, Koenig S (eds) Proceedings of the 17th international conference on autonomous agents and multiagent systems, Stockholm, Sweden. International foundation for autonomous agents and multiagent systems. pp 2171–2173

Zheng N, Waraich RA, Axhausen KW, Geroliminis N (2012) A dynamic cordon pricing scheme combining the macroscopic fundamental diagram and an agent-based traffic model. Transp Res Part A: Policy Pract 46(8):1291–1303

Zhong C, Manley E, Arisona SM, Batty M, Schmitt G (2015) Measuring variability of mobility patterns from multiday smart-card data. J Comput Sci 9:125–130

Andrew Crooks is a Professor of Geography within the Department of Geography and a faculty member in the RENEW Institute at the University at Buffalo. His research interests relate to exploring and understanding the natural and socio-economic environments, specifically urban areas, using GIS, spatial analysis, social network analysis, and agent-based modeling methodologies.

Alison Heppenstall is Professor of Geocomputation at the University of Leeds (UK) and a Fellow at the Alan Turing Institute. She is currently developing approaches within urban analytics, including detecting spatio-temporal patterns in data, quantifying uncertainty in agent-based models, and building more robust models via probabilistic programming and reinforcement learning.

Nick Malleson is a Professor of Spatial Science at the University of Leeds (UK). His research focuses on the development of agent-based models aimed at understanding and explaining social phenomena. He is also interested in how "Big Data" and smart cities initiatives can be used to understand the daily dynamics of cities.

Ed Manley is Professor of Urban Analytics in the School of Geography, University of Leeds, and Turing Fellow at the Alan Turing Institute for Data Science and Artificial Intelligence. He is Associate Editor of the Applied Spatial Analysis and Policy journal and chairs the GIScience Research Group at the Royal Geographical Society.

Chapter 47
Transportation Modeling

Eric J. Miller

Abstract Informatics are rapidly and radically transforming urban transportation in ways not seen since the introduction of the automobile over a hundred years ago. Near-ubiquitous smartphone usage, pervasive cellular and Wi-Fi connectivity, powerful and cost-effective computing capabilities, advanced GIS software and databases, advanced platforms for managing and scheduling service operations, etc., are combining to enable the introduction of new mobility services and technologies that are increasingly disrupting conventional trip-making behavior and the "rules of the game" in terms of transportation network operations and the regulation of system performance. The implications of these major informatics-driven changes for transportation modeling are equally disruptive and major. These include changes in: travel behavior; transportation system performance; the data available for model development and application; and modeling methods. Each of these broad areas of impact are discussed in this chapter.

47.1 Introduction

Use of large, computer-based models of travel demand and transportation system performance is standard practice in urban regions worldwide for transportation planning and decision-support purposes (Meyer and Miller 2013). They enable planners to estimate quantitatively the likely future impacts of a wide variety of policy options, including investment in major new transportation infrastructure (roads, transit, etc.), land-use policies, pricing/fare policies, new technologies, population and employment growth trends, etc. Detailed discussion of these models is well beyond the scope of this chapter, but the state of the art is extensively documented in the literature (see, for example, Ben-Akiva and Lerman 1985; Train 2009; Ortuzar and Willumsen 2011; Castiglione et al. 2015). Rather, this chapter explores current and emerging impacts of urban informatics on transportation modeling needs, capabilities, opportunities, and challenges1.

E. J. Miller (✉)
University of Toronto, Toronto, Canada
e-mail: miller@ecf.utoronto.ca

W. Shi et al. (eds.), *Urban Informatics*, The Urban Book Series,
https://doi.org/10.1007/978-981-15-8983-6_47

Informatics are rapidly and radically transforming urban transportation in ways not seen since the introduction of the automobile over a hundred years ago. Near-ubiquitous smartphone usage, pervasive cellular and Wi-Fi connectivity, powerful and cost-effective computing capabilities, advanced GIS software and databases, advanced platforms for managing and scheduling service operations, etc. are combining to enable the introduction of new mobility services and technologies that are increasingly disrupting conventional trip-making behavior and the "rules of the game" in terms of transportation network operations and the regulation of system performance.

The implications of these major informatics-driven changes for transportation modeling are equally disruptive and major. These include:

- Changes in travel behavior.
- Changes in transportation system performance.
- Changes in the data available for model development and application.
- Changes in modeling methods.

Each of these topics are discussed in detail in the following four sections. Looming over this discussion of technology-driven changes in the transportation system and associated modeling needs is the potential for the introduction into widespread usage within a currently ill-defined but still foreseeable future of electric vehicles (EVs) and connected and autonomous vehicles (CAVs), which may also be electrified (CAVEs). Full discussion of these technologies and their potential impacts goes well beyond the topic of urban informatics per se. But some possible impacts of eventual CAV impacts on travel behavior and transportation network performance are briefly discussed in Sects. 47.2 and 47.3.

47.2 Informatics and Travel Behavior

The primary impacts of informatics on travel behavior to date derive from two related informatics-based services:

- Real-time travel-related information.
- New mobility services and technologies.

These are discussed in the following two sub-sections. As becomes clear in this discussion, the driving technology enabling all these services are cellular- and Web-based apps running on smartphones and other computing devices, tied to centralized computing platforms that receive and send massive amounts of data and that process customer data requests for information and services, match customers with service providers, etc. The evolution and widespread adoption of smartphones among a broad segment of trip-makers, in particular, has been fundamental to the development and implementation of these various services.

47.2.1 Real-Time Travel-Related Information

A veritable plethora of Web- and smartphone-based apps exist that trip-makers can use to plan their trip destination, mode, and route choices prior to traveling and to dynamically choose their travel route during their trip. Many of these apps are provided by private companies, but public-sector apps also exist. For example, most public transit agencies provide some form of route guidance, as well as schedule and fare information.

Perhaps the most pervasive and impactful of these apps are the wide range of route-guidance apps based on the Global Positioning System (GPS) and available either on-board many automobiles or as apps for smartphones or other mobile devices such as tablets. These sense the current location of the device (and, hence, vehicle) and provide real-time estimates of current traffic conditions on the roadway being used. They also provide estimates of current travel times to a user-specified destination, along with recommended best routes to take to this destination. The definition of best route may be based either on shortest distance or shortest expected travel time, with the latter being the preferred and, increasingly, the most common option. Link and route travel times are determined based on crowd-sourced information on speeds gathered from all the users of the service, as well as possibly other information that may be available to the service provider (police/traffic center advisories, other roadway sensor data, etc.). They also depend critically on access to very precise and accurate geographic information system (GIS) representations of the road network, including speed limits and other road attributes. Huge effort over the past several decades has gone into developing such detailed maps for much of the world, particularly, in urbanized areas. Thus, these route-guidance apps represent an advanced marriage of GPS tracking and GIS mapping and analysis capabilities.

Both real-time and historical data are used in the calculations. The quality of the travel-time and route-selection calculations obviously depends on the number of users in the system at any one time, the depth and relevance of the available historical information, and, critically, the quality and accuracy of the (typically proprietary) algorithms used by the service provider to do these calculations. Machine learning methods (running on powerful cluster/cloud computing platforms) play a key role in sifting through the massive real-time and historical data to identify traffic patterns and to make short-term predictions of best routes to recommend. While these algorithms still are not 100% perfect under all conditions and in all places, their accuracy in making short-run predictions of roadway performance is typically quite impressive.

In addition to on-board route-guidance apps, conventional variable message signs on roadways and radio traffic reports have for decades provided a certain amount of high-level, real-time information concerning current travel conditions on major roadways, although these rarely provide route guidance. That is, a variable message sign might indicate that the roadway is congested ahead, but will not actually suggest or advise to take an alternative route. This is both due to legal concerns (if a driver takes a suggested alternative route and gets into an accident, who is liable?) and to minimize the potential for introducing instability into the system (what if everyone took the alternative route?).

Many apps also exist for providing static or real-time information concerning public transit routes, schedules, fares, and travel times. Most transit agencies now provide such an app, but many private and open-source public apps also exist. Such apps may provide information concerning: when the next transit vehicle is expected to arrive at a given stop; assistance for planning a trip from a given origin to a given destination at a given time of day; fare policies and payment options; service disruptions notices, etc. In addition to mobile-device-based apps, many transit agencies also provide real-time information at transit stops and stations concerning expected next-vehicle arrival times, by transit line. Various apps also exist to help bicyclists track their bike usage and routes are taken. Personal fitness apps for tracking distance walked also exist.

Although not generally thought of as being particularly travel-related, a vast array of Web sites provides information concerning every form of activity imaginable—restaurants, stores, entertainment venues, hotels, etc. These activity locations are potential destinations for trip-making that is not related to work or school, and the ubiquitous and voluminous availability of such data may well influence trip-makers' decision-making, especially regarding trip destination.

In general, most of these apps and services can be used for pre-trip planning ("Where should I go for dinner tonight"? "Should I drive or take transit for this trip?") as well as for on-route dynamic decision-making ("Accident ahead; let's get off the freeway"). While usage of these various apps is clearly very widespread, the actual impacts of this usage on travel behavior are not at all well understood. What percentage of the population are using what kinds of apps? Does this usage significantly influence choice of mode or destination, or timing of trips? Route-guidance apps must be affecting route choices, given their widespread use, but how great are the resulting deviations from the routes that drivers would have chosen in the absence of the app? To what extent is congestion being reduced (or increased?) through extensive use of these apps? These issues are discussed in greater detail below.

47.2.2 New Mobility Services and Technologies

Current and emerging information and communications technology (ICT) is not only dramatically increasing and improving the information available to trip-makers to help them in their travel decision-making, it is also revolutionizing the services available to them by which they may travel. New ICT-based mobility services and technologies are emerging virtually daily that provide new travel options for trip-makers. As with the new information services, these critically depend on smart mobile devices for communicating with potential customers of the service and on powerful computing platforms to manage the service.

As discussed in detail by Calderón and Miller (2019, 2020), a *mobility service* can be defined as an operation that enables a person to complete a trip from an origin to a destination by means of a given mode (technology) and service process. Public transit and conventional taxis are traditional mobility services. But a wide range of informatics-enabled mobility services has emerged in recent years. These take many forms, including:

- *Ridehailing*: Services such as Uber and Lyft (also conventional taxi), in which a service provider connects drivers with passengers to provide passengers with a door-to-door trip from their origin to their destination. Ridehailing can be further sub-divided into *single-user* and *shared-ride* services, with the latter involving passengers sharing the vehicle with other passengers and, as a result, experiencing some amount of trip deviation from a direct origin-to-destination trip in order to accommodate the pickups and drop-offs of the other passengers sharing the vehicle.
- *Vehicle-sharing*: These services provide short-term rentals of vehicles to customers who pick up the vehicle from where it is parked, use it to execute one or more trips, and then leave the vehicle safely parked once they are finished with it. Different services use different types of vehicles, including: automobiles (car-share), bicycles (bike-share, using both conventional bicycles and e-bikes), and, most recently, e-scooters. Vehicles usually are parked at designated stations (parking lots, bike-share docking stations, etc.), but dockless systems increasingly exist, in which the car, bike, e-scooter, etc., can be left anywhere, and is picked up by the next customer from wherever it was last left. Such dockless systems obviously depend on GPS tracking of the vehicle so that its location is known at all times. Vehicle-sharing services are usually provided by a for-profit company, but examples of peer-to-peer systems also exist in which private individuals offer their vehicle for usage by others when they do not need it for their personal use.[1]
- *Demand-responsive transit (DRT)/microtransit*: A wide variety of transit services exist (or can be imagined) that deviate from conventional fixed-route, fixed-schedule (typically large-vehicle) transit operations, including various combinations of route deviation, flexible stop location, on-demand scheduling of vehicle routing, and, usually, use of smaller vehicles that are cost-effectively matched to travel demand levels. Various forms of DRT have operated basically as long as public transit has existed. In particular, in much of the world jitney operations (along with other forms of privately operated informal transit services) are critical components of urban transportation, especially for lower-income trip-makers. In additional, DRT (often referred to as paratransit) services are a standard means of providing on-demand transit to mobility-impaired trip-makers who are unable to use conventional transit services. Platform-based informatics systems are redefining and enhancing the capabilities and potential applications of such

[1]Examples of peer-to-peer shared-ride systems also exist in which a platform connects private individuals who are willing to share rides with other individuals. A common example of such a system occurs on many university campuses, in which students offer rides to other students to travel back and forth between the university and nearby home cities during holiday weekends, etc.

services by significantly improving both the quality of service that can be offered to customers (through improved real-time scheduling and more efficient routing) and the cost-effectiveness with which the service can be provided.

While a wide diversity of mobility services exists, they all involve some combination of a generic set of operating functions (Calderón and Miller 2019, 2020). These consist of:

- *Matching* trip-maker requests for service with drivers and vehicles.
- *Rebalancing* vehicle fleets to maintain an appropriate spatial distribution of vehicles available for service.
- Trip *pricing* and payment.
- *Pooling* customers within vehicle tours for shared-ride operations.

Clearly not all operations pertain to all services. Bike-share services, for example, only provide real-time information concerning the current availability of bicycles by location, leaving customers to find their way to and rent one of these available bicycles. They do, however, have to deal with rebalancing, since usage patterns often result in large numbers of bicycles at popular destinations and too few bicycles at some origin locations. Ridehailing operators, on the other hand, primarily are concerned with matching customers to vehicles so as to both maximize the customer experience (usually meaning minimizing service wait times) and minimizing operating costs (e.g. avoiding very long dead-heading of vehicles). They may or may not engage in active attempts to rebalance the locations of the vehicles currently in service.[2] Pooling, of course, only pertains to shared-ride operations, but is a very critical component of the service, since the classical weakness of shared-ride services has been poor customer experiences: long wait times and circuitous routing (and hence long travel times relative to a more direct origin–destination journey).

Pricing levels and policies vary from one service to another and vary to the extent that prices dynamically vary with demand levels (so-called surge pricing) and, possibly, other factors (such as weather). Online payment systems based on credit cards are, however, an important feature of all new mobility systems. The convenience of this automated payment system should not be underestimated. At the end of the day, differences between a conventional taxi and an Uber are arguably not that great,[3] but the convenience of being able to simply step out of the car at the end of the trip (as well as the convenience of booking the trip with a few key-strokes on a smartphone) appears to be a significant factor in the success of new mobility services.

The role of informatics-based platforms, involving an integrated of GPS, GIS, real-time cell- and Web-based communications, combined with high-capacity computing and data processing and analytics based on artificial intelligence (AI) is fundamental

[2]Since ridehailing services currently depend on independent driver contractors, the ability of the ridehailing platform provider to influence their locations when not in service tends to be indirect at best.

[3]Although differences clearly exist, particularly, perceptual differences. Taxis, for example, are often criticized as being "dirty". Safety/security differences also exist, as do price differentials.

to all such mobility services. It is such platforms that have allowed both conventional taxi and transit services to be re-invented and for new technologies and services such as bike- and e-scooter share services to emerge.

The concept of *mobility as a service* (MaaS) generalizes mobility services by extending the platform concept to integrate two or more mobility services to provide seamless, and door-to-door mobility solutions that dynamically mix and match mobility services customer by customer to optimize their travel experience within a one-stop-shopping process. MaaS is seen by many as the future of transportation, with MaaS platforms acting as brokers that piece together different mobility services to best meet the trip-maker's needs and preferences. In such a future, a trip-maker may be picked up at her door in a suburb by a ridehailing company, taken to a commuter rail station just in time to board her train, and then have an e-bike waiting for her at her downtown egress station to complete her journey to her office, all for one fare automatically charged to her credit or debit card (perhaps with various loyalty points as well).

Such complete mobility solutions do not generally currently exist, although many companies and organizations are working toward their implementation. A particularly important policy question exists concerning the extent to which MaaS solutions can be integrated to improve the cost-effectiveness and attractiveness of public transit, so as to maintain it as a primary mass mover of trip-makers in high-density corridors. Urban areas worldwide are currently overwhelmed by auto congestion, and it is essential, however MaaS plays out, that it enables more efficient usage of transportation networks through the promotion of transit (where appropriate) and congestion reductions, while still accommodating the growth in travel that is inevitable as urban regions continue to grow. Notably, there is a growing literature that indicates that current mobility services are both adversely impacting conventional transit usage and increasing the amount of congestion (at least in central areas) in many cities (Li et al. 2019; Graehler et al. 2019; Rayle et al. 2016).

While an academic literature exists that explores the potential impact of route-guidance information on travel behavior, most of this is based on stated preference surveys or hypothetical simulation experiments rather than real-world data. A major barrier to investigating these questions is that the vast bulk of data concerning app usage and subsequent behavior is proprietarily held by private companies who are usually unwilling to share it with public agencies or academic researchers.

Enormous speculation currently exists concerning the potential impacts on travel behavior of the ubiquitous availability of fully autonomous vehicles. Exploration of this issue is well beyond the scope of this chapter. We simply note that CAVs potentially might dramatically alter auto ownership levels (people may simple rent mobility on a per-trip basis), public transit usage, and roadway congestion levels, among many possible other impacts. Transit ridership impacts are a particularly important policy question. CAVs might be used to support the use of higher-order transit by providing first- and last-mile solutions for getting to and from transit in low-density suburban neighborhoods. Or ubiquitous automated ridesharing services might decimate transit usage, likely leading to increased, rather than decreased, congestion on urban streets. In any event, increasing connectivity and automation of the transportation system will further increase the availability of massive, dynamic real-time information concerning travel and the associated need for advanced informatics methods for the storage and analysis of these data for transportation planning and operations purposes.

47.3 Informatics and Transportation Network Performance

Transportation network performance is the emergent outcome of a short-run (day-to-day, hour-by-hour, minute-by-minute) demand–supply interaction, in which the performance of a network link (road or transit line segment) depends on the volume of flow (cars, passengers, etc.) using the link at a given time. That is, the travel time required to traverse the link (and associated congestion level) depends on the level of link usage, while the number of users of the link depends (at least in part) on the travel time experienced on the link.

Route-guidance apps surely have an impact on the route choices of individual trip-makers (otherwise, why would they use them?), and, hence the distribution of flows across links and paths within the network, and ultimately on link and path travel times. Such apps are used both for pre-trip planning (What's the best way of getting there? What's a good time to leave to avoid traffic?) and dynamic on-route guidance. The actual impacts of such route-guidance apps on trip-makers' route choices, however, are typically unknown, since only the app companies usually see the data and they are generally not telling.

Note that a major impact of CAVs is likely to be to take route choice decisions largely out of the hands of the trip-maker and place them under control of the vehicle and its associated automated route-guidance system. This should help improve roadway performance since vehicles will be more likely to be spread across network paths so as to minimize overall congestion. But this may also involve an ethical issues of whether it is appropriate to impose a longer trip on one user so that other users may benefit from shorter travel times (which is usually what is required in order to reduce overall delay in the system).

Informatics-based connectivity (whether in an automated or conventional vehicle) offers the potential for ubiquitous road pricing, in that if every vehicle's location is known and local roadway congestion levels are also known at each point in the network, then usage of the road system can be dynamically priced to encourage more system-optimal route choices by trip-makers, or, at least, to charge trip-makers the actual social cost of their trip. Such a system addresses the ethical issue raised above by creating the potential of offering multiple route choices to trip-makers: for example, a quicker but more expensive route (since it involves higher social marginal costs associated with the trip) or a slower but less expensive one (in which socially beneficial behavior is encouraged or rewarded by a discounted travel cost).

Parking could be similarly monitored and dynamically charged to reduce on-street parking on congested streets, direct cars to vacant parking spaces, etc. Parking lots and garages take up an enormous amount of valuable space, on-street parking very significantly reduces the capacity of our streets to carry traffic of all sorts (i.e. bicycles, transit, etc. in addition to cars and trucks), and drivers cruising to find (cheap) parking is a major source of congestion in its own right in most urban centers. Even with conventional cars, informatics-based parking apps and usage monitoring systems in parking lots can reduce these impacts considerably, as is being demonstrated, for example, by the SF Park demand-responsive parking pricing experiment in San Francisco (https://sfpark.org/). A major asserted benefit of CAVs is that they may eliminate most on-street parking as well as significantly reduce parking lot needs, especially in urban cores. As with all aspects of CAVs, these benefits are at the moment speculative, but are the subject of considerable research (Nourinejad et al. 2018).

Informatics is also extensively (and increasingly) used in transportation network operational control. Traditionally, roadway performance (volumes, speeds, congestion levels) has been monitored by electromagnetic loop detectors embedded in roadways that detect vehicles passing over the detector by the magnetic signature of the vehicle. While useful, such loop-detector systems are expensive to install and maintain and are often subject to failure. Numerous other technologies now exist for monitoring roadway traffic, including video cameras (which require advanced image-processing methods for automated data gathering from the video images), Bluetooth detectors (which detect the unique MAC addresses of vehicles, smartphones, and other Bluetooth-enabled devices, thereby being able to trace the paths and average speeds of these vehicles as they pass a sequence of detectors within the network), and purchasing of on-board route-guidance and other passive location-detection app data from third-party providers. In the case of public transit, many agencies have automatic vehicle location (AVL) systems for tracking transit vehicles in real time and automatic passenger counting (APC) systems for measuring real-time passenger boardings and alightings per vehicle at each stop along a given transit route.

47.4 Informatics and Data Support for Travel-Demand Modeling

The informatics-based services and apps discussed in Sect. 47.2 are generating tremendous amounts of data, day after day, concerning millions of trips being made within a given metropolitan region.

Travel-demand modeling has always depended heavily on large cross-sectional surveys of trip-makers within an urban region. Such surveys are expensive and time-consuming to undertake, subject to various sampling and other biases, and often facing increasing challenges in terms of being able to generate representative samples (Miller et al. 2012; Srikukenthiran et al. 2018). While traditional large household travel surveys are likely to continue be undertaken for the foreseeable future (Miller et al. 2018), current and emerging informatics methods offer promising alternatives and complements to traditional surveys in terms of both new modes and technologies for conducting surveys and new passive (non-survey) methods for observing travel-related behavior, which are discussed in the following two sub-sections. Common to all these sources of data is the problem of imputing missing attributes of the trip or the trip-maker, which requires advanced statistical data fusion and modeling methods, which are briefly discussed in the third sub-section.

47.4.1 Informatics-Based Survey Methods

The primary two informatics-based survey methods are Web-based surveys and smartphone-app-based surveys and trackers. Web-based surveys have become a de facto standard method for undertaking travel surveys, replacing or complementing more traditional methods such as telephone interviews, self-completed mail-back surveys, and face-to-face interviews.[4] Web-based surveys can be very cost-effective since they eliminate the need to hire interviewers, and the marginal cost per survey completion is very low once the up-front cost of the survey development and implementation is accounted for. On the other hand, establishing and contacting a representative sample can be challenging, response rates can be low, and the quality of responses can also be sometimes problematic given the lack of supervision and assistance provided by an interviewer. This last problem, however, can be significantly mitigated by very careful software design to maximize the clarity of the questions being asked and to minimize respondent burden (Loa et al. 2015; Chung et al. 2020; Srikukenthiran et al. 2018).

Similarly, many custom smartphone apps exist that have been explicitly designed to track persons' trip-making and to gather information concerning trip and trip-maker attributes. These generally involve a brief up-front survey to gather key demographic and socio-economic information concerning the trip-maker (and, ideally, the trip-maker's household). The app then is designed to actively track all movements by the person over multiple days, or even possibly weeks, using the smartphone's on-board GPS and other tracking capabilities. This generates space–time traces of the person's movements while carrying the smartphone (assuming that it's turned on!). The potential to gather detailed information concerning personal travel behavior is considerable. In particular, route choice and information concerning active modes, both of which are typically challenging to gather with conventional survey methods, are readily gathered by such apps (Grond and Miller 2016; Lue and Miller 2019). Numerous technical issues, however, are not fully resolved, thus limiting their current widespread usage. These include issues of phone battery life versus the precision of the route tracking (the more precise the tracking, the greater the drain on the battery); the ability to impute travel mode and trip purpose purely from the trip trace; and the representativeness of the smartphone-based samples and sample recruitment methods (Rashed et al. 2015a; b).

[4]Even for these traditional survey modes, tablet-based Web software is being used to conduct and record the interviews. See, for example, Chung et al. (2020) and Harding et al. (2017).

Considerable processing of the raw traces also needs to be undertaken in order to identify the end (stop) point of a trip in space and time (e.g. has the person stopped for a quick shopping activity in a store or is she or he just waiting a long time at a bus stop?), the purpose of the trip (i.e. the type of activity engaged in at the trip end), and the mode of travel used to undertake the trip. Location, purpose, and mode are all essential trip attributes if these data are to be useful for travel-behavior analysis and modeling. Ideally, these attributes should be imputable from the trace data themselves, combined with additional available data, notably GIS datasets concerning land use and points of interest (POI—schools, stores, etc.) and transportation network data concerning road and transit networks. That is, the respondents are passively tracked, without having to explicitly query them concerning their trip-making. If sufficient multiple-day data for enough trip-makers are available, then machine learning methods can, in principle, be used to impute trip stop, mode, and purpose. The current state of practice, however, is such that it is generally required to actively gather at least some information concerning the trips being made, either on the fly as the trips are being detected or at the end of a day through retrospective questioning of the respondents. This active questioning allows labels to be attached to the detected trips (this trip was by car to go shopping) that greatly enhances the ability to train the automated attribute imputation models, at the price of imposing an on-going response burden on the survey participants. Thus, active questioning is often undertaken for a few days at the beginning of the survey period and then turned off with the tracking app running totally passively for the remainder of the survey under the assumption that the imputation apps can be sufficiently trained with the sample of active data obtained (Faghih Imani et al. 2020; Harding et al. 2020; Harding et al. 2016a, b).

47.4.2 Passive Trip Tracking

Numerous informatics-based methods exist to gather information concerning trip-making behavior. These include (Miller et al. 2012):

- Passive smartphone-based location trackers.
- Cellphone traces.
- Transit smartcard transaction data.
- Bluetooth sensors.
- Credit card transaction data.

Passive Location Trackers: As discussed in Sect. 47.2.1, vast quantities of information concerning trip-making are being collected by route-guidance apps, as well as other apps that track smartphone locations for a variety of purposes. In addition to facilitating route guidance, the data collected by such apps can be used to identify origin-destination trips by time of day. These data can be distinguished from the smartphone-app data discussed in the previous section in that they do not require involvement of the phone user in any way and they are completely anonymized (and generally aggregated in one way or another).

Cellphone Trace Data: Whenever turned on, all cellphones are in constant communication with their cellular network. Movements of cellphones (and, hence, their owners) can thus be tracked through time and space. These cellphone traces require significant processing in order to be useful for the analysis of travel behavior, but many analysts are working with such processed data to develop datasets on origin-destination trips by time of day in many urban regions (see Faghih Imani and Miller (2018) for a comprehensive review). The primary attraction for cellphone trace data is its ubiquity in providing massive amounts of travel data, day after day, in virtually every urban region worldwide. Also, given the very deep penetration of cellphones in today's society, these traces can likely be treated as being reasonable representative of the trip-making public. The major limitation of these, data, however, is that the spatial-temporal resolution of the traces is inherently limited by the spacing of the cell towers receiving the cellphone transmissions. Achievable resolutions vary considerably within an urban region. The relatively gross resolution generally achieved poses significant challenges with respect to imputing trip mode (which generally requires good speed measurements) and trip destination activity type (Caceres et al. 2013; Faghih Imani et al. 2018).

An interesting special use of cellphone tracking data is to identify intercity trips. When a cellphone is detected in a city other than its home city, one can impute that an intercity trip has occurred. Intercity travel is a particularly difficult travel market to survey effectively, and so use of cellphone data for this purpose is a promising avenue of research (Bekhor et al. 2013; Janzen et al. 2017).

Transit Smartcard Transaction Data: Another major informatics-enabled source of travel data are data from smartcard transactions collected by public transit agencies. Most major cities worldwide employ some form of smartcard for riders to use to pay their fares, with these cards becoming almost universal in usage. These data thus provide a near-complete record of transit usage in a city. These smartcard systems vary in technical sophistication, but they generally involve one of two primary designs: tap-on systems, in which transit riders tap into the system when they first board a transit vehicle or enter a transit station; and tap-on-and-off systems, in which riders must also again tap the card when they exit the system. These latter systems obviously provide a complete record of all trips made from a first-boarding stop or station to a last-alighting stop or station, by time of day. Tap-on systems require extensive processing to impute trip-alighting locations (typically by observing the boarding location of the next transit trip), but still provide very usable information concerning transit usage (Trépanier et al. 2007; Munizaga and Palma 2012; Parada and Miller 2017).

Bluetooth Sensor Data: As noted in the previous section, Bluetooth detectors can be used to track the passage of Bluetooth-enabled vehicles and personal devices as they pass by detectors mounted along the side of a road. Using records from multiple antennas makes it possible to derive travel times between antenna locations. Hence, depending on the setting, data could be used to derive O-D matrix and partial route choice of a sample of vehicles (cordon setting). While the available data have mostly been used to provide information on vehicle movements, it is also becoming possible to study pedestrian behavior. Malinovskiy et al. (2012) investigated the feasibility of using Bluetooth for pedestrian studies using two separate sites. Their results suggest that "given sufficient populations, high-level trend analysis can provide insights into pedestrian travel behavior."

Credit Card Transaction Data: Although not currently widely used due to lack of access to the data, credit card transaction records can provide detailed information concerning travel for a wide variety of purposes (basically any activity that involves paying with a credit card at an out-of-home location for a good or a service). It also provides expenditure data along with the activity/travel data, something which is not generally gathered in conventional surveys, but could be very useful in modeling not just time but monetary budget allocations. Further, it could provide information concerning in-home versus out-of-home shopping/recreation expenditures, again, something that is of considerable interest for understanding travel behavior. The major limitations of this data source, of course, are whether access to such data can be obtained, and the protection of the confidentiality of the data.

While each of these passive data types have their individual strengths and weaknesses, they share common strengths in terms of:

- Providing a continuous stream of data over days, weeks, and even longer periods of time, thereby permitting time-series analysis of travel trends and dynamics (as opposed to the typically one-day cross-sectional snapshots obtained through conventional surveys).
- Generating massive amounts of data, potentially for thousands or even millions of trip-makers in a large urban region (as opposed to the small samples that can typically be observed in conventional surveys); they truly are big data.
- Being total passive—they require no effort (or perhaps even awareness) on the part of the trip-maker for the data to be collected.

They also, however, share common, significant challenges in their usage in travel-behavior analysis and modeling:

- The data are inevitably anonymized to preserve confidentiality, and, thus, no personal attributes of the trip-makers are known.
- The data are individual-based, not household-based. That is, we generally know nothing about the other members of the trip-maker's household. Household interactions and constraints, however, generally significantly affect an individual's travel behavior.
- As with passive smartphone-app survey data, trip attributes beyond origin, destination, and trip start and end times are generally unknown. That is, trip mode[5] and purpose need to be imputed.
- The spatial-temporal precision of the trace data can vary considerably from one type of data source to another, and even from one trip to another within a given data

[5]Except, of course, in the case of transit smartcard data, where the travel model obviously is transit.

type. Cellphone traces are particularly problematic in this regard, often making mode and purpose imputation challenging.

47.4.3 Data Fusion and Imputation

As discussed above, there are many sources of information concerning travel behavior, ranging from traditional surveys to various informatics-based passive data streams. Virtually all such datasets are incomplete in one way or another in terms of missing one or more attributes of the trip-maker or the trip that are desirable for travel analysis and modeling purposes. This may range from trip-makers' incomes not being collected in a household travel survey to a complete lack of information concerning trip-maker characteristics in most passive datasets. Passive location-tracking data also often lack explicit information concerning key trip attributes such as travel mode and trip purpose. In all such cases, it is desirable to impute the missing information through the fusing of two or more datasets to create a new, combined dataset that contains a richer set of attributes than either original dataset. A common, relatively simple example of this is using census data to impute missing income information in a household travel survey. This is done by using the correlation between income and other household attributes observed in the census data to impute the missing incomes for households observed in the survey, based on the household attributes that are observed in both the census and survey datasets (Bonnel et al. 2009).

A wide typology of data fusion and imputation use cases exist, with many methods available for addressing these cases. Detailed discussion of these use cases and methods is well beyond the scope of this chapter, but can be found in a range of sources, including the work of Miller et al. (2012) and Srikukenthiran et al. (2018). Only two observations are included here. The first is that a particularly important type of data needed for many data fusion exercises that have not yet been mentioned herein are data based on GIS concerning the spatial distributions of people (and their attributes), jobs, and other economic and social activities (stores, schools, etc.). These may be stored at various levels of spatial aggregation (traffic zones, census tracts, etc.), but are also often available in increasingly accurate and comprehensive POI datasets from a variety of commercial and open-source providers. POI data provide information concerning land uses at the very fine level of detail of the individual building, parcel, or geocoded point in space. They thus enable highly disaggregated analysis of point-to-point travel behavior, which is increasingly the level of detail at which travel-demand models are being developed.

Second, as in virtually every sphere of data analysis today, machine learning methods are being increasingly applied to a wide variety of transportation data fusion problems (Gao et al. 2017). One such example involves the use of transit smartcard transaction data, combined with conventional household survey travel data, to train a deep neural network model to predict travel mode. This model is then applied to cellphone trace data to impute the travel mode for the trips represented by these traces (Vaughan et al. 2020).

47.5 Informatics and Modeling Methods

As noted at the beginning of the chapter, a thorough discussion of travel-demand modeling methods is well beyond the chapter's scope. A few characteristics of the current state of best practice, however, include those by Miller (2018, 2019):

- Essentially, all best-practice models are based on activities and tours, in which: (a) travel is the emergent outcome of the need to participate in out-of-home activities; and (b) individual trips are modeled within the context of the overall tours or trip-chains that people engage in throughout their daily activity pattern, so that within-tour decision-making interactions can be accounted for (e.g. if a car leaves the driveway it must eventually return home).
- Travel behavior is largely modeled using sophisticated discrete-choice models based on random utility theory, which provides a very strong behavioral foundation for operational models.
- Increasingly, these activity- and tour-based models are implemented within an agent-based microsimulation modeling framework (see Chap. 44).
- The development of such models has been based on sophisticated, but classic, econometric parameter-estimation techniques (typically maximizing log-likelihood functions).
- Even very complex model systems for large urban regions are developed based on relatively small, cross-sectional samples of a region's trip-making population.

Modern informatics is providing both challenges to the current modeling status quo and opportunities for the development of next-generation models. As noted in Sects. 47.1 and 47.2, informatics-based apps are providing enhanced information and influencing travel choices in ways that are not completely understood and that definitely are not being captured in currently operational models. However, it might also be noted that current models typically assume implicitly that trip-makers have perfect information concerning their travel options and attributes. Hence, it might be argued that these new information sources are actually bringing behavior more in line with modeling assumptions since trip-makers now do have much better information to use in their decision-making!

While the future is perhaps more uncertain than ever before, a few important, specific, and informatics-related observations concerning the current and emerging state of the art in travel-demand modeling can be made with reasonable confidence and are provided below.

First, current best-practice models definitely are not well suited for analyzing new mobility systems, let alone CAVs (Miller 2019). These models need to be redesigned and rebuilt to much better represent both demand decisions and the performance and supply characteristics of these new services (Calderón and Miller 2019, 2020). As data concerning the performance and usage of a wide variety of mobility services become available, the potential for developing improved models increases. New informatics-enabled survey methods also provide the opportunity to gather data on trip-maker preferences and attitudes that will assist in this endeavor.

Second, the increasing availability of massive and passive big data is going to profoundly change how we model travel behavior. While significant technical issues remain, they will provide the opportunity to:

- Develop dynamic models of travel-behavior evolution, freeing us from the tyranny of infrequent, cross-section survey datasets as a basis for model building.
- Establish much more comprehensive and complete representations of travel in an urban region, freeing us from dependency on small-sample surveys which, despite their richness in socio-economic information, inevitably contain significant sampling and response biases.

Third, machine learning and other AI-based methods are rapidly being applied to travel-demand modeling (Yin et al. 2016). While such methods often produce better fits to base data than conventional econometric methods, whether they actually represent improved models for policy analysis and forecasting is very much an open question. A very interesting panel session was held at the US Transportation Research Board Annual Meeting in 2017 titled "Machine Learning Is from Venus, Econometric Modeling Is from Mars: Two Different Travel Forecasting Perspectives." The very strong consensus coming out of this session was that the two modeling approaches are primarily complementary, and that travel-demand modeling needs to optimize its exploitation of both modeling disciplines if it is to meet the profession's modeling needs. In particular, the notion that the advent of big data and AI-based analysis methods will mean the death of (travel demand) models does not appear to be either a likely or attractive alternative. Longer-term, strategic forecasting requires models that can generate emergent, out-of-sample, extrapolated behavioral responses to new scenarios, policies, etc. They cannot just extrapolate current patterns. Further, the interpretability of model sensitivities, elasticities, etc., is a critical component of travel-demand modeling, something that machine learning methods are notoriously poor.

More speculatively, two final questions concerning how informatics-based data and methods might fundamentally change travel-demand modeling in the coming years are the following.

First, can the relatively rich theory of travel behavior that the field has developed over the past sixty years, combined with advanced simulation, data fusion, and machine learning methods be used to both bridge the socio-economic information gaps typical in big data and to merge complementary data sets together to create much more comprehensive representations of travel behavior? Vaughan et al. (2019) provide one example of this approach, in which cellphone traces, transit smartcard transactions, and conventional home-interview travel survey datasets are merged to create a more comprehensive representation of base-year travel than it is possible to achieve from any of the three datasets independently.

Second, is there a quantum theory of travel behavior out there? That is, is there a more explicitly statistical (as opposed to behavioral) approach to modeling that is better suited to the strengths (and weaknesses) of the new datasets? But such a theory or model would still need to be predictive to answer what-if questions. In physics, prediction is the ultimate proof of a theory: Einstein's theories of special and general relativity were accepted, not because of their elegance, but because they are capable of predicting actual behavior. And, indeed, quantum theory's acceptance rests on its ability to predict real-world phenomena (and despite the objections of Einstein on philosophical grounds). The great question facing travel behavior theorists and modelers going forward is how urban informatics-based data and methods will enable us to obtain deeper understanding of actual travel behavior, and, building on this understanding, to develop more powerful and compelling theories and models of travel behavior that enable us to better predict travel behavior in support of transportation policy analysis and forecasting.

47.6 Chapter Summary

This chapter has examined the many ways in which informatics has been changing transportation modeling. These include disruptive changes to: travel behavior, transportation system performance, the data available for model development and application, and modeling methods themselves.

Travel behavior is being influenced primarily by two types of informatics-based services. The first is travel-related Web- and smartphone-based apps that provide a wide range of real-time information, including roadway route guidance, transit service information, and information concerning alternative activity locations. This information is used in both trip preplanning and on-route dynamic decision-making. The second disruptor of travel behavior is the wide variety of new informatics-enabled mobility services that provide trip-making alternatives to conventional travel modes such as public transit, taxis, and even the privately owned car. Most notable are the Uber and Lyft ridehailing services. Other mobility service types include ridesharing (UberPool), car-sharing, bike-sharing, e-scooters, and various forms of demand-responsive transit and microtransit. The mobility service field is evolving rapidly, and the final steady state with respect to these services and their impacts on travel behavior is very difficult to predict. It is clear, however, that travel-demand models will need to evolve considerably if they are to be adequate tools for modeling these impacts and to provide the level of policy guidance needed to ensure socially beneficial outcomes with respect to these services.

These changes in travel behavior and mobility service options are also impacting transportation network performance, notably in terms of roadway congestion and transit usage. Informatics also can support improved real-time control of road and transit operations, implementation of road pricing schemes, and managing parking supply and pricing.

Informatics technologies are also dramatically changing the data available to support travel-demand modeling. Web-based and custom app-based survey methods are complementing, and increasingly replacing, conventional survey methods for collecting travel-behavior information. In addition, a wide variety of sources for passively tracking trips are available, where by passive is meant that the trip-maker is not required to interact with the tracking device or answer any questions. Passive trip-tracking data sources include: smartphone-based location-tracking apps (the route-guidance apps discussed above, but many other apps routinely track the phone's location); cellphone traces; transit smartcard transaction data; Bluetooth sensors; and credit card transaction data. All these data sources offer massive amounts of information, gathered continuously over time concerning trip-making in a given region. They also share common issues concerning lack of socio-economic information about the trip-makers, as well as lack of key trip attributes such as travel mode and trip purpose. A variety of data fusion and imputation methods (including machine learning methods), however, can often be used to augment the passive data, thereby enhancing their utility for modeling.

Given the increasing availability of large, passive datasets, travel-demand modeling will inevitably evolve to exploit these data. Continuous time-series streams of data should support the development of more dynamic (adaptive) models. The very large samples of trip-makers observable within these datasets should lead to models that are more representative and comprehensive relative to current models, which have relied on relatively small-sample survey data for their development. Machine learning and other AI-based methods will continue to play an increased role in model development and application. And, finally, it is possible that travel-demand models may adopt a more explicitly statistical approach to modeling travel behavior (as opposed to the current emphasis on a more behavioral approach) as the optimal way of exploiting the massive, passive datasets with which modelers will be increasingly working.

The challenges facing transportation modelers in the emerging informatics-enriched and informatics-enabled world are large. But the opportunities to develop significantly improved and more powerful models for policy analysis and decision support are also great. It is an exciting time to be a transportation modeler!

References

Bekhor S, Cohen Y, Solomon C (2013) Evaluating long-distance travel patterns in Israel by tracking cellular phone positions. J Advan Transp 47(4):435–446

Ben-Akiva M, Lerman SR (1985) Discrete choice analysis: theory and application to predict travel demand. MIT Press, Cambridge MA

Bonnel P, Morency C, Bayart C (2009) Survey mode integration and data fusion: methods and challenges. Keeping Up with a Changing World. Emerald, Transport Survey Methods, pp 587–612

Caceres N, Romero LM, Benitez FG (2013) Inferring origin-destination trip matrices from aggregate volumes on groups of links: a case study using volumes inferred from mobile phone data. J Advan Transp 47(7):650–666

Calderón F, Miller EJ (2020) A literature review of mobility services: definitions, modelling state-of-the-art, and key considerations for a conceptual modeling framework. Forthcoming, Transp Rev https://doi.org/10.1080/01441647.2019.1704916

Calderón F, Miller EJ (2019) A conceptual framework for modeling the supply side of mobility services within large-scale agent-based models. Submitted to Transportation Letters

Castiglione J, Bradley M, Gleibe J (2015) Activity-based travel demand models: a primer. Report S2-C46-RR-1. Transportation Research Board, Washington, DC

Chung B, Srikukenthiran S, Miller EJ, Nurul Habib K (2020) An inductive experimental approach to developing a web-based travel survey builder: developing a guideline to design efficient web-survey platform. Presented at the 99th annual meeting of the Transportation Research Board, Washington, DC, January

Faghih Imani A, Miller EJ (2018) Preliminary analysis of Montevideo cellular data records. Report 4: iCity-South: Urban Informatics for sustainable metropolitan growth in Latin America. Report to CAF, the Development Bank of Latin America. University of Toronto Transportation Research Institute, Toronto. January

Faghih Imani A, Vaughan J, Yusuf B, Miller EJ (2018) Final Project Report, Report 6: iCity-South: Urban informatics for sustainable metropolitan growth in Latin America. Report to CAF, the Development Bank of Latin America. University of Toronto Transportation Research Institute, Toronto, December

Faghih Imani A, Harding C, Srikukenthiran S, Miller EJ, Nurul Habib K (2020) Lessons from a large-scale experiment on the use of smartphone apps to collect travel diary data: the "City Logger" for the Greater Golden Horseshoe (GGH) area. Presented at the 99th annual meeting of the Transportation Research Board, Washington, DC, January

Gao D, Srikukenthiran S, Nurul Habib K, Miller EJ (2017) Data fusion: techniques and applications. Transportation Tomorrow Survey 2.0, research report. University of Toronto Transportation Research Institute, Toronto, November

Graehler M, Mucci RA, Erhardt GD (2019) Understanding the recent transit ridership decline in major US cities: service cuts or emerging modes? 98th annual meeting of the Transportation Research Board. pp 1–19

Grond K, Miller EJ (2016) Analysis of GPS smartphone data for a Toronto cycling route choice model. Presented at the 5th symposium of the European Association for Research in Transportation, Delft, September

Harding C, Srikukenthiran S, Zhang Z, Nurul Habib K, Miller EJ (2016) Smartphone app and trace processing assessment, vol 2. Transportation Tomorrow Survey 2.0 research report. University of Toronto Transportation Research Institute, Toronto, November

Harding C, Srikukenthiran S, Zhang Z, Nurul Habib K, Miller EJ (2016) Smartphone app and trace processing assessment, vol 1. Transportation Tomorrow Survey 2.0 research report. University of Toronto Transportation Research Institute, Toronto, November

Harding C, Srikukenthiran S, Nurul Habib K, Miller EJ (2017) Evaluating face-to-face surveys to augment regional travel surveys. Transportation Tomorrow Survey 2.0 research report. University of Toronto Transportation Research Institute, Toronto, December

Harding C, Faghih Imani A, Srikukenthiran S, Miller EJ, Nurul Habib N (2020) Are we there yet? Assessing smartphone apps as full-fledged tools for activity-travel surveys. Presented at the 99th annual meeting of the Transportation Research Board, Washington, D.C., January

Janzen M, Vanhoof M, Smoreda Z, Axhausen KW (2017) Closer to the total? Long distance travel of French mobile phone users. Travel Behav Soc 11:31–42

Li W, Shalaby A, Nurul Habib K (2019) UTTRI technical support for the City of Toronto vehicle for hire bylaw review, report no. 5, The relationship between PTC and Public transit: a descriptive analysis. University of Toronto Transportation Research Institute, Toronto, May

Loa P, Srikukenthiran S, Nurul Habib K, Miller EJ (2015) Current state of web-based survey methods. Transportation tomorrow survey 2.0 research report. University of Toronto Transportation Research Institute, Toronto, September

Lue G, Miller EJ (2019) Estimating a Toronto pedestrian route choice model using smartphone GPS data. Travel Behav Soc 14:34–42

Malinovskiy Y, Saunier N, Wang Y (2012) Pedestrian travel analysis using static bluetooth sensors. Paper 12–3270 in Transportation Research Board annual meeting compendium of papers

Meyer MD, Miller EJ (2013) Urban transportation planning, A decision-oriented approach. 3rd Edn. https://mtsplan.com/services.html

Miller EJ (2018) Agent-based activity/travel microsimulation: What's next? In: Briassoulis H, Kavroudakis D, Soulakellis N (eds) The practice of spatial analysis. Springer, pp 119–150

Miller EJ (2019) Travel demand models, the next generation: Boldly going where no-one has gone before. In: Goulias KG, Davis AW (eds) Mapping the travel behavior genome, the role of disruptive technologies, automation and experimentation

Miller EJ, Nurul Habib K, Lee-Gosselin M, Morency C, Roorda MJ, Shalaby A (2012) Changing practices in data collection on the movement of people. Final report. Report to the Transportation Association of Canada, Île d'Orléans, Lee-Gosselin Associates Ltd., Quebec

Miller EJ, Srikukenthiran S, Chung B (2018) Workshop synthesis: household travel surveys in an era of evolving data needs for passenger travel demand. In: Bonnel P, Munizaga M, Morency C, Trepanier M (eds) Transportation research procedia, transport survey methods in the era of big data: facing the challenges vol 32. pp 374–382

Munizaga MN, Palma C (2012) Estimation of disaggregate multimodal public transport origin-destination matrix from passive smart card data from Santiago, Chile. Transp Res Part C: Emer Technol 24:9–18

Nourinejad M, Bahrami S, Roorda MJ (2018) Designing parking facilities for autonomous vehicles. Transp Res Part B: Methodological 109:110–127

Ortuzar JdeD, Willumsen LG (2011) Modeling transport, 4th edn. Wiley, New York

Parada Hernandez C, Miller EJ (2017) Analysis of montevideo smartcard data, preliminary report. Report 3: iCity-South: Urban informatics for sustainable metropolitan growth in Latin America. Report to CAF, the Development Bank of Latin America. University of Toronto Transportation Research Institute, Toronto, September

Rashed M, Srikukenthiran S, Nurul Habib K, Miller EJ (2015a) Current state of smartphone survey methods. Transportation Tomorrow Survey 2.0 research report. University of Toronto Transportation Research Institute, Toronto, September

Rashed M, Srikukenthiran S, Nurul Habib K, Miller EJ (2015b) Current state of smartphone survey methods: annotated bibliography. Transportation Tomorrow Survey 2.0 research report. University of Toronto Transportation Research Institute, Toronto, September

Rayle L, Dai D, Chan N, Cervero R, Shaheen S (2016) Just a better taxi? A survey-based comparison of taxis, transit, and ridesourcing services in San Francisco. Transp Policy 45:168–178

Srikukenthiran S, Loa P, Hossain S, Chung B, Nurul Habib K, Miller EJ (2018) Final report. Transportation Tomorrow Survey 2.0 research report. University of Toronto Transportation Research Institute, Toronto, December

Train K (2009) Discrete choice methods with simulation, 2nd edn. Cambridge University Press, Cambridge

Trépanier M, Tranchant N, Chapleau R (2007) Individual trip destination estimation in a transit smart card automated fare collection system. J Intell Transp Sys 11(1):1–14

Vaughan J, Faghih Imani A, Yusuf B, Miller EJ (2020) Modelling cellphone trace travel mode with neural networks using transit smartcard and home interview survey data. Forthcoming, Eur J Transp Infrastrut Res

Yin M, Sheehan M, Feygin S, Paiement JF, Pozdnoukhov (2016) A generative model of urban activities from cellular data. IEEE Transactions in ITS 1–16

Eric J. Miller has been a faculty member in the Department of Civil and Mineral Engineering, University of Toronto since 1983, where he is currently Director of the University of Toronto Transportation Research Institute. Research areas include activity-based travel modeling, integrated transport–land-use modeling, and agent-based microsimulation. He is the recipient of the 2018 International Association for Travel Behavior Research Lifetime Achievement Award.

Part VI
Perspective for the Future

Chapter 48
A Final Word: The Value of Urban Informatics

Michael F. Goodchild

48.1 Introduction

The chapters of this book include a rich collection of novel forms of data acquisition, techniques of analysis and visualization, and broader concerns about such topics as privacy, urban governance, and urban planning. It is clear from this outpouring of material that urban informatics is a large and burgeoning field. In some cases, especially the chapters in Part IV, the objectives have been the traditional ones of science: the acquisition of new and general knowledge, in the tradition of the UK's Royal Society (to give it its full seventeenth-century title as devised by Isaac Newton and others: the Royal Society of London for Improving Natural Knowledge). In other cases, the objectives are more those of planning; they are normative, in the sense that they assume an ability to design and intervene according to certain principles, using established scientific knowledge. In yet other cases, the authors have been satisfied simply to report capabilities and to discuss the new kinds of data that urban informatics is generating, without any explicit statement of the objectives to which those capabilities and new data are to be applied or how value should be assessed. The finale of the book seems an appropriate place to indulge such broader issues of context.

Several chapters have been concerned with big data, which they have defined in terms of characteristics beginning with V (see, for example, Chap. 43, which cites five Vs: volume, variety, velocity, veracity, and value). Volume, variety, and velocity are central to discussions of big data: volume implying an abundance of data, variety implying a multiplicity of sources, and velocity implying near-real time. Veracity clearly refers to data quality, which big data often lack when compared to more traditional data-production programs; in a sense, then, the fourth V might be identified as an anti-V. Including value, however, begs the question of purpose:

M. F. Goodchild (✉)
University of California, Santa Barbara, Santa Barbara, USA
e-mail: good@geog.ucsb.edu

W. Shi et al. (eds.), *Urban Informatics*, The Urban Book Series,
https://doi.org/10.1007/978-981-15-8983-6_48

whose interests are served by big data? More fundamentally, we can ask the same question about urban informatics: whose interests does it serve, and whose interests are marginalized?

To what extent should specialists in urban informatics concern themselves with these issues? In the early 1990s, a number of scholars drew attention to the social implications of geographic information systems (GIS; Pickles 1995; Schuurman 2000), with the implicit or explicit suggestion that developers of GIS were ignoring such concerns. Much of the early technical development of GIS originated in Eisenhower's military–industrial complex, where its purposes could easily be seen as diametrically opposed to the immediate concerns of a civilian society (Smith 1992). GISs were being used even then to track and monitor citizens (https://www.co.pierce.wa.us/1964/Sex-Offenders-in-Pierce-County), and today geospatial technologies are an essential part of many programs of public surveillance (Chap. 32). Asking these questions about urban informatics recalls the kinds of soul-searching that occurred during and after the development of the atomic bomb, though that case is clearly more extreme. For example, it is hard to imagine anyone working in urban informatics to be driven, as Oppenheimer was on witnessing the first nuclear explosion, to quote from the Bhagavad Gita: "Now I am become death, the destroyer of worlds" (https://www.wired.co.uk/article/manhattan-project-robert-oppenheimer). Nevertheless, it seems appropriate at the end of the book to enquire about that fifth V and its implications for the future. What kind of urban world is likely to result from all of this research and development, and what can be done to ensure that the field moves in a positive rather than a negative direction? In developing and advancing urban informatics, are we headed for a future utopia, and what kinds of dystopias might emerge as unforeseen and unintended consequences? Are we, like Mark Zuckerberg and the early days of Facebook, in favor of technical disruption for its own sake (Taplin 2017), or would we rather a more considered future, a slow urban informatics if you like? In short, what constitutes value in urban informatics?

To focus this discussion of the bigger picture somewhat, the next section proposes several alternative visions of what urban informatics is about, and its corresponding form of accountability.

48.2 Visions for Urban Informatics

48.2.1 Urban Intelligence

James Clapper, who retired in 2017 as the US's Director of National Intelligence, a position in which he oversaw the activities of 17 distinct government organizations including the National Geospatial-Intelligence Agency, argues strongly in his recent autobiography (Clapper 2019) that the gathering, assembly, and interpretation of intelligence should be driven by a simple vision: the speaking of truth to power. The policy decisions that result from that intelligence are the responsibility of other leaders and branches of government to whom the intelligence community (IC) reports, and should not bias or distort the community's primary function. We could argue, then, that the value of urban informatics lies in the scientific quality of the data acquired, and the compilations, interpretations, analyses, and visualizations performed. Urban informatics should be replicable so that independent investigators should reach the same conclusions, should capture and address uncertainties, and should use terms, definitions, and practices that are as far as possible shared and standardized. The urban IC should be driven by an objective of speaking truth to urban power, whether it be city administration, elected representatives, or the urban public.

Is this a useful vision for urban informatics? It is certainly aligned with much writing on smart cities. Its ultimate goal would be the development of data acquisition programs to capture a representation of the city and its enormous complexity—as close as possible to a digital twin—that could then support the city's decision-making processes. It implies a simple kind of accountability, and a taxonomy of different kinds of intelligence somewhat comparable to the signals intelligence (SIGINT), geospatial intelligence (GEOINT), intelligence derived from social media and other social sources (HUMINT), etc., of the IC. But there are several compelling alternatives.

48.2.2 Urban Science

Many chapters, especially those in Part IV, are driven by the traditional goals of science: the acquisition of knowledge about urban systems. Such knowledge should be generalizable, since urban science looks for processes that are replicable across many urban environments. Just as physics searches for general laws and principles, it would be of little interest in urban science to discover knowledge about London, or some part of London, that cannot usefully be applied and implemented in other cities and neighborhoods, at least in those that bear some resemblance to London; and cannot be usefully applied at other times. Urban science is driven by the belief that such general principles exist, and can be discovered through the kinds of natural experiments that rely on observations, public-sector programs that gather statistical data, crowdsourcing, remote sensing, and data that can be cajoled from the private sector's enormous stocks.

Geography as a discipline has long struggled with finding a balance between the search for general principles on the one hand, and the documentation of the unique on the other, since the latter is after all what drove the Age of Discovery in Portugal and the explorations that have always captivated the human imagination. It concerned Varenius, the Polish-Dutch geographer of the seventeenth century (Warntz 1989), who wrote about what he termed Special (idiographic) Geography and General (nomothetic) Geography. It drove a debate in the 1950s between Schaefer and Hartshorne (Harvey 1969) that remains a cornerstone of graduate courses in geographic thought. The more prestigious sciences will often describe idiography using perjorative (to them) terms such as "journalism" and "mere description."

Today, this debate has become more nuanced. Techniques such as geographically weighted regression (GWR; Fotheringham et al. 2002) and local indicators of spatial association (LISA; Anselin 1995) represent a form of compromise: a set of structures whose forms can be generalized, but whose parameters are allowed to vary in space and perhaps also in time. We might term this weak generalizability, and several arguments can be presented in its favor. In the social and environmental sciences, it is hard to imagine any principle being truly deterministic, since there will always be unaccounted factors. In short, the goal of an R^2 of 1 will always be unattainable. If those unspecified factors vary spatially, then the effect will be a spatial variation in the parameters of the model. Alternatively, we might argue that processes do truly vary with location: that growing up in Detroit is fundamentally different from growing up in New Orleans, all other things being equal.

If urban science is indeed driven by curiosity, then its responsibilities end when knowledge is shared through the process of publication. Application and implementation become the responsibility of others, as in the first vision of urban intelligence, and one can imagine an applied urban science emerging that is devoted to the use of general urban knowledge—or perhaps, it would be better termed urban engineering. The value proposition is now different: instead of the abstract concepts of understanding and explanation that drive curiosity-driven science, applied urban science would be accountable through its broader impacts.

48.2.3 Urban Planning and Design

The fifth V has already taken two different meanings in these sub-sections. Value in the case of urban intelligence will be determined by policy- and decision-makers, based on the degree of support given by the information provided to them. In the case of urban science, value derives in the first instance from the production of generalizable knowledge, and less directly from its usefulness in application. But the urban planning and design that have been discussed in several chapters of this book proceed according to a prior definition of value: the extent to which plans and designs are consistent with agreed principles. In short, they are normative, unlike the previous two visions. In some cases, these principles may be at least partially embedded in software, as in Chap. 35 and in the broader area of spatial optimization, which seeks to design solutions to problems that are optimal against defined objectives.

Many issues complicate that simple vision. First, except in the simplest instances, it will be difficult to reach an agreement on the principles that drive planning and design. Will they serve the interests of a minority at the expense of the majority? Will they adequately address the needs of those whose voices are often muted or unheard? The field of multicriteria decision-making has evolved as a model of how decisions can be made in the face of conflicting goals; its tools include methods for determining consensus weights to be applied to alternative numerical criteria

(Saaty 1977). Second, while we might argue that a decision based on agreed criteria is inherently more fair, in practice any solution is bound to be seen to favor one position or another.

48.2.4 Urban Development

The value proposition for business is of course a matter of simple economics: innovations are driven in the first instance by their ability to make money. While disruptions such as Uber or dockless bikes can certainly have redeeming social value, it is their eventual profitability that ultimately drives their growth. Many businesses invite the users of their apps to allow locations to be shared and may argue that the result will be more specific information to the user. This is the case for wayfinding apps, and also for many news or weather apps. But the business case for such apps relies at least in part on the market value of those user locations to retailers, advertisers, and others. This trading of location data will be consistent with the app's terms and conditions of use, but the user is unlikely to have taken the time to read their tens of pages of fine print and to have realized what they imply.

48.3 Unintended Consequences

Although the previous section has outlined how value can be assessed under different visions of urban informatics, it is often the unintended consequences of actions and developments that determine whether outcomes will eventually be assessed positively or negatively. How, for example, should we assess the impacts of online shopping? The individual citizen benefits from having goods delivered quickly, without the time and expense of a shopping trip. New jobs are created in the city's delivery industry, and profits are made by the owners of shopping Web sites and their suppliers. But the impact on traditional shopping is severe, with significant loss of local employment and the closure of conventional retail businesses, and in some cases, wholesale abandonment of shopping centers. Supply chains may have to be reorganized, and the city's function as a regional shopping center may be undermined.

The advent of connected and autonomous vehicles (CAVs) provides a suitable case in point. Already many new vehicles are connected to the Internet, and capable of reporting details of location, driving habits, and even driver biometrics. Such data can be useful to the parents of young drivers, to insurance companies following a crash, and to mechanics when a vehicle is serviced. They have commercial value, as already noted in Sect. 48.2.4. But they also potentially have more sinister value to traffic-control systems and law enforcement, and to what has been termed automated social control (New York Times 2020).

Cities are complex phenomena, performing functions that are not only internal but also regional and global. The growth of IoT will benefit the city through the services it provides, but will also benefit employment in high-tech industries in cities that may be half a world away; and the waste created by the city will almost certainly be exported to the city's hinterland, to areas downwind and downstream, and to foreign markets for recycled material. What may be out of sight and out of mind to a city's citizens may be very real to people elsewhere in the world.

48.4 The Future of Urban Informatics

Whether as a means for gathering urban intelligence, or as a basis for new urban science, or as a tool for planning and design, or as a source of profit for developers, urban informatics is clearly destined for accelerating growth. There is little danger of it experiencing a quick death as a short-term fad. Yet it can also be a source of the future dystopia, given its potential for surveillance and control.

This short piece has drawn attention to two issues: the different ways in which parts of the urban informatics community address the value of what they are doing; and the temptation to focus on the internal complexity of the city without addressing the complexity of its external linkages.

There are obvious similarities between the emerging field of urban informatics in 2020 and the state of GIS in the early 1990s: both are growing strongly, with enormous promise. It is important therefore that the kinds of concerns for broader social impacts that emerged at that time in the GIS research community, and led to an outpouring of important research, should also become part of the agenda of urban informatics. We are the people to explore these broader impacts and to raise these issues with our governments and with the public.

References

Anselin L (1995) Local indicators of spatial association—LISA. Geogr Anal 27(2):93–115

Clapper JR (2019) Facts and fears: hard truths from a life in intelligence. Penguin, New York

Fotheringham AS, Brunsdon C, Charlton M (2002) Geographically weighted regression: the analysis of spatially varying relationships. Wiley, Chichester

Harvey D (1969) Explanation in geography. Edward Arnold, London

New York Times (2020) https://www.nytimes.com/2020/03/01/business/china-coronavirus-surveillance.html. Accessed 13 April 2020

Pickles J (ed) (1995) Ground truth: the social implications of geographic information systems. Guilford, New York

Saaty TL (1977) A scaling method for priorities in hierarchical structures. J Math Psychol 15:234–281

Schuurman N (2000) Trouble in the heartland: GIS and its critics in the 1990s. Prog Hum Geogr 24(4):569–590

Smith N (1992) History and philosophy of geography: real wars, theory wars. Prog Hum Geogr 16(2):257–271

Taplin JT (2017) Move fast and break things: how Facebook, Google, and Amazon cornered culture and undermined democracy. Little, Brown, New York
Warntz W (1989) Newton, the newtonians, and the geographia generalis varenii. Ann Assoc Am Geogr 79(2):167–191. https://doi.org/10.1111/j.1467-8306.1989.tb00257.x

Michael F. Goodchild is Emeritus Professor of Geography at the University of California, Santa Barbara, a Distinguished Chair Professor at The Hong Kong Polytechnic University and a Research Professor at Arizona State University. He is a member of the US National Academy of Sciences, and he is interested in GIScience.

GPSR Compliance

The European Union's (EU) General Product Safety Regulation (GPSR) is a set of rules that requires consumer products to be safe and our obligations to ensure this.

If you have any concerns about our products, you can contact us on ProductSafety@springernature.com

In case Publisher is established outside the EU, the EU authorized representative is:

Springer Nature Customer Service Center GmbH
Europaplatz 3
69115 Heidelberg, Germany

Batch number: 08679876

Printed by Printforce, the Netherlands